The region of the crater Kuiper, on Mercury. This photograph, which is a mosaic of images taken on 29 March 1974 by the American probe Mariner 10, shows a region of about 1400 kilometres by 1000 kilometres situated close to the equator of Mercury. The latitude of the centre of the illustration is 6 degrees south, and its longitude about 20 degrees. North is to the right, and the grazing illumination comes from the west, at the top: it is the end of the afternoon. The appearance of the surface is very like that of the Moon: it is riddled with circular depressions, which are craters derived from meteoritic impacts. The smallest details visible are less than a kilometre in diameter. Kuiper, the brightest of the mercurian craters, is located in the upper left-hand quarter of the illustration. The crater is named after Gerard Peter Kuiper, great figure of planetary astronomy and one of the scientific team for the mariner 10 mission. He died on 21 December 1973 while the probe was on its way to Mercury. The crater, about 40 kilometres in diameter, is the result of a meteoritic impact upon the walls of the more ancient crater Murasaki of about 80 kilometres diameter. This photograph, with its astonishing sameness, hardly gives a fair impression of the progress represented by this first flyby of Mercury by Mariner 10. Before 29 March, almost nothing was known about the planet apart from its mass and its orbital parameters, although the surface was known to a resolution of 500 kilometres. Our knowledge of Mercury increased immensely in only a few hours through this single probe which is about the size of a small car and has a mass of 504 kilograms. Half of the planet has now been mapped, with an average resolution of 1 kilometre; for a few per cent of the surface, the resolution is 100 metres. (NASA)

THE CAMBRIDGE ATLAS OF ASTRONOMY

THE
CAMBRIDGE ATLAS
OF
ASTRONOMY

EDITED BY JEAN AUDOUZE AND GUY ISRAËL

THIRD EDITION

CAMBRIDGE
UNIVERSITY PRESS

This English language edition published by the Press Syndicate of the University
of Cambridge
The Pitt Building, Trumpington Street, Cambridge CB2 1RP
40 West 20th Street, New York, NY 10011-4211, USA
10 Stamford Road, Oakleigh, Melbourne 3166, Australia

Original Edition
Le Grand Atlas de l'Astronomie
© Encyclopaedia Universalis France S.A. 1983, 1986, 1994

First English Language Edition
© Cambridge University Press and Newnes Books 1985
Second English Language Edition
© Cambridge University Press 1988
Third English Language Edition
© Cambridge University Press 1994

Library of Congress catalogue card number: 84–73453

British Library Cataloguing in Publication Data
The Cambridge atlas of astronomy.
1. Astronomy
I. Le Grand atlas de l'astronomie. *English*
520   QB43.2

ISBN 0 521 43438 6

Printed in France

General Editors   J. Audouze, C.N.R.S.
G. Israël, C.N.R.S.

Editor   Jean Claude Falque

Authors   M. Arduini-Malinovsky, C.N.R.S.
J. Audouze, C.N.R.S.
J.-L. Bertaux, C.N.R.S.
J.-P. Bibring, Université de Paris-XI (Orsay)
A. Boischot, Observatoire de Paris-Meudon
A. Boissé, Observatoire de Paris-Meudon
A. Brahic, Université de Paris-VII
M. Cassé, Commissariat à l'énergie atomique
A. Cazenave, C.N.R.S.
J.-P. Chièze, Commissariat à l'énergie atomique
F. Durret, C.N.R.S.
D. Gautier, C.N.R.S.
M. Gerbaldi, Université de Paris-XI (Orsay)
G. Israël, C.N.R.S.
Y. Langevin, C.N.R.S.
J. Lequeux, Observatoire de Marseille
J.-P. Luminet, C.N.R.S.
P. Masson, Université de Paris-XI (Orsay)
P. Mein, Observatoire de Paris-Meudon
J. Paul, Commissariat à l'énergie atomique
J. Roland, C.N.R.S.
F. Sanchez, Instituto de Astrofisica de Canarias
H. Sol, C.N.R.S.
P. Thomas, C.N.R.S.
C. Turon, Observatoire de Paris-Meudon
E. Vangioni-Flam, C.N.R.S.
J.-P. Verdet, Observatorie de Paris-Meudon
J.-C. Vial, C.N.R.S.
A. Vidal-Madjar, C.N.R.S.
L. Vigroux, Commissariat à l'énergie atomique

English Language Edition

Translator   Andrew King, University of Leicester
(Third edition)

Translators   S. Demers, Université de Montréal
(First and second editions)   I. T. Drummond, University of Cambridge
R. A. W. Elson, University of Cambridge
P. G. Hayman, University of Cambridge
M. F. Ingham, University of Cambridge
D. Lindley, University of Cambridge
J. McDowell, University of Cambridge
J. Mitton, University of Cambridge
L. Murdin, Royal Greenwich Observatory
N. O'Hora, Royal Greenwich Observatory
R. Pennington, University of Cambridge
M. Penston, Royal Greenwich Observatory
D. Pike, Royal Greenwich Observatory
C. H. E. Russell, Royal Greenwich Observatory
V. Samson, Royal Greenwich Observatory
T. Snijders, Royal Greenwich Observatory
E. Terlevich, Royal Greenwich Observatory
R. Terlevich, Royal Greenwich Observatory
J. Webb, University of Cambridge
R. V. Willstrop, University of Cambridge

Contents

The stars and the Galazy *Jean Audouze* 240

The extragalactic domain
Jean Audouze 332

The scientific perspective 404

Introduction

First published by Encyclopaedia Universalis, edited by Jean Claude Falque, under the scientific leadership of Jean Audouze and Guy Israël, this volume is an invitation to travel throughout the whole Universe: to planets, stars, galaxies, and, more generally speaking, to any kind of object or particles which can now be grasped by our senses.

In the first part, under the guidance of Monique Arduini-Malinovski and Jean Audouze, the reader will start by studying the Sun, a most important star in at least two respects: it is the star nearest to us, and this allows us to study it much more closely than any other star; also, the energy it pours out makes it the mainstay of all forms of life.

The exploration of the solar system forms the subject of the second part, to which Guy Israël has devoted every care. This affords an opportunity to present a number of pictures which, only a few years ago, could simply not have been obtained, let alone interpreted. Not only because space probes have revolutionized our data about Venus, Mars, the Jupiter, Saturn, and Uranus systems, and about comets, but also because these discoveries, together with the progress of Earth-based planetary astronomy, have, as it were, set the seal of approval upon the rise of a new science, born merely one or two decades ago: planetology.

With the third and fourth parts, readers, again with Jean Audouze's help, will venture further and further into outer space. Forsaking these sections of the sky with which, largely thanks to television, we are now almost familiar, they will first roam through our own Galaxy, with its two hundred billion stars. Then, leaving behind these vast regions only perceptible to us as the dim band of light we call the Milky Way, they will thrust out into extra-galactic space, where more than a billion galaxies of all types have been identified to date.

The last section brings together all the necessary information about astronomy as a science, its history, the various observation techniques now in use. It also raises its focus higher, even though, in some respects, Jean Audouze and his co-workers have been constrained by space; on such controversial issues as the origin of the universe, the beginning and enduring existence of life, they take stock of the state of the art, bring together a number of hypotheses and define the way forward.

Jacques Bersani
Encyclopaedia Universalis

Astronomy Today

We live in an age marked by enormous technological upheavals. Western societies now possess tools and machines that undertake the most varied tasks, and computer technology affects nearly every activity – whether it be scientific research, economic management or sport. These upheavals, and it is not our purpose to assess them here, are due to the progress achieved in the fundamental, or so-called exact, sciences. It is not surprising then that those who wish to continue the technical progress or, contrariwise, those who express the most serious reservations about it take a stance on science. Nobody now disputes the unpredictable influence that the apparently purest scientific research can exercise on our lives; nuclear physics is surely the most significant and widely debated example of this.

Astronomy and its sister astrophysics occupy a very special place in the sciences' hall of fame. History and archaeology teach us that there are good grounds for believing that these subjects were among the oldest preoccupations of mankind. The worries and fears occasioned by astronomical phenomena, such as eclipses and meteors, and observation of the sky and of the motions of the Sun and Moon must be as old as the most primitive religions. Direct references to astronomical events are to be found in the mythology and religious beliefs of every civilisation. The division of time into years (the period of the Earth's revolution about the Sun), into months (related to the phases of the Moon) and into days occurred in antiquity, as we know from calendars used by Assyrians and Chaldeans, but the system is still used for the regulation of our activities. It seems then that astronomy is, together with mathematics, among the oldest of the sciences and that it is one which profoundly influenced man's thinking, everywhere.

In addition, by enabling us to unravel some of the mysteries of the heavens, astronomy helps us to appreciate our environment and to glimpse something of our origin and evolution through the veil that obscures them. It permits us to explore a time span of fifteen billion years in the past and to conjecture on a series of events that there is every reason to believe will occur long after man has ceased to exist.

Finally, despite its antiquity – or perhaps because of it – astronomy benefits from the contributions of other disciplines, as if it, in some way, embraces them all because it is concerned with every aspect of the whole Universe. The most advanced techniques and the advent of the space age have contributed enormously to astronomy which today is enjoying a veritable golden age.

The difference between astronomy and astrophysics should be explained at this point. Astronomy, the older of the two disciplines is concerned with the observation and

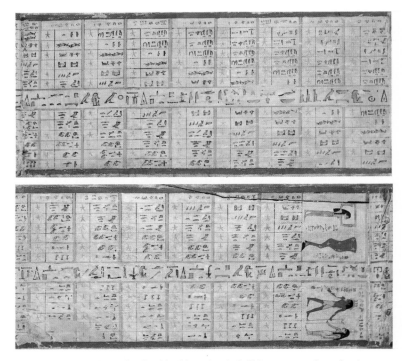

An Egyptian stellar calendar (the Idy calendar). This very complex calendar, preserved in the Tubingen Institute of Egyptology, seemingly dates from 2200 BC. It is impossible to give a detailed description of it here, but it should at least be explained that by comparing the periods of the flooding of the Nile with the earliest rising of the star Sirius – which occurs on about 19 July in our calendar – the Egyptians discovered very early on that the length of a year is 365.25 days. By observing the delays in the risings of Sirius with respect to those of the Sun they were able to determine precisely the year of the seasons. The Idy calendar is an expression of this because it shows the progress of Sirius in days and hours. Unfortunately it is not complete: it has only eighteen vertical columns, each of which covers ten days, instead of the thirty-six columns that counted the Egyptian year. It reads from right to left, the top line denotes the ten-day periods, and each column is divided into twelve compartments corresponding to the twelve hours of the day. (Ägyptologischen Instituts der Universität Tübingen. Photograph by J. Feist)

Halley's Comet in the Bayeaux Tapestry. People have always believed that astronomical events exercise an influence – either benevolent or malevolent – on their destiny. The famous Bayeaux Tapestry, which illustrates the invasion of England by William the Conqueror, shows an example of this. Near the centre of the part reproduced here a drawing of a comet can be seen: this is now known to be Halley's Comet. The appearance of this great comet in the April sky in 1066 made a profound impression; it was seen as a fatal omen for the English king, Harold, who was indeed slain during the Battle of Hastings. (Musée de la Tapisserie de la Reine Mathilde)

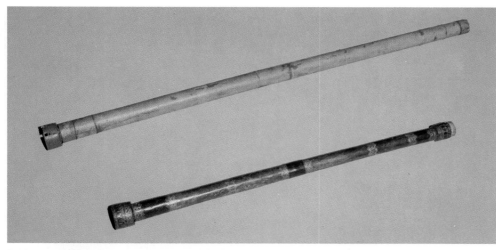

Galileo's telescopes. The top telescope has a paper-covered wooden tube 1.36 metres long; it possesses a convergent objective (a biconvex lens) of 26 millimetres aperture and 1.33 metres focal length, and a divergent eye-piece (concave lens). Its magnification is about 14. The bottom telescope has a wooden tube, covered with leather which is decorated with gold, worked in linear motifs. The convergent objective has an aperture of 16 millimetres and a focal length of 0.96 metre. The divergent eyepiece is not the original one. The magnification of this telescope is 20. It is proper to draw attention here to a point of historic importance: the name Galileo is given to this type of instrument because he was the first to make proper use of it, of astronomical interest, but he was not its inventor. In June 1609 he became aware of the existence of telescopes, which appeared in the Netherlands towards the end of 1608; he immediately set to work to build one and began to observe the sky in the autumn of 1609. (Istituto e Museo di Storia della Scienza, Florence)

The 3.6-metre telescope of the Canada–France–Hawaii (CFH) Society. This photograph, taken on the summit of Mauna Kea (Hawaii) at an altitude of 4200 metres, shows the 3.6-metre telescope which is owned by a society of which Canada and France each own 44 per cent of the capital and the University of Hawaii the remaining 12 per cent. The CFH telescope commenced operating in the summer of 1979. Some notable discoveries have been made with it, as it enjoys the best possible atmospheric conditions. (CFH)

analysis of the movements of the heavenly bodies while astrophysics, which owes its birth to the seventeenth-century invention of the telescope, is devoted to the study of the physical nature of these bodies. Neither of them should be confused with astrology, a pursuit of charlatans who claim to deduce from the motions of the planets a knowledge of the behaviour and the future of individual people; the successes they claim can only be explained in terms of the anxiety felt by every human being about their destiny. By revealing the magnificence of the observable world and the manifest splendour of the skies astronomy can – if not reassure, for such is not its aim – assist in enabling us to appreciate the beauty of the Universe that we inhabit, and perhaps increase our knowledge.

Procedures in astronomy and astrophysics

In spite of the advent of modern methods of observation we have every reason to feel overwhelmed by the vastness of the Universe. Our stature and the capacity of our intelligence seem insignificant by comparison with the objects we wish to observe, analyse and understand. Nevertheless, the resourcefulness of the human mind and imagination as well as the invention of new observational techniques and the development of new concepts and theories have enabled us to form a more detailed and exact model of the Universe and its component parts. Besides, it is only by identifying the limits of observation and of theory, and by attempting to extend them that progress can be made in any science.

At first astronomy endeavoured to look at the sky and to describe it in terms of what could be seen. Such a statement may appear redundant, but it has very important consequences, because, in spite of the body of complex mathematics, physics and even chemistry that need to be deployed to understand the Universe, astronomy and astrophysics are primarily observational sciences. Simple naked-eye observation of the stars was responsible for the notion of the luminosity and then size of a star from which came the modern concept of magnitude. The naked eye could always discern the planets (a Greek word meaning wanderers) among the stars and the great concentration of stars in the Milky Way or in the Magellanic Clouds can be observed without the aid of an instrument.

However, the human eye is not very sensitive and its 'memory' is very short. Real progress could not be achieved in astronomy and astrophysics until the telescope was introduced and until the invention of the spectroscope and spectrograph, instruments capable of analysing light. The advent of photographic techniques enabled images to be preserved and led to the institution of archives for the conservation of astronomical photographs – all of which revolutionised the methods of astronomy. It was from the rational use of these tools that astrophysics was born.

The eye is limited not only by its low resolving power and its poor capacity for detecting weak light sources but also by the restriction of its sensitivity to a very narrow band of wavelengths – a disadvantage that it shares with most photographic emulsions. Radio astronomy was born after the Second World War when, for the first time, sky radiation in the band of radio wavelengths could be received.

If man had not always aspired to flying (as expressed in the myth of Icarus, the writings of Leonardo de Vinci, the thoughts of Cyrano de Bergerac, the anticipation of Jules Verne . . .) our vision of the Universe would be confined to the so-called visible range; that is the wavelengths between 400 and 800 nanometres, and to the radio waves in the range 1 millimetre to 30 metres. This is because the Earth's atmosphere, which acts as a protective shield for biological molecules against the noxious effects of 'hard' radiation (ultraviolet, X- and gamma-rays), is equally effective in preventing observations at these wavelengths from the surface of the Earth. But by the use, first of balloons, then of rockets and finally of satellites the full range of electromagnetic

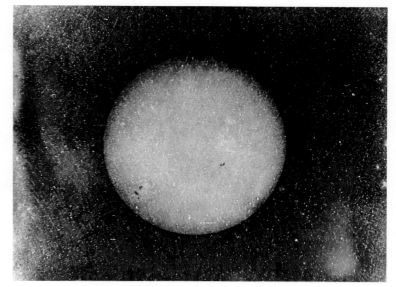

The first photograph of the Sun. This photograph is historically of great importance because it is the first daguerreotype image of the Sun, obtained by direct impression, at 09.45 on 2 April 1855, by the physicists Fizeau and Foucault. It was one of the first applications of astronomical photography. The technical quality is rather mediocre but the practised eye can discern some signs of solar granulation in this old print. (Musée des Techniques du Conservatoire National des Arts et Metiers)

Chinese tenth-century star map. This map, from a manuscript of the Tuahang collection preserved in the British Museum, shows the Great Bear (easily recognisable) and the group of constellations that were known to the Chinese as the Purple Enclosure. It is one of the earliest known star maps and the quality is striking even though it is based solely on naked-eye observations. (British Museum. Photograph by J. Needham)

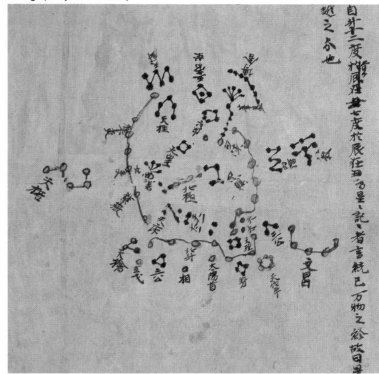

radiation emitted by the various sources in the Universe can now be observed. Already infrared, ultraviolet, X- and gamma-ray astronomy have all furnished important results.

Traditional astronomy will also benefit from space technology. A large optical telescope will be placed in orbit around the Earth by the end of the decade; it will be an extraordinary instrument, one which astronomers everywhere are awaiting with impatience because of its high resolving power, its ability to detect very weak – and consequently very distant – sources, and its capability of covering equally well the visible, the infrared and the ultraviolet ranges. It will also be able to work continuously, without interruption by the daily alteration of light and darkness or by meteorological conditions.

Astronomers are making more and more use of space probes and of photon counters of increasing efficiency for the detection of light sources as well as of other new techniques such as fibre optics, television cameras and electronic light detectors. In addition to profiting from developments in optics and space technology, they have at their disposal the whole of scientific and technical know-how and, whether they are involved in the determination of the chemical composition of a rock of lunar, terrestrial or meteoritic origin, or in the study of the bombardment of a large plastic sheet by cosmic rays, or in the examination of the infrared spectrum of certain organic molecules, or in the observation of aerosols in the atmosphere of Venus, they are all contributing to astronomy.

Today's astrophysicist is therefore a jack of all trades; he needs to have a basic knowledge of every kind of observational technique. For its devotees, the largely

The Kitt Peak millimetre radio telescope. The dish of this telescope is designed for reception in the millimetre band. Observations at such wavelengths are very difficult due to absorption of the radio waves by water vapour in the atmosphere; reception is only possible at certain wavelengths known as 'windows'. It is for that reason that this telescope is sited in an arid region, at an altitude of 2000 metres on Kitt Peak mountain, near Tucson in Arizona. The discovery of most of the known interstellar molecules was achieved with this remarkably accurate radio telescope. (NRAO)

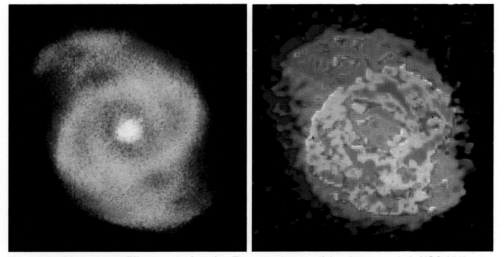

The same object at two different wavelengths. These two images of the planetary nebula NGC 6543 are very similar even though they were obtained by different means. The image on the left is a photograph taken in visible light while that on the right was obtained by radio observations in the 6-centimetre band. The striking similarity of the two enables us to establish the thermal origin of the radiation emitted by the envelope that was ejected by the planetary nebula. (Lick Observatory and NRAO)

Charles Conrad Jr, an American astronaut, on the Moon. On 20 November 1969, in the course of the Apollo 12 mission, astronauts landed on the surface of the Moon in the lunar module, which can be seen in the background about 200 metres away, and inspected Surveyor III, a space probe that had soft-landed there on 19 April 1967. The photograph shows Charles Conrad Jr scraping flakes of paint off the probe. These were returned to Earth where very refined isotopic analysis of them determined the composition of the particular radiation emitted by the Sun that is known as the solar wind. The picture is a reminder that astronomical observations are not made just by the use of telescopes, with the eye or some other kind of detector, but also by collecting information from sources as unexpected as flakes of paint off a vehicle that has been resting on the moon for two and a half years. (NASA/NSSDC)

interdisciplinary nature of astronomy is one of its strongest attractions. It is easy to understand why astronomy is making such rapid progress, benefiting as it does from all kinds of advances in science and technology.

Unlike the other observational sciences, such as physics, chemistry and biology, which are based on the performance and analysis of experiments, astrophysics, whose purpose is to analyse the whole of the observable world, is unable to arrange experiments. Besides, the uniqueness of some objects or of certain phenomena is a serious handicap for the astrophysicist: up to now only one planetary system has ever been observed – our own Solar System – and there are many bodies with unique properties, yet they too must be explained. However, the large numbers of bodies forming single families – there are about two hundred billion stars in our Galaxy and an estimated half that number of galaxies in the Universe – mean that statistical analyses of groups may be undertaken, as opposed to the study of individual members. The astrophysicist is unable to experiment directly but he can, for example, compare the properties of stars in different stages of their evolution.

The general characteristics of the observable Universe

As well as its great diversity, the observable Universe has a number of general characteristics that may be summarised as follows:
– the matter which makes up the Universe is not randomly distributed but constitutes, even on a grand scale, hierarchical entities;
– this matter appears to obey a very small number of physical laws which can be expressed by a series of unified physical laws;
– every part of the Universe appears to exert a direct influence on the rest of the Universe and to be, in turn, similarly affected by the rest of the Universe.

We are concerned with a double phenomenon, but an entirely understandable one: the simplification of the description of the Universe on a large scale and the construction of increasingly complex descriptions of the multiplicity of forms and structures observable on a small scale. This leads to the apparently paradoxical situation in which astronomical descriptions would seem easier than those dealing with small complex units such as the surface of the Sun or, to take an extreme example, a living cell. The apparent simplicity of the theories that offer global explanations of very large structures

– a planet or a galaxy for example – is nothing more than a reflection of our ignorance; it is only structures of which our knowledge is scanty that can be represented in terms of simple models.

It should be emphasised that representations of the Universe are determined by the scientific philosophy underlying the proposed models. Very different representations, depending on the points of view that the models satisfy, may be based on the same set of observations or experiments. On the whole our representations are based essentially on Greek philosophy and adhere to the notions of force and interaction, that is to say the transfer of energy between two bodies, whether they be atomic nuclei, planets, stars or galaxies. Every model of the observable Universe is, in fact, represented in terms of fields of force, and consequently, of the distribution of masses. We are so accustomed to thinking in this way that we would find it difficult to conceive other ideas that could equally well explain our observations.

The Universe as a hierarchy

The popular saying that nature abhors a vacuum is contradicted by reality. Suppose, for example, that the forces acting on the particles that constitute a man ceased to exist, then these particles would fit into a sphere with a diameter of one hundredth of a millimetre. The forces that act in matter render it very hollow so that it is composed of particles that are very small and very dense but widely separated from one another.

Let us explore the Universe, starting with bodies of small dimensions (bearing in mind that seemingly 'invisible' matter, which could be in the form of neutrinos, exists). First of all we meet the so-called elementary particles which in fact are only elementary in name – given the extraordinary complexity of the physical laws that govern them. By way of example, the nucleus of an atom, measuring a few fermi (1 fermi equals 10^{-13} centimetres) is made up of nucleons (protons and neutrons) which are themselves composed of particles called quarks. Within the nucleus, energy in the form of pi-mesons is exchanged between the nucleons. The quarks themselves interact with many of the very elementary particles – such as Higgs' gluons and bosons – in a way that cannot be discussed here. The substance in matter that is observable is in the form of nucleons. Bodies that were only recently identified, neutron stars, are composed of a kind of 'purée' of neutrons, with a density approaching that of atomic nuclei: the mass of a neutron star with a diameter of a few kilometres is comparable to that of the Sun.

Atoms for their part, with an effective mass equal to that of the nucleus (the mass of the electrons is negligible) have dimensions much greater than the nucleus because they enclose the orbits of the electrons: the dimensions of an atom are of the order 10^{-8} centimetres. White dwarf stars are related to atoms in the same way as neutron stars are to neutrons, so that their density equals the mean density of atoms and a star of this type, with the dimensions of the Earth has the same mass as the Sun.

Atoms combine to form larger structures, from diatomic molecules (composed of two atoms) up to very complex biological molecules like the double helix in genes.

Observed matter is in either a solid or fluid state. The solid state is equally well exemplified by microscopic grains of interstellar dust, with dimensions less than a micrometre, or by planets. But by far the larger portion of the matter in the Universe is fluid in form: interstellar gas – the density of which is extremely small, less than one molecule per cubic centimetre – fills such enormous regions of space that, by itself, it represents a few per cent of the mass of our Galaxy. The stars, with masses varying from a fraction of the solar mass to several hundred solar masses, are also fluid.

It is above all in the stars that observable mass is concentrated. The simplest systems are the binary (double) stars which account for more than half the stars in our Galaxy. Indeed stars are seldom born singly: they are found in associations of tens or hundreds of stars, or in galactic formations of thousands of stars gravitating in the disk of our Galaxy or again, in globular clusters of millions of stars, also gravitating around our Galaxy.

Galaxies are composed of several billion (10^9) stars, and, as with stars, they do not exist singly. Thus our own Galaxy has two satellites, the Magellanic Clouds; two wonderful galaxies, of irregular shape that are bright objects in the southern sky. Surrounding us there are about fifteen other galaxies, known as the Local Group. Galaxies are grouped in clusters of galaxies but it has not yet been established whether they combine to form clusters of clusters. However, it seems that space is not uniformly populated with galaxies and there are regions where their density is particularly small. These regions, called 'voids' (like that in the constellation of Boötes), contrast with others which are particularly dense, called 'filaments'.

Matter is therefore organised on all scales from very small to very large, in structures

Four examples of the hierarchical organisation of the Universe. This series of four astronomical photographs illustrates four stages in the hierarchical structuring of the Universe: a planet, a star, a galaxy and a cluster of galaxies. The planet is none other than our Earth as photographed by Apollo 8 from a distance of 100 000 kilometres; the Pacific Ocean is partly visible underneath the cloud cover. The Earth is a body of mass 6×10^{24} kilogrammes, with a radius of approximately 6000 kilometres. It gravitates about the Sun which, in about five billion years will evolve into a planetary nebula similar to IC 2220 (second photograph). This planetary nebula exemplifies the final stages in the evolution of the star in the centre of it, HD 65750, which is losing its shell. As planetary nebulae of this type should have a mass comparable to that of the Sun, that is, 2×10^{30} kilogrammes (including the mass of the central star), and a radius of about 10^{10} kilometres (depending on the state of its evolution). There are two hundred billion stars in our Galaxy which slightly resembles the galaxy NGC 6744 (third photograph). This particularly bright galaxy is in the direction of the Pavo Constellation. A galaxy of this kind has a mass of between 10^{41} and 10^{42} kilogrammes and dimensions of the order of 10^{18} kilometres (30 000 parsecs or 100 000 light years). Galaxies also tend to group together to form clusters of galaxies. A part of one of the most observed clusters, the Coma Cluster, is shown in the bottom photograph. A typical cluster contains about a hundred members and has a mass in excess of 10^{43} or 10^{44} kilogrammes (excluding the invisible mass they contain) and a diameter of the order 1.5×10^{20} kilometres (5 megaparsecs or 15 million light years). (NASA; Anglo-Australian Telescope Board, © 1980; Royal Observatory Edinburgh; Lick Observatory)

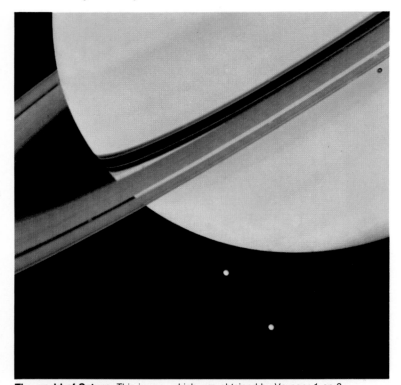

The world of Saturn. This image, which was obtained by Voyager 1 on 3 November 1980 when the probe was thirteen million kilometres from Saturn, illustrates the effects of gravitation, first on the satellites Tethys and Dione which move around the planet in elliptical orbits that conform to Kepler's laws, and then on the rings which represent the equilibrium position taken up by matter attracted at a short distance by a massive body. (NASA)

A solar eruption. This very spectacular event extending outwards over a distance comparable to the radius of the Sun was observed on 21 August 1973. The ultraviolet photograph captured this eruption which is guided by the local magnetic field in the Sun's surface. The image illustrates the effects of electromagnetic interaction in the Universe and two examples of this are evident here – the emission of photons from the Sun and the action of the magnetic field on the ejected ionised gas (or plasma). The photograph was taken by one of the two solar telescopes in Skylab, a space laboratory in orbit around the Earth. (NASA)

that vary in size from the nucleus of an atom to a cluster of galaxies. It is the task of astronomy to explain this organisation.

The laws of the Universe

Implicit in the enunciation and acceptance of the laws of the Universe is a postulate which up to now has been accepted by all astronomers: every part of the Universe is similar to our own environment. The postulate may be formally expressed by the statement that the laws of physics and chemistry that are established in the laboratory hold good throughout the whole of the universe.

It is not proposed to discuss those laws in detail here. Fortunately the physical behaviour of observable matter may be described by the four so-called fundamental interactions: gravitational, electromagnetic, nuclear and beta or weak interaction.

Gravitational interaction, which governs, for example, the motion of the planets around the Sun, is described by Newton's laws in 'normal' circumstances and by general relativity in situations where the masses and the speeds involved become much greater. Newton's laws state that the intensity of the force exercised by one mass on another is proportional to the product of the two masses and inversely proportional to the square of the distance separating them. The theory of relativity shows how the masses themselves influence the geometry of the motion of every particle, including photons (that is to say light): this theory incorporates the results of special relativity which states that nothing can travel faster than light, and that mass and energy are directly related. Gravitational force acts over distances than can be enormous – it governs interaction between galaxies – but its intensity is very feeble in comparison with the other three kinds of force.

Electromagnetic interaction controls the whole spectrum of radiation from radio waves to gamma-rays. Chemical interaction, which leads to the synthesis of molecules or ions by interchanging electrons of atoms, is governed by electromagnetic force. Like gravitational force, electromagnetic force is proportional to the product of the charges and inversely proportional to the square of the distance between them. The force is described as long-range but its intensity is much greater than that of gravitational force, being 10^{36} times more powerful.

Gravitational and electromagnetic forces directly affect the macrocosm (galaxies, stars, planets, etc.). The other two interactions, nuclear and weak, are primarily concerned with the microcosm or with particle physics. There are two kinds of 'ordinary' elementary particles: hadrons and leptons. The hadrons comprise the nucleons (protons and neutrons) and pi-mesons while leptons include electrons, mu-mesons and neutrinos. Nuclear interaction is also known as strong interaction; it is in fact the most powerful of the physical interactions, having about a hundred times the intensity of electromagnetic force. It is exercised only by hadrons, over a range measured in fermi. The nuclear reactions produced in particle accelerators or in the centre of stars are examples of this kind of interaction. Weak interaction, which is 10^{16} times less intense than strong interaction, acts over a range of only 10^{-16} centimetres, about a thousandth of that of strong interaction. It governs the interactions of leptons with one another and with hadrons. The disintegration of a neutron, resulting in a proton, an electron and a neutrino, is a typical example of weak interaction. These two short-range interactions which belong to the domain of microphysics are nevertheless prominent in the macrocosm; to take but one example the radiation energy emitted by stars is generated by the nuclear reactions taking place deep inside them.

This concise review of the fundamental laws of physics should include some reference to two important features of these interactions. Firstly, in at least two kinds of interaction – nuclear and electromagnetic – the action of the forces is accompanied by the exchange of particles (emission or absorption): pi-mesons in nuclear interactions, photons in electromagnetic interactions, intermediate bosons W^+, W^- and Z^0 for the weak interaction. All of these particles have similar statistical properties, described by

The Crab Nebula (NGC 1952 or M1). These two photographs were taken at different wavelengths; that on the left in blue light (in the wavelength range 300 to 500 nanometres) and the one on the right in red light (in the range 630 to 675 nanometres). The matter radiating in red is cooler than that radiating in blue, so it is not surprising that it is more extensive. The Crab nebula is the remnant of the best known supernova; it originated in the explosion of a star in the year AD 1054, an explosion that was triggered by a sudden enhancement of the nuclear reactions that occur in the centre of a star. (Hale Observatories)

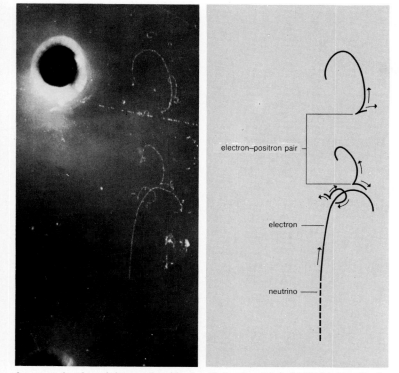

An example of weak interaction (the fourth fundamental physical interaction). Here a neutron has just collided with an electron, without change of identity, while the electron, accelerated by the impulse, radiates photons, some of which become electron–positron pairs. Such pairs are deviated in opposite directions, in curved trajectories, by the magnetic field. An event of this type is perfectly analogous to the disintegration of a neutron into a proton. This photograph was obtained with the CERN Gargamelle bubble chamber. The experiment was carried out by research teams from Aix-la-Chapelle, Brussels, CERN, Milan, Orsay and University College London.

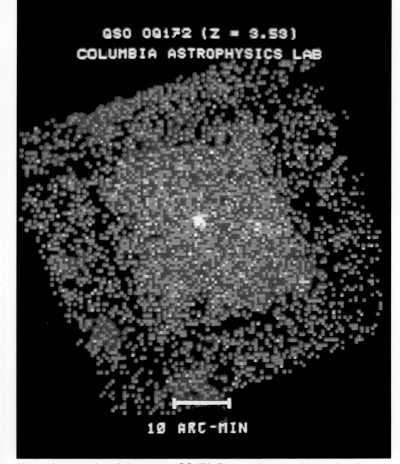

X-ray photographs of the quasar OQ 172. The popular name for quasi-stellar objects is quasars. Optically they look like stars but some have the properties of radio galaxies; they are in fact very active galaxies, apparently at a very early stage in their evolution. The quasar OQ 172 emits X-rays (indicating a radiation temperature higher than 100 000 K) and its redshift is one of the highest known – it corresponds to a velocity of recession from us equal to 0.91 times the speed of light – or alternatively 270 000 kilometres per second. Its distance from us must exceed 10^{23} kilometres, so the radiation we are now receiving from it must have been emitted ten billion years ago. This explains the very exceptional nature of this object which is situated close to the limits of the Universe and is almost as old as the Universe itself. It appears as a white spot near the centre of this X-ray photograph obtained by the Satellite HEAO–2. (NASA and Columbia Astrophysics Laboratory, courtesy of W.Hsin-Min Ku)

Einstein and Bose, hence the name bosons. In contrast the graviton, assumed to be exchanged in gravitational interaction, remains hypothetical, although emission of gravitational waves has been demonstrated indirectly through observations of a binary pulsar discovered in 1974. Secondly, physicists are on the point of unifying these four interactions, that is to say expressing them in the same general terms, at least in the domain of high energies. The unification of weak and electromagnetic interactions has been achieved following the research of S. Glashow, S. Weinberg and A. Salam, and scientists accept that the unification of these two forces with nuclear force is now practically completed. Only the unification of gravitational force with the other three still remains intractable. It can, however, be claimed that physicists now understand the essential nature of these interactions which govern not just the realm of the infinitely small but the whole of the Universe.

Recent discoveries in astrophysics

Over the past twenty years our concept of the world has been substantially modified and refined, for, as we have seen, several reasons: considerable progress in mathematics and physics, the development of new observing techniques, the ability to launch measuring instruments into space and the advent of computer technology. Every field of research has benefited from these advances.

In solar system studies the rock samples brought back by the Apollo missions have provided much information on the origin of the Earth and the Moon. The study of the samples induced geochemical laboratories to deploy great ingenuity in the development of analytical methods suitable for small quantities of materials. Most of the planets and many of their satellites have been explored by unmanned spacecraft. Equipment has been landed to make on-the-spot analyses of the soil on the surfaces of Venus and Mars. The Magellan probe has revealed the surface of Venus. The Voyager missions have revealed the wonderful variety in the satellites of Jupiter and Saturn; they also discovered Jupiter's ring and the complexity of the ring system of Saturn; and showed the rings of Uranus, first detected in 1977 from observations of stellar occultations. Such systems are therefore now believed to be a common feature of the giant planets.

In stellar astronomy great progress has been made in the study of explosive objects, novae and supernovae. Neutron stars, the existence of which had been postulated in the 1930s, were discovered in 1967 by radio astronomy. The extraordinary detection of the neutrino bursts in SN 1987a in the Large Magellanic Cloud confirms theoretical predictions about the collapse of the core of a star, and represents the birth of neutrino astronomy. The accretion of matter by certain stars provokes sudden bursts of X- and gamma-ray radiation; this discovery gives indirect evidence of objects that could well be black holes.

Our knowledge of the interstellar medium has dramatically increased over the past decade; indeed most of the interstellar molecules were discovered in the last ten years. Even more recent are the discoveries of the very hot components, revealed by observations made in the ultraviolet and X-ray domains, and of the relative composition of the gas compared with that of interstellar dust.

The first quasars were discovered in the 1960s – those objects that look like stars but radiate as much, or even more, energy than a galaxy and this with the largest known redshift, so they are perhaps the most distant sources that we can observe. In 1963 isotropic radiation at 2.7 K was detected; this is interpreted as a remnant of the big bang that marked the birth of the observable Universe.

A fossil relic of the primordial explosion which produced the observable Universe, the *2.7K cosmic background radiation*, was discovered in 1965. The COBE satellite studied it with unmatched precision at the beginning of the 1990s.

The conceptual importance of astronomy

The very great interest from which the space sciences are benefiting cannot be explained solely by the recent success of the space missions. The reasons for it are more profound and it is fitting to dwell on them at the beginning of this book. Astronomy is the science that contains all other sciences; all of them study only partial, fragmentary aspects of the real world. The ambition of astronomy – a rather bold ambition – is to describe the whole Universe and to try to explain its evolution. The historian confines himself to the study of events that have happened since writing was invented, the archaeologist seeks to unravel the history of mankind which extends over approximately a million years, the palaeontologist investigates fossils of which the oldest are about three billion years, while the astronomer describes events extending back to the origin of the observable Universe – which, according to the Big Bang theory, occurred about fifteen billion years ago. Looking to the future, the astronomer believes that the Sun will continue to evolve for the next five billion years and, guided by the particle physicist, he predicts that protons, which are the most stable form of matter, will endure for about 10^{30} years, on average. Astronomy assimilates our own history into that of the Universe and, by describing the evolution of its different components, it enables us to visualise our own situation within it: it makes us realise that we can take a detached scientific view of the Universe, even though we are deeply involved in it.

Jean AUDOUZE

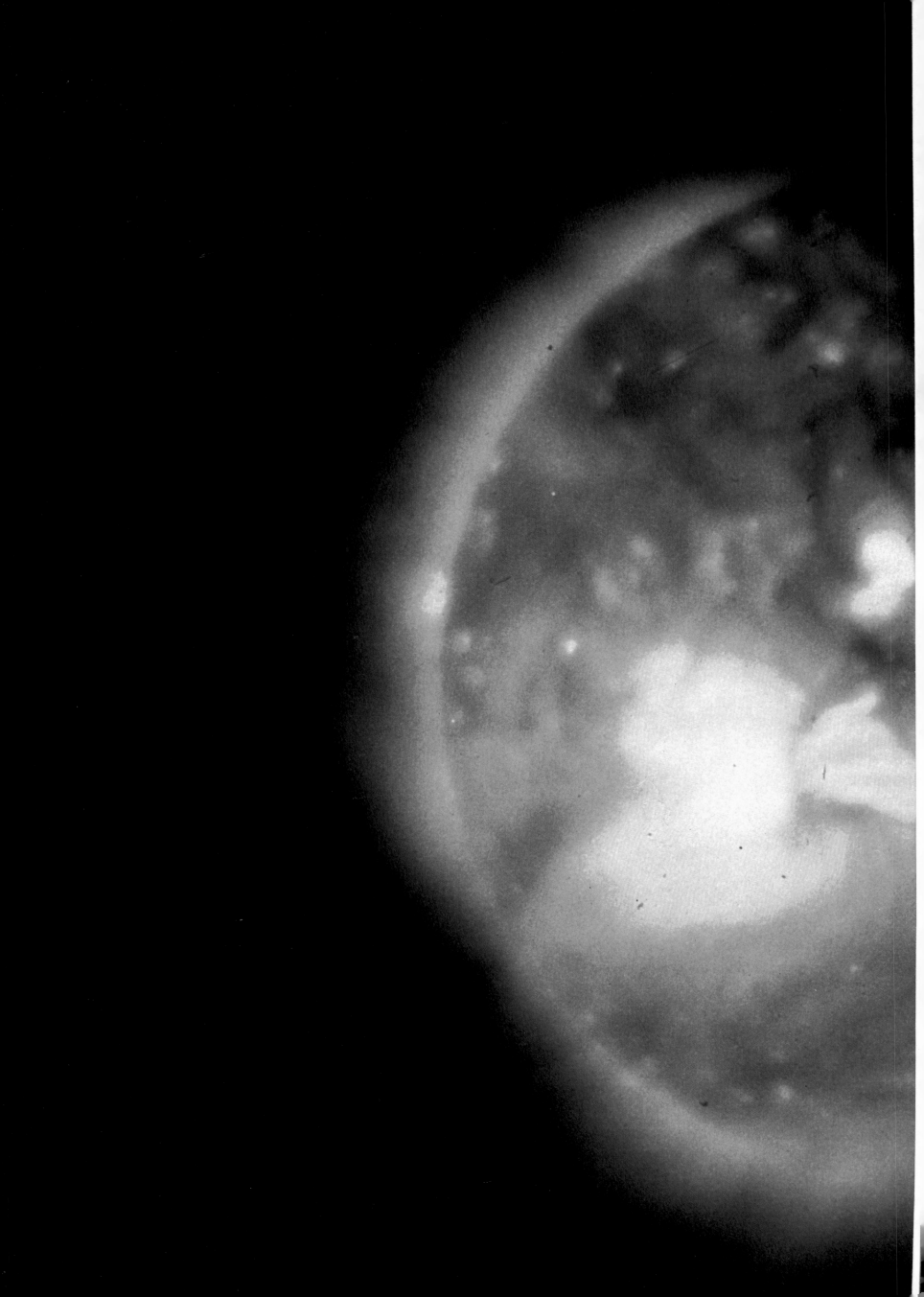

The Sun

The Sun is one of several hundred billion stars in our Galaxy. It is at an average position in a spiral arm, 8000 parsecs from the galactic centre (the radius of our Galaxy is 15 000 parsecs). The Sun participates in the rotational motion of the stars around the galactic centre, and it takes 200 million years to complete a full orbit. Inside a cube with sides of 100 parsecs centred on the Sun, there are about a hundred stars. Among these Proxima Centauri is the nearest at a distance of 1.3 parsecs. Seen from one of these stars, the Sun would appear as a luminous point. In the Hertzsprung–Russell diagram the Sun lies on the main sequence. It belongs to the category of dwarfs, in contrast to giants such as Betelgeuse and Antares, whose diameters are five hundred times larger. Other stars, much hotter or cooler, intrinsically brighter or fainter, populate the Universe. With respect to its location in the Galaxy, its size and its luminosity, the Sun appears to be a rather dull star.

Nevertheless, the Sun is unique. For us, living on Earth, it is the only star that is sufficiently near that we can study its surface in detail and, as privileged spectators, observe the grandiose phenomena happening there. The Sun's apparent diameter is 32 arc minutes, and details with an apparent size of 1 arc second are 273 kilometres across. Ground-based observations are limited to a maximum resolution of 0.3 arc second by the turbulence of the Earth's atmosphere, and those made from satellites to 1 arc second, both by stability and guidance problems and by the small space available for the instruments. Despite these limitations the observable details enable us to deduce the composition of the Sun and the characteristic features of its structure. Likewise, theories of solar evolution, and by analogy the evolution of innumerable stars, can be tested.

The Sun is a sphere of hot gas. In its centre the temperature and density reach values that are sufficient to maintain nuclear reactions. The fuel from which the Sun extracts its energy is hydrogen, whose nuclei are transformed into helium nuclei by the process of fusion. Therefore, the number of hydrogen nuclei decreases towards the solar centre. Away from the centre there is a very rapid increase in the proportion of hydrogen nuclei; beyond about a quarter of the solar radius the Sun is a homogeneous mixture of hydrogen, helium and traces of heavier elements. The latter originate from nuclear reactions that occurred either in the very first moments of the Universe, or in the interior of stars older than the Sun.

Because the Sun is gaseous, there are no abrupt discontinuities like those separating air, water and land on Earth. Even in the solar centre, where the density is ten times greater than that of any metal, the temperature of fifteen million degrees keeps matter in the gaseous state. It might therefore seem incorrect to speak of a solar surface. However, observations of the Sun in white light (all visible wavelengths together) show that the solar disk is limited by a very distinct edge, the limb, which corresponds to a true discontinuity in solar brightness. Practically all of the Sun's light is emitted from a very thin layer called the photosphere (from the Greek *photos*, light). This radiation is appreciable in visible wavelengths. Nothing can penetrate the photosphere; it is solely from this thin layer that we have been able to learn about the Sun's interior. Immediately above the photosphere lies the chromosphere (from the Greek *chroma*, colour), a layer some 2000 kilometres thick. The outermost layer is the corona, a halo of white light visible to the naked eye during total eclipses, out to about three million kilometres, but which, in the form of the solar wind, extends into the interplanetary medium beyond the Earth's orbit. The three observationally distinct layers, photosphere, chromosphere and corona, constitute the 'atmosphere' of the Sun. Even though they interpenetrate, each has its own characteristics

An unusual portrait of the Sun: a picture in X-rays. (Pages 16 and 17.) This picture was taken on 1 June 1973 by one of the X-ray telescopes aboard the space station Skylab. It reveals the corona on the whole surface of the solar disk. The clarity of the photograph is comparable to that of pictures of the photospheric disk obtained on Earth. Two important coronal phenomena are clearly visible: bright points and coronal holes. These are fundamental structures of the Sun's corona in the same way as the entanglement of magnetic loops, which can also be distinguished on this picture. (American Science and Engineering and Harvard College Observatory)

age	4.5 billion years
radius (R⊙)	700 000 km
mass (M⊙)	2×10^{30} kg
mean density (ϱ⊙)	1.4 g/cm³
luminosity (L⊙)	3.9×10^{27} kW
effective temperature	5770 K
absolute visual magnitude	+4.83 mag

The Sun: a star of spectral type G2 V.

Solar structure. The atmosphere of the Sun consists of several layers, each of which has a different structure: the photosphere (1, seen in visible light), the chromosphere (2, seen in H-alpha) and the corona (3, seen in X-rays; 4, in visible light during an eclipse). The layers below the photosphere cannot be observed. The representation of the Sun's interior is therefore the result of work done by theorists, who construct models of the internal layers from observations of the outer layers. A formidable amount of energy is generated in the Sun's central core, whose temperature and density are fifteen million degrees and 160 grams per cubic centimetre respectively. Outwards from the core the temperature and density decrease very rapidly (figure at right). At the photosphere, the visible surface of the Sun, the temperature is no more than 6000 K and the density, 10^{-6} grams per cubic centimetre. The latter is the same as the density of the Earth's atmosphere at an altitude of 50 kilometres, and is, for practical purposes on Earth, considered to be a vacuum. (Note that 90 per cent of the solar mass is inside a sphere whose radius is half of the solar radius.) The temperature decreases further until it reaches 4300 K at the base of the chromosphere. Further out, it increases again, slowly at first in the outer chromosphere, and then abruptly, rising from 10 000 to 1 000 000 K in a thin transition zone of less than 1000 kilometres, which separates the chromosphere from the corona. At the same time, the density decreases sharply to ten billion times less than that of the photosphere. In the corona, the density decreases very slowly and the temperature remains practically constant. (The figure below is the result of a photomontage using pictures from the Observatoire de Meudon, the High Altitude Observatory, and American Science and Engineering)

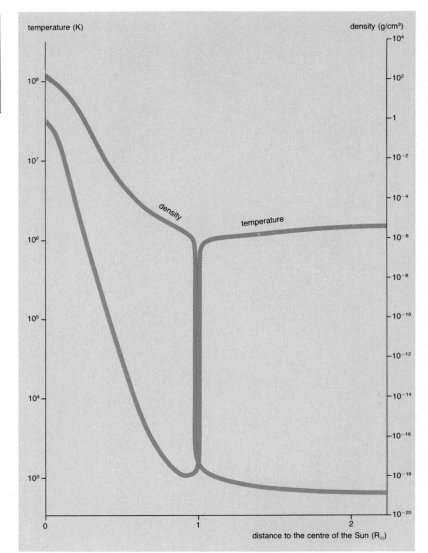

which reflect profound variations in the thermodynamics of the solar gas. The envelope outside the photosphere has a mass amounting to only a minute fraction of that of the whole Sun, and has a density comparable with a laboratory vacuum. The photosphere has a density similar to that of the Earth's upper atmosphere. The chromosphere and corona, in a progressive transition from the Sun to the interstellar medium, constitute a giant bubble of rarefied gas which the Sun blows into space without diminishing its mass appreciably.

The solar atmosphere is not homogeneous, but is highly structured; its radiation fluctuates from one point to another due to subtle variations in temperature and density. Neither is it static: it is constantly stirred by waves, currents and turbulent motions. It often undergoes perturbations such as eruptions, which are local and transitory, but violent nonetheless. The appearance of groups of dark spots on the solar surface is the most well-known manifestation of centres of activity which affect all the layers of the solar atmosphere. Because these are only local anomalies of limited duration, it is possible to define a 'static' solar atmosphere and study how it is modified, often in spectacular ways, by phenomena associated with solar activity. The number and importance of active centres shows a remarkable cyclic variation with a period of about eleven years.

The Sun emits, either continuously or sporadically, the entire spectrum of electromagnetic radiation, from X-rays, through the ultraviolet, visible and infrared, to radio waves. Radiation of different wavelengths comes from layers situated at various depths in the solar atmosphere. The path of a photon traversing the solar gas is determined partly by the absorption properties of the gas, which depend on the photon's wavelength, and partly by the density of the gas, which varies with altitude. If we observe the Sun at a wavelength at which the solar gas is very

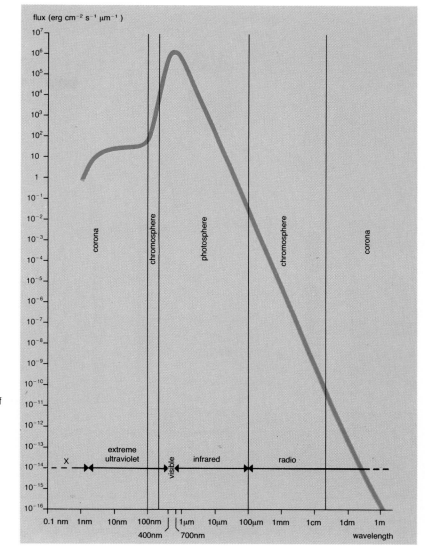

The solar spectrum. The Sun emits in all regions of the electromagnetic spectrum: 41 per cent of the energy is emitted in the visible, 52 per cent in the infrared and 7 per cent in the near ultraviolet; the energy emitted in the X-ray and ultraviolet ranges constitutes only 0.001 per cent of the total energy emitted; the energy emitted at radio wavelengths is of the order 10^{-10} per cent, and is completely negligible. Each wavelength range can be identified with a layer of solar atmosphere primarily responsible for the energy emitted. Examination of the spectrum therefore permits a study of the altitude dependence of physical conditions present in the solar atmosphere.

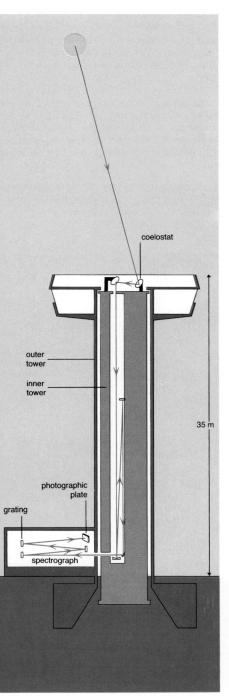

The Solar Tower of Meudon Observatory. The fixed telescope is placed in the body of a tower which allows large focal lengths while overcoming thermal perturbations due to the heating of the Sun. Furthermore, the inner tower, which supports the mirrors, is isolated from effects of the wind and vibrations of the outer tower, which has independent foundations and structure. The light rays, captured 35 metres above the ground, are relayed by a coelostat to a concave mirror at the base of the tower. The beam is then reflected by two plane mirrors before forming an image of the Sun about 42 centimetres in diameter at a spectrograph. The long focal length (14 metres) of the mirrors of the spectrograph enable very high dispersion spectra (of the order 100 millimetres per nanometre) to be produced. The coelostat (above left photograph) consists of two plane mirrors, one with a uniform rotational motion, which allows the Sun to be tracked throughout the day. (Meudon Observatory)

A solar observatory on stilts: Big Bear Solar Observatory. The large mass of water of Lake Big Bear in California stabilises the air above it, thus improving the quality of the images. The observatory houses three telescopes on the same mounting, of apertures 66, 25 and 22 centimetres. (Big Bear Solar Observatory, California Institute of Technology)

Second World War that short wavelength and radio wavelength observations began. The study of ultraviolet and X-ray radiation from the Sun, inaccessible from the ground because of atmospheric absorption, began in 1946 in the United States, using the V2 rockets recovered from Germany. Radio waves were not discovered until 1942 when British Army radars registered mysterious spurious signals whenever the Sun crossed their lobes. Since the 1940s, studies at ultraviolet and X-ray wavelengths have progressed greatly, both in the United States and in Europe, and the Sun is now under almost continuous radio surveillance at frequencies of 200, 500, 3000 and 10 000 megahertz, as well as being observed at visible wavelengths.

Solar observations, like those of all celestial objects, call for very different techniques depending on the region of the spectrum under consideration; they also require the use of special instruments such as solar towers, radioheliographs and coronagraphs. The large instruments used for observing the Sun from the ground, must satisfy two precise requirements. In the first place, high spatial resolution must be achieved. We know today that a large number of phenomena important in solar physics involve dimensions much smaller than an arc second. In order to observe them, one must minimise the effects of atmospheric turbulence on the quality of the images. A large part of this turbulence, arising from convection in the air which is heated by the ground, can be overcome by collecting light rays at the top of a tower some tens of metres high. Naturally, the choice of site is also very important. High mountains are ideal while, at certain times of year, instruments located on plains can produce equally good images. The second requirement of a large solar instrument is a high dispersion spectrograph. Even modern large diffraction gratings can achieve dispersions of 10 centimetres per nanometre only with a spectrograph about 15 metres long. Such a spectrograph is generally installed either vertically in the tower or horizontally in an adjoining building.

While the advent of modern electronics has allowed high spectral resolution radio observations, the same is not true of spatial resolution, which is strongly degraded by the effects of diffraction, exacerbated at these longer wavelengths. One can improve the angular resolution of solar radio telescopes by using interferometric techniques. At the moment we can achieve an angular resolution of 2 arc seconds at centimetre wavelengths with the VLA

opaque, we see the surface layers; if we observe it at a wavelength at which it is transparent we see deeper ones. Varying the wavelength of observation is equivalent to 'sounding' the solar atmosphere.

In addition to radiation, the Sun emits particles such as protons, electrons and helium nuclei, which it accelerates to velocities of a few hundreds of kilometres per second in the solar wind, and to velocities of a few tens of thousands of kilometres per second in solar cosmic rays.

It is by appropriate examination of these messengers of light and matter coming from the Sun, that we can discover the properties of the regions of the Sun from which they were emitted. Since the seventeenth century the Sun has been the object of numerous observations aided by ground-based telescopes; by the beginning of the twentieth century it had already yielded many of its secrets. Although visual observations of the Sun proceeded with more and more sophisticated spectrographic techniques, it was only after the

Radioheliographs. The radioheliograph of Nançay (France) consists of two large parabolic antennae (right) 10 metres in diameter relayed to a network of smaller antennae: sixteen are aligned in the east–west direction on a 3.2-kilometre base, and twelve in the north–south direction. Since March 1980, the two networks have been operated separately, and in the course of a day allow the construction by aperture synthesis of radio maps of the Sun at metre wavelengths (1.78 m, 169 MHz). The angular resolution is 1 arc minute in the east–west direction and 4 arc minutes in the north–south direction. (Paris Observatory)

There are only a few large telescopes of this type in the world: one in the United States operating at decametre wavelengths and another at Culgoora, Australia. This latter (left) contains ninety-six antennae 14 metres in diameter, regularly spaced in a circle 3 kilometres in diameter, and can make metre and decametre wavelength observations: 1.88 m, 160 MHz; 3.75 m, 80 MHz; 6.89 m, 43 MHz. (CSIRO, Division of Radiophysics)

The sky during an eclipse.
This wide-angle photograph shows an extended region of the sky during the total solar eclipse of 31 July 1981. The horizon is illuminated by the light from the chromosphere which passes beyond the lunar disk, and is reflected by layers of the Earth's atmosphere outside the Moon's shadow. Note the extensions of the outer solar corona produced by scattering off interplanetary dust grains. (Mission de l'Institut d'Astrophysics du CNRS au Kazakhstan. Photograph S. Koutchmy *et al.*)

(Very Large Array) in the United States, and of one arc minute with the large radioheliograph at Nançay, France, operating at a wavelength of 1.77 metres.

Observations of the corona pose a particular problem. Near the limb its brightness is a millionth that of the photosphere, that is, roughly that of the Full Moon. We are faced, therefore, with the problem of observing an object which is bright but difficult to see except during eclipses, since it is completely masked by a dazzling halo

10 000 times brighter, due to the photospheric light scattered by dust and molecules in the air. It is only during total eclipses, when the Moon – whose apparent diameter is, by a lucky coincidence, almost the same as the Sun's – completely occults the solar disk, that the feeble glow of the corona appears against a black sky out to about one degree from the Sun's edge. It is very inconvenient to observe a solar eclipse. Firstly, there are only one or two eclipses per year, and these rarely last more than about 5 minutes; in

addition, such observations require the organising of full-scale expeditions, and the transportation of several tonnes of equipment to the site. Site access is often difficult and, furthermore, the success of these expeditions depends entirely on weather conditions.

It is only since the invention of the coronagraph by a French astronomer, Bernard Lyot, in 1930 that the corona can be observed regularly other than during eclipses. Lyot's coronagraph is a telescope in which an artificial eclipse is created

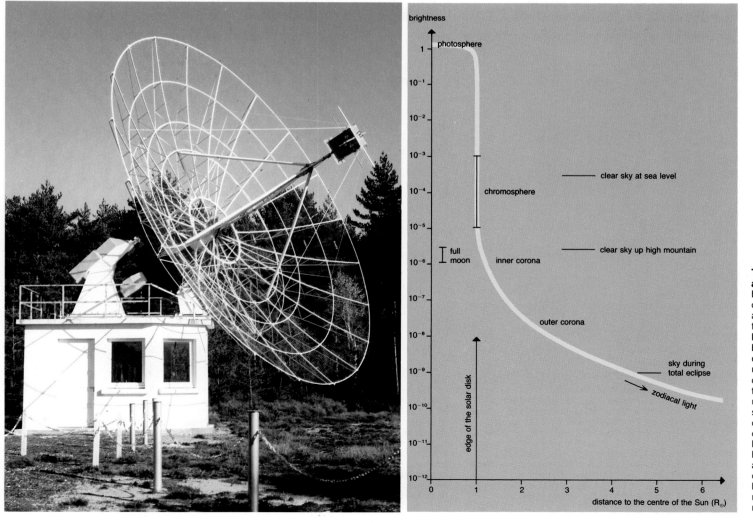

The brightness of the solar atmosphere. The comparison between the curve of brightness of the atmosphere of the Sun as a function of the distance to the centre of the star and the levels indicated for the sky brightness in various circumstances enables us to understand the difficulties of observing the upper layers of the solar atmosphere. Even under the clearest skies, the sky is too bright for us to observe the corona. It is only during total eclipses, when the light from the photosphere no longer makes the terrestrial atmosphere blue, that the corona can be seen. The sky brightness during an eclipse is no more than about one millionth of its normal value.

by occulting the image of the Sun with a disk of appropriate size placed at the prime focus of an objective lens. For optimal results the instrument must be situated where the sky is very clear: ten mountain observatories in the world are equipped with such instruments. Of these the oldest is that at Pic du Midi in the Pyrenees, and the biggest is that at Kislovodsk in the USSR. However, even with a perfect coronagraph under the clearest sky, stray white light still dominates that of the corona.

In certain narrow wavelength bands of the visible spectrum, the corona is ten to a hundred times brighter than at neighbouring wavelengths. Attaching a monochromator (spectrograph or

Lyot filter) to the coronagraph isolates these bands making it easier to separate the coronal light from the spurious light. Therefore, except during eclipses, the inner corona is most often observed in monochromatic light (green, yellow or red), out to two solar radii. Adding a polarimeter to the coronagraph separates the feeble but strongly polarised light of the corona from the intense but unpolarised direct atmospheric light, and reveals the brightest coronal structures to within one degree of the Sun, thus approaching naked-eye resolution during an eclipse.

Research in solar physics has greatly benefited from new powerful methods brought about by

the conquest of space. The launching in 1962 of the first orbiting solar observatory (Orbital Solar Observatory, OSO–1) marked the beginning of a series of eight solar observation satellites. As a result of this seventeen-year programme the Sun was, for the first time, observed at short wavelengths almost continuously for a period of 1.5 cycles of solar activity. This series of satellites was conceived with the idea, then novel, of an orbiting astronomical observatory containing several instruments simultaneously observing the Sun at different wavelengths. Priority was naturally given to instruments operating at wavelengths inaccessible from the ground, although a coronagraph aboard OSO–7 revealed

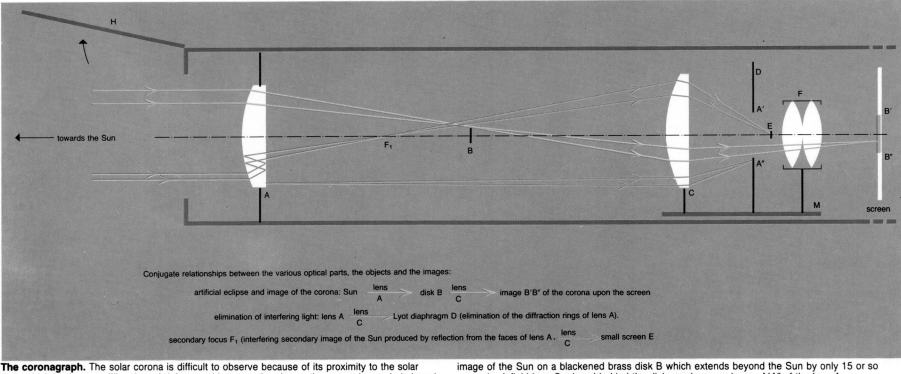

Conjugate relationships between the various optical parts, the objects and the images:

artificial eclipse and image of the corona: Sun $\xrightarrow{\text{lens} \atop A}$ disk B $\xrightarrow{\text{lens} \atop C}$ image B′B″ of the corona upon the screen

elimination of interfering light: lens A $\xrightarrow{\text{lens} \atop C}$ Lyot diaphragm D (elimination of the diffraction rings of lens A).

secondary focus F₁ (interfering secondary image of the Sun produced by reflection from the faces of lens A. $\xrightarrow{\text{lens} \atop C}$ small screen E

The coronagraph. The solar corona is difficult to observe because of its proximity to the solar disk, which is about a million times brighter, and because the observations must be made in broad daylight under very clear skies. It is necessary to block out the photosphere by setting up an artificial eclipse, and to eliminate as much as possible of the spurious light coming from the sky and the optical system.

Around 1930 Bernard Lyot perfected the artificial eclipse technique by carefully eliminating spurious light scattered by the optical system. Here is the description of his coronagraph which he gives in the *Bulletin de la Société astronomique de France*, in 1932: 'I tried to design a coronagraph which would eliminate all scattered light. The heart of the apparatus is a plane convex lens 13 cm in diameter with a focal length of 3.15 m. It was ground at the Optical Institute in specially selected Parra-Mantois optical glass. It was polished with the utmost care, and contains no grains, bubbles, or surface scratches in its central half. This lens is placed at A and forms the

image of the Sun on a blackened brass disk B which extends beyond the Sun by only 15 or so seconds. A field lens C, placed behind the disk, produces an image A′A″ of the lens A, on a diaphragm D whose centre is occupied by a small screen E. The edges of the diaphragm stop the light diffracted by the edges of the first lens. The small screen stops the light of the solar image produced by reflection on the face of this lens. Behind the diaphragm and the screen, shielded from the scattered light, an objective F, very strongly corrected, forms an achromatic image of the corona at B′B″. These optical parts are fixed to a board M which can slide in order to bring the solar image onto the screen. All the optics are contained within a wooden tube G, 5 metres long, whose interior walls are coated with thick oil; the tube is closed by a lid H which is opened only during observations. The lens and its mounting completely close up the tube in order to prevent air currents. The lenses are thus conveniently protected from dust.' The photograph shows the coronagraph installed at the Pic du Midi observatory. (Observatoire du Pic du Midi)

for the first time a white-light corona set against a black sky background. The orbiting coronagraph was in an ideal environment, free from the scattered light in the Earth's atmosphere, and furnished images of the outer corona out to ten solar radii, much further even than during eclipses. For this an outer-occultation coronagraph was used, in which the Sun was occulted by a disk placed a certain distance in front of the objective lens. Furthermore, these coronagraphs furnished observations of the corona continuously over several months, an enormous advantage over eclipse observations, which last only a few minutes.

The largest of these orbiting solar observatories, the manned station Skylab, carried a battery of eight large telescopes, including one coronagraph. From May 1973 to February 1974, in the course of three missions, the astronauts brought back from this orbiting laboratory thousands of photographs unveiling the wonders of the solar atmosphere. The most recent satellite, SMM (Solar Maximum Mission) was put in orbit in 1980 in order to examine the Sun at the maximum of its activity cycle. It was repaired by astronauts on the space shuttle in 1984.

Other satellites and interplanetary probes like Explorer, Imp, Mariner, Vela, Pioneer, Helios A (1974) and B (1976) and ISEE (1980) carried instruments for observing the solar wind, which was also studied by the Apollo missions.

Thus, in the course of the last twenty years the advent of space research, together with the use of improved telescopes has permitted increasingly precise observations of all the layers of the solar atmosphere, and, consequently, has allowed more and more sophisticated models to be tested. These observations have uncovered phenomena which were previously unsuspected; they have revealed the true and complicated nature of the Sun.

SMM (Solar Maximum Mission) was launched in 1980 with the aim of studying the Sun at the maximum of its activity.

The European Ulysses probe was launched on 6 October 1990 with the particular aim of studying the Sun's poles.

Other satellites and interplanetary probes, such as Explorer, Imp, Mariner, Vela, Pioneer, Helios A (1974) and B (1976), ISEE (1980), carried instruments for observing the solar wind, which was also studied during the Apollo missions.

Thus over the last three decades the advent of space research, together with the use of ever-improving telescopes, has allowed still more precise observations of all the layers of the solar atmosphere and hence the testing of models of growing sophistication. These observations have revealed previously unsuspected phenomena; they have shown the Sun's true nature to be extremely complex. Understanding the life cycles of the millions of stars in our Universe requires the solution of the strange riddles still posed by the Sun.

Monique ARDUINI-MALINOVSKY

Skylab, laboratory in the sky. The part of the orbiting station surrounded by four solar panels was allocated to solar observations. The electrical power required for the operation of the array of telescopes was supplied by these four panels, independently of the large rectangular panel which supplied the rest of the station. The casing containing the telescopes was 3 metres long and 2 metres in diameter: the Skylab telescopes were not scale models but full-size observatory telescopes used for the first time in space. The solar telescopes comprised two X-ray telescopes, one far ultraviolet spectroheliograph, a visible light coronagraph, and two H-α telescopes. The giant single-storey cylinder was the living quarters for three astronaut crews who arrived successively during the nine-month mission (May 1973–February 1974). The station was 36 metres long and weighed 91 tonnes (NASA).

The repair of SMM. The SMM satellite malfunctioned after ten months; it was repaired in orbit on 11 April 1984 on board the space shuttle *Challenger* by the astronauts George Nelson and James Van Hoften. The satellite was retrieved and placed in the shuttle's cargo bay by a mechanical arm. After repair, SMM was replaced in orbit. It reentered the Earth's atmosphere in 1989. (NASA)

Internal structure and evolution

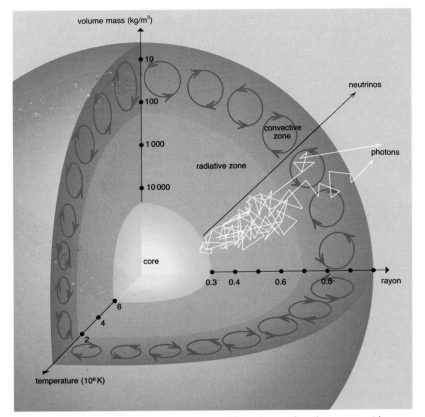

The Sun is at present the only star whose mass, size, luminosity, surface composition and age are accurately known. Further, the Sun's neutrino flux and modes of vibration are better measured than for any other star. These facts make it a special object for astrophysicists: the Sun allows us to check theories of the internal structure of the stars. The daytime star is also in the simplest and longest-lived phase of stellar evolution: it is a main sequence star, and in theory at least, quietly converts hydrogen into helium. We must give a precise picture of it by probing its deep layers by means of a quantitative physical model. Our exploration will lead us to the sources of stellar energy in the centre of the star, where the temperature is high enough that many protons reach velocities high enough to overcome their mutual electrical repulsion via the tunnelling effect, thus making the Sun a huge self-controlled fusion reactor. Some fusion reactions result in the transformation of protons into neutrons and liberate neutrinos. Observation of these particles would provide the first experimental confirmation of the nuclear origin of solar energy, and tell us the central temperature of the solar nuclear furnace with remarkable precision, of order 10 per cent.

The Sun is not only a reactor, but also a resonant cavity, just like our planet. It has been possible to measure several thousand oscillation modes whose properties are particularly sensitive to the internal structure. Their study is called helioseismology. Using a current solar model, it is possible to calculate the oscillation frequencies very accurately. The results amount to a very clear diagnostic of conditions inside the Sun. Thus helioseismology, which takes its techniques from terrestrial seismology, gives one of the best tests of the models, placing severe constraints on the internal structure, particularly the sound speed in the interior.

Neutrinos and helioseismology promise revelations in the near future about the interior of the Sun. For the moment the neutrinos probe the very centre and the oscillations the envelope, but an overlap is likely before long. What are the reactions producing solar neutrinos? How many neutrinos should we expect? What is their energy spectrum? And what confirmation can we expect of the observation of neutrinos? What are the prospects for solar seismology? These questions can only be answered after analysing the nuclear working of the Sun through a numerical model.

To simulate the solar 'machine', astrophysicists have no other means than to construct a theoretical Sun, in the hope that this model describes the Sun in space and time, at all depths, throughout its evolution. When a solar model is conceptually constructed almost all our knowledge is used: general physics, thermodynamics, nuclear physics, plasma and particle physics. Quantum mechanics is the integrating aspect of this enterprise; it is the key to understanding the interplay of particles and thermonuclear fusion which supplies the Sun's energy. The assumptions made in constructing the model are very plausible. We assume complete spherical symmetry and neglect the effects of rotation and magnetic fields; we assume hydrostatic equilibrium and energy balance (at each point of the star, thermal pressure can support it against gravity). The energy lost from the surface is replaced by new energy released in nuclear reactions in the centre. The energy escapes at a rate determined by the opacity of the solar gas. In the core, a photon cannot move more than a centimetre before interacting, and on average a

photon interacts 10^{20} times before reaching the surface, which takes several million years. The photons from central nuclear reactions (gamma-ray photons) lose energy in these interactions, essentially to the free electrons, but also to electrons bound to heavy nuclei as they approach the surface. Laboratory physics, a little extrapolated, supplies the model with the essential ingredients of *nuclear reaction rates* as functions of temperature, and *opacity coef-*

Internal structure. The standard solar model divides the star into three concentric zones: the *core*, the site of thermonuclear reactions where energy is produced by fusing hydrogen into helium, is a sphere of radius 0.3 solar radiuses; the *radiative region*, where heat is transported by photons, extends from 0.3 solar radiuses to 0.7 solar radiuses; above, in the *convection zone*, the energy is transported mainly by overturning motions of the matter. The temperature is about 15.5 million degrees at the centre, 8 million degrees at the core-radiative region interface, and 2 million degrees at the radiative-convective interface. For comparison, the surface temperature is 5780 kelvin.

About 98 per cent of the energy released in the core is removed by gamma-ray photons, the remainder by neutrinos. The Sun's visible radiation reaching us from the photosphere involves photons which were emitted as gamma rays in the centre of the Sun, and have taken several million years to reach the surface, gradually losing their energies in each scattering. In contrast, the neutrinos cross the entire Sun freely in about two seconds.

Observing the Sun from the bottom of a mine. Measuring the flux of solar neutrinos is very difficult. We can only hope to detect the most energetic neutrinos, that is those with the greatest probability of interacting with matter. The principal contribution comes from the reaction:

$$^8B \rightarrow {^8Be} + e^+ + \nu_e$$

of the cycle PP111 where ^8B designates the boron 8 nucleus (5 protons, 3 neutrons), ^8Be beryllium 8 (4 protons, 4 neutrons), e^+ the positron, and ν_e the neutrino. To detect these neutrinos, R. Davis Jr used a reaction in which the neutrino is captured by an atom of chlorine 37:

$$^{37}Cl + \nu \rightarrow {^{37}Ar} + e^-$$

where ^{37}Cl denotes chlorine 37 (17 protons, 20 neutrons), ^{37}Ar argon 37 (18 protons, 19 neutrons), and e^- the electron. Large quantities of chlorine are necessary because of the small interaction probability: 400 cubic metres of the chlorine compound ethyleneperchlorate (C_2Cl_4) are stored in a reservoir. There are other constraints: at ground level cosmic rays produce numerous reactions which transform ^{37}Cl into ^{37}Ar. To shield it from these, the detector-reservoir is housed at the bottom of a disused mine in South Dakota, 1500 metres below the ground.

Argon 37 is radioactive with a half-life of 35 days, so Davis waits about a month before recovering the few atoms of argon 37 which have formed in the reservoir. The number of these atoms is related to a flux of solar neutrinos which is then compared with the predicted theoretical flux. (Photograph furnished by R. Davis Jr, Brookhaven National Laboratory)

ficients for various depths, as the latter depend on temperature, density and chemical composition. In the outer third of the Sun's radius the temperature varies so sensitively with depth that matter becomes turbulent so as to remove the heat energy, leading to what we call convection. Below this region the most profound calm reigns: radiation filters through without perturbing the structure. This region is said to be in radiative equilibrium. The radiated energy is replaced in nuclear reactions; as a result the Sun shines in a constant and long-lived fashion.

The equations of internal structure expressing hydrostatic equilibrium, the conservation of mass and energy, and its mode of transport, form a system of coupled differential equations. These, complemented by the necessary physics (equation of state relating temperature, pressure and density, nuclear reaction rate, opacity of the matter to radiation of various wavelengths, initial abundances of the elements) are solved numerically.

Why does the Sun not explode? In its present stage, the Sun is a clever and supple star; it was not always so. It is now a self-regulating star. Let us imagine that the nuclear reaction rate begins to rise dramatically in its core; the temperature would rise, but as a direct result, the matter would expand because of the ideal gas law, which is very accurate in this region. This would cause a temperature decrease: nuclear reactions are very sensitive to this, and would resume their original moderate rate. Conversely, if the nuclear reaction rate declines, the Sun's core would contract very slightly, causing it to increase again. The Sun's good health is entirely a result of the ideal gas law, which holds right into the depths of the core (to be precise, there is a very small degree of quantum-mechanical degeneracy right at the centre because of the very high density, but this has no effect on the 'suppleness' of the core).

The Sun's structure is calculated for a star of 2×10^{30} kilograms, initially homogeneous and with given chemical composition, taken to be identical with that observed spectroscopically at the surface of the present Sun. In other words, we assume, as can be verified afterwards, that the Sun's surface is chemically virgin, and represents the composition at birth, as the nuclear reactions are so deep within the Sun that their 'ashes' have not contaminated the visible part at all. The two free parameters of the model, the initial helium abundance (unobservable at the Sun's surface) and the depth of the convection zone, are adjusted so as to obtain the right luminosity and radius at a given solar age. Thus, when the Sun became a star, we assume that it was 70.5 per cent hydrogen and 27.5 per cent helium, the sum of all the other elements, called metals, supplying only 2 per cent of the total mass. The helium abundance at the birth of the Sun, adjusted so as to get the right luminosity at 4.6 billion years (from 0.27 to 0.28 grams per gram of matter) is in good agreement with the value from the big bang (0.23 to 0.24), increased by production in stars.

For the present Sun, the model shows that the temperature and density vary with radius (15.5 million degrees and 150 000 kilograms per cubic metre at the centre as compared with 5800 degrees and less than 10^{-9} kilograms per cubic metre at the surface) and that the convection zone, although occupying the outer 26 per cent of the radius, constitutes only 1.7 per cent of the star's mass. At its base the temperature is 2 million degrees and the density 140 kilograms per cubic metre. Beyond the structural details, the model gives a wealth of information about the evolution in the form of a closely-spaced sequence of instantaneous snapshots. Viewing these successively gives an impression of continuous change at all depths, from the core to the surface, allowing one to follow the evolution of basic physical quantities (temperature, density, chemical and isotopic composition) at each point. At an age of 856 million years for example, the central temperature was 13.68 million degrees, instead of the current 15.5 million, and the luminosity were 0.76 and 0.96 of their present values. The amount of hydrogen, the primordial fuel, at the centre, drops continuously, being replaced by helium 4. We thus read an open book: the main virtue of the theoretical Sun is its transparency.

Thus the properties of an isolated star which rotates only slowly and has only a modest magnetic field can be precisely determined by calculation, given its initial mass and chemical composition and its age. To check this standard model we have to compare it with observation. Every serious model must account for the age, luminosity, radius and surface chemical composition, and also for the emitted neutrino flux and the vibration frequencies observed for the Sun.

Should we regard the neutrino flux as the first failure of the model? This question, current since 1970, has made rapid progress towards a solution. It is opportune to discuss its history.

The neutrino is an electrically neutral elementary particle with a mass which is extremely small, if not zero, which therefore moves at the speed of light or very near it, and which interacts very little with matter. There are three kinds, each associated with a lepton and sensitive to it. These leptons, electrons, muons and taus, interact only via the weak rather than the strong nuclear force. The Sun must inevitably produce some. The global reaction for transmutation of hydrogen into helium in the Sun's core can be written 4 protons → 1 helium nucleus + 2 positrons + 2 electron neutrinos + energy (25 megaelectron volts). It can be decomposed into three distinct chains.

Each time a proton is transformed into a neutron by the weak interaction, a neutrino is released in the centre of the Sun. It crosses it in 2 seconds, and if emitted in the right direction, arrives at the Earth about 8 minutes later. Because of this, neutrinos, which are incorruptible messengers, would allow astronomers to observe the Sun's core directly and examine the physical processes occurring there, unlike visible photons, which only give direct information about the complex agitated outer regions. Solar neutrino astronomy could thus, in principle, be used to check our ability to calculate the inner workings of the Sun, and by extension the stars. But this astronomy is very difficult and requires massive detectors. Moreover, these have to be placed underground to prevent confusion due the background noise caused by cosmic rays and their secondary particles. The first to be built was the chlorine detector at the bottom of the Homestake Mine in the US, and is known as the Davis detector after its constructor. The detection principle is the rare but calculable transmutation of chlorine 37 into radioactive argon 37 by electron neutrinos. This underground experiment, the size of an Olympic swimming pool, is essentially sensitive to high energy neutrinos from boron 8 (prediction 8 SNU) and beryllium 7 (1 SNU) (the SNU = 'solar neutrino unit' = 10^{-36} captures per second per target atom). Its energy threshold is 0.8 megaelectron volts. The Davis experiment has operated since 1970. Every three or four months a sample of fifteen argon atoms among more than 10^{30} chlorine 37 atoms is extracted. The low value of the detected flux (2.1 ± 0.3 SNU) compared with theoretical predictions (7.5 ± 2.3 SNU) is the 'solar neutrino problem', and has caused much ink to flow since the beginning of the 1970s. Should we blame the solar model itself or invoke some strange property of neutrinos which until now has escaped notice in experiments? The question is in the process of being decided by the advent of new detectors.

The Kamiokande detector in Japan uses a very different detection principle, Cherenkov radiation from electron-neutrino scattering, and is only sensitive to neutrinos with energies above 7 megaelectron volts, hence those from boron 8. It consists of 2140 tonnes of water surrounded by a battery of photomultiplier tubes. The directionality of this detector has shown for the first time that the neutrinos are indeed emitted from the Sun. This is, moreover, real-time detection of solar neutrinos, unlike radiochemical detectors, which require very long exposures. Analysis of the events indicates a flux about half the expected value. Thus the Kamiokande experiment provides verification of the Homestake result, while emphasising the disagreement between theory and experiment. Some researchers have thus begun to ask about the reality of the solar neutrino problem. It is therefore very important to evaluate the uncertainties associated with both experiment and calculation. For the latter the most important is connected with boron 8, as its production depends on the rate of the reaction proton + beryllium 7, which is relatively poorly known and very sensitive to the temperature. The uncertainty connected with the opactity tables used in the models is difficult to estimate. Combining all the uncertainties one finds about 40 per cent, but this value is itself uncertain. It may be that we will be unable to reach a final conclusion based on electron neutrinos from boron alone. The main hope for the future is in diversifying the modes of detection. The judgment of Solomon may be pronounced by another type of experiment, a radiochemical one like the Davis experiment, but using the transmutation of gallium 71 into germanium 71 under the impact of low-energy neutrinos produced in the basic reaction proton + proton → deuterium + positron + neutrino. The expected flux, about 120–130 (±20) SNU, is almost independent of the model: any deviation from this number would signal an unexpected property of the neutrino, and would have important consequences for particle physics. Two experiments of this type, SAGE and Gallex, have recently got under way.

SAGE stands for Soviet-American Gallium Experiment and Gallex (Gallium Experiment) is a European collaboration (Germany, France, Italy) at the Gran Sasso underground laboratory in Italy. The SAGE detector, consisting of 30 tonnes of metallic gallium at Baksan in Russia, started working in January 1990. After five

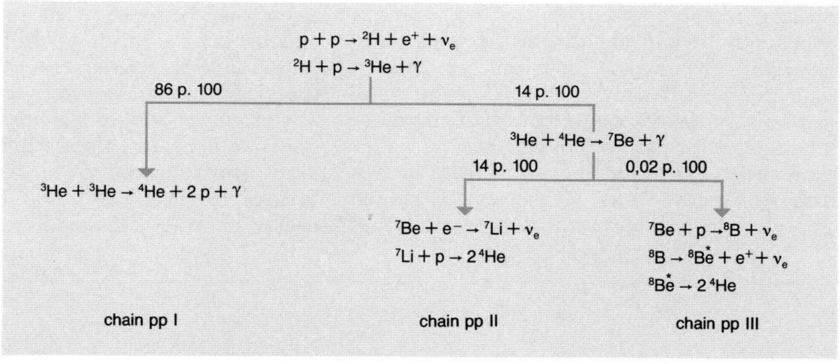

Reactions involved in converting four protons into helium. Of the three chains *pp* I, *pp* II, *pp* III, the first is the most frequent. Its energy spectrum extends from 0 to 0.420 megaelectron volts. Two protons *p* fuse to make a deuterium nucleus ${}^2H$ while emitting an electron neutrino ν_e. By capturing a proton, the deuterium nucleus transmutes into helium 3 (${}^3He$). When two helium 3 nuclei fuse, a nucleus of helium 4 (${}^4He$) is formed and two protons are released.

In 14 per cent of cases it may happen that a helium 3 reacts with a helium 4 to form an isotope of beryllium, beryllium 7. Then there are two possibilities. Usually the beryllium 7 captures an electron from the surrounding gas and transmutes into lithium 7 (${}^7Li$) by emitting a monoenergetic neutrino of 0.861 megaelectron volts in 90 per cent of cases, or 0.383 megaelectron volts in the rest. The *pp* II chain ends when lithium captures a proton to give two ${}^4He$s.

The *pp* III chain is much less fequent, but very important for observation, because it produces relatively high-energy neutrinos, measurable with Davis' chlorine detector. If instead of an electron, the beryllium 7 captures a proton, it forms boron 8, which immediately disintegrates into an excited beryllium 8 by emitting a neutrino whose energy spectrum extends from 0 to 14 megaelectron volts. The *pp* II chain ends with the fission of the excited beryllium 8 into two helium 4s.

The net result of the three chains is

$$4p \rightarrow {}^4He + 2e^+ + 2\nu_e + 25 \text{ megaelectron volts}$$

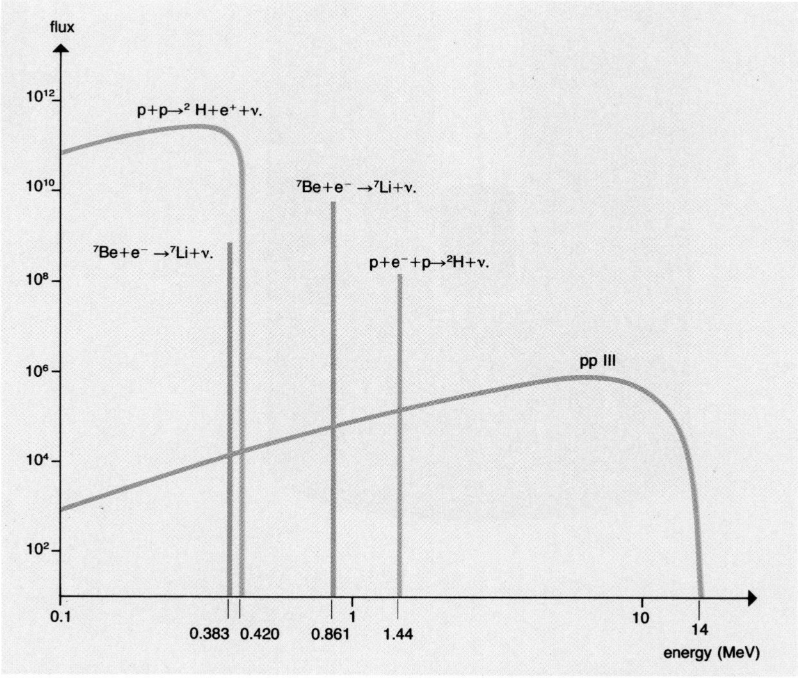

Energy spectrum of solar neutrinos according to the standard solar model. This spectrum corresponds to the reactions of the three chains *pp* I, *pp* II, and *pp* III. The flux is shown as the number of neutrinos per square centimetre per second in an energy interval of 1 megaelectron volt, at the Earth's orbit. Note that each square centimetre of our skins is crossed by 65 billion solar neutrinos each second.

One should, however, note that there is another way of starting the chains; this is the reaction $p + e^- + p \rightarrow {}^2H + \nu_e$, whose reaction rate is about 3 per cent of the *p* + *p* reaction; the emitted neutrinos are monoenergetic.

months, to the consternation of the scientific community, it had not detected a single solar neutrino. The upper limit on the flux is estimated very liberally at 79 SNU. If this result had been verified it would have had enormous implications for particle physics. But the Gallex experiment was meanwhile also taking its first data by means of its 30 tonnes of gallium chloride. The difficult part was to extract about

a few centimetres to a few metres per second, making them difficult to distinguish from the other motions. Fourier analysing the signals in space and time subsequently revealed thousands of vibration modes with frequencies closely spaced over the range from 1500 to 5000 microhertz, superimposed on background noise of various origins. Very long observations are needed in order to distinguish close frequencies.

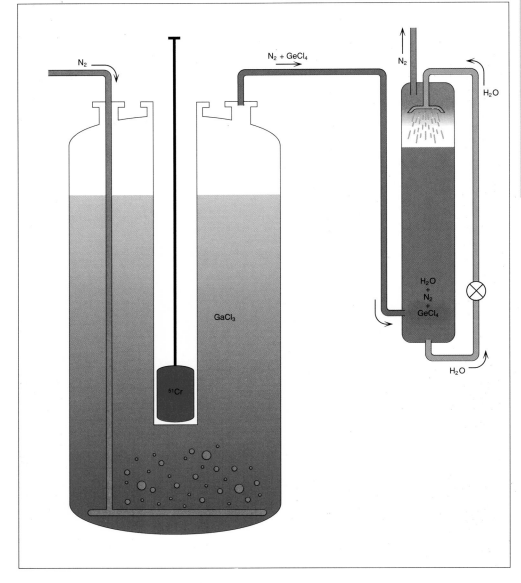

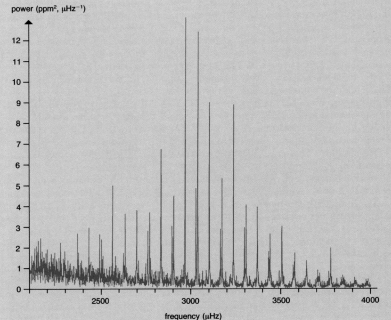

Fourier spectrum of the solar flux around 335 millimetres measured by the Iphir experiment. In 1988 the Iphir experiment on board the Soviet Phobos probe provided the first space observation without occultations of solar oscillations. Iphir measured the flux of the Sun's disc in the visible continuum for an unbroken period of 160 days, during the Earth–Mars part of the mission. The figure shows the Fourier spectrum of the integrated solar flux, giving the signal power in frequencies around 3 millihertz. We see the characteristic 5 minute p-mode (pressure) oscillations of orders $n = 16$ to $n = 26$. The discrete normal modes appear as a set of sharp peaks in the spectrum. We note that a pattern of single and double peaks repeats about every 136 microhertz for various values of n; the double peaks are the modes of degrees $l = 0,2$; the single peaks are $l = 1$ modes; the frequencies of the $l = 0$ and $l = 1$ modes are determined with a precision close to the theoretical frequency resolution of 0.072 microhertz given by the length of the observation. The resolution is poorer for the degree $l = 2$ as the peaks have smaller amplitude and a broader structure.

The low-degree modes (small l) penetrate more deeply into the Sun's interior. For the 5 minute oscillations, one mode reflects at 0.04 solar radiuses from the centre. The frequencies of these modes thus contain information about the matter structure near the Sun's core. The frequencies observed by Iphir are less than 3 microhertz from those of the most recent standard solar models, thus showing their validity. (After C. Fröhlich, T. Toutain, C.J. Schrijver, 1991)

The Gallex experiment. Electron neutrinos interact with gallium 41 to produce radioactive germanium 41, whose lifetime is 11.4 days. The reaction threshold is 0.223 megaelectron volts, much lower than the maximum energy of the neutrinos emitted by the proton–proton reaction. The expected count rate is 120–130 SNU, representing no more than one germanium atom per day.

The very rare interactions with the neutrinos transmute gallium chloride $GaCl_3$ into germanium tetrachloride $GeCl_4$, which is extremely volatile, in the presence of hydrochloric acid, which makes it easier to extract. Every three weeks several thousand cubic metres of nitrogen (N_2) are injected into the reservoir. Nitrogen entrains the germanium tetrachloride. A circulating column of water is placed at the exit to the vessel; there, the germanium chloride is captured and the nitrogen escapes. The chloride is then transformed into germanium hydride which is circulated in very sensitive Geiger counters. Hence one finds the number of interactions which have occurred in the 30 tonnes of gallium.

30 germanium atoms from 30 tonnes of gallium each month. The gamble seems to have paid off: the first results from Gallex give $80 \pm 20 \pm 6$ SNU (the first error being statistical and the second systematic). We can be reassured then: the Sun indeed emits floods of low-energy neutrinos, as expected. Moreover, the disagreement with theory is considerably reduced. At this stage of the experiment, however, the result does not have the accuracy required to implicate either the solar model or neutrinos physics. Calibration of the apparatus using an artifical neutrino source will be performed in the near future.

The Sun vibrates, and its light vibrates too, giving a rich and complex oscillation spectrum. Thus the term 'helioseismology' has entered the scientific vocabulary. Two different observational techniques are used to scan the agitated surface of the Sun for small-amplitude motions of varying degrees of regularity. The first uses the Doppler effect in spectral lines, the second measures fluctuations of light intensity. The latter approach has the advantage of simplicity, but suffers from higher background noise than the Doppler observations, largely because of the brightness fluctuations caused by the turbulent granulation of the Sun's surface. The phenomenon itself was discovered in spectroscopy. The elements of the solar atmosphere (calcium, iron, sodium, potassium) selectively absorb certain amounts of radiation at various precise wavelengths, as is shown by the dark lines seen in the solar spectrum; because of the Doppler effect these lines are displaced by an amount proportional to the velocity of the medium. In 1960 Robert B. Leighton discovered that the outer gas layers oscillated with a period close to 5 minutes. The measured displacements were small and consequently the velocities also,

The effects of day–night alternation are avoided by coordinated observations from sites at various longitudes, or from the North or South Poles, or, better still, from satellites or probes. These also have the advantage of removing the atmospheric background noise. Periodic brightness variations of the Sun have been very accurately observed by instruments on the American SMM satellite and the Soviet Phobos probe.

This plethora of waves observed at the surface bears witness to a deeper phenomenon. It reflects the global motions in the Sun's interior. In a sense these are vibrations which die out at the surface of the star, interpreted as the result of constructive interference of sound waves. The vibrations are maintained by pressure perturbations which gradually propagate as sound waves. These *acoustic waves*, called *p-type*, are excited by the granular motions of the convection zone; they are reflected at the surface because of the sharp density decrease, and refracted at depth because of the rapid increase of the sound speed in the Sun's interior. They are therefore confined within a spherically symmetrical zone which acts as a resonant cavity. The waves propagate in a series of bounces between the top and the bottom of the cavity. If the wavelength is a sub-multiple l of the Sun's circumference, constructive interference occurs and a pattern of standing waves appears, called an eigenmode. The depth of penetration depends essentially on the degree l. For each degree l there exists an infinity of vibration modes of decreasing periods, corresponding to the fundamental mode and its harmonics, just as for a stretched string. The *p-modes* of increasing degree l are confined within layers closer and closer to the surface of the Sun. The $l = 40$ mode, for example, refracts near the boundary

Pressure and gravity waves. The wave trajectories calculated from the standard solar model produce interesting patterns. The upper diagram shows two low-degree acoustic modes (very penetrating waves) and two high-degree modes (very superficial). The lower diagram shows the trajectory of a gravity wave.

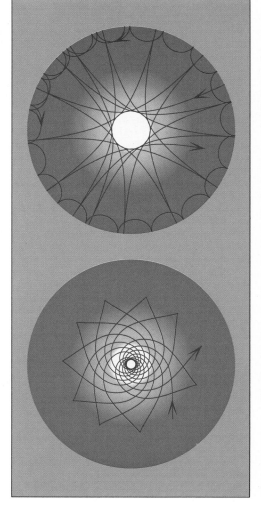

of the convection zone, i.e. at about 30 per cent of the radial distance from the surface. More generally, the frequencies, amplitudes and lifetimes of the various oscillation modes depend strongly on the density, temperature and chemical composition of the matter crossed by the wave; the values of these quantities are fixed by the solar model. The frequencies of thousands of observed vibrations have been successfully compared with the theoretical frequencies. The discrepancy between theory and observation is less than 1 per cent, confirming the globally accurate character of the solar model over all but a thousandth of its volume. Only the nuclear core, which produces the neutrinos and occupies a tenth of the solar radius, remains inaccessible to oscillation techniques. However, it will be accessible to analysis of *g-modes*, in which the restoring force is buoyancy rather than pressure. The g-modes are concentrated below the convective envelope. The resulting amplitudes at the surface of the Sun are so small that they are difficult to observe.

Considerable progress can be expected from long continuous observations. For this purpose a worldwide network of observing stations has been set up, called GONG (Global Oscillation Network Group). The data taken on the ground will be complemented by space observations which sample high-degree modes (up to $l = 4000$), and thus a detailed structure of the convection zone obtained. These observations will be carried out by instruments on the SOHO (Solar Heliospheric Observatory) satellite, a joint ESA–NASA project, ideally to be placed at the Lagrange point between the Earth and the Sun, which will allow uninterrupted observations.

We can ask if the osmosis of physics and astrophysics is about to produce a major breakthrough by means of studies of the Sun. The golden age of deep solar research is about to begin, underground (Gallex, Kamiokande, GAGE) and on the Earth's surface (GONG) as well as in space (SOHO). It is a good bet that before the end of the second millenium astrophysicists will have revealed the structure and working of the Sun's core, where for 4.5 billion years the energy needed to balance that lost in radiation from the photosphere, and prevent collapse, has been unceasingly produced.

The Sun serves today as the prototype star and a laboratory for both current and future physics. As a G2V-type star, the Sun belongs to what astronomers call the Main Sequence, a narrow band on the Hertzsprung–Russell diagram occupied by the huge number of stars burning hydrogen in their cores. Unless we can understand the Sun's internal structure there is little hope of following the workings of the other stars.

As a laboratory, the Sun allows us to check physical theories in density and temperature regimes which cannot be stably reproduced on Earth. It enables us to study complex fundamental processes in atomic and plasma physics.

Michel CASSÉ

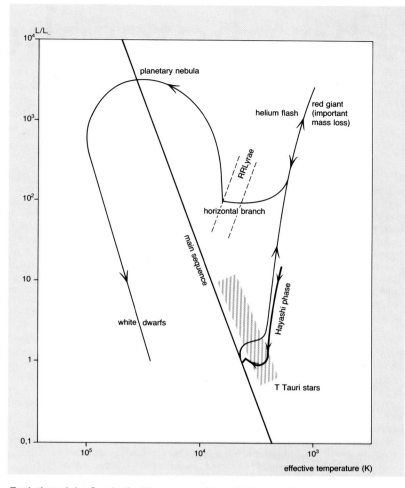

Evolution of the Sun in the Hertzsprung–Russell diagram. We believe that stars form from clouds of gas and dust present in the interstellar medium. Under certain conditions a cloud can collapse upon itself and form a protostar. The gas is heated and the pressure begins to build up in the interior of the protostar. The gravitational collapse is very rapid, of the order of a million years. The core, which is in hydrostatic equilibrium, is then completely convective; the temperature is not high enough for nuclear reactions to start at its centre and the star contracts slowly. Stars such as the Sun remain in this phase for several million years, until nuclear reactions begin. This phase is called the Hayashi phase in honour of the Japanese astronomer who constructed the first successful model. The star then passes through a phase which we believe is that observed in T Tauri stars. After about 100 million years it reaches the main sequence where it burns hydrogen in its centre. This phase started about 4.5 billion years ago in the Sun, and will last about another 5 billion years. Once the Sun's hydrogen supply is exhausted, the core, now containing only helium, will contract and its envelope will expand. Entering the red giant phase, its radius will increase as far as Mars' orbit and it will lose a lot of mass. The central temperature will be such that the nuclear reactions providing the energy will be different from those which took place on the main sequence; they will involve nuclei of carbon, nitrogen and oxygen, and are termed CNO reactions.

When the Sun reaches the end of the red giant phase it will have been a billion years since it left the main sequence. As contraction of the core continues, the Sun will have, towards the end of its life as a red giant, a central temperature of more than a hundred million K, and a correspondingly enormous central pressure. In fact this pressure will be so great that the matter at the centre will acquire special quantum properties due to the crowding of electrons. This sort of matter is called degenerate. The central density will be about a hundred thousand times that of water! The degenerate matter is a very good conductor of heat and is very difficult to compress. The contraction of the core will therefore stop, resulting in an isothermal core of degenerate helium surrounded by a shell of partially degenerate helium which can still contract. The core will have a radius of only a few Earth radii. Around the core will be a layer where hydrogen burns by the CNO cycle, this layer itself surrounded by a very extensive, rarefied convective envelope. The star will be unstable in this phase and the 'kappa' mechanism will cause it to undergo long-period pulsations.

What will happen when the central temperature continues to increase? Is an onset of helium-burning reactions possible? This does not occur readily since two colliding helium nuclei form beryllium 8 which is very unstable, and decays almost immediately into two helium nuclei. Three helium nuclei must collide simultaneously, and this requires temperatures of nearly a hundred million K. In this so-called triple-alpha reaction, the three helium nuclei are transformed into carbon 12. But this reaction takes place in a degenerate medium: the temperature increases because of this gain in energy and the high thermal conductivity of the medium. The triple-alpha reaction accelerates and a giant explosion takes place in the star's interior: this is called the helium flash.

It should be emphasised that the post-main-sequence evolution of the Sun is subject to much greater uncertainty than is its main sequence evolution. We will therefore only give a rough sketch of the Sun's fate after the helium flash. As a result of the helium flash the solar core expands rapidly and begins to oscillate. This motion is damped by the red giant's extensive envelope. The centre, in which helium is transformed into carbon and carbon into oxygen, is surrounded by a shell of burning hydrogen. After the helium flash the star moves onto the horizontal branch. In fact the star zigzags horizontally across the Hertzsprung-Russell diagram, increasing its luminosity. This phase lasts only a few hundred million years. Now we shall assume that all nuclear combustion after hydrogen leads to a convective core, leaving an inert core of oxygen and carbon surrounded by a shell of burning helium. This is in turn surrounded by a shell of burning hydrogen. The star evolves from the horizontal branch towards the asymptotic branch. During these zigzags the star may traverse an unstable band and begin to pulse like an RR Lyrae star. It is difficult to predict the evolution after the asymptotic branch: in general the star becomes very unstable and very luminous. It ejects an envelope of gas (or perhaps loses it progressively) to become a planetary nebula. The remaining core of the star is formed primarily of matter which is electron-degenerate. Consequently it cannot contract any further, and the star cools off slowly to become a white dwarf. It is estimated that the Sun will become a white dwarf having only half its present mass. The rest will have been lost in the form of violent winds and the ejection of surface layers during post-main-sequence evolution. The star cools quickly at first and then more slowly over billions of years. The white dwarf then stops shining and becomes a black dwarf: a cold mass of degenerate matter. This is our Sun's ultimate fate.

Solar oscillations. The various layers of the Sun oscillate with periods between several minutes and several hours. The observed oscillations are short-period (from 4 to 15 minutes). Using a computer one can reconstruct the oscillatory motions at various depths, as shown in this picture. Zones shown in blue are moving towards the observer and those in red are moving away (J.W. Leibacher and J.W. Harvey, National Solar Observatory).

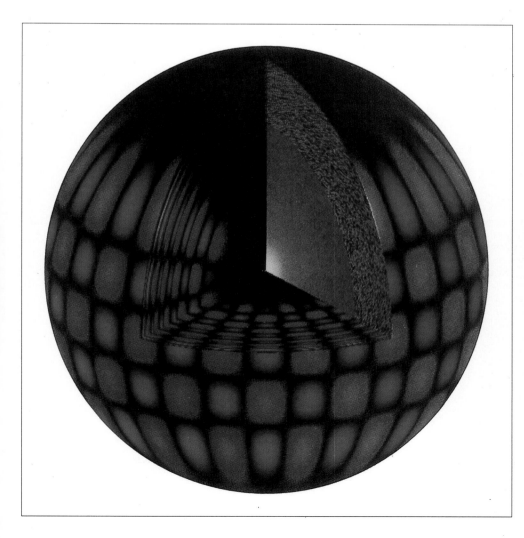

The photosphere

In a white-light photograph the Sun appears as a sharply bounded circular disk, with a brightness slightly greater in the centre than towards the edge. It is easy to interpret such a picture. The energy produced by nuclear reactions in the central regions flows in various ways, principally by radiation, towards the exterior. The radiation traverses very opaque layers and the photons are absorbed and re-emitted by the highly ionised atoms which make up the solar fluid. As the radiation travels outwards from the central regions, the layers it crosses become less and less hot and dense, since the pressure decreases as the 'weight' exerted by the outer layers lessens. Eventually the opacity becomes small enough to allow the radiation to flow freely into interstellar space, undergoing no further absorption *en route* to Earth. It thus preserves the image of the last layer which emitted it. The layer responsible for most visible radiation is called the photosphere because it appears bright and spherical.

Since the solar matter is a fluid, we should expect the opacity to diminish outwards, but this would not give rise to the observed sharp boundary of the disk. We can calculate the matter density scale-height, that is, the difference in solar altitude corresponding to a decrease in the density by a factor e (e=2.718). This scale-height is about 100 kilometres. At the edge of the solar disk, such a dimension would appear to have a size less than 0.2 arc seconds, practically invisible to ground-based observers.

The darkening from the centre to edge of the solar disk proves that the opacity decreases progressively. Light rays coming to us from the edge traverse the solar atmosphere obliquely, undergoing more absorption than those coming from the centre, and constitute an image of the Sun's outer layers. Since the temperature decreases outwards in the photosphere, these light rays come from the cooler regions, explaining why the edge of the disk appears less bright than the centre.

While a comparative study of the centre and edge of the disk can yield information about the stratification of the solar atmosphere, spectral analysis provides a much more powerful method. The photospheric spectrum straddles the visible range from the near ultraviolet (about 200 nanometres) to the infrared (about 100 micrometres). It is composed of a continuum which varies with wavelength, on which numerous dark bands, called Fraunhofer lines, are superposed. These lines correspond to the wavelengths of photons exchanged between radiation and matter during atomic excitation and de-excitation. They also correspond to an increased opacity, and therefore to cooler, outer layers of the solar atmosphere. It can easily be understood that the lines appear dark, in absorption against the continuum background. In addition, there are regions of the spectrum where the opacity is relatively small. For example, the principal absorber giving rise to the photospheric continuum is the ion H^- (a hydrogen atom which has captured an electron) which has a minimum efficiency in the near infrared, towards 1.6 micrometres. Observations at this wavelength allow penetration to a few tens of kilometres into the base of the photosphere.

Spectral lines contain information about physical properties of matter. Each line is unique to a particular element in the solar fluid, and its strength is related to the abundance of that element. In this way the chemical composition of the Sun can be determined. By number of atoms, it is 92 per cent hydrogen, 7.8 per cent helium and 0.2 per cent a mixture of other elements heavier than helium. Of these heavier elements, the most abundant appear in the same proportions as in the Earth's crust. The size and shape of a line's profile is related to the temperature, pressure and turbulent motions of the fluid. Their wavelength shift is directly related to the speed of the matter, through the Doppler effect. (Approaching corresponds to a blueshift.) Detailed observations of fluctuations in these various quantities on the surface of the solar disk allow the study of a great

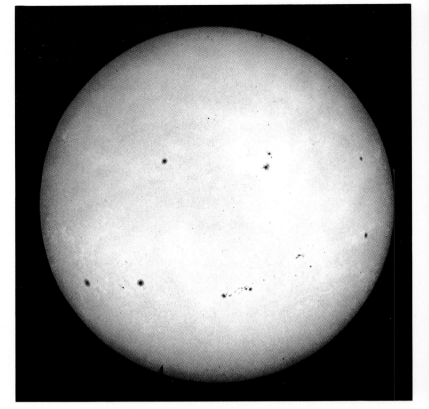

The photosphere in white light. One can clearly see a darkening towards the edge of the disk, produced by the decrease in temperature towards the outer layers: the temperature of the emitting regions is about 6400 K at the centre of the disk and 4500 K at the edge. These layers have an altitude difference of only 300 kilometres, which makes the edge of the disk appear sharp. The photograph also shows large-scale details of the photosphere, which are linked with magnetic phenomena occurring during the activity cycle: 'cold' dark spots with magnetic fields of several thousand gauss, and bright spots, visible only at the edge of the disc, where the field is several hundred gauss. The spots display an overall rotation about an axis between the poles; the rotation period varies from about 25 sidereal days for spots near the equator to 27.5 days at high lattitude. (Observatoire de Meudon)

Section of solar atmosphere in a plane passing through the Earth. In this diagram the thickness of the photosphere has been deliberately exaggerated. The white light reaching the Earth comes on average from regions whose distance from the centre of the Sun varies according to whether it comes from the centre of the solar disk or towards its edge. The light of a spectral line comes from layers nearer the surface because of the increased opacity of matter at the wavelength of the line.

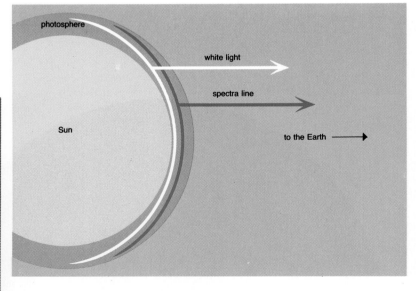

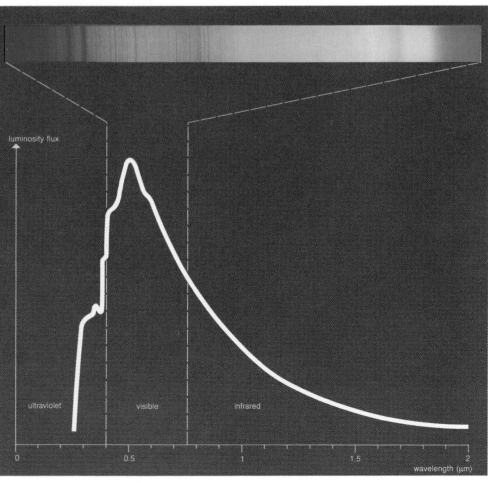

Visible solar spectrum. The luminosity flux–wavelength curve resembles that of a 5800 K black body, that is, a body that would be in thermodynamic equilibrium at this temperature. (Absorption by the Earth's atmosphere is neglected.) The dark lines visible in the spectrum (not shown in the light curve) correspond to wavelengths at which the matter is more opaque to the radiation. Several thousand lines can be enumerated in the solar spectrum. Their strength depends on the abundance of the atom in question; for a given atom the relative line strength is related to the temperature, while the line profile depends on the matter's motion, and on the pressure. (Observatoire de Meudon)

number of phenomena of diverse size and time-scale.

The surface of the Sun appears to be composed of bright grains or granules 1000–2000 kilometres in size, with rising motions of about 1 kilometre per second in their centre and lifetimes of about 10 minutes. This granulation in the photosphere is the outward sign of motion in the underlying convective zone. Superimposed on the granulation are oscillatory motions with periods of about 10 minutes. This granulation in the photosphere is the outward sign of motion in the underlying convective zone.

Study of the fractal dimension of the granules (the relation between their area and their perimeter) and Fourier analysis show that the small granules are turbulent.

The granulation is not the only phenomenon produced by the overshooting of convective motions into the photosphere. We can also see large-scale cells of the *mesogranulation* (5 000–10 000 kilometres) and *supergranulation* (30 000 kilometres). The latter is seen as mainly horizontal motions of photospheric matter from the centre to the edges of the cells. The supergranular motions are partly responsible for confining the magnetic field. There are intense magnetic field in the photosphere even outside active regions. They are in very small structures (of order 100 kilometres) only visible in very high resolution observations.

Superimposed on the granulation are oscillatory motions with periods of about 5 minutes, which ionvolve larger regions of the disk (5 000–10 000 kilometres) and which are a consequence of very low frequency sound waves. These waves, which have speeds of about 300 metres per second, are interesting for two reasons. Firstly, they can carry mechanical energy to layers outside the photosphere, and secondly, they can provide us with information about the structure of the Sun's interior. From their detailed analysis we can obtain an exact measure of the thickness of the convective zone.

The rotation of the Sun itself can also be measured by spectral line shifts, and is found to vary with both latitude and altitude. Its in depth study is of prime importance in interpreting active phenomena.

Pierre MEIN

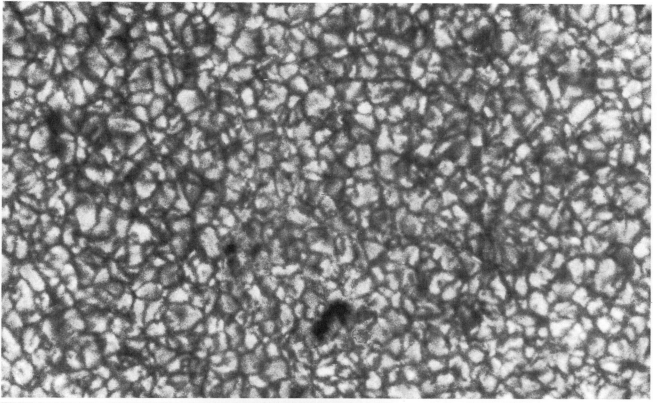

Granulation in white light. Granulation in the photosphere indicates convectively induced motions in the layers immediately below. Each bright granule corresponds to a rising flux of hot matter. The variations in temperature, at constant altitude, are of the order of 500 K. (Observatoire du Pic du Midi)

High resolution solar spectra. This image represents a small part of the solar spectrum in the near ultraviolet, around a wavelength of 393.5 nanometre. Its horizontal extent corresponds to only 0.35 nanometre in wavelength, and its vertical extent to 110 000 km on the solar disk. Variations in the radial velocity of the matter give rise to different wavelength shifts through the Doppler effect (towards the red for receding matter). This phenomenon is responsible for the zigzag appearance of the spectral lines. The velocities obtained from studying spectral line shifts indicate that the solar atmosphere undergoes complex motions: from granules which evolve on a scale of a few tens of minutes to 5-minute oscillations due to very low frequency sound waves. By observing several lines simultaneously one can study wave propagation in the solar atmosphere. In this example the narrow lines are formed in the photosphere, and the strongest line in the chromosphere. (Sacramento Peak Observatory)

Abundance of elements in the Sun. The chemical composition of the Sun can be deduced from studying spectral lines. The abundances of various elements are represented here in terms of the number of atoms, compared to the value 10^{12} for hydrogen. Note that the abundance of hydrogen is about a thousand times greater than that of the other most abundant elements, except helium. (From E. Anders & M. Grevesse, 'Abundances of the elements: Meteoritic and solar', in Geochimica Acta vol LIII, pp. 197–214, 1989.)

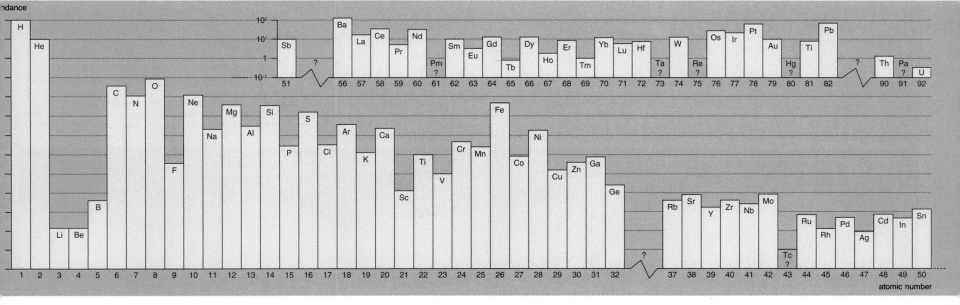

The chromosphere

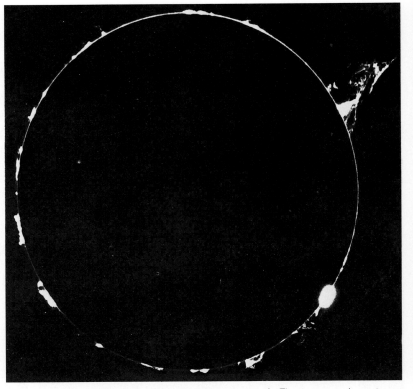

The solar chromosphere viewed through a coronagraph. The coronagraph creates an artificial eclipse of the disk of the photosphere and the chromosphere appears as a pink irregular fringe with streamers. (Observatoire du Pic du Midi)

Those who have been lucky enough to witness a total eclipse of the Sun will recall the precise moment when the edge of the Moon occults the brilliant disk of the photosphere. The phenomenon is all the more spectacular since the apparent diameters of the Moon and Sun are almost equal. The Moon appears surrounded by a ring of coloured (hence chromosphere) light, broken by prominences extending into the outermost region of the corona. Spectral analysis enables the chromosphere to be observed also when not in eclipse. Certain spectral lines are very dark, because the solar matter absorbs significantly at these wavelengths, and reveal layers further out than the photosphere. A complete picture of the Sun can be reconstructed in one of these bands using a narrow band filter or a spectroheliograph. Observed in this way, the disk has a diameter greater than that of the photosphere, which explains the appearance of phenomena during eclipses. Furthermore, its detailed structures are very different from those of the photosphere. This is particularly noticeable in the red H-alpha line of hydrogen where granules, reflecting convection cells, give way to more elongated structures. In the vicinity of sunspots, they are called fibrils and resemble the patterns formed by iron filings between two poles of a magnet. The magnetic field plays an important role in determining the structure in the chromosphere. While the density is sufficiently large in the photosphere that mechanical energy dominates magnetic energy, the reverse is true in the chromosphere.

Images obtained in calcium H and K bands (near ultraviolet) reveal slightly different structures, essentially because these bands have a different sensitivity to temperature and density fluctuations. They reveal the mesh of a large-scale network called the chromospheric network. The mesh is related to cells in the convection zone, of much greater size than granules: their diameter is of the order of 30 000 kilometres. At the edge of each mesh of the network the magnetic field is concentrated, and gives rise to spicules, jets of matter which last several minutes and escape at velocities of several tens of kilometres per second. The spicules, easily observable in H-alpha, reach a variety of heights; for every cell in the network there are about thirty 3000 kilometres high, and only one 10 000 kilometres high. It is the interplay of their number and orientation which gives the edge of the chromosphere its irregular appearance.

Other spectral bands are 'formed' higher in the chromosphere than calcium H and K bands. Examples are the H and K bands of ionised

magnesium, about 280 nanometres, and the resonance line of hydrogen, Lyman-alpha, at 121.6 nanometres, which no longer appears in absorption, but in emission against the continuum.

The prominences visible in H-alpha on the edge of the disk appear as dark filaments projected against it during the course of solar rotation. In general these filaments follow the lines of changing magnetic polarity on the disk's surface. Their inclinations with respect to the solar meridian result from the deformation of the field produced by the differential rotation already mentioned in connection with the photosphere.

The chromospheric spectral lines in the visible range are particularly dark. This does not, however, imply that the temperature continues to decrease from the photosphere towards the chromosphere. The temperature goes through a minimum of about 4300 K at the top of the photosphere, and then rises to more than 10 000 K in the chromosphere. The lines appear strong in absorption only because the density is too low to maintain the correspondence between matter temperature and radiation flux which characterises local thermodynamic equilibrium. It should be noted that millimetre wave observations of the Sun indicate that the chromosphere's radiation is approximately black body. While it is relatively easy to understand why the lines in the visible range appear in absorption on the disk despite the increase in temperature, it is harder to explain this increase. The phenomenon is accentuated beyond the chromosphere: in the few hundred kilometres separating the chromosphere from the corona, the temperature rises further to a million degrees. For a long time, the only explanation proposed involved the transport of mechanical energy, in the form of pressure waves already mentioned in reference to the photosphere. As these outbound waves encounter less and less dense layers, their amplitude increases with altitude. When this amplitude reaches the speed of sound, shock waves form, and dissipate their energy as heat. While this scenario is tenable in the chromosphere, it cannot explain the coronal temperature, because the flux of acoustic energy is already very weak at the top of the chromosphere. Astrophysicists are now leaning towards other mechanisms, taking into account the fine structure of the magnetic field. We can only hope to explain the high temperature of the chromosphere of the quiescent Sun by studying the magnetic mechanisms associated with the active Sun.

Pierre MEIN

The temperature of the photosphere and chromosphere. While the temperature decreases from the interior towards the exterior of the photosphere, the opposite is true in the chromosphere. The left part of the curve is easily explained by the transfer of radiation in a medium of decreasing opacity. The right part poses a problem. The flux of mechanical energy reaching the chromosphere in the form of acoustic waves, with periods of several minutes is insufficient to explain the very high coronal temperatures without some other source of energy, probably related to the inhomogeneous structure of the magnetic field. The circles indicate the regions of average emission of visible continuum radiation, and the Hα and K bands.

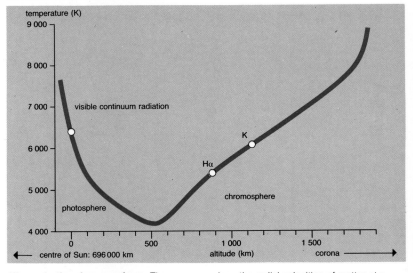

Waves in the chromosphere. These curves show the radial velocities of matter at a single point on the disk, deduced from spectral line observations. These lines correspond to different levels in the solar atmosphere separated in altitude by 200 kilometres. The small temporal separation of the two curves shows that the accoustic waves are almost stationary. The variation in altitude leads to a variation in velocity amplitude of the two lines.

Spectrum of the chromosphere. The portion of spectrum reproduced to the right of the figure was obtained by placing the slit of the spectrograph along a radius of the solar disk. The position of the slit is schematically shown on the left, with the boundaries of the photosphere and of the chromosphere (the curvature of the disk has been exaggerated in the drawing for clarity). While the continuous spectrum has an upper edge level with the boundary of the photosphere, at the wavelengths of the strongest absorption lines, light is registered beyond this edge over distances of several thousands of kilometres. The height of the chromosphere is thus demonstrated, or at least that of the highest structures in the chromosphere.

The lines emitted by the chromosphere appear in absorption interior to the disk of the photosphere, and in emission beyond it. Note that the very attenuated extension of the narrow lines and of the continuous radiation that can be distinguished upon the photograph comes from light diffused by the terrestrial atmosphere. (Observatoire du Pic du Midi)

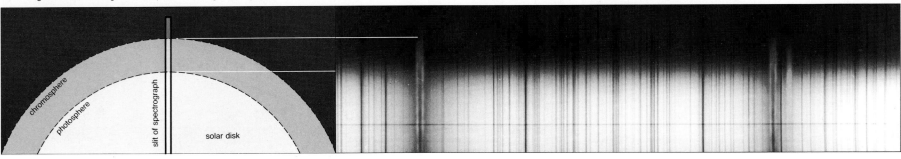

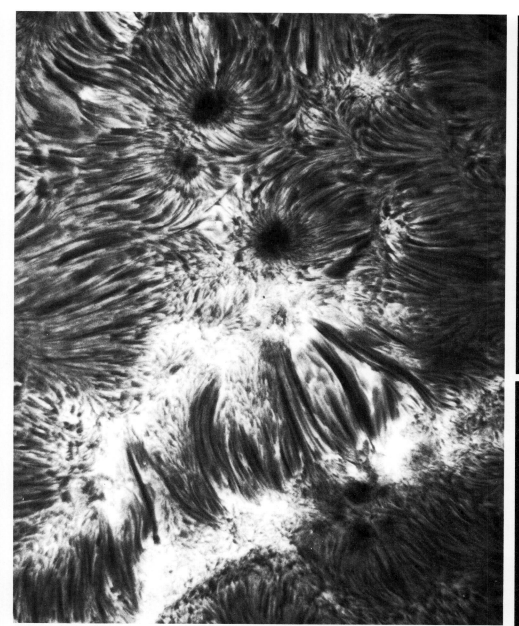

Fibrils. While photospheric granules are largely circular and display apparent convective motions, the fibrils in the chromosphere – here observed in Hα – are elongated structures aligned by the magnetic field. This alignment is particularly strong near sunspots. (Observatoire du Pic du Midi)

Spicules. Spicules are almost linear, more or less vertical, structures which vary from 5000 to 10 000 kilometres in length. Photographed here partially in Hα, they are grouped on the edges of chromospheric cells in regions of strong magnetic fields. They have rising motions (with speeds of about 25 kilometres per second) but after 5–10 minutes seem to dissolve into the surrounding corona. (Sacramento Peak Observatory)

Spectroheliograms of the chromosphere. These pictures – in Hα (above) and in ionised calcium K (below) – were taken on the same day as the white light view of the photosphere (page 26). The K line clearly shows the structure of the chromosphere. The bright regions appear so because of the higher temperatures associated with the stronger magnetic field. Note that this rule, true also for faculae, is not true for sunspots, despite their very strong magnetic fields. The dark filaments that are seen particularly well in Hα are prominences, viewed projected against the solar disk. When not directly linked to an active centre, that is a sunspot or a facula, they can persist for several months and are termed quiescent. They are composed of 'cold' matter (7000 K) somehow suspended in the corona (1 000 000 K). (Meudon Observatory)

Filaments observed in Hα. Note the alignment of filaments towards the equator and the west. This results from differential rotation. (Meudon Observatory)

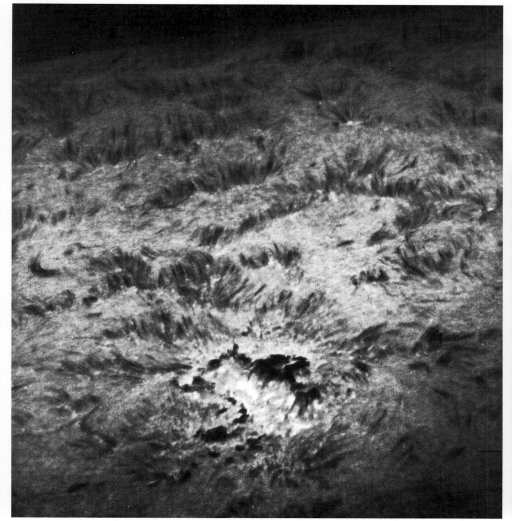

The corona

The corona is the halo of white light which appears around the Sun during total eclipses. The inner corona reaches to about 2 solar radii beyond the disk's edge, and can be distinguished, albeit somewhat arbitrarily, from the outer corona which extends further. The corona can be observed in the entire electromagnetic spectrum, from X-rays, through the visible, to radio waves.

Observations in white light

Photographs in white light, taken during a total eclipse or with a coronagraph on board a satellite, reveal that the corona, far from being a spherical homogeneous envelope, contains a large variety of structures. The most spectacular of these are large jets, each of which has a bulbous base which thins to a long neck out to 3 or 4 solar radii. In the tail, coronal matter leaves the Sun with supersonic velocities. Other more modest jets only have the bulbous base, and terminate at 2 or 3 solar radii. These bulbs consist of a multitude of coronal arches. Other characteristic structures are prominences which appear to emanate from the polar regions, and delineate the magnetic lines of force. Jets and prominences faithfully reproduce the configuration of the coronal magnetic field. Therefore the jets all have the same structure: a rounded base of closed field lines below a bundle of open field lines pulled into almost parallel lines. Like the outer corona, the inner corona, observed regularly in visible monochromatic light with ground-based coronagraphs, also appears strongly heterogeneous, with a system of loops extending to 100 000 kilometres beyond the disk's edge. Continuous observation of the corona using an orbiting coronagraph has revealed that the corona changes not only from month to month, but also from hour to hour. While the bulbous jets last several months, the big jets evolve in several weeks, and the prominences in several tens of hours. Generally, the evolution of coronal structures follows that of the magnetic fields throughout the solar cycle, and is associated with solar activity.

Extreme ultraviolet and X-ray observations

White light photographs taken during eclipses or with a coronagraph only reveal the part of the corona beyond the solar disk. However, photographs in extreme ultraviolet yield a view of the

The inner corona observed in green light. This photograph was taken on 30 October 1980, with the coronagraph of the high-altitude station Saint-Veran (High Alps). The filter attached to the coronagraph only lets green light pass, that is, light emitted by ions of Fe XIV (ionised 13 times) at a wavelength of 530.3 nanometres. These ions are at about two million degrees.
   The very fine loops, seen clearly on the right are produced by the magnetic field which structures the very hot, highly ionised coronal gas in the same way as a magnet orients iron filings. The two dark spikes as well as the bump on the left side are instrumental artifacts. The black disk is the occulting disk of the coronagraph which produces the artificial eclipse; its radius is slightly bigger than the solar disk. The bright white ring, slightly larger than the occulting disk, results from a leakage of blue light through the filter. (Observatoire de Paris)

Ultraviolet view of the corona. This striking view of the solar corona was prepared from data supplied by the Solar Maximum Mission satellite. The colours represent densities of the corona and go from blue (densest) to yellow (least dense). The blue, dense coronal regions overlie sunspot regions. When the solar flare occurred the portion of the corona above the flare was completely disrupted and changed its shape in a matter of a few minutes. (NASA)

The corona and solar activity. These white light photographs of the outer corona were obtained during two eclipses using a coronagraph on board an aeroplane flying at high altitude. Employing a special technique in which several images are used to form one composite image, the corona can be seen out to 11 solar radii. The small central disk is not the eclipsed solar disk, but an occulting disk with a diameter corresponding to 4 solar radii.

Comparison between these two photographs, one taken in 1973 during a minimum of solar activity (left), and the other in 1979 during a maximum (right), reveals the spectacular global evolution of the corona throughout the solar cycle. During the minimum the large jets appear concentrated in the equatorial regions, while during the maximum they flourish at all latitudes. (Los Alamos National Laboratory, Charles Keller *et al.*)

The corona in white light. This photograph was obtained using a neutral radial filter whose transmission increases with altitude above the Sun's edge. It is correctly exposed everywhere, and thereby reveals the entire corona in a single exposure. A special photographic procedure was used to enhance the contrast of coronal structures: large bulbous jet (1), thin jet (2), blade-like jet (3), polar streamers (4) and coronal condensation (5). Note that the shape of a jet depends on the angle at which it is viewed. This photograph was taken at Kazakhstan during the total eclipse of 31 July 1981. (Mission de l'Institut d'Astrophysique du CNRS, Paris. Photograph S. Koutchmy *et al.*)

corona projected against the disk. In this region of the spectrum the photosphere is scarcely visible, while the corona radiates most of its energy. At shorter wavelengths, we observe higher in the corona; the edge of the Sun becomes less well defined, and gives way to the edge of the photosphere which is more luminous than the centre of the disk. This brightening from centre to edge is a simple geometric effect. The opacity of the gas is small in this spectral region, so the luminosity is proportional to the number of emitting atoms. Since there are more of these along a line of sight to the edge of the disk than to the centre, the former appears brighter. Images in extreme ultraviolet monochromatic light allow us to explore the transition zone between the chromosphere and the corona. These images show that the chromospheric network becomes increasingly blurred higher in the transition zone, disappearing completely in the corona. This change of structure with altitude suggests that the magnetic field, which is tied to the network at chromospheric altitudes, gradually spreads out to near uniformity in the corona.

X-ray images of the interior of the corona show X-ray emission to be very localised on the disk, and the brightest regions to coincide with the bright regions seen near the edge in white light photographs. These X-ray images, whose resolution is equivalent to that of visible light photographs, confirmed what had already been demonstrated by photographs of the corona in monochromatic visible light obtained from ground-based coronagraphs: the corona contains loop-like structures whose size depends on the amount of activity in the vicinity. This presupposes that the magnetic field in the corona is a quasi-uniform interlacing of closed loops. These X-ray photographs have also revealed curious structures previously invisible except in the corona: dark cavities from which no light escapes, appropriately named coronal holes. Another major revelation of X-ray images is the presence of a myriad of isolated bright points speckling the surface of the solar disk. At any given moment there are about a hundred of these, each lasting a few hours.

Radio observations

At the other end of the electromagnetic spectrum, in radio waves, the corona is also visible against the disk. In this range what little radiation is emitted by the photosphere is blocked by the opaque corona. The opacity of the solar gas varies with density and wavelength in such a way that radiation coming from layers more distant from the Sun's centre, has a longer wavelength. Refraction, which bends the trajectories of radio waves, acts in the same sense as opacity, so that a wave at a given frequency comes to us from a layer of the solar atmosphere whose density is fairly well defined. As in the extreme ultraviolet range, the outer corona is sounded by varying the wavelength of observation from centimetres to metres. The 'radio Sun' is larger than the photospheric disk, and its size increases at lower frequencies. The brightening from centre to edge of the disk, observed only at centimetre wavelengths, reflects the increase in temperature in the transition zone.

The diffraction-limited angular resolution of radio telescopes at metre wavelengths is poor (1 arc minute for 1.77 metres wavelength); only intense sources of radiation can be identified, and no fine structure is revealed. These intense sources coincide with the bright zones of the X-ray corona, and with the coronal condensations of the visible corona. At centimetre wavelengths the instruments reveal structures in the inner corona down to a resolution of 2 arc seconds.

The Sun's radio emission is variable. Each centre of activity is a concentrated source of radio waves, and the contribution of these sources to the total radio flux varies with their evolution and rotation. At metre wavelengths the slowly varying radio flux arises chiefly from jets. Subtracting this component from the total radiation yields the radio emission of the quiescent Sun.

Jets, arches, loops and bright spots are coronal

manifestations of solar activity. Phenomena of the active Sun are not localised in the corona as neatly as in the photosphere and chromosphere, and this makes it very difficult to distinguish a truly quiescent corona, shaped only by phenomena of the quiet Sun.

Physical characteristics of the corona and transition zone

A study of the radiation emitted by the corona in the entire electromagnetic spectrum indicates that its temperature is much higher than that of the chromosphere; it is of the order of a million degrees. The temperature changes abruptly from 20 000 to 500 000 K in the transition zone, which is less than 1000 kilometres thick, peaks at 1.5 million K in the inner corona, and remains at 1 million K in the outer corona. It is only at large distances from the Sun, in the interstellar medium, that the temperature decreases to very low values of a few hundred degrees. The high temperature produces a large scale-height, of the order of 100 000 kilometres and is thus responsible for the extension of the corona out to a few tens of solar radii. The solar gas density, which decreases very rapidly with altitude in the photosphere, decreases slowly in the corona. The average density, which is about a hundred million (10^8) atoms per cubic centimetre in the inner corona, remains as high as a hundred thousand (10^5) atoms per cubic centimetre at 4 solar radii beyond the disk's edge. There are large deviations from these averages, revealed by the brightness of certain coronal structures. For instance, in coronal condensations, the brightest structures at all wavelengths, the density reaches ten billion (10^{10}) atoms per cubic centimetre and the temperature, more than three million degrees.

The extraordinarily high temperature and very low density of the corona are responsible for its strange properties. It has approximately the same chemical composition as the photosphere, but the atoms are highly ionised; they gradually lose their electrons as the temperature rises in the transition zone, and many lose all their electrons in the corona. Hydrogen and helium are reduced to their nuclei, oxygen loses six or seven out of eight electrons, and iron between ten and fifteen out of twenty six.

Even though the temperature is very high, the total amount of energy lost in the corona is very small. The average energy of a particle is very high, but the density is low, so the energy density is small. The energy emitted by the corona, mainly in X-rays (radio radiation is completely negligible) is almost a thousand times less than that emitted by the chromosphere.

The coronal spectrum

Coronal radiation has two components: one diffuse and one intrinsic.

The diffuse corona. Free electrons in the corona scatter photospheric light producing the K corona (K for *Kontinuum* in German) which is the dominant component of the white halo visible during eclipses out to 2 or 3 solar radii from the Sun's edge. Beyond this region, as the gas density decreases, the diffuse continuum merges gradually with a weak background produced by photospheric light scattered by interplanetary dust grains. This is called the F corona (F for Fraunhofer) and extends as far as Earth where we observe it along the ecliptic on very clear nights, just after sunset or just before sunrise, as zodiacal light. The outer corona is also responsible for the infrared radiation observed at 2.2 micrometres, between 3.5 and 10 solar radii, concentrated in the equatorial plane. Since it is non-polarised, this component due to dust can easily be separated from the polarised electromagnetic component, using a polarimeter.

The photospheric continuum is only slightly modified by scattering. The spectrum of Fraunhofer lines, on the other hand, becomes completely indistinguishable because the high speeds of the electrons in the corona enlarge the lines, through the Doppler effect, to several hundred times their original size. However, as the dust's thermal velocity is small, the spectrum of the light scattered by the dust is a faithful copy of that of the photosphere.

The intrinsic corona. The intrinsic emission of the corona consists of a spectrum of lines, mainly in the soft X-ray range, and a continuum which is very intense in the decimetre and metre radio ranges. The lines are emitted by atoms ionised to levels common at temperatures between one and two million degrees in the quiescent corona. Magnesium occurs in the form Mg X, oxygen as O VI and O VII, and iron in states of ionisation from Fe X to Fe XVII. The permitted lines of these ions, that is the lines due to the most probably transitions, are in the extreme ultraviolet and X-ray ranges (from 2.5 to 100 nanometres). On average, the shorter wavelengths come from the most highly ionised atoms, in the hottest regions of the corona. In active regions the line spectrum extends to 0.185 nanometres (Fe XXV line), and is characteristic of temperatures of three to ten million degrees.

The coronal plasma also emits forbidden lines – lines due to transitions which are quantum-mechanically improbable – throughout the visible and extreme ultraviolet spectrum. The brightest visible lines are the green line at 530.3 nanometres, the red line at 637.4 nanometres and the yellow line at 569.4 nanometres. These are due to the ions Fe XIV, Fe X and Ca V, respectively. The corona observed at these wavelengths is called the monochromatic green, red or yellow corona, and is characteristic of the inner corona. About a hundred forbidden coronal lines have been discovered at visible wavelengths.

The radio continuum of the quiet Sun is purely thermal in origin. It is characteristic of the long wavelength tail of the distribution of radiation in

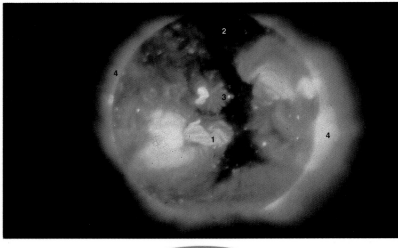

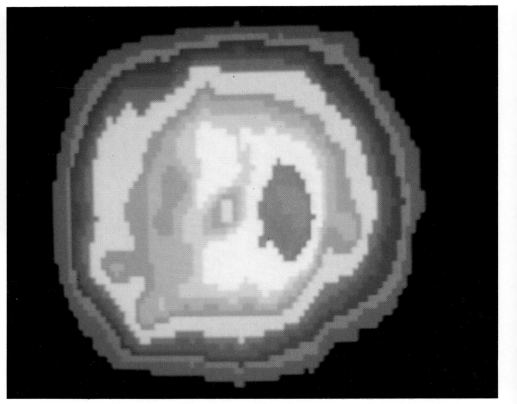

The radio corona at 169 megahertz. This radio map of the quiescent sun was obtained during the day, 1 July 1980, by aperture synthesis using the east–west and north–south channels of the radio telescope at Nançay, France. The angular resolution is 1.4 and 4 arc minutes respectively, in the two scanning directions. The brightness distribution on the disk is represented by a colour coding in which intensity decreases from brown to violet. The brown spot corresponds to a source of activity probably due to fast-moving electrons in a coronal arch. (Observatoire de Paris, Radiotélescope de Nançay)

The inner and outer corona. These simultaneous images of the corona were obtained by two of Skylab's solar telescopes; the upper image shows the inner corona against the disk and near the edge in X-rays, the other, the outer corona beyond the edge, in white light, with the first image superposed. In the inner corona one can distinguish coronal loops related to active sites (1); an extensive polar coronal hole (2), descending well below the solar equator to a latitude of 45° south (note that no loops encroach upon the hole); and bright spots speckling the surface of the disk including in the coronal hole (3). The uppermost layers of the inner corona, above the hidden hemisphere of the Sun, are visible behind the limb (4). The edge of the disk appears bright except above the coronal hole (5). In the outer corona we see the occulting disk whose radius is half that of the Sun's image (1); the shadows of the supports of the occulting disk (2); black and bright concentric rings on the occulting disk, which are not coronal structures, but are due to optical effects of the coronagraph (3) and the outer limit of the field of the coronagraph (4).

The superposition of these two images shows that the regions which are brightest in X-ray correspond to jets in the white light corona. Note also that there is no emission in white light above the coronal hole identified on the picture in X-rays. (High Altitude Observatory and American Science and Engineering)

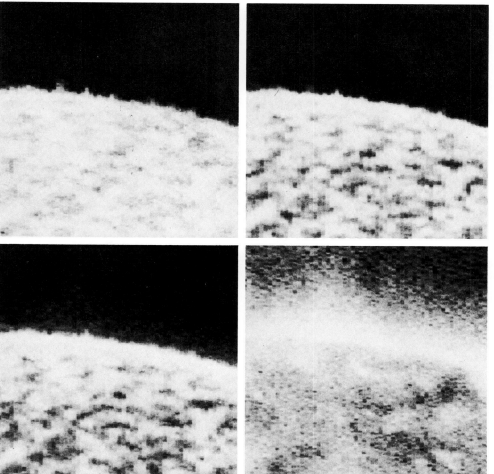

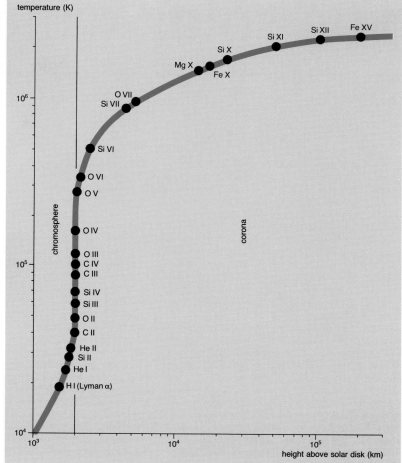

Disappearance of the chromospheric network in the corona. These monochromatic images of the edge of the quiescent Sun were obtained at wavelengths corresponding to four emission lines produced by ions at different altitudes in the solar atmosphere. The first image is taken in Lyman-alpha (neutral chromospheric hydrogen), the second in C III (twice ionised carbon), the third in O VI (oxygen five times ionised) – representative of the lower and upper transition zones respectively – and the fourth in Mg X (magnesium nine times ionised), characteristic of the corona. In this last photograph the cells of the chromospheric network can no longer be distinguished. The images were obtained with the Harvard College Observatory spectroheliometer on board Skylab. The chequered appearance is due to the scanning process used to construct the image.

Distribution of temperature in the chromosphere and the corona. In the transition zone between the chromosphere and the corona, only a few hundred kilometres thick, the temperature abruptly increases from 10 000 K to 10^6 K. The atoms successively lose their electrons in passing through higher and higher ionisation states. Each atom can only exist in a given ionisation state in a narrow range of temperature; the study of the relative intensities of the characteristic emissions of each ion constitutes a unique means of exploring the thermal structure of the transition zone.

The spectroscopic symbol A XV designates an atom A having lost 14 electrons. The point associated with each symbol represents the temperature where the abundance of the corresponding ion passes through its maximum.

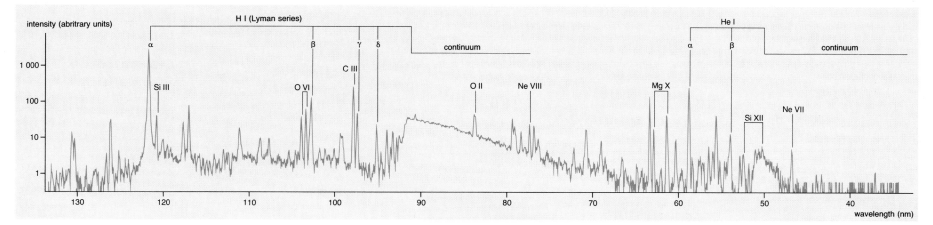

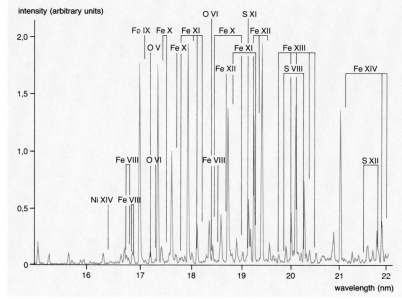

Ultraviolet coronal spectrum. This spectrum between 40 and 130 nanometres was obtained with the Harvard College Observatory ultraviolet telescope on board Skylab. It corresponds to the centre of the quiescent disk, which explains the simultaneous presence of hydrogen (H I), Si III and C III lines, formed in the transition zone, as well as coronal lines such as Mg X, Si X and Si XII. Note that in certain parts of the spectrum the lines are superposed on the H I and He I continuum (E. M. Reves, M. C. E. Huber and J. G. Timothy, 1977).

The spectrum of the solar disk below about 23 nanometres was obtained during a rocket flight in April 1979, by the Cambridge Air Force Research Laboratory. The light here comes from the entire solar disk. At these shorter wavelengths, the lines are due to ions formed in the upper transition zone and corona, and there is no continuum. Note that the intensity scale is logarithmic in the case of the first spectrum, and linear in the second.

thermodynamic equilibrium at one million degrees, which has its maximum emission in the X-ray domain. The emission is due to deceleration of free electrons in the corona; the radiated energy is borrowed from the energy of the electrons' thermal motions. At the other end of the spectrum, in the X-ray range, continuum emission is dominated by ionisation and recombination processes, but it is still negligible compared with line emission.

The line spectrum contains the most detailed information about the temperature of the electrons and protons, the electron density, the speed of ions and electrons, and the magnetic field.

Monique ARDUINI-MALINOVSKY

The solar wind

The solar wind consists of parts of the solar corona rushing into interplanetary space at supersonic speeds. This wind, blown continuously by the Sun, passes the Earth at an average speed of 400 kilometres per second, and eventually blends with the interstellar medium beyond the edge of the Solar System. During its passage it sweeps up evaporated gases from planets and comets, fine particles of meteoritic dust, and even cosmic rays of galactic origin. Its influence is felt throughout interplanetary space, and it provokes in the Earth's atmosphere polar aurorae and magnetic storms.

Awareness of the existence of the solar wind came gradually. In 1896 the Norwegian physicist Olaf Kristian Birkeland confirmed for the first time that something other than light was arriving at Earth from the Sun. He suggested it was an electrically charged corpuscular radiation which, drawn in near the pole by the terrestrial magnetic field, gave rise to aurorae boreales. It was many years before the observation of another phenomenon, geomagnetic storms, brought new proof of the emission of particles by the Sun. These storms, caused by abrupt perturbations of the terrestrial magnetic field, often interferred with radio and telephone communication. A series of solar and geomagnetic observations revealed a correlation between the appearance of these storms and the appearance, one or two days earlier, of solar eruptions. Around 1930 Sydney Chapman and V. C. A. Ferraro calculated that a cloud of ions ejected by the Sun would move at a speed of 1000 to 2000 kilometres per second and reach Earth in one or two days, perturbing the terrestrial magnetic field in the same manner as observed during magnetic storms.

The third piece of evidence for the emission of particles by the Sun appeared towards the end of the 1940s from studies of galactic cosmic rays. Scott Ellsworth Forbush discovered that cosmic rays reaching Earth had a low intensity when the sun was active, and diminished, often abruptly, during magnetic storms. Apparently something in the Sun's radiation tended to impede the flux of cosmic rays entering the Solar System, especially when the Sun was most active. It was suggested that a magnetic field, transported by a stream of charged particles from the Sun, barred the way for galactic cosmic rays.

It was clear that sporadic jets of ionised matter emitted by the Sun during very active periods explained well the fluctuations in the terrestrial magnetic field and galactic cosmic radiation.

But the most decisive proof of the existence of a corpuscular radiation from the Sun came around 1950 from the work of Ludwig F. Biermann on the tails of comets. It had been known for centuries that comets' tails were systematically oriented away from the Sun, independent of the comet's position on its orbit through the Solar System. The hypothesis then accepted, that pressure due to solar electromagnetic radiation was responsible for this orientation, had to be abandoned when Biermann showed that this pressure was much too small, and that only a flux of particles moving at speeds of several hundred kilometres per second could explain the observations. Observations of comet tails brought further evidence. No matter what the comet's trajectory was, the ionised tail was pushed away from the Sun, so the emission of particles had to be constant and isotropic. It could become more intense during periods of solar activity, but it was always present, whether or not there were spots and eruptions on the Sun.

While the analysis of comet tails furnished proof that the solar atmosphere was in a state of quasi-stationary expansion, the mechanism by which the wind could be ejected at several hundreds of kilometres per second, as deduced by Biermann, remained to be explained. A model of this expansion was calculated by Eugene Newman Parker in 1958.

At more than a million degrees, the lightweight electrons in the coronal plasma have thermal velocities greater than 5000 kilometres per second, and therefore tend to escape from the solar atmosphere. Through charge separation they create an electric field which couples the movement of electrons and ions which are dragged along by the electrons. The gravitational

Evidence for the existence of the solar wind: comet tails. In this photograph of the comet Mrkos, one can clearly distinguish two types of tail: the first has a diffuse structure, and its orientation is further from the antisolar direction than that of the second, which has finer structures. The first consists of microscopic dust and neutral molecules produced by cometary degassing; the second consists of matter ionised by solar radiation.

The ionized tail is pushed in the antisolar direction by the electrically charged plasma pouring continuously from the Sun at speeds of about 1000 kilometres per second. The neutral tail is pushed less violently by radiation pressure alone, and is insensitive to the solar wind. (Observatoire de Haute-Provence du CNRS)

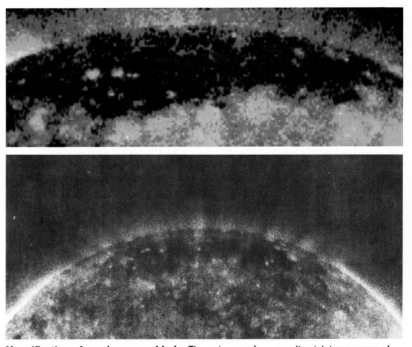

Magnification of a polar coronal hole. These images form an ultraviolet panorama of the north pole of the Sun, viewed by two of Skylab's telescopes on 28 January 1974. The upper photograph was taken at a wavelength of 62.5 nanometres, which corresponds to the magnesium X line; the lower one was taken at 36.8 nanometres, which corresponds to the magnesium IX line, that is to a temperature slightly lower than the first, but still about a million degrees. The polar streamers appear to spout from bright spots observed all over the Sun's surface, including coronal holes. The bright spots are structures where the magnetic field is concentrated. The streamers are jets of coronal gas which follow the force lines of the magnetic field implanted in these concentrated islands. The bright spots and prominences are not permanent structures but they may nevertheless last several days. At the edge of the Sun, the coronal hole is characterised by an abrupt vanishing of the bright ring observed in ultraviolet coronal emission lines. This is explained by the fact that the temperature varies differently with altitude above a coronal hole than over the rest of the Sun. (Harvard College Observatory and Naval Research Laboratory)

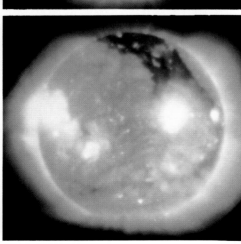

Rotation of a coronal hole observed over a period of six days. Coronal holes are quasi-permanent structures of the solar corona which are characterised by an absence of ultraviolet emission, and a particularly pronounced absence of X-rays. These structures do not appear in the photosphere or the chromosphere. Almost always present at the poles, they can extend as far as the equator, and occupy, during a minimum of solar activity, a large fraction of the solar surface. During the nine months of observation carried out from Skylab in 1973, this coronal hole, extending from the north pole, took on a curious boot shape resembling the map of Italy and covering almost a quarter of the Sun's surface. This sequence of four images, taken in X-rays at two-day intervals (19, 21, 23 and 25 August 1973), show it as it turns with the Sun. Curiously, this hole rotates as if the Sun were a solid body. Unlike the layers of the photosphere and chromosphere, which turn faster at the equator than at the poles, it is not affected by differential rotation. Note that most of the changes in the shape of the hole during the six days are due only to perspective effects.

The flux of solar wind leaving coronal holes sweeps past the Earth most rapidly when the huge coronal holes descend near the solar equator. It deforms the terrestrial magnetic field and disturbs the upper atmosphere of our planet. (American Science and Engineering and Harvard College Observatory)

field of the Sun constrains them in the corona under very high pressure, whereas in interplanetary space the pressure is very low. At a certain critical altitude in the corona their thermal velocities reach the escape velocity: above this altitude they escape radially at supersonic speeds. This is why, at this stage, the term solar wind, which suggests an expansion, is more appropriate than the term solar atmosphere, which suggests a medium at rest with respect to the Sun.

At a large distance from the Sun, where the pressure of the solar wind is of the same order as the ambient pressure, it is impossible for the coronal fluid to displace the interstellar medium. This is the limit of the heliosphere, but the way in which the solar wind and the interstellar medium meet remains mysterious. It is not yet known whether the wind is stopped abruptly across a shock or if the two mediums interpenetrate in a diffuse fashion.

Long suspected and modelled, the existence of the solar wind was not confirmed until space probes measured it *in situ*. The first probes on board Soviet Luna 2 and 3 rockets in 1959, revealed exactly what was expected from a supersonic solar wind. Since then numerous probes have furnished continuous measurements of the velocity and density of the solar wind, and all have confirmed the basic elements of Parker's model. At present the solar wind has been sounded from 0.3 astronomical units (60 solar radii) to about 10 astronomical units (about the orbit of Saturn). Emanating from the corona, it is a highly diluted mixture of electrons and ions (mostly hydrogen and helium). At Earth's orbit its density is less than about ten particles per cubic centimetre, and its velocity less than 400 kilometres per second. Although the Sun loses a million tonnes of hydrogen per second through this process, the loss is negligible. The wind would take 10^{14} years to disperse the entire mass of the Sun into interplanetary space, and the estimated lifetime of the Sun is only fifteen billion years; the Sun loses mass more rapidly through radiation. The energy of the solar wind is only a millionth of the total energy furnished by the Sun. The wind's characteristics are highly variable; its velocity can fluctuate between 300 and 1000 kilometres per second and its density between 0.1 and 30 particles per cubic centimetre. These changes reflect both coronal inhomogeneities and solar activity. The most rapid fluxes of solar wind issue from coronal holes, strange structures revealed by X-ray images, where no X-ray emission is detectable. In these regions the solar magnetic field opens into interplanetary space, permitting the outflow of ionised particles. The speed of the solar wind is therefore modified depending on whether or not it is observed above a coronal hole.

Many problems must be explained before the solar wind is understood; the mechanism for heating the corona, necessary for the expansion of the solar wind, and the mechanism for accelerating the wind are not yet known. These problems are not specific to the Sun, as a large number of stars have coronas and winds by which they lose mass. Moreover, the problems posed by the interface between the solar wind and the interstellar medium concern the evolution of the interstellar medium and therefore the Galaxy.

Monique ARDUINI-MALINOVSKY

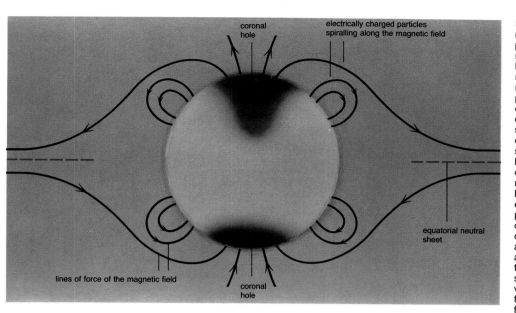

coronal hole

electrically charged particles spiralling along the magnetic field

equatorial neutral sheet

lines of force of the magnetic field

coronal hole

Structure of the coronal magnetic field in a meridian plane. The overall dipolar magnetic field of the Sun, blown by the solar wind, produces a neutral sheet in the neighbourhood of the equatorial plane, from one side to the other of which the field reverses direction. The polar zones are surmounted by holes in the corona which are the main sources of the solar wind. The lines of force in these regions diverge into the interplanetary medium while elsewhere they loop back onto the Sun. The electric charges of the coronal plasma wind around these lines of force. The electrical conductivity being quasi-infinite, all motion of matter is accompanied by a movement of the lines of force which supports this motion, and vice versa: the magnetic field is frozen in matter. The lines of force are dragged along by the plasma in the expansion of the solar wind above the holes in the corona. In contrast, the coronal magnetic structures reclosing upon the Sun trap the plasma within the corona.

The ballerina's skirt. Polar coronal holes evolve and deform, causing the neutral sheet to undulate like the skirt of a pirouetting ballerina. At a solar minimum the depths of the skirt's folds are about ±15 degrees of latitude. Because of these undulations the neutral sheet crosses the equatorial plane in several places thereby dividing the equatorial region into sectors with alternating polarity. The diagram represents a structure in four sectors. A stationary observer sees the neutral sheet, dragged by the Sun's rotation, produce the impression that the magnetic field is divided into four sections. (Artist's conception by W. Heil, scientific adviser, J. M. Wilcox)

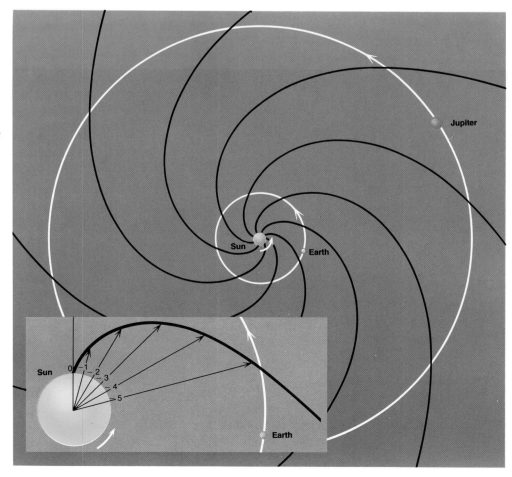

Jupiter

Sun

Earth

Sun

0 1 2 3 4 5

Earth

Spiral structure in the ecliptic plane of the interplanetary magnetic field. The structure of the interplanetary magnetic field is determined by the coronal expansion. The magnetic field, frozen in the matter, is dragged with it in its radial expansion. Because of the Sun's rotation, the magnetic field's force lines rooted in the solar corona, form an Archimedes spiral around the Sun. At the Earth's orbit the angle between the spiral and the direction to the Sun is about 45 degrees.

Details of a magnetic field line spiralling in the ecliptic plane are shown in the drawing (the Sun is not to scale). The plasma escapes radially at an average speed of 400 kilometres per second. It drags with it the magnetic field of the coronal zone from whence it issues. The diagram represents a force line rooted at point 0 and bound to the plasma emitted one, two, three, four and five days earlier by the same point on the corona which, due to the rotation of the Sun, has moved from the position −5 to the position 0.

The active Sun

Active regions

The photosphere, the chromosphere and the corona are layers of the Sun superposed like onion skins. These layers are not homogeneous, and contain diffuse structures whose variable or ephemeral character is the basis of the concept of activity.

In 1611, David Fabricius and Galileo identified sunspots already observed by the Chinese. In 1843, the amateur astronomer Samuel Heinrich Schwabe discovered solar cycles and, in 1859, Richard Christopher Carrington and Richard Hodgson discovered eruptions. These three phenomena are fundamental manifestations of solar activity.

Sunspots are the structure which seem to be the least active. They appear dark because they are 'cold', that is, cooler by about 1700 K than neighbouring regions. The temperature in the central region of the umbra can drop to 3000 K. Granulation seems to disappear, suggesting that convection, one source of heat, is almost absent. The dark core is stirred by movements of matter (hot rising structures with a diameter of about 200 kilometres), 'flashes', etc., but with their retinue of faculae and filaments, sunspots may survive for several solar rotations without undergoing any noticeable change.

In 1908 George Ellery Hale noticed that certain spectral lines were doubled when the slit of his spectrograph was aligned across a sunspot. This doubling, known as the Zeeman effect, is a 'fingerprint' of the magnetic field on the emitted light. The strength of the doubling is proportional to the intensity of the magnetic field, which reaches 2500 to 3000 gauss on the Sun's surface, six thousand times stronger than Earth's field. The solar magnetic field is certainly sufficiently intense to suppress or inhibit convection.

The higher one goes in the solar atmosphere above a sunspot, the more the magnetic field shapes the medium: magnetic pressure rapidly dominates gas pressure. This is demonstrated by the penumbra of the sunspot. Dark or bright filaments delineate the force lines; rooted in the centre of the sunspot, they close back quite quickly on the neighbouring photosphere. This is not observed in the case of flocculi, small spots in the penumbra.

The magnetic map of an active region demonstrates that strong fields are not restricted to sunspots, but that they can also occur in bright zones, faculae. The brightness of these faculae is explained by high temperatures revealed by intense radio emission. It is difficult to understand how equally strong magnetic fields can produce bright hot regions such as faculae as well as dark cold regions such as sunspots. According to Eugene N. Parker this is caused by the varying structure of the convective magnetic field.

The solar landscape has so far only been examined 'at ground level'. Higher up, magnetic pressure dominates all other types of pressure (kinetic and gaseous), which are diminished by the low density of the coronal gas, and matter, ionised by the high temperatures, is confined by the magnetic field and adopts its geometry. We therefore observe loops of all sizes (from 100 to 10 000 kilometres) and temperatures. This confinement results from the trapping of charged

A sunspot and flocculi.
This photograph, taken in white light with a 38-centimetre aperture lens shows a sunspot and solar flocculi on a granulated background. One can distinguish the umbra of the sunspot (black) surrounded by the penumbra, consisting of filaments aligned along rays which appear to emanate from the umbra. The other small spots are flocculi, structures of the penumbra. Everywhere else there is solar granulation. (Observatoire du Pic du Midi)

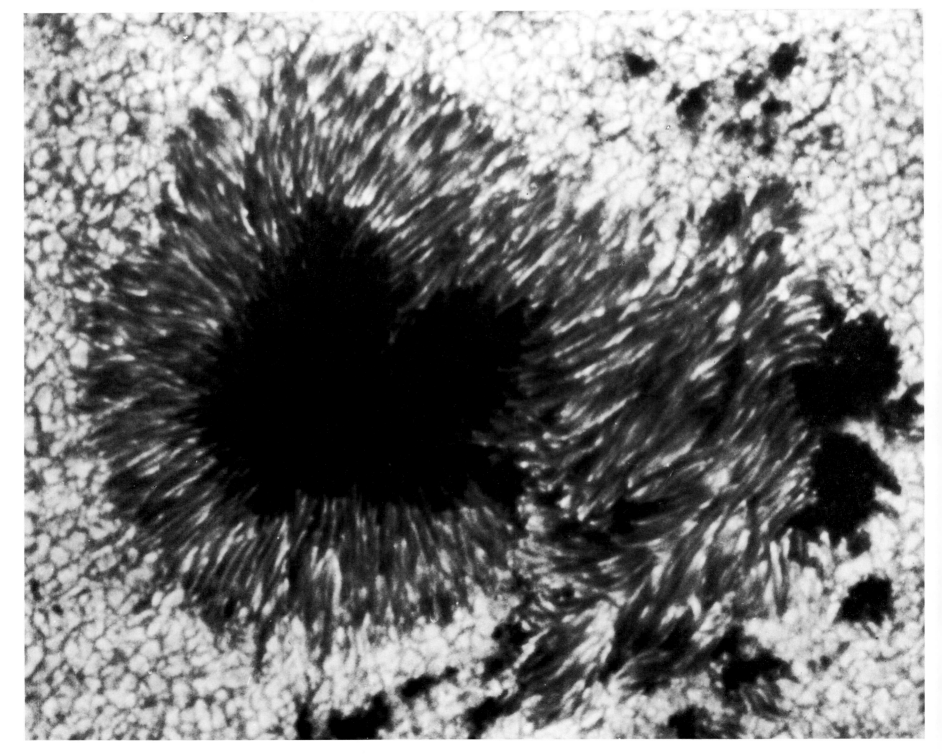

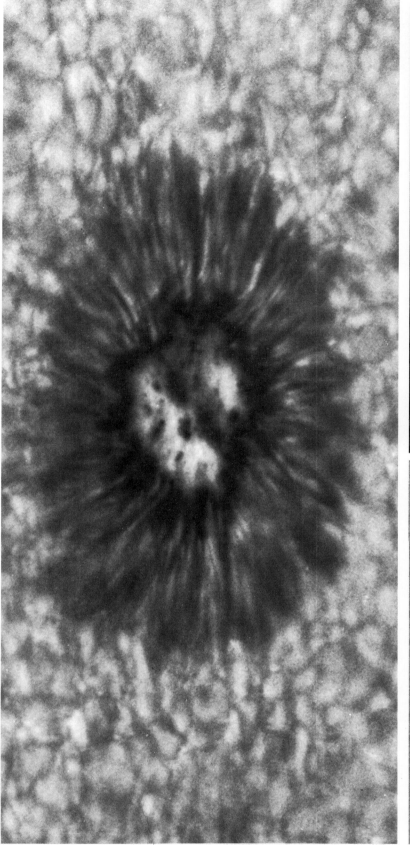

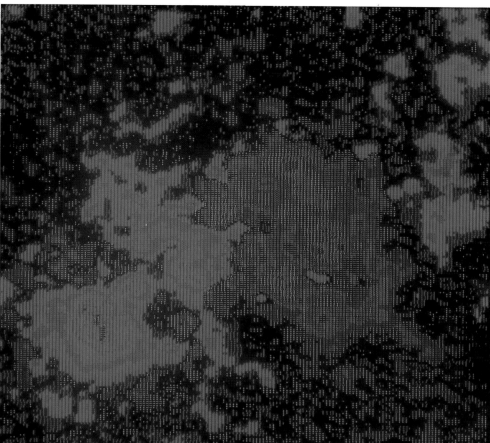

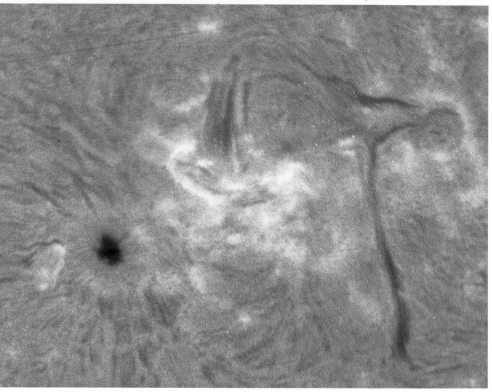

Bright points in a sunspot. In order to reveal the umbral dots, a negative and a positive of the same photograph were superposed. Small bright (hot) spots are thereby discernible in the umbra. Their diameter is less than an arc second and they exist for about half an hour. (Photographs: Sacramento Peak Observatory. Compositor: S. Koutchmy)

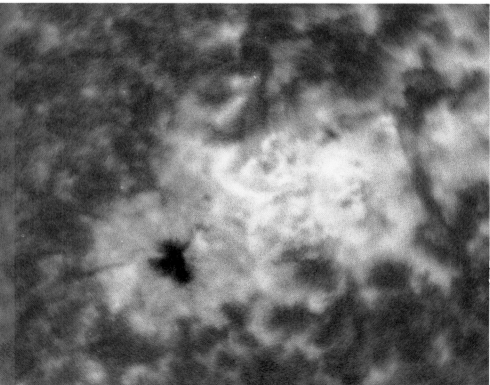

An active region. These three views represent the same active region. The first (above) is a magnetogram: the image is obtained by repeating Hale's observations point by point. Polarity and field strength are represented by the intensities of the two colours: red corresponds to southern polarity and green to northern. In the black regions the field is weak. Two zones of opposite polarity are visible.

The second image (middle) is a filtergram taken in H-alpha. Sunspots, filaments (dark) and fibrils (dark or bright threads) are visible around bright regions and faculae (bright).

The faculae are seen even better in the third image, taken in ionised calcium K.

Comparison of the three images shows that the two opposite polarities are predominantly implanted in the sunspot and the faculae. The filaments are situated along the borders between these zones. They are supported by the force lines of the magnetic field, which connect the regions of opposite polarity. (Observatoire de Meudon)

particles by the magnetic field: electrons, protons and ions spiral around the lines of force. It is remarkable that the same confinement is found in objects as different and distant from the Sun as nebulae.

X-ray images of the Sun reveal not only loops, but also bright spots, with radii less than 1000 kilometres, interpreted as small loops serving as bridges between regions of opposite magnetic polarity. Even the calmest regions are crossed by intense but localised magnetic fields; the result is an average magnetic field with an intensity about the same as Earth's.

Matter is not only contained, but is supported by the magnetic field. This explains the existence of filaments, which appear dark against the disk because they absorb sunlight, but bright beyond the limb since the light they emit is more intense than the sky. In the latter case they are called prominences. Their temperature (8000 K) is much lower than that of the ambient corona, which reaches a million degrees. The magnetic field only allows heat exchange along force lines and thereby isolates the filaments from the coronal furnace.

Under diverse influences (heating or magnetic restructuring) a prominence can be jolted out of equilibrium, or 'activated'. In some cases the matter (cold) rises at more than 100 kilometres per second to a height of about one solar radius, and vanishes. In other cases, the filament reforms after an abrupt disappearance and can thereby survive during several solar rotations.

In summary, an active region involves most of the phenomena resonsible for solar activity: sunspots, faculae, and filaments. Above all, as shown by X-ray images of the corona, it is a region of complex *closed* magnetic fieldlines. Reorganizations of this field cause spectacular and sometimes violent dynamic phenomena such as matter ejection and flares.

Jean-Claude VIAL

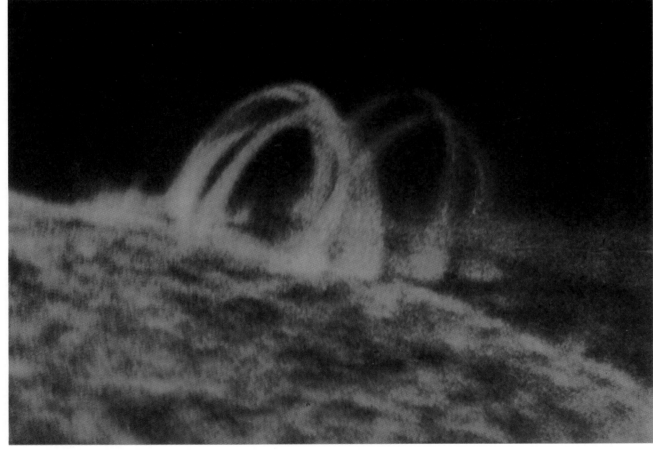

A looped prominence. Loops are observed on the edge of the Sun in the neutral hydrogen resonance line, Lyman-alpha (top). The matter is relatively cold (20 000 K) and, under the effect of collisions with ions, follows the force lines of the magnetic field. This image was obtained during a rocket flight on 3 July 1979, by the Laboratoire de Physique Stellaire et Planetaire (LPSP).

The lower picture was taken in the 46.5-nanometre resonance line of sextuply ionised neon; the matter is hotter (500 000 K). This photograph was taken on 14 August 1973 during a Skylab flight. The red colour is arbitrary. (Naval Research Laboratory)

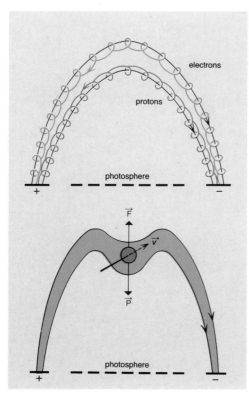

Loops and filaments. In a loop the charged particles spiral around the force lines of the magnetic field (above): positively charged particles (i.e. protons) circle clockwise, while negatively charged particles (i.e. electrons) circle anticlockwise. The high density makes these motions very slow.

A small electric current perpendicular to the plane of the figure ($\vec{v}$) is sufficient to make the Laplace force ($\vec{F}$) balance the weight ($\vec{P}$) of the matter (below). The field can thereby support filaments and, furthermore, isolates them from the hot environment.

Nebulae and magnetic fields. Photographs taken through differently oriented polaroids show that the light emitted by the Crab Nebula is polarised. The emission comes from electrons moving relativistically in a magnetic field, and is called synchrotron radiation. Maps of the field show that, as with solar loops, matter is aligned along the force lines of the magnetic field. (Lick Observatory)

A prominence in Hα. This prominence, seen here in Hα has a complicated structure. One can distinguish 'feet' and arches tens of thousands of kilometres high. The matter is fragmented into small vertical filaments with descending or, according to some researchers, ascending motions. Prominences probably form out of surrounding coronal matter, leaving a cavity. The chromosphere and spicules are visible in the foreground. (Big Bear Solar Observatory, California Institute of Technology)

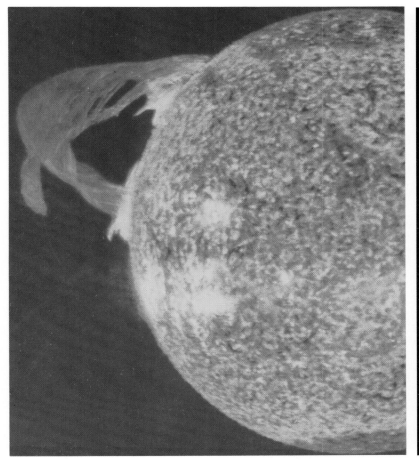

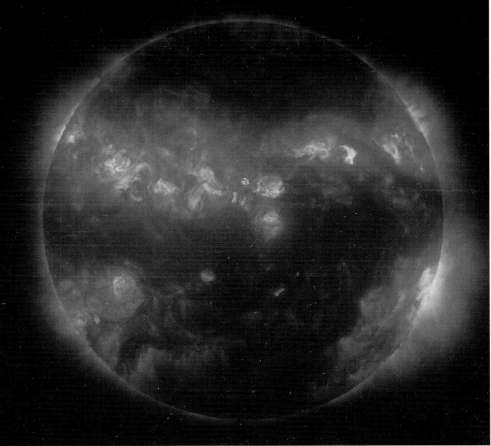

A prominence in ultraviolet light. On 19 December 1973, a giant prominence majestically took flight and escaped from the Sun under the eyes and cameras of the Skylab astronauts. One can clearly distinguish the matter 'braided' by the magnetic field. This observation was possible without a coronagraph since it was made in the ultraviolet (30.4 nanometre line of ionised helium). The brightness of the prominence and the disk are comparable; the colour is arbitrary. (Naval Research Laboratory)

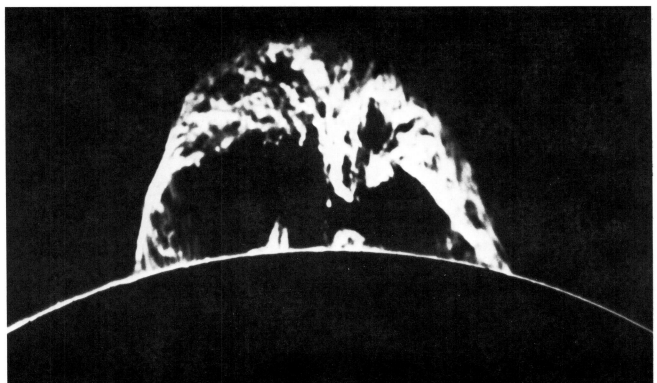

The disappearance of a prominence. This prominence, observed here in Hα with a coronagraph, will disappear as the matter accelerates upwards, starting at a few kilometres per second. Such an 'abrupt' disappearance can take several hours and often the filament reforms in the same place, suggesting that the magnetic lines of force remain intact. (Institute for Astronomy, University of Hawaii)

The active Sun

Solar eruptions

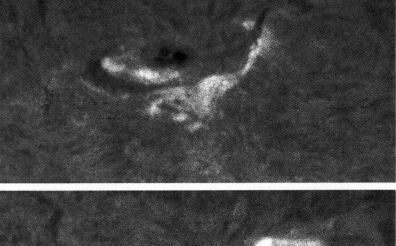

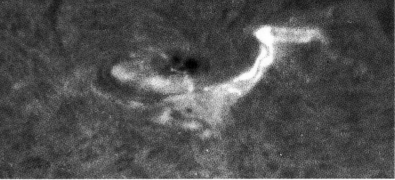

The solar atmosphere is shaken periodically by eruptions, violent phenomena whose effects can be felt as far away as Earth. They were not discovered until nearly two centuries after sunspots, because they are rarely visible in white light; the best and most frequent observations are made in H-alpha. An eruption is characterised by a large increase in brightness over a region as enormous as five million square kilometres. The H-alpha line of such a region is an emission feature rather than an absorption line as in the lower chromosphere, indicating that the emitting region is hot and dense. Deformations are also found in ultraviolet emission lines characteristic of the lower chromosphere. The entire electromagnetic spectrum changes abruptly: from gamma-rays to radio waves, the continuum and all lines are amplified or emitted. This signifies that all the external layers are perturbed in the neighbourhood of the eruption.

There is a large range of eruptive events, which are classified predominantly according to their area of emission rather than their geometry, lifetime, spectral nature, etc. Their one common feature is the abruptness of the event: in less than a minute the line intensities increase more than ten-fold: this is called the 'flash' phase. It then takes several tens of minutes or hours for the emission to return to its normal level. The solar atmosphere becomes relatively calm again, but other eruptions may follow, sometimes strictly consecutively, each in the same place as the preceding: these are known as homologous eruptions.

Such movements of matter are spectacular and deserve the name eruption. Though not always visible at the centre of the solar disk, they are evident on the limb: loops and arches form and the matter mounts and descends along them. Here again, loops are bridges of matter linking two regions of opposite magnetic polarity. Eruptions occur most frequently at the interface of two such regions and are called double filament eruptions. The magnetic field plays many roles, and may initiate eruptions as follows. A filament often straddles the neutral border between two polarities. If matter impinges on this border from opposite directions the magnetic

field may no longer be able to sustain the filament. It will disappear and the field will untwist, freeing a large amount of energy. As plausible as it may be, this scenario is only one of many which have been advanced to explain eruptions.

All the plasma particles, particularly ions and electrons, feel the effect of the liberated energy. Since the mechanism for accelerating these particles is still highly controversial, only its effects will be examined. These can be studied with X-ray detectors in space and with ground-based radio telescopes. Hard X-rays (energy greater than 50 kiloelectronvolts) are emitted by the braking effect of ions on electrons; this is bremsstrahlung, and is only observed after eruptions. For a long time it was thought that the observed X-ray spectrum was emitted by a non-thermal electron source, that is by electrons much hotter than the ambient plasma (5×10^6 K, equivalent to 0.5 kiloelectronvolts). However, it is now believed that the entire plasma is hot enough (10^9 K) to produce hard X-rays. The problem of confining such a hot plasma for a sufficiently long time still remains. The magnetic field is theoretically capable of this, since charged particles are trapped around lines of force. They emit gyrosynchrotron radiation at wavelengths from millimetres to metres. The synchronisation between X-rays and radio emission suggests that the acceleration of electrons takes place in the corona.

At radio wavelengths eruptions appear as bursts. Bursts of type III show a very rapid frequency drift which is interpreted in the following way: below a limiting frequency (called the plasma frequency) the waves can no longer propagate. This frequency is proportional to the square root of the density, which decreases rapidly away from the Sun. If the electrons move quickly enough they produce waves in the plasma with about the plasma frequency. The speed of propagation of the electrons and hence that of the drift in frequency can thus be deduced. The latter is greater than 100 000 kilometres per second. Bursts of type IV are observed throughout the radio spectrum. Some are due to shock waves which move at more than 1000 kilometres per second. Very rapid pulsations, lasting several

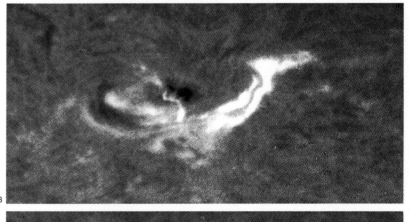

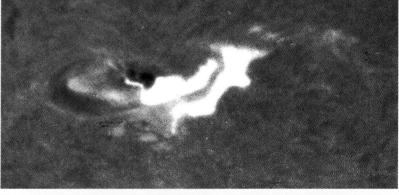

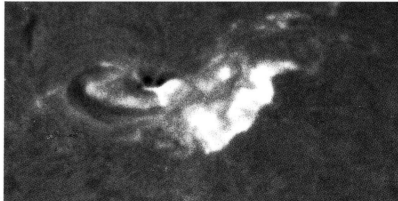

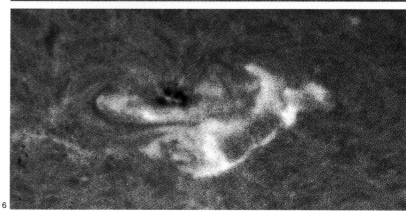

Evolution of an eruption. Surveillance of eruptions is carried out in a routine manner in Hα. This series of photographs, taken on 7 September 1973, shows the evolution of an eruption in a region 15° south and 46° west. The field is about 5′ by 4′.
1. At 11 h 38 min universal time (UT), the active region is in a pre-eruptive phase: one can distinguish near the sunspot a filament broken in two. To the left the remaining piece delineates the border between two oppositely polarised regions; to the right, the other piece is in the process of rising.
2. At 11 h 42 min UT bright spots appear along a thin dark ribbon in the place of the vanished filament. Such an eruption is called a double filament. One can distinguish knots in these bright patches.
3. At 11 h 50 min UT the eruption develops: the bright zones begin to lengthen and rise.
4. At 12 h 02 min UT is the moment of maximum emission.
5. At 12 h 32 min UT the bright matter (right) reaches its maximum altitude and takes on an irregular form, probably imposed by the magnetic field.
6. At 14 h 22 min UT it is almost calm. Note that a filament has reformed in the place of the double filament. The piece of the filament at the left survived the eruption. (Photographs: Observatoire de Meudon)

Evolution of line profiles during an eruption. During an eruption, line profiles are modified considerably.
a. The resonance line K of ionised calcium, usually an absorption line, becomes an emission line. Note that the far wings and the few absorption lines are little reinforced.
b. An emission line (the resonance line K of ionised magnesium) usually inverted, becomes very intense, large, and has a single peak. Note the presence of some other emission lines on the wings.
c. The same phenomenon, only more pronounced, occurs in hydrogen Lyman-alpha. The profile of the quiescent Sun (white line) has been amplified by a factor of ten; otherwise it would barely be discernible. An increase in brightness by a factor of twenty is observed.
d. Light curves of three lines, Ca K, Mg K, and Lyman-alpha. The energy contained in each of these profiles was calculated, and the evolution in intensity was traced. Note that the flash phase begins in less than 20 seconds, and lasts several minutes; the time required to return to normal depends on the line: 300 seconds for Lyman-alpha, 500 seconds for Mg K, and more than 1200 seconds for Ca K. These three atmospheric lines were measured simultaneously by the multichannel spectrometer of the Laboratoire de Physique Stellaire et Planetaire (LPSP) on board the satellite OSO–8.

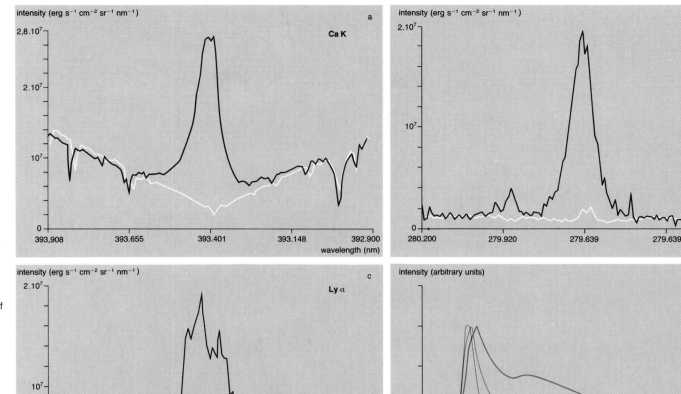

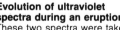

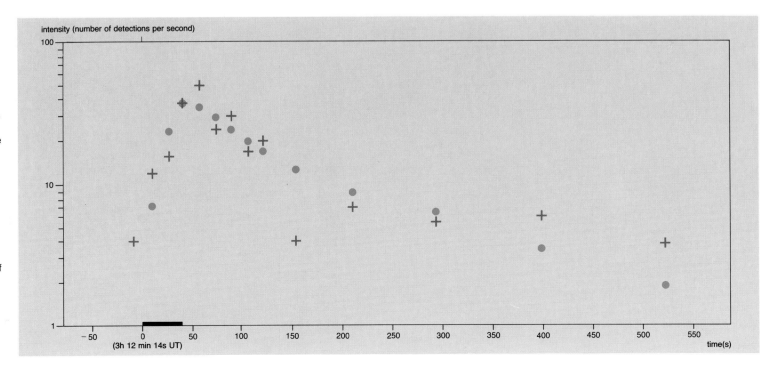

Evolution of ultraviolet spectra during an eruption. These two spectra were taken in the ultraviolet (around 152 nanometres) on board Skylab. As on a negative, the bright regions are black. The spectrum stretches (horizontally) over 3 nanometres.
Above, the solar region observed was quiescent; below, undergoing an eruption. The emission lines (chromospheric and coronal) are intensified and enlarged in the latter case. A multitude of lines usually drowned in the grain of the film (above) appear during the eruption. (Naval Research Laboratory)

Type IV burst. This observation was made with the decametric radioheliograph at Nançay (169 MHz). The signal intensity at this frequency was maintained as a function of time. Note the abrupt variation every second, with no well-defined period. (Observatoire de Meudon)

Gamma-radiation during an eruption. Detection of gamma-rays from a nuclear reaction 150 million kilometres away is difficult. This observation was made on board the SMM (Solar Maximum Mission) satellite during a big eruption on 7 June 1980. The line observed at 2.223 megaelectron volts results from the capture of emitted neutrons and is produced by the photosphere, bombarded by rapid neutrons from the eruption. The curve shows the line intensity as a function of time. The crosses are measurements and the circles are points calculated assuming that the production of neutrons lasts 40 seconds (blackened line).
Note that the light curve is similar to curves observed at more easily accessible wavelengths (or energies). (From E. L. Chupps *et al.*, 1981)

minutes with periods of the order of a second, are also observed.

In addition to electrons, protons are accelerated. Through collisions with helium nuclei they produce neutrons which lose energy, are captured by hydrogen, and emit a gamma-ray at 2.2 megaelectron volts. This ray has only been observed quite recently, mainly with the SMM (Solar Maximum Mission) satellite. The light curve of the ray is similar to that of hard X-rays or radio waves.

These energetic particles can also be detected outside the Earth's atmosphere by satellites: flares in which protons are detected are call proton flares. Some particles can even be detected by means of the atmosphere: as they travel through the upper atmosphere, relativistic solar particles (protons with energies above 0.5 GeV) produce secondary particles detected by neutron monitors. The latter part of 1989 was particularly prolific in such events and it is thought that some particles of energies around 20 GeV were detected. Two large solar events in September and October 1989 each lasted about a week, and injected far more particles than the total detected since 1986. Flares of this violence are dangerous to human beings exposed to the radiation in space (in orbiting space stations or shuttles) or even in the stratosphere.

It is clearly important to understand how flares work. The SMM satellite, lost in December 1989, made an important contribution to this. But it is also important to try to predict them, in view of the perturbations they cause to the Earth's environment. Marie-Josèphe Martres and Zadig Mouradian of the Meudon Observatory have shown the existence of unusual solar magnetic regions which seem to rotate rigidly, as if attached to deeper layers. These are the preferred sites for flares, and are therefore to be monitored carefully. The American researchers Bai and Clever have discovered a periodicity of 54 days in the appearance of proton flares between 1955 and 1986 (i.e. over three solar cycles).

Finally we should mention that stars of most spectral types show flares, some of them thousands of times more violent or extended than solar flares.

Jean-Claude VIAL

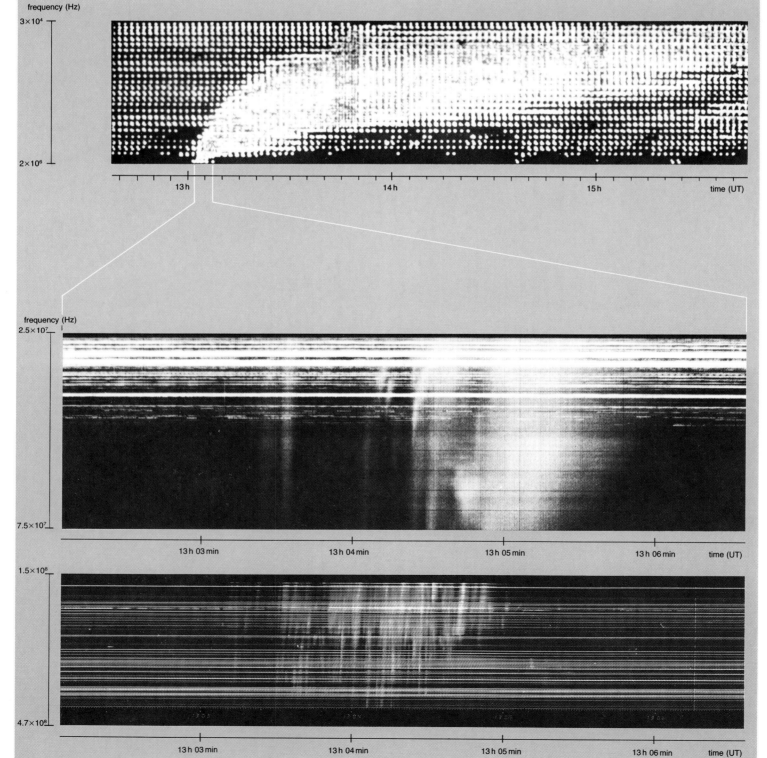

Dynamic spectrum of solar bursts.
This figure shows the dynamic spectrum of type III solar bursts in three frequency ranges, observed by three different instruments: a multi-channel radiospectrograph (24 channels from 30 kilohertz to 2 megahertz) on board the probe ISEE–3 (above); a wide channel decametric frequency scanning spectrograph (from 25 to 75 megahertz) (middle); and a multichannel digital radiospectrograph (from 150 to 470 megahertz) (below).

One line (one frequency) represents the intensity of the radio signal as a function of time.

A group of bursts takes off at a high frequency (470 megahertz), which it maintains for about one minute (above). The frequency then decreases and the individual bursts merge (middle). Still later one observes a single burst lasting more than an hour with a frequency less than 30 kilohertz (below). This is the radio visualisation of subrelativistic electron packets which, accelerated in the base of the corona, propagate into interplanetary space beyond Earth's orbit. Each packet stretches considerably while propagating so that the initially distinct groups are transformed into a single homogeneous packet. (Above: ISEE–3; middle and below: Observatoire de Meudon, Station de Nançay)

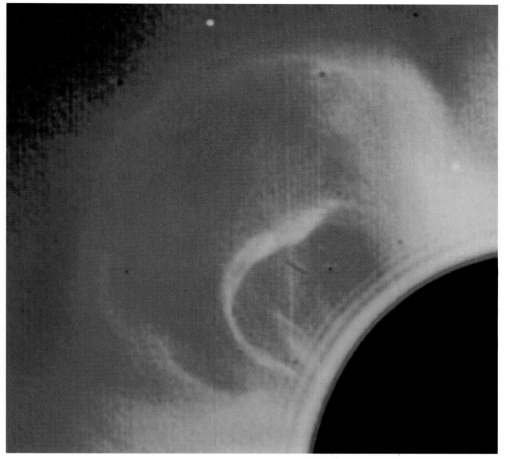

A coronal transient (opposite). This coronal transient was observed by the coronagraph on board the SMM satellite. This instrument allowed observation of coronal structures up to 20 solar radiuses above the limb. On 14 April 1980 plasma was ejected in a flare and moved through the corona as an enormous loop of several hundred thousand kilometres with a speed of the order of 1000 km/s. A shock wave formed ahead of the loop, and the gas behind was rarefied. Transients are often associated with radio outbursts (types II and IV) and an increase in the solar wind speed. (High Altitude Observatory, National Center for Atmospheric Research)

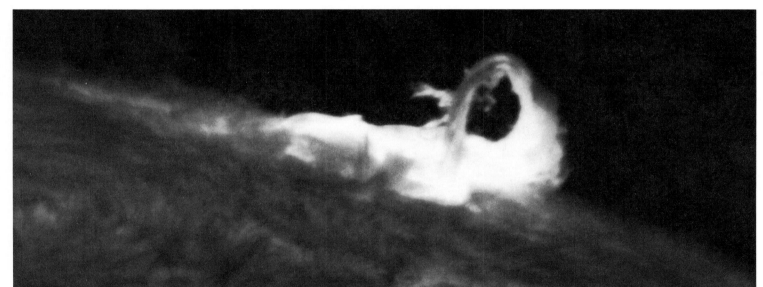

An eruption in Hα. At the edge of the Sun it is possible to observe the vertical extension of an eruption. An eruption of 9 July 1974 is seen here in Hα. Matter is ejected and visibly follows magnetic loops. The mass involved reaches 10^{13} kilograms, and speeds can surpass 1000 kilometres per second. The energy dissipated is of the order of 5×10^{24} joules, almost equal to the thermal energy of the eruption, which is about 10^{25} joules. (Big Bear Solar Observatory, California Institute of Technology)

Post-flare loop system. A montage of two Hα photographs by the large coronagraph of the University of Wroclaw on 25 October 1989 at 9.44 UT. The artificial moon of the coronagraph was moved between the two exposures. The passband width of the filter (0.3 nanometres) allows one to see the photosphere. The structures clearly visible around the limb are matter loops which form after flares and disappear as they rise in the corona. The picture of the disk shows the photospheric active centres responsible for the flares (sunspots, faculae...). Some of the sunspots show the *Wilson depression* (no visible penumbra on the side opposite the limb). Similar montages on the disk away from the limb are difficult to obtain and interpret because of projection effects. (B. Rompolt)

The active Sun

The solar cycle

On the Sun we observe sunspots, pores, faculae, regions of intense magnetic field. This field becomes increasingly important as we consider higher layers of the solar atmosphere, and is without doubt involved in triggering flares. But how and where is it generated? Obviously below the atmosphere, within the enormous rotating mass of gas (2×10^{30} kilograms); as they move, electrons and protons create an electric current which induces a magnetic field. The Sun is a giant dynamo.

Let us examine the working of this dynamo more closely, particularly its origin in the Sun's rotation. This rotation has a period of about twenty-seven days (the synodic period, measured from the Earth), but the equatorial regions rotate more rapidly than the other regions: twenty-six days compared with thirty-one at a latitude of 60 degrees. For this reason the rotation is called differential. The rotation periods are measured by observing the appearances and reappearances of surface features (spots, prominences) under rotation, or by measuring the Doppler shifts of spectral lines at the Sun's limb: the surface rotation velocity of about 2 kilometres per second shifts a line at 600 nanometres by about 0.004 nanometres; the shift is to the blue (shorter wavelengths) if we observe the East limb, and to the red if we observe the West limb.

This differential rotation transforms a weak poloidal field (from pole to pole) into a strong toroidal field (parallel to the equator): the rapid equatorial rotation contorts the field lines. Because of convection the field lines twist around each other and form magnetic 'ropes', in the same way that a strong rope can be made by twisting fine threads together. Where the magnetic field is strong the magnetic pressure is also strong, and the density lower (assuming there is equilibrium with neighbouring regions). Buoyancy forces overcome the weight of the

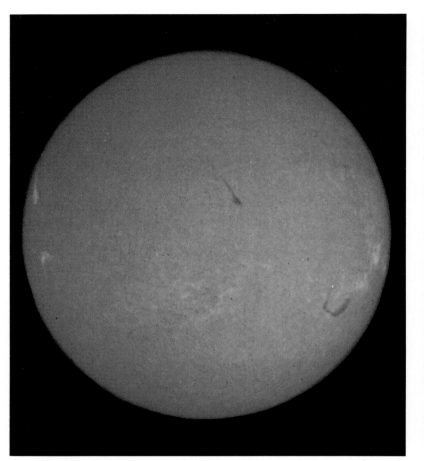

From minimum to maximum. These photographs of the chromosphere in the hydrogen Hα line show the solar disk during a minimum and a maximum of its activity: dark filaments – prominences – and bright facular plages have increased. (Observatoire de Paris-Meudon)

The Maunder diagram. If the sunspot number is counted each day and a mean – the Wolf mean – is constructed and plotted annually one notes the presence of minima and maxima with a period of around eleven years (in fact, the period varies between 9 and 12.5 years). However, one also notes various irregularities such as the fact that the decline is less abrupt than the rise, and alternations of strong and weak cycles. Further, if one plots the latitude of the observed sunspots as a function of time one obtains a so-called *butterfly* diagram. The diagram shown here combines the data collected since sunspots were observed systematically. Each sunspot (or group of sunspots) is marked by a vertical 1° line. We note the eleven-year period, with two characteristic times: a maximum (with sunspots at various latitudes, e.g. in 1969) and a minimum which can last two or three years (e.g. in 1975–6): one moves from one butterfly to the next. This diagram is very rich in information. First, the wings do not extend above 45°, i.e. sunspots do not form above this latitude, called the royal zone. Second, the wings are V-shaped: the sunspots first appear at latitudes of 30°–40° and descend towards the equator. Finally the equator seems no more favoured than the poles. The total area covered by the sunspots is also shown: it is expressed in thousandths of the solar surface, a quantity which differs little from the sunspot or Wolf number. We note here too that the eleven-year period is disturbed by cycle-to-cycle variations (compare 1958 and 1969) and that within a cycle the rise is quite abrupt and the decline more gentle. One would like to know what happened before 1880, and what will happen 'afterwards'... (Science Research Council, Royal Greenwich Observatory)

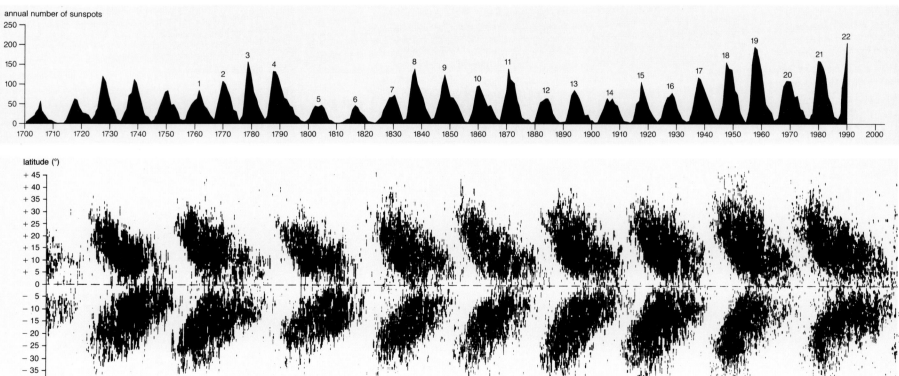

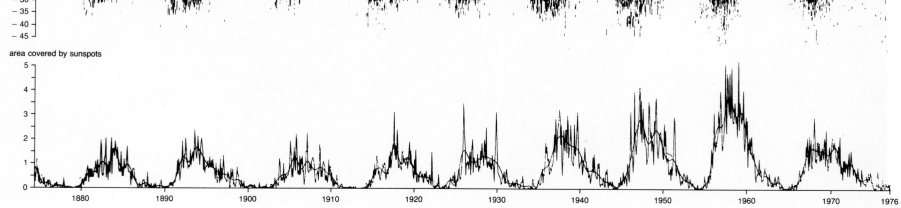

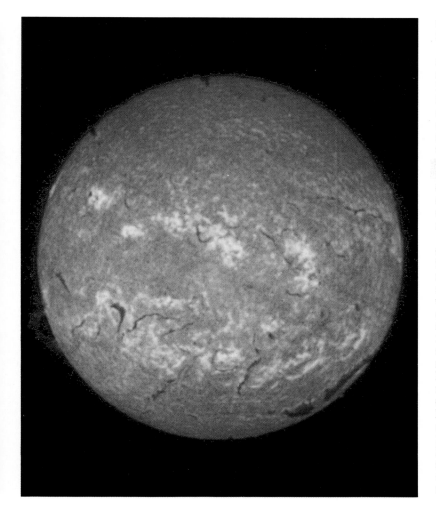

structure: magnetic regions rise and emerge, forming sunspots among other things.

In reality the Sun is more complex. The dynamo shows periodic changes: roughly every eleven years (the period varies between 9 and 12.5 years) the field changes its orientation, giving the twenty-two year cycle revealed by long and patient observation of the sunspots. The moment when the sunspots appear in the largest numbers is called solar maximum. Sunspots are very numerous at this time, as also are polar aurorae on the Earth. The terrestrial magnetic field is disturbed; radio communications and even electrical power supplies may be disrupted, and there is danger for astronauts. The lifetimes of low orbit satellites are considerably reduced at solar maximum.

This influence is clearly very old, and ancient observations of terrestrial phenomena allow us to reconstruct the Sun's history. Direct solar recording, such as sunspot counts or measurements of the solar diameter, really began only with Galileo's telescope, and was carried out by different methods (and different observers). The near-absence of sunspots between 1650 and 1700 (called the Maunder minimum) is nevertheless well established, and it is interesting to note that it coincided with a colder climate in Europe. To go back further in time one has to use cosmic isotopes. Protons, neutrons and other cosmic ray nuclei continually bombard the Earth's atmosphere, especially the abundant atoms of nitrogen and oxygen, forming carbon-14 and beryllium-10. These isotopes have limited lifetimes (with half-lives of 5730 and 1.5 million years respectively), and their present abundances allow one to measure the time since they were formed. Certain conditions are never-

theless required: the object from which they are sampled must not have been exposed to any other irradiation since then (this is particularly true for tree trunks, frozen carrots and sea sediments) and the production mechanism must have stayed constant over time. The latter condition probably holds as the cosmic radiation is unlikely to vary except as a result of supernova explosions. The target, the atmosphere, is certainly sensitive to solar activity, but also to the Earth's magnetic field (particularly at the Equator). The Sun's activity itself can play contradictory roles: the solar wind can inhibit the penetration of cosmic rays but it can also contribute isotopes in strong flares. If, on the other hand, we know the elapsed time, we can work out the atmospheric (and thus solar) conditions under which the isotope was formed. Isotopic analysis of carbon-14 is complicated by its stay of around thirty years in the atmosphere, but analysis of tree trunks in the Caucasus shows eleven-year cycles even in the Maunder minimum. Beryllium-10 sampled from glacial layers is even more useful: its lifetime allows one to go further back, and its shorter stay in the atmosphere (two years) provides better time resolution. The influence of solar activity on the Earth's climate is seen in a correlation between stratospheric heating and solar maxima, but we should be cautious: after all we have not found a correlation between the average global temperature and the solar cycle in about a century.

In the present epoch magnetograms giving the intensity and orientation of the magnetic field are decisive: magnetic maps obtained at two maxima separated by eleven years show that the field completely changes direction in the two

The solar dynamo. How do we make a strong toroidal field from a weak poloidal field? We start from a poloidal field restricted to the polar regions, limited by the dashed curves (a). The field lines cross the surface layers of the Sun. Here the North geographical pole is a South magnetic pole. Because of the differential rotation the field line segments at the equator are pulled along more rapidly than the rest. This produces an equatorial stretching and the appearance of a toroidal field (b). The 'porthole' (c) allows us to see the strong concentrated fields and the opposing orientations of the two hemispheres. We note their slight inclination towards the equator, explaining the lower latitude of the leading sunspots compared with the trailing sunspots. In the following eleven-year cycle all of the arrows are reversed except for the Sun's rotation.

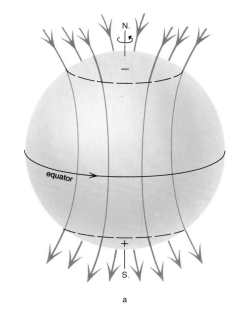

a

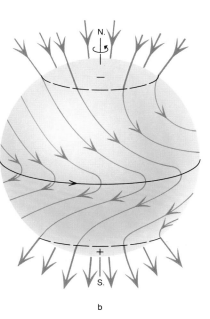

b

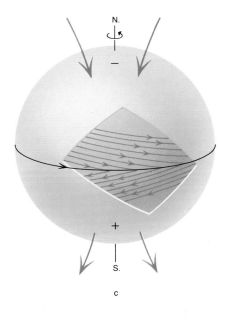

c

Formation of bipolar regions. At the beginning of the cycle (left) the sunspots result from the emergence of field lines under buoyancy forces. In general they appear in pairs: the leading sunspot has the same polarity as the hemisphere to which it belongs, while the trailing one has the opposite polarity, but it is situated at a slightly higher latitude. Because of the drift of the field towards the poles the magnetic flux of the trailing sunspot will arrive before the flux of the leading spot, and the combined effect of several sunspots will reverse the field at the poles. At the end of the cycle (right) the new poloidal field creates a new toroidal field which begins to oppose the initial field, until completely annihilating it at solar minimum.

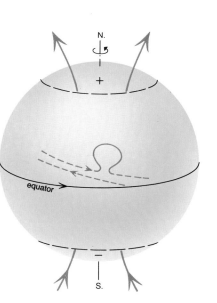

hemispheres. A closer study reveals that the field at the poles changes sign right at the solar maximum. Can we understand the migration of the sunspots and their recurrence over periods of eleven and twenty-two years? There is an explanation. The sunspots appear in pairs with opposing signs: at the beginning of the cycle the sunspot which leads in the direction of rotation has the same polarity as the hemisphere in which it is situated; it moves towards the equator, and under the effect of the differential rotation its distance from the following sunspot increases and its field progressively weakens. In contrast the field of the following sunspot, whose polarity is opposite to that of the hemisphere, slowly moves towards the pole under the effect of a slight but effective meriodional circulation. At maximum the sum of all these fields cancels and then reverses the established poloidal field, implying also a reversal of the creation of the toroidal field, which also eventually vanishes and reverses. The Sun thus reaches a minimum with few sunspots. At this point the strong

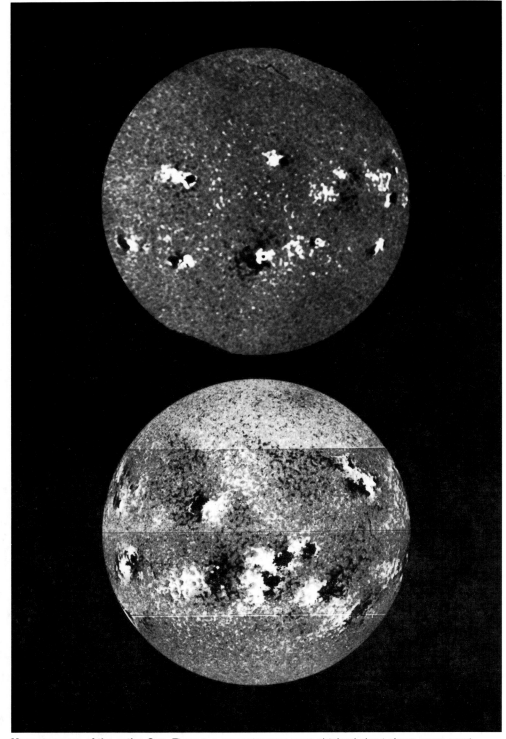

Magnetograms of the active Sun. These two magnetograms were obtained about eleven years apart, near solar maxima. The black and white areas correspond to two field polarities; grey represents a weak magnetic field.

We note the association of black and white pairs corresponding to bipolar regions (pairs of sunspots). Between the two pictures each hemisphere changed its system of polarities (the field reverses every eleven years). We note finally the slight inclination of the magnetic structures with respect to the equator. This inclination is also seen in the filaments visible on Hα spectroheliograms. The other differences between these two pictures result from the improved quality of the lower heliogram; the horizontal and sloping lines result from the observational technique. (Kitt Peak National Observatory)

Relative variation of the amount of carbon-14 measured in tree trunks in the Caucasus. This variation is representative of the rate of carbon-14 formation between 1600 and 1730, before and during the Maunder minimum. Note the increase and a periodicity coinciding with solar activity minima. (after G. Kocharov)

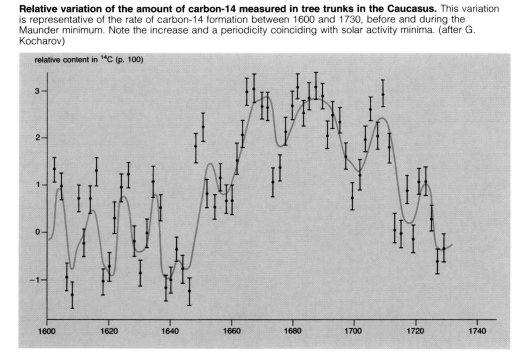

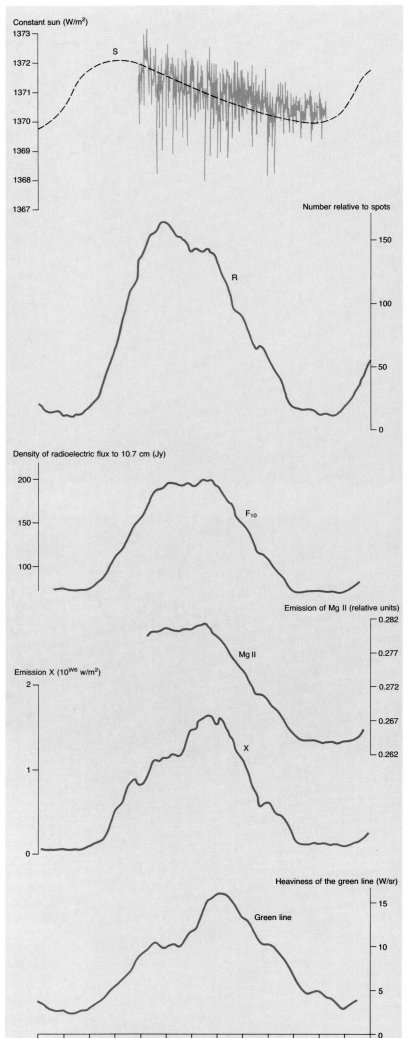

Some indices of solar activity. There are many indices of solar activity, each corresponding to a particular phenomenon. The range of these indices has increased considerably in the space age. In cycle 21, which extended from 1975 to 1986 it was thus possible to measure: the solar constant S, the total luminous flux (essentially of photospheric origin) crossing a one metre square surface at normal incidence at one astronomical unit from the Sun, i.e. at the Earth's orbit, above the atmosphere (result obtained by the Nimbus-7 satellite: the dashed curve is an average): the relative number R of sunspots: the flux density at 10.7 cm wavelength at one astronomical unit from the Sun, F_{10}: the emission in a chromospheric line of singly-ionised magnesium Mg II (result obtained by an instrument on board Nimbus-7): the emission in the soft X-ray region from 0.1 to 0.8 nm from the corona (result obtained by an instrument on board the GOES satellite); the intensity of the 'green line' (line of thirteen-times ionised iron at 530.3 nm), also formed in the corona.

All these indices show a maximum at about 1980–2, but as we go down in the figure (and rise in the solar atmosphere) the form of the maximum changes considerably. The abrupt rise of R between 1977 and 1979 for example corresponds to a slower growth in X-rays and the green line, so that the maximum moves from 1979–80 for R to 1981–2 for the X-rays and the green line. (After R.F. Donelly, 1990)

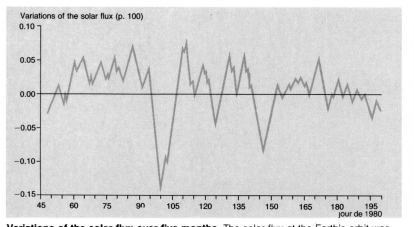

Variations of the solar flux (p. 100)

jour de 1980

Variations of the solar flux over five months. The solar flux at the Earth's orbit was measured by a radiometer on board the SMM satellite. The relative variations over the 153 days after launch in February 1980 are expressed in percentages of the mean detected flux (1368 W). The two minima at days 100 and 145 in 1980 correspond to the passage of large sunspots across the Sun's disk. The maximum deviation is 0.15 per cent of the mean.

No doubt because of perturbation of the Earth's atmosphere such effects have not previously been observed despite thirty-five years of effort by Charles Abbot at the beginning of the twentieth century. (After R.C. Wilson, C.H. Duncan and J. Geist, 1980)

Variations of the solar constant over ten years. These measurements of the solar constant were made by instruments on board the Nimbus-7, SMM and ERBS satellites. They begin before the solar maximum of the years 1980–1 and end at the start of the maximum of the next cycle, which ended in 1990.

One might be surprised by the systematic differences between the three series of measurements; however, we should note that the differences are of order 3 or 4 W/m², or 0.3 per cent of the value of the solar constant, which is excellent for absolute photometry (the vertical bars show the absolute accuracy).

The three instruments all show a minimum of the constant at the time of minimum solar activity (in 1986); as the passages of active regions are then less frequent, the variations of the constant on the scale of months are weaker. The 'valleys' correspond to numerous extended sunspots. The modulation of the Sun's emitted luminous flux between solar maximum and minimum is of the same order as the seasonal variation of the received solar flux at Earth. (After R.B. Lee, III, 1990)

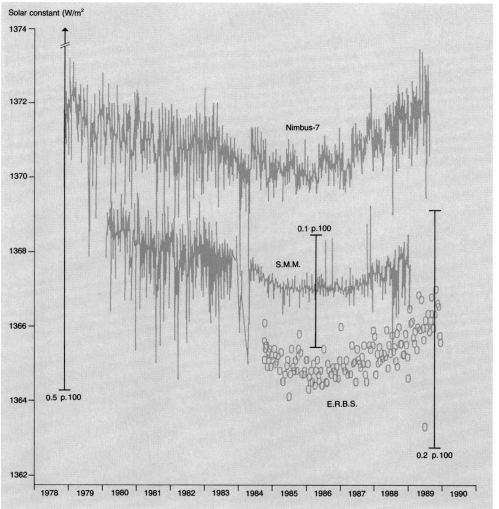

Solar constant (W/m²)

Nimbus-7

0.1·p.100

S.M.M.

0.5 p.100

E.R.B.S.

0.2 p.100

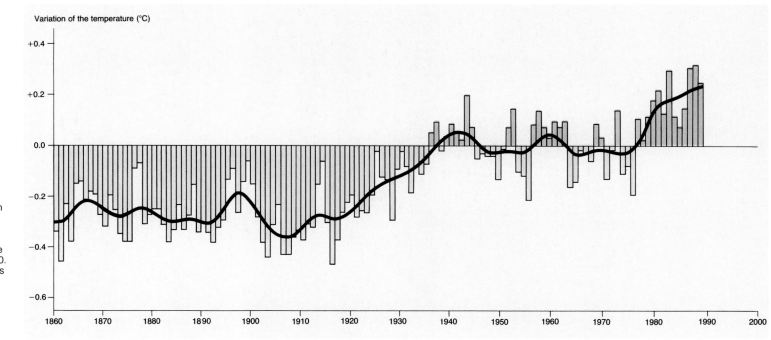

Variation of the temperature (°C)

Variation of the average global temperature. This histogram shows the variation of the global average temperature (above the continents and oceans) expressed with respect to the average of the years 1951–80. If we compare the form of this variation with the sunspot diagram there is no correlation: there is nothing resembling the periodicity of the solar cycle. (After J. T. Houghton *et al.*, 1990)

polarity of the pole spreads widely where the toroidal field is weakest and organises the open field lines. These diffuse regions are the coronal holes.

The Sun is, however, so complex that many aspects of it remain unknown. Where exactly is the dynamo situated? Should not the concentrated magnetic fields be rapidly destroyed by matter motions? What is the exact role played by turbulence? How does the differential rotation react to the creation of the field? Does the Sun rotate more rapidly when the number of sunspots decreases? How do we simultaneously explain the whims of the solar cycle, such as the Maunder minimum, and the phase loss observed

between the periods before and after this minimum? Why do the coronal holes rotate more or less rigidly? Does the solar cycle have a chaotic origin?

Systematic varied observations are required to answer these questions: measurements of velocities and magnetic fields, studies of the formation and disappearance of structures, and precise continuous study of the full spectrum of emitted radiation. The first results obtained by the SMM satellite surprised astronomers: the emitted solar radiation decreased systematically each time a large sunspot passed over the disk. Nine years of activity allowed one to study the phenomena in greater depth: in fact the solar flux decreases at

solar minimum, a result confirmed by other satellites (Nimbus-7 and ERBS). Another tool has recently appeared in the shape of observations of solar oscillations. Study of these over the last cycle shows that the oscillation frequencies seem to increase slightly (by about one part in ten thousand) with solar activity. The variation is stronger at low latitudes and shallow depths. This information is particularly valuable for understanding the source of solar activity.

Jean Claude VIAL

The Solar System

The Solar System

Only a few years ago, the Solar System could not have been given a sumptuous pictorial treatment. The richness of the pages which are devoted to it today demonstrates the importance of scientific results obtained from exploration using an unprecedented technology. Well before artificial satellites were orbiting the Earth, scientists realised how much they would increase our understanding of the environment of our planet. The years which followed the launch of the first Sputnik by the Soviets in 1957 have been marked by a large number of discoveries. Among these, the realisation that there is a dense population of highly energetic charged particles surrounding the Earth – the Van Allen belts – was without doubt the most remarkable. The desire of scientists to apply measuring techniques to study the environments of other planets was made possible by rapid improvements of space vehicles and the mastery of the techniques of very long distance communication. Of course the first target was our own satellite; we will have to wait a long time before our knowledge of the planets reaches the same level as that which we have of the Moon. The greatest scientific contribution, and the most spectacular feature, of the Apollo missions came from the work of the astronauts on the lunar surface. Nevertheless, an equally rich scientific harvest, from the analysis of the samples of lunar rocks in laboratories all over the world and their comparison with studies of meteorites, has truly opened up a new discipline – planetary science.

During the Soviet Luna missions, the collection of lunar samples was made entirely automatically. In addition, the exploration of the Moon required a series of scientific platforms to be placed in orbit around it. In this way it was shown how useful remote sensing could be as a method for the geological analysis of planetary surfaces, a method whose efficacy has been largely confirmed during the Viking missions to Mars. In this particular case it was also shown that a good analysis of the properties of a planetary surface could be carried out using extremely complex instruments on board an automatic station placed on the Martian surface.

As for Venus, it shows the determination of engineers to develop apparatus to overcome a hostile environment as soon as the scientific rewards merited the effort.

This mysterious planet is similar to the Earth in mass and size but has a higher surface temperature (470 °C) and a denser atmosphere consisting almost entirely of carbon monoxide. The exploration of Venusian soil was pioneered by Soviet scientists with their highly successful seven Venera and two Vega landing probes (the latter being the Venusian component of the

mission to Halley's Comet). These probes were able to transmit signals, sometimes for over an hour, Global studies of the planet topography were achieved through satellites and radar images (from Pioneer Venus and Venera 15 and 16). Such methods, as for the Moon and Mars, advances our understanding of the thermal evolution of the planet.

Comparative analysis of the geological process which shaped both the planetary surfaces and those of the satellites of the giant planets have opened a new chapter in Earth sciences, called comparative planetology. It effectively began in 1974 with the Pioneer 10 flyby of Jupiter. A few years later, the Voyager probes explored the frozen and colourful worlds which make up the systems of Jupiter and Saturn. The mission is not finished: Voyager 1 continues to transmit information about the frontiers of the Solar System and Voyager 2, after observing the Uranus system in 1986, is now travelling through interplanetary space towards Neptune.

Perhaps the impact of technology on planetary research has been greater than in other areas of the astronomical sciences. However, it is obvious that the scientific demands of this discipline have directly influenced the rapid development of new technologies.

Microelectronics would not have experienced its prodigious expansion nor reached its present level of reliability if there had not been, from the outset, the need to transport ever more complex and precise instruments over ever greater distances. The most often quoted example in this respect is the comparison between the weight of scientific instruments – the payload – of two probes built by engineers at the Jet Propulsion Laboratory. In 1962, after a four month journey to Venus, Mariner 2 flew by the planet at a distance of about 3000 kilometres; the probe weighed 656 kilograms and carried 18 kilograms of scientific instruments. Launched in 1977, the Voyager probes underwent an ordeal of several years which brought them to the neighbourhood first of Jupiter, then of Saturn. Slightly heavier than Mariner 2, these probes carried 113 kilogrammes of extremely sophisticated apparatus. This difference in payload shows the extent of the progress made with the auxiliary equipment in spacecraft, and with the techniques of communication and data transmission. These techniques have advanced to such an extent that it is possible to apply to the Earth Sciences methods of remote sensing developed mainly for planetary exploration.

Planetologists have used the most modern techniques of *in situ* chemical analysis to study the chemical composition of the telluric planets and their atmospheres. This has stimulated the

The Pioneer missions to Jupiter and Saturn. When the pictures of Jupiter obtained since 1974 by Pioneer 10 were published, NASA engineers proved that a probe could cross the asteroid belt unhindered and transmit data after being bombarded by high-energy particles around the giant planet. One year later, Pioneer 11, by passing 43 000 kilometres from Jupiter, repeated the feat. In fact, pictures such as the one shown, in which the southern hemisphere and the Great Red Spot are 'seen' by Pioneer 11 at a range of about one million kilometres, are reconstructed from the readings of a photopolarimeter, using the probe's own rotation as a scanning system. Subsequently, Pioneer 11 reached Saturn in August 1979 and perfectly fulfilled its mission after surviving the passage through the plane of rings and diving towards the planet, which it flew past within 21 000 kilometres. (NASA)

Saturn seen by the Hubble Space Telescope. Despite the handicap of the spherical aberration of its mirror, the American space observatory has provided images of a resolution unattainable from a ground-based telescope. This picture was acquired by the Wide Field Camera on 26 August 1990, with a resolution of 670 kilometres per pixel. Features in the atmosphere near the North Pole are revealed which could not be resolved by the Voyager probes, because of their trajectories around the giant planet. The ring system is very clear, with the bright A and B rings separated by Cassini's division and the very tenuous inner C ring all visible. Encke's division, near the outer edge of the A ring has never been photographed from the Earth (NASA)

On the preceding double page: *Saturn observed by Voyager 2.* The system which comprises Saturn and its rings is without doubt the most beautiful spectacle which space exploration has disclosed to us. Taken on 4 August 1981 at 21 million kilometres distance by the probe Voyager 2, this picture reveals all the features of this immense concentration of material lost in the depths of interplanetary space: the coloured bands marking the disk of Saturn, the structure of the rings with, in shadow, Cassini's Division, and also, looking like fingerprints on a gramophone record, the marks known as 'spokes'. In contrast to what we see around Jupiter there is no large satellite near Saturn. Three little icy satellites appear as minute white dots in the lower part of the picture (from left to right, Tethys – which casts a shadow on the southern hemisphere of the planet, Dione and Rhea). If the system of Saturn, with its outer A ring extending to 270 000 kilometres, were placed in the space between the Earth and the Moon, it would occupy two-thirds of the distance which separates us from our satellite. (NASA/JPL)

The Moon seen by *Galileo.* The trajectory chosen for the Galileo probe launched by NASA put it close to the Moon (and then the Earth) in December 1990. The probe provided images of the Moon from an angle impossible from Earth.

On this picture of the Western hemisphere we see Mare Orientale in the centre, Oceanus Procellarum at upper right, and at right the small near-circular 'sea' Mare Humorum (NASA).

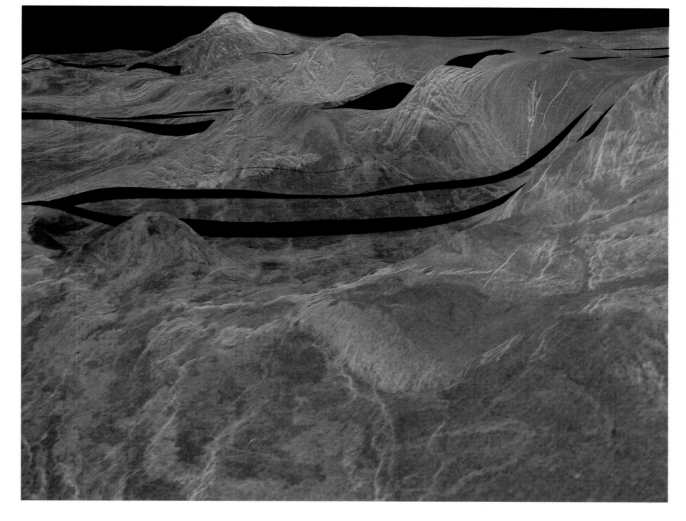

Venus revealed by *Magellan*. The aperture synthesis radar of the American orbiter *Magellan* established the topography of the Earth's sister planet with unmatched precision. This picture is a spectacular three-dimensional perspective reconstitution of the topography of Lakshmi Planum. This plateau rises above the continent of Ishtar Terra, in places by 2.5–4 kilometres. In the centre of a series of tortured valleys and fracture zones, Danu Montes (on the 'horizon') rises 1.5 kilometres above the main relief. (JPL/NASA)

Chandor Chasma. Chandor Chasma is a depression belonging to the system of canyons called Valles Marineris on Mars. This picture of it shows the quality of remote sensing techniques applied to the planets. This is a mosaic of twenty-nine pictures taken by the Viking 1 and 2 orbiters, treated by combining high-resolution (59 metre) monochrome images and low-resolution (260 metre) colour images. The colour contrasts have been exaggerated. The region shown is approximately 200 kilometres square. (A. McEwen, USGS)

development of instruments such as mass spectrometers or gas-phase chromatographs in ever more miniaturised form, and required extremely complex data analysis. Moreover, data processing has become an essential tool of space research, not only for the precise correction of space probe trajectories but also for modelling planetary environments.

Along with planetary exploration other innovations and technological improvements have kept pace with the imagination of the scientists. Sadly, we know today that we have to allow for the unforeseen in such a high-technology adventure. The accident which cost the lives of the Challenger crew also wrecked the timetable for planetary missions. Several probes have nevertheless been successfully placed in orbit by the space shuttle after a delay of several years. Among these were the Magellan and Galileo probes, which were injected into interplanetary trajectories by the shuttle Atlantis in May and October 1989. The aims of Galileo are on the one hand to place a scientific platform in orbit around Jupiter in December 1995, and on the other hand to send a probe into the Jovian atmosphere to analyse its various layers.

In the near future, two missions will use a new space vehicle, Mariner Mark II, developed by NASA. The first mission, called CRAF (Comet Rendezvous/Asteroid Flyby) will involve observing the comet Wild II and the asteroid Eunonia. The second mission, Cassini–Huygens, a joint NASA/ESA project, is aimed at studying Saturn. The descending probe Huygens, built by ESA, will allow a special study of the atmosphere of Titan.

Guy ISRAËL

Characteristics and dynamics

The Solar System is a well-structured unit consisting of a star – the Sun – around which orbit nine planets: Mercury, Venus, Earth, Mars, Jupiter, Saturn, Uranus, Neptune and Pluto, in order of increasing distance from the Sun.

Around these planets revolve numerous satellites: more than fifty are actually known, but it is probable that there exist others not yet discovered, especially around the most distant planets. As well as the Sun, the planets and their satellites, the Solar System contains another category of objects called the 'minor bodies'. These can be divided into the asteroids, thousands of small objects in orbit about the Sun between Mars and Jupiter, the comets, the ring systems of Saturn, Jupiter and Uranus, and lastly the interplanetary dust.

A good model of the Solar System is a flat disk, with the Sun at its centre and a succession of nearly concentric circles in the plane of the disk representing the orbits of the planets. Only the orbits of the nearest planet to the Sun – Mercury – and the furthest – Pluto – are appreciably inclined to the plane of the disk. The planetary orbits are well separated from each other; moreover, the separation increases in approximately geometrical progression with distance from the Sun. Compared with the sizes of the planets or even of the Sun, the distances between the orbits are so great that the kilometre is no longer a useful unit: instead, distances in the Solar System are expressed in astronomical units (AU) defined as the mean distance between the Earth and the Sun, that is 149.6 million kilometres. Thus Pluto lies between 30 and 50 AU from the Sun, while Mercury is less than 0.5 AU.

Halley's Comet. The Solar System, besides its planets, satellites, asteroids, meteors, etc., also contains comets. There are about 700 known, of which a hundred or so have orbital periods of less than 200 years. These are called the short period comets; their aphelia are situated at distances comparable to the orbit of the outer planets. Halley's Comet, shown in this photograph obtained at Calar Alto station in Spain on 10 January 1986, is a short period comet. The other comets, numbering about 600, are called long term or 'new' comets. The orbital elements of some of these show that they are making their first passage close to the Sun and their aphelia lie at very great distances (from 50 000 to 100 000 times the Earth–Sun distance). These observations strongly suggest that the comets belong to our Solar System and that they are stored in a vast reservoir called the Oort Cloud, whose dimensions are only slightly less than interstellar distances. It is thought that from time to time the passage of a star in the vicinity of the Sun 'perturbs' the cometary cloud, causing some comets to escape from the Solar System and sending others towards the Sun. Thus the so-called new comets appear at intervals.

The Outer Solar System. A hypothetical traveller 6.7 billion kilometres from the Sun and 5 degrees above the plane of the Earth's orbit (the ecliptic) would have this view (also hypothetical since the trajectories are not real objects) of the orbits of the planets Jupiter, Saturn, Uranus, Neptune and Pluto; the planes of these orbits coincide for the most part with the plane of the ecliptic. The orbits of two asteroids, 944 Hidalgo and 2060 Chiron, are also shown: Chiron lies between the orbits of Saturn and Uranus. This diagram shows that the orbits of Hidalgo, Chiron and Pluto are very eccentric and highly inclined to the ecliptic plane (different colours distinguish the parts of the orbits lying above and below the ecliptic). Note also the very eccentric and highly inclined orbit of Halley's Comet. (Courtesy J. Kelly Beatty, from *The New Solar System*, 1981)

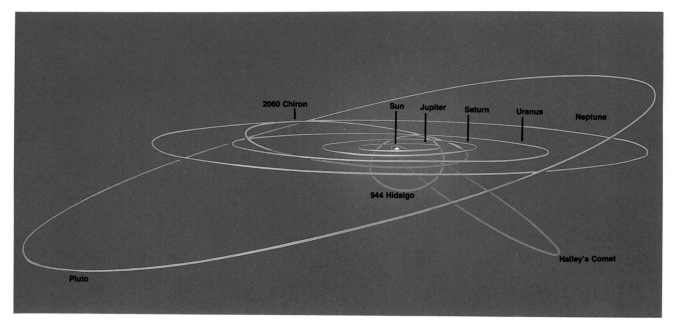

The relative sizes of the planets. This plate (right) shows eight planets of the Solar System: the relative sizes are correct, as are, roughly, the colours, but not the albedos. The ninth planet, Saturn, is shown on the title page of this section to exactly the same scale. It is not posssible to show the whole disk of Jupiter, the largest of the planets, and on the same scale, the Sun would be about 3 metres in diameter. The picture of Jupiter was obtained by the Voyager 2 space probe (NASA/NSSDC). From top to bottom are shown Mercury, photographed by Mariner 10 (NASA), Venus photographed by Pioneer Venus (NASA), the Earth seen from Apollo 8 (NASA), Mars viewed from the Earth with the 200-inch (5 metre) telescope on Mount Palomar (Hale Observatories); the image of Uranus was obtained by Voyager 2 (JPL/NASA); that of Neptune in the red, from Earth (Courtesy B. A. Smith). The dimensions of Pluto are still uncertain.

It is also known that Uranus and Jupiter, like Saturn, have a system of very thin rings, probably solid dust grains for Jupiter and Uranus, and ice in the case of Saturn. For all three, the plane of the rings and that of the orbits of most of the satellites, coincide with the equatorial plane of the planet.

The physical properties of the planets. The table below summarises some of the physical properties of the planets, separated into inner planets out to Mars and outer beyond.

		inner planets				outer planets			
	Mercury ☿	Venus ♀	Earth ⊕	Mars ♂	Jupiter 21	Saturn h	Uranus ♀	Neptune —	Pluto —
mass relative to the Earth	0.553	0.8150	1.0000	0.1074	317.833	95.159	14.54	17.204	(0.0022)
mass in kg*	3.303×10^{23}	48.70×10^{23}	59.74×10^{23}	6.419×10^{23}	189.9×10^{25}	56.86×10^{25}	8.66×10^{25}	10.30×10^{25}	$(6.6 \times 10^{21} - 1.6 \times 10^{22})$
equatorial radius relative to the Earth	0.382	0.949	1.000	0.532	11.19	9.44	4.10	3.88	(0.15 – 0.21)
equatorial radius in km	2439	6050	6378	3393	71.492	60.268	25.554	24.769	(1000 – 1350)
flattening	0.0	0.0	0.0034	0.0052	0.0648	0.107	(0.024)	0.0259	?
mean density in g cm^3	5.43	5.24	5.52	3.94	1.314	0.69	(1.3)	1.76	(0.6 – 1.7)
acceleration of gravity at the equator in m s^2	3.79	8.869	9.798	3.72	22.88	10.59	7.77	11.61	(4.3)
velocity of escape at the equator in m s^2	4.3	10.4	11.2	5.0	59.5	35.6	21.22	23.3	(5.3)
sidereal poeriod of rotation (d = days. h = hours)	58.65 d	243.01 d	23.9345 h	24.6229 h	9.841 h†	10.233 h‡	17.9 h	14.2 h	6.3874 h
inclination of equator to the plane of the orbit	0°	177.3°	23.44°	25.19°	3.08°	26.73°	97.92°	28.80	122.46

The values in parentheses are uncertain by more than 10%

* Not including the mass of the satellites

† The internal period of rotation of Jupiter (system III) is 9.925 hours
‡ The internal period of rotation of Saturn (system III) is 10.675 hours

The Solar System

Compared with the distance between the Sun and the nearest star, Alpha (α) Centauri, situated at about 250 000 AU, the dimensions of the Solar System, of the order of 50 AU, are very modest: seen from nearby stars, the orbits of the planets would appear very close to the Sun.

The greater part of the mass of the Solar System is concentrated in the Sun, which, with a mass of 1.9×10^{33} grams, represents 99.867 per cent of the whole. Among the planets, Jupiter is by far the most massive (318 times the mass of the Earth). Next comes Saturn (95 times), then Neptune and Uranus. Because of differences in size, mass and composition, the planets fall into two distinct groups:
– the inner planets: Mercury, Venus, the Earth and Mars;
– the outer planets: Jupiter, Saturn, Uranus, Neptune and Pluto.

The inner planets – also called the terrestrial planets – are essentially rocky bodies. Of this group, the Earth is the largest, with a radius of 6378 kilometres; then comes Venus ($R = 6050$ km), Mars ($R = 3398$ km) and Mercury ($R = 2439$ km). One usually adds the Moon to this category because its composition is closer to that of the terrestrial than to that of the outer planets. Mercury, Venus and the Earth are the densest bodies in the Solar System (with a density of about 5.5 grams per cubic centimetre). The Moon and Mars, with densities of 3.3 and 3.9 grams per cubic centimetre, respectively, reveal a composition a little less rich in heavy elements – such as iron – than the first three planets.

The terrestrial planets are bodies which can be termed solid in contrast to the outer planets which appear gaseous. Their mineral composition consists mainly of silicates, iron and magnesium. Venus, the Earth and Mars have an atmosphere, while Mercury and the Moon are devoid of one. These atmospheres are called secondary because they are the result of the degassing of compounds and light elements (carbon dioxide, water vapour, nitrogen, …) trapped in the interior of these planets since their formation. In addition, the Earth's atmosphere has had a very complex history resulting from the development of life.

The outer planets are very different from the terrestrial ones; firstly because of their size (Jupiter has a diameter ten times greater than the Earth's), but especially because of their composition. These planets are made up mainly of very light elements: hydrogen, helium, methane and ammonia. Their density is about 1 gram per cubic centimetre. Jupiter and Saturn consist almost entirely of hydrogen and helium in a proportion very similar to that observed in the atmosphere of the Sun, while Uranus and Neptune are richer in methane, ammonia and water ice.

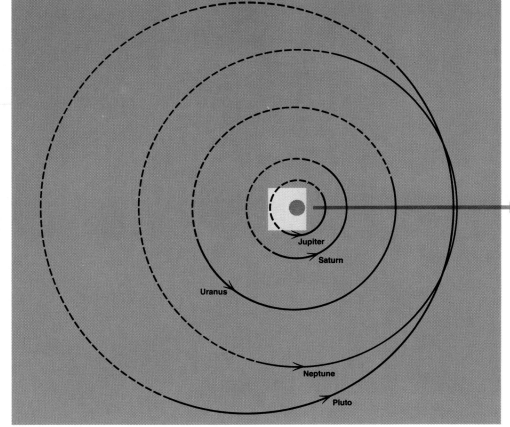

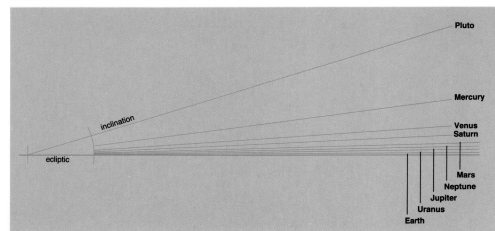

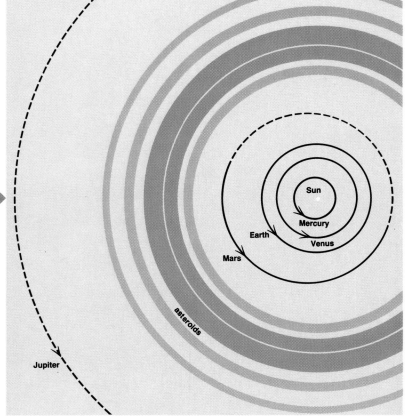

The planetary orbits. The orbits of the planets of the Solar System are shown, seen from vertically above the ecliptic plane (the plane of the orbit of the Earth about the Sun). Those in the outer Solar System (the giant planets) are shown on the left, those in the inner on the right. The parts of the orbits lying below the ecliptic plane are shown by broken lines. The chief asteroid belts also appear (in blue) on the diagram for the inner Solar System.

Note the highly eccentric orbit of Pluto, which intersects the orbit of Neptune. Luckily, thanks to the phenomenon of orbital resonance, there is no risk of Neptune and Pluto ever colliding. The figure opposite shows the respective inclinations of the planetary orbits with respect to the ecliptic.

Mercury, the planet closest to the Sun, and Pluto, the most distant, have the largest inclinations (7 and 17 degrees respectively). Their orbits are also the most eccentric in comparison with the other planets.

The orbital elements of the planets are summarised in the table below.

Orbital parameters of the planets

	Mercury	Venus	Earth	Mars	Jupiter	Saturn	Uranus	Neptune	Pluto
aphelion distance									
in AU	0.466 7	0.728 2	1.016 7	1.666 0	5.454 1	10.042	20.083	30.291	49.351
in millions of km	69.817	108.94	152.10	249.23	815.92	1 502.3	3 004.4	4 531.5	7 382.8
perihelion distance									
in AU	0.307 4	0.718 4	0.983 3	1.381 4	4.952 5	8.990 0	18.247	29.715	29.655
in millions of km	45.986	107.47	147.10	206.65	740.88	1 344.9	2 729.7	4 445.3	4 436.3
mean distance from the Sun									
in AU	0.387 1	0.723 3	1.000 0	1.523 7	5.203 3	9.516 2	19.165	30.003	39.503
in millions of km	57.909	108.20	149.60	227.94	778.40	1 423.6	2 867.0	4 488.4	5 909.6
sidereal period of revolution									
in tropical years	0.240 85	0.615 21	1.000 04	1.880 89	11.862 22	29.457 7	84.022 1	164.771	247.698
in days	87.969	224.701	365.256	686.980	4 332.59	10 759.2	30 688.4	60 181.3	90 469.7
synodic period of revolution (days)	115.88	583.92	–	779.94	398.88	378.09	369.66	367.49	366.73
mean orbital velocity (km/s)	47.89	35.03	29.79	24.13	13.06	9.64	6.81	5.43	4.74
eccentricity	0.205 6	0.006 8	0.016 7	0.093 4	0.048 2	0.055 3	0.047 9	0.009 6	0.249 3
inclination to the ecliptic(°)	7.004	3.394	0.000	1.850	1.308	2.488	0.774	1.774	17.148

	diameter or dimensions (km)	mass (kg)	medium density	relative size
Earth				
Moon	3476	7.3483×10^{22}	3.34	
Mars				
Phobos	26.8 × 22.4 × 18.4	9.6×10^{15}	(2.2)	
Deimos	15.0 × 12.2 × 10.4	2.0×10^{15}	(1.7)	
Jupiter				
Metis	(40)	9.5×10^{16}	?	
Adrastee	25 × 20 × 15	1.9×10^{16}	?	
Amalthee	262 × 146 × 134	7.22×10^{18}	1.1	
Thebé	110 × ? × 90	7.6×10^{17}	?	
Io	3630	8.886×10^{22}	3.55	
Europe	3138	4.785×10^{22}	3.04	
Ganymede	5262	1.481×10^{23}	1.93	
Callisto	4800	1.075×10^{23}	1.83	
Leda	16	5.7×10^{15}	?	
Himalia	186	9.5×10^{18}	?	
Lysithee	36	7.6×10^{16}	?	
Elara	76	7.6×10^{17}	?	
Ananke	30	3.8×10^{16}	?	
Carme	40	9.5×10^{16}	?	
Pasiphae	50	1.9×10^{18}	?	
Sinope	36	7.6×10^{16}	?	
Saturn				
Atlas	40 × ? × 20	?	?	
Promethee	148 × 100 × 68	?	?	
Pandore	110 × 88 × 62	?	?	
Epimethee	138 × 110 × 110	?	?	
Janus	194 × 190 × 154	?	?	
Mimas	392	4.5×10^{19}	1.2	
Encelade	500	7.4×10^{19}	1.1	
Tethys	1060	7.4×10^{20}	1.0	
Telesto	34 × 28 × 26	?	?	
Calypso	30 × 16 × 16	?	?	
Dione	1120	1.05×10^{21}	1.4	
Helene	36 × 32 × 30	?	?	
Rhea	1530	2.50×10^{21}	1.3	
Titan	5150	1.353×10^{23}	1.88	
Hyperion	410 × 260 × 220	1.7×10^{19}	?	
Japet	1460	1.9×10^{21}	1.2	
Phœbe	220	4.0×10^{17}	?	
Uranus				
Cordelia	50	?	?	
Ophelia	50	?	?	
Bianca	50	?	?	
Cressida	60	?	?	
Desdemona	60	?	?	
Juliet	80	?	?	
Portia	80	?	?	
Rosalind	60	?	?	
Belinda	50	?	?	
Puck	170	?	?	
Miranda	480	1.7×10^{20}	1.3	
Ariel	1158	1.6×10^{21}	1.6	
Umbriel	1172	1.0×10^{21}	1.4	
Titania	1580	5.9×10^{21}	1.6	
Oberon	1524	6.0×10^{21}	1.5	
Neptune				
1989 N 6	(54)	?	?	
1989 N 5	(80)	?	?	
1989 N 3	(180)	?	?	
1989 N 4	(150)	?	?	
1989 N 2	(190)	?	?	
1989 N 1	400	?	?	
Triton	2705	1.3×10^{23}	(2.0)	
Nereide	(340)	2.1×10^{19}	(2.6)	
Pluto				
Charon	1186	?	(2 ?)	

The satellites of the planets. Of the inner planets, only two (the Earth and Mars) have satellites. The orbits of the satellites are in general almost circular and not inclined to the equatorial plane of the parent planet. Such satellites are termed regular. The inner satellites of Jupiter, Saturn and Uranus fall into this category; these systems are often likened to miniature solar systems. The largest regular satellites of Jupiter are the Galilean satellites, Io, Europa, Ganymede and Callisto. The sizes of Io and Europa are similar to that of the moon, while Ganymede and Callisto have sizes comparable to that of Mercury. Jupiter also has satellites termed irregular, circling far from the planet in highly inclined and very eccentric orbits. It is generally considered that their origin differed from that of the regular satellites.

The regular satellites of Saturn are Mimas, Enceladus, Tethys, Dione, Rhea, Titan and Hyperion. Titan, one of the largest satellites in the Solar System, is comparable to Mercury in size. It is the only satellite known to have an atmosphere. Beyond Hyperion revolve Iapetus and Phoebe in irregular orbits.

Numerous small satellites have been discovered in recent years during the space exploration of the planets. These are of quite modest size and revolve either very close to the planet or in the same orbit as a satellite already known. Uranus has fifteen known regular satellites. In addition we know at present of two satellites of Neptune, Triton and Nereid, and it has recently been discovered that Pluto also has a companion, Charon. Some asteroids probably also have satellites.

The planets in motion

All the bodies of the Solar System revolve about a central body (the Sun in the case of the planets, comets and asteroids, or a planet in the case of the natural satellites and the rings), and also rotate on their own axes.

The orbital motion is a consequence of gravity. To a very good approximation, the motion of the planets about the Sun can be described by simple laws enunciated by Johannes Kepler at the beginning of the seventeenth century and later explained by Newton using his theory of universal gravity. Kepler's laws, three in number, are as follows: the planets describe ellipses with the Sun at one focus (first law); the radius vector Sun–planet describes equal areas in equal times (second law); lastly, the orbital periods about the Sun and the semi-major axes of their orbits are connected by the relation $a^3/P^2 = \text{constant}$, where a is the semi-major axis and P the period.

The size of the orbit is given by the semi-major axis of the ellipse, its shape by the eccentricity. The plane of the Earth's orbit is called the ecliptic. The planes of the orbits of all the other planets (except Pluto) are very slightly inclined to the ecliptic. The eccentricities of the orbits are close to zero; the orbits are thus nearly circular. The duration of an orbital revolution of a planet about the Sun, according to Kepler's third law, increases with distance from the Sun. Thus the 'year' for Mercury lasts 88 days, that for Venus 224 days and that of Earth 365 days.

Although the path of each planet about the Sun is quite well described by Kepler's laws, the presence of the other planets causes perturbations. These disturbances, proportional to the masses of the disturbing planets and inversely proportional to the square of their mutual distances, cause the path to oscillate about a mean ellipse. They also cause a rotation of the line of apsides of the orbit in its own plane (this motion is called the precession of perihelion). The periods of oscillation of the trajectory are very long, of the order of 100 000 years. However, since these perturbations are small compared to the attraction of the Sun, we can represent the orbits of the planets as Keplerian ellipses whose size, shape and inclination are constantly changing with time.

If the planets were much more massive (for example 100 times greater) or much closer to one another, their motion round the Sun could not be described even to a first approximation by Kepler's laws (two-body approximation), but only by what is called N-body theory. Very complex motions would result.

All the planets revolve about the Sun in the same sense. When the rotation of the planet is also in this direction we speak of direct rotation. In the opposite case the rotation is termed retrograde. Among the planets, Venus and Uranus have retrograde rotations.

Compared to their orbital periods, the sideral periods of rotation of the planets are usually short, less than 25 hours. This is also true of the asteroids. Of the planets, Jupiter and Saturn have the most rapid rotation, of the order of 10 hours; Uranus and Neptune rotate in about 15 hours; the Earth and Mars in 24 hours. In contrast, Mercury and Venus have much slower sidereal rotation periods, 59 and 243 days, respectively. It is generally agreed that these are not the original rotation periods of these two planets, but that they have been progressively braked throughout the history of the Solar System by the effects of solar tides.

The inclination of the equator of a planet to the plane of its orbit determines the obliquity. For the planets these are in general moderate – less than 30 degrees – with the exception of Venus (177°), Uranus (98°) and Pluto (122°).

The obliquity, so long as it is not zero, is the cause of the phenomenon of the seasons. But it is not well known what originally determined the velocity of sidereal rotation and the obliquity acquired by the various bodies of the Solar System. It is, however, assumed that, during the final phase of formation, the last collisions undergone by the bodies in the course of their formation were responsible for their rotational characteristics. There is as yet no complete theory of this problem.

Like the Moon, practically all the known natural satellites of the planets have sidereal rotation periods equal to their orbital periods. This rotation is said to be synchronous. The synchronism is the result of rotational braking by tidal forces, a phenomenon which is especially efficient in slowing down the rotations of the natural satellites of the planets to the equilibrium value corresponding to synchronous rotation. Thus the satellites always present the same face to their parent planets.

The rotation of a planet can be perturbed by oscillations acting on a time-scale much more rapid than tidal braking. These oscillations affect the position in space of the planet's axis of rotation and its sidereal period. The axis of rotation of a planet can thus execute small oscillations on its surface. These movements are entirely unobservable, except of course on our own planet, the Earth.

The rotation axis of the Earth turns about the pole of the ecliptic (with which it makes an angle of 23° 27′) in 26 000 years. This motion, called precession, is the result of the attractions of the moon and Sun on the Earth's equatorial bulge. On this slow movement are superposed the oscillations called nutations, also caused by the Moon and Sun.

The variations in the speed of rotation of the Earth and the oscillations of its axis at the surface (called polar wandering) exist because the Earth is not a rigid body and because it has an atmosphere and oceans.

There are, in the Solar System, a great number of examples of orbital resonances corresponding to exact commensurabilities. The best known is that which affects the first three Galilean satellites of Jupiter; Io, Europa and Ganymede. Orbital commensurabilities also exist among the satellites of Saturn: Mimas and Tethys are in resonance, so are Dione and Enceladus and also Titan and Hyperion.

There are many other commensurabilities as well: for example within the asteroid belt there are divisions corresponding to an absence of material where the orbits would be commensurable with that of Jupiter. An analogous phenomenon is shown in the rings of Saturn where the principal divisions – Cassini's Division in particular – occupy positions corresponding to a commensurability with the satellite Mimas.

The orbital periods of the planets themselves are almost commensurable, but, except for the

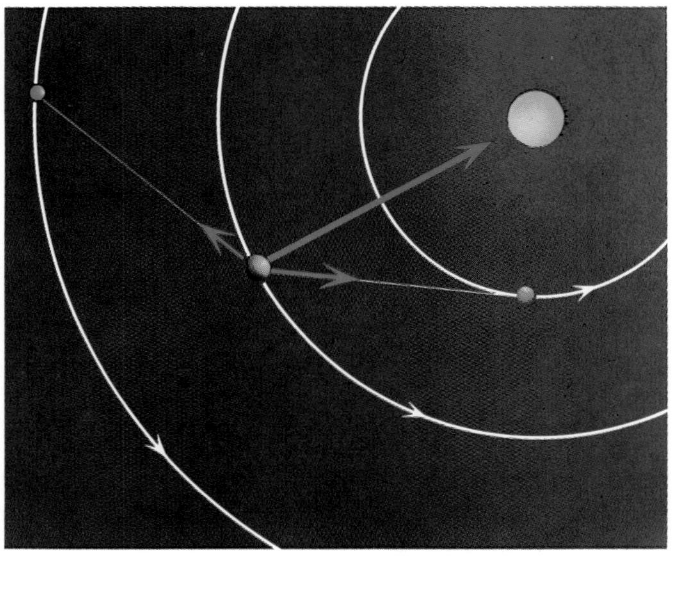

The orbital perturbations. Since the Sun contains almost the whole mass of the Solar System (more than 99.85 per cent), it effectively controls the movements of the planets. However, their mutual attractions cause small deviations in their orbits. Even so, compared with the Keplerian ellipse which a planet would describe if it were alone with the Sun, these deviations are very small. The gravitational forces which produce them are called perturbations. We recall that it was the observation of the perturbations of the motion of Uranus which led to the prediction of the existence of Neptune, discovered in 1846 thanks to the calculations of the astronomer Urbain Jean Joseph Le Verrier.

The orbital resonances. Among the manifold periodic motions of revolution and rotation which are exhibited by the collection of bodies of the Solar System, we observe very many commensurabilities. approximate or exact. In most cases the ratio between the orbital periods of two planets or of two satellites of the same planet is a simple fraction, for example ½, ⅔, ¾, . . . That implies that the two bodies periodically find themselves in the same configuration with respect to the Sun or the planet. The mutal gravitational interaction between the two bodies is increased and we then have an orbital resonance. This resonance is stable if, despite external perturbations on the system – for example, the gravitational attraction of more distant bodies or tidal forces – the commensurability of the orbital periods does not change with time. The above figure shows an example of an orbital resonance ½. The outer satellite describes exactly one revolution while the inner one accomplishes two. Thus each time they come closest together the two satellites are in the same orbital configuration with respect to the central body, so that the mutual perturbations are amplified.

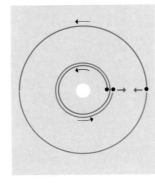

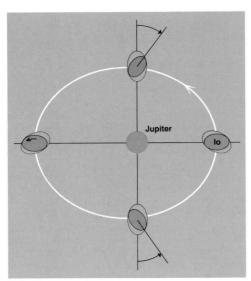

pair Neptune–Pluto, the remainder do not exhibit the phenomenon sufficiently exactly to be able to speak of orbital resonance.

The existence of numerous orbital commensurabilities between pairs of bodies in the Solar System is certainly not due to chance; it is usually explained as the result, on the one hand, of dynamical conditions during its formation 4.5 billion years ago and, on the other, of the slow subsequent evolution caused by tidal effects; these, in causing the orbits to evolve, act so as to 'trap' certain bodies in commensurable orbits. If the proportionality is exact, the amplification of the mutual gravitational interactions maintains the stability of the configuration and prevents further change.

The tides

The law of gravity is the dominant force in the Universe. When two bodies approach one another closely, their mutual gravitational attraction modifies their motion. No body in the Universe is perfectly rigid, however, not even the terrestrial planets which are mainly composed of rock. Because of this, the force of gravity also gives rise to what are known as tidal effects when two bodies approach each other closely. Each body attracts the different particles which make up the other body, and the two bodies are consequently deformed, more or less elastically depending on their composition. If the bodies are very close to one another the least massive may fragment into pieces if the forces holding it together are less than the tidal forces exerted by the other body.

In the Solar System the bodies are in general very far from each other; as a result, the tidal deformations are insignificant compared to their

dimensions. These deformations are periodic because they are associated with periodic configurations of bodies in motion about each other.

Under the attraction of the Moon, the Earth is periodically deformed; each point on its surface undergoes its greatest displacement when the Moon is at its closest. The ocean tides are known to everyone; changes in sea-level of several centimetres to many metres occur regularly. A part (about 1/5) of the ocean tides is due to the action of the Sun. Like the oceans, the solid Earth is deformed to the same rhythm, but with a weaker amplitude: thus the surface upon which we stand 'rises' and 'falls' by about 30 centimetres every twelve hours. The displacements of the surface are accurately measured using special instruments such as gravimeters, extensometers and also horizontal pendulums.

The tidal deformations are not perfectly elastic since, because of friction, energy is dissipated as heat (by friction on the ocean floor and between different layers of the Earth's mantle). Consequently, the distortions lag somewhat behind their cause, that is the passage of the Moon or the Sun. This lag is a few minutes for the tides in the solid Earth, but can reach many hours in the case of the ocean tides.

This dissipation of energy has two important consequences for the long-term evolution of the Earth's rotation and the lunar orbit; the energy dissipated as heat is taken from the rotational energy of the Earth which results in a slow and progressive retardation of its rotation (the Earth turned on its axis more rapidly in the distant past or, what is the same thing, the length of the day was less).

A three-body orbital resonance. Three of the Galilean satellites of Jupiter, Io, Europa and Ganymede, are involved in a three-body resonance: their mean orbital motions (respectively, n_1, n_2, and n_3) are connected exactly by the equation $n_1 - 3n_2 + 2n_3 = 0$. It is called Laplace's relation because it was he who, in the last century, drew attention to it and showed that it led to a stable resonance. This resonance is such that when Europa and Ganymede are in conjunction with Jupiter, Io is 180 degrees from the direction of the conjunction, thus preventing the three satellites from ever being aligned on the same side of the planet.

In addition, the conjunction of Io with Europa takes place when Io is close to its periapsis and Europa is near its apoapsis. On the other hand, Europa and Ganymede are in conjunction when Europa is close to its periapsis and Ganymede is anywhere in its orbit. This three-body resonance is remarkable in that, not only is it unique in the Solar System, but it is also responsible for a very special phenomenon: the volcanoes on Io. In the absence of Europa and Ganymede, Io would have an extremely small eccentricity, of the order of 0.000 01. In that case the tides raised by Jupiter on Io, rotating in synchronism with its orbital motion, would exert no dissipative effect on the interior of the satellite: in fact, for a satellite in synchronous rotation in a circular orbit, the tidal bulge is always aligned with the planet–satellite direction whatever the position of the satellite in its orbit. That is no longer true if the orbit is eccentric; there are parts of the orbit in which the satellite has an orbital velocity more rapid than its velocity of rotation about its own axis, others where it is less. In consequence, the tidal bulge raised by the planet on the satellite is sometimes in front and sometimes behind the planet–satellite direction so that Jupiter exerts a couple on the bulge. The diagram above (after C. Yoder, 1979) shows this effect. The couple should tend to alter the rotation of the satellite, but cannot because it is in synchronous rotation; consequently it acts to 'extract' orbital energy. This loss of orbital energy is balanced by an increase of thermal energy in the interior of the satellite which raises its temperature.

This is just what happens to Io. Because of the orbital resonance with Europa and Ganymede, the eccentricity of Io increases significantly (we speak in this case of a forced eccentricity). The existence of a forced eccentricity then gives rise to the effects just described. Theoretical estimates have shown that the production of heat in the interior of Io by this means is of the order of 215 calories per gram of material ever since the resonance was established. If there had been no loss of heat the temperatures would have risen by 900 K, enough to partially melt Io and account for observed volcanoes. This mechanism which had been predicted before the flyby of Io by the Voyager 1 and 2 probes, is now thought to be the most likely source of energy to explain the intense volcanic activity on Io. A mechanism of this kind is perhaps also responsible for the particularly 'young' appearance of the surfaces of Enceladus and Iapetus, satellites of Saturn. The picture opposite shows two volcanic eruptions in Io: a mushroom-shaped eruption behind the edge of the disk, and a large diffuse ring surrounding Pele volcano. This diffuse ring is produced by ejections of sulphur dioxide falling back and condensing around the caldera (Courtesy Alfred McEwen, USGS).

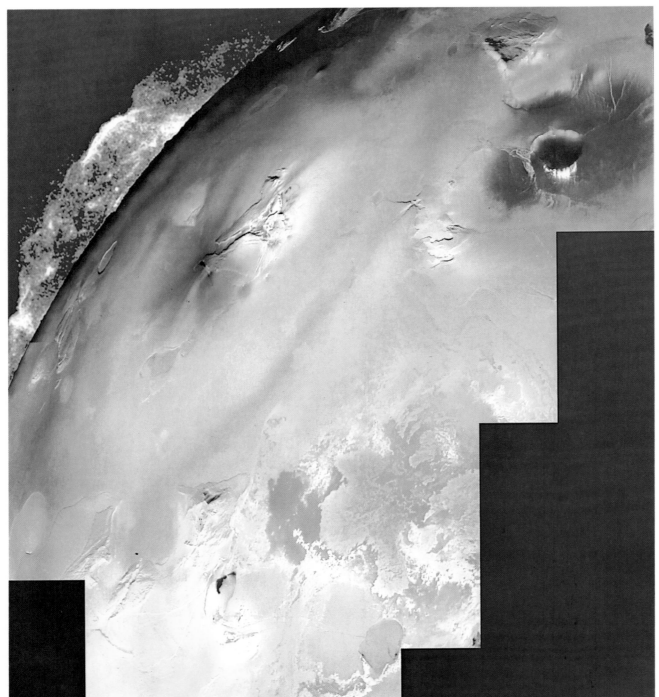

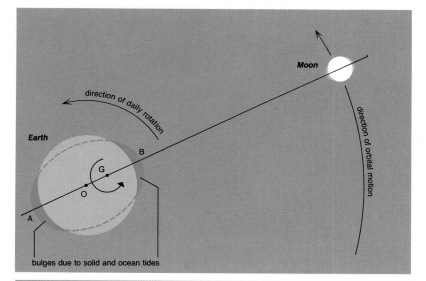

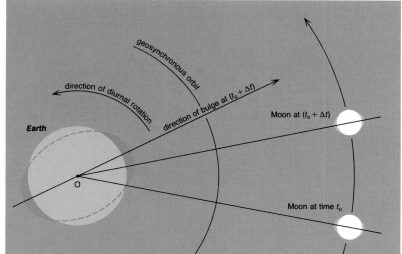

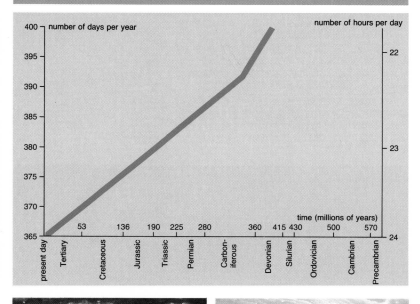

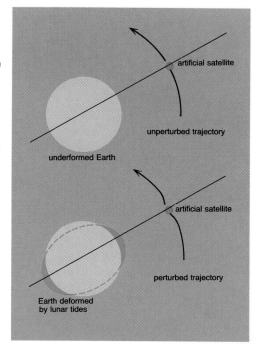

The tides in the Earth–Moon system. The Earth and the Moon revolve around point G, centre of gravity of the two bodies, which is very close to the centre of the Earth, point 0 (upper left). The Moon exerts a force upon every particle of the Earth, or, what comes to the same thing, an acceleration which is the greater the nearer the particle is to the Moon. The acceleration of each particle in the reference frame of the Earth is a differential acceleration: it is the difference between the acceleration produced upon a particle and that of the centre of mass of the Earth. It results from this that two diametrically opposite points located on the Earth–Moon axis, such as A and B, are subject relative to the centre of mass of the Earth to accelerations of the same magnitude but opposite in direction. Two diametrically opposed tidal bulges thus appear. It is the same upon the Moon. As the Earth turns upon its axis in 24 hours, all the points of maximum deformation such as A and B make a circuit of the Earth in 24 hours giving a tide every 12 hours. The ocean tides are well-known phenomena but there are also solitudes in involving deformation of the terrestrial mantle. Because of internal friction, the deformation is retarded (by a time Δt) with respect to the passage of the Moon, located after the instant t_0 (opposite). Dragged along by the rotation of the Earth, more rapid than the motion of the moon in its orbit, the tidal bulge is displaced in relation to the Earth–Moon direction: the moon exerts a couple upon the bulge which tends to pull it back to the Earth–Moon direction thus braking the rotation of the Earth. The bulge exerts an equal but opposite couple upon the Moon which decelerates the moon in its orbit: the radius of the lunar orbit increases.

The measurement of the tidal deformations. The periodic deformations of the Earth caused by the tides give rise to measurable perturbations of the orbit of an artificial satellite. Thanks to the analysis of these perturbations, one can deduce several parameters which characterise the tidal deformations, in particular the amount of energy dissipated within the Earth. There is a direct relationship between the tidal dissipation of energy and the rate at which the Moon's distance increases.

As the Earth's rotation is retarded the Earth–Moon distance increases. Why? Two laws govern all mechanical phenomena:
– the law of the conservation of energy
– the law of the conservation of momentum.

The retardation of the Earth's rotation due to the tides leads to a reduction in the angular momentum of the Earth, but the total angular momentum of the Earth–Moon system must be conserved and so the orbital angular momentum of the moon about the Earth must increase. Thus the radius of the lunar orbit grows by about 3 metres a century.

This progressive recession of the moon is revealed by various methods and observational techniques:
– astronomical observations of the movement of the Moon;
– techniques of laser ranging to the moon (since 1970);
– perturbation of the orbits of artificial Earth satellites.

These observations verify the present phenomenon. However, there exist indirect proofs of its existence in the geological past: we know from the study of fossil marine organisms that about 400 million years ago the length of the day was 21 hours instead of 24 hours as at present. It was discovered twenty or so years ago that corals exhibit diurnal, monthly and annual concentric rings; by examining fossil corals many hundreds of millions of years old, it is apparent that during one annual cycle the number of daily growth rings very clearly averages more than 365. Because the length of the year has not changed since the formation of the Solar System (since the Earth–Sun distance could not have changed), we conclude that the number of days in the year was

greater in the past and hence that the day was shorter than it is now. Four hundred million years ago there were a little over 400 days in the year and the day lasted 22 hours. The Moon must have been closer to the Earth than it is today.

We have seen that the tides raised by the Moon on the Earth retard its rotation. In the same way, and more efficiently, the tides raised by the Earth on the Moon have braked the rotation of our satellite. This retardation was very rapid, lasting only a few million years (the time-scale of the retardation is directly proportional to the mass of the body responsible for the distortions). The tidal braking ceased when the rotation reached a state of stable equilibrium in which its period equalled that of the orbital revolution. We then speak of synchronous rotation. In the case of the Moon this synchronism is easily observed; the Moon rotates on its own axis once in 27 days in the same time it makes one revolution of the Earth. That is why, from the Earth, we always see the same face of the Moon.

Almost all the satellites of the planets are in synchronous rotation and thus always keep the same face towards their parent planets. As with the Earth and Moon, the synchronism of rotation of the natural satellites must have become established very soon after the Solar System was formed.

On the Earth, the solar tides are weaker than the lunar ones; although much more massive, the Sun is much more distant than the Moon and so its effect is less. But the planets Mercury and Venus are nearer to the Sun than the Earth and it is probable that the solar tides have significantly affected their rotations. While the majority of the bodies in the Solar System rotate in less than 25 hours, Mercury and Venus have very slow rotations; 59 days for Mercury, 243 days for Venus. It is generally agreed that the rotations of these planets have been braked. However, Mercury has not yet reached synchronous rotation (equal to 88 days, the length of the year on Mercury). The planet is in fact 'trapped' in a resonance such that, in two revolutions around the Sun, it makes exactly three rotations on its axis. This situation is stable, preventing further retardation. As for Venus, its rotation appears by contrast to have 'overtaken' the synchronous rotation (with a 224-day period). That might have been because Venus has a 'retrograde' rotation (it rotates in the opposite sense to its orbital motion). We do not at present know if Venus has always had a retrograde rotation or if, in the beginning, its rotation was direct, like the majority of the other members of the Solar System, and if a progressive tilting of its axis of rotation has transformed direct rotation into retrograde. Tides in the dense atmosphere of Venus might account for such a tilting.

Anny CAZENAVE

The length of the day. Over the ages the length of the day has changed (opposite left); this variation in the period of the Earth's rotation is due to the effects of the lunar tides. The length of the day in the Carboniferous Era (22.2 hours) and in the Devonian Era (21.9 hours) has been measured by the annual growth of corals. The days were then shorter, but the length of the year remains constant because the tidal interactions between the Earth and the Sun are too weak to modify the year. Corals and certain other marine organisms grow by forming concentric rings according to the diurnal rhythms, monthly and annual so that we have marks for the days and the seasons. By examining the fossil corals dating from the Devonian (about 400 million years ago), a very interesting discovery was made: it was found that, in the course of an annual cycle, the number of diurnal rings was close to 400, suggesting that the number of days in a year in this epoch was greater than it is today (it is generally agreed that the length of the year has not changed). This implies that the length of the day was shorter and that the Moon was closer to the Earth, which is in accord with what is known of the phenomenon of tides in the Earth–Moon system. The photograph on the left represents a Devonian fossil coral from the Moroccan South, magnified a dozen times; the rings are readily distinguishable. The right hand photograph shows a present day coral (Macinia) from Barbados; the magnification is the same as for the fossil. (From the collection of the Museum National d'Histoire Naturelle; photograph J. C. Recy).

The primordial nebula

The most striking feature of the Solar System is its extraordinary diversity. Understanding how it was formed and how it has evolved is like piecing together a huge puzzle. The fact that it is the only planetary system we know of does not make the task any easier. The cosmogony of the Solar System (that is the study of its formation) has been rejuvenated since the 1970s thanks to the great space exploration enterprise; to ever more precise laboratory analyses of lunar rocks and the 'poor man's space probes', the meteorites; to numerous new observations from the Earth and also to the recent development of new theories.

The problem of the origin of the Solar System was hardly considered before the seventeenth century. Before then most philosophers thought that the planets and stars had existed from eternity and would be there for ever; in Western Europe dogma placed the Earth at the centre of the Universe. With Copernicus, who placed the Sun at the centre of the Solar System; with Kepler, who stated his three famous laws; and with Newton, who discovered the law of universal gravitation, the foundations of a better understanding of the origin of the planets were laid. The fact that all the planets are confined to the same disk, the plane of the ecliptic, close to the equatorial plane of the Sun, indicates that the planets grew by condensation of material in a protoplanetary disk. For a long time it was debated whether the formation of the planetary disk was caused by the close passage of a star past the Sun (catastrophic theories), or if the Sun and the planets were formed from the same nebula of which the Sun would then be just the central condensation (nebula theories).

Descartes, in 1632, was the first to introduce the idea of evolution and to propose a scientific theory of the origin of the Solar System. He can properly be considered as the father of theories of cosmogony. The father of catastrophic theories was Buffon – in 1749 he put forward the idea that the Sun had been struck by a comet which had torn out a filament of material from which the planets formed. A better understanding of the nature of comets soon led to the abandonment of this theory, though keeping the idea accepted today, that one and the same physical phenomenon caused the formation of the Solar System in its entirety. Most catastrophic theories suppose that a star once passed close by the Sun and pulled out a tidal filament from which the planets were formed. This model, which was quantitatively developed by James Jeans and Harold Jeffreys between 1915 and 1929, has now been discarded because the filament pulled from the Sun would not extend far enough and the solar material drawn out would be much too hot to condense into planets (it would disperse instead) and also because the abundance of deuterium and lithium on the Earth suggests that the planets were formed at a relatively low temperature.

Kant in 1755 and Laplace in 1796 put forward the hypothesis that a primordial nebula was the common ancestor of both Sun and planets. This idea seems now to be accepted by the majority of astrophysicists to describe the stages of the formation of the planets. After the fragmentation of an interstellar cloud, the primordial nebula was rotating and contracted under the influence of its own gravity. Because of centrifugal force and viscosity, the contraction produced a disk whose parts condensed and accreted to give birth to the planets in the cooler regions while the central part contracted to become the Sun. It is still a problem to know whether the Sun formed just before or just after the planetary disk. By studying elements with long half-lives such as rubidium 87, thorium 232 or uranium 238 we can estimate the age of the planets to within about a hundred million years. It is found that 4.55 billion years have elapsed since solidification.

The Sun condensed in the central part of the nebula about 4.56 billion years ago. It is thought that the protoplanetary nebula was strongly heated by the gravitational contraction; this nebula would be sufficiently opaque to prevent the infrared radiation from escaping from its central regions. Near to the luminous proto-Sun the temperature would be very high, although further out it would be no more than a few tens of kelvins. As the nebula cooled, then various minerals would be condensed. Refractory compounds of calcium, aluminium, magnesium and titanium would appear below 2000 K; at about 1000 K silicates and metal oxides would condense, while iron sulphide probably forms at about 700 K. At around 180 K water vapour turns to ice and between 50 and 20 K solid grains of methane appear.

During this sequence of events, various chemical reactions took place, and produced the minerals which fill the Solar System today. Thus the density and chemical composition of the planetary system are a direct consequence of the variation of temperature in the protoplanetary disk. The abundance of the elements and molecules (i.e. the content of volatile elements) can be used to measure the temperature at which the accretion occurred. Organic molecules are formed from reactions between carbon monoxide (CO) and molecular hydrogen (H_2), Fe_3O_4 and hydrated silicates acting as catalysts.

After the 'ignition' of thermonuclear reactions in the centre of the Sun and the chemical condensation of grains from a cooling nebula we must ask how the planets formed in the granular disk. Chemical condensation alone cannot produce bodies whose dimensions are more than a few centimetres. The growth of large objects like the planets or satellites from such small debris was for a long while an enigma. A few years ago, the study of the development of gravitational instabilities in a thin disk of condensed debris showed that this process could allow objects several kilometres in size to accumulate in a very short time. Collisions between these 'planetoids', after some breaking up and re-forming, led to the formation of the planets by accretion.

It thus seems that the first four stages in the evolution of the Solar System were the formation of a protoplanetary disk from the contraction of a nebula, the chemical condensation of grains when the gas in the nebula cooled, the formation of planetoids by means of gravitational instabilities in the granular disk and finally the formation of planets and satellites by accretion. It should be remembered that these stages seem to have lasted in all less than a few hundred million years – soon after its formation the Solar System probably already looked very much as it does today.

This scenario is the most generally supported, but it is far from being universally accepted. in fact, the problem of the origin of the Solar System is peculiar: on the one hand we observe today the result of processes which took place more than four billion years ago and not the processes themselves and, on the other hand, as already stated, we know of only one planetary system; our own. That is why the only useful approach consists of gathering facts.

As far as astronomical observations proper are concerned, the study of young stars, molecular clouds or infrared sources (where it is thought new stars are forming) and T Tauri-type stars (the proto-Sun probably passed through this stage) raises hopes of observing the birth of a star at first hand. Theoretical studies of stellar formation are progressing rapidly; the classical methods of solving the equations of hydrodynamics allow us to simulate the first phases of the contraction of stars from the interstellar medium.

We must not forget that cosmochemistry is particularly active with its laboratory analysis of meteorites and lunar rocks. It can even be claimed that a large part of the firm data (especially the

A cradle of stars. Lying in the constellation of Sagittarius about 1400 parsecs from us, the Lagoon Nebula (M8, NGC 6523) is a cloud of hydrogen and dust which extends to about 10 parsecs. The bright parts are due to gas excited by the ultraviolet radiation from very hot young stars embedded in it. Towards the centre of the nebula can be seen the star cluster NGC 6530, whose age is estimated at only two million years.

Also visible are dark areas formed by dust which absorbs the radiation from the gas and stars. Some lines and circular dark patches called Bok globules may be clouds of dust and gas in process of collapsing to give birth to stars. One of these globules, called the Dragon (visible at the lower left), is being exhaustively studied. The very bright region includes the star Herschel 36 which is less than 10 000 years old. (Royal Observatory, Edinburgh)

time-scales) which we have on the origin of the planets came from meteorites. The problem of the origin of the meteorites, whose population differs from that of the asteroids, is not altogether solved. In spite of the turbulence prevailing at the beginning of the history of the Solar System, it does not seem likely that the system was homogeneous. We must ask whether the population of the meteorites is representative of the initial state of the system?

Some research work centres on the isotopic anomalies and the role of supernovae. Despite their rarity (among the hundred billion stars in a galaxy, there are on average only about three per century) supernovae are very important; most of the elements heavier than hydrogen and helium found in the Sun and the planets were ejected by a supernova more than five billion years ago. Isotopic anomalies revealed in some primitive meteorites suggest that a supernova exploded in the neighbourhood of the proto-Sun about a million years before its formation. Some astronomers at first thought that the explosion of this supernova played an important part in initiating the process of formation of the Solar System and in the early evolution of the solar nebula. A different picture is now developing: the Sun probably formed in the midst of a cluster of stars. The most massive of these stars evolved very rapidly and some ended their lives in supernova explosions while the less massive Sun had scarcely finished contracting. Numerous isotopic anomalies found in the meteorites probably resulted from multiple supernova explosions in the cluster within which the Sun was born. These supernovae were not the cause of the birth of the Solar System, but rather the witnesses of its birth. A veritable firework display marked the early life of the Sun.

The minor bodies of the Solar System (asteroids, comets, rings and even small satellites of the planets) are still poorly understood inasmuch as their apparent angular diameters are significantly less than one arc second, which makes them particularly difficult to study from the Earth.

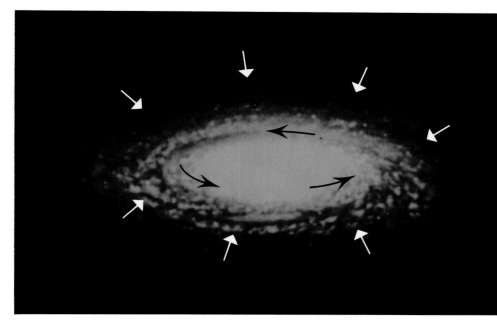

The primordial nebula. The primordial nebula would have had the appearance of a cloud of material which collapsed upon itself and gradually took the shape of a flattened rotating disk.

Infrared observations have shown that MWC 349 is a massive young star (about 30 solar masses) surrounded by a disk of gas. Its luminosity has decreased by 88 per cent in nearly forty years of observations, which is attributed to the rapid dissipation of disk material. Some astronomers believe that this disk is a primordial nebula like that which gave birth to the Solar System.

They are nevertheless of great interest since they should furnish unique information about the early history of the Solar System. The comets and the asteroids are the most primitive bodies in the Solar System; their chemical composition reflects that of the protosolar nebula (close to the Sun for the asteroids, at the edge of the system for the comets). They are probably the remains of the original planetoids from which the planets were formed and they are too small to have undergone any thermal or chemical evolution as did the planets. They thus contain firsthand information about the protosolar nebula. Whatever their origin, the planetary rings resemble in some ways the protoplanetary disk before the planets formed. The existence of material making inelastic collisions around the planets probably represents an intermediate phase in the formation of the planets just before accretion prevailed over fragmentation. The discovery of very many concentric rings around Saturn suggests that the mechanisms for the confinement of the material are actually at work around this planet, the large particles confining the small ones in thin rings. This intermediate stage probably occurred before the formation of the Earth.

Research into the origin of the Solar System is particularly well developed at present and is a truly multidisciplinary field. Physicists, mathematicians and chemists, astrophysicists, geophysicists, meteorologists and biologists; each makes a vital contribution towards the understanding of our past.

André BRAHIC

The Allende meteorite. Did the explosion of a supernova trigger the formation of the Solar System?

Isotopic anomalies were first discovered in a meteorite which fell in the north of Mexico on 8 February 1969 near the village of Pueblito de Allende (hence its name). It appears that, when it was formed, the meteorite contained a radioactive isotope of aluminium, ^{26}Al. Now aluminium has only one stable isotope, of mass number 27 (the nucleus is composed of 13 protons and 14 neutrons), and the aluminium found on the Earth consists only of this isotope. Aluminium 26, with 13 protons and 13 neutrons, has a half-life of 720 000 years; after this time half its atoms have disintegrated radioactively. A positron and a neutrino are emitted when a proton in the nucleus is transformed into a neutron; the mass number is conserved, but the new nucleus has 12 protons and 14 neutrons and is a nucleus of the stable isotope magnesium 26. Suppose that there was a certain amount of aluminium 26 in the protosolar nebula just before the Allende meteorite condensed. After 4.6 billion years – the age of the Solar System – all the aluminium 26 has been transformed into magnesium 26. Magnesium has three stable isotopes, of mass numbers 24, 25 and 26; the normal isotopic composition of magnesium is 78.99, 10 and 11.01 per cent, respectively. All the magnesium 26 formed by the radioactive disintegration of aluminium 26 will be detected as an excess over its normal abundance. That is just what has been found in certain inclusions in the Allende meteorite, where the proportion of magnesium 26 is 11.5 per cent. Obviously it is necessary to make sure that the excess of magnesium does indeed result from the radioactive disintegration of aluminium 26; that is done by comparing the percentages of magnesium 25 and magnesium 26 and by measuring the magnesium/aluminium ratio.

This discovery of an excess of magnesium 26 shows that the radioactive isotope aluminium 26 was incorporated into the meteorite less than a few million years (i.e. a small number of half-lives of aluminium 26) after its creation, otherwise all the aluminium 26 would have been transformed into magnesium 26 and mixed unhindered with the other isotopes of magnesium. Aluminium 26 is produced by thermonuclear reactions in the interiors of stars. It was probably not formed in the protostellar nebula itself, otherwise other isotopic anomalies would have arisen, which are not observed. The only tenable explanation is thus that a supernova exploded just before the birth of the Solar System. However, it is generally agreed that this explosion was not the origin of our Solar System. The supernova would have been an onlooker at its birth. (Collection of the Museum National d'Histoire Naturelle; photograph Les Ateliers MS)

The formation of the planets

Scientists today seem to agree in thinking that planets are a by-product of the formation of stars and that, in the case of the Solar System, the Sun and the planets were formed nearly simultaneously, 4.55 billion years ago. Many hypotheses have been advanced to account for the existence of a planetary system around the Sun and all must answer a vital question: was the Solar System formed as the result of a rare accident, or during a normal stellar evolution, or in the course of a peculiar stellar evolution? In the first case planetary systems would be extremely uncommon, in the second very widespread and in the third, without being very rare, they would result from relatively infrequent initial conditions. Although there can be no certainty in this matter, the third scenario is at present preferred; the formation of a planetary system around a star requires special conditions, counteracting the natural tendency to form double or multiple stellar systems. By the same token, the first possibility can definitely be discarded, inasmuch as clouds of dust are observed around stars in the process of formation.

To understand the process of formation of the planets, we must first explain how the Sun condensed from a cloud of gas and dust and how a very small part of the material (less than 1 per cent) was gathered into planets instead of falling into the Sun. In fact, modern theories of the formation of the planets separate the two problems and assume at the outset the existence of a central condensation – the Sun – formed by the gravitational contraction of a cloud of gas – hydrogen, helium and other elements – and that the rest of the envelope, made up of gas and dust, formed rapidly into a flattened disk rotating with the central condensation. This disk is often called the solar nebula. From there, two types of mechanism are proposed to explain the formation of the planets.

The first involves the formation of large gaseous protoplanets, but with a solid central core, as a result of gravitational instabilities in the solar nebula. Just as the initial protostellar cloud collapsed inwards to give birth to a star – the Sun – this theory proposes that the disk of gas and dust collapsed and caused the very rapid formation of giant protoplanets.

The second mechanism involves the rather slow aggregation of solid particles of ever larger sizes from the dust initially present in the nebula.

At present there are no compelling reasons for preferring one theory or the other. However, the first encounters serious difficulties, especially in accounting for the formation of the terrestrial planets. In practice, the protoplanetary disk would have to be very massive – about the mass of the Sun – or the collapse would not occur. Now once the planets have formed, the total mass of planetary material is less than 1 per cent of the mass of the Sun, which implies that, if we adhere rigidly to the theory, an enormous quantity of material has to be expelled from the Solar System. The difficulty then is that we do not know of a mechanism efficient enough to remove such an excess of material. A second difficulty arises because, in the region of the terrestrial planets, that is, close to the Sun, a giant protoplanet is not stable because of the tides raised by the Sun. There are other difficulties as well, connected with the chemical composition of the planets.

On the other hand, a study of the various geophysical data on the terrestrial planets seems for the moment to favour the second theory, which gives a good account of the formation of those bodies. The process of formation of the planets by the gradual accumulation of larger and larger solid objects has been the subject of quite elaborate theories. In particular, the theory of the formation of the planets by means of collisional accretion, developed in the 1970s by the Soviet astronomer Safronov, is a major contribution in this field.

According to this theory, the planets were formed in three stages:
– a very rapid (about 1000 years) initial phase

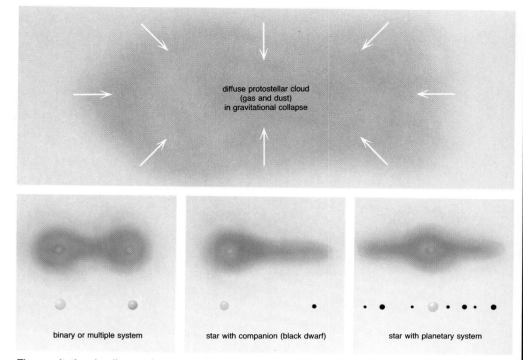

binary or multiple system

star with companion (black dwarf)

star with planetary system

The gravitational collapse of a protostellar cloud. Two factors govern the gravitational collapse, the mass of the initial cloud, and how much material is left in the gaseous envelope surrounding the first nucleus of condensation to form a star. Either a multiple (or binary) star system may be formed, or a single star surrounded by a disk-shaped nebula. Under certain conditions, a system of planets may form from this disk.

during which small solid planetoids were formed, between 1 and 5 kilometres in size, as a result of small-scale gravitational instabilities in the solid dust grains present in the nebula;
— an intermediate phase giving rise to embryo planets about 1000 kilometres in size. During this phase, the growth of these objects was controlled by two opposing mechanisms; rapid destruction by collision and slow accretion by encounter;
— a final phase, characterised by the growth of the embryos into planets; this phase is dominated by the mutual gravitational interactions between the embryos.

The last two phases would have taken place slowly, over about 100 million years.

At the end of the first phase, the proto-planetary disk was composed of a relatively homogeneous mass of little planetoids a few kilometres in size revolving round the Sun in Keplerian orbits. These objects encountered each other frequently. When a pair of planetoids make a close encounter, but without colliding, their orbits are slightly altered. Close encounters without collision usually make the orbits more eccentric. If two objects pass very close to one another, a collision can take place.

There are then two possibilities. If the relative velocity of approach is large, the collision completely destroys the objects; the fragments are added to the planetoids. If, on the other hand, the velocity of approach is small, the two bodies fuse and remain together, as a result of their mutual gravitational attraction. A single body is thus created.

During this intermediate phase, the evolution of the planetoids is directly linked to their changing relative velocities. If the encounter velocities are too small some embryo planets will grow quite quickly. But, once they have swept up

those planetoids whose orbits lie in their immediate neighbourhood, these embryos will find themselves in almost circular orbits, isolated from each other. Since collisions are rare, growth ceases. The result is then a planetary system with a large number of small planets. If on the other hand, the encounter velocities are too large, growth is equally hindered by the catastrophic collisions.

Clearly our Solar System has escaped these two misfortunes, and during its development the relative velocities reached equilibrium, thus allowing the embryos to grow. Theoretical models show that growth depends mainly on the relative velocities always being of the order of the escape velocity from the largest objects.

To describe the final phase in the formation of the planets, it is necessary to use numerical simulations on a computer to take very precise account of the mutual gravitational interactions between the embryos. When some objects have grown to a size of from 1000 to 2000 kilometres, their subsequent growth no longer depends upon chance encounters. Because of their size and mass, their gravitational spheres of influence extend far beyond their mere diameters. If a planetoid enters such a region, its orbit will be strongly perturbed, with a strong probability that it will be 'swallowed' by the embryo.

The present characteristics of the Solar System, namely the small number of planets, their sizes, masses and rotations, the distances between them, etc., result from gravitational competition between the embryos during this final stage of their growth. However, and in spite of recent progress, all these features are not yet well understood.

Anny CAZENAVE

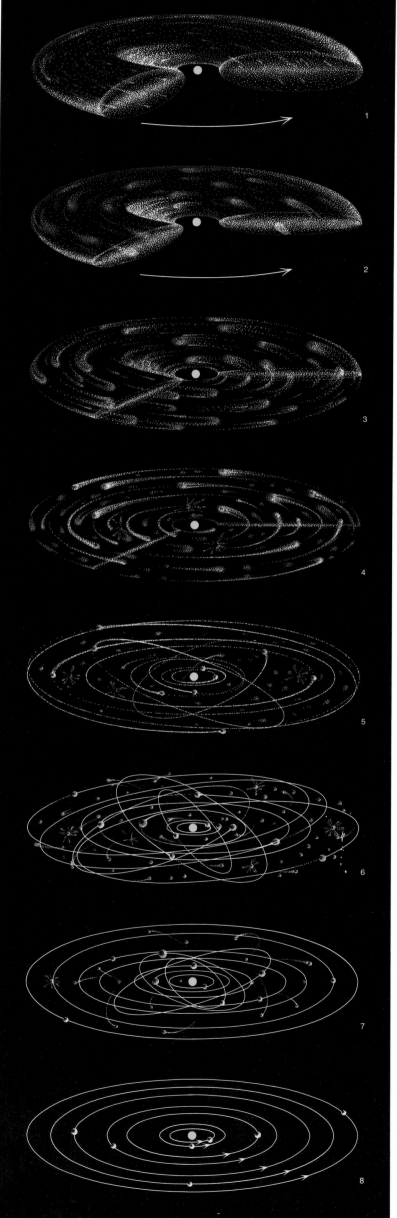

The successive phases in the formation of the planets according to the theory of collisional accretion. The rotating disk of gas and dust (1) develops a thin concentration of dust and small bodies in its central plane (2) which, as a result of small-scale gravitational instabilities (3), produces embryo planets, the planetoids, which begin to collide (4). Two kinds of event can occur during the growth of the planetoids by collisional accretion. Either a close encounter between two of them alters their orbits by increasing their relative velocities, or a collision produces a single planetoid in a less elliptic orbit. On average, collisions reduce the relative velocities of the planetoids, which, moreover, undergo fewer and fewer collisions since the probability of encountering other bodies gets less as the orbit becomes more circular (5, 6, 7). This model assumes that the initial relative inclinations of the orbital planes were small, since the planetoids emerged from a thin disk. However, gravitational concentration allows a large planetoid to capture small objects, even if their orbits do not lead to a direct collision, and also to displace planetoids, which do not collide with it, out of the initial plane. This is what is shown in stages 5, 6 and 7. The inclination of the orbits reduces the frequency of collisions. Gravitational concentration has played an important part in the final phase of the formation of the planets by accretion, in allowing the growth of large embryos. Stage 8 shows the state of the system (which is not necessarily our own) at the end of the accretion process.

The interiors of the terrestrial planets

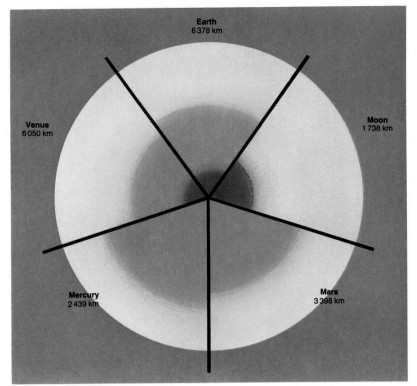

Comparative sizes of the mantles and cores of the terrestrial planets and the Moon. The values of the radii of the mantles and cores are well known only for the Earth. In the case of the Moon and Mars some observations allow limits to be placed on the sizes of mantle and core. For Venus and Mercury these values are very uncertain and are based only on the mean densities of these planets.

Before the era of space exploration very little was known about the terrestrial planets and their structure. The only available data obtained from observations from the Earth were their mean radii, densities and periods of rotation. The exploration of space has enabled great strides to be made; many planetary missions to the Moon, Mars, Venus and Mercury have revealed the main characteristics of these bodies and clearly demonstrated their great diversity.

The five principal objects in the inner Solar System are composed essentially of rocks and metallic compounds. However, differences of composition are revealed by their mean densities, which (in units of grams per cubic centimetre) are 5.42 for Mercury, 5.25 for Venus, 5.52 for the Earth, 3.33 for the moon and 3.94 for Mars. These differences of composition have been confirmed by many kinds of measurement made during planetary missions.

Thanks to these observations, we can now say that, like the Earth, the bodies of the inner Solar System do not have a homogeneous composition, but on the contrary are made up of layers of material of ever greater density as we go towards their centres. This layered structure consists, in the simplest case, of a light outer crust, a more or less rigid mantle and a very dense central core, either solid or liquid. Of course the reality is more complex but, in the present state of our knowledge, filling in the details of this simple scheme is quite difficult enough.

Of all the members of the inner Solar System, the best understood, after the Earth, is undoubtedly the Moon, thanks to the numerous Apollo missions, among others. Models of the internal structure of the Moon – at least to a depth of 1000 kilometres – have now reached a consensus. The lunar crust is thinner on the side visible from the Earth than on the hidden side. Its thickness varies from about 60 to about 100 kilometres. The whole of the lunar surface is riddled with vast impact craters. These have been filled by basalt flows which have formed the lunar maria.

The density of the surface crust can be deduced from analyses of samples of lunar rocks. The results, about 2.9 to 3.0 grams per cubic centimetre, are very similar to those of terrestrial rocks, with the exception of certain sedimentary rocks.

The lunar mantle extends from the base of the crust (about 100 kilometres in depth) to at least a depth of 1100 kilometres. It is in the deepest parts of the mantle (between 700 and 1100 kilometres deep) that the lunar quakes originate. These have been revealed by the small seismological stations left on the surface by the Apollo missions. It is now known that these quakes are periodic and that the observed periodicities are connected with variations in the configuration of the Earth–Moon system; it is the tidal deformation of the Moon caused by the Earth which is responsible. Apart from these, the Moon has almost no surface seismic activity, in agreement with the absence of tectonic motions on its surface.

The existence of a lunar core is still a matter of controversy. The presence of a core was suggested by a single seismic measurement, which indicated a maximum radius of about 500 kilometres. Other geophysical considerations suggest that the likely size of the core is less.

The great Martian volcanoes, the desert plains, the canyons and valleys, these last bearing witness ot the existence of rivers in the past, have been discovered by the Mariner 90 planetary probe and especially by Viking. The surface of Mars has largely been photographed, but the interior has hardly been directly examined. A seismometer was indeed placed on the Martian surface but only one seismic event was detected during its operation. Our understanding of the interior of Mars is thus largely based on models.

Measurements of the overall gravitational field suggest that Mars has a fairly thick rigid outer

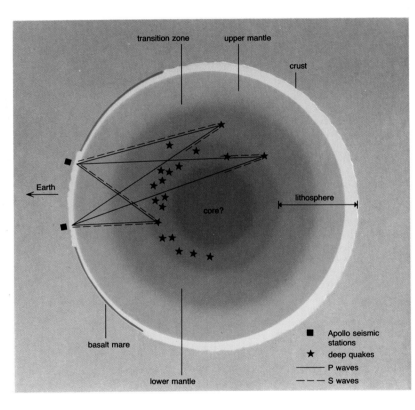

Schematic section of the interior of the Moon. The layered structure shown is inferred from numerous geophysical measurements made during the Apollo missions, especially the seismic experiments. The division into crust, mantle and core is based on differences of chemical composition, while the division between the lithosphere and aesthenosphere is based on differences in rigidity (and hence of temperature).

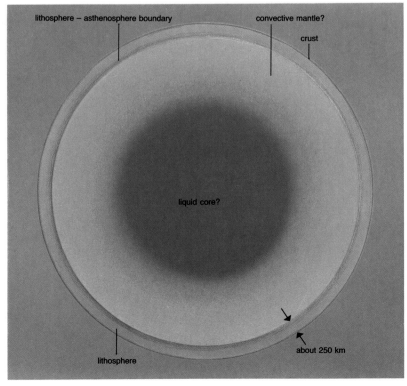

Schematic internal structure of Mars. The thicknesses of the crust and lithosphere are inferred from some sparse geophysical measurements made during the Viking mission in 1976. The deep structure of Mars has not been investigated and is thus deduced largely from simple models based on the mean density of the planet, its moment of inertia and the equations of state appropriate to the most plausible composition of the Martian mantle. It is not known if there are convective motions in the mantle, and the size of the core depends on the thermal model used. Some observations suggest that the Martian mantle is enriched in iron in comparison with that of the Earth.

layer – the lithosphere – about 200 kilometres thick, which excludes surface tectonic activity of the terrestrial kind. The Martian crust is estimated to be about 50 kilometres thick. The size of the core is very uncertain and, depending on the model, varies between 1300 and 2000 kilometres. There is no proof of the existence of convection currents in the Martian mantle. Nevertheless, the large recent volcanoes in the Tharsis region are evidence of the existence of thermal anomalies in the interior of the planet.

The study of the atmosphere of Venus was for long the only object of space missions to the planet. It is only very recently that we have been able to gather data about its surface, thanks to the Soviet Venera missions and the American Pioneer Venus.

Indirect analysis of the composition of the soil shows that the Venusian crust is composed of silicates like the Earth, indicating that the planet is differentiated. The most recent measurements made by the Soviet probes placed on the surface have also revealed the basaltic nature of the landing sites. A complete radar mapping of the surface has been made by Pioneer Venus, thus providing for the first time a relief picture of Venus of adequate resolution. It seems that the altitudes are, on average, small, since 70 per cent of the surface does not exceed 1 kilometre in altitude (with respect to the mean radius). Even so, certain limited areas (Maxwell Montes, Beta Regio, Aphrodite Terra) have very significant reliefs, reaching 11 kilometres. These areas are very similar to the terrestrial continents, while those which are less elevated are more like the oceans – if the Earth had no water. Some long canyons have also been detected on the surface of Venus, but the general relief has nothing to compare with the great tectonic structures on the Earth (mid-ocean ridges, subduction zones, faults, etc.) so that it is very difficult to say whether or not there is a system of moving plates on the surface of Venus. Consequently it is not known if there are convection cells in very slow motion within the planet's mantle. Measurement of the gravitational field of Venus reveals a very close correlation between the gravitational anomalies and the relief, which indicates either that the interior of Venus is very rigid, enough to support the topographic loads on a geological time-scale, or that the main relief is relatively recent and has not yet been compensated isostatically. Another possibility is that the reliefs are supported dynamically by rising currents resulting from convective motions in the mantle. Finally we note that measurement of the gravitational field of Venus shows an absence of flattening, which agrees with the fact that the planet rotates very slowly, in 243 days.

Anny CAZENAVE

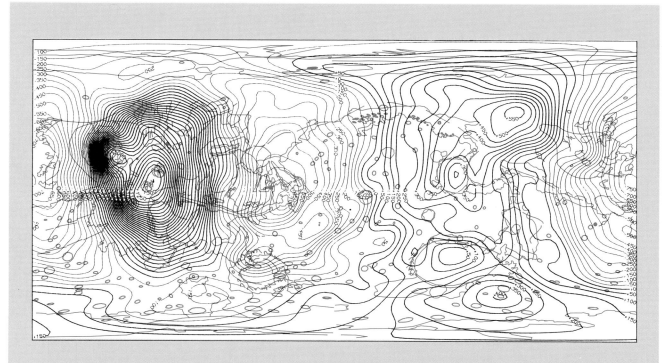

The Martian geoid. The geoid is the equipotential surface everywhere normal to the Martian gravity. The heights of the Martian geoid, shown here as contours at 50-metre intervals (in black), have been calculated from the perturbations of the orbits of the probes Mariner 9 and Viking 1 and 2. Note the very large positive anomaly conspicuously correlated with the region of large volcanoes, Tharsis. There are other correlations between anomalies in the geoid and the Martian topography – shown in red – especially at the level of the Hellas and Isidis basins. The amplitude and horizontal extent of these anomalies allow the internal rigidity of the planet to be estimated, particularly that of the outer layer called the lithosphere which should be rigid enough to give elastic support to the loads imposed by the topography. (Courtesy G. Balmino, GRGS/CNES)

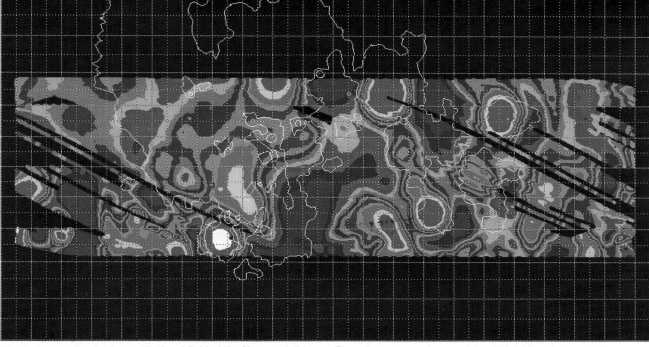

Chart of anomalies in the gravitational field of the Moon – visible face. These anomalies, shown as accelerations due to gravity and expressed in milligals, were measured by Apollo 15. The effects of the lunar flattening and of the elliptical shape of the lunar equator have been subtracted from the anomalies. The dull colours (violet, indigo, blue-green, dark green, orange-yellow) indicate negative anomalies (between −40 and −5 milligals) corresponding to a mass deficiency. The colours orange, brown, pink and dark blue show anomalies between 5 and 15 milligals. Finally the bright colours (blue, green, yellow) represent positive values from 15 to 40 milligals and red indicates anomalies of between 40 and 100 milligals. Higher anomalies are shown in white. These positive anomalies represent an excess of mass. The seas and oceans are outlined in white. (After W. L. Sjogren *et al.*, 1971; courtesy L. A. Soderblom)

type of measurement	giving information on	planets measured by these means
magnetic field (in orbit)	the existence of a metallic core	Mercury, Venus, Moon, Mars
gravitational field (in orbit)	the rigidity of the body; the existence of internal convective motions	Venus, Moon, Mars
heat flux (in orbit)	the thermal state of the interior; the thermal evolution of the body	Moon
imaging: measurement of the surface minerals by remote sensing (in orbit)	the chemical composition of the surface	Venus, Moon, Mars
optical and radio imaging of the surfaces (in orbit)	the topography of the surface: evidence for tectonic activity; information on the presence of internal convective motions	Venus, Moon, Mars
seismic measurements (*in situ*)	internal homogeneity; layered structure, existence of a liquid core; relative sizes of mantle and core	Moon

The study of the internal structure of the terrestrial planets. The table summarises the principal measurements made with planetary probes. These measurements form the basis for the models of composition and internal structure of the bodies observed (the Earth is not included).

The origin of the atmospheres

Of the four inner planets only Mercury, on account of its weak gravitational field, has not been able to retain an atmosphere. Venus, the Earth and Mars, the terrestrial planets, have atmospheres which recent space missions have made it possible to describe and compare.

Mars is ten times less massive than Venus or the Earth, whose masses are similar. The atmospheric pressures at surface level are very different, as also are the chemical compositions. Carbon dioxide (CO_2) is the main constituent on Venus and Mars, whereas nitrogen (N_2) and oxygen (O_2) are the most abundant on our planet, probably because of the existence of life on Earth. Why should three planets so similar in size and distance from the Sun today have atmospheres so different that only one of them supports life? The answer should perhaps be looked for in different initial conditions and in their having evolved along different paths.

The way in which the dust grains of the primordial nebula were gathered together into planetismals or planetoids eventually to form the planets 4.6 billion years ago is now well described by quite a simple theory. As for the origin of the atmospheres, there were until recently three competing theories, but a comparison of the primordial content of rare gases of the three planets now allows two to be discarded.

Neon, primordial argon 36 (as opposed to argon 40, a radioactive decay product) and krypton are especially useful for this purpose since these gases are too heavy to escape from the atmosphere and do not combine with the solid matter on the surface, so that the present amounts of these gases reflect the original amounts. The abundance ratio between neon 20 and argon 36 is about the same for the three planets, but very dfferent from the solar ratio, which also reflects that of the primitive nebula. We can therefore exclude the hypothesis that the atmospheres were captured by the solid bodies direct from the primitive nebula.

On the other hand, the absolute abundance of primordial argon (i.e. the mass of argon relative to the mass of the planet) is sixty times less on the Earth than on Venus, and much less still on Mars than on the Earth. Since Venus and the Earth have had the same chance of being bombarded by asteroids and comets, one would expect to find equal abundances of primordial argon if it is assumed that these small bodies were the source of their atmospheres.

By a process of elimination there remains the accretion hypothesis; the atmospheric constituents were already present in the dust grains (accounting for 0.01 per cent of their mass) before they were gathered first into planetismals and then into planets. For example, water could have been present in hydrated minerals such as serpentine; strongly heated in the interior of the planet, these minerals would give off water vapour finally to be released at the surface by volcanic activity.

The laws of mineral chemistry tell us that minerals condensed out of the primitive gaseous nebula at different temperatures and hence at different distances from the Sun. The planets could therefore have been made of different minerals and released different gases. This distance factor may be seen to play a part in comparing the compositions of Mercury, the Earth and the satellites of Jupiter (lying 5 AU from the Sun), for example. However, its importance is less obvious when comparing the Earth with its two planetary neighbours, Venus and Mars. For example, it is estimated that comparable amounts of carbon dioxide (CO_2) and nitrogen (N_2) have been degassed into the atmospheres of Venus and the Earth. But while

CO_2 has remained in the atmosphere of Venus, it has been largely fixed in the water and rocks on the Earth as calcium carbonate Ca CO_3.

Mars was able to form with absolute abundances of nitrogen, carbon and water comparable with those on the Earth. However, because of its smaller mass, its interior was heated less and outgassing has not yet finished, whereas in the case of Venus and the Earth it was certainly rapid (less than a billion years) because of the higher internal temperatures.

The chemical composition of the gases released into the atmosphere depends upon this temperature and the state of oxidation of the rocks. In all probability the gases released were mainly water vapour (H_2O), carbon dioxide (CO_2), carbon monoxide (CO) and nitrogen (N_2). Oxygen (O_2) and hydrogen (H_2) were also produced by the decomposition of water while hot, but oxygen was immediately re-incorporated into the rocks. Consequently there was originally no free oxygen in the atmosphere; the chief reducing gases were hydrogen and carbon monoxide.

The gases thus released experienced various processes of transformation and loss so that the atmosphere of each of the terrestrial planets has undergone very important but very different evolutions. It is the distance from the Sun which is the most important factor in these divergent developments. However, a common feature was that water condensed on the Earth, on Mars and no doubt on Venus, to form oceans and glacial deposits. Some carbon dioxide gas was fixed in the form of limestone. The very light hydrogen was able to escape from the atmosphere.

In this way, in a primitive atmosphere consisting mainly of nitrogen and carbon dioxide, with a little hydrogen, water vapour and carbon monoxide, there began, under action of solar ultraviolet radiation and lightning, the first stages of organic chemistry which led to the appearance of life on the Earth some 3.7 billion years ago.

But how can we elucidate the subsequent development of our atmosphere up to the present? Firstly, we understand and can estimate quantitatively the mechanisms of evolution actually at work and can calculate their influence in the past. Secondly, we have two precise pointers to the history of the oxygen in the atmosphere. Uranium oxide (UO_2) and galena (PbS) have been found in sediments dating from 2 to 2.5 billion years ago, which shows that there was very little free oxygen (less than 1 per cent of the present amount) in the atmosphere at that time, otherwise the products would have been the oxides UO_3 and $PbSO_4$. In contrast, the existence of quite extensive beds of red rocks (the colour being due to Fe_2O_3) dating from 1.8 billion years ago, implies that oxygen existed then.

Further, we know that life invaded the continents only 420 million years ago, which implies that there was probably enough oxygen and equally important ozone (O_3), produced from oxygen, to prevent dangerous ultraviolet radiation from the Sun reaching the ground. There then resulted a rapid, almost explosive, increase in the biomass, which led in its turn to the production of oxygen by photosynthesis and thus to a positive feedback process.

As for Venus, so like the Earth in mass and size, it is a little nearer the Sun than our planet (the semi-major axis of its orbit is 0.7 astronomical unit) and it is that which accounts for the extraordinary difference in the evolution of its atmosphere. Formed from the same material, the two planets should have vented similar amounts of water, carbon dioxide and nitrogen. Today we find similar quantities of carbon dioxide and nitrogen in the two planets (including the carbon

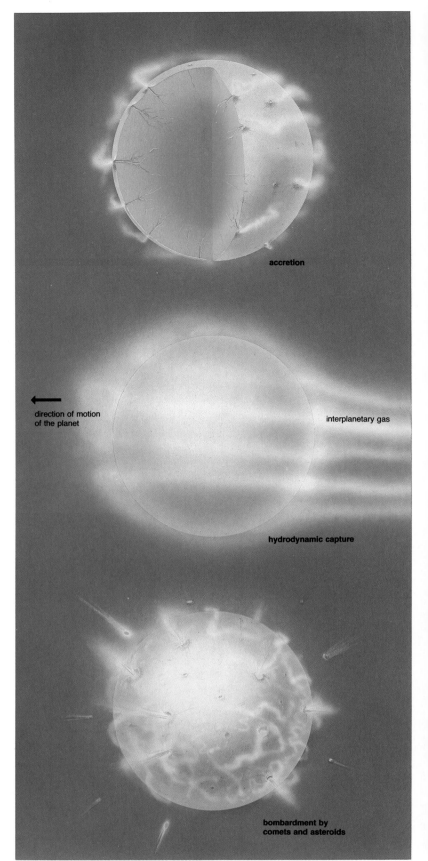

accretion

direction of motion of the planet

interplanetary gas

hydrodynamic capture

bombardment by comets and asteroids

The various hypotheses on the formation of the atmospheres. There are three main hypotheses as to how the atmospheres of the terrestrial planets were formed. According to the accretion hypothesis (top), the volatile compounds were already contained in the dust grains before they formed into planetismals, and they were degassed after these combined to make the planet, mainly as the result of internal heating and the differentiation of the solid mass. According to the primitive atmosphere hypothesis (centre) the grains did not contain volatile compounds, but the planet, once formed, captured some of the gas of the primitive nebula by a hydrodynamic process. In a third hypothesis (bottom) the planet acquired its atmosphere after an intense bombardment by bodies rich in volatile substances, themselves coming from other parts of the Solar System, and by asteroids and comets. The latest knowledge of the present state of the atmospheres leads us to believe that the accretion hypothesis is the most plausible with, perhaps, a small contribution from asteroids and comets.

mineral	gas released
C, N adsorbed in iron–nickel alloys	N_2
olivine (Fe,Mg)$_2$ SiO$_4$ pyroxine (Ca,Fe,Mg) SiO$_3$	CO, CO$_2$, NO, NO$_2$
troilite FeS	COS, S, SO$_2$
tremolite Ca$_2$Mg$_5$Si$_8$O$_{22}$(OH)$_2$ serpentine (Fe, Mg)$_3$Si$_2$O$_5$(OH)$_4$ talc (Fe,Mg)$_3$Si$_4$O$_{10}$(OH)$_2$	H$_2$O, NH$_3$, PH$_3$, CH$_4$ HCl, HF, H$_2$S, H$_2$

Some minerals which may have made up the grains of the primitive nebula and the gases which they could have released. The gases could have been released either by 'devolatisation' (the mineral changed its nature) or by the desorption of a gas previously adsorbed in the nebula material as is the case with the nitrogen in iron–nickel alloys, or even by chemical reactions.

dioxide fixed as carbonates in the terrestrial rocks) but in contrast to the 3 kilometre thick layer of water in the Earth's oceans, there is scarcely 10 centimetres of water vapour in the atmosphere of Venus. If there really was once water on Venus, where has it gone?

It is the greenhouse effect in the atmosphere which is the cause of the scorching temperature (450°C) on the surface of Venus. The solar radiation heats the ground, which emits in the infrared; a part of this radiation is returned to the surface after being blocked in the atmosphere by carbon dioxide, water and the clouds, as if by the walls of a greenhouse.

Sometime in the history of Venus, the greenhouse effect generated by the first gases to be released was sufficient to prevent water condensing. Water vapour then accumulated in the atmosphere which accentuated the greenhouse effect; this positive feedback phenomenon has been called the runaway greenhouse effect. It remains to be discovered whether this process has been at work since the formation of Venus or if it only began later, at a time of increased solar radiation. We can speculate as to the existence of liquid water on Venus in an initial phase, which would have allowed life to develop until the overheating by the Sun unleashed the runaway greenhouse effect and caused all the water to be vapourised into the atmosphere.

The opposite fate has certainly befallen the water on the surface of Mars, where it is now to be found only in the polar caps and probably frozen in the ground. Here also the greater distance of Mars from the Sun (on average 1.5 astronomical units) has been an important factor. However, its smaller mass and lower internal temperature when compared with the Earth have also played a part in that they have allowed only a partial outgassing (about 20 per cent). Nevertheless, it would have been possible to release the equivalent of a layer of water 100 metres thick, 2 bars of carbon dioxide and 100 millibars of nitrogen. To prevent all the water being fixed as ice, carbon dioxide must have been present in the atmosphere to create a greenhouse effect and maintain a temperature greater than 0°C. However, as on the Earth, carbon dioxide would progressively disappear from the atmosphere to be fixed in limestone, but, whereas on the Earth plate tectonics causes these rocks to sink and melt in the magma and release the carbon dioxide again via volcanoes, the weak geological activity of Mars does not allow a similar recycling. The carbon dioxide diminishes, the greenhouse effect weakens, the temperature falls and water becomes ice in a runaway glaciation effect. Almost all volatile substances were removed from the atmosphere and fixed in surface reservoirs; polar caps, fine grained regoliths and the soil itself.

We conclude that the distance of a planet from the Sun plays a crucial role in the evolution of its atmosphere. By simulating on a computer the evolution of the terrestrial atmosphere, Michael H. Hart has calculated that, if the Earth had been in an orbit less than 0.95 astronomical unit from the Sun, a runaway greenhouse effect would have started and the Earth would have gone the way of Venus. In contrast, if the Earth had been at more than 1.01 astronomical units, there would have been as runaway glaciation as on Mars. Similarly, in any solar system we can find a very narrow habitable zone within which liquid water could condense on a planet. Life could appear and persist for billions of years in this extremely narrow zone.

Jean-Loup BERTAUX

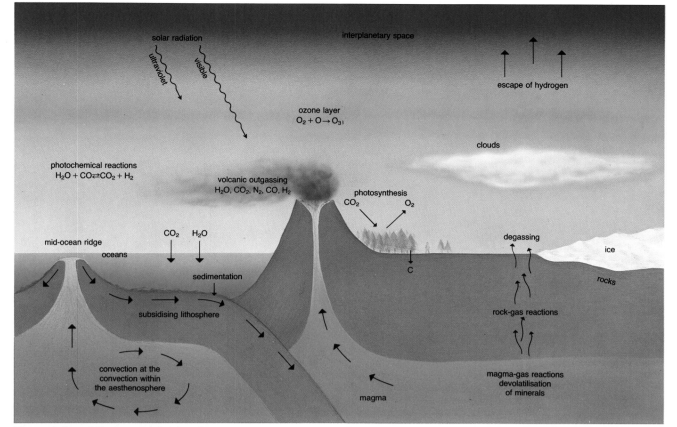

The evolution of the terrestrial atmosphere. A great variety of mechanisms acted on the different gases, so that each one (except the rare gases which accumulated in the atmosphere) followed cycles whereby they were in turn present in the atmosphere, condensed (as is water) or dissolved in the oceans, trapped in the rocks, then released anew by volcanoes or by effusion into the atmosphere. A state of equilibrium was thus established, but has, however, been able to evolve slowly during the 4.6 billion years which is the age of the

Evolution of the atmosphere of Venus (lower left). Although the amounts of nitrogen and carbon dioxide on Venus and the Earth are comparable (counting the carbon dioxide fixed in the terrestrial rocks), there is now practically no water on Venus; only the equivalent of 10 centimetres, as vapour in the atmosphere. Supposing that to begin with these planets had similar amounts of water, we need to explain how Venus has been able to get rid of the equivalent of 3 kilometres of water. The runaway greenhouse effect explains why this water does not condense, but remains as vapour in the atmosphere. Being so light, the hydrogen has almost entirely escaped. As for the oxygen, we have to suppose that it is removed from the atmosphere by oxidation of the surface rocks. In the evolutionary model shown here, the amount of carbon dioxide is constant at its present level of 100 bars, and water disappears very rapidly in 200 million years with the production of an equivalent amount of oxygen. It then requires four billion years (note the change of scale on the time axis) to remove all

Evolution of the Earth's atmosphere (right). It is possible to attempt a computer simulation of the evolution of the Earth's atmosphere from its formation up to the present day (and even beyond), taking into account a wide spread of evolutionary processes: outgassing from the interior, condensation, glaciation, photodissociation of water, oxidation of surface material, the fixing of carbon dioxide in the rocks, the presence of life and the evolution of the biomass, photosynthesis and the decay of organic sediments, and climatic factors (changes in the solar radiation, cloud cover, greenhouse effect). Obviously the simulation leads to the present state of the atmosphere.

The changing chemical composition of the atmosphere is shown as a function of time, for a model in which the original composition of the outgassed material was: 84% H_2O, 14% CO_2, 1% CH_4 and CO, and 0.2% N_2. In this picture, the water very soon condenses, the total

Earth, though for different reasons. Hydrogen is light enough to have largely and permanently escaped. The composition of the minerals, which was at first that of the dust grains of the primordial nebula, changed bit by bit, as did that of the gases which resulted from the 'devolatilisation' of these minerals. The water released by volcanic degassing condensed to form the deep oceans, where the dissolved carbon dioxide could form calcium carbonate ($CaCO_3$) and return to the rocks.

the accumulated oxygen from the atmosphere, supposing that newly formed rocks, not yet oxidised, reach the surface of Venus at a rate equal to the present rate on the Earth.

A recent measurement of the quantity of deuterium (a heavy isotope of hydrogen) in the atmosphere of Venus has lent strong support to this evolutionary scheme. Thomas M. Donahue, of the University of Michigan, has shown that, on Venus, the ratio of heavy water (HDO) to normal water (H_2O) is 1 per cent, that is, a hundred times greater than on the Earth. Deuterium, twice as heavy as hydrogen, has much more difficulty in escaping from the atmosphere and in the course of time the atmosphere becomes enriched in deuterium. There was once at least 10 metres of water on Venus and probably a lot more, as on the Earth. Only its greater proximity to the Sun, by setting off the runaway greenhouse effect, prevented Venus from later having an atmosphere like the Earth's (After J. C. G. Walker, 1975)

pressure varies between 0.6 and 1.3 bars, and the chemical composition is radically altered. The carbon dioxide is dissolved in the water and fixed in the rocks and the methane (CH_4) and carbon monoxide (CO) are oxidised by oxygen, itself produced by the photodissociation of water. As soon as methane and carbon monoxide were exhausted, two billion years ago, oxygen reappeared (in agreement with what we know about the history of oxygen) and slowly increased, until there was a thick enough protective layer of ozone (O_3) for life to emerge from the sea and overrun the continents, 420 million years ago. This increase in the biomass then led to an explosive increase in the amount of oxygen liberated by photosynthesis. The extrapolation of the curves beyond the present takes no account of future changes resulting from human activity.

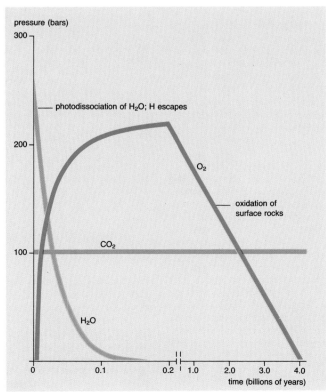

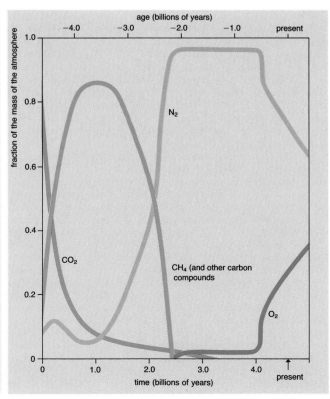

Planetology and climatology

The exploration of the Solar System has shown that on the majority of planets and several of their satellites there are atmospheric phenomena which are in numerous ways comparable to some of those observed in the Earth's atmosphere. Several examples will illustrate the analogies between the mechanisms giving rise to the most characteristic structures of planetary atmospheres, particularly the processes leading to the formation of clouds.

Most terrestrial atmospheric motions on the synoptic scale, such as cyclonic or anticyclonic systems, have been detected in the atmospheres of other planets. The Viking 1 orbiter thus observed two cloud formations resembling terrestrial cyclonic disturbances in Mars's northern hemisphere. The shapes of these clouds also resembled those in the westerly current forming part of the general circulation of our atmosphere. This shows that baroclinic instabilities occur in the Martian atmosphere just as in our own. As another example, the pictures transmitted by the Voyager probes allowed a determination of the rotation sense – anticlockwise – of the three ovals and the Great Red Spot in Jupiter's southern hemisphere. This sense corresponds to a tropical anticyclonic phenomenon on Earth, supporting one of the explanations of these turbulent structures.

In the Earth's atmosphere, cloud formations of average scale, from several to a hundred kilometres, are easily distinguished by meteorological satellites. This allows one to study the clouds according to their mode of formation, and thus to classify them within the general circulation or as a function of the terrestrial relief. Several cloud types have been identified in planetary explorations, especially in the Martian atmosphere, where in general we find the same classification: orographic clouds, formed as a result of forced rising motions of the air; convection clouds, resulting from atmospheric instability; and short-lived clouds such as fogs.

Another important feature of several planetary atmospheres is the formation of large dust clouds. On our planet, these clouds of dust and aerosols can grow to sizes where they change the local albedo of the Earth's surface. In Mars's atmosphere, where many dust clouds are seen, the effect can sometimes extend over the whole planet and mask its surface completely.

Similarly in the upper atmosphere of the earth, Venus and Titan, we see the formation of photochemical hazes. A thin layer of photochemical aerosols is permanently present on a global scale in the terrestrial stratosphere. The haze layer observed at a level 25 kilometres above the base of the Venus clouds has properties similar to those of the terrestrial stratospheric layer. The yellowish disc of Venus is particularly bright at the poles, where there are spots like caps. Their shapes and brightnesses are variable, and they result from an increased concentration of the global aerosol layer over the poles. These particles consist of aqueous solutions of sulphuric acid and are comparable to the stratospheric aerosols on Earth, although they are larger (of the order of 1 micrometre). The same phenomenon of photochemical haze formation occurs at high altitude on Titan: the opaque layer consists of aerosols, which are thought to be organic polymers, hydrocarbons, and nitrides, formed from nitrogen and methane, which are important constituents of the lower atmosphere of Titan. Strangely, an aerosol polar cap is only seen at Titan's north pole. Generally, the whole northern hemisphere is clearly redder and darker than the southern hemisphere. As yet there is no explanation for this asymmetry.

Finally, exploration of the planets and their satellites has revealed evidence for strong surface activity. The most spectacular example is without doubt the gigantic plumes, probably of sulphuric anhydride, ejected at certain places on the surface of Io in what we must call volcanic eruptions. The surface of Triton shows similar activity, in the form of geysers of several kilometres in height. But unlike terrestrial geysers, these plumes consist of liquid nitrogen.

Beyond this purely phenomenological approach, meteorologists and climatologists rapidly realised the advantage of setting their research in the context of comparative planetology. The planetary circulations of the atmospheres of Mars, Venus, and the Earth show that the general wind pattern has the same properties and seems to obey the same laws. For these three planets it is the dissimilar solar illumination at the equator and the poles which drives the general atmospheric circulation. This dissimilarity is different for each of the planets, and its influence on the atmospheric dynamics and radiative balance depends on the properties of each planet. The mass of the atmosphere, proportional to the surface atmospheric pressure, is clearly an important parameter; its value fixes the timescale for responding to radiative changes. Thus, because of its high thermal inertia and very long day – 117 Earth days – the temperature does not vary strongly with longitude in Venus's lower atmosphere. This effect is seen on the Earth, although the atmosphere is less dense. However, here the effect lasts the length of the day, which is much shorter. In contrast, an atmosphere reacts to dynamical changes according to the strength of the Coriolis forces, and thus to the rotation rate of the planet. In the dense Venus atmosphere the *dynamical* response time is very short compared with the *radiative* response time, so heat transport to the dense interior is entirely controlled by dynamical processes, while Mars's very rarefied atmosphere is very close to radiative equilibrium.

Comparison of planetary meteorology has gone futher. Geophysicists such as Gareth P. Williams, of Princeton University, have adapted their very detailed models of the Earth's general atmospheric circulation by changing a small number of parameters: the rotation period of the planet, the inclination of its rotation axis and the length of its day. The results are extremely encouraging, as they do reproduce the main features of planetary atmospheric circulations. Further, the models apply equally well to the giant planets of Jupiter and Saturn, whose rotation periods are of order 10 hours, as to Venus, which rotates once every 243 days. In this case parametrising the diurnal variation of the solar heating is the most important part of the modelling.

The properties of the upper atmosphere of Venus (above 50 kilometres) are now well known, in particular after the infrared measurements made by Pioneer Venus. These show the near-permanent presence of a wave caused by the atmospheric tides. As for the Earth, this type of atmospheric wave must play an important role in the transport of heat and kinetic energy. Some theoreticians even suggest that the atmospheric tides could have been what tipped over Venus's rotation axis, and thus caused the retrograde motion of the planet. These tides may also be the origin of the superrotation of the Venus atmosphere. Finally, on Mars, the atmospheric tides certainly amplify the dust storms on the planetary scale.

Photochemical generation of aerosols in the upper atmosphere of the planets. There is a clear correspondance between the fog layers which form in the stratosphere of our planet and those discovered at altitude, above the dense Venus clouds. Despite important differences between the temperature and the chemical composition of these atmospheres, a photochemical process is the origin of the formation of sulphuric acid aerosols in both cases. The gas–particle conversion mechanism is in Venus caused by the photo-oxidation of sulphur dioxide (SO_2) by sunlight ($\lambda < 217$ nanometres). Further, in models of the thick fog layers covering the whole of the satellite Titan, the assumption is made of aerosols made of polymers derived from hydrocarbon molecules, after complex photochemistry of methane gas, which with moelcular nitrogen constitutes the atmosphere of this satellite of Saturn.

In the upper photograph, transmitted on 12 November 1980 by Voyager 1 when the probe was 4000 kilometres away, one sees several fog layers above Titan's limb. The outer layer is about 300 kilometres above the surface of the satellite (NASA). The lower photograph was taken in the Earth's atmosphere from a balloon at an altitude of 31 kilometres. The camera photographed the limb of our planet below the horizontal line of sight. The image was obtained at several wavelengths for a solar depression of less than 5 degrees. The white-orange part is the stratospheric aerosol layer, called the Junge layer and situated at about 17 kilometres altitude. A finer layer is also seen at 22 kilometres. The summit of the clouds is at 12 kilometres and delimits the lower dark region. (Photograph by M.Ackerman, of the Belgian Institute of Spatial Aeronomy)

History of the climates of the telluric planets

Comparative planetology has reached a stage where it is possible to check climatic models whose object is to describe the past evolution of the Earth's climate, and thus make a link between the present environmental conditions and those prevailing more than 4.6 billion years ago. Clearly one's confidence in these models is increased the more one can apply them successfully to other telluric planets. Using the same

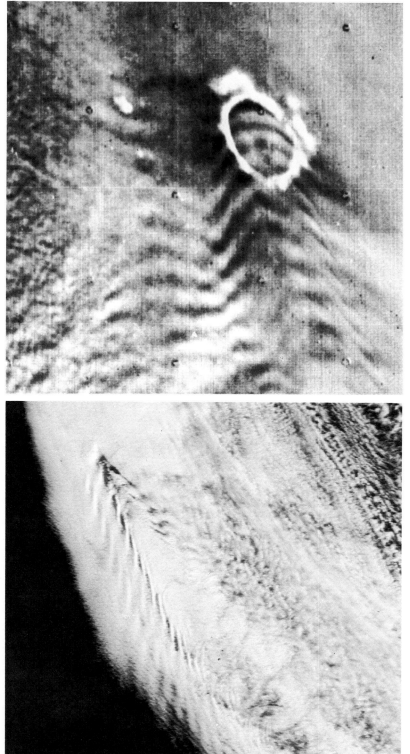

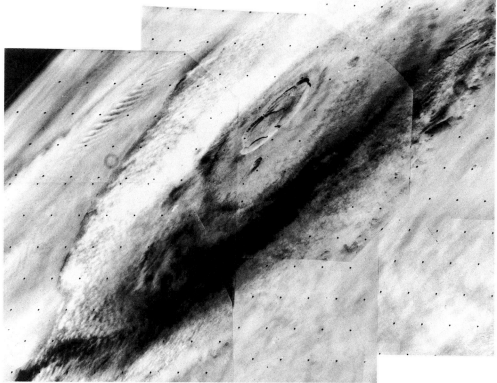

Orographic clouds around Olympus Mons, the highest Martian relief (above). Taken at a distance of 8000 kilometres, this photograph of the Martian volcano Olympus Mons was transmitted by the Viking 1 probe on 31 July 1976. It clearly shows the phenomenon of orographic clouds which form as the atmosphere cools in contact with the slopes of the volcano which rises to 24 kilometres. The clouds which completely encircle the relief reach a height of 19 kilometres. Formed of water-ice particles, they are densest on the western slopes. In the centre, overhanging the cloud belt, we see the system of calderas completely uncovered by clouds. The scale of the photograph is such that the clouds which form mainly in the hot seasons can be observed from Earth. Above left, one sees cloud waves characteristic of orographic formations. (NASA)

Clouds associated with lee waves (opposite). When an air current is forced to rise because it encounters an obstacle such as a mountain range, gravity waves are created within it. One thus sees a wave train in the lee of the obstacle. The lower picture was transmitted in the visible (0.6–0.7 micrometres) by the NASA satellite NOAA-5 on 1 September 1976, and received by the Department of Electrical Engineering and Electronics, University of Dundee. NOAA-5 was launched in July 1976 and worked until 1979.

The upper picture shows the same phenomenon caused by the edges of a Martian crater in the region of Mare Acidalium. The Mariner 9 probe, orbiting about Mars, provided the best observations of these cloud waves, in winter over the northern hemisphere. The white circle corresponds to the frost-covered edges of the 90 kilometre wide crater. The wavetrains extend over several hundred kilometres. (NASA)

Orographic Karman turbulence. In the Earth's atmosphere one sometimes observes small-scale turbulence in the lower cloud layers downstream from a mountainous island. The clouds resemble the eddies produced by a round obstacle in a purely turbulent laboratory flow. The Tiros-N satellite detected these characteristic formations (vortex streets) formed by Jan Mayen Island on 12 April 1979, shown on the image at right. (DEEE, University of Dundee)

The left-hand image is a spectacular picture of the cloud layers observed by Voyager. It is striking that despite the difference of scale, there is a certain similarity in the form of the turbulence in the clouds forming in the wake of Jupiter's Great Red Spot. This can be understood if, as certain theoretical meteorologists working on Jupiter think, the Spot is a cyclonic system created deep in the atmosphere by enormous convection cells. These are an anomaly in the general circulation and play the same role as a mountain, and as a result, produce a Karman vortex street. (NASA)

model, can one understand the evolution of the atmospheres of the Earth, Mars, and Venus from the epoch of formation from the primitive nebula?

We know the origin of the various volatile elements found today in the environments of the planets. These elements were those incorporated, either in grains (not vaporised) of the interstellar cloud giving rise to the protosolar nebula, or in planetesimals and protoplanets. Their chemical properties depend mainly on the distance of the accretion zones of the planets from the centre of the nebula. The Earth, Mars, and Venus were formed at similar distances from the proto-Sun, so the temperature differences between them were too small for there to be much difference in the volatile elements retained. The differences we now see in the composition of their atmospheres must be explained by different evolutions.

The main conclusion of this comparative analysis of climates is that the Earth's environment has been relatively stable throughout its evolution, with moderate climatic conditions. This has not been the case for Venus or Mars. The most complete version of this theory was worked out by two American scientists, James F. Kasting and James B. Pollack. Their work was stimulated by the strange results discussed in 1978 by Carl Sagan and Michael Hart. They showed that the Earth's atmosphere several million years after the formation of the planet could not have been the same as that of today. In fact the Sun has not always had its present luminosity: stellar evolution theory shows that in its youth it had a luminosity 25 to 30 per cent less than now. The solar constant has increased gradually ever since the formation of a star inside the primitive nebula. Its current rate of increase is of the order of 1 per cent per 100 million years. Now, when we compare the energy balance now and 4.6 billion years ago, we find that the simplest assumption is that the Earth has benefitted from an atmospheric greenhouse effect. Similar to but more powerful than the present effect, this must have been able to compensate for the lower solar input. If this is the case, what do we have to assume about the properties of the atmosphere immediately after the Earth's formation, and in particular its composition?

It is probable that in the very first accretion phase, the Earth acquired by direct capture of nebular gases a primitive atmosphere whose composition reflected that of the Sun. We assume however that most of this atmosphere has disappeared through the escape of light gases into space. Only the remainder, mainly the rare gases of the primitive atmosphere, could appear in the present atmosphere. But apart from this exception, the atmosphere which originated the present one was a secondary atmosphere, which formed and evolved by various processes over the Earth's long accretion phase lasting an estimated 50 million years or more. The main formation process of this secondary atmosphere was the volatilisation of the planetesimals as a result of numerous impacts in the accretion phase. In this case, water, the most volatile of the elements retained inside the bodies participating in the accretion, will be the major constituent of the atmosphere. What were the other constituents of this atmosphere and how did it evolve until the end of the Earth's formation period? Once the atmosphere became sufficiently opaque to infrared radiation, a greenhouse effect appeared which tended to raise the surface temperature. This thermal effect differs from the runaway greenhouse effect nowadays discussed for our atmosphere, as the energy comes from below. Thermal energy is supplied not by the Sun but by the impacts occurring during accretion. This energy is a component of that directly deposited in the surface at the moment of impact, and that stored inside the planet and then reinjected into the planet. The resulting effect is thus called 'thermal blanketing'. A critical threshold is reached once the heat flux transmitted through the surface exceeds that which the atmosphere can radiate into space in the infrared. The atmospheric opacity increases as more volatilised water is released in impacts, which become more and more numerous. The thermal blanketting effect feeds itself, like the runaway greenhouse effect. The temperature rises rapidly, reaching values higher than the solidification temperatures of silicate rocks. In the model of Kasting and Pollack, the surface temperature is of order 1500 K. At this stage of the accretion epoch the atmospheric pressure is of order 30 bars (3 megapascals) and the surface is in a state of fusion, in equilibrium with the atmosphere. After a long period of 35 million years an important event occurs during which the internal layers of the planet liberate large amounts of water. While part of the impact energy of the planetesimals in the accretion process raises the surface temperature, some of the heat is also transported to deeper layers, where significant convection then begins. Once the surface water is exhausted, internal convection dredges up magma which cools at the surface and releases its water content. This results in the formation of a very dense atmosphere whose pressure can reach 250 bars (25 megapascals: the Earth's oceans contain 1.4×10^{21} kilograms of water, equivalent to an atmosphere of 27 megapascals).

Towards the end of the accretion phase, as the collisional energy decreases, the environmental properties of the Earth change completely. The surface heat flux is no longer enough to maintain the irreversibility of the greenhouse effect. The surface temperature falls rapidly, and water vapour condenses, the precipitation leading to the formation of vast expanses of water. Once the Earth is completely formed, the surface temperature is 300 K and the atmospheric pressure 2 bars (0.2 megapascals). The oceans progressively form. Outgassing processes, either at the moment of impact of planetesimals or after gas is released by internal layers, increase the atmospheric content of water vapour and other gases. Kasting and Pollack studied the infrared opacity of an atmosphere rich in water vapour but dominated by molecular nitrogen (N_2) and to lesser degree by molecular oxygen (O_2) and carbon dioxide (CO_2). They calculated that to compensate a greenhouse effect which was originally stronger, the energy deficit resulting from the lower luminosity of the young Sun, one has to assume a carbon dioxide concentration corresponding to partial pressures of order several tenths of bars. This is a thousand times greater than that observed today (about 300 parts per million). Of all the gases considered, carbon dioxide is most sensitive to the input of the mathematical models. One can show that the deficit from the fainter Sun can be exactly compensated by modifying its concentration over time. Above all, its presence has a stabilising effect and the concentration changes are not uncontrolled. They fluctuate automatically as a function of the surface temperature variations. The permanent recycling of carbon dioxide between the lower atmosphere and the Earth's crust guarantees relative stability of climatic conditions. When the surface temperature rises, the atmospheric carbon dioxide content decreases, and conversely. For the Earth, the processes involved in the geochemical cycle of about 500 000 years mainly cause the solution of carbon dioxide in rainwater to form carbonic acid (H_2CO_3). This reacts with rocks containing calcium and silcates and releases calcium (Ca^{2+}) and bicarbonate (HCO_3^-) ions into underground rivers which flow into rivers and oceans. Other factors participate in the process, such as the presence of living organisms, fractures of the ocean floor, subduction and volcanic eruptions. Thus, if the Earth has enjoyed moderate climatic conditions throughout its lifetime, this probably did not result from a fortuitous coincidence between the variation of the carbon dioxide content of the atmosphere and the increase of the Sun's luminosity. The stability of the climate results from correlated fluctuations of the carbon dioxide content, i.e. from the damping effect of its geochemical cycle .

The climatic model applied to Mars and Venus

Assuming that the atmospheres of the Earth, Venus and Mars evolved from similar types of secondary atmospheres, we still have to understand the differences observed today. From space probes we know that Mars's surface was covered in liquid water for several hundred million years. For temperatures to have remained above freezing for this time appears to have only one explanation according to current theoreticians, in the form of a greenhouse effect heating the surface. Such a process could well have occurred if Mars's atmosphere had originally been dense and composed of carbon dioxide (at pressures of 0.5 to 1 megaparsecs). A carbon geochemical cycle very similar to that on Earth would then have maintained the required environmental conditions for almost a billion years. But, unlike what happened on our planet, the Martian cycle stopped working in a closed loop. Because of the smaller mass and size of Mars, the internal heat and tectonic activity became too weak for the surface to release enough carbon dioxide into the atmosphere.

The Venus atmosphere originally had similar properties to that of the Earth after the condensation of its secondary atmosphere and the formation of the oceans. Slightly denser (a little more than 0.2 megaparsecs) and hotter because of its proximity to the Sun, the Venus atmosphere was nevertheless cold enough (350 K) for oceans to exist on its surface.

Kasting and Pollack analysed the consequences of an incease in the Sun's luminosity for the evolution of the environment which the Earth and Venus must then have possessed: atmospheres with high water vapour content, oceans, and surface temperatures between 350 and 450 K. They showed that once the surface temperature reached 500 K, a very small increase in the Sun's luminosity would make this temperature rise to 1400 K. This effect is connected with the infrared opacity of an atmosphere of increasing density, and becomes more and more effective if it is not balanced by the cooling resulting from the reradiation into space of the infrared radiation emitted by the planet and its atmosphere. Within this narrow range of variation of the solar luminosity, we can say that the planetary environment experiences a *humid atmosphere greenhouse effect.* This runaway effect causes surface temperatures of more than 1400 K. Then the surface and the lower atmosphere start to radiate in the near infrared and visible, where water vapour opacity is lower. The greenhouse effect recedes.

Although the secondary atmosphere considered by Kasting and Pollack is hot and humid, their analysis makes no assumptions about the presence of clouds. The threshold at which sunlight is able to initiate the greenhouse effect is $1.4S_0$ (S_0 is the present irradiation at the Earth's orbit). Now, as the young Sun was 25 to 30 per cent less luminous, a planet at Venus's orbit received a solar irradiation of between $1.34S_0$ and $1.43S_0$. This range is too narrow to be sure that Venus could have lost a hypothetical ocean.

Kasting and Pollack thus tried to test the sensitivity of their model to the presence or absence of clouds. For a hot and humid atmosphere like that considered here, the clouds play a dominant role: their presence causes an increase in the reflection of sunlight rather than an increase in the opacity of the atmosphere to the infrared radiation emitted by the surface. The surface tends to become cooler; consequently the value of the solar irradiation at which the runaway of the greenhouse effect starts becomes larger. The threshold is at $4.8S_0$ (rather than $1.4S_0$) if one assumes that the clouds are high (corresponding to a pressure of several megapascals) and cover the surface completely. Thus theoretically, oceans could have survived on the surface of Venus almost until the present. But this is clearly in disagreement with observation. The reason is that the properties of the atmosphere considered for Venus (secondary atmosphere hotter than the Earth's) particularly favour the rapid outflow of hydrogen into space, even without a runaway greenhouse effect. The solar irradiation need only reach $1.1S_0$ for very large amounts of water vapour (corresponding at least to the same mixing ratio as that of the lower atmosphere, i.e. 20 per cent) to exist in the upper atmosphere, at about 100 kilometres altitude. With the specific properties assumed for the primordial Venus atmosphere, large quantities of water vapour would reach the upper layers of the atmosphere. At these altitudes hydrogen escapes from the water molecule and leaves the Venus environment through hydrodynamic escape. At altitudes of 100 kilometres, water is photodissociated by the solar ultraviolet radiation. Hydrogen thus becomes a major constituent of the upper atmosphere. The atoms collide often enough for them to acquire the energy to escape from Venus's gravity. This escape has the properties of a hydrodynamical flow of a fluid (the energy needed to expand a dense atmosphere to high altitude is supplied by

Temperate cyclone in the Martian atmosphere. Two cloud formations with appearances similar to cyclones forming in temperate latitudes on Earth were discovered by the Viking mission. This picture shows the formation observed in summer by the Viking 1 orbiter at 81° north and 160° longitude. The spiral cloud, composed of water ice crystals, has a size of 200 kilometres, and its altitude is estimated at 7 kilometres. (G. Hunt/NASA)

the heating of the gas by the absorption of solar ultraviolet rays). This hydrodynamic escape process explains the absence of liquid water on Venus: it would take only a few million years to lose an amount of water equivalent to a terrestrial ocean.

Carbon dioxide plays no role in this phase of the evolution of Venus; it is in contrast the constituent mainly responsible for the very high temperature measured now at the surface of the planet. After the disappearance of the oceans from Venus the geochemical cycle of carbon dioxide was completely changed, and it gradually replaced water vapour as the major constituent of the atmosphere (its pressure is about 9 megapascals today).

If after its formation the Earth had received solar radiation about 10 per cent greater, i.e. if it had been formed slightly nearer the Sun, at 0.95 astronomical units, it would rapidly have lost its oceans, like Venus. Life could not have developed here. The idea of a region of habitability of a planetary system has to be considered in estimating the probability of the appearance and survival of life for planets around other stars. In the Solar System this region is not very restricted, as its outer boundary is about at Mars's orbit.

Guy ISRAËL

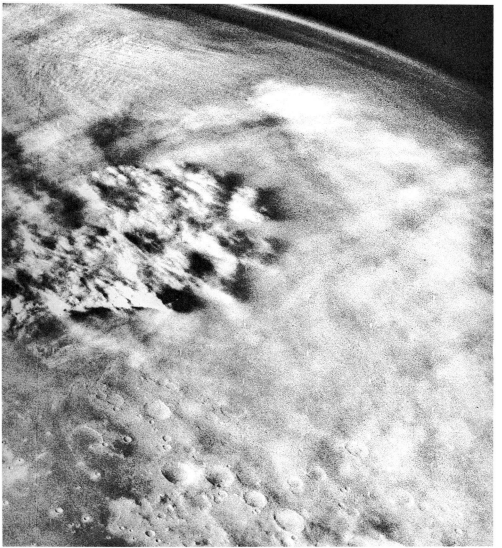

Dust clouds. Observing the disc of Mars from the Earth, one can discern spots of orange colour which today can be identified with dust clouds. These form episodically when relatively strong surface winds – of order 30 to 60 metres per second – raise the ground dust. Particles of 5 to 10 micrometres diameter can remain in suspension in the low-density Martian atmosphere, as there is no rain to wash out the atmosphere. Exceptionally the phenomenon can involve the whole planet; thus at the end of 1971 Mars was completely covered by a 'curtain' of dust occulting its surface. At this time of the Martian year the planet was at its closest point to the Sun, and because of its fairly eccentric orbit, the solar irradiation was 40 per cent higher than usual (this effect is only 3 per cent for the Earth). This cannot be the only parameter responsible for the phenomenon on a planetary scale, for in February 1977 the Viking 2 orbiter observed a dust storm involving almost the whole of the southern hemisphere, at a time when the the planet was still far from periastron.

This picture shows a more localised storm observed by the Viking 1 orbiter when surveying the site of the Viking 1 station. The black spots are the shadows of each cloud, the light coming from the left. (NASA)

Radiation fog. Meteorologists are familiar with the phenomenon of radiation fogs, which are common in cold seasons. They are characterised by the formation at sunrise of a relatively thin featureless layer of condensates which dissipates during the morning. The right hand photograph was transmitted by the Meteosat 1 satellite on 14 May 1979, at midnight. We see a vast fog outlining the southwest coast of England and the south coast of Ireland (Meteosat, ESA). Sometimes the fog forms just above the surface. This is the surface fog phenomenon clearly seen above certain regions of the Martian surface. The left-hand photograph shows two images of the same region taken 30 minutes apart, on the morning of 24 July 1976 by the orbiter Viking 1, which was surveying Mars from about 10 000 kilometres altitude. Contrasts can be seen on the second picture between the crater floors and the channels; the thickness of the condensates (water) is estimated at several microns. (NASA)

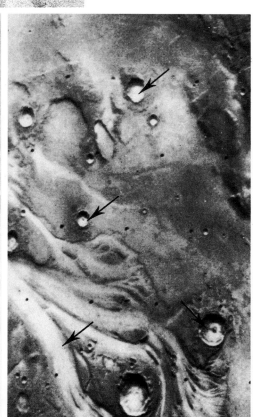

Mercury

Mercury is a very bright object because of its particular position in the Solar System; at perihelion, it receives ten times as much solar radiation as our Moon. The flux of thermal energy emitted in the infrared from the surface was measured very precisely during the first and third encounters of Mariner 10 with the planet. This confirmed that the ground was subject to large temperature changes during a Mercurian 'day' of duration 176 terrestrial days. Close to perihelion, the temperature may reach 430 °C in the vicinity of the subsolar point (where the Sun is at the zenith), while the minimum night side temperature is about −170 °C.

One of the great discoveries of the Mariner 10 mission was that there is an intrinsic magnetic field, which is much stronger than would have been expected from the slow rotation of the planet. Measurements made by instruments on Mariner 10 of the magnetic field close to the planet show that the intensity at the surface is relatively large, corresponding to a field equal to about 1 per cent of the magnetic field of the Earth. This is, however, sufficient for a magnetosphere to form.

Mercury, like Venus, does not have a natural satellite. Compared to the other planets, Mercury exhibits two important peculiarities from the point of view of dynamics. Its orbit has a very large eccentricity ($e = 0.206$), and the orbital plane is inclined to the ecliptic by the large angle of 7 degrees. Furthermore, the planet is very difficult to observe from Earth because of its apparent path in the heavens. Mercury oscillates from one side of the Sun to the other every four months and is never more than 28 degrees away. Because of the great brilliance of the Sun, the weak contrasts which appear on the planetary disk could never have been defined well enough to determine the rotation of the planet. Hence the technique of radar echoes has been used in order to establish that the rotation period is 58.65 ± 0.25 Earth days, whereas it was previously thought to be equal to the revolution period (87.97 days). The hypothesis that the rotation period of Mercury was exactly two-thirds of its revolution period was thus confirmed. This commensurability is interpreted to be the consequence of a dynamical coupling due to the gravitational forces acting upon a non-spherical body. The form of Mercury deviates only slightly from that of a sphere; this was, however, sufficient for the gravitational action of the Sun to result, after the formation of the planet, in a slowing down of its rotation. Analysis of the perturbations shows that stabilisation upon a given value of the rotation period is possible when there is a commensurability with the revolution period.

Theoretically, the mass of Mercury is too low for an atmosphere to have been able to form and persist. The escape velocity is only 4.3 kilometres per second while that of the Earth is 11.2 kilometres per second. Furthermore, the daytime temperatures are high because of the proximity of the Sun. The result is that the thermal velocities acquired by molecules degassed by the surface are much greater than the escape velocity. The very low atmospheric pressure (10^{-9} millibar), determined by observations in the ultraviolet during the passage of Mariner 10 1000 kilometres from the surface, corresponds to values in the exosphere of the Earth at an altitude where collisions between molecules are practically non-existent. The only definitely known atmospheric constituent is helium. The existence of this light gas in the atmosphere may be explained by the beta-decay of uranium and thorium in the crust of Mercury, or by the accretion of particles emitted by the solar wind.

Guy ISRAËL

The approach of Mariner 10 to Mercury. A mosaic of eighteen photographs, taken at 42-second intervals on 29 March 1974, six hours before the flyby at the minimum distance, shows part of the illuminated side of Mercury. The probe was then 200 000 kilometres from the planet. A large number of craters were discovered; some are 200 kilometres in diameter. In particular, towards the centre of the image, the brightest crater of Mercury, named Kuiper, may be distinguished. (NASA)

Rotation and revolution of Mercury. The left-hand drawing shows the path of Mercury around the Sun. The red point indicates successive positions of the same region of the planet during two revolutions around the Sun. It is necessary to note that the Mercurian 'days' are twice as long as the Mercurian 'year'.

The diagram illustrates the position of the Sun with respect to various equatorial regions of Mercury in the course of the revolution of the planet. As a result of studying the relative heating of a given point of longitude with respect to another, it is convenient to represent the apparent motion of the Sun with respect to the centre of Mercury in polar coordinates which turn with the same movement of rotation as the planet. The right-hand drawing shows the apparent path of the Sun in the Mercurian sky. The position of the Sun every eleven days is shown. The longitude 0 degrees has been arbitrarily chosen to represent one of the two subsolar points at perihelion. The presence of loops shows that the period of heating is rather long in certain regions. Near perihelion, the angular velocity of Mercury in its orbit is very large, and may compensate for, or even exceed, the angular velocity of the planet's rotation on its axis. In this way, for an observer upon the surface of Mercury at the equator, there will be a period of about eight days at perihelion during which the path of the Sun is very peculiar. During this short period, which occurs around 'midday' with respect to the Mercurian day of 176 days, the observer will see the Sun move very slowly, then stop, and set off again in the opposite direction. (After S. Soter and J. Ulrichs, 1965)

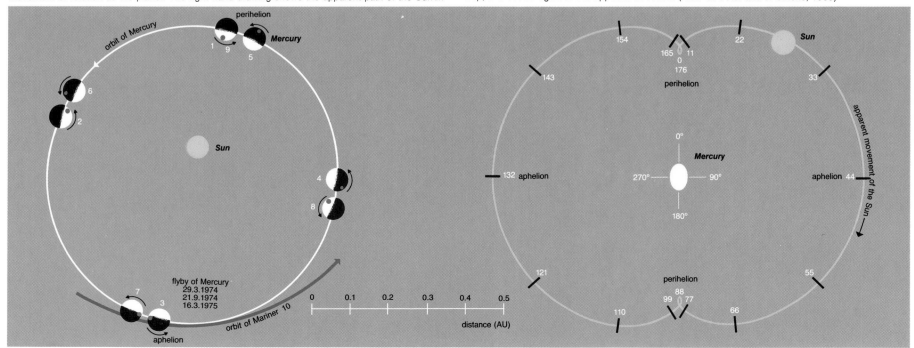

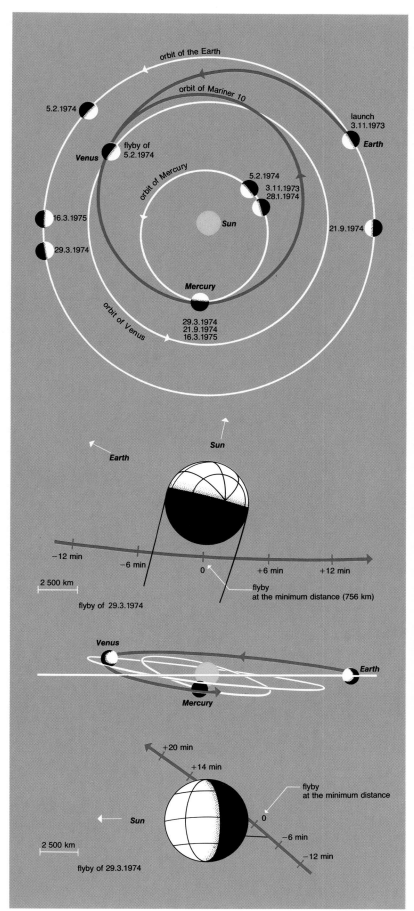

Mariner 10 flies away from Mercury. This photograph (above) was constructed from eighteen images taken at a distance of 210 000 kilometres, six hours after the flyby at the minimum distance of 29 March 1974. The Caloris basin is near to the centre of the terminator. It is thus named (*calor*, Latin for heat) because it is located in the vicinity of one of the two points which alternately face the Sun when Mercury is at perihelion, and because of this reach the highest temperatures. The Caloris basin represents a mass anomaly on the scale of the planet; the axis of the minimum moment of inertia crosses it. (NASA)

A determination of the rotation period of Mercury. It is possible to measure the rotation period of the planet accurately by comparing high resolution photographs taken during the three flybys of Mariner 10.

The photographs below illustrate the method used. The shadow projected by the walls of a crater of diameter 65 kilometres located at 1° north and 17° east upon its flat bottom at the time of the first flyby (left) is compared with the shadow which was projected at the time of the third flyby 352 days later (right). The altitudes of the Sun were respectively about 7.6 and 7.1 degrees, and the lengths of the shadows about 18.0 and 19.2 kilometres. During the 352 days, the planet had revolved six times on its axis and two Mercurian days had passed. The rotation period resulting from the comparison of the two pictures is 58.644 1 ± 0.004 5 days. (Paper kindly communicated by the Jet Propulsion Laboratory; after Kenneth P. Klassen, 1976)

The Mariner 10 mission. The only probe to have flown past Mercury, Mariner 10, was launched from Cape Canaveral on 3 November 1973. Two months before the first flyby of Mercury, Mariner 10 bypassed and photographed Venus on 5 February 1974. NASA experimented with a new ballistic technique, consisting of using the gravitational deflection of a vehicle caused by the pull of a planet, the goal being to use a rocket inferior in performance to the rocket which would normally be required for a direct approach to the planet. Mariner 10 was in this way first aimed at Venus, and came sufficiently near to the planet (at 5800 kilometres from its surface) for the gravitational pull to reorientate its trajectory considerably. After the probe had passed behind Venus, the modulus of its heliocentric velocity vector was smaller: the energy necessary to bring the probe onto a trajectory closer to the Sun, and to reach the orbit of Mercury, was reduced in proportion. In fact, thanks to the particularly happy choice of the ballistic parameters for the Mariner 10 mission, a triple resonance was established between the orbits of the probe, of Mercury, and of the Sun. After its first encounter with the planet, the probe was in an orbit around the Sun with a period equal to twice the revolution period of Mercury. In that way, once every two revolutions of the planet, hence at intervals of two Mercurian 'years', the probe and Mercury were located at the same positions in space relative to the Sun. As the planet turns on its axis exactly three times while performing two revolutions around the Sun, the result was that the same regions of the surface of Mercury were photographed at each encounter with the same conditions of illumination, although with the resolution varying considerably.

The top diagram shows, viewed from the north pole of the ecliptic, the trajectory of Mariner 10 and the positions of the Earth, Venus and Mercury at various dates. Immediately below, the first flyby is shown as it would have appeared from the north pole of the trajectory of the probe. The next diagram shows a panoramic view of the trajectory from a far distant point located in the plane of the ecliptic. (The vertical scale is three times greater than the horizontal scale.) Finally, the first flyby is shown as it would have been seen from the direction of the Earth. Mariner 10 ceased to be operational after its third flyby, but it will continue to revolve around the Sun indefinitely. The minimum distances for the three encounters were respectively 756, 48 069 and 327 kilometres. (Diagrams after James A. Dunne, 1974)

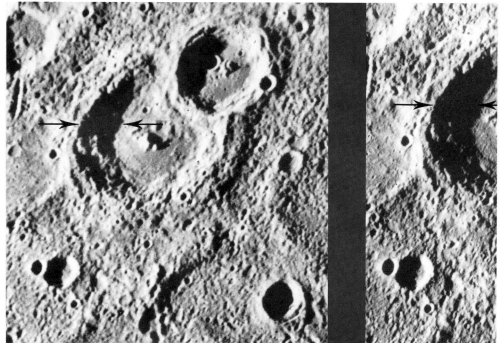

Whatever the resolution of the images, the dominant morphological features shown to exist by the Mariner 10 mission are the craters. The largest, the Caloris basin, has a diameter of 1300 kilometres; on the very high resolution photographs, a multitude of craters with diameters around 100 metres can be seen. Indirect measurements such as those of the polarisation of the light re-emitted by Mercury, show that the ground is riddled with craters whose diameters range down to 1 micrometre.

Morphological studies of the craters reveal an underlying unity despite the many forms. Craters of diameters less than 10 kilometres have the characteristic appearance of shell holes: they are bowl-shaped, with depths between a fifth and a tenth of the diameter, surrounded by a ring of debris, termed ejecta. As the crater diameter increases from 10 to 20 kilometres, the form gradually changes until the bottom is more or less flat. For diameters upwards of 20 kilometres, the craters are all flat-bottomed with, at the centre, the shape of a peak. Between 20 and 150 kilometres diameter, the ring of ejecta develops and the peak becomes more and more well defined and prominent. In the craters of diameters from 150 to 200 kilometres, the peak widens and tends to become a ring. Above 200 kilometres diameter, craters are termed basins; they exhibit one or more concentric rings unless they have been filled with volcanic material subsequent to their formation. All these characteristics (ejecta, variation of form as a function of the diameter, etc.) are found in the least detail on the Moon and in the hundred or so terrestrial craters for which studies *in situ* (which allow, for example, meteorite fragments to be found) have shown the craters to be impact craters. The change from the simple bowl form to the more complex forms can be explained by recoil phenomena immediately after the impact.

Craters are not subject to any kind of erosion upon Mercury as there is a total absence of water and virtually no atmosphere. A crater can only be covered by more recent terrain, or obliterated by the impacts of further meteorites. This 'eternity' of the craters explains why they are so abundant. (Note that as many meteorites fall upon the Earth as upon Mercury, and that the Earth's atmosphere is totally unable to halt the largest; however, a terrestrial crater will undergo erosion, and may be covered with sediment, etc.) Upon all of the planets without atmospheres, a chronology

of craters according to their state of 'freshness' may be established. The recent craters are intact, while the very old craters have been degraded by other impacts. Estimates of the numbers of craters as a function of their age show that the frequency of impacts has diminished considerably in the course of time: meteorite falls, very rare today, were very common in the past.

The craters may be used to date the terrains of Mercury: the old terrains, exposed to the bombardment for a long time, will be extremely cratered. On the other hand, there will be very few craters on the recent terrains. By applying this method, it is found that 70 per cent of the surface of Mercury consists of very old terrains. These are termed intercrater plains. Ten per cent of the surface consists of averagely cratered plains, called intermediate. The rest of the planet consists of smooth, not very cratered plains which are therefore relatively recent. By comparing counts of craters on Mercury with counts made upon the Moon where the ages of the terrains are known by measuring the radioactivity of the rocks, the ages of the young plains are estimated to be between 3.8 and 3.9 billion years, and the ages of the intercrater plains are estimated to be at least 4.1 to 4.2 billion years. The plains were therefore formed very early in the history of Mercury. The planet was very active up to 4.2 billion years ago; this activity almost completely ceased 3.9 billion years ago. An intense but short period of activity took place 3.8 billion years ago. Since then, there appears to have been no significant production of new terrains upon Mercury.

The chemical and geological nature of the terrains upon Mercury are not completely elucidated, but two different arguments indicate that the surface is probably volcanic. The morphological argument is as follows: even if active volcanoes are not in evidence (contrary to what is seen upon Mars) it is extremely probable that numerous structures such as ridges and domes are of volcanic origin. The second argument is that photographs showing the details of the contacts between the smooth plains and the intercrater plains indicate that the smooth plains cover the intercrater plains, invading the lower regions and transforming the high parts into 'islands'. In just the same way, upon Earth, layers of lava cover an uneven terrain. The arguments from spectroscopy are as follows: the reflection spectrum of Mercury contains certain spectral

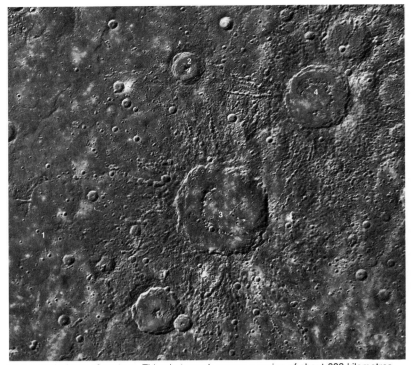

The morphology of craters. This photograph covers a region of about 600 kilometres across. North is at the top; the light comes from the south-east at lower right. The diversity of forms of the craters of Mercury are shown. The simple form (1) is produced by the impact of 'small' meteorites from 10 000 to 100 000 tonnes. Craters with diameters between 20 and 150 kilometres have a flat bottom and a central peak (2). These craters are termed complex craters and result from the impact of 'large' meteorites from 1 to 100 billion tonnes; they were the site of recoil phenomena leading to the formation of the central peak. Something of the kind may be seen when a raindrop falls into a puddle. These craters are surrounded by a ring of radial debris, the ejecta, expelled by the shock. In the craters of larger dimensions, termed basins, the recoil phenomena have given rise to one or several rings which replace the central peak; the Strindberg basin (3) of 165 kilometres diameter, and the Ahmad Baba basin (4) of 115 kilometres diameter, are particularly good examples; the ejecta are very plentiful and show radial chains of small secondary impact craters created by the heavy debris ejected from the basin at the time of the original impact. (NASA)

Young and old craters. This photograph covers a region of about 800 kilometres by 1200 kilometres. North is at the top; the light comes from the south-east at lower right. Craters of various ages are seen. The craters Degas (1) and Brontë (2) exhibit a very fresh morphology: pointed form, well-defined central peak, without any more recent craters being superposed; the ring of ejecta forms bright lines or rays which extend over several hundreds of kilometres. These craters are called ray craters. The rays are formed of finely pulverised ejecta and are relatively rapidly darkened by the solar wind. They disappear completely after a billion years. Degas and Brontë are therefore young craters. On the other hand, the crater Chŏng Ch'ŏl (4) does not have a central peak or a ring of ejecta, being riddled with tens of smaller craters. It is a very old crater which has undergone considerable erosion by meteorite falls subsequent to its formation. All intermediate forms between the young craters and the very old craters are found: thus crater (3) still has a central peak but no longer has a ring of bright ejecta. (NASA)

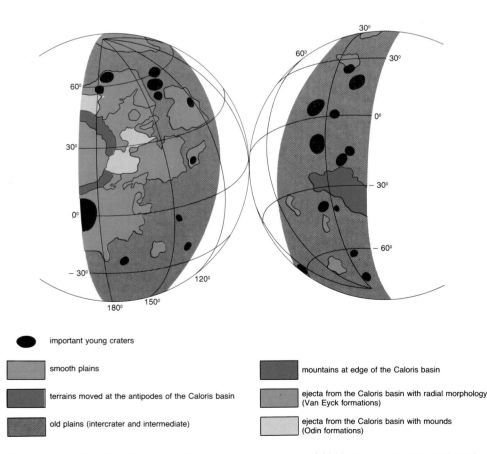

- ⬤ important young craters
- ▨ smooth plains
- ▨ terrains moved at the antipodes of the Caloris basin
- ▨ old plains (intercrater and intermediate)
- ▨ mountains at edge of the Caloris basin
- ▨ ejecta from the Caloris basin with radial morphology (Van Eyck formations)
- ▨ ejecta from the Caloris basin with mounds (Odin formations)

Formations and terrains. It was possible to make a geological map for only 35 per cent of the surface of Mercury. Fifty per cent of the planet was in shadow during the flybys of Mariner 10 and 15 per cent of the surface was photographed at too oblique an angle. Each of the two parts of this simplified geological map corresponds to photomosaics of hemispheres figuring in the preceding pages. Note the relatively large size of the Caloris basin, and also the smooth plains which fill it and partially cover its ejecta. The so-called young craters are those whose formation is later than both the impact forming the Caloris basin and the volcanic eruptions which have given rise to the smooth plains. (After N. J. Trask and J. E. Guest, 1975)

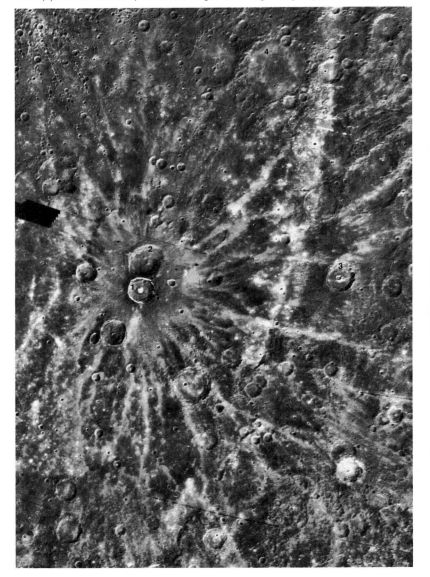

Map of Mercury. This map shows the known part of Mercury, that is, slightly less than 50 per cent of the surface of the planet. Principal reliefs are shadowed in the drawing such that they appear to be illuminated by a fictitious Sun located to the west.

The projection is based on a sphere of radius 2439 kilometres. The lattitudes and longitudes have been defined by adopting the hypothesis that the axis of rotation of the planet is normal to the orbital plane. A new reference meridian has been defined for convenience. The new point of reference that has been chosen is a very well-defined small crater about 1500 metres in diameter, located at 0.2° south at approximately 20° of longitude. A new system of longitude has been adopted, defining the 20° meridian to pass exactly through the centre of the crater. The crater is named Hun Kal, the word for twenty in the language of the Mayas, people who used that number as the base of their number system. This new system of longitude differs by about 0.5° from the previous one.

The accuracy with which the topographical formations are located upon the map varies between 10 and 25 kilometres. Six kinds of place names have been adopted by the IAU for the six principal types of morphological features. The craters carry the names of artists of all disciplines and nationalities (Rodin, Mahler, Botticelli, Ahmad Baba . . .). There

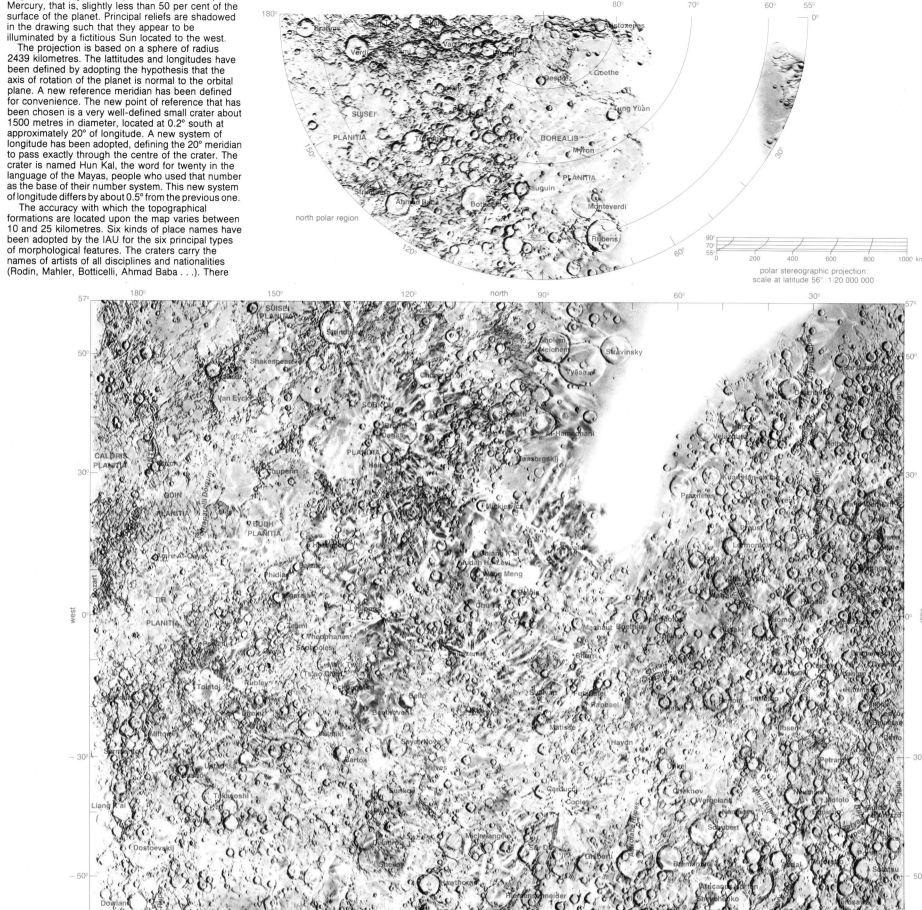

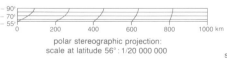

polar stereographic projection:
scale at latitude 56°: 1/20 000 000

equatorial Mercator projection;
scale at latitude 0°: 1/35 474 000

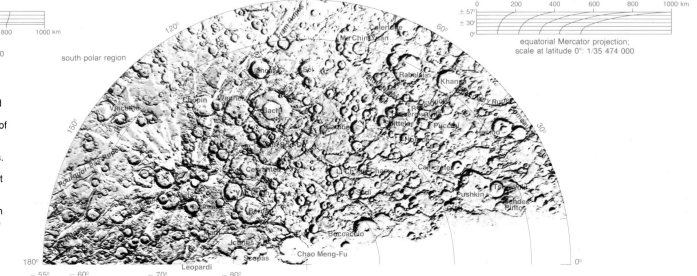

are only two exceptions to this, Hun Kal, obviously, and Kuiper, which is named after a contemporary astronomer. The plains (in Latin, *planitia*) are named after divinities corresponding to Mercury in various mythologies and religions, with the two exceptions of Caloris Planitia and Borealis Planitia. The largest mountains discovered by Mariner 10 border the Caloris basin and have been named Caloris Montes. The valleys (*vallis*) are named after radio-astronomical observatories to which was owed most of our knowledge of the relief of Mercury before Mariner 10. The ridges (*dorsum*) honour astronomers who have been closely associated with the study of Mercury; the names of famous vessels of exploration were given to escarpments (*rupes*).

bands which are characteristic of volcanic materials. Detailed studies of the spectra and of the albedo show that the surface is remarkably homogeneous and that its properties are very close to those lunar basalts that are the least rich in iron and in titanium. (Those samples were obtained during the Apollo 12 and Apollo 15 missions.) The surface of Mercury therefore seems to be entirely composed of very old volcanic terrains. These terrains are totally moved around by meteorite impacts in the case of the intercrater plains and the intermediate plains, an intense but short volcanic episode having produced the smooth plains. The smooth plains have changed very little because the impacts have become rare since then.

One of the characteristic features of the geology of Mercury is the existence of long cliffs called lobate scarps. These cliffs, from 50 to 3000 metres high and 50 to 500 kilometres long, occur here and there over the entire surface of the planet, just as much in the intercrater plains as in the smooth plains. The cliffs often cut across craters and, in some cases, modify their dimensions: a diameter of a crater measured perpendicular to a cliff may be 15 kilometres less than a diameter measured parallel to the cliff. This phenomenon, and the morphology of the cliffs, leads to an interpretation of the lobate scarps as compressive faults and overlappings. The origin of this compression would be a cooling of the interior of the planet – in particular of the relatively large iron core – giving rise to a contraction of the planet, that is, a reduction of its radius. This cooling is indirectly confirmed by the rapid decrease in volcanic activity. By studying the shrinkages of the craters upon all of the scarps, the reduction in the radius of Mercury is found to be 1.5 kilometres. The lobate scarps, of which many are more or less orientated towards the Caloris basin, are the sole indisputable evidence of large-scale tectonic activity upon the surface of Mercury. The surface does not, in particular, exhibit plate tectonics (perhaps because there is no internal convection, or because the lithosphere is too thick). Additionally, here and there old faults can be seen which are the evidence of some internal activity.

The Caloris basin, the largest impact formation upon Mercury (its diameter is 1300 kilometres) is the dominant feature of the planet. Incidentally, only the eastern half of the basin is known, the other half being immersed in night during the flybys of Mariner 10. Its rampart of mountains and its ejecta give it the morphology typical of large impact basins. Its present internal structures (concentric rings) are totally hidden by the smooth plain which completely fills the basin. The same smooth plains also partially cover the ejecta and form a vast ring around Caloris. A count of the craters indicates that they were formed just after the impact. Moreover, it is necessary to note that 80 per cent of the smooth plains upon Mercury are located inside or around the basin. It therefore seems that there is a causal relation between the gigantic impact which gave rise to the Caloris basin, and the sudden fresh outbreak of volcanism which produced the smooth plains.

The basin was the site of important tectonic events. The peripheral ejecta have been transported and now form more or less parallel 'piano keys'. Moreover, the interior plain is smooth only in name. Its periphery is covered by approximately concentric ridges, which are the folds formed by a compression directed towards the centre of the basin. The central region of the basin is not folded, however, but is on the contrary broken up, which indicates an expansion. The origin of these movements is not wholly elucidated. A local explanation may be put forward: a sinking of the basin under the weight of the basalt gave rise to the compression, a central lifting then causing the expansion. However, the size of the Caloris basin, whose diameter is one-quarter of that of the planet, enables a global explanation to be put forward. At the epoch of the impact, between 3.8 and 3.9 billion years ago, the lithosphere (rigid superficial part) was under compression. The almost total cessation of volcanic activity and the formation of lobate scarps are evidence of this. It was in such a context that the basin was formed; its floor was completely crushed and perhaps more or less melted by heat released by the impact. The peripheral lithosphere, under compression, was 'attracted' towards the centre of the basin, which was not under constraint. This movement transported the outer part of the basin, assisting the fresh outbreak of volcanism begun by the shock, and causing the radius of the basin to be reduced. This reduction of radius gave rise to a compression of the outer part of the floor, producing ridges, and causing the central part to bulge. The central part fractured. Moreover, this overall movement of the lithosphere was able temporarily to orientate the formation of lobate scarps towards the basin.

Mariner 10 has thus revealed Mercury to be a planet whose geological evolution is old and relatively simple. Its formation about 4.6 billion years ago was followed by a meteoric bombardment and an intense period of volcanic activity. After this, the planet cooled and slowly contracted while the bombardment considerably lessened. A gigantic impact 3.8 billion years ago temporarily reactivated geological activity. The surface of Mercury has not evolved for at least 3.5 billion years. This apparently dead planet has undergone no more than some rare impacts. Today, only the magnetic field shows that there is activity in the core.

Pierre THOMAS

Intercrater plains and smooth plains. This photograph shows a region near the north pole. (The light comes from the south at lower right.) Two quite distinct morphological domains are visible. To the west (lower left), the terrains are riddled with craters of all sizes, some on top of others, with morphologies ranging from fresh to very degraded. This is a very old intercrater plain; the crater Mansart, indicated by an arrow, has a diameter of about 80 kilometres. To the east (upper right), the terrains are much flatter and show only some rare craters, all of rather fresh morphology. This is the young, smooth plain Borealis Planitia. (NASA)

A volcanic structure, Mirni Rupes, and a compressive structure, Discovery Rupes. Mirni Rupes is a ridge 350 kilometres long (A–B), 10 kilometres wide and 500 to 1000 metres high. The mean slope of its flanks is 5 to 15 degrees, that is, comparable with the slopes of Hawaiian basaltic volcanoes. This principal ridge is oriented north-west to south-east (north is at the top and the light comes from the west, at the left). The ridge is associated with small ridges running north–south. It ends to the south-east with a series of pits and overlapping craters situated close to a very old crater. The ridge and all of the region to the south-east are averagely cratered, indicating an age intermediate between that of the intercrater plains and the smooth plains. All of these characteristics suggest an extrusion of lava along a pre-existing fault; the pits in the south-east would be volcanic craters. The slopes of the flanks of the ridge show that the lava had a fluidity close to that of basalt. This ridge would have given rise to the entire plain in the south-west.

The escarpment Discovery Rupes is a typical example of a form of relief which is very common over the entire surface of Mercury. The escarpment is a cliff about 500 kilometres long (C–D), facing the south-east. Measurements of the shadows cast indicate a height of 2000 to 3000 metres. This cliff cuts across the entire region, and in particular across two craters, numbered 1 and 2. The second, called Rameau, is about 45 kilometres in diameter. Crater 1 no longer has a perfectly circular form: its north-west–south-east diameter appears to have shrunk. A comparison of this cliff with terrestrial and lunar structures, and this shrinkage, lead us to interpret the cliff as an overlapping. A north-west–south-east compression has produced a displacement of the entire north-west region towards the south-east; the cliff represents the front of the overlapping which has resulted from this displacement. The reduction of the diameter of crater 1 enables the shrinkage due to the displacement to be estimated at about 10 kilometres. Such tectonics in compression, which are found over the entire surface of Mercury, are unique in the Solar System. (NASA)

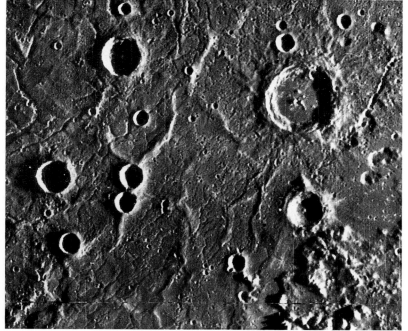

The interior plain of the Caloris basin. This plain, of which a detail is seen here at high resolution (of the order of 500 metres), contains few impact craters. The largest such crater is about 60 kilometres in diameter. The plain is classified as a smooth plain. However, despite its name, the plain is very uneven. Close to the outer rampart, part of which may be seen to the south-east at bottom right, there is a series of reliefs in the form of ridges. (The light comes from the east, to the right.) These ridges are the folds which have resulted from a compression of the floor of the basin towards its centre. In the central part of the basin at top left, the ridges give way to cracks and fractures, showing that this region has undergone an expansion. These ridges and fractures are symmetrically placed with respect to the basin. (NASA)

The Caloris basin. This mosaic of photographs taken at sunrise with the light coming from the east, at right, shows the Caloris basin and its immediate environs. More than 1200 kilometres lie between the terminator, located at around 190 degrees of longitude, and the extreme right-hand edge of the picture.

The Caloris basin is dominated by a circular mountain chain made up of large, juxtaposed blocks about 2 kilometres high. This rampart corresponds to the outer limits of the original formation. Beyond these mountains, particularly to the north-east and to the south-east, extend terrains which form a series of hills and valleys, which are elongated and parallel to one another (Van Eyck formations). These are the ejecta of Caloris. (NASA)

The antipodes of the Caloris basin. This picture shows the region of the crater Petrarch, which lies at the antipodes of the Caloris basin. (North is at the top; the light comes from the west, at left). The terrain is extremely chaotic, consisting of juxtaposed hills 5 to 10 kilometres across and 1.5 kilometres high, covered by fractures orientated mainly north-west–south-east, like Vallis Arecibo, which joins Petrarch, 170 kilometres in diameter, to a crater about 80 kilometres in diameter. This valley is more than 100 kilometres in length and about 7 kilometres wide. The entire ramparts of the old craters have moved. This tectonic movement would have been the seismic consequence of the enormous impact which produced the Caloris basin. After this fracturation, volcanic eruptions have filled the important topographic depressions, like the old crater Petrarch. (NASA)

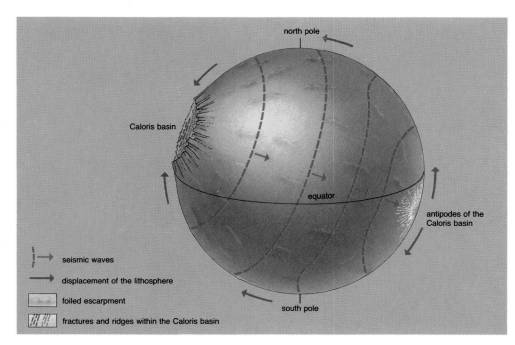

seismic waves

displacement of the lithosphere

foiled escarpment

fractures and ridges within the Caloris basin

The tectonic consequences of the impact which produced the Caloris basin. The tectonic events which took place after the formation of the basin are of two quite distinct kinds, and may be summarised thus:
– at the moment of the impact, the shock waves produced by the shock waves were focussed towards the antipodes, completely shifting the antipodes of the basin;
– after the impact, the lithosphere, which was under compression because of the cooling of the core of the planet, was slowly set in motion towards the centre of the basin; this movement brought about the formation of ridges and fractures within the basin in the same way as the peripheral movements, and temporarily oriented the formation of lobate scarps. (After P. H. Schultz and D. E. Gault, 1975; L. Fleitout and P. G. Thomas, 1982)

Venus

Venus moves around the Sun in a nearly circular orbit which has a mean radius of 0.723 astronomical units, or 108 million kilometres. The planet receives nearly twice as much radiation from the Sun as reaches the orbit of the Earth. Venus has a dense atmosphere, consisting almost entirely of carbon dioxide. While 50 per cent of the surface area of our planet is, on average, covered by clouds made up of water droplets or of water crystals, the surface of Venus is completely hidden by clouds, the upper layers of which are made of droplets of sulphuric acid in aqueous solution. The strong reflection of sunlight by the clouds of Venus makes this object, after the Sun and the Moon, the brightest object in the whole sky. When observed in visible light by telescope from the Earth Venus looks like a very bright, slightly yellow disk without a single distinctive feature. However, when the light reflected by Venus is received by a detector sensitive to the ultraviolet, dark bands on the image are revealed, whose semipermanent presence must be attributed to phenomena associated with atmospheric waves. The existence of these bands observed in the ultraviolet is the fundamental factor in comprehending how very different Venus is from the Earth, although the planets differ little in size. Indeed, by following the temporal variations of these features, it has been established that the dense atmosphere of Venus at the level of the upper cloud layer, which is situated at a height of 50 to 70 kilometres, rotates sixty times faster than the solid planet.

This discovery of *superrotation* is one of the most important in the exploration of the solar system. In spite of the thick cloudy atmosphere, the dynamics of the solid body of Venus have been determined through radar observations. As a matter of fact these have the advantage of using waves which can penetrate the layers of clouds. The planet executes its rotational movement around its geometrical axis very slowly – the period is 243 days – in a retrograde sense. The effects connected with rotation, such as the Coriolis force, which contribute to the extreme complexity of the movements of the Earth's atmosphere, must, in the case of Venus be negli-

The veiled planet. Venus has long been regarded as the Earth's sister planet; its size, mass and mean density are similar. It nevertheless differs in its gaseous envelope: the surface pressure is about 95 times higher than that at sea level on Earth. This atmosphere, 105 times as massive per unit surface as the Earth's, represents about one ten-thousandth of the planet's mass (as opposed to a millionth for the Earth and a ten-millionth for Mars). Further, a thick layer of opaque clouds prevents any observation of Venus's surface in the visible. Only radar techniques and space probes – from Mariner 2, in 1962 to Magellan – have revealed the characteristics of Venus.

Launched in 18 October 1989, the *Galileo* probe (whose main purpose was to explore the Jovian system) used the gravitational fields of Venus and the Earth to reach Jupiter in December 1995, passing close to Venus in February 1990. This flyby provided the occasion to observe the planet and its environment. This picture, taken on 14 February 1990 at a distance of 2.7 million kilometres at a resolution of order 70 kilometres, shows Venus in the spectral range centred on 418 nanometre wavelength. For this reason the monochromatic original has been coloured purple, corresponding to this region. In contrast, the other images in this chapter restore the pale yellow colour of the planet. North is up, and the terminator at left.

One can see the main markings of the cloud tops: the Y-shape (partially visible), the cellular structure of the illuminated region near the subsolar point (local 'midday') near to the limb, and the bright polar caps (NASA)

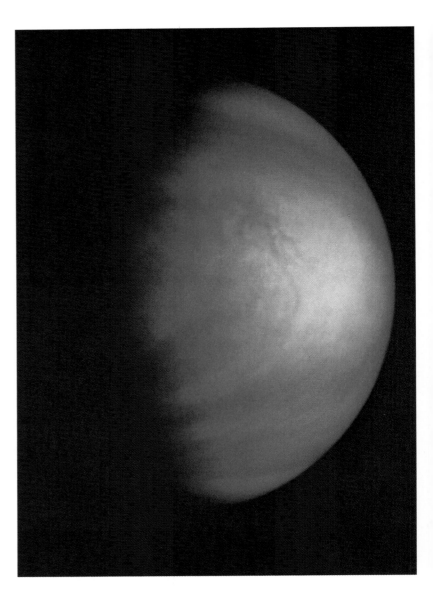

The Pioneer Venus mission. The Pioneer Venus mission was executed in two stages: one part was the placing in orbit around Venus of a satellite, the Pioneer Venus Orbiter (secondary project) and the other part was the ejection from a carrier vehicle of four probes which would penetrate the dense atmosphere of the planet (main aim). Launched on 20 May 1978, the Pioneer Venus Orbiter was inserted in orbit around Venus on 4 December 1978. The carrier and its four probes were launched later than the Orbiter, on 8 August 1978, and entered the atmosphere of Venus on 9 December 1978. The sounder carried seven scientific instruments, each of the three little probes contained three and the carrier vehicle itself was equipped with two instruments, mass spectrometers earmarked for the analysis of the upper atmosphere during the flight over the planet.

The Pioneer Venus Orbiter was placed in a very eccentric orbit (perigee was at 200 kilometres and apogee at 66 000 kilometres). Its period of revolution around the planet was twenty-four hours and the inclination of the orbit was very large (105 degrees). It was equipped with twelve instruments. An ultraviolet polarimeter was used to construct this image of the disk of Venus in which the characteristic contrasting patterns of the atmosphere and the clouds show up. This image was obtained on 11 February 1979 by the Pioneer Venus Orbiter, which was then vertically above a point 30° south. This made it possible to study the southern polar region, which is difficult to observe from the Earth, and in particular the bright polar caps. These are manifestations of a veil of thick mist of variable brightness, located higher than the Venusian cloud layers, and formed by small aerosols with a radius of 0.25 micrometres.

The bright filaments that break away from the polar region are a result of the presence of these small aerosols, which are probably made up of droplets of sulphuric acid. (NASA)

The observations of Venus.
The progress made in the observations of Venus during the last two decades is clear from these two pictures (left), each obtained with a detector sensitive to ultraviolet radiation.

Using observations made from the Earth, one can just detect a sort of dark belt on the photograph, described by the astronomers as a 'Y' on its side (upper picture, photograph taken 24 July 1966).

In February 1974 the first probe, Mariner 10, flew past Venus. The distance at which the observations were obtained, 720 000 kilometres, was near enough for a large number of contrasting features to become visible (lower picture, image obtained on 10 February 1974). In particular the features which, in the equatorial regions, are the source of the Y-shaped form, observed from the Earth, are perfectly distinguished. Because the contrasts appear only in ultraviolet light, the clouds or the atmosphere of Venus must contain a constituent which absorbs in that region. It cannot be sulphuric acid (H_2SO_4), which is a weak absorber in the ultraviolet. The presence of sulphur dioxide (SO_2) makes it possible to understand only part of the ultraviolet albedo; the chemical and physical nature of a second agent has still to be discovered and the positions of both components with respect to the clouds have to be established. Probably hidden under the layer of clouds, they follow the vertical movements of the planetary waves whose existence seems to be demonstrated by the permanence of the large-scale contrasting pattern. (NASA)

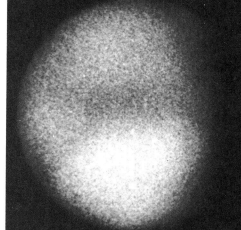

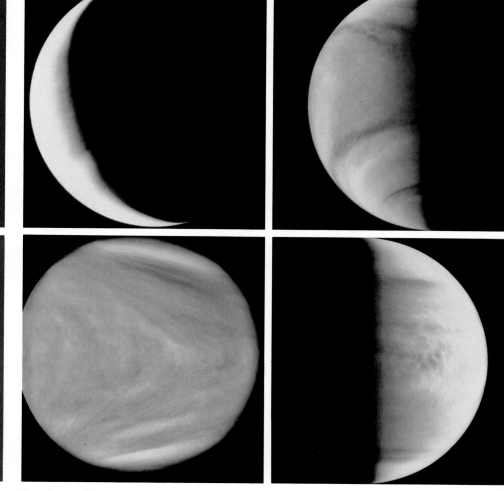

gible. In addition the duration of the solar day on Venus, which equals 117 terrestrial days, is sufficiently long that all effects connected with the relative movement of the Sun with respect to the surface of Venus are diminished. Finally the orbit is virtually circular, but above all, the very small inclination of the rotational axis of Venus with respect to the vertical to its plane of revolution around the Sun (the obliquity of the planet is 177.3 degrees) must lead to nearly undetectable seasonal effects.

Exploration of Venus has shown that its atmospheric dynamics are as complex as Earth's,

The phases of Venus. Venus, a very bright planet when viewed from the Earth, is always located in the neighbourhood of the Sun. Venus is an evening object after sunset and in the morning object before (it is never more than 48 degrees away from the Sun). Observed through binoculars in visible light, Venus is dazzling white without a single visible spot. Venus shows the same phases as shown by Mercury and the Moon, as shown in this series of four images taken by the Pioneer Venus Orbiter in the ultraviolet region of the spectrum:
– the top left image, which shows a crescent Venus, is the first image obtained by satellite (5 December 1978);
– the next image (first quarter) was obtained on 25 December 1978 when the mission of Venera 11 ended. That Soviet satellite is located near the equator in the illuminated part of the planet, close to the limb in the extreme left of the image;
– 9 February 1979, the disk is completely illuminated and the main ultraviolet pattern has become visible;
– 10 April 1979 (last quarter), one can see spots at the equator, which are a consequence of convective phenomena connected with the absorption of solar radiation. (NASA)

and that the properties of the lower atmosphere change from place to place.

The planet dynamics

How do we explain the very slow rotation of Venus? The other planets, except for Mercury, have rotation periods of hours. Apparently

Relative movements of the Earth and Venus (right). The analysis of the relative movements of the Earth and Venus around the Sun, during one synodic period (which is equal to 583.92 days), shows that there exists a pseudo-synchronisation between the rotation of Venus and that of the Earth. The same meridian, identified by the black dot, apparently points at each inferior conjunction towards the Earth. A real synchronisation would require a rotational period for Venus equal to 243.16 days. In reality the period is 243.01 days, which differs from the synchronisation value. Consequently no resonance exists between the rotational movements of the Earth and Venus.

The structure of the high atmosphere (left). A comparison between the structure of the high atmospheres of Venus and the Earth makes it obvious that there is an essential difference. For the Earth, a profile of the temperature as a function of pressure or altitude exhibits a strong maximum due to the absorption of solar ultraviolet radiation by a layer of atmospheric ozone. This absorption shows up as a characteristic knee in a temperature profile of the terrestrial atmosphere at an altitude of around 45 kilometres. It marks the frontier (the stratopause) between the stratosphere and the mesosphere. There is no such effect in the case of Venus; its high surface temperature is a consequence of a much stronger greenhouse effect than that which heats the surface of our planet. This temperature decreases progressively and follows an adiabatic curve very closely up to a height of about 50 kilometres, where the temperature is 345 K. In the region which extends from this level up to an altitude of 90 kilometres, a temperature minimum is observed, which has its equivalent in the terrestrial mesosphere.

The correspondence is also valid for still higher regions, where the temperature again increases. At the highest levels we can classify the layers as the exosphere. However, for Venus the exosphere begins at 135 kilometres instead of 600 kilometres as is the case for the terrestrial atmosphere. Furthermore, this region is curiously subject to strong daily temperature variations from 300 K on the day side to 130 K on the night side; the term cryosphere has been used to designate the high atmosphere on the night side of Venus. (After F. W. Taylor and D. W. Hunten, 1980)

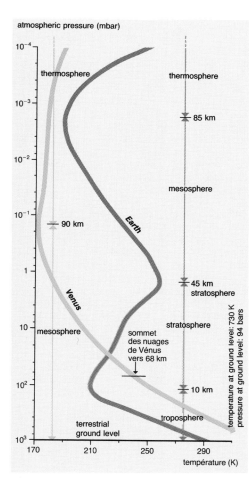

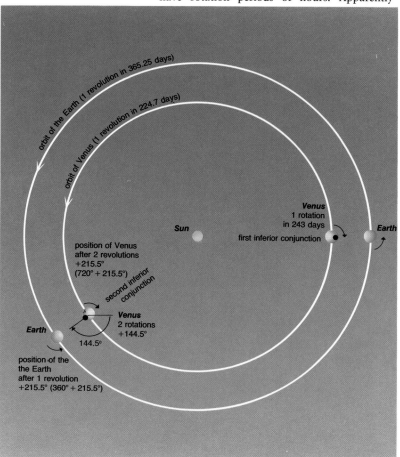

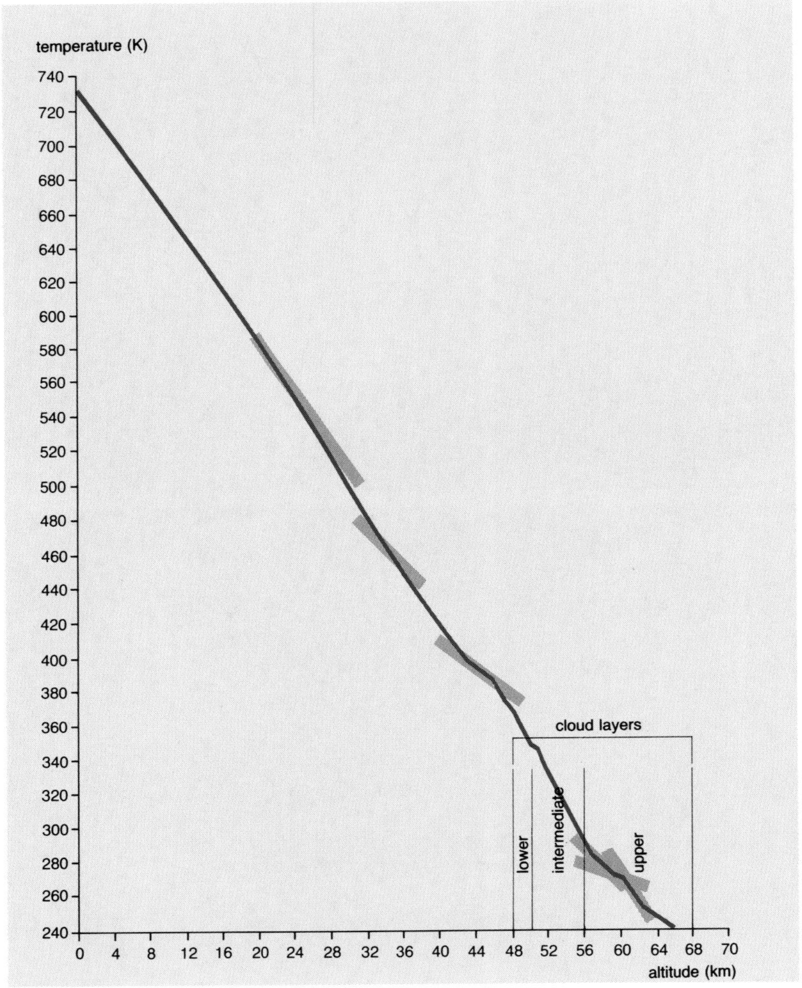

Vertical temperature profile of the lower atmosphere. The landing probe of Pioneer Venus (the sounder), transmitted the results which had the best temperature resolution. The measured values apply to the layers between 12 and 64 kilometres high, at which the temperatures are respectively 636 and 247 K. At lower altitudes the thermometers of the four probes of Pioneer Venus stopped working. The temperature profile can, nevertheless, be extrapolated using the results of the Soviet probes which just kept working until they reached the surface of Venus. The general profile below 60 kilometres corresponds to an average temperature gradient of 8 K per kilometre. This value is close to that for the adiabatic gradient, for which the calculated value for the atmosphere of Venus specifies the criterion for hydrostatic stability. When a mass of air rises, its pressure decreases, which lowers its temperature. If the vertical velocity is not too small radiative effects and the exchange of matter by turbulence or heat loss from the element can be ignored. It amounts to treating the movement of the air mass as adiabatic. If one assumes that the pressure changes with altitude according to the equation of hydrostatic equilibrium, then the adiabatic temperature gradient can be calculated, and we can compare the results with the experimentally determined temperature gradients (which are indicated by the short red lines in the figure). With the exception of a layer between 50 and 55 kilometres, the temperature profile inside the clouds deviates from the adiabatic profile in a way which enhances the stability of the atmosphere. Below the clouds, down to an altitude of 30 kilometres, there is a very extended region of the atmosphere with a high degree of stability. At the altitude of the cloud tops, around 64 kilometres, the net temperature gradient decreases (to 3.5 or 4 K per kilometre), which indicates that this region is also stable. Two unstable regions exist: in the middle of the cloud layers, between 50 and 55 kilometres and in the cloud-free region between 20 and 30 kilometres. It appears highly likely that the atmosphere at ground level (below 10 kilometres), also forms a stable region. (From A. Seiff *et al.*, Measurements of Venus thermal structure, *Journal of Geophysical Research*, 1980, p. 7924)

Venus has had a rather different dynamical evolution from that of the Earth, whose size and mass are however similar.

Venus is thought to have acquired its dense gaseous envelope early in its evolution. One could imagine that the direction of the rotation of Venus was initially retrograde, but with a much shorter period of about 10 hours. Two tidal forces would act upon such a planet; forces due to the gravitational attraction of the Sun on the solid body (solid tides), which in the case of the Earth cause the diurnal tides in the oceans and crust; and, thermal forces caused by diurnal variations in atmospheric temperature. The latter force occurs in the Earth in the vicinity of the stratopause and generates a diurnal thermal wave on a planetary scale. Thus, it is referred to as an atmospheric or thermal tide.

In the case of Venus, the diurnal temperature changes of its very dense atmosphere, generate a strong atmospheric tide which would increase rotational velocity. The solid tides acting on Venus exert a torque in a direct sense which tends to slow down the retrograde rotation of the planet. In the absence of an opposing force this torque would cause a progressive reduction of the initial rotation of the planet, until after about 100 million years. It would be in synchronisation with its orbital motion, with the consequence that the same side of the planet would always be facing the Sun. The fact that this is not the case indicates that thermal tides have always been present, brought on by the dense atmosphere.

It is however possible to show that the gravitational potential of the Sun acts over the semi-diurnal component of the thermal tide producing a retrograde torque in the atmosphere which is subsequently translated to the planet. The model is simple; during the early dynamical evolution of Venus, the solid tides were dominant; later, when the rotational velocity of the planet had decreased, the effect of the atmospheric tides was accentuated. Calculations of the rotational period from current observations reveal that the forces are of comparable strength.

The dynamics of the lower atmosphere are not sufficiently well known to be able to decide on the details of this mechanism. Nevertheless it has the advantage of providing a plausible explanation for one puzzling characteristic: the apparent synchronisation between the rotation of Venus and Earth. At each inferior conjunction, that is, when Venus crosses the line from the Earth to the Sun at the point closest to the Earth, the same meridian on Venus always points in the direction of the Earth. If the two main forces acting on Venus were in equilibrium, the weak gravitational pull of the Earth could become important, and Venus could be 'captured' in a synodic resonance; the theoretical period corresponding to such a resonance is 243.16 terrestrial days. In fact precise radar measurements, covering a period of nearly ten years, have revealed a value of 243.01 days. Although the discrepancy is small, it is enough to reject the theory of 'capture'. The apparent quasi-stability could exist due to the equilibrium of the tidal forces acting on Venus. It is possible that the dynamical evolution will lead to a real capture in the end, or, conversely, that such a stage has already been passed and the capture forces have been overcome.

The atmosphere and the clouds

The atmosphere of Venus has two special characteristics. The first is the very large atmospheric pressure, between 92 and 95 bars at ground level and associated with it, a surface temperature of the order of 735 K. The second is the presence of clouds at altitudes of 48 to 68 kilometres. These clouds have a particle density comparable to thick mist observed at the surface of the Earth. The principal components of the thermosphere are carbon dioxide gas (CO_2), oxygen atoms (O), carbon monoxide (CO) and molecular nitrogen (N_2). In contrast, the lower atmosphere of Venus essentially consists of carbon dioxide (96.5 per cent) and molecular nitrogen (3.5 per cent). All the other constituents of the lower atmosphere have relative concentrations which are counted in parts per million (p.p.m.). The determination of the concentration of argon and its isotopic ratio is an important confirmation of the hypothesis of evolution of the atmosphere of Venus. The isotope ^{40}Ar that comes from radioactive decay of ^{40}K is slightly more abundant than the primordial ^{36}Ar. The ratio $^{40}Ar/^{36}Ar$ is 1.03, notably different from the ratio of 294 measured in the Earth's atmosphere. There is a large analytical error associated with the determination of concentrations of highly reactive chemical compounds. Nonetheless, it has been established that sulphur dioxide (SO_2) concentrations are low (1 p.p.m.) above the clouds, and higher below them. Water vapour concentration also varies with altitude, from a few p.p.m. at the cloud tops to 200 p.p.m. in the upper cloud layers. Water vapour and sulphur dioxide react to form sulphuric acid droplets. The absolute concentration of such reactive gases is best measured by spectroscopic techniques. Using special equipment onboard the Vega probe, it was possible to obtain new values for the SO_2 concentration; 20 p.p.m. at 40 kilometres altitude. The reaction of this gas with the surface rocks affects the chemical composition of the lower atmosphere. Spectrophotometric equipment onboard Venera 13 shows that H_2O concentrations decline as one approaches the surface (40 p.p.m.). Similarly, the concentration of carbon monoxide (CO) increases monotonically from above the clouds towards the lower atmosphere. Direct measurements by gas-phase chromatography contradict the spectroscopic measurements of oxygen (O_2) concentration, recording 20 to 30 p.p.m. at the top of the clouds and 0.1 p.p.m. underneath the clouds. Finally, while hydrochloric acid (HC1) has been detected above the clouds by infrared spectroscopy from ground-based observations, no chlorine compound has been found in the lower atmosphere.

Results from the Pioneer Venus probe indicated that SO_2 was the dominant sulphur compound. The latest information obtained by the Vega probes, and Venera 13 and Venera 14 indicates the presence of atomic sulphur, and has shown that the gases H_2S (80 p.p.m.) and COS (40 p.p.m.) are at least as abundant as SO_2. These results are in agreement with the predictions of models assuming thermodynamic equilibrium between the surface and the atmosphere of Venus.

Compared with the clouds in the terrestrial troposphere, which are formed through the condensation of atmospheric water vapour, the clouds of Venus show several peculiarities:

– the mass density of the aerosols is never very high and does not exceed a few milligrams per cubic metre of Venusian atmosphere, while the hydrometeors of the terrestrial clouds reach values of the order of 400 milligrams per cubic metre of air;
– the majority of the Venusian aerosols have a diameter which rarely exceeds 10 micrometres, while the size of the hydrometeors varies from a few micrometres to a few millimetres;
– The Venusian clouds occur mainly in three strata, located between altitudes of 47 and 65 kilometres and covering the entire planet. Thus the large optical depth associated with the clouds is not due to particle size or concentration, but to the 20 kilometre zone where they are present. Above the clouds, between altitudes of 70 and 90 kilometres, there is a layer with an optical depth one hundred times smaller.

There now exists a number of observations which show that the droplets which make up the majority of the aerosols of the upper layer are watery solutions of sulphuric acid, 85 per cent acid by weight, and their diameters are about 2 micrometres. One can explain their presence by a photochemical process which leads to the formation of acid molecules ***in situ***: the sulphur dioxide of the atmosphere of Venus is photodissociated, at a height of about 70 kilometres, by solar ultraviolet radiation with a wavelength less than 218 nanometres; the liberated oxygen atom then reacts with sulphur dioxide in the presence of carbon gas to form sulphur trioxide (SO_3), after which the combination of this unstable gas with the water vapour present in the atmosphere of Venus leads to the formation of molecules of sulphuric acid. Then the aerosol particles are formed through condensation out of the gas phase. This is a phenomenon little different from that which accounts for the formation of acids in the haze of urban pollution.

A layer of mist is present between altitudes of 70 and 90 kilometres. It is made of small particles of sulphuric acid, with diameters of about 0.25 microns.

By their size, the method of formation and the composition of their particles, and also by their stratification in layers at a planetary scale and in particular through the mist observed at high altitude, the clouds of Venus suggest a comparison with what is observed in the stratosphere of the Earth. We know that there exists around the Earth, at a height of about 18 kilometres, a permanent layer of aerosols of which the rather small mass density increases from time to time after a volcanic eruption, to reach values of the order of 40 micrograms per cubic metre of air. In comparison, the mass density of the aerosols in the upper clouds of Venus is of the order of 2 milligrams per cubic metre of the atmosphere. The stratospheric aerosols are comparable to very concentrated solutions of liquid acids, 75 per cent by weight, and their mean diameter (0.1 micrometre) is slightly less than that of the small particles in the upper layer of thick mist.

The upper and lower limits of the three layers which characterise the basic structure of the clouds of Venus are well known. The base is always at around 47 kilometres altitude. However, uncertainties remain about the aerosol content inside the different layers. It has been confirmed by the Vega mission that the upper layer, between 56 and 68 kilometres of altitude could be slightly richer in aerosols, but uncertainties remain about the size distribution of the particles and the presence and absence of solid particles with a chemical compositions completely different from the sulphuric acid droplets. The Venus probe did not detect large particles which, if present, could consist of chlorine substances such as monohydrate perchloric acid ($HClO_4$,

H_2O). There is no photochemical model however, capable of producing them in a reliable cycle.

The global energy balance

The ensemble which Venus and its atmosphere form in interplanetary space emits as much energy as it receives from the Sun. The use of the Orbiter of the Pioneer Venus mission as an observing platform has resulted in numerous photometric measurements of the high atmosphere, made at various longitudes and latitudes. The northern hemisphere of the planet has a radiation temperature, integrated over all wavelengths, of 230 K. This corresponds to an energy output into space of 158 watts per square metre.

At the level of the Venus orbit, the electromagnetic energy received from the Sun is 2621 watts per square metre, which corresponds to nearly double that reaching the top of the terrestrial atmosphere. Nearly 80 per cent of this energy is sent back to space by the Venusian atmosphere and by the clouds, due to the high albedo (0.77). The solar energy absorbed by the planet is only 150 watts per square metre (as opposed to the 240 watts per square metre absorbed by the Earth). More than 70 per cent of the available energy is absorbed by the clouds and by the upper atmosphere above 68 kilometres. The remaining energy is absorbed equally by the lower atmosphere and by the surface of the planet. The thermal energy which reaches the surface of Venus from the interior is estimated to be 0.06 watts per square metre. This is three orders of magnitude less than the energy from the Sun which succeeds in penetrating through the thick atmosphere of the planet. Consequently the very high temperature at the surface is due only to solar energy (which is roughly 20 watts per square metre).

How can one understand the surface temperature of 735 K, which was first estimated and later confirmed thanks to the series of Soviet Venera landing probes? The heating mechanism is now well proven and is based on a particularly powerful atmospheric greenhouse effect. A crucial event during the evolution of Venus has produced a fundamental modification of the properties of the atmosphere and, in particular its chemical composition. It is mainly the infrared opacity of the water vapour that contributes to this greenhouse effect. In the case of the Earth, the complete absence of any atmospheric effect would lead to a surface temperature 31 °C lower. (−18 °C rather than the actual 13 °C.) In the case of Venus this phenomenon is produced on a larger scale. Due to its proximity to the Sun, at some point in time the lower atmospheric layers reached a critical temperature above which all the water was present in a vapour state. The greenhouse effect became irreversible as almost all the water and all the carbon dioxide locked in the planet were now back in the atmosphere. Assuming thus that an amount of water equal to the Earth oceans content (1.39 × 10²¹ kg) was transformed into gas one would find that, after such a runaway greenhouse effect, the atmosphere of Venus contained water vapour with a partial pressure of 265 bars. On the other hand, one could assume that the Venusian atmosphere had always contained the same amount of carbon dioxide gas as it has today, with a partial pressure of 90 bars. Taking into account the present properties of the atmosphere, this means that considerable quantities of hydrogen would have escaped from the planet.

The actual time when the surface temperature was enough to trigger off this runaway greenhouse process remains uncertain. It depends on various hypotheses about variations in the solar flux, which at the present, orbit of Venus is 1.91 $S_\odot$ ($S_\odot = 1360$ W.m⁻² is the solar constant). It has been demonstrated that solar luminosity has increased over the 4500 million years since the formation of the Solar System. Taking this into account, the effective solar flux received by Venus originally was between 1.34 $S_\odot$ and 1.43 $S_\odot$. This is a very narrow window, because in an atmosphere saturated with water vapour but without cloud layers, the critical threshold of the runaway greenhouse effect is estimated at 1.40 $S_\odot$. Thus one cannot be sure that the threshold was reached. Thus theoreticians now prefer the idea that the increase in the surface temperature has been very modest. We can envisage Venus evolving very similarly to the Earth, at least in a first relatively long phase, during which oceans formed. According to the models, the surface temperature would not have exceeded 80 °C. Well after formation, Venus would have had oceans like the Earth's — because of the identical accretion conditions of the two planets — and would have had the following properties: an atmosphere of base pressure 2000 hectopascals, composed mainly of molecular hydrogen and water vapour (20–30 per cent), with very low carbon dioxide content (0.03 per cent). These are the characteristics of a

The super-rotation of the upper atmosphere. In spite of the progress which has been made, the dynamics of the atmosphere of Venus are not satisfactorily understood; in particular the nature of the mechanism which maintains the super-rotation of the upper layers is not well explained. This last phenomenon is shown by the permanent presence of a large-scale pattern in the form of a horizontal 'Y', which executes a rotational movement with a period of four days. The zonal outflow from this super-rotation corresponds to wind velocities of the order of 100 metres per second at altitudes between 50 and 70 kilometres. These winds are also strong at lower altitudes but, after a very abrupt inflection height, the winds become gradually weaker; at the surface of the planet their velocity is only about a metre per second.

Detailed analysis of the images shows in every case the existence of a meridional circulation, at the level of the upper cloud layers where the meridional winds reach 10 metres per second. Moreover, it is clearly indicated that the upper atmosphere is crossed by horizontal waves on a planetary scale. These propagate very slowly in a movement above the mean zonal outflow and superimposed upon it.

It was through the images obtained during the Pioneer Venus mission that clear evidence for the existence of the super-rotation was obtained:
− on the first image, taken on 15 February 1979, one can easily distinguish the Y-shaped pattern;
− on the second image, obtained on 16 February, 24 hours later, the pattern has been displaced towards the left and one can only see its tip;
− the third image, taken on 17 February, shows the face of the planet on the opposite side to where the 'Y' pattern is visible; the atmospheric circulation here is more symmetric and shows a series of longitudinal bands;
− on the last image, obtained on 18 February, the stripes curve progressively, announcing the approach of the first contours of the Y-shaped dark pattern. (NASA)

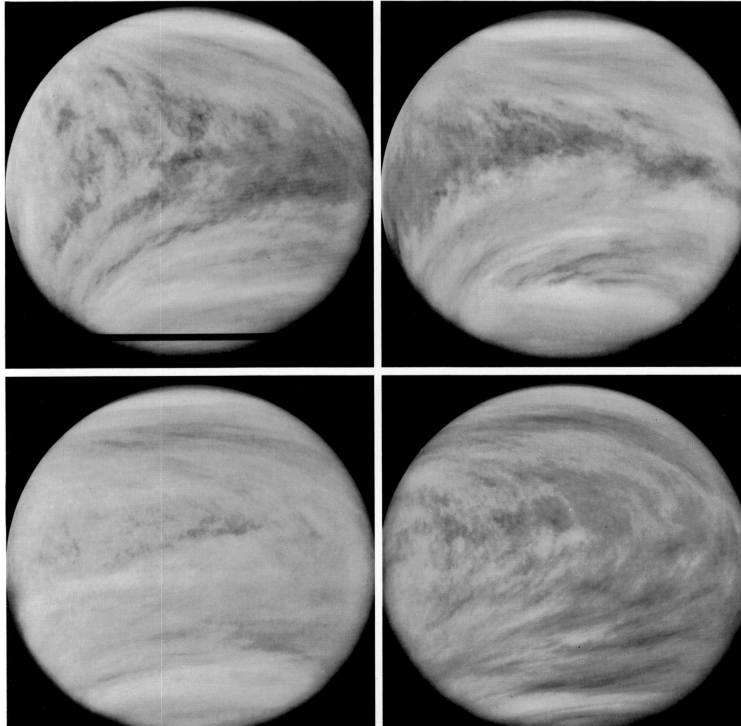

hot and humid atmosphere whose main source is the progressive surface outgassing of the ocean. They differ from those of the greenhouse effect in dry vapour (with no ocean present on Venus). But whatever the initial conditions and the type of greenhouse effect, large amounts of water, liquid or vapour, have been lost. Almost all of the water on Venus at the moment of accretion–corresponding to about 265 000 hectopascals – has disappeared; only a few hectopascals are observed today in the Venus atmosphere. The mechanism responsible is the photodissociation of the water molecule and the escape of hydrogen into space. But this requires the thermal structure of the atmosphere to be such that water vapour is again an important constituent (at least 20 per cent). In this case one can envisage several escape processes for hydrogen. The most powerful is hydrodynamic escape, during which, despite the high altitude, collisions are frequent enough that the flow remains a (hypersonic) fluid. The more satisfying of the two theories is the idea of a humid atmosphere greenhouse effect associated with the presence of an ocean, revealed by a low carbon dioxide concentration in the atmosphere. It explains the complete loss of the Venus ocean in less than 600 million years. Further, it explains the evolution of the atmosphere. The carbon dioxide gas released can maintain itself in the atmosphere without being dissolved in an ocean. It then becomes the main constituent of the atmosphere, and the pressure reaches the 92 000 hectopascals necessary for the surface temperature to reach 735 kelvin.

Atmospheric circulation

In 1960 Charles Boyer and Henri Camichel observed a periodicity pattern in the dark patches over the planet disk. This confirmed the theory of zonal circulation of the Venusian atmosphere. The atmospheric zone between 50 and 70 kilometres altitude rotates sixty times faster than the solid planet. The presence of zonal winds which produce a vertical gradient below the clouds is known from the landing probes. The super-rotation of the atmosphere starts at an altitude of 10 kilometres with velocities, on the order of a few metres per second. At the cloud top, located at an altitude of 68 kilometres, velocities are on the order of 150 metres per second. The super location extends to at least 95 kilometres altitude, by which height the wind velocity has decreased to nearly zero. In order to explain the rapid rotation of the upper atmosphere, one has to assume a vertical transfer of momentum through it. How is it possible to maintain the super-rotation if, as observed, the vertical gradient of the local winds has the same orientation everywhere? In fact, small scale vertical turbulent diffusion would cancel the differential rotation in a few months. It can be shown, however, that an axially symmetric meridian circulation (Hadley cell) in which angular momentum decreases with latitude could compensate for the effect of turbulent diffusion when applied to the atmosphere of Venus. Whirl-like mechanisms would need to be invoked at the same time to balance losses due to meridional transport generated by the Hadley cell.

The problem of maintaining the zonal circulation in the clouds region (between 48 and 68 kilometres altitude) that received almost all of its energy from solar radiation, is different from that for the lower atmosphere (below 48 kilometres) where radiative transfer occurs in large time scales.

From a comparison of Venus with Earth it is not at all surprising that a Hadley meridian circulation exist at cloud level. Both planets show important horizontal temperature gradients at the middle latitudes, but because the rotational velocity of Venus is so low, the Coriolis forces are too weak to generate the baroclinic instability observed above the temperate regions of the Earth. It is to this instability, associated with the propagation of planetary waves, that the atmosphere of our planet owes the persistence of the large cyclonic depressions and the anticyclones which are characteristic of the temperate regions. These are the processes which assure the transport of thermal energy to the polar regions. In the case of Venus, the forces generated by the horizontal pressure gradients are in equilibrium, not because of horizontal forces (a geostrophic equilibrium), but because of the centrifugal forces generated by the super-rotation of the atmosphere (a cyclostrophic equilibrium). One can show that such an equilibrium favours the maintenance, at the altitude of the highest cloud layer, of a north–south atmospheric circulation, characterised by a Hadley cell which extends as far as the polar regions.

During the long mission of the Pioneer Venus Orbiter it was possible to follow the thermal variation in the upper atmosphere as a function of latitude. Above 96 kilometres the analysis

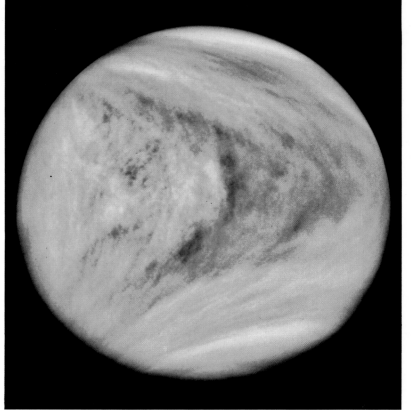

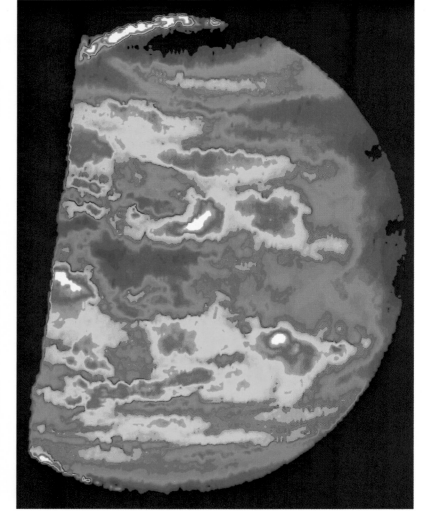

The clouds and the dynamics of the atmosphere. The top picture was taken by *Mariner 10* on 6 February 1974 from a distance of 720 000 kilometres. The probe, on its way to Mercury, was at its minimum distance from Venus of 5800 kilometres one day earlier. The contrasts or brightness variations which are of the order of 30 per cent observed over the disk of Venus, betray the complexity of the dynamics of the cloudy atmosphere. For certain contrasts located at the subsolar point (on the left), the cellular structure is a consequence of the convective movements at the altitude of the cloud tops, between 60 and 70 kilometres. The cells have dimensions of the order of 100 to 1000 kilometres. The polar regions are also characterised by bright caps, of eliptical form, resulting from the very strong vorticity of the atmosphere above the poles and the gigantic scales of the eddies. Infrared observations made by the Pioneer Venus Orbiter have confirmed these discoveries.

The entire disk of Venus is visible in the middle image, obtained from photometric measurements, made from a distance of 65 000 kilometres by the Pioneer Venus Orbiter. One sees again most of the large-scale pattern already noticed in 1974 in the Mariner 10 images and in particular the semi-permanent 'Y'. With the exception of a few, so far unexplained, periodic, disappearances, the pattern evolves on the same time-scale as the rotating atmosphere. During the entire Pioneer Venus mission, the super-rotation with its four-day period at the altitude of the clouds, was maintained over the whole planet at all latitudes and at approximately the same velocity. Nevertheless, the analysis of the Mariner 10 images has made it possible to prove clearly the existence, four years earlier, of a jet stream, localised above the regions at middle latitudes. (NASA)

The lower cloud layer. One of the *Galileo* probe instruments, NIMS (Near Infrared Mapping Spectrometer), a near infrared spectrophotometer, was extremely useful in exploring Venus. In particular, it obtained images of the unilluminated hemisphere at 2.3 micrometres: radiation at this wavelength is emitted by a layer of the atmosphere between 48 and 53 kilometres altitude, below the summit of the sulphuric acid cloud layer. The pressure is of order 500 hectopascals (half of the atmospheric pressure at sea level on Earth) and the temperature is about 75 °C.

This false colour image from measurements made on 10 February 1990 at a distance of 100 000 kilometres, shows the dayside appearance of Venus at this altitude. The colours show the relative transparency of the clouds to the thermal radiation from the base of the atmosphere. White and red correspond to thin clouds, and black and blue to fairly thick clouds. Near the equator the clouds are fluffy; further north they are drawn out into east-west filaments by winds of more than 250 kilometres/hour; the poles are capped by thick clouds.

We note that the structures in this altitude range are completely different from those previously reconstructed from Venus probes whose instruments operated in the visible or ultraviolet. The clouds are distributed far less uniformly than hitherto thought. (JPL/NASA)

showed the existence of a heat deficit for regions at high latitudes and a heat excess for tropical regions. To re-establish the equilibrium, heat must be transported towards the poles. This can be easily achieved by a pair of Hadley cells situated each side of the equator. Each loop consists of a rising current of air at the tropics, then a high altitude flux towards the poles followed by a descent to the surface at the polar regions and a return to the equator parallel to the surface. Such a system of cells could achieve the transfer of momentum to the upper atmosphere and the compensation for the loss of momentum due to turbulent diffusion, if one is considering only the upper layers of the atmosphere (between 48 and 75 kilometres altitude). At the same time, the meridian circulation linked to the Hadley cells would transfer momentum from the maximum intensity equatorial regions to just above the poles. This would result in a loss of momentum above the equator that has to be compensated for by some other mechanism of transfer from the pole to the equator. Thermal tidal waves at a planetary scale are capable of producing horizontal whirls and produce a uniform angular velocity. The models can be confirmed experimentally. A semi-diurnal thermal tide has been clearly shown by infrared radiometry performed from the orbital probe Pioneer Venus. The propagation of waves at planetary scale is also suggested by the persistent dark patches detected in ultraviolet light. In particular the recurrent horizontal 'Y' pattern that forms with a 4.2 days period as a consequence of planetary waves interacting with the zonal flux at the level of the highest clouds. If albedo changes in the ultraviolet are due to abundance inhomogeneities of the absorbers (particles or gas), these absorbers can be used to trace the horizontal motions at a large scale (Y-pattern) or small scale (convection cells).

The meridian circulation in the lower atmosphere, consists of two Hadley cells. One retrograde and positioned just below the clouds at 48 kilometres altitude, the other in direct sense and just above the planet's surface. The meridian circulation extends from the equator to the poles, although it is weaker than that formed at cloud altitude. Gravity waves excited by convection at the surface, might be responsible for the compensation of the momentum deficit due to turbulent diffusion in the high density region of the atmosphere.

This vertical atmospheric structure of Venus fits well with a two zone dynamical model. A highly stable zone between 48 and 30 kilometres altitude where the atmospheric waves are trapped, has been found by direct measurements.

Polar circulation

The interest in the measurments made by the Pioneer Orbiter lies in the fact that the satellite passes vertically over the polar regions. By operating at a wavelength of 11.5 microns, where a narrow window in the infrared absorption spectrum of carbon dioxide gas exists, the temperature at the top of the clouds can be measured. It has in this way been shown that the previously held theory of a uniform distribution of the highest clouds around the planet, at a height of 68 kilometres, was incorrect. Over the north pole a very distinct atmospheric depression, covering a large area, has been discovered. It shows up on infrared images in two ways: as a relatively dark circumpolar region lying between 53 and 70 degrees north latitude, and experiencing low temperatures (205 to 225 K); and, as two bright connected zones, 2000 kilometres long by 1000 kilometres wide, centred within a few degrees on the north pole, and forming a polar dipole. The brightness temperature wich corresponds to the 'eyes' of the dipole is about 250 K. At the level of the highest cloud layer, these temperatures are the highest ever observed at a wavelength of 11.5 microns. The dipole is not stable in time. It has a very rapid rotational movement, with a period of 2.7 days. Moreover, the circumpolar collar defines a region of the upper atmosphere wich has a large temperature inversion, a difference of 50 K in comparison with much hotter equatorial regions. These polar phenomena are a consequence of a substantial local lowering of the altitude of the ceiling of the clouds; the difference amounts to about 15 kilometres for the cloud layers where the 'eyes' of

the dipole are found. Such a change in the relative uniformity of the clouds can only be attributed of phenomena associated with the dynamics of the atmosphere; this is also strongly suggested by the rapid rotation of the dipole. The images transmitted by Mariner 10 in 1974, enabled us to prove the importance of vorticity in the Venusian atmosphere above the polar regions (polar vortex). Furthermore, the strong response of the general circulation is illustrated by the extreme variability of the bright caps created by the thick mist of small sulphuric acid particles.

Guy ISRAËL

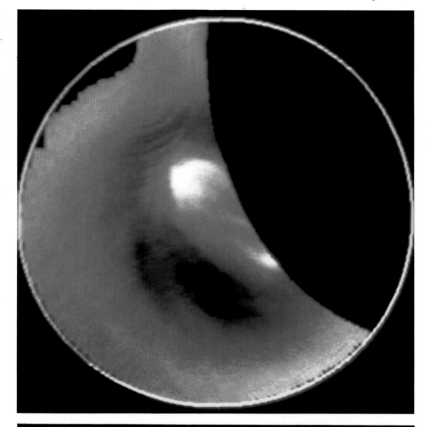

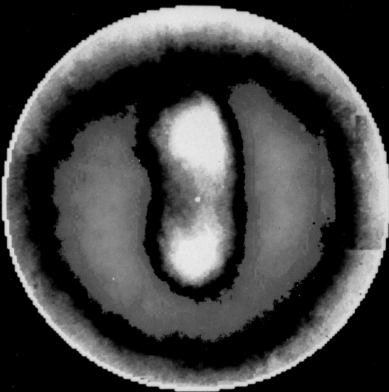

Atmospheric circulation in the polar regions. These two images are maps, in polar stereographic projection, showing the dipole structure of thermal emission in the atmosphere of Venus above 50°N. They represent infrared at 11.5 microns.

The upper image, at high resolution, was acquired by the Pioneer Orbiter in a single orbit on 11 February 1979. Subsolar longitude (local midday) is given below. White areas show dipolar emission at a brightness temperature of 250 K. The darker circumpolar collar has a brightness temperature below 210 K. The bright dipole is evidence for a layer of cloud well below the general cloud ceiling. Its form varies; here, two very hot patches are joined by a hot filament. The circumpolar collar demonstrates a temperature inversion.

The lower image is the result of observations acquired during 72 orbits. Using average values and a system of coordinates to match the rotation of the dipole, it can be shown that the dipole makes one revolution of the pole in 2.7 days. Blue corresponds to 215 K and white to 254 K. (Kindly communicated by Tim Schofield, Department of Atmospheric Physics, University of Oxford.)

Venus

Topography and surface

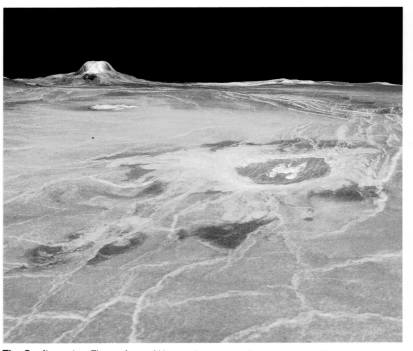

The Cunitz crater. The surface of Venus shows many impact craters. However, there are many fewer of them than on the Moon or Mars. The Venus atmosphere acts as an effective shield, shattering the smallest meteorites. This explains the preponderance of large craters, such as Cunitz, which is about 50 kilometres in diameter, and whose floor appears little modified.

The surface Radar observation techniques could not fail to be applied to Venus, whose atmosphere prevents all observations from outside in the visible part of the electromagnetic spectrum. The available photographs only showed the sites where some of the Soviet Venera spacecraft landed. On the global scale, only altimetric radars mounted on orbiting probes could provide topographical maps. The first was produced by the American mission Pioneer Venus. It covers most of Venus with an altitude precision of about 200 metres. This map was completed for regions near the north pole by the altimetric radars of the Venera 15 and 16 orbiters, with an altitude precision of about 50 metres. However understanding the morphology of the terrain and its geological implications requires radar images. Using the spin of the orbiter about its own axis, the radar of Pioneer Venus, used in imaging mode, gave a surface resolution of 70 kilometres in 1979.

Images with a better surface resolution have been regularly obtained by the most powerful ground-based radars at each inferior conjunction of Venus. The progressive improvements of the sounding techniques have allowed the picking out of details of the order of a kilometre (1.3 kilometres in 1986 for the Goldstone station, from 1.5 to 2 kilometres over summer 1988 for the Arecibo radio telescope). However, these techniques can only be used for equatorial regions.

The images obtained by the aperture synthesis radars of the Venera 15 and 16 probes have resolutions of 1 to 2 kilometres over much more extended regions above 30° north. These radars, associated with altimetric radars, made maps between October 1983 and July 1984. Regions near the north pole were observed for the first time.

In September 1990 the American Magellan probe began to transmit aperture synthesis radar images. With a resolution of order 120 metres, these have overthrown our ideas about Venus by revealing hitherto unsuspected features of its surface. Because of these observations we now have a very precise view of almost all of the surface of Venus.

The level corresponding to the mean radius of the planet – 6051.4 kilometres – is taken as the reference level. The surface is extremely flat: for about 70 per cent of the surface there are no deviations of more than 500 metres. This first type of terrain is called *undulating plains*; it may be compared with the ancient plains of Mars or the Moon. Then, more than 1500 metres high, there are several regions of high relief – the *high ground* occupying almost 10 per cent of the surface. Two main formations are observed there, with dimensions comparable to those of terrestrial continents: these are Aphrodite Terra and Ishtar Terra. Less extended and situated each side of the equator, two mountainous formations, Beta Regio and Phoebe Regio, also belong in this category. These regions were the subjects of the first ground-based radar observations in 1967. A third category of terrain, the *low ground*, covers about 20 per cent of the surface. Its altitude is about 1000 metres below the reference level, but we distinguish the *planitiae*, large depressions which can reach 3000 metres in depth.

The geological analysis of the surface from radar images shows evidence on Venus for the three main mechanisms which formed the Earth's surface: tectonics, vulcanism, and impacts by large meteorites.

Many forms of relief are attributed to fractur-ing, and show that the crust of Venus, like that of the Earth, has fragmented. It seems that one can identify in Beta Regio and other high terrains such as Phoebe Regio the characteristics of the terrestrial tectonic systems of the East African type, showing a succession of deep depressions bounded by parallel crests. In regions to the west of Aphrodite Terra there appear also linear deformations similar to the oceanic fracture zones associated with terrestrial plate tectonics.

With an oval shape, Beta Regio is about 2000 kilometres wide and overhangs the average level of the undulating plains by about 5000 metres. Two volcanic structures, Theia Mons and Rhea Mons, emerge from the tormented structure forming Beta Regio; the higher, Theia Mons, extends for about 300 kilometres and rises to 2500 metres above this elevated region. It is bounded by deep canyons within which there appear to be traces of lava flows. Theia Mons is indisputably linked to the general structure of Beta Regio. The sequence of the two events – tectonics then vulcanism – is confirmed by analysis of the radar reflectivity from the ground, whose great brightness shows that the lava flows are rough, and thus relatively new. At the eastern end of Aphrodite Terra, another region, Atla Regio, has similar characteristics to Beta Regio. There again, structures resembling volcanic ones are superimposed on the tectonic system and strongly reflect radar waves.

The geophysical study of Venus is also very interesting. It has provided information about the link between tectonics and vulcanism, as well as on the type of vulcanism. However, it does not lead to the same conclusions as before as to the dynamical characteristics of tectonics. Geophysicists have established a correlation between the topography and the global distribution of density variations in the crust, as deduced from measurements of the gravitational field (determined to high precision by analysing the orbital perturbations of the Pioneer Venus probe). This correlation – clearly more marked than for the Earth – implies on the global scale a crust which is thicker and more rigid, which cannot be affected by plate movements as on the surface of our own planet. Further, the gravimetric map shows very strong anomalies in some places, principally Beta Regio and Atla Regio, which confirm the relatively recent origin of the volcanic structures associated with them, and constrain hypotheses about the type of volcanism. The process must be specific to Venus, as on the Earth mountains are not systematically associated with gravitational anomalies: the positive anomaly caused by a mountain chain is compensated by the lower density of the rocks of the upper mantle within

Venus revealed by Magellan. These two mosaics of images were acquired by the aperture synthesis radar of the American probe Magellan and show the morphology of most of the surface of Venus. Some regions have not yet been observed by Magellan; they are usually mapped by using the older data from the Pioneer Venus probe. The surface colours are based on an analysis of the surface colours found by the Soviet landers Venera 13 and 14.

The upper picture shows one hemisphere of the planet centred on longitude 270 degrees; the equator runs horizontally halfway up.

The lower picture centres on the north pole, not yet observed. Longitudes 0, 90, 180 and 270 degrees are at 6 o'clock, 3 o'clock, 12 o'clock and 9 o'clock respectively. We clearly see Maxwell Montes with Cleopatra Patera (the yellow spot at longitude 0 degrees), and Lakshmi PLanum to its left. These reliefs belong to Ishtar Planum. (JPL/NASA)

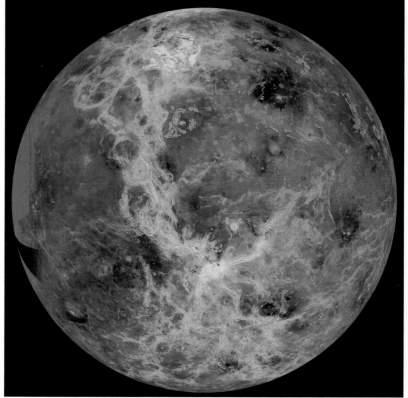

which the foundations of the reliefs are laid. This model of isostatic compensation results from the hypothesis of hydrostatic equilibrium in the Earth's interior down to a certain depth, the compensation depth. Because of the very high surface temperature, the lithosphere of Venus should be less dense than that of the Earth. Isostatic compensation capable of supporting the reliefs of Venus would thus have to apply over a depth of at least 100 kilometres, requiring a differentiation leading to the formation of a much thicker crust than the Earth's continental crust (30 kilometres).

The most likely hypothesis thus consists of assuming that the volcanos of Venus are supported not by isostatic systems, but by dynamical mechanisms, implying a very slow rise of the rocks of the upper mantle and the existence of hot convention zones. On our planet similar dynamical processes are invoked to explain the formation of cold point vulcanism. If the surface of Venus is actually a single rigid plate, hot points are the only places where the internal thermal energy from radioactive decay can be liberated.

Radar images of Ovda Regio (the western part of Aphrodite Terra) acquired by Magellan do not show any evidence for a horizontal displacement of the crust down to a resolution of 120 metres. However, the images obtained by the Pioneer Venus and Venera 15 probes seem to show that this region is the seat of fractures and faults signalling plate tectonics of the terrestrial type. The fact that such characteristics were not confirmed by the Magellan images would reinforce the hypothesis of single-plate tectonics. This was the type of tectonics prevailing on the Earth three billion years ago. Nevertheless, no theory of the thermal evolution of Venus can be definitely dismissed in view of the great diversity of observed terrain.

Topographical studies raise many questions. Ishtar Terra, the highest continent of Venus, is surmounted by a mountainous formation called Maxwell Montes, the highest structure on the planet, 6000 metres above the level of the plateau. Its volcanic origin is not proven: the circular crater of 100 kilometres diameter observed in its centre – Cleopatra Patera – could as easily be a caldera as an impact crater. It is 1500 metres deep and has at its centre an crater embedded 1000 metres deep over a diameter of 55 kilometres. If doubts remain over the origin of the crater at the centre of Cleopatra Patera, there are none for the oval depression Colette (60 kilometres by 120 kilometres) and Sacajawea (100 kilometres by 200 kilometres), which resemble some calderas of Martian volcanos. As on the images of Theia Mons and Beta Regio one observes descending from Colette a system of radial striations of strong radar reflectivity, resembling lava flows. The complexity of the relief of the continent Ishtar Terra makes it difficult to identify the two types of terrain there. The first, localised in the eastern and western regions, is made up of giant circular structures of 200 to 600 kilometres in diameter, called *crowns*. These formations interest researchers as they are 10 to 20 times as large as the equivalent structures on Earth. However, their morphology suggests that they may have originated from mantle material associated with hot points. They would thus not be linked to plate tectonics. The interpretation of the second type of terrain is not as simple. To the west of Cleopatra Patera, a succession of parallel crests spaced at 5 to 10 kilometres extends over several hundred kilometres. The region surrounding the central crater forms an oval 400 kilometres from north to south, and 200 kilometres from east to west. About 200 kilometres north of Cleopatra Patera is the highest point of the planet: it rises to 11 000 metres above the reference level. This relief dominates the high plateau of Lakshmi Planum, to the west of Ishtar Terra. This region is of great interest, as the system of folded parallel chains seen near to Lakshmi Planum is transformed as one moves away into regularly overlapping systems which Soviet scientists baptised with the Greek word *tessera* (tiles) because of their shapes. Unlike other comparable morphological features, which can be explained by vertical tectonics (domes are common in the region of Beta Regio), these special structures may have been formed in horizontal tectonic motions.

About 850 impact craters, all of large size, have been identified on the Magellan radar

images. Their number and distribution correspond to established models for other objects in the Solar System. The relative absence of small craters is explained by the shield provided by Venus's very dense atmosphere. The rate of cratering allows one to estimate the age of the surface at less than one billion years; the surface is thus relatively young.

We should note that Beta Regio is characterised by the relative absence of impact craters; the volcanic activity giving rise to Theia Mons and Rhea Mons is therefore still more recent.

The nature of the surface

The Soviet landers made direct measurements of the surface of Venus; these have added considerably to our knowledge of the geological processes which formed its surface by giving information about the chemical composition of its rocks. These come from gamma-ray spectrometer analyses of radioactive elements in the rocks – uranium, thorium and potassium – and chemical analysis through X-ray fluorescence, of other elements such as silicon, calcium, iron, magnesium and sulphur. Complementary information came from analysis of the atmospheric gas composition at the interface with the surface (particularly water, hydrofluoric acid, hydrochloric acid, sulphur dioxide and hydrogen sulphide) and considerations of cosmogony and thermodynamic equilibrium. Interpretation of photographs of the Venus surface taken by various landers (Venera 9, 10, 13 and 14) has also been very valuable.

The natural radioactivity of rocks is a signature of their content of radioelements uranium, thorium, and potassium. These elements, dispersed through the planet in the course of its accretion, have contributed to chemical differentiation through the resulting heat. After partial fusion of the rocks, they are concentrated in the magmas, whence subsequent volcanic activity transported them to the surface. The enrichment of the surface rocks in radioactive elements thus shows the effectiveness of these processes over the history of the planet. Well protected from cosmic radiation beneath a dense atmosphere, the surface lends itself well to gamma-ray spectrometric analysis of its natural radioactivity. Thus one can show that the thermal evolution of Venus and the Earth were similar, particularly as the mean density of Venus (5.25) is only very slightly less than that of the Earth (5.52). The difference between these two densities reduces to only 3 per cent if one takes account of the different gravitational compressions resulting from the slightly different masses.

The landers Venera 9, 10, 14, Vega 1 and 2 all landed on the Venusian plains. Their analyses showed a certain analogy with magmatic rocks such as tholeiitic basalts and gabbros, found on the Earth's surface. Everything suggests that on

Venus too the mineral content of some rocks has been changed by recent volcanic activity. Further, the landers Venera 8 and 13 also analysed matter similar to terrestrial rocks of high alkalinity. These rocks, which could be nephelinic or potassium syenites, are rare on the Earth's surface, where they are mainly found on ancient terrain.

The original contribution of the Venera 13 and 14 missions was an X-ray fluorescence analysis of the chemical composition of the Venus surface. More than fifteen years after the first Soviet attempt (Venera 4) and twenty years after the launch of the American probe Mariner 2, the Venera 13 and 14 missions showed the progress made in exploring this inhospitable planet. Surface samples were irradiated by a plutonium source (^{238}Pu) and two iron sources (^{55}Fe). The doses – 50 and 250 millicuries respectively – were enough to excite fluorescent X-ray emission from elements between magnesium and atomic mass 35. Each instrument provided more than twenty spectra over its active lifetime – 127 minutes for Venera 13 and 55 minutes for Venera 14. These showed the diverse nature of the rocks analysed. The Vega 1 and 2 stations landed in the region of Aphrodite Terra, to the northwest of Diana Chasma, where they worked for almost 20 minutes. Only Vega 2 was able to carry out an analysis of a surface sample.

Examination of photographs of the landing sites clearly led to geochemical interpretations. Despite the thousands of kilometres between the sites, one is struck by the extreme similarity of the view. However, the stations transmitting pictures landed in Beta Regio and Phoebe Regio, and it would not be surprising for much of these regions to be characterised by lava flows, when we recall that they are thought to be volcanic. This would explain the similarity between the landscapes. However there are differences: the Venera 13 site is smooth but segmented; it is dotted with debris of various sizes which might be the result of the impact of the lander; the smooth parts of the surface suggest a crust of fine particles stuck to the surface through chemical processes linked to the special composition of the very hot and chemically active atmosphere. Over the whole of the Venera 14 site we observe a series of rocky plates in overlapping layers, and there is no covering of fine particles. The distinct angular edges of the plates suggest a relatively intact surface, perhaps of consolidated rocks; the resemblance of these plates to those sometimes left by terrestrial lava flows is striking. A volcanic explanation is, however, not the only possibility: changes through impact, mechanical effects or chemical effects by the atmosphere are equally possible.

The Venus surface has suffered numerous erosion processes comparable to those which have modelled the Earth's surface: chemical and aeolian erosion, but not erosion by rivers. The two stations landed at different altitudes; the

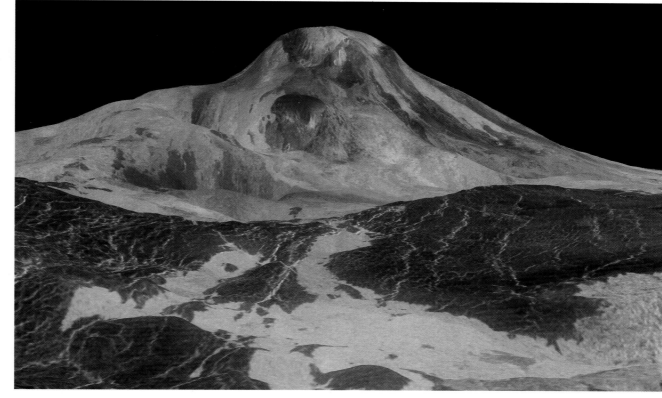

Maat Mons. Rising to an altitude of 8000 m above the Venus reference level, the volcano Maat Mons is seen in three-dimensional perspective from a distance of 560 kilometres. Lava flows of several hundred kilometres extend over the fractured planes of the foreground at the base of the volcano. (JPL/NASA).

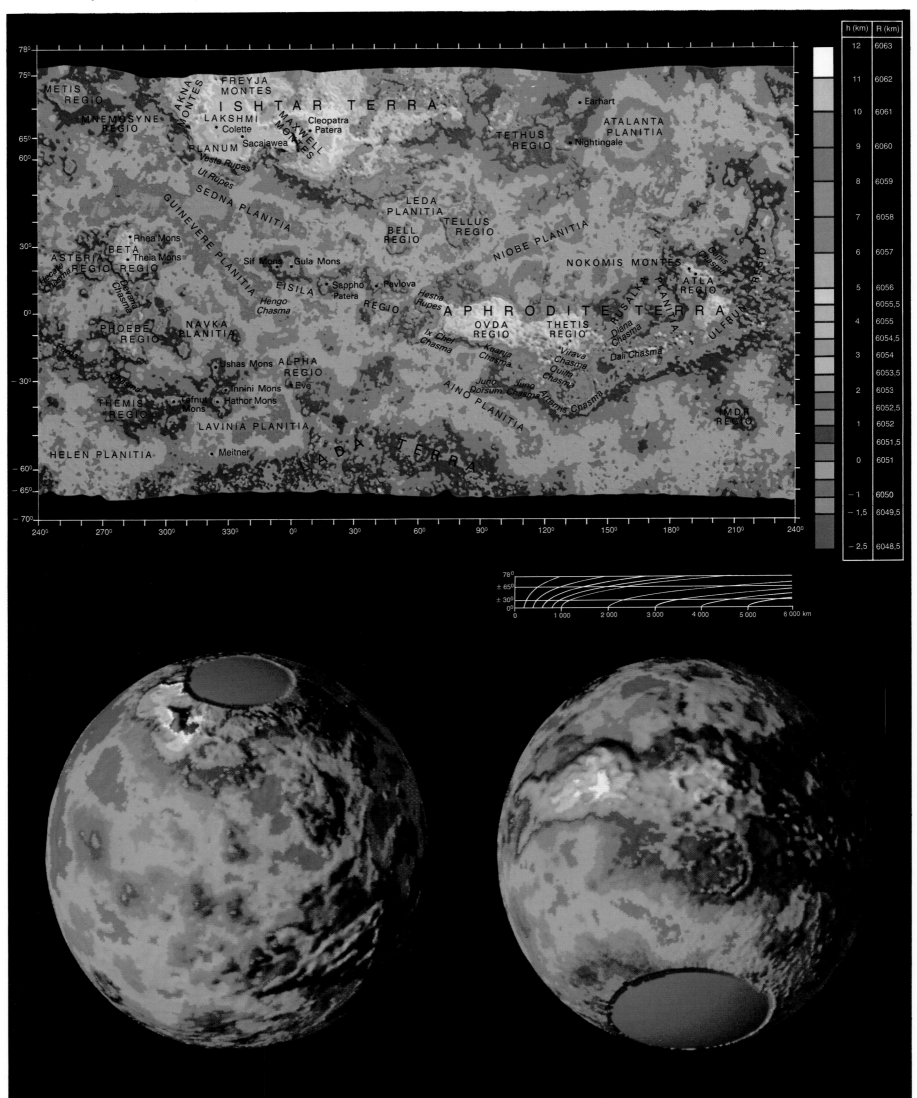

The topography of Venus. Many months of operation of a radar in orbit around Venus, with an accuracy of 200 metres in altitude, have resulted in the topographical information for nearly the whole surface area of the planet (93 per cent, between 75° latitude north and 63° latitude south). Most of the surface, about 70 per cent, appears to have a more or less flat homogenous relief. In this region, similar to the lunar plains and shown in pale blue and blue–green, are located many zones called planitia, whose level corresponds to the mean radius of the planet, 6051.4 kilometres. The zones which are clearly higher, on average, than this level are shown in red, yellow and light green. With its edges some kilometres higher than neighbouring planitia, Ishtar Terra is a vast plateau, or continent, as large as Australia. It stands out strongly against the plains. Near the centre of the 'Terra' is Maxwell Montes, the highest observed formation on the surface of Venus, 11 kilometres above the base level. To the west, Lakshmi Planum forms a plateau 2500 kilometres in diameter, three times more vast than the Tibetan plateau. It is 3 or 4 kilometres higher than the mean level of the plains. The other Terra, Aphrodite Terra, is as big as Africa. It is more elongated than Ishtar (although this cannot be seen from this Mercator projection). Of a very tortuous shape, it contains a more or less circular formation 2400 kilometres in diameter. It is crossed by many steep sunken trenches, some of which are hundreds of kilometres wide and up to 3 kilometres deep. They can be up to 1000 kilometres long.

Other regions do not have the continental dimensions of Ishtar Terra and Aphrodite Terra but are also higher than the plains. Among them, Beta Regio is the most notable because a volcanic formation of very recent origin can be found there, the summit of which is 4 kilometres high. The other parts of the terrain, shown in dull blue, make up about 20 per cent of the surface area. These are depressions which are relatively less noticeable but the bottoms of which can be 3 kilometres below the mean level of the plains. It is in these zones, and in those where the depressions are not so deep that circular structures, reminiscent of impact craters, have been found. The diameter varies enormously from one crater to another, from 10 to 140 kilometres. (Documents supplied by E. M. Eliason/NASA, USGS, MIT)

Venera 14 site is at the altitude of the plains, while that of Venera 13 is 500 metres higher. Chemical processes operating in atmospheric conditions which differ because of the altitude, have perhaps modelled the two sites differently. Chemical erosion would have been more effective at higher altitude, as the wind velocity is higher and the atmospheric density and temperature lower. If this erosion powdered the Venus surface into very fine grains, we can imagine that winds would transport this towards regions at lower altitude, where they would be chemically modified and agglomerated in compact layers. This transport mechanism favours particular chemical reactions, as minerals formed by chemical erosion at high altitude can react with the very hot atmospheric gas of the low-altitude regions where they are transported.

Although we have very little information about the atmospheric sulphur compounds near the surface, we can make several assumptions about the chemical reactions which would change the rocks. Thus a geochemical cycle catalysed by the modification of pyrite under the action of water vapour and CO_2, with production of H_2S and COS gas would play a key role in atmospheric cycles which are the main origin of aerosols. But we do not know if these reactions are stable and rapid enough to explain the high concentration of SO_2 observed in the atmosphere below the clouds. Further, the mineralogical analysis of the rocks appears to show that calcium appears as an oxide in silicates and carbonates. The absence of the anhydrite $CaSO_4$, which would have permitted a regeneration of pyrite by reacting with iron oxide, shows that another origin is needed to explain the strong concentration of SO_2 compared with the other sulphurated gases. Injections of SO_2 of relatively recent volcanic origin might episodically animate the atmospheric sulphur cycles. The injections would have to be frequent enough to explain the persistence of a high ratio of the observed SO_2 concentration at 20 to 50 kilometres altitude to that required by thermodynamic equilibrium. In 1985, measurements of the ultraviolet absorption by sulphur compounds were performed during the descent of the Vega 1 and 2 landers. These measurements were compared with those made in 1978 by the Pioneer Venus probes using mass spectroscopy. The SO_2 concentrations between 20 and 50 kilometres in 1985 were half those of 1978. If the accumulation of SO_2 in the lower atmosphere is an episodic phenomenon of volcanic origin, this means that an eruption occurred before the arrival of Pioneer Venus. This idea is supported by another SO_2 detection, this time in the upper atmosphere, just above the clouds. The atmospheric properties above 68 kilometres altitude were studied over a long time by the orbiting probe Pioneer Venus from 1978. This shows that the SO_2 content of the atmospheric layer between 70 and 90 kilometres altitude as well as the optical depth of the upper layer of haze decreased smoothly by a factor of 10 until 1983. Analysis of this shows that there must have been an extremely powerful volcanic explosion just before Pioneer Venus arrived, comparable to that on Tambora (Indonesia) in 1815. On the Earth such events occur only about once per century, and there is nothing to suggest that volcanic eruptions are explosive on the surface of Venus, given the absence of water.

Another atmospheric phenomenon supports the idea of comtemporary volcanic activity. In December 1978, during the descents of the Venera 11 and Venera 12 probes through the atmosphere, one instrument detected radio emission which was attributed to atmospheric lightning. At the same epoch the instrument on board the orbiter Pioneer Venus designed to detect radio emission in the frequency range from 50 to 50 000 hertz episodically located low-frequency bursts coming from the atmosphere or the surface. The characteristics of these bursts suggested an interpretation as whistlers, usually associated with atmospheric lightning. A first analysis of the cloud composition showed that the particle concentration was not high enough for accumulated charges to cause such lightning. Moreover, the emission seemed to come from a source above 30 kilometres in altitude, and thus from a region of the atmosphere with no clouds. This suggested that these signals could only be radioelectric signals associated with volcanic events: on the

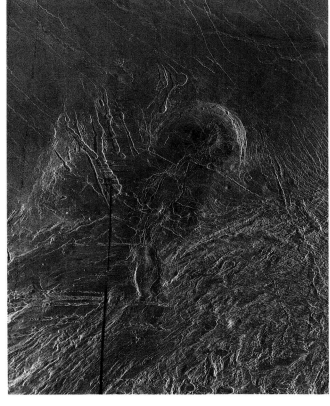

Lavinia Planitia. This picture from Magellan shows a surface of almost 500 square kilometres to the north of Lavinia Planitia, in the southern hemisphere of the planet. Near the centre is an impact crater of 40 kilometres diameter. To the south of the crater are two fans of lava flows which are extremely bright, showing that the surface is rough. All the region to the east of the crater is surrounded by a large dark zone which probably results from debris from the impact. The lines seen on the dark zone resulting from the displacement of various sediments give an idea of the winds at the surface of Venus, in a region where the relief is marked by impact. The surface resolution is about 120 metres. (JPL/NASA)

The tessera. The relief of Lakshmi Planum, to the south of the continent Ishtar Terra, appears without radar contrast in the upper part of this mosaic of images from Magellan. In contrast in the south of this region we see signs of extremely tortured reliefs which are the tessera. Above, the circular shape probably corresponds to a volcanic caldera; this caldera, called Siddons, has a diameter of 64 kilometres. (JPL/NASA)

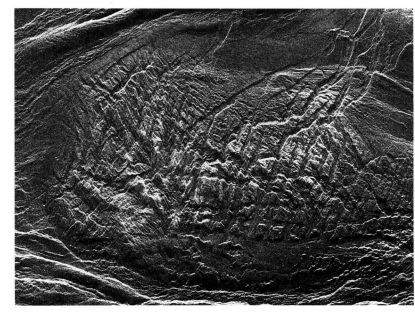

Relief of Freyja Montes, to the northwest of Ishtar Terra. Magellan tranmitted this picture, revealing a surprising aspect of the surface of Venus. The observed area corresponds to a region about 125 kilometres square and is on the western slopes of Freyja Montes. The central part is characterised by what must be a dome crossed by deep cracks. The shape, resembling a tortoiseshell, probably results from two opposing motions of the crust of Venus which have created a system of overlapping cracks. The study of these reliefs should explain how mountains were formed on the planet. (JPL/NASA)

Impact craters. Three large craters dominate this perspective picture of a region northwest of Lavinia Planitia: the crater Howe, in the foreground, has a diameter of 37 kilometres; Danilova, in the left background, a diameter of 48 kilometres, and Aglaonice, at right, 63 kilometres. No small impact craters are visible. (JPL/NASA)

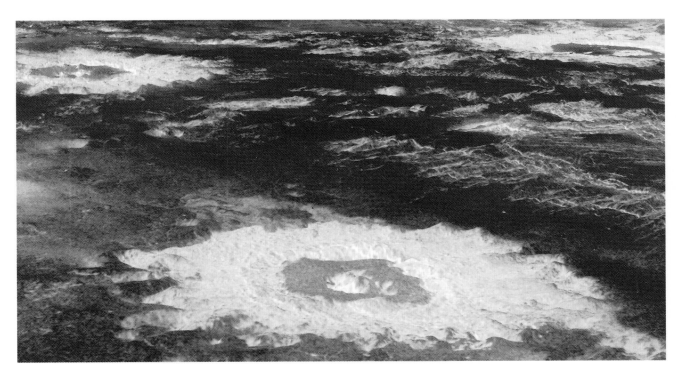

station	date	latitude (°)	longitude (°)
Venera 7	15 Dec. 70	— 5	351
Venera 8	22 July 72	—10	335
Venera 9	22 Oct. 75	+31.7	290.8
Venera 10	25 Oct. 75	+16.0	291.0
Venera 11	25 Dec. 78	—14	299
Venera 12	21 Dec. 78	— 7	294
Venera 13	1 Mar. 82	— 7.5	304
Venera 14	5 Mar. 82	—13.25	313
Vega 1	11 June 85	+ 8.1	176.9
Vega 2	15 June 85	—7.5	179.8

The Soviet landers which transmitted from the surface. After 1967, with Venera 4, Soviet scientists were able to use every launch window for Venus. The landing sites of the ten stations which were able to transmit scientific data from the surface are shown.

With the exception of Venera 9 and 10, which landed to the east of the mountainous Beta Regio area, and the Vega 1 and 2 stations, which landed on Aphrodite Terra, the stations landed in a relatively low-lying area near to the equator and to the east of Phoebe Regio.

Earth, plumes of dust and volcanic ash are often the sites of electric discharges. This interpretation was challenged by other scientists, who regarded the signals as random, resulting from instabilities of the local ionospheric plasma, like noise in telemetry. Another analysis in 1988 suggested the simplest explanation of all, that the signals were lightning discharges between clouds, and thus at about 60 kilometres altitude. The debate began again in 1990: on 10 February an instrument on the Galileo probe detected nine radioelectric bursts, of which six could only be lightning at low altitude. Although this supports the idea of volcanic activity, in the absence of direct proof we should suspend judgement.

Guy ISRAËL

elements	oxides	Venera 13	Venera 14	Vega 2	tholeitic basalt
magnesium	MgO	11.4 ± 6.2	8.1 ± 3.3	11 ± 3.8	6.3
aluminium	Al_2O_3	15.8 ± 3.0	17.9 ± 2.6	16 ± 1.9	14.1
silica	SiO_2	45.1 ± 3.0	48.7 ± 3.6	45.6 ± 3.2	50.8
potassium	K_2O	4.0 ± 0.63	0.2 ± 0.1	4.7 ± 1.5	0.8
calcium	CaO	7.1 ± 0.1	10.3 ± 1.2	0.1 ± 0.08	10.4
titanium	TiO_2	1.6 ± 0.5	1.3 ± 0.4	7.3 ± 0.7	2.0
manganese	MnO	0.2 ± 0.1	0.2 ± 0.1	0.2 ± 0.1	0.2
iron	Fe_2O_3	9.3 ± 2.2	8.8 ± 1.8	8.5 ± 1.3	9.1
sulphur	SO_3	1.6 ± 1.0	0.9 ± 0.8	4.7 ± 1.5	

Chemical composition of the Venus rocks. The chemical composition of the surface rocks of Venus, determined by X-ray fluorescence spectroscopy, is compared with that of a terrestrial magmatic rock. The values are given as fractions by weight (per cent). The rocks analysed by the Venera 13 station are distinguished by very strong alkalinity: the potassium oxide (K_2O) content is higher than at the Venera 14 site. The rocks analysed by Vega 2 are comparable with those of the Venera 14 site. Their mineral content is close to that of terrestrial magmatic rocks of tholeitic basalt type, which make up vast deposits on the continents, the present ocean floors, and some archipelagos.

The Venera 13 and 14 sites. These two photographs reconstruct for each of the stations, the view opposite to that shown in colour. Black-and-white reproduction seems to give a better contrast. The 1975 pictures already showed that despite the very dense atmosphere, the illumination of the Venus surface was sufficient to show detail and contrast. In these two pictures, we see the articulated arm of the instrument designed to measure the density of the surface. That of Venera 14 unfortunately deployed over the cap of the camera. (USSR Academy of Sciences)

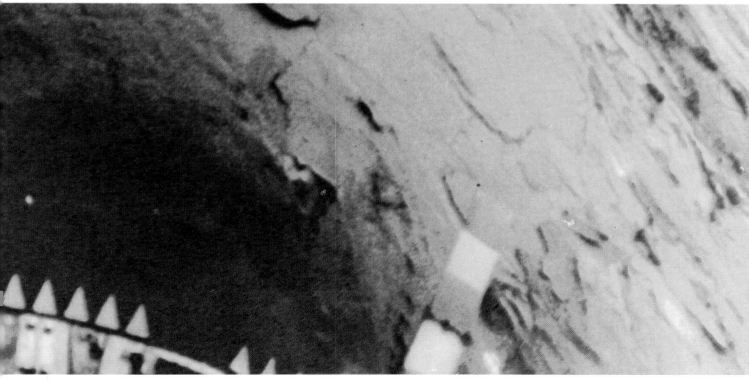

The Venera 13 and 14 sites. These 180 degree views were transmitted by the Venera 13 (above) and 14 landers (below). We see the circular base of the shock absorber, a belt of triangular projections which had a stabilising effect during the descent through the atmosphere. The rectangular appendage, deployed after landing, is for calibrating the colours. In the centre of each picture one can see the camera cap which was ejected on landing. Venera 13 landed on 1 March 1982 at 20 degrees to the east of Phoebe Regio (7° 30′ S, 304°). The site shows a fairly smooth but not homogeneous surface. It appears to have a sandy base, and is strewn with rocky debris. Around the station these may result from its impact. The regions of smoothest appearance were probably created by the formation of a surface crust made of an agglomeration of very fine grains. This could not have resulted from a process of cementation with liquid water as the eroding agent. Chemical erosion mechanisms using the atmospheric gases have been proposed to explain the cementation. Venera 14 landed on 5 March 1982, further to the east of Phobe Regio (13° 15′ S, 313°), 950 kilometres from the Venera 13 site. In many ways the Venera 14 site is exceptional. We do not see, as on the Venera 13 site, a swarm of pebbles and stones. The rocky plates are scattered, and the site is practically covered in 'flagstones' made from the superposition of distinct layers. In this case too, chemical erosion followed by a cementation process is suspected. Whatever the mechanism of formation, their angular and distinct appearance and the lack of granular deposits show that the surface is relatively recent, formed about 10 million years ago. The cameras had three filters, red, blue and green, which were used in succession. The lower atmosphere of Venus totally filters out blue, because of its chemical composition. The surface and the sky seen at the horizon at the edges of the pictures, do not have the orange colour seen here. The surface is supposed to be dark grey. (USSR Academy of Sciences)

The Earth

The mechanics of the Earth

The Earth is the third planet, in terms of distance from the Sun, after Mercury and Venus, and appears to have the form of a solid sphere. It has a mass of 5.976×10^{24} kilograms. In actual fact it has long been known that the Earth is not exactly spherical: it is close to an ellipsoid of revolution in shape, being flattened at the poles and distended at the equator. The mean equatorial radius is 6378 kilometres while the mean polar radius is 6357 kilometres; the difference is 21 kilometres, which corresponds to a flattening of 1/300. This deformation is due to the Earth's rotation about its own axis. The centrifugal force generated by this rotation draws the Earth's matter out along the equatorial plane, giving rise to the flattening at the poles and the bulge at the equator.

The mean density of the Earth (5.52 grams per cubic centimetre) is greater than that of the surface rocks (less than 3 grams per cubic centimetre). Moreover, the Earth's moment of inertia about its axis of rotation is appreciably less than that of a homogeneous sphere of the same mass and radius. A reasonable conclusion, therefore, is that the Earth's density increases from the surface towards the interior.

As a result of many measurements drawn from different branches of geophysics, and by means of physical models, it has been possible to determine the physical and chemical properties of the Earth's interior, at least in broad outline. However, the most precise information on the variation of density with depth, and on the existence of clearly distinguished layers from the surface crust to the central core, has been obtained through seismic soundings of the Earth's interior. Earthquakes give rise to acoustic waves, as do artificially created explosions such as nuclear explosions. These waves propagate through the Earth, undergoing reflection, refraction and diffraction due to the existence of the different internal layers, and after following a complicated path are detected by measuring instruments and seismographs at various points on the Earth's surface. Since the velocity of the waves depends on the elastic properties of the rocks through which they travel it has been possible, by observing their progress at different distances from the epicentres, to gather information on the properties of rocks situated at depth.

There are two main types of seismic wave: the faster P waves, which can travel through both solid and liquid matter, and the slower S waves. These S waves are transverse – that is to say the displacement is perpendicular to the direction of propagation. They cannot occur except in a

region	constituents	percentage of the Earth's mass
mantle	SiO_2	32
	MgO	23
	FeO, Fe_2O_3	7.5
	Al_2O_3	2
	CaO	2
	other constituents	1.5
core	iron (Fe)	24
	nickel (Ni)	3
	sulphur (S)	5

The chemical composition of the Earth. The chemical composition of the Earth's interior is not directly accessible. It is deduced from thermodynamic models which describe the variation of density and temperature as a function of depth. These models take into account the mechanical and chemical properties of the surface rocks as well as information from geography, geology and geochemistry. The Earth's crust, which is the only part which can be observed directly, is made up of aluminosilicates of calcium, potassium and sodium. The continental crust which is less dense than the oceanic crust (2.2 compared with 2.8 grams per cubic centimetre) has a predominantly granitic composition while that of the oceanic crust is mainly basaltic. Models of the Earth's mantle indicate the presence of ferromagnetic silicates of which the principal constituent is olivine. Certain secondary discontinuities at depths of 400 and 650 kilometres are associated with phase transitions.

sufficiently rigid medium, and therefore cannot pass through fluids. Since the beginning of the century, seismological observatories have been operating all over the world, and have made it possible to demonstrate the existence of several internal discontinuities, at about 30 kilometres below the continents and 10 kilometres below the oceans. The celebrated Mohorovicic discontinuity (named after the Yugoslav geophysicist who discovered it in 1909) is a case in point. This discontinuity, called the Moho, is the boundary between the Earth's crust and the mantle. At a depth of about 2900 kilometres the speed of P waves falls sharply and S waves no longer propagate. It is this observation which reveals the existence of a fluid core. Furthermore, certain measurements indicate the presence of a solid core within the fluid core. This solid core has a radius of approximately 1250 kilometres. The study of seismic waves reveals in addition the existence of secondary discontinuities, at depths of about 400 and 650 kilometres, the latter being the boundary between the inner and outer mantle.

The plate tectonic theory describes the evolution of the outer solid skin of our planet: the lithosphere. This includes the crust and part of the upper mantle. This lithosphere is made up of a mosaic of rigid plates of various sizes which move relative to each other. A dozen or so of these plates have been identified, some of which carry the continents along by their motion. The zones of seismic activity are concentrated along the boundaries of the plates. Some of these zones are not very deep and are the site of submarine basaltic volcanism corresponding to the ocean ridges. Each ridge corresponds to a narrow zone of fracture between separating plates, along the length of which molten rock (magma), rising up from the mantle, fills the fractures and brings to the Earth's surface additional material which is subsequently pulled out along the ocean floor. The plates separate at the ridges by sliding on a more plastic layer (the asthenosphere) at the base of the lithosphere. This increase of the lithosphere is only possible because an equivalent quantity of material disappears elsewhere.

The plate tectonic theory was preceded by the idea of the expansion of the ocean floor (continental drift). This theory was worked out in detail by F. Vine and D. Matthews during the early 1960s, following the discovery that periodic inversions of the Earth's magnetic field were registered in the suboceanic crust. According to their theory, when the deep lying magma arrives at the surface, the ferromagnetic minerals contained within it become polarised along the direction of the Earth's magnetic field. As the magma cools and solidifies, the direction and polarisation of the magnetic field are fossilised in the volcanic rock. The successive inversions of the Earth's magnetic field are recorded as

The blue planet. Photographs such as this one, taken during an Apollo mission, allow us to grasp the relative importance of the oceans and the diversity of cloud layers. The Earth is a huge machine in full operation. (NASA/NSSDC)

Diagram of the Earth's interior. Our knowledge of the Earth's interior derives from observations of the propagation of seismic waves. In this diagram the thickness of the outer shell is considerably enlarged. (From *Les Volcans. Aux sources de la connaissance de la Terre,* CNRS, 1981)

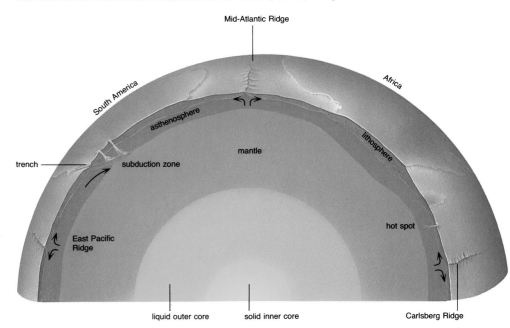

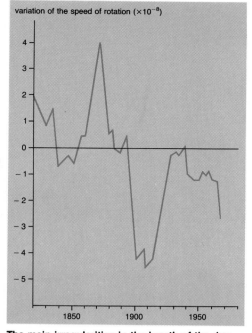

The main irregularities in the length of the day.
This figure shows the long-period variations (those with a time-scale of several decades) which have been observed from about 1820 onwards. The periods when the Earth's rotation has slowed down (such as 1870 to 1900) and when it has speeded up (1900 to 1940) are very obvious. Between 1820 and 1900 the length of the day increased by 10 milliseconds. Annual and semi-annual variations of the length of the day have also been detected since 1955.

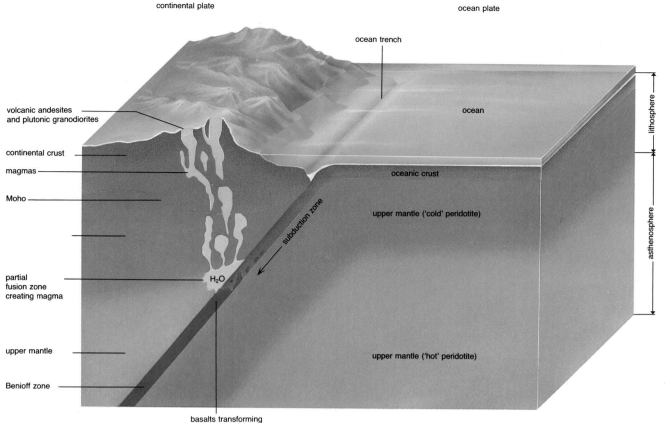

magnetic stripes parallel to the axis of the ridge. The chronology of magnetic field reversal has been independently established by absolute dating of continental rocks. Since the edges of the stripes correspond to these reversals, their width is a measure of the rate of expansion of the ocean floor; this expansion occurs symmetrically about the axis of the ridge.

A reconstruction of the movement of the plates throughout geological time indicates the temporary existence, about 250 million years ago, of one single continent – Pangea.

It is generally accepted that the driving forces causing the movement of the lithospheric plates are the convection currents in the Earth's mantle, which over a period of hundreds of millions of years behaves like a viscous fluid, in which convection currents are maintained by internal heat sources.

The Earth's rotation

As seen by an observer situated at the North or South Pole, the stars appear to describe concentric circles, the common centre of which defines the celestial pole, that is the prolongation of the Earth's axis of rotation.

Nowadays, the north celestial pole is close to the star Polaris, but precise observations over many years have revealed that the position of the celestial pole is not fixed relative to the stars. This fact was first noted by Hipparcus in 120 BC. The terrestrial axis of rotation describes a cone of opening angle 23° 27' about the ecliptic pole and it takes about 26 000 years to make a complete revolution. Some 5000 years ago the 'polar' star was Alpha Draconis, and in 5500 years Alpha Cephei will become the new 'polar' star. This slow movement of the Earth's axis of rotation in space is called precession.

The precession is the result of a double phenomenon: the existence of the Earth's equatorial bulge and the substantial inclination of the Earth's equator to the ecliptic. Because of the Earth's flattened shape, the resultant gravitational attraction exerted by the Sun does not pass through the Earth's centre of mass. This gives rise to a couple which tends to twist the equator back into the plane of the ecliptic. However, because of the angular momentum generated by the Earth's rotation, the effect of this couple is to cause the precession of the Earth's axis described above.

As well as solar precession there is a lunar precession caused by the gravitational pull of the Moon on the Earth's equatorial bulge. The other planets act similarly but the effect is extremely weak compared with that of the Sun and the Moon. The terrestrial orbit round the Sun is very slightly eccentric which causes a periodic varia-

tion of the solar couple. The same is true of the lunar couple. More generally, all the various irregularities in the Earth's movement around the Sun and in the Moon's movement round the Earth give rise to small periodic oscillations of the Earth's axis of rotation which are superimposed on the secular motion of the precession. They are known as forced nutations. The principal nutation has a period of 18.6 years with an amplitude of 9 arc seconds. This oscillation is linked to a variation, of period of 18.6 years, of the plane of the Moon's orbit relative to the ecliptic. This nutation was first detected by the astronomer James Bradley in 1747.

Precise observations of these movements of precession and nutation are of importance in geophysics because their magnitudes depend on the Earth's flattening and its internal distribution of mass.

Close observation shows that the celestial pole is not fixed relative to the Earth but seems to describe a small circle with a period of approximately one year. This is an effect different from precession and nutation and reflects the movement of the axis of the Earth's rotation around its axis of inertia. It amounts to a circular movement of the Earth's pole. Although this movement was predicted by Euler in 1765 it could not be detected until the end of the nineteenth century.

The movement of the Earth's pole is revealed by observations of the stars. The trajectory of the pole in a terrestrial frame of reference is like a spiral. This has been confirmed by observations over a period of nearly eighty years. A spectral analysis of the motion reveals two components: one oscillation of period 14 months and another of period 12 months, both having an amplitude of about 0.1 arc second. This means that the trajectory described by the moving pole lies within a region on the Earth's surface no more than a few metres across, which underlines the difficulty of observing the phenomenon. The period of the Earth's revolution can be determined very accurately by using very accurate clocks to time the passage of the stars across the meridian. These astronomical observations have revealed that the length of a day thus measured is not constant, but varies perceptibly by as much as one millisecond. Measurements have been carried out since 1820, but it is since the appearance of atomic clocks that the observations have attained a precision sufficient to make a satisfactory inventory of the various fluctuations, of which only a few, from a few years to a few decades, have been detected since 1955. At present, fluctuations on time-scales of a few days are known.

Astronomical observations of various aspects of the Earth's rotation have shown the existence of an impressive number of irregularities whose

interpretation involves various Earth sciences. Numerous phenomena are potential candidates for perturbing the rotation of the Earth, such as the redistribution of mass associated with the atmosphere and the oceans, movements of the core and the associated electromagnetic effects, global deformations caused by the tides, the displacements of the lithospheric plates and the induced seismic activity.

Two laws govern phenomena in mechanics: the conservation of energy and the conservation of angular momentum. As the Earth–oceans–atmosphere system can be considered as isolated in space, its total angular momentum is conserved. Thus changes in moment of inertia and movements of particles within the Earth or upon its surface (e.g. winds and currents) induce variations in the Earth's rotation in such a way as to conserve the total angular momentum in the Earth–oceans–atmosphere system. It is also through conservation of total angular momentum in the Earth–Moon system (orbital angular momentum of the moon and rotational angular momentum of the Earth) that the phenomenon of tides simultaneously gives rise to the secular slowing of the Earth's rotation (by dissipation of the energy of the tides) and an increase in the radius of the Moon's orbit around the Earth.

Large variations in the rate of rotation with a time-scale of several decades appear to originate in variations of the magnetic field and associated movements within the core.

On these low frequency variations are superimposed more rapid oscillations of the length of the day, which are seasonal in character, as well as fluctuations of even shorter period which last for several months or even only a few days. The atmospheric origin of these fluctuations now seems to be well established.

The amount of solar energy which enters the atmosphere varies with latitude in the course of the year. This gives rise to an annual variation of zonal winds, greatest in the stratosphere and at the tropopause and opposite in phase in the two hemispheres. The asymmetric distribution of the continents in each of the two hemispheres modifies the radiative properties of the lower atmosphere and therefore the properties of the zonal winds, so that the annual variation of the winds is not completely antisymmetric with respect to the equator. The angular momentum of the atmosphere therefore exhibits an annual variation (which would not exist if the continents were uniformly distributed) directly transferred to the rotation of the Earth. In this way, the yearly cycle of the rotation of the Earth is a measure of the imperfect equilibrium between the atmospheric circulation of the northern hemisphere and that of the southern hemisphere with a frequency of one year. As for higher frequencies

Subduction of the lithosphere. The material of the lithosphere returns to the mantle in the depths of the large ocean trenches. These regions, called subduction zones, are characterised by intense seismic and volcanic activity which is very different from that associated with the mid-ocean ridges. The earthquakes associated with these trenches are usually very deep (down to 700 kilometres) and occur on a plane inclined along the direction in which the tectonic plate is being driven into the mantle. A third type of boundary is characterised by faults, such as the San Andreas fault in California, along which two plates can slide against each other. When two continental masses come in contact along a subduction zone, the collision gives rise to mountain chains. (From *Les Volcans. Aux sources de la connaissance de la Terre*, CNRS, 1981)

(variations occurring from a few days to a few weeks), it has been possible to establish their meteorological origin with certainty: The majority of these irregularities reflect meteorological anomalies taking place in the atmosphere.

The motion of the Earth's axis is characterised principally by two components of periodical nature:
– an oscillation with a period of 14 months which is in fact a free oscillation of the axis of terrestrial rotation predicted by mechanics, whose period is lengthened because of the complex structure of the Earth;
– a forced oscillation with a period of 12 months, of atmospheric origin (it is caused by seasonal redistributions of air masses in the atmosphere).

At lower frequency, the axis appears to exhibit a drift of three thousandths of an arc second in the direction 70° of longitude west. The existence of this drift has long been controversial but it could be the effect of variations in sea-level.

The 14-month oscillation is called Chandler's oscillation, after its discoverer. It has a characteristic feature: observations over the last hundred years show that while the amplitude of the motion has varied, the period itself has fluctuated very little. Fourier analysis suggests that it is a dampened oscillation, maintained by a source of energy not yet identified with certainty, but which could well be connected with seismic activity.

This history of uncertainty is due partly to a lack of precision in astronomical measurements, and partly to a lack of information on the important earthquakes of past decades. An earthquake gives rise to a redistribution of mass in the Earth, and consequently modifies the position of the axis of rotation relative to the axes of inertia. But in order to determine this relative displacement of the axes, and hence the change in the direction of the pole's trajectory, a large number of parameters characterising the earthquake must be known with precision. These parameters are only known for the more recent big earthquakes, and it will be necessary to accumulate a record of a large number of events before being in a position to associate a detectable change in the trajectory with seismic activity. This approach, although promising, will not be conclusive for several years.

The gravitational field

The strength of the gravitational field varies from one place to another over the surface of the Earth. This variation reflects the asymmetric distribution of mass in the Earth's crust and mantle. A notable departure from symmetry is the polar flattening. Every particle inside the Earth is subject to both the gravitational attraction of the rest of the planet and also, because of

the daily rotation, a centrifugal force which acts perpendicular to the axis of rotation. This latter is greatest at the equator, where it opposes the gravitational attraction, and vanishes at the poles. A fluid Earth subject to these forces would attain an equilibrium shape of a flattened ellipsoid very close to what is observed. This suggests that the Earth's material indeed behaves as a fluid when subject to forces acting over a long period of time.

A more precise description of the shape is given by the term *geoid*, the shape that the Earth would have to be for the effect of gravity to be constant at all points. Because of anomalies in the distribution of mass in the Earth, the geoid departs in shape from the ellipsoid of the ideal fluid. It exhibits various dips and bumps. The dips are about 100 metres deep in some places. When scientists talk about the shape of the Earth, it is usually the geoid which they have in mind. It is only at sea that the geoid is directly accessible and its shape can be deduced from gravitational observations on the surface. On land the geoid can also be determined from gravitational measurements on the topographical surface.

These departures or anomalies of the geoid from the shape that the Earth would have, were it in hydrostatic equilibrium, are directly related to lateral inhomogeneities of the mass distribution in the interior. The aim in interpreting the associated gravitational anomalies is to determine whether, at the depths studied, the material of the Earth is rigid enough to behave elastically, or whether, if it is plastic in its behaviour, a dynamical support is necessary (for example by means of convective motion).

The large-scale undulations of the geoid show no obvious correlation with surface topography – continental or oceanic – suggesting that they reflect more the presence of deep lying density anomalies, several hundred kilometres down in the Earth's mantle.

In the last few years many geophysicists have tried to relate the large-scale gravitational anomalies to the possible existence, within the mantle, of convection cells. Such large-scale convection currents, with speeds of the order of 1 centimetre per year, have already been postulated to explain the movement, proposed in the well-known theory of plate tectonics, of the lithospheric plates. In spite of much effort, both theoretical and experimental, convection in the Earth's mantle remains a very badly understood phenomenon. Each model of convection, be it of the outer mantle (down to 700 kilometres), the whole mantle or even, on two scales, of the outer and inner mantles, has its supporters and its opponents. It seems very clear that small-scale laboratory experiments cannot be extrapolated to account for what is actually happening inside the Earth. Furthermore, numerical models of convection have, up till now, a natural limitation,

The terrestrial geoid over the oceans from satellite altimetry. In recent years a very precise map of the terrestrial geoid has been obtained by a new technique: satellite altimetry. This technique involves a direct measurement of the satellite's altitude above the ocean surface. If the satellite's orbit is already known – by means of classical tracking methods (laser, Doppler) – the height of the geoid relative to a reference ellipsoid can be obtained with very high precision over the oceans. This map shows the average topography of the ocean surfaces derived from measurements carried out over a period of eighteen months by the satellite GEOS–3; the Seasat satellite measurements are also shown. (Document kindly communicated by R. E. Cheney, NASA Goddard Space Flight Center)

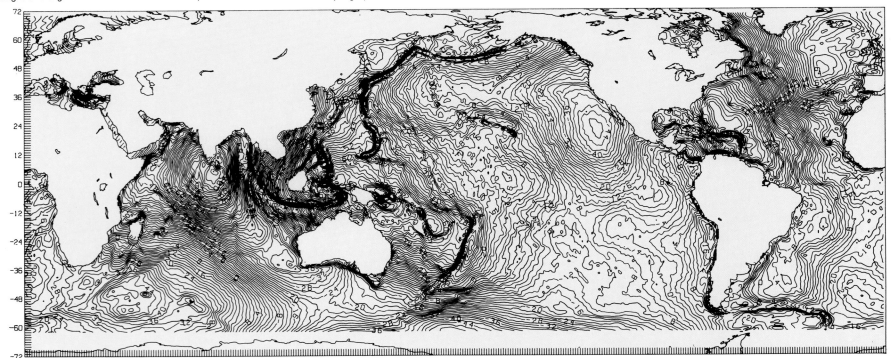

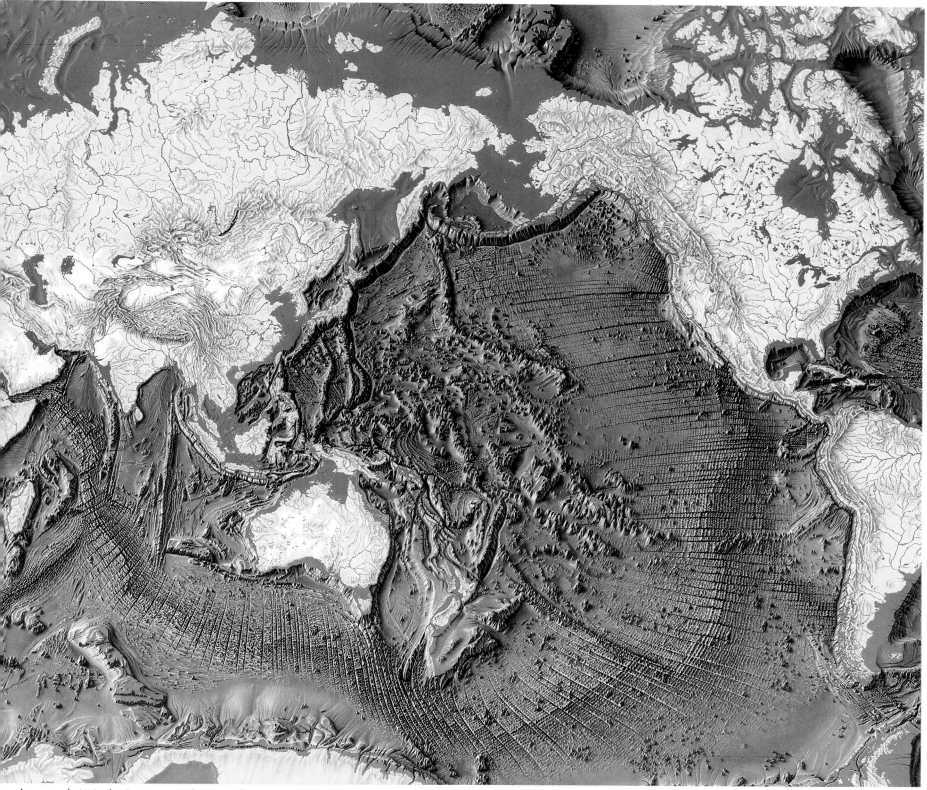

Large-scale tectonic structures on the ocean floor. Each of these structures has a characteristic effect on the geoid, that is, on the mean sea-level. These effects occur, for example, above subduction zones, near ocean ridges and at the edges of continents. (From the map of the floor of the oceans by Tanguy de Rémur, © Hachette-Guides Bleus).

in that they cannot deal with the phenomenon in three dimensions. This explains why attempts to relate large-scale gravitational anomalies to convection in the mantle have so far proved unsuccessful. In spite of these difficulties it seems clear that these anomalies are indeed the result of dynamical processes occurring within the mantle.

By contrast, the link between short-scale (less than 1500 kilometres) gravitational anomalies and topography, notably undersea topography, is particularly remarkable.

There is a very striking correlation between medium wavelength undulations (from 100 to 2000 kilometres) of the geoid, and the large tectonic structures in the ocean depths. Each

structure has a characteristic effect on the geoid.

Above the ocean trenches, where the lithospheric plates plunge down into the mantle (subduction zones), the geoid typically exhibits a depression of 15 to 20 metres throughout a region about 200 kilometres across. Over the ocean ridges, where the lithospheric plates are being formed, the geoid is raised up by a few metres throughout a region which extends for about 1000 kilometres. The height profile of the geoid above fracture zones (transforming fossilised faults, at right angles to the direction of the mid-ocean ridge) shows a sharp jump of several metres over about 100 kilometres. At the continental margins (transition zones between the

continental and ocean crusts) the geoid shows a discontinuity of the order of 5 metres over 100 to 200 kilometres. Anomalies of the geoid associated with undersea volcanoes is another striking instance. In most cases, the associated anomaly is positive with a size of 5 to 10 metres, extending for a distance of 100 to 500 kilometres.

The study of these anomalies will lead to a determination of the state of isostatic compensation in the outer mantle, and ultimately to an understanding of the mechanism by which this occurs. This information is extremely important for the creation of models of the oceanic lithosphere.

Anny CAZENAVE

The Earth

The fluid environment

If one sets aside the extraordinary phenomenon of life, then the most remarkable feature of the Earth, if one views it simply as one of the planets in the Solar System, is the interaction between the solid surface and the fluid environment – atmosphere and oceans. The conditions which held during the emergence and evolution of the volatile substances acquired by the Earth at its formation have led to an atmosphere relatively rich in oxygen. They also brought into being the oceans which cover large parts of the Earth's surface, as well as the polar ice caps. A machine was set in motion, each part of which has played or continues to play an important role.

The precise chemical composition of the Earth's atmosphere is due to the dynamic mechanism of life. However life would not have survived were it not for the presence of vast stretches of water which protected it from the hostile primordial environment. On all Earth-like planets, water and carbon dioxide are the most prominent volatile substances. But for the Earth itself the most important point is that the evolution of its atmosphere has resulted in the accumulation of the water in the oceans and ice caps, and the trapping of the carbon dioxide, in the form of carbonates, in the sediments of the ocean beds. This can explain why, unlike on Mars or Venus, the surface of the Earth has a marked influence on the way its atmosphere responds to the Sun's radiation.

This is well illustrated by the following example. During the last fifty years the burning of coal and oil for industrial and domestic purposes has increased the concentration of carbon dioxide in the atmosphere. Between 1958 and 1980 the concentration of carbon dioxide rose from 315 to 338 parts per million. The consequences are a greater opacity of the lower atmosphere to infrared radiation and a limitation on the ability of the Earth to radiate into space the heat stored in its land surface and oceans.

The dynamical mechanisms in the atmosphere which maintain its thermal balance are also strongly influenced by the Earth's surface. Excess energy near the equator is transferred horizontally towards the poles. At latitudes of less than 35 degrees the absorption of heat is greater than the emission, with a resultant warming of the atmosphere. There is a comparable cooling at latitudes of more than 35 degrees. The temperature gradient in the atmosphere is, however,

The blue planet as seen from space. Not much more than twenty years ago the spectacle of our planet in space could only be glimpsed in the imagination. The above is one of many pictures of the Earth taken by astronauts during the Apollo series of missions. The principal meteorological features show up clearly. Firstly, it is obvious that a large fraction of the surface of the blue planet is liquid and that significant quantities of water can change phase as a result of favourable atmospheric temperatures. This gives rise to layers of cloud maintained in the atmosphere and to such meteorological phenomena as rain and snow. The cloud cover of the Earth is about 50 per cent which in this respect puts our planet half way between Venus (100 per cent cloud cover) and Mars (10 per cent). Secondly, the large-scale trends of the general circulation are easily made out as are motions on a range of scales down from the synoptic, with its characteristic turbulent formations, to small-scale convective cells. (NASA)

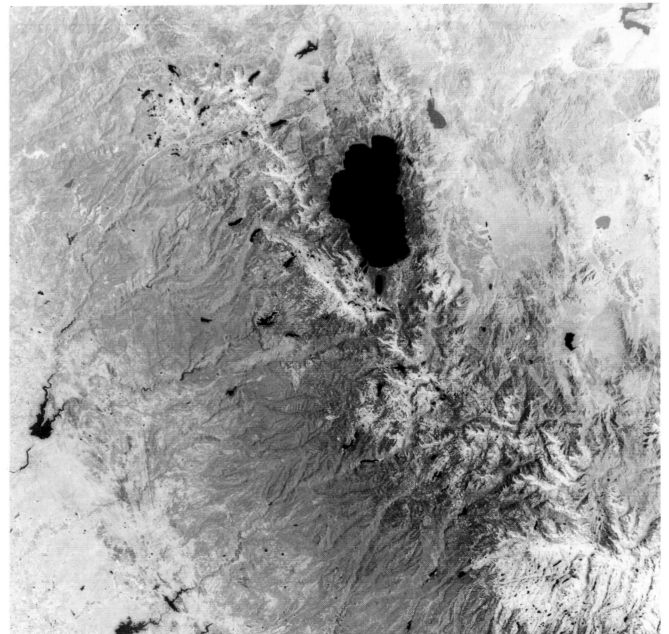

Study of snow cover. The observation satellite Landsat 1 was launched on 23 July 1972 and continued to function until 1979. This picture, taken on 29 May 1976, covers the region of Lake Tahoe in the Sierra Nevada. The importance of this type of photograph is the record it provides from one year to the next of the extent of the snow cover, with the aim of revealing its effect on the water reserves of the region. The appearance of the snow depends on its age and the nature of the terrain. At this time of year the snow covering the Sierra Nevada shows up white and is in retreat. It gives place to zones of vegetation, in red. Desert areas are grey but here and there irrigated regions with vegetation show up as red. The resolution of the Landsat images is 80 by 55 metres (NASA)

Mean surface albedo of various cloud and terrestrial zones from satellite photographs. (From Conover, 1965)

surface	albedo
cumulo-nimbus (thick and large)	92
cumulo-nimbus (small, top ± 6 km)	86
cirro-stratus (thick with lower clouds)	74
cumulus and strato-cumulus (above land*)	69
strato-cumulus (above land*)	68
stratus (thick up to about 500 m above the ocean)	64
sands (White Sands, New Mexico)	60
massed strato-cumulus (within a cloudy layer above the ocean	60
snow (3 to 7 days old covering all the mountains in a woody area)	59
stratus (thin above the ocean)	42
cirrus (isolated above land)	36
cirro-stratus (isolated above land)	32
fair-weather cumulus (above land*)	29
sand (in valleys and on plains and inclines)	27
sand and brushwood	17
conifer forests	12
lake (Great Salt Lake)	9
ocean (Gulf of Mexico)	9
Pacific Ocean	7

* cloud cover greater than 80 per cent

insufficient to explain the necessary heat transfer.

In fact this heat flow is brought about by large-scale movements of the atmosphere, which transport the thermal energy from the equator towards the poles, both in the form of latent heat of water vapour and in the form of sensible heat. (Latent heat is the heat taken up or released at a fixed temperature during a change of phase, while sensible heat is freely available and detectable through changes in temperature.) Thus when ocean water is evaporated into the atmosphere at equatorial latitudes, it absorbs a certain quantity of heat (latent heat of evaporation) which is released at more northerly latitudes when the water vapour, transported by large-scale atmospheric motion, condenses. Ocean circulation provides another significant contribution to the movement of heat towards the poles.

Atmospheric dynamics are also involved in vertical heat exchange between the Earth and the lower atmosphere, where non-radiative heat transfer is partly due to thermal convection. In this case the effect is the result of small-scale atmospheric turbulence. Meteorologists refer to large-scale convection as advection, reserving the term convection for small-scale transport.

Recent research into the fluid environment of the Earth is aimed at clarifying the relative importance of the different types of energy flow. It draws upon several disciplines – meteorology, climatology, oceanography and the study of the Earth's resources.

Radiative phenomena

The electromagnetic radiation in the neighbourhood of the Earth has an important effect on the environment of our planet. In order to understand it better, geophysicists use methods similar to those employed by astronomers. There are two main components of this radiation: namely solar radiation and radiation from the

Three Meteosat images of the whole terrestrial disk, 15 November 1981. These three pictures transmitted by the European satellite Meteosat 1 illustrate the technique of teledetection which makes use of observations in different spectral bands. The first image (Vis 2) was obtained in the band 0.4–1.1 micrometers (approximately the visible) with a resolution of 2.5 kilometres. Variations of brilliance depend on three factors: the intensity of solar radiation, the height of the Sun and lastly the reflectivity of the body. The reflectivity depends on the type of cloud and the nature of the terrain or ocean which gives rise to very different values of the albedo (see table on page 86). It is possible to study the contribution to the albedo of the land mass which was almost without obvious cloud cover on the day the picture was taken; the brilliant sands of the Sahara contrast sharply with the cultivated ground in the valley of the Nile, the vegetation of South Africa appears markedly darker.

The second image (IR1) was taken in the thermal infrared, within a window between 10.5 and 12.5 micrometres, with a resolution of 5 kilometres, thus providing a thermal map (200–300 K) of the Earth's surface and the tops of the clouds, in which weak emission corresponds to low temperature. During the day the warmer land is darker while the colder sea shows up as a lighter grey.

The third image (WV) was obtained from infrared emission in the spectral band 5.7–7.1 micrometres, with a resolution at ground level of 5 kilometres. In this band, the absorptivity of water vapour is so strong that emissions from the Earth's surface and the lower layers of the atmosphere are usually totally absorbed. The dark grey and black correspond to higher temperatures; that is, regions where the radiation is coming from low altitudes. Regions of high humidity show up lighter because of the lower radiation temperature. Measurements in this 'water vapour' band essentially indicate the humidity level of the upper layers of the troposphere. (Atlas Photo/ESA Meteosat)

Solar illumination at the top of the atmosphere (averaged over a day, expressed in calories per square centimetre per day). The figure exhibits the variation of this quantity as a function of latitude and time of year. The inflow of energy is a maximum in the polar regions at the summer solstice in the northern hemisphere and at the winter solstice in the southern. Furthermore, since the Earth is nearer the Sun in January (winter in the northern hemisphere) the distribution of solar energy is slightly asymmetric. In the southern hemisphere this weak seasonal effect of the eccentricity of the Earth's orbit comes into play during the summer, thus reinforcing the seasonal effect of the obliquity of the Earth's axis. As a result the maximum thermal energy received by this hemisphere is greater than the maximum energy received in the northern hemisphere.

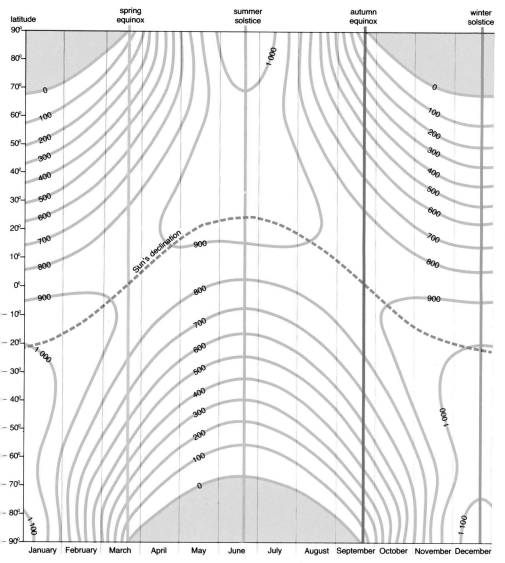

Earth itself. This latter comprises radiation emitted by the Earth's surface, including the oceans, and that emitted by the cloudy atmosphere.

As a result of measurements using rocket probes and satellites, the spectral distribution of solar radiation is known sufficiently well for most geophysical applications. It is subject to fluctuations induced by solar activity. However, in the visible spectrum these variations of intensity are less than 2 per cent, which is comparable with the precision with which they can be measured at the top of the Earth's atmosphere (an altitude of about 600 kilometres). The mean energy flux over the whole spectrum outside the Earth's atmosphere is called the solar constant, and is equal to 0.132 watts per square centimetre. This flux is made up of 42.4 per cent visible radiation, 48.4 per cent infrared and 9.2 per cent ultraviolet.

The intensity of solar radiation at the top of the atmosphere depends upon the angle at which the Sun's rays strike the surface. This varies with latitude, season of the year and time of day. The figure on page 87, taken from List (1958), shows how the daily irradiation depends on these variables. Maximum irradiation occurs over the poles at the summer or winter solstice, depending on the pole, as a result of the 24-hour day.

Only 49 per cent of the available solar radiation reaches the surface of our planet, 5 per cent through direct radiation, 22 per cent through clouds and 22 per cent by downwards scattering in the atmosphere. The greater part of the radiation is absorbed.

This attenuation of the Sun's radiation by the atmosphere is due to two processes, genuine absorption and scattering. Absorption leads to a permanent loss of radiant energy which is converted into heat when it interacts with atmospheric particles – not only the atoms and molecules of the solid, liquid and gaseous substances, but also the fine particles of aerosol suspensions which have diameters in the range 0.1 to 20 micrometres. By contrast scattering leads to some radiant energy being sent in all direction by each particle, which therefore acts like a new source of radiation. Attenuation occurs particularly in the infrared and visible parts of the spectrum. Water vapour, carbon dioxide and ozone are the substances most heavily involved. A recent estimate by the meteorologist Wittman suggests that a cloud-free atmosphere would absorb 22 per cent of the incident solar radiation.

In the main, clouds reflect radiation. Nevertheless, for certain types of cloud true absorption can be important, giving rise to a contribution to the total atmospheric absorption of the order of 4 per cent.

The total albedo is the ratio of the energy returned to space, by reflection or upward scattering, to the energy incident on the top of the Earth's atmosphere. Again a distinction can be made between the effects of the cloud-free atmosphere on the one hand, and the clouds themselves and the Earth's surface on the other. Scattering by the cloudless atmosphere ranges from 8 to 9 per cent depending on the water vapour content and the aerosol concentration. It has been estimated that pure dry air would reflect 5 per cent of incident radiation, while scattering by water vapour would contribute 2.5 per cent to the albedo. Aerosols provide a yet smaller contribution. Clouds reflect 17 per cent of the incident energy. Since the Earth's surface reflects only 6 per cent it follows that clouds provide much the largest part of the total albedo, for which Wittman has given a mean value of 31 per cent.

Whatever the climatic variations of the past there has been no detectable cumulative heating or cooling. This means that the Earth, together with its cloudy atmosphere, constantly re-emits just as much radiant energy as it absorbs from the Sun. This radiation, apart from a small contribution from atmospheric luminescence, is thermal radiation. That part of it which emerges directly from the Earth's surface corresponds to black-body radiation at a temperature of 225 K. This is infrared radiation with a distribution of intensity which has a maximum at a wavelength of about 12 micrometres and which stretches down to about 2.5 micrometres. In the same way a cloud (or indeed any other object) will emit black-body radiation, with an energy distribution determined by its temperature. Finally, cloudless air emits infrared radiation mainly from the consti-

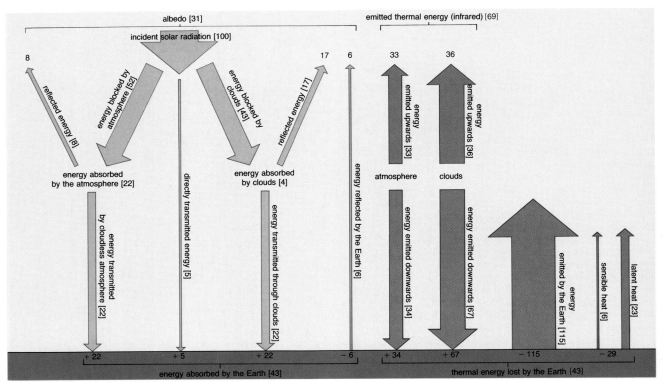

Thermal balance sheet for the ground and atmosphere. The numbers placed at the beginnings and ends of the arrows indicate contributions to the vertical transport of energy expressed as a percentage of the incident solar energy. It can be seen that the system ground plus atmosphere has a mean global albedo of 31 per cent and that a radiative equilibrium holds (69% + 31%). If the transfer of radiation through the subsystem 'atmosphere' is calculated taking into account only solar energy and the radiation from the Earth's surface, then a deficit is revealed; the atmosphere emits more radiation than it receives. Correspondingly the Earth's surface picks up an excess of energy. Necessarily therefore there is a compensating transport process for the Sun's energy (29%). The necessary vertical transport is guaranteed by air movement. The relevant non-radiative mechanisms are thermal convection, the release of latent heat of evaporation at sea-level and heat recovered when water vapour condenses in the atmosphere. (After G. D. Wittman)

tuents water vapour, carbon dioxide and ozone.

Near the surface of the Earth the radiation received and absorbed produces a number of thermal effects before being re-emitted back into space. The net result is positive or negative according to whether the thermal emission from the surface is greater or less than the radiation incident from above. The precise importance of these effects depends on the thermal conductivity of the land surface, the ocean surface and the very lowest layers of the atmosphere.

The main thermal effects of radiation in the atmosphere are radiative cooling of clouds and the transfer of heat through clear atmosphere by 'radiative conduction'. The radiative cooling of clouds occurs mainly from their upper parts, and

Terminology for the structure of the atmosphere.
– The *troposphere* is a region of negative temperature gradient; it extends up to 12 kilometres. Thermal equilibrium is maintained by radiative transfer and convection. The temperature in the tropopause is about 200 K.
– The *stratosphere* is a region where the atmosphere is in radiative equilibrium. The temperature is initially constant then rises slightly towards the top. The lower stratosphere, up to 20 kilometres, is heated by infrared radiation from the Earth's surface but because of the low density of water vapour at these altitudes this effect diminishes, with the temperatures stabilising at 200 K. The temperature then rises progressively because of the absorbtion by oxygen and ozone of ultraviolet solar radiation. Ozone has a characteristic vertical distribution with maximum concentration at about 25 kilometres, and plays an important role in the stratosphere.
– The *mesosphere* is a region where the temperature falls again to 180 K at an altitude of 80 kilometres (mesopause). The lower boundary of the mesosphere is at a level above which turbulence again plays a dominant role.
– The *thermosphere* above 120 kilometres is a region of strong heating as a result of photodissociation of molecular oxygen and photoionisation of molecular oxygen and nitrogen, and atomic oxygen, produced by solar radiation in the far ultraviolet. At the top of the thermosphere the temperature seems to level out at about 1300 K.
– The *exosphere* is defined as that region of the atmosphere where the density is so low and the probability of collision so small that the particles follow ballistic trajectories in the Earth's gravitational field. The base of the exosphere may be taken to be 700 kilometres where the density of particles is a million per cubic centimetre. The upper limit must correspond to a density equivalent to that of the interplanetary medium, namely about a hundred particles per cubic centimetre. This upper limit of the atmosphere is at about 5000 kilometres.

It should be noted that within this general structure maxima of temperature occur at those levels where the absorbtion of solar radiation is greatest. Starting with the maximum at the terrestrial surface where 50 per cent of the visible solar radiation is absorbed, a second maximum is encountered at 50 kilometres due to absorption by ozone (at wavelengths in the 200 to 300 nanometre range), and a third maximum at the top of the thermosphere where ultraviolet radiation is absorbed mainly by oxygen. At each altitude the observed temperature is the result of a particular balance between the heating due to absorption and thermal emission in the infrared by the constituents of the atmosphere.

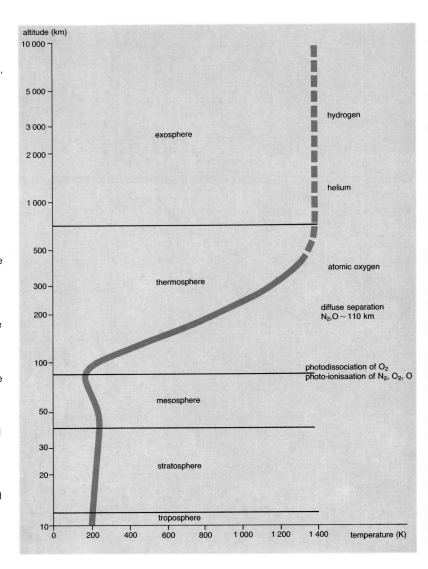

is responsible for a permanent cooling of the troposphere. Radiative conduction is a vertical transfer of energy from one layer of atmosphere to another. It occurs mainly when the radiation emitted by one horizontal layer of air is precisely such as can be absorbed by water and carbon dioxide. This phenomenon is limited to the troposphere.

Clearly, nowadays, radiative phenomena and their thermal effects in the atmosphere are well understood, and it is possible to establish with precision a comprehensive thermal balance sheet.

One piece of information which is important in drawing up this balance sheet is that the Earth's surface receives, in addition to the solar energy (visible) which penetrates the atmosphere and clouds, a considerable amount of thermal energy (infrared) radiated or scattered downwards from the low lying layers of hazy atmosphere. This is the reason why the surface has a global mean temperature, T_s, greater than the theoretical value it would have if only the solar energy and the global albedo, A, were taken into account. This equilibrium temperature, called the effective temperature, is determined by the equation $\varepsilon\sigma T_{eff} = (1-A)S$, where S is the solar constant, ε the infrared emissivity of the terrestrial surface and σ is the Stefan–Boltzmann constant. In the absence of any additional contribution from atmospheric heating or an internal source of heat, the Earth would have an effective temperature, T_{eff}, of 255 K (or $-18\,°C$), while the actual temperature is 288 K. The additional heating caused by the atmosphere is due to its opacity to infrared radiation, and is referred to as the greenhouse effect. In fact this additional heating of the Earth's surface is less effective than would be suggested by calculations based on radiative processes alone. This is because of the big heat loss brought about by atmospheric convection. It is therefore a reduced greenhouse effect which operates at the surface of the Earth.

When extended to the high atmosphere, the analysis of radiative and non-radiative processes provides an explanation for the structure of the whole atmosphere. The diagram on page 88 (bottom right) illustrates the different regions: troposphere, stratosphere, mesosphere, thermosphere and exosphere.

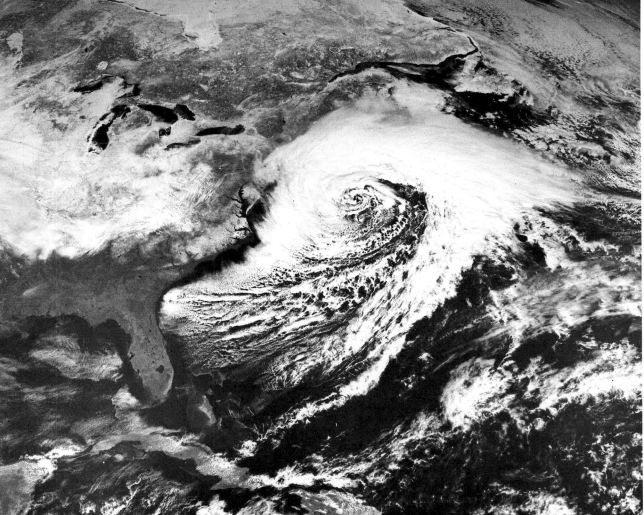

Schematic representation of the global circulation. The ultimate consequence of the planetary circulation is the transport of heat from low latitudes towards the poles. This circulation is driven by solar radiation. The solar energy received per unit area decreases with increasing latitude. This dependence on latitude brings about temperature gradients in the atmosphere: differential heating with respect to latitude. The effect of gravity on the fluid atmosphere, subject to horizontal temperature gradients, is to bring about a displacement of the air mass because warmer air is less dense than colder air. Cold air descends and hot air rises. In order to maintain continuity of fluid flow horizontal movements arise at the same time as the vertical movements. Although solar radiation is the prime cause, gravity is the final controlling factor in atmospheric circulation in which horizontal motion predominates.

Were the Earth not to have a large angular velocity, then the basic structure of the atmospheric circulation would comprise wind movement towards the equator at the surface of the Earth and towards the poles at high altitude: a gigantic convection cell in fact. Because of the Earth's rotation, a fluid particle undergoes an additional acceleration equal to twice the vector product of the angular velocity Ω, and the wind velocity of the particle, U. This is the Coriolis acceleration, which results in a deflection of the particle trajectory. The same mechanism operates in the large-scale motion of the Earth's atmosphere, but because of the curvature of the Earth's surface the dynamical effects are much more complicated. The direction of the Coriolis force is perpendicular to the axis of rotation. At the equator it is perpendicular to the surface of the Earth so its horizontal component vanishes, and therefore produces no meteorological Coriolis effect. Away from the equator a horizontal component does exist which produces a Coriolis effect which increases with latitude. In general the horizontal Coriolis force tends to balance the horizontal component of the pressure gradient. Horizontal displacements of the atmosphere are referred to as geostrophic motion, and the above balance is called geostrophic equilibrium (by analogy with hydrostatic equilibrium, in which a vertical pressure gradient balances the Earth's gravity). When this hypothesis is taken into account in dynamical theories of the atmosphere (geostrophic approximation), a close correlation between the distributions of wind and pressure shows up in the atmospheric circulation described by these models. Since this circulation refers only to long-term average displacements it is called general circulation.

This figure represents a general circulation with wind distribution forming six systems: north-east and south east trades, middle latitude winds of the western sector, and the winds of the eastern sector of the polar regions. The systems are separated by the 'intertropical convergence' (equatorial regions of low pressure or 'equatorial depressions'), and by the two extratropical convergence zones (low pressure zones around 60° latitude in each hemisphere).

For the planet's energy balance sheet, atmospheric circulation compensates by moving energy towards the winter pole. This imposes a meridian circulation shown in the diagram: six cells which ensure this transport by horizontal currents requiring in turn vertical currents to complete the cycle. There are two equatorial cells with direct circulation (Hadley cells) and two cells with inverse circulation (Ferrel cells) and also two polar cells with direct circulation. This model, too schematic in view of recent observations, can only be applied to transient periods around spring and autumn. Ferrel cells must be replaced by a meridian circulation at planetary scale regulated by a complex system of pressure waves, especially during winter; cyclonic and anticyclonic depressions, largely contributing to horizontal heat transfer, are caused by these waves.

An extratropical perturbation. Synoptic-scale extratropical perturbations have similar origins to large-scale perturbations at middle latitudes: cyclones on the polar side and anticyclones on the equatorial side of the westerly air stream. Superimposed on the Rossby waves, smaller synoptic-scale waves can be observed which also have a similar cause. Air masses which are thermally distinct are separated by frontal surfaces. For example in the northern hemisphere in winter the Atlantic polar front divides the polar air mass controlled by the large Icelandic depression and the tropical air mass controlled by the anticyclone centred on the Azores. It is along such fronts that the synoptic oscillations of the westerly air stream encounter the strongest vertical wind shear. This is produced by meridional thermal gradients which become important at middle latitudes. In the presence of the baroclinic instability so produced, the oscillations become amplified and distort the cloud front. This can then wrap itself in a spiral round a vortex which gives rise to the cyclonic system which is an extratropical perturbation.

A rapid westerly air stream from the Pacific to Europe draws a succession of cyclones which form over North America and become fully developed over the warm waters of the Gulf Stream. The diagram shows the tail of the preceding perturbation on the right, and on the left the first cirrus clouds of the following one. In the centre is a fully developed perturbation. The cirriform and stratiform clouds characteristic of warm air are arranged in spiral bands around the vortex while the cold air penetrates under the warm air to its south and west. The terrain of Canada and the United States is covered with snow up to the 32nd parallel. (Doc. DEEE, University of Dundee)

Large-scale movements in the Earth's atmosphere

The atmospheric motion which gives rise to the most rapidly varying cloud formations is that which occurs over mesoscale and synoptic scale distances. These motions are recorded by meteorological stations all over the world so that short-term forecasts may be made. They are however merely minor disturbances or perturbations which take place against the background of planetary-scale circulation of the atmosphere.

In order to make long-term forecasts and lay the basis for climatology it is essential to be able to describe atmospheric movement on the very largest scales and to construct a model of planetary circulation. Moreover, it is just those synoptic cloud systems which can easily be identified, such as cyclones and jet streams, which are determined by this planetary circulation.

Complex mathematical models are needed to describe the average movements constituting the general circulation, which is strongly influenced by the Earth's rotation and in which horizontal motion is much more important than vertical. Because the Earth drags the atmosphere up to a high altitude round with it horizontally, the horizontal component of the Coriolis force is very important.

The planetary circulation exhibits a marked annual variation. In summer, the meridian circulation is disorganised and weak at all latitudes. In addition the planetary circulation undergoes various changes of which the most rapid have the time-scale of only a day. In spite of these fluctuations, cloud formations on a planetary scale can be linked directly to certain trends in the circulation as a whole. Three types of cloud band can be identified: namely, the extratropical bands associated with cyclones which appear along the polar front, intertropical cloud bands near the equator and cloud bands of the upper subtropical troposphere.

The extratropical bands have been the subject of the most intensive study. In fact it is in the temperate zone of each hemisphere that the general circulation shows up particularly clearly as a west–east flow. These westerly currents each carry several large-scale undulations varying between three and eight, which are called planetary or Rossby waves, after the meteorologist who showed, in 1939, that they are the consequence of the dynamical stability of the currents. As they travel these planetary waves can, on encountering certain meteorological conditions, split up and give rise to secondary circulations. This is how Rossby explained the origin at high altitude of the large-scale atmospheric motions such as anticyclones and large depressions. The crucial conditions, which are referred to as baroclinic instability, arise when there is a temperature gradient across the horizontal (north–south) oscillations, which is so large that it gives rise to a vertical shear in the zonal wind pattern.

The same theory may be applied to extratropical perturbations on the synoptic scale.

The tropical cyclone Gloria. The picture above, transmitted by Meteosat 1 on 13 September 1979, is the last of a sequence following the formation and evolution of a cyclone called Gloria. The cyclone is at its peak and the eye shows clearly at the centre of the tropical cloud mass. It was the first to appear in the Atlantic in 1979. Gloria arose from a perturbation above the west African coast on 3 September. Tropical cyclones arise between 5 and 20 degrees of latitude on either side of the equator. They only attain the full strength of a tempest (cyclone, typhoon or hurricane according to region) over the ocean. (Atlas Photo/ESA Meteosat)

Formation of vortices. A geostationary satellite like Meteosat is particularly suited to the analysis of meteorological events because its wide-angle pictures provide a synoptic-scale map of great value to forecasters. The picture opposite, taken in visible light, shows cloud formations surrounding a vortex. A large cloud band can be seen leaving Ireland and splitting into narrower strips at its western edge. It is a continuous cloud formation about 4000 kilometres long and 400 kilometres wide. Another cloud band encircling a large part of the Iberian Peninsula has a more clearly pronounced curvature. (Atlas Photo/ESA Meteosat)

Perturbations and variations of the general circulation

Cyclonic systems and jet streams show up as perturbations of the general circulation. They may be distinguished according to the type of cloud band which they perturb. Thus when an isolated extratropical cloud band wraps itself in a spiral round an eddy, an atmospheric wave is produced at the boundary between the polar and tropical air – that is, on the polar front. Meteorologists think that Rossby's explanation of the formation of large depressions and anticyclones at moderate latitude is equally applicable on the small scale. When a baroclinic instability appears the atmospheric wave can grow extremely rapidly and give rise to cyclonic activity. Satellite observations are particularly useful for following the evolution of this mechanism right up to the formation of the extratropical cyclone.

In the same way perturbations of the inter-tropical cloud bands lead to tropical cyclones. These disturbances, infrequent but the most violent of all, appear in each hemisphere almost exclusively in the hot season, and nearly always over a region of the equatorial sea where the temperature of the surface water is greater than 26 °C. These are on the one hand the typhoons of the west Pacific and the Indian oceans, and on the other the hurricanes of the west Atlantic above the equator and the eastern Pacific along the coast of Mexico. Finally jet streams are regions of violent winds in the high subtropical troposphere. In their vicinity the vertical and horizontal air movements have a readily detectable affect on the appearance of high-level clouds.

All these atmospheric movements are on the synoptic scale, of the order of 100 kilometres. There are also numerous other cloud formations associated with air movement on the synoptic scale such as anticyclones, vortex formations and frontal clouds.

The undulations of the westerly currents of the general circulation, Rossby waves, are not the only large-scale oscillatory motions. The Earth's atmosphere as a whole shows oscillations which are approximately periodic, the predominant motion being semi-diurnal. Such oscillations are very weak below 50 kilometres and generally have a negligible effect. They have periods of the order of an hour while Rossby waves have periods of the order of a day. The amplitude of the semi-diurnal oscillation depends only on latitude being a maximum at the equator and diminishing towards the poles. The amplitude is approximately proportional to the cube of the distance from the Earth's axis.

The analogy with the semi-diurnal tides in the oceans produced by the Sun's gravity is obvious. These air movements are often referred to as atmospheric tides. As suggested by this analogy there is a corresponding lunar effect but it is very much weaker, the opposite of the situation in the case of ocean tides.

The theory of the solar tide in the atmosphere shows that it is a thermal wave due to the diurnal temperature variation of the air near the stratopause, hence the name 'thermal tide', which is occasionally used. Above 30 kilometres its amplitude grows until at 90 kilometres it is a hundred times greater than at the surface of the Earth. In equatorial regions the variation of pressure at sea-level follows the semi-diurnal thermal wave, and has a maximum two hours before the Sun crosses the local meridian. This timing of the wave gives rise to a very weak acceleration of the Earth's rotation. In the case of Venus, such atmospheric tides have resulted, over a long period of time, in the overturning of the axis of spin thus explaining that planet's retrograde rotation. In the case of Mars, astmospheric tides may be a contributory factor to the process which triggers off intermittent dust storms of a planetary scale.

A complex frontal system over the North African continent, pictures taken by satellite NOAA–7 in visible light and infrared. This is a typical satellite picture, taken on 17 February 1982, and shows a frontal band. Images obtained using visible light alone are normally insufficient to distinguish between different types of cloud. Used in conjunction with infrared pictures, however, they reveal the multilayered structure of the band and identify the active zones. The two pictures were taken simultaneously at 13.18 Universal Time. Note that on the infrared picture the African continent appears darker than the sea because of the lower temperature of the latter. This would not be the case at greater latitudes were the land areas are cold. (Doc. DEEE, University of Dundee)

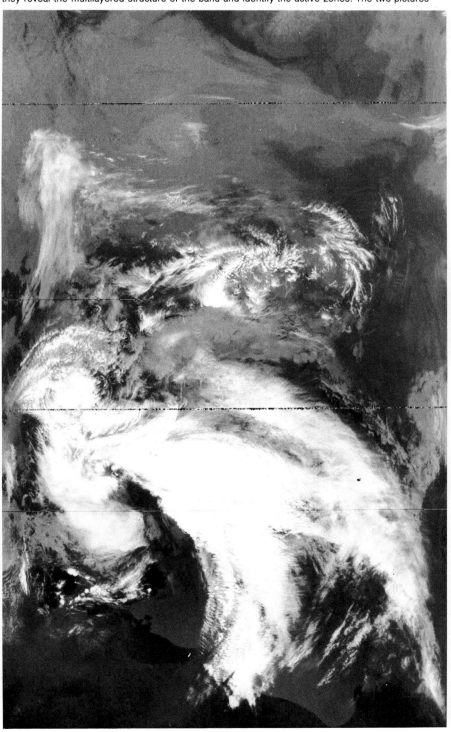 

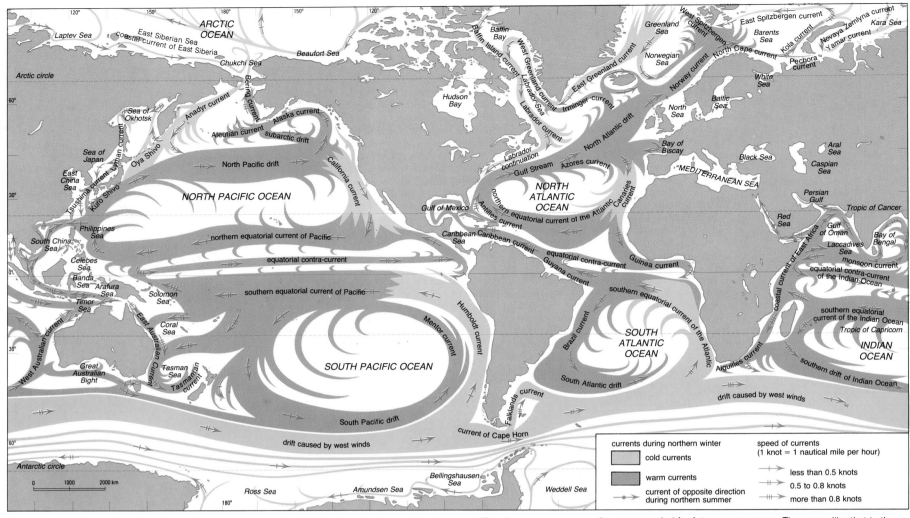

Legend:
currents during northern winter
☐ cold currents
☐ warm currents
→ current of opposite direction during northern summer

speed of currents
(1 knot = 1 nautical mile per hour)
→ less than 0.5 knots
⊣⊢ 0.5 to 0.8 knots
⊣⊢ more than 0.8 knots

The general surface circulation. Part of the energy accumulated by the atmospheric winds in the form of kinetic energy is passed on to the seas. This gives rise to the marine currents of the general circulation and to waves. Two other forces besides those of the winds perturb the sea: terrestrial gravity and the Coriolis force. Under the action of gravity alone, the water of the seas would be at rest. The form of the surface which would be produced if the oceans were at rest upon the globe is called the *geoid*. Its contours have been precisely measured by the radar of the oceanographic satellite Seasat. If, by the way, one measures directly the three components of the vector describing the large-scale motion of the seas, one can estimate the size of the deviations of the ocean surface from the geoid. The deviations determined experimentally are very small, at most 1 or 2 metres. Although measuring these deviations is very difficult, knowledge of them is of great interest for oceanographers because it enables them to analyse the relative importance of the relevant mechanical forces and to calculate the large-scale trends for the circulation of the oceans. The resolution demanded by oceanographers is of the order of 100 km. This is far from being possible to obtain unless appeal is made to the most modern techniques of *teledetection* by

satellite; these are consequently recommended for future programmes. The maps, like that in the figure, that specialists can draw today only give a global view of the marine currents. The circulation is evidently linked to the dominant winds. The general currents of the oceanic circulation have a zonal direction; from east to west at low latitudes, they correspond to trade winds from tropical and subtropical regions. At higher latitudes, the marine currents are from west to east in conformity with the atmospheric circulation. In general, the circulation takes the form of large eddies, anticyclonic and cyclonic. However, they are not of equal intensity, nor distributed homogeneously. The speeds of the currents are greater and the widths of the currents are less in the west of the oceans than in the east. To quote just one example: the Gulf Stream, to the west of the North Atlantic, along the East coast of the United States. Besides, the currents of the Gulf Stream, in transporting the warm waters of the equatorial currents, counterbalance the role of thermal sinks played by the polar and subpolar regions. The effect is appreciable upon the meteorology of the British Isles where changes are frequent.

Utilisation of aperture synthesis radar. Oceanographers expect that knowledge of ice floes will be considerably expanded by satellite radar techniques. This is not only important for maritime safety. Each year ice covers an area four times that of the American continent. The effect on the Earth's albedo is so marked that the importance of ice cover for the climate cannot be doubted. The rapidity with which the ice cover varies means that it must be continually monitored however cloudy the atmosphere. Radar imaging makes this possible. The best results are obtained by aperture synthesis radar (ASR) such as that used by Seasat. Like all radar, ASR uses the Doppler effect and the transit time of the signal to fix the position of the target, the final image being reconstructed numerically. The originality of ASR is its use of the motion of the satellite to scan a strip of fixed width parallel to the orbit. The coherence of the radar beam makes it possible to synthesise echo

signals numerically thus effectively increasing the area of the antenna. ASR is just as useful in oceanography for measuring wave heights and interference phenomena in coastal regions.

The photograph shows a radar image obtained by these means from the orbiter Columbia during a preparatory mission of the Space Shuttle in November 1981. According to the Jet Propulsion Laboratory who constructed the radar system, the effective width of the antenna is 2.5 kilometres as opposed to 9 metres apparent width. The picture, taken on 14 November at 3.00 a.m. local time, shows a region of the Mediterranean of 50 × 120 kilometres, to the east of Sardinia. The resolution, of the order of 40 metres shows up the white horses generated by the winds. It is comparable to resolution in pictures obtained with visible light by Landsat whose resolution is 80 × 55 metres. (M. Kobrick, JPL/NASA)

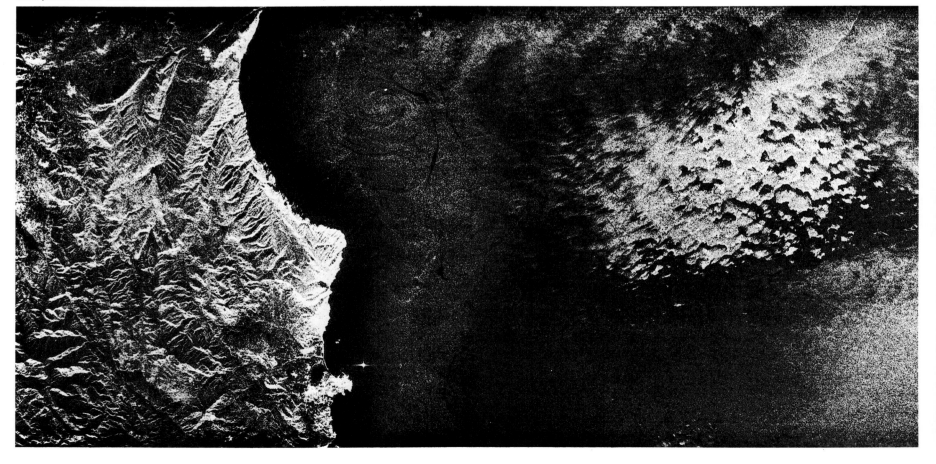

Global distribution of three parameters characterising the inter-relation of ocean and atmosphere. The ocean surface of our planet covers 360 million square kilometres. The new satellite observation techniques gives us a global view of the fluid system of the Earth: the general circulation of the oceans, the Gulf Stream, etc. During the last twenty years these satellites have been equipped with passive instruments, such as television cameras, visible spectrum radiometers (such as the Landsat multispectral scanner (MSS)) or infrared radiometers (such as the high resolution infrared radiometer (HRIR) of the Nimbus 2 satellite) and more recently with radio frequency radiometers (such as the scanning multifrequency microwave radiometer (SMMR) on Nimbus 7). The novelty of the latest NASA observation satellite, Seasat, launched in 1978, is that it was directed towards measuring ocean characteristics by means of radio frequency techniques, either passive (SMMR) or active (altimetry and radar). These techniques allow the problem of cloud cover to be overcome. Seasat's mission was curtailed by a short circuit and lasted only three and a half months. Nevertheless it strikingly demonstrated what could be done. A group working at the Jet Propulsion Laboratory in California used the results of Seasat's radar altimeter and scanner to draw up the three maps on the right. They illustrate the global distribution of atmospheric water vapour, surface wind speed and wave height. The highest concentration of water vapour is in the tropics but also unexpectedly over the Antarctic, presumably because of the sublimation of ice. Very fine altimetric measurements of the order of 5 centimetres are necessary to determine the speed of the winds above the oceans (with a precision of 2.5 metres per second), and to thus deduce the height of the waves and the currents. Map 2 shows wind speeds; most of the winds are well known such as the sub-tropical trade winds separated by the equatorial calm. The strongest winds are in the North Atlantic. Map 3 provides a global account of wind-induced oscillations – the first ever measurements for the southern hemisphere in winter – and thus demonstrates the power of 'active' radar observations. In the western Atlantic and in the Pacific wave heights are usually less than 1.5 metres but may be as much as 5.5 metres to the south-west of Australia. (JPL/NASA)

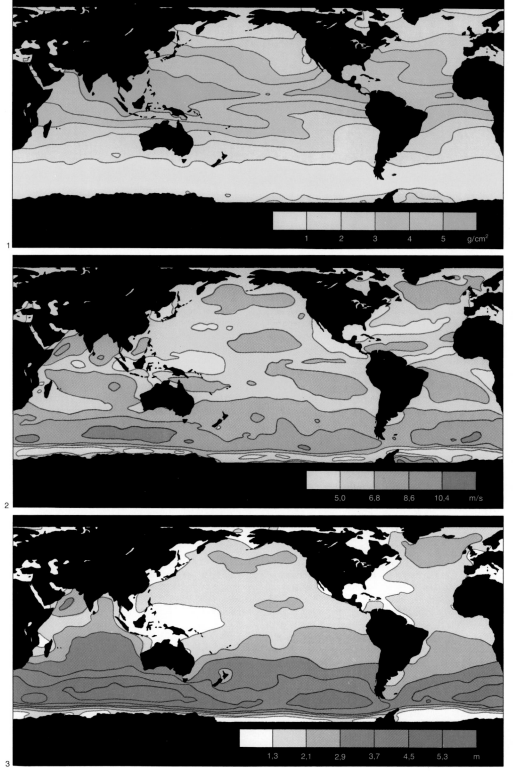

Interaction between the atmosphere and the oceans

Unlike the oceans which receive their thermal energy from above, the atmosphere receives the bulk of its energy from below. As a result, the atmosphere exists in a state of very active convection. In fact the kinetic energy inherent in wind motion is an important part of the energy available in the atmosphere. It is from this source that the oceans draw the energy necessary to sustain marine circulation, and in particular the Gulf Stream and the equatorial currents.

As a result of many satellite observations of the Earth from polar orbits there is sufficient information to determine the local net balance of radiation at the top of the atmosphere and to record its variation with latitude and time of year. The net radiation (measured in watts per square centimetre) is the result on a local scale of the balance between energy sent in by the Sun (mainly in the visible range), and the sum of the reflected (or albedo) radiation and the thermal radiation emitted by the Earth and atmosphere (in the infrared). There is a marked annual variation, but on a global scale and in the long term the net radiation must be zero, otherwise the Earth would progressively heat up. A horizontal redistribution of the locally available thermal energy is therefore necessary. At the surface of the Earth, the oceans which are the best absorbers of the Sun's heat, are very effective in bringing about its redistribution horizontally. Because the upper layers are constantly stirred up by the winds, the ocean can absorb heat on a time-scale of the order of six months, without its temperature changing by more than a few degrees. The mixing of ocean water by the winds can be effective down to depths of a hundred metres.

Furthermore, there is an intimate connection between the cooling of the oceans by evaporation and the heating of the atmosphere by release of the latent heat of water vapour. As a result, the thermal stability of the oceans is imposed on the lower atmosphere whose temperature cannot depart from the surface temperature of the sea for any length of time. This influences the atmospheric circulation which is responsible for the transport of heat released by the surface or in the atmosphere. The ocean currents themselves transport a significant quantity of heat. Clearly then, the two systems, ocean and atmosphere, cannot be understood separately. For this reason the most promising techniques of analysis in climatology are those made possible by satellite observations, which provide a global picture of the important atmospheric and oceanic parameters – surface winds, water vapour concentration, wave heights and the movement of ice floes.

Guy ISRAËL

The Earth

The magnetosphere and radio waves

The electromagnetic radiation and particles emitted by the Sun cause some of the atoms and molecules in the Earth's atmosphere to become ionised. X-rays and ultraviolet radiation are particularly effective. The free electrons and ions thus created diffuse throughout the space surrounding the Earth forming what can loosely be called the ionosphere.

In that part of the ionosphere nearest the Earth, that is at altitudes of less than a few hundred kilometres, the density of charged particles is less than the density of neutral atoms and molecules. The motion of the charged particles is therefore dominated by scattering and ionic recombination, and the Earth's magnetic field has only a secondary influence. At higher altitudes the atmosphere is completely ionised forming a plasma whose density decreases with distance. It has a maximum of several million particles per cubic centimetre at altitudes of 200 to 300 kilometres. This is the true ionosphere. The decrease is relatively slow out to distances of three or four Earth radii, where the density is about 100 particles per cubic centimetre. Beyond this distance the density falls sharply to values of about one particle per cubic centimetre. This boundary, the plasmapause, is the transition region beyond which lies the external magnetosphere where the motion of particles is greatly influenced by the variable effects of the solar wind on the Earth's magnetic field.

Below the plasmapause, the plasma has a low average temperature, maintained by the solar radiation. By contrast the external magnetosphere is mainly populated by high-energy particles which have undergone substantial acceleration in certain parts of the magnetosphere, in particular the tail and the auroral regions. Here, therefore, one is dealing with a hot plasma.

The first studies of the magnetosphere were carried out on the basis of measurements of the magnetic field at the surface of the Earth. As is well known, the motion of ionised particles in the magnetosphere and ionosphere give rise to electric currents. They modify the static dipole field of the Earth. The variations so induced are known as magnetic storms and can easily be detected at ground level by means of magnetometers.

There are various types of magnetic storm corresponding to the different kinds of electric current in the magnetosphere. In principle a knowledge of the variations of magnetic field over the whole surface of the globe, and of the physical conditions of the high atmosphere are sufficient to determine the electric currents. In practice, the measurements are often incomplete. So, although the study of magnetic storms has provided a great deal of useful information particularly on the internal magnetosphere, it is space techniques which have, over the last twenty years, enabled a precise picture to be drawn up.

In fact the magnetosphere is easily accessible to satellites which can measure, *in situ*, the characteristics of the plasma: density, degree of ionisation, composition and energy distribution of the particles, etc. By these means it has been shown that the most important currents occur in the ionosphere where the charged particles are intermingled with the neutral atmosphere. There are also currents both within the external magnetosphere, and between it and the ionosphere. Knowledge of them is essential for a good understanding of particle dynamics in these regions.

Three regions are of particular interest:
– the interface with the solar wind (shock waves, magnetic sheath, magnetopause), where it is abruptly slowed down, heated up and deflected from its original trajectory by the Earth's magnetic field;
– the plasmasphere comprising the ionosphere and the internal magnetosphere;
– regions of high particle acceleration, that is the magnetospheric tail and the auroral regions.

The auroral regions lie, one in each hemisphere, between latitudes 60 and 75 degrees and play a special part in the running of the magnetospheric 'machine', that is, in the formation and discharge of electric currents between different regions of the magnetosphere. The appearance of visible aurora, which also emit in the ultraviolet, shows up the presence of powerful electric currents comprising highly accelerated downward moving particles. The auroral regions are also sources of intense low-frequency radio emission: terrestrial kilometric radiation. This is further evidence that these regions are traversed by jets of ionised particles which travel down the magnetic lines of force from the outermost parts of the magnetosphere, particularly the tail. These showers of particles occur mainly when variations in solar activity give rise to changes in the solar wind together with consequential disturbances in the Earth's magnetic field.

André BOISCHOT

Polar aurorae. These spectacular phenomena occur in two 'auroral arcs' lying between 60 and 75 degrees of latitude in both hemispheres. At times of heightened solar activity they can be observed at much lower latitudes.

There are several types of aurorae of various forms and colours. The most spectacular are the sheet aurorae which result from the discharge of electrons along the magnetic field lines which link the tail of the magnetosphere to these regions of high latitude. These large aurorae are associated particularly with big eruptions in the solar chromosphere. One of the effects of these eruptions is to increase the intensity of the solar wind which disturbs the Earth's magnetic field thus bringing about the acceleration of electrons in the magnetospheric tail through a series of complex mechanisms. These are the particles which descend on the auroral zones exciting certain molecules in the high atmosphere thus giving rise to these luminous phenomena. The photograph shows a magnificent aurora borealis observed above a forest near Fairbanks, Alaska. (© Jack Finch/SPL/Cosmos)

The Earth's magnetic field and the Van Allen belts. In the first approximation the Earth has a dipole magnetic field, that is its magnetic field is like that of a bar magnet with its south pole in the northern hemisphere. This field originates in the movements of rotation and convection of the Earth's fluid interior, which is a conductor. The axis of rotation and the magnetic axis are inclined to one another at an angle of about 11 degrees so the magnetic and geographical poles do not coincide. Moreover, the magnetic poles are not exactly stationary. Studies of the magnetism of rocks have shown that the magnetic field has reversed several times in the course of some ten thousand years.

The dipole field has superimposed on it other components of lesser strength arising from local 'anomalies' due to the permanent magnetism of certain rocks and from magnetic storms due to electric currents in the ionosphere and magnetosphere.

The first satellite launched by the Americans to study the Earth's environment, Explorer 1, discovered regions of space which contained concentrations of high-energy particles: electrons with several hundred kiloelectronvolts to several megaelectronvolts and protons of up to 100 megaelectronvolts. These particles are trapped in a region of the magnetic field near the Earth, extending for five or six Earth radii in the equatorial plane, the field being sufficiently strong for the containment to be stable. The precise extent of the regions depends on the energy of the particles.

Two main theoretical problems are posed by the existence of the Van Allen belts are how the particles got there (very likely as the result of an acceleration during a magnetic storm) and how they disappear (by interaction with waves and particles and by collision with neutral molecules in the auroral regions of the lower atmosphere).

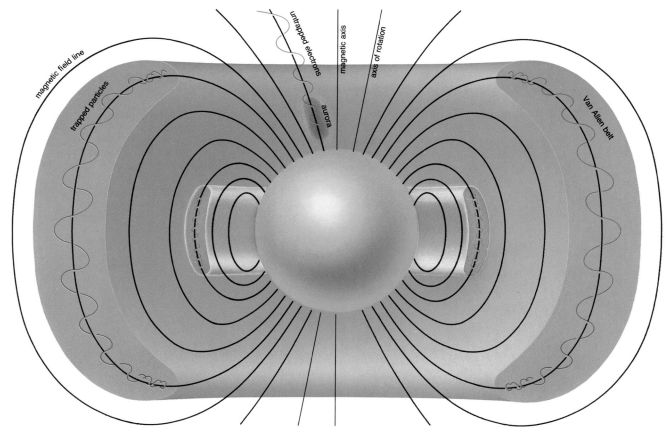

The Earth's magnetosphere. The Earth's dipole field and dense atmosphere give rise to a well-developed magnetosphere surrounding the Earth. The various regions identified in the diagram are:
– the shock wave which accompanies the Earth as it travels supersonically through the solar wind; its peak is at about 12 Earth radii in the direction of the Sun;
– the magnetopause, the effective boundary to the range of influence of the Earth's field; its peak is at about 10 Earth radii in the direction of the Sun and in the opposite direction it forms a long tail of plasma;
– a region of unstable confinement in which charged particles follow the magnetic field lines but frequently make transitions to neighbouring regions, particularly in the 'polar horns' through which the particles of the solar wind have direct access to the internal magnetosphere;
– a zone of stable confinement (Van Allen belt) extending up to 5 Earth radii on the equatorial plane.
 The diagram shows on the same scale the magnetosphere of Venus. The magnetosphere of Jupiter is a hundred times, and that of Saturn twenty times, the size of the Earth's.

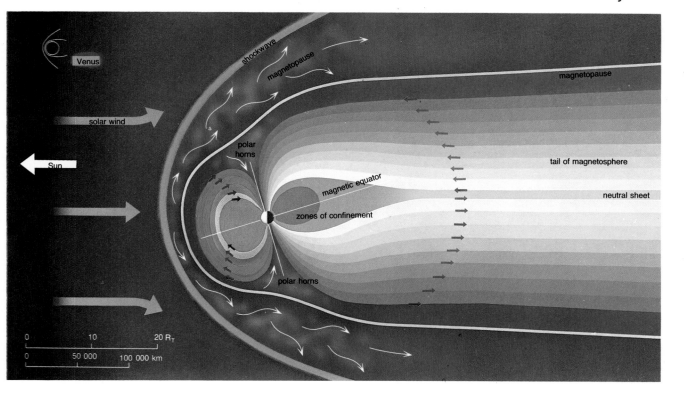

NASA program Dynamic Explorer. In 1980 NASA began the Dynamic Explorer program aimed at studying the dynamic structure of the Earth's magnetosphere with two satellites. One, in a highly eccentric orbit, was able to take these photographs of the Earth using light in the oxygen lines at 130.4 and 135.6 nanometres.
 The auroral arc of the northern hemisphere where the electrons descend, is clearly visible. The picture on the far left shows two regions of high intensity, one on the night side and one on the day side. The first corresponds to electrons coming from the neutral tail of the magnetosphere and the second probably corresponds to solar particles which penetrate directly to the lower atmosphere through the polar horn. These pictures are of a calm day. During a disturbance of the magnetosphere following solar activity, the auroral arc is more spread out and more complex. (NASA/NSSDC)

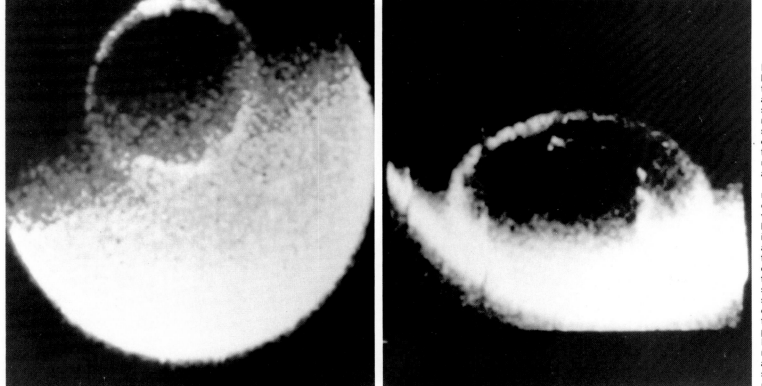

Electromagnetic radiation from Earth. Like Jupiter and Saturn, the Earth is an intense source of long (kilometre) wavelength radio waves. These waves were discovered by satellites above the ionosphere. They cannot penetrate the ionosphere and so cannot be observed at ground level.
 This radiation is intermittent and intimately connected with polar aurorae. Its source lies in regions of high altitude with a position determined by the Sun and having a maximum in the quadrant 18.00 hours to 0 hours. Its intensity increases with the occurrence of polar aurorae since the particles which are accelerated in the tail of the magnetosphere and precipitated onto the auroral zones, cause both phenomena. The mean intensity of radio waves emitted by the Earth in the kilometric range is 10^8 watts and hence several orders of magnitude greater than radio broadcast transmitters. The figure shows a meridional cross-section of the magnetosphere with the position of the sources of the radiation and the regions where they can be observed. On the right is a diagram illustrating the fraction of time during which a terrestrial satellite experiences kilometric emission as a function of distance. The appearance of a maximum at 18.00 and 0 hours is obvious. The three shades correspond to three levels of intensity. (D. L. Gallagher & D. A. Gurnett, 1979, and J. K. Alexander & M. L. Kaiser, 1976)

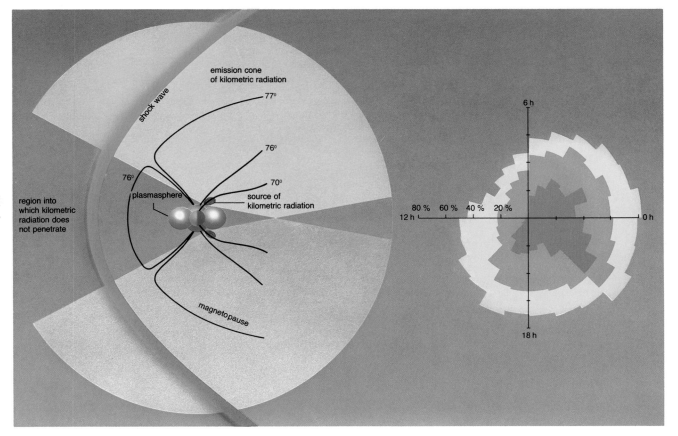

The Earth–Moon System

The Moon gravitates around the Earth, of which it is the only natural satellite, at a mean distance of 384 402 kilometres. Its mass, 7.4×10^{22} kilograms, is 1.23 per cent of the Earth's mass; its equatorial radius is some 1740 kilometres. The mean density is 3.34 grams per cubic centimetre and the acceleration due to gravity on its surface, 162 centimetres per second per second, is low compared with that on the surface of the Earth (981 centimetres per second per second).

From the distant past, down through the centuries, astronomers have always observed the moon. At first the purpose of their observations was more utilitarian than scientific; marine navigation in particular necessitated precise tables of the Moon's motion.

This motion is very complex since the Moon is not just subject to the gravitational pull of the Earth but also to that of the Sun and of other planets. If the Sun's influence did not exist and if the Earth were a homogeneous spherical mass then the Moon's orbit would be elliptical. As it is the orbit approaches ellipticity with an eccentricity of 0.054 and with the orbital plane inclined at approximately 5 degrees to the ecliptic. However, this ellipse is permanently deformed by the perturbations of the Sun and, to a smaller degree, by the flattened form of the Earth.

The principal perturbations were known in ancient times, long before the advent of the telescope. They were first explained by Isaac Newton but it was mainly in the eighteenth century that the foundations of a complete lunar theory were first established by the works of Leonhard Euler, Alexis Clairaut, d'Alembert, Joseph Lagrange and Pierre Simon de Laplace. In the nineteenth century a great advance in lunar theory was made by Peter Hansen but a completely new theory of great mathematical skill and complexity – it contains several hundred terms each relating to separate periodic perturbations – was advanced in 1860 by Charles Eugene Delauney. In 1920 E. W. Brown, a young Englishman who settled in the USA, published his famous tables of the Moon which were based on his own theory involving 500 separate terms. The tables enable the position to be obtained without the enormous labour of computing each separate term. From 1924 to 1983 the definitive positions of the Moon as published in astronomical almanacs and ephemerides were computed from Brown's Tables.

At the end of the seventeenth century, following a long series of observations of the rotation of the Moon, Giovanni Domenico Cassini derived three empirical laws that bear his name:
– the Moon rotates with constant angular speed and with a rotation period of a sidereal month;
– this inclination of the axis of rotation to the ecliptic is fixed and equals 5 degrees 8 arc minutes;
– the planes of the lunar orbit and the lunar equator intersect in the ecliptic, this last plane being between the other two.

The first of these laws expresses the fact that the period of rotation of the Moon is equal to its period of revolution about the Earth. However, this only holds good on average. In the course of a revolution about the Earth the orbital speed of the moon varies slightly because of the eccentricity of the orbit; the rotation of the Moon is consequently sometimes in advance and sometimes in arrear of its orbital motion and this has the effect of revealing periodically portions of the Moon's surface that are not normally visible. This oscillation is known as geometrical libration in longitude. There is also a geometrical libration in latitude due to the fact that the axis of rotation of the moon is not exactly perpendicular to the plane of its orbit. Because of libration, up to 59 per cent of the Moon's surface is visible from the Earth.

In addition to the geometrical librations there are physical librations caused by oscillations in the rotation of the Moon arising from variations in the gravitational attraction of the Earth (associated with fluctuations in the lunar orbit) on the body of the Moon which itself is not perfectly spherical – being slightly flattened at the poles and having an elliptical equator. These deformations are attributed to asymmetric mass distribution inside the Moon. The variations in the attraction of the Earth on the equatorial bulge of the Moon are responsible for the physical librations. Observations of the motions of the Moon about its centre of mass reveal the physical librations and from these we can deduce the basic parameters of the internal global structure; these parameters are indicative of how mass is distributed within the body of the Moon. Even though they are difficult to discern because of their minute amplitudes, a knowledge of these parameters is of fundamental importance in investigations of the physics of the Moon's interior.

As a result of tidal action by the Earth on the Moon the same face of the Moon is always turned towards us; this action has slowed down the moon's rotation until its period is equal to that of its orbital revolution. The orbit has also been irreversibly affected by terrestrial tides: the tidal energy dissipated in the body of the Earth and in the oceans is responsible for the progressive increase in the distance of the Moon – the current rate of this increase is about 4 centimetres per year. The time-scale of the orbital evolution depends directly on the amount of tidal energy dissipated since, if this remained constant in the past and equal to its present value, it can be shown that the Moon was contiguous to the Earth 2 billion years ago. However, we know the Moon's age to be 4.6 billion years, its crust solidified at least 4.3 billion years ago and formation of the lunar seas occurred between 3.9 and 3.3 billion years ago. The hypothesis that the Moon has moved away from close proximity to the Earth over the past two billion years can therefore be rejected; such proximity would have generated tremendous tidal activity, traces of which should still be evident in both lunar and terrestrial rocks. In fact it is now known that the dissipation occurs mainly in the oceans and is critically dependent on the distribution of the continents on the Earth's surface. Over geological time this distribution has changed enormously so that the problem of the time-scale no longer seems insoluble.

Many suggestions have been advanced to explain the origin of the Moon. In fact, these fall into one of three categories each of which invokes a basic mechanism: fission, capture or accretion in orbit about the Earth.

Despite numerous space missions to the Moon, the origin of our satellite still remains an enigma. We are not yet in a position to adopt one or other of the mechanisms because of dynamical or chemical discrepancies. It remains quite possible that a variant of one of these scenarios will enable us to overcome existing difficulties.

Anny CAZENAVE

View of the Earth from the Moon. On 23 August 1966 the unmanned space probe Lunar Orbiter I transmitted this photograph, our first view of the Earth above the lunar horizon. The eastern coast of the USA can be seen in the upper left of the crescent; lower down the Antarctic is visible. Scarcely two and a half years later the crew of Apollo 8, were able to view this scene directly. (NASA)

The Earth and the Moon. This exceptional photograph of the crescents of the Earth and the Moon was obtained on 18 September 1977 by Voyager 1 at a distance of 11.66 million kilometres from the Earth, thirteen days after its launch on a voyage towards the outer planets. Everest is in darkness close to the terminator but on our planet the Pacific Ocean, Eastern Asia and the Arctic can be distinguished. The Earth has a much higher albedo than the Moon so that the image intensity of the latter was increased by a factor of three to render it clearly visible. (NASA/NSSDC)

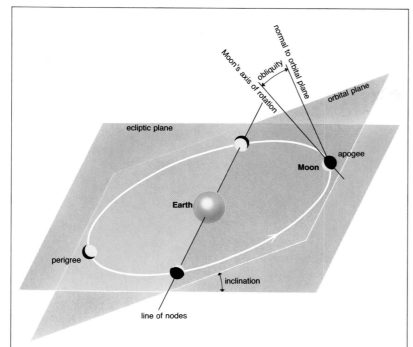

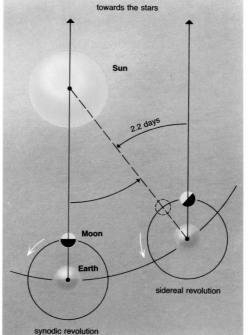

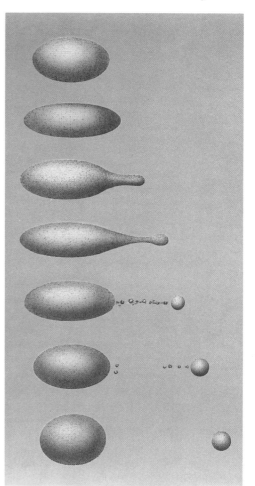

The Moon's orbit. The plane of the Moon's orbit is inclined at 5 degrees 8 arc minutes to the ecliptic (above left). The Sun perturbs the orbit; the principal effect of its action is to cause the line of apsides (the major axis of the orbit) to turn in the plane of the orbit, completing a revolution in 8.85 years. This motion, known as the advance of perigee, is similar to but in the opposite sense to the regression of the nodes which is the revolution of the line of nodes (the line of intersection of the plane of the orbit with the ecliptic) in a period of 18.6 years. The different motions of the Moon have several effects, particularly on its orbital period. Depending on how they are measured, there are several kinds of lunar month (above right). A month defined by the revolution of the Moon with respect to the stars is called a sidereal month, its duration is 27.32 mean solar days. In this time the direction of the Sun changes because of the orbital motion of the Earth so that the interval between two identical lunar phases, the synodic

month, is 29.53 days, longer than the sidereal month. A month may also be defined in terms of the time it takes the Moon to return to the same point in its orbit; this is the anomalistic month of 27.55 days, slightly longer than the sidereal month because of the advance of lunar perigee. Finally, because of the retrograde motion of the Moon's nodes, a month based on successive passages of the Moon through the same node is 27.21 days, shorter than the anomalistic month.

Furthermore, because of secular movements of the nodes and of perigee there are a large number of periodic oscillations in the orbit, due to the same causes. Among the most important of these is the evection, a fluctuation in the eccentricity consonant with the solar attraction and the variation caused by interference between solar and terrestrial attractions. Then there is the anomaly associated with the eccentric orbital motion of the Earth.

The librations. The three diagrams illustrate the libration in longitude (left), libration in latitude (top right) and diurnal libration (lower right) which causes us to see slightly different aspects of the Moon in mornings and evenings. The two photographs show the effect of libration in longitude: the apparent rotation of the Moon is due to changes in speed in the course of an orbit. These variations periodically enable us to see a small portion of the 'hidden' face. The photograph on right shows the face that is normally turned towards the Earth. (Lick Observatory)

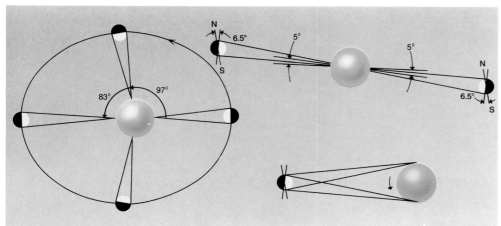

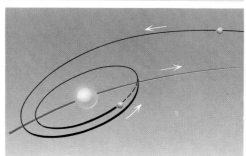

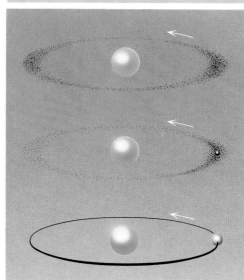

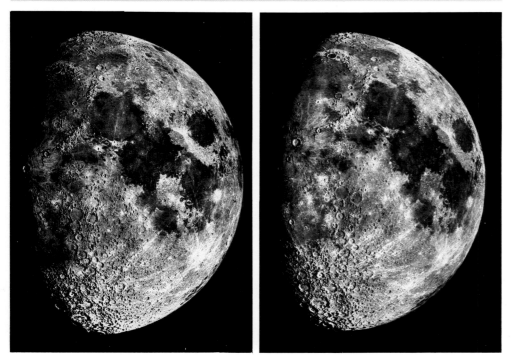

The origin of the Moon. Three hypotheses may be advanced to explain the origin of the Moon. The fission theory (top) supposes that shortly after its formation the Earth was rotating very rapidly (in a period of 2 to 3 hours), in consequence of which polar flattening developed, to such a degree that the equatorial bulge became unstable and broke up, allowing material from which the Moon is formed to escape. This theory has the advantage of explaining the chemical differences between the Earth and the Moon, but it clashes with dynamical theory. Fission would require the angular momentum of the Earth–Moon system to be much greater than its current value, yet there is no way of explaining how it could have decreased in the past.

The capture theory (centre) suggests the Moon was formed in some distant part of the Solar System and that it happened to approach close enough to the Earth to be captured by it. This theory also poses difficulties: the probability of such an event is very small especially since the transformation, at the moment of encounter, of the Moon's parabolic orbit about the Earth into a geocentric elliptical orbit would require an enormous retardation in orbital speed. No known mechanism is adequate for this.

There remains the possibility that the Moon was formed at the same time as the Earth by the accretion of debris and dust gravitating about the Earth (bottom).

The Moon

It has been known since 1609, when Galileo first examined the sky with a telescope, that the Moon is a solid body on the surface of which plains and mountains, valleys and peaks can be seen. In the eighteenth century there were wild claims of sightings of settlements. Improvements in telescopes proved these to be false. The surface features of the Moon have now been explored in much detail with very large modern telescopes which have a resolving power equivalent to 200 metres at the distance of the Moon. Telescopic observations made on the Earth first revealed the dual nature of the lunar landscape and discerned the innumerable craters. Up to 35 per cent of the visible face of the Moon is quite dark and smooth in appearance; since the seventeenth century regions of this type have been known as 'seas' (*maria* in Latin), although we now know they contain no water. The other 65 per cent of the surface is brighter terrain that is clearly mountainous and, by contrast with the seas, this type of surface was divided into 'continents' that are usually called highlands. Both seas and continents are riddled with circular depressions called craters; these are surrounded by mountains that are sometimes very high.

Although the lunar topography has been charted for a long time experts could not agree on the origin of the craters until the middle of the twentieth century. But from the 1960s, with the advent of space exploration, enormous advances have been made in our knowledge of the moon. Mention of only the more important stages in the exploration can be included here. In September 1959 a Russian probe, Luna 2, crashed on the far side of the Moon. A month later, Luna 3, sent back to Earth photographs of the hidden face. In 1964 and 1965 probes of the American Ranger series relayed detailed pictures as they approached the surface in crash landings. Then in

February 1966 Luna 9 made a soft landing following which it sent back photographs of its immediate surroundings. Between August 1965 and August 1966 a systematic photographic survey, with a resolution of about 2 metres, of the whole of the lunar surface was completed by five American satellites known as Lunar Orbiters. From 1968 to 1972 nine manned American space craft, Apollo 8, 10, 11, 12, 13, 14, 15, 16 and 17 travelled in different orbits around the Moon, improving photographic coverage and increasing our geophysical knowledge of it. Geological analyses of its surface soils were made following the six landings made by Apollo 11, 12, 14, 15, 16 and 17 as a result of which a total of 382 kilograms of lunar samples were brought back to Earth. Between 1970 and 1976 the Soviet Union also placed satellites in lunar orbits and landed three unmanned vehicles, Lunakhods, to explore the surface. These also returned with lunar samples, but only 0.3 kilogram. It was the Lunar Orbiter missions and above all Apollo which have contributed the most to our knowledge of the Moon.

The most prominent features in all lunar photographs, whether obtained by terrestrial telescopes or by spacecraft are the craters which are found on the hidden and visible faces. As on most planets and satellites the craters vary greatly in form; the smallest, those with a diameter less than 10 kilometres are known as simple craters – they are shaped like a bowl and surrounded by a ring of ejected debris. Complex craters with diameters in the range 20–200 kilometres are completely different morphologically. The floor is flat except for a peak at the centre and the inner side of the surrounding ring of debris is stepped or terraced. The ring is quite substantial and is often perforated by smaller craters lying on lines radiating from the centre. Then there are craters

The Moon viewed from the Earth. This photograph of the Moon at first quarter was taken from an observatory on Earth; it is representative of the state of our knowledge of the Moon before its exploration by automatic probes and the Apollo missions. It shows clearly the difference between the maria, the dark terrains which are flat and tend to be circular in shape, and the highlands which are bright and mountainous. The grazing illumination close to the terminator highlights the large number of craters in the highlands as well as their characteristic morphology. The two quasi-circular maria situated in the upper half are Mare Serenitatis, close to the terminator and Mare Crisium, near the limb. The somewhat irregularly-shaped maria in the centre are, moving downwards, Mare Tranquillitatis, where the first astronauts landed, Mare Fecunditatis and Mare Nectaris. In the eastern extremity Mare Marginalis (top) and Mare Symthii may be discerned. These differences in morphology have long been recognised and even a century ago very good maps of the visible face of the Moon were available. (Lick Observatory)

The region of the crater Euler. This photograph, taken during the Apollo 17 mission, shows the crater Euler, located at about 25° north and 29° west together with other smaller craters and the maria that surround them. North is at the top. The 'small' crater between Euler and the bottom of the photograph has a diameter of 4 kilometres. It is a typical simple crater; it has a perfect bowl form and is surrounded by a ring of debris. The crater Euler, 25 kilometres in diameter, has a more complex form; its internal slopes exhibit terraces and steps. At the centre there is a peak. The bottom between the terraces and the peak is almost flat. Euler is surrounded by a large ring of debris thrown out from the centre – the ejecta – which are perforated in places by chains of small secondary craters caused by large debris originating in the principal crater. A few highland 'islands' rise above the almost uncratered plain of the surrounding mare, Mare Imbrium. (NASA/NSSDC)

The crater Langrenus. Located at about 8° south and 61° east, Langrenus has a diameter of 140 kilometres. With its very extended terraces, its flat bottom and fairly complicated central peak, it is a perfect example of a complex crater. Older than the crater Euler, Langrenus is slightly more delapidated; in particular, its ejecta have a less characteristic morphology. (Apollo 8 photograph; NASA/NSSDC)

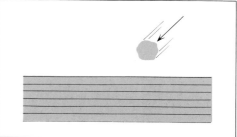

The formation of impact craters. The series of four diagrams illustrate the formation of impact craters on the Moon as well as on other planets and their satellites. The different kinds of crater are not drawn to scale.
1. Craters with a diameter less than 10 kilometres are described as simple. They are shaped like a shell-hole or a bowl, with a depth of between a fifth and a tenth of the diameter.
2. and 3. Craters with a diameter between 20 and 150 kilometres have complex properties. Rebound phenomena lead to the formation of a central peak which can be somewhat complex where the impacts are capable of forming craters of diameter greater than 30 kilometres. Widespread slipping at the crater rim creates terraces below the rim inside the depression.
4. Craters with diameters exceeding 200 kilometres are known as basins. The primary rebound is large enough to provoke secondary rebounds that transform the central peak, formed initially, into a ring and lead to the formation of one or more concentric structures.
(After J. B. Murray, 1980)

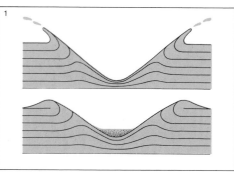

The Moon as seen from Apollo 16. The crews of the Apollo missions were able to photograph the Moon from unusual angles. In this picture, obtained by Apollo 16, North is at the top, but only a third of the image, on the left side, is visible from Earth. Mare Crisium, left, is recognisable as well as Mare Marginalis and Mare Smythii, close to the centre. The right of the photograph shows part of the hidden face, a feature of which is the almost complete absence of maria and the high density of craters of all sizes. (NASA)

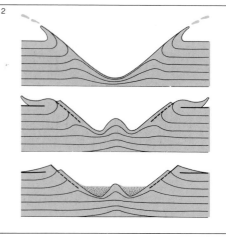

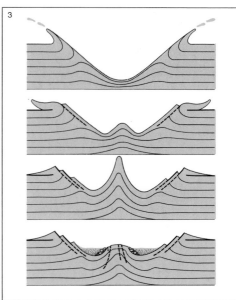

with diameters exceeding 200 kilometres, known as basins; they are like complex craters except in the centre where instead of a peak they have a small raised ring.

However, this variety of forms conceals a unity in origin, because all of the intermediate forms exist, passing gradually from the simple craters to the complex craters and to the basins. For a long time scientists were divided over the origin of these craters, some arguing that they were formed by volcanic activity, others holding

that they were created by the impact of meteorites. Comparisons have been made with terrestrial craters formed naturally by volcanoes and meteorites and with artificial ones created by explosions, including atomic bombs, and also with many kinds formed in laboratory experiments. These comparisons indicate that lunar craters were caused by the impact of meteorites and this evidence is supported by the presence of craters in regions of the Moon that are clearly free from volcanic activity and also by the discovery

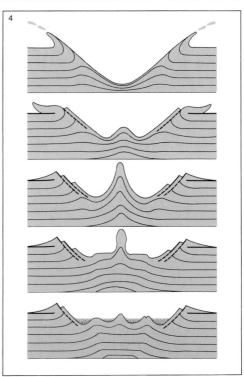

The Schrödinger basin. Situated close to the South Pole which is marked by the cross, the Schrödinger basin is an example of a typical basin with a central ring. Its diameter is 350 kilometres and that of the ring about 120 kilometres. The terraces are very well developed. The radial valleys were carved out by secondary projectiles ejected from the basin at a very low angle of incidence. Around Schrödinger there are a large number of ordinary complex basins. Note that in spite of its small size (140 kilometres, the same as Langrenus), the crater Antoniadi (marked by the arrow) has a central ring. The increase in the complexity of craters is not therefore uniquely related to their size; it depends on other factors which still remain something of a mystery. (Lunar Orbiter IV photograph; NASA)

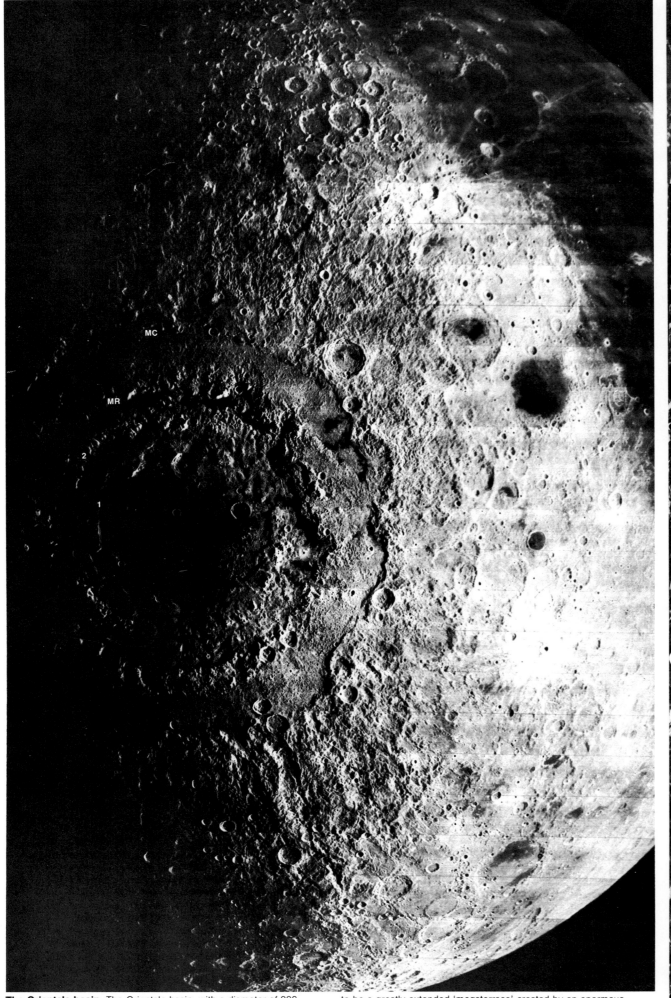

The Orientale basin. The Orientale basin, with a diameter of 900 kilometres is one of the two great lunar basins with multiple ring systems. Moving out from the centre, there are first the two interior rings (1, 2) which are equivalent to the central ring that is typical of less massive craters. At 600 kilometres from the centre are the Rook Mountains (MR) which seem to define the edge of the impact cavity. Beyond these are the Cordillera Mountains (MC) which would appear to be a greatly extended 'megaterrace' created by an enormous surface slip towards the inside of the basin. The basin is completely surrounded by ejecta that are very extensive, covering 'normal' highlands to the north and the south. The Ocean of Storms (Oceanus Procellarum) is the very dark area in the north-east. A few 'recent' craters are visible, inside the basin and in the ejecta. (Lunar Orbiter IV photograph; NASA)

Detail of ejecta from the Orientale basin (opposite). This mosaic of four Lunar Orbiter photographs shows details of features that are characteristic of the Orientale basin, that is, three of the concentric rings (1, MR, MC) and the ejecta. Just outside the edge of the original excavation (MR), the ejecta, having fallen almost vertically are chaotically strewn without any structure. Further out they have a well-defined radial structure with lines of hills and valleys. These terrains, known as a Hevelius formation, are created by the oblique fall of more-or-less molten debris, melted by the heat generated during the impact and splashed outwards in a blazing spray. The thickness of the ejecta continues to diminish further out, and to the south, at a distance of 1000 kilometres from the centre of the basin, the continental substratum becomes visible again. In places the ejecta have a transverse structure; these features are interpreted as accumulations of ejecta where obstacles impeded the normal flow of the 'flood'. (Lunar Orbiter IV photograph; NASA)

The Imbrium sculptures. The Imbrium sculptures (4° S, 3° W) are easily visible from Earth, with a telescope. They comprise a series of hills and valleys, almost parallel, directed towards the basin, but 600 kilometres south of its edge (which in this photograph is beyond the horizon and 300 kilometres from the foreground). The sculptured landscapes are attributed to the ejecta from Imbrium and should therefore be a somewhat delapidated equivalent of the Hevelius formation. Moving inwards from the sculptures towards the basin all traces of the ejecta disappear as they are covered by the maria: Mare Vaporum, Sinus Aestuum and Sinus Medii. Some craters, like Herschel, to the edge of the photograph, on the right, are superposed on the ejecta and were therefore formed subsequent to the Imbrium impact. Others have their edges 'sculpted' and consequently ante-date the basin. The antenna of the Apollo 16 craft, visible to the right of centre, should not be confused with a morphological feature! (NASA/ NSSDC)

on the spot of traces of the shock waves created in the terrain by the impact of meteorites. All the evidence shows conclusively that practically all craters were formed by meteorites; the peaks or rings in the centre are due to rebounding material following particularly violent impacts which also cause the surrounding rims to slip and form terraces.

The two largest basins, Orientale and Imbrium, exhibit the principal morphological conditions of the Moon. Their structure is unusually complex with three or four concentric features the origin of which remains a puzzle, and the ejecta from both these basins extend over several hundred square kilometres, lying in furrows and ridges running more or less radially from the basins.

The variety of basins and craters is enhanced by the obliteration of some by later impacts, subsequent to their formation, which completely changed their original appearance. Such obliteration is of importance in establishing the sequence in which craters were formed; obviously the newer craters are less affected by subsequent meteoritic bombardment than the older ones. Also, the most recently formed craters have systems of bright rays, composed of finely pulverised ejecta, radiating from them. The rays centred on the crater Tycho can be seen from the Earth with small binoculars. Such rays are gradually darkened by the solar wind.

In a landscape where there is no atmosphere and consequently no erosion physical features are preserved free from change almost for ever, except when altered by bombardment by more meteorites. Consequently the relative age of a surface can be determined by counting the craters in it because the older a geological formation is, the longer it has been subject to meteoritic impacts and the more it will be cratered. Evaluating crater density (number of craters per unit surface area) gives a method of relative dating of different parts of the surface. In the case of the Moon this method enables us to compare the ages of the two basic kinds of terrain; the maria are only slightly cratered and must therefore be younger than the densely cratered highlands. The fact that the Moon has two different kinds of surface – as was first shown by telescopic views of bright and dark regions – is confirmed by this method of relative dating.

Standard methods of analysis using radioactive dating techniques that were highly refined for processing lunar samples have furnished values for the age of the Moon and of the surface soil where the samples were collected. The results give, for the age of the Moon 4.55 billion years, for that of the highlands 3.8 billion years and for the maria 3.2 billion years. Counting craters in areas where samples were collected gives a relationship between the age of the terrain and the crater density. The relationship is valuable for two reasons, firstly the ages of all parts of the surface, including those that have never been visited, may be determined, and secondly because it reveals that in the past the intensity of bombardment did not remain constant. Prior to 3.8 billion years ago the rate of impacting was catastrophic but between 3.8 and 3.5 billion years ago it gradually decreased until it reached the very low level of intensity that has persisted ever since.

Pierre THOMAS

A ray crater: Giordano Bruno. This smaller crater (its diameter is about 20 kilometres) situated on the hidden side is surrounded by a system of brilliant rays extending outwards 400 kilometres from the crater. For a long time such features were a puzzle but it is now known that the rays are composed of a sprinkling of finely pulverised ejecta which, in a few hundred million years, will become darkened by the action of external agencies such as the solar wind and microbombardment. The scarcity of ray craters is a good indicator of how seldom large meteorites now strike the Moon.

Relationship between the age and the density of craters on a surface. Comparisons between landscape ages as determined by radioactive dating methods and crater density (i.e. the number of craters with diameters exceeding a certain value per unit area of surface) give a relationship between the two parameters. The relationship confirms that the older a terrain is, the more highly it is cratered. The curve also shows that the decrease in the density of cratering is linear over the past three billion years, this indicates that the impact frequency has remained constant over that time and equal to its current value. It also establishes that the impact frequency was very high prior to four billion years ago and that from then until three billion years ago it decreased sharply. The full line represents the observed density and the dashed line the theoretical density that would obtain if the current impact frequency had remained constant in the past. This relationship, established by analyses of samples returned to Earth by different Apollo missions, gives a method for dating terrains that have never been visited by spacecraft. For example, the density of craters of a particular diameter in the ejecta of Orientale, as evaluated by counting in photographs, amounts to forty in terms of the adopted unit of area, and this corresponds to an age of 3.8 billion years. Approximate dating of the whole of the lunar surface may be achieved in this way.

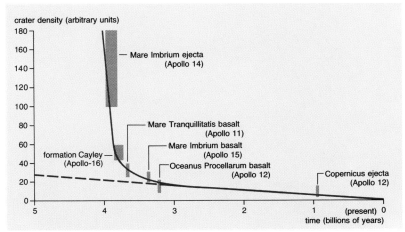

crater density (arbitrary units)

— Mare Imbrium ejecta (Apollo 14)

— Mare Tranquillitatis basalt (Apollo 11)

— Mare Imbrium basalt (Apollo 15)

formation Cayley — (Apollo-16)

— Oceanus Procellarum basalt (Apollo 12)

— Copernicus ejecta (Apollo 12)

180
160
140
120
100
80
60
40
20
0

5   4   3   2   1   (present) 0
time (billions of years)

The Moon *Topography*

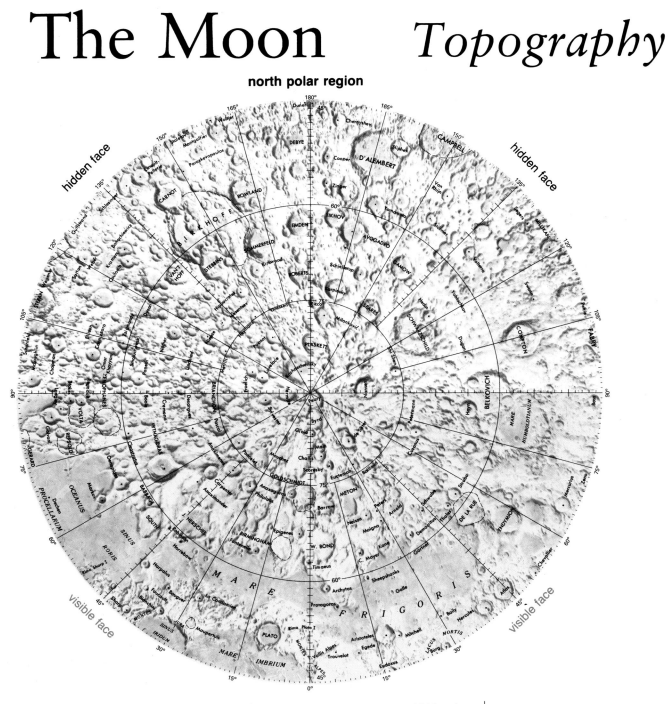

There are two ways in which the whole of the Moon can be represented in a map. In the first method the visible face, as it is seen from Earth, and the hidden face are represented separately, using the same projection. Such maps are convenient for the amateur astronomer whose map mirrors the view that his telescope presents. Although it has been in use since the seventeenth century when it was adopted for showing the only face known at the time, this method has two major disadvantages: firstly, morphological features close to the limb are seen at an angle so that small detail cannot be discerned and secondly, this kind of map emphasises the notion of two different faces even though such a distinction only depends on the synchronisation of the moon's rotation and revolution and not upon some fundamental difference in the nature of the two faces. The use of a planisphere to represent the moon is preferred nowadays; as with the planets, including Earth, a Mercator-type projection is used to show the equatorial and intermediate zones and a stereographic projection for the polar regions.

The whole of the lunar surface has been accurately mapped except for a small zone, close to the South Pole, that has yet to be photographed from a low altitude. The map shown here is based on the photography of Lunar Orbiters I, II, III, IV and V. The maria of which there are many in the north and west of the visible face are beige in the map. The highlands dominate the hidden face and the south-west of the visible face; they are shown in blue. The smallest details shown have dimensions of about 10 kilometres.

The coordinate reference system is based on the principal axes of inertia of the Moon. The poles are defined by the axis of rotation; the pole on the same side of the ecliptic plane as the terrestrial North Pole is conventionally the North Pole of the Moon. The equatorial plane is perpendicular to the axis of rotation. Note that the image of the Moon given by a telescope is inverted with north at the bottom and the right and left sides reversed. The definition of lunar east and west was ambiguous for a long time; before the space age some maps showed east on the left side while others had it on the right. In

◀ hidden face | visible face ▶

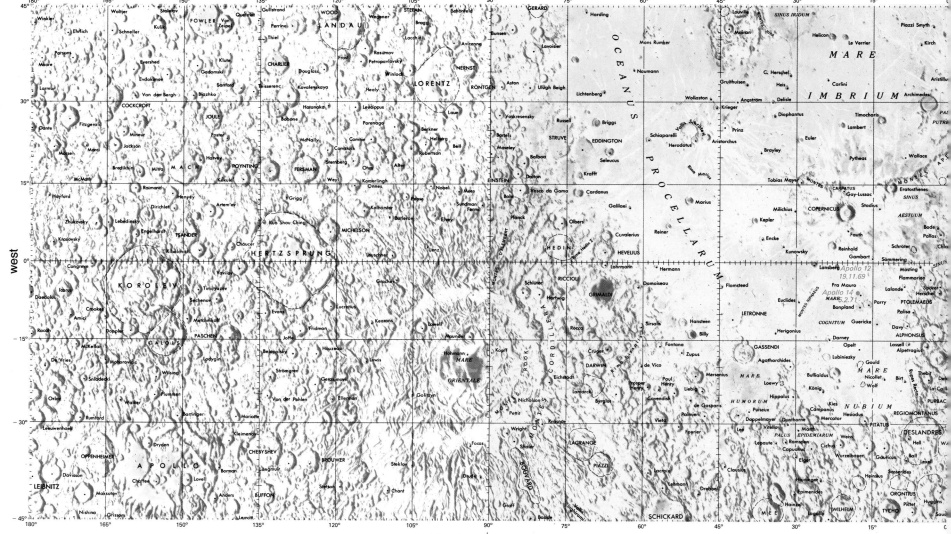

◀ hidden face | visible face ▶

1961 the convention known as astronomical was adopted; in this the cardinal points correspond to the naked-eye view of the moon with north at the top and west on the left (for an observer in the northern hemisphere). Using this convention an observer on the Moon would see the Sun rising in the east and setting in the west, just as on Earth. Astronomers chose the meridian passing through the centre of the visible face as the prime meridian. Historically lunar positions were first referred to the small bright crater Mösting A. Precise determination of the coordinates of this crater with respect to the new axes has furnished the absolute coordinates of all the other features.

The naming of lunar features has many origins. On the visible face it owes much to Johannes Hevelius and even more to Riccioli and Grimaldi. These astronomers gave to the maria the names of human qualities or meteorological phenomena (Mare Serenitatis, Mare Vaporum . . .). Mountains received the names of great mountain ranges on Earth (Montes Carpatus, Montes Apenninus . . .). Craters were given the names of scientists, artists, philosophers and writers (Copernicus, Ptolemaeus . . .). Latin, which was used for a long time in naming lunar features, was officially adopted for this purpose in 1964.

The naming of formations on the hidden face is much more recent and dates from discoveries made by space probes. The most prominent features of the far side were first seen in 1959 in very poor resolution images that were transmitted by Luna 3. This was a Soviet probe and the features it revealed were given Russian names (Mare Moscoviense, Tsiolkovsky . . .). Less visible features that were subsequently discovered received the names of great scientists of every nationality (Schrödinger, Fleming . . .).

The landing sites of the Apollo missions and of the unmanned Soviet vehicles that brought back samples are shown on the map.

Pierre THOMAS

south polar region

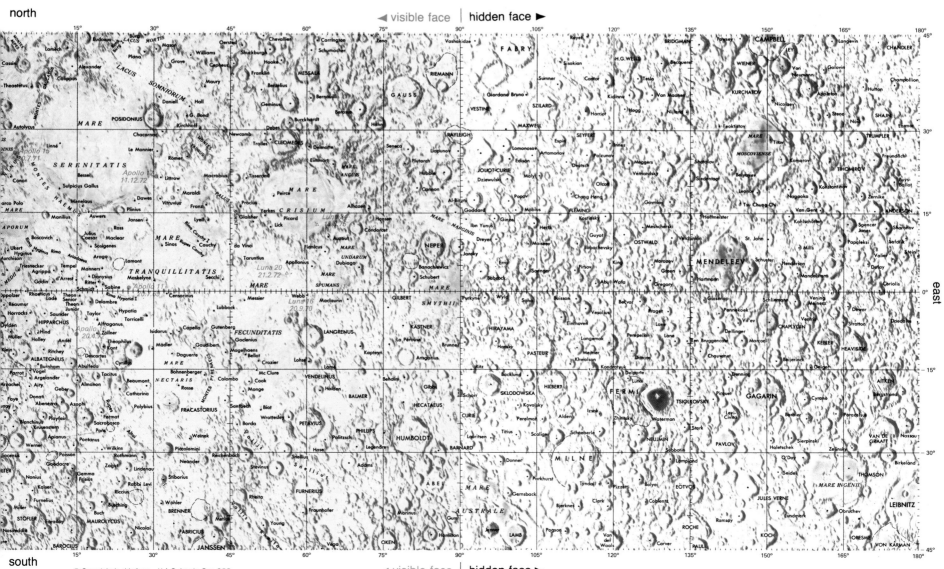

north

◄ visible face │ hidden face ►

south

◄ visible face │ hidden face ►

The Moon

The highlands

The highlands (sometimes called continents and known as *terrae* in Latin) are recognisable by their bright colour and mountainous nature; they cover 80 per cent of the Moon's surface. Morphologically they appear as the juxtaposition or even superposition of a myriad of craters of all sizes, and of their ejecta. In this confused scene some basins play a dominant role because of their size. There are in fact twenty-nine basins with diameters exceeding 300 kilometres; they are equally distributed on the two sides, with fourteen on the visible face, twelve on the hidden face and three straddling the two faces. The two largest, Imbrium (1100 kilometres diameter) and Orientale (900 kilometres diameter) are also the youngest because ejecta from them have covered all the others.

Apart from a few 'recent' impacts the craters and ejecta of the highlands are severely defaced, forming mountains that are very rounded in relief, between which may sometimes be found relatively flat stretches, the bright plains.

The Apollo and Luna missions enabled chemical and petrological examinations to be carried out at a few sites in the highlands. In every case the samples returned to Earth mainly consisted of anorthositic breccia, associated with a little norite or other associated rocks as well as very rare basalts: the KREEP basalts which derive their name from the constituent elements – K (potassium), REE (rare earth elements) and P (phosphorous). The particles of the breccia are completely crystallised. Anorthosites are rich in lightweight metals such as aluminium and calcium that appear to have surfaced in a primeval molten crust in which heavier elements sank to the bottom. Norite on the other hand is rich in heavy radioactive elements; it is to be found all over the Moon, scattered by impacts in the few areas in which it is concentrated. These minerals are representative of the chemistry of highland 'soil' but their severe fragmentation and the nature of the breccia are indicative of fractionation by countless impacts. The 'bright plains' on which samples were collected by Apollo 16 are also composed of breccia but why these plains are perfectly horizontal remains a puzzle. Anorthosite and norite occur in the Earth but they differ significantly from the samples found in the lunar highlands; the latter contain no water and are richer in rare earth elements.

Dating lunar samples using radioactive dating methods indicates that the era of transformation into breccia extended from the time of the moon's origin up to 3.8 billion years ago, with the exception, very locally, of examples close to the extremely rare 'recent' impacts. The crystallisation of minerals is dated between 4.6 and 4.4 billion years, which corresponds to the formation of the Moon. The two youngest large basins, Imbrium and Orientale are aged about 3.9 and 3.8 billion years. Dating by counting craters shows that the highlands' surface is nowhere younger than 3.8 billion years old.

Pierre THOMAS

The crater Heaviside and surroundings. This region on the hidden face around the crater Heaviside, which is 150 kilometres in diameter, (near the top of the photograph and at 10° S, 167° W) represents a typical view of the highlands: a juxtaposition of craters of every size, and of their ejecta. Heaviside is perforated with a multitude of younger, smaller craters, but itself covers an older crater of equal size, the shape of which may be discerned in the south-east. The crater in the foreground is an example of a crater with a fractured floor. It is not known why such fracturing occurs. (Apollo 17 photograph; NASA/NSSDC)

The crater Anderson and surroundings. A photograph of the highlands taken from a low altitude only covers a small field so it is unlikely to contain a large crater. No crater that has been officially named appears in this photograph of the region south-east of the crater Anderson, at 12° north and 170° east: it is an area full of randomly spaced craters that are rounded with blunted edges. The very high crater density is responsible for the soil in the highlands being reduced to breccia; there is no material that could remain intact and coherent under such bombardment. The crater in the foreground on the right is 10 kilometres in diameter. (Apollo 16 photograph; NASA/NSSDC)

The crater Orlov and surroundings. This region on the far side of the Moon around the crater Orlov (in the centre of the photographs and at 26° S, 175° W) is typical highland terrain. However, in the crater Leeuvenhoek, which has a diameter of 120 kilometres, (near the horizon, top right) the floor is very flat, dark and only slightly cratered; it is composed of maria-type material. Although there are no large maria on the hidden face, a large number of craters, including some quite small ones, were filled by maria-type volcanic discharges after their formation. At the same time other craters were not filled and there is no explanation for this. (Apollo 17 photograph; NASA/NSSDC)

Sample of anorthosite breccia from the highlands. This sample (no. 6119516) was obtained at the Apollo 16 landing site. It is made up of bright angular fragments of every size which are composed of a feldspar known as anorthite. They are tightly packed and bonded together by a dark 'cement'. The whole of the highlands is made up of such breccia-type rocks, that is rocks formed by the cementing together of a deposit of fragments laid down when the parent material was smashed and splintered. In this sample the reunion of the fragments, to form breccia, took place between 3.9 and 3.8 billion years ago. (Jean-Pierre Bibring)

The crater Descartes and surroundings (above). This photograph, covering an area 120 by 120 kilometres shows the region where Apollo 16 landed (see arrow). This site was chosen so that rock samples of 'normal' highland terrain (south-east) could be collected as well as some from the bright plain, known as the Cayley formation, in the north-west. This bright plain, which fills more than a quarter of the area in the photograph, occupies low lying regions and especially the craters Dolland B and Dolland C in the north-west. The plain, of which the origin is unknown, is perfectly horizontal although, like the rest of the highlands, it is composed of anorthositic breccia. (Apollo 16 photograph; NASA/NSSDC)

Simplified geological map of the Moon (below). This map, showing separately the visible face (left) and the hidden face (right) illustrates the two kinds of lunar terrain: areas in blue, pink and violet represent the anorthositic highlands while the beige coloured regions are the maria, great plains of basalt. The highlands are everywhere covered by a disorderly mass of craters and basins. The twenty-nine basins with diameters greater than 200 kilometres, are shown; their random distribution is evident. The two largest, Orientale and Imbrium are conspicuously marked because of their predominance. Over the highlands, subsequent to the era of intensive cratering, maria-type basalts were discharged, covering indifferently some basins and their ejecta. These basalts are distributed very unequally, 80 per cent are to be found in or around the Imbrium basin where they have also filled all the neighbouring basins (Serenitatis, Humorum) and even some areas where no large impact had occurred (Oceanus Procellarum). Other discharges happened here and there, far from Imbrium, filling some pre-existing craters and basins while other craters and basins (Hertzsprung) were unaffected by volcanic action. The reason why some craters were affected rather than others is not known.

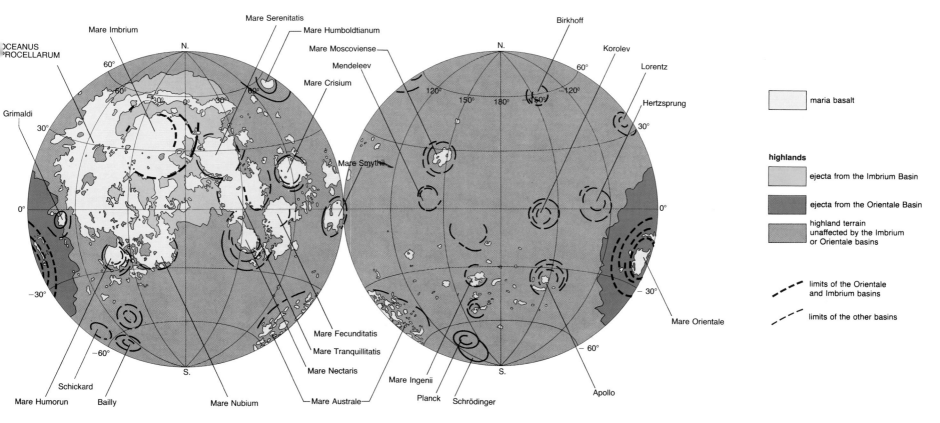

The Moon

The maria

The maria (plural of mare, the Latin for sea – and in the context of the Moon the English word is seldom used) are prominent features of the Moon. They are the dark areas that make the Full Moon resemble a human face. In fact they are large plains composed of a black material that is only slightly cratered. They are mostly circular in outline because they fill basins and craters that they have covered over. On the visible face where they are quite numerous they occupy 35 per cent of the surface but on the hidden face they are very rare. Their surface is horizontal and they are located close to each other in low lying regions. The transition from maria to highlands is everywhere the same: the maria encroach on the land but here and there, close to the 'shore', elevated land pushes up through the flat surface to form islands, capes and peninsulas. All these features indicate that the maria were formed from a molten liquid that filled the low lying areas with layers of very fluid lava.

Even before the acquisition of lunar samples, reflection spectra obtained from Earth and the presence in the surface of the maria of structures that are indubitably volcanic (the fronts of lava flows, domes and some volcanic craters) strongly suggested the maria were volcanic in origin. By analogy with equivalent terrestrial features, other structures that for a long time were even more puzzling are now understood; these are the sinuous rilles that represent the channels in which the flow of molten lava ended. The lunar samples collected in the maria have confirmed their volcanic origin and have provided an understanding of the nature of this volcanism. It is now known that the maria are composed of basalts, quite similar to those occurring in the Earth but differing slightly in the relative abundance of some elements. Lunar basalts are richer in ferrous iron, titanium and magnesium and poorer in alkali and volatile elements; they contain neither water nor ferric iron. The global chemical composition of the basalts is completely different to that of the highlands; the basalts contain much more iron and magnesium and less silicon and aluminium, and therefore cannot be derived from the melting of the highlands.

The age of the basalt samples brought back by the Apollo and Luna vehicles is in the range 3.2 to 3.8 billion years. Crater counts indicate that the expulsion of basalts commenced at about 3.7

The Hills of Marius (left). This region, situated in Oceanus Procellarum, (12° S, 51° W) typifies the general morphology of the lunar maria; it is flat, slightly cratered and marked here and there by fairly low-altitude features of two kinds: the first are the hills and domes of volcanic origin, known here as the Hills of Marius after the large 30 kilometre crater in the background; the second are the elongated sinuous wrinkles which are tectonic in origin. Even though they appear young in comparison with the highlands the maria are aged between 3 and 3.9 billion years. (Lunar Orbiter II photograph; NASA/NSSDC)

The great lunar maria. This photograph (right, opposite) covering more than 2000 kilometres from north to south, shows the whole of the west side of the visible face of the moon, with, in the north-west and west of centre Oceanus Procellarum, the largest of the maria, in the north-east Mare Imbrium, just south of centre Mare Humorum and in the south-east Mare Nubium. Highlands are visible in the south and south-west. The large ray crater in the north-east is Copernicus. Mare Humorum, which is 400 kilometres in diameter, is perfectly circular; it fills a very ancient basin the ejecta of which can no longer be identified. The basin is perforated by many later impact craters such as Gassendi to the north, Hippalus to the east and Puiseux and Doppelmayer to the south. The maria-type basalt which fills Mare Humorum is much more recent – 500 million years, at least, elapsed between the impacts and the filling with basalt. This basalt partially covers the craters Gassendi, Hippalus, Puiseux and Doppelmayer. Only the upper parts of the crater Puiseux rise above the basalt plain with the result that it resembles a crown. In Oceanus Procellarum also, the heights of the underlying highlands pierce the surface; in many instances these are craters in the form of more or less complete crowns – like Sirsalis E, Lansberg C, Flamsteed P, etc. (Lunar Orbiter IV photograph; NASA/NSSDC)

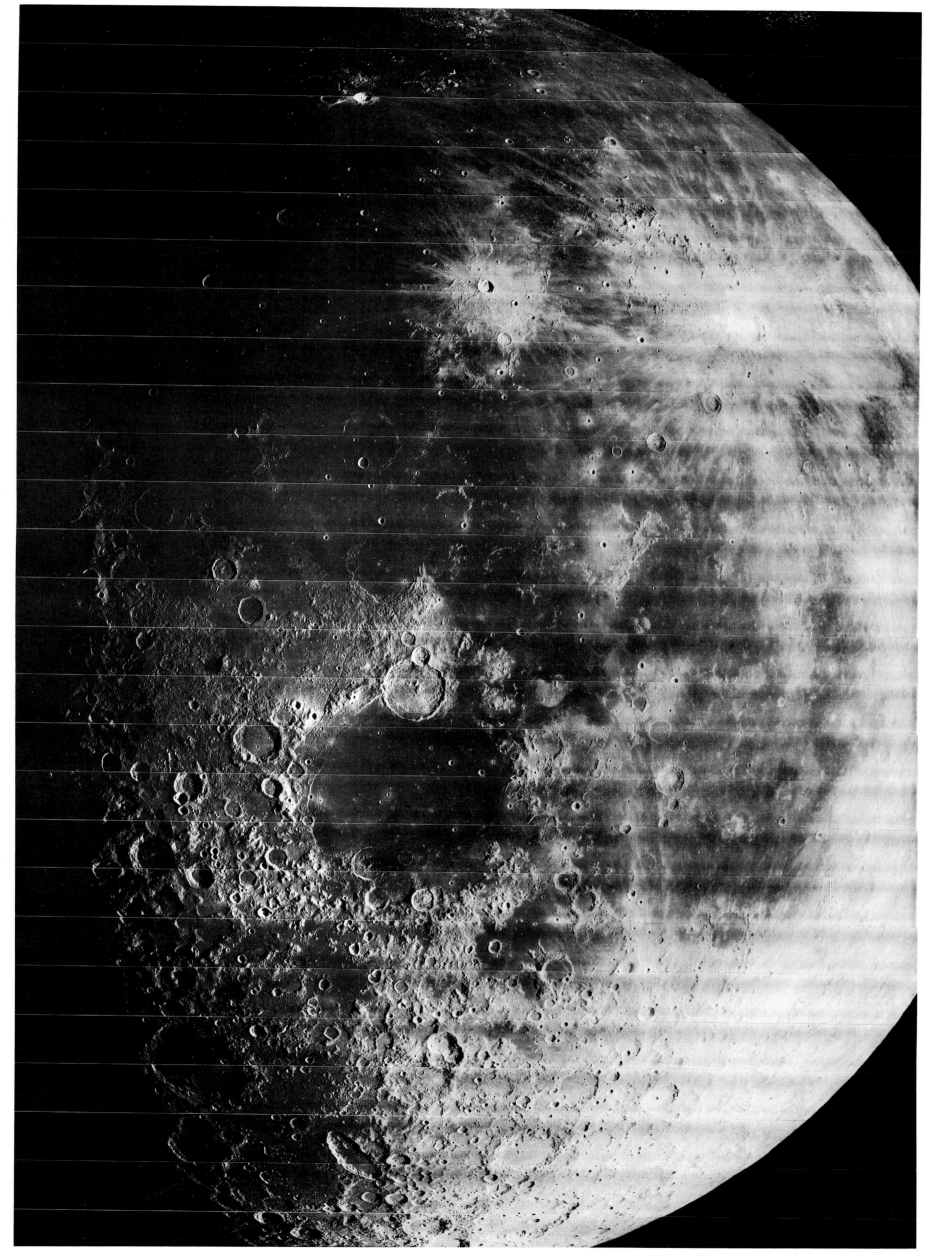

The crater Tsiolkovsky. This crater in the hidden face (20° S, 131° E), with a diameter of 180 kilometres, is completely filled with maria-type basalt the dark colour of which contrasts with the brightness of the highlands. The basalt is contained by the crater terraces which are clearly visible in the upper part of the photograph and its surface is broken by the emerging central peak. The surface boundaries are perfectly level – this explains why they resemble an indented coastline, with islands, capes and gulfs. All this proves that the maria-type material was extremely fluid at the time it filled the crater. (Apollo 15 photograph; NASA/NSSDC)

A lava flow in Mare Imbrium (below). This very high resolution photograph shows a lobate denticulated escarpment situated at about 35° north, 20° west. The southern part of the escarpment (top) is higher than the northern plain by about 25 metres. The structure corresponds to the wave front of a lava flow coming from the south. The crater density in the flow is the same as in the layer underneath it (an earlier flow) – which shows that the two flows are almost contemporary; they are dated at 3.4 billion years.

Such flows can be several hundreds of kilometres in length (only the last 12 kilometres are shown here) on a slope of less than 1 per cent, which demonstrates the very high fluidity of the lava. The white spots are due to defects in the transmission equipment of the unmanned vehicle. (Lunar Orbiter V photograph; NASA/NSSDC)

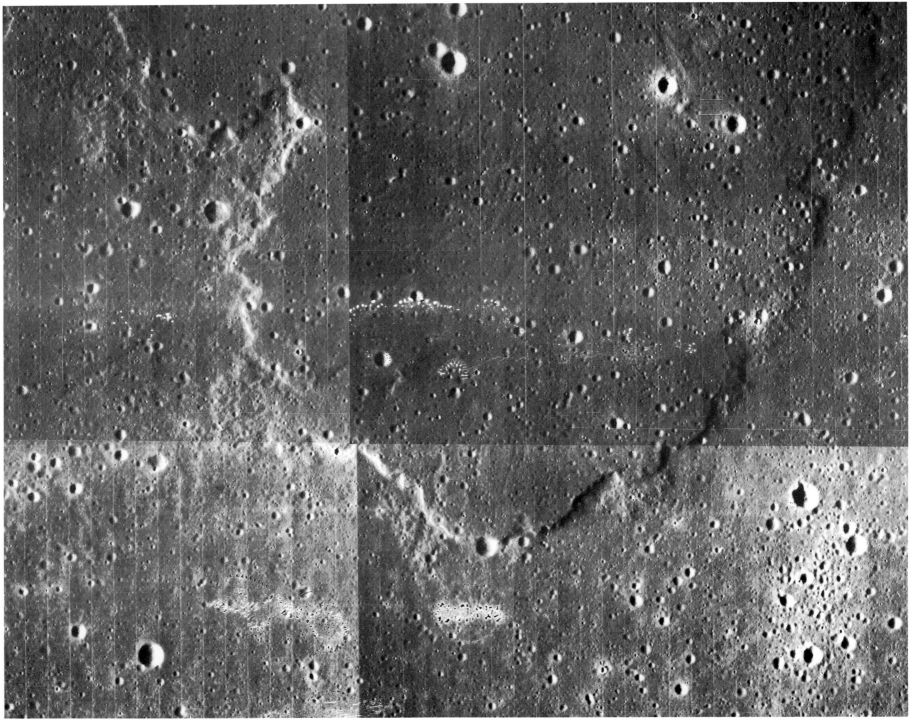

billion years ago (Mare Tranquillitatis), the largest surfaces (Mare Imbrium, Oceanus Procellarum) are aged between 3.3 and 3.5 billion years and the youngest about 3 billion years.

There is clearly a geographical relationship between the large impact basins and the basaltic volcanism which generally filled them. It should however be noted that some basins contain no basalt (Hertzsprung, for example, which is 450 kilometres in diameter) and that the largest of the maria (Oceanus Procellarum) is not associated with any basin. Moreover, it was long after their formation that the basins were invaded by the basalt; it is estimated that 500 million years elapsed between the Imbrium impact and its filling with lava. Furthermore, the distribution of maria is very asymmetric while that of the basins is very uniform. Unlike what happened on Mercury, impacts on the Moon do not appear to have been directly responsible for volcanism; the basin depressions just happened to trap the basalt. It is however possible that the thinning and fracture of the crust as well as other profound alterations caused by the impact could have facilitated the subsequent upsurge of magma.

Pierre THOMAS

The volcanism of Lacus Veris. Although 20 per cent of the lunar surface is volcanic in nature, features that are wholly volcanic are very rare and unobtrusive. This photograph shows details of a lunar volcano in Lacus Veris, a small mare that is isolated in the Orientale basin, between the Rook Mountains (top right) and the interior ring I2 (bottom left). This volcanic system includes a volcano with an extinct crater (near centre) and with a slope very nearly the same as in volcanoes in Hawaii; its diameter is about 6 kilometres. Immediately south of the volcano there is a fissure about 10 kilometres long which is flanked by plateaus, also with volcano-type slopes. The two structures could be explained in terms of lava accumulations that were much less viscous than the 'ordinary' basalt of the maria; the extinct volcano and the fissure are seen as the sources of the discharge. Similar accumulations can be found on all sides of the central structure as well as 20 kilometres north of the volcano – but in this case the mouths of the volcanoes cannot be seen. (Lunar Orbiter IV photograph; NASA)

Two volcanic rilles, on the Moon and on the Earth. The region of the crater Aristarchus at 24° north, 48° west in Oceanus Procellarum is marked by flexuous meandering channels or rilles (*rima* in Latin) in the basalt plain (left). For a long time the nature of these features remained enigmatic but the puzzle was solved by comparisons with similar terrestrial features, such as the Snake River basalts in Idaho, USA (right).

When a layer of lava is almost completely solidified some flows can persist in channels underneath the solid crust. When the source dries up the flow in the channels continues until they are empty, leaving long hollow tubes within the lava and these are apt to be flexuous where the terrain is nearly flat. The rilles we observe are these tubes which have been exposed by the collapse of the lava covering them.

However, lunar rilles are very different from terrestrial ones. The photograph on the right was taken from an altitude of only a few hundred metres (part of the wing of the aircraft can be seen); it shows the largest terrestrial volcanic rille which is 5 kilometres long and 20 metres wide. The picture on the left was obtained by Apollo 15 from an altitude of 100 kilometres; here the rilles are 100 kilometres long and 1 or 2 kilometres wide. The reason for the difference in size is not clear – it may be due to differences in the viscosity or in the mass of the lavas. (NASA/NSSDC)

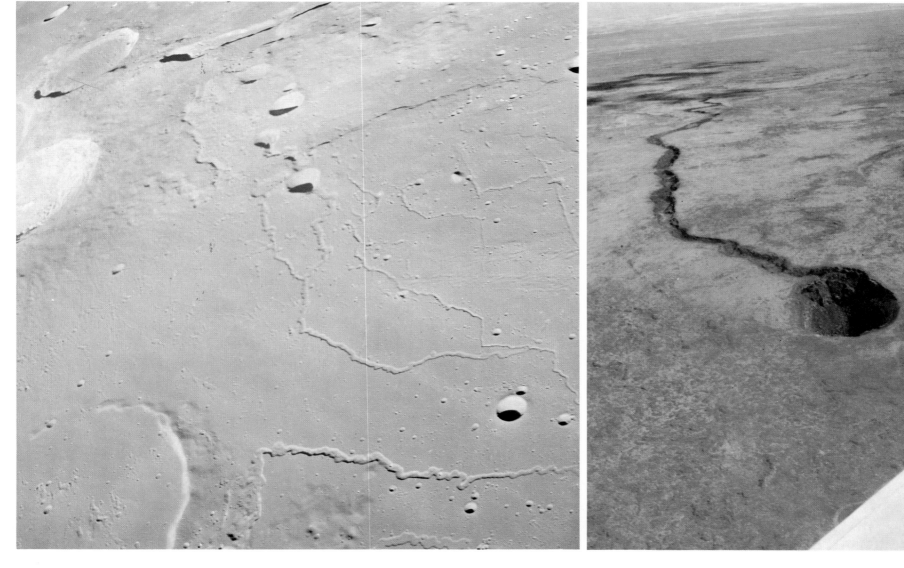

The Moon

Tectonics and structure

Throughout its whole history, the Moon has been noteworthy for its almost complete lack of significant tectonic movements of the kind observed on Earth and Mars. Nevertheless, three types of moderate tectonic activity can be discerned, in addition of course to the ground movements accompanying impacts, such as the recoil phenomena or landslipping on the rims of craters, known as slumping.

The first kind of activity is associated with depressions which have a significant filling of basalt. These volcanic plains are sometimes affected by undulations in the form of ripples, which are interpreted as creases originating from a compression of the surface; the same ridges can cross the borders between several regions of lava and even lightly encroach upon neighbouring highlands, thus certifying to their tectonic and not volcanic nature. These ridges are usually arranged concentrically around the centre of the depression, which is often an old crater. The outside of these same basins also shows a system of grabens and stress-induced crevices which are more or less concentric. These two kinds of phenomena are understood to be the indirect result of vertical movements: under the weight of the basalt which has filled it, the bottom of the depression has bowed, thus generating the compression which will form the ridges. The periphery of the depression has then risen up again thus leading to straining and fracturing.

In addition to these undulations, the whole surface of the Moon is affected everywhere by faults and stress-induced rilles, certifying to a tendency of the moon's surface to be under stress at the very moment when the volcanism of the seas was in full development. These rilles are in the main directed radially or tangentially to the Imbrium basin, whose presence seems to have

'orientated' the Moon's expansion. Such a phenomenon is reminiscent of the Caloris basin on Mercury. This expansion is, however, very small and any consequential increase in the Moon's radius would in any case be less than 1 per cent.

For a long time astronomers have noted that all the Moon's topographical features are statistically orientated in three dominant directions: north–south, north-east–south-west and north-west–south-east. This arrangement which has been called the 'lunar grid' is the reflection of former dislocations caused by the slowing down of the moon's rotation under the effects of the tides.

If past tectonic activity has been very weak compared with that of Earth, the present activity is almost nil. Seismometers left by the Apollo missions have shown that the seismic energy released by the Moon is a hundred billion times weaker than that released by Earth (remember that the masses of these two bodies are in a ratio of about 100 : 1). The epicentres of the rare quakes recorded on the Moon are situated in the main either between 600 and 900 kilometres deep or very near to the surface. The first kind are relatively periodic and are correlated with the cycle of lunar tides which suggests that gravitational interactions are responsible for these quakes. The surface events are very well synchronised with the risings and settings of the Sun; they are due to stress phenomena and thermal contractions. Only 1 per cent of lunar quakes have characteristics similar to earthquakes (tectonic origin) but they could not be located with precision for they were all situated far from the seismometers.

In addition to locating natural moon quakes, the seismometers left on the Moon have also allowed the signals emitted by natural or artificial

The ridges of the Imbrium basin. The origin of the ridges has been enigmatic and the majority of researchers used to attribute them to a magmatic origin (extrusion of viscous lava along the fissure for example). The tectonic origin of these ridges is demonstrated by this photograph, which shows the region around 28° north and 23° west. In fact a complex of lava flows crosses the whole region from south-west to north-east over a length of more than 200 kilometres. These flows 'cross' the ridges and the escarpment reappears at the edge of the flow 'perched' at the summit of the ridge, which proves that an upheaval of the ground and not its swamping by a viscous lava is responsible. (Apollo 15 photograph; NASA/NSSDC)

Mare Serenitatis. This mare fills the very old basin of the same name, and has a diameter of about 500 kilometres. On this photograph (below left) taken from Earth, the grazing illumination shows up a number of ridges crossing the surface of the mare. These ridges are more or less concentric with, and often parallel to, the edges of the basin. This arrangement is also found in all former basins which have been filled in with basalt and in particular in the Mare Humorum. At the extreme left of the photograph is the edge of the Imbrium basin. (Lick Observatory)

The southern edge of the Mare Serenitatis. This detailed photograph (below right) which covers a square of side 120 kilometres (towards 18° north and 25° east) shows the southern edge of the Mare Serenitatis. The ridges have quite a complex structure and a sinuous shape. They are understood to be the result of the buckling of the surface of the basalt by tectonic compression a long time after its solidification. The south of the basin is affected by a series of linear fissures parallel to its rim. These fissures are grabens limited by faults and indicate the presence of surface stress. (NASA/NSSDC)

Moon tremors to be analysed. The artificial tremors were stimulated by the impacts of the third stage of the Saturn V rocket or of the lunar module (LM) from the Apollo missions. This analysis has allowed seismic wave propagation in the Moon's interior to be studied. Since the variation in speed of these waves is a function of the density and viscosity of the material through which they pass, the internal structure of the moon can be deduced. The main discontinuity is situated between 50 and 75 kilometres deep; it separates an anorthosite crust from a denser material (peridotite) forming the mantle. The basalt of the seas covers this crust locally with a thickness of a few kilometres, twenty at the most. The crust is moreover a little thinner on the Moon's visible face than on the hidden face, which explains the relative rarity of volcanism on the latter. This asymmetry in the mass distribution (an excess of crust on the hidden face) can, furthermore, be related to the synchronism of the rotation and revolution of our satellite. A second discontinuity exists at a depth of about 1000 kilometres. It does not correspond to a chemical discontinuity but to a rheologic discontinuity; it separates the rigid mantle (lithosphere) at the base of which are situated the rare deep lunar quakes, from a slightly more viscous mantle (asthenosphere) which blocks the seismic shear waves. There is still insufficient data to confirm or otherwise the presence of a metallic core. In any case, this core, if it exists, is small (with a radius of less than 500 kilometres). The determination of the moon's moment of inertia, which is very close to that of a homogeneous sphere, allows us to set an even lower upper-limit (300 kilometres) to the value of any radius of any possible care.

The determination of the Moon's gravitational field by the precise study of the trajectory of artificial satellites in lunar orbit has shown that it is not regular; the gravity is stronger in the vicinity of certain large impact basins (Mare Imbrium, Mare Serenitatis, Mare Crisium, Mare Nectaris and Mare Humorum). These gravitational anomalies are explained by the presence, at the base of each of these five basins, of excesses of mass, which have been called 'mascons'. This excess of mass is the result of the existence of a layer of basalt (density = 3.3 grams per cubic centimetre) about 10 kilometres thick on top of the normal lower density (2.9 grams per cubic centimetre) anorthosite crust. The persistence to the present epoch of such an anomaly shows that the lithosphere was, even at the time of filling 3.6

Rima Arideus. This 'ditch' of which only 120 kilometres are seen here, located at about 6° north and 15° east, has a total length of 300 kilometres, and a width of about 5 kilometres. The bottom, of the same nature as the surrounding terrain, seems to be sunken between two parallel faults. When this graben crosses a relief, it widens as may be seen at the centre of the photograph. This fact, welll known in terrestrial geology, proves that these faults are due to an expansion. (Apollo 10 photograph; NASA/NSSDC)

billion years ago, too thick to allow isostatic equilibrium to be completely restored, despite the slight sinking movements shown to exist by the tectonics.

Lunar exploration has demonstrated the absence of any contemporary magnetic field on the Moon. On the other hand, a fossil magnetism has been discovered in certain rocks, but the origin of this ancient field (dating back 4 billion years) has not been determined. Various hypotheses have been put forward: an ancient lunar magnetic field, a solar magnetic field, or possibly magnetic phenomena accompanying impacts?

The thermal flux of the Moon was determined by measurements taken from the boreholes drilled during the Apollo 15 and 17 landings; this flux is equal to half of the mean terrestrial flux, showing that the production of heat per unit of mass is twice as much on the moon as on the Earth, indicating a relatively strong abundance of radioactive elements on the Moon.

Pierre THOMAS

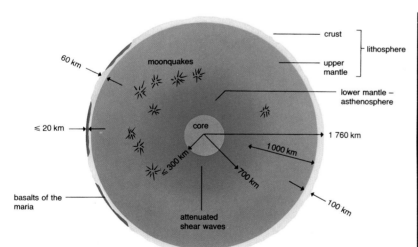

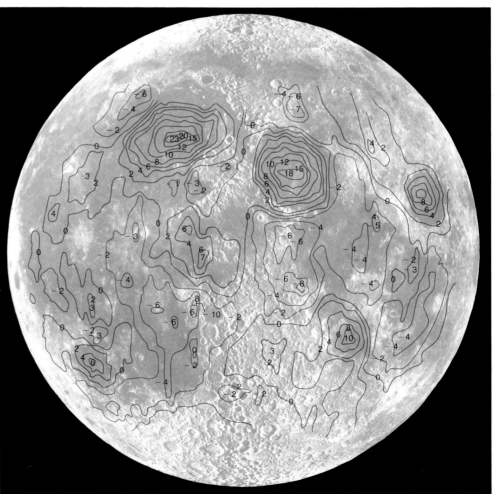

Schematic structure of the Moon, according to the data of seismic movements (above). The external envelope constitutes the anorthosite crust, about 70 kilometres thick, thinner on the visible than on the hidden side. This crust which is not shown to scale can be locally covered by basalt. Situated under the crust, the mantle forms the majority of the lunar mass. It is subdivided by a progressive decrease in rigidity, into the upper mantle (lithosphere) and the lower mantle (asthenosphere). The majority of moonquakes emanate from this region, but because of its changing characteristics, the exact position of this boundary has no real significance and it is alternatively located according to different authors either above or below the moonquake zone. Insufficient data exist to deny or confirm the presence of a core, which, in any case, would have a radius of less than 300 kilometres.

The gravitational field anomalies of the Moon. This map (opposite right) of the visible face shows the differences between the real gravitational acceleration deduced from the study of the orbits of lunar satellites and the average theoretical value of 162.7 centimetres per second per second, equal to a sixth of the terrestrial gravity. The differences are recorded in units of 0.01 cm/sec/sec and are shaded red for a positive difference and blue for a negative difference. The differences are mainly positive; they are due to the concentrations of mass, called 'mascons' which coincide with the basalt-filled basins. It is the great thickness of basalt (dense rock) in these basins which causes these excesses of mass. (Photograph Lick Observatory, map by P. M. Muller and U. L. Sjogren, 1968)

The Moon
Chemistry, origin and history

Chemistry

The morphological and petrographical descriptions of the Moon have enabled us to investigate the external manifestations of its evolution. However, it is geochemical studies which have contributed most to our knowledge about the Moon's origin and internal structure. The main source of chemical information is provided by the analysis of samples brought back by the Apollo and Luna missions. These analyses relate not only to the most abundant elements but also to trace elements (like rare earths) and to the isotopic composition of different elements. These extremely careful analyses, using quite small quantities of materials, have, furthermore, necessitated the development of new analytical techniques and have resulted in very important advances in the entire field of geochemistry.

Thorough chemical analysis of a rock sample allows us, through application of the laws of thermodynamics, to understand the physical and chemical conditions of the place where it was produced. For example the analysis of lava allows us to deduce the exact nature of the body whose melting gave birth to it and to know the temperature and the depth at which this melting took place. It was thus possible to determine, from the return of the first samples, that the lunar basalts had formed at a depth of between 300 and 600 kilometres at 1500 K to the detriment of the partially melted mantle, which in chemical composition is slightly different from that of the Earth's mantle. The chemical models of this mantle which have been suggested show likewise that the latter has undergone an earlier melting

which had removed certain elements from it, like europium. Subsequent Apollo missions have confirmed all these deductions; in particular the europium deficiency in the mantle has been 'rediscovered' in the highlands.

Furthermore, although the surface composition of the Moon is only known with accuracy at the locations of the nine landing sites, it has, however, been possible to extrapolate these results to the whole surface using the geochemical measurements carried out in lunar orbit by the Apollo 15 and 16 modules:
– from the study of the spectrum of gamma-rays in the region 0.55 to 2.75 megaelectronvolts (natural radioactivity) emitted by the crust, we may deduce its content of radioactive elements (uranium, thorium, potassium 40);
– the neutrons of cosmic rays which strike the nuclei of the atoms in the crust generate gamma-radiation of a very precise energy. Studies of this radiation reveal the abundance of various elements (iron and titanium, for example);
– certain elements re-emit by fluorescence a part of the X-ray radiation which they receive from the Sun; the study of the spectrum of this X-ray fluorescence tells us about the surface chemistry, in particular about the concentrations of silicon, aluminium and magnesium.

The internal chemical composition is deduced not only from surface petrographic analyses, but also from the global density of the moon, from the relative importance of the different layers revealed by geophysics and by the flux of thermal energy. The latter gives direct information on the

global abundances of radioactive elements and indirect data on the abundances of elements having similar physical and chemical properties. All these data, internal and external, show that the Moon has an overall chemistry close to that of the terrestrial mantle. However, perceptible differences do exist; with respect to the Earth's mantle, the Moon is especially poor in volatile elements like alkalines or lead and rich in refractory elements like titanium or zirconium.

Origin and history

The origin of the Moon has always posed a problem and three very distinct scenarios have been proposed to explain it:
– The hypothesis of fission supposes that the Moon is a piece of terrestrial mantle which detached itself, for example under the effect of tides. However, notable differences in chemical composition between the Moon and the supposed 'Mother Earth' completely invalidate this model.
– The second hypothesis – that of the double planet – envisages the accretion of two bodies (Earth and Moon) in the neighbourhood of each other. This hypothesis is also contradicted by lunar chemistry; in fact two bodies whose accretion took place in the same zone of the Solar System ought to have the same iron – silicate

Sample of lunar basalt. This small sample of lunar basalt collected by the Apollo 15 astronauts shows greenish minerals (olivine and pyroxine) and white minerals (feldspar). There is a small quantity of glass. It is the extremely careful chemical analysis of such samples – from which the abundance of the rare earths has been deduced for example – which has allowed the internal history of the Moon to be retraced. The history of this sample has thus been reconstructed: it is the result of the crystallisation of a basalt magma resulting from the melting of a peridotite mantle, which itself originated from the fractionated crystallisation of the 'ocean' of magma resulting from the global (or very nearly) melting of the Moon. (Jean-Pierre Bibring)

Comparative chemical compositions of the Earth and Moon. The table opposite gives the approximate contents of the main cations constituting the Moon, the Earth and various parts of their envelopes (oxygen being always the main anion). The global differences in content between the Earth and Moon, particularly in iron, excludes the hypothesis of the Moon originating in the neighbourhood of the Earth. The differences in content between the Moon and the primitive terrestrial mantle, in particular in alkalines and in titanium, excludes the hypothesis of fission.

An example of 'orbital' geochemical cartography. The command modules of Apollo 15 (inclined orbit) and Apollo 16 (almost equatorial orbit) recorded the intensity of the gamma-radiation emitted by the surface in the spectral band 0.55–2.75 megaelectronvolts, a consequence of the natural radioactivity of the crust. The intensity of the radiation is represented on a lunar planisphere, where the limits of the main maria are indicated. The strongly radioactive zones are represented in red and yellow, the weakly radioactive zones in pink and purplish-blue and the intermediate zones in blue or in green. Note that the highlands are weakly radioactive, which indicates a low content of uranium, thorium and potassium, while the maria are plainly more radioactive. A more accurate analysis shows that the basalts of Oceanus Procellarum, Mare Imbrium and Mare Nubium are clearly more radioactive than those of the Mare Serenitatis. Mare Crisium, Mare Fecunditatis and Mare Orientale; there are thus two kinds of basalts on the Moon with respect to the elements uranium, thorium and potassium. (L.A. Soderblom/USGS)

element	Moon	Earth	original terrestrial crust	lunar highland crust	terrestrial continental crust	mean of the lunar maria	terrestrial oceanic crust
silicon	20	14	21	21	27	22	23
magnesium	19	16	24	4.1	2.1	6	4.6
iron	10.6	33	6.2	5.1	5.8	14	8.2
calcium	3.2	1.2	1.9	11	5.4	7	8
aluminium	3.2	1.2	1.7	13	10	5	8
titanium	0.18	0.06	0.09	0.3	0.5	3	0.9
sodium + potassium	0.07	0.17	0.26	0.36	3.8	0.25	2.2
uranium	0.033	0.010	0.018	0.24	30	0.2	0.6

The numbers given are percentages by weight, except for uranium, where they are expressed in parts per million.

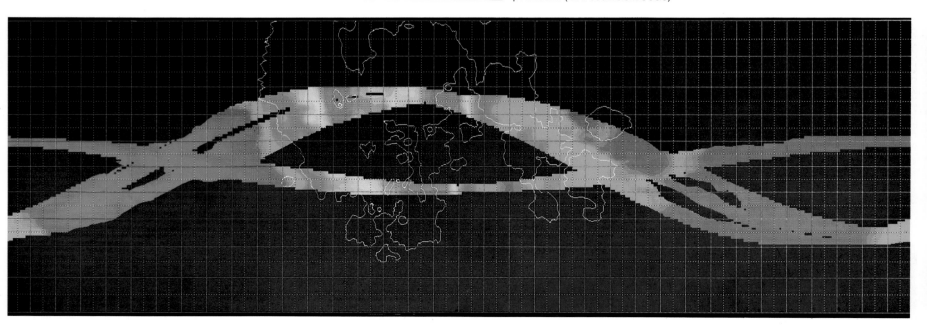

ratio. Now, in the Earth, iron represents 30 per cent of the total mass, mainly contained in the core, while in the Moon it is three times less abundant.

– The hypothesis of capture supposes that the Moon, initially in solar orbit, was captured by the Earth. However this capture could only have occurred very early on and it could only have taken place before the crystallisation of the crust 4.4 billion years ago, because the variations in speed and kinetic energy caused by such a capture would have led to the reheating and total melting of the Moon. In addition, celestial mechanics shows that the probability of capture is negligible unless the orbit of the captured body around the Sun has a radius and an eccentricity close to that of the Earth, which again poses the problem of the difference in iron content.

None of the three classical models is therefore compatible with all the data currently available and the origin of the Moon still remains a problem. Another proposal for explaining all the available data was made in 1975: that the Moon formed by coalescence of debris from a glancing collision between a proto-Earth and another proto-planet. As this was a purely qualitative idea, it was largely ignored by the scientific community. However, since 1984, high-speed computers have been able to simulate such collisions. The simulations show that this idea is plausible and could indeed have produced an Earth–Moon system with the present characteristics.

An oblique collision between the Earth and a planet with a mass close to that of Mars is simulated by computer. The two bodies are assumed differentiated, thus having iron cores and silicate mantles. Just before the collision, tides rapidly deform the incident planet. After the collision, part of the debris torn off the two bodies can form a disk around the Earth and produce a single satellite by accretion.

One can simulate several types of collision as a function of the mass of the incident planet. If the mass exceeds 0.17 Earth masses most of the debris falls back onto the Earth or is ejected into space; the debris remaining in Earth orbit is too small to form a body with the mass of the Moon. If the incident planet mass is less than 0.12 Earth masses, it provides almost all the debris, and very little mass is lost into space. It can thus form a satellite with the mass of the Moon, but not iron-poor, as we know the Moon is. These two scenarios are therefore in conflict with the data.

By contrast, if the mass of the incident planet lies between 0.12 and 0.17 Earth masses, the simulations predict a complex series of events which can give rise to the Moon as we know it. The silicate part of the incident planet is almost completely stripped off, and only about one Moon mass remains in Earth orbit. After a complicated trajectory the core of the incident planet hits the Earth again and is completely absorbed. All this occurs in less than twenty-four hours. The silicates remaining in Earth orbit are heated by the impact sufficiently to lose a large part of their volatile elements, hence becoming enriched in refractory elements.

These simulations, with judicious choices of masses, speeds, and directions of motion for the proto-Earth and incident planet, show that this scenario is possible, which is not the same as saying that it must actually have occurred. However, this model is the only one currently conceived of which produces a realistic Moon. The very low probability of such a collision with the 'right' parameters explains why the Earth–Moon system is unique in the Solar System. But whether or not this hypothesis represents reality, we know quite definitely that the Moon is unique as a body isolated for 4.5 billion years, with a crust which crystallised 4.3–4.4 billion years ago. The postulated collision must therefore have been contemporaneous with the end of the accretion of the Earth.

If the origin of the Moon is very poorly known, at least an outline of its history is understood. After its formation as a separate body was completed about 4.5 billion years ago, the heat released by the accretion and the radioactivity (in particular due to elements with a short half-life aluminium 26) melted the Moon to a depth of at least 400 kilometres. This ocean of magma cooled and crystallised slowly. The densest of the minerals (olivine and pyroxene) settled to the bottom of the ocean and formed the mantle, while the lightest elements (the feldspars) floated to the surface to give an anorthosite 'scum' which gradually solidified and formed the crust, which represents around 10 per cent of the volume of the Moon. These floating feldspars carried with them certain chemical elements which today are present in very small proportions in the mantle, like potassium (K), rare earths (Rare Earth Elements) and phosphorous (P) from which is derived the acronym KREEP given to certain rocks of the crust. The continental basalts (KREEP basalts) thus represent the last evidence of this ocean of magma. This differentiation and crystallisation happened between 4.5 and 4.4 billion years ago, an age during which the newly formed crust was completely dislocated by a catastrophic bombardment of meteors which only ended 3.8 billion years ago. It was at the end of this bombardment that the biggest meteorites fell giving rise to the great basins like Orientale and Imbrium. About 3.9 billion years ago, the mantle, which was probably completely crystallised about 4.4 to 4.2 billion years ago, began partially to melt again under the action of the heat released by the radioactivity of the long-lived elements. This melting involved the release of basalts from the seas, which although the process lasted for nearly a billion years involved only 1 per cent of the Moon's volume. Since the last eruption of basalt between 3 and 2.5 billion years ago, the Moon appears internally to have ceased any geological action, and only its surface is evolving as a result of very light residual meteoritic bombardment.

Pierre THOMAS

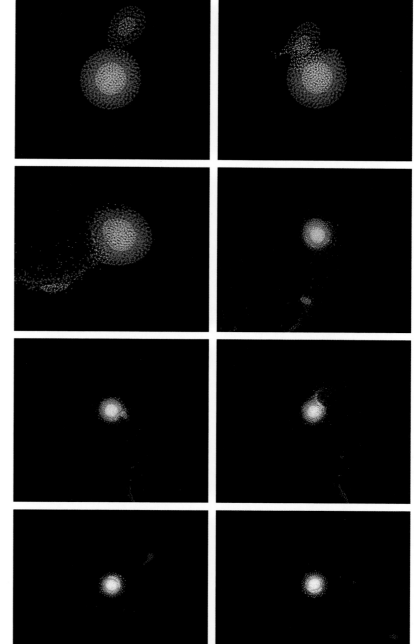

Computer simulation: this shows a possible mechanism for the formation of the Moon as a result of a giant impact between the Earth and a large asteroid. In this simulation the Earth is struck tangentially by an object of 0.14 Earth masses at a relative velocity of 11 km/s. The Earth is shown here with its nuclear core coloured rose and the mantle is red. The first two images, at the top, show the actual collision (t = 1.2 min and t = 11 min), in which the incoming object is destroyed. In the third frame a plume of the mantle streams into space, in an unstable orbit round the Earth. Most of it falls back, but a small nucleus (fourth frame) remains in orbit. The last four frames (t = 2.3 hours, to t = 24 hours) show debris (faint red object) with the mass of the Moon in Earth orbit. (W. Benz, W.L. Slattery and A.G.W. Cameron).

The history of the Moon. The first illustration represents the Moon 3.8 billion years ago. The anorthosite crust, long since solidified, is completely covered by craters of all sizes. One recognises, top left, the Imbrium basin with its characteristic concentric structures reminiscent of the Orientale basin (which has not been subsequently modified).

The second illustration represents the Moon a billion years later, 2.8 billion years ago. Numerous lower lying parts, in particular certain depressions and large craters, have been flooded by the discharges of basalt from the maria, represented here in dark grey. The high unsubmerged parts within the same craters are almost unchanged, which shows that the bombardment had virtually ceased at this time.

The third photograph, which is a mosaic of photographs, represents the Moon at the present time. There have not been any additional volcanic eruptions. Only a few new craters are conspicuous by their ray systems. (After D. E. Wilhelms and D. E. Davis, 1971; documents kindly provided by D. E. Wilhelms)

The Moon

The surface

A simple observation of the full Moon will show two kinds of regions on the surface of our satellite: the dark 'seas' or maria which surrounds the lighter 'continents' or highlands. With the advent of the first astronomical telescopes, Galileo discovered circular formations which riddle its surface: these are the lunar craters. The progressive improvement in the resolution of telescopes has allowed an ever increasing number of craters to be observed, as well as other, much rarer, structures such as sinuous valleys and escarpments. The maria appear much less rich in craters than the highlands, from which we deduce that they are much 'younger', that is to say, they formed later. Furthermore, it has been discovered that a few of the largest craters (Tycho, Copernicus) project 'rays', light, narrow bands capable of reaching 1000 kilometres in length. However, for a long time the scientific interpretation of all these observations remained scant. In particular it was not known whether the craters had a volcanic origin or whether they were the result of giant impacts. In addition, nearly half of the surface of the Moon remained unobservable from Earth.

The space age was destined to produce a revolution in our knowledge of the Moon. In 1959, Luna 3 revealed that the hidden side of the Moon is very different from the visible side: there are hardly any maria. The observation, in 1967 by the first artificial Moon satellites, of flat-bottomed craters demonstrated that the lunar surface is completely covered by a layer of dust, some 5 to 10 metres thick – the regolith. Subsequently, from 1969, the Apollo and Luna missions allowed multiple experiments to be carried out *in situ* and brought back nearly 400 kilograms of rock samples from nine different sites, both from the maria and highlands. It is the laboratory analysis of these samples which mainly has been responsible for the advance in our understanding of the moon's history and its interactions with the interplanetary environment.

The Moon has been dead for about three billion years. All internal activity has stopped. Only the interaction with its environment has modified its surface.

The incessant impacts of meteors of all sizes have eroded the surface rocks, producing the regolith. The finest meteoritic particles have unceasingly reduced the size of the regolith grains so that today they consist of a very fine dust (less than 0.1 millimetre on average). As a result of the violence of the impacts, a fraction of the grains has melted, and been projected over the surface in the form of fine droplets, giving rise to a great variation in the distribution of partially or totally vitreous material: spheroids, vitreous aggregates and splashes on the lunar rocks.

The formation of craters in the regolith produces a continual mixing of the regolith, exposing those grains at the surface to direct

The visible face of the Moon. Taken at Full Moon, this photograph clearly shows the light highlands surrounding the dark maria: the albedo is on average twice as high on the highlands as on the maria. Overall, however, the Moon is an extremely dark body, with a mean albedo close to 10 per cent. The brightest regions on the surface are those which are covered by the debris from very young craters. This debris can spread out far from the crater in the form of rays. (In the case of Tycho, which is prominent at the bottom of the photograph, these rays are longer than 1000 kilometres.) The contrast in albedo which they present with the surrounding material is explained by the difference in exposure times on the lunar surface: the action of the interplanetary environment induces an ageing of the exposed grains, which results in a progressive darkening. In a few tens of millions of years, the rays will merge into their environment. In the maria, the depth of the largest craters, such as Copernicus (towards the centre of the photograph) exceeds that of the layer of basalt; the debris made up of the underlying highland material, presents an even more marked contrast to the basalts of the maria. The smallest features discernible in this photograph taken from Earth are several kilometres across. (Hale Observatories)

contact with space. The smaller the grain, the less time it remains on the surface: a 1 micrometre grain only stays there a few thousand years, while a 100 micrometre grain would be exposed for hundreds of thousands of years. At any single location on the surface, the accumulation of layers of debris from neighbouring craters forms a stratified structure which is observed in the core samples taken by drilling to a depth of more than 3 metres. These samples are of great interest: in fact the deepest grains in these samples were on the surface and exposed to space during a distant epoch which can only be dated approximately. The core samples thus provide a mechanism by which we may follow the evolution of the interplanetary environment.

The flux of charged particles

No atmospheric or magnetic shield protects the Moon, which as a consequence receives the full force of all particles crossing interplanetary space.

The solar wind is by far the most intense source of these particles; every second, 100 million protons impact upon each square centimetre of the lunar surface, at speeds ranging from 350 to 650 kilometres per second, corresponding to an energy of the order of 1 kiloelectronvolt per particle. With such an energy, the particles only penetrate to a depth of a few tens of nanometres and as a result, only the grains on the top surface of the regolith are exposed to the solar

The collection of samples. An astronaut gets ready to collect samples around an enormous rocky boulder at the time of the last Apollo mission, 13 December 1972. Note that the surface of the regolith is strewn with rocky fragments. The majority are small and they only constitute a very small fraction of the total volume of the regolith. The antenna on the lunar vehicle which appears in the foreground allowed the astronauts to communicate directly with the control centre at Houston, even when they found themselves out of sight of the lunar landing module. (Apollo 17 photograph; NASA/NSSDC)

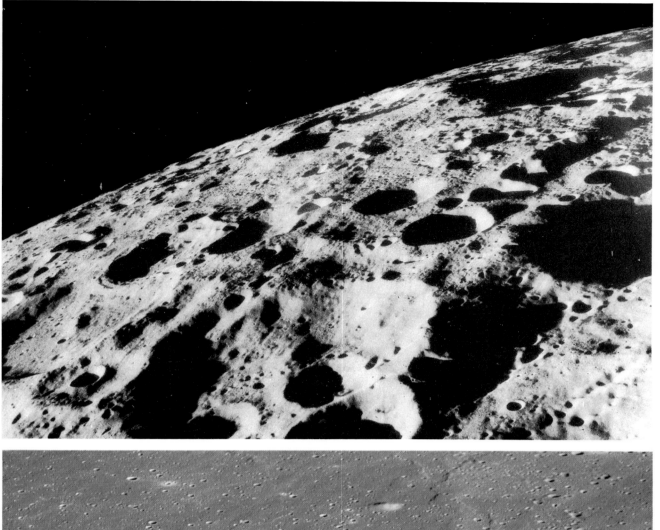

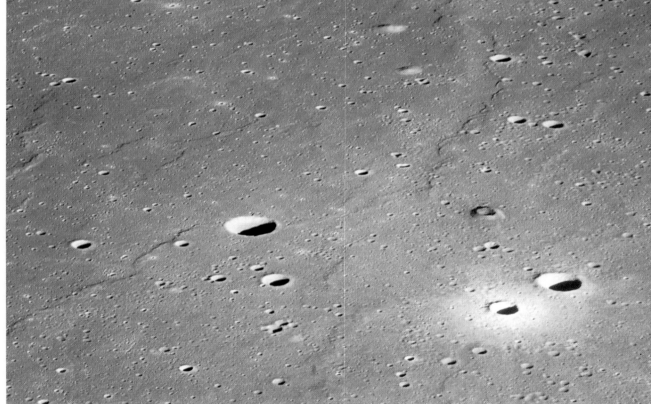

wind. The properties of the surface layers of these grains are, therefore, changed significantly.

Solar cosmic rays are a million times less intense, but have an energy a thousand times greater (about 1 megaelectronvolt per nucleon). They, therefore, penetrate up to a few tens of micrometres into the regolith. The tracks left by the heavy ions (of the iron group) as they traverse the material can be seen under a microscope. By counting the tracks it is possible to measure the length of time for which the grains were exposed on the surface. The characteristics of the tracks permit the determination of the energy spectrum and the relative abundances of the many elements in cosmic rays.

Galactic cosmic rays, one hundred times less abundant than solar ones, are sufficiently energetic (about 1 gigaelectronvolt per nucleon) to penetrate to a depth of more than a metre into the regolith and there to induce nuclear reactions in the atoms which constitute the grains. These effects in particular allow us to trace the history of the formation of successive strata which are observed in each sample of lunar soil.

The ageing on the surface

The combined effects of the meteoritic bombardment and the flux of charged particles show up as a progressive alteration of the macroscopic properties of the surface, or ageing. First of all, the ceaseless erosion by the smallest meteorites progressively smooths the surface features. A newly formed impact crater with an initial size of 100 metres will be completely eroded in a few hundreds of millions of years. The contrasts of light and dark observed on the surface of the Moon have several causes. Maria and highlands have a mineralogically different composition: the highlands are almost entirely made up of feldspars (anorthite) while the maria, arising from outflows of basalt, contain a high proportion of pyroxenes, as well as olivine and ilmenite, which make them on average twice as dark as the highlands. However, the gradual darkening of the surface produced by the ageing process affects both the maria and highlands equally. Although this phenomenon is still poorly understood, it can be interpreted as an increase in the glass content where the iron abundance is reduced and by the appearance on a large fraction of the grains of a layer badly damaged by the ions of the solar wind. Thus, a ray produced by a large continental crater initially presents a very strong contrast with the maria which it crosses. Gradually its constituent soils darken, the meteoritic bombardment mixes it with the surrounding material and it disappears after a few hundreds of millions of years.

Highlands and maria. Observed from lunar orbit, the highlands (above) appear a lot richer in large craters than the maria (below). This is explained by the fact that the maria solidified nearly a billion years after the highlands. The latter therefore underwent the intense meteoritic bombardment (which followed the accretion of the planetary bodies) of which traces are found throughout the Solar System from Mercury to the satellites of Saturn. During that period, certain giant impacts created the large circular basins which, when filled, formed the basaltic maria. Then and for more than three billion years, the highlands and maria underwent an identical, but less intense, bombardment and therefore display a similar density of small craters. (Apollos 17 and 16 photographs; NASA/NSSDC)

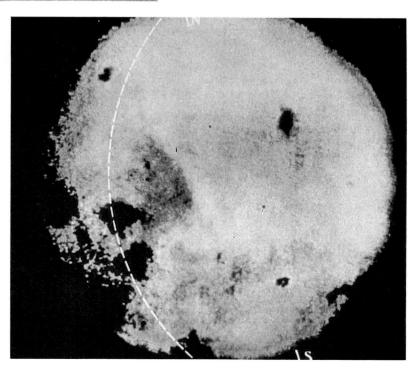

The hidden face of the Moon. It was in 1959 that the hidden face of our satellite was revealed for the first time by the Soviet Luna 3 mission. It was discovered that this face was almost completely devoid of dark maria. It is thought today that this asymmetry arises from the fact that the lunar crust is thicker on the hidden than on the visible side. The large basins formed on the hidden side have therefore not been filled up by the outflow of basalt as they have on the visible face. (USSR Academy of Sciences)

The surface of the Moon. This panorama of the western extremity of the Descartes mountains was taken on 23 April 1972 by the astronauts John W. Young and C. M. Duke on the Apollo 16 mission. On the left, the vehicle used by the astronauts to explore up to nearly 5 kilometres from the descent module is visible. In front of that is a crater of some tens of metres in diameter. This

The regolith. The lunar vehicle used by the astronauts from Apollo 15, 16 and 17 had a top speed of 12 kilometres per hour. Even at this low speed, because of the small average size of the regolith grains and their very weak cohesion, it raised a cloud of dust. (Top left; Apollo 16; NASA)

The wheel tracks of the vehicles (below left), like the foot prints left by the astronauts, will only disappear after several million years. The site of the last Apollo mission can be seen here in the valley of Taurus-Littrow. At each lunar landing site, the astronauts used instruments which were intended to study the Moon and its environment over a period of several years. Some of these instruments can be seen at the centre of the photograph (Apollo 17 photograph; NASA).

The seismometer put in place at the time of the Apollo 14 mission (above; it is also visible in the middle distance of the photograph below left) has measured the Moon's response to the explosion of grenades released by the mortar situated in the foreground; this has enabled the local thickness of the regolith – close to 10 metres – to be determined. The detection of moonquakes has, moreover, allowed astronomers to determine the internal structure of the Moon. (NASA)

crater has already been smoothed by the erosive action of the impacts of small particles which constitute the sole means of erosion; it will be completely filled up in a hundred million years. The hills which appear on the horizon are not unduly steep escarpments; even in this mountainous area, the average slope rarely exceeds a few degrees. (NASA/NSSDC)

The lunar surface and the history of the interplanetary environment

While the lunar surface has remained practically unchanged for more than three billion years, it is directly in contact with the interplanetary environment. It is therefore an ideal medium through which to study the history of this environment. Each sample taken from the regolith contains grains which have registered the effects of all the fluxes of particles and matter over several hundreds of millions of years. Among the major results already obtained, we may cite the variations in the isotopic composition of nitrogen in the solar wind – and therefore in the solar corona – which was progressively enriched in isotope 15 over a period of two billion years. The average energy of the ions of the solar wind also seems to have increased during this period. From these studies it appears that the composition and energy spectrum of cosmic radiation has evolved very little.

The future of Moon exploration

Despite the immense progress made in the last fifteen years, a lot remains to be done in the study of our satellite. In particular, only the regions near to the equator have been flown over, and mapped with good resolution from the point of view of their mineralogical and chemical compositions. The feasibility of a lunar satellite in polar orbit is currently being studied. This would extend the coverage to the high-latitude regions. Such systematic observations would enable scientists to test the hypothesis which proposes that volatile compounds, like water, should be trapped in certain polar craters which, being always in the shade, maintain a temperature of less than 200 K. Such a result would be essential for the long-term exploitation of our satellite.

Jean-Pierre BIBRING and Yves LANGEVIN

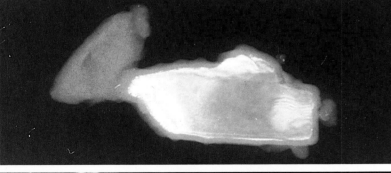

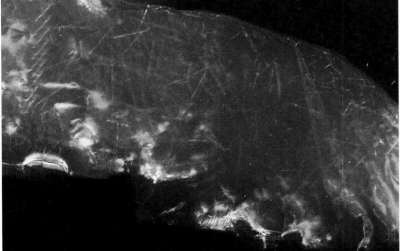

The effect of the solar wind and cosmic rays. Grains, a few micrometres in size, observed by a high-voltage electron microscope (above left) display the effects of their exposure at the surface of the regolith. The ions of the solar wind which travel at 400 kilometres per second only penetrate into the grains a few tens of nanometres. At the end of several thousand years, the dose accumulated by a grain is such that the crystalline structure is completely destroyed in the surface layer which then becomes amorphous. This appears as a grey surface completely surrounding the grain for, as a result of the micrometeoritic bombardment, the grain is frequently turned during its time on the Moon's surface.

More highly energetic ions emitted at the time of solar eruptions (solar cosmic rays) completely cross these grains of a few micrometers in size, leaving in their wake a linear track of destruction which is visible in the electron-micrograph (below left). The analysis of these effects in the grains of lunar dust allows us to retrace the history of the regolith grains and also to retrace the evolution of the solar wind flux over a period of two billion years. (Laboratoire d'Optique Électronique du CNRS, Toulouse)

Grains and lunar spherules. When grains of more than 100 micrometres are extracted from a sample of lunar dust (right), crystals of feldspar (light), pyroxine and olivine (yellow-green) and dark spherules can be observed. At the time of the impact of a micrometeorite, a fraction of the substratum is in fact liquefied; certain droplets thus ejected vitrify in flight and then assume a spherical shape: meanwhile another fraction percolates down between the grains and forms vitreous aggregates, which can constitute up to 50 per cent of certain soils. (Jean-Pierre Bibring)

Meteorites

Meteorites are 'stones fallen from the sky', which appear mysteriously in fields and gardens, following luminous phenomena ('shooting stars'), and intense explosions. It was Ernest Chladni, in 1794, and Jean-Baptiste Biot, in 1803, who demonstrated the extraterrestrial origin of these strange stones. A systematic search then started for these objects, which we can classify in two groups depending whether their fall has, or has not, been observed.

For nearly two centuries, meteorites were the only source of extraterrestrial matter at the disposal of scientists. Since 1969, Moon samples have been brought back from Apollo and Luna missions and meteorite science has been completely altered by the extremely delicate and precise analysis methods applied to lunar samples. Today, more than 3000 meteorites have been recorded on different falls; they are usually named after the place where they have fallen. This number is increasing rapidly; indeed, it has been recently discovered that there are some very rich 'meteorite fields' in the Antarctic ice. These meteorites are even more interesting as they have been preserved intact for thousands of years.

The number of preserved objects in the museums of the world greatly outnumbers the 3000 meteorites already catalogued. Indeed, each meteorite is usually considerably fragmented on entry into the atmosphere and sometimes dispersed over hundreds of square kilometres. Because of the atmospheric shield, the size distribution of objects collected on the ground is very different from that in interplanetary space. It is estimated that hundreds of tonnes of meteoritic matter of all kinds fall each year, about 10 000 tonnes penetrating the Earth's atmosphere with a velocity of 11 to 70 kilometres per second. 'Micrometeorites' of a smaller size, about one-tenth of a millimetre, are continually falling slowly through the stratosphere without getting overheated because of their higher surface to volume ratio; after reaching the low stratosphere, they are descending at a slower velocity of the order of one centimetre per second which allows them to be recovered. If the initial size of the object is from one-tenth of a millimetre to a few centimetres it is completely destroyed. Bigger meteorites, up to a few tonnes, heat up very rapidly on the outer surface; nevertheless, after ablation, that is, melting and stripping away of material, a fraction of their mass falls to the ground, at a speed of one hundred metres per second. Finally, the very large objects (many metres in diameter) do not slow down, and are the origin of big craters such as the meteor crater in Arizona. Luckily such events are very rare.

Shooting stars or meteors result from the volatilisation of intermediate-sized objects in the atmosphere. The frequency of these events varies during the year. In fact, at the same times each year, we can see especially intense meteor showers during which all shooting stars seem to come from the same part of the celestial sphere, centred on a point called the radiant. All the objects coming from a given radiant in fact belong to the same swarm, made up of the debris of a comet during its successive journeys in the neighbourhood of the Sun. However, as we will see, most meteorites probably come from another

Distribution of meteorites. The diagram below shows the number N of objects with masses greater than a given mass m which each year penetrate the atmosphere over the entire surface of the Earth with velocities of several tens of kilometres per second. This distribution is schematically represented by three power functions which are negative powers of the mass for m varying from 10^{-15} to 1 kilogram. In the case of the smallest objects (less than a milligram) the function is determined directly by exposing metallic foils placed in orbit around the Earth to the meteoritic bombardment. We note that it is the smallest objects weighing less than 0.1 milligram which constitute most of the mass hitting the terrestrial atmosphere. Only 10 000 objects weighing more than 1 kilogram, which are the origin of meteorites, penetrate the atmosphere each year.

The photograph above shows two meteor trains photographed in the star field in the constellation of Scutum. (Yerkes Observatory)

Extraterrestrial dust. The particle below – its greatest dimension is approximately 18 micrometres – is representative of a large fraction of the dust of extraterrestrial origin collected in the stratosphere (photograph kindly supplied by D. E. Brownlee). To collect these particles small plates covered in very viscous oil are deployed under the wings of aircraft flying at an altitude of about 20 kilometres, (right, NASA photograph kindly supplied by U.S. Clanton). The trapped particles are then cleaned and sorted to separate matter of extraterrestrial origin from terrestrial contamination. More than half of these extraterrestrial particles are aggregates measuring less than 100 nanometres, whose composition approximately reflects that of carbonised chondrites. Chemical and physical analysis of these samples leads us to postulate that they are of cometary origin.

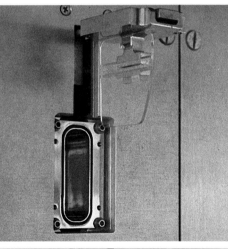

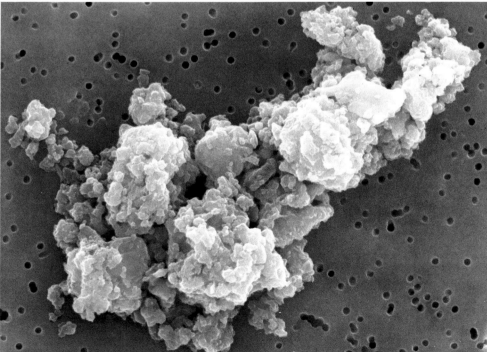

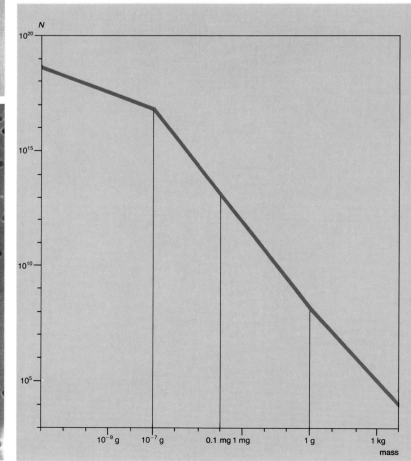

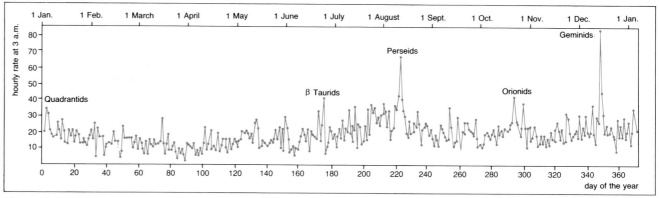

source, the asteroids.

Mineralogical and chemical studies of meteorites have rapidly shown us their extreme diversity. Some are nearly completely metallic (iron and nickel), others are like terrestrial rocks, some others, very friable, are rich in carbon. Moreover, we discover inclusions in a large fraction of these objects of a more or less well-preserved spherical shape, the chondrules. Finally, the size of the constituent crystals, which in terrestrial rocks characterises slow or fast cooling, varies greatly among meteorites of the same chemical and mineralogical composition. It is on the basis of these three criteria that we have been able to classify the best known meteorites.

Meteorites are thus divided into three groups: iron meteorites, lithosiderites and stony meteorites, depending on the metal content. Stony meteorites are by far the biggest group (accounting for more than 90 per cent of observed falls. We can distinguish among them the chondrites – containing chondrules – and the achondrites – containing no chondrules. The chondrites are the most abundant type of stony meteorites. We can distinguish the carbonaceous *Text continued on page 122*

Meteor showers. The average hourly number of shooting stars, of magnitude greater than 2, that can be seen varies little during the course of a year (above, from D. W. Hughes, 1978) with the exception of the peaks corresponding to meteoric showers, when the Earth encounters a cometary stream. On average, the most intense are the β Taurids at the end of June, the Perseids at the beginning of August, the Orionids at the end of October, and the Geminids in December. During a shower all the shooting stars seem to come from the same region of the sky, the radiant (right; this shows the great Leonid shower of 1966). This phenomenon is caused by the fact that the trajectories of objects in the same swarm are nearly parallel and thus appear to be diverging when viewed from Earth. (© 1974, Charles Capet)

The main showers. The table gives the periods over which the shooting stars of the main showers can be observed and the comet which is associated with each of these showers; the period of heliocentric revolution of each of the comets is also indicated.

shower	dates	associated comet	period (years)
Lyrids	10–20 April	Thatcher	
η Aquarids	1–8 June	Halley	76
Orionids	18–26 Oct.	Halley	76
β Taurids	24 June–6 July	Encke	3.3
Perseids	25 July–17 Aug.	Swift–Tuttle	120
Draconids	9–10 Oct.	Giacobini–Zinner	6.4
Andromedids	2–22 Nov.	Biela	
Leonids	14–21 Nov.	Tempel–Tuttle	33.2
Geminids	7–15 Dec.	*	*
Ursids	17–24 Dec.	Tuttle	13.6

* Nucleus recently discovered (1983) by IRAS observations. Period unknown.

The Meteor Crater. The most famous crater made by the impact of a giant meteorite on the Earth, is in Arizona. With a diameter of 1.2 kilometres and 150 metres deep, it was formed about 40 000 years ago, and shows a marked similarity to lunar craters. It has been estimated that the size of the metallic meteorite responsible for the crater was of the order of 25 metres (that is a mass of 65 tonnes). However, because of the violence of the impact, only a very small fraction of this mass escaped destruction and could be recovered; this is the Canyon Diablo meteorite. Such an event only happens about once every 25 000 years on land above sea-level. (The impact of the largest objects (of the order of 1 kilometre) only occurs about once every 100 million years. However, such events could have played an important role, in particular by throwing into the atmosphere hundreds of cubic kilometres of matter, which is nearly one hundred times more than the greatest volcanic eruptions. Such a quantity of dust encircling the whole Earth for many years would profoundly alter the climate by intercepting the light from the Sun. An event of this type could have been responsible for the disappearance of a large number of species at the end of the Cretaceous Period. (Yerkes Observatory)

The streams. The streams responsible for three of the main meteoric showers are represented here in relation to the orbits of the Earth and Jupiter. The most clearly visible part is situated below the plane of the ecliptic. It is the tangent to the trajectory of the stream at the point it cuts the Earth's orbit which determines the radiant of the shower. The orbits of these streams seem distributed at random in space, unlike those of the planets which are all situated near the ecliptic plane. It has been found that some of these streams, such as the Perseids, are associated with long-period comets, and others such as the Geminids are associated with short-period comets. On each perihelion passage the comet loses a large number of particles. Under the effect of solar radiation pressure, the smallest grains are put into a hyperbolic trajectory and finally leave the Solar System. Only particles bigger than a few micrometres remain in the same trajectory as the mother comet and thus renew the associated stream. (D. W. Hughes, 1978)

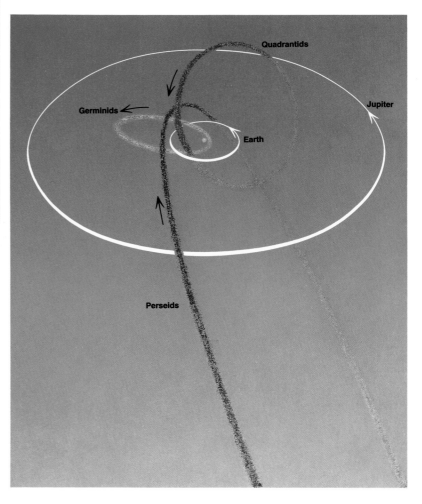

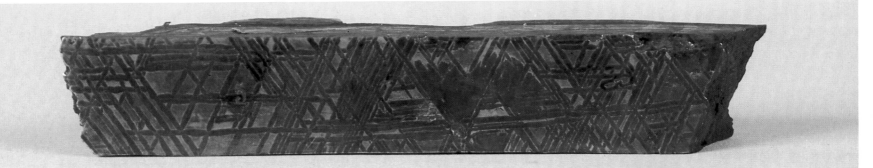

Metallic meteorites. Metallic meteorites consist almost entirely of iron and nickel in two main groups: kamacites, where the nickel content is less than 8 per cent and which crystallises in a body-centred cubic lattice and taenites, where the nickel content is over 20 per cent and which crystallises in a face-centred cubic lattice. The hexahedrites, which are nickel-poor (containing less than 6 per cent) consist of large cubic crystals of kamacite. When acid-etched a delicate series of parallel bands known as Neumann lines appear, most probably the direct result of the shock of the impact resulting in the fragmentation of the parent meteorite. The more numerous octahedrites, richer in nickel (content about 12 per cent), consist of kamacite layers bordered by taenite surrounding plessite domains (mixtures of kamacite and taenite crystals). These are in a geometrical arrangement known as the Widmannstäten pattern, arranged in an octahedral shape, which gives this class its name. The higher the nickel content, the thinner the kamacite layers. This particular structure can be explained either by extremely slow cooling or, on the contrary, by very sudden solidification of a relatively nickel-rich phase. These shapes clearly appear on the faces of three octahedrites: the Carthage meteorite (above top), the Roebourne (left) and the Canyon Diablo (above right), which comes from the Arizona Meteor Crater.

The lithosiderites. The lithosiderites consist of a metallic and a silicate phase, closely imbricated. Two groups can be distinguished: the mesosiderites, such as the meteorite Vaca Muerta (below), in which the silicate phase mainly consists of feldspar and pyroxenes; and the pallasites, such as the Springwater meteorite (left), where olivine prevails. In this last group a metallic matrix, similar to that of the octahedrites surrounds the olivine crystals, but very often the opposite is true for the mesosiderites (that is a silicated matrix with metallic inclusions). The pallasites would have been formed by the invasion of a metallic liquid phase at the interface between the olivine-rich mantle and the core of the parent asteroid, whereas the metallic inclusions of the mesosiderites have most probably been incorporated when in a solid phase.

Carbonaceous chondrites. Carbonaceous chondrites are made up of a matrix of very fine carbon-rich crystals which can contain inclusions of olivine, pyroxine, glass or metals. These are the only meteorites where silicate sheets (of mica) are found, most frequently in the hydrated form: the water content can be more than 20 per cent in carbonaceous chondrites. They are classed in types I, II or III according to the decreasing carbon, water and iron sulphide (troilite) content. The Orgueil meteorite (opposite) of the very friable type C I, is made up of very small particles. The carbon content is more than 3 per cent, and that of iron sulphide 15 per cent. The Allende meteorite of type C III (above) is one of the most massive known: its total weight was 2 tonnes. It contains very many light coloured refractory inclusions in which isotopic anomalies were first observed. These meteorites, which have no equivalent among terrestrial rocks, are generally considered to be those which have been least altered since their formation.

The table opposite gives the original abundances in numbers of atoms in the Orgueil meteorite (top line) and in the solar atmosphere (bottom line) taking the silicon content as unity.

hydrogen	carbon	oxygen	sodium	magnesium	aluminium	silicon	sulphur	calcium	chromium	iron
6.2	0.7	8.2	0.07	1.1	0.1	1	0.5	0.06	0.01	0.9
30 000	12	21	0.06	1.1	0.09	1	0.5	0.07	0.01	0.8

Note. The twelve meteorites and meteoritic fragments in the colour photographs appearing on these two pages and the following page come from the collection of meteorites of the Muséum National d'Histoire Naturelle (National Natural History Museum) in Paris, and have been photographed by kind permission of P. Pellas .

Ordinary chondrites. Ordinary chondrites are divided into H, L, and LL chondrites in order of decreasing iron content (H, L, and LL standing for high, low, and low-low). In each of these three groups degrees of crystallisation from III to VI are attributed which could indicate the increasing depth at which they were buried in the parent body or bodies. The main constituent minerals are olivines, pyroxenes and feldspars, just as in terrestrial or lunar rocks, often with a metallic phase. The chondrules are spherical inclusions of approximately 1 millimetre diameter, perfectly recognisable in certain chondrites, such as the Saint-Mesmin (below), but nearly completely incorporated in the mineral matrix in others, such as the Meso-Madaras (opposite). Chondrules have only been observed in some meteorites and their origin is still not well understood. In the Saint-Mesmin meteorite a dark iron-rich material appears. This inclusion became incorporated in the meteorite about 1.3 billion years ago. This event therefore took place much more recently than the formation of most meteorites, which reflect the origin of the Solar System some 4.5 billion years ago.

chondrites, the ordinary chondrites, composed mainly of olivine and pyroxene, and the chondrites of enstatite, where nearly all the iron is in a metallic or sulphide state. The achondrites are the meteorites most similar to certain lunar and terrestrial basaltic rocks.

Almost half the meteorites are breccias, compressed materials of the same composition (monomict breccias) or of different compositions (polymict breccias). Among these breccias some are notably rich in concentrations of rare gases produced by the solar wind, and some produced by irradiation by solar cosmic radiation.

The great diversity of meteorites is directly linked to their origin and their history in the parent body. Their mineralogic composition especially reflects different degrees of thermal evolution; most of the meteorites, including the carbonaceous chondrites, are 'primitive', in the sense that they have not been submitted to temperatures high enough to alter the main minerals. The ordinary chondrites would have undergone such phase changes. Finally, the achondrites, the lithosiderites and the siderites are the result of a complete fusion followed by a re-crystallisation (or differentiation). Among the possible parent bodies the comets have kept very low temperatures (under 300 K) since their formation. If they were the origin of 'primitive' meteorites, especially the very small ones, they cannot be the source of differentiated meteorites, achondrites, siderites and lithosiderites. Therefore, it is now thought that the majority of meteorites come from the asteroids, whose diversity amply covers all classes of meteorites. The achondrites could come from the 'crust' or 'mantle' of different asteroids, the siderites from the cores, and the lithosiderites from the mantle/core interface. However, this plausible scenario requires a major source of heat to melt the asteroid; this cannot be provided by free energy during accretion or by the release of heat from long-period radioactive elements (uranium, thorium, potassium) since the small sizes of these bodies (some hundred kilometres for the biggest) would allow efficient radiative cooling. On the other hand, heat from a 'short'-period radioactive element (some million years at most), such as aluminium 26, would be sufficient to melt bodies of a size greater than ten kilometres. The cometary nucleus and the smaller asteroids would have escaped fusion.

To explain the breccia structure of a large proportion of meteorites, we usually think of a violent impact on the surface of the main mass able to fracture and to agglomerate the substratum by the compression waves generated. Among these breccias those formed from exposed particles on the surface of a regolite would correspond to gas-rich meteorites, as they would have been richer in volatile elements produced by the solar wind. In contrast to cometary origin, asteroid origin poses the dynamical problem of the transfer of the meteorites from the main belt to the Earth. One of the possibilities proposed is that by collision in the belt, then by gravitational perturbations, the special Apollo–Amor family is formed, whose orbit passes the neighbourhood of the Earth. This family would constitute one of the main meteorite reservoirs.

The time to traverse the distance between this last reservoir and impact on Earth may be determined by the degree of irradiation of these meteorites by cosmic rays, which induces nuclear reactions. Thus, it appears that siderites have very often stayed many hundreds of millions of years in space. Stony meteorites are generally exposed for a much shorter time, less than a hundred million years. This striking difference is still not well understood. Moreover, these exposure times are often found to lie in groups for meteorites of the same class. We can postulate that reservoirs of different families of meteorites are re-stocked following a small number of events, such as the complete fragmentation of a nearby asteroid.

The systematic study of the chemical composition of meteorites shows us an essential result: that the vast family of chondrites show a remarkable similarity in their composition. Also, the proportions of the elements are close to those observed in the Solar System. The only exceptions concern the very volatile elements (such as hydrogen, carbon, oxygen and the rare gases), very rare in chondrites, and lithium, burned by the nuclear reactions of the Sun. This finding reinforces the hypothesis that the chondrites represent a condensate, little modified chemical-

The achondrites. The achondrites consist of similar materials to those found in lunar rocks and terrestrial basalts; that is pyroxenes and plagioclases. We can distinguish aubrites, consisting almost entirely of enstatite, magnesio-pyroxene, and achondrites composed of pyroxenes and plagioclases: eucrites, howardites and diogenites, classified by decreasing iron and calcium content. The eucrites (Serra de Mage, bottom) are the meteorites which most resemble terrestrial basalt. The howardites (Binda, top, and Pavlovka, centre) are polymict breccias resulting from initial fragmentation and subsequent aggregation of different types of rock, whereas the diogenites are monomict breccias, where the fragments are of a single type of rock.

ly, of the matter constituting the primitive solar nebula. The achondrites, on the other hand, present some very varied chemical compositions due to the differentiation they have undergone.

The isotopic composition of meteorites is generally similar to that of terrestrial rocks and lunar samples. Some important differences have been observed in certain elements, which we can explain by the influence of three processes: the fragmentation which occurs during all phase changes, chemical fusions and reactions; nuclear reactions induced when the particles are irradiated by cosmic rays; and the decay of radioactive elements, which enriches the isotopic particles. This last process has a very important application: the radioactive dosage produced by these isotopes allows us to date the main stages in the history of meteorites.

The main 'radioactive series' used are rubidium 87 and its daughter isotope strontium 87, potassium 40 and argon 40, and uranium 238 and lead 206. Knowing the disintegration times, this isotopic measurement of the (daughter) elements gives us the date when the meteoric reservoir was created, that is to say, the last time that the original isotope was mobile. We then find the time at which the constituent material was isolated from the primitive solar nebula, that is, the time of the last crystallisation during the last important metamorphosis event. Furthermore, it is possible to determine the isotopic composition of the mixture from which the material is condensed. The most important results are as follows. The constituent material of the meteorites condensed 4.55 billion years ago, at the same time as the accretion of the Earth and the moon. The constituent minerals of most meteorites were formed very early in the history of the Solar System. Among the meteoritic breccias, the majority have been formed by being compacted more than four billion years ago.

Up to 1970, the main point of isotopic analysis of meteorites was in the context of a model of the condensation of the Solar System from a protosolar nebula of homogeneous composition. The differences observed were explained through the later history of meteorites.

The possibility of achieving measurements with increased accuracy on very small samples (less than a milligram) has completely upset this conception. Indeed, some meteorites have revealed themselves very inhomogeneous on a small scale. This is especially the case of the carbonaceous Allende chondrites, which fell in 1969 and weighed more than two tonnes. Within the dark matrix are inclusions made up of very refractory minerals rich in calcium and aluminium. The isotopic analysis of oxygen in these inclusions has given an important result: this result is that up to 5 per cent of meteoritic particles contain only oxygen 16, though most are of 'normal' isotopic composition. Such an 'isotopic anomaly' can only be explained by the incorporation of particles condensed from a primitive solar nebula in the vicinity of a star containing only the oxygen 16 isotope, very probably a supernova. This now refutes the hypothesis of the homogeneous nature of the primitive solar nebula, which would have implied the volatility of all presolar particles. Following this discovery, the systematic study of the analyses of isotopic oxygen content in the different families of meteorites, has shown that the differentiated meteorites, and also the terrestrial and lunar samples, are formed from a common isotopic source. On the other hand, the H chondrites on one side, and L and LL on the other are each produced from different isotopic sources.

Isotopic anomalies in other elements have been similarly observed in meteorites. We shall mention here only observations made in the isotopic analysis of magnesium. Indeed, some minerals contain a superabundance of magnesium 26, correlated with an abundance of aluminium in the sample. This excess of magnesium 26 very probably results from the radioactive decay of aluminium 26. The high quantity of magnesium 26 could constitute the necessary heat source for the differentiation of a body of more than 10 kilometres diameter.

As the half-life of aluminium 26 is approximately 700 000 years, the period between the formation of this isotope (by nucleosynthesis in a supernova) and the condensation of meteoric particles cannot exceed some millions of years. Thus, a supernova explosion could only have preceded the condensation of the primitive solar nebula by a relatively short time. This could not possibly be coincidental but must be a general phenomenon: the most massive stars, exploding violently would generate a shock wave during their final explosion, which would induce the formation of a new stellar system. Thus, the progress of analytical techniques has brought to the study of meteoric samples an important result in planetology and in astrophysics. Meteorites are today shown to be unique witnesses to the origin and to the first stages of the evolution of the solar system; to the stellar environment of the primitive solar nebula; to possible survival of pre-solar particles; to the dynamic condensation of planetary bodies; and to the thermal evolution of the solar nebula.

Jean-Pierre BIBRING and Yves LANGEVIN

Irradiation tracks observed with an optical microscope in a meteoritic olivine crystal. These tracks are formed when a heavy ion from solar or galactic radiation passes through the particle. The crystal matrix is then damaged along a fraction of the ion's course. This is made visible by the preferential action of an appropriate reagent. It is then possible by counting the number of these tracks on the surface (their surface density) to determine the period of time for which the particle has been exposed to cosmic radiation, whether at the surface of the parent body or in transit between the point of fragmentation of the parent body and the Earth. Another type of track is produced by spontaneous fission of uranium and thorium nuclei, which are present in some minerals. When observing the densities retained by neighbouring particles, it is possible to trace the kinetic cooling of stony meteorites, due to the differences of thermal stability between these different minerals. For the lithosiderates and metallic meteorites other techniques allow thermal history to be retraced, by study of nickel diffusion between the kamacite and taenite boundaries when in the metallic phase. The deeper the meteorite was buried within the parent body, the slower the cooling rate. (D. Lal, 1972)

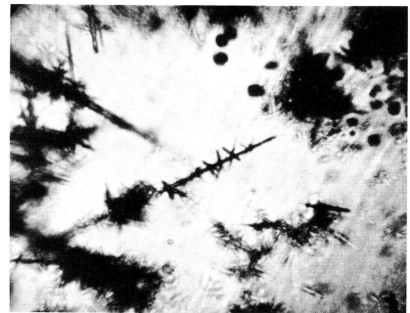

Graph of the rubidium–strontium evolution of the eucrites. One of the most commonly used isotopic series for the dating of events many billions of years ago is the decay series from rubidium 87 to strontium 86 and 87. Indeed strontium 87 results from decay of rubidium 87 with a half-life of 47 billion years. The $^{87}Sr/^{86}Sr$ ratio increases with time, in proportion to the time elapsed, until the original ratio of $^{87}Rb/^{86}Sr$ is reached. In a graph where the $^{87}Sr/^{86}Sr$ ratio is plotted as a function of the $^{81}Rb/^{86}Sr$ ratio, two samples of different initial Rb/Sr composition, but formed at the same time, are situated on a line whose gradient is a function of their common age. Thus the points which represent the different eucrites lie upon a line where the gradient corresponds to an age of 4.56 billion years. This age represents the time when the parent body of these objects was differentiated, only a very short time after the formation of the Solar System. (From C. J. Allegre et al., 1975, and J. L. Birck et al., 1975)

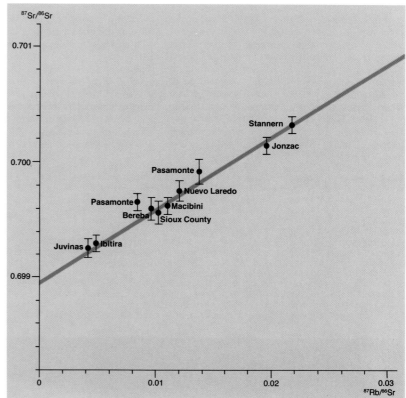

Anomalies in isotopic oxygen. Terrestrial oxygen possesses three stable isotopes of atomic masses 16, 17 and 18 in the following approximate proportions: oxygen 16, 99.76%; oxygen 17, 0.04%; oxygen 18, 0.20%. However, the isotopic ratios of oxygen are not strictly constant in all terrestrial matter, whether in the atmosphere, in the oceans or in various rocks. During chemical reactions and phase changes such as vapourisation or condensation, the various isotopes do not behave in exactly the same manner. This isotopic differentiation occurs with a common rule: that all variations of the ratio of oxygen 16 to oxygen 17 (separated by one atomic mass unit) correspond to a double variation of the ratio of oxygen 18 to oxygen 16 (separated by two atomic mass units). In a graph where the $^{17}O/^{16}O$ ratio is plotted as a function of the $^{18}O/^{16}O$ ratio, this is expressed by the fact that all points corresponding to terrestrial samples are situated on the same line of gradient 1/2, called the line of isotopic fractionation. A schematic graph is shown in the left-hand diagram. We note that points representing lunar samples similarly fall on this line. On the other hand points representing oxygen samples from the refractory inclusions of the Allende meteorite are situated on a line of gradient 1 called the isotopic mixture line, which means that in these mineral

phases the ratio of oxygen 17 to oxygen 18 is constant with only the absolute quantity of oxygen 16 varying. No known processes of radioactive decay, irradiation or mass fractionation can account for this exclusive enrichment of isotope 16. This isotopic anomaly, discovered in 1973, could be interpreted in the following manner. These refractory inclusions have in varying proportions (up to 5 per cent) some particles of normal isotopic ratio (shown by the black circle) and some particles having only oxygen isotope 16.

An enlargement of the region of the graph in the neighbourhood of the intersection of the mixture and fractionation lines is schematically shown on the right. It presents evidence of the spatial inhomogeneity of the primitive solar nebula. Lunar and terrestrial samples form one line. The ordinary H chondrites, the L and LL types of chondrite, and the carbonaceous chondrite matrices are each situated on different parallel lines of gradient 1/2. Each line corresponds to a different fractionation. However, processes of mass fractionation would not allow samples to pass from one line to another. This would imply that the parent bodies of different meteoritic families have condensed from different reservoirs, that is, containing different proportions of pre-solar particles. (From R. N. Clayton et al., 1973).

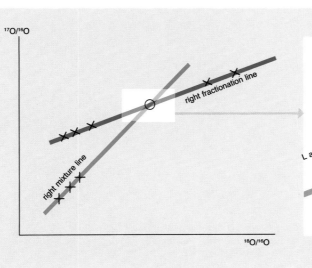

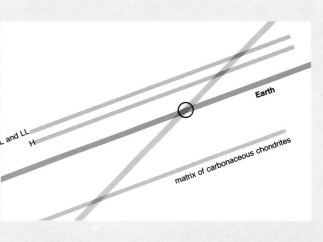

Mars

It is not surprising that astronomers have always considered Mars as a second Earth; in fact these two planets have many features in common and planetary exploration has done nothing to change this idea: the Red Planet is always held to be the most 'terrestrial' of the other objects of the Solar System.

Mars is the fourth planet counting outwards from the Sun; its equatorial radius – 3398 kilometres – and its mass – 6.418×10^{23} kilograms – make it the second smallest in size (after Mercury) of the inner planets. Mars is a solid, differentiated body whose fairly low average density – 3.9 grams per cubic centimetre – shows that iron represents 25 per cent (by mass) of the composition of the planet. By way of comparison, the corresponding figure on Earth is 33 per cent. No radiation belt has been detected, but a very weak magnetic field (its intensity is equal to about 2 per cent of the Earth's magnetic field) suggests the possible existence of a metallic core which may form the centre of a dynamo mechanism. The planet's moment of inertia also reinforces the idea that Mars has a dense core, probably rich in iron, both in metallic (Fe) and

sulphurous (FeS) form. The density of the mantle – 3.45 grams per cubic centimetre – shows that it contains a high proportion of forsterite and a low abundance of metallic iron.

The formation of Mars by accretion must have occurred at sufficiently high temperatures to enable it to be enriched in metallic iron and depleted in volatile elements, but also at low enough temperatures to allow a degree of oxidation of the planet. The major differentiation of Mars into a crust of about 50 kilometres in thickness, a mantle rich in olivine and ferrous oxide (FeO) and a metallic core (Fe–FeS) of 1500 to 2000 kilometres radius, implies that processes involving melting of the interior occurred at least once in the history of the planet.

It was necessary to await the images transmitted by the American probe Mariner 9 in 1971 and especially those taken by the Viking missions since 1976, to discover the multiple aspects of the surface of Mars. Previously, the planet had been briefly explored by three American space probes: Mariner 4 in 1965 and Mariners 6 and 7 in 1969. The Soviet probes, Mars 2 and 3, launched in 1971, and Mars 4, 5, 6 and 7 launched in 1973,

Mars during the approach of Viking 1. These two pictures of Mars were taken by the orbiter Viking 1 during its approach to the planet on 17 (below) and 18 June 1976 (opposite).

The surface of Mars can be roughly divided into two:
– the northern hemisphere is occupied by relatively recent plains, containing few impact craters; these plains appear light in the two pictures. The giant volcanoes, the single Olympus Mons and the Tharsis Montes chain dominate the plains of the northern hemisphere from a height of 20 kilometres (below).
– the southern hemisphere is made up of very ancient cratered lands, which appear dark. In these regions are found large basins, such as the Argyre Planitia (internal diameter 900 kilometres) clearly visible at the terminator on the picture opposite and at the lower right edge of the image below; on the latter the edges of the basin are highlighted in white by deposits of hoar frost (CO_2).

On the equator, the large canyon of Mars, Valles Marineris, spreads out over a length of 5000 kilometres. A part of this canyon is visible at the terminator on the image opposite.

The northern and southern ice caps do not appear on these two images. The light shading visible at the limb is produced either by deposits of hoar frost, or by cloud formations. (NASA)

were designed to place a satellite in Martian orbit and deposit an automatic station on the surface of the planet; unfortunately, technical difficulties cut these missions short before all the scientific objectives had been achieved.

The aim of the Mariner 9 mission was to photograph 70 per cent of the surface of the planet using two television cameras, one with a wide field and a resolution on the ground of about one kilometre, the other with a narrow field giving a resolution of about a hundred metres; in fact, more than 7000 images covering virtually all of the surface were obtained. Ultra-violet and infrared spectrometers studied the composition, temperature and density of the atmosphere. The infrared spectrometer was also designed to measure the temperatures and physical properties of the surface.

The tens of thousands of images transmitted by the Viking orbiters complemented (with a higher resolution) the general coverage of the planet achieved by Mariner 9 and increased the extent of our geological knowledge.

The topography of Mars has been established from radar observations carried out from Earth and also with the help of the study of radio occultations of Mariner 9. Further observations carried out in the infrared and ultraviolet added to our knowledge of this topic. The planet is not a perfect sphere since its polar radius measures 18 kilometres less than its equatorial radius, so that the degree of flattening is of the order of 1/192. Differences in altitude of more than 30 kilometres exist on Mars. The highest area is that of the Tharsis plateau where the giant volcanoes are situated and where the height of the crust reaches about 6 kilometres; this plateau extends over 4000 kilometres from north to south and over 3000 kilometres from east to west. The Elysium Planitia region overlooks the surrounding plains from a height of 4–5 kilometres; as in the case of the Tharsis plateau, it consists of a large dome of 1500 kilometres diameter, within which are numerous volcanoes. The Thaumasia plateau, situated at an average altitude of 4 kilometres, is noteworthy for the intense scarring of the surface.

The surface of Mars is characterised by a very pronounced morphological and topographical asymmetry between the northern and southern hemispheres. This is most evident either side of the great circle inclined by about 35 degrees to the equator. From the morphological point of view,

Ancient terrains and recent plains. The boundary between the ancient, strongly cratered terrains situated mainly in the southern hemisphere and the recent plains of the northern hemisphere constitutes one of the predominant morphological features of the Martian surface. Depending on the region, this boundary is seen in various guises. In numerous places (as illustrated by this image showing an area of about 700 × 800 kilometres situated to the south of the Amazonis region) the boundary is marked by the obvious 'flooding' of the ancient terrains by the sparsely cratered formations of the plains. The ancient terrains represent the primitive crust of the planet which has been intensely bombarded since its formation. In these lands two major populations of meteorite impact craters can be observed: firstly, a population of craters with an average diameter greater than 20 kilometres, which show signs of erosion and, secondly, a population of smaller, better preserved, craters, whose formation has contributed to the erosion of the craters of the other population. The first population of craters is older; it was produced right at the beginning of the planet's history, at the time of the intensive bombardment which marked the end of the period of accretion (4.6 to 4.5 billion years ago). The most recent craters date from the end of the period of intensive bombardment about 4.0 to 3.9 billion years ago.

The ancient terrain shows networks of channels which generally end abruptly at the boundary of the regions covered by the formations of the recent plains. These channels probably result from liquid flows on the surface. The young plains have quite a smooth, sparsely cratered surface. Formed after the end of the general meteoritic bombardment (3.9 billion years ago), their origin is still poorly understood. (NASA/NSSDC)

Formations and terrain. The polar formations correspond to the different features situated on the periphery of the northern and southern ice caps, the recently stratified aeolian deposits and the thick, irregular, wind-eroded deposits.

The volcanic systems comprise the recent volcanic structures, such as the giant volcanoes of Olympus Mons or Tharsis Montes, and the great ancient volcanic plains situated around these volcanoes and on both sides of the canyon of the Valles Marineris; here it is possible to distinguish:
– volcanoes;
– sparsely cratered recent volcanic plains;
– moderately cratered and ancient volcanic plains;
– heavily cratered volcanic plains displaying ancient eroded volcanic shapes, such as the surface ripples which are similar to those of the lunar seas.

The remodelled terrains are in general situated in the northern hemisphere and correspond to recent terrains which have been modified by erosion (wind, flow). They consist of:
– chaotic terrains made up of huge boulders, in general associated with channels or situated in closed depressions;
– eroded terrains with a high relief, situated on the boundary between the ancient, very cratered terrains of the southern hemisphere and the young plains of the northern hemisphere;
– terrains with a high relief, isolated in the plains of the northern hemisphere and unconnected to the channels or the boundary between ancient terrains and young plains;
– deposits at the bottom of the channels composed of materials transported by wind and liquid or produced by the collapse of the sides of the channels;
– lightly cratered plains containing channels and irregular wrinkles, sometimes covered by powdery deposits brought by the winds;
– terrains situated on the periphery of Olympus Mons, with irregular reliefs containing depressions and quasi-rectilinear mountainous alignments.

The ancient terrains are in the main situated in the southern hemisphere and are characterised by their relief due essentially to the presence of numerous meteoritic impact craters of all sizes; represented are:
– very cratered terrains;
– mountainous terrains situated at the periphery of the large basins and made up of materials ejected at the time of the formation of the basins.

The landing sites of the Viking 1 and 2 stations are shown. (Cartography after T. A. Mutch and J. W. Head, 1975)

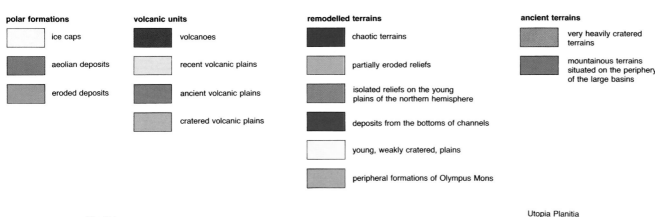

polar formations	volcanic units	remodelled terrains	ancient terrains
ice caps	volcanoes	chaotic terrains	very heavily cratered terrains
aeolian deposits	recent volcanic plains	partially eroded reliefs	mountainous terrains situated on the periphery of the large basins
eroded deposits	ancient volcanic plains	isolated reliefs on the young plains of the northern hemisphere	
	cratered volcanic plains	deposits from the bottoms of channels	
		young, weakly cratered, plains	
		peripheral formations of Olympus Mons	

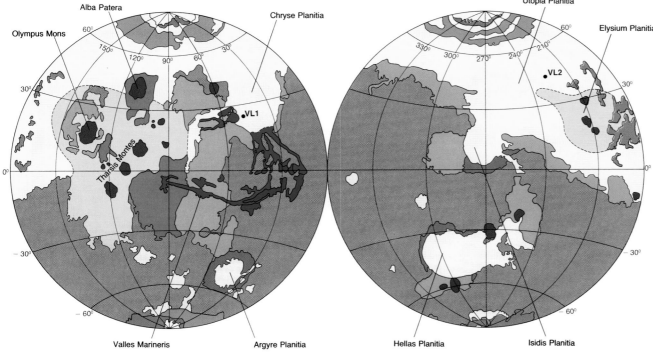

Geological evolution of the planet. These schematic maps (left) show the geological evolution of Mars. It is the eastern hemisphere which is represented. Mars was formed 4.5 billion years ago by the accretion of numerous small bodies in a relatively short time, probably not exceeding a few hundreds of thousands of years. During the first billion years of the planet's existence, the flux of meteorites and asteroids decreased to a rate which has remained relatively constant during the last 3.5 billion years. The final stages of the meteoritic bombardment which lasted for the first billion years affected the whole primitive crust of Mars (1). Similar surfaces have been preserved on the Moon and on Mercury; on Earth this type of surface must also have existed at the beginning of its history, but it is no longer visible nowadays due to its subsequent evolution and erosion. Shortly after the period of accretion, the interior of the planet became differentiated – probably as a result of the kinetic energy acquired during the period of accretion – into a crust, a mantle and a central core. The crust's thickness is variable – thin in the northern hemisphere, thick in the southern hemisphere. During this period of change, the first great fractures (2) affected part of the planet, particularly in the future Tharsis area, and a discontinuity appeared between the plains of the northern hemisphere and the terrains of the southern hemisphere. This first great period of tectonic activity would correspond to an expansion phase in the mantle; at this time convective movements would be responsible for the upheaval of the Tharsis dome.

It is generally admitted that a dense atmosphere was formed quite early in the planet's history, probably by the intense outgassing which followed the period of accretion. Originally the atmosphere must have been relatively hot and it would certainly have contained large quantities of water. This must have condensed at the time the surface cooled and caused torrential rains which would have resulted in the flow erosion whose effects are observed in many places on the surface. As long as the upper layers of the ground were at temperatures higher than the freezing point of water, the water would be able to penetrate into the soil. When the temperature dropped below freezing, subsurface ice formed in the upper layers. The growth in thickness of the subsurface ice increased the water pressure beneath, which could have induced artesian eruptions on the surface. Local melting of the ground-ice, perhaps caused by volcanic activity, as well as any artesian phenomenon, could have produced local subsidences in the crust and thus catastrophic floods which would produce the networks of channels. As a result of increased cooling of the atmosphere, part of the remaining water and carbon dioxide migrated towards the polar zones. This whole period is therefore particularly important in the geological history of the planet.

Volcanic activity developed subsequent to the period of intense erosion (3). Huge discharges of lava covered the major part of the northern hemisphere plains at that time. This volcanic activity was almost contemporaneous with that which formed the lunar maria; it continued intermittently and would only have finally ceased some hundreds of millions of years ago. The uplifting of the Tharsis dome continued (4), becoming more pronounced and causing new generations of faults and fractures. It was at this time that the Valles Marineris canyon must have opened up. The formation of these networks of faults probably favoured the release of water, which channelled out the bottoms of valleys and created new channels. The volcanism of the Tharsis region (5) covered a great many of the faults in this area with lava flows. The four giant volcanoes – Olympus Mons, Ascraeus Mons, Arsia Mons and Pavonis Mons – date from this time.

In recent times, wind activity has played an important role in the evolution of the surface. It is possible that a large fraction of the material of the equatorial regions was eroded and carried towards the poles. The stratified deposits observed in the polar regions do not have a precise age, but they certainly indicate important climatic changes which coincided with the formation of layers of carbon dioxide and frozen water. The ice caps must have been formed at the end of the great period of erosion (2). (After T. A. Mutch *et al.*, 1976)

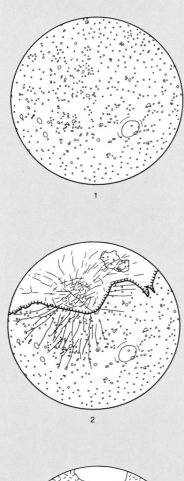

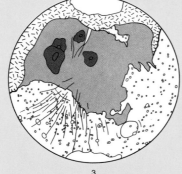

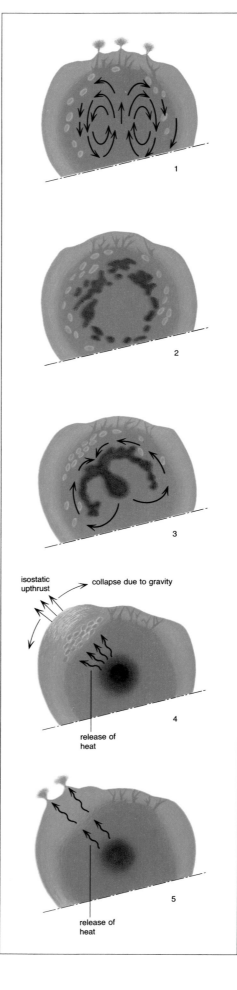

isostatic upthrust → collapse due to gravity

release of heat

release of heat

An evolutionary model of the internal structure of Mars. This model comprises five principal stages here represented schematically (opposite left).

The convective cells cause thinning and erosion of the underside of the crust, as well as a migration of lighter elements (1).

After the disappearance of these convective motions, the densest elements of the mantle segregate and begin to form layers of the future core (2).

A concentration of core elements appears underneath the future Tharsis site, and leads to lateral movements in the mantle which will generate an isostatic upthrust and the beginning of local topographical collapse (3).

The formation of the core is completed with the creation of an asymmetric thermal anomaly under the Tharsis region (4).

Finally, the transfer of heat from the deep zones will generate a long period of volcanic activity in the Tharsis area (5). (After D. U. Wise *et al.*, 1979)

this asymmetry is due to the presence of old terrains strongly marked by numerous meteoritic impacts, which make the southern hemisphere of the planet resemble the lunar highlands (it is in the southern hemisphere that the great basins analogous to the lunar basins are found: i.e. Hellas, Argyre and Isidis) and in the existence of young, sparsely cratered plains in the north. From the topographical point of view, this asymmetry appears as a difference in altitude between the two hemispheres, the northern being lower than the southern.

The large plains situated around the northern polar cap are in general 2 or 3 kilometres lower than the ancient terrains situated in the southern hemisphere, and the difference in altitude can reach 2–3 kilometres over a distance of a few hundreds of kilometres: the slope between the two surfaces resembles that which exists between the continents and the terrestrial oceanic basins. This asymmetry between the Martian hemi-

spheres could correspond to a difference in thickness of the crust, similar to that observed on Earth where the continental crust has a depth of from 30 to 50 kilometres, while the oceanic crust only attains 6 to 8 kilometres.

Gravity at Mars equator is some 371 centimetres per second per second. Comparison of the topography with variations in gravitational field allows regional variations in the thickness of the crust to be spotted which correspond to anomalies in the isostatic equilibrium.

Mars has a tenuous atmosphere: atmospheric pressure on the ground is of the order of 6 millibars. Owing to the inclination of the axis of rotation of the planet with respect to the plane of the ecliptic, the Martian year, which is almost twice as long as the Earth's, is subjected to the rhythm of the seasons. Because of the eccentricity of the orbit, summer is hotter and shorter in the south than it is in the north.

Philippe MASSON

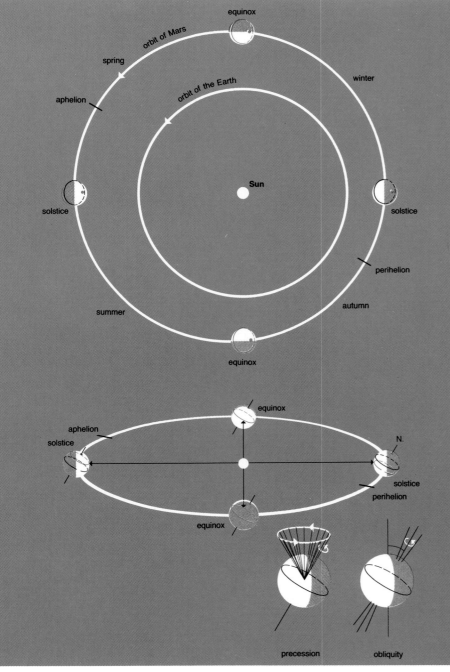

The Martian winter. In the course of their missions, the Viking stations observed changes in the Martian landscape linked to the climatic conditions. Thus, at the Viking 2 site in the Utopia Planitia, each morning during the winter white hoar frost deposits were observed, situated generally in the shade of rocky boulders (above). The formations of hoar frost, produced by the low nocturnal temperatures (−113 °C), disappear gradually as the temperature rises during the day (up to −98 °C). Given that, in the conditions prevalent on Mars, carbon dioxide would only condense into hoar frost at a temperature below −122 °C, it is probable that the hoar frost which is observed is composed of a mixture of one part carbon dioxide and six parts water.

The size of the rocks visible on this image, taken on 25 September 1977, towards the end of the first Martian winter experienced by the station, varies between about ten centimetres and a little less than a metre. In the foreground, various measuring instruments situated on the Viking 2 station are visible. These include the colour charts with which the images taken through different filters were calibrated after first having been transmitted in black and white and then combined with the help of a computer in order to give images in true colour. These images confirmed that the ground of the planet was a reddish-ocre colour owing to the presence of iron oxides in the surface materials and revealed a pink sky caused by the permanent presence, even in calm weather, of fine dust particles suspended in the atmosphere. (NASA/NSSDC)

The seasons. Nineteenth-century observers had noticed that the Martian surface showed dark zones whose shape and size varied according to the season and the year. They had even observed variations in colour ranging from grey to green, this latter shade occurring during the local spring; it was in these observations that some observers saw the proof of the existence of vegetation. It is known today that the seasonal variations in colour and albedo owe nothing to vegetation but are the consequences of quite simple physico-chemical processes.

Martian seasons are determined, in the same way as on Earth, by the obliquity, the angle between the axis of rotation of the planet and the plane of its orbit around the Sun (opposite). Measured with respect to the normal to the plane of the orbit, this inclination is some 23.5 degrees in the case of the Earth and some 25.1 degrees in the case of Mars. The seasons on Mars are defined in the same way as on the Earth, that is to say, as a function of the successive positions of the planet in relation to the Sun. As Mars' period of revolution is about twice as long (669 Martian days, or 687 terrestrial days) as that of the Earth, the seasons are themselves twice as long. But owing to the large eccentricity of the Martian orbit, they are of unequal duration: the northern hemisphere's spring, summer, autumn and winter last respectively 194, 178, 143 and 154 Martian days (corresponding to 199, 183, 147 and 158 terrestrial days; recall that, on Earth, the lengths of the seasons are respectively 92.9, 93.6, 89.7 and 89.1 days). The northern hemisphere's spring thus lasts 52 days longer than autumn. Moreover, during spring in the northern hemisphere, Mars is close to aphelion, the point at which it is furthest from the Sun. Because of this, summers in the southern hemisphere will be hotter but shorter than in the northern hemisphere (154 Martian days instead of 178), a consequence of the fact that the further a planet is from the Sun, the slower its orbital motion.

The seasons cause the changes in the distribution of the dark and light zones seen on the surface. These changes of colouring are caused by the shifting of dust linked to seasonal changes in the wind pattern. Likewise, the size of the ice caps observed at the poles of the planet vary differently as a function of the seasons; thus, in summer time in the southern hemisphere, the southern cap shrinks more than the northern cap because of the proximity of the planet to the Sun at that time. So it is now clear that the surface phenomena attributed by certain nineteenth-century observers to seasonal changes in vegetation are now explicable in terms of physical processes.

Finally, perturbations produced by the other planets cause slow variations in the eccentricity and obliquity of Mars (and also of the Earth). The phenomenon of precession also leads to changes in the lengths of the seasons. The two diagrams (below right) illustrate the phenomena of precession and variation in obliquity. These variations in the cycle of the seasons have important geomorphological repercussions. The photograph (above left) is a mosaic of the three images acquired by Mariner 9 on 7 August 1972; it shows a large part of the northern hemisphere of Mars. It is the end of spring in this hemisphere and the northern polar cap has already contracted, revealing a complex system of sedimentary deposits. (NASA)

Mars

Topography

The Red Planet has always intrigued observers. The first maps of Mars – based on telescopic observations – go back to 1659 and are due to Christiaan Huygens. More elaborate documents, such as the maps of Richard A. Proctor (1867) or Giovanni Schiaparelli (1877) are planispheric representations based upon the interpretation of variations in albedo.

Space missions to mars (Mariner 4 in 1965, Mariner 6 and 7 in 1969, and especially Mariner 9, which was operational from 13 November 1971 to 27 October 1972) gave the first detailed maps of the surface. It was, however, the Viking mission which was responsible for a decisive step forward in our knowledge of the planet's topography.

The Viking mission

A main objective of the Viking mission was to land carefully, on the surface of Mars, two automatic stations, one of the basic scientific objectives being to try to answer the fundamental question: 'Is there or has there been a form of life on Mars?' Initially the two Viking orbiters were placed in elliptical orbits quite close to the planet (perigee distance: 1520.5 kilometres for Viking 1 and 1519 kilometres for Viking 2). Subsequently the orbits were modified so that at the end of the mission the two probes at perigee were only 290 kilometres from the surface of Mars. In the course of the first month after the Viking 1 Orbiter was put into orbit, it transmitted numerous images designed to study the projected landing site for the automatic probe; the landing took place in the Chryse Planitia region on 20 July 1976. The orbit was then modified in such a way that the Orbiter could record very high resolution photographs of the surface and pass as close as possible to one of the two natural satellites of Mars, Phobos. The Viking 2 Orbiter was placed in a very inclined orbit which allowed it to fly relatively low over the polar zones. Later the Orbiter was redirected to allow it to pass very near to Deimos, the second natural satellite of Mars. The second automatic Viking station was placed in position on 3 September 1976 in the Utopia Planitia region. After four years of operation, the Viking 1 Orbiter was stopped on 7 August 1980; the Viking 2 Orbiter had ceased to function on 25 July 1978. The automatic Viking 2 station continued to transmit until February 1980. The Viking 1 station stopped transmitting information in Mid-November 1982, due to battery failure. It must be stressed that, overall, the Viking experiment has greatly exceeded its initial lifetime expectancy. Both Viking orbiters were equipped with two television cameras, with an infrared spectrometer for use in producing a map of the distribution of water vapour and with a radiometer for thermal cartography of the surface. The resolution of the images obtained at apogee (distance: 1500 kilometres) is of the order of a hundred metres; the most recent images obtained at 300 kilometres distance have a resolution of less than ten metres. During their descent towards the surface, the two automatic stations studied the ionosphere and the high atmosphere. Once on the ground, they analysed the atmosphere and carried out meteorological observations (measurements of temperatures, winds, etc.). However, the most important experiments performed by the ground stations were, firstly, the systematic observations of the landscape by the two cameras which gave panoramic views of the stations' vicinities and, secondly, analysis *in situ* of ground samples while simultaneously studying their physical characteristics, their mineralogical composition and also

looking for organic compounds which could have indicated the existence of a form of biological activity, either past or present. Besides the devices used for these experiments, the two stations were each equipped with a seismograph to record any possible internal activity of the planet.

The tens of thousands of images transmitted by the Viking Orbiters completed the global mapping begun with Mariner 9, but at a better resolution, and also contributed to our knowledge of the planet's geology.

The cartography of Mars

A cartographical atlas of the planet has been produced by the United States Geological Survey using Mariner 9 and Viking images. The maps consist of rectified photomosaics; the kinds of projection utilised depend on the latitude: Mercator for the maps included between 30° north and 30° south, Lambert between 30 and 65° both north and south and polar stereographical for the two polar zones (between 65 and 90° north and south).

These maps are produced in shaded relief in which the relief is represented by shading with a uniform illumination coming from the west. The topographical relief maps have, in addition to the representation of the relief, kilometric contour lines and toponymy details. Altitudes are defined with respect to a reference level (level 0) which, in the absence of seas, is established both from the gravitational field, and from a reference surface corresponding to an atmospheric pressure of 6.1 millibars defined by studies of radio occultations at the equator. The basic reference surface is a triaxial ellipsoid with two semi-major axes of 3394.6 and 3393.3 kilometres and a semi-minor axis of 3376.3 kilometres. The contour lines result from the combination of altitudes determined from terrestrial radar observations and measurements made with Mariner 9 and Viking. The toponymy is that which has been approved by the International Astronomical Union. Each of these maps has a name, a number and a standardised alphanumeric code, allowing recognition of the planet, the scale of the map, its location and its type (topographical map or other). Each of these basic maps is useful in establishing geological maps on which appear the different geological formations and structures identified by colours and labels. The identification of the terrain and surface structures is based on the photo-interpretation of the Mariner 9 and Viking images.

The Mariner 9 images, 7300 in all, have a resolution of 1 to 3 kilometres on the ground over more than 98 per cent of the Martian surface. The high-resolution images (100–500 metres) only cover about 1 per cent of the planet. The Viking orbiters have transmitted more than 50 000 images. Large zones in the equatorial region are covered at a resolution of the order of 7 to 30 metres and at least 90 per cent of the planet is covered by images having a resolution of 100 to 150 metres. The equatorial zone and certain parts of the polar regions have a stereoscopic cover which gives a topographical cartography with contour lines spaced from 500 to 1000 metres. A small number of areas have stereoscopic cover sufficient to yield contours spaced from 20 to 100 metres. A certain number of images have been obtained through calibrated coloured filters, to give geochemical information. They have been used to create colour photomosaics.

Because the Viking missions were able to obtain better quality images at a better resolution than those of Mariner 9, a new series of maps at

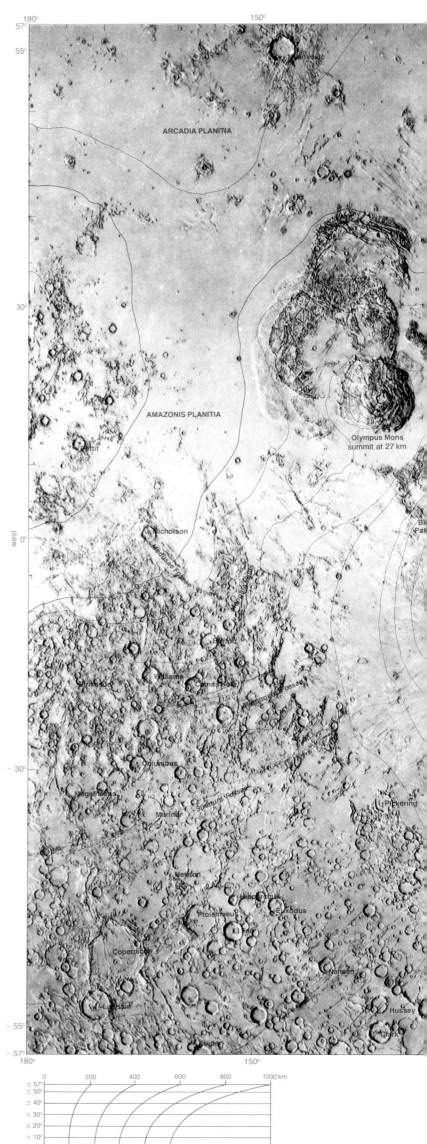

equatorial Mercator projection; scale at the equator 1:30 000 000

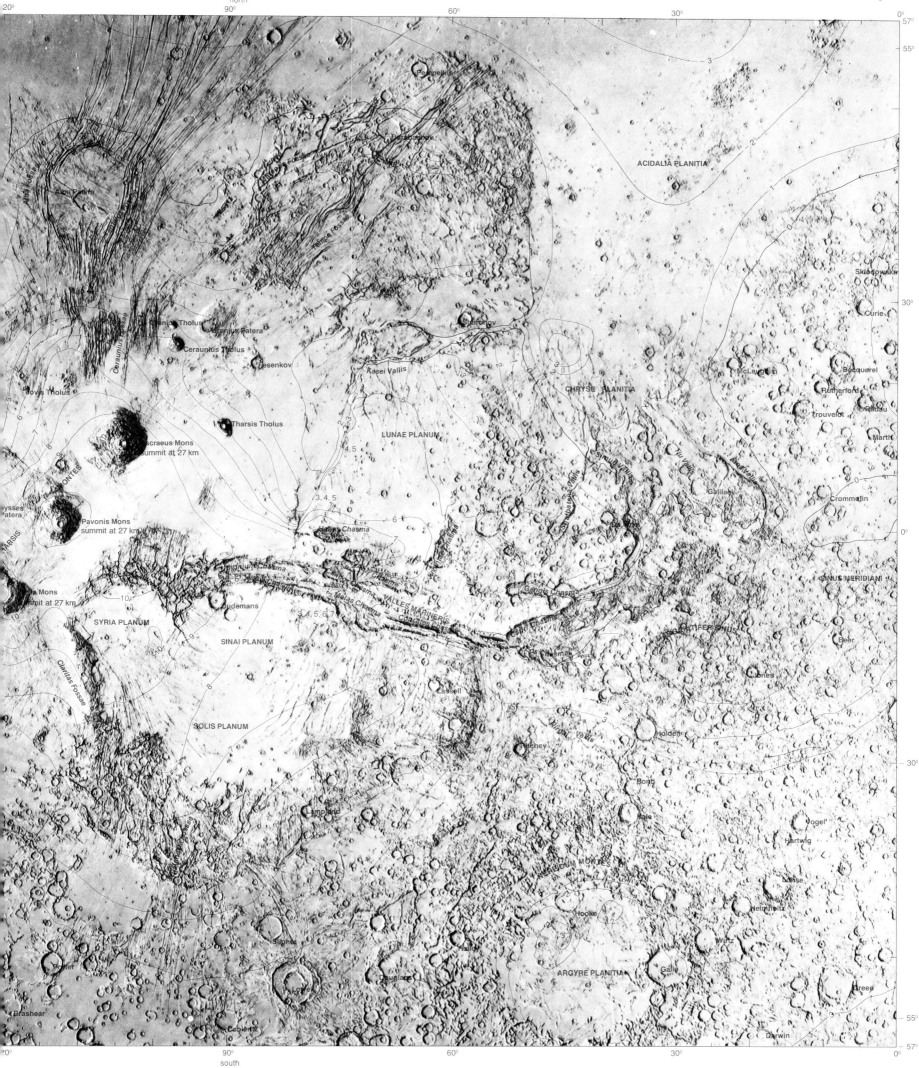

The toponymy of Mars. The nomenclature commission of the International Astronomical Union (IAU) is responsible for the nomenclature used for the regions and the features (craters, basins, volcanoes, channels, valleys, etc.) of the planetary surfaces. In the case of Mars, the names used for the large regions of the planet are generally those adopted by nineteenth and twentieth-century astronomers to describe the variations in albedo of the Martian surface. Giovanni Schiaparelli, from 1877 to 1899, and Eugene Antoniadi, in 1930, are responsible for the choice of these names. Thus, in 1879, Schiaparelli borrowed from *The Odyssey*, the name of Nix Olympica (snows of Olympus) to name one of the giant volcanoes of Mars, now called Olympus Mons. In 1909 Antoniadi gave the name of the goddess of the night (Nox) Noctis Lacus, to the network of canyons situated in the region of the Tharsis volcanoes; this network has been renamed Noctis Labyrinthus. The names of the valleys and Martian craters are those of scientists. Thus, the Polish astronomer,

Nicolas Copernicus, gave his name to the crater Copernicus, in the planet's southern hemisphere. The names of rivers, (Parana Valles comes from the Brazilian Rio Parana) towns, villages and famous places on Earth (for example Bordeaux, and Kourou) and equally the translation of Mars into different languages (for example Kasei Vallis comes from the word Kasei, which means Mars in Japanese) serve to complete the Martian toponymy.

The latinised terms used to define the principal forms of relief are coded in the same way for all the planets. Thus, *catena(ae)* is used for a chain of craters, *dorsum(sa)* for a wrinkle, *fossa(ae)* for a ditch-like linear depression, *mensa(ae)* for little plateaux, *mons(montes)* for a mountain, *planitia(ae)* for a low plain, *plan(na)* for a great plateau, *rupes* for a cliff, *tholus(li)* for a little hill, *vallis(les)* for a valley, etc.

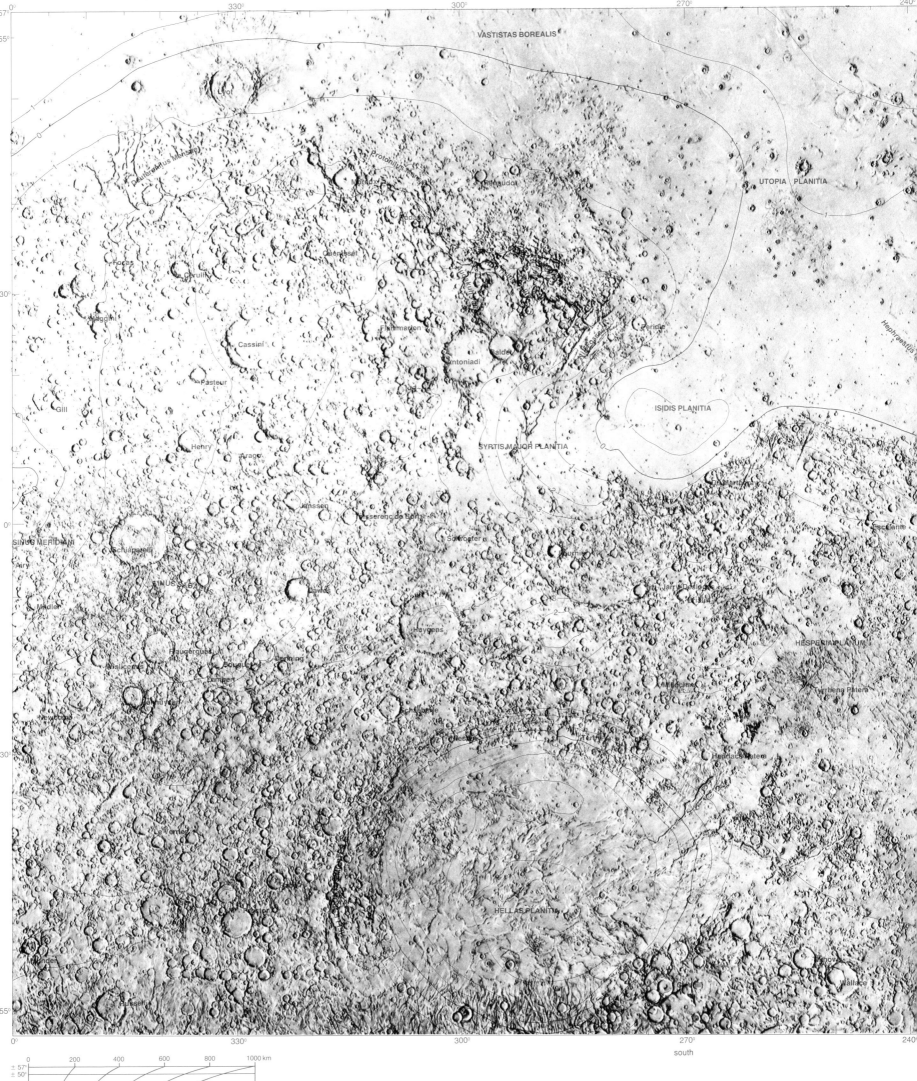

equatorial Mercator projection; scale at the equator 1:30 000 000

Note. This map is based on the topographical map of Mars published in 1982 by the United States Geological Survey at a scale of 1:15 000 000. The contour lines indicate the altitude (in kilometres) above or below the reference spheroid, which has an equatorial radius of 3393.4 kilometres and a polar radius of 3375.7 kilometres. The reliefs are represented as they would be seen illuminated by an imaginary Sun situated to the west. S. Davis and B. Peacock are responsible for this map, which has been kindly communicated by R. M. Batson. (USGS)

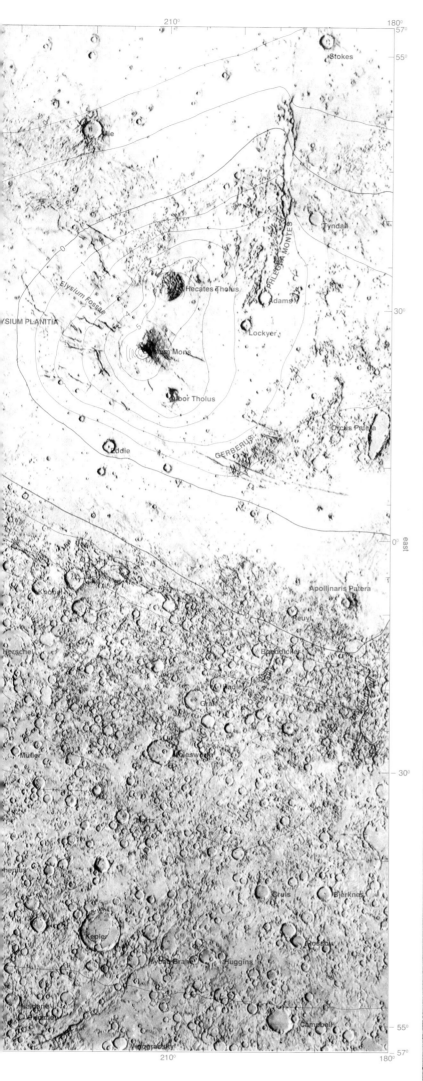

the scale of 1 : 2 000 000 has been produced. This series will include 140 sheets and will cover the whole of the planet, with the exception of the poles (only fifty-one photomosaics and nine provisional topographical maps have been published up until now). From the complete set of thirty maps at a scale of 1 : 5 000 000, a general map at 1 : 25 000 000 has been prepared by combining the topographical and geological maps. Besides the maps at a scale of 1 : 5 000 000, special sections at 1 : 1 000 000 and at 1 : 250 000 have been produced to cover the needs of the Viking experiment, in particular in relation to the choice of landing sites for the two automatic stations.

Philippe MASSON

Chryse Planitia. The automatic Viking 1 station landed on 20 July 1976 in the Chryse Planitia region, in the northern hemisphere of Mars. The carefully chosen landing site, indicated by an ellipse (top image) had been selected in a relatively flat, sparsely cratered region made up of great volcanic discharges (to the left of the image), hollowed out in places by large stream channels.
The surface of the volcanic discharges displays wrinkles analogous to those found on the lunar maria (bottom image). These wrinkles might have been formed when the volcanic discharges occurred, or they could be the result of later deformations caused by local tectonic movements. These wrinkles are affected by erosion. (NASA)

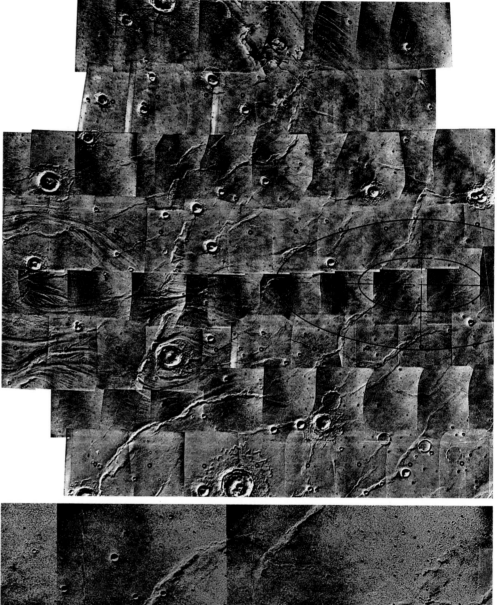

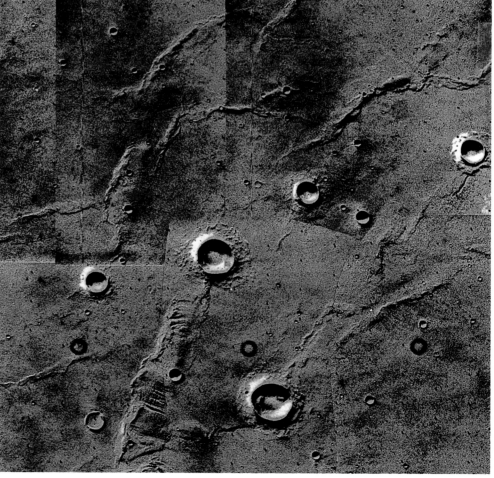

Mars

Volcanism and tectonics

In addition to the large volcanic discharges found all around the Tharsis region, Mars has volcanoes which are inactive today. The most spectacular among them are the great shield volcanoes whose diameter at the base reaches several hundred kilometres: Olympus Mons, Elysium Mons, Ascraeus Mons, Pavonis Mons and Arsia Mons. There also exist numerous little volcanoes whose diameters are only a few hundred metres. The majority of the great volcanoes are localised in two main areas: Tharsis, on the equator, and Elysium, in the plains of the northern hemisphere. Tharsis is situated at the centre of a wide, raised area whose average altitude is about 10 kilometres and which extends over about 6000 kilometres. Towards the summit of this dome there are three volcanoes in a row – Arsia Mons, Ascraeus Mons and Pavonis Mons – which reach a maximum height of 27 kilometres and form the Tharsis Montes chain. Olympus Mons, situated to the north-west of the summit of the Tharsis dome, also reaches a height of 27 kilometres; the diameter of its base is 600 kilometres. To the north of the dome, an old volcano, Alba Patera, with a diameter of 1500 kilometres, is surrounded by a large network of faults. By way of comparison, the largest terrestrial volcanoes which show similarities in shape to the shield volcanoes of mars are those of the Hawaiian islands; however, they only reach 120 kilometres in diameter and only stand some 9 kilometres above the ocean bed.

The volcanoes of the Tharsis region followed a common evolution. The formation of these shield volcanoes was brought about by successive eruptions of liquid lava, probably basaltic, which has accumulated on the summits and sides. The central calderas situated at the summits were formed by post-eruptive collapses; they measure several tens of kilometres in diameter. During the course of the final episodes of activity, great quantities of lava escaped onto the slopes of the shield volcanoes via secondary craters and fissures; these lava flows covered large areas in the vicinity of the volcanoes and engulfed the lower parts of the mountains themselves.

The great size of some of the Martian volcanoes is probably the consequence of a stable and thick lithosphere. Indeed, it seems that plate tectonics is absent on Mars; so that the volcanoes always remain above the source of magma and their development continues as long as the magma is available. By using dating techniques based on a count of the numbers of impact craters which are found on the slopes and in the calderas, the volcanoes of the Tharsis region seem relatively young and their activity appears to have been spread out over a very long period, lasting perhaps several billion years. How can such volcanic activity be explained? The rising of magma is due to a large extent to hydrostatic pressure created by the difference in density which exists between the magma and the rocks which it traverses during the course of its journey towards the surface. The fact that the three volcanoes of the Tharsis chain have the same altitude suggests that this corresponds to the limit beyond which the magma cannot rise: a comparison with terrestrial data indicates that the source of magma must be situated at a depth of 200 kilometres while, in the case of the Hawaiian

Text continued on page 145

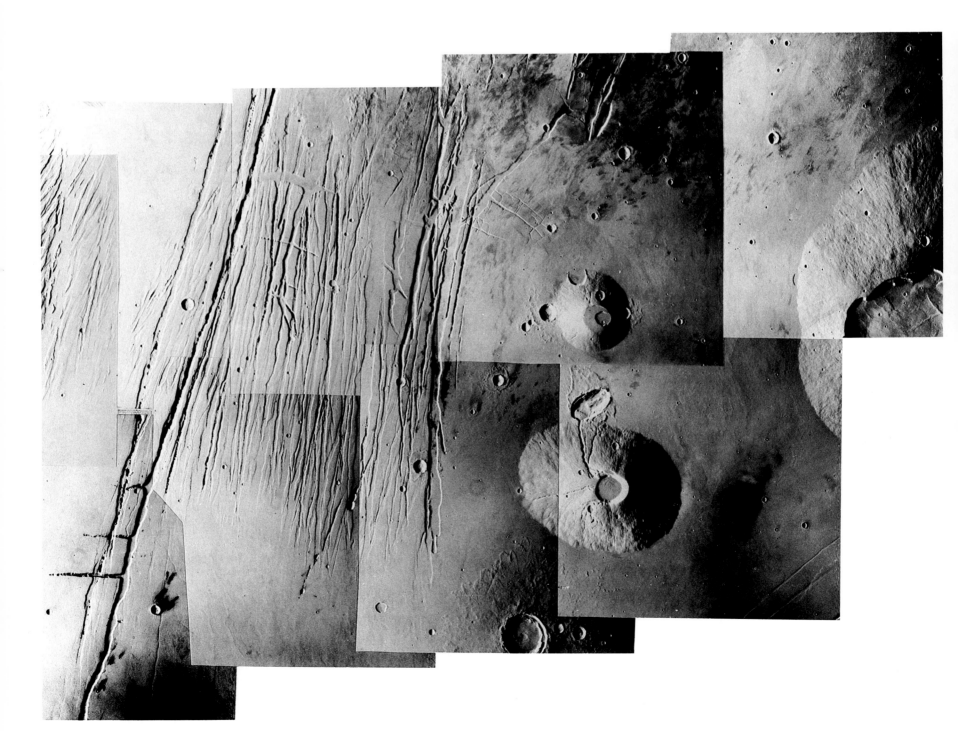

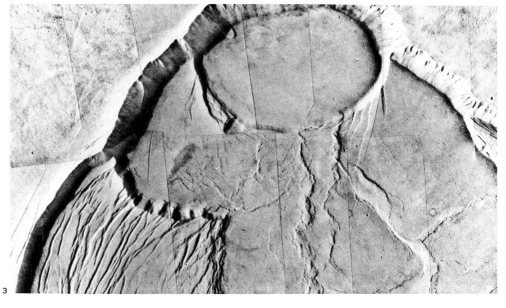

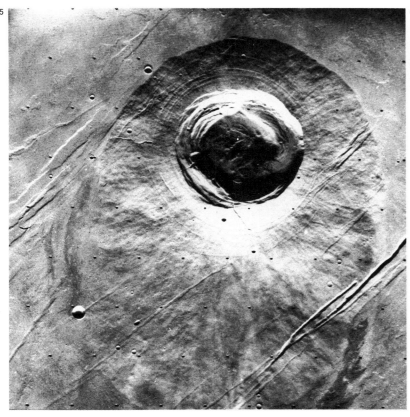

The Martian volcanoes. The volcanic edifices are concentrated essentially in two areas of the planet: the Tharsis dome, to the west of the Valles Marineris, and Elysium Planitia, to the east. All the volcanoes are situated in the northern hemisphere of the planet.

In the Tharsis region, three giant volcanoes – Ascraeus Mons, Pavonis Mons, and Arsia Mons – form a chain orientated north-east–south-west. Another large volcano – Olympus Mons – is situated 1500 kilometres to the north-west of the Tharsis volcanoes. These four volcanoes are true giants. Olympus Mons (image 2) measures 600 kilometres in diameter and soars to a height of about 27 kilometres. The summit caldera has a diameter of 90 kilometres. An escarpment 6 kilometres wide separates the base of its cone from the surrounding plain which is composed of great volcanic discharges.

Olympus Mons and the three Tharsis volcanoes have identically shaped cones at their summits, in which are found systems of calderas, or concentric craters, formed by the successive collapse of their floors in the course of eruptions. Image 3 shows a part of Olympus Mons' summit caldera. Image 4 displays the summit of Arsia Mons with its 120-kilometre diameter caldera. The slopes of these volcanoes are covered by successive flows of liquid lava which spread to great distances on the surrounding plains

Volcanoes of lesser dimensions also exist in the neighbourhood of the giant volcanoes; there are, for example, (image 1) Ceraunius Tholus (115 kilometres diameter) and Uranius Tholus, situated to the north of the Tharsis volcanic chain, or Biblis Patera (image 5), whose caldera measures 50 kilometres in diameter, to the west of Pavonis Mons. The slopes of Ceraunius Tholus (at the lower part of image 1) exhibit a kind of channel which connects the summit caldera (22 kilometres in diameter) to a meteoritic crater situated at its base. The presence of this channel could indicate that the last eruptions of this volcano might have been caused by the impact of large meteorites. The base of Uranius Tholus (image 1) shows an impact crater partially filled by volcanic flow, which suggests that the activity of these volcanoes ended relatively recently in the history of the planet. The great fault system seen to the west of these two volcanoes belongs to the structural network encircling another volcano, situated more to the north – Alba Patera.

The high resolution view of the Ascraeus Mons caldera (image 6) demonstrates the concentric faults formed around and at the crater's interior in the course of its successive collapses. The main caldera, of which only a part is visible here, is 40 kilometres wide and about 4 kilometres deep. (NASA)

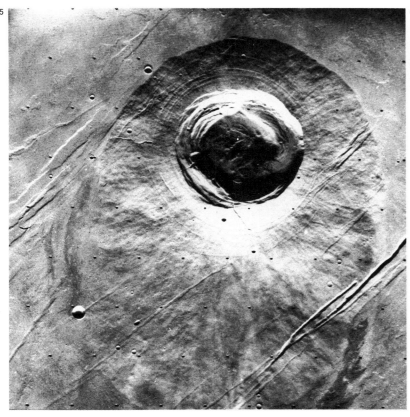

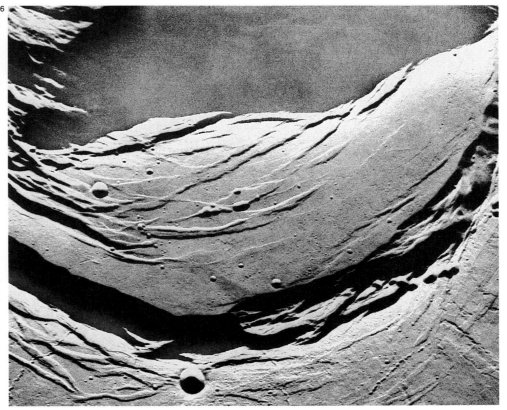

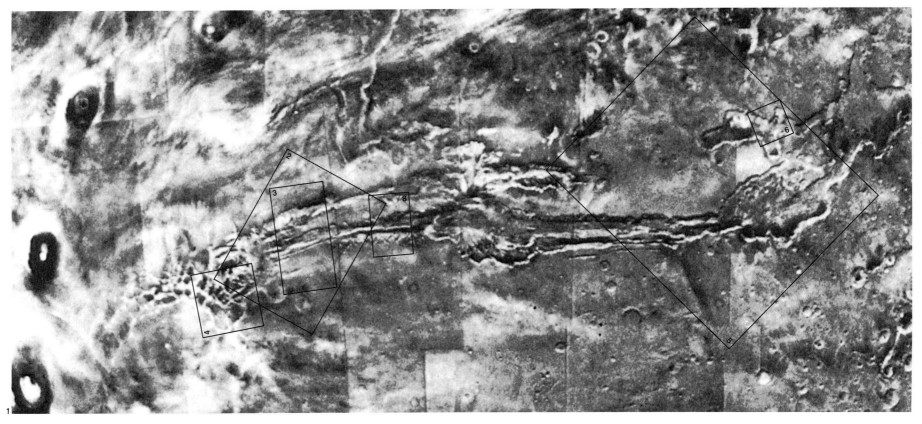

Valles Marineris. The great equatorial canyon of Mars, Valles Marineris, is one of the most outstanding morphological features of the planet. This system of valleys extends from east to west for about 5000 kilometres at an average depth of 6 kilometres (see image 1, which also shows the locations of the other images; one centimetre represents about 200 kilometres). It commences in the west in the Noctis Labyrinthus region by a complex system of graben (valleys of very steep-sided faults), cutting up the surface into regular polygons (image 4). In its central part, Valles Marineris is made up of several parallel canyons (images 2 and 3) aligned with large faults. These canyons join up at the centre of the system with Melas Chasma, a vast depression, which is 160 kilometres wide and which links the valleys together. To the east, Valles Marineris, emerges into a vast depression, Capri Chasma (image 5), littered with huge boulders and debris. The flat bottom of

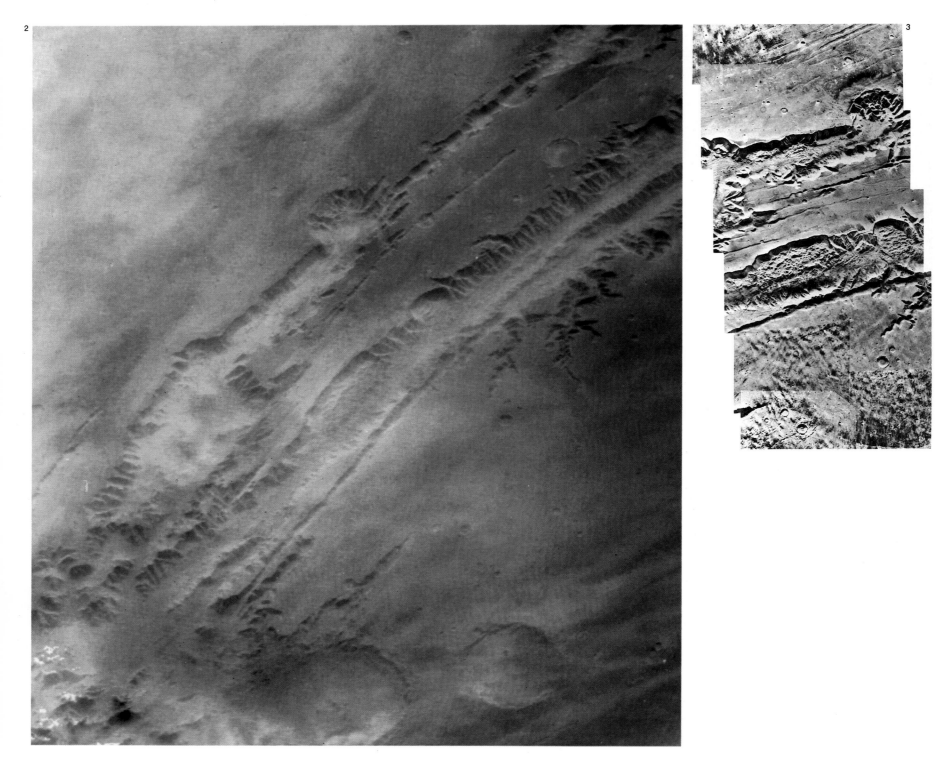

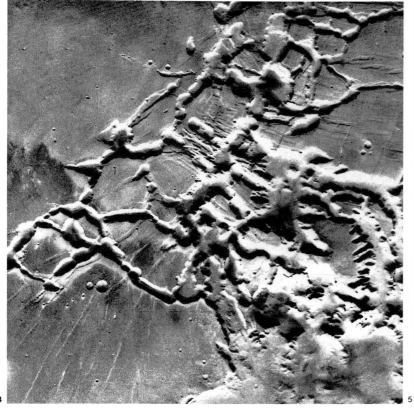

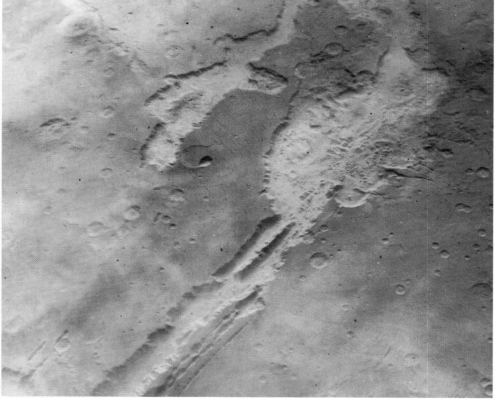

4

5

the canyon is often covered by crumbling material which comes from the erosion of the slopes or from the ancient action of flows which formed deep gorges along the slopes (image 8). The erosion of the slopes, under the action of gravity, causes spectacular collapses (image 6: the width of the canyon is of the order of 10 kilometres). This erosion allows us to observe that the high part of the plateau, dissected by the canyon, is made up of a succession of horizontal layers (image 7).

The Valles Marineris system resulted initially from a collapse of the high part of the plateau, along the great parallel faults formed at the time of the upthrust of the Tharsis dome to the west. The valleys have since been enlarged by stream erosion which grooved the slopes and by a sequence of landslides which have occurred recently and which perhaps are still happening. (NASA)

6

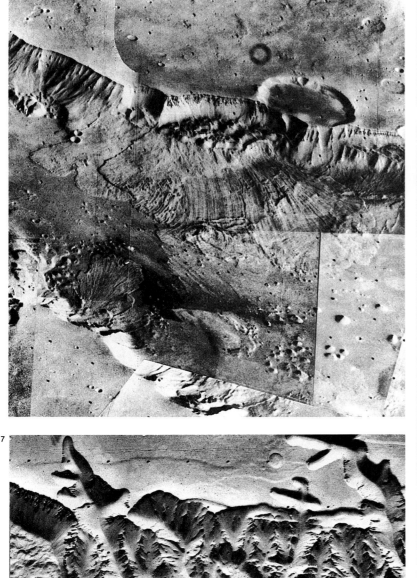

8

7

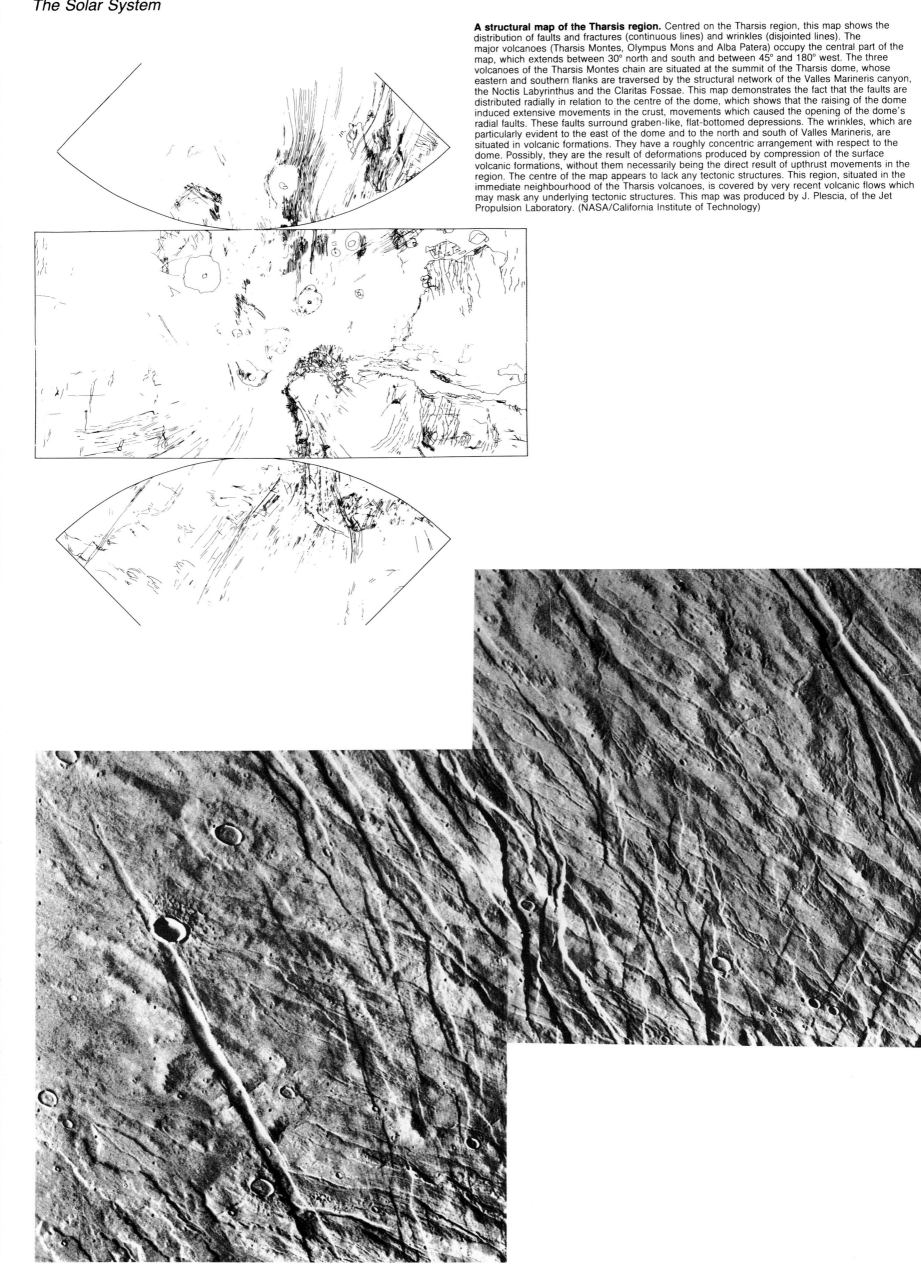

A structural map of the Tharsis region. Centred on the Tharsis region, this map shows the distribution of faults and fractures (continuous lines) and wrinkles (disjointed lines). The major volcanoes (Tharsis Montes, Olympus Mons and Alba Patera) occupy the central part of the map, which extends between 30° north and south and between 45° and 180° west. The three volcanoes of the Tharsis Montes chain are situated at the summit of the Tharsis dome, whose eastern and southern flanks are traversed by the structural network of the Valles Marineris canyon, the Noctis Labyrinthus and the Claritas Fossae. This map demonstrates the fact that the faults are distributed radially in relation to the centre of the dome, which shows that the raising of the dome induced extensive movements in the crust, movements which caused the opening of the dome's radial faults. These faults surround graben-like, flat-bottomed depressions. The wrinkles, which are particularly evident to the east of the dome and to the north and south of Valles Marineris, are situated in volcanic formations. They have a roughly concentric arrangement with respect to the dome. Possibly, they are the result of deformations produced by compression of the surface volcanic formations, without them necessarily being the direct result of upthrust movements in the region. The centre of the map appears to lack any tectonic structures. This region, situated in the immediate neighbourhood of the Tharsis volcanoes, is covered by very recent volcanic flows which may mask any underlying tectonic structures. This map was produced by J. Plescia, of the Jet Propulsion Laboratory. (NASA/California Institute of Technology)

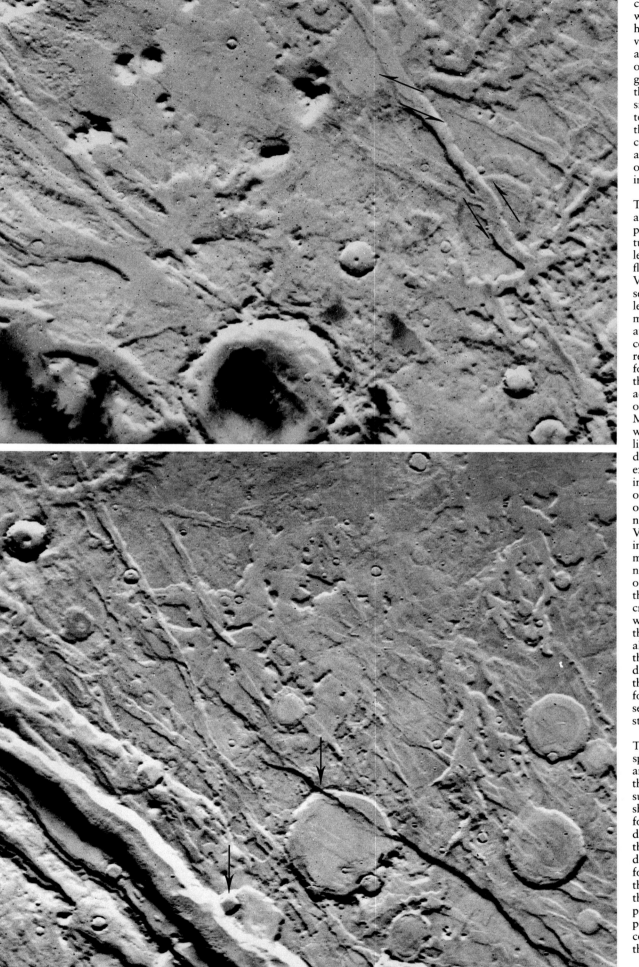

volcanoes, this source would only be at a depth of 60 kilometres. In other words, the Martian lithosphere in this region seems to be thicker than that of the Earth, which agrees with the structural models constructed for the planet. The history of volcanic activity on the planet Mars can be established from dating based on a census of impact craters. The great discharges which covered the plains occurred during the first half of the planet's history; in the second half, the volcanic activity has been limited to the Tharsis and Elysium regions. For the last billion years, the only obvious volcanic activity has been in the giant volcanoes, that is, Olympus Mons and those of the Tharsis Montes chain. This progressive decline in volcanic activity is probably linked to the general cooling of the planet and to a thickening of the lithosphere. The mineralogical composition of the Martian lava could not be analysed directly but the measurements carried out by the Viking automatic stations gave information on their petrological nature.

The great volcanoes are associated with the Tharsis dome, whose formation was obviously an important event in the planet's history. At the periphery of this dome, numerous radial fractures emanate in most directions, affecting at least a third of the planet. Situated on the eastern flank of the Tharsis dome is the great canyon, Valles Marineris. It consists in fact of a system of several parallel canyons which extends over a length of about 5000 kilometres. The depth of the main canyon is from 6 to 7 kilometres and its average width is some 200 kilometres; where it connects up with other canyons, this width can reach 600 kilometres. These canyons seem to be, for the main part, of tectonic origin; many of their walls resemble collapsed valleys. Tectonic activity has also led to gigantic landslides, a few of which extend over more than 100 kilometres. Moreover, the deposits observed on the canyon walls are channelled as under the effect of flowing liquid; these slopes would have been subjected to different degrees of surface erosion. At its eastern extremity, the Valles Marineris canyon emerges into a vast depression littered with a chaotic array of huge boulders. Numerous large channels originate from this region and head towards the north, to the Chryse Planitia region. To the west, Valles Marineris terminates in a veritable labyrinth of graben – Noctis Labyrinthus – which are a multitude of collapsed valleys aligned on a network of faults and fractures. These valleys often continue in the form of crater chains, but these must not be confused with impact volcanic craters: they are circular or elliptical depressions which are due to collapse of the surface during the degassing of the crust. Deposits observed along the slopes of the graben of Noctis Labyrinthus could be due to gravitational collapses during periods of humidity. Further to the south, the Noctis Labyrinthus system continues in the form of a regular network of parallel great faults separating caved-in valleys and arranged in stepped stages.

What is the origin of these tectonic structures? The rise of the Tharsis dome again appears to be responsible. This region is the centre of a gravitational anomaly which has been precisely measured by the Mariner 9 and Viking probes. Calculations of surface forces, based on topography and gravity, show that the main directions of the compression forces are orientated perpendicularly to the directions of the fractures. These fractures would therefore result from the very presence of the dome and not from the stresses generated by its formation. The ancient networks of fractures and the most ancient lava flows seem to indicate that the dome was formed quite early on in the planet's history, four billion years ago. One of the possible causes of this formation could be convective movements in the mantle, caused by the segregation of the elements of the core.

Philippe MASSON

Faults. Mars has been subjected to significant tectonic activity which resulted in the formation of great networks of faults along which collapses occurred (for example, on image 3, two impact craters, indicated by arrows, are affected; the diameter of the larger of the two craters is about 40 kilometres).

These faults, which extend over several hundreds of kilometres, are in the main situated on the periphery of the Tharsis region (Valles Marineris, Noctis Labyrinthus, Claritas Fossae, Alba Patera...). In this region, the faults are linked (in part) to the rise of the Tharsis dome which caused stress movements in the crust. These stress movements caused collapses along the vertical faults which would be how the Valles Marineris canyon and the Noctis Labyrinthus network came into being.

In the ancient terrains, the faults dissected or affected the impact craters and are themselves distorted by more recent craters (image 3) which suggests that the tectonic movements which produced these faults occurred relatively early on in the planet's geological history.

Around the volcano Alba Patera, the faults form a parallel network bordering flat-bottomed troughs, which have a depth of about 100 metres and are orientated north–south (image 1, on which 1 centimetre represents about 15 kilometres); these faults are probably connected with the region's volcanic activity. If the majority of these faults have experienced vertical movements, some of them must equally have suffered a horizontal shift which could have caused the dislocation or the displacement of some craters or channels (image 2). To set the scale, the large crater situated at the bottom of the photograph has a diameter of the order of 50 kilometres. (NASA)

Mars

The surface relief

The process of erosion

The plains of Mars' northern hemisphere display many networks of riverlike channels. The presence of these channels poses a problem, since the existence of water in liquid form is incompatible with the present physical conditions at the planet's surface. Three main categories of channel have been discerned: runoff channels, fretted channels and outflow channels.

Runoff channels resemble terrestrial rivers and seem to have been formed by the slow erosion of running water. They increase in size distally and have numerous tributaries; the majority of them measure less than a kilometre in width and a few tens of kilometres in length. They generally end abruptly, as if the water had disappeared into the soil. These channels are observed exclusively in the ancient, heavily cratered terrains, which seems to indicate that a climatically humid period must have existed quite early on in the history of Mars. Nevertheless, they seem to have remained at quite a primitive stage of evolution; the effects of their erosion have been quite limited.

The fretted channels resemble runoff channels. However, in contrast to the latter, they have wide, flat bottoms. They were probably formed from runoff channels which were enlarged by erosion of the sides.

The outflow channels are generally broadest and deepest at their head and they emanate from regions cluttered with great boulders, to the east and west of the Valles Marineris. Comparison with analogous terrestrial formations suggest that these channels are due to catastrophic floods whose origin is still very controversial; it could be a question of ice melting under the influence of volcanic activity, artesian eruptions of water under pressure, liquefaction of fine sediments saturated in water or additional slow erosion by glaciers. These outflow channels criss-cross the sparsely cratered plains, which rarely show runoff channels. They are therefore younger than these runoff channels. A possible explanation of this fact is that the majority of the water – or other liquid – which hollowed out the runoff channels, supplied a great artesian system; a general cooling of the planet would have brought about the creation of subsurface ice and with it a tendency of the water trapped in the soil to remain under the surface. The water had then only two avenues of escape, either violently by causing catastrophic floods, or by slowly leaking away through glaciers.

The surface of Mars has also been exposed to wind erosion; this is still evident from the numerous dust storms, observed both from Earth and by space probes. The effects of this erosion are quite small; it has not been able for instance to erode away craters with diameters of about 100 metres. Nevertheless, formations associated with wind erosion are visible in various places, for example, to the south-west of Olympus Mons or in the high latitudes of the northern hemisphere. In the latter regions, the presence of friable materials on the surface demonstrates well the effects of wind erosion. The most tangible effect of this erosion consists of a redistribution of fine debris and particles. Stretches of dunes are quite common, especially around the northern ice cap. The appearance of some impact craters shows evidence of the mobility of fine surface materials. Their appearances are stable on the time-scale of a year, but they are modified or destroyed by major storms and they reform during calm periods, thereby indicating the direction of the prevailing winds.

The polar formations

In the vicinity of the poles, we can observe some of the youngest features of the planet. Strata, at least 1–2 kilometres thick, are visible at the periphery of the polar caps. These strata lack any impact craters, and are therefore relatively young. The succession of strata indicates variations in the conditions during the deposition and, in particular, in the climatic conditions of recent

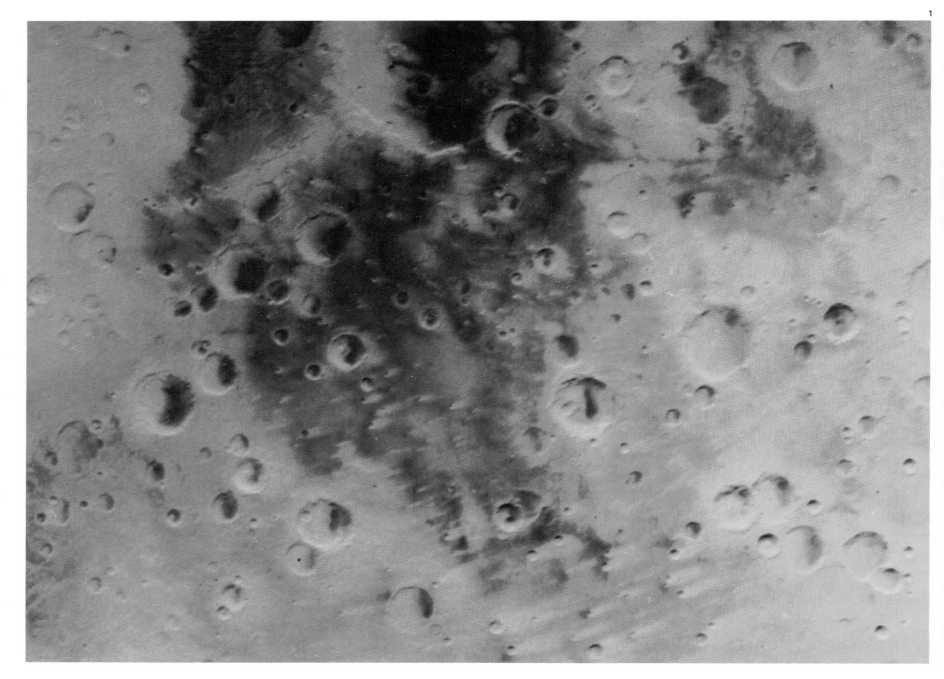

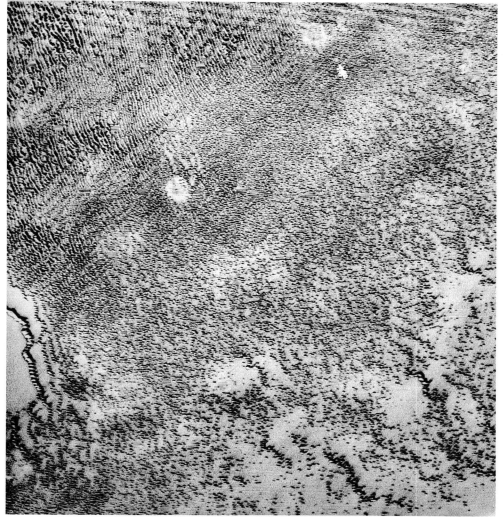

2

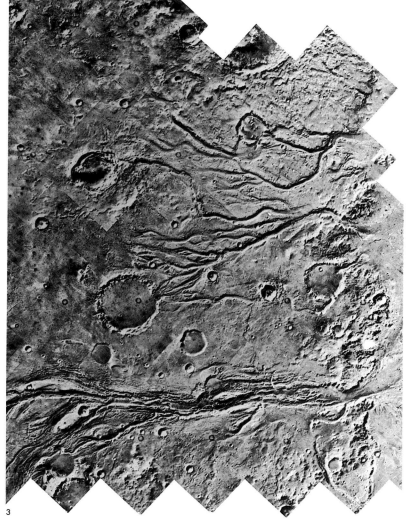

3

Different forms of erosion. The variations in albedo on the Martian surface can be attributed to aeolian phenomena as well as to seasonal phenomena causing deposits and sublimation of hoar frost.

The comparison between the seasonal variations in albedo observed by the Mariner 9 and Viking probes shows that the winds are relatively weak in the northern hemisphere in summer: on the other hand, their activity increases in the southern hemisphere in the same season.

Traces of aeolian erosion (image 1) and deposits behind the impact craters give the best indications of the variations in direction and intensity of the Martian winds. On the images acquired by the Viking Orbiters, the dark zones correspond to the places on the surface where the fine materials have been cleared away by the wind. The lighter zones correspond, in contrast, to the places where materials carried by the wind are deposited and accumulate. To illustrate the scale, the double crater, bottom right on the image, extends over about 110 kilometres. In the northern polar region, the winds fashion huge stretches of dunes (as in image 2, where 1 centimetre represents about 3 kilometres) analogous to the terrestrial desert ergs. In these ergs, the dunes are most often grouped in parallel lines. The individual dunes have a crescent shape and the direction of their elongation corresponds to the prevailing wind direction.

Another form of erosion frequently encountered in the planet's northern hemisphere consists of networks of channels (image 3; which shows Maja Vallis to the south and Vedra Vallis, to the north). The diameter of the large crater near the centre of the photograph is of the order of 75 kilometres. These anastomotic channels (that is to say those whose courses become confused) extend over several hundreds of kilometres and have a distribution similar to that of the terrestrial fluvial networks (principal flows and tributaries). In spite of the current total absence of liquid water on the surface of Mars, it is probable that at some time, when the climatic conditions were different, some flowing water did exist. The water would have subsequently disappeared (perhaps under the influence of degassing on the planet). A small quantity of this water is still found in the polar caps. Some more could exist in a frozen state in the soil (permafrost).

Note the existence in a few places of almost filled-in depressions (image 4: the maximum width of a depression is about 40 kilometres) whose bases are littered with chaotic material and fallen debris. These depressions progress via a sort of channel at the bottom of which are observed deposits which seem to have been violently hollowed out. This kind of depression could result from the sudden melting of the ground-ice which would have caused a local collapse of the soil and the liberation of a large quantity of water. The water could then have hollowed and channelled out the surface, as would happen if a dam ruptured catastrophically.

The 'mouth' of the Ares Vallis (image 5) shows 'islands' fashioned by the current. These islands, each about 40 kilometres long, seem to result from the erosion of a plateau. The current, flowing from bottom left, has skirted around two craters. (NASA)

5

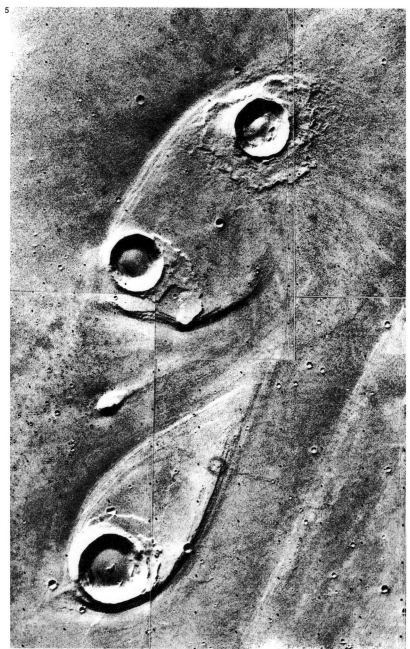

4

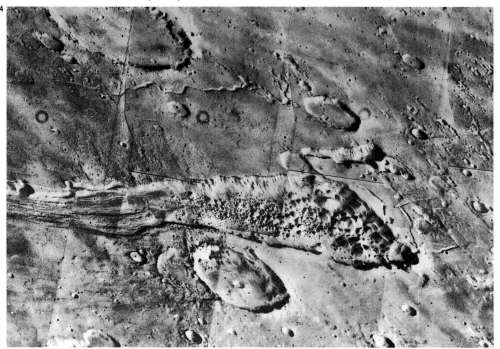

times. The partially eroded deposits show that the periods of deposition have alternated with periods of erosion.

Under contemporary climatic conditions, dust storms occur in the southern hemisphere while the northern polar cap is in the course of expansion. As a consequence, dust probably finds itself trapped in the ice. On the other hand, when the southern polar cap is forming, the atmosphere is relatively clear and, owing to this, little dust finds itself trapped in this cap. Calculations have shown that the rate of accumulation is some 0.04 centimetres per year. At this rate, it would have taken 100 000 years to form the strata which are seen to be 30 metres thick in places.

The 50 000-year cycle of the precession of Mars' axis of rotation produces climatic changes between the two hemispheres. The caps alternate in the accumulation of dust, which could explain the alternation of the strata. The real situation is a lot more complicated since the eccentricity of the orbit and the obliquity change with time. The variations in eccentricity affect the degree to which the climates in each hemisphere differ: when the eccentricity is small, the two hemispheres have identical climates, while, during periods of high eccentricity, the differences are very important.

The variations in obliquity influence the intensity of solar radiation which different latitudes receive. These effects are not only important with regard to the stability of volatile elements and their effect on the general atmospheric circulation, but can also influence the absorption or evaporation of volatile elements in or from the regolith. It is thought that these changes of the volatile elements can cause variations in atmospheric pressure of the order of a factor of ten. Thus, the deposits observed in the neighbourhood of the poles could result from the depositions of fine grains of dust and the condensation of volatile elements. Their stratification and their irregular shapes are the result of a combination of several factors which affect the general conditions. They include the frequency of the dust storms, the stability of the volatile elements near the surface and the circulation of the winds.

Craters

The southern hemisphere of the planet Mars is made up of ancient terrains which were subject to intense bombardment after the planet's formation 4.5 billion years ago. The meteorite craters resulting from this bombardment are of all sizes, ranging in diameter from the limit of resolution of photographs to a few tens of kilometres. Only two large basins (Argyre Planitia and Hellas

The ice caps. The appearance of the Martian poles is radically different from that of the rest of the planet. This is because the terrain is very different from that encountered elsewhere.

The two poles are covered by ice (water and carbon dioxide) the extent of which varies as a function of the seasons. In summer, as shown by the two montages made from images acquired by the Mariner 9 probe, the residual caps are of very different sizes; the northern cap (image 3) is clearly bigger than the southern cap (image 4). Each cap has a spiral appearance (image 2, which displays the northern residual cap) emphasised by successive layers of ice and by a few steep-sided valleys. This unique appearance could be correlated with the displacement of masses of air from the poles towards the equator under the influence of the Coriolis force. This hypothesis does not seem applicable to the southern pole, for the spiral formation is orientated in the opposite direction to that which would result from motion driven by the Coriolis force.

The terrain at the periphery of the ice caps is made up of thick strata, reaching depths of several tens of metres. These deposits rest on a sparsely cratered substratum to the north, and upon very ancient formations to the south. The deposits themselves show few impact craters and are therefore relatively recent. These deposits are considered to be the result of the accumulation of fine materials carried by the wind from the equatorial regions. Their stratified appearance (image 1) seems to indicate that these deposits are laid down cyclically owing to periodic variations in the eccentricity of Mars' orbit.

In the northern polar region, the strata are surrounded by great stretches of dunes which, in places, completely obscure the substratum. In certain regions, the dunes are sparser and are organised in crescent-shaped lines. The summer view in colour (1) of a small part of the northern polar cap of Mars (60 × 30 kilometres) taken by the Viking 2 Orbiter at a distance of about 2200 kilometres shows the details of the strata of the wind-blown materials in contact with the ice cap. Above a height of about 500 metres the deposits have been etched by erosion. (NASA)

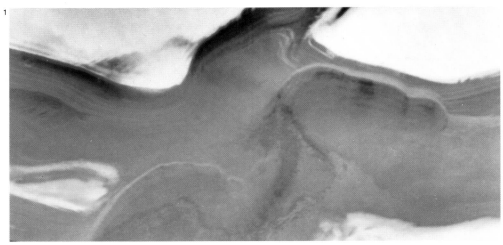

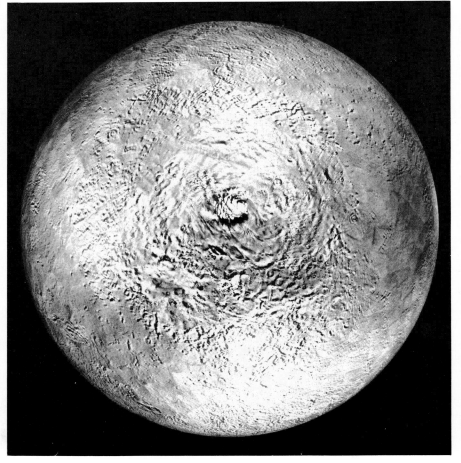

Planitia), are visible on Mars and are analogous to the lunar basins.

In general, Mars' meteorite craters display the same morphological characteristics as those of the Moon, with quite steep rims and a central peak. On the other hand, important differences in the level of the ejecta blankets of many Martian craters, such as that of Arandas, are observed. The ejecta blanket is made up of several superimposed layers of material which seem to have 'flowed out' in successive sheets from the edges of the crater over relatively short distances. It would seem that there has been a softening or liquefaction of the soil under the effect of the impact, causing the ejection of a viscous material instead of the pulverisation of the soil.

This phenomenon could support the hypothesis of the presence of frozen water at a depth of a few tens of metres. The frozen soil would be suddenly reheated at the moment of the impact and would cause the liquefaction of the soil, generating the viscous ejecta blanket. This phenomenon is observed around certain meteorite craters in the northern hemisphere of the planet, where numerous runoff channels are also observed.

Philippe MASSON

The Argyre basin. The internal diameter of Argyre is 900 kilometres (above), but taking into account the blanket of materials ejected at its periphery at the time of its formation, the overall diameter is some 1400 kilometres. The bottom of the basin is relatively flat and is covered in powdery material. The immediate surroundings of the basin are occupied by quite dramatic terrain which has been re-shaped by less important meteoritic impact, more recent than the basin itself. (NASA)

The crater Arandas. Situated in the planet's northern hemisphere (43° north, 14° west) the crater Arandas measures 28 kilometres in diameter. It displays the typical morphology of a meteorite impact crater, with a bowl shape whose centre is occupied by a peak. By contrast with the observed craters on the Moon and Mercury, this crater, like many others situated on Mars above latitudes 30° or 40° north, is surrounded by a thick ejecta blanket which forms a kind of slope with scalloped edges. The surface of the slope, in places, displays limited accumulations of deposits. The general appearance of this ejecta blanket is comparable to that of mud flows or very viscous lava flows. The presence of such forms at the periphery of a number of Martian craters seems to indicate that the surface could have been abruptly softened by meteorite impacts, which could be an argument in favour of the presence of frozen water in the soil of the planet. (NASA)

Mars

The soil

The two automatic Viking stations which in 1976 landed in the planet's northern hemisphere on the Chryse Planitia and Utopia Planitia sites, analysed the soil *in situ* and observed the landscape for several years.

The images transmitted by these machines reveal to us a landscape similar to the terrestrial stony deserts. In the Chryse Planitia area, the landscape is fairly flat, ochre-coloured, strewn with rocky boulders and sprinkled with small dunes. Trails of finely granulated material are visible in the shelter of the rocky boulders. These trails, just like the dunes, show that the effects of the wind in transporting material and eroding the surface occur in quite constant directions. Parts of the soil which are laid bare show a hardened and vitrified crust which has resulted from an encrustation by mineral salts caused by the evaporation of water originally contained in the soil. In the Utopia Planitia region, the surface appears to form a vast, stony plain. The relief is essentially made up of depressions in the form of channels which cut up the terrain into polygons. No outcrop of any rocky layers is observed here, but boulders and pebbles are seen which originate perhaps from a meteorite crater situated not far away. Small depressions are also visible. The

origin of some of them is probably connected to the freezing and thawing of the soil, while others could arise during the desiccation of the surface.

Soil samples from Mars were taken and analysed by fluorescent spectrometry. This technique allows an abundance determination for those elements with an atomic number greater than 11. These analyses showed that the overall composition of the soil is identical at the two sample sites. The composition of Martian soil is not comparable to any mineral or type of known rock. It must therefore comprise a complex mixture. This soil could have originated from rocks produced from a molten mixture rich in magnesium and iron. Compared to terrestrial rocks, Martian rocks are rich in magnesium, iron and calcium. They are poor in potassium, silicon and aluminium. This composition would correspond to that of materials originating during a partial melting of Mars' mantle, that is to say of the deep hidden layer under the crust. It is probable that the soil analysed is a compound of argillaceous minerals rich in iron and in iron hydroxides, as well as minerals rich in sulphur and carbon. The experiments performed with a mass spectrometer and a gas chromatograph measured the quantities of water vapour and

The site of the Viking 1 station. Chryse Planitia, the site of Viking 1, is an undulating plain, strewn with angular-shaped rocks. The large rock situated at the centre of the photograph and nicknamed 'Big Joe' is about 8 metres distant from the station and measures 3 metres wide and 1 metre high. The salmon-pink colour of the sky is caused by fine particles of dust suspended in the atmosphere. Judging by the circular arrangement of the small rocky boulders which surround Big Joe and its own proximity to other large boulders, it is probable that this great boulder is not in its original location. It was probably thrown aside by a meteorite impact in the region. (NASA/NSSDC)

Geochemical Map of Mars. As it approached Mars in August 1976, Viking Orbiter 2 took over 50 pictures of the planet to make this computer-generated mosaic illustrating variations in the planet's surface chemistry. Red depicts rocks that contain a high proportion of iron oxides; dark blue signifies fresher, less oxidised materials, probably volcanic basalts; bright turquoise shows surface frosts and fogs made from carbon dioxide or water ice; brown, orange and yellow represent different types of sand and dust deposits. This snakeskin-like swath covers the entire 360° of planetary longitude from a maximum of 30° latitude north to 60° south latitude. (NASA)

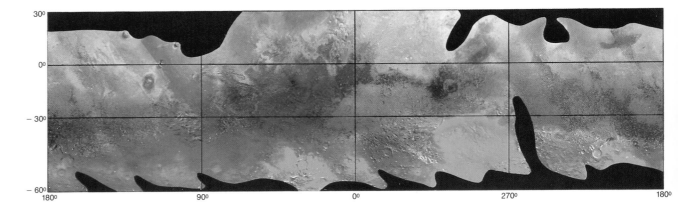

Utopia Planitia. The relief of Utopia Planitia (the inclination of the horizon is due to the fact that one of the station's feet settled on a boulder and inclined it at an angle of 8 degrees) is marked by depressions in the form of channels similar to that visible in the left foreground. This depression, a few centimetres deep, is filled with fine debris and dust accumulated by the wind. These depressions could have originated during the desiccation of the soil. The boulders are probably pieces of volcanic rocks, and their porous appearance is due to degassing at the time the lava

carbon dioxide released when the soil samples were heated. The soil contains about 1 per cent water by weight, of which part is probably contained in hydrated minerals. On the Earth, minerals of the same kind which chemically react with water give rise to clay soils rich in iron. The same kind of phenomenon could have happened on Mars at a time when water was in abundance. In addition, the soil on Mars contains significant quantities of magnetic minerals. It is probable that these minerals are iron oxides (magnetite or maghemite) and metallic compounds of iron and nickel. The maghemite has a reddish-yellow colour and would be responsible for the colour of the Martian soil. Both the abundance of magnetic minerals and their composition are as would be expected with the ultrabasic origin of rocks like the basalts, volcanic rocks which seem to be widespread over the surface of the planet.

Philippe MASSON

The site of the Viking 2 station. At first sight, Utopia Planitia, the Viking 2 site, resembles Chryse Planitia with its rock-strewn surface. However, these boulders, of which the nearest to the station measure about 20 centimetres across, have more rounded shapes and many of them have a spongy appearance. In addition, we note the presence of more massive, different coloured boulders, as well as small scattered pebbles. The high-gain antenna visible in the upper part of the photograph is pointed towards the Earth; it provided direct links with the receiving network (Deep Space Network) of NASA. The Viking 2 station is situated nearly 7500 kilometres from the Viking I station. (NASA/NSSDC)

The collection of samples. Besides making observations of the landscape and performing climatological studies, the two Viking stations were designed to obtain and analyse *in situ* samples of the Martian soil. This was in order not only to determine the main mineralogical constituents, but also to look for traces of organic constituents resulting from some kind of biological activity. For this task, the two stations were each equipped with a remote-control arm, operated from Earth, which held a scoop thus allowing the surface to be scratched to a depth of a few centimetres. The material was then introduced into the measuring apparatus. Because there were no positive results from the search for organic materials in the first samples analysed, some scientists decided to take samples from under rocks where any organic material would have been protected from the dangerous solar ultraviolet radiation. Because of this, it was necessary to clear the terrain by displacing some boulders with the help of the articulated arm and the shovel, before hollowing out the trench; just as on Earth a mechanical shovel is used before navvying begins! The photographs show the arm of the Viking 2 station during that operation, on the afternoon of 8 October 1976; the porous boulder has been displaced towards the left by several centimetres. (NASA)

cooled. This heterogeneous collection of boulders is probably debris projected from a giant meteorite crater situated about 200 kilometres from the landing site. This photograph is in fact a mosaic of several images which covers an angle of more than 120 degrees; the illumination was slightly different at the time of each photograph, since they were taken many months apart (the two-thirds on the right show the surface in February 1977, the left third in September 1976). North is to the left and east is towards the centre (NASA/NSSDC)

Mars

Martian activity

In deciding to prolong the Viking mission beyond its planned duration, NASA provided geophysicists with the opportunity to study the dynamics of the atmosphere and the variability in the appearance of the Martian surface. For more than a century this variability, which was first shown by observations of global changes, was at the heart of the hypotheses which postulated the presence of a vegetation which changed with the seasons. The cameras of the Viking probes transmitted images which have allowed this hypothesis to be refuted. Nevertheless, they did show the existence of Martian activity which is most apparent at the poles; however, it quickly became apparent that, on a smaller scale, changes in the surface's appearance occur at all latitudes. By comparing photographs of the Viking landing sites, minute changes could be discerned over long time intervals, typically greater than a Martian year.

The atmosphere participates in a fundamental way in this Martian activity, firstly because clouds form and evolve there, and secondly because it is the site of strong winds, which are capable of picking up particles from the soil. The atmosphere's opacity can thus vary considerably

gas	concentration
carbon dioxide (CO_2)	95.32%
nitrogen (N_2)	2.7%
argon (Ar)	1.6%
oxygen (O_2)	0.13%
carbon monoxide (CO)	0.07%
water vapour (H_2O)	0.03%
	(variable)
neon (Ne)	2.5 p.p.m.
krypton (Kr)	2.5 p.p.m.
xenon (Xe)	0.08 p.p.m.
ozone (O_3)	0.3 p.p.m.
	(variable)

Composition of the Martian atmosphere. The relative concentrations are expressed in percentages or in parts per million (p.p.m.).

isotopic ratios	Mars	Earth
$^{12}C/^{13}C$	90	89
$^{16}O/^{18}O$	500	499
$^{14}N/^{15}N$	165	277
$^{40}Ar/^{36}Ar$	3000	292
$^{129}Xe/^{132}Xe$	2.5	0.97

Isotopic ratios of the gases of the atmospheres of Mars and Earth. Geochemists compared the isotopic ratio of nitrogen, carbon and oxygen, and other elements present in the Martian atmosphere. Molecular nitrogen N_2 is a relatively inert gas which does not display chemical affinity with the Martian rocks; being relatively light, it was able to escape from Mars' high atmosphere after ionisation mechanisms dissociated it into single atoms. Escape from the atmosphere is selective; the lightest atoms escape most easily. The isotopic ratio $^{14}N/^{15}N$ measured in the Martian atmosphere is clearly lower than that observed in the terrestrial atmosphere. This lower proportion of the lighter atoms implies a very active mechanism of gas emission. Because its mass is ten times greater than that of Mars, Earth has obviously emitted greater quantities of gas; nevertheless, it can be concluded from the comparison of the two isotopic ratios that the Martian atmosphere today only represents 1 per cent of the total quantity of gas liberated by the planet. Compared to molecular nitrogen, carbon dioxide (CO_2) and water vapour (H_2O) would have had more chemical reactions with the carbon and oxygen-rich Martian rocks and as a result would have participated much less in the gaseous exchanges, which would suggest that the isotopic ratios $^{12}C/^{13}C$ and $^{16}O/^{18}O$ are not anomalous. They are, in fact, within about 5 per cent, equal to the ratios observed in the terrestrial atmosphere. (After T. Owen *et al.*, 1977)

during the course of the year. In fact, the observed changes mainly appear to be a consequence of interactions of the surface materials with the Martian atmosphere, that is physico-chemical interactions with the gases of the atmosphere and dynamical interactions with the dust particles in suspension.

When a terrestrial planet reaches the final stages of its formation, it has acquired practically all its share of volatile substances. Because of volcanic activity or other forms of emission, Mars gradually became surrounded by an atmosphere during the four billion years following its formation. As in the case of our planet, a large part of the substances produced by this emission of vapour are no longer present today in the Martian atmosphere. Water, in particular, has been hidden in great quantities in the form of ice inside the polar caps, or incorporated into the Martian subsoil in the form of ground-ice; besides it is known also that the Martian regolith acts as an equally good reservoir for water molecules which are chemically trapped in the structure of the minerals making up this regolith. On the surface of the Earth, carbon dioxide is, with water vapour, a major constituent of volcanic gases. It is therefore convenient by analogy to estimate the importance in the case of Mars of the emission of water and carbon dioxide. We must also evaluate the quantities of volatile substances which, after having reacted with the Martian surface, are today, in one form or another, hidden in different 'reservoirs'.

After the Viking mission, there was much discussion – the firmest arguments were based on the geological considerations we have already discussed – as to the effectiveness of the surface of Mars as an important reservoir of volatile substances, in particular of carbon dioxide and water. During the course of Mars' evolution, the mechanism of the interaction of the carbon dioxide gas and water vapour in the Martian atmosphere with surface materials must have been a very effective selective adsorption of these gases. This interaction still exists today and varies with time. It shows up in particular through variations which are sensitive to the relative concentration of water vapour, which can change from 1 to 10 per cent depending on the place, season or altitude. In winter, above each polar region, there is an even more distinct decrease in the relative concentration. However, in the northern hemisphere in summer, the values measured above the polar regions are close to those which correspond to a saturated atmosphere; the atmosphere is then the scene of intense meteorological activity.

The Viking mission scientists extended the chemical analysis of the Martian atmosphere to its non-gaseous constituents. For this, they attempted to interpret images transmitted by the two stations landed on the Martian soil. The optical thickness of the lower atmosphere depends almost totally on the attenuation of the light caused by aerosols. These are made up of dust in suspension, and of ice particles produced by the condensation of water molecules and carbon dioxide. Almost certainly, water ice crystals of 1 or 2 micrometres diameter cause the morning mists which form above specific areas of the Martian surface. As for the particles lifted from the soil – mainly at the time of dust storms affecting a whole hemisphere – they have a radius of between 1 and 10 micrometres and their size distribution is similar to that of terrestrial aerosols. This information on the physics of the atmospheric particles was obtained by studying the variations in the intensity of the Martian sky as a function of the angle of illumination. In the same way, their distribution as a function of altitude can be studied. This depends on the state

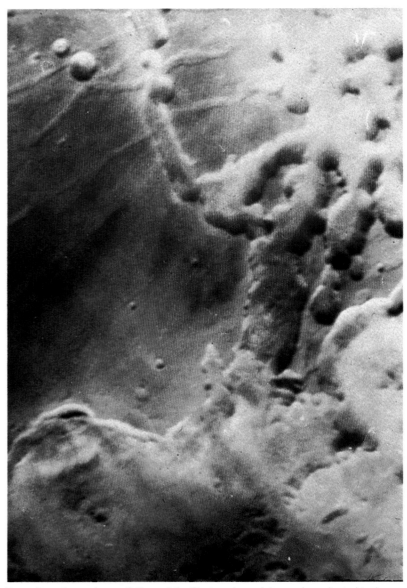

Formation of morning radiation mists. The troughs and depressions situated to the east of Coprates Chasma are here completely hidden under layers of mist, which give them a very brilliant appearance. This kind of mist, which affects the regions of low relief and which seems to 'touch' the soil, is an effect of the solar radiation which appears in the early hours of the morning; meteorologists give it the name of radiation mist. The mist forms because particles of water ice which condensed on the surface of Mars in the course of the night sublimate into water vapour under the effect of the first rays of the Sun. This water vapour recondenses in the cold atmosphere, where it forms a mist made up of ice particles. This image of Coprates Chasma was transmitted by the Viking 1 Orbiter in July 1976. In September 1977 the Viking 2 station recorded a view of the landing site at around midday, in which it can be seen that the base of some rocks are covered with a whitish deposit. This is proof that, at least during the Martian winter, deposits of water ice can exist in the soil in particular areas. From the onset of winter, hoar frost formations begin to cover the surface as the water vapour of the Martian atmosphere is transferred from the southern to the northern hemisphere wherein is situated, on the Utopia Planitia region, the site of the Viking 2 station. (NASA)

Atmospheric ozone. Unlike Earth's lower atmosphere, that of the Red Planet is not protected against solar ultraviolet radiation by a layer of ozone. This radiation can photodissociate the carbon dioxide (CO_2) molecules (into carbon monoxide, CO, and atomic oxygen) and water vapour (into the radical OH and the hydrogen atom). The abundances of carbon dioxide and water have consequences for the chemistry of the whole atmosphere. In particular, the products of the photodissociation of water vapour react with other constituents in such a way that they reduce the efficiency of the formation of ozone. Maximum concentration of this gas is thus observed above the winter pole when the water vapour content is the weakest. An analogous phenomenon causes variations in the concentration of ozone in the terrestrial stratosphere above a height of 40 kilometres.

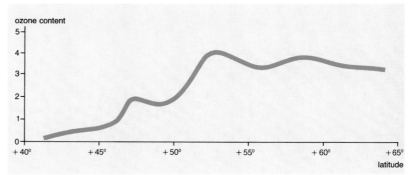

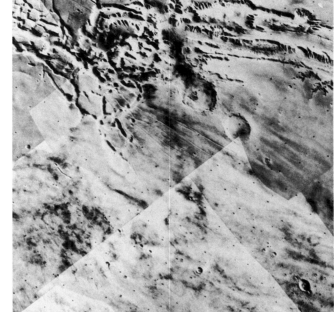
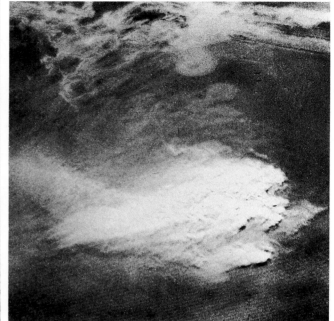

Formation of a vast cloud of dust above a region situated to the north of Solis Planum. The birth of this characteristic phenomenon of Martian activity is shown well by the comparison of these two images showing the same region including Noctis Labyrinthus (upper part of the image): to the left is a mosaic constructed with the aid of several images transmitted by the Viking 2 Orbiter on 25 March 1977. The central region of the image on the right is covered by a cloud of dust extending over about 600 kilometres and belongs to Solis Planum, a region of high Martian relief which, in places, reaches a height of 10 kilometres. (NASA)

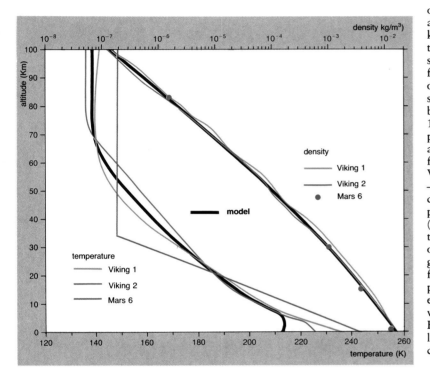

of the atmosphere; during periods of storms, aerosols can be observed above an altitude of 50 kilometres; even in calm periods, aerosols are transported up to an altitude of 20 kilometres, sometimes higher. Even more surprising is the fact that the wind speed at the Martian surface – of the order of 10 metres per second – does not seem sufficient to lift the dust above the planetary boundary layer, whose thickness reaches about 100 metres. Information on the chemical composition of the aerosols is provided by spectral analysis of the light from the sky, determined from colour photographs transmitted by the Viking stations. This study suggests the presence – in combination with particles of water ice – of dust in suspension which could contain a small proportion (1 per cent) of grains of magnetite (Fe_3O_4). These measurements must be compared to the much more precise ones which have been obtained from infrared studies at wavelengths greater than 10 micrometres of the emissions from the dust. When, in 1971–72, the Mariner 9 probe, in orbit around Mars, observed the extraordinary dust storm which raged planet-wide, its infrared radiometer analysed the dust. By comparing the results to those obtained in the laboratory for different minerals, a definite correlation was found between the infrared properties of the Martian dust emission and those of the grains of a clay rich in iron, similar to montmorillonite. Taking into account the extreme complexity of the Martian soil's composition, it is not impossible that grains with such different chemical natures should occur in the form of aerosols.

The meteorology of Mars

The most obvious aspect of Martian activity is the very great variability in the meteorological parameters – temperature, pressure and wind speed and direction – measured at the surface of the planet. The temperature on the surface varies considerably in the course of the Martian day: it varies for example between −100 °C and 0 °C in summer at the site of the Viking 1 station; the cycle repeats itself regularly from one day to another. There also exists a diurnal variation of pressure and wind; however, there is also a semi-diurnal variation superimposed on the diurnal variation of these parameters. The changes in pressure are caused by forces acting globally and these fluctuations are attributed to the atmospheric thermal tides.

Text continued on page 155

Temperature structure of the atmosphere (above). The extreme variability of the properties of the Martian atmosphere makes it difficult to define its temperature structure, that is to say the variation of temperature with altitude. Thus, the presence of dust in suspension modifies the opacity of the atmosphere and as a result governs the heating and therefore the thermal structure. Temperature and circulation are directly controlled by the concentration of aerosols and by their vertical distribution. These parameters vary considerably as a function of the season. Nevertheless, it is possible to build a model which is valid for a given season – summer in the model proposed by A. Seiff – and for the middle latitudes – lower than 60 degrees. Very obviously such a model will not be representative of conditions prevailing during dust storms, especially when they affect the entire globe of Mars. Moreover, it does not take into account oscillations of temperature created by the atmospheric tides; this undulatory phenomenon, whose effects are not only dynamic but also thermal, manifests itself on a planetary scale. The temperature profile shown is an average one established from actual measurements of temperature, obtained in the course of the descent of the two Viking stations.

From the measurement of the temperature and from the vertical profile of the atmospheric pressure (also measured on board the Vikings) a profile of the density has been determined.

On the ground, the mean pressure is about 7.3 millibars, while the density is approximately equal to 1.7×10^{-2} kilograms per cubic metre. Using the observed chemical composition, these data give a mean molecular weight for the atmosphere of 43.5

Dusk at the site of the Viking 1 station. After landing in the region of Chryse Planitia, the Viking 1 station transmitted a great number of colour photographs of the soil. Because of the particularly favourable illumination, this image, taken 15 minutes before sunset on 21 August 1976, highlights the surface details well. Thus a little trough is visible just in front of one of the station's feet. The rocks overhanging the trough have a size of about 30 centimetres. The colour of the sky betrays the properties of the aerosols in suspension in the Martian atmosphere. (NASA)

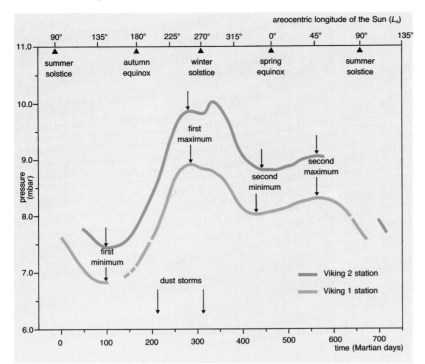

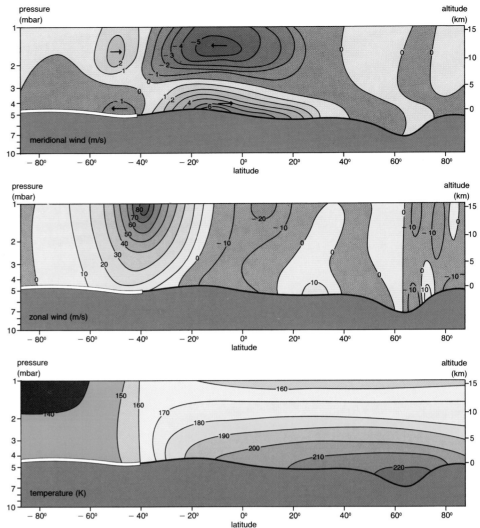

Seasonal variations in atmospheric pressure.

Today one can trace the variation in atmospheric pressure over a time interval greater than a Martian year, equal to 669 Martian or 687 Earth days (a Martian day lasts 24.62 hours). In the figure the diurnal oscillations and fluctuations due to the effects of meteorological disturbances have been smoothed and one observes the kind of seasonal variations corresponding to the physical process already discussed, that is the condensation and sublimation of carbon dioxide at the polar caps. The season parameter, L_s, is the longitude of the Sun in a system of coordinates, which is centred on Mars (areocentric) and moves with respect to the fixed system of stars. $L_s = 0°$ corresponds to the spring equinox of the southern hemisphere; each season then begins at 90° intervals. Perihelion, regardless of the season, occurs at $L_s = 250°$. The interval between the minimum of atmospheric pressure near $L_s = 135°$ and the first maximum towards $L_s = 250°$ is an indication of the importance of the deposit of carbon dioxide sublimated by the southern polar cap. The interval between the second minimum and the second maximum is less, showing that the quantities of carbon dioxide liberated by the northern polar cap are probably compensated for by a non-negligible condensation of this gas at the southern polar cap, during the same period. It is possible that, contrary to what happens in the northern regions, the southern polar cap is a permanent reservoir for carbon dioxide, for the whole of the year. The other phenomenon which the figure illustrates is the extreme rapidity of the response of the Martian atmosphere to the occurrence of a 'planetary' dust storm towards $L_s = 280°$. This is not surprising if one remembers that the temperature and circulation depend on the dissipation of solar heat and that this is strongly influenced by the aerosol content. It is probable that the phenomenon of the global sand storm owes its existence to the 'planetary' nature of the dynamic effects of the atmospheric tides. If locally the atmosphere is sufficiently charged with aerosols, the dynamic effects can be amplified and the winds become strong enough to raise more dust; after which the mechanism can accelerate and extend to still vaster regions of the atmosphere.

Zonal winds, meridional winds and temperatures at the time of the summer solstice in the southern hemisphere.

The mean variations of the winds and temperature are deduced from a model of the general Martian circulation established by J. Pollack in 1981. Positive values indicate an easterly direction for the zonal winds, and a northerly direction for the meridional winds. The model is a variant, using the characteristic conditions of the atmosphere of Mars, of a model established by the Department of Meteorology of the University of California to simulate the general circulation of the Earth's atmosphere. The modifications take into account the particular topography and the variable albedo of the Martian surface. The strong gradients of the temperature and winds displayed near the surface in the course of the Martian day have been incorporated by introducing into the model an additional atmospheric layer. In the figure, the area coloured white in the representation of surface levels corresponds to the extent of the polar caps containing carbon dioxide.

By comparing the model with that of the Earth's atmosphere, we note a certain resemblance between the circulation in the Martian winter hemisphere and the winter circulation in the Earth's atmosphere. It is also clear that baroclinic waves develop, as in the case of Earth, in the regions for which a strong temperature gradient in the direction of the pole exists. These waves are the cause of the spiral cyclonic clouds observed in the Martian atmosphere, above the northern polar regions in summer.

Lee waves.

Prior to the Viking mission, the Mariner 9 probe had, in 1972, transmitted photographs of the Martian surface, with the help of which the main properties of the planet and of its atmosphere had been defined. The image shows a system of wave-like clouds comparable to those which form in the Earth's atmosphere and which are called lee waves. Here the obstacle creating the disturbance is the edge of a crater 100 kilometres in diameter. The waves extend over a distance of 800 kilometres. The crater, called Milankovic, is situated in the northern hemisphere (53° north, 148° west). It was photographed by Mariner 9 at $L_s = 344°$. The winds blow generally towards the east with speeds, near the surface, of the order of 10 to 20 metres per second. (NASA)

Extratropical cyclone during the Martian summer.

From the Viking 1 Orbiter a spiral cloud formation was observed, which has been identified as an extratropical cyclonic system. At the centre of the zone flown over, which is near the polar regions (66° north and 227° west), the cloud covers an area 600 kilometres wide. The observation date corresponds to the Martian summer $L_s = 126°$. The altitude of the cloud is estimated at about 7 kilometres. In the Earth's atmosphere, the appearance of an atmospheric wave on a polar front can set the frontal cloud bands into spiral motion when, encountering a baroclinic instability, it will increase in amplitude. Such an atmospheric instability is an east-west asymmetry. It creates a shearing in the vertical distribution of the winds and is produced when the temperature variation as a function of latitude becomes too great. (NASA photograph, kindly communicated by G. Hunt)

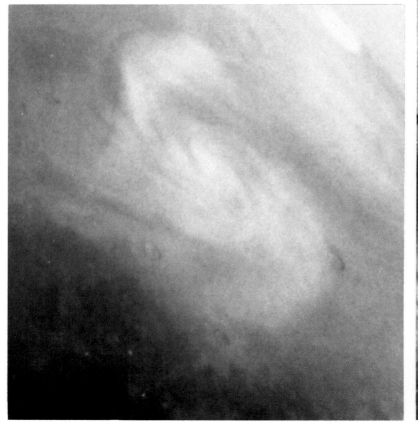

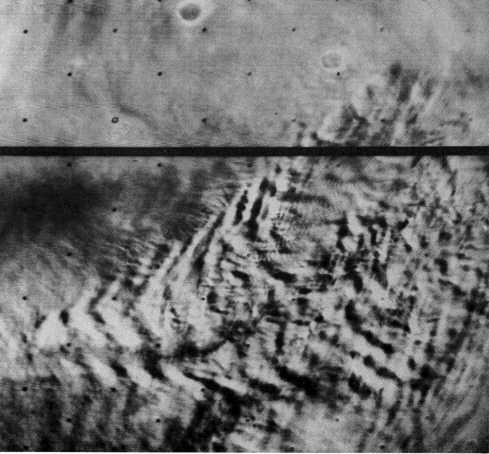

Because of the similarity in the inclinations of their rotation axes and their periods of rotation, models of the Martian atmospheric circulation can be derived from those which have been established for the Earth. These allow the most important aspects of the Martian meteorology to be understood. Thus, instabilities in the general circulation give rise to atmospheric phenomena, which are well known to meteorologists. Spiral cloud patterns, similar to those photographed in summer by the Viking 1 Orbiter above the northern polar regions of Mars, generally occur at the boundary between air masses originating from the polar and tropical regions, that is to say, on the polar front. There is good reason for thinking that the spiral clouds are condensation clouds made up of water ice particles. If they have the same origin as the terrestrial extratropical cyclones, this would prove that systems of baroclinic waves caused by instabilities in the Martian atmosphere do exist.

In winter, above each polar region, meteorological conditions are particularly favourable for the formation of condensation clouds. In summer, the conditions which prevail above the zones of high relief are more conducive to their formation. A great variety of cirrus-type clouds or ice mists thus cover the Martian surface. During winter, the north pole is hidden by a veil of mists composed of water ice particles (and also of dust) which creates the so-called 'polar hood'. This phenomenon also occurs at the south pole, but to a lesser extent. The polar hoods occasionally extend to the middle latitudes (40 degrees for the northern hood). The Viking 2 station observed the polar cloud as it passed overhead, resulting in a darkening of the sky in the course of a few hours. Moreover, in some regions of the planet, the condensation of carbon dioxide can also give rise to clouds or ice mists, though admittedly of short duration. Instances have been recorded where the temperatures of the atmospheric layers have dropped to $-120\,°C$ a few hours before sunrise; these temperatures are sufficiently low to allow carbon dioxide ice to form at the Martian atmospheric pressure.

Compared to the clouds observed in the terrestrial atmosphere, those which form in the Martian atmosphere are less spectacular; no elements resembling the very dense and well-defined terrestrial cumulus are observed. All the Martian clouds have, however, characteristics analogous to the terrestrial clouds. Four categories have been identified: convection clouds, orographic clouds, wave clouds and fogs.

Convection clouds, caused by the rising of atmospheric layers heated near the soil, have the characteristic of forming into very distinct layers covering a wide area and situated at an altitude of between 4 and 6 kilometres.

One of the most spectacular phenomena of Martian meteorology is the seasonal formation of a huge ensemble of white clouds, which develop mainly in the course of the afternoon. This phenomenon, which can be observed from Earth, occurs over the Tharsis region where Mars' four highest volcanoes are located. The clouds are wispy and situated at a high altitude, sometimes several kilometres above Olympus Mons, which itself reaches 27 kilometres in height. These clouds undoubtably have an orographic origin, and are caused when the hot air cools while rising along the volcanoes' slopes and the water vapour it contains condenses.

Wave clouds associated with the phenomenon of lee waves have often been observed in the Martian atmosphere, in particular at the time of the Mariner 9 mission. This type of cloud develops when rapidly moving air masses encounter an obstacle. In the regions of Mars where these wave clouds form it has been established that the surface winds blow from the west with speeds of 10 to 20 metres per second near the surface and more than 100 metres a second at a height of a few kilometres.

The finest example of the strong interaction of the Martian soil with the atmosphere is furnished by the radiation fogs. The colour photograph given on page 144 shows this phenomenon in the bottom of Coprates Chasma's depressions. The particles of water ice condensed on the surface sublimate into water vapour under the influence of the first rays of the Sun, from which it follows that a deposit of hoar frost is formed during the night in these regions.

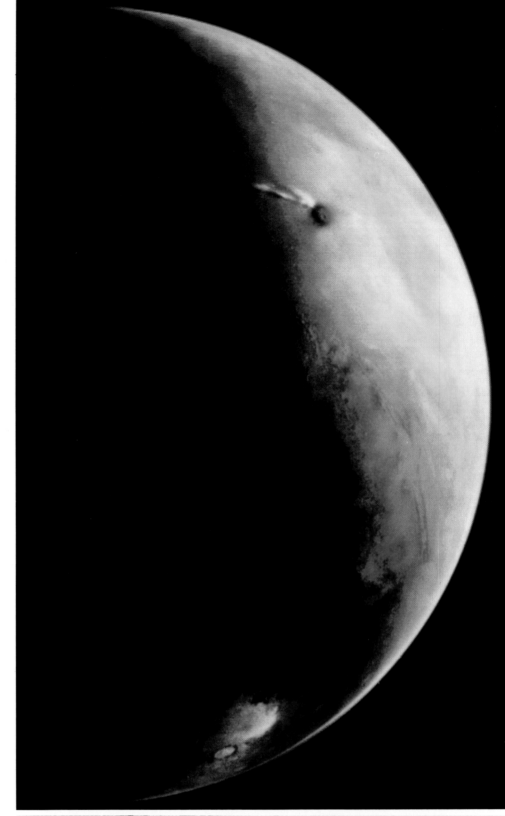

Orographic clouds. Some Martian cloud systems are directly related to the topography. The region of the planet most affected by very extended formations of orographic clouds is obviously the Tharsis plateau, where the great Martian volcanoes are situated. Upwind of the summit of a volcanic mountain, the air is forced to rise and, in so doing, cools, giving rise to this kind of cloud. The image of the crescent Mars shown above was transmitted by the Viking 2 probe in August 1976, when it was approaching the planet. It allowed two plumes forming a 'V' to be discovered around the Ascraeus Mons volcano. The clouds are made up of water ice crystals. In the northern hemisphere, they form especially right at the beginning of summer. The other reliefs are the Argyre basin, towards the south pole in the lower part of the disk, and the immense fractured region, Valles Marineris, at the centre. During the course of its revolution around the planet, the Viking 2 Orbiter flew very near the Ascraeus Mons region. On the lower image, obtained during the 225th orbit ($L_s = 210°$) an orographic cloud is observed, closely associated with the volcano; its surface area is less than 5000 square kilometres, and its morphology was seen to change during the course of a day. (NASA)

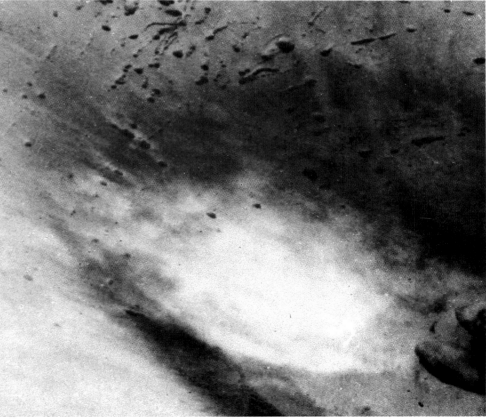

Origin of the rocks observed on the Viking 2 site (image above obtained on 18 February 1977). A gully has been hollowed out in the middle of the cover of rocks and stones which are scattered over the immediate neighbourhood of the Viking 2 station. On its edge (top right of the image) is a rock one metre in diameter. The majority of the rocks are pitted with alveoli whose dimensions are no greater than a centimetre. One cannot avoid the comparison with certain, porous, volcanic terrestrial rocks. (NASA/NSSDC)

Layered structure of the polar relief. Observations of the polar regions reveal a very peculiar relief which must be the consequence of the strong surface–atmosphere interaction. This photograph (below) of the south pole, obtained by the Mariner 9 probe, shows a superposition of very distinct horizontal layers. Each profile corresponds to the edge of a plate made up of several layers whose thickness is of the order of 50 metres. The topography where the whirlpool-shaped contours are seen corresponds to the lowest relief; so that in the picture a range of tones in the structure is observed from left to right. This tiered structure was progressively built up in the course of time by numerous deposits of an aggregate of dust and water condensates. The tiered shape must be associated with periodic variations of the atmospheric parameters, such as the temperature and dust content. The process which controls precipitations is similar to that which has been seen, on a smaller scale, by examining the variations of the surface properties on the Viking 2 site. The periodic changes of the properties of the atmosphere and in particular of the relative concentrations of the very volatile constituents, are the result of the regular variations in the eccentricity of the Martian orbit, but especially of the tilt of the planet's rotation axis, which varies between 15 and 35 degrees with a period of 120 000 years. (NASA)

Changes in the appearance of the soil in the course of the Martian year. The scientists in charge of the analysis of photographs of the sites of the two Viking probes tried to investigate even the most trivial traces of change of the soil in the course of time. One of the most obvious features of the panorama transmitted by Viking 1 was a large rock 2 metres long, named 'Big Joe'. It appears on the three images above, obtained successively on 5 December 1976, 22 March 1977 (239 Martian days later) and 11 December 1977 (583 Martian days later). Even though the lighting conditions are different on the second image, a change is apparent. A slight collapse, evidenced by the appearance of a surface deposit resembling a collection of very fine grains at the base of the rocks, is ascertained. This may be a direct effect of the wind or the result of thermal forces. However, the most likely explanation is that at the time of the landing of the station, the soil was disturbed in the region where Big Joe is located. Subsequently an event occurred – sometime between 74 and 183 Martian days afterwards – which led to subsidence, followed by the re-covering of that part of the soil by a thin film of sediments, whose thickness is little greater than a centimetre. It is possible that this event was of seismic origin, since examination of the third image shows that the appearance of the soil does not seem to have changed. Several indications of minor changes in the surface have been found on other images of the Viking 1 site. On some images, it is noted that two dark spots appear in the local panorama. An explanation proposed for this is that, in each case, a high-albedo deposit originally in that position has been covered by wind-borne dust of lower albedo. (NASA/NSSDC)

Variability of the surface

The geomorphological analysis undertaken, the aid of images of the Martian surface transmitted by the cameras of the Viking orbiters allowed scientists to confirm the existence of two phenomena which have been observed for a long time and which manifest themselves by seasonal or secular changes in the albedo of specific areas.

The first seasonal phenomenon is the dilation in the winter and contraction in summer of the polar caps; this is without doubt linked to the process of condensation and sublimation of carbon dioxide (today we have evidence that the deposit which permanently covers the polar regions is made up of water ice). The other phenomenon is the result of aeolian erosion. Traces of aeolian activity have been observed in the regions which flank the three volcanoes of the Tharsis dome, and in particular Arsia Mons. Moreover, a comparison of images taken in 1972 by Mariner 9 and in 1976 by the Viking Orbiters, shows dark radial striae associated with the craters in particular on the eastern edges of Hellas Planitia. But, on the whole, the effects of aeolian erosion are a lot less important even after a global dust storm than those which were expected based on the analysis of the Mariner 9 results. This is equally true for the chemical erosion of the soil, which today seems non-existent. If in the recent past significant changes have appeared in the morphology of the Martian surface, they can only have been due to the formation of craters.

The two Viking stations observed the Martian landscape for several years. Scientists have tried to make use of the excellent resolution of the images of the soil transmitted, to discover barely perceptible modifications of the site. Attempts were made in the first place to discover any changes over intervals of a few Martian days. Secondly, pairs of photographs of specific areas taken many Martian months apart, but under identical illumination, were studied. Finally, the camera was programmed so that it could be used to detect movement, in particular to chart any movement of dust.

The detailed analysis of variable phenomena on a short time-scale is very instructive, for it can give clues about the long-term evolution – both past and future – of the Martian surface. It also contributes to a better understanding of the planet's climatology. It is probable that the mechanism which leads to the formation of a complex mixture of dust and ices (water and carbon dioxide) in the Martian atmosphere is also responsible for the tiered structure of the polar regions. The successive layers arranged in tiers would consist, after sublimation of carbon dioxide, of a mixture of dust and water ice. The differences characterising the layers would then be indirectly connected to the oscillations of Mars' rotation axis, oscillations which in particular result in substantial changes in pressure – and therefore of the dust content – and changes in temperature of the polar surfaces – and therefore of the water content.

Guy ISRAËL

Analysis of changes in the appearance of the Viking 2 site.
Among the variable phenomena seen at the Viking 2 site, some are of greater extent than all those detected on the Viking 1 site. They are connected with the movement, across the site, of the north polar cap. The appearance upon the soil, in the course of the first winter, of a covering of condensates, constitutes the most obvious proof of this movement. The deposits which covered the whole site remained for a third of the Martian year. Then, after the condensates had dispersed, the soil curiously did not recover its former appearance. The enhanced red colour testified to the existence of a new, very fine layer of shiny dust. The correlation between, on the one hand, the process of formation of condensates and, on the other hand, the mechanism of transport of the dusts into the atmosphere, has been shown by analysing three images taken at long time intervals. These three photographs of a specific area of the Viking 2 site were obtained under almost identical lighting conditions, the first 36 Martian days after the landing date of the Viking 2 station, the second after 503 Martian days and the third after 1050 days. In this way, their analysis highlights the possible changes occurring during the course of two Martian winters. The sidereal year is equal to 669 Martian days. Examination of the first two images clearly shows an increase in the brilliance of the soil. On the first image very light spots are discerned in contrast to the darker surface of the soil, while in the second image they have disappeared under an accumulation of highly reflective dust. The date at which the shiny deposit began to form corresponds to 22 days after landing. The process ceased when the Viking station manoeuvred its articulated arm to collect the Martian soil to carry out biochemical measurements. That is the reason why the two imprints retained a dark appearance. The temporal relationship of the phenomenon to the occurrence of the second global dust storm observed at the time of the Viking mission (275 days after landing) has been well established. One can thus try to identify the chemical nature of the surface coating and make hypotheses as to the origin of the mechanism responsible. At the time of a sand storm, dust particles whose diameter is estimated at 2 micrometres are carried from lower latitude regions. As they are swept northwards, they increase in diameter because water vapour in the atmosphere condenses on them. When the temperature decreases further, the same condensation phenomenon intervenes, but this time with carbon dioxide, which increases their size even further.

The third photograph, in colour, acquired at the end of the second winter, shows traces of white-coloured condensates which have accumulated periodically at the base of the rocks. The metallic object ejected at the foot of Viking is a protective cap of the articulated arm used for taking soil samples. (NASA/NSSDC)

Mars

Phobos and Deimos

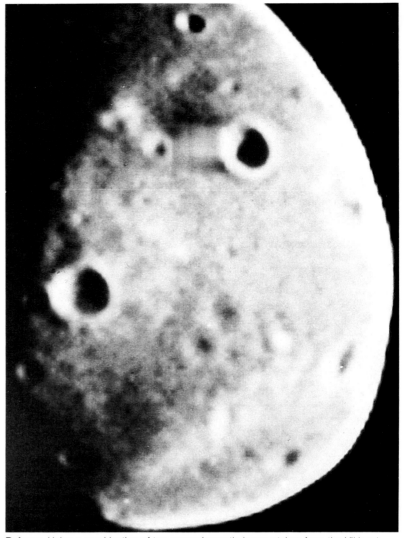

Mars has two satellites: Phobos ('Terror' in Greek) and Deimos ('Panic'). They are very small, very dark bodies, covered with craters. They are also very near the planet, and are difficult to observe from Earth. They were only discovered in August 1877 by Asaph Hall on the occasion of a very favourable opposition of Mars, which was only 0.57 astronomical unit from Earth. Hall was using the new 66-centimetre diameter telescope of the US Naval Observatory at Washington. It is amusing to note that the two Martian moons were mentioned in literature long before their discovery: convinced of the harmony of the cosmos, Johannes Kepler believed that the Earth had one moon, Mars two and Jupiter four (only the four Galilean satellites were known at that time); Jonathan Swift in *Gulliver's Travels* and Voltaire in *Micromegas* also speak of the two Martian moons!

Phobos rotates about Mars more rapidly than the planet spins. Because of tidal forces it is inexorably spiralling in towards Mars, and will crash into it in about 30 million years, a very short time in astronomical terms. In contrast Deimos orbits Mars almost synchronously and is almost undisturbed by tidal forces.

Phobos and Deimos were observed by the American Mariner 6 and 7 probes in July and August 1969 respectively, then by Mariner 9 at the end of 1971. These were the first satellites of another planet to be photographed from close up. But most of what we know was discovered by the Viking 1 and 2 orbiters in 1977. In 1989 the Soviet Phobos 2 probe provided some data before malfunctioning.

These two irregularly-shaped bodies are to a first approximation ellipsoids of revolution: the long axis of Phobos measures 27 kilometres and that of Phobos 15 kilometres. Like all bodies of the Solar System without atmospheres or geological activity, the surfaces of Phobos and Deimos are saturated with craters and covered in a regolith. Analysing the diameters and depths of the craters reveals the history of impacts on the two satellites and gives a measure

of the depth of the regolith. Using models of the cratering rate at Mars's orbit, one can estimate that their surfaces are 2.5 to 3 billion years old.

While Mars is very red, Phobos and Deimos are grey-black, and very dark, as dark as the blackest maria on the Moon; their surfaces reflect less than 6 per cent of the incident sunlight. This colour and reflectivity show that the satellites are composed of matter similar to that of the Orgeuil meteorite (carbonaceous chondrite) or the Ceres asteroid.

On 18 February 1989 Phobos 2 flew by Phobos twice at less than 190 kilometres; precise measurements of the perturbations of its trajectory give the mass and density of the satellite (its volume is relatively well known), which turn out to be smaller than previously thought: $1.08 \pm 0.01 \times 10^{17}$ kilograms and 1.95 ± 0.1 grams per cubic centimetre. This density is noticeably lower than that of meteorites found on Earth. It is estimated that Deimos's density is similar, but its volume is too poorly known to give a precise estimate. The low density of Phobos suggests that it consists of matter rich in light elements similar to those of carbonaceous chondrites, and that its interior is very porous or contains low density material (such as water ice). The spectra of Phobos and Deimos seem to show that their surfaces are very dry: if there is water on Phobos it is deeply buried.

All this suggests that Phobos and Deimos are primitive objects, perhaps formed from the same body which fragmented in a collision after being captured by Mars. Some researchers think that the matter making up the two satellites is close to the 'primordial dust' and the primitive planetoids from which the planets and satellites of the Solar System formed. Thirty-seven high-resolution images of Phobos were acquired by Phobos 2. Contrary to what was thought at the start of the exploration of the Martian system, Phobos is not uniformly covered in grey dust: close comparison of visible and infrared observations reveals zones which are fairly blue or red; this surface heterogeneity may cor-

Deimos. Using a combination of two monochromatic images taken from the Viking 1 Orbiter, one through a violet filter, the other through an orange filter, the image above contains to give an idea of the colour of Deimos' surface. The resolution is about 200 metres. Deimos is uniformly greyish, the shades of orange and blue are in fact only artefacts introduced by the image-processing.

Phobos and Deimos are very dark, even darker than some asteroids of the inner belt. Their spectra have similarities with those of type C asteroids – present in the outer belt – and with Trojan asteroids. The material of which they are composed is probably very close to that of the carbonaceous chondrites. (NASA/NSSDC)

respond to a body whose interior is a conglomerate of cosmic dust, Martian material and carbonaceous chondrites.

The origin of Phobos and Deimos is still unknown. Many astronomers think that they may be asteroids captured shortly after Mars's formation, or instead remnants of the formation process. An answer may come from two lines of research: chemical analysis of the surfaces and the study of the dynamics of the orbits. Chemi-

Phobos. These two images of the inner satellite of Mars were obtained by the Soviet Phobos 2 probe on 28 February 1989, at a distance of 439 kilometres (left) and the American Viking 1 orbiter, on 19 October 1978, at a distance of 612 kilometres (right). Most of the characteristics of Phobos are visible: the irregular shape, the Stickney crater (clearly visible at the bottom of the right-hand image), chains of craters and parallel grooves, probably created in the impact producing Stickney, which almost split up the satellite. These grooves are about 100–200 metres wide and have a depth of about 20 metres. (Left, IKI/CNES; right, JPL/NASA)

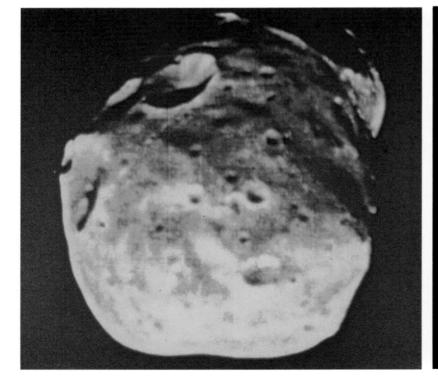

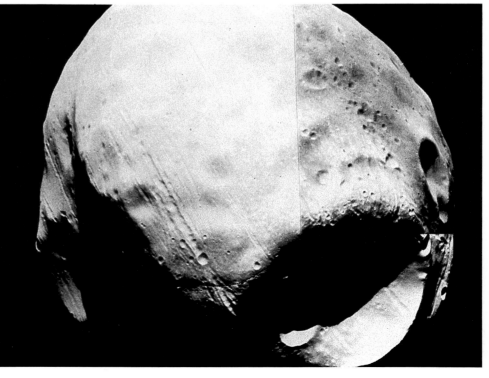

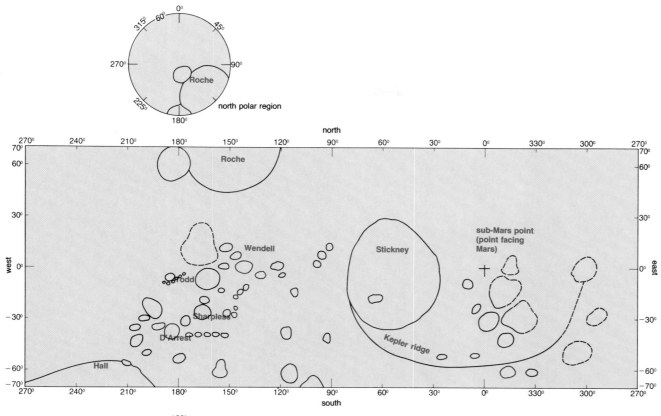

Orbital characteristics. Deimos revolves around Mars in a period of 30 hours 18 minutes at a distance of 23 490 kilometres from the planet. Phobos, at a distance of 9354 kilometres from Mars, has a period of revolution of 7 hours 39 minutes, which is much shorter than the rotation period of Mars, 24 hours 37 minutes. An observer situated on the surface of Mars would therefore see Phobos moving in the opposite direction to Deimos, the stars and the planets; he would see it rise in the west, set in the east, and from nowhere would it remain in the sky for more than 3 hours 10 minutes. In contrast, Deimos moves through the Martian sky very slowly at a rate of about 2.8 degrees per hour from east to west. On the figure, the angular distances travelled by Mars and Deimos during one revolution of Phobos are indicated. The two satellites are so close to the planet's surface that they cannot be seen from the polar zones; an observer must be situated between latitudes ±82 degrees to see Deimos and ±69 degrees to see Phobos.

Map of Phobos. It is difficult to draw up maps of Phobos and Deimos; these two objects are in fact roughly represented by ellipsoids of 28 kilometres by 22 kilometres by 18 kilometres and (16) kilometres by 12 kilometres by (10) kilometres, respectively, whose major axes point towards Mars. (Values in brackets are uncertainly by more than 10 per cent.)

A system of triaxial coordinates has, however, been developed by T. C. Duxbury to draw up a 'map' of Phobos. The two biggest craters have been named Hall and Stickney (Mrs Hall (née Stickney) being honoured for her encouragement of her husband, particularly on the night of the discovery of the Martian satellites). The maximum diameter of Stickney is about 10 kilometres, that of Hall about 6 kilometres. Note that Stickney's diameter is more than a third that of Phobos' major axis.

No cartography of Deimos has yet been drawn up; however, one crater has been named Voltaire, another Swift. (After T. C. Duxbury, 1974)

The surfaces. Phobos (top) and Deimos (bottom) are saturated with craters and covered by a regolith. On a scale of a few hundred metres, Phobos is homogeneous, while Deimos shows regions that are about 30 per cent brighter than the surrounding area. In contrast with Phobos, there are no striae on the surface of Deimos. The image of Phobos on the left was acquired by Viking 1 from a distance of 120 kilometres. It covers an area of 3 by 3.5 kilometres and details down to a few metres in size can be distinguished. The two large craters have diameters of 0.9 and 1.3 kilometres; the smallest craters visible have diameters of 10 metres. The large striae are between 100 and 200 metres wide and 20 metres deep. The image of Phobos on the right has a comparable resolution and shows a surface saturated with craters similar to the most ancient terrains of the Moon. Crater counts enable us to estimate that the greater part of Phobos' surface is of the order of three billion years old. The two images of Deimos were taken by Viking 2 from a distance of about 60 kilometres. The mosaic on the right covers 2 kilometres from top to bottom; its upper part is reproduced at a larger scale on the left-hand image, which covers a region 1.2 by 1.5 kilometres. The smallest details visible are about 3 metres across. At the top of these two images, a large dark crater is discernible.

These high-resolution images enable the differences in appearance between the two satellites to be understood. Deimos' surface, although appearing smoother than that of Phobos, has in fact the same surface density of craters; the smallest are hidden by a regolith about ten metres thick. Isolated rocks are seen which are not visible on Phobos. The thickness of the regolith covering Phobos is estimated at about 300 metres. (Laboratory for Planetary Studies, Cornell University)

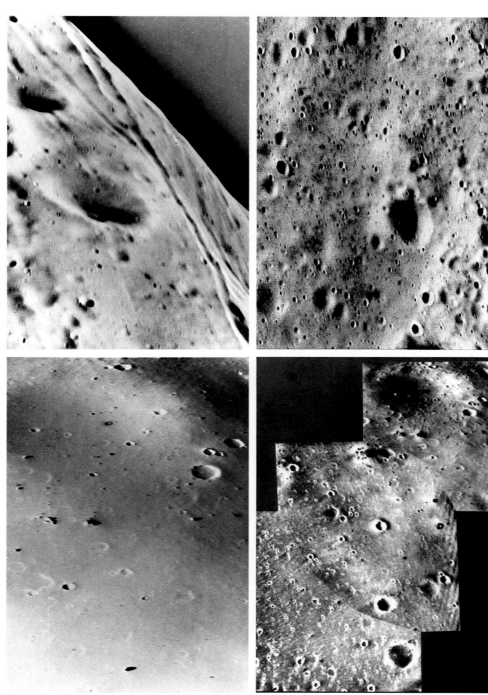

cal analysis is difficult as the surfaces are complex. The spectacular difference in the surface composition of Mars and its satellites, however, makes improbable the formation of the latter at the same time as Mars by agglomeration of material in orbit about the protoplanet. It seems more likely that Phobos and Deimos were initially formed in the outer asteroid belt, then later captured by Mars. This would not be surprising; perturbations by Jupiter as well as encounters and collisions between asteroids strongly affect their orbits. Some of them cross the orbits of the planets and crash into them, explaining the many impact craters in the Solar System. Nevertheless, the capture mechanism for asteroids is complex and still poorly understood (arrival of a body at very low velocity, tidal effects, presence of a third body, collisions . . .?)

Calculating the evolution of the orbits of Phobos and Deimos in the past shows that a capture was conceivable. It is even possible that the two orbits crossed and that the two bodies are the result of a collision. If Phobos and Deimos are indeed captured asteroids, they are for the moment the only well-studied members of the family of primitive bodies which formed the Earth and planets through the interplay of their mutual collisions.

André BRAHIC

The Asteroids

It was on 1 January 1801 that the Sicilian Guiseppe Piazzi discovered Ceres, a minor planet with a diameter of 1000 kilometres, at a distance of about 2.8 astronomical units from the Sun. Until that time, the geometric progression of the major planets' distances from the Sun (Titius–Bode law) had been embarrassed by the absence of any object between Mars and Jupiter. The orbit of Ceres corresponded exactly to that of the 'missing planet' predicted by that law. However, Ceres' mass was much smaller than that of Mercury, the smallest of the planets. In addition, it was quickly shown that Ceres was not the only object with an orbit between those of Mars and Jupiter: three other asteroids, Pallas, Juno and Vesta, were discovered at the beginning of the nineteenth century. A number of advances in observational techniques followed these initial discoveries so that today several thousand asteroids are known. However, the total mass of all these objects is barely twice that of Ceres.

The orbital parameters of asteroids cover a wide range of semi-major axes, eccentricity and angles of inclination to the ecliptic, which can reach 30 degrees. The vast majority of asteroids have semi-major axes between 2 and 4 astronomical units and constitute the main belt. However, no asteroid has an orbit whose period is related in a simple way (3/1, 5/2, 7/3, 2/1) to that of Jupiter. These zones where orbits are unstable are known as the Kirkwood gaps. There are also about a hundred very small asteroids (a few kilometres in diameter) whose orbits cut across or approach close to that of the Earth. They make up the Apollo–Amor group which originates essentially from asteroids of the main belt, perturbed from the Kirkwood gaps by the effects of Jupiter's gravity. Some others of the group could be the nuclei of former short-period comets.

Physical characteristics

In the course of time, the magnitude of an asteroid varies, sometimes considerably. There is a slow variation depending on the distance to Earth, and also on the phase angle (that is to say, the angle between the Earth–asteroid and Sun–asteroid directions). This 'slow' variation is accompanied by a variation of much shorter period, produced by the rotation of the object. Rotation periods for numerous asteroids have thus been determined and are typically between 2 and 24 hours. Some light curves suggest that certain asteroids, such as Camilla or Eunomia, are very elongated; others are very probably made up of two asteroids in contact (e.g. Hektor). Furthermore, some recent results from stellar occultation studies seem to indicate the presence of satellites around a few asteroids such as Metis.

The size distribution of the asteroids can be determined by two different optical methods. The most direct method, only recently developed, consists of evaluating separately the energy emitted and absorbed by the object in the visible and infrared regions respectively. The sum of these two terms is equal to the solar flux received by the asteroid, which is proportional to its projected surface area and therefore to the square of its size. A more indirect method uses an empirical relationship between the variation of the polarisation of light emitted as a function of the phase angle on the one hand, and the percentage of solar light reflected (the albedo) on the other. The measurement of the apparent visual magnitude then allows the size of the object to be determined. These measurements bring two essential characteristics to light; first of all, the number of asteroids grows very rapidly with diminishing size. A decrease in size by a factor of ten increases the numbers by a factor of one hundred. In addition, two classes of asteroids may be distinguished on the basis of their albedo alone: 'light' asteroids with an albedo greater than 0.1 and 'dark' asteroids with an albedo of less than 0.05. These latter objects are thus the darkest in the whole Solar System (by way of comparison, the lunar maria, themselves very dark, have an albedo of 0.07).

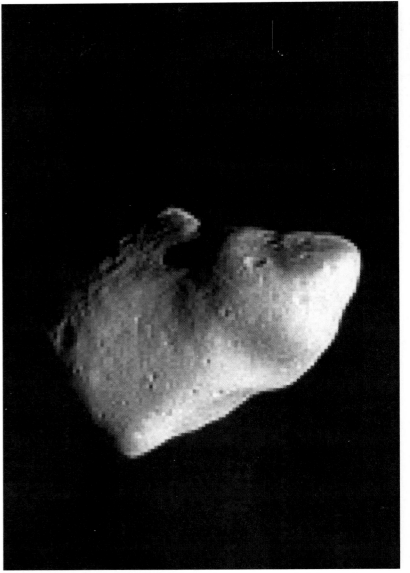

The asteroid Gaspra. The first high-resolution (160 metres per pixel) image of an asteroid – 951 Gaspra, in the Flora family – was taken by the space probe *Galileo* on 29 October 1991, from a distance of 16 200 kilometres. The North ecliptic pole is at the top and the Sun at right. The illuminated part of Gaspra is about 16 × 12 kilometres. The size of the asteroid is estimated at 20 × 12 × 11 kilometres; Gaspra is thus comparable in size to Deimos or the nucleus of Comet Halley. It rotates anticlockwise with a 7 hour period; its North Pole is near the upper left corner of the illuminated part.

The surface shows many impact craters, whose diameters vary between 1500 metres and the resolution limit of the image.

The irregular shape of Gaspra suggests that it is a fragment of a larger asteroid broken in collisions. (JPL/NASA)

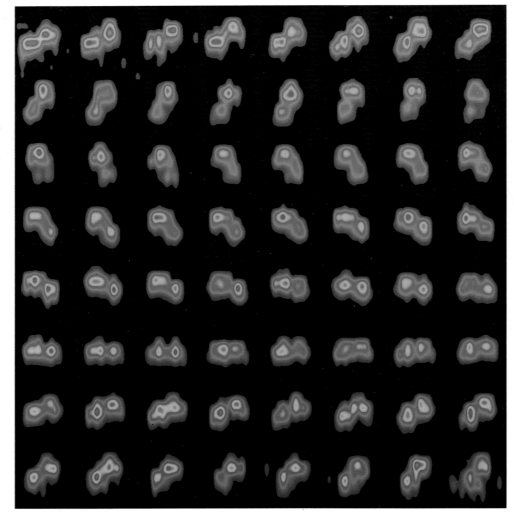

Radar images of the asteroid 4769 Castalia. These images were acquired by Steven J. Ostro on 22 August 1989 using the Arecibo radiotelescope. They reveal the structure of 4769 Castalia (initially called 1989 PB), which seems to consist of two objects of about 900 metres diameter in contact with each other.

It is likely that the two objects had separate origins and slowly collided to form a double asteroid. This sequence (left to right and top to bottom) shows the anticlockwise rotation of Castalia around an axis pointing at the observer, with a period of order 2 hours 30 minutes (S.J. Ostro, JPL/NASA)

Mineralogical characteristics and classification

For a few years now, advances in infrared astronomy have yielded information on the surface composition of asteroids. Numerous minerals conveniently display characteristic absorption bands in the near infrared, from 0.6 to 4 micrometres. In particular, the relative abundance of the large groups of silicates – olivines, pyroxenes and feldspars – can be determined. Also, the presence of hydrated minerals can be deduced thanks to the strong water absorption band at 3 micrometres. In this way, it is possible to deduce that the two classes distinguished by their albedo are in fact composed of very different minerals: light asteroids are made up of silicates analogous to those observed in the lunar rocks. Hydrated minerals are very abundant in the dark asteroids, whose low albedo is attributed to the presence of a few per cent of carbon on the surface.

Spectral characteristics thus allow the asteroids to be classified into different groups:
– the C asteroids are the darkest asteroids, rich in hydrated silicates and carbon; they are the most numerous (about 60 per cent) particularly in the outer regions of the main belt;
– the S asteroids have the spectral characteristics of rocky bodies and are mainly made up of pyroxenes and olivines as well as metals (iron and nickel); very abundant among the Apollo–Amor group and in the inner regions of the main belt,

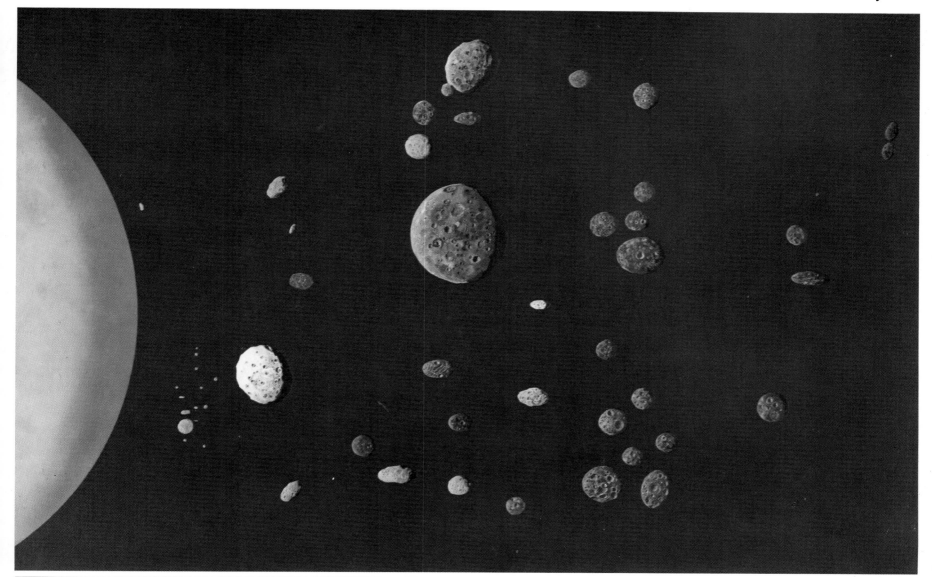

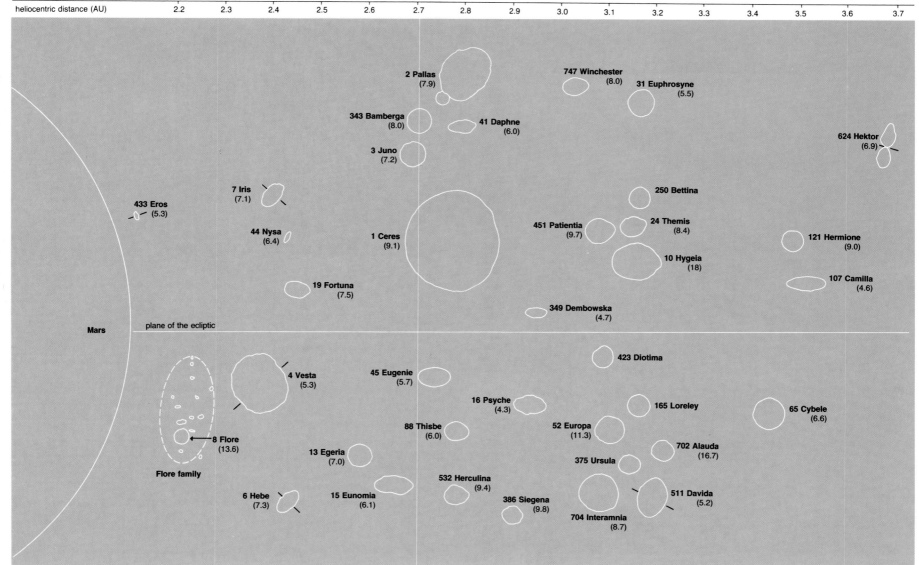

| heliocentric distance (AU) | 2.2 | 2.3 | 2.4 | 2.5 | 2.6 | 2.7 | 2.8 | 2.9 | 3.0 | 3.1 | 3.2 | 3.3 | 3.4 | 3.5 | 3.6 | 3.7 |

2 Pallas (7.9)

747 Winchester (8.0)

31 Euphrosyne (5.5)

343 Bamberga (8.0)

41 Daphne (6.0)

624 Hektor (6.9)

3 Juno (7.2)

7 Iris (7.1)

250 Bettina

433 Eros (5.3)

451 Patientia (9.7)

24 Themis (8.4)

44 Nysa (6.4)

1 Ceres (9.1)

121 Hermione (9.0)

10 Hygeia (18)

19 Fortuna (7.5)

107 Camilla (4.6)

349 Dembowska (4.7)

Mars

plane of the ecliptic

423 Diotima

45 Eugenie (5.7)

4 Vesta (5.3)

16 Psyche (4.3)

165 Loreley

88 Thisbe (6.0)

52 Europa (11.3)

65 Cybele (6.6)

8 Flore (13.6)

13 Egeria (7.0)

702 Alauda (16.7)

375 Ursula

Flore family

532 Herculina (9.4)

511 Davida (5.2)

6 Hebe (7.3)

15 Eunomia (6.1)

386 Siegena (9.8)

704 Interamnia (8.7)

Physical properties of the larger asteroids. The thirty-three asteroids whose largest dimension exceeds 200 kilometres are shown in this graphic representation, which illustrates several physical properties of the larger asteroids. The distance from the Sun of each of the objects is shown. The relative sizes of the asteroids are to scale, and the shapes are those deduced from their light curves. The very small size of these objects compared to Mars, whose limb is represented to scale, and the great difference in size between the largest of the asteroids, 1 Ceres, and the second largest, 4 Vesta, should be noted. In the upper figure, the contrasts of colours reflect the differences in albedo observed among the asteroids. It is apparent that the dark asteroids (C) predominate in the outer part of the belt, while the light asteroids (S, M) are abundant nearer the

Sun. Some of these objects have very similar orbits and make up asteroid families (for example, the Flore family of objects, whose members with a size greater than 15 kilometres are shown) which are probably the debris resulting from the fragmentation of a single parent body. The asteroids situated towards the top or towards the bottom of the figure have relatively eccentric or inclined orbits with respect to the plane of the ecliptic (or simultaneously possess both these characteristics). However, the orbits of the asteroids shown near the plane of the ecliptic are relatively circular and have small inclinations. In the lower diagram, the periods of rotation, when they are known, are shown in hours below the name of each asteroid, which itself is preceded by the asteroid's sequence number. (Figures from Andrew Chaikin and J. Kelly Beatty.)

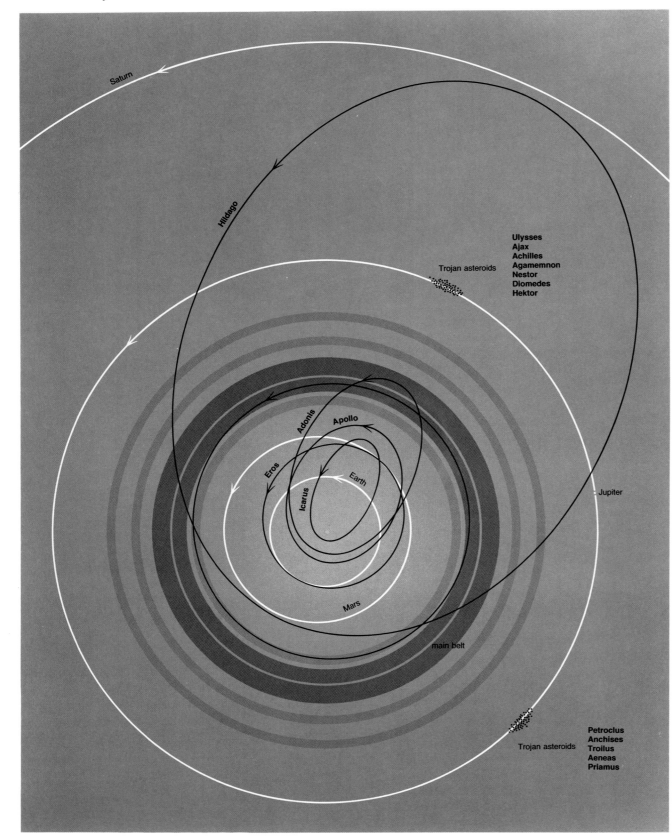

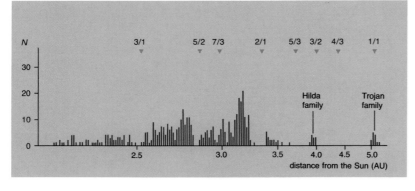

Trojan asteroids

Ulysses
Ajax
Achilles
Agamemnon
Nestor
Diomedes
Hektor

Jupiter

main belt

Trojan asteroids

Petroclus
Anchises
Troilus
Aeneas
Priamus

Orbits of some asteroids. This representation of orbits projected onto the plane of the ecliptic shows the main belt (tinted grey) which contains nearly all the known asteroids, the majority of which have small eccentricities, and the characteristic orbits of a few particular asteroids, like Hidalgo, whose aphelion is situated in the neighbourhood of Saturn's orbit. Only a few asteroids have orbits which penetrate into the inner regions of the Solar System: they consist of the Amor group, which cross Mars' orbit, and the Apollo group, which cross that of the Earth. In this diagram, in order to facilitate comparison, the major axes of the orbits are aligned, although obviously this is not the case in reality. The names of a few Trojan asteroids are listed.

they represent about 30 per cent of the total number of known asteroids;
– M asteroids appear entirely metallic (iron and nickel).

Some asteroids do not fall into these categories. For example, Vesta has a particularly high albedo of 40 per cent and a spectrum dominated by the bands of pyroxenes and feldspars.

Asteroids and meteorites

The similarities between the absorption spectra of asteroids and meteorites have been investigated and from these studies it seems that asteroids are the most likely parent bodies of meteorites. It thus appears that carbonaceous chondrites originate from C asteroids and some stony meteorites from the S asteroids. The metallic meteorites could originate from M asteroids, while metallic–stony meteorites could result either from M asteroids or S asteroids. However, the ordinary chondrites, which represent the majority of the collected meteorites, do not seem to have any counterpart among the known asteroids. This could be due to these meteorites not being representative of the surface of asteroids but rather of their interior. The achondrites, themselves also relatively abundant, only seem to have links with a very small number of asteroids, such as Vesta.

Origin and evolution of the asteroids

Two major hypotheses have been put forward to explain the origin of the main-belt asteroids. The first involves the existence of a mother planet, orbiting between Mars and Jupiter, which itself was fragmented. It is now considered more likely that their origin lies in a population of objects whose accretion into a planet was interrupted at an intermediate stage by gravitational perturbations brought about by the newly-formed Jupiter.

In this hypothesis, the asteroids are the sole survivors of the myriad of small objects from which the terrestrial planets themselves formed. After their accretion, objects with a size greater than a few kilometres are heated to high temperatures by the decay of short-lived radioactive elements; in some of these bodies, a metallic core thus forms, surrounded by a silicate crust. All of these processes take place in a few million years, right at the beginning of the history of the Solar System. After this brief period of internal activity, only collisions between objects of the main belt are able to modify their characteristics, fragment them and, together with planetary perturbations, evolve their orbits. The relative speed at the time of these collisions is of the order of a few kilometres per second.

Bombardment by the smallest objects pulverises the surface rocks and leads to the accumulation of a surface layer of debris, or regolith. Because of the intensity of the bombardment and of the weak gravitational field, the thickness of this layer reaches nearly a kilometre. It is in this bed of dust that some meteoritic holes are formed.

Impacts between bodies of similar size lead to the complete fragmentation of the two objects. The life span of an asteroid *vis-a-vis* this process is short compared to the age of the Solar System, for asteroids less than about a hundred kilometres in size. Thus the majority of the asteroids are fragments of primordial bodies of greater size, metallic asteroids originating, for example, from the core of a larger, differentiated body.

Distribution of the asteroids. The diagram opposite represents the distribution of the semi-major axes of the orbits of those asteroids of more than 80 kilometres diameter which orbit between Mars and Jupiter. Each vertical bar represents the number (N) of asteroids orbiting around the Sun at that mean distance. Notice first of all that the densest concentrations of asteroids are found between 2.5 and 3.5 astronomical units (AU). Apart from this, it is clear that the distribution is very irregular: beyond 3.5 AU the asteroids are contained in only two families, the Hildas (at 4 AU) and the Trojans (at 5 AU), the latter following or preceding Jupiter by 60 degrees in its orbit. Nearer to the Sun, discontinuities in the distribution separate regions rich in asteroids. These are the Kirkwood gaps, situated at heliocentric distances (2.5 AU, 2.83 AU, 3.3 AU) corresponding to orbital periods that are simple ratios of that of Jupiter: 7/3 signifies, for example, that the asteroid orbits the Sun seven times in the time it takes Jupiter to orbit three times.

Principal characteristics of the ten largest asteroids. The nine largest asteroids are of type C (for carbonaceous) or display a complex spectrum (type U, for 'unknown'). Only the tenth, Eunomia, is a stony type (S, for 'stony'). S objects, the most similar to lunar or terrestrial rocks, are generally small. Originally asteroids were named after female figures from mythology. Because of the proliferation of discoveries, it was decided to give them names that were female but not exclusively mythological. This system too had to be abandoned; asteroids are today designated by the year of their discovery followed by two letters; when their orbital elements are determined, they receive successive numbers, but their discoverer can always give them a name. Thus 1988 BJ was later named 4635 Rimbaud.

asteroid	type	apparent magnitude	diameter (km)	semi-major axis (AU)
1 Ceres	C	7.5	1025	2.768
2 Pallas	U	8	583	2.773
4 Vesta	U	6.5	555	2.362
10 Hygeia	C	10	443	2.386
704 Interamnia	U	11	338	3.061
511 Davida	C	11	335	3.181
65 Cybele	C	12	311	3.428
52 Europa	C	11	291	3.095
451 Patientia	C	11.5	281	3.065
15 Eunomia	S	9.5	261	2.642

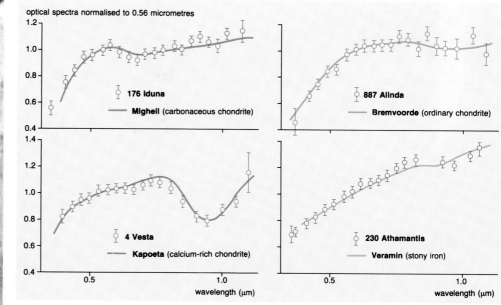

optical spectra normalised to 0.56 micrometres

176 Iduna
Mighei (carbonaceous chondrite)

887 Alinda
Bremvoorde (ordinary chondrite)

4 Vesta
Kapoeta (calcium-rich chondrite)

230 Athamantis
Veramin (stony iron)

wavelength (μm)

wavelength (μm)

Asteroids and meteorites. The spectral characteristics of the principal classes of meteorites (solid lines) are very similar to those of some asteroids (shown as open circles with error bars). This similarity suggests that the asteroids are the most plausible parent bodies of the meteorites. (After C. R. Chapman, 1976)

Some of the fragments resulting from a collision will end up in zones where orbits are unstable such as the Kirkwood gaps. These fragments are the Apollo–Amor asteroids. Their number will always remain very small, for they 'rapidly' (after a few hundreds of millions of years) collide with the terrestrial planets. In the case of the Earth, objects smaller than about a metre supply our meteorite collections. The very rare collisions with asteroids several kilometres in diameter have formed vast impact craters and have projected gigantic quantities of dust or water vapour (oceanic impact) into the atmosphere. Such catastrophes were capable of profoundly modifying the evolution of life on Earth.

Because of the smallness of the asteroids, our knowledge can hardly progress much further with astronomical techniques alone, whether from the ground or with the future Space Telescope, for which a large asteroid 300 kilometres in diameter will still remain a point source. Only a space mission would allow this family of objects to be studied in detail. One would expect them to display characteristics in common with the smaller rocky bodies already observed in detail such as Phobos, observed during the Viking mission, and Amalthea, the closest of Jupiter's satellites. Such an exploration would have economic repercussions, for it is possible to envisage in the medium term the exploitation of small asteroids, which could be placed in Earth orbit, as reservoirs of minerals (in particular the metallic ones).

Jean-Pierre BIBRING and Yves LANGEVIN

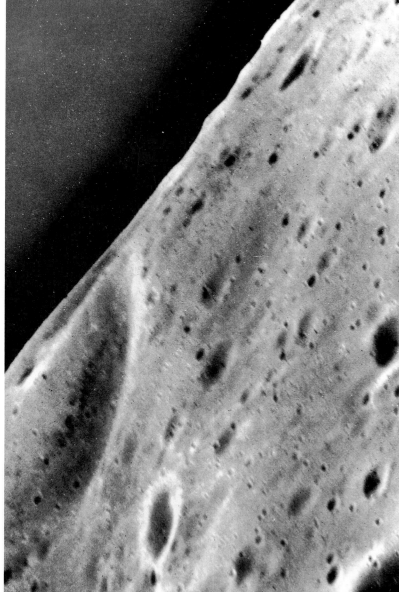

Close-up view of Phobos.
The closest satellites of Mars and Jupiter, Phobos and Amalthea, may be regarded as the closest relatives of the asteroids. They have very irregular shapes, as seems to be true of many asteroids. These small bodies have masses too low for gravitational forces to give them the quasi-spherical shape of the telluric planets or Galilean satellites for example. This very high-resolution photograph was taken by the Viking 1 Orbiter at the time of a close approach to Phobos. Top to bottom of the photograph covers less than 2 kilometres at the surface of the satellite. (Laboratory for Planetary Studies, Cornell University)

Mineralogical characteristics of the asteroids. The spectral characteristics in the near infrared of the principal constituent minerals of the asteroids are represented on the left: pyroxene (a) displays a narrow absorption band at 1 micrometre and a wide one at 2 micrometres, while olivine (b) produces an asymmetric band between 1 and 1.5 micrometres. Feldspar (c) shows absorption at 1.25 micrometres. An iron–nickel alloy (d) is characterised by a smooth increase of albedo with wavelength. Finally, the hydrated minerals (e) are characterised by the water absorption band at 3 micrometres. The near infrared spectra of typical asteroids (right) reveal profound differences in mineralogical composition. In particular, the characteristic bands of the silicates appear clearly in Vesta's spectrum, while Ceres' spectrum is dominated by opaque hydrated minerals. (After Gaffey and McCord, 1979, and Larson and Veeder, 1979)

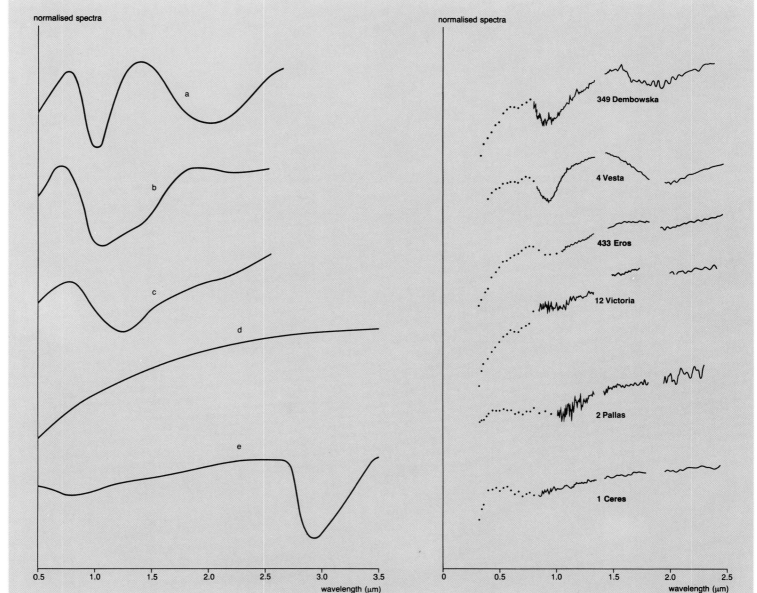

normalised spectra

a

b

c

d

e

wavelength (μm)

normalised spectra

349 Dembowska

4 Vesta

433 Eros

12 Victoria

2 Pallas

1 Ceres

wavelength (μm)

Jupiter

Jupiter, like the other giant planets of the Solar System, is a very different object from the terrestrial planets, Mercury, Venus, the Earth and Mars, which are characterised by a solid surface a few thousand kilometres in diameter, and a thin atmosphere, which is non-existent in the case of Mercury. In contrast, Jupiter is an enormous ball of gas composed essentially of hydrogen and helium, like the Sun and other stars. The splendid images which we see in telescopes and which are transmitted by space probes are images of the outer layers of cloud. These clouds hide the deep structure of the planet, but the modern techniques of measurements of electromagnetic rays reflected and emitted from the planet, the exact form of the trajectories of space probes passing close to it, and the application of the laws of physics, allow us to form a surprisingly accurate idea of the interior of the planet.

The analysis of the planetary radiation in the ultraviolet, visible and infrared regions of the spectrum both observed from terrestrial observatories and with apparatus on board the Pioneer and Voyager space probes has enabled us to determine the temperature and the chemical composition of the outer layers of Jupiter over a thickness of about 2000 kilometres (which is obviously very small compared with the 70 000 kilometre radius of Jupiter). What then would an observer see as he descended into Jupiter, equipped with the necessary – and indestructible – means of investigating?

Coming from interplanetary space, and heading towards the centre of the planet, our traveller would first of all meet an extremely tenuous upper atmosphere, essentially consisting of hydrogen at a temperature of the order of 1500 K. He then arrives at the levels where the pressure is of the order of a millionth of the pressure of the terrestrial atmosphere at ground level, a zone below which the turbulence is sufficiently strong for the various constituents of the atmosphere to be continually mixed. The temperature here is at most about 370 K; it continues to fall as the traveller descends. From then on, the atmosphere is composed of about 90 per cent molecular hydrogen (H_2) and about 10 per cent helium. A small amount of methane (CH_4), of the order of 0.1 per cent, is added to this, and even smaller quantities of acetylene (C_2H_2) and ethane (C_2H_6). The last two gases are produced in the upper atmosphere by solar ultraviolet radiation breaking the methane molecules into pieces which subsequently recombine into more complicated molecules, the hydrocarbons. Acetylene and ethane are the only molecules which have been detected so far, but it is probable that other hydrocarbons are present in very small quantities. According to recent analyses of data from the Voyager probes, ethylene (C_2H_4), benzene (C_6H_6) and methylacetylene (C_3H_4) might also be present.

As the traveller continues to descend, at levels where the pressure is some thousandths of an atmosphere, he finds ammonia (NH_3) in minute quantities but sufficient for it to be detected from astronomical observation satellites in orbit around the Earth. He also begins to discover a thin mist composed of small particles less than a micrometre in diameter, whose nature is not yet known (they could be small crystals of ammonia or particles of hydrocarbons in the solid or liquid state). On arriving at a level where the pressure is about one tenth of an atmosphere, the traveller finds himself in a region called the tropopause where the temperature is about 120 K. From here on, the temperature begins to rise, and increases continuously until the centre of the planet is reached. From the tropopause downwards, the quantity of ammonia increases very rapidly with depth. (This continues until it reaches a few parts

The king of the planets. In this photograph, taken by Voyager 1 on 1 February 1979 at a distance of 30 million kilometres, the resolution is approximately 600 kilometres. We can see clearly the structure in elongated bands parallel to the equator of Jupiter. The bands correspond to cloud layers which are probably at different heights in the atmosphere. The upper cloud layers are almost certainly made of crystals of solid ammonia, but the nature of the constituents giving rise to the colours is as yet unknown. The darker bands correspond to regions where the upper layers of cloud are at a lower height than elsewhere. The clouds move at high speed with respect to the rotation of the rest of the planet, some from east to west and some from west to east. The west–east movement exceeds 100 metres per second in the Equatorial Zone. Numerous eddies are also clearly visible, especially in the southern hemisphere, the largest and most celebrated being the Great Red Spot. (NASA)

per ten thousand, at a level where the pressure is 0.6 atmosphere). A gas called phosphine (PH_3) also appears, which, although in modest quantity (less than 1 part per million), considerably absorbs infrared radiation, as does ammonia. At pressures of about 0.3 to 0.5 atmosphere, the traveller discovers a layer of white clouds resembling the cirrus clouds in the Earth's atmosphere; these white clouds are made of ammonia crystals whose dimensions may be as much as 100 micrometres. This cloudy layer is optically thin so that it does not prevent the coloured clouds which are deeper down (in all probability at pressures of 2 to 3 atmospheres) from being seen from Earth. On the other hand, the ammonia 'cirrus' clouds absorb strongly in the infrared and block such radiation from the warmer clouds situated lower down. The ammonia layer is not, however, homogeneous, and in various regions of Jupiter, particularly in the equatorial zone, it is very thin or non-existent, enabling infrared radiation at 5 micrometres to reach us. The coloured clouds, in contrast, are opaque to the infrared, like the visible. Their nature is still unknown: the question is which of ammonium hydrogen sulphide (NH_4SH), compounds of phosphorus, or even organic compounds are involved? An answer to this question could well require awaiting the descent of a probe into the atmosphere of Jupiter.

At pressures of about 3 or 4 atmospheres, the traveller begins to detect other components of the atmosphere such as water vapour, germane (GeH_4) and carbon monoxide (CO). Other minor components, which have not yet been detected, are undoubtedly present in very small quantities. From pressures of 4 to 5 atmospheres, at about 270 K, visible and infrared radiation can no longer give us information, but the electromagnetic radiation emitted by these layers can still be detected from the ground with large radio telescopes. Beyond pressures of about 40 atmospheres, at about 320 K, we no longer have direct information at our disposal. We enter into the realm of the internal structure, which is the subject of complex theories. We will say a few words about them before going more deeply into the Jovian mystery.

Three kinds of information give us constraints on the theories of the internal structure of Jupiter. Firstly, there are the respective proportions of the two major constituents of Jupiter, hydrogen, and helium; these proportions have been measured precisely by the Voyager probes in the upper atmosphere. Secondly, measurements in the infrared have shown that Jupiter emits 1.7 times more energy than the planet receives from the Sun; in other words, there is a source of energy at the centre of Jupiter which produces about 70 per cent as much energy as the planet receives from the Sun. The presence of this internal energy source constrains the value of the central temperature. Finally, as with all massive bodies, there is a gravitational field around the planet; this field is not symmetric, and its variations perturb the trajectories of space probes. The departures from symmetry of the gravitational field thus deduced give information on the distribution of mass within the interior of the planet.

Let us therefore return to our imaginary traveller. As he sinks below the visible clouds of Jupiter, he undoubtedly finds more complex clouds. Besides this, with the temperature rising more and more, he begins to find various compounds, which become volatile, of carbon, nitrogen, silicon, magnesium, sulphur and other elements. These compounds are always in very small quantities compared with hydrogen and helium, which remain uniformly mixed. The pressure becomes greater and greater, attaining values well above those attainable in terrestrial laboratories. The components nevertheless remain fluid and not solid because of the relatively high temperatures. However, at about two million atmospheres and 10 000 K, a radical change takes place. Hydrogen becomes monatomic and metallic; its density and conductivity suddenly become much higher. The local density consequently increases very sharply. It is believed that helium remains mixed with metallic hydrogen in this region of Jupiter as a result of the high temperatures, as opposed to what takes place in Saturn. For the same reason, metallic hydrogen occurs in the liquid form and not as a solid.

Continuing his descent, the traveller reaches the fantastic level of 45 million atmospheres and 20 000 K at a distance of about 57 000

kilometres below the visible clouds of Jupiter. It is thought that the upper limit of the solid core of the planet is here, constituted at its origin by accretion of grains and dust immersed in the primitive nebula. This core would be composed of silicates, metals and ices (of water, ammonia and even methane). While the accretion was taking place, the core would have been considerably heated. The remainder of this primordial heat is the origin of the internal energy source observed in Jupiter.

Studies of the internal composition of Jupiter are important. Gaseous molecules of planetary atmospheres tend to escape as a result of their own agitation. This Brownian motion is greater as the temperature is increased. Opposing this tendency to escape is the gravitational field of the planet. In the case of Jupiter, gravity is strong (about three times that of the Earth) and the temperature of the planet's outer layers is much less than for the terrestrial planets, so much so that even the lightest molecules cannot escape from the atmosphere. The composition of the atmosphere of Jupiter must still be the same as it was at the time of the formation of the planet, about 4.5 billion years ago. In other words, by determining the present composition of Jupiter, one may find out the composition of the primitive nebula, from which the entire Solar System is thought to be derived. The composition of the interstellar medium in this part of the Galaxy 4.5 billion years ago may thus be found.

Among the elements which make up the interstellar medium, two of them, which are measurable in Jupiter, hydrogen and deuterium, are particularly interesting from the point of view of cosmology. In fact, the theory of the Big Bang predicts that these two gases are essentially made during the first three minutes of the existence of the Universe. Subsequently, additional helium is produced inside stars in the course of their evolution. Some of these stars end their evolution by exploding: these are the supernovae. They thus enrich the interstellar medium in materials which they have manufactured, notably helium. The proportion of helium in the interstellar medium therefore increases constantly with time. The measurement of the helium abundance of Jupiter therefore gives an upper limit to the primordial abundance of helium. This upper limit, determined by the Voyager mission, is of the order of 24 per cent by mass, which agrees well with the upper limits deduced from observations of very old galaxies.

The measurement of deuterium in Jupiter is even more important. This element, also essentially formed during the Big Bang, is destroyed in stars. Supernova explosions therefore enrich the interstellar medium in all of the elements except deuterium. It follows that the relative proportions of deuterium with respect to other elements such as hydrogen decrease continually with time.

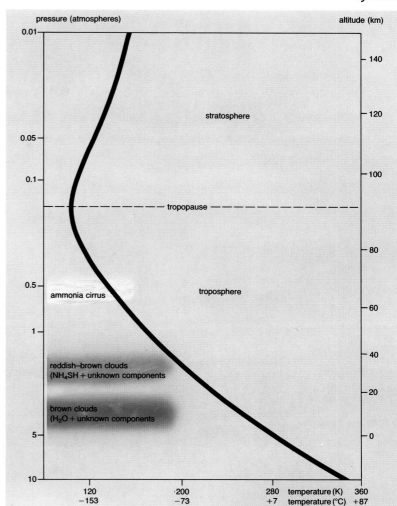

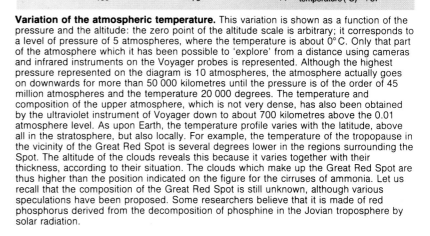

Variation of the atmospheric temperature. This variation is shown as a function of the pressure and the altitude: the zero point of the altitude scale is arbitrary; it corresponds to a level of pressure of 5 atmospheres, where the temperature is about 0°C. Only that part of the atmosphere which it has been possible to 'explore' from a distance using cameras and infrared instruments on the Voyager probes is represented. Although the highest pressure represented on the diagram is 10 atmospheres, the atmosphere actually goes on downwards for more than 50 000 kilometres until the pressure is of the order of 45 million atmospheres and the temperature 20 000 degrees. The temperature and composition of the upper atmosphere, which is not very dense, has also been obtained by the ultraviolet instrument of Voyager down to about 700 kilometres above the 0.01 atmosphere level. As upon Earth, the temperature profile varies with the latitude, above all in the stratosphere, but also locally. For example, the temperature of the tropopause in the vicinity of the Great Red Spot is several degrees lower in the regions surrounding the Spot. The altitude of the clouds reveals this because it varies together with their thickness, according to their situation. The clouds which make up the Great Red Spot are thus higher than the position indicated on the figure for the cirruses of ammonia. Let us recall that the composition of the Great Red Spot is still unknown, although various speculations have been proposed. Some researchers believe that it is made of red phosphorus derived from the decomposition of phosphine in the Jovian troposphere by solar radiation.

Jupiter in the infrared and in the visible. On the left is an infrared image at 5 micrometres made with the 200-inch telescope at Mount Palomar on 22 May 1979. The right-hand photograph was taken simultaneously by Voyager 2. The flux of infrared radiation is proportional to the temperature of the region of the atmosphere that is emitting. In the left-hand image, the coldest regions are black, the regions of intermediate temperature are red and the hottest are white. Since the temperature of Jupiter increases as one descends deeper into the atmosphere, the regions appearing white correspond to the deepest layers of the planet. That enables us to locate at which level the clouds are. The Great Red Spot appears black in the infrared image (lower right, on the limb); it is therefore relatively high in the atmosphere, and in particular higher than the orange–brown clouds of the Equatorial Zone, which appear red in the infrared image – and are therefore hotter. (JPL/NASA)

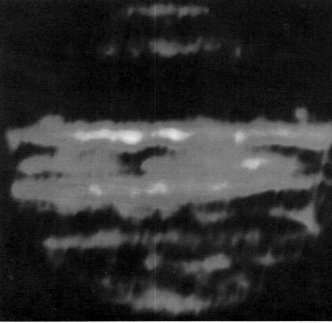

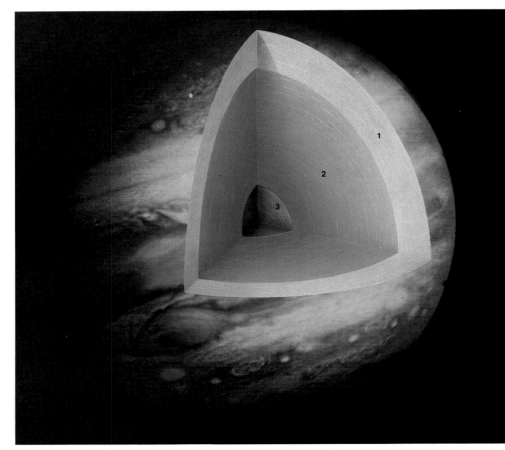

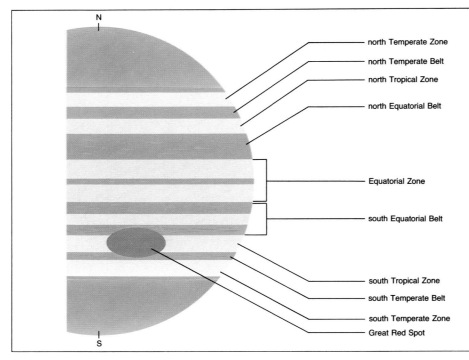

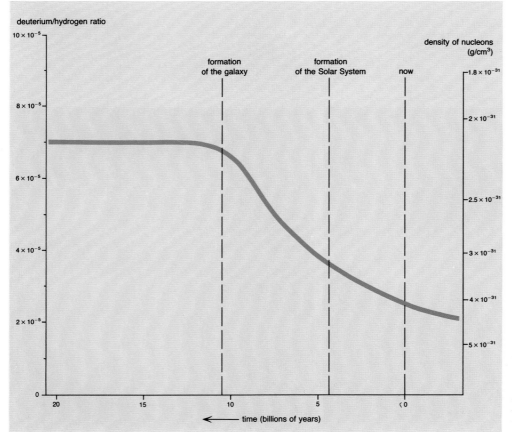

Internal structure. Jupiter is essentially made of hydrogen and helium. These gases are compressed under their own weight. At a pressure of two million atmospheres, at about 8500 kilometres below the level of the visible clouds, hydrogen probably becomes metallic and the density suddenly jumps from 1.1 times that of water to 4 times that of water. The solid core is only reached at about 57 000 kilometres below the visible clouds, at pressures of the order of 45 million atmospheres, and at temperatures of 20 000 to 30 000 degrees. The core is probably composed of iron and silicates and perhaps ices of water, ammonia and methane. The radius of Jupiter (R_J), calculated from the centre of the planet to the level where the pressure is 1 atmosphere, is 71 400 kilometres at the equator and 66 750 kilometres at the pole. The mass of the planet is 1.90×10^{27} kilograms, 318 times that of the Earth, and its volume is 1317 times that of the Earth. The density of Jupiter is therefore 1.34 times the density of water while the density of the Earth is 5.6 times that of water.

The permanent structure of the bands. The relative permanence of the banded structure of the atmosphere of Jupiter has led astronomers to adopt a nomenclature which distinguishes light zones delimited by darker belts. The South Equatorial Belt is often divided in two by a narrow zone, and the belt which is shown here is not a permanent feature of the Jovian physiognomy. Perhaps it is not truly a belt. As for the Great Red Spot, it was observed for the first time by the English astronomer Robert Hooke in 1664. Its appearance is very changeable both in size and in colour, and a remarkable fact must be noted: no observations of the Spot were reported for about fifty years at the end of the eighteenth century. However, the Great Red Spot has been the dominant feature of Jupiter since 1840.

Variation of the ratio of deuterium to hydrogen in the interstellar medium. Deuterium was essentially produced during the Big Bang; the amount in the interstellar medium has subsequently decreased. The various possible models of the variation of the deuterium/hydrogen ratio must be adjusted in such a way as to be in agreement with the observations which we have at our disposal, which, however, correspond to only two epochs. The first is now (from the measurement of the interstellar medium); the second epoch is 4.5 billion years ago at the moment of formation of the Solar System (from the present-day measurements in Jupiter). One can thus use a model to extrapolate backwards from these measurements to find the value of the deuterium/ hydrogen ratio at the moment of the Big Bang. This quantity depends on the density of nucleons, that is, protons and neutrons, which constitute the essential mass of the Universe in the first phase of the expansion. One can therefore deduce the density of nucleons from the primordial value of the ratio of deuterium to hydrogen. If this ratio were equal to 7×10^{-5}, the density of nucleons would be about 2.2×10^{-31} grams per cubic centimetre, which is 20 to 80 times less than the value necessary for the Universe to be closed. If neutrinos have a mass, however, the density of the Universe could be several tens of times greater and it would be closed.

Now the abundance of the interstellar deuterium can at present only be measured in our Galaxy, that is, we can only measure the quantity of deuterium at the present epoch. The measurement of deuterium in Jupiter is therefore extremely valuable since it gives us a second point on the curve of evolution of deuterium, located 4.5 billion years ago, and a lower limit for the primordial abundance.

The measurements of the deuterium abundance obtained by the Voyager mission seem to confirm that the deuterium/hydrogen ratio has decreased slightly since the birth of the Solar System, in a manner conforming to the prediction of the model of evolution of the abundance of deuterium as a function of time.

In making use of such a model, one can also extrapolate backwards to the abundance of the deuterium that was produced during the Big Bang. The theoretical model of that primordial explosion enables us thus to deduce the density of protons and neutrons (which are called nucleons or baryons) in the universe. This value of the density, according to cosmological models, has fundamental consequences for the structure of the Universe; the Universe will be open, that is, it will continue to expand for ever. This result will, however, be called into question if the elementary particle called the neutrino should have a mass. Some experiments already carried out with large particle accelerators suggest this. As neutrinos are much more abundant than protons and neutrons, the total density of the Universe would then be much greater, and the universe would thus be closed, that is, after having continued its present expansion for some further time, it would contract until it reached its initial size.

Two scenarios are envisaged at present for the formation of Jupiter. In the first scenario, it is supposed that in the region of Jupiter and of the other giant planets, some sufficiently large fragments of the primitive nebula (of the order of several thousand times the present radius of Jupiter) condensed and formed giant gaseous protoplanets. A core would subsequently be formed from grains of iron and of silicates already occurring in the nebula and falling towards the centre of the protoplanet. In this scenario, the composition of the atmospheres of the giant planets must be similar to that of the Sun, if we admit that the primitive nebula had the same composition at its centre and at its periphery. in particular, carbon, nitrogen and oxygen, which are the most abundant elements in the universe after hydrogen and helium, must be in the same proportions with respect to hydrogen in the atmosphere of Jupiter and in the Sun. This is not what is observed; the carbon/hydrogen ratio in all of the giant planets, and, it seems, the nitrogen/hydrogen ratio in Jupiter and Saturn are higher than in the Sun.

In another scenario, the planets are considered to have formed at two times. In the first phase, a nucleus was formed by concentration of grains floating in the primitive nebula. These grains were composed of iron and silicates, and, because of the low temperatures existing in the nebula at its periphery, also ices of water, ammonia and methane. The nucleus grew until it attained a certain critical mass, of the order of ten times the mass of the Earth. The heat released during this process would partially revapourise the ices. When the nucleus attained the critical mass, it attracted the surrounding materials of the primitive nebula, made essentially of hydrogen and of helium which had not been able to condense because that would have required extremely low temperatures. Thus, the atmospheres of Jupiter and the other giant planets would be constituted in the second phase, enriched in carbon, nitrogen and oxygen compared to the Sun as a result of ices in the atmosphere being revapourised.

Future missions of exploration of the giant planets are planned with probes being sent to the interior of their atmospheres, the most important being the Galileo mission to Jupiter. These probes, descending slowly suspended on parachutes, will make very precise measurements of the atmospheric composition up to pressures of 15 or 20 atmospheres, enabling considerable clarification of the exact scenario of formation of these planets.

Daniel GAUTIER

Clouds and colours

The forms and contours which, in the visible, are displayed on the disk of Jupiter correspond to strong contrasts in brilliance; the former are due to the particular distribution of more or less opaque layers of cloud forming at various altitudes. The extreme singularity of the spectacle of Jupiter (and Saturn) lies above all in its being extremely symmetric about one axis: a dozen bright and darker bands alternate in lines parallel to the equator. The regular geographical distribution of contours is easily discernible from Earth, and makes it clear that there are permanent atmospheric dynamics, which lead to zonal winds blowing alternately from the east and from the west. This has resulted in the adoption of a nomenclature which establishes a distinction between two kinds of bands: bright bands which are conveniently called zones and dark bands which are called belts. Observations in the visible of the disk of Jupiter also reveal another peculiarity, a very great variety of colours. The information obtained on the chemical composition of the atmosphere and on its structure – which was emphasised earlier – narrows down the substances that could be envisaged as capable of constituting the mass of the aerosols of the clouds of Jupiter. Among these substances, the best candidates are the solid particles, the crystals of ammonia and of ammonium hydrogen sulphide. These substances are colourless and their presence can only account for the white parallel bands and for such other contrasts as the white plumes observed below the equatorial regions. In consequence, there must be minor constituents which are the origin of the colouration of the clouds, in all probability thanks to local changes of the chemical equilibrium. The identification of these chromogeneous (colour-producing) constituents is not easy, in view of the variety of colours. Besides the colour white, roughly four tints are seen: red, red-brown, maroon and blue-grey. Analyses of infrared images have revealed that each colour must more or less occur at a given altitude. For example, the spots of blue colour are associated with regions with the strongest infrared emission. This shows that, for these regions, the hottest layers of the atmosphere, therefore the deepest, emit without the clouds at higher altitude attenuating the infrared radiation. If one can observe in the visible the blue spots characteristic of clouds at depth, it is

because there are 'windows' in the upper clouds. The colour of these clouds varies with the altitude; first maroon, then yellow tending to red–brown, then white at high altitude. It is because of the geographic variability of the opacity of the upper cloud layers that it is possible, in observing in the visible, to see zones, more or less prominent, of the underlying clouds. Speculations concerning the nature of the chromogeneous constituents are not lacking; it is thought that the maroon and yellow colourations may originate from various polymerised forms of sulphur. It is very probable that elementary sulphur (S_n) is present among the gaseous constituents of the lower atmosphere. Sulphur, in condensing, may be transformed into a large number of allotropic forms which are potential colouring agents. Elementary sulphur is itself created by photodissociation of hydrogen sulphide (H_2S) under the action of ultraviolet solar radiation down to relatively low altitudes. Although it has not been definitely identified, hydrogen sulphide is, with water and ammonia, a constituent of all models of the lower atmosphere of Jupiter.

The colour of the Great Red Spot is an enigma, because, taking into account the level of the atmosphere where it is located, it should be associated with white clouds. The latter resemble terrestrial cirrus clouds; they are made of ammonia crystals formed at 150 K and are of a purity which evidences the absence of colouring agents. Now, from the infrared measurements, the Great Red Spot is without doubt a phenomenon of the very cold upper atmosphere, higher than even the white clouds.

It is in any case clear that the relationship between colour and stratification has a physical basis and that the analysis of the morphology of the clouds, according to their colour, is not arbitrary. It is thus that the visible markings have been classified into various groups. A first group consists of the blue clouds, a second of the red spots and a third of the brown or black clouds of elongated form and a fourth consists of the marks of blue–grey colour. Whereas the parallel banded structure and the Great Red Spot are easily discernible from terrestrial observatories, it was the pictures transmitted in 1974 and 1975 by the Pioneer 10 and Pioneer 11 probes which have really revealed the other markings characteristic

of Jovian clouds. Subsequently the Voyager probes have furnished yet more precise data. The separation between clear zones and dark bands seems to be much less distinct than one had imagined. For example, as far as the white clouds are concerned, the prevailing impression is rather that the planet is entirely covered by an upper mantle of more or less uniform cirrus clouds. As for the white plumes, they would originate from vertical atmospheric movements transporting gas saturated in ammonia vapour up towards high altitude where the condensation of the characteristic white crystals would occur. The stretched-out form of the plumes must then result from the horizontal high altitude winds. Regarding the other category of white clouds, the ovals of large size, such as those in the southern hemisphere which accompany the Great Red Spot, there is no doubt that they have the characteristics of stormy formations of the anticyclone type, like the Spot. The Great Red Spot, which belongs to the second group of clouds, is in fact generally interpreted as the consequence of a meteorological phenomenon of the upper atmosphere. It is definitely joined to the inside of the Tropical Zone of the southern hemisphere. Of oval form, its width has hardly changed since the epoch of its discovery almost three hundred years ago. Its expanse – its width is today 26 200 kilometres – and its colour do however show small variations. Other spots of smaller dimensions are joined to the insides of the bright zones; the colours of these spots are also a mixture of red and orange. Probably due to the same cyclonic phenomenon as the Great Red Spot, they do not have its permanent character. To mention one example, a sort of replica of the Spot was identified in the Tropical zone of the northern hemisphere. After Pioneer 10 had photographed it in 1973, it was estimated that it must have appeared eighteen months earlier; one year later, when Pioneer 11 scrutinised the same region, the spot had disappeared. In the other hemisphere, in the South Tropical zone where the Great Red Spot is immersed, a long dark spot was identified in 1919, called the Great Southern Perturbation. It has not been seen since 1939.

Observations of the brown ovals, which are classed in the third group, also have a long history. A large brown elongated semi-permanent spot, 10 000 kilometres long, has often been seen between the North Equatorial Belt and the

A riot of colour. The left-hand image, rich in colour, was transmitted by Voyager 1 on 1 February 1979. The probe was 28.4 million kilometres from the planet. Besides the parallel banded structure, most of the characteristic coloured markings can be seen. The North Temperate Zone and the North Tropical Zone form a single white band within which the North Tropical Belt is barely discernible. Slightly to the north of the Equatorial Zone a very bright plume can be seen; it is a type of cloud formation already observed during the Pioneer missions. The Great Red Spot is bordered to the west by a turbulent zone which gives an eddy-like aspect to the clouds; below the Great Red Spot, we distinguish one of three white ovals which 'sailed' near it during the entire period covered by the two sequences of observations by the Voyager probes. Io is in transit in front of Jupiter. (JPL/NASA)

In the right-hand image, taken by Voyager 2 on 9 June 1979 at a distance of 24 million kilometres, we distinguish particularly well the bright plumes to the north of the Equatorial Zone, and the eddies situated to the west of the Great Red Spot. Io is visible to the right, and the shadow of Ganymede on the left. (NASA/NSSDC)

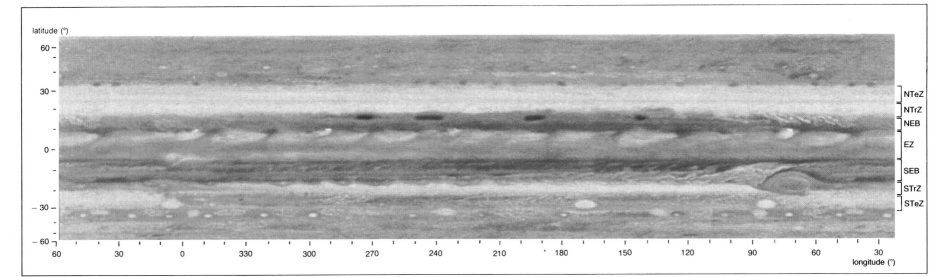

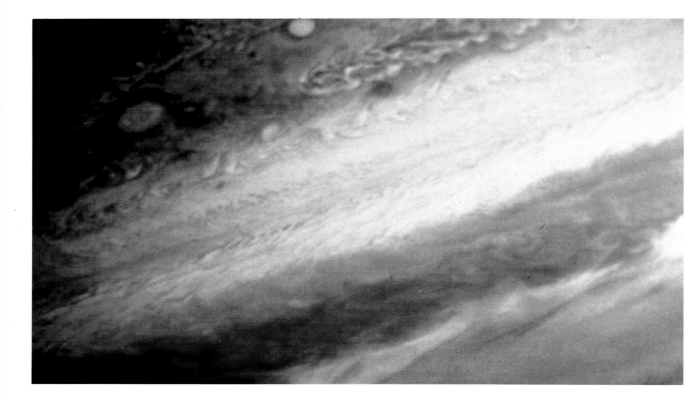

Cylindrical projection of Jupiter (above). This cylindrical projection, made of ten pictures transmitted by Voyager 1 on 1 February 1979, enables us to reproduce one entire rotation of the planet. Each of the pictures is built up with the help of three monochromatic pictures taken through orange, green and blue filters. There are brown ovals in the North Equatorial Belt (NEB) and white ovals in the South Temperate Zone. (STeZ). At the time of the flyby of Voyager 1, the Temperate Zone and the Tropical Zone of the northern hemisphere (NTrZ and NTeZ) form only a single bright band. The phenomenon is less marked in the southern hemisphere. In contrast, the South Equatorial Belt (SEB) stands out very easily above the South Tropical Zone (STrZ) within which lies the Great Red Spot. (JPL/NASA)

Fine structure of the atmospheric streams within the North Tropical Zone. The Voyager 1 probe was 14 million kilometres from Jupiter when this photograph (left) was taken on 19 February 1979. In the upper part of the picture may be seen the very wide, bright band which is the North Tropical Zone, and also the stripes which correspond to small-scale movements in this region of the atmosphere. Note also the contrast in colours between the North Equatorial Belt and Equatorial Zone. Analyses of infrared spectra show that for these regions, the visible emitting clouds are located deep in the atmosphere. (NASA)

The large brown oval. The picture on the left, transmitted by Voyager 1 in March 1979, presents a striking view of a contrasting feature, dark in colour, and of oval form, whose size is comparable with the diameter of the Earth. Other examples of this kind of contrast have been found; they occur because of 'windows' in the veil of upper clouds. The orange band in the interior of which is immersed the brown oval spot is the North Equatorial Belt. Observations of the same region in the course of the Voyager 2 mission (see right-hand picture) have enabled us to confirm the stratification of layers of various colours. The upper edges of the oval spot are less well defined and, at one point, there is an intrusion in the form of a tongue of orange clouds. (NASA/NSSDC)

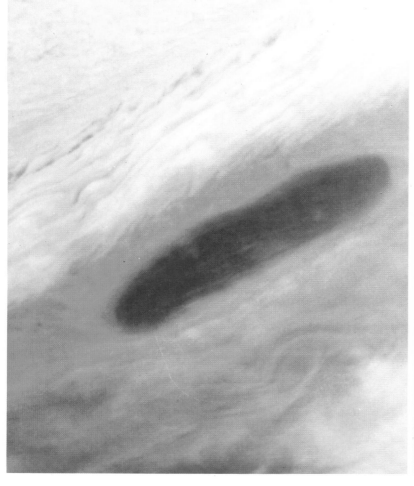

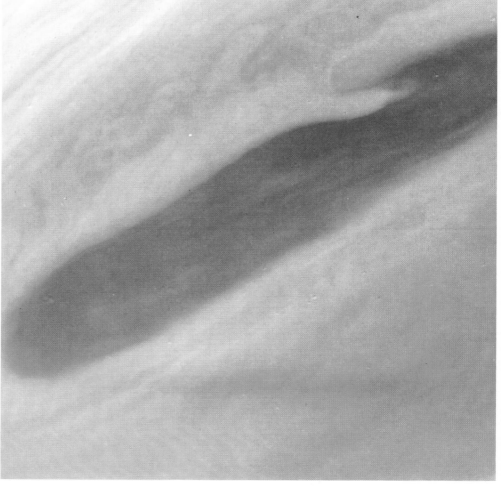

The Great Red Spot and the white oval. The left-hand picture was transmitted by Voyager 1 as it flew by the planet at a distance of less than a million kilometres. The visible area is 25 000 kilometres across. Both the Red Spot and the white oval acccompanying it have the characteristics of an anticyclonic storm. The right-hand picture was transmitted four months later; the white oval is more to the south of the Spot and it has therefore moved with respect to the latter. (NASA/ NSSDC)

The variety of colours. This picture taken by the probe Voyager 2 on 28 June 1979 has been processed by a computer to enhance the contrasts between the colours. Traces of the deepest layers of the atmosphere which have been probed now show up very much better in white instead of a very pale blue–grey. The dark belt crossing the top part of the picture is the North Equatorial Belt. On its upper edge, in the centre, there is a very dark brown oval. More to the south, owing to the great contrast, the white plume forms a gigantic slightly blue-shaded streak in the middle of the equatorial region. Even further south can be seen the chaotic region which borders the west regions of the Great Red Spot. This region can be easily picked out because of the numerous eddies and loops which form there. The probe is 10.3 million kilometres from the planet, and the resolution is 190 kilometres. (NASA)

North Tropical Zone. Cameras mounted on the Voyager probes have photographed it at four-monthly intervals. Other brown ovals are always found in the same region near the North Equatorial Belt. These observations have confirmed that the oval features are seen because there are windows in the intermediate layer of cloud whose colour tends towards red–brown, the latter being situated underneath the white clouds. It is, however, curious to note that the windows which exist within the red–brown cloud cover always lie at 13° north latitude. They are never observed in the equatorial regions. Observations over many decades have shown this. And the question still remains unanswered as to the colouring agent of the brown ovals, which one perceives through the windows and the holes, particularly as the spots of brown colour are never observed in the equatorial regions.

The fourth group of clouds, whose colours are less well defined, are definitely manifestations of phenomena appearing very deep within the atmosphere of Jupiter. Some people have suggested that the blue–grey spots are solely due to Rayleigh diffusion of gas from the very dense atmosphere. They therefore correspond to the fact that there are places where the atmosphere will be cloudless, even at very low altitudes.

Guy ISRAËL

Dynamics of the atmosphere

The most pronounced cloud formations in Jupiter's atmosphere have been used as tracers of the general circulation to demonstrate its very strong axial symmetry. In the same way, it has been possible to follow the recurrent movement with a period of 6 to 10 days of the small spots which appear in the vicinity of the Great Red Spot. This demonstrates the permanence of the winds which always blow anticlockwise around the Great Red Spot. Although only showing in an approximate way the real movement of masses of air, this method has shown itself capable of establishing the main features of the general circulation as well as other atmospheric movements.

Unlike Earth's atmosphere where two general currents are observed – the jet stream of the intermediate latitudes blowing towards the east, and a weak general current from east to west for latitudes close to the equator – Jupiter's atmosphere is traversed by several jet streams. The speed attributed to these jets is measured relative to the planet's rotation, whose period (9 hours 55 minutes) is determined by measurements of its magnetic field. The tight correlation between, on the one hand, the winds blowing in an easterly direction and those blowing towards the west and, on the other hand, the alternation of zones and belts is well established. However, the temporal permanence of the meridional circulation of the winds is surprising. Their relative stability has been demonstrated using observations obtained over a period of eighty years; the wind speeds reach 130 metres per second south of the Equatorial Zone.

In order to agree with the observations, models of Jupiter's atmospheric circulation must predict the alternation of the east and west winds and demonstrate the permanence of this phenomenon; they must in addition take account of the two characteristics of Jupiter's atmosphere, in which it differs significantly from that of the Earth. Firstly, Jupiter's atmosphere receives a flux of heat, nearly half of which originates from the planet's interior; the existence of an internal source of thermal energy has been demonstrated by measurements in the infrared. Secondly, the temperature of the upper atmosphere changes very little between the equatorial and polar regions; the difference is only 3 degrees at most.

Without attempting to be exhaustive, it is possible to present the principles on which at least two kinds of models of the general circulation are based. The first is due in particular to Gareth P. Williams of the University of Princeton; his studies concentrate on the search for a unification of the parameters governing the circulation of the atmospheres of the principal planets. Despite the peculiarities mentioned above, Williams takes as a type of unified model his three-dimensional model of the Earth's atmosphere. It is obvious that, as far as Jupiter is concerned, the dominant parameter is the speed of the planet's rotation. This can explain the extreme stability of the general features of the circulation. In the Earth's atmosphere, the baroclinic waves, which arise in the mid-latitudes because of the strong thermal gradients existing between the equator and the poles, have a destructive effect on the cellular Hadley-type circulation. In Jupiter's atmosphere, the planet's very high speed of rotation completely overcomes the effects of the baroclinic waves. The results of numerical simulations applied to Jupiter's atmosphere do show the alternating profile of the east and west winds.

One of the criticisms which can be made of the Williams model is that it assumes that the atmosphere situated below the upper layers which absorb solar radiation has negligible effect upon the general circulation. The great depth of internal fluid of a planet like Jupiter is a parameter which has stimulated other meteorologists to suggest an entirely different kind of model. In the model constructed by Friedrich H. Busse of the University of California at Los Angeles, it is re-emphasised that the zones and belts could be the surface manifestation of a bundle of convection cells which are very deeply rooted in the atmosphere. Several theoretical and experimental considerations have shown that convection at the interior of a rotating sphere is divided into long columns which dovetail into one another, and have their axes parallel to the axis of rotation. Opposite ends of the same column emerge at the surface and are visible in each hemisphere in zones of opposite latitude. In order that this mechanism can be applied to Jupiter, the fluid layers at the interior of the planet must have an adiabatic temperature gradient. There are few experimental results concerning the physics of layers situated below the summit of clouds, in particular concerning their wind speed. The model nevertheless has the advantage that it explains the permanence of the jet streams. A life span of the phenomenon of the order of eighty years suggests a strong imbalance between, on the one hand, the major fluid mass involved in the movement of the jets and, on the other hand, the much smaller mass involved in the movement of the jets and, on the other hand, the much smaller mass involved in the ovals, plumes and eddies, which are the other characteristics of the dynamics of Jupiter's atmosphere. In this way, the effects of the eddies and other baroclinic instabilities which are confined to levels where the pressure is about 5 bars only have a negligible influence on the organised movements linked to the jets, which extend down to pressures of about 1000 bars.

Whatever the model adopted, it must also allow an understanding of the nature and dynamics of the phenomena which cause not only the Great Red Spot, but also the other long-lived ovals. For example, the three white ovals which border the Great Red Spot appeared in 1938. They are also part of an anticlockwise atmospheric circulation which indicates that they, like

Cloud movement. Long-lived clouds which are observed in the form of different contrasts in the atmosphere do not always have exactly the same position on the disk. Winds blowing in different directions and with different speeds at different latitudes enable cloud formations to move relative to one another. This is quite obvious on examining the two photographs above. These were taken by Voyager 1 on 24 January 1979 (top) and by Voyager 2 on 9 May 1979 (bottom). The most important changes occur in the vicinity of the Great Red Spot. During the interval of four months, one of the white ovals which had been observed below and slightly to the left of the Spot, changed position. Another oval which, on the first photograph, was to the left of the Great Red Spot is, four months later, situated directly beneath it. Note also the white tongue-shaped plume situated above the Red Spot on the Voyager 1 photograph, which four months later interacts with a bright cloud formation which had appeared on this part of the disk. (NASA)

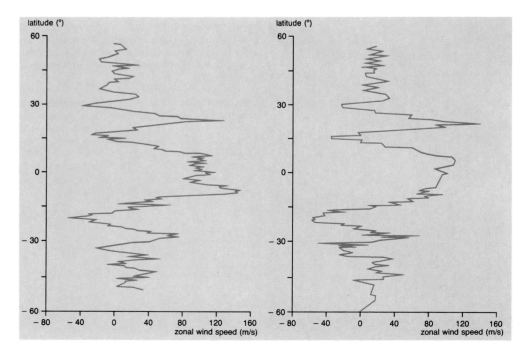

The zonal winds. The distribution of the zonal wind speeds (from east to west) along a meridian was established by measuring the displacement of the smallest discernible features. Several thousand measurements have been necessary to investigate a portion of the disk with a large coverage in longitude (150°) centred on the Great Red Spot. The profile established by Voyager 1 (left) relates to the longitude interval 330 to 120 degrees; that established by Voyager 2 (right) relates to the longitude interval 0 to 150 degrees. There are only very minor differences between these two profiles; for example, in the Equatorial Zone, the speeds measured at a latitude of 7° south changed from the 150 metres per second, measured in the course of the Voyager 1 mission, to the 95 metres per second measured four months later. In comparing the wind speed distribution in latitude to the features observed upon the disk it is clear that the strongest winds occur at the boundaries between zones and belts. (After B. A. Smith et al., 1979)

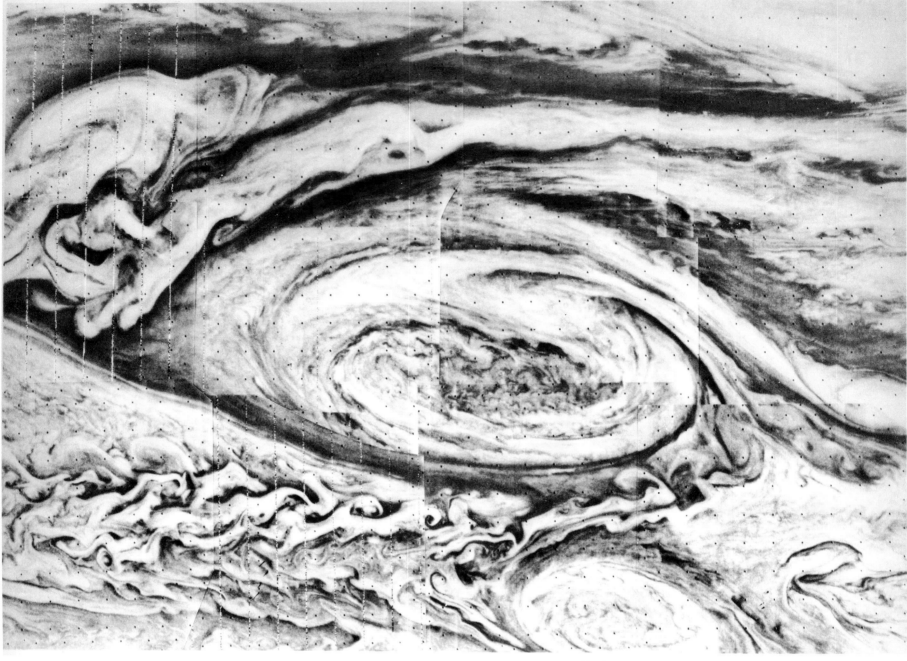

Meteorology of the Great Red Spot. The image (above), transmitted by Voyager 1, is a demonstration of the extreme complexity of the dynamic flow of the atmosphere around the Great Red Spot (the resolution of this image is less than 30 kilometres). However, to judge from the long life of the Spot (it has been observed for more than 300 years), a certain order must reign amid this chaos. A large number of theories has been put forward, but none is entirely satisfactory. Gareth P. Williams' model, based on terrestrial models, provides a good description of the circulation of Jupiter's atmosphere with a dominating tropical jet stream, extratropical jets and associated eddies. It also has the advantage of reproducing the distribution of the atmospheric currents which surround the Great Red Spot. According to Williams, in order to explain the equatorial jet stream, there must exist, under the thick layers of cloud, an intermediate layer acting as a thermal exchanger, similar to the role played by the Earth's surface in tropical meteorology. Williams then shows that a specific disturbance in the layer at the level of the Great Red Spot, introduced into his model of Jupiter's circulation, gives rise to a long-lived, anticyclonic eddy. The numerical model also reproduces numerous other disturbances which show up on the images transmitted by the Voyager probes. (NASA/NSSDC)

the Great Red Spot, are centres of high pressure. White in colour, the ovals must be composed of clouds of ammonia crystals which shows that they are a high altitude phenomenon. In fact, they are formed at a level just below the Great Red Spot, which is the highest formation above the mean upper layer of the clouds. If one considers now the horizontal structure of the markings, one observes that the Great Red Spot moves very slowly, at a few metres per second westwards, while in an environment where the zonal circulation is maintained at a speed of 100 metres per second. The drift of the white ovals is greater but their displacement clearly remains less rapid than the zonal winds. Moreover, each oval has in addition its own rotation; the interaction of the matter making up the Great Red Spot with the east and west winds causes a rotation with a period of six days.

Yet none of the models is completely satisfactory in explaining both the origin and the permanence of these eddy currents. Considered as meteorological formations, they certainly have the properties of a cyclonic system; in view of the direction of their rotation, the Red Spot and the white ovals of the southern hemisphere are anticyclonic formations. When Williams' general circulation model is used with the hypothesis that the Great Red Spot originates as a small, local disturbance, it reproduces the general properties of the phenomenon: that is, its anticyclonic

rotation and its stable track and associated vorticity. Nevertheless, the model assumes the presence of an intermediate layer between the layers of cloud and the very dense gases. A 'topographical' singularity directly below the region where the phenomenon occurs would then explain the permanence of the Great Red Spot; but how do we reconcile the anomaly of the internal structure thus suggested by the theory?

Another hypothesis has been established which considers the very peculiar position occupied by the Great Red Spot between the two opposite currents which border it. The permanence of its shape and the morphology of its interaction with the atmospheric circulation are characteristics which are reminiscent of the propagation properties of special waves called solitons. This is the name that specialists in fluid mechanics have given to waves which display a single boundary, without crests or waves, and which move without altering their shape. They may arise from arbitrary initial disturbances and can mutually interact without changing their structure. The analogy is quite extraordinary when one makes the hypothesis that the Great Red Spot is a soliton sustained by the horizontal shearing observed in the atmospheric circulation of the South Tropical Zone. Nevertheless, this does not explain how the eddies originated in the first place.

Guy ISRAËL

The model of coaxial cylinders. The persistence of the properties of the circulation of Jupiter's atmosphere, and in particular of the strong axial symmetry, must be associated with the existence of an internal source of thermal energy. The zones and belts are only the surface manifestation of a very-large-scale convective system which extends deep down into the interior of the planet's fluid envelope. It has been shown in the laboratory that the internal convection of a rotating sphere takes the form of co-axial cylinders, with each cylinder having its own rotation period. A configuration like that shown in the figure forms the basis of models to explain the circulation of the atmospheres of Jupiter and Saturn.

Note that such an internal convective system will occur if the temperature of the deep layers varies with depth so as to have an adiabatic gradient. (After F. H. Busse, 1976, and A. P. Ingersoll, 1981)

Jupiter

Magnetosphere and radio emission

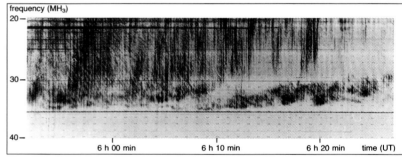

In 1955 scientists using radio telescopes made the discovery that Jupiter is a very powerful source of radio waves. For the first time there was proof that the Earth was not the only planet with a magnetic field. The characteristics of Jupiter's magnetic field and of the resulting magnetosphere are well known today from radio astronomical observations made on Earth and from measurements made near the giant planet by the Pioneer and Voyager probes.

One very important difference between the Jovian and terrestrial magnetosperes is the presence of a source of charged particles in the inner part of the former. Whereas the terrestrial magnetosphere is essentially populated by particles of solar origin, or by particles accelerated under the influence of the Sun's activity, the Jovian system is itself a very important source of ions and electrons, arising from volcanic activity on Io. Io is the only satellite on which such activity has been observed. The volcanoes eject large quantities of gas and dust (in particular sulphur compounds) into the atmosphere of Io. These are ionised by solar ultraviolet radiation and are then distributed into an enormous plasma torus surrounding Jupiter at the distance of Io's orbit. (A torus is the shape of a ring doughnut). Complex systems of electric currents certainly exist between Io's torus and the iono-sphere of Jupiter, making the magnetosphere into an immense particle accelerator. These high-energy particles – electrons and ions – are observed throughout the interior of the magnetosphere and often outside it as well. Jupiter appears to be one of the main sources of cosmic rays of medium energy observed in the interplanetary medium in the vicinity of the Earth.

Jupiter is known to be a powerful emitter in several wavelength domains. The sources of these emissions are all in the interior of the magnetosphere and studying them gives precise information on the dynamics of energetic electrons in those regions.

Studies of the different radio emissions have enabled the structure of the magnetic field to be determined, and a precise value of the axial rotation period of the planet to be determined. Observing the movement of the details visible upon the disk of Jupiter in fact only gives a rotation period for the clouds (and this varies with latitude). However, radio emissions in the decimetre to decametre range vary with a period equal to the rotation period of the magnetic field. This period, equal to 9 hours 55 minutes 23.70 seconds, represents the true period of rotation of the planet, that of its interior where the magnetic field is generated.

André BOISCHOT

Synchrotron radiation from the Van Allen belts. The large modern radio telescopes enable astronomers to take pictures of radio sources with angular resolutions of a few seconds. The Very Large Array in New Mexico was used to map the synchrotron emission from Jupiter at a wavelength of 21 centimetres by Imke de Pater of the Lunar and Planetary Laboratory, University of Arizona. At the observing frequency of 1415 megahertz, the resolution is 1.5 arc seconds. White indicates the strongest emission.

As can be seen, the emission does not come from the surface of Jupiter but is concentrated in two zones either side of the planet: these are the Van Allen belts. Symmetrical in relation to the magnetic equator, they extend out to 4 or 5 Jupiter radii. (NRAO)

Dynamic spectrum of a decametre burst. The decametre bursts of Jupiter are very complex. Their dynamic spectrum (i.e. the variations in intensity according to the frequency and to time) presents a great number of structures which are due either to the emission process itself, or to a modulation of the signal during its propagation between the source and the Earth. The figure shows the dynamic spectrum of a Jovian 'storm' taken on 4 December 1978. This spectrum was taken at the radio astronomical station at Nançay in France. The large dark structures are emissions from Jupiter. Their great complexity is noticeable and in particular the sudden disappearance of radio emission towards high frequencies. The straight lines parallel to the time axis (therefore indicating fixed frequencies) correspond to transmissions from radio stations on Earth which the antenna has picked up. The different sources identifiable on LΦ diagrams (see below) have dynamic spectra that differ greatly from one another. (Figure from Observatoire de Paris)

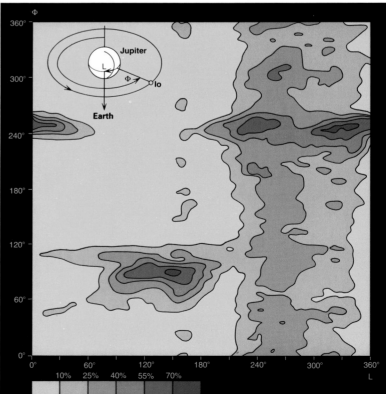

Various radio emissions. Thermal radiation from the cloud layer predominates at millimetre and centimetre wavelengths. The decimetre emission is synchrotron emission from relativistic electrons. Very high resolution radio 'images' show how the electrons are trapped close to Jupiter by the planet's magnetic field, like the particles in the Van Allen belts around the Earth. Another emission, at decametre and hectometre wavelengths, reaches its maximum at about 10 megahertz. This emission was closely studied both from the Earth and from the Voyager probes. It is very irregular, consisting of sudden bursts lasting from a few seconds to several minutes and storms lasting for a few hours. The occurrence of these depends both on the rotation of Jupiter and on the position of Io in its orbit. No emission has ever been observed at frequencies above 40 megahertz. On the other hand, the spectrum extends far down into the range of kilometre wavelengths. The decametre emission is caused by jets of electrons which travel along the length of the lines of force of Jupiter's magnetic field, exciting emission at the gyrofrequency. The process by which these electrons radiate is not very well understood but it is certainly very efficient since Jupiter is a stronger source than the Sun at these wavelengths. The decametre emission was the first to be discovered, in 1955. Finally two other types of emission were discovered by the Voyager probes at kilometre wavelengths. One comes from a source situated in the high-latitude regions close to the planet, the other from the outer face of the plasma torus which surrounds the orbit of Io.

The LΦ diagram. The radio emission of Jupiter at decametre wavelengths is only observable from certain regions of the planet and only when the satellite Io is at certain well-defined points in its orbit in relation to the Earth. The regions of the planet are specified by the longitude L of the central meridian facing the observer at the moment of observation. The position of Io is specified by the phase angle Φ. (See inset below). This LΦ diagram gives the probability of observing a radio burst as a function of L and of Φ expressed as a percentage. This diagram can show the influence of the satellite on radio bursts. Clearly, a burst is much more probable when the phase of Io is close to 90 degrees with the longitude of the central meridian around 150 degrees, and when the phase of Io is close to 250 degrees with the longitude of the central meridian around 240 or 330 degrees. For the Earthbound observer, those phase angles of 90 and 250 degrees correspond to Io being east and west of the planet, respectively.

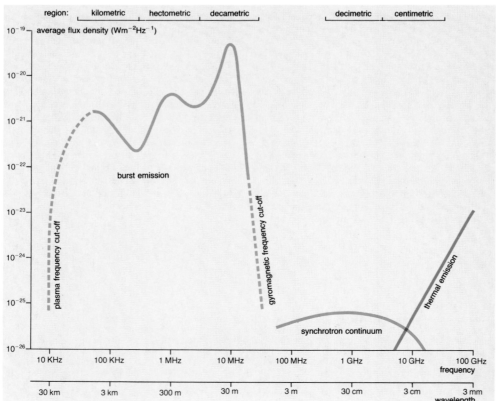

The magnetosphere (right). Although the magnetic fields of the Earth and of Jupiter are both dipolar, the sizes and shapes of the two magnetospheres are quite different.

The Jovian magnetosphere is about a hundred times larger in extent. Two factors contribute to this. Firstly, the magnetic field of Jupiter is much greater than that of the Earth; secondly, the density of the solar wind, and hence its pressure, is about twenty-five times less at Jupiter's orbit than at the orbit of the Earth. If Jupiter's magnetosphere were luminous, it would look from Earth like another sun, larger than the Moon or Sun, with a tail longer than the most spectacular comet. The extent of the magnetosphere changes considerably with the variations in pressure of the solar wind but the length of the tail is not known. Some observations indicate that it stretches at least as far as the orbit of Saturn which would give it a length of more than 600 million kilometres.

The inner regions of the magnetospheres of Jupiter and the Earth are also quite different. The giant planet rotates upon its axis two and a half times faster than our planet, and since its radius is twelve times greater the particles inside the magnetosphere are subject to a very much larger centrifugal force. Being electrically charged, they will be led on the one hand to follow the lines of force of the magnetic field, and on the other hand to leave the planet under the effect of the centrifugal force. The particles will therefore tend to concentrate around the magnetic equatorial plane, forming a large disk of plasma crossed by strong electric currents which will themselves interact with the magnetic field. This disk of current is not completely flat. Close to the planet, where the magnetic forces dominate the centrifugal forces, the disk is located in the plane of the magnetic equator. Further out centrifugal forces predominate and the disk curves to become parallel to the equatorial plane of Jupiter.

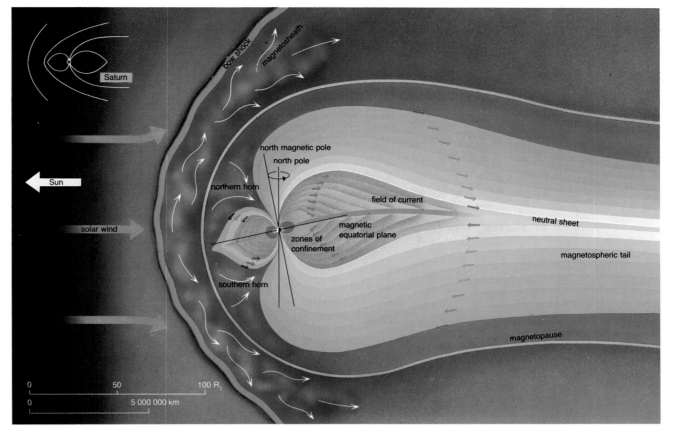

The magnetic field at the surface (left). The magnetic field of Jupiter is, to a first approximation, that of a dipole whose axis is inclined 11 degrees to the axis of rotation, and slightly offset from the centre of the planet (the offset is 0.1 Jupiter radius). The result is that the intensity of the magnetic field at the surface is far from uniform. In particular, it is not the same in the northern hemisphere as in the southern hemisphere, and the magnetic equator does not coincide with the equatorial plane of Jupiter. The magnetic moment of this dipole is 4000 times greater than that of the Earth. The radius of Jupiter being twelve times that of the Earth, the mean intensity of the field in the equatorial regions close to the planet is of the order of 4 gauss, much greater than the field at the surface of our planet. The diagram shows the distribution of the intensity of the field (in gauss) at the surface of the visible cloud layer. As Io revolves around Jupiter in its orbit, it crosses lines of magnetic force. When it crosses such a line, this line itself intersects the visible surface of Jupiter at two points. The red lines represent the coordinates of these two points (the 'feet' of the magnetic lines of force) as the position of Io varies in its orbit. The black dots mark points at equal intervals of time; those ringed correspond to longitudes of Io of 0 and 180 degrees.

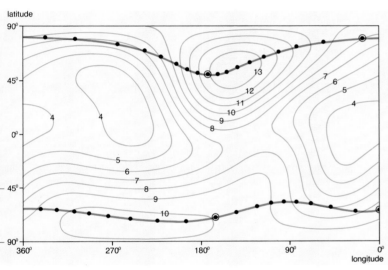

The ionosphere of Jupiter (right). Like the Earth, Jupiter is surrounded by an ionosphere. Studies of the refraction of telemetry signals from the Pioneer and Voyager probes, as the probes were entering and leaving occultation by Jupiter, have enabled us to determine the principal characteristics of Jupiter's ionosphere. The variation of electron density as a function of altitude is like that of the terrestrial ionosphere. Jupiter's ionosphere is linked to Io's plasma torus by a complex system of currents which must cause acceleration of electrons. These electrons are involved in the origin of aurorae and decametre radio emissions. (After V. R. Eshleman *et al.*, 1979)

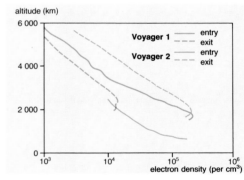

The origin of Jupiter's radio emissions. The radio emissions are due to electrons which spiral around the lines of force of the Jovian magnetic field. Near the planet, these electrons are trapped and radiate by a synchrotron effect, with a maximum at decimetre wavelengths. Further out, the presence of Io and of its plasma torus prevents the electrons from being trapped. The influence of Io on the decametre emission of Jupiter was well known before the observations made by the Voyager probes in 1979 led to an understanding of the phenomenon. The dimensions of the torus were determined; the density at its centre can be as high as many thousands of electrons and ions per cubic centimetre. This makes a more important reservoir of charged particles than the ionosphere. Made up of electrically charged particles, the torus lies in the plane of the magnetic equator of Jupiter, inclined at 11 degrees to the plane of the orbit of Io. Interactions between this torus and the planet's ionosphere through complex systems of currents along the lines of magnetic force give rise to radio emissions in the decametre and hectometre ranges, and to polar aurorae which have been detected by the instruments of the Voyager probes. Close to Io, electrons are subject to an additional acceleration hurling them along the lines of force passing by the satellite, as far as the high-latitude regions of the planet. During their journey they emit very strongly at a frequency close to the gyrofrequency. The highest frequencies (40 megahertz or 7.5 metre wavelength) come from regions close to the planet while the lowest frequencies, in particular kilometre wavelengths, are emitted much further away.

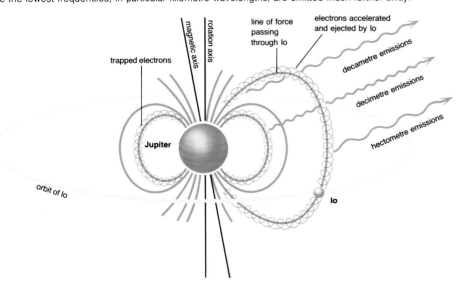

A polar aurora. When the probe Voyager 1 passed into the Jovian night it photographed an auroral arc in the high-latitude regions. This arc, which appears doubled, is highlighted against the dark background of the sky; the north pole of Jupiter is roughly in the middle. The brilliant points are lightning storms illuminating the clouds. (JPL/NASA)

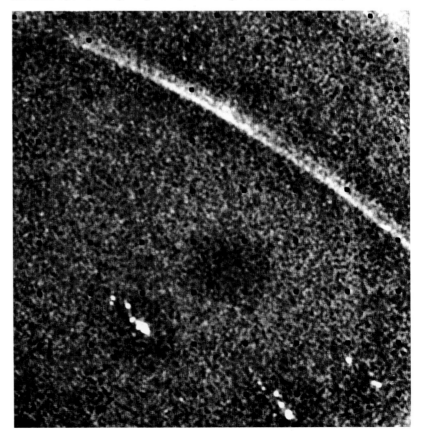

Jupiter

The small satellites and the rings

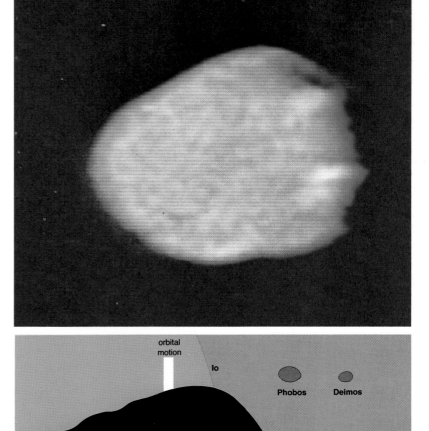

A rich family of satellites surrounds Jupiter. Sixteen were known at the end of 1985 but it is likely that several small satellites have evaded detection. In addition to the four large satellites discovered by Galileo (Io, Europa, Ganymede and Callisto), there are four small satellites close to Jupiter with circular orbits in its equatorial plane. There are also eight small, distant satellites that move in inclined, eccentric orbits. Other small bodies, the Trojan asteroids, are located at the Lagrangian points in Jupiter's orbit (see diagram below). These asteroids, all named after heroes of the Trojan war, are probably very similar to primitive planetoids. It is striking that the eight outer satellites of Jupiter have many features in common with the Trojan asteroids. Their surfaces are very dark, reflecting less than 5 per cent of the incident light, and their spectra of reflected sunlight are similar. These satellites are probably asteroids that were captured by Jupiter, but the mechanism by which they were captured is not yet well understood. Their dimensions vary from 16 to 186 kilometres and their orbits are strongly perturbed by the gravitational attraction of the Sun.

Discovered in 1892 by Edward Emerson Barnard, Amalthea was for eighty-seven years considered to be the satellite closest to Jupiter, until in 1979 the Voyager probes discovered three new satellites of which two are the closer to Jupiter. Amalthea is a very dark body, reflecting less than 5 per cent of the light. It is reddish in colour and of an irregular shape. Its surface is heavily marked by craters but it seems very different from the other known asteroids, parti-

Amalthea and the inner satellites. Amalthea (upper right) is a dark object, of a reddish colour, heavily cratered and of an irregular shape, of dimensions 270 × 170 × 150. Amalthea is very difficult to see with a terrestrial telescope, even as a point source of light. This picture, on which details of the order of 8 kilometres can be seen, was taken by Voyager 1 on 4 March 1979 at a distance of 425 000 kilometres. The major axis of this satellite points permanently towards Jupiter in the course of its movement round the giant planet. At the moment when this picture was taken, Jupiter was on the left, the right side of the satellite being plunged in darkness. The north pole is at the top. The surface of Amalthea is probably contaminated by the bombardment of sulphur compounds ejected from Io and by its interaction with the magnetosphere of Jupiter.

The figure at lower right shows the relative dimensions of Jupiter's inner satellites, from Io at 3630 kilometres in diameter down to the small Thebe (1979 J2), Adrastea (1979 J1), and Metis (1979 J3) which have maximum dimensions between 110 and 24 kilometres. Amalthea is represented here by the view from below (in silhouette) and the view from the side (below); this last drawing shows the forward side of the satellite in its movement round Jupiter, as well as its chief craters. The two Martian satellites Phobos and Deimos are also depicted for comparison. (NASA)

cularly in its spectrum of reflected sunlight. Perhaps this satellite is covered with sulphurous compounds ejected from the volcanoes of Io and circulating in orbit around Jupiter.

The environment of the satellites closest to Jupiter seems very hostile. Their surfaces are exposed to hordes of energetic particles moving within the magnetosphere of Jupiter and to

The orbits of Jupiter's satellites. Besides the Trojan asteroids which orbit the Sun in the same orbit as Jupiter, located 60 degrees behind and in front of Jupiter in the stable zones of the three-body problem, the sixteen satellites of Jupiter can be classified into four groups:
– four inner satellites (Adrastea, Metis, Amalthea and Thebe), close to Jupiter and in circular orbits in the equatorial plane of the planet; three of them were discovered in 1979 thanks to the pictures from Voyager probes;
– four large satellites (Io, Europa, Ganymede and Callisto) discovered in 1610 by Galileo and in circular orbits in the equatorial plane of the planet at distances not more than two million kilometres from Jupiter;
– four small satellites (Leda, Himalia, Lysithea and Elara) revolving in the same direction in orbits inclined in relation to Jupiter's equator at distances between eleven and twelve million kilometres from Jupiter;
– four small satellites (Ananke, Carme, Pasiphae and Sinope) in retrograde motion in very eccentric orbits that are considerably inclined in relation to Jupiter's equator at distances between 21 and 24 million kilometres from the planet. The inclinations of the orbits of the outer satellites with respect to the orbital plane of the planet, are given in the figure.

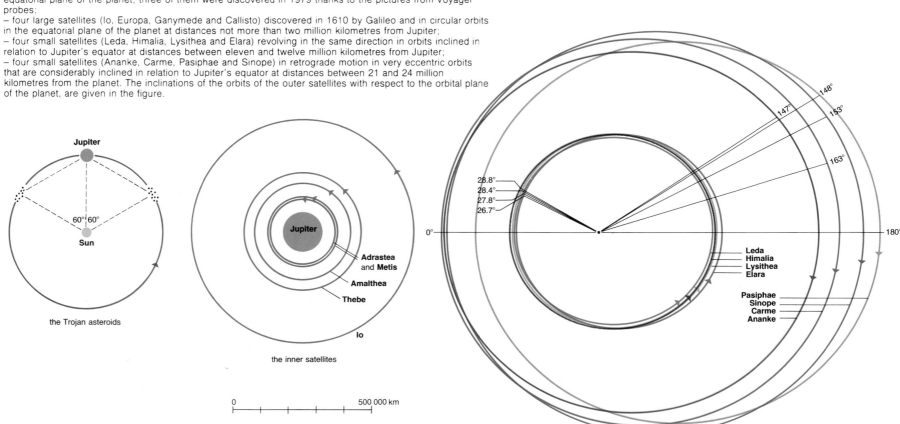

the Trojan asteroids

the inner satellites

0   500 000 km

the outer satellites

0   10 000 000 km

The discovery of Adrastea (near right). The Voyager 2 probe took this picture as it passed through the plane of Jupiter's rings. The exposure time was 96 seconds. Because of the movement of the probe, the ring, seen in cross-section, appears as a wide, bright band in the centre of the picture, and the satellite appears as a short line on the left, due to its displacement during the exposure. The stripe to the right in the plane of the ring is the trail left by a star in the field. It is noticeable that, because of the movement of the satellite, the trails of the star and the satellite have neither the same length nor the same inclination. During the exposure, Adrastea travelled about 3000 kilometres. (JPL/NASA)

The discovery of the rings (far right). The rings of Jupiter were discovered on 4 March 1979 by Voyager 1. This photograph was taken as the probe was crossing the equatorial plane of Jupiter. The exposure time was 11 minutes. Stars in the field appear as zig-zag lines because of the motion and slight oscillation of the probe. Six exposures of the thin rings appear in this photograph, which covers an extent of about 8 000 kilometres. The outer edge of the rings can be seen to the right. The black points are camera calibration points. (JPL/NASA)

The small satellites. Values in brackets have errors greater than 10 per cent. The inclination of the inner satellites is given with respect to Jupiter's equator; and for the outer ones, with respect to the planet's orbit. The sizes are the semi-axis of the elipsoid which best resembles the shape of the satellite.

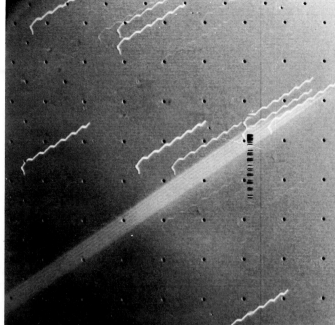

Name	orbital semi-major axis (km)	(R_J)	sidereal period of days (j)	eccentricity	inclination (°)	radius or dimensions (km)
Metis	127 960	1.789 8	0.294 780	<0.004	(0)	(20)
Adrastea	128 980	1.804 1	0.298 26	(0)	(0)	12.5 × 10 × 7.5
Amalthea	181 300	2.536	0.498 179 05	0.003	0.40	131 × 73 × 67
Thebe	221 900	3.104	0.674 5	0.015	0.8	55 × ? × 45
Leda	11 094 000	155.18	238.72	0.147 62	26.07	8
Himalia	11 480 000	160.58	250.566 2	0.157 98	27.63	93
Lysithea	11 720 000	163.93	259.22	0.107	29.02	18
Elara	11 737 000	164.17	259.652 8	0.207 19	24.77	38
Ananke	21 200 000	296.5	631	0.168 70	147	15
Carme	22 600 000	316.1	692	0.206 78	164	20
Pasiphae	23 500 000	228.7	735	0.378	145	25
Sinope	23 700 000	331.5	758	0.275	153	18

Jupiter's rings. These spectacular photographs taken by the probe Voyager 2 on 10 July 1979 help us to understand the rings of Jupiter. Their outer edge is 57 000 kilometres above the upper layer of clouds. The top mosaic is composed of six pictures taken at the moment when the probe was 2 degrees above the plane of the rings, at a distance of 1 550 000 kilometres. The finest details have been blurred in these long exposure pictures because of the motion of the probe, particularly in the photograph at extreme right. The mosaic at the bottom consists of four pictures taken at the moment when the probe entered the shadow of Jupiter. The rings are particularly bright in this mosaic because of the scattering of light by small particles. The outline of the planet is also well defined because of the scattering of light through small particles in its upper atmosphere. On each side of the planet, the portion of the rings on the same side as the probe seems broken; it is in fact in the shadow of Jupiter. The night side of Jupiter here appears completely dark; a more sharply contrasted version would show the polar aurorae and the emission which is always present in Jupiter's atmosphere. (JPL/NASA)

bombardment by gas and dust particles originating in the volcanoes of Io, as well as by micrometeorites.

Adrastea, Thebe and Metis were discovered in 1979 and 1980 on pictures taken several months before by the Voyager probes. Adrastea and Thebe are situated on the outer edge of the rings of Jupiter and have maximum dimensions of about 24 and 110 kilometres, respectively. These two satellites are intimately linked with the rings; their gravitational action probably limits the outer edge of the rings and the hypothesis has also been advanced that the continual erosion of their surfaces feeds the rings with an uninterupted supply of small particles. Metis has a diameter of about 40 kilometres; its surface is dark red similar to that of Amalthea. All three satellites are irregularly shaped.

The rings of Jupiter were discovered on 4 March 1979 by the cameras of the Voyager 1 probe. The density of these rings appears to be about a billion times less than that of Saturn's rings. This explains why, situated so close to the bright disk of the planet, they had never been observed from the Earth: detecting them is as difficult as distinguishing the feeble light of a candle from the powerful beam of a lighthouse next to it. If observations are made in the infrared at a wavelength of 2.2 micrometres (the methane abounding in Jupiter's atmosphere is then almost opaque), the ratio of the luminosity of the rings over the luminosity of the planet is greatly increased and the rings can be detected from the Earth, which was, in fact, done five days after their discovery by Voyager 1. This discovery explained why, while flying over the planet five years before, Pioneer 11 had observed sudden variations in the number of charged particles orbiting round Jupiter at certain distances from the planet. Some scientists suggested that Jupiter had satellites not then discovered, or rings at the points where the number of high-energy particles decreased; five years later this hypothesis was proved to be correct!

The discovery of the rings of Jupiter, coming two years after that of the rings of Uranus, showed that the existence of rings round the giant planets was quite natural. Like those of Saturn and Uranus, the rings of Jupiter have distinct edges and nearby satellites: they are, however, much thinner and quite different. We can distinguish four components: a bright ring, about 6000 kilometres wide is continued outwards in a very bright rim of about 800 kilometres width. On the inside more-dispersed material stretches right to the top of Jupiter's cloud layer; the whole is enclosed by a very thin halo. As for the particles making up the rings, for the time being, we know neither their nature nor their dimensions; being situated inside the magnetosphere of Jupiter, they are probably charged.

André BRAHIC

Jupiter

The Galilean satellites

Among the sixteen satellites of Jupiter known at the present day, the four large satellites discovered in January 1610 by Galileo, and afterwards named Io, Europa, Ganymede and Callisto by Simon Marius, form a distinct group because of their sizes, their physical characteristics, their compositions and their orbits. Jupiter's Galilean satellites have dimensions similar to those of the Moon (Io and Europa) Mercury (Ganymede and Callisto). Their masses (from 4.87×10^{22} kilograms for Europa to 1.49×10^{23} kilograms for Ganymede) and densities (from 1.83 for Callisto to 3.55 for Io) also approach those of the moon and Mercury. All four Galilean satellites revolve in almost circular orbits practically in the equatorial plane of Jupiter, at distances from 420 000 to 1 900 000 kilometres. They are therefore subject to important effects from the Jovian magnetosphere.

Until March 1979, the date of the Voyager 1 encounter with the Jovian system, the Galilean satellites were known only as small points of light which were difficult to observe because of the brilliance of the giant planet. Only the principal physical characteristics (diameters, densities, distances and orbital periods, albedos and surface compositions) were known, thanks to observations with terrestrial telescopes and the space missions of Pioneer 10 and 11 (1973 and 1974). The appearance of the surfaces of the satellites were completely unknown. The observations made by Voyager 1 and 2 have revealed the astonishing diversity of their surfaces showing that the four bodies have experienced very different evolutions and geological histories. Leading researchers think of them as small planets rather than as simple satellites. Launched in October 1989 by the space shuttle *Atlantis*, the Galileo probe should from July 1995 be able to study the Galilean satellites much more deeply in several flybys.

Philippe MASSON

The Galilean satellites. On this photomontage, obtained from pictures taken by the probes Voyager 1 and 2, the four Galilean satellites are represented on the same scale as Jupiter, indicated on the left. Their distances from, and their positions relative to, the giant planet are not taken into account. Europa is slightly smaller than the moon; Io is slightly larger. Ganymede, the largest of the four, is a little larger than Mercury, Callisto a little smaller.

The colours and superficial appearances of these four satellites are very different. Io shows a great diversity of tints (red, yellow, orange, white and black) caused by the presence of sulphur and sulphur compounds on the surface. Europa has a very bright surface composed of almost pure ice shot through with large dark strips. Ganymede presents a fairly wide diversity of bright and dark tints which correspond to regions covered with ice of different ages. The bright circular marks correspond to impact craters made by meteorites. Callisto has a surface of very dark ice (probably a mixture of ice and silicate debris) completely pock-marked with impact craters. The surfaces of these four satellites are not only very different in their colours and compositions but also in their ages. The surface of Callisto, riddled with impact craters, seems to be the most ancient. Very few, if any, impact craters can be seen on Io and Europa. The latter two satellites, although very different, have been the scene of processes that have rejuvenated their surfaces. Ganymede represents an intermediate stage.

	Io	Europe	Ganymede	Callisto
orbit:				
semi-major axis				
(km)	421 600	670 900	1 070 000	1 883 000
(R_J)	5.90	9.40	24.97	26.33
sidereal period (days)	1.769	3.551	7.155	26.689
eccentricity	0.004	0.009	0.002	0.007
inclination (°)	0.04	0.47	0.21	0.51
radius (km)	1815	1569	2631	2400
mass (10^{20} kg)	892	487	1490	1075
density (g/cm³)	3.55	3.04	1.93	1.83
albedo	0.6	0.6	0.4	0.2
surface composition	S,SO_2	ice	ice	dirty ice
minimum distance				
of observation (km)				
Voyager 1	21 000	734 000	115 000	126 000
Voyager 2	1 130 000	206 000	62 000	215 000
maximum resolution				
of pictures (km)				
Voyager 1	1	33	2	2
Voyager 2	20	4	1	4

Principal orbital and physical characteristics of the Galilean satellites and the circumstances of their observation by the Voyager probes. The inclination is given with respect to the the equatorial plane of the planet.

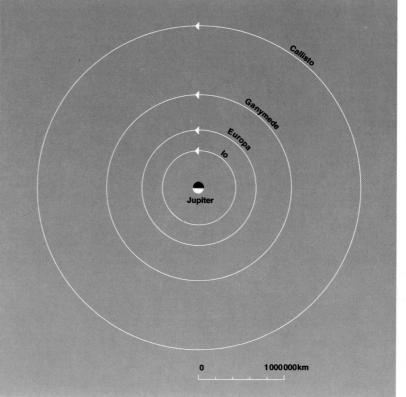

The orbits of the Galilean satellites. The Galilean satellites follow almost circular orbits in the normal (direct) direction and always keep the same face turned towards Jupiter because of their synchronous rotation. The orbits are depicted here as from the north pole of Jupiter (out of the page). Because of their respective distances from the planet. Callisto takes a little more than two weeks to make a complete revolution round Jupiter whereas Ganymede takes only seven days. Europa takes half the time required by Ganymede, and Io only forty two hours.

Jupiter, Io and Europa. This spectacular picture of Jupiter, with two Galilean satellites – Io and Europa – crossing in front of it, was taken by Voyager 1 on 13 January 1979 from a distance of 20 million kilometres. The picture gives an idea of the ratio of the sizes of the two satellites (with diameters about that of the moon) to the size of the giant planet (whose diameter is 11.2 times greater than that of the Earth). The angle of the photograph makes Io and Europa seem very close to Jupiter. Io is in fact 350 000 kilometres above the Great Red Spot and Europa is 600 000 kilometres above the tops of Jupiter's clouds. To give an idea of the scale, the distance from the Earth to the Moon is 384 000 kilometres, and the Red Spot is larger than the Earth. In this picture, where the resolution is about 400 kilometres, few details of the surfaces of the satellites are visible. Only the different colours and some dark zones can be made out. (NASA)

Callisto

Voyager 1 obtained the best pictures of the surface of Callisto. The resolution achieved was around 2 kilometres. The surface is relatively dark – Callisto is the darkest of the four Galilean satellites – but it is still lighter than that of the Moon.

The dominant morphological feature revealed by the pictures from the Voyager probes is the presence of innumerable meteoritic impact craters. The surface is almost entirely covered by craters with an average diameter of the order of 100 kilometres; there are very few craters more than 150 kilometres across. No other body in the Solar System has such a distribution of the diameters of its craters. Like all the other solid bodies in the Solar System Callisto must have received meteorites of all diameters at the beginning of its existence but the ice covering has not preserved their imprints which must have been covered over by more recent layers of ice. These layers have, in turn, been bombarded by smaller meteorites, whose impacts we can see today.

In addition to these impact craters the surface of Callisto shows a great basin – Valhalla – 600 kilometres in diameter whose origin is probably meteoritic, identical to that of the lunar maria. Similar to those, its morphology is relatively flat and it is marked by a series of concentric rings extending to a radius of 1500 kilometres and spaced 50 to 200 kilometres apart. Unlike basins on the Moon and Mercury, Valhalla has no mountainous relief at its periphery and there is no

Callisto. Voyager 1 passed Callisto in February 1979 at a distance of 126 000 kilometres. Voyager 2 approached no nearer than 216 000 kilometres in July 1979. The best pictures of the satellite obtained by the two probes have resolutions of 2 and 4 kilometres respectively. However, because of the different relative positions of the satellites and the probes and the alteration of the angle of viewing between the two encounters, the pictures obtained by the two probes do not cover the same regions of Callisto's surface and are therefore complementary even though they do not have the same definition.

Seen at a great distance (2 318 000 kilometres) on 7 July 1979 by Voyager 2, the dark surface of Callisto appears to be peppered with numerous bright spots which are actually meteoritic impact craters. The surface is probably composed of 'dirty' ice, that is to say a mixture of ice and silicates. (NASA/JPL)

radial ejecta. These differences probably arise because at the moment of the gigantic impact that created Valhalla the ice crust was not sufficiently rigid to preserve the morphology of a great basin; on the contrary the crust would have reacted to the shock as a soft material and the movement would have created deformations like a series of waves from the interior of the basin. These deformations would have become solid very quickly giving rise to the concentric circles. After the accretion phase and at the beginning of the differentiation phase of the satellite, its silicate core must have been the site of a radioactive source from which the heat flux was not sufficiently strong to be able to melt the surface crust. On the contrary, any convectional movements must have begun in the mantle, causing the crust to expand and favouring its renewal by migration of the materials in the mantle towards the surface. Arriving there, these materials (water or ice contaminated with silicates) could cause an 'oceanisation' erasing all traces of the large meteoritic impacts and intense bombardment which immediately followed the accretion. This surface, which was not completely consolidated, was itself subject to the final phase of the meteoritic bombardment which formed the Valhalla basin.

The process of renewal of the crust slowed down and then completely stopped as evidenced by the numerous small craters that can be observed there. These craters correspond to the end of the intense bombardment that took place 4.6 to 4.3 billion years ago. This period corresponds to the end of the internal differentiation of Callisto. Given that the surface seems not to have evolved since this epoch, it is probable that compared with the ice surfaces of two other Galilean satellites, Europa and Ganymede, Callisto represents the most primitive stage of evolution of the three bodies. Strangely, Callisto, which possesses about the same dimensions as Ganymede and the same type of internal structure, does not show the same type of structure on its surface, that is to say the flutings of light terrain. There are two reasons for this: firstly, Callisto may have differentiated more slowly than Ganymede and it may have expanded less; secondly, its crust must be noticeably thicker (about 100 kilometres) and it would not have been possible for it to be fractured by giant impacts and broken up into polygons.

Philippe MASSON

Internal structure. Callisto is slightly smaller than Ganymede but has about the same density (1.8 g/cm^3). The global compositions of the two satellites are probably rather similar. Callisto must be composed of a mixture of ice and silicate materials. Its internal structure is probably differentiated, with a solid silicate core surrounded by a mantle 1000 kilometres thick. The upper part of the mantle is a thick crust of ice (100–200 kilometres) mixed with silicate rocks. The rest of the mantle consists of ice or of water in a liquid state. In the latter case, convection currents would arise in the midst of the mantle. The recent craters expose the ice at the surface but the large impact basins are progressively reduced in extent.

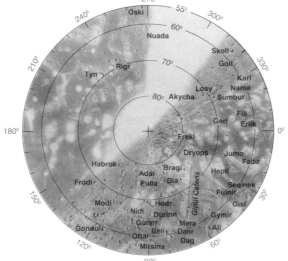

north polar region

Map of Callisto. This map is based on the 'Preliminary Pictorial Map of Callisto' composed from the pictures obtained by the probes Voyager 1 and 2 and published by the United States Geological Survey. Based on calculations of the trajectories of the probes, the positions of the morphological features must be considered as provisional. In particular, relating and joining the pictures taken by Voyager 1 to the pictures taken by Voyager 2 is a very delicate process and errors of position of as much as 10 degrees, or in some cases more than this, are probable. Only the craters and a chain of craters (Gipul Catena, 60° west and 65° north) have been named. Note the principal basins of Callisto, Valhalla and Asgard. This airbrush representation was made by P. M. Bridges under the direction of R. M. Batson. The original map was kindly communicated by R. M. Batson (USGS).

polar stereographic projection;
scale at latitude 56°: 1/41 940 000

equatorial Mercator projection;
scale at latitude 0°: 1/75 000 000

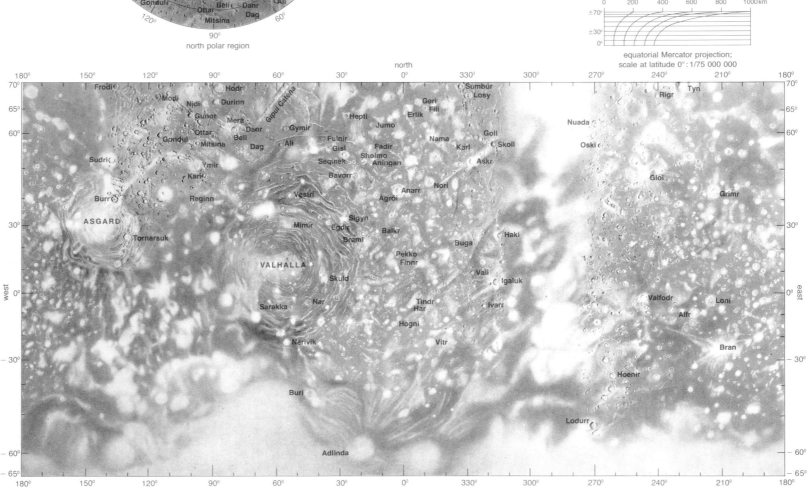

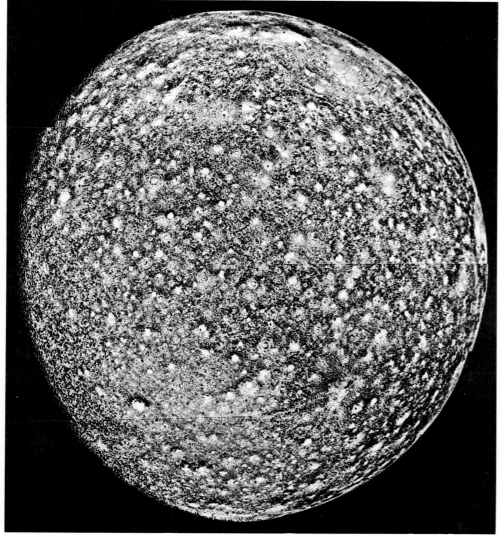

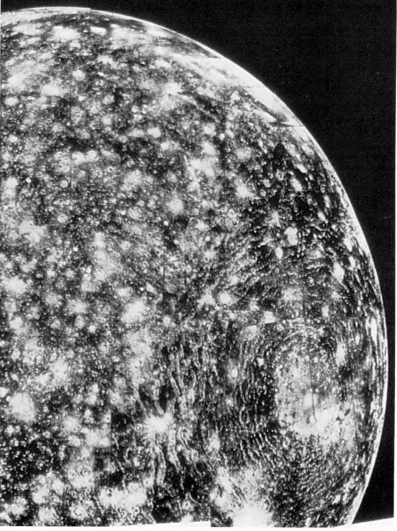

The surface of Callisto. Seen close up, from 390 000 kilometres, the surface of Callisto appears to be riddled with impacts. It is perhaps the most cratered object in the Solar System; consequently it must be very old. The almost total absence of all other surface structures and of any parts without impacts or with a lower density of craters indicates that the ice crust has changed little since its formation. If the criterion of relative dating by the density of impact craters, established for the Moon, can be applied to Callisto, the surface of this satellite must be around four billion years old. Callisto would thus possess the most primitive surface of the four Galilean satellites. Unlike other bodies in the Solar System, in particular the Moon, the craters of Callisto all have more or less the same dimensions, with diameters around 100 kilometres; craters of a larger size are rare. Some recent craters with radial ejecta can also be seen. Close to the limb at upper right is the only important structure of the satellite, the great basin of Valhalla. This picture is a mosaic formed from nine pictures taken on 7 July 1979 by Voyager 2. (NASA/JPL)

The Valhalla basin. The Valhalla basin is a large circular structure with a diameter of about 600 kilometres. It is lighter than the rest of the surface of Callisto. The basin is surrounded by about fifteen concentric rings spaced at distances between 50 and 200 kilometres. The furthest is 1500 kilometres from the centre of the basin. The rings were caused by the giant impact which gave rise to the Valhalla basin; they are ripples arising from the plastic deformation of the ice crust. Unlike the other great basins known elsewhere in the Solar System, Callisto has no important relief, probably because of the consistency of the crust at the moment of impact. The centre of the basin and the rings which surround it are marked by numerous small impact craters. There are fewer craters, however, in Valhalla than in the rest of Callisto's surface. These photographs were taken by Voyager 1 at 202 000 kilometres (above) and at 350 000 kilometres (below). (NASA)

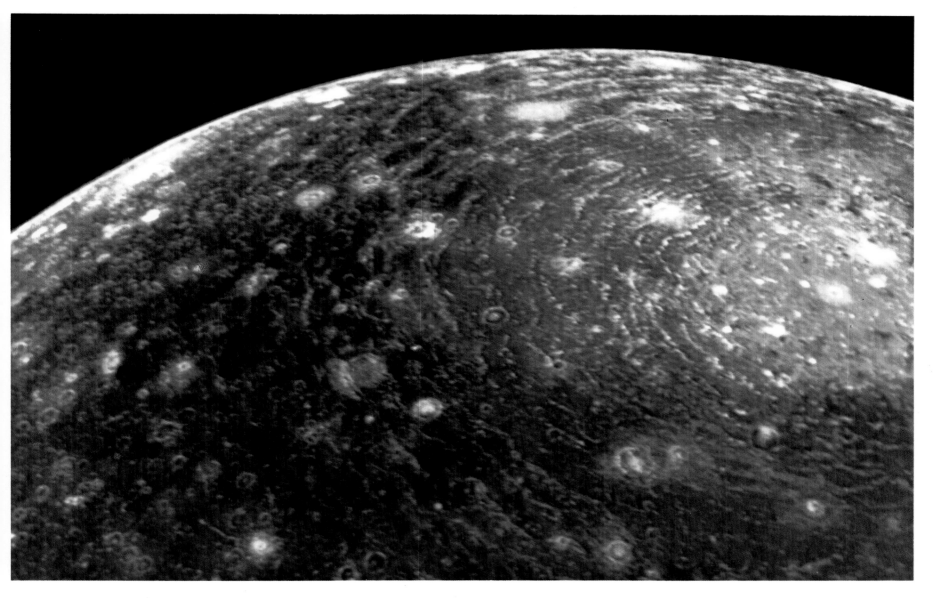

Europa

Europa appears as a white, highly reflective body in a telescope, probably covered with ice. It was necessary to wait for the Voyager probes in 1979 to discover the appearance of its surface. Voyager 1 approached no closer than 734 000 kilometres. At this distance, Europa appeared to be an object without particular forms, without a single visible meteoritic crater or any recognisable geological structure. However, the photographs of Voyager 1 did show numerous thin, dark lines criss-crossing over almost the whole surface, some over 3000 kilometres in length. When Voyager 2 passed close to Europa the photographs transmitted confirmed that the surface was furrowed with dark lines whose length varied from several hundred to several thousand kilometres, and width from several to 70 kilometres. Most of these lines are straight, but some are curved or irregular. They are situated in regions without any visible relief or morphology, where the surface is polished and bright, except for certain isolated dark points most of which have a diameter of less than 10 kilometres. In addition to these polished, shining areas with these networks of dark lines, Europa has darker regions whose surface is less even. These could show small meteoritic impact craters, but their average dimensions are close to the 4-kilometre limit of resolution of the best photographs of Voyager 2, and this makes their identification difficult. In fact, only five meteoritic impacts have been identified, in the 10 to 30 kilometre diameter range; this very small number of impact craters seems to indicate either that Europa is very young or that the craters are not well preserved in the material that constitutes the surface. Although the dark lines give Europa a cracked appearance, they may not all be fractures. They have no particular morphology. On the other hand there are lines that are brighter and thinner (about 10 kilometres wide) but much more regular and which show a pronounced vertical relief (several hundreds of metres high).

Radioactive elements within the silicate core of Europa would release a large amount of heat which could considerably modify the crust or the mantle. The heat flow might give rise to convection currents in the mantle. These convectional movements would have been able to bring about alterations in the ice crust which, linked with its expansion, would have produced the great fractures observed on the surface. These fractures would have been filled in with ice that originated in the mantle.

In spite of the little information that we have about the internal structure or the state of the surface of Europa, we can summarise the satellite's history in three periods:

– An 'ocean', perhaps itself covered by a thin crust of ice, must have covered the whole of the primitive surface after its formation. This ocean masked the traces of the intense meteoritic bombardment that must have affected the primitive silicate surface at the end of its accretion about 4.6 billion years ago. The water of this primitive ocean could have come from dehydration by heating of the hydrated silicates in the core during its accretion.

– The frozen surface of this ocean would have grown thicker, thus forming the surface crust, which developed fractures on expanding.

– The fractures formed in this way would then have been filled with materials, water or ice, coming from the depths of the mantle.

This fractured crust would be relatively recent because it must have undergone rejuvenation thanks to the material coming up from the mantle which covered all traces of its ancient history such as meteoritic impacts. According to some calculations, the ice crust would, by its expansion, have increased its surface area by between 5 and 15 per cent. The disposition of certain fractures that mark the edges of the networks of polygons on the side away from Jupiter could indicate that the cracking of the crust is not solely due to its expansion in it but is also brought about by the tidal effects of the giant planet.

Philippe MASSON

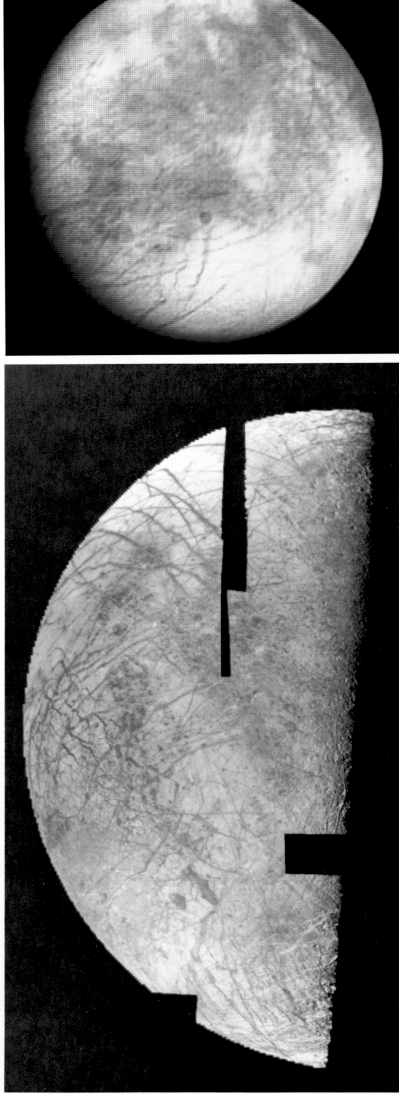

Europa at high resolution. The great fractures or cracks extend for several thousands of kilometres in every direction. Their widths range from several to seventy kilometres. Apart from these fractures, the surface of Europa presents no other structures, nor a great variety of terrain.

The fractures divide the ice crust into a mosaic of polygons without any particular morphology. The surface shows almost no impact craters: only five have been identified. One of them was identified with certainty because of the network of structures radiating out from its periphery and its bright, circular interior. It can be seen in the picture below. Such a scarcity of meteoritic impacts might indicate that the surface of Europa is relatively recent, or that it has known processes of rejuvenation that have been able to erase most of the craters. A third possibility is that the ice crust has not been able, because of its consistency, to preserve the morphology of impacts. (NASA)

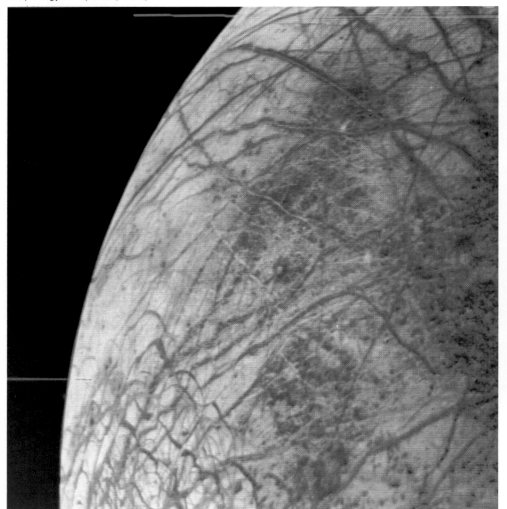

Europa. From 2 million kilometres Voyager 1 showed Europa as a bright disk marked with darker lines (top picture). These lines in fact cover a large part of the surface, as a mosaic of five photographs taken by Voyager 2 shows (bottom picture). When Voyager 2 flew past, Europa was only 206 000 kilometres from the trajectory of the probe. This more favourable position enabled more detailed pictures of the satellite's surface to be taken, with a maximum of resolution of 4 kilometres. The surface of Europa appeared to be almost entirely covered by a complex network of more or less rectilinear fractures, generally darker than the ice crust. They could result from the expansion of the crust under the effect of convectional movements in the mantle and they could have been filled with ice from below, charged with darker silicate materials. (NASA/NSSDC)

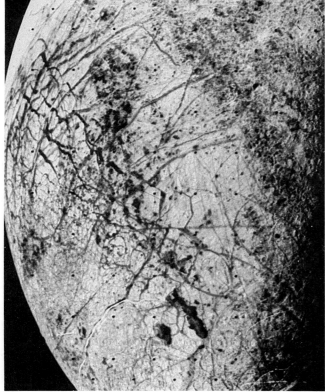

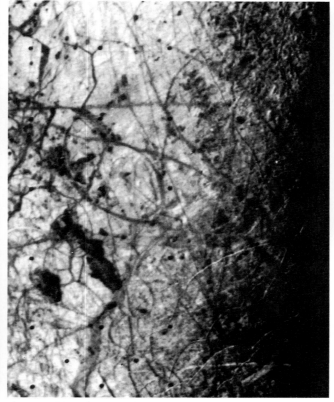

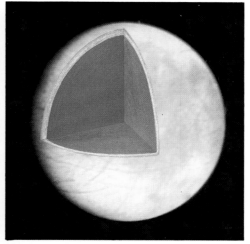

Internal structure. The density (3.0 g/cm³) of the smallest Galilean satellite indicates that it is probably composed of silicate rocks. Observations made by the Voyager probes have confirmed that the surface is entirely covered by a crust of ice whose thickness is not precisely known (at least 50, perhaps 100 kilometres), resting on a mantle of ice with a quite diffferent density and consistency, or of water in the liquid state. This rather thin mantle surrounds a silicate core with a radius of about 1450 kilometres.

Ridges and bands. The left-hand picture was taken by Voyager 2 from a distance of 241 000 kilometres. It shows the region around longitude 180 degrees. Some of the Voyager 2 pictures had a resolution of 4 kilometres, such as the picture on the right: this shows a detail from the other picture, and covers a region about 600 by 800 kilometres. Taken near the evening terminator, it shows the topography of Europa clearly, throwing into relief an entanglement of almost rectilinear fractures. Two types of puzzling structures are also seen: bright curved ridges of width 5 to 10 kilometres usually about 100 kilometres long, and very dark bands 20 to 40 kilometres wide which extend for a few hundred kilometres. These structures are probably connected with movement of the crust, but there is no plausible explanation for them as yet. The large band at lower left is Thrace Macula: to its left is Thera Macula. (NASA/NSSDC)

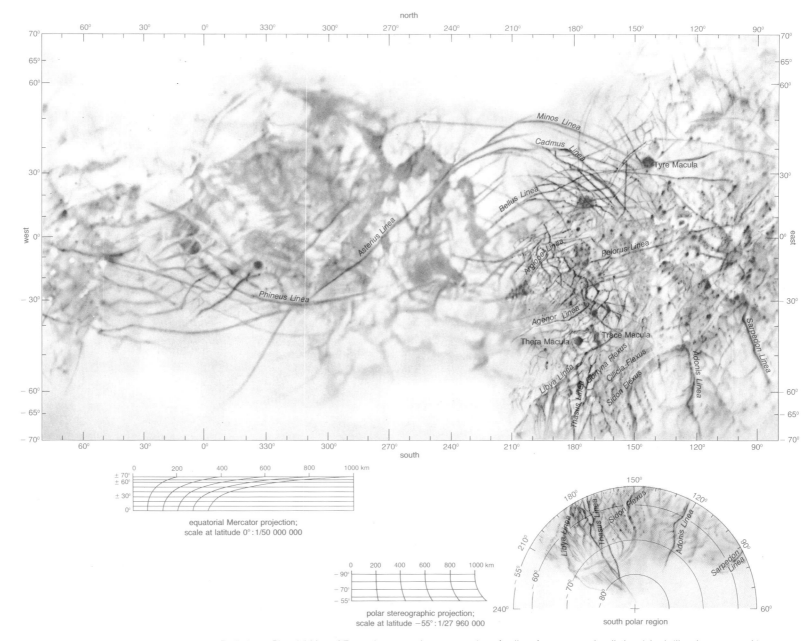

Map of Europa. This map is based on the 'Preliminary Pictorial Map of Europa' using photographs taken by the probes Voyager 1 and 2 and published by the United States Geological Survey. Based on calculations of the trajectories of the probes, the position of morphological features must be considered as provisional. On Europa three categories of salient features can be distinguished: 'linea' – pattern of long dark streaks; 'flexus' – great curving ridge, and 'macula' – dark mark. The airbrushed picture was made by J. L. Inge under the direction of R. M. Batson; the original map was kindly communicated by R. M. Batson (USGS)

Ganymede

Like the other Galilean satellites, Ganymede was almost unknown until high-resolution pictures were transmitted from the probes Voyager 1 and 2 in 1979. The best pictures (resolution of 1 kilometre) were taken by Voyager 2 which flew twice as close as Voyager 1. These pictures revealed a very complex morphology of a type not known elsewhere in the Solar System.

At first sight the icy surface of Ganymede appeared to exhibit two types of terrain, dark and light, each type covering about half of the total surface area. The dark areas form polygonal regions, separated by broad bands of light terrain. In the polar regions, deposits of frost cover both types of formation. In the high-resolution pictures, the light and dark terrains appear like distinct geomorphological units. The light terrain is generally made up of broad parallel channels while the dark terrain is essentially confined to meteoritic impact craters. This distinction between the two types of terrain is purely morphological and does not depend on their constituents which appear to be the same: both types are made of ice.

The dark terrain has a fairly shallow relief and is heavily cratered. The crater population shows the whole gradation from well-preserved craters, as upon the Moon, to circular marks without relief and only visible because of variations in the albedo. This variety in appearance must be due to a slow relaxation of their morphology caused by the rejuvenation of the crust of ice. The significantly large number of craters seen in the dark regions shows that this ice surface is very old. If the meteoritic bombardment dates from the end of the period of accretion of the Solar System, the dark patches must be at least four billion years old and they have hardly evolved at all since that epoch. In addition to these craters, large furrows, parallel but not deep, are visible in these regions. They stretch for more than 1000 kilometres and are often damaged by impacts. Possibly these furrows are ancient remains of giant impacts, now invisible, which caused the formation of huge concentric rings similar to some lunar basins surrounded by mountain ranges.

The large channels seen in the light areas form interlinked networks; they outline depressions a few kilometres wide and several hundred kilometres long. These networks terminate either by gradually disappearing under the surface or by running into other differently aligned channels. The channels are not always rectilinear, and may have winding paths which cross and re-cross, forming 'skeins'. The boundaries between the different terrain, or between the light terrain with channels and the dark cratered terrains, are in general marked by more prominent channels. Certain regions of light terrain lack channels and exhibit a relatively flat topography. The density of impact craters observed in the light terrain is much less than in the dark terrain; it therefore seems that the light terrain would have been formed after the dark terrain, beginning just at the end of the period of accretion, and continuing for a few hundreds of millions of years. The craters in the light terrain are deeper and better preserved than those in the light terrain, which appears to indicate that the viscosity of the ice has increased since the formation of the channels.

The origin of the channels is even less well explained, but they were probably formed by movements in the crust of ice. Various theories have been put forward to explain their formation by the rejuvenation of the surface ice with less dense ice from the depths, under the influence of convectional movements of the mantle. Such movements would have caused distortions of the crust and the filling of open areas with new ice. Actually the process of formation and the origin of these structures depends on what you believe to be the internal structure of Ganymede. However, whatever model is adopted for its internal structure, the interior of Ganymede would have been the seat of convection movements. These movements would have had the effect of bringing a magma of ice and water to the surface in zones where the crust is weaker or ruptured by expansion movements, and thus creating the channels and the light areas. However, calculations have shown that the expansion of the crust (of the order of 6 or 7 per cent) is insufficient to explain the importance of the light areas. It is therefore possible that the primitive crust of Ganymede (dark terrain) was partially displaced by terrific impacts into polygons of ice which would have moved apart and drifted

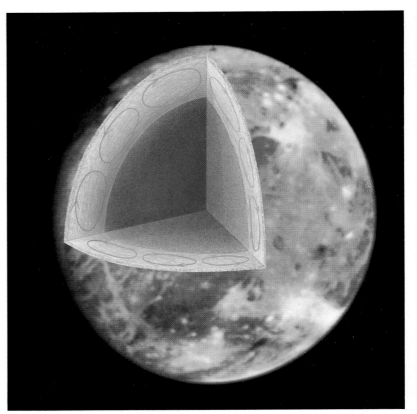

Internal structure. This is the largest of the Galilean satellites, but it has a relatively low density (1.9 g/cm^3) showing that it is half ice and half silicates. Some think that the satellite has a differentiated structure with a solid silicate core 1800 to 2200 kilometres in radius, a mantle of ice or liquid water 400 to 800 kilometres thick, and a thin crust less than 100 kilometres thick. This model is shown above. Others think that Ganymede is not differentiated but is a mixture of ice and silicates.

Ganymede. The Voyager 1 and 2 probes bypassed Ganymede at 115 000 and 62 000 kilometres respectively in 1979. As the probes approached, they first transmitted general views of the satellite from a great distance. The picture below left was taken by Voyager 1 on 4 March from 2 600 000 kilometres. In such pictures the surface shows a variety of light and dark shades and numerous bright spots. As the probes gradually moved on and the angle of observation changed, the surface appeared to consist of two types of terrain as shown below right. The picture is a mosaic of six photographs taken by Voyager 2 from 310 000 kilometres. In the top right of the picture can be seen the south-west quarter of a large dark roughly circular area called the Galileo Regio. This area, in the northern hemisphere, has a diameter of 3200 kilometres. It is crossed by a network of light bands which are parallel and regularly spaced. The rest of the satellite's surface is covered with lighter terrain in bands or in more or less regular polygons. All these areas are dotted with light spots of various sizes which are meteoritic impact craters. In the southern hemisphere some are surrounded by a system of radial ejecta analogous to that found around some recent lunar craters. The crater Osiris, at the bottom of the picture, is a good example. The density of impact craters is higher in the dark terrain of the Galileo Regio than in the light terrain; the dark terrain is therefore the older, showing that the crust of ice which constitutes the surface of Ganymede must have undergone important modifications during its evolution. (NASA/NSSDC)

Toponymy of the Galilean satellites. Because the giant planet had been given the name of the most powerful of the gods of Greco-Roman mythology in antiquity, its four large satellites (Io, Europa, Ganymede and Callisto) discovered in 1610 by Galileo, were given the names of four of the numerous lovers and mistresses of Jupiter by Simon Marius (an astronomer, contemporary with and a rival of Galileo).

Following the encounters of the probes Voyager 1 and 2 with the Jovian system, the commission for nomenclature for the outer planets of the International Astronomical Union (IAU) decided to choose names for the different regions of each satellite partly from the mythology directly associated with the person who gave her name to the satellite, and partly from different cultures such as Indian, African, South American, etc. In the latter case the names were grouped by region and continent, by lists of people, places and objects having reference to the story concerned. In this way Europa took names from mythologies of southern Europe, Callisto from northern Europe and Ganymede from other regions, in particular India. The discovery of active volcanoes upon Io led the commission to choose names of mythological deities having something to do with fire, the Sun or volcanoes (e.g. Prometheus and Pele). Certain regions with a low albedo were given the names of famous astronomers who discovered or studied the Galilean satellites of Jupiter (Galileo, Marius, Barnard, etc.).

To the morphological terms required for the inner planets, it was necessary to add new ones to describe specific forms on the Galilean satelites, like 'macula' for dark spot, 'linea' for a long dark streak or 'flexus' for a long curved ridge on Europa.

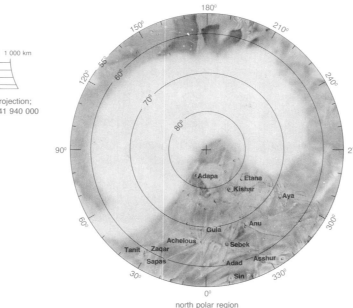

polar stereographic projection;
scale at latitude 56°: 1/41 940 000

north polar region

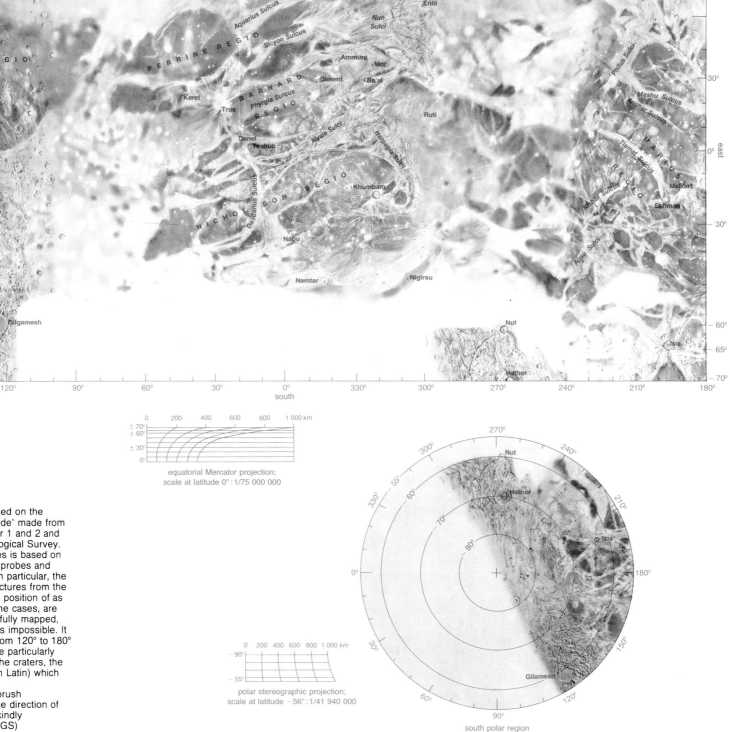

equatorial Mercator projection;
scale at latitude 0°: 1/75 000 000

polar stereographic projection;
scale at latitude −56°: 1/41 940 000

south polar region

Map of Ganymede. This map is based on the 'Preliminary Pictorial Map of Ganymede' made from pictures taken by the probes Voyager 1 and 2 and published by the United States Geological Survey. The position of morphological features is based on calculations of the trajectories of the probes and must be considered as provisional. In particular, the process of relating and joining the pictures from the two probes is very delicate: errors in position of as much as 10 degrees, or more in some cases, are probable. There is one region, doubtfully mapped, where cross-checking is more or less impossible. It extends from 45° to 55° south and from 120° to 180° longitude. Three types of features are particularly prominent on Ganymede. They are the craters, the 'regiones' and the 'sulci' ('furrows' in Latin) which are parallel channels.

The figure was taken from the airbrush representation by J. L. Inge under the direction of R. M. Batson; the original map was kindly communicated by R. M. Batson. (USGS)

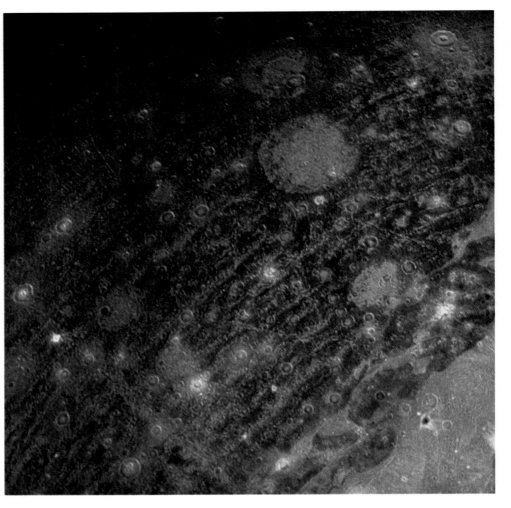

Ancient terrains. The very detailed picture below left of Galileo Regio was taken by Voyager 2 at a distance of 85 000 kilometres. The surface, probably constituted of a mixture of ice and silicates, shows a great variety of impact craters. Taking account of their state of conservation, these appear to be of very different ages. The craters that are apparently the most recent are the brightest. Of small size, tens of kilometres in diameter or more, and with a well-marked morphology, they are often surrounded by bright, more or less radial ejecta. The oldest craters, on the other hand, have almost totally lost their original morphology and are often only visible thanks to the difference in albedo of the crust of ice. (See picture at top right, taken by Voyager 2 at 310 000 kilometres). These craters are large, with diameters of several hundreds of kilometres. They are often altered by more recent craters, which are of moderate size, with diameters of about 100 kilometres. By far the most numerous, they have a morphology typical of impact craters (circular borders and a central peak).

Recent craters also affect the parallel structures which cut across the dark Galileo Regio. These structures consist of hundred-metre-high ridges about 10 kilometres wide, alternating with spaces about 50 kilometres wide. The ridges are not rectilinear but sketch huge arcs of concentric circles whose focus is located in the light region to the south-west of Galileo Regio. The great Valhalla basin on Callisto shows similar structures at its edge. By analogy, it is generally thought that the ridges on the Galileo Regio were due to another giant impact whose traces are no longer visible. They would have disappeared from the surface of Ganymede because of processes which have subsequently altered the crust, particularly in the light terrain, where the ridges are not visible. (NASA)

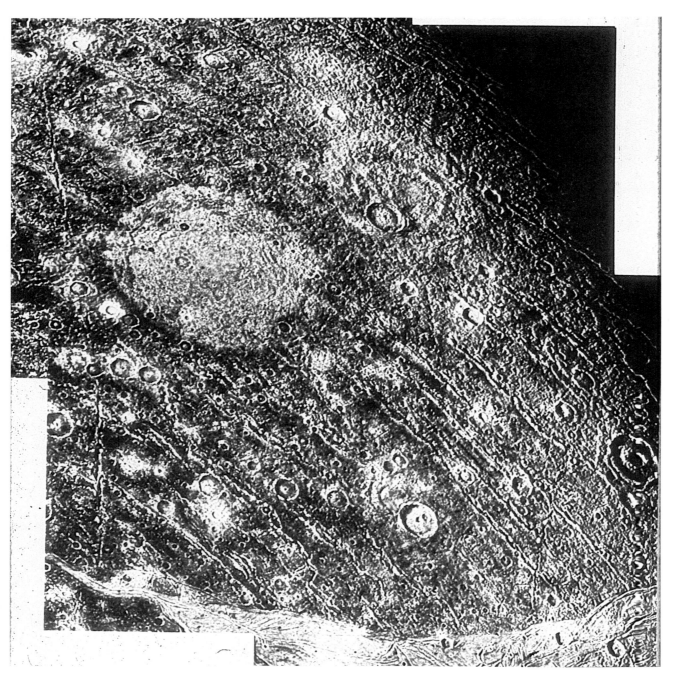

because of the expansion of the crust and the convection in the mantle. Being constituted of denser ice than the mantle these polygons would have sunk into it: the space thus left would then be fitted with less dense ice coming up from the mantle. This displacement and subsidence of dense ice would have made extensive deformations in the light ice from the depths of the mantle and in this way would have formed the channels.

Ganymede must thus have experienced an intense meteoritic bombardment four billion years ago. This bombardment, which probably affected the whole primitive surface, must have triggered a fracturing of the ice which, helped by other factors such as convection and expansion, led to the less dense ice from the mantle, or a more or less liquid magma, coming to the surface and re-freezing immediately. The displacement of these polygons in the primitive crust (which continued for some hundreds of millions of years after the end of the period of accretion, as proved by the impact craters in the light areas) split the clear ice, forming great channels. When the light areas had formed and had been split by the channels, the surface of Ganymede did not evolve any further. The interior differentiation had ceased, bringing to a stop the convectional movements in the mantle and expansion in the crust. Consequently, the surface of Ganymede is ancient and has probably remained in the same state for two to three billion years.

Philippe MASSON

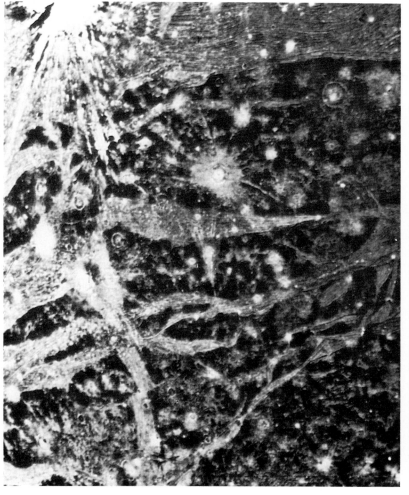

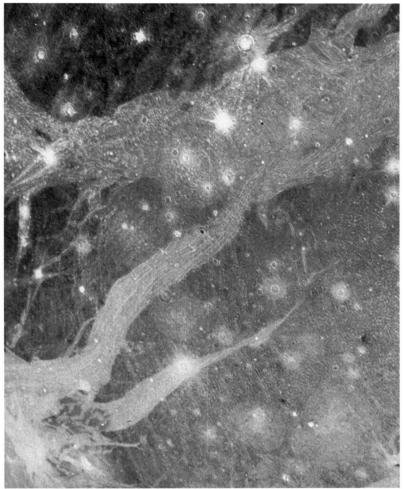

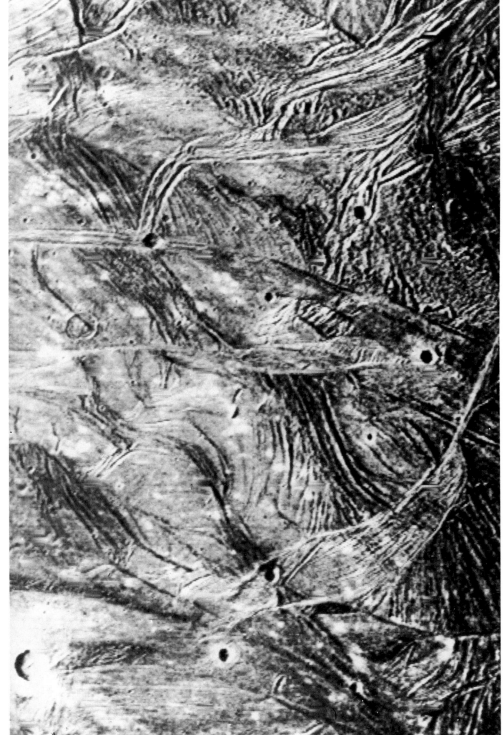

Recent terrains. The bright terrains which make up the major part of Ganymede's surface show many more meteoritic craters than the dark formations of Galileo Regio. These bright terrains are divided up into polygons of great size, often separated from one another and divided by wide very light, bands. These bands, made of ice, have no preferential orientation. They vary greatly in size (several thousand kilometres long by several hundred kilometres wide) and frequently cut back on themselves, as in the pictures above and above left. They are sometimes shifted laterally and appear to be affected by faults and great fractures (top right).

On high-resolution pictures such as that on the right, where the resolution is 3 kilometres, these bands show a very peculiar morphology; they appear to be constituted of huge parallel channels, on average about 1000 metres deep and about 10 to 15 kilometres apart. Inside the bands, the network of channels twists and overlaps and entangles itself in a very complex way. These bands of channelled terrain contain many fewer impact craters than the other formations. They must have been formed when the ice crust fractured under a great impact which created the ridges now observed in the Galileo Regio. The displacement of pieces of the crust would have led to deformations in the ice, thus creating the channels in the great bands of light terrain. (NASA)

Io

After its discovery by Galileo in 1610, and before the Voyager probes had observed it in 1979, Io was only seen from the Earth as a tiny point of light. It was difficult to observe because of its size, which is comparable to that of the moon, and because of its proximity to Jupiter. Only a few things were known about this satellite: its period of revolution – about 1.8 days; its density – 3.5 grams per cubic centimetre – comparable to that of the Moon, leading us to expect a composition based on rocks. Spectra of the surface of Io indicated the presence of sulphur compounds, and the absence of ice, in contrast to the other Galilean satellites.

Before the flybys of the Voyager probes, two other American probes, Pioneer 10 and 11 flew past in 1973 and 1974 at distances of 130 000 and 34 000 kilometres respectively, without obtaining good pictures of the surface of Io. However, the probes made measurements in the ultraviolet which increased our knowledge of the environment of the satellite.

In passing by Io at 21 000 kilometres, Voyager 1 took high-resolution photographs (resolutions of the order of a kilometre) at last enabling the surface to be examined. Voyager 2 flew by at 1 130 000 kilometres obtaining only pictures of medium resolution (20 kilometres) and made complementary observations.

Io is the most spectacular of the Galilean satellites of Jupiter. Even in pictures of low resolution its vivid colours set it in a class apart from all the other solid bodies in the Solar System.

A detailed study of the surface does not reveal even the smallest meteoritic impact crater. This appears to indicate that the satellite is very young and has had a very rapid evolution. On the other hand, the surface does show a great many volcanic centres which appear in the form of dark spots with diameters of some tens of kilometres. About 5 per cent of the surface is taken up with very dark volcanic calderas which reflect less than

5 per cent of the light. They are often encircled by irregular haloes. At the edges of these calderas, large lava flows can be seen starting in the dark part and extending for several hundreds of kilometres. They meander and look like long tentacles.

The equatorial regions of Io are relatively flat. There are some high escarpments, and valleys a few hundred metres deep can be seen. These structures appear to correspond to ruptures in the crust like fractures and terrestrial grabens. In the polar regions the morphology is more irregular. There are only a few volcanic centres and piles of successive layers give the appearance of cliffs formed by erosion and also by fracture. The cliffs would be the place where liquid sulphur and liquid sulphur dioxide (SO_2) are emitted, which would have eroded the surface by suddenly gushing up from artesian wells.

The most remarkable features upon Io are the rings or ovoids which mark the great calderas. These rings which can reach 700 kilometres in diameter (and even 2000 kilometres in the case of the volcano Pele), probably consist of sulphur and sulphur dioxide fallout thrown up by volcanic eruptions. Through temperature measurements made by spectrometry and infrared interferometry, a number of hot regions were discovered: one dark spot has a temperature of 17 °C, while the surrounding surface is at −146 °C; it could be a lake of molten lava. The bright colours, red and yellow, suggest a composition rich in sulphur. (Heated to different temperatures, then suddenly re-frozen, sulphur has a great variety of tints from black through shades of red and orange to brilliant yellow). The white spots seen in a few regions are solid sulphur dioxide.

One of the most spectacular aspects of Io is obviously its volcanic activity, discovered by Voyager 1. Io is, apart from Earth, the only body in the Solar System known to be volcanically active. When Voyager 1 flew by eight volcanic eruptions were detected. These eruptions were evident from giant plumes, probably composed of molten sulphur dioxide thrown up to altitudes

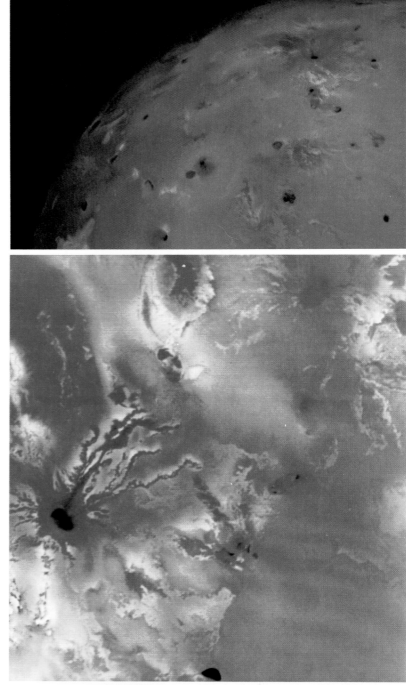

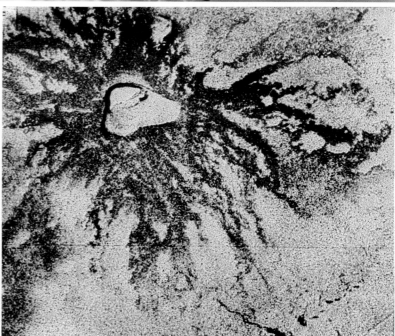

The surface of Io. This view of the whole of Io is the result of combining several pictures taken on 4 March 1979 by Voyager from a distance of 862 000 kilometres. The surface is vividly coloured in red, orange and white with various spots and dark marks. The different tints are regions of different concentrations of sulphur and sulphur dioxide (SO_2) from the volcanic activity of Io. The surface reveals no meteoritic impact craters which indicates that the volcanoes were formed recently. In contrast, there are numerous circular structures of various sizes such as the one in the centre of the picture. It revealed itself to be an active volcano, and was subsequently named Prometheus. (NASA/NSSDC)

Volcanic morphology: lava flows and craters. As it made its approach, Voyager 1 obtained pictures with a resolution of almost one kilometre, from which we can examine the volcanic morphology of Io. Four months later when Voyager 2 entered the Jovian system Io was less favourably placed and the resolution of the pictures taken was therefore much lower (20 kilometres). These pictures did not produce many new observations but did supplement the discoveries made by Voyager 1.

The top picture was taken by Voyager 1 at a distance of 377 000 kilometres. The resolution is 10 kilometres. The dark circular spots are calderas, or volcanic craters formed by collapse. Next to some of these calderas are shapes like tentacles. Some, vividly coloured, extend for several hundreds of kilometres. An example is shown in the centre picture, taken by Voyager 1 at a distance of 128 500 kilometres; an area about 1000 kilometres across is shown. These forms are probably very fluid lava flows, rich in sulphur, spreading out from the calderas. Some very high-resolution pictures like the one at the bottom show that calderas upon Io may enclose one another, as is seen in the summits of Hawaiian volcanoes upon Earth or in the giant volcanoes of Mars. In the bottom picture, taken at a distance of 30 800 kilometres, the caldera has a diameter of about 50 kilometres, and the dark flows issuing from it only extend over a relatively small area about 50 by 100 kilometres. This indicates a difference of viscosity and hence of composition (which is basaltic) compared with the flows of lighter colour. (NASA/NSSDC)

between 70 and 280 kilometres (in the cases observed). To attain such altitudes, matter must be ejected at speeds from 300 to 1000 metres per second. Each eruption ejects about 10 000 tonnes of material per second; over the whole surface of Io, over 100 million tonnes is ejected per year, equivalent to 10 metres of deposits in a million years. Together with the quantity of material produced by lava flows from the calderas, an estimated 100 metre-thick layer would be laid down in a million years. It is therefore not surprising that there should be no trace of meteoritic impact craters because the surface is continually rejuvenated.

For Io to be the scene of such intense volcanic activity, there must be a particularly strong source of energy. in the case of the terrestrial planets the source of heat and energy lies in the decay of radioactive substances such as thorium and uranium. If this were to be the energy source of Io, it would be necessary to have a hundred times more thorium and uranium than the satellite ought to contain. An explanation has been put forward involving the effects of tides caused by the enormous mass of Jupiter; these tides deform Io. The deformations would be without effect if the distance between the planet and the satellite did not vary because of gravitational perturbations caused by the other Galilean moons (particularly during conjunctions). The energy thus produced in the interior of Io, estimated at 10^{13} watts, appears in the form of intense internal heating, which provokes volcanic eruptions. This volcanic activity must have existed since the formation of Io more than four billion years ago. With a molten interior and fierce surface vulcanism, Io has had enough time to alter its composition, to lose the volatile components of its atmosphere (water and carbon dioxide), and to rejuvenate its surface continually with the permanent recycling of the lava flows of sulphur and sulphur dioxide.

Philippe MASSON

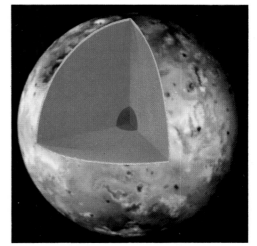

Internal structure. Io is the densest of all the Galilean satellites (3.5 gcm^{-3}). Several models of the internal structure have been made which are based on a solid silicate crust surrounded by molten silicates. Io could have a solid core which, in the model shown above, has a radius of about 600 kilometres.

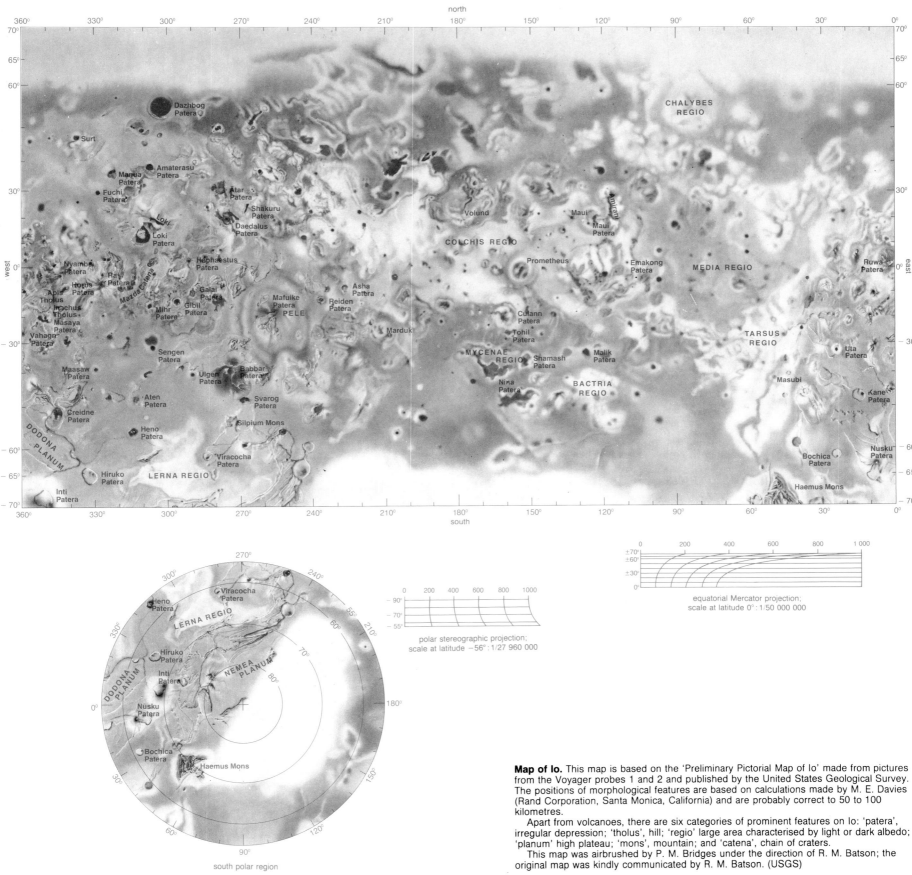

Map of Io. This map is based on the 'Preliminary Pictorial Map of Io' made from pictures from the Voyager probes 1 and 2 and published by the United States Geological Survey. The positions of morphological features are based on calculations made by M. E. Davies (Rand Corporation, Santa Monica, California) and are probably correct to 50 to 100 kilometres.

Apart from volcanoes, there are six categories of prominent features on Io: 'patera', irregular depression; 'tholus', hill; 'regio' large area characterised by light or dark albedo; 'planum' high plateau; 'mons', mountain; and 'catena', chain of craters.

This map was airbrushed by P. M. Bridges under the direction of R. M. Batson; the original map was kindly communicated by R. M. Batson. (USGS)

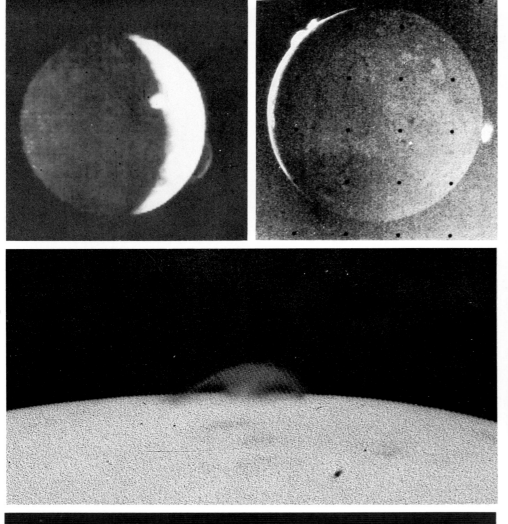

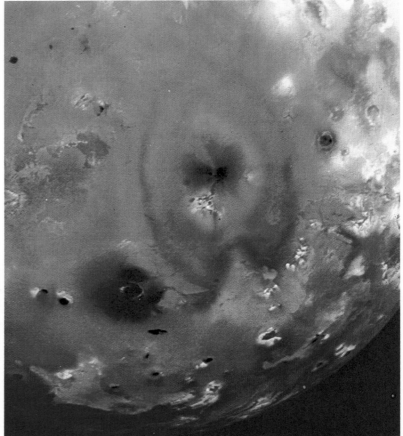

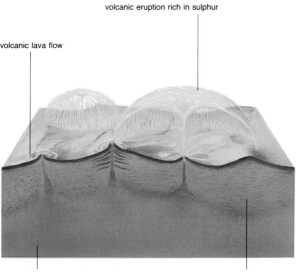

Volcanic activity. The lava flows visible around the calderas are not the sole indications of volcanic activity on Io. At the edges of some calderas one can see haloes of more or less circular form, which may be 1000 kilometres in diameter (as around the volcano Pele, just above the centre of the photograph above). These haloes are made of sulphur dioxide which has been projected above the surface and has condensed and fallen back like snow, forming a ring around the caldera.

Apart from the Earth, Io is now the only solid body in the Solar System to exhibit volcanic activity. This activity was theoretically predicted when taking account of the effect of tides caused by the mass of Jupiter. The confirmation of this hypothesis was obtained when a picture (above left) taken by Voyager 1 at a distance of 45 million kilometres was processed by computer to reveal a clear plume like a mushroom in the lower right region of the limb. This plume rose to about 260 kilometres from the surface, vertically above Pele. The spot clearly visible on the same picture in the region of the terminator is also an eruption but in this case it is seen from above; it is the volcano Loki. Voyager 1 observed eight eruptions. Four months later, Voyager 2 observed only six. The most important in Voyager 1's flyby, that of Pele, had stopped. On the other hand, two others (Prometheus was one), had increased in intensity between the approaches of the two probes. Three of these eruptions are visible on the limb in a picture taken from a distance of 1.2 million kilometres by Voyager 2 (top centre). On this picture the plume on the right (Loki) is one and a half times larger than it was at the approach of Voyager 1 (185 kilometres high, 325 kilometres wide). The high-resolution picture (opposite centre) showing the plume of Prometheus, enables us to study the structure of eruptions of which the upper part appears to be made up of fine particles of sulphur dioxide, condensing rapidly. The central part would be richer in heavier materials (silicates and sulphur). This volcanic activity is very violent, and the speed of ejection in the midst of the plumes is estimated at 900 metres per second for the largest. (NASA/NSSDC)

Models of the crust of Io. Several models have been suggested to explain the relationships between the internal structure of Io, its surface, and its volcanic activity. Some researchers are in favour of a model showing what is called silicate volcanism. In this model a silicate magma rich in sulphur passes through the solid silicate crust by means of volcanic chimneys, also rich in sulphur. The variations in physical conditions (temperature and pressure) and the chemical composition of the magma can explain the different forms of volcanic activity (high-altitude eruptions and lava flows) observed on the surface.

Other researchers believe that pockets of molten sulphur (sulphur and sulphur dioxide) exist at the boundary between the surface crust and the lower silicate crust. This model shows sulphurous volcanism. The pockets of sulphur are reservoirs of magma and they feed the eruptions and the surface flows. An 'aquifer' level of liquid sulphur and sulphur dioxide is situated at the boundary between the pockets of molten sulphur and the surface crust. This level is the origin of discharges which occur along the fault escarpments and materialise in whitish deposits. On its arrival at the surface the liquid sulphur dioxide becomes unstable and vapourises or suddenly condenses into clouds and frost. In this process the pressure is sufficiently strong to throw the sulphur dioxide 50 kilometres up, and to erode the sides of neighbouring hills.

volcanic eruption rich in sulphur

volcanic lava flow

silicate magma, rich in sulphur   silicate crust

model of silicate volcanism

eruption of SO$_2$

surface crust of sulphur and solid SO$_2$

sulphur flows

caldera

molten silicate interior

molten sulphur

lower silicate crust

model of sulphurous volcanism

deposits of frozen SO$_2$

crust of sulphur and solid SO$_2$

fog and cloud of SO$_2$

fault

'aquifer' level of sulphur and liquid SO$_2$

molten sulphur

The atmosphere of Io

Alone among the Galilean satellites, Io has an ionosphere extending out to 700 kilometres. This was revealed by the flyby of Pioneer 10 before high-resolution photographs were taken by Voyager 1 and 2. A peak of high electron density (nearly 100 000 electrons per cubic centimetre) located at an altitude of 100 kilometres has been identified. This indirectly confirmed the presence of a neutral atmosphere around Io, shown to exist by a stellar occultation in 1971. The observations were made during the latest planetary missions, and in particular a careful analysis of the interaction of the satellite with the magnetosphere of Jupiter has made it possible to specify the nature of this atmosphere.

More or less comparable to an exosphere, the atmosphere of Io is in fact directly controlled by the phase equilibrium of volatile substances frozen onto the surface, and almost exclusively composed of sulphur dioxide (SO_2). The result of this control is that at the subsolar point, on the equator, the pressure reaches a maximum of about 10^{-3} millibars (the temperature is about 140 K) whereas for elongations exceeding 80 degrees in longitude, when the temperature is below 100 K the pressure is exospheric – less than 2×10^{-8} millibars. Indications of the presence of sulphur dioxide are not marked, although considerations of phase equilibrium seem to rule out the permanent presence of all other atmospheric constituents, with the possible exception of argon. Thus, instruments for infrared analysis on the Voyager probe have clearly demonstrated the existence of absorption at 7.6 micrometres due to sulphur dioxide gas, above the hot volcanic areas of Io. Moreover, certain spots on the surface of Io are attributed, because of their colour, to deposits of ice. The process of formation proposed is the following. Sulphur dioxide is ejected during volcanic eruptions; the gas, after its condensation in the cold atmosphere of the satellite, falls back to the surface in the form of droplets whose concentration in certain parts of the surface gives a characteristic blue colour. It was possible to confirm this hypothesis through observations made on Earth: an analysis was made of the infrared spectrum of the light reflected from the surface of Io. The spectral signature was like that of absorption by solid sulphur dioxide. The best evidence, however, lies in the chemical composition of the plasma torus encircling Jupiter along the orbit of Io.

The Voyager probes were very easily able to detect the characteristic emissions of ionised atoms of sulphur and oxygen in the far ultraviolet. The strongest intensity detected is that from doubly ionised sulphur which emits at 69 nanometres. The Voyager probes were also equipped with an instrument for analysis *in situ* of the composition of the plasma in Jupiter's magnetosphere. The ions identified are those of the atoms of oxygen and sulphur and of the molecules of sulphur (S_2) and sulphur dioxide (SO_2). Their concentration is highest (5000 particles per cubic centimetre) in the midst of the plasma torus which was crossed by the Voyager 1 probe.

Although the planetocentric configuration of the plasma torus round Io's orbit leaves no doubt about the origin of the particles of which it is composed, it still remains to pinpoint the region where the atoms and molecules become charged before being trapped by the rotating lines of force of the magnetic field of Jupiter. The reservoir which constitutes the ionosphere of Io is not sufficient to explain the particle density of the torus by a mechanism of direct injection. On the contrary, the observations indicate that the atoms and particles that populate the torus are ionised in its midst. We may well ask how neutral particles ejected during volcanic eruptions acquire enough energy to leave Io. Two hypotheses have been advanced. In the first, it is supposed that the dust particles which compose the plumes of the volcanoes are sufficiently charged, following their bombardment by electrons from Jupiter's magnetosphere, to leave the environment of Io under the influence of Jupiter's magnetic field. Thus freed, the particles would be broken into atoms and molecules which are ionised by the environment of the magnetosphere. In the second hypothesis, atoms on the surface of Io would be freed by a phenomenon of ionic pulverisation following the impact of very energetic ions from the Jovian magnetosphere (sputtering effect). This latter process had already been invoked to explain another peculiarity of Jupiter's environment, one that could be identified by an observatory on Earth: the presence of a narrow belt of neutral sodium atoms around the giant planet itself and also along the orbit of Io. It was possible to photograph from Earth traces of intermittent jets of sodium atoms. This gas has a very readily recognisable spectral signature at 589 nanometres, leading us to think that the process of ionic pulverisation frees other atoms that are more difficult to identify from their spectra. This mechanism is sufficiently powerful to give atoms the energy to populate the space occupied by the torus within which they will be ionised.

Guy ISRAËL

An eruption of Loki. This picture in false colour was constructed from four pictures taken in ultraviolet, violet, green and orange on 4 March 1979 by Voyager 1 from a distance of about 500 000 kilometres. The photographs were processed by computer to show the structure of the erupting plume which reveals two components. The first, which is seen in the visible, is represented by the asymmetrical yellow region, rising to an altitude of about 100 kilometres. It is surrounded by an envelope which scatters ultraviolet radiation. This is the symmetrical blue region rising to an altitude of 200 kilometres. This envelope could be the result of the scattering of light by very fine particles. (NASA)

The Jovian Sulfur Nebula. The Jovian Sulfur Nebula is represented in false colour images depicting the nebula at two rotational phases separated by three hours. Regions populated predominantly by S^{+2} appear with purple hues, those predominantly S^+ appear green. The yellow curve marks the magnetic confinement equator associated with the orbital path of Io, Jupiter's volcanically active satellite. Each image is a composite assembled from snapshots of Jupiter (for spatial reference) and 10 minute exposure of the nebula in characteristic [SII] and [SIII] emissions while blocking the Jovian disk. The images were obtained by J. T. Trauger (Caltech) at Las Campanas Observatory on 4 May 1983.

Jupiter's plasma torus. Observations made by the probes Voyager 1 and 2 in the extreme ultraviolet have enabled the dimensions and composition of the plasma torus to be measured. Its radius is 5.9 ± 0.3 Jovian radii; the radius of its cross-section is 1 ± 0.3 Jovian radii. The torus is in the plane of the magnetic equator, which is inclined to the orbit of Io by 10.6 degrees. The torus therefore lies 'above' and 'below' the orbit of Io, as shown. The predominant emitting species are the ions of sulphur (S^+, S^{2+}, S^{3+}) and of oxygen (O^+ and O^{2+}). These ions are derived from material torn from the surface of Io by the processes of eruption and ionic pulverisation. The ions rotate in the same sense as the Jovian magnetosphere.

Four months passed between the approaches of Voyager 1 and 2. During this period, the radiation from the torus increased by a factor of about two, accompanied by a decrease in the temperature of the plasma (100 000 K to 60 000 K).

The auroral zones of Jupiter are associated with the lines of force of the Jovian magnetic field that passes through Io.

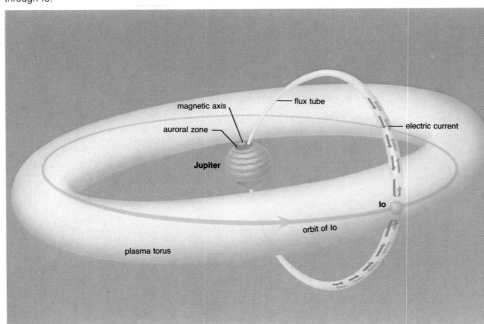

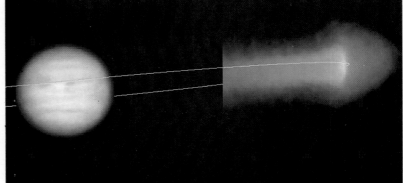

Saturn

The rings encircling Saturn give it an appearance which is unique in the Solar System. The planet itself, though, is remarkably similar to Jupiter. Its dimensions are almost the same; its equatorial radius of 60 000 kilometres is 9.4 Earth radii compared to 11.3 for Jupiter. Its mass, however, is of the order of 95 times that of the Earth, rather than Jupiter's 318 making its average density only 0.7 grams per cubic centimetre; in other words, if you could drop it in an ocean of water, Saturn would float to the surface like a buoy. This suggests that Saturn, like Jupiter, is formed mostly of hydrogen and helium, the elements that made up the primordial solar nebula. We will see, though, that these two main constituents do not stay mixed together throughout the interior of the planet and that the internal structure of Saturn is qualitatively different from that of Jupiter. It is nevertheless true that Saturn is basically a huge ball of gas compressed by its own weight, just like Jupiter, and all that we see of it are clouds at the top of its atmosphere. These clouds are made of chemical compounds that can condense out at the low temperatures prevailing at the edge of this gaseous globe.

Like Jupiter, Saturn turns on its axis very rapidly; it has a rotation period of 10 hours and 32 minutes. The marked polar flattening is more pronounced than Jupiter's, with the equatorial radius being 11 per cent larger than the polar radius, rather than a figure of 6 per cent for Jupiter. Another similarity is that Saturn also emits more energy in the form of radiation than it receives from the Sun, so it must have an internal source of energy. However, it appears that the energy sources of Jupiter and Saturn are different.

The space probes Pioneer 11 and Voyagers 1 and 2 gathered a harvest of data on the composition, thermal structure and dynamics of the outer atmosphere of Saturn. Combined with data from Earth observations, both ground-based and airborne, these results now allow us to describe the planet in far more detail than was previously possible.

Let us imagine an observer coming from interplanetary space and heading for the centre of the planet. What would he discover?

A 'cloud' of atomic hydrogen and (perhaps) molecular hydrogen surrounds Saturn in the shape of a torus (or ring). This torus lies in the equatorial plane and extends from 8 to 25 Saturn radii (480 000 to 1 500 000 kilometres) and has a thickness of about 14 Saturn radii (840 000 kilometres). It is thought that this cloud, whose density is of the order of twenty atoms per cubic centimetre, comes from hydrogen which has escaped from the atmosphere of Titan, and then been drawn into orbit around Saturn by the planet's gravitational attraction. It is possible that the torus also contains molecular hydrogen, possibly with even higher density than the atomic hydrogen.

The exosphere is the outer atmosphere of Saturn, above the zone where the different gaseous constituents stay well mixed due to the effects of turbulence. It has a temperature of about 400 K. The density of molecular hydrogen increases rapidly below 61 400 kilometres from the centre of the planet, or about 1300 kilometres above the 1 atmosphere pressure level. There is probably also some methane in this zone.

The homopause is the region below which those constituents which cannot either condense or be dissociated by sunlight are uniformly mixed. It is at a temperature of 200 K and lies 1150 kilometres above the 1 atmosphere level. Below the homopause, the relative proportions of the two major constituents, helium and hydrogen, are 7 per cent by volume (14 per cent by mass) and 93 per cent, respectively. On Jupiter, the proportions of the same elements are 10 and 90 per cent. In the stratosphere, the region between the homopause and the tropopause (which is at the 0.1 atmosphere level), the

The exploration of Saturn. Taken by Voyager 2 on 20 July 1981 from a distance of 34.7 million kilometres, this picture shows the rings – Cassini's Division is clearly visible – and also displays the bands running parallel to the equator, on the planet's disk. However, there is much less contrast in the colours than on Jupiter. Two large cloud formations can be made out in the northern hemisphere. The exact nature of the particles which make up the clouds is still unknown. Just as for Jupiter, they could be composed of ammonium hydrosulphide (NH_4SH), of phosphate compounds, or of complex organic compounds. This difference in appearance from Jupiter remains an enigma. One of Saturn's satellites, Enceladus, appears as a bright point beneath the rings. (NASA)

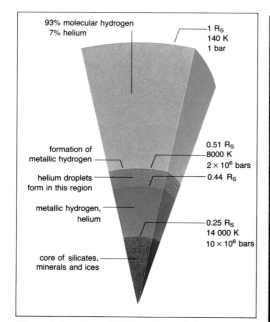

93% molecular hydrogen
7% helium

1 R_S
140 K
1 bar

formation of metallic hydrogen

0.51 R_S
8000 K
2×10^6 bars

helium droplets form in this region

0.44 R_S

metallic hydrogen, helium

0.25 R_S
14 000 K
10×10^6 bars

core of silicates, minerals and ices

Internal structure of Saturn. Beneath the clouds, whose tops we see, is an extremely deep atmosphere more than 30 000 kilometres thick, made up of hydrogen and helium. Numerous other elements are also present in very small quantities. Towards depths of 30 000 kilometres, at pressures of 2 to 4 million bars, the hydrogen becomes monatomic and metallic. Within the first 5000 kilometres of this metallic hydrogen, helium condenses in droplets which fall towards the centre of the planet. This process, which has been going on for perhaps two billion years, is responsible for the internal energy produced by Saturn. A deeper layer about 11 000 kilometres thick is thus permanently enriched in helium, which once again becomes uniformly mixed with hydrogen in this hotter zone. Underneath this is a solid core, which takes up one-quarter of the planet's radius, but only 1.2 per cent of its volume. The core consists of silicates, metallic oxides, and perhaps ices of water, ammonia and even methane.

atmosphere also contains one part in one to two thousand of methane. Methane is broken up by ultraviolet solar radiation and products of this break-up are also present in very small quantities, including acetylene (C_2H_2), ethane (C_2H_6), and probably propane (C_3H_8) and methylacetylene (C_3H_4). Other more complex molecules may also have been formed. Phosphine (PH_3) has been detected at a level of a few parts per million, up as far as the 5 to 10 millibar level (0.005 to 0.01 Earth atmospheres). The hydrocarbons formed in the stratosphere are thought not to be present in the troposphere (the layer below the tropopause), in contrast to phosphine, which comes from the interior of the planet.

Working inwards, the temperature decreases as far as the tropopause where it is only 85 K, and then increases again continuously as one descends deeper into the interior of the planet. Ammonia condenses at temperatures below 145 K and is found in the proportion of a few parts per ten thousand, below the 1 atmosphere level. The coloured clouds we observe are probably also below this level. We can gain information about the temperature of the deeper tropospheric levels because the radio waves emanating from Saturn originate there. At a wavelength of 21 centimetres, the emission is coming from the 10 to 20 atmosphere level where the temperature is of the order of 250 K.

At greater depths, Saturn's structure, like that of Jupiter, can only be deduced from theoretical models, which are constrained by three pieces of information: the value of the hydrogen–helium ratio in the outer atmosphere; the strength of the internal energy sources; the departures from symmetry of the gravitational field that the planet creates around itself. These three quantities have been measured precisely by the Voyager probes.

Measuring the gravitational field tells us about the distribution of mass in the planet's interior. Saturn is deduced to have a dense, solid core, principally composed of silicates and metals, and perhaps of water ice, ammonia ice and methane ice. However, this core must be small (about 15 000 kilometres in radius) and its mass cannot be more than 10 to 20 Earth masses.

The internal energy source is 1.76 times as intense as the solar radiation absorbed by the planet. One hypothesis postulates that this energy is a relic of the heat stored in the planet at the time of its formation. As in a heater that is cooling after the power supply has been switched off, energy flows from the centre of the planet to the surface, where it is converted into the observed radiation. But evolutionary models indicate that, because Saturn's mass is smaller than Jupiter's, it should have lost all of its initial heat several billion years ago. There is another, more plausible suggestion: at pressures of two or three million atmospheres, hydrogen atoms are broken up into their constituent protons and electrons, which considerably increases density and conductivity of the hydrogen so that it behaves essentially as a metal. Thermodynamic calculations tell us that helium is no longer soluble in metallic hydrogen if the temperature is low enough; instead, drops of liquid helium are formed and make their way towards the centre of the planet, releasing gravitational potential energy as they fall. This process could neatly account for the observed release of internal energy and, if it is true, we ought to see less helium in Saturn's outer atmosphere than in Jupiter's. As we can see from looking at the figures quoted earlier for the troposphere, this is exactly what is found by the Voyager experimenters: 7 per cent on Saturn versus 10 per cent on Jupiter. Jupiter's temperature is higher in the relevant region and so the metallic hydrogen and helium can coexist as a mixture. The formation of helium droplets within Jupiter will only take place when the planet has cooled down enough.

To summarise, travelling inwards towards the centre of the planet, one encounters successively:
– a layer about 30 000 kilometres thick basically made of 93 per cent molecular hydrogen and 7 per cent helium. All the other minor elements that made up the primordial solar nebula are probably also present in as yet unknown quantities at depths where the temperature is high enough;
– an inhomogeneous layer 5000 kilometres thick containing metallic hydrogen in the midst of which drops of helium continue to form and fall as 'rain' towards the centre of the planet;
– a layer from 10 000 to 12 000 kilometres thick of metallic hydrogen and helium, the latter in a higher proportion than in Jupiter or the Sun.
– finally, a core made of silicates, metals, and perhaps ices, about 15 000 kilometres in radius.

However, it must be kept in mind that this scenario may have to be drastically revised as our knowledge of this giant planet is increased.

Daniel GAUTIER

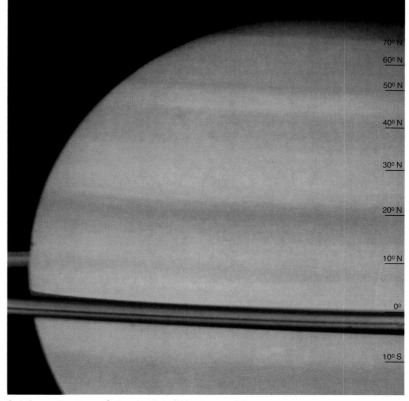

Subtle contrasts on Saturn's disk. This picture of Saturn's northern hemisphere was taken by Voyager 1 on 11 November 1980 from a distance of 1.75 million kilometres, and has a resolution of about 50 kilometres. A latitude scale is overprinted on the right. The rings are immediately below the equator. A bright spot is just visible below the rings, and a brown one can be seen at about 40° north. In spite of the lack of contrast, Saturn's atmosphere has complicated structures and in particular has many more alternating zones and belts than Jupiter. The north equatorial belt begins at about 20° north, and the very active north temperate belt begins at 40° north. (NASA)

The structure of the clouds. On 11 August 1981 Voyager 2 was only 13.9 million kilometres from the planet and two weeks away from its flyby (opposite left). The cloud formations at middle latitudes have become more visible. The ribbon-shaped formation in the white cloudy zone is a very fast stream at 47° north where the west-to-east winds reach speeds of 150 metres per second. Notice that the striped pattern extends all the way from one pole to the other. (JPL/NASA)

Night on Saturn. Voyager 2 flew by Saturn sixteen months after the vernal equinox, the beginning of spring in the planet's northern hemisphere. Because of this, the rings are now inclined to the planet–Sun direction and so are well lit. In contrast, Voyager 1's flyby was in November 1980, just after the equinox when the rings were edge-on to the Sun and so relatively dark.

On 6 November 1980, four days after closest approach, Voyager 1 passed 5.3 million kilometres 'behind' Saturn and transmitted this image of the night side of Saturn, forever invisible from Earth, with the shadow of the giant planet on the rings. The crescent Saturn shows through part of the rings, which themselves cast a shadow on the planet. (NASA/NSSDC)

Saturn

Atmospheric dynamics and clouds

At first sight, Saturn looks like a toned-down version of Jupiter. It has the same axial symmetry and alternating light and dark bands, albeit much less sharply contrasted. The colours of the patterns are less vivid, as if all the different clouds had been mixed together to produce various shades of yellow–brown. The atmospheric circulation is also very similar to that of Jupiter; there are many of the long-lived atmospheric phenomena known as ovals, and several atmospheric currents alternating in latitude between eastward and westward. However, the currents are faster on Saturn and the remarkably strong eastward equatorial current has no Jovian counterpart.

The chemical composition of the clouds is no better known than for Jupiter. The topmost cloud layer, which is probably no more than a mist, is almost certainly made up of ammonia crystals. Lower down may be a layer of ammonium hydrosulphide droplets (NH_4SH) or alternatively a mixture of ammonia and water. All of these chemicals are colourless, so the colouring of the clouds and spots must be due to other, unknown constituents. Just as in the Jovian case, the detection of phosphine (PH_3) in the troposphere leads us to suspect that phosphorus may be one of these constituents.

We do not know the altitude of the various cloud layers very well either. Ammonia condenses at temperatures below 145 K, so the base of the ammonia cloud layer must be somewhere around the 1 atmosphere pressure level. Infrared observations, showing absorption by clouds in the troposphere, above the 0.5 atmosphere level, then lead us to conclude that the ammonia cirrus clouds lie in the region between 0.5 and 1 atmosphere, which covers an altitude range of about 30 kilometres. Unfortunately we have no data that would let us determine the altitude of the coloured clouds which could be at the 2 or 3 atmosphere level.

We can study the global rotation of Saturn by observing radio emissions from the planet modulated by its magnetic field, which rotates with the core. Comparing this with the apparent motion of the clouds, the wind speeds on Saturn may be found. It turns out that eastward flowing currents have much higher speeds than westward ones and that, in general, the wind speeds are higher than on Jupiter. In particular, the equatorial current extending from 30° south to 30° north latitude flows from west to east at very high speeds, reaching 500 metres per second at 10° latitude, which is two-thirds of the local speed of sound. On Jupiter the highest speeds are only 150 metres per second and the currents correspond to the visible coloured bands, which is not the case on Saturn where there seems to be no correlation between the visible markings and the winds. There is, however, a good correlation between the speeds of the high-altitude clouds (above 0.3 atmosphere) and the wind speeds calculated theoretically from the thermal structure of the atmosphere under certain assumptions. This suggests that the thermal energy that comes out from the interior is transformed into energy of atmospheric motion when it reaches the upper

Spots and a ribbon. This picture of Saturn's northern hemisphere was taken by Voyager 2 on 19 August 1981 from 7.1 million kilometres, and has a resolution of about 125 kilometres. The colours have been altered to increase the contrast and bring out details. In particular, three yellow–green spots can be seen whose speed has been measured as a 15 metres per second westward drift. The westernmost spot shows anticyclonic motion, indicating that it is a high-pressure region. It is about 3000 kilometres across. The ribbon-shaped pattern in the north is a strong current where the wind speed reaches 150 metres per second. (JPL/NASA)

Flow patterns around a brown spot. The circulation patterns around a large brown spot in the atmosphere of Saturn can be clearly seen in these two pictures taken ten hours apart by Voyager 2 on 23 and 24 August 1981 at distances of 2.7 and 2.3 million kilometres, respectively. The resolution on the pictures is about 50 kilometres, and the one at top right. This huge spot, 5000 kilometres long, lying at 42.5° north latitude, is turning clockwise – that is, anticyclonically. This appears to indicate that the spot is in a region of high pressure. (JPL/NASA)

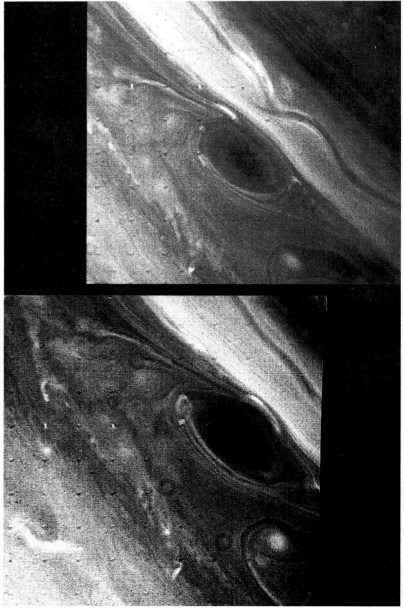

Wind speeds on Saturn and Jupiter. We know the speed of the planet's rotation by observing the rotation of its magnetic field, and, by comparing this with the speed at which the clouds appear to move, we can deduce the speed of the winds in the upper atmosphere which are responsible for the clouds' motion. Superimposed on the photograph at lower left is a graph of the wind velocities on Saturn as a function of latitude. Negative values of velocity correspond to currents travelling from east to west; we can see that almost all wind currents are west winds. These strong eastward currents reach almost 500 metres per second at the equator, which is two-thirds of the speed of sound on Saturn.

There does not seem to be any correlation between the wind pattern and the visible cloud belts, in contrast to the case for Jupiter (portrayed at lower right), where the light zones correspond to strong eastward currents. Jupiter and Saturn, it would appear, have qualitatively different atmospheric circulation. (After B. A. Smith, *et al.*, 1981)

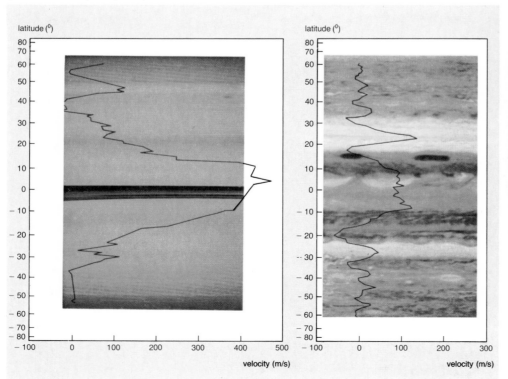

The northern hemisphere of Saturn. This picture of Saturn's northern hemisphere, taken by Voyager 2 on 15 August 1981 at a distance of 10.7 million kilometres (resolution 100 kilometres), shows that the structure of cloud bands extends into the polar regions. Note the ribbon or sinusoidal structure which extends over an enormous distance at a latitude of 47° N. This structure doubtless corresponds to a high-velocity westerly current, but no phenomenon of this type has ever been observed either on the Earth or on Jupiter. We also see two parallel dark bands to the north of this ribbon; these may be regions without clouds. (JPL/NASA)

The Great White Spot. This perturbation was discovered by amateur astronomers on 25 September 1990. It rapidly grew to a gigantic size, affecting virtually all equatorial regions of the planet as it rose through the atmosphere. This picture, by the Hubble Space Telescope, shows it on 9 November 1990, near its maximum size. These bright clouds are mainly composed of ammonia ice.

Perturbations of this type occur once about every 30 years in Saturn's atmosphere, i.e. with a period similar to that of the planet around the Sun, at an epoch coinciding with mid-summer in the northern hemisphere, suggesting a seasonal origin for them. Four other Great White Spots have been observed earlier: in 1976, 1903, 1933 and 1960; none of them reached the size of the 1990 event, however. (NASA)

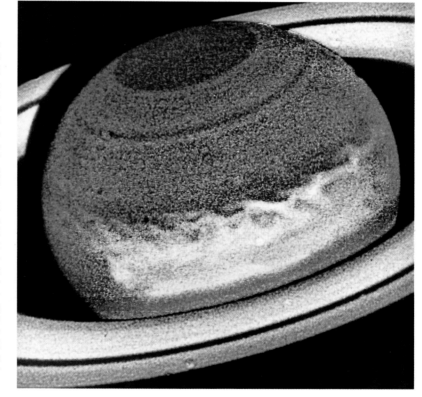

troposphere and this kinetic energy maintains the horizontal movement of the atmosphere – in other words, the winds. Below the 0.3 atmosphere level and, in particular, in the layers containing the coloured clouds, this correlation between wind speeds and thermal energy supply disappears. Chemical reactions are very important in these layers and in certain respects dominate dynamical and thermal effects.

Three kinds of very long-lived atmospheric phenomena that are important features of Saturn's meteorology. Firstly, the symmetrical ovals, which are stable and come in various colours. These include Big Bertha at 75° north, Brown spots 1, 2 and 3 at 42° north, UV Spot at 27° north, and Anne's Spot at 55° south. These ovals, which move rather slowly, appear to revolve like ball-bearings between adjacent eastward and westward currents. They are much bigger than terrestrial cyclones; for example, Brown Spot 1 is 5000 kilometres along its short axis. A second group are the 'convective features', similar to the Jovian white plumes, travelling with the westward jet at 39° north. The individual cloudlets are bright, white, irregular and short-lived formations. The physical origin of these features, and the reason for their presence at this latitude, is not well understood. A third atmospheric feature is the distinctive 'ribbon' at 47° north. This is a dark wavy line, flowing east with the surrounding current, with each crest and trough being about 5000 kilometres across in the east–west direction. On the north side of the ribbon – in the 'troughs' if you take north to be at the top – are cyclonic eddies made up of clouds, which spiral around in an anticlockwise direction. There are no comparable long-lived phenomena in the Earth's atmosphere.

The dynamics of the interactions between the ovals are also shrouded in mystery. On Jupiter, when two spots have been seen to meet each other they merge, but on Saturn, on at least one occasion, two spots have approached each other and then one spot has rolled around the other as if the wind systems of the two spots had acted on each other. This behaviour seems to exclude the theory that spots are examples of solitons, discrete travelling wave packets, which would pass through one another without affecting each other.

The dynamics of Saturn's atmosphere seem to be even more difficult to understand than those of Jupiter. The problems that face us can be divided into two basic types. The first, common to Jupiter and Saturn, concerns the modelling of the general atmospheric circulation and the explanation of the eddies and the various dynamical features that are observed. At present there are two models of the general atmospheric circulation in

vogue and there is, as yet, insufficient data to decide which theory is most plausible. These models were described in the section on Jupiter; here we will recall that Gareth Williams' model, in many ways similar to a model of the Earth's atmosphere, assumes that the lower atmosphere, below the depths where solar radiation can penetrate, has a negligible effect on the dynamics. Friedrich Busse's model, on the other hand, assumes that the observable currents in the upper atmosphere are the visible manifestations of a rotational structure involving the entire fluid interior of these planets.

The second problem is peculiar to Saturn and comes from the apparent absence of correlation between the appearance of the visible clouds and the observed winds at a given level and latitude. It will undoubtedly be necessary to analyse carefully the physical and chemical processes occurring in the clouds to resolve this problem. It is unlikely that we can do this without the help of *in situ* observations made from a probe descending through the atmosphere, at least as deep as the 10 atmosphere level. The Galileo probe will carry out this mission for Jupiter later this decade, and back-up hardware could be modified to perform the same task at Saturn in the 1990s if money were made available.

Daniel GAUTIER

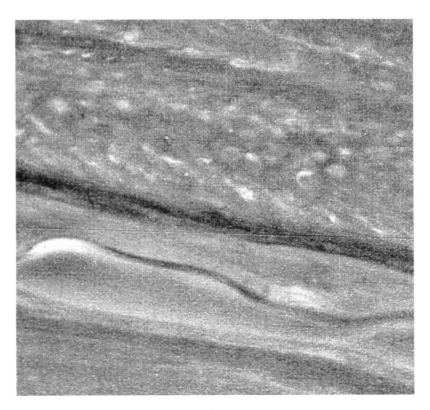

The north polar region. This high-resolution (60 kilometre) image of Saturn was obtained by Voyager 1 on 10 November 1980 from 23 million kilometres away. Studies of this image have revealed alternating zonal currents; west–east at 60° north, east–west at 55° north, then west–east again at 50° north.

Saturn

The magnetosphere and radio emissions

The electric currents flowing in the layer of metallic hydrogen in Saturn's interior produce an intense magnetic field whose properties have been determined by the Pioneer and Voyager flybys.

Saturn's magnetic field is a dipole in the first approximation. Its moment is 550 times greater than the dipole moment of the Earth and thus ten times less than that of Jupiter. Since Saturn is much bigger than the Earth, the magnetic field intensity at the height of the cloud layers is actually slightly less than that at the Earth's surface. As on Jupiter, the north magnetic pole is in the northern hemisphere, whereas on Earth it is in the southern hemisphere. There is one difference from the Jovian and terrestrial fields which is of great importance for theorists: the axis of the magnetic dipole and the axis of rotation of the planet coincide to within a degree; on Jupiter and Earth these axes point in directions 10 degrees apart. Theories of magnetic field generation must take this into account and explain the lack of misalignment on Saturn.

Saturn has an ionosphere which was detected by the distortions it caused in the telemetry signals from the Voyager and Pioneer probes. This ionosphere extends several thousand kilometres above the cloud layers. It is at a temperature of about 1000 K, is principally made up of ionised hydrogen atoms, and is slightly less dense than the terrestrial ionosphere.

Approaching Saturn, the Voyager probes detected an intense source of radio static with their radio astronomy receivers. This emission occurs for the same reasons as the low-frequency radiation from Jupiter; electrons fall down along the field lines in high-latitude regions. They emit radiation at the gyrofrequency which is proportional to the field intensity, so the Saturnian radiation is at slightly lower frequencies than terrestrial kilometric radiation, between 50 kilohertz and 1 megahertz. This intense emission comes from a source with a power of about 10^{10} watts at latitude 80° north in the north polar region on the sunward side of the planet, and from a similar but weaker source in the southern hemisphere. The sources, while fixed in position with respect to the Sun, are strongly modulated in intensity by the planet's rotation, which is puzzling given that the magnetic dipole is practically aligned with the rotation axis. The radio emissions repeat with a period of 10 hours 39 minutes and 24 seconds, which is assumed to be the period of the magnetic field and thus the rotation period of the inner regions of Saturn where the field is produced. The modulation reveals the existence of magnetic anomalies near the surface and localised in longitude, which could not be detected by the magnetometers on the flyby probes. The origin of the electrons that produce the radio emission is not yet clear. It could be that they are electrons from the solar wind finding their way into the magnetosphere, or electrons accelerated within the magnetosphere itself in magnetic storms of the kind occurring on Earth. The radio emissions also seem to be correlated with the appearance of polar aurorae, which can be detected by ultraviolet observations, and with the pressure of the solar wind.

André BOISCHOT

Particles in the rings. The above view from Voyager 1 (12 November 1980; 720 000 km) shows from top to bottom the thin, bright, F ring, A ring and Encke's division, the B ring and the dark C ring. (NASA)

The diagram below shows the density of particles detected by Pioneer 11 (September 1979). They disappear at the A ring.

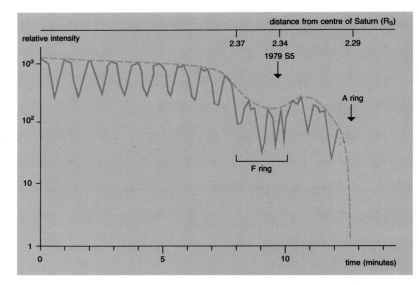

Particles in the magnetosphere. The rings and satellites are not the only matter in the inner magnetosphere of Saturn. A neutral hydrogen cloud has been detected around the planet, with a density of a few particles per cubic centimetre. Between the orbits of Dione and Enceladus is a torus of gas, mainly consisting of oxygen ions (O⁺) which have probably been torn from the surfaces of Dione and Tethys. Beyond this there is a much more tenuous torus which extends beyond Titan's orbit, made of neutral and ionised hydrogen, and more energetic deuterium and tritium ions coming from Titan's atmosphere and Saturn's ionosphere.

The zones of trapped particles are very different from those of Earth and Jupiter. In principle,

Saturn's magnetic field would be strong enough to trap a large population of charged particles – electrons and ions – in the inner magnetosphere, which could produce enough synchrotron radiation to be detectable from Earth. However, the rings prevent this trapping from occurring; as they drift in longitude, the electrons and ions will collide with dust in the rings and lose their charge. So there cannot be any trapped particles within the outer edge of the rings, which is why radio astronomers have failed to detect synchrotron radiation from Saturn. For the same reason, the presence of a satellite corresponds to a distinct drop in the number of particles trapped on lines of force passing through its orbit.

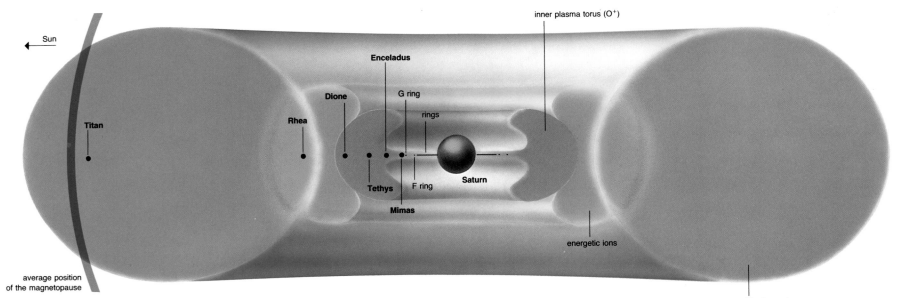

Saturn's magnetosphere. The magnetosphere is well developed, intermediate between that of Jupiter and that of the Earth. It has a bow shock, a magnetosheath, a magnetopause, and an elongated tail in the antisolar direction. The shape is very regular due to the alignment of the magnetic and rotational poles of the planet. The position of the bow shock varies between 18 and 25 Saturn radii, that is, 1 to 1.5 million kilometres, in the direction of the Sun, as the solar wind pressure varies. Because of the planet's rapid rotation, a disk of current is formed in the equatorial plane, which modifies the magnetic field in the outer magnetosphere. The inner magnetosphere is without stable trapped radiation zones due to the presence of the rings which absorb the charged particles.

On the diagram the Earth's magnetosphere is shown to the same scale.

At large distances from Saturn, the dipole field (parallel to the rotation axis) is deformed by the solar wind, which compresses it on the sunward side of the planet and stretches it into a long tail on the night side. The size of the charged particle population (electrons, protons and ions) is between that of Earth and Jupiter. The outer boundaries of the magnetosphere, the bow shock and the magnetopause are fairly unstable. They reflect the rapid pressure variations in the solar wind, and because of this the Pioneer and Voyager probes crossed them several times while they approached the planet. For example, in the three days between the arrival and departure of Voyager 2, the size of the magnetosphere had increased by more than 70 per cent.

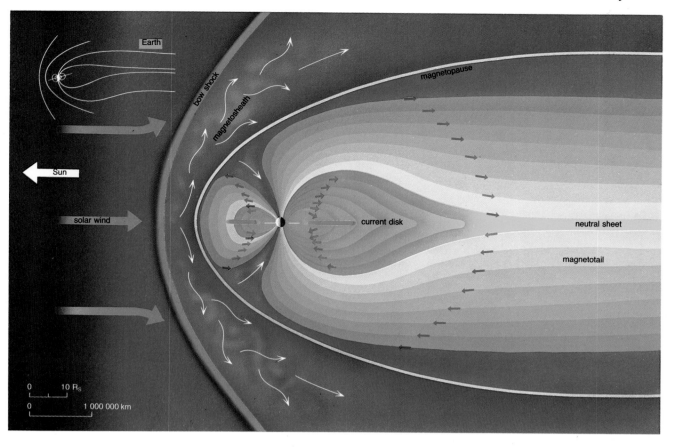

Time variation of the spectrum of radio static from Saturn. The radio noise from Saturn occurs at kilometre wavelengths, so it cannot be observed by terrestrial radio telescopes. The diagram shows a recording from Voyager 1's radio astronomy receiver made as the probe pased the planet on 13 November 1980. The variations in intensity (the darkest regions corresponding to the strongest emission) are shown as a function of frequency and time. Note the arc-shaped structure which is also present in dynamic spectra of Jovian radio emission.

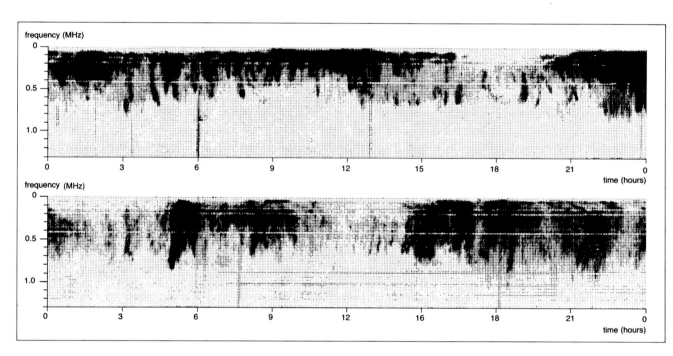

Measuring Saturn's rotation period. A spectral analysis of observations of Saturnian kilometre radiation at 174 kilometres over a period of 367 days has allowed a precise determination of the period with which the strength of the emission varies. The narrow peak in the diagram shows that this period is very stable at 10 hours 39 minutes 24 seconds ±7 seconds. It is noticeably different from the rotation period deduced from the movements of the clouds. It is accepted that the variations in the radio emission are due to effects of the magnetic field on the source, varying with the planet's rotation. Hence the period of the radio emission corresponds to the rotation of the magnetic field, and thus ultimately to the rotation of the planet's interior, where this field originates.

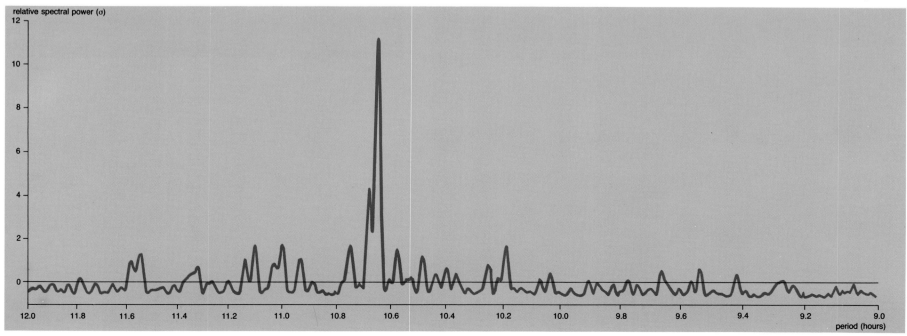

Saturn

The rings

First observed by Galileo in 1610, the rings of Saturn are probably one of the most beautiful sights to be seen in the sky with the aid of a simple pair of binoculars. The flybys of the rings by the Voyager probes in November 1980 and August 1981 revealed a magnificent system of rings made of countless billions of 'pebbles' in orbit around Saturn, forming thousands of astounding patterns. The ring system is not only one of the most striking objects in the sky, but also one of the most scientifically interesting, and currently one of the least well understood.

During the summer of 1610, Galileo, who was one of the first people to use a telescope to observe the sky, reaped a harvest of discoveries. In particular, he discovered 'something around Saturn'. At first he believed that he had found two big satellites on either side of the planet, but he noticed that these two companions of Saturn failed to show any movement with respect to the planet, which greatly intrigued him. He was even more amazed when, two years later, he noticed that these two companions had apparently disappeared. Astronomers were puzzled by the changing appearance of the planet for the next forty years. Some people saw two satellites, others a flattened planet, still others complex structures, and rival observers proclaimed the quality of their instruments and cast aspersions on the eyesight of their colleagues. It was only in 1655 that Christiaan Huygens found the solution: Saturn is surrounded by a bright ring lying in the planet's equatorial plane; in the course of Saturn's twenty-nine year revolution around the Sun, the rings are seen sometimes edge-on, and sometimes at a more oblique angle, whence their changing appearance in the crude telescopes of the day, which were much inferior to a modern pair of binoculars.

Giovanni Domenico Cassini, the first director of the then newly established Paris Observatory, discovered a gap in the rings which now bears his name, thereby showing that the rings were not homogeneous. He suggested that they were made of a large number of small satellites. Despite this, many seventeenth and eighteenth-century astronomers believed that the rings were solid, and it was not until 1785 that Pierre Simon de Laplace proved that a solid ring would be unstable and would be destroyed by the planet's tides. Laplace then suggested that the rings were made up of a succession of thin concentric solid rings. In 1857, James Clerk Maxwell proved theoretically that the rings were made of independent solid particles in differential rotation around the planet. In 1895, James Edward Keeler obtained a spectrum of Saturn and its rings and showed, by measuring the radial velocity of the rings by the Doppler effect, that the rings were indeed in differential rotation around Saturn, just as would be expected for a large collection of small independent satellites obeying Kepler's laws. The particles closest to Saturn revolve around the planet in less than 8 hours (more rapidly than the planet itself) and the ones farthest out take more than 12 hours. Maxwell's theoretical calculation was thereby confirmed. In 1911, Henri Poincaré underlined the importance of the mutual collisions of the particles in the rings and noted that the kind of collision phenomena now going on in Saturn's rings must have played a crucial role in the early stages of the Solar System. It was not until the 1970s and 1980s that quantitative theoretical studies of the role of collisions in such a ring were undertaken. It is interesting that, from Galileo to Poincaré via Huygens, Laplace and Maxwell, some of the greatest names in physics have been associated with the study of Saturn's

The night side of the rings of Saturn. The rings of Saturn are made of countless small particles covered in ice which continually undergo collisions as they move around Saturn, and which are contained in a disk whose thickness is less than one kilometre. Seen from Earth, the rings are lit by the Sun and appear very bright. These two images show an unusual view of the rings; the night side, photographed by Voyager 1 on 12 November 1980 from 740 000 kilometres (top) and by Voyager 2 on 29 August 1981 from 3.4 million kilometres (bottom). The false-colour picture at the top was produced by combining black and white photographs taken through filters; the processing reveals details which would otherwise be invisible. The B ring, normally so bright, appears here to be very dark; sunlight has difficulty in passing through this dense ring. These pictures are a bit like a negative of the usual pictures of the rings. The Voyagers were not only able to observe the rings at a resolution which was thousands of times higher than that possible from Earth, but also to see them from all angles. The relative amounts of light reflected, transmitted or scattered obviously depends on the size and optical properties of the ring particles. Comparing images taken from different angles allows us to reconstruct the size and properties of the particles. This cannot be done from Earth since the angle at which we see the rings varies little. (JPL/NASA)

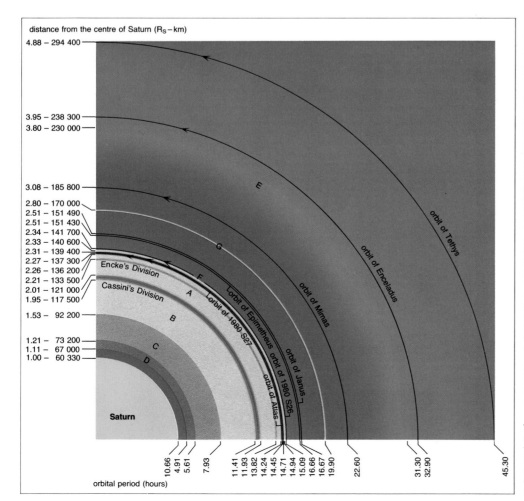

The rings of Saturn. Starting at about 7000 kilometres above the cloud tops, the main rings extend over more than 74 000 kilometres. Going outwards from Saturn, we come to the D ring, very tenuous; the C ring, just visible from Earth; the B ring, very bright; Cassini's division; the A ring, also bright; the F ring; the G ring; and the E ring which stretches well beyond the Roche limit out to Rhea's orbit at more than 9 Saturn radii. The diagram gives distances in both Saturn radii (R_S) and kilometres, and orbital periods in hours.

rings. The discovery of the rings of Uranus (1977), and of Jupiter (1979), and the flybys of Saturn's rings by space probes (1979–81), which revealed surprising structures in these rings, have led to a new surge of interest in the physics of planetary rings.

Why are there rings around the giant planets? Why is the study of the rings so important for astrophysics? Only a few years ago we wondered why Saturn was the only planet with rings; now we find the presence of rings around three giant planets. Very close to a planet, tidal forces break up all large bodies into little pebbles, and the interplay of mutual collisions between these tiny bodies leads to the formation of a disk in the planet's equatorial plane.

Consider a satellite orbiting a planet. Every particle making up the satellite is subjected to a gravitational pull from the planet. This pull depends on the distance of the particle from the centre of the planet; in fact it is inversely proportional to the square of that distance. But since two neighbouring particles are at very slightly different distances from the planet, the force they each feel is very slightly different. The differential force – a 'tidal' force – tends to push the particles apart. Other forces hold them together: the mutual gravitational force between the particles making up the satellite, and the structural strength of the solid satellite. Comparing the relative sizes of the disruptive and cohesive forces for a satellite in orbit around a given planet defines a distance range within which any large body will be torn apart into smaller pieces. Beyond this distance, a satellite can survive. If the Moon, for example, were less than 18 000 kilometres from the Earth instead of being at 400 000 kilometres it would be broken into pieces about 200 kilometres across. No large satellite can orbit Saturn at less than 140 000 kilometres from its centre. The distance range within which large satellites cannot survive is called the 'Roche limit' after the French mathematician Edouard Roche who first derived the mathematical expression for it in 1850.

Within the Roche limit, then, we expect large numbers of small bodies instead of a few large satellites. These small bodies will collide with each other as they orbit the planet. In these often violent collisions the bodies lose energy and this quickly leads to the formation of a thin disk in the equatorial plane of the planet. Calculations indicate that a spherical cloud of particles around Saturn would turn into a magnificent ring after less than a year of such collisions. Obviously, this is a very much shorter time than the age of the Solar System.

The rings of Saturn are extremely useful to astronomers, who use them as a nearby laboratory for studying processes which, although very widespread in the Universe, are difficult to study due to the immense distances of the objects in which they take place. Many celestial objects, although very different from Saturn in nature and size, have the same basic form of a disk around a central body or concentration of matter. The rings of Saturn, studied for centuries, are the nearest such disk system to us (excluding the rings of Jupiter which are poorly known and less rich in material) and should give us important information on the dynamics of flattened systems in general, which can be applied to much more inaccessible systems like spiral galaxies, accretion disks around black holes or neutron stars, or the protosolar nebula just before the formation of the planets.

Seen from the Earth, the rings of Saturn appear to us as a series of concentric zones of different brightness separated by dark divisions. Going outwards from the innermost ring, we first encounter the very faint D ring which starts only a few thousand kilometres above the cloud tops (which are 60 000 kilometres from the centre of Saturn), then the C ring, known as the crepe ring because of its transparency. It is about 20 000 kilometres wide. This is followed by the brightest ring, the B ring, which stretches over 25 000 kilometres. The B ring is separated from the A ring by Cassini's Division, 5000 kilometres wide. The A ring reaches out a further 15 000 to 136 000 kilometres from Saturn; this is the edge of the rings as seen from Earth. Two thin rings, F and G, are at 140 000 and 170 000 kilometres from the centre of Saturn, respectively. A very diffuse ring, the E ring, goes out to more than 550 000 kilometres, well beyond the Roche limit. The E ring can only be seen on Earth when the rings are seen edge on. As seen from Earth, Saturn

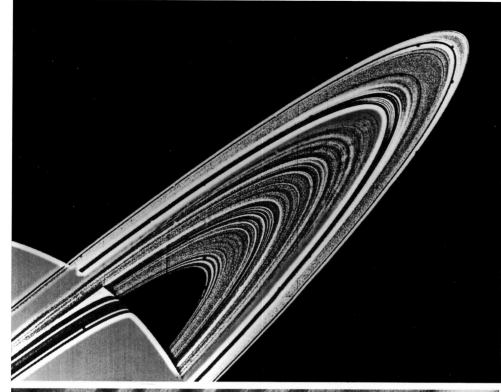

The grooved rings. While Saturn's rings look relatively homogeneous seen from the Earth, Voyager pictures reveal a remarkably complex environment. On the top picture, taken by Voyager on 6 November 1980 while it was more than 8 million kilometres away, the rings of Saturn look like a long-playing record with hundreds of grooves. Ninety-five concentric rings can be seen on this picture but more than 10 000 concentric structures appear at higher resolution. For example, the lower picture represents a small part of the B ring. It was taken on 25 August 1981 by the Voyager 2 probe from a distance of 743 000 kilometres. It shows about 6000 kilometres of the ring and enables details about 10 kilometres across to be distinguished. The very thinnest rings of the top picture resolve into even finer structures, which can themselves be split up into even thinner rings on images taken at even higher resolution. The alternation of light and dark zones across the rings is not simply related to the presence or absence of a ring of matter, but is determined by the mix of rings with dark particles or light particles, by spiral waves and undulations on the surface of the rings, as well as by real gaps in the rings.

'Ringlight' at night on Saturn. This picture, which was taken by Voyager 1 on 13 November 1980 as it receded from the planet at a distance of 1.5 million kilometres, shows the rings and their shadow on the planet. At lower left, the bright limb of the planet is clearly visible through the rings; in particular, Cassini's Division and the C ring are transparent, while the outer part of the B ring is opaque, which shows that it contains a much higher density of matter. In the foreground the F ring can be seen and Encke's Division can clearly be seen in the outer part of the A ring. At top left, from bottom to top, can be seen the B ring, light scattered by the C ring, the night side of Saturn, and a bright area which is illuminated by sunlight scattered and reflected by the rings just as moonlight lights terrestrial nights.

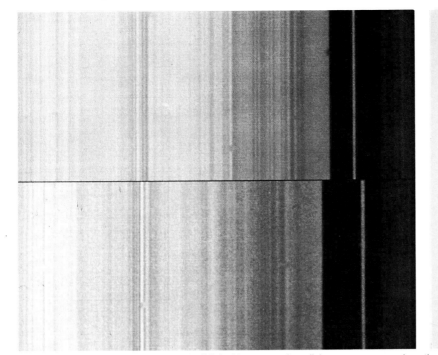

Elliptical and irregular rings. Not only did the Voyager probes discover many more rings than expected, they also revealed an extremely complex small-scale structure with rings which are elliptical, twisted or irregular. As a whole, the rings are a very stable dynamical structure, but on the small scale they undergo variations and oscillations all the time. The spectacular composite picture above, at left, shows views of the region where the B ring ends and Cassini's Division begins, taken at diametrically opposed points on the rings on 25 August 1981 by Voyager 2 from a distance of 610 000 kilometres. While most of the structure lines up well, there are many differences in the position of the individual fine details, and the edge of the ring differs by 50 kilometres in the two positions. The distorted form of this edge is thought to be due to the satellite Mimas.

The picture at top right was taken by Voyager 2 with a resolution of about 15 kilometres on 25 August 1981. It shows the discovery of a small thin ring in the middle of Encke's Division, whose wiggly, intermittent form oscillates from one side of the division to the other. The ringlet appears in the middle of Encke's Division at far right; in the middle picture, corresponding to a point further along the ring, it is near to the inner edge. The picture below left, shows the structure of the small ring with a resolution of the order of 100 metres, as deduced from the δ Scorpii occultation. As the rings passed in front of δ Scorpii, as seen from Voyager 2, the variation in the star's light allowed scientists to probe the radial structure of the rings with a resolution which is ten times better than could be obtained with the best photographs. Here we can see a very complicated structure of concentric rings within Encke's Division. The other photographs all show the F ring; as seen by Voyager 1 on 12 November 1980 (below right) and Voyager 2 on 15 August 1981 (bottom left), and as reconstructed from the occultation experiment (bottom right). The F ring was discovered on 1 September 1979 by Pioneer 11, and shows braiding, oscillations, clumps of matter and splitting. It seems to be confined by two guardian or 'shepherding' satellites which are visible on the Voyager 2 picture. (JPL/NASA and NASA/NSSDC)

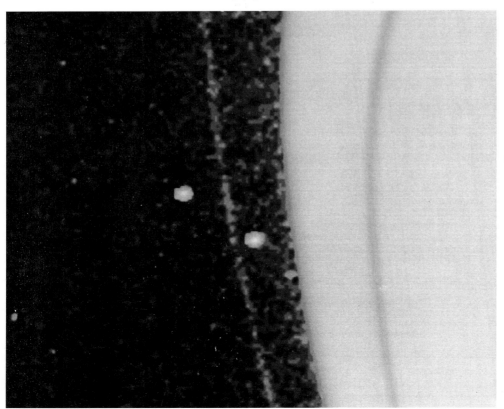

and its rings cover 48 arc seconds on the sky, and it is difficult to distinguish any structure on scales less than several thousand kilometres because of the turbulence of the Earth's atmosphere.

The ring particles orbit around Saturn according to Kepler's laws, the innermost ones having a period of revolution of 7 hours and 46 minutes, and the outermost ones 14 hours 27 minutes. While studying the rings in visible light reveals zones that are darker or lighter according to the amount of material present, observations in the infrared show the presence of absorption lines characteristic of water – so the particles are covered with frost mixed with some impurities. The rings are a very weak source of radio waves; in contrast, they reflect radar at 3 and 12 centimetre wavelengths very well, which implies that many of the particles are at least tens of centimetres in size.

During the flybys of Saturn by the Voyager probes, the main surprises came from observing the rings; the Voyager 2 spacecraft was in fact reprogrammed to give a higher priority to observing them after the Voyager 1 flyby in 1980. Despite their appearance from the Earth, the rings are not made up of large, relatively homogeneous zones, but are in fact composed of thousands of thin concentric rings giving the whole system the look of a gramophone record. The finest details that Voyager 1 discovered were observed at higher resolution by Voyager 2 and turned out to be subdivided into even finer structures. The traditional image of a ring was that of a set of particles in orbit around a planet undergoing mutual collisions and making up a structure with perfect circular symmetry and ill-defined edges, but the Voyager probes revealed ringlets, which were non-circular, eccentric, twisted, wavy, or clumpy, as well as ringlets with sharp edges from which waves were propagating in the ring. All these structures pose a challenge to physicists to construct new theories which then may be applied to other astronomical objects.

Voyager 1 discovered radial structures like immense fingerprints 20 000 kilometres long across the 'grooves' of the 'record'. Voyager 2 saw these fingerprints form in less than ten minutes. They are probably due to magnetic storms in the rings – small particles about a micrometre in size are lifted up by the action of Saturn's magnetic field and form blisters above and below the plane of the rings.

As Voyager 2 crossed the ring plane at about 2.8 Saturn radii, it detected more than ten thousand impacts by small micrometre-sized particles in the five minutes it took to cross a region 2000 kilometres thick either side of the ring plane, and the steering mechanism of the scan platform broke down at that time. The environment of the rings seems to be very hostile.

The fact that the rings have very sharp edges rather than diffuse ones, as well as the presence of thousands of narrow rings cleanly separated from each other, indicates that some mechanism that confines the particles is at work in the rings. It has been suggested that the sources of this confinement are small satellites or large particles (about a kilometre in size) which have created dark divisions free of matter and repel neighbouring ringlets. The theory of the interactions of a satellite and a ring of small particles undergoing mutual collisions shows that the two repel each other. Because of collisions, an attractive force like gravitation can lead in particular cases to a repulsive effect!

The ring plane is seen edge-on by terrestrial observers once every fourteen years. The occurrence in March 1980 prompted many observations with big telescopes and modern detectors, which gave an indirect measurement of the thickness of the rings of about one kilometre. This measurement is affected by diffuse light, the satellites, the largest particles, any possible warping of the rings, etc. and it is probable that the thickness is locally even less. It can be estimated from Voyager measurements at a few tens of metres. At over 300 000 kilometres across and less than one kilometre thick, the rings of Saturn are the thinnest disk relative to their diameter yet found in the Universe. A razor blade would have to be less than a micrometre thick to compete with them! It is quite possible that a very tenuous halo of small particles with a much larger thickness surrounds the rings.

Studies of the colour of the rings shows that, although they are all essentially covered with ice and frost, the rings at different distances from Saturn have different properties. The Voyager

The rings in colour. These false-colour pictures were prepared from three black-and-white photographs taken through ultraviolet, green and orange filters. The contrast has been highly exaggerated. For example, the C ring is here shown as blue, when it is really a grey colour much like that of dirty ice. These false-colour representations allow us to distinguish quite distinct zones within the rings; the C ring and Cassini's Division are bluer than the A and B rings, and there are three thin yellow rings within the blue rings that make up the C ring (lower image). These colour differences correspond to different chemical compositions of the surfaces of the ring particles. it was already known from Earth-based observations that the ring particles were covered in ice; it seems that other elements are mixed in with the ice in different proportions in the different zones of the rings. The origin of this variation is not yet understood. The top was obtained on 17 August 1981 by Voyager 2 from 8.9 million kilometres; the lower on 23 August 1981 from 2.7 million kilometres. (JPL/NASA)

Saturn seen through the rings (below). This high-resolution Voyager 2 image of the rings was taken on 23 August 1981 while the probe was 2.3 million kilometres from the planet. In the foreground is the A ring, then above that Cassini's Division, darker than its surroundings, going diagonally from bottom centre to top right, and the B ring. At top right the planet and the shadow of the rings on the planet can be seen through the rings. The B ring can be seen to be much more opaque in its outer regions. The succession of bright and dark regions corresponds in various cases to the absence or presence of matter, to brighter or fainter particles, or to small ripples in the ring. (Illuminated at grazing incidence by the Sun, these ripples will produce a pattern of lighted regions interspersed with regions cast in shadow.) Such pictures, compared with those of the rings seen at other angles of illumination, and to optical and radio occultation experiments, allow astronomers to deduce the density of matter in the rings and the nature and size of the particles. It will take many years to complete the analysis and interpretation of all the data that have been gathered. (JPL/NASA)

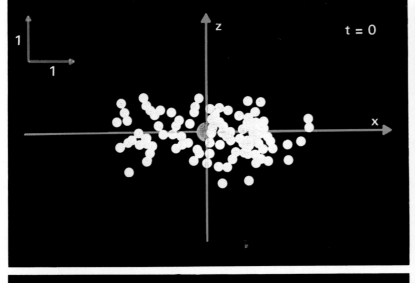

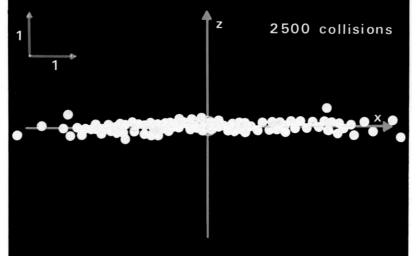

Why are there rings around Saturn? The detailed structure of Saturn's rings is far from being completely understood, and before it is there will have to be a detailed study of the interaction between the satellites and the rings, of the propagation of waves in the rings, of the development of instabilities in a disk, and of the role played by collisions. However, the presence of rings around a giant planet now seems natural, especially since the discovery of rings around Jupiter and Uranus. The presence of such rings is a natural consequence of the differential attraction by the planet (near Saturn, tidal stresses do not allow a large satellite to survive – it is either torn in pieces or cannot form in the first place) and inelastic collisions between the particles; the loss of energy due to collisions of a system of particles whose momentum is conserved leads to rapid flattening of the system and to the formation of a disk in the equatorial plane of the planet. The drawings above represent the results of a numerical experiment carried out by Andre Brahic (the x axis is in the planet's equatorial plane and the z axis is the rotation axis of Saturn and its rings). The top diagram shows the initial conditions and the lower diagrams shows the final flattened state.

Faint rings. There is matter in the equatorial plane of Saturn going right down to the cloud tops. This is the D ring as recorded by Voyager 1 on 13 November 1980 from a distance of 250 000 kilometres. The very tenuous D ring seems to be composed of separate bands in which the material is concentrated. Saturn's limb is at top right and the inner edge of the C ring at lower left. The shadow of the planet interrupts the picture of the rings at top left. (NASA/NSSDC)

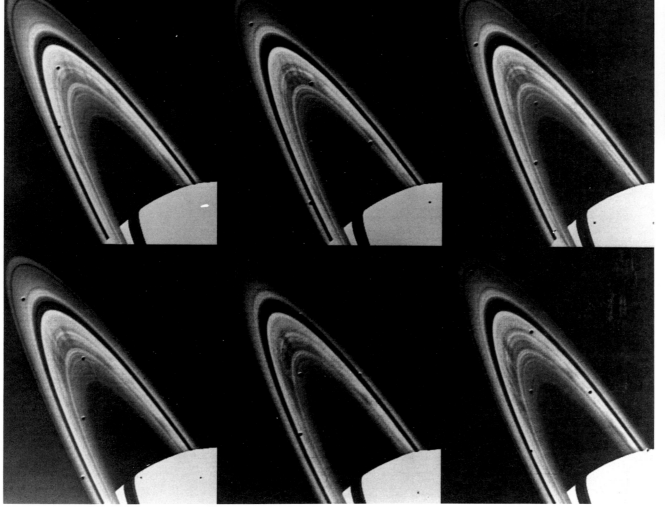

The 'spokes' in the B rings. A series of pictures taken on 25 October 1980 at a distance of 24 million kilometres by Voyager 1 shows radial structures which have been compared to giant fingerprints on the grooved 'long-playing record' of the rings, or to the spokes of a wheel. The spokes follow the rotation of the rings. Their evolution in time can be followed on the pictures; they are more than 20 000 kilometres long, have sharp edges less than 60 kilometres wide, appear in less than ten minutes and last one or two rotations, less than one terrestrial day. They continually form, disintegrate and reform; the formation preferentially occurs as the particles leave the planet's shadow, and the spokes are essentially at the synchronous altitude, where the Keplerian orbital rotation period is equal to the rotation period of Saturn's magnetic field. When they form, the spokes are radial, but as shown on the diagram above, they are subsequently slowly drawn out by differential rotation. (JPL/NASA)

probe transmitted a radio signal to Earth through the rings; analysis of the effect of the rings on this signal gives information on the nature of the ring particles, as does the study of Voyager observations of the star δ Scorpii seen through the rings. Voyager also observed the rings from all angles before and after the near encounter with Saturn, comparing reflected, scattered and transmitted light. From a detailed comparison of the 'difficulty' that radio waves and light rays have in passing through the rings, invaluable information can be deduced about the number, density and size of the particles. It will take years to complete the analysis of these observations but the first results show that there are bodies of all sizes from a micrometre to a kilometre in the rings.

The smaller particles, less than a centimetre across, would be fairly quickly removed from the Saturnian environment by the pressure of solar radiation causing them to lose speed and fall onto the planet. The presence of many small particles in the rings must therefore mean that they have been supplied to the rings comparatively recently – well after the formation of Saturn – and that there is still some source of small particles. This is most likely to be fragmentation from larger particles or satellites due to collisions. The total mass of Saturn's rings is about that of a typical satellite; if the material could condense despite the Roche limit, it would form a moon comparable to Dione or Enceladus.

The origin of the rings is of course unknown. We still do not know if they are made of primordial material which, present at the epoch of Saturn's formation, was never able to collect into a moon because it was within the Roche limit, or if they are the remains of a body that was captured by Saturn at some later time and broken up by the tidal forces within the Roche limit. Astronomers at present favour the former hypothesis, but even if the rings are recent, they are very interesting for cosmogony because the protosolar nebula once passed through a phase where particles of different sizes orbited the Sun, colliding with each other, before collecting into planets, and where confinement mechanisms also played an important role in the formation of the planets and the satellites.

The interplay between the mutual gravitational attraction of Saturn, its satellites and the ring particles, the interparticle collisions and the influence of Saturn's magnetic field and solar radiation on the smaller ring particles, make the rings of Saturn a wonderful natural laboratory for the study of many astronomical processes which will be of absorbing interest to physicists and astronomers for a long time to come.

André BRAHIC

The 'spokes' by reflected and scattered light. The spokes appear dark in reflected light and bright in scattered light (opposite and below, respectively.) This indicates that they are made of small particles comparable in size to the wavelength of light (about 1 micrometre). Such particles reflect light badly, but appear very bright when observed in scattered light. It is thought that the spokes are 'blisters' above and below the ring plane. They are probably created by the magnetic field, a by-product of magnetic storms in the ring plane which can lead to this phenomenon of magnetic 'levitation' of small particles. (NASA/NSSDC)

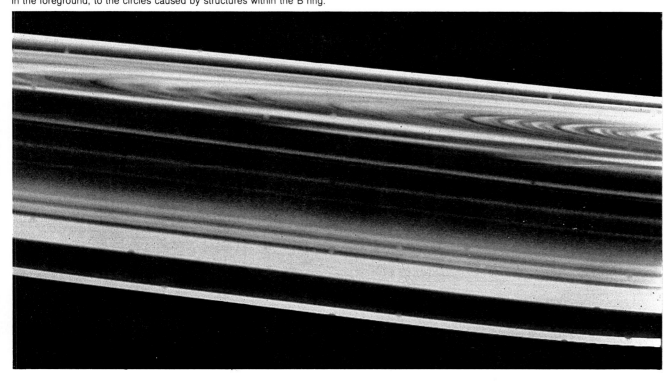

Voyager 2 passes through the ring planet. On the evening of 26 August 1981, in Saturn's shadow, out of sight of Earth, the Voyager 2 space probe crossed the ring plane at 2.86 Saturn radii out, near the G ring. This striking image, taken at that moment, was stored on board and relayed to Earth the following day. It shows the entire ring system a few seconds before the passage through the plane, from the F ring in the foreground, to the circles caused by structures within the B ring.

The bright streaks above the B ring are radial structures lying above the ring plane which appear very bright in scattered light. This was the last picture to be taken before the breakdown of Voyager 2's scan platform a few seconds later. This breakdown seems to have been due not to the ring plane passage but to a fault in lubrication, and it did not affect the observations of Uranus in 1986. (NASA)

Saturn

The satellites

With its array of rings and retinue of satellites, the Saturn system is in some ways like a miniature Solar System. The existence of only ten satellites had been established before the Pioneer encounter in 1979, but we now know twenty-one. Observations from both Earth and space probes led to the discovery of a further eleven satellites by 1981. Before 1979, these bodies only appeared as tiny points of light, which were very difficult to observe. Our knowledge of them has been revolutionised by the Voyager probes' observations; these moons are now real worlds to us, the biggest, Titan, being half as big again as our Moon. The Moon was the only satellite seen with good resolution in the 1960s but the Saturnian satellites have now been studied at a resolution comparable to that achieved for the Moon with the best Earth-based telescopes. These satellites teach us much about the behaviour of matter in the outer Solar System and the conditions in which the protosolar nebula formed.

Besides the enormous Titan, 5150 kilometres in diameter, there are six satellites with diameters between that of the Jovian satellite Amalthea and that of the Moon; they are between 400 and 1500 kilometres in diameter with surfaces essentially made of water ice. Bodies of this size have not been studied before, and their detailed analysis will fill a gap in our knowledge. There are also fourteen smaller satellites, mostly irregularly shaped, and they could all be either captured asteroids or fragments remaining after inter-satellite collisions at the beginning of the evolution of the Saturnian system. All these satellites are heavily cratered but, surprisingly, it is easy to distinguish them by their appearance; they are all different and have each had a unique history, despite expectations that they would be all much the same. Three of them are truly exceptional: Titan, with its nitrogen atmosphere; Enceladus, which shows evidence of extensive geological activity; and two-faced Iapetus, one side bright and the other dark. Expectations of what the satellites would be like were governed by two things. First, the more massive a body, the greater is its internal temperature and pressure and the more pronounced is its internal and surface geological activity such as volcanism and faulting. Second, all bodies in the Solar System underwent an intense bombardment during the 4.6 billion year history of the Solar System, principally at the beginning, when the planets were still forming and many small objects were wandering 'freely' through the Solar System. The legacy of this bombardment is a dense covering of craters, which has only been partially or completely erased when geological activity has arisen. It was thus expected that the Saturnian satellites would be saturated with craters, with no trace of geological activity because of their small size, except perhaps for the very biggest ones. As often happens, reality confounded the predictions. Just as with Jupiter's satellites, Saturn's have revealed themselves to be much more complicated and interesting than we had anticipated.

The six icy worlds, which are, in order of distance from the planet, Mimas, Enceladus, Tethys, Dione, Rhea and Iapetus, move in circular orbits, the first five lying in the planet's equatorial plane at less than 530 000 kilometres from Saturn, while Iapetus has an orbit with an inclination of 14.7 degrees at more than 3.5 million kilometres from Saturn. These satellites fall into three groups according to their size: Mimas and Enceladus, about 400 and 500 kilometres in diameter, respectively; Tethys and Dione, about 1000 kilometres across; and Rhea and Iapetus, which are about 1500 kilometres across. Their densities are between 1.0 and 1.45

times that of water, which correspond to those of bodies half rocky and half ices (of water or other compounds). In spite of its small size, Enceladus is geologically active; it is not clear at present where its internal source of energy comes from, it being too small to retain its initial heat of formation and not massive enough for internal heat to produce a molten core. Further proof that Enceladus is geologically active is that it appears to lose material to the E ring – the density of the ring is greatest in the neighbourhood of the satellite's orbit. Dione and Tethys retain some traces of geological activity post-dating the intense bombardment phase that followed their formation. Rhea and Mimas are heavily cratered and seem to have the most ancient surfaces of any of Saturn's satellites. Iapetus, with its dark and bright hemispheres, is one of the most enigmatic objects in the Solar System.

The smaller satellites of Saturn (those with a diameter of less than 300 kilometres) are mostly irregular in shape; their mass is insufficient for their self-gravity to make them spherical. Two of them, Hyperion and Phoebe, at 1.5 and 13 million kilometres from Saturn, respectively, have been known since the nineteenth century but eleven others, close to Saturn, were only discovered between 1979 and 1981, being very difficult to see in the glare of the planet and rings. When the Earth passed through the ring plane in March 1980 and the edge-on rings scattered the minimum possible amount of light towards us, improved detectors on terrestrial telescopes revealed five small new satellites, which Voyager 1 observed six months later. The Voyager flybys resulted in the discovery of six other satellites, bringing the number of known Saturnian satellites to twenty-one. The existence of three of these satellites has not yet been confirmed and some more unconfirmed satellites have since been discovered on Voyager images and more recent observations from Earth. The new discoveries all orbit within six Saturn radii of the planet and are usually illuminated by light scattered from the rings. The fourteen small satellites of Saturn have sizes ranging from 300 kilometres for Hyperion to 30 kilometres for the smallest. Their densities and masses are not known; however, they are probably made of ice. The two small outer satellites, Phoebe and Hyperion, were observed by the Voyager 2 probe. The other small satellites of Saturn are not well known, but they pose a number of interesting problems in relation to their orbits. It is no great thing in itself nowadays to discover a new satellite, but the presence of so many small bodies so close to Saturn leads us to

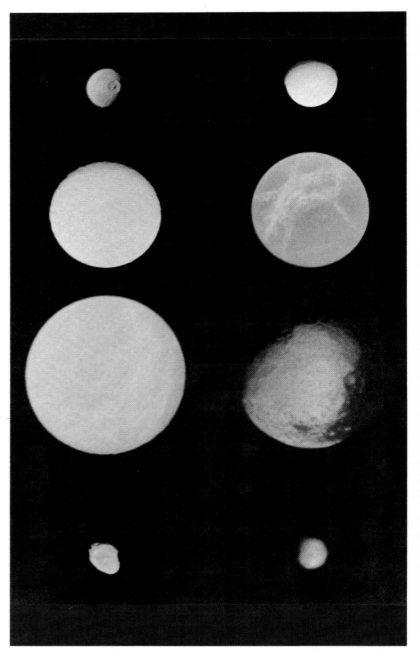

Eight Saturnian satellites compared. This photomontage shows the relative sizes of eight of Saturn's satellites: Mimas, Enceladus, Tethys, Dione, Rhea, Iapetus, Hyperion and Phoebe, in order from left to right and from top to bottom. The radius of Titan, the largest satellite of Saturn, is 3.4 times larger than that of Rhea. (After a montage prepared by J. A. Mosher, JPL Image Processing Laboratory)

The satellites of Saturn. The satellites of Saturn can be classified into five groups:
– Titan, the only satellite so far known to have its own atmosphere. At over 5000 kilometres in diameter, it is larger than the planet Mercury.
– Five satellites of intermediate diameter, between 400 and 1500 kilometres (Mimas, Enceladus, Tethys, Dione and Rhea) at between 200 000 and 500 000 kilometres from Saturn.
– One outer satellite, Iapetus, whose nature is not well understood, of 1500 kilometres diameter, moving in an elongated, inclined orbit, at more than 3.5 million kilometres from Saturn.
– Two small outer satellites, Hyperion and Phoebe, less than 400 kilometres in diameter, which may well be former asteroids that have been captured by Saturn.
– A collection of small satellites close to the planet, mostly discovered in 1980. Some of these satellites travel on the same orbit as other satellites. Eight of them have been observed several times since their discovery. Some of them have only been seen on a single image, and their existence will have to be confirmed by other observations.
The table below summarises the orbital data and sizes of the principal satellites (values in brackets are uncertain by more than 10 per cent.)

name	semi-major axis (km)	semi-major axis (R_s)	sidereal period (days)	eccentricity	inclination (°)	radius or semi-axis (km)
Atlas	137 700	2.276	0.602	0.002	0.3	(19) × ? × (13)
inner guardian of F ring (1980 S27)	139 400	2.310	0.613	0.004	0	70 × (50) × (37)
outer guardian of F ring (1980 S26)	141 700	2.349	0.629	0.004	0.1	(55) × (42) × (33)
Epimetheus	151 400	2.510	0.694	0.009	0.3	(70) × (57) × (50)
Janus	151 500	2.511	0.695	0.007	0.1	110 × 95 × 80
Mimas	186 000	3.08	0.942	0.020	1.52	196
Enceladus	238 000	3.95	1.370	0.004	0.02	250
Tethys	295 000	4.88	1.888	0.000	1.86	530
Calypso (Tethys Lagrangian)	295 000	4.88	1.888	?	?	? × (12) × (11)
Telesto (Tethys Lagrangian)	295 000	4.88	1.888	?	?	(15) × (12) × (8)
Dione	377 000	6.26	2.737	0.002	0.02	560
Dione Lagrangian (1980 S 6)	377 000	6.26	2.737	0.005	0.2	(18) × ? × (< 15)
Rhea	527 000	8.73	4.518	0.001	0.35	765
Titan	1 222 000	20.3	15.945	0.029	0.33	2575
Hyperion	1 481 000	24.6	21.277	0.104	0.43	175 × 117 × (100)
Iapetus	3 561 000	59	79.331	0.028	14.7	730
Phoebe	12 954 000	215	550,4	0.163	175	110

reflect a little on the role of aggregation and accretion in the formation of planets and satellites. It was thought in the past that each object swept up the matter within its sphere of influence and that this process led to well-separated satellites. The fact that there are satellites so close to the rings of Saturn leads to a continuing process of interaction, which probably explains some of the complicated structures in the rings. An even more interesting discovery is that of satellites that follow each other around the same orbit like cyclists chasing each other on a racing track. These satellites, however, never catch each other; despite the perturbations due to all the other satellites, they move on perfectly stable common orbits. There are three satellites in Dione's orbit: Dione itself, and smaller satellites 60 degrees behind it and ahead of it. Tethys also has small companions in its orbit, Telesto and Calypso, leading and trailing it by 60 degrees. This situation is reminiscent of the Trojan asteroids which similarly lie 60 degrees behind and ahead of Jupiter in its orbit. Their positions correspond to stable points predicted by Joseph Louis Lagrange in his study of the famous three-body problem, before the discovery of these asteroids. The satellites lying at the Lagrange points of Dione and Tethys are between 20 and 30 kilometres in size. Two other satellites, Janus and Epimetheus, play out a unique dance as they travel around the same orbit 2.51 Saturn radii from the planet. They each travel round Saturn in 17 hours and alternately approach and recede from each other without ever colliding. When photographed from Earth at their discovery in March 1980, they were 170 degrees away from each other while Voyager 1 found them separated by 110 degrees in November 1980. Both objects move in circular orbits round Saturn, but each describes a horseshoe-shaped trajectory with respect to the other. It has just been discovered that such relative orbits can be stable over long periods of time. Janus was first seen in 1966 by Audoin Dollfus and is a pear-shaped body 160 by 220 kilometres in size; Epimetheus is smaller. The three satellites closest to Saturn are each in separate orbits, but probably play an essential role in confining the outer edge of the rings.

André BRAHIC

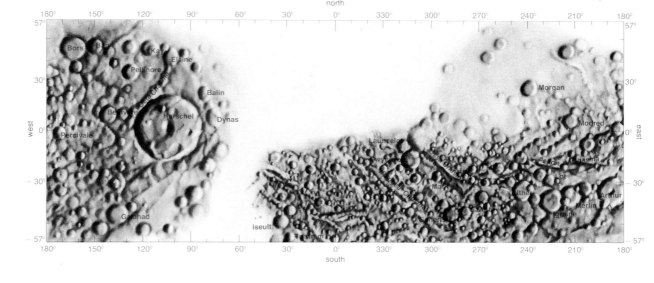

equatorial Mercator projection;
scale at latitude 0°: 1/8 000 000

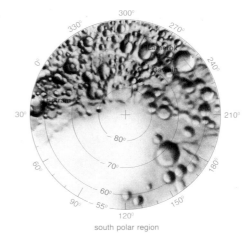

south polar region

polar stereographic projection;
scale at latitude −56°: 1/4 473 600

Mimas

Mimas takes less than 23 hours to orbit Saturn. Since it is so close to the planet, it is very difficult to observe it with a telescope from the Earth. Voyager 1 flew within 88 000 kilometres of this satellite and was able to obtain images with a resolution of about 2 kilometres. The most obvious feature is a huge crater, Herschel, one-third the size of the satellite at 130 kilometres diameter, 10 kilometres deep and with a central peak 6 kilometres high. This crater was probably created by the impact of a body about 10 kilometres across. An impact by an only slightly larger body would have smashed Mimas into several pieces. The surface of Mimas bears the scars of this devastating collision; fractures which were probably formed as the shock wave from the impact travelled around the satellite are found all over its surface. The rest of Mimas' surface is saturated with craters of all sizes.

Note. The maps of Saturn's satellites are based on maps drawn up from Voyager images and published by the United States Geological Survey. The half-tone airbrush drawings were made by Jay L. Inge (Rhea, Mimas, Iapetus) and Patricia M. Bridges (Tethys, Dione, Enceladus). Data preparation and preliminary image processing were done by K. F. Mullins, Christopher Isbell, E. M. Lee, H. F. Morgan and B. A. Skiff. The cartographic project as a whole was directed by R. M. Batson, who kindly supplied the necessary original documents. (USGS)

The surface of Mimas. In spite of its small size, Mimas is spherical, which suggests that it is sufficiently plastic to have readjusted itself after each violent collision. Herschel, Mimas' largest crater, is clearly visible on the two images at left which were taken on 12 November 1980 by Voyager 1 from a distance of 500 000 kilometres. The lower image shows the Saturn-facing hemisphere. The impact that produced Herschel nearly destroyed the satellite. Other craters ranging from 10 to 45 kilometres in diameter are visible. The image at right was taken on the same day by Voyager 1 when it was only 129 000 kilometres from Mimas; craters only 2 kilometres across can be seen. The surface of Mimas seems to be very ancient; there is no evidence for significant geological activity since its formation. Most of the craters must have been formed over four billion years ago. (NASA/NSSDC)

Enceladus

The images obtained by Voyager 2 with a resolution of 2 kilometres reveal a young surface with diverse features, and suggest that the satellite was still geologically very active until recently. Since Enceladus and Mimas are comparable in size, scientists expected to find two similar objects. Instead, while Mimas is highly cratered and inactive, the surface of Enceladus shows a variety of geological formations. Some regions are devoid of craters, having been covered by ice flows or geological processes which have remodelled the surface. Even the most heavily cratered regions are far from saturated; the oldest craters there have also been covered over by geological activity. The surface of Enceladus has lost all traces of the intense bombardment it underwent at the beginning of the Solar System's history. The fact that certain areas contain no craters means that the process which remodelled the surface was active not long ago; a large part of the surface has been remodelled in the past 100 million years, which is a very short time compared to the 4.6 billion year ago of the Solar System. The variations in the observed surface morphology suggest a continuous process rather than an isolated catastrophic event. Enceladus is a 'thermal machine' beneath whose crust the mantle must remain plastic or even liquid at less than 10 to 20 kilometres deep in order to generate these geological formations. No volcanic features like those on Io have been seen, but because the E ring has a density maximum in the neighbourhood of Enceladus' orbit, some scientists think that the satellite must occasionally eject material to augment the E ring, perhaps by a volcanic mechanism. It is very surprising that a small body like Enceladus has a source of internal heat, and there is currently no satisfactory theory of its origin. The heat that Enceladus had at its formation must have long since been radiated away from such a small satellite; heating by radioactive decay does not seem to be a good solution, provided that the satellite does not contain abnormally large quantities of uranium! Tidal heating due to the central planet, as in Io, seems not to be applicable here. Enceladus is too light to have deflected Voyager's trajectory by a measurable amount, and its mass – and hence its density – is poorly known, which does not make it any easier to model the internal structure and heat source.

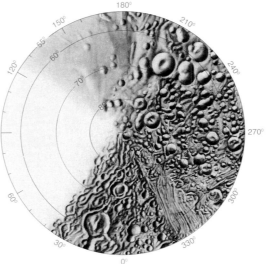

polar stereographic projection;
scale at latitude 56° : 1/4 473 600

north polar region

equatorial Mercator projection;
scale at latitude 0° : 1/8 000 000

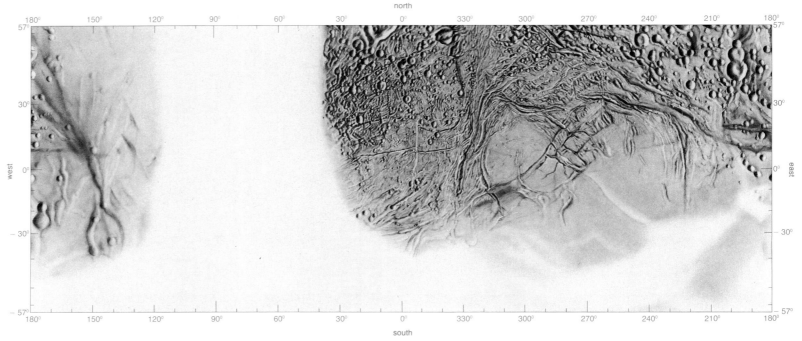

The surface of Enceladus. The picture below left is a mosaic of images acquired by the Voyager 2 probe when it was 119 000 kilometres from the satellite; details of the order of 2 kilometres in size are visible. The largest craters that can be made out are about 25 kilometres in diameter. However, a large part of the surface is craterless, which indicates that recent geological activity has wiped out older formations. The huge fault at lower left and the numerous striations all over the surface are probably the result of a thin plastic crust lying on top of a liquid interior. In many ways, the surface of Enceladus is reminiscent of that of the Solar System's largest satellite, Ganymede.

This is strange, since Enceladus is ten times smaller than Ganymede and should have long since lost its initial heat. An impressive ice flow can be seen at top left, and is shown in close-up on the right-hand image, which was obtained by Voyager 2 on 25 August 1981 when it was 112 000 kilometres from Enceladus. The age of this feature can be estimated from crater counts to be less than 100 million years, making it a very recent event on the astronomical scale. Other traces of older flows and other activity can also be seen on the pictures. (NASA/NSSDC)

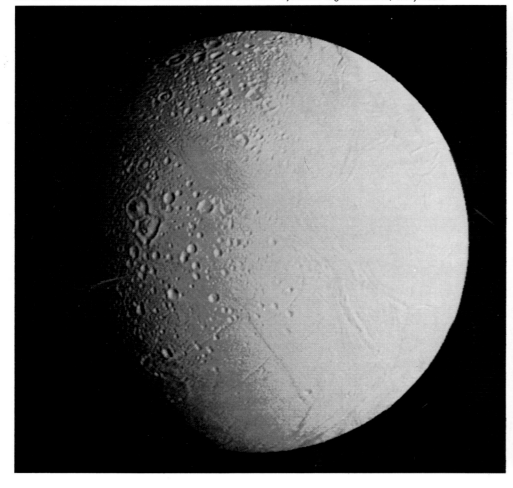

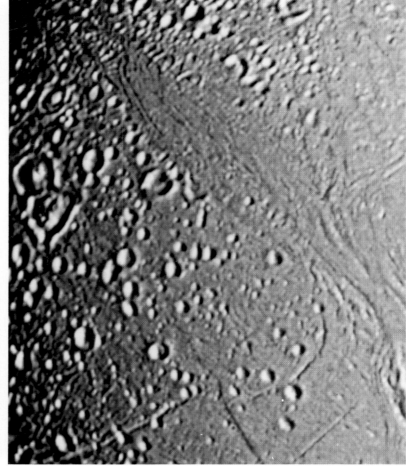

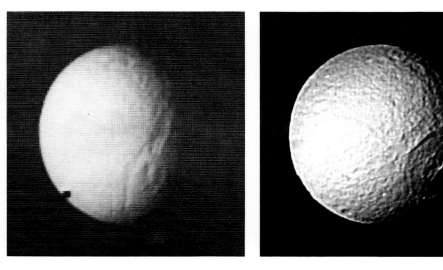

This satellite, about 300 000 kilometres from Saturn and over 1000 kilometres in diameter, seems to be an intermediate object in Saturn's family of satellites as far as both geological activity and chemical composition are concerned as well as in size. Voyager 1 passed 416 000 kilometres from Tethys and Voyager 2 passed 93 000 kilometres away. Unfortunately, Voyager 2 was unable to observe Tethys at high resolution because of the breakdown of its scan platform as it crossed the ring plane. The best images we have were taken by it from a quarter of a million kilometres and have a resolution of the order of 5 kilometres. As can be seen on the image to the left (taken by Voyager 2 on 25 August 1981) Tethys' surface is completely covered with craters. The density of these craters is as high as on Mimas in many places, so the surface must be very old. There are, however, some regions where there are fewer craters, which means that geological processes leading to re-surfacing occurred well after the satellite formed.

The right-hand image shows a particularly impressive crater at lower right. It is called Odysseus, and is 400 kilometres in diameter, larger than the satellite Mimas. While it is very large, it is shallow, its floor having been flattened from its original bowl shape by ice flows; the impact which formed the crater must have nearly shattered this large satellite.

The central image shows another remarkable geological formation, Ithaca Chasma, a huge valley or depression more than 2000 kilometres long cutting a swath three-quarters of the way round the satellite. It is 100 kilometres wide and several kilometres deep. Although very different in detail, the canyon makes one think of Valles Marineris on Mars. It is not yet known whether Ithaca Chasma is a fracture caused by the Odysseus impact or whether it is the result of a slight expansion of the satellite shortly after it formed when the liquid water in the interior slowly expanded as it froze. The picture was taken by Voyager 1 on 12 November 1980. (NASA/NSSDC)

Tethys

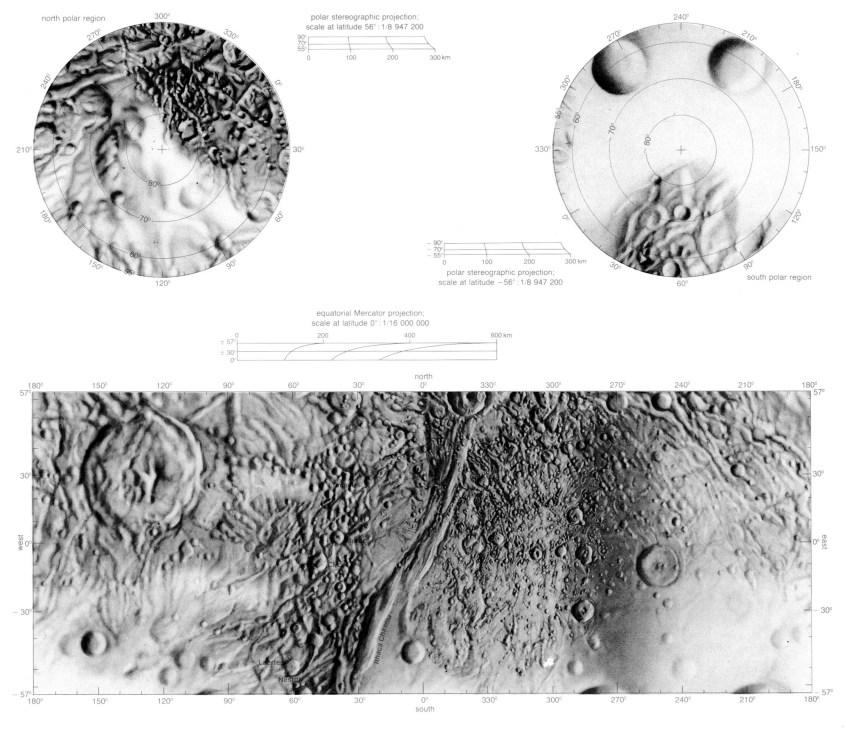

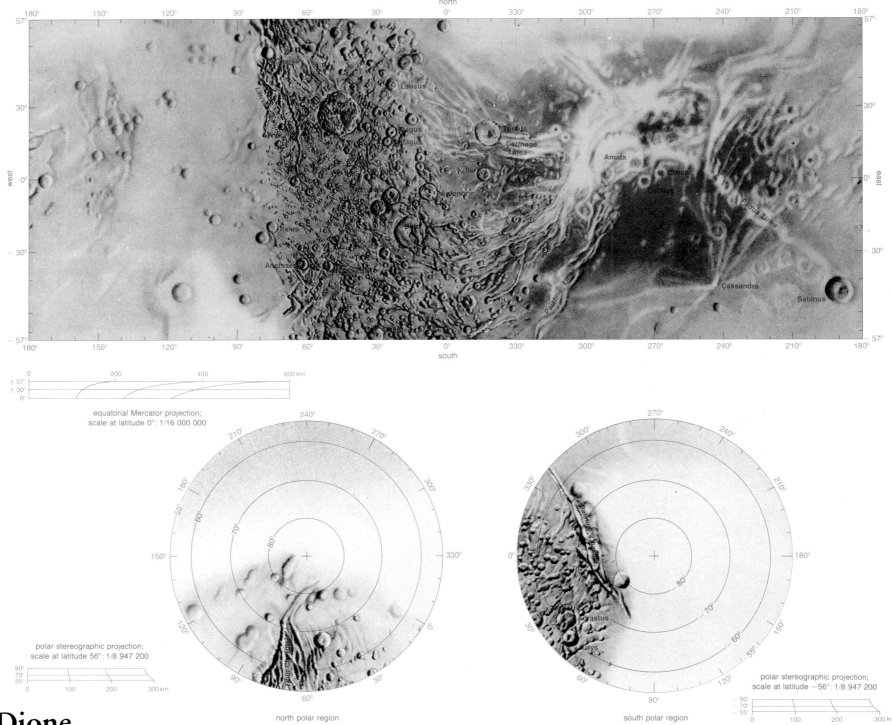

equatorial Mercator projection;
scale at latitude 0°: 1/16 000 000

polar stereographic projection;
scale at latitude 56°: 1/8 947 200

north polar region

polar stereographic projection;
scale at latitude −56°: 1/8 947 200

south polar region

Dione

Dione has practically the same diameter as Tethys (to within 60 kilometres) but cannot be considered as its twin sister. Its density of 1.4 grams per cubic centimetre (compared with 1.2 for Tethys) is the largest density of any of the icy satellites. Dione probably has a larger proportion of rocky materials and should therefore have a larger amount of internal radioactive heat. Dione displays a variety of geological features, including faulting, valleys, bright spots and depressions. The fractures seem to be linked to the outgassing of water and perhaps of methane. The density of craters varies from one place to another, showing that the surface was remodelled over several periods during the first couple of billion years of its history. Dione seems to have been more geologically active than Tethys, and this is probably because it had more radioactive heat. Like most satellites, Dione always keeps the same face towards the planet that it orbits. The face that leads in its motion around Saturn is brighter and more heavily cratered than the trailing hemisphere.

Many impact craters can be made out on the left-hand image, which was taken by Voyager 1 on 12 November 1980 from a distance of 162 000 kilometres. The largest craters are about 100 kilometres across and have a well-developed central peak. The winding valley that can be seen at top left of the picture, near to the terminator, is probably a fault or crack in the icy crust. The bright streaks represent material ejected from the impacts which formed the craters. The image at upper right was taken by Voyager 1 on 11 November 1980 from a distance of 377 000 kilometres; this very spectacular picture shows a transit of Dione in front of the orange-coloured globe of Saturn. The difference in brightness between the leading hemisphere (on the right) and the trailing hemisphere (on the left) is very apparent. The rear hemisphere consists of dark material covered with bright streaks, while the forward hemisphere has a relatively uniform and bright surface covered with many craters. The image at lower right shows a close-up of the trailing hemisphere of Dione; it was taken on 12 November 1980 by Voyager 1 from a distance of 790 000 kilometres. (NASA/NSSDC)

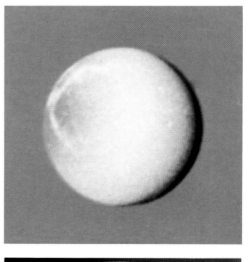

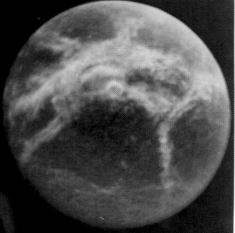

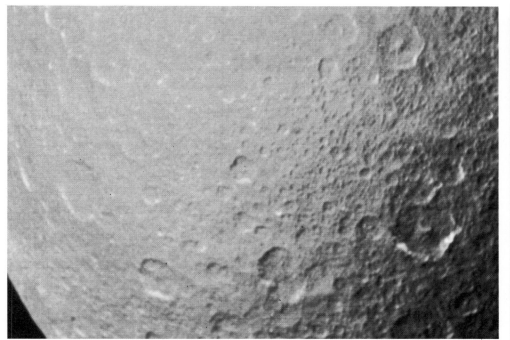

Rhea is, after Titan, the largest satellite of Saturn, but it displays fewer traces of surface geological activity than smaller bodies like Dione, Tethys and Enceladus. Voyager 1 flew within 74 000 kilometres of Rhea and provided images with a resolution of the order of one kilometre. These images reveal a surface saturated with craters resembling the highland areas of the Moon; however, the material covering the surface of Rhea is made of white, reflective ices and not of dark brown rocks.

The surface of Rhea, together with that of Mimas, is probably one of the oldest in the Saturnian system. In contrast to Ganymede and Callisto, for example, Rhea's craters show almost no trace of erosion by glacial flows, which suggests that it has a more rigid crust and a lower surface gravity. Between day and night, the surface temperature varies from 99 K to 73 or 53 K, according to the region. At these temperatures, ice behaves like hard rock as far as impacts by external bodies are concerned. The picture on the right was taken on 12 November 1980 by Voyager 1 from 128 000 kilometres. Notice that in some regions there are many craters more than 40 kilometres in diameter, while other regions are also saturated with craters, but of much smaller size. Some scientists have suggested that this is a relic of two distinct populations of projectiles which have bombarded Rhea. The first consisted of the original material from which Rhea was formed. The second, which contained no large objects, was probably debris left over from the formation of the Saturnian system which impacted the satellite later on. (NASA/NSSDC)

Rhea

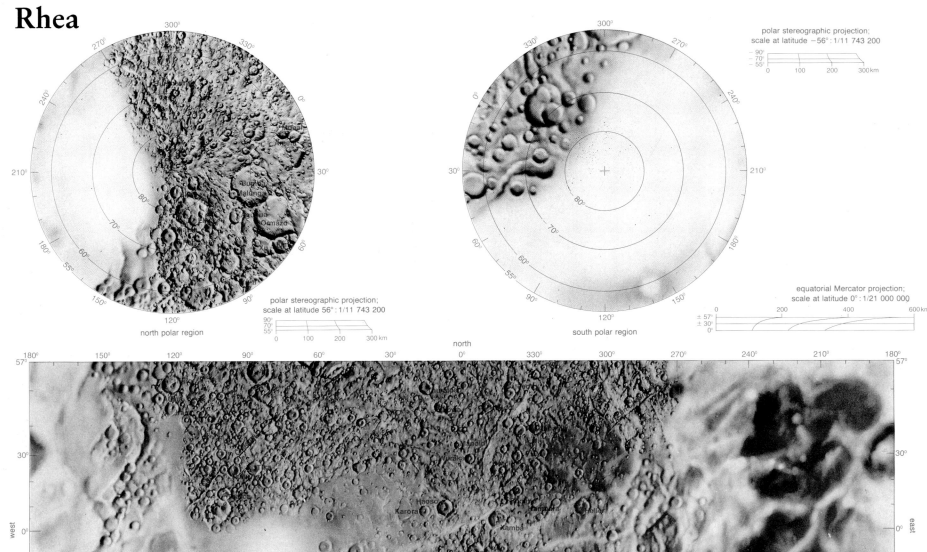

polar stereographic projection;
scale at latitude −56°: 1/11 743 200

polar stereographic projection;
scale at latitude 56°: 1/11 743 200

north polar region

south polar region

equatorial Mercator projection;
scale at latitude 0°: 1/21 000 000

Iapetus

The eccentric and inclined orbit of Iapetus is on the borders of the Saturnian system, at nearly four million kilometres from the planet. While it is similar in size to Rhea (to within 70 kilometres). Iapetus is very different in nature. It is one of the strangest and least understood objects in the Solar System. With a density of 1.2 times that of water, Iapetus probably has methane and ammonia ices in its make-up; these are too volatile to condense closer to Saturn, where the temperature would have been higher when the Solar System formed. Giovanni Domenico Cassini, who discovered Iapetus at the end of the seventeenth century, immediately noticed an unusual characteristic of this satellite: it was only visible on one side of Saturn. It was just as if Iapetus, which always shows the same face to Saturn, had a dark leading hemisphere and a bright trailing hemisphere. The Voyager probes confirmed Cassini's hypothesis. The bright rear face is cratered and probably covered with ice, like the surface of Rhea, while the front face is uniformly covered with a material which is six to ten times darker. It is not known whether this material is the result of an external collision or whether it comes from the interior of Iapetus. Because some of this dark material is found on the floor of some of the craters in the bright hemisphere, some scientists are convinced that the material is of internal origin and is due to eruptions of methane from the interior. There is no explanation for the fact that these eruptions must only have occurred in one hemisphere. The best photographs from the Voyager 2 flyby have a resolution of 20 kilometres; Iapetus remains an astronomical enigma.

polar stereographic projection;
scale at latitude 56°: 1/11 743 200

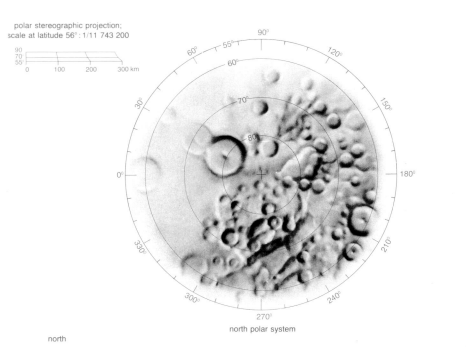

north polar system

equatorial Mercator projection;
scale at latitude 0°: 1/21 000 000

north

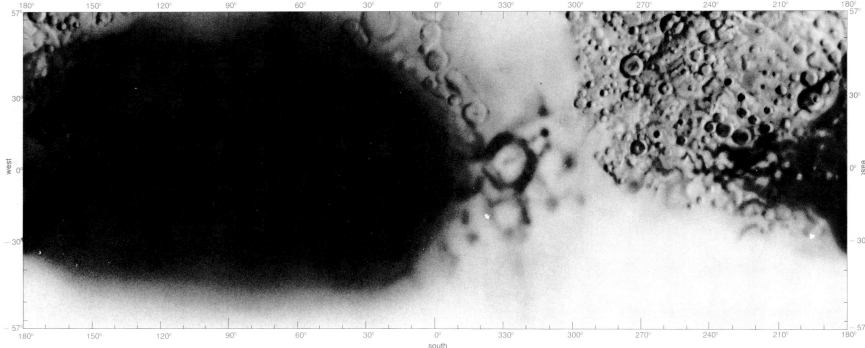

south

The two faces of Iapetus. The image at lower right, taken by Voyager 1, shows the striking difference between the dark and bright hemispheres of Iapetus. The left-hand image was taken by Voyager 2 on 22 August 1981 from a distance of 1.1 million kilometres. It shows the northern hemisphere, the north pole being on the right near to the crater with the central peak. A patch of dark terrain is visible on the left in the equatorial region. (NASA/NSSDC)

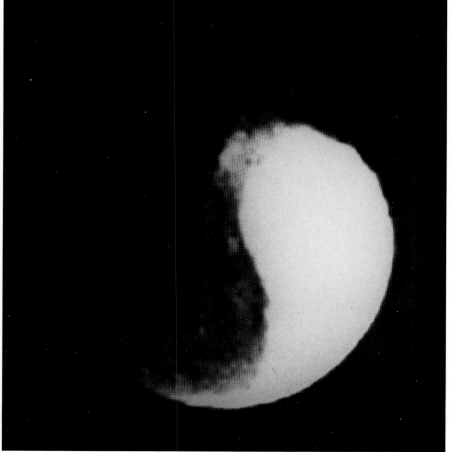

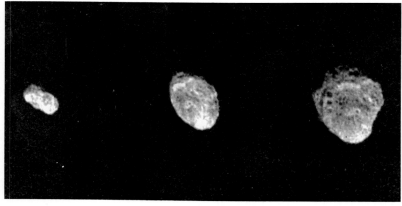

Hyperion. Located almost 1.5 million kilometres from Saturn, Hyperion revolves between Iapetus and Titan and takes 21.3 days to complete one orbit. Voyager 2 sent back images with a resolution of 9 kilometres as it flew past at a distance of 500 000 kilometres. This satellite is also made of ice, but the fact that its albedo is slightly lower than that of Saturn's other icy satellites suggests that its surface contains some impurities. Depending on the side from which you look at it, Hyperion resembles a peanut or a hamburger 350 × 234 × (200) kilometres in size. Its highly irregular shape contrasts with the immaculate sphericity of Phoebe and makes one think of a fragment of a violent collision.

While it is as large as Mimas, it has a very different appearance. This object was apparently never sufficiently heated for its surface to be remodelled and take on a spherical form. It may be a primitive object which has never undergone differentiation. It is interesting to note that the orbital periods of Titan and Hyperion are in the ratio of 3 : 4. Because of tidal effects with Saturn, Titan, like the Moon from the Earth, is slowly receding from Saturn, and drags Hyperion along with it because the latter has an orbit which is in resonance with that of Titan. The images above were taken by Voyager 2 on 23 and 24 August 1981 while the probe was between 500 000 and 1.2 million kilometres from Hyperion. (NASA/NSSDC)

Phoebe. The satellite that is furthest from Saturn, Phoebe, has a retrograde circular orbit inclined at 15 degrees to the equatorial plane of Saturn, 13 million kilometres from the planet. It takes 550 days to complete an orbit. Voyager 2 went within 1.5 million kilometres and obtained the image above on 4 September 1981; it has a resolution of 38 kilometres. This image shows that Phoebe is a spherical body 220 kilometres in diameter. Unlike the other satellites of Saturn, which are mostly made of water ice, its surface is very dark. It reflects less than 5 per cent of the incident sunlight. Phoebe resembles the most primitive asteroids, which are thought to be made of matter that condensed directly from the protosolar nebula. It may be one of these primitive bodies, having been captured later by Saturn, and if so would be the first such object to have been photographed by a space probe. (NASA/NSSDC)

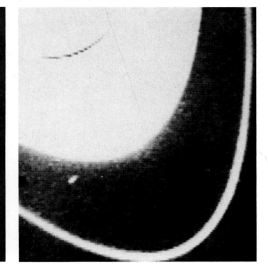

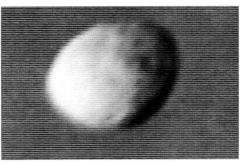

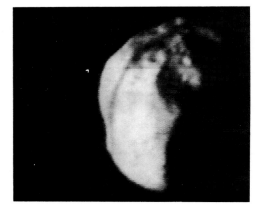

The small inner satellites of Saturn. Eleven new small satellites of Saturn were discovered in 1980 and 1981. Five were discovered from the Earth when the rings of Saturn were seen edge-on in March 1980. They are all near the planet, and all of them have unusual orbits. In particular, some of them move together on the same orbit, a phenomenon which was previously unknown in the Solar System. Eight of the inner satellites have been observed several times, and their existence is now well established. All of the information we have about their physical nature comes from the Voyager probes. The rest of the satellites have only been observed once, and their orbits are therefore poorly known. Several more satellites have been tentatively identified, but their existence has not yet been confirmed. The composite photograph above shows the eight satellites that were observed by the Voyager probes in November 1980 and August 1981, and reproduces accurately their relative sizes. Their dimensions range from 16 to 220 kilometres; all these objects are irregular and probably have icy surfaces. From left to right, these small satellites have been arrranged in order of their distance from Saturn. First is the small satellite Atlas, less than 500 kilometres from the outer edge of Saturn's A ring; then two satellites which lie on either side of the F ring and probably confine it; two co-orbital satellites, Janus and Epimetheus; two satellites at the Lagrangian libration points of Tethys, Telesto and Calypso; and one satellite, Dione B, at one of Dione's Lagrange points.

Atlas, the closest satellite to Saturn, was discovered on an image (above right) taken by Voyager 1 on 7 November 1980. On this overexposed image, the satellite can be seen between the outer edge of the A ring and the F ring. This was the first time that a satellite had

been seen between two rings. Atlas probably plays a very important role in confining the outer edge of Saturn's rings. The images opposite, of the two co-orbital satellites Janus and Epimetheus, were taken on 12 November 1980 by Voyager 1. These two bodies are elongated; the shadow of the F ring can be seen on tooth-shaped Epimetheus (lower picture). These satellites were in fact first seen in 1966, the last time that the Earth passed through the plane of the rings. It was not then possible to determine their orbits, and a controversy arose over the reality of their existence. When the rings were again edge-on and the glare from them was therefore at a minimum in March 1980, these two satellites were again observed from Earth. They were moving in the same orbit, 170 degrees from each other. When Voyager 1 observed them in November 1980, they were only 110 degrees from each other. Amusingly enough, the pear-shaped Janus was being pursued by Epimetheus, which is shaped like a tooth! The tooth will never bite into the pear, though, since the two satellites oscillate with respect to each other. Their distance from Saturn varies very slightly and hence, by Kepler's laws, they alternately speed up and slow down. Such orbits, which take the form of a horseshoe in a frame of reference in which one of the satellites is at rest, had been postulated by mathematicians at the beginning of this century, but this is the first time that they have been found in nature.

The rings and inner satellites of Saturn constitute an extremely complex dynamical system. All these bodies continually perturb each other and yet the whole system has remained perfectly stable for more than four billion years. (NASA/NSSDC)

Titan

Titan, Saturn's largest satellite, was discovered in March 1655 by Christiaan Huygens, who noted that it orbited in the same plane as the rings, with a period of about sixteeen days. It was the sixth satellite in the Solar System to be discovered, after the Moon and the Galilean satellites.

Titan is similar in many ways to the terrestrial planets. It is intermediate in size between Mercury and Mars. Of all the satellites in the Solar System, except possibly for Triton, it is the only one to have a substantial atmosphere. The pressure is five hundred times greater than that of the Martian atmosphere, and ten times greater than Earth's. As is the case for our planet, molecular nitrogen (N_2) is the principal constituent of this atmosphere but it does not contain any oxygen. Methane plays an important role because of the photochemical reactions it gives rise to and it makes the atmosphere a strongly reducing environment. The surface temperature allows the existence of oceans of liquid methane and, potentially, the formation of methane clouds and methane rain in the lower atmosphere.

In 1944, Gerard P. Kuiper detected gaseous methane on Titan through spectroscopy, thus demonstrating the presence of an atmosphere. A thick layer of haze entirely cloaks the satellite. Most of our basic knowledge comes from Voyager 1, which passed less than 500 kilometres from its surface on 12 November 1980. In particular, the radius of Titan was determined by the occultation of radio waves emitted from the probe, and the gravitational deflection of the probe's trajectory revealed the mass, and hence the mean density, of the satellite. This density turned out to be equal to 1.92 grams per cubic centimetre and implies that the internal composition is a mixture of rock (about 3 grams per cubic centimetre), water ice, methane ice and perhaps ammonia ice.

Three facts show that Titan was formed at the same time as Saturn: the orbit is circular, almost exactly in the ring plane, and the satellite orbits in the same direction as Saturn rotates. In a similar way to the formation of the planets around the Sun from a flattened gaseous nebula, it is imagined that a flattened disk of dust and gas orbited around a central concentration which would one day contract to form Saturn. The gravitational energy liberated by the contraction of the proto-Saturn was radiated outwards and the inner parts of the disk were, therefore, hotter than the parts that were more distant from the proto-Saturn. After a time, the intensity of the radiation from the proto-Saturn lessened and the temperature of the disk decreased, allowing the least volatile ices to condense first: water ice, then ammonia ice and finally methane ice. Clathrates,

mixtures of these ices, were also formed; these clathrates have the interesting property of being able to trap gases like argon, nitrogen, and carbon dioxide.

The accretion of these ices and rocky materials into a single body, a proto-Titan, involved the release of enough internal heat to allow differentiation in Titan's interior, the rocky matter (mostly silicates) forming a heavy central core and the different ices settling in concentric layers, according to one model still under discussion. Any good model, which should describe the internal distribution of pressure, temperature, density and composition, must account for the observed mass, radius and density of Titan, which limits the possibilities. It is already clear that there is a rocky core making up about 55 per cent of Titan's mass and 1700 kilometres of its radius of 2575 kilometres. A recent model proposes that this core is surrounded by concentric layers of solid clathrates, followed by a liquid layer of ammonia and methane hydrates ($NH_3.H_2O$ and $CH_4.H_2O$, which are combinations that are easier to liquefy than the pure ices) containing solid chunks of water ice, and finally, at the surface, an ocean of liquid methane.

Molecular nitrogen is the principal constituent of Titan's atmosphere. Titan is the only object in the Solar System, apart from the Earth, for which this is true. There are two possible sources for this nitrogen: it may have been trapped in the form of molecular nitrogen in the clathrates of the disk that turned into Titan before being released during the accretion phase, or it may be the relic of an enormous ammonia atmosphere that has slowly been destroyed by solar radiation splitting up the ammonia molecules and letting the hydrogen atoms escape to space. For this latter hypothesis to work, the temperature must have stayed above 150 K to prevent the ammonia condensing. This would have been possible due to the greenhouse effect, by which the heat released by the planet is trapped within the atmosphere by the ammonia. Whatever the truth, there is no ammonia in Titan's atmosphere today and the most active gas present is methane. Solar ultraviolet photons with wavelengths shorter than 160 nanometres photodissociate methane molecules into hydrogen atoms and methyl radicals (CH_3), both of which are very chemically reactive. This leads to the formation of hydrocarbons like ethane (C_2H_6), acetylene (C_2H_2) and ethylene (C_2H_4), which have been detected at a proportion of a few parts per million, and possibly polyacetylene chains ($C_{2n}H_2$). A few nitrogen atoms produced by photodissociation of molecular nitrogen allow the formation of hydrogen cyanide (HCN), one of the fundamental compounds of organic chemistry which must have been important in the formation of life on Earth.

The hydrocarbons and the more complicated molecules that form from methane and nitrogen

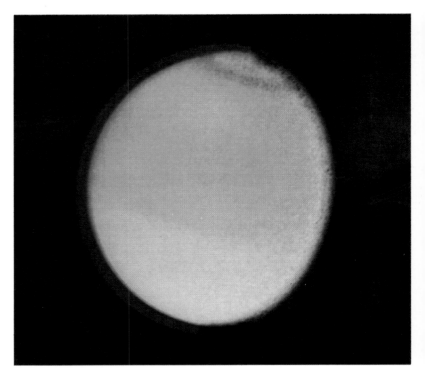

Seasons on Titan. This photograph was taken by Voyager 2 on 23 August 1981 from a distance of 2.3 million kilometres. The surface is hidden by a continuous covering of cloud. Bluish haze layers are visible above the limb.

The contrast of the image has been exaggerated, revealing that the northern hemisphere is slightly darker and redder than the southern hemisphere; a dark 'collar' can also be seen towards 70° north latitude. These markings provide indirect confirmation that Titan's rotation axis is parallel to Saturn's. Because the latter is inclined at 27 degrees to the normal to Saturn's orbital plane, the Sun shines preferentially on alternate hemispheres for 15 years at a time – half of Saturn's orbital period. When Voyager 2 flew past, spring was just beginning in Titan's northern hemisphere.

This difference in solar illumination may initiate more active photochemistry in the summer hemisphere, leading to a difference in the composition of the aerosols which give Titan its colour.

It is assumed that this planet-sized satellite always keeps one face towards Saturn, as the Moon does to the Earth, and the Galilean satellites do to Jupiter. A day on Titan would then be equal to its 15.9 Earth day orbital period around Saturn. (NASA/JPL)

The atmosphere of Titan. On this image, acquired by Voyager 2 from the night side of Titan on 25 August 1981, from a distance of 907 000 kilometres, the orange crescent fades into a bluish ring which surrounds the planet and reveals the existence of an atmosphere. The colours are exaggerated. The orange crescent is produced by sunlight reflected off an aerosol layer, a layer of tiny liquid or solid particles in suspension in the atmosphere at some 200 kilometres altitude. The bluish halo is due to scattering of sunlight by haze layers several hundred kilometres high. The polarisation of the reflected light, its colour, and its distribution across the disk provide information about the size of the particles; some are about 0.3 micrometres in size, while others are larger, especially lower in the atmosphere.

The destruction of methane (CH_4) and the formation of hydrocarbons of higher mass probably leads to the formation of small particles. Accreting slowly, they grow and eventually fall to the surface.

Titan, Mars and Earth compared.

	Titan	Mars	Earth
mass (g)	1.35×10^{26}	6.4×10^{26}	5.98×10^{27}
equatorial radius (km)	2575	3395	6378
surface gravity (cms^{-2})	144	374	981
density (gcm^{-3})	1.92	3.95	5.52
mean surface temperature (K)	95	215	288
rotation period (days)	15.9(?)	1.03	1
atmospheric pressure (kgcm^{-2})	10.9	0.02	1
surface pressure (mbar)	1600	8	1000
atmospheric composition	N_2, Ar, CH_4	CO_2, N_2	N_2, O_2

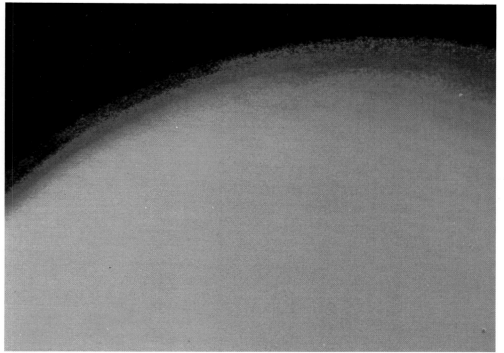

Haze layers. On 12 November 1980, Voyager 1 was only 435 000 kilometres from Titan and the resolution of its pictures was close to 10 kilometres. The colours in this photograph have been exaggerated. Above the uniform orange haze layer are some separate layers, bluish in colour, which are visible on the satellite's limb. (NASA)

Titan's atmospheric structure. The temperature at ground level, measured by the infrared spectrometer, is 96 K (−178 °C). The temperature distribution as a function of altitude and pressure (white curve) was determined from analysis of the attenuation of Voyager 1's radio signal as it passed through Titan's atmosphere on its way to Earth. The surface pressure is 1600 millibars. The temperature decreases as the altitude increases up to 40 kilometres high where it reaches a minimum of 74 K (−199 °C). The region below this point is the troposphere. The temperature then rises in the stratosphere until it reaches 175 K (−98 °C) at about 250 kilometres, after which it stays constant up to the exosphere, which begins at an altitude of 1500 kilometres. A layer of light haze at about 300 kilometres is clearly detached, at least in the southern hemisphere, from the main haze layer, whose top is at 230 kilometres. Made of small aerosol particles which conceal the surface, this layer plays the role that ozone does in the Earth's atmosphere. The dots outline a layer which absorbs ultraviolet radiation.

The temperature and pressure in the troposphere, where convection is· effective, will almost certainly allow clouds of methane to form and even precipitate as methane rain. As liquid methane may be present on Titan's surface, a complete methane meteorology is possibly analogous to terrestrial meteorology for which the ubiquitous presence of liquid water is the controlling factor.

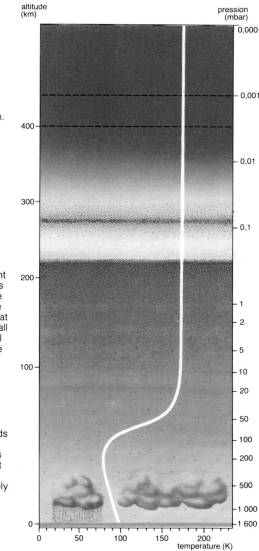

The extreme ultraviolet spectrum. (left). The extreme ultraviolet spectrum of the upper atmosphere of Titan as measured by Voyager 1 reveals the presence of atomic hydrogen (Lyman-alpha line at 121.6 nanometres) and nitrogen in three forms – molecular (N_2), atomic (N) and ionic (N^+). These emission lines are excited by either solar radiation (in the case of the hydrogen Lyman-alpha line) or by Saturnian magnetospheric electrons (in the case of nitrogen) precipitating in Titan's upper atmosphere. These electrons are tied to Saturn's magnetic field and turn with the same angular velocity as Saturn, one rotation in about ten hours, and so move much faster than Titan does in its orbit. Nitrogen makes up about 99 per cent of Titan's atmosphere with about 1 per cent methane and traces of the hydrocarbons from methane's breakup. (Diagram from Broadfoot *et al.* 'Extreme Ultraviolet Observations from Voyager 1 Encounter with Saturn' in Science, vol. CCXII no. 4491, p. 206, 1981)

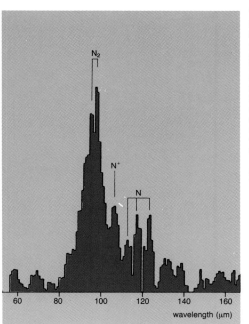

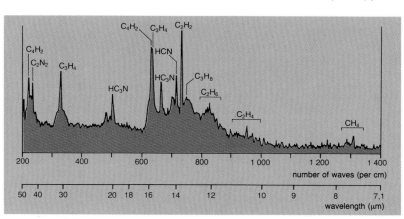

Infrared spectrum of Titan. This Voyager 1 spectrum shows the infrared emission of Titan's limb in the region of the north polar hood (after V. G. Kunde, A. C. Aikin, R. A. Hanel, D. E. Jennings, W. C. Maguire and R. E. Samuelson, C_4H_2, HC_3N and C_2N_2 in Titan's Atmosphere, *Nature* vol. 292, p. 686, 1981). The peaks in the spectrum are at wavelengths characteristic of particular molecules, allowing several gaseous compounds to be identified and have their abundances measured. All the gases that have been identified in the spectrum are products of the break-up of methane followed by other chemical reactions. They include hydrocarbons like ethane (C_2H_6), propane (C_3H_8), acetylene (C_2H_2), and organic compounds containing nitrogen like hydrogen cyanide (HCN), cyanogen (C_2N_2) and cyanoacetylene (HC_3N). These chemicals are the simplest building blocks from which much more complex chemical reactions could lead to pre-biotic molecules.

In Titan's atmosphere, some of these molecules will form polyacetylenes, ($C_{2n}H_2$) which will condense into small particles making up the orange-red haze and then fall to the surface.

condense in the coldest region of the atmosphere, within 200 kilometres of the surface and ultimately form particles between 0.2 and 1 micrometre across, which constitute the thick, uniform, orange layer of haze that covers the entire surface of Titan.

These organic aerosols fall slowly towards the surface and settle, either on a solid surface or sink to the bottom of liquid methane oceans, to form a layer that must have reached 1 kilometre in thickness over the 4.6 billion year lifetime of Titan, while the hydrogen atoms escape and form an immense torus around Saturn in Titan's orbit.

The formation of the organic aerosols from the methane in the atmosphere would only take a few million years to use up all the methane. Methane must, therefore, be supplied continuously to the atmosphere from the surface, which is at a temperature of 95 K. This temperature is determined by the partial pressure of methane at the surface and implies the presence of liquid methane. Whether this liquid is in puddles, lakes or huge oceans will remain uncertain until another space probe explores Titan.

Jean-Loup BERTAUX

The hydrogen torus (opposite right). Voyager 1 discovered a gigantic torus of atomic hydrogen around Saturn, by observing its characteristic ultraviolet Lyman-alpha emission line at 121.6 nanometres. Lyman-alpha is the equivalent for hydrogen of the characteristic yellow-orange glow of sodium, the colour of the light emitted by sodium street lights. This torus surrounds the orbit of Titan, between 8 and 25 R_s from the centre of Saturn, and is 14 R_s thick (1 R_s = 1 Saturn radius = 60 330 kilometres) and engulfs the orbit of Rhea. Its existence was postulated as long ago as 1973 from simple theoretical considerations. The upper atmosphere of Saturn is always losing hydrogen atoms and molecules, which have enough speed to escape Titan's gravity, but not enough to escape Saturn's gravitational field. Those atoms leaving Titan in the direction of its motion go into the outer part of the torus, while those leaving in the backwards direction populate the inner part of the torus.

The concentration of hydrogen atoms in the torus has been measured at twenty atoms per cubic centimetre, and ionisation losses are compensated by the escape of one kilogram of hydrogen every second from Titan's atmosphere. Hydrogen molecules should also escape from Titan, and may be present in the torus in ten times greater numbers, but they are much more difficult to detect.

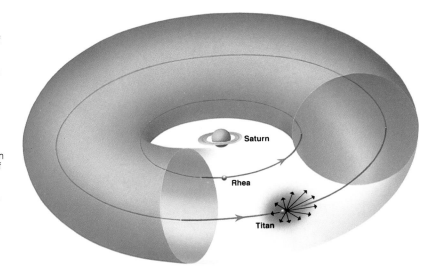

Uranus

Almost two and a half times smaller than Saturn, and twice its distance from the Sun, the planet Uranus is just visible to the naked eye. Unknown to the Ancients, for whom Saturn marked the edge of the Solar System, Uranus was not discovered until 13 March 1781 by the musician and amateur astronomer William Herschel, who, observing by chance in the constellation Gemini with a telescope of 16 centimetres diameter, noted an object which was not as point-like as a star. He thought that he had discovered a new comet, but calculation of its orbit quickly showed that the object was in fact a planet revolving in an orbit more than 3 billion kilometres from the Sun.

An experienced observer can see Uranus on a very clear night as a very faint star. Uranus was in fact plotted several times on maps of the sky between 1690 and 1780, a fact that was later exploited in determining the orbit of the planet. Seen from the Earth, Uranus subtends an angle of 4 arc seconds (about a thousandth of a degree). Because of atmospheric turbulence, it is difficult to resolve points in the sky separated by less than an arc second, and so Uranus appears, even with the largest telescopes, as a small greenish disk on which no details may be seen.

With a diameter of 52 000 kilometres, Uranus is fifteen times more massive than the Earth. Smaller than Jupiter and Saturn, and with about the same density as Jupiter, it is similar to Neptune. Like the other giant planets, Uranus is 99 per cent hydrogen and helium. Its rotation period is 17 hours 12 minutes. Unlike the other planets, the axis of Uranus lies almost in the plane of its orbit; thus the equatorial plane of Uranus and the orbital planes of its five known satellites are almost perpendicular to the plane of its orbit around the Sun.

After the Voyager 2 flyby on 24 January 1986 Uranus became the most distant object ever to be explored by a robot. For 205 years it has been only a small point of bluish light. In a few hours it revealed itself as a particularly rich world surrounded by amazing rings and by satellites which are much more active than had previously been predicted, including among them the smallest satellite, Miranda. The bulk of our knowledge of Uranus comes from that brief encounter; it is important therefore to realise the need to complement Voyager 2 observations with ground-based ones. The study of the rings by stellar occultations observed from Earth, for example, provided data on their structures that could not be obtained by the probe. On the other hand, small particles in the rings detected by Voyager are invisible from Earth.

Seven thousand images of the Uranian system, 2 000 of which are from the minimum distance; thousands of infrared and ultraviolet spectra; and millions of radio and magnetic measurements were transmitted to Earth. They revealed clouds, bands parallel to the equator and layers of mist. In particular, the pole presently facing the Sun is covered by a mist cap. Studies which have followed the cloud motions have allowed the calculation of the rotation period of the atmosphere to be determined.

The helium to hydrogen ratio is similar to that of the Sun (about 15 per cent); it has been estimated by infrared spectra and radio occulations by the atmosphere. Thus it appears that the Uranian atmosphere is primordial and not the product of evolution of the planet. Temperature variations with latitude and depth have also been measured with infrared instrumentation. These results have been followed with great interest due to the peculiar orientation of the rotation axis. Everything about the magnetic field of Uranus was unknown before the encounter; radio radiation from Jupiter is detected from Earth; that of Saturn was received by Voyager at large distances, more than a year before the encounter; but the veil was not lifted from Uranus until a few hours before the nearest approach of the probe.

The atmosphere of Uranus has a temperature of about 50 K. Spectroscopic study has revealed the presence of methane and molecular hydrogen. The Voyager detected an extended atmosphere of molecular hydrogen and an even more extended corona of atomic hydrogen.

Uranus is much more different from Jupiter or Saturn than it was previously thought, due to its seasonal rhythm, the complex dynamics of its atmosphere, the heating processes and the chemical reactions that they cause.

André BRAHIC

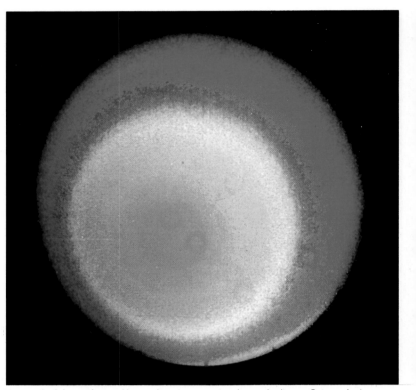

Uranus in false colours. Uranus is not as spectacular as Jupiter or Saturn. An image (such as the one on page 51), modified in order to represent how it would be seen by the naked eye by a passenger in a spatial probe, does not show any detail. It is composed of images obtained on 17 January 1986 through the blue, orange and green filters of the narrow field camera on board the Voyager 2, when it was 9,1 million kilometres from the planet's surface. The blue-greenish colour results from methane present in the atmosphere; it absorbs the red portion of the solar radiation and lets through the blue.

On the other hand, the false colour picture presented above, is composed of images taken on the same day at the same distance, through ultraviolet, violet, blue, green and orange filters. The contrast is enhanced and more details are revealed. The south pole, to the left of the disk and pointing towards the camera, is covered by mist. The Uranian atmosphere, like those of Jupiter and Saturn, has fringes. Other features, in particular the small rings that correspond to specks of dust accumulated over the objective of the camera, are enhanced by image processing. (JPL/NASA).

The 'seasons' and climatology of Uranus. The motions of Uranus upon its axis are very peculiar: the planet, its rings, and its attendant satellites roll upon an axis of rotation which is in fact located practically within the orbital plane. The north and south poles point alternately towards the Sun in the course of each revolution around the Sun lasting 84 years. A day, or what comes to the same thing, a season, of length 42 terrestrial years follows a night – of length 42 years at the poles. This must have striking consequences for the climate of Uranus and for its atmospheric circulation. Important temperature differences can be expected between the pole illuminated for more than 20 years and the dark one. On the contrary, the temperature of the upper atmosphere is almost the same at the poles as at the equator, as shown by the right-hand diagram. This confirms the model advanced by Bezard and Gautier at the Paris Observatory, one year before the encounter. The dark pole is even slightly hotter than the illuminated one, and the coldest area (about 2 K) is situated between 20 and 40 degrees latitude, in a region that could qualify as tropical at about 60 K. Seasonal temperature variations at the poles are less than 5 K. Temperature differences between the two hemispheres correspond not to differences in light but to complex dynamical processes. Uranus could be compared to a refrigerator; heat is extracted from the illuminated pole and energy is provided as the equator. Temperature variations are limited by the large atmospheric thermal inertia, mainly due to the very low temperatures. The Uranian atmosphere is as complex and interesting as that of Jupiter and Saturn, though on a less spectacular scale.

Due to the peculiar inclination of the rotation axis, the Voyager, travelling along the orbital planes of the planets, approached Uranus perpendicularly to the equator (and therefore also to the rings and satellites). Voyager's speed had also been accelerated by Jupiter and Saturn, and therefore, the encounter was brief.

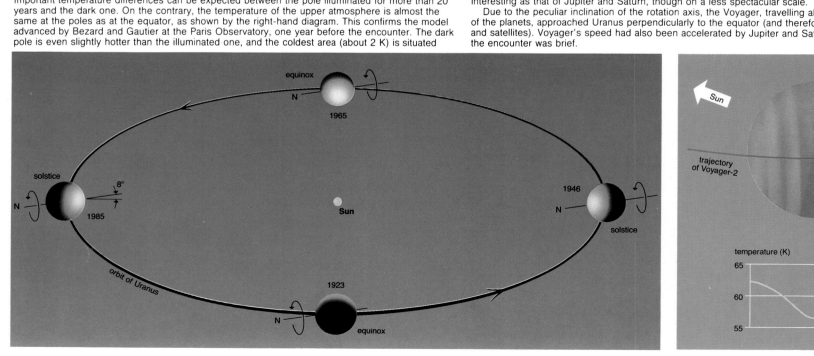

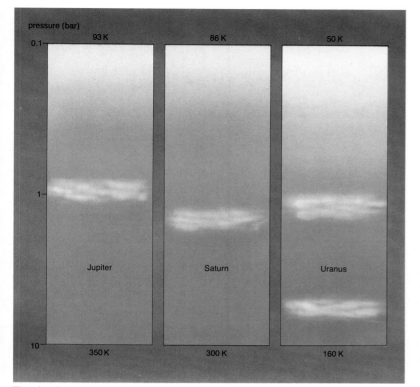

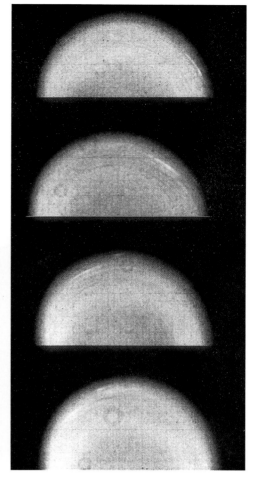

Atmospheric rotation. The false colour image represents a superimposition of images taken on 14 January 1986. A cloud can be seen near the edge, top right. The black and white images were taken through an orange filter. The south pole is at the centre of the disk. Between the recording of the top and bottom images a total time of 4 hours 36 minutes elapsed. The rotation of two clouds can be observed. The atmosphere rotates anticlockwise and faster than the solid planet; all the winds come from the west. The rotation period of the largest cloud (at 35 degrees latitude) is 16 hours 18 minutes; and the small one, at 27 degrees latitude, has a period of 16 hours 54 minutes. Thus, the upper atmosphere shows differential rotation, being faster towards the poles than at the equator, which is contrary to what happens on Saturn. At 25 degrees latitude, the period is 17 hours, and at 40 degrees latitude, it is 16 hours. Atmospheric bands concentric to the pole can also be seen in the figure (JPL/NASA).

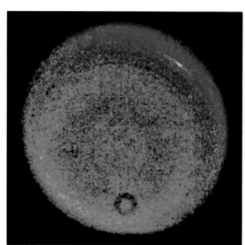

The clouds of Uranus. The schematic figure represents the relative altitude of the clouds in Jupiter, Saturn and Uranus, as a function of pressure. The clouds of Uranus are covered by a mist layer, much thicker than that of Jupiter. This explains the sharply contrasting colours of the two planets, Saturn is in an intermediate position.

The magnetic field and the magnetosphere. One of the biggest surprises of the Voyager 2 mission was the discovery of an inclination of about 60 degrees between the axis of the magnetic field and the rotation axis of Uranus. The magnetic field is fifty times stronger than Earth's. The surface intensity is therefore weaker, after considering the difference in sizes of the two planets. The magnetic field is probably generated by a dynamo effect inside the liquid mantle, which contains large numbers of ionised atoms. This strong magnetic field interacting with the solar wind drives zones comparable to the van Allen belt around the Earth. The rotation period of the magnetic field, that corresponds to the period in the interior of the planet where the magnetic field is generated, is determined by variations in radioelectric emissions. The period is 17 hours 14 minutes, which is different from the rotation period in the upper atmosphere. The unexpected orientation of the magnetic field has an important consequence. Ultraviolet emissions due to atomic (Lyman α) and molecular hydrogen have been observed from Uranus since 1984 by the ultraviolet satellite, IUE (International Ultraviolet Explorer), in orbit around the Earth. Polar aurorae (like those observed on Earth, Jupiter and Saturn, have been thought to cause the observed emissions. Charged particles falling into the atmosphere along the magnetic field lines cause the auroral lights. They are expected to appear day and night, near the magnetic pole. However, these emissions, surprisingly, are not concentrated, but are distributed over the disk; similar to those of Jupiter and Saturn, linked to low energy electrons interacting with the ionosphere independently of the magnetic field. This phenomenon, called electroluminescence, is only evident in the day and can eventually obscure any aurora. The discovery of the magnetic poles has shown that we have been looking for aurorae in the wrong places. A faint aurora has been observed on the dark side of the planet. They might also exist on the bright side, but be masked by the electroluminescence phenomenon.

The magnetosphere of Uranus extends up to 18 Uranian radii towards the Sun, with a long tail on the opposite direction. An electromagnetic environment much more astonishing than the one predicted was discovered during the 46 hours that Voyager 2 travelled through it. High energy electrons and protons were detected, but very few heavy ions of helium, carbon or oxygen. These ions are quite common in the solar wind, therefore the Uranium magnetosphere must be well isolated from it. Protons with kinetic temperatures of about a hundred thousand degrees are found everywhere; but it was only outside Miranda's orbit that highly energetic protons with kinetic temperatures of around ten million degrees were found. These protons could come from rings, or from the surfaces of satellites that are inside the magnetosphere, or even from the upper atmosphere (after JPL/NASA documents).

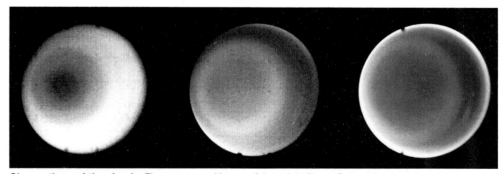

Observations of the clouds. The camera on Voyager 2 has eight filters. Several images taken through different filters can be superimposed to give a composite colour image. Images taken at different wavelengths reach layers at different depths into the atmosphere. Ultraviolet radiation is absorbed by particles (fog looks dark) and diffused by gas (a clear atmosphere looks bright). On the other hand, at the wavelengths were methane absorbs (red), gas absorbs and particles diffuse the light; therefore areas of fog look bright, and clear areas look dark. Uranus, seen through violet, orange and red filters is shown in this figure (from left to right). The images were obtained on 23 January 1986. Clouds deep in the atmosphere are detected through the red filter (0.619 microns). This wavelenght corresponds to a methane absorption band. The atmosphere is too opaque to the violet light, and the polar zone – covered in mist – looks dark. The individual clouds observed through the red and orange filters are probably the lightest ones of a deeper complex of clouds (JPL/NASA).

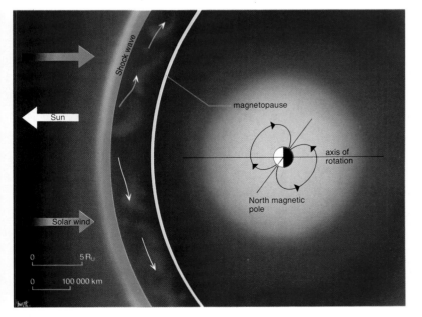

The internal structure of Uranus. Uranus is internally very different from Jupiter and Saturn. Its central temperature is on the order of 7000 K and its pressure, some twenty million times the atmospheric pressure on Earth. Starting at the centre, one probably encounters a rocky core of about 7500 kilometres in radius, hot, solid or liquid, and mainly formed by silicates and iron; then a mantle, more than 10 000 kilometres thick, formed by ices of water, methane and ammonia; and finally a gaseous envelope of hydrogen and helium that forms the atmosphere, observed from Earth, which also contains some minor compounds such as methane.

The pressure in the centre is not high enough for the hydrogen to become liquid and thus conduct electricity. Uranus is denser than Jupiter and Saturn although its mass is less. This means that Uranus has relatively less hydrogen and helium internally, even if the atmospheres of the three planets have very similar chemical compositions. This probably reflects the conditions at the time of formation. The magnetic field of Uranus is generated by ionised elements in the mantle. Uranus does not seem to show an important source of internal energy, as do Jupiter, Saturn and Neptune.

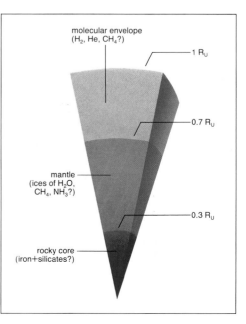

Uranus

The rings

One of the most important discoveries of recent years was the detection on 10 March 1977, of rings around Uranus during observations of the occultation of a star by the planet. Until then, only one planet – Saturn – was known to possess a ring system, and the discovery less than two years later of rings around Jupiter showed that the existence of rings around the giant planets was a common phenomenon. The rings of Uranus have many of the properties already noted in those of Saturn. Nine rings encircle the planet between 42 000 and 52 000 kilometres from the centre of Uranus. Compared to their circumference of 250 000 kilometres they are particularly narrow; eight of them are less than 10 kilometres wide. Three rings are circular and six are elliptical and of variable width. The characteristics of these six rings are well illustrated by the outer ring, which is the biggest. Its distance from Uranus varies by more than 800 kilometres and its width varies between 20 and 100 kilometres, linearly with the distance from Uranus.

The structure and size of these nine rings have been determined from Earth by stellar occultations, that is, by observing a star when Uranus passes in front of it. The presence of the rings around Uranus produces modulations to the stellar light just before and after the occultation. The position and structure of the rings were determined with a precision of a few hundred metres using this technique.

The existence of the rings was confirmed by the images taken by Voyager 2, and new ones were also discovered. Two of the new ones are at distances of 45 736 and 50 040 kilometres from the centre of the planet. Voyager 2 took images of the rings under very different light conditions. The angle between the Earth, Uranus and the Sun is always very small because the Earth is twenty times nearer to the Sun than Uranus; thus the angle between the probe, Uranus and the Sun, changed by more than 180 degrees during the encounter. Before the encounter, the probe had the Sun at its back and the images reproduced the rings found from Earth. After the encounter, the Sun was in front of the probe and the rings showed a very different structure. Small particles, with sizes on the order of the light wavelength (less than 1 micron) projected diffuse light forward, which was obscured by the probe after the encounter. These particles form a disk around Uranus and are not visible from Earth. This disk, observable in diffuse light, has a different structure from the previously known rings which have relatively few small particles.

Information about the rings has also been obtained by observing two stars (σ Sagitarii and β Scorpii) through the rings one after the other. Using this method a resolution of less than 10 metres for the rings δ and ε, and one hundred metres for the rest has been attained. Radio-occultation is another technique which has been employed. It consists of sending radio signals of known characteristics and frequency towards the Earth, and measuring the difficulty experienced by the signals in passing through the rings. The distribution of particle size in the rings is

The rings of Uranus in colour. This colour image of the dark and narrow rings was obtained by combining six images taken by Voyager 2 on 21 January 1986, ε is a neutral colour, while the other eight rings show clear difference; the rings δ, γ, and η have colours varying in shades of blue-green; α and β are lighter and 4, 5 and 6 are white. These colour differences are probably due to physical or chemical differences in the particles forming the rings (JPL/NASA).

estimated by comparing signals at wavelengths 3.6 and 13 centimetres, sent by Voyager 2 through the rings. These signals have been found to travel in a comparable manner, contrary to signals passing through the rings of Saturn. A 3.6 centimetre signal is stopped by particles around 10 centimetres in size, but a 13 centimetre signal is not. This is what occurs in Saturn's rings. Since both signals pass through the rings of Uranus it appears that the Uranian particles are longer and not as 'dusty' as those comprising the rings of Saturn. A different interpretation could be obtained if the observations are the result of

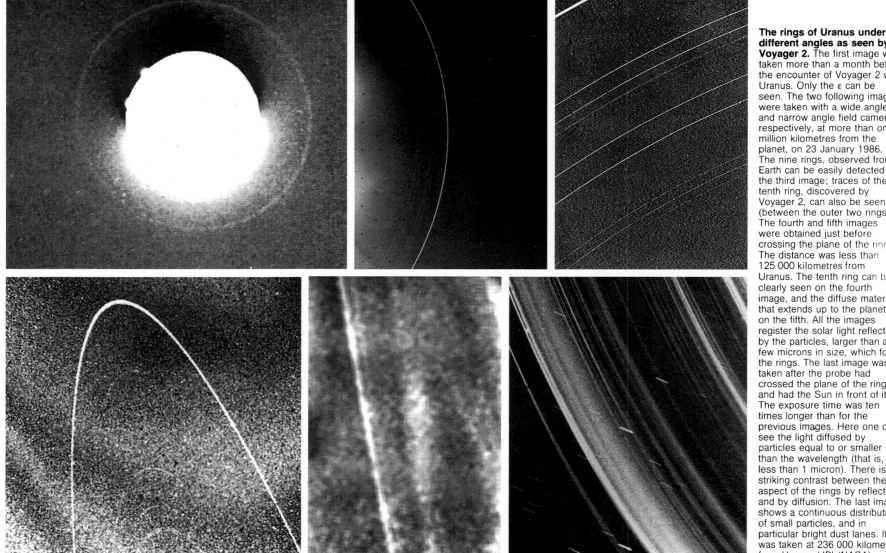

The rings of Uranus under different angles as seen by Voyager 2. The first image was taken more than a month before the encounter of Voyager 2 with Uranus. Only the ε can be seen. The two following images were taken with a wide angle and narrow angle field camera, respectively, at more than one million kilometres from the planet, on 23 January 1986. The nine rings, observed from Earth can be easily detected on the third image; traces of the tenth ring, discovered by Voyager 2, can also be seen (between the outer two rings). The fourth and fifth images were obtained just before crossing the plane of the rings. The distance was less than 125 000 kilometres from Uranus. The tenth ring can be clearly seen on the fourth image, and the diffuse material that extends up to the planet, on the fifth. All the images register the solar light reflected by the particles, larger than a few microns in size, which form the rings. The last image was taken after the probe had crossed the plane of the rings, and had the Sun in front of it. The exposure time was ten times longer than for the previous images. Here one can see the light diffused by particles equal to or smaller than the wavelength (that is, less than 1 micron). There is a striking contrast between the aspect of the rings by reflection and by diffusion. The last image shows a continuous distribution of small particles, and in particular bright dust lanes. It was taken at 236 000 kilometres from Uranus (JPL/NASA).

Comparison of the rings of Uranus and Saturn. The drawings show the relative sizes of the rings; the two planets are represented with the same radii; the radii are actually 26 145 km and 60 000 respectively. The distances in each figure are represented in units of the planetary radii. The Roche limit is indicated both for satellites which have the same density as the planet (red line) and a density of 1 gram per cubic centimetre (green line). The typical size of the particles which make up the rings of Uranus could be on the order of centimetres or metres, with a surface density between 10 and 100 grams per cubic centimetre, leading to a total mass for the rings of between 10^{15} and 10^{17} kilograms.

The table summarises the principal properties of the rings of Uranus. Rings 6, 5 and 4 have the smallest integrated optical depth and are the most difficult to detect. Ring α has a double structure: perhaps it is 'plaited'? In the case of ring β the occultation profile is flat-bottomed. Ring η is wide with a narrow dense component. Ring γ has much sharper edges than the others; the occultation profiles show the largest amplitude interference fringes, the same contrast as those which would be obtained from a sharp razor blade. There is probably a diffuse halo in the vicinity of ring δ. The width of ring ε varies between 20 and 100 km. Its profile is sharp-edged and similar at all points whatever its width. Its width varies linearly with distance from Uranus.

Small and large particles in the rings. Images of the rings taken before and after the encounter with Voyager 2 are shown here together. Rings α, β and δ, can be seen in both images, as can the excentric ε ring. The others are more difficult to recognise in the lower image; the bright dust lanes are relatively absent in the rings themselves.

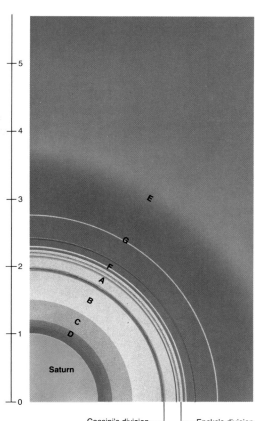
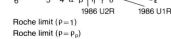

Roche limit (ρ = 1)
Roche limit (ρ = ρ_p)

Cassini's division —— Encke's division

surface characteristics of the particles, and not of their size.

The ring system is richer than previously envisaged, judging from the comparison of the images sent by Voyager 2, and by the radio and stellar occultations. The new rings – narrow, thin and sharp-edged – are shrouded in a disk of small particles. The outermost ring, ε, is about twenty metres wide, as ascertained from the occultation of σ Sagitarii. For a ring which is more than one hundred thousand kilometres in diameter, it is extremely thin.

The rings are quite dark in comparison with those of Saturn. The satellites of Uranus are also covered in dark material. This material reflects about 5 per cent of the light at all wavelengths. Its composition is not known. The somewhat lighter material found on the satellites could be a mixture of the dark material with water ice. One theory postulated to explain the dark material suggests that high energy protons have blown off the lighter hydrogen molecules from methane, leaving a dark layer of carbon. Alternatively, this dark material could be a mixture of carbon, opaque minerals and organic matter of the kind found in some carbonated chondrite meteorites. The first explanation implies that methane ices are plentiful around Uranus. The second is in agreement with observations of the relatively high densities of the Uranian satellites.

André BRAHIC

ring	mean optical depth	mean width (km)	radius (km)	radius (R_U)
6	0.2–0.3	1–3	41 850	1.60
5	0.5–0.6	2–3	42 240	1.62
4	0.3	2–3	42 580	1.63
α	0.3–0.4	7–12	44 730	1.71
β	0.2	7–12	45 670	1.75
1986 U2R	0.001–0.0001		45 736	1.75
η	0.1–0.4	0–2	47 180	1.81
γ	1.3–2.3	1–4	47 630	1.82
δ	0.3–0.4	3–9	48 310	1.85
1986 U1R	0.1	1–2	50 040	1.91
ε	0.5–2.1	22–93	51 160	1.96

Transit of the rings over the disk of Uranus. The left image shows the transit of the rings over the bright disk of Uranus, taken 27 minutes before crossing their plane. The right-hand figure shows the field of the camera and the geometry of the image; rings 5 and 6 show slight inclinations with respect to the equatorial plane of the planet (JPL/NASA).

The rings as seen by stellar occultations. The different structures of the rings are shown by computer reconstruction; γ ring is very narrow (around 600 metres wide and more than 100 000 in diameter), β is double the width of γ, ε has an extremely complex structure and δ is triple the width of γ. Some small thin rings (with widths of about 50 metres) seem to exist in addition to those already known.

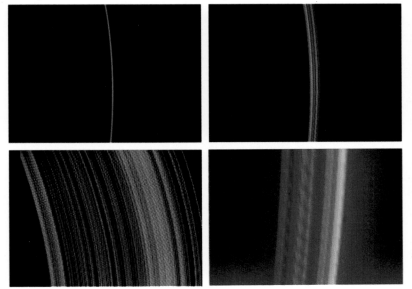

The guardian satellites. This image, taken 4.1 million kilometres from Uranus, shows two satellites on either side of ring ε. These satellites, known as 1986 U7 and 1986 U8, seem to be the 'guardians' of ring ε. Their radii are about 20 to 25 kilometres. This 'guardian' system resembles that of Saturn's F ring. The narrowness of the rings can be explained by the existence of such satellites: collisions, resonances and disturbances among the particles and between them and the satellites, produce an energy exchange between the ring and the satellites so that they repel each other. No guardian satellites have been found for the other rings, and there exists as yet no explanation for their narrowness (JPL/NASA).

Uranus

The satellites

Before the Voyager 2 flyby five satellites of Uranus were known. All follow direct circular orbits around the equatorial plane of the planet and are very difficult to see from Earth. Titania and Oberon, the two largest, were discovered by William Herschel in 1787; Ariel and Umbriel, were discovered by William Lassell in 1851; and the smallest, Miranda, was first observed by Gerard Kuiper in 1948. Absorption lines characteristic of ice can be seen in the spectra of Ariel, Titania and Oberon. A mixture of water and other ices and silicates probably make up these satellites; their temperatures and central pressures are probably too low for them to have a melted core.

Ten new satellites, all between Miranda and the planet, have been discovered through analysis of the images produced by Voyager 2. Thus, Uranus is currently known to have fifteen satellites, and there are probably more. This confirms that the environments of the giant planets are much more 'cluttered' than those of the terrestrial ones. The Uranian satellites, like those of Jupiter and Saturn, have proven to be more active and varied than was previously thought. Important traces of geological activity were unexpectedly found on the small frozen satellites; large numbers of impact craters had been expected. The diversity of the satellites and in particular their geological activity are indeed striking. The richness of phenomena increases as one approaches the planet, culminating with Miranda, the jewel of the Voyager 2 encounter. Before this encounter, study of the five largest satellites was limited to their orbits and global photometry and spectroscopy. Through the better spatial resolution obtained by Voyager it was verified that the satellites always show the same face to Uranus, and their diameters were accurately determined. The results have to be interpreted cautiously as only the illuminated southern hemisphere could be observed by Voyager 2. The northern hemisphere could be different, as is Iapetus in Mars. Major geological phenomena, if present in the northern hemisphere, would have escaped observation.

The satellites of the giant planets are formed by ices (of water, methane, ammonia and carbon dioxide) mixed with rocks. The lower the density, the higher the relative quantity of ices. The Uranian satellites are denser (around 1.6 grams per cubic centimetre) than those of Jupiter with the exception of Titan, whose density is 1.2. This indicates that they have less ices and that heating by the natural radioactivity of the rocks might have been important in the cores of the Uranian satellites.

Important geological activity on the surface of terrestrial planets or satellites is generally shown when they are sufficiently massive and have a hot core. The most massive bodies should show signs of such activity on their surfaces, as is the case with Venus and Earth, still geologically active, while Mercury and the Moon are inert. The satellite, Ganymedes shows many traces of geological activity while small Mimas is covered in craters. There are nevertheless, several exceptions to this rule: some small bodies, like Enceladus, have a complex surface. The energy source for this geological activity has yet to be indentified. Important energy sources can be provided by the natural radioactivity of the rocks, or even by gravitational perturbations by other satellites, when a body is already under the effect of tidal forces from the central planet; such as on Io with its striking volcanoes. In cases where there is so much geological activity or erosion, the surface of the planet accumulates the scars of meteoric impacts, and the craters cover the entire planet, thus allowing the study of the bombardment history.

Craters about 100 kilometres in diameter (population I) are probably produced by debris in orbit around the Sun since the beginning of the Solar System. On the other hand, craters of less than 50 to 60 kilometres in diameter (population II) are thought to have been produced by secondary debris due to collisions inside the system of satellites. A third class (population III), could be associated with comets. It is estimated that Uranus might 'capture' 600 to 700 times more comets than Jupiter, and around 100 times more than Saturn. These numbers are obtained both from theoretical models of the motion of comets and by extrapolating the number of comets passing in the vicinity of the Earth. The comets captured by Uranus were in fact in long period orbits around the Sun; these orbits have been transformed by gravitational perturbations due to Uranus, into short period orbits passing regularly by the planet. Eventually some comets could hit the satellites causing some of the craters observed on the surface. Nevertheless this process cannot explain all the craters observed, the majority of which were formed more than 4000 million years ago by the bombardment of the planetoids which formed Uranus and Neptune (large population I craters), and by small debris in orbit around Uranus (population II small craters).

The presence of Uranus in the centre of the system drives the bombardment; the nearer to the planet the more intense it is. It is possible that the inner satellites have been broken many times by the collisions, and later formed again. These collisions could also have provided the material that form the rings.

Small satellites covered by craters and with few traces of geological activity were expected by astronomers – the opposite was found.

Oberon and Umbriel contain an important population of craters between 50 and 100 kilometres in diameter, similar to those observed on the Moon's oldest terrains. Titania and Ariel, on the other hand, have very few such craters and the numbers of small ones increases rapidly with their size. Thus it appears that Oberon and Umbriel's surfaces retain the more ancient traces of population I, while Titania and Ariel show signs of having been remodelled by subsequent geological processes.

Scientists collect and try to understand thousands of pieces of information gathered from the neighbourhood of Uranus in order to compare them with theories of the formation and evolution of the planets, and possibly imagine new theories. Scientists hope that this information will be both a test of theories of planetary formation and evolution and a source of ideas for new scenarios. As the journey continues the Solar System will seem to us a more complex and amazing machine. Nature will always have more imagination than the brightest theoretician.

André BRAHIC

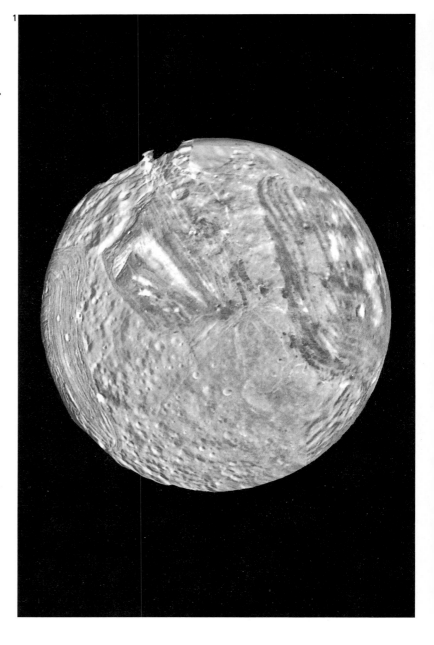

The satellites of Uranus. This photographic montage shows the relative sizes and albedos of the five largest satellites of Uranus: Miranda, Ariel, Umbriel, Titania and Oberon, from left to right. They are much darker than Saturn's satellites except for Phoebe and the dark face of Iapetus. Umbriel is the darkest. Their albedos are on the order of 30 per cent, 40 per cent, 15 per cent, 27 per cent and 25 per cent, respectively. It is not known what the dark material at the surfaces of the satellites and rings is composed of. It could be primaeval, and similar to that found in carbonated chondrite meteorites; or, it could be a surface layer of carbon uncovered by the removal of hydrogen atoms from methane as a result of bombardment by high energy protons (JPJ/NASA).

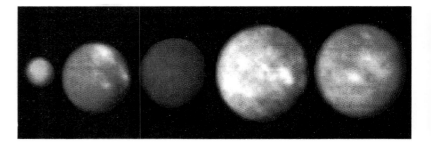

	semi-major axis		sideral period (days)	excentricity (days)	inclination (°)	radius (km)
	(km)	(R_U)				
Cordelia	49 700	1.90	0.33	0.0	0	(20)
Ophelia	53 800	2.06	0.37	0.0	0	(25)
Bianca	59 200	2.26	0.43	0.0	0	(25)
Cressida	61 800	2.36	0.46	0.0	0	(30)
Desdemona	62 700	2.40	0.47	0.0	0	(30)
Juliet	64 600	2.47	0.49	0.0	0	(40)
Portia	66 100	2.53	0.51	0.0	0	(40)
Rosalind	69 900	2.67	0.56	0.0	0	(30)
Belinda	75 300	2.88	0.62	0.0	0	(30)
Puck	86 000	3.29	0.76	0.0	0	85
Miranda	129 783	4.96	1.412 5	0.017	3.4	242
Ariel	191 239	7.31	2.521	0.002 8	0	580
Umbriel	265 969	10.17	4.146	0.003 5	0	595
Titania	435 844	16.67	8.704	0.002 4	0	805
Oberon	582 596	22.28	13.46	0.000 7	0	775

R_U is the radius of Uranus inclination is given with respect to the equatorial plane of the planet. The values in parentheses have uncertainties greater than 10 per cent.

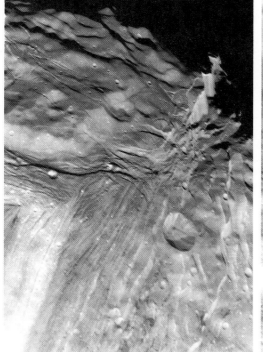

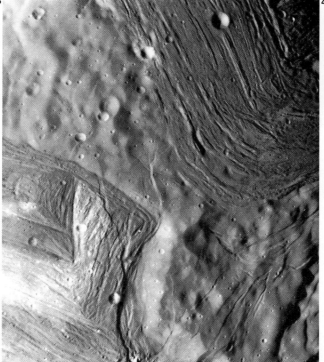

Miranda, the jewel of the Uranian system. The Voyager 2 probe passed by Uranus after being accelerated by Jupiter and Saturn, and therefore at a much higher speed. In addition, as Uranus receives 400 times less solar light than the Earth, longer image exposure times are needed, with the consequent risk of losing resolution. Technicians succeeded in reprogramming the onboard computers to allow the cameras to compensate for the motion of the probe, by changing the orientation so as to keep the object under observation fixed. Image quality is improved 100 times with the use of this facility, which was not available at the time of the probe's launch. Thus, details with a resolution of 500 metres could be observed on Miranda; otherwise, the resolution would have been more than 56 kilometres. In fact, the images of Miranda are the best ones obtained by the Voyager probes. All the images shown here were obtained on 24 January 1986. The first (on the opposite page) is a mosaic of images obtained at distances between 30 160 and 40 310 kilometres; the resolution varies between 560 and 740 metres. Miranda contains the most extraordinary show of the Voyager 2 encounter with the Uranian system. An unexpected variety of landscapes is observed in neat contrast: valleys, fractures, faults, cliffs, canyons, plateaux, craters, etc.; it is as this small satellite of less than 500 kilometres in diameter, has assembled over its surface a summary of the geological formations found in the Solar System. Miranda seems to be made out of bits of Mimas, Ganimede, Enceladus and Iapetus, with a crust from Mars and Mercury. Two clearly different types of terrain can be distinguished: old regions of relatively uniform albedo with many craters, and young regions with few craters, showing complex structures and characterised by dark and light fringes. The small number of impact craters indicates that these structures formed well after the formation of the satellite. Miranda is too small and light to have developed such an amazing structure and variety of geological formations. A speculative model of the origin has been put forward as follows. The satellite was formed by the accretion of small planetoids. Before developing a core, mantle and crust, it suffered a violent collision that broke it into rocks and ices. The pieces stayed in orbit around Uranus, to reassemble in a new body. Readjustment tensions cause the geological structures observed, when buried ices tried to emerge to the surface, and the rocks tried to submerge.

Image 2, was obtained at a distance of 36 000 kilometres; the resolution is on the order of 800 metres. The crater in the lower right quadrant is about 25 kilometres in diameter. Two different kinds of terrain are superimposed: a high one full of accidents, and a low one showing numerous alignments, judging from the number of craters the high terrain is older than the low one. A cliff with an estimated height of 5 kilometres can be seen in the top right corner.

Image 3, with a resolution of about 600 metres, was obtained at a distance of 42 000 kilometres. Different kinds of terrains can be identified, as well as a brighter 'rafter'; sinuous steep slopes, probably faults, can be seen. Craters of less than 5 kilometres in diameter pock mark the surface of these terrains.

Image 4, acquired at 30 600 kilometres, clearly shows the accidental character of Miranda, particularly on the limit, where 'mountain' profiles can be identified (JPL/NASA).

Oberon (left). This image was taken at a distance of 660 000 kilometres by Voyager 2 on 24 January 1986; it has a resolution of 12 kilometres. This outer satellite displays numerous impact craters surrounded by bright ejecta. A large crater is visible near the centre of the disk; it has a bright central outcrop and a background which is partially covered by dark material. A mountain, or perhaps the central peak of a crater, appears on the lower left edge. Its height is about 20 kilometres (JPL/NASA).

Umbriel (right). Umbriel remains an enigma. It is much darker than the other four large satellites and seems to be uniformly covered by craters, with the exception of a bright ring 40 kilometres in diameter. The ring can be seen in this image taken on 24 January 1986, at a distance of 557 000 kilometres. Numerous large craters indicate that the surface is old, although the presence of dark material and bright rings suggests recent changes. One can assume that the surface has recently been covered by dark material that erased the ancient topography, with the exception of the craters. We do not know if the dark material comes from Umbriel's interior or from a diffuse ring around Uranus, similar to the E ring around Saturn (JPL/NASA).

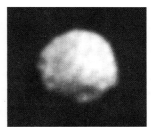

The new satellites. Ten new satellites all located between Miranda and the planet, were discovered by Voyager 2. 1985U1 (above) was discovered on 30 December 1985, just in time to reprogramme the probe and observe it under better conditions on 24 January 1986, at a distance of 500 000 kilometres. This satellite is very small and dark: it has a diameter of 170 kilometres and reflects less than 7 per cent of the solar light. Some craters can be seen on its surface (JPL/NASA).

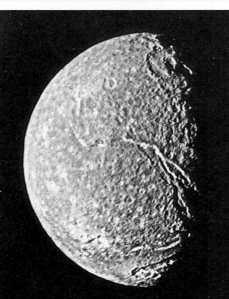

Titania (left). This is the largest satellite of Uranus. It resembles Oberon in size, colour, density and albedo, but has fewer large craters. More differences become apparent when it is observed at high resolution. Oberon and Titania are examples of satellites that, despite having similar size and density, have undergone quite different physical and geological evolutions. Numerous small craters, some with flat bottoms, and regions containing few craters suggest that the process of re-shaping the surface took longer on Titania than on Oberon. Numerous valleys, faults and fractures are scattered over Titania's surface. Some cut the large craters in half and seem little modified by the more recent small craters; they are probably the youngest geological formations on Titania. A grill of faults could be the result of crust fractures, caused by the expansion of water freezing underneath the crust. The image was taken at a distance of 368 000 kilometres, on 24 January 1986. A crater cut by a fault can be seen at the bottom of the picture (JPL/NASA).

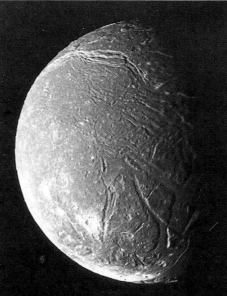

Ariel (left). Ariel is a spectacular satellite which appears to contain few craters. Its surface is much younger than those of Oberon, Titania and Umbriel. The oldest region is covered by craters with diameters less than 60 kilometres. Old large craters have disappeared, either due to viscous relaxation, or to eruptions from the inside. Ariel displays traces of considerable geological activity, in spite of its small size; several faults, fractures, valleys, depressions and cliffs can be observed. Material seems to have leaked to the surface. As in the case of Titania, we do not know whether the ices are water, methane ammonia or a mixture of them. It is possible that ice that leaked at the bottom of the valleys can be seen in this image, obtained at a distance of 130 000 kilometres on 24 January 1986 (JPL/NASA).

Neptune

Gravitating at about 4.5 billion kilometres from the Sun on a near-circular orbit, Neptune takes 165 years to complete one revolution. Its equator is inclined at almost 30 degrees to the plane of its orbit. Although three times smaller than Jupiter, Neptune is a giant planet which is 99 per cent composed of hydrogen and helium. With a diameter of 49 520 kilometres, Neptune is scarcely smaller than Uranus. However, its mass is slightly larger, of the order of 17.2 times that of the Earth (as opposed to 14.5 times for Uranus), making its density the largest of the giant planets (1.64 grams per cubic metre). As Neptune is significantly less massive than Jupiter or Saturn, and thus less compressed by gravity, it contains a larger proportion of elements heavier than hydrogen and helium.

The discovery of Neptune created a great stir in the nineteenth century. It was a landmark in the history of science because it marked the triumph of celestial mechanics: calculation led to the discovery of a celestial body more than 4 billion kilometres from the Earth. Since the end of the eighteenth century, astronomers had had difficulty in reconciling observations of Uranus with its calculated position. Alexis Bouvard, an astronomer at the Paris Observatory, was one of the first to notice these 'irregularities' in the motion of Uranus. Especially because of Francois Arago, the idea of an unknown body perturbing its orbit achieved currency. The Englishman John Couch Adams in 1843 and the Frenchman Urbain Jean Joseph Le Verrier in 1846 independently calculated the position and mass of the body with sufficient precision to allow its discovery in the constellation of Aquarius. Little notice was taken of Adams's prediction: Cambridge University did not possess up-to-date charts of Aquarius, and Adams's colleagues did not offer much help to the new scientist, whom they considered too young to make such a prediction. In contrast, on 23 September 1846, the same day he received Le Verrier's letter, Johann Gottfried Galle discovered the new planet at the Berlin Observatory, less than 1 degree from the predicted position. By a curious chance of history, 233 years before its discovery, Neptune had been close to Jupiter in the sky during the winter of 1612–13. Galileo, observing Jupiter on 28 December 1612 and 22 January 1613, had indicated Neptune in his sketches, thinking that it was a star. Neptune is an eighth magnitude object, and thus invisible to the naked eye, It appears in a telescope as a blue-green disc with an apparent diameter of 2 seconds of arc; some marks in its atmosphere can be discerned with difficulty.

Before the space age only two satellites were known: Triton and Nereid. By observing occultations of stars by Neptune in independent observations, French and American observers simultaneously detected at least two 'arcs' of material around the planet in 1984 and 1985.

But most of our knowledge about Neptune, its environment, satellites and rings, came from observations by the space probe Voyager 2 in 1989. Initially designed to explore Jupiter and Saturn, this probe was improved and repaired remotely with the aim of studying Uranus and Neptune. The exploration of the Neptune system was particularly delicate for two reasons however: the rings and the satellites are intrinsically dark, and the intensity of the Sun's radiation is 900 times weaker at Neptune's orbit than near the Earth. It was therefore not straightforward to take pictures from a probe which would, moreover, pass the Neptune system on 25 and 26 August at more than 27 kilometres per second. The engineers did nevertheless succeed in programming the movements of Voyager 2 so as to compensate for the motion during each exposure. In a few days the probe collected several thousand pictures and spectra, as well as millions of measurements of radio emission, magnetism, particle fluxes, and so on. It discovered six new satellites, a system of complete rings, and showed the complexity of Neptune's atmosphere, which is much more active than previously expected for so cold a body. The greatest surprise was the images of Triton, revealing a satellite with a complex geological history which still shows traces of activity.

The existence of violent winds, the persistence of large oval structures like immense eddies, as well as the great variability of smaller markers, was completely unexpected for an atmosphere receiving 20 times less radiation from the Sun than Jupiter, or 350 times less energy than the Earth. The large structures near the Equator move at a speed of 325 metres per second with respect to the interior of Neptune, while the small structures move twice as fast. Along with Saturn, Neptune is the planet with the fastest winds in the Solar System. As for Uranus, and unlike in Jupiter and Saturn, Neptune's atmosphere rotates less rapidly near the equator than at high latitude. The upper atmosphere has bright white clouds of methane ice in a very transparent atmosphere which overlies a cloudy layer containing ammonia and hydrogen sulphide ices. During the six month approach of the probe, many cloud-like structures appeared and disappeared in a few hours. However, three of them remained stable: the Large Dark Spot, the Small Dark Spot, and a a third, lighter spot called the Scooter.

The extremely rapid changes of the bright clouds (sometimes in less than 40 minutes) have greatly intrigued astronomers; some have suggested that they are actually the tops of convection cells; as it rises, the gas condenses in solid crystals in the cold regions of the atmosphere. Others think that they are the crests of atmospheric waves; these waves would have to be quite high and cold so that methane solidifies. Voyager 2 probed the atmosphere by transmitting radio waves through it, complementing visible, ultraviolet and infrared observations.

Like Uranus, Neptune possesses a reducing atmosphere, rich in hydrogen (unlike the Earth, which has an oxidising atmosphere, rich in oxygen); it is about 25 per cent helium and at least 1 per cent methane. The blue colour of the planet is in large measure caused by the absorption of red light by methane. In the upper atmosphere, at a pressure of order several hectopascals, methane molecules (CH_4), dissociated by solar radiation, recombine to form hydrocarbons such as ethane (C_2H_6) and acetylene (C_2H_2), which were detected by Voyager 2. Lower down, at a level of order 1300 hectopascals, methane condenses into ice crystals. Still lower, the presence of an opaque layer of hydrogen sulphide (H_2S) is suspected. It is not impossible that ammonia (NH_3) is also present at this level. Voyager 2's infrared detectors measured an average temperature of $-214\,°C$ (59 K). The equatorial and polar regions have approximately the same temperature, while the intervening regions are several degrees colder. Where the Sun's radiation is currently a maximum, that is at intermediate latitudes, the gas rises and cools, as on Uranus. Towards the equator and the poles it falls back, and is compressed and reheated. Computing Neptune's energy budget shows that the planet radiates 2.7 times as much energy into space as it receives from the Sun. The origin of this excess energy is not yet understood.

A week before the Neptune flyby, Voyager 2 detected bursts of radio emission at regular intervals, the first signs of the planet's magnetic field. This results from electric currents at great depth, allowing the scientists to deduce the period of internal rotation as the interval between adjacent bursts (16 hours 3 minutes). Neptune is slightly flattened by its rotation.

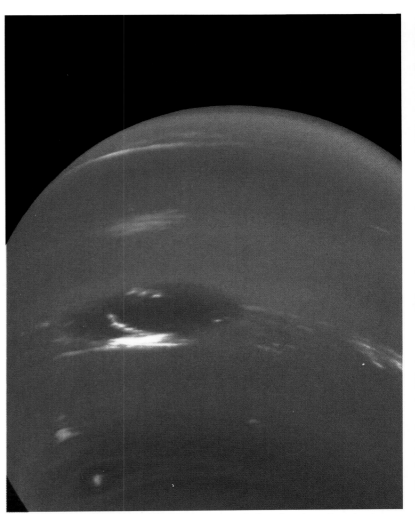

Neptune's upper atmosphere. The blue colour results from the absorption of red light by the small quantity of methane present in an atmosphere mainly composed of hydrogen and helium. This image, acquired by Voyager 2 on 21 August 1989 at a distance of 6.1 million kilometres, shows most of the marks visible in the upper atmosphere of this giant planet. The Large Dark Spot, near the centre, revolves in 18 hours 20 minutes at a latitude of 22° south; the Small Dark Spot, or D2, at lower left, near 55° south completes a revolution in only 16 hours, comparable to the internal rotation period of Neptune (16 hours 7 minutes). Finally, above and a little to the left of the latter spot, is a cirrus-like cloud which has been called the 'Scooter' because its rotation period of 16 hours 50 minutes is very short for a cloud. The white clouds probably consist of crystals of methane ice. Winds blowing towards the west at 2200 kilometres per hour with respect to the interior of the planet have been detected: these are the fastest observed in the Solar System.

The Large Dark Spot has a size similar to that of the Earth. Extending on average about 38 degrees in longitude and 15 degrees in latitude, it is proportionally as important as Jupiter's Great Red Spot, which extends over 30 degrees in longitude and 20 degrees in latitude. (JPL/NASA)

The Large Dark Spot. This spot has many similarities to Jupiter's Great Red Spot. Like the latter it shows an anticlockwise (i.e. anticyclonic) rotation, with a period of 16 days; this must be a high-pressure region. The two spots are also at similar latitudes: 22° south on Neptune, and 20° south on Jupiter. However, the Large Dark Spot seems to affect its environment less than the Great Red Spot. Further, the speeds of the two spots with respect to the planetary interior (towards the west in both cases) are very different: 300 metres per second for the Large Dark Spot, and only 3 metres per second for the Great Red Spot. (JPL/NASA)

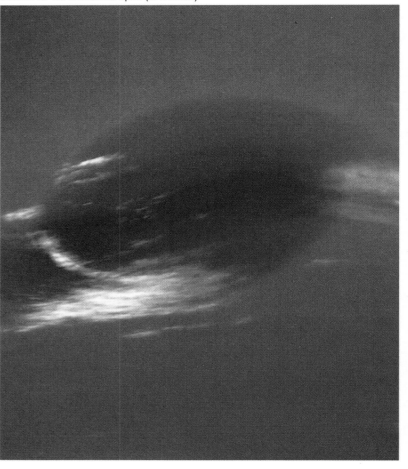

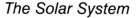

Evolution of the Large Dark Spot (left). This sequence shows the changing appearance of the spot over about four and a half days (the time lapse between each image is a Neptune rotation period, about 18 hour). A dark extension to the west gradually changes into a line of small lighter marks. (JPL/NASA)

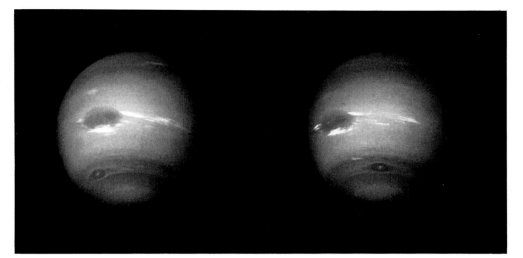

Differential rotation of the spots (right). Over the 17.6 hours separating the left-hand picure from that on the right, the Large Dark Spot performed a little less than one revolution about Neptune, while the Small Dark Spot performed more than one complete turn: the relative velocity of the two spots is of order 100 metres per second. (JPL/NASA)

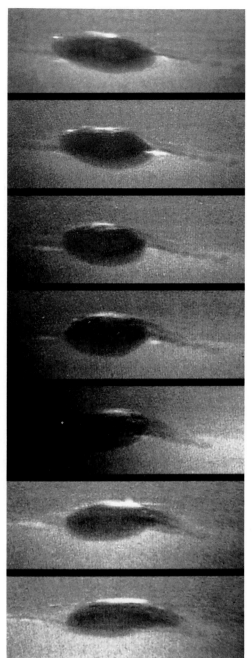

The south polar region (right) This high-resolution image (38 kilometres) shows that inside the Small Dark Spot there is a nucleus apparently consisting of many bright clouds, whose shapes and sizes varied rapidly (on scales of hours) during the Voyager 2 flyby. These bright clouds might form at the top of ascending matter in the centre of the nucleus. Unlike the Large Dark Spot, the Small Dark Spot would rotate clockwise, making it a cyclonic depression.

In the lower part of the image, near latitude 71° south, one can see an arc with bright ends extending about 90 degrees in longitude and less than 5 degrees in latitude. The intensity and brightness distribution of this arc changed sigificantly over a rotation period of Neptune. (JPL/NASA)

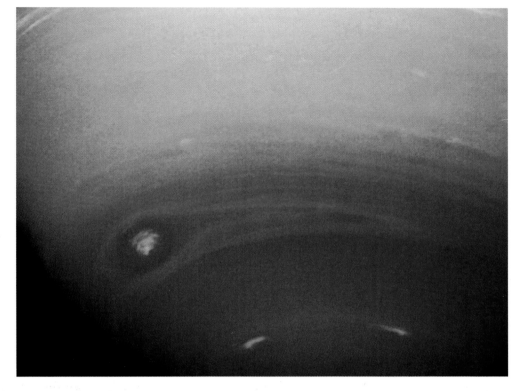

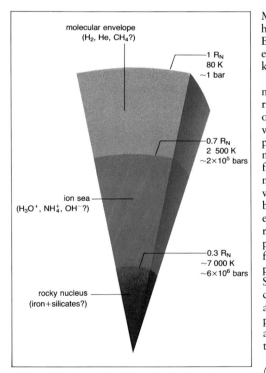

Schematic view of the internal structure of Neptune. A three-layer model appears currently to work as well for Neptune as for Uranus. The distance to the centre (in Neptune radii R_N), the temperature and pressure are shown. Unlike Uranus, it appears that Neptune's cooling has not yet finished.

molecular envelope
(H_2, He, CH_4?)

1 R_N
80 K
~1 bar

0.7 R_N
2 500 K
~2×10^5 bars

ion sea
(H_3O^+, NH_4^+, OH^-?)

0.3 R_N
~7 000 K
~6×10^6 bars

rocky nucleus
(iron + silicates?)

Measured at a depth where the pressure is 1000 hectopascals (the same as at sea level on the Earth) the polar radius is slightly less than the equatorial radius, at 24 340 and 24 764 kilometres respectively.

Neptune has a magnetosphere. The axis of the magnetic dipole is inclined at 47 degrees to the rotation axis; moreover, it is shifted: the source of the magnetic field is not in the core, but half-way between the centre and the boundary of the planet. When Voyager 2 flew by Neptune, the magnetic pole was pointing less than 20 degrees from the Sun, and the probe entered Neptune's magnetosphere via the polar funnel, where solar wind particles can penetrate most deeply before being repelled. This was the first time, with the exception of the Earth, that a polar magnetic region of this type had been explored by a space probe. These observations are very important for a better understanding planetary magnetospheres. Neptune's is the 'emptiest' in the Solar System: along the magnetic equator, where charged particles are most concentrated, Voyager 2 found only 1.4 protons or heavier particle per cubic centimetre, three times less than around Uranus and three thousand times less than around Jupiter.

The ultraviolet instrument detected an aurora (much weaker than those observed around other giant planets) and a faint diffuse luminescence on Neptune's night side.

André BRAHIC

Cirrus on Neptune. The shadows of narrow 'cirrus' bands over the terminator, around 27° north, allow an estimate of their 'altitude': they reach their highest point about 50 kilometres above the uniform cloud cover. The resolution of this image is 11 kilometres and the cirrus bands have a width varying between 50 and 200 kilometres (JPL/NASA)

Neptune

The Rings

Like the other giant planets, Neptune has rings. However, they are very unusual: they involve arcs of matter. The discovery of these arcs from the ground through observation of stellar occultations in 1984 and 1985 led to a change in the observing programme of Voyager 2, so as to study Neptune's environment better. The probe revealed that the planet is surrounded by a complete system of tenuous rings studded with bright arcs.

Astronomers had long wondered why Saturn was the only planet surrounded by rings. The discovery in one decade of rings around Jupiter, Uranus and Neptune showed that such systems are natural around giant planets. However, these four ring systems are quite different from each other: whether we look at rings, satellites, or planets, the Solar System shows a stupefying diversity.

The story of the discovery of Neptune's arcs is worth telling. Certain astronomers had long believed that rings could not exist around this planet because of the perturbations produced by the two irregular satellites Triton and Nereid. However, the presence of these abnormal satellites seemed at least to show that Neptune's environment was unusual. But observing/photographing such rings from the Earth was out of the question, as all dark material in the planet's immediate neighbourhood is drowned by its scattered light in a telescope. Only observations of the occultations of stars would permit the detection of such material from the ground: as a planet passes between a star and the Earth the starlight is refracted and then absorbed by the upper atmosphere of the planet.

The two main rings, the arcs, and the satellite Larissa. On this image acquired by Voyager 2 on 23 August 1989 at a distance of 2 million kilometres we can see the narrow rings 1989 N1R and 1989 N2R. In the lower part of the image the three bright arcs of 1989 N1R are clearly visible. They extend over 10, 4 and 4 degrees in longitude, and are separated by 14 and 12 degrees respectively. The average optical depth of the matter of 1989 N1R between the arcs is of order 0.01 to 0.02, compared with 0.04 to 0.09 for the arc material.

The satellite Larissa appears outside these rings in the upper part of the image, deformed by its orbital motion over the exposure time of 111 seconds (the second bright object is a star). The respective radii of the rings 1989 N2R, 1989 N1R and Larissa'a orbit are 2.15, 2.53 and 2.97 Neptune radii. (JPL/NASA)

The ring system. This view of the rings is actually composed of two images acquired by Voyager 2 on 26 August 1989 at a distance of 280 000 kilometres. The rings are illuminated from below, and the large white area partly blocked by the black band is the overexposed disc of Neptune.

One can easily see the two main narrow rings: outside, 1989 N1R (with three bright arcs, not visible here) which has a width of order 20 kilometres; inside 1989 N2R, at least 15 kilometres in width. A broad ring between 1989 N2R and Neptune is more difficult to see, this is 1989 N3R, with a width of about 1700 kilometres.

Finally one can make out a diffuse structure extending from the inner principal ring up to halfway to the outer main ring 1989 N1R: this is the ring 1989 N4R, called a 'plateau'; its width is about 5800 kilometres.

These rings are in direct circular orbits in the equatorial plane of the planet. (JPL/NASA)

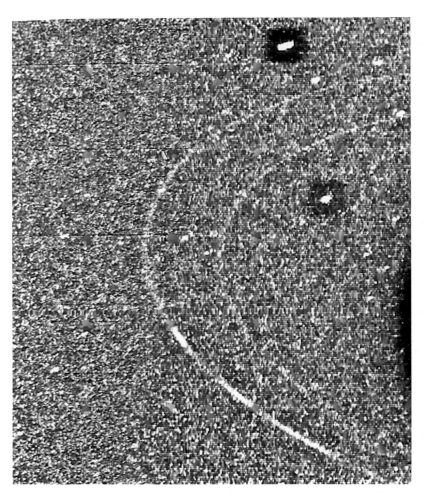

Main characteristics of the rings. R_N, Neptune's equatorial radius, is measured from the level of pressure 1000 hectopascals; it is 24 760 kilometres.

ring	mean optical depth	radius (km)	radius (R_N)	
1989 N3R	< 0.000 1	38 000	1.5	inner edge of 1989 N3R?
	0.000 1	41 900	1.69	40 to 70 per cent dust
	< 0.000 1	49 000	2.0	outer edge of 1989 N3R?
1989 N2R	of 0.01 to 0.02	53 200	2.15	40 to 70 per cent dust
1989 N4R	0.000 1	53 200	2.15	inner edge of 'plateau'
	0.000 1	59 100	2.4	outer edge of 'plateau'
1989 N1R	of 0.01 to 0.1	62 930	2.53	contains three bright arcs

The variation of refractive index allows one to measure the temperature of the atmosphere at different depths. Further, if the planet possesses rings, the starlight is reduced just before and just after the occultation, because it is blocked by the ring material. The rings of Uranus were discovered in this way. This type of observation is more delicate for Neptune, as this planet is further away from the Earth, and moves slowly across the sky with a surface area only half that of Uranus. Stellar occultations by Neptune are thus rarer. Further, the rings of Uranus are presently seen face-on, while those of Neptune are seen almost edge-on. However, because of its methane-rich atmosphere, Neptune (like Uranus) is very dark in the infrared, at 2.2 micrometres, and does not pollute the signal of the star passing behind the rings very much. An occultation, even of a relatively faint star, may thus be easily observed at this wavelength. As an example, a typical star a hundred times fainter than Neptune in the blue part of the spectrum is a hundred times brighter than it at 2.2 micrometres. In the blue, the occultation is almost undetectable; in the infrared it is easily seen. But not every weakening of the stellar signal is due to rings: rapid variations of atmospheric absorption (through turbulence, winds, etc), errors in guiding the telescope, small fluctuations in the electrical power supply or even the passage of clouds, aeroplanes or birds would all produce the same effect. One can, however, keep track of these extraneous effects as they occur at all wavelengths and are observed at any given time by one telescope only. One therefore has to observe an occultation at various wavelengths with different telescopes. Moreover, as the star can be regarded as being at infinity, the distance between two telescopes corresponds to the same distance at Neptune. Each telescope sees the star on a different apparent trajectory with respect to the planet. With several observations one can thus survey the environment of the planet. Until recently, a ring was regarded as detected if there were two interruptions of the signal corresponding to two intersections of the apparent trajectory of the star and the ring on each side of the planet. Now each detection is regarded as significant even if it is single, if it was seen by at least two telescopes, so as to eliminate extraneous effects.

A stellar occultation by Neptune was observed for the first time on 7 April 1968 from Australia. Ten years later, after the discovery of the rings of Uranus, some astronomers claimed to have seen the star disappear just before the occultation by the planet. Unfortunately this was not pointed out at the time, and the original data were not recovered.

The first systematic campaign to observe the neighbourhood of Neptune began on 10 May 1981. On 24 May two observers claimed to have observed a secondary occultation with two telescopes six kilometres apart in Arizona. It was realised only later that by an amazing chance they had actually observed the occultation of a star by the Neptune satellite 1989 N2. The most important observing campaign was organised from 15 June 1983 over the whole Pacific basin. However, from Hawaii to Australia, from China to California, no-one observed the slightest secondary occultation. After this setback, only a few astronomers decided to continue the work. Two teams observed the occultation of 22 July 1984: a French team led by André Brahic and Bruno Sicardy, at the European Southern Observatory, with two telescopes, and an American team led by William B. Hubbard, 100 kilometres further south in the cordillera of the Andes, at the Cerro Tololo Inter-American Observatory. Both teams detected a 35 per cent diminution of the stellar signal over 1.2 seconds, with 0.1 seconds difference between the telescope at the two sites. This occultation corresponded to an object of order 10 kilometres in width and at least 100 kilometres long, situated at least three Neptune radii from the centre of the planet, in its equatorial plane; but this could not be a continuous ring, as the star was not occulted on the other side of the planet. The same type of observation was performed a year later, in August 1985, by the same astronomers. At the Canada–France–Hawaii telescope, André Brahic and Bruno Sicardy observed a secondary occultation confirmed by a neighbouring NASA telescope. But in the Andes cordillera, William B. Hubbard saw no secondary occultation. This interruption of the signal from one side of the planet only caused the astronomers to conclude that the ring around Neptune was fragmented, and that one or several 'arcs' of matter were gravitating around the planet.

A little earlier, on 7 June 1985, the occulted star was actually double, and the observers detected a single secondary occultation (not two!) without being able to say which of the two stars had been occulted.

Between 1981 and 1989 almost a hundred stellar occultations by Neptune were observed. Seven of them showed the presence of matter around the planet. So as to avoid a possible collision between Voyager 2 and this matter, the Jet Propulsion Laboratory decided to increase the flyby distance slightly. And images taken by the probe from 11 to 26 August 1989 showed the rings of Neptune. This discovery provides a magnificent example of what can result from a collaboration between observatories and scientists of different countries; it emphasises again the complementarity of space- and ground-based observations.

At least four tenuous rings surround Neptune. The one furthest out contains three arcs of denser material extending over 4, 4 and 10 degrees; these were responsible for the secondary occultations observed from the ground. The names L, E and F were proposed for these names (for Liberty, Equality and Fraternity in the bicentenary year of the French Revolution). The rings are so tenuous that they cannot be observed from the ground; the space probe cameras only obtained the images after the longest exposures attempted in the course of the entire mission (up to 600 seconds without motion, as compared with a fraction of a second for Jupiter and Saturn).

We still have to understand how the matter around Neptune can be confined not only

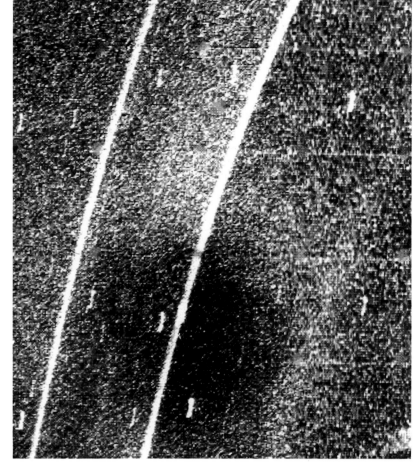

Close up of the rings. The inner ring 1989 N2R appears brighter than the outer ring 1989 N1R at this longitude. This image shows the broad diffuse ring 1989 N4R. (JPL/NASA)

radially (like the thin rings around Saturn and Uranus), but also azimuthally in the arcs, which appear to be stable. The explanation probably involves the interaction of close satellites with these arcs. The currently known satellites, however, are insufficient to explain this stability, but theoreticians are studying other hypotheses.

It is unlikely that a space probe will return to Neptune in the next 50 years. Until one does we must observe stellar occultations from the ground to follow the evolution of the arcs. With a little patience we may see a magnificent new ring form, when Triton breaks up as it reaches Neptune's Roche limit, in a little less than 100 million years!

André BRAHIC

The arcs at high resolution. The rings are seen here lit from behind, as Voyager 2 moved away from Neptune, unlike the image at the top of the preceding page, where the light was backscattered by the rings (the probe was approaching Neptune and the Sun was below it). The narrow rings as well as the three arcs of 1989 N1R are very bright for this viewing angle, showing that they contain many small particles, many more than the rings of Saturn or Uranus. The large white area at the bottom of the picture is the overexposed image of Neptune. (JPL/NASA)

Neptune

The satellites

Before the Neptune flyby of Voyager 2, only two satellites, Triton and Nereid, were known. They are called irregular as their orbits are strange: Triton has a retrograde orbit highly inclined to Neptune's orbital plane, and Nereid moves on a very eccentric orbit. In 1989, the images taken by Voyager 2 revealed six new satellites forming a regular system moving on direct circular orbits almost in the orbital plane.

Triton was discovered on 10 October 1846 by William Lassel, only seventeen days after the discovery of the planet itself. With a diameter of 2705 kilometres, it is one of the largest satellites in the Solar System. Because of tidal effects with Neptune (analogous to those between the Moon and the Earth) and its retrograde orbit, Triton is moving inexorably closer to the planet. In less than one hundred million years, when it is no more than one to two thousand kilometres from Neptune, it will reach the Roche limit of the planet and break into fragments of a few kilometres which will collide with each other: some will add to the rings, while others will smash into the planet.

Triton appears as a bright body whose surface is geologically very young; it has recent volcanic craters and a nitrogen atmosphere, like the Earth and Titan. The size, mass, tenuous atmosphere, and many other characteristics of Triton strongly resemble Pluto, the only planet not yet visited by a probe. For several decades Triton will offer us probably the best idea of Pluto that we can have.

Triton's orbit suggests that is was captured long ago by Neptune, for example after a collision with a satellite of Neptune which it destroyed, or by being braked in the cloud of gas and dust which probably enveloped the planet at its formation. Triton would then have moved in an eccentric orbit around Neptune. Tidal effects would then have slowed it and circularised its orbit after about a million years. In this energy exchange process Triton would have been heated and differentiated, the heavy elements

falling into the centre, forming a rocky core, while the lighter and more volatile elements would have condensed in the mantle and crust.

The images transmitted by Voyager 2 on the night of 24–25 August 1989 showed that Triton, the coldest body in the Solar System (−235 °C, or 38 K), is much more active than expected. With the same size as the Moon, Triton is far from being a dead world: like the Earth and Io it possesses active volcanos. At least four geyser-type eruptions were detected in Voyager 2 images (but of much larger scale than those on Earth). Columns of dark material several tens of metres to a kilometre in diameter rose vertically up to 8 kilometres altitude, where they form dark clouds swept along in winds over more than 100 kilometres. The cause of these eruptions is not yet understood. However, their proximity to the subsolar point suggests that the energy source has a solar origin. One model appeals to a greenhouse effect under the surface of Triton, assuming that this has very low conductivity. Just below its transparent surface, the Sun would heat the nitrogen ice, which would sublime and compress, exploding and throwing ice and dark particles into the atmosphere. A temperature difference of 4 K would be enough for the material to be ejected to an altitude of 8 kilometres. Every second, tens of kilograms of dust and several hundred kilograms of nitrogen would thus be injected into the atmosphere. An eruption could last a year or more, through sublimation of about a tenth of a cubic kilometre of ice. Other hypotheses have been proposed: some scientists assume that an internal heat source powers the geysers. Others wonder if these phenomena may be purely atmospheric, analogous to 'dust devils' which appear on Earth under clear skies, where instability conditions lead to the formation of spectacular vortices. In deserts, around midday, when the ground temperature exceeds that of the air, dust may be swept up by wind vortices. However, it does appear that Triton's geysers

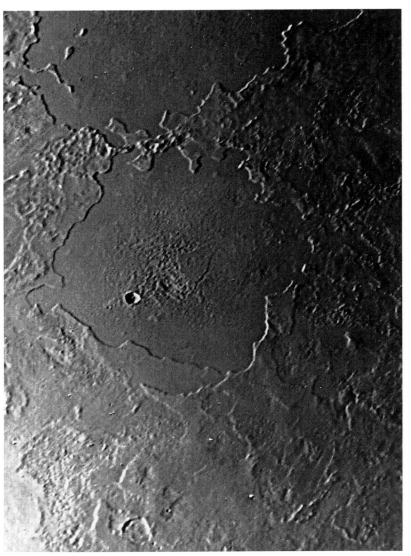

The plains of Triton. This image, with resolution of order 900 metres, shows the complex geological history of the satellite. Two large depressions (the second only partly visible) show a flat surface. This probably results from the flooding of ancient meteoritic impact basins by water, or a mixture of water and ammonia, coming from below, and subsequently freezing. Terraces show that several episodes of filling took place. The diameter of the plain at the centre of the picture is about 175 kilometres; it has only one impact crater, of about 15 kilometres diameter.

Triton's black marks. One of the strangest morphological features of Triton is the still enigmatic existence of very dark spots surrounded by aureoles of bright material. There are several impact craters in and around them, but no relief can be seen. These marks seem to be relatively recent; however, they are situated in the most heavily cratered, and thus the oldest, region observed by Voyager 2. The resolution of this picture is of order 3 kilometres; the largest aureole, at left, stretches over about 320 kilometres. (JPL/NASA)

Triton. This mosaic of a dozen pictures taken by Voyager 2 on 25 August 1989 shows a part of the southern hemisphere of Neptune's largest satellite. The south polar cap (below) is currently in sunlight, and has a high reflectivity. It seems to consist of a crust of nitrogen and methane ices deposited during the previous 'winter' eighty years ago. This crust is slowly subliming.

This satellite of rock and ices (of nitrogen, methane, water and ammonia) has a complex geology: one can see faults, valleys, cliffs, depressions filled with ice, meteoritic impact craters, and so on. Its surface is probably young (several hundred million years). One can see geysers of blackish gas rising to about 10 kilometres altitude.

Triton has a very tenuous nitrogen and methane atmosphere: the surface pressure is of order 0.015 hectopascals. (JPL/NASA)

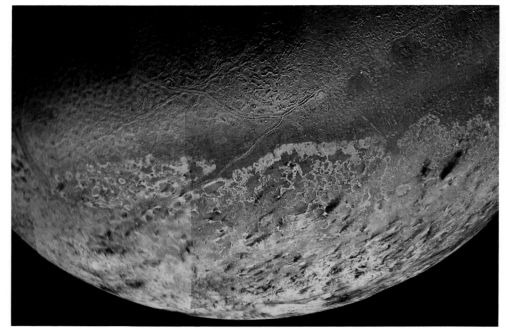

Two morphological units of Triton. The picture on the left has the best resolution – 750 metres – of any taken of Triton. It shows a region about 220 kilometres wide whose very complex surface has been described as 'canteloupe' after the rough-surfaced melon of that name. The region is uniformly covered by depressions which are probably not impacts craters but regions where a local fusion effect has caused many circular collapses. A dense network of folds and furrows gives evidence of periods of fracture and deformation of the surface of Triton. This type of terrain is unlike any other discovered in the Solar System.

The right-hand picture, with about 2.5 kilometres resolution, shows a zone near the terminator: the Sun is below the horizon, and the long shadows emphasise the relief. The lines could be collapsed ditches with ice crests in their centres. The lines have widths of 15 to 20 kilometres and are several hundred kilometres long. Few impact craters are visible, suggesting that the region is geologically young. (JPL/NASA)

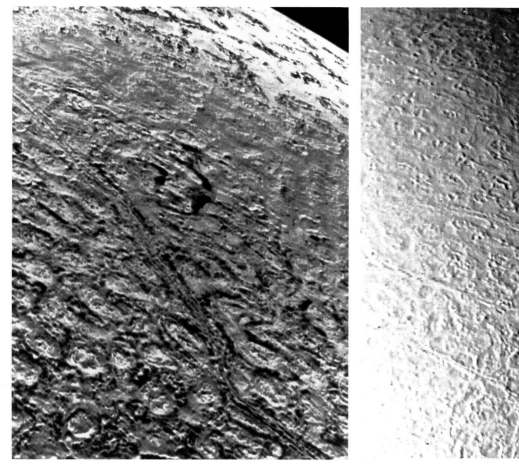

	orbital semi-major axis		sidereal period of revolution (days)	eccentricity	inclination (°)	radius or semi-major axis (km)
	(km)	(R_N)				
Naiad (1989 N6)	48 000	1.938	0.296	(0)	(0)	27 ± 8
Thalassa (1989 N5)	50 000	2.019	0.312	(0)	(4.5)	40 ± 8
Despina (1989 N3)	52 500	2.120	0.333	(0)	(0)	90 ± 10
Galatea (1989 N4)	62 000	2.503	0.429	(0)	(0)	75 ± 15
Larissa (1989 N2)	73 600	2.972	0.554	(0)	(0)	95 ± 10
Proteus (1989 N1)	117 600	4.748	1.121	(0)	(0)	200 ± 10
Triton	354 800	14.327	5.877	0.00	157	1352.5 ± 3
Néréide	5 513 400	222.637	360.16	0.749	29	170 ± 25

Neptune's satellites. R_N is the radius of Neptune; the inclination is measured with respect to the planet's orbital plane. The values in parentheses are uncertain by more than 10 per cent. Only Triton has a retrograde orbit. 1989 N2 was probably detected in 1981 by Harold Reitsema and William B. Hubbard during a stellar occultation observation.

are eruptive. They are, moreover, fundamentally different from those detected on Io.

The young active surface of Triton, possessing few impact craters, has recently been subject to ice fusion phases. Large plains and calderas seem to have been flooded by 'lava' of water, ammonia, and methane. Triton's crust should contain much water ice, which at 38 K behaves like a hard rock.

Measurements of perturbations of the orbit of Voyager-2 show that the mass of Triton is of order 2.14×10^{22} kilograms, with a density 2.07 times that of water.

The atmospheric pressure at the surface is very low, of the order of 0.016 hectopascals, i.e. about 100 thousand times less than at the surface of the Earth. Triton should have a tropopause at a height of order 25 to 50 kilometres. Nitrogen molecules are transported from the south pole, currently in sunlight, to the north pole, which is in shadow.

Discovery of Neptune's second satellite took another century. Nereid was found in 1949 by Gerard P. Kuiper; even today we still have little information about it, as Voyager 2 passed 4.7 million kilometres from it, and could only provide a low-resolution image.

Of the six small satellites found by the probe, only Proteus and Despina could be photographed so as to reveal surface details: they seem to be irregular and covered in craters.

With the exception of Triton, all Neptune's satellites have very dark surfaces: Nereid reflects only 14 per cent of the sunlight its receives, and the small satellites about 6 per cent.

Some satellites of Neptune, particularly the smallest, may be the fragments of a larger primitive satellite which was broken in a collision.

André BRAHIC

The orbits of Triton and Nereid. Triton has a retrograde orbit highly inclined to the planet's equatorial plane. The orbit of Nereid is direct, but is also very unusual: it is the most eccentric in the Solar System.

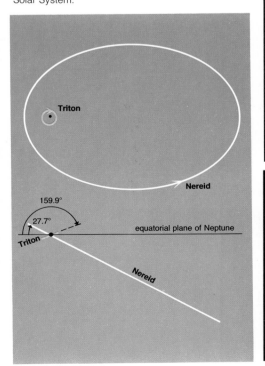

Proteus and Larissa. These two small satellites were discovered by Voyager 2. Proteus (opposite) is very dark – its reflectivity is about 6 per cent. This picture has a resolution of about 2.7 kilometres; furrows and numerous impact craters are visible, the largest (near the terminator) being about 150 kilometres in diameter.

Larissa (below) has several craters of 30 to 50 kilometres diameter on this picture of about 4.2 kilometres resolution; its reflectivity is about 5 per cent. (JPL/NASA)

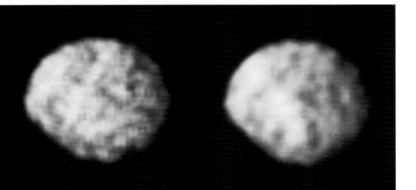

Pluto

The last planet to be discovered in the Solar System is still something of a mystery, and will probably remain so for several decades, as no space probe to it is planned. In telescopes it appears as a faint point of light without any discernible surface features.

However, several important discoveries have been made in the last two decades. In 1976 spectroscopic observations showed that the surface of Pluto is at least partially, and perhaps even entirely, covered in frozen methane. In 1978 a satellite, Charon, was discovered. Between 1979 and 1985 the diameter and mass of Pluto were accurately measured: with a diameter of 2300 kilometres, this planet is smaller than the Moon, and has a mass 500 times less than that of the Earth. In 1988, the observation of a stellar occultation by Pluto showed the existence of an atmosphere. In 1989, the exploration of Triton by the Voyager 2 probe revealed an object which so closely resembles Pluto as to be something of a twin brother.

Pluto makes one revolution around the Sun every 247.7 years in an unusual orbit for a planet: it is neither circular nor in the plane of the ecliptic. The orbit is inclined by 17° with respect to this plane, so that the planet rises 1.25 billion kilometres above it, a distance similar to that between the Sun and Saturn. The orbital eccentricity (0.25) is by far the largest among the planets. It is large enough that since 1979 Pluto has been closer to the Sun than Neptune. In 1989 it passed its perihelion, and it will not become the most distant planet in the Solar System again until 1999. In 2113 Pluto will reach aphelion: its distance from the Sun will be more than 7.5 billion kilometres.

One might think that the orbit of Pluto could intersect that of Neptune, and that one day these two planets might collide. In fact this cannot happen. Even though viewed 'from above' the orbits of Pluto and Neptune appear to intersect, the inclination of Pluto's orbit is such that these orbits are never close to each other. At perihelion Pluto is above Neptune's orbital plane by more than one quarter of the Neptune–Sun distance. In fact Pluto's orbital inclination and motion in its orbit leave the maximum safety margin against collision. Pluto and Neptune never get closer than 2.5 billion kilometres from each other, although the closest points on their orbits are much nearer than this; it is interesting to note that this minimum distance between Pluto and Neptune is well above the minimum distance between Pluto and Uranus, which is only 1.6 million kilometres. Not only is the perihelion of Pluto far away from Neptune's orbit, the motions of Pluto and Neptune on their orbits are synchronised in such a way that at the moment that Pluto reaches its perihelion, Neptune is more than 60° away from the Sun on its orbit. The mean period of Pluto's motion about the Sun is exactly 1.5 times that of Neptune, which means that every 495 years (after Neptune has performed three revolutions about the sun and Pluto two) the two planets are in the same relative positions. At the moment when Neptune 'overtakes' Pluto in its orbit, Pluto is at aphelion. This synchronisation is clearly not accidental: this is a dynamical phenomenon known as a 'stable resonance'. The resonance between Neptune and Pluto was discovered using powerful computers which allow one to reconstruct the motions of the two planets over the last five million years. Stable resonances of this kind play an essential role in the motions of the planets and satellites of the Solar System, and they explain in large measure the present configuration of the bodies within it; the planets and their main satellites move so that there is no danger of collision with each other. If a small perturbing force acted on Pluto to change its motion about the Sun, the resonance with Neptune would return it to the present resonant state. If in the distant past Pluto had been in a non-resonant orbit near its present orbit, the perturbing force of Neptune's gravity would have changed its orbit until it was permanently locked in the present resonant orbit.

Pluto appears anomalous when compared with either the four terrrestrial planets or the four giant planets. This frozen world is more like a large asteroid or Triton; some people have indeed suggested that Pluto is an escaped satellite of Neptune.

Infrared observations show the presence of methane on the surface of Pluto. These observations seem to show that the surface is dark and reddish at the equator, while the poles are covered in methane ice. The brightness of the polecaps varies with the distance from the Sun. Seasonal phenomena linked to the sublimation of ices heated by the Sun occur on Pluto's surface. In contrast, the surface of Charon does not seem to possess any methane, but is covered in water ice.

Observation of an occultation of a star of twelfth magnitude by Pluto on 9 June 1988 showed the presence of an atmosphere around Pluto. As the star disappeared behind Pluto its light decreased gradually rather than suddenly, which would have happened if the planet had had no gaseous envelope. Pluto's atmosphere seems to contain methane and a heavier gas (perhaps carbon monoxide or nitrogen). The surface atmospheric pressure is less than one hundred thousandth of that of the Earth. The characteristics of Pluto's atmosphere probably vary greatly with the seasons; the pressure is a maximum when Pluto is closest to the Sun, and may decrease by a factor of ten near aphelion.

In 1978 it was noticed that enlarged photographs of Pluto showed a slight elongation: a satellite had been discovered. It was called Charon.

The orbit of Charon is inclined at 118° to that of Pluto about the Sun. If, as is likely, Charon moves in Pluto's equatorial plane, the planet must have its rotation axis close to the plane of

Neptune and Pluto. The orbits are seen from the North ecliptic pole and 'from the side' (the line of sight is the line of nodes). Upper picture: orbits projected on the plane of the ecliptic; the part of Pluto's orbit below this plane is shown dashed. An orbital resonance keeps the two planets far enough from each other that there is no danger of a collision. Their positions at the moment of closest approach to each other are shown; this occurs close to conjunction, and their separation is about 18 AU. Another approach occurs at points not near conjunction, but the distance is of the same order (25 AU).

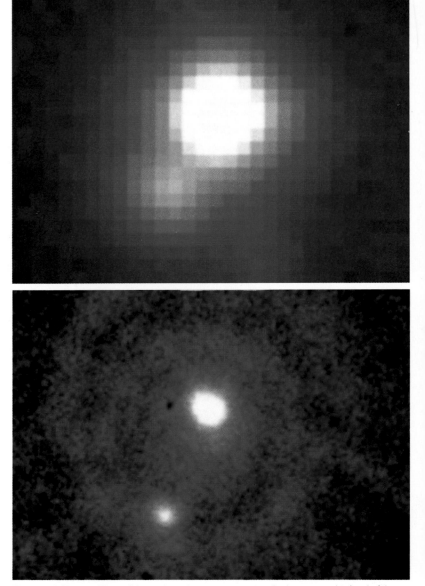

Pluto and Charon. These false colour pictures are among the best ever obtained of the Pluto–Charon system. The upper one was taken on 20 May 1990 with a ground-based CCD detector sensitive in the near infrared, mounted on a high-resolution camera built by the Dominion Astrophysical Observatory, Canada, at the prime focus of the Canada–France–Hawaii telescope. The aperture was restricted to 1.2 metres. The resolution of this picture is an exceptional 0.48″ over an exposure of 90 s. The angular separation of Pluto and Charon (appearing as an outgrowth at the lower left) was then 0.88″.

The lower picture was taken by the European Space Agency's Faint Object Camera on board the Hubble Space Telescope. The angular separation of Pluto and Charon was about 0.9 here also. The two objects are perfectly distinguishable, however. Charon is fainter than Pluto, partly because of its smaller size, and partly because its surface is thought to be covered in water ice while that of Pluto is thought to consist mainly of methane frost or 'snow', which has a higher reflecting power. (top, D. Salmon, CFHT; bottom, ESA)

The discovery of Pluto. Even near to perihelion, Pluto has an apparent diameter of 0.2″ less than the resolution limit of ground-based telescopes. On most photographic plates it therefore appears as a faint point of light lost among the twenty million brighter stars in the sky. Pluto is revealed only by its motion: when we compare two photographs of the same star field taken 24 hours apart (as is the case with the photographs above), anything that moves in relation to the stars in the field will be noticed. When Pluto is at opposition its motion can be as much as one arc minute per day. By using this method of comparison, Clyde William Tombaugh identified Pluto. Encouraged by the success of Urbain Jean Joseph Le Verrier and John Couch Adams in the prediction and discovery of Neptune, at the beginning of the nineteenth century, numerous astronomers began looking for a ninth planet by analysing the as yet unexplained perturbations of the orbit of Uranus (the orbit of Neptune was not at the time known with sufficient precision to be used). Percival Lowell and William Pickering, in particular, at the beginning of this century predicted the existence of a transNeptunian planet. Beginning in 1905, photographic research was undertaken at the Lowell Observatory in Arizona and the Mount Wilson Observatory in California. Percival Lowell put into operation an intensive photograph search but it was not until 1930, several years after his death, that the ninth planet was discovered after very long and systematic photographic searches that covered a large part of the sky. In 1929, Tombaugh began to search using a new telescope specially made for this purpose, capable of photographing a wide field. After a year of minutely examining hundreds of plates, he discovered a tiny object of magnitude 15, and about 5 degrees from the predicted position. The date was 18 February 1930. This planet, like the others, was given a name from Greek mythology: Pluto (the first two letters preserving the memory of Percival Lowell). It was believed that the story of the discovery of Neptune had repeated itself. Recent determinations of the mass of Pluto throw this idea into question, for the mass is too small to disturb the motions of Uranus and Neptune noticeably. Furthermore, observational errors seem to have led to an overestimation of the perturbations. The discovery of Pluto therefore seems to be not a matter of precise mathematical calculations, but rather the fruit of systematic searching. In other words, the calculations certainly led to the discovery, but themselves were mistaken! (Hale Observatories).

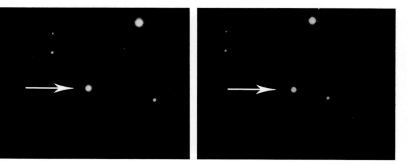

its orbit, like Uranus. With Venus and Uranus, Pluto may be the third planet to spin in a retrograde fashion.

As it moves about Pluto, Charon passes in front of and behind the planet. These eclipses and occultations, observed since 1985, have allowedmeasurement of the diameters of both bodies as well as their reflecting power. Pluto's albedo is about 50%, and that of Charon about 37%. Both objects have a density of about 2. Pluto probably has a rocky core and a mantle containing ices of water and methane. Its surface is thus much brighter than assumed when Pluto was thought to be a rocky body.

A tenth planet?

In the last two centuries we have discovered three of the nine planets, and it is natural to ask if there may be a tenth. It is doubtful that one will again discover a large planet by accident, as in the case of Uranus: if the object is very distant, much further than Pluto, it will be indistinguishable from a star, while if it were nearer it would already have been discovered long ago through its perturbations of the motion of Neptune. A tenth planet, if it exists, must either be a small object of the size of Pluto situated a little further out, or a giant planet like Jupiter much further out than Pluto. This far from the Sun, a tenth planet would have such small

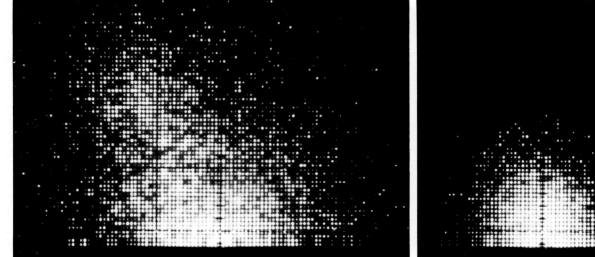

Observation of Charon by speckle interferometry. In 1970 the French astronomer and optics expert Antoine Labeyrie suggested the method of *speckle interferometry*. The idea of this method originates in the similarity between the grainy nature of stellar images at the focus of a large telescope and the coherent granulation produced by laser light which has passed through a diffuser. In optical astronomy the Earth's atmosphere acts as the diffuser; local temperature fluctuations cause fluctuations of the refractive index of the air, so that the atmosphere behaves as a non-uniform optical medium. Starlight can be strongly coherent if we observe an unresolved star in a narrow spectral range. In all cases an individual granule can be light or dark depending on whether the random phases of the many wavefronts reaching it from the diffuser interfere constructively or destructively. A stellar image thus appears instantaneously as a spot of several seconds angular diameter studded with bright granules. Speckle interferometry involves following the image in a narrow wavelength range so as not to

decrease the contrast of the figure of interference. The exposure time is currently between 1/50th and 1/1000th of a second so as to freeze the atmospheric turbulence. As an object is observed several hundreds or thousands of images are therefore recorded on photographic film using a very sensitive television camera. The photographic images are usually treated by Fourier transform optics, i.e. by recording the diffraction pattern obtained by illuminating each image with a laser beam.

The use of photon-counting television cameras has allowed speckle interferometry of faint objects. Thus in 1980 Pluto and its satellite Charon (discovered in 1978) were observed by Daniel Bonneau and Renaud Foy using the Canada–France–Hawaii telescope. The picture on the left, obtained on 23 June 1980, is typical of the Pluto–Charon system: the satellite is seen as an outgrowth at the top left; the picture on the right, typical of an unresolved star in the field of Pluto, acts as a reference object. (material kindly supplied by R. Foy)

	Pluto	Charon
distance to Sun in 1991 (AU)	29	30
sidereal rotation period (days)	6.3867 (retrograde)	5.9 (retrograde)
diameter (km)	2 300	2 700
mean density	2.1	2.05
albedo (%)	50	70
surface temperature (K)	50	37
atmospheric pressure (atm) .	≃ 0.01	0.01
surface composition	ices of methane and nitrogen organic ices	ices of nitrogen and methane nitrogen, methane
atmospheric composition	methane+(nitrogen, carbon monoxide, neon)?	

Comparison of orbital and physical characteristics of Pluto and Neptune's largest satellite Triton.

The nature of Pluto's surface. An infrared spectrum of Pluto (black dots sometimes extended up or down by an error bar) is compared with a laboratory spectrum of methane ice (CH_4) (continuous red line). The comparison shows that Pluto's surface is covered with methane ice (revealed by strong absorption features at 1.7 and 2.3 micrometres wavelength), confirming observations showing that the surface is very reflective. (from B.T. Soifer, G. Neugebauer and K. Matthews, 1980)

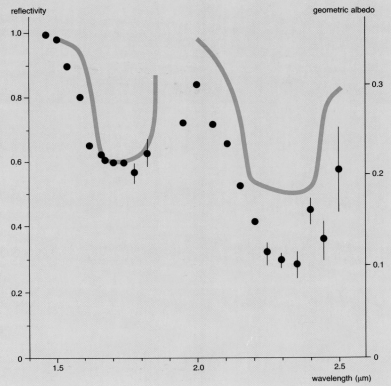

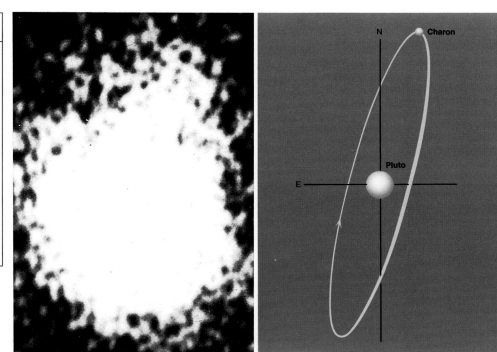

A double planet. Neither the usual images nor those obtained by speckle interferometry give a good idea of the real dimensions of Pluto and Charon or their separation. This historic photograph, taken on 2 July 1978 by James W. Christy, using the 1.55 metre aperture telescope of the US Naval Observatory, nevertheless allowed the discovery of Charon, which is seen as an extension to the top right of the main image, which is that of Pluto. The grain of the photographic emulsion is clearly visible on this considerably enlarged photograph.

The diagram shows the true relative sizes and positions of the two objects at the moment the photograph was taken. It is not on the same scale as the photograph. Charon orbits at about 19 000 km from Pluto and has a diameter of order 1190 km. In comparison to the size of the central planet (of diameter 2300 km) this is the largest satellite in the Solar System. Following the example of the Earth–Moon system, Pluto can be regarded as a double planet: Pluto and Charon are separated by only eight Pluto diameters, and the Earth and Moon by thirty Earth diameters. The rotation period of the satellite (6.39 days) is probably equal to the period of Pluto. Pluto and its satellite thus form the only synchronously rotating and spinning pair in the Solar System: other satellites generally have a spin period equal to their orbital period about the planet – so that they always present the same face to the planet – but the planet's spin period is different. (from *The New Solar System*, 1990)

luminosity and apparent motion among the stars that detecting it would be extremely difficult. All studies of bodies moving slowly with respect to the stars have been fruitless, but have led to the discovery of many asteroids and comets. The Pioneer and Voyager probes offer faint hopes: if by chance they pass close to a new planet, the deviation of their trajectories should be detectable.

In any case, the orbit of Pluto definitely does not mark the edge of the Solar System. Billions of comets gravitate around the Sun and move almost 1500 times futher away from it than Pluto; the nearest known star is more than 6000 times further from us than Pluto. We are far from the frontier!

André BRAHIC

Comets

Comets which are feared as much as thunder,
Stop terrifying the people of the Earth,
In an immense ellipse complete your course,
Arise, descend, close to the daytime star,
Start up your fires, fly away, and return without
ceasing,
Rekindle the old age of exhausted worlds.

Voltaire

Written in 1738 by Voltaire to his friend the Marquis de Châtelet, this verse dramatically illustrates a complete revolution that had recently taken place in the general understanding of comets. Until that time, these 'stars', which were set on an apparently erratic course and whose spectacular and rapidly changing appearance was quite unpredictable, had been regarded with fear and superstition as fearful omens heralding great catastrophes. But in the seventeenth century it was finally realised, due mainly to the work of Johannes Kepler, Isaac Newton and Edmund Halley, that the apparently strange motions of comets on the celestial sphere actually obey the same laws of motion as do the planets – namely, they travel in elliptical orbits with the Sun at one focus. In the case of comets, the ellipse is simply more elongated – more eccentric – than the ellipses that are traced out by the planets, which, with the notable exception of Mars and Pluto, are almost circular.

What is a comet? What is its physical nature? An analysis of the observations accumulated over recent decades leads us to believe that all the observed cometary phenomena can be explained in terms of what happens to a solid body, the nucleus, with a diameter of a few kilometres (which is too small to have ever been seen directly) composed mainly of ice or compacted snow, closely mixed with solid dust particles of different sizes that are, in general, of the order of a micrometre. This hypothesis was brilliantly confirmed during the encounter of the European and Soviet probes Giotto and Vega with Halley's Comet in March 1986.

It is the motion of the solid nucleus that defines the orbit of the comet. This motion obeys Kepler's laws in the first approximation. This nucleus is cold and naked when it is at aphelion but when it approaches the Sun the temperature of the surface rises, the ice evaporates, and the resulting gas escapes rapidly into the surrounding space trailing all the particles of dust in a gigantic tail many tens of millions of kilometres long. This dust cloud is the main source of the luminous phenomenon observed, and was baptised 'comet' after the Greek word 'kometes', meaning head of hair.

At each perihelion passage the nucleus thus loses a substantial part of its mass; one can calculate that a comet is 'consumed' in a limited number of perihelion passages – between a hundred and a thousand. It has been established that the brightness of Comet P/Encke, whose period is 3.3 years, is gradually diminishing.

Even for a comet of long period, the lifetime, defined as the total time for all orbits during which it becomes active each time it gets close to the Sun, does not exceed some tens of millions of years at maximum, which is very short relative to the age of the Solar System (4.6 billion years). There is every reason to believe that the nuclei of comets – Fred L. Whipple has called them 'dirty snowballs' – were formed at the same time as the planets. Since the formation of the Solar System they have been 'preserved' at great distances from the Sun in the immense 'refrigerator' that constitutes the outer Solar System. From time to time, a chance event, for example an infinitesimal perturbation caused by the relatively close approach of another star to the Sun, hurls a nucleus towards the interior of the Solar System where it becomes a visible comet and evaporates relatively rapidly in at most some hundreds of

orbits in a magnificent and spectacular swansong.

With this picture, modern physicists no longer consider comets as messengers of the gods, as was once believed, but as messengers of the past, suddenly appearing, intact and virginal, presenting direct evidence of the physical conditions existing at the epoch of formation of the Solar System. It is thus clear why the study of comets has generated so much interest; all theories of formation of the Solar System clearly must include an explanation of the formation of cometary nuclei.

By studying the various phenomena that appear at each perihelion passage of a cometary nucleus, astronomers try to find out about the physical nature of nuclei, which have been inaccessible to direct observation until space probes approached the nucleus of Halley's Comet. The nucleus, a relatively fragile ball of snow and dust, gives rise, when it approaches the Sun, to the three main components; the coma (or head), the hydrogen cloud and the plasma tail. These three components, which are between 1 and 10 million times larger than the nucleus, are extremely tenuous and are only observable by their interaction with solar radiation. Their combined mass is only 300 000 tonnes, compared with the billion to 1000 billion tonnes of the nucleus.

Photographs of comets generally show two very distinct sorts of tail: a dust tail, which appears pale yellow because it reflects sunlight, and a plasma tail (of ionised matter), which appears blue because the principal emission is from the CO^+ ion (ionised carbon monoxide) at a wavelength around 420 nanometres and thus in the blue part of the spectrum.

Jérôme Frascator, a medical doctor from Verona, first remarked in 1538 that 'the tails of three comets observed recently were in the

Comet Bennett. Shooting stars can often be seen, particularly on clear moonless nights. They are a spectacular, very short lived phenomenon, lasting about a second. The bright trail is produced by the disintegration in the Earth's upper atmosphere (at an altitude of about 90 kilometres) of solid pieces of extraterrestrial material (meteoroids) heated by atmospheric friction.

People who have never seen a comet often confuse them with shooting stars. A comet looks completely different; the luminous trail appears fixed in the sky, and follows the slow daily movement of the stars. However, by observing the comet day after day it is clear that it moves relative to the stars and the Sun. Comets are often seen close to the Sun in the sky since they are at their brightest when they are close to the Sun in space.

This photograph of comet Bennett was taken by Claude Nicollier on 26 March 1970 from the summit of Gornergrat, in the Swiss Alps, near Mount Cervin. The Sun is below the horizon and the snow is illuminated by the Moon. (C. Nicollier)

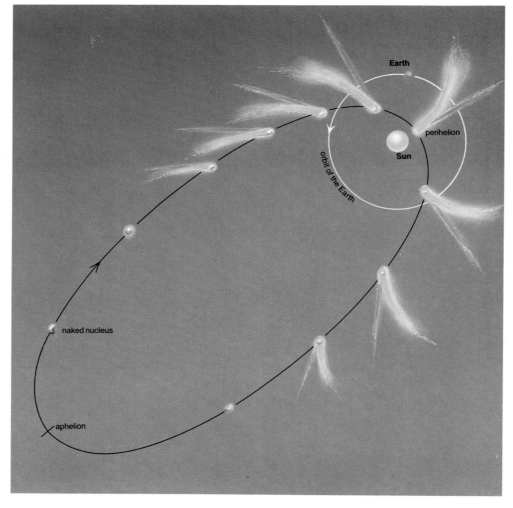

The evolution of a comet in its orbit. As the nucleus approaches the Sun, the tails of dust and ionised gas begin to develop and they attain their maximum size in the neighbourhood of perihelion. The tail of ionised gas is straight and away from the Sun while the dust tail is curved and behind the nucleus of the comet with reference to its orbital motion. Cometary tails may be very long and sometimes cross the Earth's orbit or the Earth itself. This happened on 19 May 1910 when Halley's Comet passed directly between the Sun and the Earth. Seen from the Earth, perspective effects can modify the apparent length of the tail. The grains of dust that make up the tail move around in elliptical orbits themselves, in the same plane as the orbit of the comet, and this is in general different from the plane of the Earth's orbit. In the illustration the size of the nucleus has been greatly exaggerated.

The cloud of hydrogen. This is a photograph in visible light of comet Bennett (1970 II) taken at the observatory of Haute Provence on 1 April 1970, on which have been superposed the contours of intensity (isophotes), of the Lyman-alpha line of hydrogen obtained simultaneously by a photometer on board the satellite OGO–5 (Orbital Geophysical Observatory–5). The Lyman-alpha line at 121.6 nanometres wavelength, in the far ultraviolet, is characteristic of atomic hydrogen, H. The isophotes have intensities increasing towards the centre. Notice the immense extent of the hydrogen cloud, which demonstrates that hydrogen is a major constituent of the cometary nucleus, arising from the molecule H_2O. (After J.-L. Bertaux, J.-E. Blamont and M. Festou, 1973)

direction opposite to that of the Sun'. At the beginning of the seventeenth century, Johannes Kepler first suggested that the pressure of sunlight might be responsible for a comet's tail. From modern physics we can calculate the strength of this repulsive force, the radiation pressure exerted by sunlight on the grains of dust, acting in a direction opposite to the gravitational attraction of the Sun. After leaving the nucleus, a grain of dust follows a trajectory in the Solar System that is slightly different from that of the nucleus (the radiation pressure is negligible on the nucleus). Thus, the dust is seen in an extension of the Sun–nucleus line yet slightly curved away from the direction of motion of the comet. The length of the tail may attain one to ten million kilometres for the most spectacular comets.

Actually, the strength of the radiation pressure depends on the size of the grain and its chemical composition. By careful examination of the tail of comet Arend-Roland it was determined that the grains were of the order of 0.05 to 5 micrometres in size and that some of them contained silicates. This was corroborated by observations in the infrared. This 'new' comet produced 75 tonnes of dust per second at its perihelion passage whereas the short-period comet P/Encke only loses about 200 kilograms of material per second at each perihelion passage.

Sometimes a tail is observed that is said to be anomalous or called an antitail as it is directed towards the Sun, for example as was observed in comets Kohoutek and Arend-Roland. An anomalous tail consists of fairly large grains (from 0.1 to 1 millimetre) for which the radiation pressure is insignificant. These tails have a tendency to remain on the same trajectory as the nucleus yet are dispersed along this trajectory. A simple effect of perspective can then make the tail appear to point towards the Sun, particularly when the

Earth is situated in the plane of the orbit of the comet.

More rarely, structures are noticed in the dust tail that can be explained by a discontinuous ejection of material from the nucleus, with, for example, sudden outbursts, resembling explosions.

What is the ultimate fate of these dust grains? The smallest, those that can be seen in the tail, are dispersed by radiation pressure and end up feeding the zodiacal cloud, the cloud of interplanetary dust that can be seen in the shape of a spindle stretched out along the plane of the ecliptic. The largest grains generally disperse all along the cometary orbit. When the Earth meets such a stream of dust, a shower of meteorites results, with each particle of more than about 1 millimetre in diameter producing a shooting star. The first association between a meteor shower and its parent comet was made by Giovanni Schiaparelli who was able to associate Comet Swift-Tuttle of 1862 with the well-known Perseid shower, which peaks between August 9 and 14 each year.

The physical characteristics of the bright tails imply that the mean density of the constituent particles is of the order of 1 gram per cubic centimetre, while the density of silicates is of the order of 3 grams per cubic centimetre. This is almost certainly explained by the fact that a cometary particle consists of a very porous aggregate of very much smaller grains, perhaps of the order of 0.01 micrometre.

Calculations show that the solar radiation pressure is insufficient to explain the rectilinear plasma tails, which are bigger than the dust tails and can reach a hundred million kilometres. In 1951, the German astronomer Ludwig Biermann suggested that the action of a solar plasma, moving radially outwards from the Sun at great

Visual appearance of a comet. A large well-developed comet has a diffuse, nebulous head, from which emanates a luminous tail, directed away from the Sun. Under favourable conditions, two tails can be seen, one straight and formed of plasma, the other denser and more curved, made up of dust. If a comet is examined with a powerful telescope it is often possible to make out a central brighter zone within the head, which by custom is called the nucleus, the diffuse coma and nucleus together forming the head. However, this terminology, based on naked-eye observations, is now considered out of date, and only partially represents the actual physical structure of a comet. In particular, the term nucleus is now used more specifically for the solid body from which the cometary phenomena arise, and not for the very bright condensation which is seen at the centre of the head and whose size is about a thousand kilometres. The visible part may be a thousand times larger than the solid nucleus which, as yet, has never been directly observed. For certain very bright comets the angular dimension of the tail can be several tens of degrees. This can be compared with the size of the Moon which is only half a degree or, as here, with that of the constellation of Lyra.

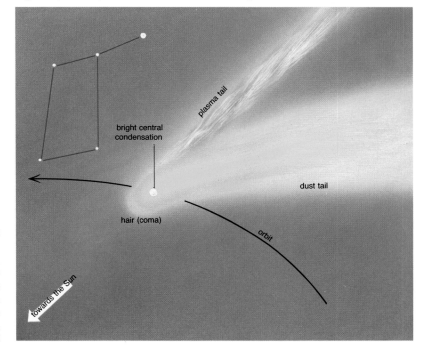

velocity and violently repulsing the ions formed in the atmosphere of the comet, was responsible for the formation of the plasma tails. Thus, the solar wind was discovered, which has since been studied 'in situ' by space probes. It consists essentially of a mixture of protons and electrons leaving the Sun at 400 kilometres per second and contains between one and ten particles per cubic centimetre at the distance of the orbit of the Earth. This plasma, which carries the solar magnetic field with it, does not interact with the neutral gas of the coma but interacts strongly with the cometary plasma, which is composed mainly of CO^+ and CO_2^+ but contains other ions, such as H_2O^+, OH^+, CH^+, N_2^+ and single electrons. The cometary plasma seems to originate very close to the nucleus of the comet in a zone about 1000 kilometres across. The ions are produced either by photo-ionisation of the neutral cometary molecules by solar ultraviolet radiation, or through the action of the solar plasma by the phenomenon of charge exchange in which a proton tears off an electron from a neutral cometary molecule and thus ionises it.

The plasma tails often show streaks or nodules whose velocity away from the nucleus can reach 50 kilometres per second. Sometimes the whole plasma tail detaches itself from the head; this is undoubtedly due to a sudden inversion in the solar magnetic field. In 1970, atomic hydrogen was discovered in comets Tago-Sato-Kosaka and Bennett by means of observations of Lyman-alpha at 121.6 nanometres by the satellites OAO–2 and OGO–5. French astronomers showed that an enormous cloud of hydrogen (more than ten million kilometres in diameter) surrounding comet Bennett implied a production of ten atoms of hydrogen per second, well above the production of other constituents observed until then. Combined with the observation of the production of the hydroxyl radical, OH, this discovery gave considerable support to the hypothesis of Fred L. Whipple in which the principal constituent of the cometary nucleus is water ice. Ever since then, all the comets observed by space vehicles have shown the Lyman-alpha emission of atomic hydrogen and it has been possible to measure the rate of production for a good many of them.

Sometimes a comet does not have a visible tail. However, comets always have a coma (or head), a bright circular region whose brightness increases towards the centre and whose radius may be a hundred thousand to a million kilometres. Spectroscopic study of the light of the coma, effectively the atmosphere of the comet, shows that it is made up partly of dust and partly of gaseous molecules, or rather fragments of molecules (free radicals such as OH, C_2, C_3, CH, etc.) which are leaving the nucleus at about 0.5 kilometre per second. This gas flow, evaporated at the surface of the nucleus, drags the cometary dust away with it.

Free radicals are produced in the coma by the influence of ultraviolet radiation from the Sun, which breaks up the molecules of gas – the so-called parent molecules – which are escaping from the nucleus. The radicals may be split up again into atoms or ions, defining the size of the coma when the comet is very close to the Sun.

Which are the most important parent molecules? Water, is the principal cometary gas; it was only directly detected for the first time in Halley's Comet in 1986. Since water is so difficult to detect the general picture has been established by inference from many other observations, Water (H_2O) begins by being dissociated into H and OH. The OH is then dissociated into H and O. The quantities of O, H and OH observed are consistent with a massive initial production of H_2O by the nucleus. The H_2O^+ ion has also been detected in certain plasma tails.

It is evident that parent molecules containing carbon must also exist because C, C_2, C_3, CH, CN and CS are observed. Molecules of CO_2, HCN and CO were recently detected in Halley's Comet. They are trapped in the ice of the nucleus, but if there are no other more complex molecules, the presence of the radicals C_2 and C_3 could be explained by chemical reactions in the central regions. They could come, nevertheless, from partial evaporation of organic compounds composed primarily of silicates (with silicium, sodium, magnesium, iron and calcium) that were found on the grains of Halley's Comet.

Several metals are also detected in the spectrum of a comet when it approaches the Sun. The dust is overheated in this case, vaporises and releases the metals.

The visible parts of a comet (coma, tail, hydrogen cloud) have very little mass – at most 300 000 tonnes; although they occupy an enormous volume, compared with the Sun, they are extremely tenuous. The source of this matter, the nucleus, is a piece of dusty ice of just several kilometres in diameter, with a mass between 1 are 1000 billion tonnes. The spectacular phenomena are the tail and coma, but it is the nucleus that forms the essential part of the comet; it contains all the information about the nature, origin, and role played by comets in our general description of the formation of the Solar System.

The nucleus is so small that we have very little direct information about it. Using the giant radio dish at Arecibo on the island of Puerto Rico, a radar echo of the comet P/Encke was obtained in 1980 and a total radius of between 0.4 and 4 kilometres was determined. When a comet is far from the Sun it reverts to being a solitary nucleus. One sees then only a single point of light but the amount of light received can be related to the size

Comet Humason. Although it was visible for four years, comet Humason (1962 VIII) did not get closer than 2.13 astronomical units (AU) to the Sun. Its spectacular blue colour was due to emission from CO^+ ions. The spectrum of this comet was peculiar. The continuum revealed the presence of dust, but no emission from the neutral gases C_2 or CN was detected although the unusually strong CO^+ emission was detected out to a distance of more than 5 AU. As water (H_2O) does not vapourise further than 3 AU from the Sun, the parent molecule of CO^+ (perhaps the CO molecule itself) must have controlled the gaseous evaporation of this comet and must then be a principal constituent. (Hale Observatories)

Comet Arend-Roland. This photograph of comet Arend-Roland (1957 III) which was taken on 25 April 1957 shows a rather rare phenomenon: a streak of light or antitail seems to point in the direction of the Sun, and therefore opposite to the normal tails which are directed away from the Sun. This is actually due to particles of dust ejected from the nucleus more than a month before this photograph was taken, and when the comet was at least two weeks before perihelion. As the comet moves on in its trajectory, the dust particles, which remain in the plane of the comet's orbit, get left further and further behind relative to the Sun–comet line. If the Earth is in a suitable position, it may happen that, by a perspective effect, this dust appears to be directed towards the Sun. The fact that this antitail is still close to the head of the comet indicates that it contains large particles of the order of a millimetre in size for which the effect of radiation pressure is small, while the dust grains of the normal tail are some micrometres in size. (Lick Observatory)

Physical model of a comet. In order to represent the different parts of a comet on the same diagram, a logarithmic scale has been used for the distance, allowing the nucleus (with a diameter of less than 10 kilometres) and the enormous hydrogen envelope (extending to more than ten million kilometres) to be shown together. Gas and dust are escaping from the rotating nucleus. The gases that make up the inner region of the coma are the sublimated ices from the nucleus, H_2O with probably some HCN, CO_2 and CO. These parent molecules are broken up and ionised by the ultraviolet solar radiation giving birth to the free radicals observed throughout the coma and to ions which are carried at great speed by the solar wind in the ionised tail. The neutral dust grains, which are illuminated by the Sun, are pushed away by the action of radiation pressure exerted by the solar photons. Their curved trajectories (represented in linear coordinates in the figure on the right) are contained within a parabolic envelope. The plasma of the solar wind meets the cometary obstacle head-on in a shock wave.

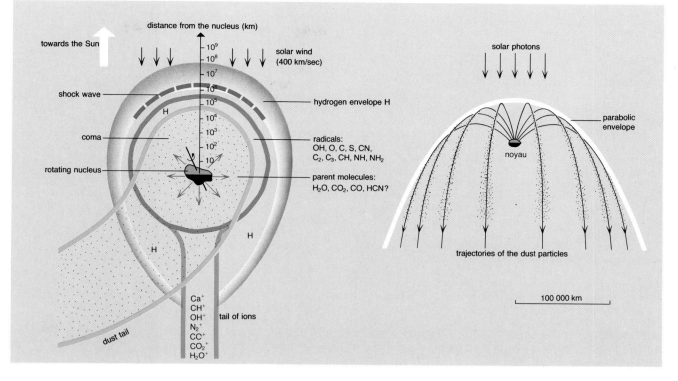

Comet IRAS-Araki-Alcock. The left-hand photograph was taken on 9 May 1983 with the Herstmonceux 26-inch telescope. The right-hand false colour image produced from IRAS (Infrared Astronomical Satellite) data. It shows the dust cloud surrounding the comet; warmed by solar radiation, this dust emits in the far infrared. (Courtesy (left) Royal Greenwich Observatory, Herstmonceux; (right) Rutherford Appleton Laboratory)

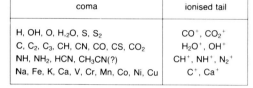

coma	ionised tail
H, OH, O, H_2O, S, S_2	CO^+, CO_2^+
C, C_2, C_3, CH, CN, CO, CS, CO_2	H_2O^+, OH^+
NH, NH_2, HCN, $CH_3CN(?)$	CH^+, NH^+, N_2^+
Na, Fe, K, Ca, V, Cr, Mn, Co, Ni, Cu	C^+, Ca^+

Atoms, molecules and ions observed in comets.

The solar wind and the ionised tail. These photographs show comet Mrkos (1957 V) on 17, 22 and 27 August 1957. During this time the curved dust tail remained homogeneous and stable, while the straight ionised tail, consisting primarily of CO^+ ions which give it a blue colour, evolved rapidly and showed filaments and nodules for which the velocity of displacement can be measured.

The measured velocities are very high and are of the order of 50 kilometres per second, which cannot be explained by the pressure of the solar radiation. In studying the behaviour of comet tails, Ludwig Biermann demonstrated the existence of the solar wind, the continuous flow of a plasma of electrons and protons escaping radially from the Sun at high velocity (of the order of 400 kilometres per second). The solar magnetic field, carried by the solar wind, plays a vital role in the interaction of this wind with the cloud of ions produced in the coma. The lines of magnetic field move around the ionised coma, which is an obstacle in its path, and the ions flow back along the lines of the coiled-up field. (Hale Observatories)

of the nucleus by assuming that the nucleus reflects from 10 to 30 per cent of the incident solar radiation. By this method one finds that the radii of the nuclei are between 1 and 10 kilometres. However, in this calculation it is necessary to assume a total absence of any coma around the nucleus although it is known that comae do exist for certain comets out to 7 astronomical units from the Sun! It was not until 1986 that a comet nucleus (Halley's) was directly observed by special probes: its dimensions are around 15 kilometres by 8 kilometres. Its albedo is only 4 per cent, smaller than previously thought. As for the mass, this is estimated by assuming that the nucleus is nearly spherical and has a density of the order of 1 gram per cubic centimetre. One finds a mass of forty billion tonnes for a radius of 2 kilometres. At the surface of such a nucleus, the pull of gravity would be very weak and the velocity of escape would only be about 1 metre per second; this means that all matter – gas or dust – escaping from the nucleus at a velocity greater than 1 metre per second would be lost for ever into space. Thus, a layer of material some metres thick is lost at each perihelion passage.

The nucleus itself is rotating. Sometimes a dust coma is not homogeneous and one can see structures such as spirals or haloes, which are more or less concentric, moving about hour by hour. This phenomenon can be explained by making the hypothesis that the surface of the nucleus is not uniform; certain regions are more active than others and when these are exposed to the Sun they eject much more gas and dust than

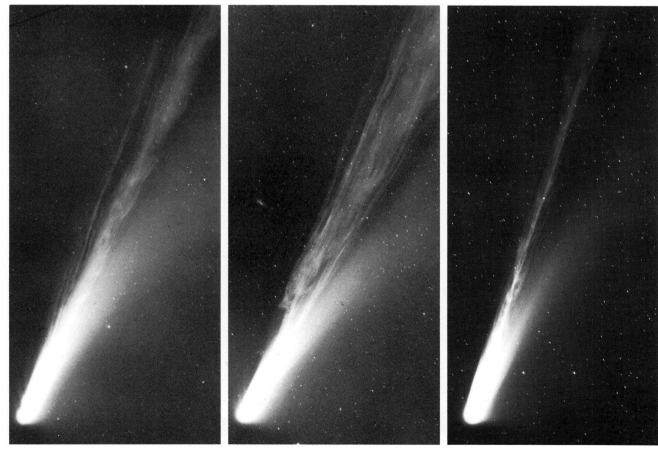

neighbouring regions. A jet of material is shot out from the nucleus, like water from a swirling garden hose, and this is responsible for the observed structures. Thus, it is possible to determine the periods of rotation of some comets (which are between 4 and 30 hours) and the orientation of the axes of rotation (which are found to be uniformly distributed). These jets have an opposite reaction on the nucleus and have a small perturbing effect on its orbit. Fred L. Whipple and Zdenek Sekanina analysed fifty-nine passages of comet P/Encke and showed that its axis of rotation was very variable. According to the direction of this axis, the effect of the reaction of the jet can slow down or accelerate the comet in its orbit. The secular variations of the period of P/Encke are actually explained very well by the calculated effects of variations in the axis of rotation.

This reasoning has been behind the general acceptance over the past thirty years of the idea that comets have solid nuclei and has been to the detriment of the model of a 'flying sand bank', which was ardently defended by Raymond A. Lyttleton in the 1950s. According to his model, there was no solid nucleus but only a cloud of small individual particles but this fails to explain how a persistent jet could could be maintained over a number of revolutions.

What is the stability of the nucleus of a comet? Some succeed in surviving a close approach to the Sun in which they could be expected to be broken up by the tidal force induced by the Sun. Others literally blow up and break into many pieces, although their distance from the Sun is too large for tidal effects to be the sole cause. The breaking up of some twenty comets has been observed and in particular that of comet West in 1976. In this case, it is possible that the nucleus became cracked and that a pocket of gas formed deep in the nucleus which eventually resulted in an explosion. On this occasion it was established that the coma rigorously maintained the same chemical composition after the explosion. The fresh matter exposed in the centre of the nucleus was just the same as that of the surface.

How can one explain the fact the surface of the nucleus is inhomogeneous (since one observes the presence of jets) while the interior is undoubtedly homogeneous? It seems likely that when the ice evaporates, some grains are too big to be carried away by the gas; they could then agglomerate one with another to form a relatively solid crust free of any ice. There would then be no more evaporation in this crusty outer region of the nucleus. Observation of Halley's Comet appear to confirm this. Sometimes comets show sudden increases in brightness by a factor of 100 or so. Many of these can be explained by the existence of particularly active zones that are suddenly exposed to the Sun as a result of the rotation of the nucleus. In this case the flashes are periodic. Other flashes appear spasmodically and could then be simply due to the explosion of pockets of gas formed under the crust of dust.

Thus, the nucleus is a mixture of dust and ice, or, in other words, volatile substances which are sufficiently cold to be in a solid state. There is about as much dust as volatile elements – the ratio of masses varying slightly from one comet to another, between 0.2 and 1. The composition of the dust grains is now known thanks to the exploration of Halley's Comet in 1986. Beside silicates, they contain slightly, volatile organic compounds, judging by the presense of atomic hydrogen, carbon, oxygen and nitrogen.

In contrast, the chemical composition of the coma clearly reflects that of the cometary ice from which it is formed by the process of sublimation. As H and OH are observed to be the major constituents of the coma, the ice must principally be water ice. Another argument towards this conclusion is drawn from observations of the light curve of the comet as it approaches the Sun. The rate of sublimation of an ice depends very strongly on its nature and the temperature. The temperature of the nucleus depends on its distance to the Sun. Thus one can calculate that, if dry ice (ice of CO_2) were present in the nucleus, it would begin to sublimate when the comet was at 8 astronomical units, and methane (CH_4) ice at 38 astronomical units from the Sun. For water, the calculated distance is 2.5 astronomical units, which actually corresponds to the distance where the coma begins to develop. However, nothing is known about the exact state of this water ice, which constitutes about 80 per cent of the volatile matter; is it crystalline, amorphous or formed of compacted snow?

The theory must also explain the origin of the other observed volatile compounds in the coma; CO, nitrogen and carbon compounds, HCN and sulphur. According to Armand Delsemme, the cometary material could be a clathrate hydrate, that is to say, a mixture of water-ice crystals containing many cavities where other gases (CO_2, CO, N_2, CH_4) are trapped and which are released when the water evaporates.

If the structure of the nucleus, as well as its chemical and isotopic composition, were perfect-

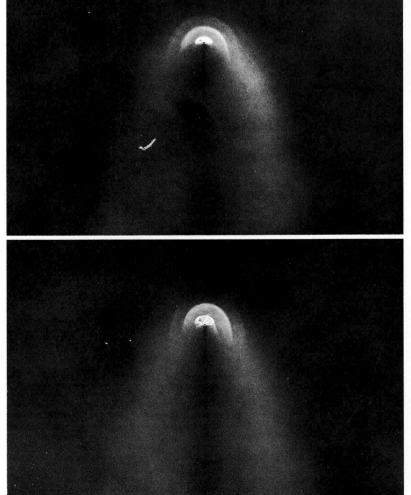

Comet West. Easily observable to many before dawn in the spring of 1976, comet West (1976 VI) was one of the most spectacular of recent comets. The bluish colour of the plasma tail and the pale yellow of the dust tail could be perfectly distinguished.

An ultraviolet spectrum, obtained by a rocket-borne experiment, showed for the first time, evidence of the presence of carbon and carbon monoxide. The nucleus of this 'new' comet is believed to have broken up in March 1976 into four pieces. (© Treugesell-Verlag/Photeb)

The rotation of the nucleus and non-gravitational forces (opposite right). Certain comets show concentric haloes which progressively move away from the centre, as can be seen in these drawings of the comet Donati (1858 VI) made by G. P. Bond on 4 and 5 October 1858. In this case the haloes followed each other with a period of 4.6 hours. Their origin can be understood by imagining that the nucleus is in rotation with the same period and that the surface of the nucleus has a particularly active region: each time this region is exposed to the Sun a jet of gas and dust is shot out of it producing another halo of dust. The reaction of this jet on the nucleus changes its orbit slightly by providing a supplementary non-gravitational force and this explains, for example, why the orbital period of comet Encke decreases by about two hours in each orbit. (Harvard College Observatory)

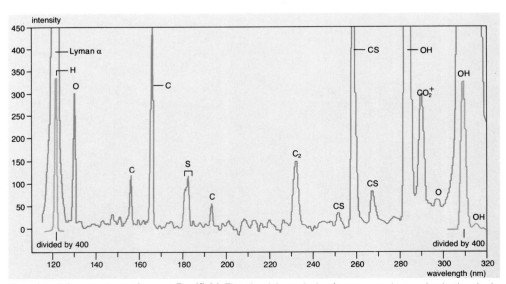

The ultraviolet spectrum of comet Bradfield. The ultraviolet emission from comets is completely absorbed by the Earth's atmosphere and one can only study it using rocket probes and satellites. This ultraviolet spectrum of comet Bradfield (1979 X) was obtained by the International Ultraviolet Explorer (IUE) space telescope. The intensity of the emission, expressed in rayleighs per nanometre, is plotted against wavelength. The strongest lines are those of OH and H indicating that the principal constituent of the nucleus is water ice. The atoms O, C and S were detected for the first time in the ultraviolet on this occasion as was the radical CS. The visible C_2 and the ion CO_2^+ were already known in the visible spectrum. (After H. A. Weaver, P. D. Feldman, M.-C. Festou, M. F. A'Hearn and H. V. Keller, 1981) About ten comets have been observed with IUE, and they all show the same spectrum, indicating that their chemical composition is very similar.

Conceptual representation of the nucleus. The nucleus of a comet was photographed for the first time in 1986; it was Halley's Comet. These photographs have confirmed previous theories based on indirect evidence. The shape is irregular, the surface rather dark. Gas and dust may be flowing from relatively small zones in the nucleus. Some whole pieces can become detached from the nucleus. Pockets of gas accumulate under the surface and give rise to sporadic explosions. Large grains of dust can remain on the surface and form a solid crust which would be ejected at the following perihelion passage by the pressure of the underlying gas.

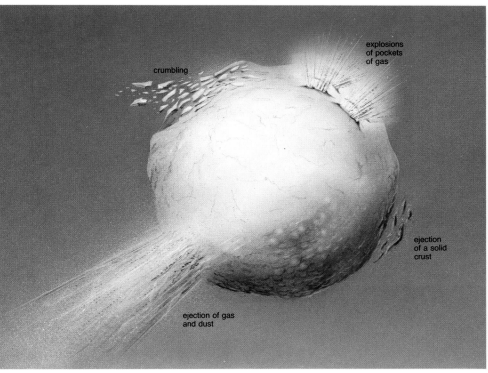

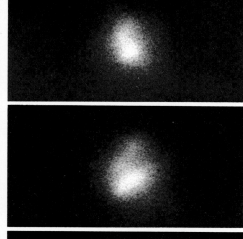

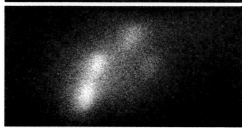

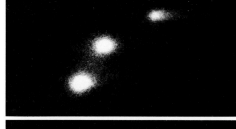

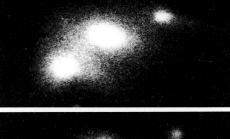

name of the comet	perihelion distance q (AU)	eccentricity e	date of perihelion passage
1957 III	Arend-Roland	0.316	1.00017	8.04.1957
1957 V	Mrkos	0.355	0.9994	1.08.1957
1962 VIII	Humason	2.133	0.989	14.05.1962
1963 I	Ikeya	0.632	0.9934	21.03.1963
1965 VIII	Ikeya-Seki	0.0078	0.9999	21.10.1965
1969 IX	Tago-Sato-Kosaka	0.473	0.9999	21.12.1969
1970 II	Bennett	0.5376	0.9962	20.03.1970
1973 XII	Kohoutek	0.142	1.000008	28.12.1973
1975 IX	Kobayashi-Berger-Milon	0.426	1.0001	5.09.1975
1976 VI	West	0.197	0.9997	25.02.1976
1977 XI	P/Encke (period 3.3 years)	0.34	0.84	17.08.1977
1986	P/Halley (period 76.1 years)	0.59	0.967	9.02.1986

Elements of the orbits of some recent comets (left). The orbits of these comets all have an eccentricity close to 1. Three of them whose eccentricity is greater than 1 will have left the solar system for ever in a hyperbolic trajectory. Two wellknown periodic comets, Encke and Halley, have been added. The period, T (in years), perihelion distance q (in AU), semi-major axis of the elipse, a (in AU), aphelion distance, Q (in AU), and the eccentricity, e, are related by the equations:
$q + Q = 2a$, $a^3/T^2 = 1$;
$q = a(1 - e)$; $Q = a(1 + e)$.

Visible spectrum of the comet Kobayashi-Berger-Milon (right). The light of the comet has two components which are distinguished by their spectra. The dust diffuses the light of the Sun and displays a continuous spectrum identical with that of the Sun. The gas emits a line spectrum and the wavelengths of the emission lines are characteristic of its constituents.

Along the ordinate one has the distribution of light along a segment 300 000 kilometres long. (Susan Wyckoff, 1981)

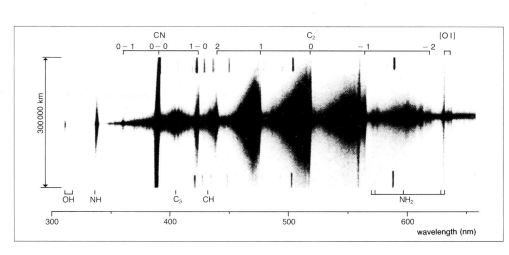

ly known we would understand where, when and how the comets are formed and where they stand in the general picture of the formation of the Solar System. The rings of Saturn and most of the satellites of the giant planets contain, like the nuclei of comets, a large amount of water ice showing that at 10 astronomical units from the Sun it is possible for ice to have formed and remain in existence until the present time. Fred L. Whipple has gone further in proposing that comets are indeed the elementary 'bricks' from which Uranus and Neptune were formed, following a parallel theory of the formation of the terrestrial planets by V. S. Safronov. In this model, the proto-Sun was formed at the centre of the solar nebula, while the residual dust and gas became concentrated in a flat disk. The dust agglomerates bit by bit, as a result of collisions, to form bodies whose dimensions are of the order of a kilometre, the so-called 'planetesimals'. A planet of terrestrial type forms by the accretion of all these 'planetesimals' situated at a common distance from the Sun.

Likewise, beyond 10 astronomical units some 'cometesimals' will have been formed differing only from their terrestrial counterparts by the massive presence of ice, permitted by a very much lower temperature. Meanwhile, only part of these cometesimals would have been used in the formation of Uranus and Neptune by accretion, the rest being dispersed throughout the Solar System very early in its history, by the effect of gravitational perturbations by the major planets.

There is a striking similarity between the molecules detected in the interstellar clouds and the radicals and elements found in comets. In particular, it is known that interstellar clouds also contain solid grains of carbon covered with a coating of ice. It is believed that the Solar System was born out of such a cloud of gas and dust and it seems likely that the elements contained in our planets were contained in the interstellar cloud before its condensation but, at present, this remains an important uncertainty, the key to which may be found in cometary nuclei. Indeed, one can imagine two variants to the scenario proposed above: either the comets are formed directly from the grains and ices of the interstellar cloud or the Solar System has passed through a hot phase in which the ices and grains have been vaporised and then have condensed to form ices and grains again during a cooling phase. It is now believed that the second variant is the most likely explanation of the formation of the inner planets because of their proximity to the Sun. However, a fundamental question still remains to which only

Disintegration of comet West (right). Sometimes comets disintegrate or break up into many pieces. This spectacular series of photographs of comet West (1976 VI) taken in March 1976, showed that the nucleus broke up into four pieces of similar sizes which each later developed a separate tail. The four pieces had identical spectra indicating identical composition. This demonstrates that a solid nucleus exists (since it is liable to break up), that the nucleus is relatively fragile and that its structure is homogeneous: the chemical composition of the exposed interior was the same as that of the surface. The first five images of this sequence (from top to bottom) are photographs in the yellow–green taken on 8, 12, 14, 18 and 24 March. The five images at the bottom were taken using a red-sensitive detector and show the three fragments that were visible from 31 March. These were taken on 31 March and 1, 2, 3, and 7 April. (Photographs by C. F. Knuckles and A. S. Murrell, New Mexico State University Observatory)

the comets can one day provide an answer: to what distance from the Sun has the interstellar material been thus vaporised?

At the present time at least a dozen comets appear each year. Some are already known and are the expected reappearances of periodic comets but others are new discoveries. Amateur astronomers contribute to these discoveries and are recognised by the attribution of the name or names of the discoverers to the said comet. Observations of the trajectory of the comet allow the elements of the orbit to be determined; in particular the period, T, its eccentricity, e, and its semi-major axis a.

The astronomical annals of Chinese astronomers contain the earliest information on comets that were visible some twenty-three centuries ago. Alexandre-Gui Pingré made a most comprehensive study of comets which was published in his *Cométographie ou Traité historique et théorique des comètes* (2 vols., Paris, 1783–84). The most recent catalogue (Brian G. Marsden, *Catalogue of Cometary Orbits*, 1983) lists 1105 cometary appearances to the end of 1982, corresponding to 706 different comets. Of these, 117, all with periods of less than 200 years, qualify by convention as periodic (the letter P is added in front of their name). 74 comets have already been observed at least twice.

Among the 589 long-period comets, 316 have parabolic orbits ($e = 1$) while 169 have elliptical orbits ($e<1$) and 104 have hyperbolic orbits ($e>1$). However, the distinction between an elliptic orbit and a hyperbolic one is undoubtedly artificial, and is evidence of the uncertainty in the determination of the eccentricity.

A statistical study of cometary trajectories reveals some very valuable information. For example, it is established that the largest known eccentricity, that is corresponding to the most hyperbolic orbit, is only 1.057 for comet Bowell (1980 b). What is more, it is established that no comet comes in towards the Sun in a hyperbolic orbit. However, during this inward journey a few comets may undergo a small planetary perturbation which can augment the eccentricity to greater than one in which case they swing away past the Sun never to return: in other words comets as a group are gravitationally bound to the Sun; they all belong to the Solar System.

In general, orbits of comets with short periods have low inclinations to the plane of the ecliptic, and they travel in the same sense around the Sun as do the planets. Most of these comets have an aphelion distance, Q, close to the orbit of Jupiter indicating that this planet has played a major part in the establishment of the present orbits by capturing comets from among the population of long-period comets.

On the other hand, the inclination of the orbits of the long-period comets to the plane of the ecliptic can have any value. In 1950, Jan Oort examined a statistical sample of forty-six long-period comets and proposed the existence of an enormous spherical cloud of comets with aphelia of the order of 50 000 astronomical units (by way of comparison one of the nearest stars to the Solar System α Centauri is found at 250 000 astronomical units). This Oort Cloud would contain about a thousand billion comets, representing a mass equal to that of the Earth. The passage of a star close to the Sun (within one parsec or so) is a rare event occurring about once every 100 000 years. The gravitational effect of this close approach disturbs the stability of the orbits of the comets, some of which may be torn away from the Solar System altogether, while others have their perihelion moved closer to the Sun to give the spectacle we observe, which are the death-throes of cometary nuclei, since they cannot survive many orbits. These stellar perturbations are responsible for a permanent rain of new comets towards the interior of the Solar System.

It is questionable whether the comets were formed at 50 000 astronomical units from the Sun, where the density of matter would be extremely small. As proposed by Fred L. Whipple it is more likely that they were formed beyond the orbit of Saturn in the region occupied by Uranus and Pluto. A large number of the original bodies formed these planets by accretion; others were ejected outwards by the planetary perturbations to make up Oort's cloud. Others may have been projected towards the interior of the Solar System where, by crashing into the terrestrial planets, they could have been able to play an important role in the formation of planetary atmospheres, and in addition have been able to supply the original contribution of organic molecules necessary for the appearance of life on Earth.

Jean-Loup BERTAUX

Comet Ikeya-Seki (1965 VIII) grazed the Sun at the time of its perihelion passage. It was part of the Kreutz group, a family of comets with orbits that all passed very close to the Sun. The dust released by the nucleus was vaporised by the solar radiation and metallic elements were detected in it by means of optical spectroscopy. This comet is responsible for all that is known about the existence of metals in comets until the measurements of Halley's Comet obtained by the spacial probes in 1986. (Lick Observatory)

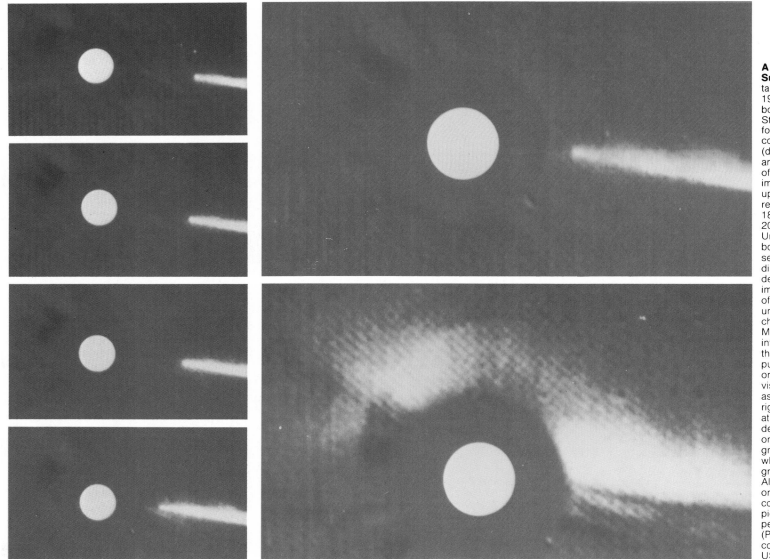

A comet absorbed by the Sun. These photographs were taken on 30 and 31 August 1979 using a coronagraph on board a satellite of the United States Air Force used normally for observing solar flares. The coronagraph occults the Sun (drawn in on the photographs) and its corona out to a distance of 2.5 solar radii. On the four images on the left and on the upper right picture, taken respectively on 30 August at 18 h 36 m, 19 h 15 m, 19 h 35 m, 20 h 32 m, and 20 h 49 m Universal Time (from top to bottom and left to right) one can see a comet and part of its tail directed away from the Sun. A detailed analysis of a series of images showed that the nucleus of this comet, which was unknown until then and was christened Howard-Koomen-Michels (1979 XI), had fallen into the solar atmosphere. Only the dust tail, which had been pushed outside the cometary orbit, survived and remained visible on two sides of the Sun as seen in the image at bottom right taken on 31 August 1979 at 8 h 21 m UT. The orbit thus determined is very similar to the orbits of the so-called Kreutz group – of nine other comets which have the peculiarity of grazing the Sun at perihelion. All these comets would have originated in the same parent-comet which broke into many pieces during an earlier perihelion passage. (Photographs and details kindly communicated by D. J. Michels; US Naval Research Laboratory)

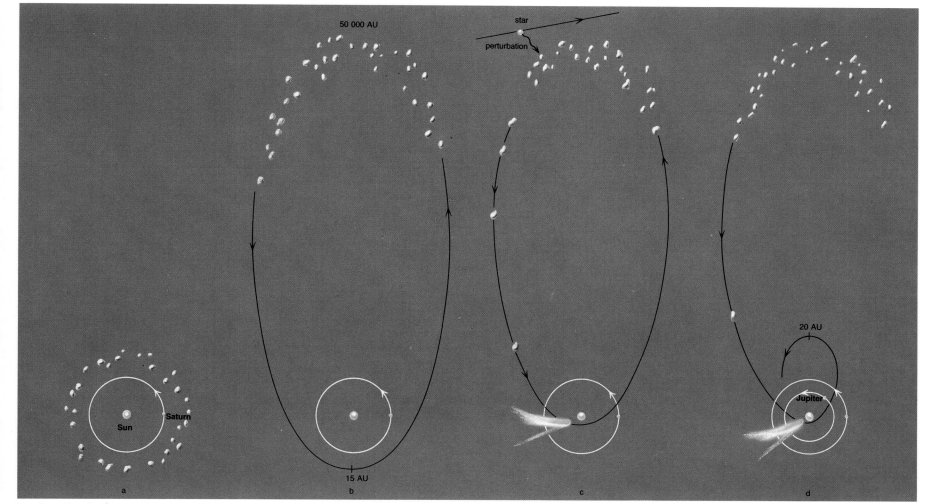

The Oort Cloud. The comets we see today are thought to come from a gigantic 'reservoir', the Oort Cloud, at the very edge of the Solar System but nevertheless still bound to it. The nuclei of comets, made of ice and dust, were formed 4.6 billion years ago, at the same time as the planets. According to a well-accepted theory, these nuclei were formed in a circular orbit in the region of Saturn and Uranus, about 15 astronomical units (AU) from the Sun (a). The perturbations brought about by the giant planets rapidly transferred these nuclei into very eccentric orbits with perihelia at 15 AU and aphelia at 50 000 AU where they spend most of their time (b).

There they are subjected to the perturbations of stars passing in the neighbourhood of the Sun. It only takes a small perturbation to transfer a nucleus into a new orbit taking it closer to the Sun (perihelion of the order of 1 AU, aphelion remaining at 50 000 AU) and this produces a 'new' observable comet of very long period (c).

If this nucleus passes close to a giant planet, its trajectory could be modified again and changed into an orbit of short period (perihelion about 1 AU, aphelion about 20 AU) such as in the case of Halley's Comet (d). Further perturbations could shrink the orbit further and shorten the period.

The cataclysm of Tunguska.
In the morning of 30 June 1908, a fantastic explosion occurred in central Siberia in the valley of the river Podkamennaya Tunguska, 800 kilometres to the north-west of Lake Baikal. Witnesses described an enormous meteoric bolide or fireball which was visible in the sky for a few seconds. Other witnesses at a distance of 60 kilometres from the point of impact were knocked over by a resounding shock wave at the time the explosion was heard. Seismic shocks were registered over the whole world. Scientific experts suspected that a giant meteorite had fallen but political instability at the time prevented them visiting the presumed area of the impact. Finally, in 1927 the first expedition, directed by the Soviet scientist Kulik, surprised the waiting world by announcing the absence of any crater in the area. However, as the photograph shows, trees were torn up in an area 30 to 40 kilometres across, their trunks aligned in a direction away from the centre of the explosion.

In 1930, the Englishman Francis Whipple (not to be confused with the American Fred L. Whipple) proposed an explanation which is now largely accepted: that the event was due to the collision of a comet with the Earth. Later it was calculated that it must have been a block of ice 40 metres in diameter weighing 30 000 tonnes, which exploded at an altitude of some kilometres and released energy equivalent to that of a thermonuclear bomb of 12 megatonnes. A calculation of the trajectory revealed that it was likely to have been a part of the nucleus of comet Encke. An unusual night sky brightness persisted for nearly two months after the event due to the presence of a large quantity of cometary dust in the atmosphere. (TASS)

Comets

Halley's Comet

The most famous comet is undoubtedly Halley's Comet. Its last approach, in 1986, was greeted with a salvo of space probes which studied it in detail and decisively advanced our knowledge of comets. Observations of Halley's Comet go back to 240 BC, according to Chinese records; since then it has been observed on each of its appearances.

The English astronomer Edmond Halley was twenty-six when he was intrigued by the spectacular appearance of a comet in 1682. Until then comets had been thought of as random events without periodic recurrence. It occurred to Halley to consult Isaac Newton who was working on his theory of gravity and planetary orbits. He strongly urged Newton to publish this new theory, paying the costs of publication himself although he was not rich. The *Philosophiae naturalis principia mathematica* appeared in 1687. Halley used Newton's theory to calculate the orbits of twenty-four comets; he noticed that the orbits of the comets seen in 1531, 1607 and 1682 were very similar and deduced that they were one and the same comet. He predicted that it would return in 1759. The comet actually reappeared on 25 December 1758 when it was seen by an amateur astronomer, Johann Palitzsch, a farmer near Dresden. This rediscovery caused a great stir: it showed that comets orbit round the Sun like planets, but in very elongated orbits, and it confirmed Newton's theory of gravitation.

Astronomy was not Edmond Halley's only occupation. Between 1698 and 1701 he directed scientific expeditions on board his ship the *Paramour Pink* from which he made a survey of the tidal currents in the English Channel, mapped the trade-winds, and measured the magnetic variations over the whole Atlantic Ocean, north and south.

Physical studies of Halley's Comet began with the work of Friedrick Bessel in 1835. In 1910 an international effort produced a series of photographs which have provided the basic material for planning the 1986 space missions.

The orbit of Halley's Comet has a perihelion of 0.59 astronomical unit, an aphelion of 35 astronomical units and a mean period of 76 years. The precise period varies by as much as two and a half years either side of this because of perturbations by the planets. Now we know that the diameter of its nucleus is about 10 kilometres and at each passage it exhibits the complete range of cometary phenomena: coma and tails of dust and of plasma. In 1986, Halley's Comet did not look spectacular. At perihelion it was in the least favourable configuration for 2000 years: opposite to the Earth with respect to the Sun. If as a visual show it has been disappointing, the unprecedented scientific effort that has been devoted to it, has on the contrary, rendered numerous results. Coordinated ground-based observations under the International Halley Watch scheme, combined with the international collaboration that was established to launch five probes to meet the comet, harvested an amazing wealth of results. This success was only clouded by the Challenger catastrophe in January 1986 which meant the cancellation of the NASA mission, planned for March 1986. The European Space Agency launched a space probe in July 1985 called Giotto (because Giotto painted the comet in his famous fresco *The Adoration of the Magi*, at Padua). It weighed 750 kilograms and passed the nucleus at a distance of 600 kilometres on 14 March 1986 at 0003h. The Soviet Union sent two probes, Vega 1 and Vega 2, each weighing almost 2 tonnes and carrying about 10 instruments. After passing close to Venus these probes surveyed the nucleus from a distance of 9000 and 8000 kilometres on 6 and 9 March 1986. Japan put up more modest probes, only 140 kilograms in weight, called Sakigake and Suisei. The scientific instruments on board there probes have taken pictures of the nucleus and analysed the nature of the gas using mass spectrometry and ultraviolet, visible and infrared spectroscopy. They have determined the number of dust grains, their dimensions, their chemical composition and studied the interaction between the comet and the solar wind. While the probe Giotto passed as close as possible to the comet at the risk of being destroyed, the two Vega probes maintained less dangerous distances. They pointed a television camera and two optical spectrometers permanently at the nucleus using three-axis stabilisation.

The Vega probes photographed the nucleus of the comet for the first time ever on 6 and 9 March 1986; they obtained the precise location of the

Halley's Comet after its perihelion passage. The comet became impossible to observe when it approached the Sun, shortly before perihelion (9 February 1986). A dust tail had just started to develop at the end of January. When the comet reappeared after the perihelion, a large fan-shaped tail with complex structure was displayed. Several jets extended in various directions: from the anti-Sun direction of the plasma tail, whose twisted structure can be recognised (top, right), to the anti-tail with sharp edges formed near the solar direction (bottom, left). The dust along the jet was ejected by a single violent event lasting only a few hours, which occurred some days before.

This photograph was taken by K. S. Russel from Australia on 22 February 1986; the field covered is about 2 degrees by 2 degrees. The knots seen in the plasma tail, imply a 'disconnection event': when the polarity of the magnetic field carried by the solar wind reverts, whole portions of the ionised tail separate from the comet. A chain of small telescopes was operated by amateur astronomers all over the world under the framework of the International Halley Watch; these observations, when put together, provide a continuous 'film' of the development of the comet's tail, that is, presuming that the sky is not cloudy everywhere at the same time. (Royal Observatory, Edinburgh)

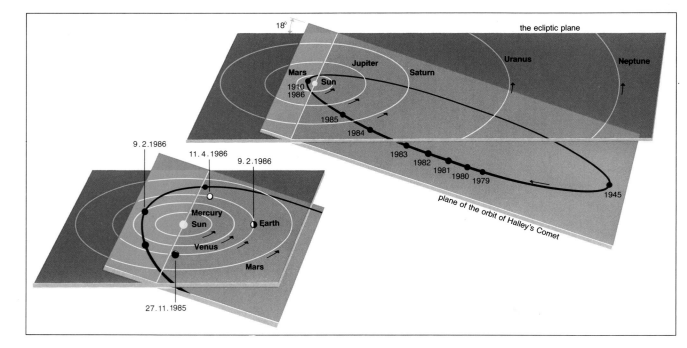

The orbit of Halley's Comet. The path of Halley's Comet is shown in relation to the orbits of the planets of the Solar System. The path is slightly inclined relative to the plane of the ecliptic. The comet's orbit is retrograde, that is in the opposite direction to the planets, as its inclination of 162 degrees suggests. This characteristic puts severe constraints on space probes, which must keep to the plane of the ecliptic because they are launched from the Earth whose velocity lies in this plane. For this reason all the space missions in 1986 encountered the comet between 6 and 14 March at the time when it crossed the ecliptic from north to south, after passing the perihelion. In addition, the probes themselves are revolving in the Solar System in the same direction as the planets and their relative speeds in relation to the comet were very high: 68.4 (Giotto) and 79.2 (Vega 1) kilometres per second. So the encounters only lasted a few hours.

The orbital parameters of Halley's Comet are as follows: present period 76.09 years, perihelion distance 0.58720 astronomical units, aphelion ditance 35.33 astronomical units, eccentricity 0.96727, inclination to the ecliptic 162.239 degrees (retrograde motion), perihelion passage on 9 February 1986, at 0640 hours Universal Time.

Because of the Earth's orbital motion, the comet makes two close approaches, one before perihelion, and one after perihelion, on 27 November 1985 and 11 April 1986. The respective positions of the comet and Earth are shown at perihelion and on the dates of the close approaches.

The nucleus as seen by Vega 1. The first images of the nucleus of Halley's Comet were taken by Vega 1 at a distance of 9000 kilometres, on 6 March 1986. These three prints were taken at intervals of about one minute, from different angles. The colours are artificial and correspond to an arbitrary luminosity coding. The resolution of the camera was about 100 metres, and the nucleus contour corresponds roughly to the boundary between turquoise and blue.

In the second print, the Sun is approximately at the back of the observer, at 20 degrees below. The light comes from the left on the first image, and from the right on the third. Here we only see the face of the nucleus illuminated by the Sun, but the details of the surface are masked by the dust that escapes abundantly. The size of the illuminated area is 10 kilometres. Images obtained by Vega 2 and Giotto, on the other hand, show an elongated nucleus, 15 kilometres long. This indicates that the camera in Vega 1 observed the nucleus from the thickest end. An appendix, first seen to the right and then to the left of the nucleus, is an extension illuminated from below. (Intercosmos/CNES, Laboratory of Space Astronomy/Aeronomy Service)

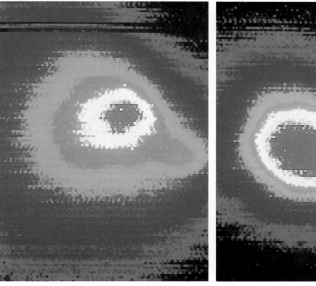
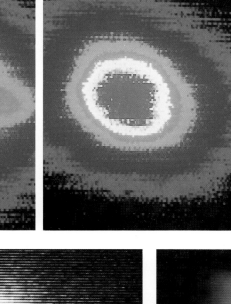
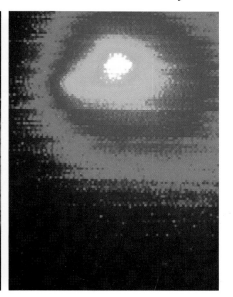

The nucleus as seen by Vega 2. Vega 2 was 8200 kilometres from the nucleus on 9 March 1986. By that time the nuclear activity had grown weaker and therefore the amount of dust was smaller. Thus the nucleus can be seen much more sharply in this black and white photograph. It has a very irregular shape; the bright face measures 15 kilometres by 8 kilometres. The image in false colours has been produced by mathematical transformation of the black and white photograph: the contrast has been increased in the nucleus, and decreased outside. The largest axis is seen here almost horizontally. The Sun is shining from behind the camera, slightly to the left. Dust jets, driven by the sublimation of ice from the nucleus, originate in selected regions, whose surfaces seem to be strongly inhomogeneous. The nucleus appears bright in these images with contrast enhancement. It is nonetheless darker than a piece of charcoal, with an albedo that is at most only 4 per cent. Such a small value (three times smaller than the Moon's) is found in the Solar System only in the type C asteroids, and in the Martian satellite, Phobos. (Intercosmos/CNES, Laboratory of Space Astronomy/Aeronomy Service)

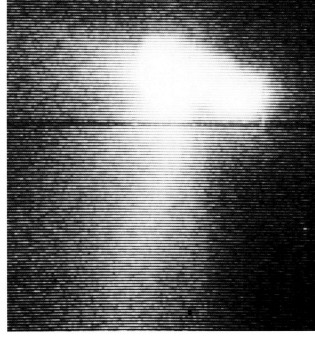
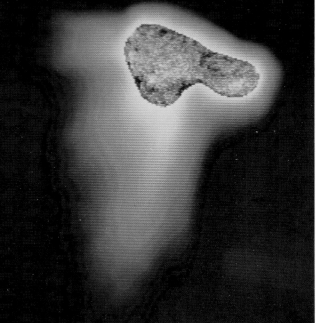

The nucleus as seen by Giotto. These three false colour images of the nucleus of Halley's comet were taken by the Giotto camera from distances (respectively from left to right and from top to bottom) of 57 000, 18 000, and 10 700 kilometres, with a spatial resolution of 450 metres for the picture at lower left. The fourth picture, at lower right, is a composite of 60 images combined so as to bring out the maximum morphological detail of the nucleus and the dust jets. The resolution varies from 800 metres (near the top left) to 80 metres near the lower jet. The nucleus appears as a solid object of irregular shape. The face not illuminated by the Sun stands out in shadow against a lighter background of light scattered by dust above the nucleus. The brightest parts of the image are two dust jets from the face illuminated by the Sun (whose light comes from the right) and which masks the exact shape of this face from the Sun. As in the image obtained by the Vega-2 probe the nucleus has an elongated shape with maximal dimensions 14.9 × 8.2 square kilometres. The topography of the nucleus has a tortured appearance, as shown by the complicated appearance of the terminator (the curve separating the day side from the night side). Visible on the lower images is an elliptical depression with major axis 2 kilometres long, whose lower face appears to be illuminated by the Sun. A crater like this could be the result of a collision at the time of formation of the nucleus, or possibly of the explosion of a pocket of accumulated gas suddenly liberated as the comet passed close to the Sun. (H.U.Keller, Halley Multicolor Camera Team, Max-Planck-Institut für Aeronomie, ESA)

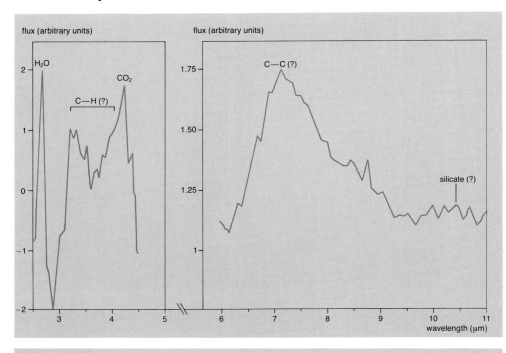

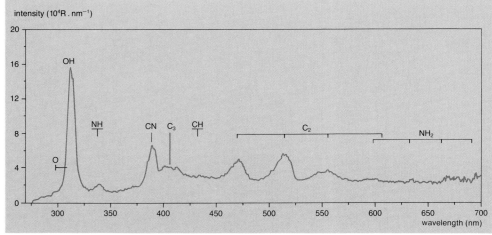

Infrared and near ultraviolet–visible spectrometry. The emission spectrum of the comet in the circumnuclear region was observed simultaneously in infrared, near ultraviolet and visible light by the Vega probes. It is not possible to obtain these data from ground-based observations. The most intense emission peaks in the infrared (top figure) correspond to water vapour (H_2O) (at 2.7 microns) and to carbon dioxide (CO_2) (at 4.5 microns). The presence of this gas has been suspected for a long time, but this is the first direct detection. CO_2 is thirty times less abundant than H_2O. The emissions at 3.3. and 7.5 microns are attributed to compounds containing the chemical bonds C—H and C—C, respectively; they are superimposed over a 350 K black-body spectrum. They might come from dust or even from the surface of the nucleus, as they are too strong to be due to gas. These emissions coincide with some already identified in the interstellar medium, thus reinforcing the hypothesis of the relation between comets and the interstellar medium from which the Solar System formed. A water feature was also identified at the 1.38 micron wavelength (not visible in the figure); without any doubt, this is the source of the radical OH, that shows an intensive emission at 310 nanometres. The classical emissions of the radicals OH, NH, C_3, CN, C_2 and NH_2 can be identified in the ultraviolet–visible region of the spectrum that is shown in the bottom figure; this confirms previous observations from Earth. The increase in these emissions at 4000 kilometres from the nucleus, nevertheless, was much faster than what was expected on the basis of models of the distribution of radicals. When a molecule (for example H_2O) is dissociated by a solar ultraviolet photon, it produces a radical OH* in an excited state. This radical can get 'de-excited' by spontaneously emitting a photon either in the ultraviolet (310 nanometres) or in the near infrared (1.5 to 1.9 microns). The spatial distribution of the gas that escapes from the nucleus can thus be reproduced by this indirect means, independently of the dust jets that it drives along as it expands. (*Nature*, Vol. CCCXXI, p. 266, 1986)

of protons and electrons. It leaves the Sun at a velocity of 400 kilometres per second and reaches the cometary cloud. Molecules and atoms in the comet are ionised, either by ultraviolet solar radiation (which can then be important up to large distances from the nucleus) or by charge exchange between a solar wind proton and an electron from a cometary molecule.

Cometary ions mixed with the solar wind were detected by the probes approaching the comet at distances larger than one million kilometres from the nucleus. The solar wind gradually became more and more charged with cometary ions, until it lost its identity; however, there is a vast area where the two plasmas mixed. Finally the solar plasma disappeared at a distance of 4700 kilometres from the nucleus; only cometary plasma was found inside that distance.

This complex interaction is not yet fully understood. It is present up to large distances, and is noticed by variations in the magnetic and electric fields, associated waves, and the distribution of the ionic velocity field. Opinions are still divided about the existence of shock waves in this interaction.

Water in a comet has been detected directly for the first time. It was previously suspected to be the major constituent of the volatile phase of the nucleus. It was first detected directly by infrared observations from a plane, then by infrared spectrometry from the Vega probes, and finally by mass spectrometry from the Giotto probe. H_2O represents at least 80 per cent in volume of the gas escaping from the nucleus, while CO_2 is thirty times less abundant. Cyanhydric acid (HCN) has also been detected convincingly for the first time by radioastronomic observations. The observed quantity of HCN is, nevertheless, about ten times smaller than that needed to explain the amount of radical CH, so easily observed from the Earth. It seems therefore reasonable to expect to find other parent molecules by a more refined analysis of the mass spectrometric data; CO in particular is seen in the ultraviolet spectra.

Large quantities of small dust particles (smaller than 1 micron) were found by the space probes; these escaped to ground-based astronomical observations as well. The bulk of the mass is nevertheless formed by large particles, escaping at a rate of 4 to 10 tonnes per second. This is about four times lower than the gas. The most important new information provided by the space missions has no doubt been the rather basic chemical composition of the dust grains. A small percentage of grains show the kind of silicates found in carbonated conditions of the type Cl (C, O, Na, Mg, Si, Ca, Fe), similar to solar abundances. Others also contain a large component of light elements such a C, H, O and N. There is a third class of particles which contain only light elements. They are refractory organic molecules, undoubtedly similar to those pro-

Neutral mass spectrometry. The purpose of placing a mass spectrometer into the heart of the gaseous coma at only 600 kilometres from the nucleus, was to determine the nature of the parent molecules directly escaping from it. Only the products of the dissociation of these molecules were known before the Giotto experiment. The predominant molecule is water; together with its sub-products of OH and O it represents more than 80 per cent of the gaseous composition, while carbon dioxide (CO_2) is thirty times less abundant. The inset in the figure below shows a histogram of the relative proportions of H_2O, OH and O. With decreasing distance from the comet, the concentration of water vapour increases as the inverse square of the distance. An expansion velocity of 900 metres per second was measured for the vapour cloud, with a production rate of 15 tonnes per second.

Many other chemical species were found as Giotto approached the nucleus. Metallic ions of sodium (Na^+) and iron (Fe^+), carbon (C), radicals C_2^+, CH^+, O^+ and sulphur were detected by the mass spectrometer. For sulphur in particular, it was possible to measure the isotopic ratio $^{34}S/^{32}S$; its value, 0.045, is equivalent to ratios measured on the Earth. A sharp fall in the ionic temperature is encountered at the 'contact surface', at about 4700 kilometres from the nucleus. It corresponds to the boundary between the area of interaction of the gaseous coma and the solar wind, and the zone, nearer to the nucleus where the solar wind cannot penetrate, (*Nature*, vol. CCCXXI, p. 326, 1986)

nucleus which was used to make a last minute correction to the trajectory of Giotto. This allowed the mass spectrometer to analyse the composition of the gases escaping from the nucleus.

What was previously a theory has now become a certainty; there is a nucleus at the centre of Halley's Comet – and probably at the centre of all the other comets – a solid block formed by ice and dust. This 'dirty snow ball' as it is called, was proposed by Fred L. Whipple in 1950. The model of a nucleus as a 'sandbank', a cloud of independent particles, proposed by Raymond A. Lyttleton is now considered wrong. The nucleus is irregular and elongated, like a potato, not spherical as previously imagined, and much larger: 15 kilometres by 8 kilometres, instead of the 5 to 6 kilometres estimated from observations made at large distances, when the nucleus was still naked and inactive and only its surface reflected the solar light. The albedo was previously estimated as 15 per cent, but the albedo measured by the cameras of the Vega probes was 4 per cent; one fact compensates the other; larger surface, smaller albedo.

The nucleus rotates over itself with a period of around 52 hours; this was confirmed by the periodic variations in Lyman α emission, observed in the hydrogen cloud. Therefore the nucleus is not presenting the same face to the Sun at the time of the three encounters with Vega 1, Vega 2 and Giotto. This relatively slow rotation implies that the side facing the Sun

becomes very hot. Temperatures of 350 K have been measured. The other side remains much cooler. The ice therefore evaporates exclusively from the 'day' side, as testified to by the dust jets that escape only from the face that is exposed to the Sun.

Several questions arise: how can ice be darker than coal? Besides, at 350 K ice should sublimate at a rate that is 10 to 100 times faster than the observed rate of 15 tonnes per second. This paradox can be resolved if one imagines that the surface layer (a few centimetres thick) of the comet is formed, not by ice, but by a mantle of extremely dark and porous solid material, formed by grains that have not escaped from the nucleus, or are too heavy, and fall back. It is this mantle's external temperature that is being measured. The ice, situated some centimetres deeper, stabilises as a much lower temperature. Only a fraction of the surface of the nucleus – less than 10 per cent – is active, as can be seen from the images. Only certain regions of the surfaces of the nucleus eject material. Two extemely bright jets are shown by Giotto images. The jets seem to have their origin in 'hollow' regions: faults or craters are indeed needed to channel the escape of gas by a nozzle effect.

The space probes felt the proximity of the comet in the solar wind plasma, allowing us to study the interaction between two very different plasmas in a giant natural laboratory. This subject is of great interest to fundamental physics. The solar wind is essentially composed

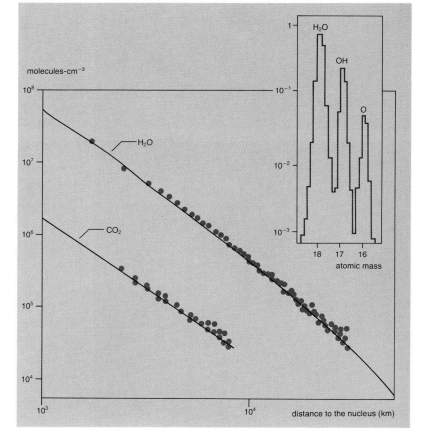

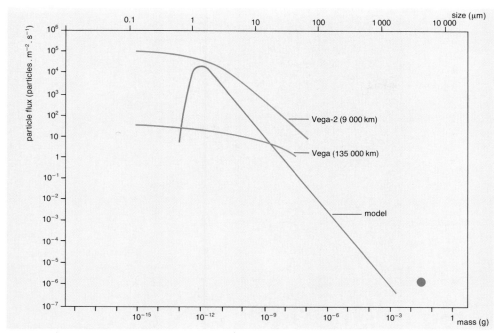

Particle distribution by size. Ground-based observation made it possible to estimate the size of cometry dust grains and to construct a model (see red curve). Satellite dust detectors have found quite a different distribution pattern (see blue curve), with slightly more bigger grains and a very large quantity of tiny grains measuring less than 1μm. These had escaped detection by being too small to diffuse the Sun's light. We do not yet know whether these originate directly from the nucleus or from the fragmentation of larger grains.

The size and flux distribution of grains vary greatly across the coma. Images from the Vega-2 satellite show the coma to be composed of two jets originating from the side towards the Sun; they comprise a fairly narrow cone (half-angle c45°, 9th March 1986).

The first grain was detected by Vega-1 at a distance of 320 000 km from the nucleus. The largest particle encountered by Giotto was 40 mg (see green point). The flux was as high as 10^5 particles $m^{-2} s^{-1}$, to the detriment of the space satellites. Four instruments were destroyed and there was 50% degradation of the solar panels of both Vegas. As for Giotto, 14 seconds before closest passage to the nucleus (610 km) it met a ferocious hailstorm of particles which badly damaged the camera mirror and destabilised it. This caused a partial halt in communications which lasted for 32 minutes.

The quantity of dust spewed out from the nucleus has been estimated at about 10 tonnes per second on 6th March and 4 tonnes per second on the 13th. That is only a fraction, 10%–25%, of the total mass ejected, the rest being in the form of gases – mainly water vapour.

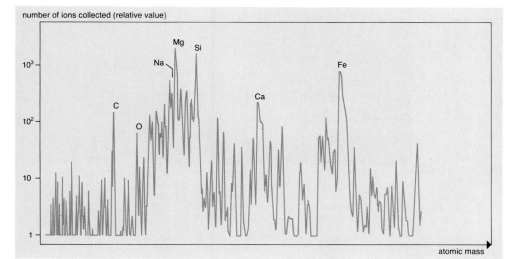

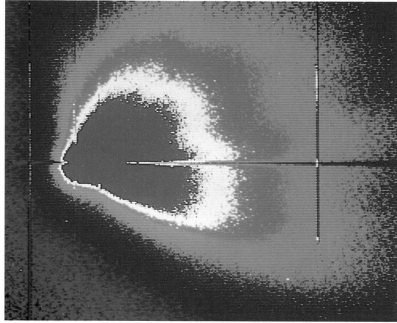

The dust coma. The distribution of particles in the dust coma is shown in the black and white image obtained by Vega 2 before the encounter, at 20 000 kilometres from the nucleus. By colouring-coding the information in the black and white image one obtains a coloured image which shows the differences in dust densities. Dust jets are distributed in a cone-like pattern towards the Sun. (Intercosmos/CNES, Laboratory of Space Astronomy/Aeronomy Service)

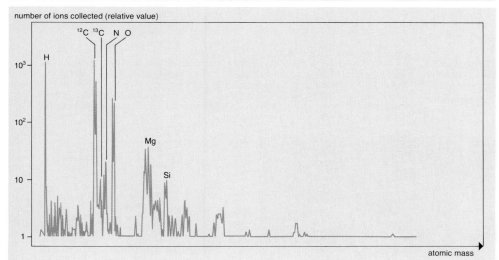

Chemical composition of the dust. The only way to arrive at the chemical composition of cometry dust was to go and measure it on the spot. Once again a means of measurement was needed which could travel at 70km a second. It was precisely this great speed which was put to advantage with both the Giotta and Vega satellites.

When a grain hit a metal target (silver or platinum) it was completely vaporised, its molecules were destroyed and its constituent atoms ionised. These ions were analysed by a mass-spectrometer. In this way, the composition of over 1500 grains was analysed. The grains were very small, of the order of 10^{-16} g, and did not all have the same composition, but can generally be divided into three categories.

The upper graph shows a distribution of elements such as C, O, Na, Mg, Si, Ca and Fe – the obvious sign of a silicate-type of mineral composition, with which sulphur and chlorine are sometimes associated. In contrast, the lower graph shows predominantly light elements – H, C, N and O.

The third, and most numerous, category is simply a combination of the first two. The presence of these light elements is perhaps the most important discovery of the cometary space missions. In fact, the H, C, N, O atoms cannot have come from molecules of CO_2, HCN or H_2O ice, which would have had time to evaporate during the journey from the nucleus to the probe. We have to look to organic composites which are only slightly volatile – this would support the notion that comet cores are made up of small interstellar grains. Mayo Greenberg has shown in the laboratory how a composite grain can be formed in interstellar space. At very low temperatures (c10°K) H_2O, CO_2 (?) and HCN ices condense around a silicate particle. Over several hundreds of millions of years ultra-violet irradiation breaks the molecules into radicals and ejects the O atoms. The radicals recompose to form a shield of complex organic material which is resistant to temperature. The total size of the composite particle, grain and shield, is about half a micrometre. These particles can survive passage near the Sun. At several dozens of astronomical units they are once again covered in ice. This in turn condenses and acts as a cement when they eventually assemble to form the nuclei of comets. (After *Nature* vol. 321, p. 280, 1986.)

duced in a laboratory by Mayo Greenburg to simulate the evolution of interstellar grains.

This supports the hypothesis that comets are formed by the accretion of interstellar grains, covered by a refractory organic component, water, ice and CO_2. But these grains also come from the primeval nebula as do the Sun and the planets: they belong to our Solar System. Analysis of the sulphur and carbon isotopes reveal no anomalies; there is no signature of a foreign origin.

Many mysteries still remain unsolved about comets. For example from the volume of the nucleus – around 800 cubic kilometres – estimates of the mass and the density can be deduced. The mass is determined only by indirect and hazardous methods, by estimating the decrease in the comet brightness through the centuries. The density was 20 kilograms per cubic metre, 50 times lower than the density of ice!

New space missions are being designed to approach a comet at even shorter distances, and during longer periods of time, and to bring back samples so that they can be analysed in the laboratory. They will help to unravel the mysteries of the comets and the enigma of the formation of the Solar System. It would not have been possible to even conceive these missions, without this first indispensable stage: the coordinated survey of Halley's Comet during its passage in 1986.

Jean-Loup BERTAUX

The interplanetary medium

The Solar System consists primarily of planets, satellites, comets and asteroids orbiting around the Sun, but the interplanetary space between these bodies is not a complete vacuum, as is spectacularly shown by meteors, or shooting stars. These luminous objects are produced in the upper atmosphere when the Earth intercepts a small fragment of interplanetary material. Such fragments are known as meteoroids. Detailed study of these luminous trails yields important information about the meteoroids; their direction and velocity (which allows the orbit in the Solar System to be determined), and their mass, density, cohesion and sometimes material content.

Objects which fall to the ground are called meteorites. The smallest meteorites, with a mass less than 10^{-7} gram, fall slowly through the atmosphere without being destroyed. Their size is typically a few tens of micrometres. Some can be collected by stratospheric aircraft fitted with air filters.

The meteorites whose mass is between 10^{-7} gram and 1 kilogram are completely destroyed by heating and vaporisation when they are suddenly braked by the atmosphere. Those whose mass is less than 10^{-2} gram can be detected only by radar, through the ionised trail they leave behind. Above 10^{-2} gram the meteorites produce a luminous trail whose intensity increases with the mass. These trails can be studied statistically by a network of cameras on the ground, providing a permanent nocturnal surveillance.

Meteorites of more than 1 kilogram survive their passage through the atmosphere and fall to the ground, after losing about four-fifths of their mass, at a rate of between two and ten per day over the whole surface of the Earth. They produce very bright, luminous trails. Occasionally, it is possible to work out the probable place where a meteorite fell and find it on the ground, newly arrived from interplanetary space.

Where do meteorites come from? It is now thought that fallen meteorites, over 1 kilogram in mass, are fragments of asteroids torn off by collisions. Their arrival in the atmosphere is completely unpredictable. On the other hand, it is

believed that the fragments causing meteor showers are friable, of low density (1 gram per cubic centimetre) and that none reach the ground. Moreover, a reconstruction of their orbits shows clearly that they are fragments of comets, gradually spread out in a swarm around the orbit.

Micrometeorites, a few tens of micrometres in size, form an enormous spread of at least 600 million kilometres, the zodiacal cloud. This cloud reveals itself by the Sun's light, which it scatters, creating a faint glow (like the Milky Way) extending along the ecliptic and called the zodiacal light. The zodiacal light has been observed for a long time and was correctly explained for the first time by Gian Domenico Cassini in 1683 Space probes have been equipped with photometers to measure the zodiacal light more precisely than from the ground (for example the French satellites D2-A Tournesol and D2-B Aura) and with impact detectors to measure *in situ* the mass and numbers of the particles.

Luckily for space vehicles, the zodiacal cloud is a very tenuous medium, whose total mass is certainly less than 10^{-10} of the total mass of the planets. At the Earth's orbit its mean density is 10^{-23} grams per cubic centimetre: roughly speaking, a grain two micrometres in diameter is alone in a cube of space measuring a hundred metres along its sides, and travels at several tens of kilometres per second. Concentrated near the ecliptic, the plane of symmetry of this dust cloud differs slightly from the ecliptic plane and coincides more nearly with that of the orbit of Jupiter.

The particles, in elliptical orbits around the Sun, are influenced by the Poynting–Robertson effect, according to which the scattering of solar photons produces a deceleration. Each particle thus follows a spiral path and gradually approaches the Sun, taking about 10 000 years. In the immediate vicinity of the Sun, the particles are eroded by collision, evaporation and the action of the solar wind. When their size becomes less than one micrometre they either sublime entirely or are expelled by solar radiation pressure towards the outer Solar System.

If the zodiacal cloud is not a transitory

The zodiacal light. Soon after evening twilight or shortly before dawn, on a dark, moonless night a faint luminosity called the zodiacal light may be seen above the horizon. The name stems from the fact that the band of light, centred on the Sun below the horizon, has the shape of an ellipse elongated along the plane of the ecliptic. The projection of this plane in the sky is the Zodiac. The intensity of the light decreases away from the Sun and from the plane of the ecliptic. It is, however, present in all directions and it is even possible to notice a slight increase in the zodiacal light in the antisolar direction; this is the *Gegenschein*. This phenomenon is caused by the optical properties of the interplanetary dust, which is concentrated in the plane of the ecliptic and scatters more light backwards than to the sides.

The zodiacal light can best be observed when the ecliptic is highly inclined to the horizon at the appropriate time, as it always is in the tropics, but which happens in northern temperate latitudes in February and March after dusk in the west and in September and October before dawn in the east.

This photograph of the zodiacal cloud was taken on 30 June 1982 as part of the night-sky photography experiment on the Soviet space station Salyut 7 during the flight of the French astronaut Jean-Loup Chrétien. The station was at an altitude of 300 kilometres in the Earth's shadow. The bright diagonal band at lower right is luminous emission from the terrestrial ionosphere. (Courtesy S. Koutchmy, Institut d'Astrophysique du CNRS, CNES-Intercosmos)

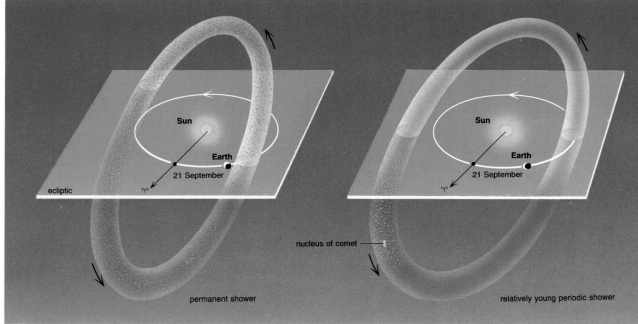

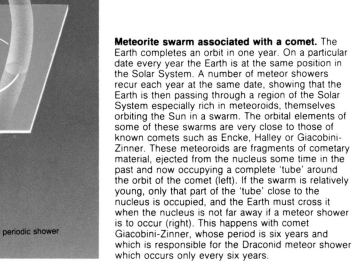

Meteorite swarm associated with a comet. The Earth completes an orbit in one year. On a particular date every year the Earth is at the same position in the Solar System. A number of meteor showers recur each year at the same date, showing that the Earth is then passing through a region of the Solar System especially rich in meteoroids, themselves orbiting the Sun in a swarm. The orbital elements of some of these swarms are very close to those of known comets such as Encke, Halley or Giacobini-Zinner. These meteoroids are fragments of cometary material, ejected from the nucleus some time in the past and now occupying a complete 'tube' around the orbit of the comet (left). If the swarm is relatively young, only that part of the 'tube' close to the nucleus is occupied, and the Earth must cross it when the nucleus is not far away if a meteor shower is to occur (right). This happens with comet Giacobini-Zinner, whose period is six years and which is responsible for the Draconid meteor shower which occurs only every six years.

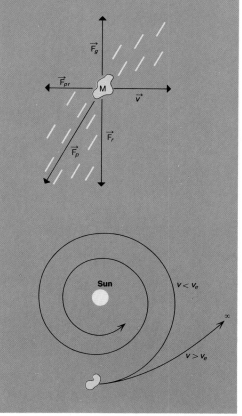

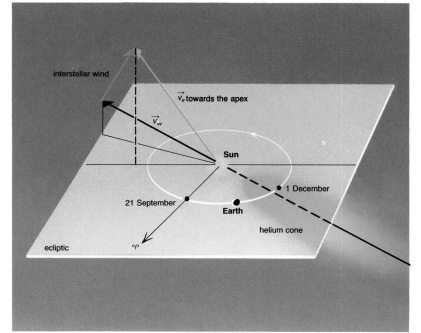

Hydrogen and helium in interplanetary space. The Sun moves through the interstellar material at 23 kilometres per second, in the direction indicated by the vector $\vec{V_w}$. Atoms of hydrogen and helium in the interstellar medium thus pass through the solar system in the opposite direction. The helium atoms are focussed by the solar gravitational field; their paths are hyperbolae whose planes contain the half-line opposite to $\vec{V_w}$. The atoms concentrate along this gravitational focal axis and form a luminous cone, as a result of resonance scattering of solar photons of wavelength 58.4 nanometres emitted by helium atoms at the surface of the Sun.

The hydrogen atoms are also attracted by the Sun, but that attraction is counterbalanced by the radiation pressure of the solar Lyman-alpha photons, which are strongly scattered. The hydrogen atoms are not focussed, and they become ionised when they approach the Sun. They thus produce a backward-directed ionisation cavity. In the direction of $\vec{V_w}$ the concentration of hydrogen atoms decreases towards the Sun since the intensity of the solar Lyman-alpha has increased. Thus, at about 2.5 astronomical units, there is a cone of maximum Lyman-alpha emission from the interplanetary hydrogen. In the figure the trajectories are shown parallel. In fact, since the interplanetary medium is very hot (about 10 000 K), there is a dispersion of velocities which tends to dilute the cone of helium and to fill the back of the ionisation cavity carved out in the interstellar medium by the Sun.

phenomenon but a permanent member of the Solar System, this continual loss of material must be compensated by an equivalent supply, of several tonnes per second. The asteroids are incapable of supplying such a mass (besides, the dust density does not increase at all at the middle of the asteroid belt). However, it is known that comets produce swarms of meteorites, by ejecting dust at each perihelion passage. In course of time, as a result of the effects of collisions, gravitational forces and the solar radiation, the swarms disperse. The largest particles give rise to the meteor showers, while the smallest are gradually incorporated into the zodiacal dust.

The comets are, therefore, the most likely source of replenishment for the zodiacal cloud. However, the question is not completely solved, for although the parabolic comets produce plenty of dust, they depart to infinity and are lost to the Solar System, while the short-period comets produce much less dust, the exact amount of which it is hard to calculate, but some astronomers think it quite inadequate to maintain the zodiacal cloud: 2 kilograms per second for comet Encke. With only sixty or so periodic comets at present known, this source fails to compensate the losses estimated above by a factor of ten. Perhaps the space exploration of the Solar System especially the data from Halley's Comet obtained *in situ* in 1986, will allows us to resolve this paradox.

Photons and charged particles also fill the interplanetary medium. Their principal source is the Sun, which emits photons in the radio, infrared, visible, ultraviolet and X-ray regions of the spectrum and particles as a hot plasma, the solar wind, composed mainly of protons, electrons and helium nuclei (alpha-particles). The solar wind flows outwards from the Sun at 400 kilometres per second with a flux of 3×10^8 particles per square centimetre per second at the Earth's orbit. Very energetic particles and gamma-rays are also emitted by the Sun during sporadic eruptions, which generate the solar cosmic rays. These mingle with the galactic cosmic rays, whose origin is still obscure (supernovae, perhaps). They are composed mainly of relativistic particles with an average flux of one particle per square centimetre per second.

The interplanetary medium also contains neutral gas, consisting of atomic hydrogen (H) and helium (He). At first, this is surprising since these atoms cannot survive for very long subject to the ionising effects of the Sun (charge-exchange with the plasma of the solar wind, photo-ionisation by the extreme solar ultraviolet). The lifetime is only about twenty days for an atom at 1 AU from the Sun. These atoms come

from the interstellar medium close to the Solar System. Like other stars, the Sun moves through the interstellar material. This being so, a continual stream of this material, composed mainly of atoms of hydrogen and helium, flows through the Solar System. In its wake the Sun forms an ionisation cavity where there are no more hydrogen atoms, while the helium aroms, much less easy to ionise, distribute themselves along the wake by gravitational focussing; the attraction of the Sun on these atoms has much the same effect as a lens focussing parallel rays of light.

It was only in 1969 that the origin of these hydrogen atoms was understood. They can be observed indirectly by the solar photons of wavelength 121.6 nanometres (Lyman-alpha) which they scatter in all directions; these photons were detected by a French experiment mounted on the American satellite OGO-5. By studying the 58.4-nanometre emission from helium as well, it is possible to determine the characteristics of the interstellar medium through which the Solar System is passing at present; its temperature is 10 000 degrees and the concentrations are 0.05 hydrogen atoms and 0.01 helium atoms per cubic centimetre.

The high temperature and low concentrations are typical of what is called the 'intercloud medium', that is the medium between the stars and outside the interstellar clouds, defined as having a concentration of more than one hydrogen atom per cubic centimetre and a much lower temperature, less than 100 K.

During its existence, the Sun should have encountered many interstellar clouds, some with a concentration of a thousand atoms per cubic centimetre and containing dust particles as well. These encounters would surely have modified the terrestrial climate in important ways, by various processes that acted in opposite directions (raising or lowering the surface temperature) and whose relative importance has not yet been worked out. The gravitational energy of the dust falling into the Sun increases its luminosity but, on the other hand, the interplanetary medium becomes more opaque. The solar wind does not reach further than the Earth. Hydrogen penetrates the upper atmosphere to combine with oxygen and the resulting water condenses on the dust and small particles of ice which finally form a partial screen against sunlight.

By studying the interaction of our Sun with the interstellar medium, at present tenuous, we can reconstruct the possible climatic effects of past encounters with very dense clouds, and look for evidence of such events in the history of our climate.

Jean-Loup BERTAUX

Radiation pressure and the Poynting-Robertson effect. This diagram shows the action of light on a particle M moving with velocity $\vec{V}$ in the Solar System. Solar photons travel with the velocity of light, but appear to the particle M to come from a direction slightly different from that of the Sun, just as raindrops falling vertically seem to a walker to come from in front. The photons intercepted by the particle give up their momentum to it and exert a force $\vec{F_p}$ on it. This force can be resolved into a radial component $\vec{F_r}$, the radiation pressure, opposing the force of gravity $\vec{F_g}$, and a tangential component $\vec{F_{pr}}$ in the direction opposite to $\vec{V}$, which causes the Poynting–Robertson effect. The total radial force $\vec{F_r} + \vec{F_g}$ is smaller than $\vec{F_g}$ and its effect is as if the Sun's attraction was reduced, with a critical velocity of escape v_e. If the velocity of the particle is greater than v_e, it will leave the Solar System along a hyperbolic path. If it is less than v_e, the Poynting–Robertson effect reduces the orbital radius and the particle gradually spirals towards the Sun. This process takes about 10 000 years for a 1 micrometre particle starting 1 astronomical unit from the Sun.

The interstellar wind. It has long been known that the Sun moves with respect to the nearby stars at a velocity of 19 kilometres per second in the direction of the solar apex (vector $\vec{V_a}$) inclined at 60 degrees to the plane of the ecliptic. If the interstellar medium were fixed with respect to the nearby stars, the motion of the Sun relative to it should be in the same direction. But it is nothing of the sort; observations of the hydrogen and of the luminous cone of helium have shown that the Sun is moving, relative to the interstellar medium, in the direction of the vector $\vec{V_w}$ very close to the ecliptic plane (latitude 7 degrees). This means that the interstellar medium is itself in motion with respect to the nearby stars along another vector (in red) $\vec{V_a} - \vec{V_w}$. A real interstellar wind thus blows through our Local Group of stars in a direction lying in the galactic plane. From the direction of the interstellar wind it would therefore seem that, like all the stars, the interstellar material revolves around the galactic centre, but at a different speed. Note that each year the Earth passes close to the helium cone on about 1 December. The position of the Earth on 21 September is shown; the arrow passing through this position points to the First Point of Aries ♈.

The stars and the Galaxy

The stars and the Galaxy

Having carried out a detailed exploration of the Solar System, which consists of the Sun and the procession of planets and other bodies in orbit around it, we shall now review the general properties of stars and the interstellar gas from which they are formed. We have one great advantage in this investigation in that we have a very detailed knowledge of the nearest star, the Sun. However, we must be content with present-day observational methods, which cannot resolve any other star into anything but a point image (except for speckle interferometry, which offers a means of measuring indirectly the diameters of a few, relatively close giant stars).

The proximity of the Sun is a great advantage, not only because it has allowed life to develop on Earth, but also because it facilitates detailed analysis of a star of average characteristics, comparable with more than 80 per cent of the stars of our Galaxy. There are more massive stars (the most massive being about a hundred times the mass of the Sun) and stars at the other extreme, which have only a tenth of a solar mass. Its radius of 700 000 kilometres is also intermediate between those of the stars called supergiants, which have radii a thousand times larger, and the radii of neutron stars (or pulsars), which are less than 10 kilometres. Lastly, the Sun's surface temperature, close to 5700 K (which gives it its orange–yellow colour), is between those blue-white stars which have a surface temperature of several tens of thousands of degrees kelvin, and those of 'cool' stars, whose temperature is as low as 2000 K. Detailed study of the Sun therefore allows us to understand many of the characteristics of the stars. Nevertheless, as we shall see in the illustrations in this part of the atlas, even the closest stars are difficult enough to observe: fluctuations in the atmosphere blur the images and it is impossible with even the best traditional methods to see even a very close star as anything but a twinkling point. It is therefore not possible to determine a star's radius directly. Only interferometric methods will allow us to overcome this very great problem and to start to resolve star images.

Despite this major drawback, we can use photometric and spectroscopic observations of stars to deduce a very large amount of information about their physical state and their evolution. By photometry, using filters of various colours, we can locate a star in a graph, known as the Hertzsprung–Russell diagram, in which total luminosity is plotted against surface temperature.

The total luminosity of a star is measured with an instrument called a bolometer, which is sensitive to all the radiation coming from the star. To determine the surface temperature, we measure the luminosity in different wavelength ranges such as the blue, the yellow or the red. The dominant colour of the star is a measure of its temperature: a very hot star, whose surface temperature is greater than 10 000 K, radiates mostly in the violet, while the coolest stars (with temperatures of the order of 2000 to 3000 K) are red. We shall see in this Atlas that stars are not distributed uniformly throughout in the Hertzsprung–Russell diagram: most lie on a diagonal strip in the diagram, going from hot, luminous stars to cool, faint stars. The point representing our Sun lies on this diagonal, which is called the main sequence. The way that the points for particular groups of stars fall on the diagram gives the best clue we have to the evolution and history of stars. The stars called red giants, which are situated in a region of the diagram indicating relatively high luminosities and low surface temperatures, are at a later stage of evolution than main sequence stars.

Spectroscopy, that is analysis of how the intensity of a star's emitted radiation varies with wavelength, also allows us to obtain a large amount of data: analysis of the characteristic emission and absorption lines allows us to determine the chemical composition of the outer layers of the star. Then we can compare this chemical composition with that of the Sun. We notice, for example, that stars that belong to the dense, spherical globular clusters (see page 290) have a lower abundance of elements such as nitrogen, oxygen and iron than is observed in the surface of the Sun. Thanks to spectroscopy, we can determine other parameters besides the chemical composition, for example, the rotation of the star, which causes a broadening of the lines because of the Doppler effect; magnetic fields, which cause a doubling of the lines by the Zeeman effect; the temperatures of emitting regions, which govern the relative abundances of different ionisation states and excitation levels of each chemical element, etc. Astrophysicists can thus in principle, find from observation practically all the characteristics necessary to determine the physical state and stage of evolution of stars.

One might be surprised that many astrophysicists spend their efforts in the study of stars, when it might seem more interesting to explore the Solar System or to speculate on the origin of

The North America Nebula (NGC 7000). This nebula in the constellation Cygnus takes its name from its shape, which resembles the North American continent. Ionised hydrogen gives the brilliant pink colour. To the lower right, the shape of the 'Gulf of Mexico' is created by a cloud of dust, which may be recognised by the fact that the density of observable stars is much lower than at the top and on the left of the photograph. The hydrogen in this nebula is ionised by hot (and therefore massive) stars which are close to this gas (in particular HD 199579 and Deneb= α Cygni). Some blue parts are due to reflection of the stellar radiation. Short wavelengths are preferentially reflected, according to Rayleigh's law, and this law applies to light reflected by gaseous nebulae in the insterstellar medium. This nebula is at a distance of 2300 light-years and has an angular diameter of 1 degree. (Hale Observatories)

Two globular clusters. This photograph shows two globular clusters NGC 6522 and NGC 6528 which are close to the centre of our Galaxy in the direction of Sagittarius. This photograph was taken with the 4-metre telescope at Kitt Peak National Observatory in Arizona. These two globular clusters are each a collection of hundreds of thousands of stars whose age must be about fifteen billion years: their Hertzsprung–Russell diagrams, on which luminosity is plotted against surface temperature for the member stars, show a very short main sequence because stars having masses comparable to that of the Sun have already evolved into their later stages (see page 254). These stars, and therefore these clusters, were formed very rapidly during the very first stages of the evolution of the Galaxy, before the formation of the disk. The stars are all of low mass since they are old and they give a yellow–red colour to these clusters. The abundance of elements heavier than helium is much lower in globular clusters than in the Solar System. They contain some of the oldest stars which were created at the birth of our Galaxy. (D. F. Malin, Anglo-Australian Telescope Board)

On the preceding double page:

The star S Monocerotis and the nebula NGC 2264. The region around the fifth magnitude star S Monocerotis (on the left of the photograph) is a nebulous complex composed of hydrogen (identified by the orange–red colour) and dust (which obscures the part on the right of the photograph). NGC 2264 is at a distance of 2700 light-years from the Earth; this complex contains about 250 listed stars. (D. F. Malin, Anglo-Australian Telescope Board, © 1981)

the Universe or the way in which large structures like galaxies were formed. The reason for this is quite simple: the stars are the main sources of energy in the present Universe (with the possible exception of giant black holes at the centres of some galaxies, which would explain their very high emissivity: this problem is discussed in several chapters of this book).

The stars produce their energy from nuclear reactions in their central regions. Main sequence stars radiate energy liberated in the transformation of hydrogen into helium in their centres. Stars like the Sun can radiate for times as long as ten billion years. The oldest stars have ages comparable with that of the Universe, between fifteen and twenty billion years. We must remember that the Sun could shine for only ten million years if the only energy source it had was its own gravitational contraction. The processes of thermonuclear fusion make most stars into nuclear power stations constituting one of the principal energy sources on the scale of the Universe. In the course of time they modify its chemical composition.

The single most important parameter governing the physics and the evolution of a star is its mass. Stars of high mass (ten times that of the Sun and above) are luminous stars with high surface temperatures when they are on the main sequence. They evolve in much shorter times (of the order of millions of years) than less massive stars like the Sun, because the luminosity of such a star is roughly proportional to the fourth power of its mass, while the available energy is only proportional to the mass of nuclear fuel. They are capable of synthesising greater quantities of heavy elements. They finish their evolution by exploding; this is the supernova phenomenon

which is thought to occur about every thirty years in our Galaxy. Stars of mass comparable with that of the Sun have a lower surface temperature and a lower luminosity when they are on the main sequence. They evolve in times of the order of ten billion years and end their lives by losing their outer envelope, that is, by producing a planetary nebula. The heart of the star then becomes what is conventionally called a white dwarf, that is a star with a very high surface temperature but a very low luminosity. This very low luminosity is due to the fact that the radius of these stars is very small (comparable with that of the Earth, i.e. about 0.01 of a solar radius for a white dwarf of one solar mass). A white dwarf consequently has a density of 10^6 to 10^8 grams per cubic centimetre. The capacity of these stars to synthesise new chemical elements is much less than that of massive stars.

The stars therefore have several very important roles: they are particularly efficient energy sources; they may provide favourable conditions for the establishment of processes as complex as life in some of the planetary systems which may well orbit round many of them; they are factories where chemical elements such as carbon and all

the elements of higher atomic mass are formed. Looked at like that, this Atlas and ourselves are nothing but the ashes of stars whose evolution finished before that of the Solar System began.

Stars are being formed continually by the gravitational contraction of interstellar clouds; stars evolve at different rates and die. At any time, stars in various stages of evolution can be observed. The astrophysicist attempts to deduce the stages of evolution of the stars that he is studying.

Stars are not formed singly: more than half of the stars making up our Galaxy (the Milky Way) are members of double or multiple systems. The oldest clusters are the globular clusters, which are tightly packed collections of several hundreds of thousands of stars. The globular clusters lie within a spherical volume of space surrounding the whole Galaxy, called the galactic halo. Open clusters, which are loose and irregular, containing as many as several thousand members, can be seen up to 3 kiloparsecs from the Sun close to the galactic plane. Further away, they merge into the starry background. The youngest stellar groups are the O and B associations, named after the spectral types of the young stars therein. They

The Crab Pulsar. This photograph of the Crab Pulsar (below) shows its great optical variability. The Crab Pulsar is a neutron star, and is the remnant of the core of the star that underwent the supernova explosion observed in the year 1054. The central parts, which this pulsar, contracted very dramatically and so the density there is as high as 10^{15} grams per cubic centimetre, that is, the density of the material of the nuclei of atoms. At such densities, the protons of the nuclei absorb the surrounding electrons and are transformed into neutrons, which are stabilised by the conditions of high density and pressure. The outer parts of the star were blown away by the explosion and mixed with the surrounding interstellar medium. The pulsation of the electromagnetic radiation emitted by the star occurs in all wavelength ranges including radio, visible, and X-rays. The very short period, of the order of 33 milliseconds, is due to the very small radius (less than about 10 kilometres) of the neutron star. It is thought that the emitting region is very localised on the surface of the star, which makes the star resemble a lighthouse. During its evolution, the pulsar will slow down (its period will increase) over a period of the order of a hundred thousand years. (Lick Observatory)

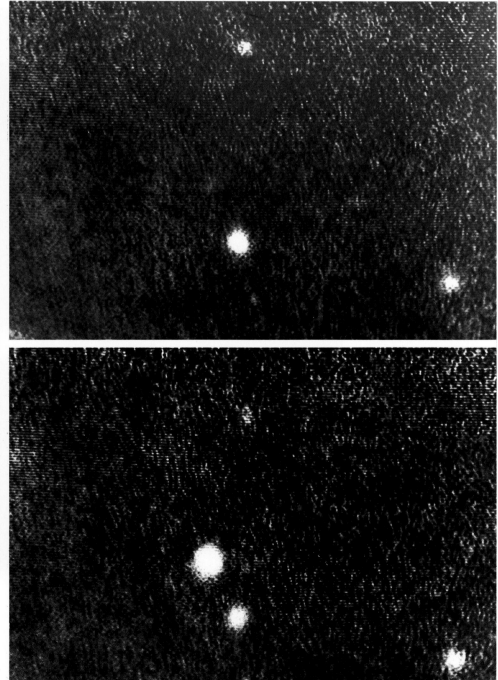

An open cluster (above). This cluster of stars (NGC 5897) was photographed with the 200-inch telescope on Mount Palomar. It is in the constellation of Libra and is of the 'open' or 'galactic' type consisting of only a few thousand stars which are all in the disk of the Galaxy. Their relative velocities with respect to the galactic plane are much less than those of stars in globular clusters. Open clusters are much younger than globular clusters and their abundance of heavy elements is comparable with that of the Sun. Some open clusters, of which several have ages of less than a billion years (that is less than 20 per cent of the age of the Sun) are relatively blue compared with the yellow–red colour of globular clusters, showing the presence of massive stars which have not yet finished their evolution. (Hale Observatories)

243

Supernova SN 1987A. On 24 February 1987, the brightest supernova since the one observed by Kepler in 1604 appeared near the Tarantula Nebula in the Large Magellanic Cloud, 170 000 light years from the Solar System.

The upper photograph was taken in 1984; the star indicated by an arrow, called Sanduleak − 69° 202, is the precursor of the supernova. The latter is visible on the lower photograph, taken about two weeks after the explosion.

The total energy emitted in the course of this type of event is of the order of 10^{44} joules, or about 10 per cent of the output of a typical galaxy in a year: i.e. a supernova emits as much energy as ten million stars like the Sun over a year. In a given galaxy one observes two or three supernovae per year.

The phenomenon represented by a supernova explosion is truly fundamental. Despite its rarity it is actually one of the most important energy sources of a galaxy: through the cosmic rays it accelerates, a supernova heats the interstellar medium and contributes to ionising it; the compression of the interstellar gas around the star may create favourable conditions for the formation of new stars.

Only the most massive stars (more than $5M_{\odot}$) explode in this way; these are also the stars which evolve most rapidly (on timescales less than ten million years, while the Sun's evolution will last about ten billion years) and contribute to the enrichment of the interstellar medium in heavy elements. (D. Malin, Anglo-Australian Telescope Board, 1984 and 1987)

The Rosette Nebula. This superb nebula, also called NGC 2237, is visible in the constellation of Monoceros. It is a spherical object made of hot ionised hydrogen gas. In the centre, in the 'cavity', we see several brilliant stars, which form the cluster NGC 2244. At first sight, the Rosette Nebula could be mistaken for a planetary nebula, which it strongly resembles. A planetary nebula is a low mass star that is ending its life by losing its outer layers, which expand away from the centre with a velocity of the order of 50 kilometres per second, leaving a central star, which becomes a white dwarf at the end of the process. In fact, the Rosette Nebula is a much more massive object which is excited and heated by more than one star, and the stars that are associated with it, which ionise the cloud of hydrogen, are very young (less than a million years old). It is thought that the central cavity, from which the ionised hydrogen has been blown away, was created by radiation pressure from these young stars pushing the gas outwards.

A large quantity of interstellar dust persists despite the energy liberated in this object. This dust is pushed, like the gas, and acquires velocities of the order of several kilometres per second.

To sum up, the Rosette Nebula is a structure of energetic ionised gas also containing dust, which has to disperse at velocities of several thousand kilometres per second because of the pressure exerted from its centre. (D. F. Malin, Anglo-Australian Telescope Board)

contain several tens or hundreds of members. It is of interest to study the stars in a cluster which, having a common origin, are of the same age: the stage of evolution of each star thus depends only on its mass.

It was said at the start of this introduction that most stars are similar to the Sun. Nevertheless, we see a very large number of 'exotic' stars in the sky, of which certain types have already been mentioned: stars such as the Cepheids whose brightness varies on short time scales from less than a day to over a hundred days; eruptive and explosive stars such as novae and supernovae; 'remnant' stars such as white dwarfs and neutron stars. The neutron stars reveal themselves by the rapid periodicity of the radiation they emit (which gave them the name of pulsars); they are extremely dense objects (up to 10^{14} grams per cubic centimetre compared with a density of 10^6 to 10^8 grams per cubic centimetre in white dwarfs). The density of neutron stars is that of nuclear material; neutron stars are just gigantic atomic nuclei. With a mass of the order of one to two solar masses, they have a radius of a few kilometres. A neutron star is the remnant of a supernova explosion which violently ejects the outer layers of its parent star at velocities from 10 000 to 20 000 kilometres per second. A neutron star revolves rapidly and may be observed as a radio pulsar for a hundred thousand years.

The most massive stars, which undergo a strong gravitational contraction, can eventually end their lives in the form of black holes; such objects are so dense that they 'trap' the radiation, which they would otherwise emit, purely by the force of gravity. Black holes therefore interact with the rest of the Universe only by their dynamic effects. However, they are, indirectly, intense energy sources. They are able to attract the matter that surrounds them. As this matter falls into them, it is strongly heated and can radiate intensely in wavelength ranges as energetic as X-rays before disappearing inside the black hole.

Although the idea of black holes or pulsars may seem extremely bizarre, detailed observations of stars almost certainly prove that stars exist with physical properties that have no parallel on Earth or in its immediate neighbourhood.

We see, then, how much the study of stars achieves: we can understand the origin of most of the available energy in the Universe and the processes that have created the various chemical elements. The very peculiar stars extend our knowledge of the most extreme physical states of matter.

Stars come from what are, by convention, known as interstellar clouds in the interstellar medium. There is, therefore, an intimate connection between the stars and the interstellar medium that surrounds them. This gas is very tenuous, since the density of the densest clouds does not exceed a million particles per cubic centimetre, this corresponds to a vacuum far better than we can produce on Earth. The interstellar medium is an extremely heterogeneous entity: these cold, dense clouds (with temperatures of about 100 K or −173 °C) bathe in an inter-cloud medium which is very rarefied (its density is of the order of one particle per 10 cubic centimetres) but hot (with a temperature of about 10 000 K). It is because this medium fills extremely large volumes (for example our Galaxy is a disk 1000 light-years thick and 100 000 light-years in diameter) that its mass reaches about 5 to 10 per cent of that of the stars (in the case of our Galaxy).

The interstellar medium is heterogeneous not only in its morphology but also in its nature. It is a gas in which we find a sizeable quantity of dust, representing 1 per cent of the total mass of the interstellar medium. Although the relative mass of dust appears to be low, it efficiently absorbs the visible radiation coming from the central regions of the Galaxy. We think that most of the interstellar heavy elements (iron, silicon, aluminium, etc.) are in interstellar grains rather than in the gas. The gas is made up of atoms (in particular the hydrogen atom, characterised by its radio radiation at a wavelength of 21 centimetres), ions (we observe, for instance, five-times-ionised oxygen in certain interstellar regions) and molecules (some of which are quite complex, like those of ethyl alcohol). To begin to understand the interstellar medium, we need to look across the whole range of wavelengths in the electro-

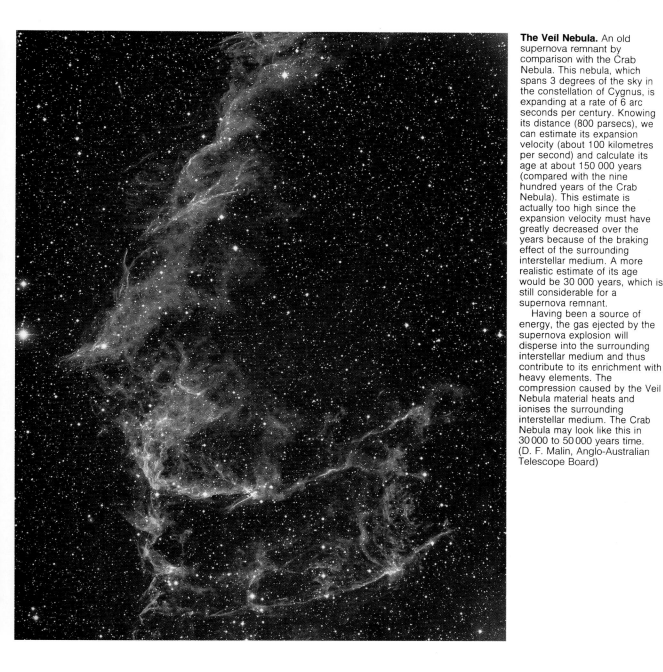

The Veil Nebula. An old supernova remnant by comparison with the Crab Nebula. This nebula, which spans 3 degrees of the sky in the constellation of Cygnus, is expanding at a rate of 6 arc seconds per century. Knowing its distance (800 parsecs), we can estimate its expansion velocity (about 100 kilometres per second) and calculate its age at about 150 000 years (compared with the nine hundred years of the Crab Nebula). This estimate is actually too high since the expansion velocity must have greatly decreased over the years because of the braking effect of the surrounding interstellar medium. A more realistic estimate of its age would be 30 000 years, which is still considerable for a supernova remnant.

Having been a source of energy, the gas ejected by the supernova explosion will disperse into the surrounding interstellar medium and thus contribute to its enrichment with heavy elements. The compression caused by the Veil Nebula material heats and ionises the surrounding interstellar medium. The Crab Nebula may look like this in 30 000 to 50 000 years time. (D. F. Malin, Anglo-Australian Telescope Board)

nature of our Galaxy and of the multitude of galaxies cannot be understood without correctly analysing the physics of stars and of the media which surround them.

A little more effort is perhaps required on the reader's part in this section of the book, as one cannot base it on such spectacular pictures as in other chapters. Nevertheless a knowledge of the mysterious world of the stars and their surroundings is indispensable to an understanding of modern astrophysics and its constant progress.

Jean AUDOUZE

The Rho Ophiuchi complex. This cloud near the star Rho Ophiuchi is very large, with an angular size of 2 degrees and a distance of about 200 parsecs.

The dark appearance of this cloud is caused by the large amount of dust, which obscures the radiation emitted by the stars inside it. This cloud has been observed by radio astronomers who believe that it must be a site of star formation like the Orion Nebula. Indeed, inside this cloud, a maser radiation source has been seen. (A maser emits in the same way as a laser but the radiation is produced as radio waves: in this phenomenon the most energetic energy levels are more populated than less energetic energy levels, as opposed to vice versa, which is the normal arrangement.) We associate these masers with star formation regions. (Photograph courtesy of Ronald E. Royer)

magnetic spectrum from gamma-rays to radio waves. Its study is, therefore, very recent since it requires mastery of observational techniques in wavelength ranges most of which are accessible only from in space.

The interstellar medium, especially its densest regions, acts as a birth place for stars. Stars, both during their evolution (by the winds of particles which they constantly emit) and at the end of their lives (by the ejection of their envelopes or by supernova explosions), restore to the interstellar medium a fraction of the matter that they took up at their formation. The matter that returns to the interstellar medium can be enriched in the chemical elements manufactured in the interiors of stars. The interstellar medium is thus intimately bound to the stars: indeed, its ionisation state and its heating depend on energy sources that have, directly or indirectly, a stellar origin. We notice, for example, that the hottest stars are capable of ionising the hydrogen of the interstellar medium which surrounds them; cosmic rays, which are very energetic particles probably accelerated during or after supernova explosions, are the principal source of heating of the interstellar medium. We cannot therefore independently understand the stars and the interstellar medium in which they are bathed.

The Sun, the stars and the interstellar medium which surrounds them constitute our Galaxy, which is seen as a shining band across the sky, called the Milky Way. Our Galaxy is a disk with a bulge in the middle (similar in cross-section to two fried eggs, back-to-back) surrounded by a very rarefied medium called the halo. The sizes of these different components are given later. The central regions are mostly composed of low-mass stars with very little gas, while if we observe the outside of the disk, we note the presence of more gas and, in proportion, more massive stars. The

Observation of the stars

For centuries, astronomers studied only the positions of stars. They had to contend with the impossibility of performing experiments *in situ* that would allow them to determine the nature of the objects. This led Auguste Comte to make an unfortunate prediction in the mid-nineteenth century: that we would never know the chemical composition of the Sun, for its temperature is too high for us to succeed in approaching it. But despite that, today we can describe the surface of the Sun with a detail unimaginable a century ago. This revolution is due to the fact that we have learned to analyse the light of stars.

The fact that matter absorbs or emits energy in the form of radiation was demonstrated by Gustav Kirchhoff and Robert Bunsen in 1859. However, the interpretation of the laws they discovered in terms of physical properties could be fully realised only when the quantum nature of light was discovered in the twentieth century. So, the analysis of the intensity of radiation as a function of its wavelength became an important tool for the determination of the fundamental parameters not only of stars, but of the greater part of the observable Universe. The various processes of absorption and emission of radiation by matter are a function of the state of the matter and give rise to specific characteristics in radiation, which can be recognised from spectroscopic analysis; physical parameters, which describe the structure of the star, are then deduced from these characteristics.

Apparent magnitudes

Straightforward naked-eye observations of the sky show that some stars appear brighter than others. The oldest known star catalogue was compiled by Hipparchus in 150 BC, who gave positions for 1080 stars, enabling them to be located in the sky. Estimates of brightness were also included in this catalogue, which remained the standard reference work for observers for nearly sixteen centuries. The brightest stars were called stars of the first magnitude and the faintest stars were called stars of the sixth magnitude. The principle of this classification is still used in the quantitative magnitude system used today.

This scale, proposed by Norman Pogson in 1856, is based on the following observational fact: we receive a hundred times more light from a first-magnitude star than from a sixth-magnitude star. A difference of five magnitudes between two stars corresponds to a brightness ratio of a hundred between them. Thus stars of magnitude 1 have a brightness about two and a half times those of magnitude 2. The number corresponding to the magnitude of a star will be correspondingly smaller if the star is brighter. The magnitude scale has been established so as to match that of Hipparchos. There are a set of definite standard stars whose magnitude values have been carefully measured and defined by agreement between astronomers and which are used as a reference for any new measurement. Magnitudes can also be negative numbers for the brightest stars, since the measurement scale has an arbitrary, but nevertheless precisely defined, origin.

The determination of a magnitude consists of a measurement of the brightness of a star. The energy radiated by a star is not constant through the whole spectral range, so the brightness measured depends on the spectral sensitivity of the detector. We always precede the term 'magnitude' with a qualifier defining the spectral range of the observation. Magnitudes are determined by observation, either photographically or photometrically. In the case of photographic observations, the darkening of the photographic plate in each stellar image is converted into a magnitude measurement by comparison with an appropriate calibration. This way we obtain a photographic magnitude, written m_{pg}. If we measure the brightness with a photoelectric detector, a filter defines a spectral range or passband whose width varies from a few tenths to several tens of nanometres; in the first case, the measured magnitudes are classed as 'monochromatic'. Nowadays magnitudes are commonly measured in the UBV system, consisting of three passbands, in the ultraviolet, blue and visible regions of the spectrum. This system was

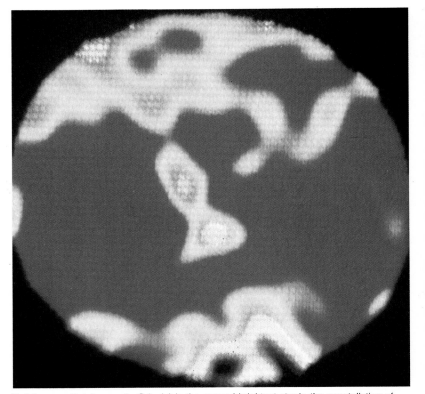

Betelgeuse. Betelgeuse (α Orionis) is the second brightest star in the constellation of Orion. It is a red supergiant about 650 light-years away, which has one of the largest stellar radii known: about 800 times the solar radius. Despite that, its image in telescopes is still point-like. By means of speckle interferometry, the astronomers, C. R. Lynds, S. P. Worden and J. W. Harvey at Kitt Peak, have been able to see the image of Betelgeuse. Different coloured zones appear in the image (the contrasts are exaggerated here), which are interpreted as being of different temperatures: the blue, the purple and the black correspond to the coolest regions. Spots such as the two at the centre have a size of about 300 million kilometres, that is of the same order of size as the Earth's orbit.

Betelgeuse is a variable star and according to its spectrum it has a strong stellar wind. Stellar evolution theories predict the existence in such stars of a deep surface convection zone. The spots observed on the surface of Betelgeuse are interpreted as being a sign of this convection. (Association of Universities for Research in Astronomy, Inc., Kitt Peak National Observatory)

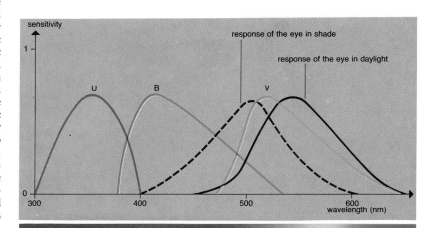

Johnson–Morgan UBV system		Strömgren uvby system	
U (ultraviolet)	330–400 nm	u (ultraviolet)	330–367 nm
B (blue)	390–490 nm	v (violet)	400–420 nm
V (visible)	505–595 nm	b (blue)	462–478 nm
		y (yellow)	538–562 nm

Spectral ranges of the most common photometric measurement systems.
Photometric magnitudes are represented by symbols that signify the spectral ranges considered: U for an apparent magnitude measured in the ultraviolet spectral range, B for a measure of the apparent magnitude in the blue; the V magnitude stands for 'visible' because it corresponds to the range of maximum sensitivity of the eye.

The diagram shows the response curves of the filters of the UBV system, compared with those of the eye.

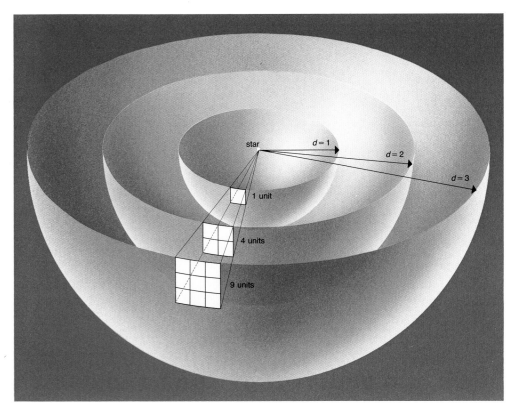

Flux and luminosity. The luminosity of a star is the total amount of energy emitted as radiation per unit time. The determination of this luminosity requires the measurement of the energy radiated in all spectral ranges.

The observable quantity is the flux F: this is the amount of energy collected in unit time by a detector of unit surface area placed at right angles to the line of sight; it will be less if the detector is placed further from the star. The flux (F) varies as the inverse square of the distance (d) to the object ($F = L/4\pi d^2$). Consequently, measurement of the flux from a star allows us to determine its luminosity (L only if we know its distance. For example, the luminosity of the Sun is equal to 3.82×10^{26} watts. (Diagram from *Contemporary Astronomy*, J. Pasachoff).

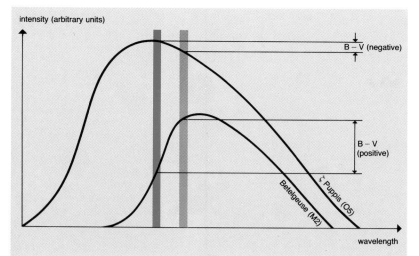

The colour index. A hot star radiates more energy in the blue spectral range than in the yellow. As a result, its B magnitude will be smaller than its V magnitude (the larger a magnitude is, the smaller the luminosity of the star in the spectral range considered). This result is reversed for a cool star.

A magnitude expresses, on an arbitrarily decreasing scale, the integrated flux over the spectral range in which it is measured. The difference B−V is negative for a hot star and positive for a cool star. This parameter, known as the colour index, has the property of being independent of the surface area of the star and of its distance. The colour index thus depends only on the nature of the radiation from the stellar atmosphere. Colour index can be calibrated in terms of temperature.

spectral type	effective temperature (K)	bolometric correction $m_{bol} - m_v$
O5	35 000	−4.6
B0	21 000	−3.0
B5	13 500	−1.6
A0	9 700	−0.68
A5	8 100	−0.30
F0	7 200	−0.10
F5	6 500	0.00
G0	6 000	−0.03
G5	5 400	−0.10
K0	4 700	−0.20
K5	4 000	−0.58
M0	3 300	−1.20
M5	2 600	−2.10

Bolometric correction.

introduced in the 1950s. The passbands are wide, of about 8 nanometres width at half maximum response, and peak at 360, 420 and 520 nanometres. From two magnitudes measured at different wavelengths we can derive a colour index by subtracting one magnitude from the other. The U-B and B-V colour indices, in particular, are closely related to stellar temperature and luminosity. For several years we have been able to measure magnitudes in other spectral ranges thanks to observations from space: catalogues of X-ray, ultraviolet or even infrared magnitudes are starting to become established. Observations from the ground in radio wavelengths have even given radio magnitudes for a few stars.

As the magnitude of a star is a function of its distance we frequently use the qualification 'apparent'. If two identical stars are at different distances we see a lower brightness, therefore a higher magnitude, for the further star. Whichever method is used, the brightness measurement must be corrected for the effects of absorption of the radiation by the Earth's atmosphere, as well as for the sensitivity of the measuring instruments used, so that the results from one observer are comparable with those from another.

Because of the limited spectral sensitivity of measuring instruments, the apparent magnitude of a star is not a measure of the total brightness of the star, but only of what is an almost monochromatic intensity. For the same star, apparent magnitudes in different spectral ranges will therefore have different values.

The apparent magnitude determined taking account of the energy radiated by the star in all spectral ranges is called the apparent bolometric magnitude, written m_{bol}.

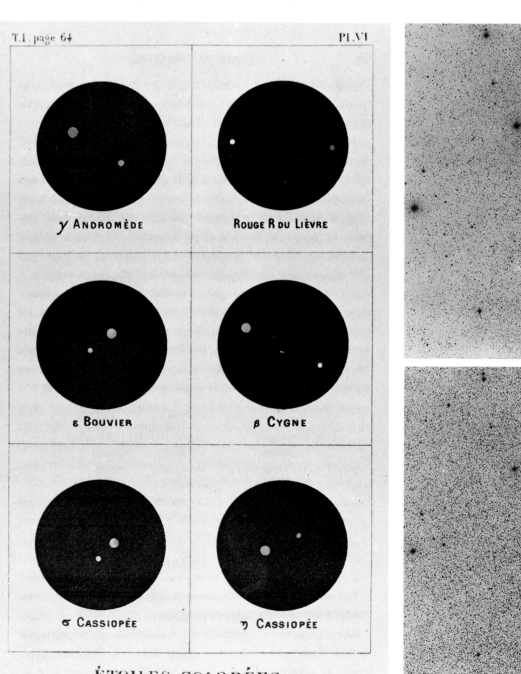

ÉTOILES COLORÉES

instrument	limiting apparent magnitude
eye	6.5
binoculars	10
15 cm (6 inch) telescope (observation by eye)	13
5 m (200 inch) telescope (observation by eye)	20
5 m (200 inch) telescope (photographic magnitude)	24

Limiting apparent magnitude. The smallest flux that can be measured with a given telescope depends on the size of the telescope and the exposure time.

Although stars appear as points, stellar images in photographs have a finite size due to scattering in the emulsion. The size depends on various factors: the star's apparent brightness, the telescope, the photographic emulsion, the exposure time and the weather conditions during the observation, and in particular of the atmospheric turbulence. All images on the same photographic plate are affected in the same way by these parameters, so it is possible to calibrate each photographic plate in terms of photographic magnitudes m_{pg} by establishing a correlation between the blackening of the plate and the known magnitudes of standard stars.

The ratio of fluxes corresponding to an apparent magnitude difference of 20 is 100 million. With the Space Telescope, we hope to reach an apparent visual magnitude of 29.

The colours of stars. The brightest stars can be seen with the naked eye to have different colours. However, our eyes lose the ability to distinguish colours for faint objects. The range of magnitudes for which colours can be distinguished can be extended by using a pair of binoculars or a small telescope.

These colours indicate the temperatures of the stellar atmospheres. The radiation from a star corresponds approximately with that of a black body. Of course, the colour perceived is weighted by the spectral sensitivity of the eye. As an example, Antares (α Scorpii) appears red to us while Rigel (β Orionis) is blue. The colour temperature of Antares is 3000 K, that of Rigel is 20 000 K.

At sunset, or at dawn, when the Sun is low above the horizon, it appears very yellow or even red. This effect is not due to a change in the Sun's surface temperature, but results from the interaction of photons emitted by the Sun with particles in the Earth's atmosphere, which absorb part of the radiation and scatter part as well. This scattering affects blue photons preferentially. When the Sun is low

above the horizon, the solar radiation travels through a much larger thickness of the Earth's atmosphere than when the Sun is high in the sky. As a result, a greater proportion of blue photons are scattered; the lack of blue photons makes the solar disk appear red. These blue photons are scattered by the Earth's atmosphere in all directions and give the blue colour to the sky.

The left-hand picture shows the colours of some double stars; it is a plate of coloured drawings from a work by Angelo Secchi.

Two photographs of the same star field, taken with two emulsions of differing spectral sensitivity, can show a completely different appearance (right): the upper photograph was taken with an emulsion sensitive mainly in the blue, the lower one with a red-sensitive emulsion. Some stars appear the same in both pictures. Others are brighter in the 'blue' picture: these are hot stars; while others are brighter in the 'red' picture and are thus relatively cool stars. (S. Counil, Observatoire de Paris (left); Palomar Observatory Sky Survey (right)

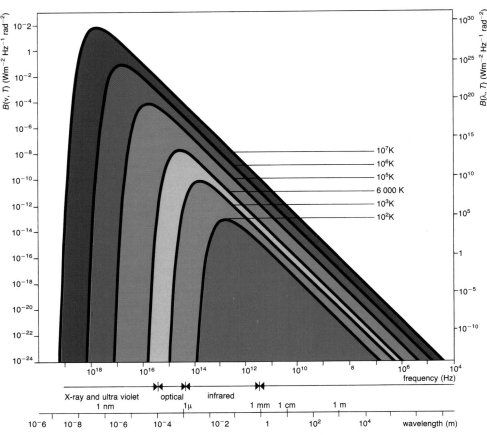

The difference between the apparent visual magnitude, m_v (i.e. that corresponding to a measure of the flux in the visible region only) and the apparent bolometric magnitude is smaller or larger according to the characteristics of the star's radiation. This radiation is essentially a function of the surface temperature of the star. Tables give the difference between these two magnitudes, called the bolometric correction, as a function of the surface temperature of stars, or more precisely as a function of the effective temperature. This correction will be very important for stars with a high surface temperature and also for those of low surface temperatures since they emit their maximum energy respectively in the far ultraviolet or the infrared. Stars like the Sun have a very small bolometric correction because the maximum energy radiated is in the visible region.

The measurement of luminosities

To determine the luminosity of a star from its apparent bolometric magnitude, we need to know its distance from us. The apparent visual magnitude of the Sun is -26.5 while that of Sirius, the brightest star in the sky apart from the Sun, is -1.5. Knowing that Sirius is five hundred thousand times further away than the Sun, we deduce, knowing the difference in apparent bolometric magnitudes between the two stars, that Sirius is about twenty-five times more luminous than the Sun. The fact that the Sun has a flux ten billion times that of Sirius results purely from its closeness.

The closest star to us (apart from the Sun) is similar to the Sun, but has an apparent visual magnitude of -0.1. The Sun lies in the middle of the observed range of stellar luminosities; there are many stars whose luminosity is ten thousand times larger and others ten thousand times fainter than the Sun. One of the most luminous stars that has been detected is η Carinae, visible in the southern hemisphere, whose luminosity is three million times that of the Sun. It is obvious that one cannot give a value for the lowest luminosity actually detected, this being of course a function of instrumental sensitivity: each time we push back the limiting observable apparent magnitude, we expand our domain of exploration, but that allows us also to observe stars which are not necessarily distant, but intrinsically fainter.

The apparent magnitudes of stars are affected by absorption of the radiation by interstellar matter. When photons cross a cloud of interstellar matter, absorption takes place, which is more important for blue photons than for red photons. The star appears to us as if it were emitting radiation relatively more intensely in the red than

Black-body radiation. The atoms in any body are in motion, vibrating around their mean positions. If a body is heated while perfectly insulated from the ambient medium, it reaches a state of thermodynamic equilibrium. In this state, the atoms have the same average kinetic energy at every point in the body. This energy characterises the state of the body. If we heat the body further, the vibrations of the atoms increase, and this average kinetic energy increases. We can specify the state of the body by another parameter, the kinetic temperature, which is a number proportional to the average energy of the atoms in the body. This parameter is normally said to be the temperature of the body. All bodies with temperatures above absolute zero emit electromagnetic radiation. For a body in thermodynamic equilibrium, the properties of its radiation are perfectly defined: in particular, the energy distribution is given by Planck's law. This body is then known as a black body.

Planck's law expresses the quantity of energy radiated in unit time by a unit surface area of black body at temperature T in a given direction into a unit solid angle, per unit wavelength interval at a given wavelength. This quantity, called the spectral luminance, depends only on the temperature T and on the wavelength λ considered, because the radiation of a black body is isotropic.

The curves that describe the variation of the spectral luminance $B(\lambda, T)$ as a function of λ do not cross; the luminance is always greater for a higher temperature at a given wavelength.

For each of these curves, there exists a value of the wavelength, λ_{max}, for which the spectral luminance is a maximum. The relationship between λ_{max} and temperature is known as Wien's law: from observation of the wavelength of the maximum of the spectral luminance of a black body, we can deduce its temperature. At high temperatures, the maximum is at short wavelengths and for low temperatures the maximum is in the infrared or the radio region.

It is common experience that the colour of a heated body changes with temperature. This observation is consistent with Planck's law and Wien's law. A heated body emits photons over a range of wavelengths. The colour the body appears to us depends on the proportion of photons emitted in the visible region and the response of the eye. If we heat a piece of iron, at first we do not see it radiating. When the iron is cool, its maximum radiation takes place in the infrared, a range to which the eye is not at all sensitive, and the number of photons emitted in the visible region is far too small for us to see them. As the temperature rises, the energy radiated in the visible rises. The piece of iron will look red because red photons predominate. As we raise the temperature further, the piece of iron becomes yellower. By raising the temperature, we have raised the number of emitted photons: photons corresponding to other colours have become numerous enough so that mixed together, the dominant colour is yellow. For higher temperatures still, the iron appears white; in fact, when the maximum luminous intensity is in the range of sensitivity of the eye, we receive in almost equal proportions photons of all visible wavelengths; this mixture of photons gives us the impression of white light. If we could heat it still further, the piece of iron would appear blue.

To calculate the total energy emitted by a black body, it is necessary to add up the spectral luminance for all wavelengths. We can show that the total power radiated by a black body – its luminosity – is proportional to the surface area of the body and to the fourth power of the temperature. This relation between luminosity and temperature, called Stefan's law, shows that if we know the luminosity of a black body and its surface area, we can deduce its temperature. If we know the wavelength of the maximum spectral luminance of a black body and its luminosity, it is then possible to deduce its radiating surface area.

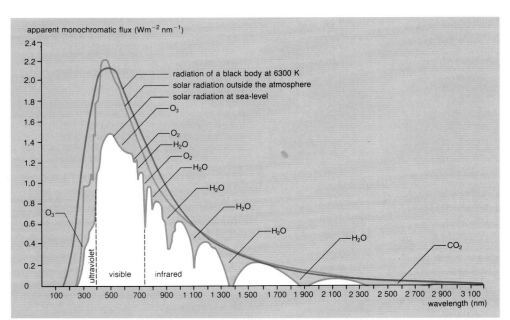

Temperature of the solar atmosphere (below left). The spectrum of continuous radiation from a star may be approximated by that of a black body. However, the atmosphere of a star is obviously not a black body, because lines are seen in stellar spectra. Supposing that the surface of a star radiates like a black body at temperature T, the monochromatic power radiated by the star is expressed by the product $\pi \times B(\lambda, T) \times S$, where S is the surface area of the star and $B(\lambda, T)$ is the spectral luminance of the corresponding black body at wavelength λ and temperature T. It is this quantity that can be measured effectively when the distance to the star is known. Generally, we know only the monochromatic flux $\pi \times B(\lambda, T) \times S/(4\pi d^2)$ where d is the distance to the star. The variation of this quantity with wavelength is the energy distribution of the star. These two expressions are proportional to $B(\lambda, T)$ and they take the same form as the Planck law. As a result, the determination of the wavelength of the maximum of the curve representing the stellar energy distribution allows us, thanks to Wien's law, to determine a temperature characteristic of the atmosphere of the star.

This temperature is often called the radiation temperature. It is relevant to the continuum radiation, but does not take account of the conditions in which lines are formed. However, it is a useful quantity. The diagram shows the solar energy distribution at the Earth's orbit, outside the atmosphere; the position of the maximum of this curve near to 460 nanometres corresponds to a black body at 6300 K. The third curve represents the solar radiation as measured at sea-level: we notice how much the atmosphere of the Earth affects the solar radiation and therefore stellar radiation in general. The principal absorption bands, mostly due to water vapour, are shown.

If we compare the spectral luminance of the Sun with that of a black body at 6300 K for a wavelength other than the maximum, we find that they are different: the Sun and the stars do not radiate exactly like black bodies.

Several procedures have been defined, each allowing us to define a temperature. If a star were really a black body, all these methods would give the same numerical result. The comparison of the monochromatic power radiated by the star at two different wavelengths with the power emitted by a black body at the same wavelengths allows the definition of another temperature, called the colour temperature. Another method, more often used, consists of finding which black body emits the same total power as the star. The temperature of this black body defines the effective temperature of the star. For the Sun, it is 5800 K.

We see, therefore, that it is incorrect to speak of *the* temperature of a star or even the temperature of the 'surface', which is an undefined region. In the case of a black body, the temperature appearing in the Planck function has the same value as the temperature describing the average speed of the atoms in the body. These two temperatures are not the same in stellar atmospheres, and can differ considerably.

None of the methods described here allowing us to define a temperature parameter is fully satisfactory, but by combining them with the laws of physics describing the matter–radiation interactions, the astrophysicist can determine more precisely the structure of stellar atmospheres.

if there were no interstellar matter. This phenomenon is known as interstellar reddening. If one does not correct apparent magnitudes for interstellar reddening, the temperature of the star that is inferred will be lower than its actual value.

The absolute magnitude

Comparisons between the apparent bolometric magnitudes of stars cannot give any information about their relative luminosities, for the apparent bolometric magnitudes depend on the distance. A scale of magnitudes has therefore been defined,

called absolute magnitudes, these being the magnitudes that stars would have if they were placed at a distance of 10 parsecs; the absolute magnitudes of stars are therefore directly comparable to their luminosities. Just like apparent magnitudes, absolute magnitudes are a function of the spectral range over which they are measured and absolute bolometric magnitude can be defined, written M_{bol} (absolute magnitudes are always written with a capital M to distinguish them from apparent magnitudes written with a small m).

If we can estimate the absolute magnitude of a given star, for example from particular spectro-

The formation of stellar spectra (below). The stellar radiation we observe is usually made up of a continuum spectrum and absorption lines.

In the most simplified model for the formation of this radiation there is an incident continuum at the base of the atmosphere and the interaction of this radiation with the gas of the atmosphere is responsible for the formation of lines. Actually, there are not two separate zones and at each point in the atmosphere the processes of absorption and emission take place simultaneously.

The continuum radiation has been represented here at the base of the atmosphere by a collection of photons of different colours. Only certain photons of particular energies can be absorbed. However, an atom de-excites very rapidly, re-emitting a photon whose energy is equal to the difference in energy between the two successive states of the atom. This de-excited state is not necessarily the same as the initial state of the atom before the absorption of the photon, but in the simplified model considered here, we shall suppose that the state is the same, so that the re-emitted photon has the same energy and colour as the incident photon. The direction of this re-emission is random. Let us then consider blue photons. They are capable of interacting with certain atoms in the atmosphere. Some are re-emitted in the same direction as the incident photon and others in random directions. It is the same for red photons. For the sake of argument, we shall suppose that green and yellow photons are not susceptible to interactions with atoms in the atmosphere.

Let us suppose that the observer can observe the star in two directions. Direction 1 corresponds to an observation directly towards the centre of the stellar disk, and direction 2 to an observation of the edge of the disk (possible only during eclipses). For stars, only the observation corresponding to direction 1 is possible; their apparent diameters are too small for it to be possible to aim in direction 2. In direction 1, the observer will collect photons that have crossed the atmosphere without undergoing interaction, as well as those that have interacted and have been re-emitted in the same direction. The number of these is less than the number of incident photons: the observer will therefore see an absorption line. Observing in direction 2, the observer collects photons that have been re-emitted in that direction after absorption, and only those. He will therefore see a spectrum of emission lines. That is what can be seen during a solar eclipse.

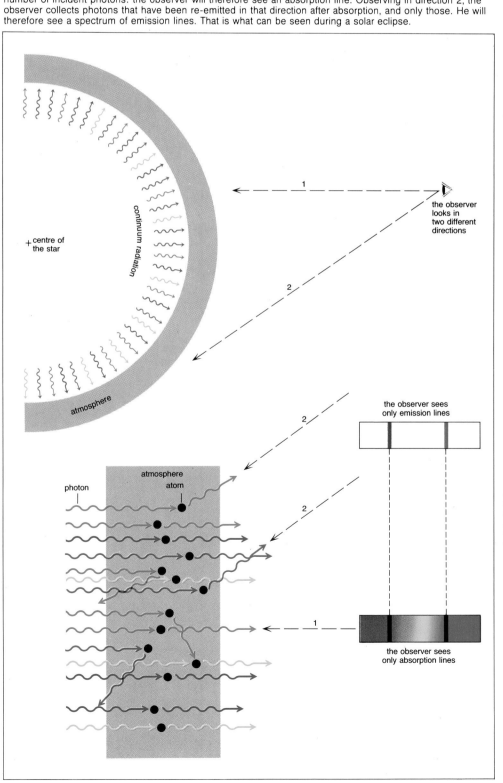

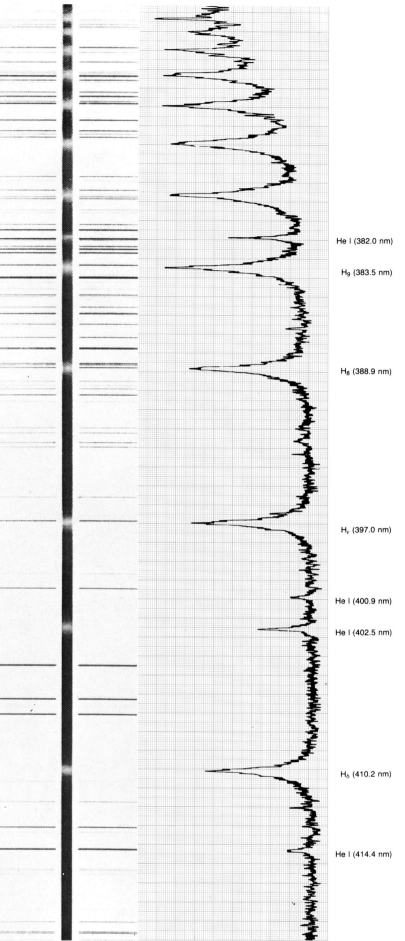

He I (382.0 nm)

H$_9$ (383.5 nm)

H$_8$ (388.9 nm)

H$_ε$ (397.0 nm)

He I (400.9 nm)

He I (402.5 nm)

H$_δ$ (410.2 nm)

He I (414.4 nm)

Analysis of a stellar spectrum (above). How do we identify the wavelengths of lines in a spectrum? On a photographic plate (left) used to record a stellar spectrum (the dark stripe in the middle), the spectrum emitted by an iron arc (or other comparison source) is also exposed, on both sides of the star's spectrum. This is an emission-line spectrum with known wavelengths. These reference lines are used to determine the wavelengths of the lines in the stellar spectrum, and thus identify the elements that caused the lines. The spectrum shown here is that of ν Orionis (spectral type B3 V); it was obtained at the CNRS Observatory in Haute-Provence, France. The density of the blackening of the plate is analysed to show the relative intensity of the spectrum as a function of wavelength (right). Analysis of the line spectrum can give data about the region of the star's atmosphere where the lines are formed. Analysis of the shapes of the lines – their 'profiles' – is extremely complex. These profiles are compared with theoretical profiles calculated from models of stellar atmospheres. The parameters of the theoretical models may be adjusted until the calculated profiles of the lines correctly represent the observed profiles.

The quantities that can be determined in principle in this way are: the chemical composition, the density, the pressure and the temperature. These last three parameters are not constant throughout the atmosphere, but vary with depth.

Stellar temperatures obtained in different ways can differ strikingly from each other if the gas is far from being in an equilibrium state. Temperature parameters associated with the states of particles are different from those obtained from the spectral luminance.

Other facts can also be estimated from the study of spectra: radial velocity of the star, the rotation speed of the star about its own axis, the strength of its magnetic field and whether it is ejecting matter. (Photograph F. Warin, Institut d'astrophysique de Paris)

The stars and the Galaxy

Measurement of radial velocities. The velocity V of a star can be determined if the two components of its velocity (top diagram) can be measured. The transverse velocity V_t causes an apparent displacement of the image of the star in two photographs taken on different dates. This displacement is in general very small and difficult to determine because of the star's great distance. The radial component V_r can often be measured easily thanks to the Doppler effect.

The radial velocities of several thousand stars have been measured. They are between −400 kilometres per second and +400 kilometres per second, the + or − sign referring to the fact that the star may be receding from or approaching the Sun. However, stars whose absolute value of radial velocity is more than 100 kilometres per second are rare, most having radial velocities between −50 and +50 kilometres per second. To determine the radial velocity of a star with respect to the Sun, it is obviously necessary to take account of the movement of the Earth around the Sun.

The photographs show the spectra of two stars (the dark strips in the middle). These pictures are negatives: the stellar lines, which are absorption lines, appear white and the reference lines of the iron arc (at the extreme left and the extreme right), which are emission lines, are black. The star HD 17971 is a star in our galaxy which is approaching the Sun at a radial velocity of −35 kilometres per second while the star HDE 271182, is receding at a radial velocity of +300 kilometres per second. This star does not belong to our Galaxy, but to a neighbouring galaxy, the Large Magellanic Cloud.

By reference to the iron arc spectrum the Doppler shifts of the two stellar spectra can be compared. Three lines of neutral iron (Fe I) have been identified. Notice that the spectrum of HD 17971 shows a shift towards shorter wavelengths: the Fe I lines are shifted upwards compared with those of the iron arc; in the case of the star HDE 271182, these lines are shifted downwards, that is towards longer wavelengths. (Observatoire de Haute-Provence, F. Warin)

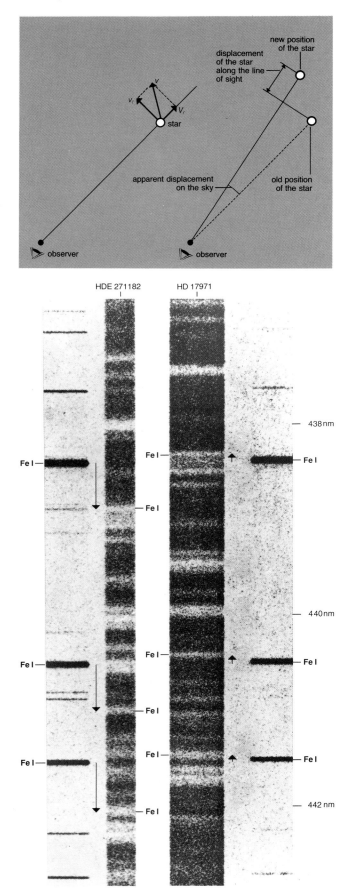

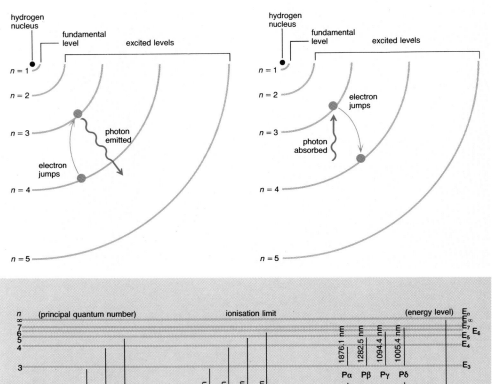

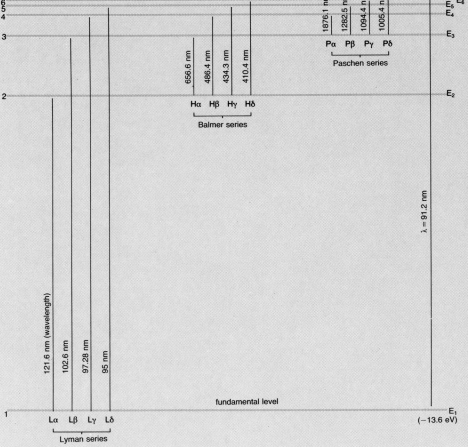

The Bohr model of the atom. An atom is composed of a nucleus, which is small and dense, surrounded by electrons. The hydrogen atom consists of a proton and one electron. The motion of the electron may be represented by a circular orbit (top diagram). According to the laws of quantum mechanics, the electron can exist only in certain well-defined orbits. Any movement from one orbit to another is called a transition and corresponds to a change in the energy of the system. The closest orbit to the nucleus corresponds to the minimum energy and is called the ground state or fundamental level. Any other situation is called an excited level. When the electron goes from one orbit to another closer to the nucleus (top left), its energy decreases. According to the law of conservation of energy, the energy must appear in another form, here in the form of a photon whose energy is equal to the energy difference between the two orbits. There is emission of radiation. The electron can also jump from one orbit to another further out if it gains energy, for example by absorbing a photon (top right), but the energy of the absorbed photon must correspond exactly to the energy difference between two possible orbits. The wavelengths of emitted or absorbed photons are correspondingly shorter as the energy associated with the photon is larger. The radii of the possible orbits in theory extend to infinity; however, when the electron ceases to be subject to the direct influence of the nucleus the energy of the electron can have any value. We say that the atom is ionised. To ionise an atom enough energy is needed to tear the electron from the attraction of the nucleus. The absorption of any photon – as long as it gives enough energy to the atom to permit ionisation – is possible. The reverse phenomenon can happen: a free electron can re-combine with an ion to re-form the atom with the emission of a photon. This model of the atom of hydrogen, proposed by Niels Bohr in 1913, is now recognized as a gross oversimplification. The model is nevertheless very convenient for the rapid and schematic understanding of the processes and the notion of quantised energy levels remains fundamental (lower diagram). The possible energy levels are represented schematically in a Grotrian diagram. The levels are placed on a scale proportional to their energy. Lines associated with the fundamental level, which involve the largest energy differences, are in the ultraviolet (lines of the Lyman series) while those which involve lesser energy differences are in the infrared (lines of the Paschen series). The lines observed in the visible involve transitions from the second level. These are the lines of the Balmer series. The higher levels are close to one another and are squeezed together right up to the limiting energy value above which the atom is considered as ionised.

An atom may be excited (i.e. jump to a higher energy level) either by collision with another particle (the energy is then taken from the kinetic energy of the particle) or by absorption of a photon whose energy corresponds to a possible transition. An atom does not remain in an excited state for long. It is de-excited spontaneously and returns to the fundamental level. It may also de-excite following a collision.

scopic characteristics, we can deduce the distance
of the star by comparing the absolute and
apparent magnitudes. However, if the apparent
magnitude has not been corrected for interstellar
reddening, the distance so determined will be
overestimated.

The radiation of stars

For a given star, measurement of its monochromatic apparent magnitudes at a series of
wavelengths allows us to determine the flux from
this star as a function of wavelength. If the
distance to the star can be estimated, we can then
deduce the energy distribution as a function of
wavelength (i.e. the quantity of energy radiated
by the star in a second into unit spectral intervals
at each of the wavelengths considered). An ideal
situation is to be able to carry out, on the same
object, measurements for the entire electromagnetic spectrum, that is from the gamma-ray
region to the radio region. We can be content in
the case of stars, to observe their radiation from
the ultraviolet to the infrared since outside this
region they emit very little energy, but these
observations must call on space techniques for
the study of radiation that is normally absorbed
by the Earth's atmosphere.

The transformation from a magnitude to a flux
is not easy. We have seen that the magnitude scale
does not refer to any real measurement of energy;
consequently, in order to transform a series of
observations of monochromatic magnitudes into
an energy distribution of the star's radiation, it is
necessary to calibrate in terms of energy. This
calibration must take account of energy losses
not only in the Earth's atmosphere but also in the
instruments used for the observation. The very
simple principle of an absolute calibration consists of the comparison of the radiation of a star
with that of a standard source. The standard
source is a black body, that is an opaque, isolated
body whose temperature is kept constant. The
spectrum of the radiation from such a body
depends only on the temperature: its intensity as
a function of the wavelength is described mathe-

matically by the Planck law. The materials
generally used are copper or platinum, placed in a
thermostatically-controlled enclosure whose
temperature is raised to 2000 K; the observation
with the telescope plus photometer of the
radiation emitted by the black body is translated
into a magnitude and we can thus establish for a
black body a correlation between magnitude and
energy flux. It remains only to observe, with the
same combination of telescope and photometer,
a star whose apparent magnitude can then be
converted into an energy flux by reference to the
calibration curve. This measurement, simple in
theory, is not easy to put into practice so
secondary standards are generally used. For some
stars, their energy distributions are determined
by reference to the black body. For other
observations, these stars replace the black body
and become in turn calibration sources. Vega (α
Lyrae) is one of these standard stars; it is actually
the primary standard.

Stellar spectra

Fortunately, extremely important data concerning the structure of stars can be obtained simply
by spectroscopic analysis, without needing an
absolute energy calibration. The study of the
intensities of the lines in a stellar spectrum
relative to the continuum on either side allows us
to gain access to the physical properties of the
atmosphere of the star.

While magnitudes can be determined either by
photography or photometry, detailed observation of stellar radiation – and more particularly
the lines – requires the use of a spectrograph.
Photometry allows us to determine the intensity
of stellar radiation from a measurement of the
average of the intensity of the radiation over a
passband several tens of nanometres wide.
Spectroscopy determines the intensity of the
radiation from a star, over a large spectral range,
using a much smaller passband, perhaps as little
as a hundredth of a nanometre. In this case, the
analysis of the variation of the intensity of the
radiation as a function of wavelength is therefore

much more detailed. We notice, however, that
photometric analysis allows observation of much
fainter stars (i.e. of much larger apparent magnitude) than can be studied by detailed spectroscopy. Obviously, these two types of analysis are
complementary.

Stellar atmospheres

The radiation that we observe from a star is
emitted by the outer layers of the star, called the
stellar atmosphere. There is no precise limit to the
atmosphere, and it is impossible to define
rigorously the 'surface' of a star. The atmosphere
is a transition zone between the condensed gas of
the star and the very rarefied gas of the interstellar
medium. The density of the gas decreases as one
goes away from the centre of the star.

The atmosphere of a star can be thought of as
several zones with different properties. For
example, the solar disk that we see is called the
photosphere. This outer layer extends for a
thickness of about 500 kilometres. The two terms
'photosphere' and 'atmosphere' are frequently
confused. The photosphere merges into other
regions, the chromosphere and the corona, which
have totally different properties. It is possible to
make direct measurements of the geometrical
extent of the chromosphere and corona for the
Sun, but not for other stars. The existence of the
chromosphere or a corona can be shown only by
particular features of the star's spectrum. The
energy emitted from the central regions of the star
by thermonuclear reactions leaves those regions
in the form of radiation. Passing from the centre
to the surface of the star, photons undergo
numerous interactions with matter, which greatly increase the time taken for the journey. Thus,
energy created at the centre of the Sun will not
leave the surface until, on average, ten million
years have passed. The radiation that we observe,
therefore, represents the interactions of these
photons with the outer regions of the star (i.e. the
atmosphere): these are the last interactions that
the photons undergo before travelling more or
less freely through the interstellar medium and
coming to Earth.

star	apparent diameter (″)	distance (parsecs)	radius ($R_\odot$)
Betelgeuse (α Orionis) (variable star)	{ 0.034 / 0.042	150	{ 250 / 375
Aldebaran (α Tauri)	0.020	21	22
Arcturus (α Bootis)	0.020	11	12
Antares (α Scorpii)	0.040	150	320
Scheat (β Pegasi)	0.021	50	55
Ras Algethi (α Herculis)	0.030	150	250
Mira (o Ceti)	0.056	70	210

The measurement of the radii of stars (table opposite left). In principle, the radius of a star can be determined if we know its apparent or angular size and its distance. Actually, the direct image of a star in a telescope cannot be used to determine its apparent radius because of the perturbing effect of the atmosphere, which blurs the images. The space telescope due to be launched in 1986 will make such measurements possible. On photographs, stars have images of different sizes; this effect is purely instrumental being a function of the telescope, the photographic emulsion used, and also the weather conditions during the observation.

We have, however, been able to measure effectively the apparent diameters of several dozen fairly close stars, whose distances we know, by using interferometric techniques (see table opposite, $R_\odot$ = solar radius).

Another method that can be used to estimate the angular diameter of a star involves observing an occultation of the star by the Moon. This technique can obviously be applied only to stars close to the Moon's path and it is limited by our lack of knowledge about the irregularities of the lunar surface.

Analysis of stellar radiation is another way of estimating the radius of a star. We suppose that the star radiates like a black body; we know then how to determine the temperature of the atmosphere of the star. The measurement of the apparent flux of the star allows us, if we know the distance, to deduce its luminosity. From Stephan's Law the relationship between luminosity, radius, and temperature can be derived; luminosity is proportional to the square of the radius and the fourth power of the temperature. Hence, knowing luminosity and temperature, the radius can be found.

The only method for which the star's distance is not needed is the observation of eclipsing binary stars. Observation of the relative movements of the two stars and its interpretation through the law of gravitation allows, in some cases, the direct determination of the radii of each star of the pair (see table below). Few measurements are obtained by this method, but they are of great interest because they give independent verification of the other methods described.

Radii of the stars of eclipsing binary star systems (table opposite right). The values of the radii measured for stars range from a few tenths of a solar radius ($R_\odot$) to about a thousand $R_\odot$. The radii determined is that of the luminous disk called the photosphere. In the case of the Sun, we know that other regions extend beyond the photosphere – the chromosphere and corona – but these regions are relatively transparent. The spherical shape of a star can be distorted, for example by rapid rotation which causes a flattening at the poles; this effect is observed in the case of the Sun. A star in a binary system undergoes tidal effects due to the pull of its companion. If the two stars are very close to each other, the spheres can be considerably deformed and can in certain cases come into contact with each other. However, these shapes predicted by the law of gravity have not yet been observed directly. The values of the radii of stars belonging to the main sequence are of the order of 0.1–15 $R_\odot$. Giant and supergiant stars can have very extended atmospheres, as large as 2000 $R_\odot$. At the other extreme, white dwarfs have radii less than 0.01 $R_\odot$.

star	period (days)	radius of each component ($R_\odot$)	
WW Aurigae	2.5	1.9	1.9
YY Geminorum	0.8	0.6	0.6
U 356 Sagittarii	8.9	4.9	12.7
32 Cygni	1140	353	4
ε Aurigae	9898	1278	716

The masses of stars (table below right). The mass is the fundamental property that wholly determines the structure and evolution of a star of given chemical composition. The only method of direct determination of masses is the study of the movement of binary stars. However, the mass of each component of a binary system can be determined only in the case where we have either an eclipsing spectroscopic binary or a visual spectroscopic binary.

These measurements, although few in number – about fifty altogether – are interesting because from them we have been able to establish a few results that furnished the key to the evolution of stars. Certain stars have luminosities and masses that obey a mass–luminosity relation. These stars are those which lie on the main sequence of the Hertzsprung–Russell diagram. This mass–luminosity relation thus gives us a means of estimating the mass of a star that does not belong to a binary system from its luminosity, on condition that it belongs to the main sequence, which can be discovered from spectroscopic criteria.

We have also found for the same group of stars a mass–radius relation. This relationship results from the physical processes governing the behaviour of gas in a star. The range of stellar masses is relatively small – from about a tenth to sixty solar masses – while the range of luminosities is much wider: from a hundredth to more than a hundred million times the solar luminosity.

If we combine the values of the masses and the radii so determined, we can estimate a mean density. Thus, we find for the Sun a mean density of the same order as that of water. Of course, the density actually varies in a star's interior: it is high in the central regions – nearly a hundred times this mean value – and much lower in the atmosphere.

Some stars have densities considerably greater than that of the Sun. Sirius, the brightest star in the night sky, is a binary star. Its companion, whose mass has been estimated as 0.98 times the mass of the Sun by a study of the movements of the system, has a luminosity 400 times lower than that of the Sun. Taking account of the surface temperature of the companion of Sirius, the value of its radius has been estimated as 10 000 kilometres. Its mean density is therefore of the order of a hundred kilograms per cubic centimetre. This extremely condensed star is a white dwarf.

Is there a limiting mass for stars? This complex question is still the subject of study. Very massive stars have been observed, perhaps 120 times the mass of the Sun such as the star HD 93250, of spectral type O3, and extremely luminous stars two million times as luminous as the Sun. The stars with the smallest observed masses are Ross 614 B, with a mass of 0.08 solar mass and Luyten 726–8B at 0.04 solar mass, but most stars have a mass somewhere between 0.3 and 3 solar masses.

luminosity class	mass ($m_\odot$)			Radius ($R_\odot$)			effective temperature		
	V	III	I	V	III	I	V	III	I
spectral type:									
B0	17.5	20	25	7.4	15	30	30 850	29 050	26 000
B5	5.9	7	20	3.9	8	50	15 400	15 050	13 600
A0	2.9	4	16	2.4	5	60	9500	10 100	9700
G0	1.1	1.0	10	1.1	6	120	6000	5850	5550
K0	0.79	1.1	13	0.85	15	200	5250	4750	4400
M0	0.51	1.2	13	0.60	40	500	3850	3800	3650

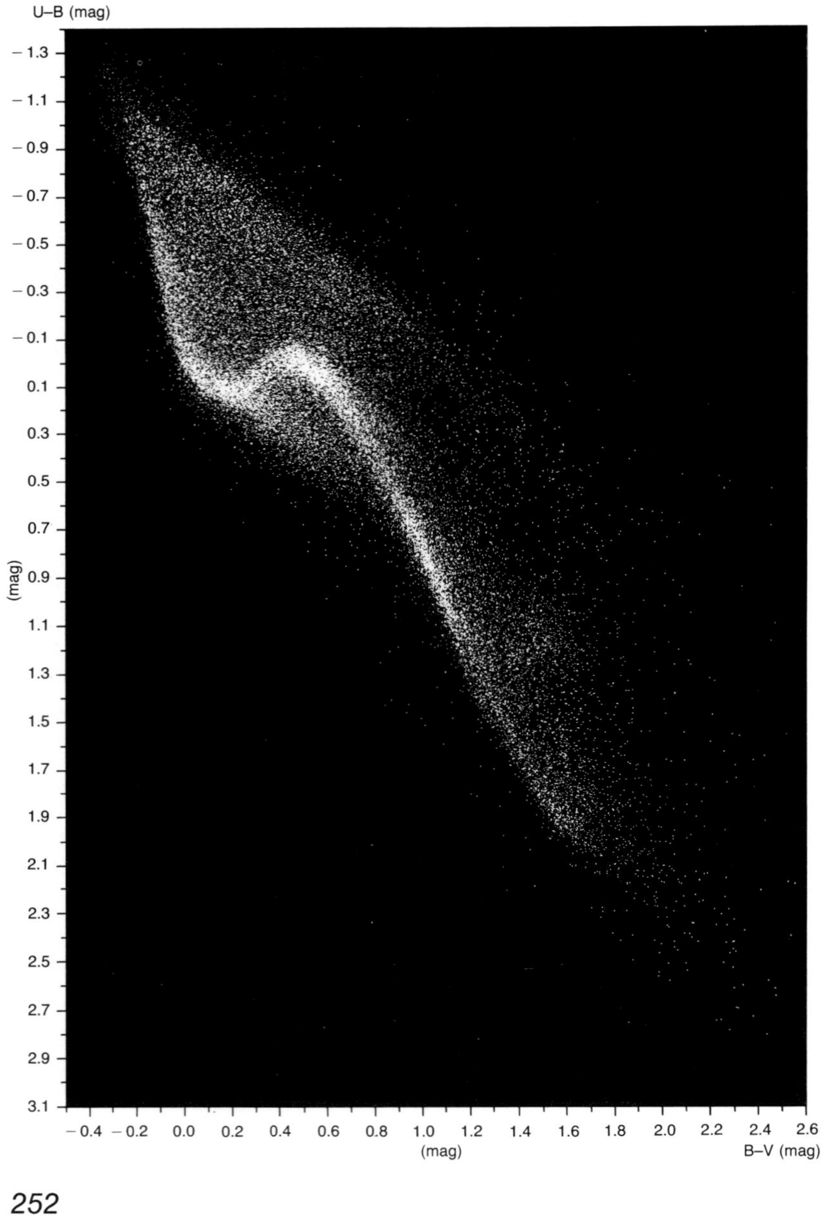

(U–B)–(B–V) diagram (opposite). The UBV photometric system, explained on page 234, gives two colour indices, U–B and B–V, which are correlated with the spectral type and luminosity class. These correlations are not one-to-one: each of the indices is sensitive to both parameters at once. This double sensitivity comes essentially from the fact that each of the filters defining the system measures the intensity of the stellar radiation over a large spectral range. The diagram of U–B as a function of B–V is a diagram of the same type as the Hertzsprung–Russell diagram, but it is much less easy to interpret in terms of stellar evolution, because the luminosity classes are separable only with difficulty. On the other hand, this diagram is a powerful tool for understanding the energy distribution of stars. This diagram allows us to detect stars with a particular energy distribution.

Photometric measurements are affected by interstellar extinction. The interstellar space is not empty of all matter; even if the density is very low, matter exists between the stars in the form of gas and dust. This matter causes an absorption of the stellar radiation, an absorption which affects different regions of the spectrum unequally. Interstellar extinction is greater on the ultraviolet side of the spectrum than in the red. The star appears to us as if it were emitting a relatively more intense luminous flux in the red than if there were no interstellar matter. We call this phenomenon 'interstellar reddening'. Such an absorption can be attributed only to grains of dust mixed with the interstellar gas. It is necessary to determine the intensity of this absorption in order to interpret correctly the energy distribution observed for a star. The (U–B)–(B–V) diagram is an essential tool for this determination.

Suppose that we observe, by means of UBV photometry, a star of spectral class O. Interstellar absorption affects the colour U more than the colours B and V. Thus, the B–V index as measured will have, for this star, a value which will correspond, for example, to that of an A-type star, while the line spectrum of the star does not resemble at all that of an A star. Interstellar absorption reduces the intensity of the continuum radiation, but does not modify the line spectrum at all: it does not make lines appear or disappear. This is why spectral classification based on the presence or absence of lines is not affected by interstellar reddening, while photometric classifications suffer from interstellar reddening. An O-type star of a certain luminosity class will have well-defined colour indices B–V and U–B. The difference between the measured index and the index the star should have allows us to measure the intensity of the interstellar reddening in the direction of the star. The reddening is not the same everywhere in the Galaxy and such measurements have allowed us to find out about the distribution of dust. If we limit ourselves to the study of stars close to the galactic plane, we find that a cloud of interstellar matter 1000 parsecs across causes an absorption of about 2.2 magnitudes. We can see that clouds of interstellar matter are obstacles to the study of distant objects, but their distribution is fortunately far from uniform.

Interstellar reddening explains the distribution of stars in the (U–B)–(B–V) diagram. The diagram shown here presents the results of observations on 46 084 stars. A line corresponding to main sequence stars of class V is well defined by the thickly clustered points. These are unreddened stars. Points representing stars which are reddened are shifted relative to this line. This shift is down and to the right, in a direction which makes an angle of about 30 degrees to the horizontal axis. The size of the shift depends on the amount of the interstellar reddening. The main difficulty lies in estimating this shift from the point it would have occupied had there been no interstellar reddening. The degree of interstellar reddening is a function of the distance to the star: the further the star is from the Sun, the more dust its radiation is likely to meet on its path and the greater the absorption will be. However, a star close to the Sun, but surrounded by a cloud rich in dust, will also be observed with a strong reddening. Mapping the interstellar reddening therefore requires a knowledge of the distance to the stars. In general, distant objects are very difficult to observe, because of the cumulative absorption caused by all the dust on the line of sight. This absorption has made the understanding of the structure of our Galaxy difficult. (Diagram by B. Nicolet, Geneva Observatory)

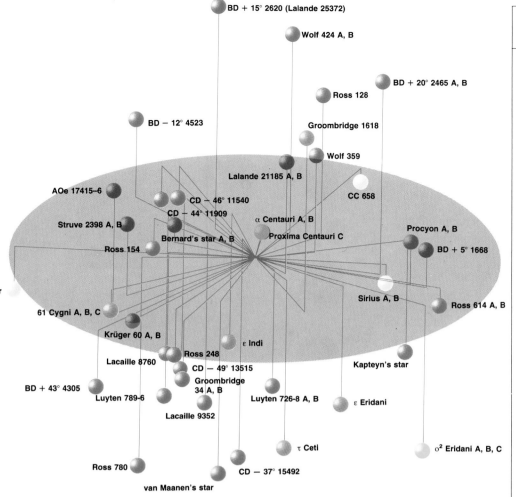

The forty nearest stars to the Solar System (figure above and table opposite). All these stars have had their parallax measured. For those that are binary or multiple systems, the components are indicated by the letter A, B or C. The colours correspond to spectral types.

distance (parsecs)	star or system	components in the case of a system	apparent visual magnitude (m_v)	absolute visual magnitude (M_v)	spectral type
0.0	Sun		− 27	5	G2
1.3	α Centauri	A	0	4	G2
		B	1	6	K5
		C	11	15	M5
1.8	Barnard's star	A	10	13	M5
		B	?	?	?
2.3	Wolf 359		14	17	M8
2.5	Lalande 21185	A	8	10	M2
		B	?	?	?
2.7	Sirius (α Canis Majoris)	A	−1	1	A1
		B	9	12	white dwarf
2.7	L726-8	A	12	15	M5
		B	13	16	M6
2.9	Ross 154		11	13	M4
3.2	Ross 248		12	15	M6
3.3	L789-6		12	15	M7
3.3	ε Eridani		4	6	K2
3.3	Ross 128		11	14	M5
3.4	61 Cygni	A	5	8	K5
		B	6	8	K7
		C	?	?	?
3.4	ε Indi		5	7	K5
3.5	Procyon (α Canis Minoris)	A	0	3	F5
		B	11	13	white dwarf
3.5	Struve 2398	A	9	11	M4
		B	10	12	M5
3.5	Groombridge 34	A	8	10	M1
		B	11	13	M6
		C	?	?	K?
3.6	Lacaille 9352		7	10	M2
3.6	τ Ceti		4	6	G8
3.7	BD + 5° 1668	A	10	12	M5
		B	?	?	?
3.8	Lacaille 8760		7	9	M0
3.9	Kapteyn's star		9	11	M0
4.0	Krüger 60	A	10	12	M3
		B	11	13	M4
		C	?	?	?
4.0	Ross 614	A	11	13	M7
		B	14	16	M7
4.1	BD − 12° 4523		10	12	M4
4.2	van Maanen's star		12	14	F0
4.3	Wolf 424	A	13	14	M5e
		B	13	14	M5e
4.5	Groombridge 1618		7	9	K6
4.6	CD − 37° 15492		8	10	M3
4.7	CD − 46° 11540		9	11	M3
4.75	BD + 20° 2465	A	9	11	M3e
		B	9	11	M3e
4.8	CD − 44° 11909		10	12	M5
4.8	CD − 49° 13515		9	10	M3
4.85	AOe 17415−6		9	11	M4
4.85	Ross 780		10	12	A5
4.85	BD + 15° 2620		9	10	M1
4.9	CC 658		11	13	M1
5	Altair (α Aquilae)		1	3	A5
5	o² Eridani	A	5	6	G5
		B	9	11	B9
		C	11	12	M4e
5	BD + 43° 4305		10	12	M5e

The seventeen brightest stars to the naked eye.

star	apparent visual magnitude	spectral type	luminosity class	distance (parsec)
Sun	− 26.7	G2	main sequence	0.0
Sirius (α Canis Majoris)	− 1.4	A1	main sequence	2.7
Canopus (α Carinae)	− 0.7	F0	supergiant	60
Rigil Kentaurus (α Centauri)	− 0.1	G2	main sequence	1.33
Arcturus (α Bootis)	− 0.1	K0	red giant	11
Vega (α Lyrae)	0.0	A0	main sequence	8.1
Capella (α Aurigae)	0.1	G0	red giant	14
Rigel (β Orionis)	0.1	B8	supergiant	250
Procyon (α Canis Minoris)	0.4	F5	main sequence	3.5
Achernar (α Eridani)	0.5	B5	main sequence	39
Hadar (β Centauri)	0.6	B1	supergiant	120
Altair (α Aquilae)	0.8	A5	main sequence	5
Betelgeuse (α Orionis)	0.8	M2	supergiant	200
Aldebaran (α Tauri)	0.8	K5	red giant	21
Acrux (α Crucis)	0.9	B1	main sequence	80
Spica (α Virginis)	1.0	B2	main sequence	80
Antares (α Scorpii)	1.0	M1	supergiant	130

The stellar atmosphere is in a gaseous state: the atoms are separated and free to move with respect to one another, which is not the case in a liquid or a solid. In the atmospheres of the coolest stars, we even find a few molecules. The atoms and molecules in the atmosphere are in fact responsible for the final interactions with the radiation. Let us take the example of the Sun: the average density of the solar photosphere is 5×10^{-8} grams per cubic centimetre; outside the photosphere are the chromosphere and the corona, with densities ranging from 10^{-9} to 10^{-14} grams per cubic centimetre. In such a rarefied medium, interactions between photons and matter are rare and the radiation from these zones, observed during solar eclipses, is due to other mechanisms: this radiation follows the excitation of the gas by the dissipation of waves generated in the deepest regions of the photosphere, which cross the photosphere without losing the energy they carry. To understand this mechanism, we can make an analogy with the swell of the sea and the energy that is dissipated when a wave hits a rock.

Thus, the analysis of radiation allows us to determine the physical conditions in the atmosphere: temperature, pressure and chemical composition.

energy levels forms what we call the line spectrum of that element. The wavelengths of these lines are perfectly defined: they are characteristic of the atom because they depend on its structure. This is why its line spectrum is a signature unique to each atom. An ionised atom has a line spectrum totally different from the neutral atom. As the wavelengths of photons re-emitted or absorbed by atoms are determined in the same way, from differences between the discrete energy levels, we can understand why absorption lines and emission lines are produced at the same wavelengths.

Spectral lines can be identified by looking them up in reference tables to find out by which atom each line could have been formed. From that, we then deduce the presence of that element in the body and we can thus find out the chemical composition of the atmosphere.

This process of matter–radiation interaction is not the only one that can occur, but it is the one mostly responsible for the stellar radiation.

Michèle GERBALDI

The matter–radiation interaction

The energy produced by thermonuclear reactions in the centre of a star propagates towards the surface by different processes, but when it arrives at the outer layers it is essentially in the form of photons.

When matter interacts with radiation it can reflect a part of it, transmit a part without altering it, and absorb the rest. The result of these interactions is radiation such as we observe.

Stellar radiation resembles, to a first approximation, that of a black body. But a more precise analysis shows us that a star's radiation cannot be described by the Planck function. We see, in particular, absorption lines characteristic of a deviation from equilibrium. For a given atom, the set of possible transitions between its discrete

Positions, distances and apparent motions of the stars

The place of the Earth in space and time is one of humanity's permanent questions. To find the positions and relative motions of various observable objects at a given epoch, as well as their distances, their shapes and sizes, has been and remains a major problem, discussed on every continent and in every century for as long as human thought can be traced.

The first attempts to give a coherent description of the Universe containing an explanation of the various observed motions dates from about the fifth century BC. The first phenomenon clearly and daily visible, the motion of the sphere of fixed stars, to which the stars were assumed to be 'attached', already caused controversy. Two explanations are, in fact, possible. Either the motion is real, and the stars revolve with the sphere of fixed stars about the unmoving Earth, or the motion is merely apparent, produced by the Earth's rotation on its axis. Both theories were put forward from this epoch onwards by Greek astronomers and philosophers, but the observations possible with the instruments of the time could not decide the question in favour of either of them.

The other heavenly bodies whose motions are easy to study are the Sun and planets. Mercury, Venus, Mars, Jupiter and Saturn, as well as the Moon were regularly observed and their motions were known. In the fourth century BC Eudoxus of Cnide and then Aristotle developed geometrical models of concentric spheres whose motions combined to explain the positions and their variation in time. In these models the Earth was at the centre of the system and the other bodies revolved around it in the following order: the Moon, Mercury, Venus, the Sun, Mars, Jupiter, Saturn, and, finally, the stars. Hipparcus of Nicea in the second century BC, and then Ptolemy in the second century AD, 'improved' this system so as to account better for the observed motions, without fundamentally changing it. This cosmological system, with variants, was to prevail for almost two millenia, although in the third century BC the physicist and astronomer Aristarcus of Samos had proposed that the 'fixed' stars and the Sun were motionless, and that the Earth revolved around the Sun. A very strong argument against this theory was that if the Earth revolves around the Sun one should see a slight parallactic displacement of the stars during the year. An enormous underestimate of the distances of the stars, allied of course to very strong ideological resistance to the idea that the Earth was not at the centre of the Universe, thus refuted what was in fact a correct description of the cosmos.

Reference systems and their realisation. The idea of a reference system is quite theoretical and abstract (three axes 'fixed' in space) if it is not made marked by the positions of one or more easily observable objects, and the variations of these positions in time.

The catalogue of Hipparcus, containing the ecliptic coordinates of more than 800 stars, (with an accuracy of about ten minutes of arc), as well as their 'magnitudes' (apparent luminosities), was the first realisation of a dynamical reference system, defined by the apparent orbit of the Sun on the celestial sphere, i.e. the ecliptic plane. The origin of the angles in this plane is the intersection of the planes of the celestial equator and the ecliptic, i.e. the line of equinoxes. The compilation of this catalogue allowed Hipparcus to discover and interpret the precession of the equinoxes, which causes a regular change

of the position of the origin of angles in the ecliptic plane.

The dynamical reference system is still used, realised partly by the motions of the major and minor planets, and partly by the accurate positions and proper motions of 1535 stars of the 'fundamental' catalogue FK5 (*Fünfter Fundamentalkatalog*, Astronomisches Rechen-Institut, Heodelberg, 1981). The density of this catalogue (about one star per 25 square degrees) is clearly inadequate, and extensive observational programs have been instituted, using transit instruments and astrolabes. These programs have provided a much denser realisation (one star per square degree) and are being continued to improve the accuracy of the data. The space astrometry program carried out by the European Space Agency using Hipparcos (High Precision Parallax Collecting Satellite), launched on 8 August 1989 has greatly improved this realisation. The positions of 108 thousand stars (2.9 stars per square degree) are known with a precision of 0.002 arcseconds at the mean observing epoch (about 1992).

The next step is to define the reference system by use of the apparently unchanging positions (certainly at the level of the best current observations) of compact extragalactic sources simultaneously emitting in the radio and the visible, such as quasars.

The positions of these objects are observed in the radio by VLBI (Very Long Baseline Interferometry) with an accuracy of order one milliarcsecond or better.

In the optical these objects are unfortunately too faint, with one exception, to be observed by Hipparcos, and the connection between the Hipparcos observations and the extragalactic system must be made indirectly. One observes with Hipparcos some stars which also emit in the radio (about 200 are included in the program), or one observes stars close to quasars (in projection on the sky) simultaneously with Hipparcos and the Hubble Space Telescope or ground-based astrometric instruments, which can also observe objects as faint as quasars. The set of stars observed by Hipparcos will thus realise the extragalatic system, as their positions are defined with respect to it.

The accuracy of proper motions obtained by Hipparcos is, however, inadequate to retain the precision of this reference frame for long. A second Hipparcos-type satellite launched some decades after it, would solve the problem completely.

The determination of distances in the Universe. Despite Aristarcus's extraordinary premonition, man had to wait for Copernicus and Kepler in the sixteenth and seventeenth centuries for a convincing demonstration of the central position of the Sun and the positions, distances and motions of the planets. Similarly, it was the absence of reliable criteria for distance determinations which made understanding the nature and form of our Galaxy so difficult, as well as comprehending the position of the Sun in this collection of stars, even though the latter had long been identified.

Lacking distance criteria, Willaim Herschel at the end of the eighteenth century underestimated the size of our Galaxy by about a factor of 10, and placed the Sun at its centre. For the same reason, at the beginning of the twentieth century Jacobus Cornelis Kapteyn and Harlow Shapley proposed structures of a size

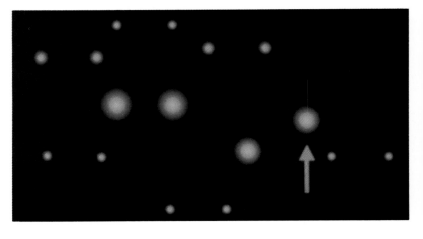

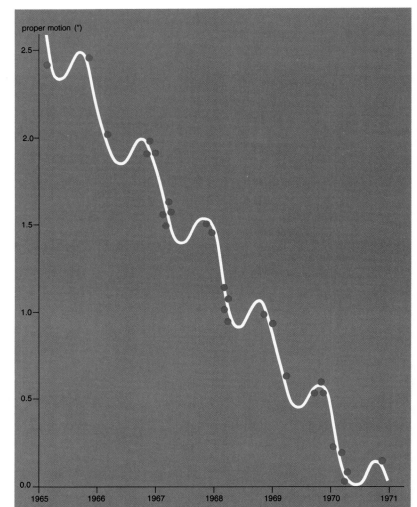

Proper motion. The Sun, like the nearby stars, moves with the global rotation of our Galaxy. In addition to this global motion, each star has its own motion. The relative motion of a star with respect to the Sun is revealed partly by an angular displacement on the celestial sphere, its *proper motion*, and partly by a *radial velocity*, away from or towards us, which can be measured by the Doppler effect. Seen from the Earth, the angular motion combines with the parallactic motion to produce an oscillatory motion.

Superposing two photographs (with a horizontal shift) taken at two epochs in the field of the star AD Leonis (marked by an arrow) shows that the latter has a larger proper motion than the other field stars. The schematic figure shows the angular displacement of AD Leonis over six years. The red dots denote the observations; the white curve running through them shows the proper motion of the star and the oscillations caused by the Earth's motion around the Sun.

The largest proper motion known is that of Barnard's star, at 10.2 arcseconds per year. Only seven stars have a proper motion larger than 5 arcseconds per year; it exceeds 1 arcseconds per year for several hundred.

With ground-based measurements it is necessary to measure the position at an interval of at least ten years to determine the proper motion of most stars. With the Hipparcos satellite an accuracy of 0.002 arcseconds per year can be obtained in three years. To fix the velocity vector of a star we have to know its radial velocity, its proper motion and its distance, as the tangential velocity is obtained from the proper motion and the distance. (above, Royal Greenwich Obsrevatory; below, ESA)

Evolution of the accuracy of stellar position measurements. Small field astrometry has the main aim of localising a star with respect to the other stars present within the field of the instrument. Global astrometry compares the positions of stars very far from each other on the celestial sphere.

Since the time of Hipparcus of Nicea the accuracy of angular measurements has improved by a factor of more than 10 000. The instruments of classical astrometry have been gradually automated, the positions becoming more and more precise and the observed stars fainter and fainter. As an example, the transit instrument of the Bordeaux Observatory gives positions to 0.10 to 0.15 arcseconds down to a limiting magnitude 13; this instrument, the Anglo-Spanish-Danish one on the island of La Palma in the Canaries, reaches stars of 15th magnitude with an average precision of about 0.15 arcseconds. The teams running these two instruments are studying the possibility of improving their sensitivity, speed and precision, with the aim of a limiting magnitude of 16–17, two million observation per year (currently 30 000 and 100 000 observation per year respectively) and an accuracy of 0.04 arc seconds.

Determining stellar distances. Because of perspective, certain stars appear to us to be grouped in constellations. In reality, the stars of any constellation are at very different distances. The distances of various stars in the constellation of Leo are shown in this figure. They vary from 12 to more than 700 parsecs. The distances of α, β, δ and μ are obtained by trigonometric parallaxes, those of the other stars by spectroscopic parallax.

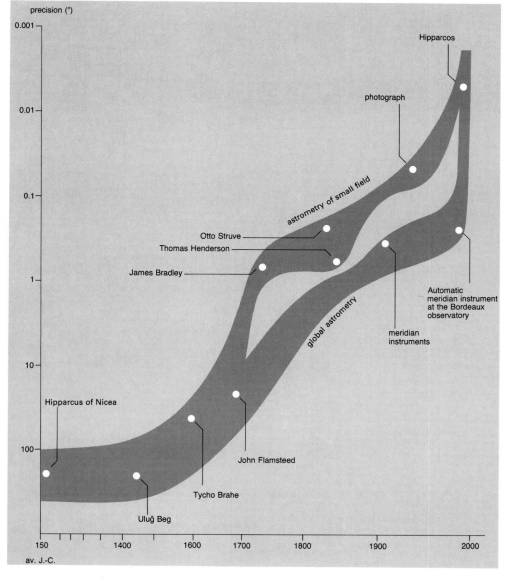

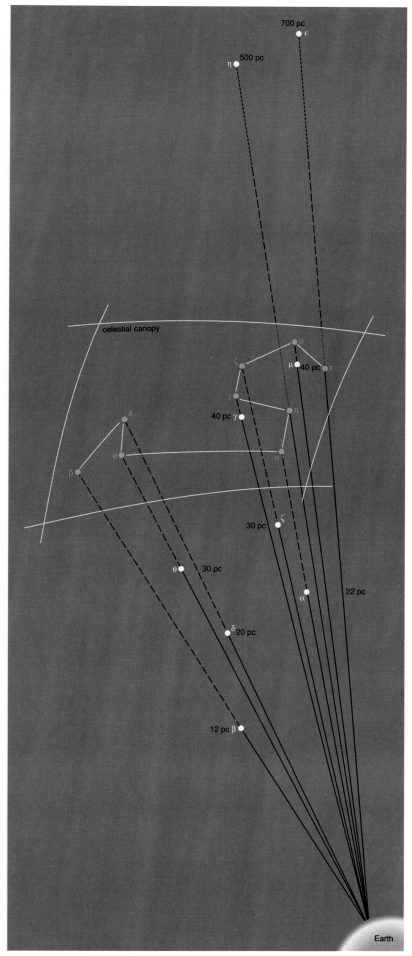

respectively very close to Herschel's and about 100 times larger! Similarly, although the distinction between nebulae and galaxies dates from the nineteenth century, the distances and sizes of the latter were not correctly estimated until about 1930.

The direct measurement of the distance of the stars by trigonometric parallax is only possible within an extremely small volume around the Sun. From ground-based measurements, one can even today scarcely hope to determine precise distances beyond about 20 or 30 parsecs, particularly because of systematic errors which are very difficult to estimate. Other, indirect, methods have been developed using observational data which are easier to obtain than trigonometric parallaxes: apparent magnitudes in different wavelength ranges and spectral types.

There are established relations between the intrinsic luminosity of stars and their colour or spectral type, allowing photometric or spectroscopic parallaxes. Another method is to study the variation of apparent luminosity of certain stars, for example the Cepheids and RR Lyrae stars. The brightness of these stars varies in a regular characteristic way, allowing them to be easily recognised. By studying the variability of the Cepheids in the Magellanic Clouds (one may reasonably assume that all these stars are at approximately the same distance from the Sun), it is possible to establish a relation between their average intrinsic luminosity and their period of variation. Making the hypothesis that the Cepheids of our Galaxy and those observable in neighbouring galaxies are the same as those in the Magellanic Clouds, one can obtain their intrinsic luminosities from their periods, and thus their distances if one measures their apparent magnitudes. This assumes, however, that the period–luminosity relation can first be calibrated by using stars whose distance can be determined by another method. A third method uses the stars of open and globular clusters.

The measurements of trigonometric parallaxes available at the end of the 1980s did not

allow one to calibrate securely distance determinations by photometric and spectroscopic parallax. With the Hipparcos satellite one gets direct distance determinations out to distances of order 100 parsecs, which greatly extends the range of spectral types for which such direct determinations exist. This, in turn, widens the observational base for calibrating other distance determination methods. Instrumental projects exist for attaining precisions of 10^{-5} to 10^{-6} seconds of arc. One would then obtain excellent calibrations for the period–luminosity relations within and outside our Galaxy, as well as colour–luminosity relations for various types of star clusters. The distance scale, first of all in our Galaxy, and then beyond, depends on these. It is the key to understanding the place of the Sun in the Galaxy and the Galaxy in the Universe.

Catherine TURON

The Hertzsprung–Russell diagram

In 1911, the Danish astronomer Ejnar Hertzsprung compared the colour and the luminosity of stars belonging to several open clusters. He traced the curve representing the variation of their apparent visual magnitude as a function of their colour. He observed that the points were not scattered at random in the diagram. In 1913, the American astronomer Henry Russell came to the same conclusion for a different sample of stars. The work of Hertzsprung and Russell showed empirically the existence of a relation between the luminosity and the effective temperature of stars. The diagram that, for a group of stars, represents the variation of one of these parameters as a function of the other is called the Hertzsprung–Russell diagram (HR diagram).

The location of each star in the HR diagram depends on its stage of evolution and, for that reason, the HR diagram is a basic tool for the study of the structure and history of our Galaxy. The fundamental quantities defining the HR diagram can be measured by different parameters, thus giving different forms of the HR diagram. The classical HR diagram uses the two quantities spectral type, which is a qualitative determination of the effective temperature, and absolute magnitude. When a colour index (e.g. B−V) is used instead of the spectral type (surface temperature) the resulting graph is called a colour–magnitude diagram.

The spectral type

Our only source of information about the nature of the atmospheres of stars is the analysis of their spectra. Two approaches are possible: one quantitative, the other qualitative.

Quantitative analysis is based on a study of the spectrum, which ends in determination of physical parameters describing the stellar atmosphere. Such a study is generally very long and it is applicable only to a limited number of brighter stars.

Qualitative analysis rests on the fact that a simple visual inspection of a large number of spectra shows that they can be grouped into a number of families: this is spectral classification, which considers only the appearance of the spectrum in the visible region, and which is founded on the morphology of the absorption line spectrum without considering, a priori, the physical causes of this appearance. The first classification of spectra, for example, that by Angelo Secchi in 1863, included only four classes, but already Secchi himself noticed that his classification was not a chance result, but had a physical cause, each class grouping together stars with similar effective temperatures. The basis of the present classification of spectra was defined in 1901 at Harvard College Observatory, where a collection of several tens of thousands of stellar spectra was used by Antonia Maury and Annie Cannon, who ordered the stars into seven principal classes designated by the letters O, B, A, F, G, K and M.

Within each class, it was soon necessary to make a decimal subdivision to take account of differences in appearance between spectra of the

Examples of the MK spectral classification. Spectral classification is done by visual comparison of the spectrum of a star with a set of spectra of standard stars. The Sun is a G2 V star, G2 designating the spectral type and V the luminosity class.

The spectra of twelve stars belonging to different spectral types and luminosity classes are shown, and on them some of the hydrogen lines of the Balmer series, Hβ. Hγ. Hδ, have been identified.

It may seem surprising that the letters used for the spectral types are not in a logical order – for example alphabetical order. In fact, the first classifications, defined at the turn of the century, adopted an alphabetical order, but it soon became apparent that some classes were spurious because of the poor quality of photographic plates. When the physical mechanisms behind the formation of spectra were better understood, it became obvious that the spectral classes had to be ordered according to the temperature. Some letters designating the original classes were kept, and we finished with the order O B A F G K M. It appears that it was Henry Norris Russell himself who invented the mnemonic for remembering the order of these classes: 'Oh Be A Fine Girl, Kiss Me'. (Photograph, observatoire de Haute-Provence du CNRS)

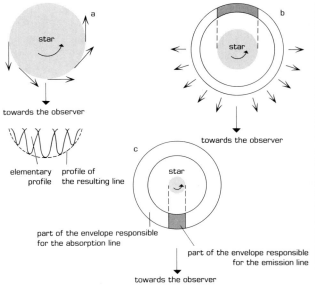

Spectra of Be stars. Two spectra of Be stars are shown, between spectra of an iron arc used for wavelength calibration. These spectra are in the region near the Hβ line of the Balmer series of hydrogen.

The upper spectrum shows a broad diffuse absorption line, on which is superposed a much narrower emission line. The absorption line is due to the star itself and its width results from various effects, such as the temperature.

Another cause of broadening is the rotation of the star on its axis (a). The part of the star that is approaching the observer contributes to the formation of an absorption line which the observer sees, because of the Doppler effect, at a shorter wavelength. On the other hand, the parts of the star that are going away from the observer contribute to the formation of an absorption line seen at a longer wavelength. The whole disk therefore contributes a series of lines, differently shifted which, seen together, make up a broad line.

Stars of the spectral types, O and B, have high rotational velocities, of the order of 200 to 250 kilometres per second. Cool stars show much lower rotational velocities: G stars have rotational velocities of the order of 20 kilometres per second.

The envelope around the star, responsible for the emission line, would be formed from gas ejected from the star, probably by a centrifugal effect caused by the rapid rotation (b). Be stars in fact have rotational velocities of the order of 300 kilometres per second, which,

taking into account their masses and radii, is close to the limit of dynamic stability of these objects. The emission line is shifted to the shorter wavelengths by the Doppler effect due to the expansion of the gas: in fact, the observer never really sees the part of the envelope going away from him because it is hidden by the star. The width of the emission line is essentially due to the velocity of expansion of the envelope.

The lower spectrum shows a situation where the gaseous envelope is thicker than in the first case. The absorption line from the star disappears. A very narrow absorption line is present due to absorption of light from the star by the part of the gaseous envelope that is in front of the star as seen by the observer (c). This line is narrow, since the variation in the radial component of the velocity, whether from an expansion or from a rotation of the whole envelope, is small in this limited region of the envelope. The absence of absorption due to the star and the presence of the absorption line due to the envelope constitute two proofs of the thickness of the envelope, which is thick enough to absorb light coming from the star. Physical conditions in the envelope can vary in the course of time. Thus at certain times the envelope may become transparent and the radiation from the star can then be observed. (Photograph, observatoire de Haute-Provence du CNRS)

Principal spectroscopic characteristics of the spectral classes.

spectral class	example	effective temperature (K)	
O	10 Lacertae	25 000 and above	few absorption lines are visible in the spectrum: some lines of helium singly ionised, nitrogen doubly ionised, silicon triply ionised;
B	Rigel (β Orionis) Spica (α Virginis)	11 000–25 000	intense lines of neutral helium (maximum intensity at class B2); lines of silicon ionised singly or doubly; lines of oxygen and magnesium ionised singly; hydrogen lines begin to appear but are very weak; the lines of ionised helium have disappeared;
A	Sirius (α Canis Majoris) Vega (α Lyrae)	7500–11 000	very intense hydrogen lines: they dominate the spectrum (maximum intensity at class A0); lines of elements ionised singly: magnesium, silicon, iron, titanium, calcium, etc.; very weak lines of neutral metals;
F	Canopus (α Carinae) Procyon (α Canis Minoris)	6000–7500	intensity of hydrogen lines decreasing; lines of calcium, iron and chromium ionised singly are still present, as well as lines of neutral metals, whose intensity is increasing;
G	Sun Capella (α Aurigae)	5000–6000	lines of ionised calcium are the most remarkable spectral characteristic because of their intensity; numerous lines of neutral and ionised metals; appearance of the molecular bands of CH and hydrocarbons;
K	Arcturus (α Bootis) Aldebaran (α Tauri)	3500–5000	lines of neutral metals dominate the spectrum; the molecular band of CH is still intense;
M	Betelgeuse (α Orionis) Antares (α Scorpii)	3500 and less	intense lines of neutral metals and molecular bands of titanium oxide very developed.

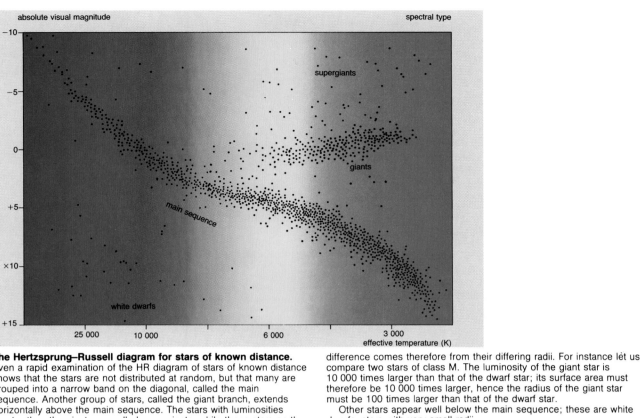

The Hertzsprung–Russell diagram for stars of known distance.
Even a rapid examination of the HR diagram of stars of known distance shows that the stars are not distributed at random, but that many are grouped into a narrow band on the diagonal, called the main sequence. Another group of stars, called the giant branch, extends horizontally above the main sequence. The stars with luminosities greater than the giants are called supergiants, while those stars on the main sequence are called dwarfs.

These qualifications, 'dwarf' and 'giant' make sense only for a given spectral type. If we consider two stars of the same spectral type, one belonging to the main sequence and the other to the giant branch, the two stars show a large difference in luminosity. The luminosity of a star is proportional to the surface area ($4\pi R^2$) and to the fourth power of its effective temperature (T_e^4). If these two stars have the same spectral type, they have the same effective temperature. Their luminosity difference comes therefore from their differing radii. For instance let us compare two stars of class M. The luminosity of the giant star is 10 000 times larger than that of the dwarf star; its surface area must therefore be 10 000 times larger, hence the radius of the giant star must be 100 times larger than that of the dwarf star.

Other stars appear well below the main sequence; these are white dwarfs, stars with very small radii.

The points that form the main sequence show some dispersion about the average. This dispersion does not all come from uncertainties in the absolute magnitudes. In the HR diagram for stars in a cluster, the absolute magnitude on the vertical axis can be replaced by the apparent magnitude because all the stars are essentially at the same distance. The points in the HR diagram for stars of the same cluster also show scatter, which can be interpreted in terms of changes taking place with stellar evolution.

same class. Therefore the spectral type of a star is represented by one of the symbols: B0, B1, B2, . . ., B8, B9, A0, A1, etc., where a spectrum of type B9 has characteristics closer to those of type A0 than to those of type B0.

From 1911 to 1924, Annie Cannon classified, according to these spectral classes, the 225 000 stars of the Henry Draper catalogue. But it was only in 1925 that a physical interpretation of this classification was possible thanks to the discovery, made in 1920 by Meghnad Saha, of the laws of the ionisation of atoms.

The characteristics of the stellar spectrum used to establish the spectral classification are the presence or absence of the lines of certain elements. This presence or absence is not due to differences in chemical composition between the atmospheres of stars, but reflects only the differences in temperature of these atmospheres. Thus hydrogen, which is the most abundant element in the Universe and which has nearly the same abundance in all stars, predominates in the line spectra of stars with an effective temperature near to 10 000 K, because the conditions favour the excitation of hydrogen atom at this temperature. In the atmospheres of the hottest stars (of spectral type O) hydrogen is almost all ionised and therefore does not produce a significant spectrum of absorption lines. Because of the mechanisms of ionisation, there exist some neutral atoms in the medium, despite everything, which will produce an absorption line spectrum, but it will be very weak.

In the atmospheres of cool stars (e.g. of spectral type K) hydrogen atoms are in the neutral (not ionised) state and practically all in the unexcited or ground state, and the spectrum of lines produced belongs mostly to the ultraviolet range, not observable from the Earth; the lines of hydrogen observable in the visible are very faint.

Thus O-type stars, which are the hottest, show in their spectra lines of ionised helium, but no hydrogen lines. Going from type B0 to type A0, the intensity of helium lines decreases as the temperature conditions are not favourable for their formation, while the intensity of hydrogen lines increases progressively so as to reach a maximum around type A0. The intensity of hydrogen lines then decreases while those of metals increase for the spectral types corresponding to lower effective temperatures. For the coolest stars, lines of neutral metals become more and more intense while bands characteristic of molecules appear.

The luminosity class

It became apparent that the early classifications were inadequate as spectrum lines can show different characteristics in the same spectral class; thus Rigel (β Orionis) and Regulus (α Leonis) are both of spectral type B8, but Rigel has narrow lines and Regulus has broad lines. To take account of this difference, it was necessary to introduce a second parameter into the spectral classification.

Since 1913, thanks to the work of Hertzsprung and Russell, we have been able to understand that this difference is explained by the difference in luminosities between stars of the same effective temperature, and therefore reflects a difference in radius. This second parameter has allowed us to define a luminosity class which can be interpreted in terms of physical conditions in the atmosphere of the star. Although temperature is the predominant factor in the determination of the characteristics of a spectrum, other causes have a non-negligible effect, for example density. Thus the degree of ionisation is a function of temperature but it also depends on the density of the gas: if the density of the gas is high, the particles are close to one another, and recombinations between ions and electrons are facilitated, so at a given instant the number of ionised atoms is smaller than in a medium at the same temperature but with a lower density.

However, the density of a gas is proportional to its pressure, and this results from the weight of the atmosphere, that is from the strength of gravity in the stellar atmosphere. The gravity is proportional to the mass of a star, but inversely proportional to the square of the radius of the star. Stellar radii vary over a much larger range than stellar masses, and it is this that makes all the difference. Thus, in the atmosphere of a star with

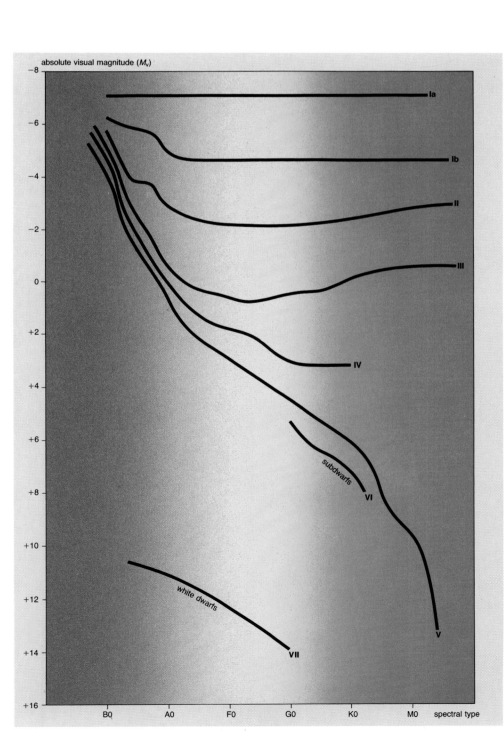

The luminosity classes. This HR diagram shows different luminosity classes. Five principal classes have been defined. Class I is supergiant stars; it is subdivided into class Ia, (the most luminous stars) and class Ib (stars of the same spectral type as stars of class Ia, but a little less luminous). Class II is giant stars, slightly less luminous than class I but much more luminous than class III. Class III is giant stars. Class IV is subgiant stars, intermediate between stars of classes III and V. Class V is dwarf stars, that is, stars on the main sequence. Some stars in the HR diagram are below the main sequence; these are subdwarfs, to which we give a luminosity class VI.

The spectra of white dwarfs do not come into this classification and are identified by their own characteristics. The luminosity class of these objects is sometimes given as VII.

Stars on the main sequence are still called dwarfs by comparison with other stars, this term 'dwarf' being applied even to the massive stars at the top of the main sequence – which have large luminosities and radii several times larger than that of the Sun. This is because the luminosity classes distinguish only stars that have the same spectral type, and therefore the same effective temperature. To each stellar spectrum, we can therefore attribute a spectral type and a luminosity class.

a large radius, elements are more easily ionised and the spectrum resembles that of a star with a higher effective temperature but a smaller radius. This resemblance, which could cause confusion, is fortunately not complete: it varies as a function of the element considered. Thus visual analysis of the lines of several elements in a stellar spectrum allows a spectroscopist to tell whether it is a giant star with an extended atmosphere or a more compact star – a dwarf star – showing a higher effective temperature.

The MK classification

A spectral classification based on the spectral type and the luminosity class is called two-dimensional since it uses two independent physical parameters. The spectral type is a qualitative estimate of the temperature of the atmosphere of the star, usually defined by the effective temperature. The luminosity class reflects the density of the atmosphere, that is the size of the star or its surface gravity. The spectral classification used most frequently today is the MK classification, which was defined by W. W. Morgan and P. C. Keenan in 1943. This classification attributes to each star a spectral type and a luminosity class. The spectral classes are similar to those of the Harvard classification and the same names have been kept to designate them. This classification is entirely empirical in that only spectroscopic characteristics are used. So that this classification remains homogeneous, it is necessary that it is made with spectra having the same dispersion, from 6 to 13 nanometres per millimetre. The spectral type and the luminosity class are attributed to a star by reference to a set of stars, called standard stars, which characterise the system. Thus the classification remains unchanged when the interpretation of spectral classes in terms of physical parameters is modified, as the models of stellar atmospheres improve.

Peculiar stars

There are catalogues that list all classifications made with the MK system, but these classifications are not necessarily homogeneous. A large collection of these data and analyses is kept at the

Centre for Stellar Data, at the Strasbourg Observatory. At the University of Michigan, N. Houck and A. P. Cowley have undertaken to re-classify all the stars of the Henry Draper catalogue in the MK system with two parameters. Ninety per cent of all stars can be classified in the MK system. We generally call stars peculiar if their spectra cannot be compared with one of the standard stars defining the MK system. The other stars are called normal stars.

At the beginning of the century, the Harvard classification was extended by introducing the three spectral classes R, N, and S after class M.

The classes R and N contain stars with effective temperatures close to those of stars of classes G, K and M, but whose spectra show very intense molecular bands due to carbon. Today, these stars are collected into a single class, class C, or the class of carbon stars. We now know more than 3000 of these stars, which are giants in which the molecular bands of CN, C_2 and CH are very intense, showing that the abundance of carbon with respect to that of oxygen is four or five times larger than in normal stars of the same effective temperature. Attempts are being made to explain the origin of these stars in terms of stellar evolution. It is possible that these were once massive stars of a few solar masses in which helium burning took place to produce carbon in a layer around the centre. Turbulent movements due to thermal instabilities could then develop and carry elements produced by nuclear reactions to the surface, where they become observable, in particular carbon. This mechanism would explain qualitatively the appearance of the spectra but it cannot be compared with the quantitative measurements.

Stars of class S are giant stars, whose effective temperatures are close to those of type M, but whose spectra show strong bands due to the oxides of zirconium, yttrium and barium. It is very probable that the origin of these elements is the same as that of the elements of carbon stars, that is, nucleosynthesis. These heavy elements would be produced in massive stars – a few solar masses – by neutron addition near the region where helium is burned in a layer around the core.

A great variety of stars ranging from type B5 to F5 and showing a whole series of peculiarities are classified as Ap and Am stars. These stars are on the main sequence and their anomalies are explained in terms of their atmospheric structure

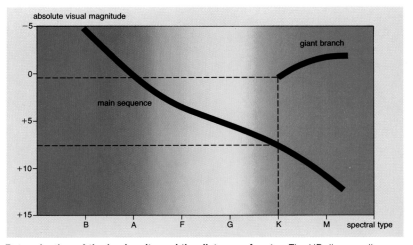

Determination of the luminosity and the distance of a star. The HR diagram allows us to find the luminosity of a star. Suppose that we have a reference HR diagram (spectral type–absolute visual magnitude) constructed from all the stars whose distance we have been able to measure. If we observe any star and determine its spectral type (K0 in this case) and its luminosity class (the luminosity class of a star allows us to decide whether the star belongs to the main sequence or the giant branch), the reference diagram will allow us to find its absolute visual magnitude, therefore its luminosity, and also its distance.

and not in terms of evolution. Ap stars are those whose spectra show the appearance of those of normal stars from types B5 to A5, but they have, in addition, intense metallic lines, belonging usually to the elements manganese, mercury, silicon, chromium, strontium and europium, and corresponding to an overabundance of these elements compared with normal stars. An important characteristic of stars of this group is the presence of an intense magnetic field in the photosphere. The flux from some of these stars is variable, and this variation is interpreted as resulting from an inhomogeneous distribution of the elements at the surface of the star. As it rotates on its axis, the star shows the observer regions of different structure and therefore of different luminosity. The redistribution of certain elements into vast regions called spots, very similar to sunspots, at the surface of the star is certainly favoured by the presence of a magnetic field. The overabundance of certain elements, generally observed in the spots, is attributed to processes of selective diffusion in the atmosphere of the star, causing some atoms to rise to the surface of the star and dragging others into the deep regions of the atmosphere, thus making them invisible to the observer. Am stars are those whose spectra look like those of normal dwarfs of types A0 to F5, but with very weak lines of calcium and scandium, while the lines of the elements of the iron peak and the rare earths are reinforced. These intensities correspond respectively to an underabundance and to an overabundance of these elements. These stars have rotation speeds much lower than those of normal stars of the same temperature. We think that this characteristic allows the diffusion to occur in the atmosphere of the star, creating the observed abundance anomalies.

The spectra of certain O and B stars also show emission lines. The suffix 'e' is then added to the spectral type of the star. More than 15 per cent of the stars of these hot types possess such a spectrum. These emission lines are attributed to a tenuous envelope of gas surrounding the star. These stars show large variations in the intensities of the lines, with, for example, complete disappearance of the emission lines; the variations are attributed to changes in the structure of the envelope.

Stars which are placed below the main sequence in the HR diagram are called subdwarfs. These stars are underluminous compared with those of the main sequence for a given spectral type. However, their chemical composition is totally different from that of main sequence stars: they are underabundant in metals; these are population II stars.

The term 'symbiotic stars' is applied to certain stars whose spectra show both absorption and emission lines. The absorption line spectrum resembles that of a cool giant star, for example of type M. The emission line spectrum requires physical conditions leading to a strong excitation of the atoms producing these lines. Some of these lines have also been observed in the solar corona, which could point the way for research into mechanisms that produce these emission lines.

Masses of some stars in the Hertzsprung–Russell diagram. Along the main sequence, the masses of stars are related to position as shown. This relationship arises from the mechanisms governing the internal structure of the stars. Elsewhere, the masses are mixed, since stars of different initial masses evolve towards the giant branch. Moreover, important phenomena of mass loss occur during stellar evolution, which markedly change the initial masses. The masses are expressed in solar masses.

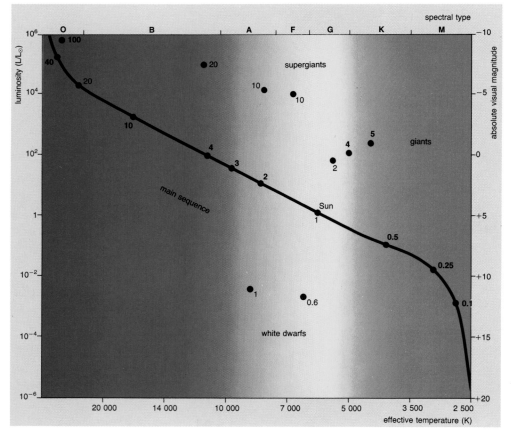

We know about a hundred stars belonging to this group, whose membership is defined by spectroscopic criteria. These criteria are much looser than those that define the MK spectral classification and the spectra of stars in this group exhibit considerable variety. The understanding of the mechanisms leading to the formation of these spectra is important, because it gives a clue to certain stages of stellar evolution.

Many spectra of this type have been shown to belong to binary stars, but it is not yet proven that all such spectra are due to symbiotic pairs. In the case of binary stars, the spectrum is interpreted as the result of the combination of the radiation of a normal giant star of type K or M and a close companion, which is a very hot star whose radiation excites a gaseous envelope which must have formed, by the mechanism of mass loss, during the evolution of the system. Thus certain nova stars appear among the list of symbiotic stars. Symbiotic stars could therefore be a stage in the evolution of certain binary stars and the question of the relationship between symbiotic stars and planetary nebulae is raised.

The peculiar stars which have just been defined offer theorists concerned with stellar atmospheres a greater variety of observations than do normal single stars, thus leading to a much better understanding of the structure of stellar atmospheres and the physical phenomena that can occur there.

Quantitative spectral classification

From analyses of the physical meaning of spectral type and luminosity class, it is possible to define a quantitative spectral classification.

A quantitative spectral classification measures some spectroscopic characteristics – for example the intensity of the continuum radiation, or the ratio between the intensities of certain lines – which are then translated into quantities representing the effective temperature and the absolute magnitude of the star, or its luminosity.

The principal difficulty lies in the choice of criteria to use and in their calibration in terms of effective temperature and absolute magnitude. There are no criteria that are valid for all stars, whichever spectral class they belong to; this is why several quantitative classifications are necessary, each referring to a well-defined set of stars. Photometric measurements are most usually used for this calibration. Quantitative spectral classifications are indispensable for the plotting of observational HR diagrams which will be interpretable in terms of physical parameters deduced from stellar models. A theoretical stellar model provides, among other things, the distribution of energy radiated, that is, a synthetic spectrum, but the spectral type and luminosity class of the theoretical star cannot be derived from the synthetic spectrum.

Michèle GERBALDI

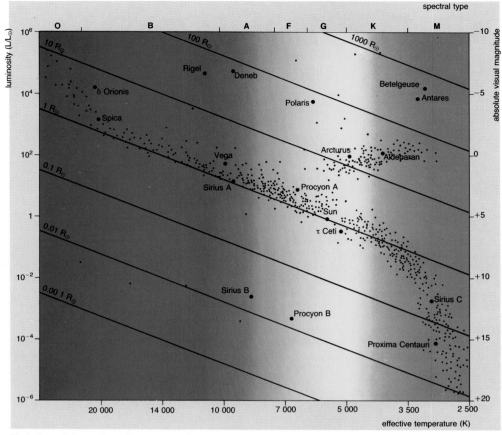

Variation of the radius of a star as a function of its spectral type and its luminosity. In the HR diagram, the straight lines each correspond to stars of the same radius (expressed in solar radii R$_\odot$). The positions of some well-known stars are indicated.

We note that a star whose radius is 10 R$_\odot$ is a dwarf star on the main sequence if its spectral type is O, but is a giant star if its spectral type is G.

The stars of Populations I and II. This HR diagram (right) includes points for stars belonging to both Population I and Population II. These two populations are distinguished by their distribution in the Galaxy, their age and chemical composition. Stars of Population I have a chemical composition similar to that of the Sun. The stars belonging to open clusters are typical examples. The stars of Population II represented in this diagram are typical of globular cluster members in the halo of the Galaxy, and are poor in metals.

The positions of the sequences in the HR diagram differ between the two populations. Models of the internal structure of stars have shown that the initial chemical composition fixes the positions of the main sequence and of the giant branch.

The Population II stars considered here are all as old as the galaxy, that is, about ten billion years. They were formed from the primordial interstellar medium, poor in heavy elements. The Population I stars are much younger.

The main sequence of Population II is much shorter than that of Population I, because the high mass Population II stars have already left the main sequence and have become either giants or degenerate stars (below).

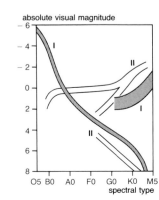

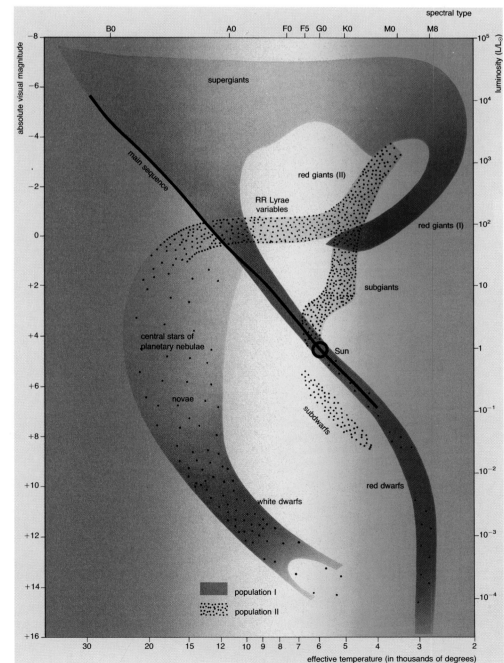

Star formation

It is known that the stars of our Galaxy show quite a range of ages: globular cluster stars, with an age of the order of ten billion years are among the oldest, while at the other extreme we observe giant stars exciting nebulae of ionised hydrogen, called H II regions. At ten to twenty million years old these are examples of the youngest stars. It was via infrared observations, however, that stars in the process of being formed were detected.

All evidence points to clouds of atoms and molecules (mainly hydrogen) as the cradle of new generations of stars in our galaxy. The average density of the interstellar medium in the Galaxy is near 1 atom per cubic centimetre. The formation of a star thus requires a mechanism able to raise the local density by a factor of nearly 10^{24}! Only the force of gravity – which can act at large distances and hence plays an essential role here – can lead to such a high compression factor. On the other hand, the thermal motion of molecules and the turbulent motion of the interstellar gas produce a pressure which prevents a runaway contraction being imposed by the gravitational force.

It is when this equilibrium is broken in favour of gravity that a star or a group of stars is able to form. In very general terms, this happens when the mass of a cloud of interstellar gas surpasses a critical mass fixed by its internal energy (as indicated by its temperature). Higher temperature clouds have a higher critical mass. A cloud will collapse if, for example, its mass increases following collisions with smaller clouds but its average temperature only increases slightly, or if the mass of a cloud stays constant but its temperature drops. A third example could involve the two previous scenarios simultaneously. In any case, computations show that for masses greater than about 2000 solar masses gravity always wins out over the pressure forces as long as the cloud is homogeneous. The cloud becomes

gravitationally unstable and contracts faster and faster. Because we know that the mass of a typical star is about a thousand times smaller, we must conclude that a process of fragmentation of the original cloud takes place. Without stopping the collapse, this process isolates the fragments of the cloud.

Radiation is the most efficient cooling agent of the interstellar medium. When two atoms or two molecules collide, a fraction of their kinetic energy is temporarily transformed into excitation energy. This populates the excited electron energy levels. The atoms de-excite themselves by spontaneously emitting a photon or a cascade of photons. Collisions thus cool off the gas which becomes less and less able to resist the gravitational collapse. The inhomogeneities in the cooling process could contribute to the fragmentation of the cloud as previously mentioned.

The very cold giant molecular complexes – with temperatures between 10 and 90 K – are the recognised sites of stellar formation. Their masses are very large, reaching up to a million solar masses. Their structure is very inhomogeneous, however and we often observe shells of gas of increasing densities boxed one inside another. In the Rho Ophiucus cloud and the Orion molecular complex a few dozen stars of intermediate masses have been detected. These stars are invisible to the optical astronomer because of the dust in the cloud. They can, however, be seen in the infrared.

It seems that these grains of dust play an important role in the formation of stars. They protect the sites of stellar formation from the ultraviolet radiation emitted by nearby stars, which could otherwise heat up the cloud and prevent its collapse. Furthermore, they serve as nuclei to facilitate the condensation of the molecules of the gas.

The mechanism for the formation of massive stars seems to be somewhat different. The giant O and B stars, which excite the H II regions, form

Star formation in Barnard 5. A recently discovered newborn star (dark patch pointed out by arrow) is seen embedded in a cloud of gas and dust called Barnard 5 in this image produced from IRAS data. The protostar is believed to be in an evolutionary stage similar to that of the Sun when it formed 4.6 billion years ago. (NASA)

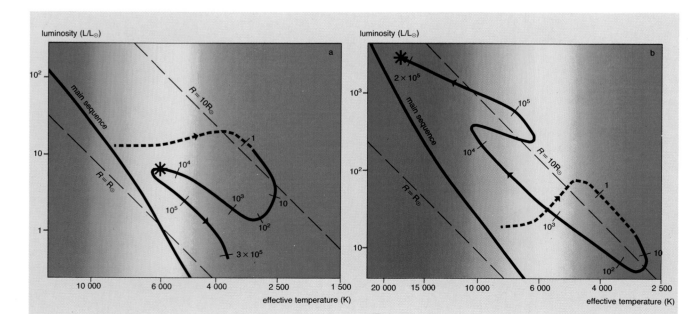

Evolutionary tracks on the Hertzsprung–Russell diagram of the stellar core of protostars, 0.25 M$_\odot$ (left) and 10 M$_\odot$ (right). On the theoretical HR diagram the abscissa represents the logarithm of the effective temperature of the stellar surface, T_e, expressed in kelvin; the ordinate represents the logarithm of the luminosity of the star, L, expressed in terms of the solar luminosity (L$_\odot$ = 3.86 × 10^{26} joules per second). The times in years elapsed since the formation of the stellar core are indicated on the tracks.

The initial temperature, radius and density of the gaseous mass just before the collapse are, respectively: 10 K, 4.1 × 10^{16} cm; 1.8 × 10^{-18} g/cm^3 (a); and 20 K, 8.2 × 10^{17} cm; 8.8 × 10^{-21} g/cm^3 (b). After a first contraction phase, not shown here, a stellar core soon appears and the figures show its evolution. The dashed lines indicate how the radius of the stellar core evolves. The dashed curves represent the phase during which all the emitted radiation from the core is absorbed by the collapsing envelope.

The solid line starts when 90 per cent of the radiation emitted by the core escapes in the form of infrared radiation. When the protostar reaches the symbol *, half of the total mass has been accreted by the core. (From R. B. Larson, Evolution of spherical protostars, in *Monthly Notices of the Royal Astronomical Society*, vol. 157, p. 121, 1972)

NGC 2264. Situated at 800 parsecs from the Sun, this cluster was formed about two million years ago. Only a few massive O and B stars have had time to reach the main sequence. The majority of the stars seen on the picture are rapidly variable stars of the T Tauri type. The red streaks are due to Hα emission from hydrogen; the dark spur is part of the cloud of gas and dust from which the stars were formed.

T Tauri stars, young irregular variable stars, are found in the neighbourhood of young (one million to ten million years old) OB associations made up of hot stars, near Herbig–Haro objects, and in opaque clouds. The properties of the radiation received from these stars (continuous emission, Balmer lines, infrared excess) reveal the presence of a dense and hot envelope (10^9 to 10^{12} atoms per cubic centimetre). Most of these relatively low mass (less than 2 solar masses) stars have a strong stellar wind and significant mass loss. On the H–R diagram they reach the main sequence along a Hayashi track. (D. F. Malin, Anglo-Australian Telescope Board, © 1981)

preferentially on the perimeter of the molecular complexes. It is probable that the formation of these stars sustains the chain reaction according to the following scheme: a first generation of massive stars, formed near the surface of the cloud, emits a strong ultraviolet flux. This radiation produces an H II region which, with the help of radiation pressure, enters into the cloud by compressing the matter with a shock wave. This compression triggers the formation of a new group of massive stars which, in turn, continue the process. This is apparently seen in the great Orion Nebula, a large H II region on the surface of a molecular complex.

It is very difficult to tell at present whether or not molecular complexes are in gravitational contraction. Recent developments in millimetre astronomy, more specifically the interferometry of millimetric waves, will bring new insights to the dynamical state of the molecular clouds and to the details of the processes of fragmentation.

There is a class of a small molecular clouds

called Bok globules, some of which have been found to be in gravitational contraction from observations of the 26 millimetre line of carbon monoxide. Their collapse velocity is about half a kilometre per second and their radius is of the order of half a parsec. If nothing slows down their collapse, these globules will condense into stars in a million years.

These isolated objects (seen as black patches on photographs of nebulae or against the background of the Milky Way) illustrate the theoretical models of star formation. The central region, highly compressed and much denser than the periphery, attracts the surrounding matter. The core slowly becomes opaque to the infrared radiation emitted by the grains. The temperature rises progressively and the pressure becomes sufficiently high to momentarily stop the collapse of the core. Little by little, however, all of the matter in the envelope eventually falls onto the protostar. When its temperature exceeds about ten million degrees, thermonuclear reactions

begin. Put on the Hertzsprung–Russell diagram, these stars approach the main sequence following a path called the Hayashi track.

The observation of stars being formed or of very young stars along with their environment provides important contributions to the theory of stellar formation. In the scheme sketched previously the formation of stars is directly related to the evolution of molecular clouds. Even though it is the most thoroughly studied case, it is not the only one. One way to learn more about stellar formation is to investigate it in nearby galaxies. It is in spiral galaxies like our own that we observe O and B stars associated with H II regions. These young stars are formed in spiral arms where dust and gas are found. This fact suggests the existence of a stimulating factor leading to the formation of stars on a galactic scale. The density wave theory invoked to explain spiral arms predicts that these waves should be preceded by shock waves capable of triggering the formation of stars along their path. Nevertheless, the idea of self-prop-

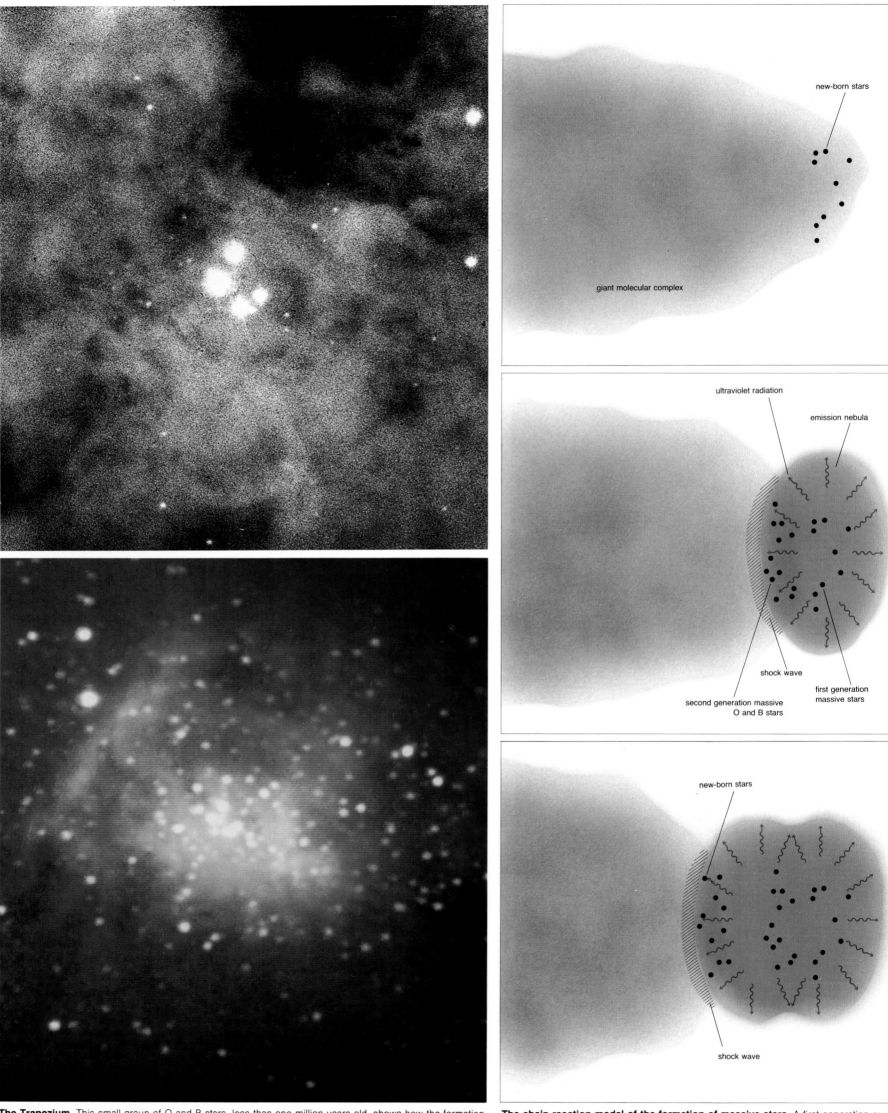

The Trapezium. This small group of O and B stars, less than one million years old, shows how the formation of massive stars can be propagated towards the interior of a dense cloud. (The top photograph was taken in blue light; the bottom photograph is an infrared composite photograph at 1.2 and 2.2 micrometres.) Behind the Trapezium group there is a large molecular cloud called Orion A. This 10^5-solar-mass cloud has been torn up by the ultraviolet radiation emitted by the Trapezium stars. They are, at the present time, exciting the optical emission of the Orion Nebula. Two maxima of radio emission due to the CO and HCHO molecules are detected in this molecular cloud. One of them. OMC 1, is located near and behind the Trapezium. This dense region, most probably in gravitational contraction, contains many OH and H_2O masers and several infrared sources. The strongest one, called the Becklin–Neugebauer object, is almost certainly a star in formation still surrounded by its cocoon. Theory and observation both suggest that the formation of massive stars can be propagated from star to star by the compression of the gas produced by the passage of an ionisation front that was triggered by the ultraviolet flux coming from the previous generation of stars. (D. F. Malin, Anglo-Australian Telescope Board, © 1981 and D. A. Allen, Anglo-Australian Telescope Board © 1984)

The chain-reaction model of the formation of massive stars. A first generation of massive stars forms in a cluster near the surface of a giant molecular cloud. After about one million years the ultraviolet radiation emitted by the group of stars heats and ionises the surrounding gas. The ionisation front enters into the molecular cloud and compresses a layer of gas inside which a new generation of massive stars then forms. The new ionisation front they produce will, in turn, trigger the formation of new stars deeper inside the cloud.

agating stellar formation on a galactic scale, carried along by explosions of successive generations of supernovae, can explain the spiral structure of galaxies similar to M101. We are then dealing with a common, large-scale phenomenon.

Irregular galaxies are usually poorer in heavy elements (carbon, nitrogen, oxygen, etc.) than our Galaxy. Therefore, one finds very little dust in them. The stellar formation in the Large Magellanic Cloud causes some problems: in a region called 30 Doradus one observes fifty or so O and B stars associated with a cloud of fifty million solar masses of neutral hydrogen. There is no dust in this region and no molecular cloud has been detected. This clearly shows that the stellar formation theory based on molecular clouds does not explain all stellar births.

Elliptical galaxies, do not show signs of contemporary star formation. The morphology of these galaxies can be understood only if we suppose that their very old stars were formed nearly simultaneously with the galaxy.

Jean-Pierre CHIÈZE

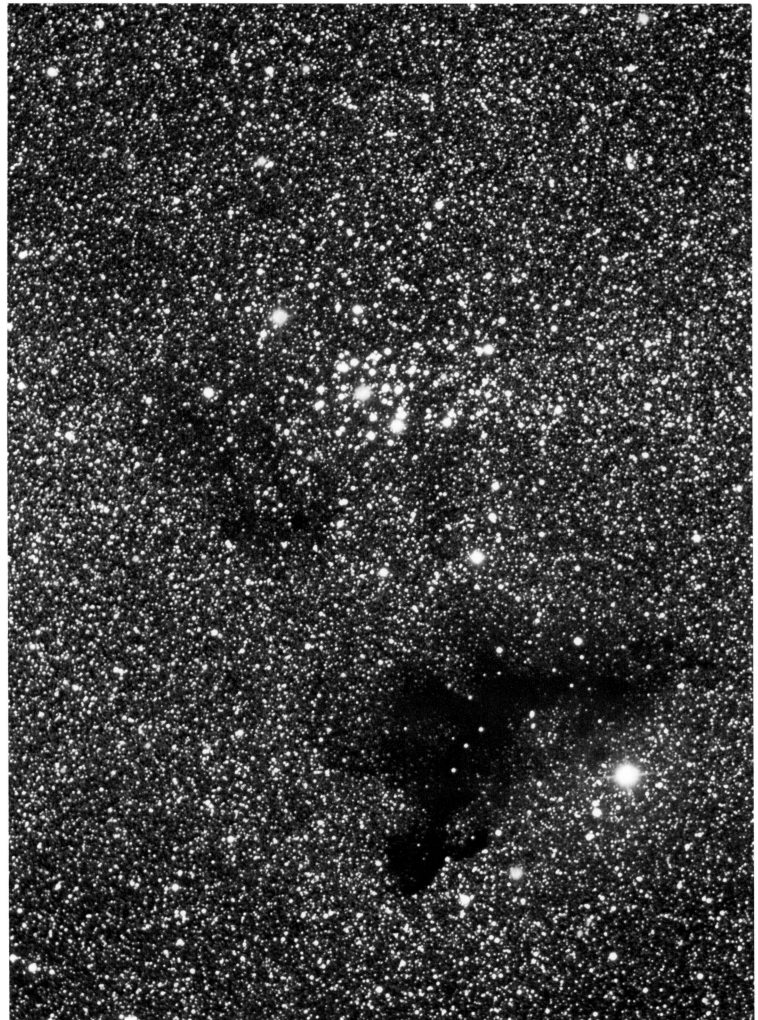

The Bok globules. These small dark clouds, called Bok globules, can be seen against the stellar background of the Milky Way or bright areas of gaseous nebulae (NGC 6520). Because of the large amount of dust they contain, they are opaque to visible light. Rich in all sorts of molecules, they are very similar to molecular clouds. Radio observations of Bok globules, which are more or less spherical in shape, reveal that some of them are in gravitational contraction. Their radius is less than 1 parsec and their mass is usually between 1 and 200 solar masses. Theories predict that the core of the globule collapses more rapidly than the envelope. After one million years its thermal energy could be high enough to trigger the first nuclear reactions which characterise the beginning of the life of a star. The radiation emerging from the star can then disperse the gaseous shells surrounding it. Eventually, fragmentation processes can take place in the contracting globule, complicating enormously this oversimplified scenario. (D. F. Malin, Anglo-Australian Telescope Board, © 1980)

Protoplanetary systems

Astrophysics is advancing rapidly in many fields, such as cosmology, the study of the Universe in its entirety, and the analysis of particular objects such as quasars, galaxies, stars and planets. However, one area is stagnating in comparison; this is cosmogony, the study of the formation of planetary systems, so named when it was still thought that the solar system constituted the entire cosmos. This stagnation comes from the fact that only the Solar System is truly accessible to observation, while other candidate planetary systems all pose real problems.

There are two motives for studying other planetary systems. One is comparative, the other statistical. In the first case, if we wish to understand how the Solar System formed 4.5 billion years ago, we have either to find clues in the organisation, composition, and structure of its constituents (from dust grains to the Sun itself, including meteorites, comets, asteroids, planets and their satellites) or to observe other planetary systems in different stages of evolution. In this way we could try to reconstruct the formation history of a planetary system in general, and our own in particular without having to work back in time. For example, we have arrived at an understanding of stellar evolution by comparing millions of stars, observed 'instantaneously' at various stages of their lives, rather than observing a single star for millions of years. Comparing these observations with theories of stellar evolution has led to a plausible picture of the life of a star. Confronting theories of the formation of planetary systems with observations has already given some insight into the various mechanisms involved.

From the statistical point of view, we would like to know if the formation of a planetary system is a rare event or not. Here, the most generally accepted star formation theories suggest that planets are often if not always formed around young stars condensing from an interstellar cloud. If this is correct, planetary systems should exist around many stars. If it is wrong, planetary systems will be rare. We should note in passing that the answer to this question would have immense repercussions for the chances of life appearing elsewhere in the Universe than the Solar System.

The possible detection of other planetary systems is difficult however. There are two methods: studying either systems similar to ours, or systems in a different evolutionary stage, between the primordial interstellar cloud and the final system.

Systems like ours might reveal themselves by the presence of planets, but these are small faint objects, several billions times fainter than the central star, and very near to it. In consequence they remain inaccessible to direct observation at present. However, their gravitational perturbations should slightly perturb the motion of the central star. This perturbation, although extremely small, has led to several possible detections. These measurements are, however, very difficult, as they require estimates of tiny separations, and only appear over relatively long timescales, related to the periods of the putative planets, i.e. several years. The apparent displacement of the central star can be at most a few hundredths of a second of arc, which must be observed over several years, equivalent to detecting a displacement of 1 millimetre at a distance of 1 kilometre over the same time! Nevertheless, with great patience, Peter van de Kamp has succeeded in doing this for several stars near to the Sun. These observations seem to have revealed the presence of giant planets (masses several times that of Jupiter) around several nearby stars. However, these results are

at the limit of credibility and require confirmation.

In its motion, a planet forces the central star to rock back and forth. This rocking can be detected by spectroscopy, using the Doppler effect to measure the star's radial velocity. For planets of the size of Jupiter velocity variations of several tens of metres per second are expected. These measurements, at the limit of feasibility, are very difficult as they too must be sustained over several years. Since the beginning of the 1980s Bruce T.E. Campbell has been carrying out such measurements at the Canada–France–Hawaii telescope, and appears to have discovered several planetary systems.

The search for planetary systems in a different evolutionary state from our own requires knowledge of a formation scenario. Comparing earlier observations with theoretical models, this scenario has to be adjusted continuously in the light of new theoretical and observational insights. The most widely accepted picture is as follows: the planetary system forms, at almost the same time as the central star, from the contraction of a large interstellar cloud. Many studies have been aimed at observing this last contraction phase, which should produce a vast disc of gas and dust. These studies are made in the infrared and the radio, as it is only at these wavelengths that light can emerge from these dense opaque regions. Then, when the star has formed and lit up, many phenomena seem to appear: formation of an accretion disc, violent gas jets, strong stellar activity, and so on. All these effects are observed and incorporated into the picture. This is the phase of observation of young stars. Then, while the star stabilises and reaches the main sequence (i.e. begins to burn hydrogen stably), the circumstellar disc evolves, and dust condenses there, forming comets and planets, and the gas is dissipated. Study of this phase thus amounts to observing traces of this gas and circumstellar dust. Finally the system evolves to a planetary system like our own, or deviates along other paths that we must predict or discover. For example we know that 70 per cent of stars belong to binary systems. We can thus look for companion stars of lower and lower mass, which in the limit could amount to the transition to the largest planets. These small cold stars have been called brown dwarfs. Discovery of them would be very important as it would demonstrate the lowest mass that a star can have, and show if a small star can in fact be a large planet. Obviously the formation picture of planetary systems would be strongly affected by such a discovery.

The detection of planetary systems in other evolutionary stages has made notable progress. To quote only a few spectacular cases, many discs have been observed around the very young stars called T Tauri stars. Many observers have shown that these discs seem to feed the central star, to the point that a region of intense friction appears at the limit where the disc is in direct contact with the star's atmosphere. The rotation of these discs has been directly observed by millimetre-wave radio interferometry (by Anneila I. Sargent and Steven Beckwith), and the discs themselves may contain a mass equivalent to that of the central star itself, extending to sizes several thousand times the Earth–Sun distance. Some discs expel bipolar jets along their rotation axes. The interaction of the jets with the ambient interstellar medium has been observed. Sometimes, some of these young stellar systems show violent eruptions making the system suddenly more than 100 times brighter. This has occurred in several systems whose prototype is the star FU Orionis, studied for

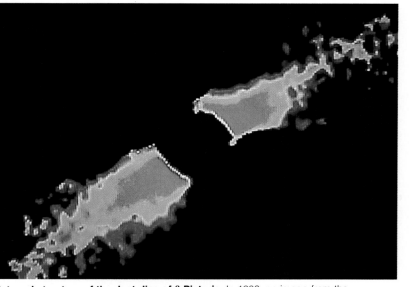

Internal structure of the dust disc of β Pictoris. In 1992, an image from the European Southern Observatory's 2.2 metre telescope showed, perhaps for the first time, the internal structure of the disc near β Pictoris supporting the idea that planets had already formed around the star. The yellow bar shows a length of 100 AU, corresponding to the smallest distance (given by the radius of the occulting disc) on the original image by Bradford A. Smith and Richard J. Terrile; we thus observe the disc structure directly at less than 30 AU from the star, where planets ought already to be formed. (A. Vidal-Majdar et al, C. Buil, CNES; F. Colas, Bureau des Longitudes; A. Lecavalier des Etangs and G. Perrin, Institut d'Astrophysique de Paris)

Some observations delimiting the 'classical' formation scenario for planetary systems. The observations of Bruce T.E. Campbell are indisputably the most convincing in the search for planetary systems resembling our own. He has surveyed eighteen stars spectroscopically and given evidence for radial velocity variations in nine of them; these variations are generally compatible with the presence of planets with masses between 1 and 10 Jupiter masses.

Also importantly, Campbell has not detected any companion in the range occupied by the brown dwarfs; these objects must then be much rarer than thought, at least in orbit around another star.

star	minimum and maximum masses (unit: Jupiter mass)	
ε Eridani	1.0	4.5
36 Ursae Majoris	1.6	12.7
β Virginis	0.9	9.6
β Comae Berenices	1.0	9.0
61 Virginis	0.8	7.2
β Aquilae A	1.0	12.7
η Cephei	1.1	19.4
61 Cygni A	0.7	4.0
γ Cephei	1.6	

many years by George H. Herbig. It seems that the events result from the sudden accretion of about a hundredth of a solar mass onto the central star. Whatever the cause, these observations show that the evolution of a disc is certainly very complex and that a non-negligible part of the matter continues to fall on to the star via the disc. After these fairly violent evolutionary phases the dust condenses rapidly near the equatorial plane and the residual gas starts to dissipate. This evolutionary phase of a planetary system was discovered by H. Aumann and his collaborators using infrared observations by the IRAS satellite. The dust present in the discs is cold and emits in the far infrared, while the central star is much hotter and radiates in the visible. Circumstellar dust has been found around several tens of nearby stars through an excess of infrared emission. Among them, the star β Pictoris provides a spectacular example. Its disc has been directly imaged by Bradford A. Smith and Richard J. Terrile. This is a circumstellar disc in the most advanced evolutionary phase currently known to us outside the Solar System. Within it, planets may be forming from dust grains; it is even possible that they have already formed. This system is thus probably in the last stage of evolution of a disc before it becomes a real planetary system.

Evolution to various planetary systems probably involves brown dwarfs. Study of these requires several techniques. In particular the 'discovery' in 1984 of a brown dwarf around the star Van Biesbroeck 8 made much ink flow. This 'non-discovery', which was not confirmed several years later by more accurate observations, stimulated many studies, with very positive results. Theoretical studies thus showed that these objects, with mass less than 0.08 times the mass of the Sun, are not true stars as nuclear reactions cannot start in their centres. However, the evolution of their luminosity as they cool is extremely difficult to predict. It is thus impossible to say whether we do not see them today because they are too faint or because they do not exist. Here the observations of Bruce E. Campbell are extremely important as they show that such objects are rare, since they were not detected in any of the eighteen nearby stars studied.

This appears to suggest that stars and planets are fundamentally different and that there is more than a simple difference of mass between them. Stars are formed from condensations of parts of interstellar clouds, in a mass range from 8 per cent to a hundred times the Sun's mass, while planets are much smaller objects formed in circumstellar discs by accumulation of dust grains. Their masses do not exceed a few tens times the mass of Jupiter, i.e. 0.1 per cent of the Sun's mass. Thus in the range of celestial objects there seems to be nothing between the largest planets and small stars, and these two classes seem to be fundamentally different.

The future of this field of astrophysics is very promising, as everything is still to be discovered, and several different approaches seem about to bear fruit.

Alastair G.W. Cameron believes that powerful computers are needed for this research area to evolve, as the modelling of such complex situations is limited by the capacity of current computers. The next generation of computers will allow new theoretical studies.

From the observational point of view, many existing methods will continue to produce results. New β Pictoris stars may be discovered, and real brown dwarfs may at last be found. In any case new telescopes will completely transform the observational situation.

Clearly, now the spherical aberration of its mirror is corrected, the Hubble Space Telescope will undoubtedly have a major effect, not in directly detecting planets around other stars (this remains very difficult, and at the limit of the telescope) but in a much finer analysis of the circumstellar discs recently discovered around young stars. This telescope also has astrometric capabilities which will perfectly complement the remarkable performance of the European Space Agency's Hipparcos satellite, launched in 1989. Many planetary companions of nearby stars should be discovered. The European Space Agency's ISO infrared satellite, the successor to IRAS, should allow the study of circumstellar discs in various evolutionary stages and probably discover many more. It will also allow more detailed study of phases in the formation of stars, particularly the very beginning of the process, which is very unclear.

Some large ground-based telescopes such as the VLT (Very Large Telescope) of the European Southern Observatory (ESO) will be operational. Their great sensitivity and angular resolving power (using optical interferometry) will allow them to probe the fine structure of many circumstellar accretion discs.

Finally, there are projects for ground or space-based telescopes designed to perform stellar coronography using superpolished mirrors. These telescopes will be able to reduce the scattered light from the central star by a factor of a 1000 compared with standard telescopes, thus allowing the direct detection of planets around nearby stars. The study of planetary systems will then truly begin.

It is easy to convince oneself of our immense ignorance in this area of astrophysics. This ignorance results from the extreme complexity of the possible physical processes, which are almost unconstrained by current observations. It is, however, clear that the study of the formation

Dust disc around β Pictoris. In the southern constellation of Pictor, this star was noticed in 1983 by the IRAS satellite as having an abnormal infrared excess. Since 1984 Bradford A. Smith and Richard J. Terrile have tried to image the circumstellar dust assumed to be responsible for this excess by stellar coronography. Masking the emission from the central star, they discovered an immense dust disc around the star, extending to more than a 1000 times the Earth–Sun distance. If this disc is a protoplanetary system, it is more than 20 times the size of the Solar System, which is usually assumed to be bounded by the orbit of the last planet, Pluto.

In fact the Solar System is certainly much larger, since we have long observed comets coming from much further out, from a region called the Oort cloud at 10 000 times the Earth–Sun distance. To explain the long life of this cloud requires that it be constantly fed. One source of comets coming from regions further in has been proposed by Alastair G.W. Cameron, who assumed that the comets come from a vast disc beyond Neptune, i.e. between a 100 and a 1000 times the Earth–Sun distance. It is interesting to note that this disc closely resembles that of β Pictoris.

Alfred Vidal-Majdar, Hervée Beust Roger Ferlet and Anne-Marie Lagrange-Henri, from the Institut d'Astrophysique in Paris, in collaboration with Lew Hobbs, from Yerkes Observatory in the US studied the gas in the disc from the European Southern Observatory, using the signature that it leaves in absorption of the central star (the disc is seen edge-on). These observations show the presence of a large amount of gas in the disc (with density 100 000 atoms per cubic centimetre), suggesting that the system has probably not evolved yet.

A surprise came from the fact that absorption lines appear sporadically in the spectrum, showing that large quantities of gas are falling on the central star. This result is confirmed by observation with the IUE ultraviolet satellite, made in collaboration with Magali Deleuil and Céecile Gry, from the Laboratoire d'Astronomie Spatiale in Marseilles. These show that the gas falls on to the star with velocities reaching 400 kilometres per second. This is exactly the velocity that would be reached by a body falling freely onto the star's surface. All of this suggests that these events can be interpreted as due to comets which become volatile as they fall on to the star. We can estimate their diameters at several kilometres, which seems to be compatible with what we know of Solar System comets.

If this intepretation is correct, it would show that small bodies have already been able to form in the disc around β Pictoris, and that they suffer large perturbations to fall onto the central star at such a rate. What could cause these perturbations? Could this be the gravitational interaction of a massive body like Jupiter, which has already formed?

Many other questions remain unanswered, but we are indisputably seeing a protoplanetary system where planets have doubtless already formed. (B.A. Smith and R.J. Terrile)

and evolution of planetary systems is about to mature and will make spectacular advances because of the observational revolution which is occurring. It is a good bet that in a few years several planetary systems will have been indisputably and directly detected, and that between now and the end of the twentieth century the foundations of this fascinating subject will have been laid. With a good idea of the origin of the Universe and of the Solar System, we shall be ready for the great adventure of the twenty-first century: the search for life in the Universe, in other planetary systems, or conceivably in other more exotic locations.

Alfred VIDAL-MAJDAR

The evolution of stars

The concept of stellar evolution

Our ideas on the life history of stars have been modified at the same time as our ideas of their nature. The history of astrophysics can be divided into two great periods: up until the seventeenth century the view was that the Universe did not evolve; nothing was more immutable than the Heavens. After 1687, with the publication of the *Principia* by Isaac Newton, it was demonstrated that the same law of physics – the law of gravity – governed both the motions of the planets in the Heavens and the fall of objects onto the Earth. From that time Heaven and Earth could no longer be considered as two fundamentally different entities. The idea of evolution, at least that which results from the laws of mechanics, was accepted but it was necessary to await the second half of the nineteenth century and the development of thermodynamics before the true problems of stellar evolution were realised.

The idea of evolution does not come naturally to us on observing the Heavens. On the contrary it seems as if the stars have not changed during the whole history of humanity. Copernicus, in his *De Revolutionibus Orbium Caelestium* described the Universe thus 'First of all and everywhere is the sphere of the fixed stars which contains itself and everything and is for this reason immutable.'

This dogma of immutability was believed for more than 2000 years until 1572 when Tycho Brahe published his observation of the discovery of a new star (a supernova) in the constellation of Cassiopeia. Tycho Brahe showed that the star was located well beyond the orbit of the moon and therefore belonged to the sphere of the stars. Other observations like his contributed to the downfall of the myth of immutability. But astronomical observations were not the only influence on the evolution of ideas. There was a general trend of thought, re-questioning the theories of Plato and Aristotle and the dogma of the Scriptures.

In 1842 Julius Mayer formulated the principal of conservation of energy. Now the existence of both animal and vegetable fossils implies that the intensity of the solar radiation has remained practically constant for several million years. Consequently, in order to supply this radiation the Sun must have a tremendous energy reserve. However this reserve is not infinite, one day it will be exhausted.... Realising the importance of the Sun's rays for life on Earth man wanted to know for how much longer the Sun would shine. Answers to these questions had to wait until 1919. Until then, the various theories proposed could not account for the age of the Solar System, estimated, on essentially geological grounds, to be 4.6 billion years. In 1919 Jean Perrin suggested that thermonuclear fusion reactions were the energy source of the Sun and stars.

In 1920 Arthur Eddington issued a challenge to all physicists: to determine the energy source, central temperature and density of the Sun.

M16 (NGC 6611). M16 is a young star cluster, which was formed about two million years ago from a cloud of dust and gas. It bathes in a nebulosity whose pink colour is characteristic of hydrogen gas. In the dark regions near to the centre of the cluster, new stars will soon be born. The future evolution of each star will depend on its mass and chemical composition. (D. F. Malin, Anglo-Australian Telescope Board, © 1980)

Models of the internal structure of stars on the main sequence. The internal structure of a main sequence star depends on whether its mass is above or below 1.5 solar masses ($M_\odot$). The diagram on the left represents the current state of the Sun. In the central regions, where some of the hydrogen has already been converted into helium, the chemical composition is as follows: hydrogen 36 per cent, helium 62 per cent, other elements 2 per cent. Elsewhere the composition has remained fixed at the initial values: hydrogen 73 per cent, helium 25 per cent, other elements 2 per cent. The energy produced by nuclear reactions in the interior of the Sun is transported outwards by photons, except near the surface where convection predominates. The diagram on the right shows the structure of a massive star of 9 solar masses. The central temperature exceeds that of the Sun, but the central density is less. Energy transport is by convection in the core and by radiation in the envelope. (For greater clarity the diagrams are not drawn to scale.)

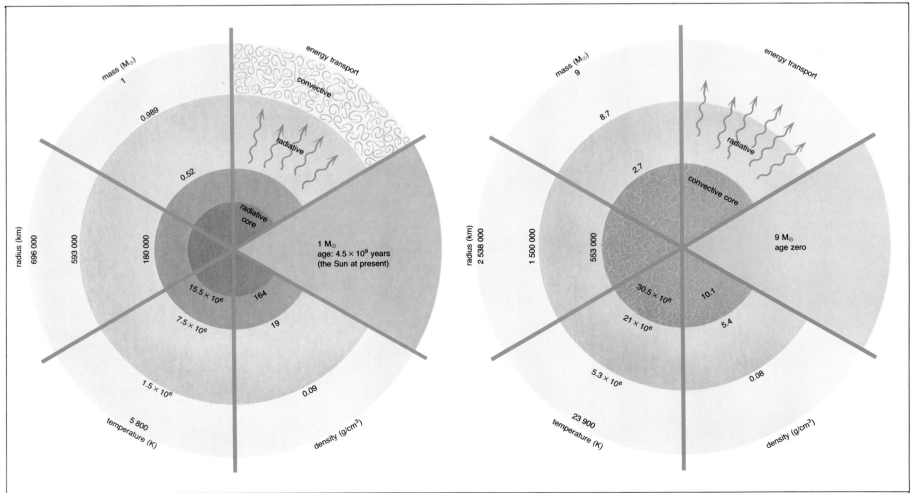

However, it was only in 1939 that Von Weizsäcker and Bethe described the fusion reaction that converts hydrogen into helium (the proton–proton chain and the carbon cycle) and is capable of producing the radiant energy of the stars.

The fact that the Sun (or indeed any star) radiates implies that its structure will change. Evolution takes place. However, this evolution is very slow and is imperceptible to us, even in a time comparable to the history of man. How, under these conditions can we predict the future evolution of the Sun? Stars are born, evolve and die. How can we determine the age of a star when the life of a man is infinitely brief in comparison? For Sir William Herschel, an astrophysicist was like a botanist who, in order to study the growth of trees, was only allowed to observe a forest for one hour. This period is too brief to yield any noticeable growth on a single tree. It is, in compensation, sufficient to observe, during the course of a walk, the trees of the forest at different stages in their life. The astronomer is in the same position as this botanist: he sees the stars at different stages in their evolution but he cannot observe this evolution. On the other hand he can predict theoretically how a star will evolve, using the laws of physics to construct a model of a star. It might seem presumptuous to wish to describe the internal structure of stars when only their outermost regions can be observed. However, the analysis of the radiation emitted by a star gives its luminosity, that is the total energy radiated in all wave bands, as well as the temperature of the outer regions of the star. Observations of the motions of binary stars in some cases enable the mass and radius of these stars to be calculated. One of the major successes of astrophysics has been to show that just a limited number of parameters that can be observed suffice to describe the internal structure of a star. The laws of physics then allow us to establish by calculation what is called a model of the internal structure. Such a model is fundamental in determining the rate at which energy is produced at the centre of the star. This energy is transported to the surface by multiple radiation–matter interactions. We can calculate the amount of energy radiated by this theoretical model of a star. Comparing this with the amount of radiation actually observed provides a test of the validity of the model. If the agreement is poor, a new model is constructed by changing the values of the parameters; this is repeated until the agreement between the model and the observations is satisfactory. It now remains to describe and understand the evolution of stars.

Observations enable us to construct the Hertzsprung–Russell diagram, which shows that there is a correlation between the spectral type (or effective temperature) of a star and its luminosity. The laws relating the values of these parameters to the state of evolution of the star can be derived, thanks to the models of internal structure. It is in fact simple to evolve a star theoretically. One determines how the physical structure of the initial model changes as soon as thermonuclear reactions occur. A new model is then constructed. A succession of models is thus obtained, for each of which the effective temperature and luminosity of the associated theoretical star can be determined. The points corresponding to each of these models may be plotted on a theoretical Hertzsprung–Russell diagram. The points will trace out a path in the diagram, each region of which corresponds to a well-defined phase in the life of the star, giving us a method of determining its age. The Hertzsprung–Russell diagram is the principal tool in the study of stellar evolution.

A gas sphere in equilibrium

The laws of physics used in the theory of the internal structure of stars are mathematical expressions of the following facts: a star is a gas sphere in equilibrium, and the energy produced in its centre by thermonuclear reactions is propagated towards the surface where it escapes as radiation.

Terrestrial physics has accustomed us to the fact that a mass of gas tends to dissipate and to occupy all of the available volume. However, the Sun behaves differently. Its gas does not disperse into space but remains within a limited volume. This happens because the mass of gas involved is considerable and has a gravitational field strong enough for cohesion.

The constituent particles of the gas move more quickly the higher the temperature. This movement, known as thermal agitation, gives rise to a pressure force, which is directed towards the exterior as a result of the decrease in temperature from the centre to the surface of the star. In regions where the temperature is high, photons also exert a significant force, called radiation pressure. The magnitude of this force is proportional to the fourth power of the temperature. Now imagine a small volume of gas within a star. It will be in equilibrium under the action of two forces, which act in opposite directions: one, due to the gravitational attraction, tends to pull it towards the centre of the star; the other, due to pressure, tends to push it towards the exterior. When these forces are no longer in equilibrium the star becomes unstable.

A stable thermonuclear reactor

The gravitational contraction of a protostar causes the temperature of the gas to increase to about ten million kelvins in the central region. At such temperatures hydrogen exists in the form of a nucleus (the proton) with a positive charge and a free electron, of the opposite charge. All of these particles move at very high speeds, high enough to allow protons to fuse on colliding despite the electromagnetic repulsive force which acts between charges of the same sign. The fusion of the protons finally gives rise to the formation of a helium nucleus accompanied by the liberation of a vast amount of energy. This transformation only occurs where the temperature is sufficiently high, in the central regions of stars. For the Sun, only 12 per cent of the mass is involved. The energy produced escapes from the star as the electromagnetic radiation which we observe.

The fusion of hydrogen nuclei has been realised by man in the explosion of the H bomb. However, he is unable to control the energy liberated and therefore cannot use it for peaceful ends. We do not know how to build vessels that can withstand the tremendous temperatures that occur in nuclear fusion reactions. Such vessels already exist in the Universe: the Sun and the stars are sufficiently massive to control the stability of the thermonuclear reactor in their centre and to produce a regulated supply of energy. The stars are stable nuclear reactors.

The transport of energy from the centre towards the surface

How does the energy produced in the central regions of the star propagate outwards? The simplest way of visualising this transport would be to say that a particle is able to move from the centre to the surface without being stopped on the way. However, the gas density in the star is so high that the particle is subject to many collisions. During a collision the more energetic particle loses some of its energy to the other. If electrons are present then the transport of energy is described as conductive. In the case of multiple photon–matter interactions the transport is said to be radiative. Energy can also propagate

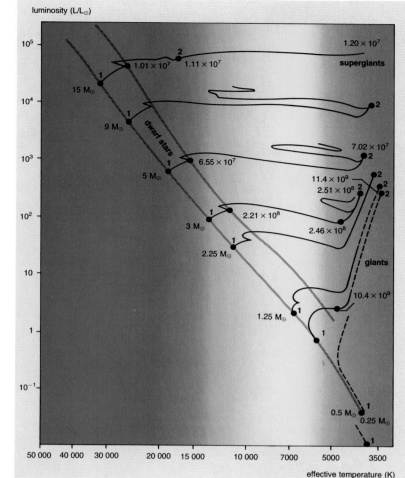

Evolutionary tracks of stars in the Hertzsprung–Russell diagram. The life of a star may be represented theoretically by a series of models of its internal structure. The time interval between two models (i.e. two points on the track of the star in the HR diagram) depends on the stage of evolution: certain very short phases profoundly alter the observed properties of the star, while other very long phases produce very little change. In the theoretical HR diagram (above) the tracks of seven stars of various masses are traced. The starting point, point 1, corresponds to the birth of the star when hydrogen combustion begins in its core. The two grey lines enclose the region of the HR diagram where stars (of any mass) are found while they derive their energy from the combustion of hydrogen. This is the main sequence. Point 2 corresponds to the onset of helium combustion. The least massive stars will just reach this phase and will not go beyond. Massive stars, on the contrary, will go on to synthesise heavier elements. The durations of the principal stages in the life of a star are a function of its mass and may vary considerably. For some of the tracks the time taken (in years) to reach certain points is shown, starting from the main sequence at age zero.

These evolutionary tracks in the HR diagram form the basis for a large number of observational tests, which allow the verification of their validity. The evolutionary tracks shown as broken lines correspond to models of the internal structure that have been determined less precisely. The initial chemical composition of these Population I models is as follows; in 1 gram of matter there is 0.708 gram of hydrogen, 0.272 gram of helium and 0.020 gram of all other elements. There are three groups of stars in the diagram: the dwarf stars, the main sequence stars, the giants and supergiants.

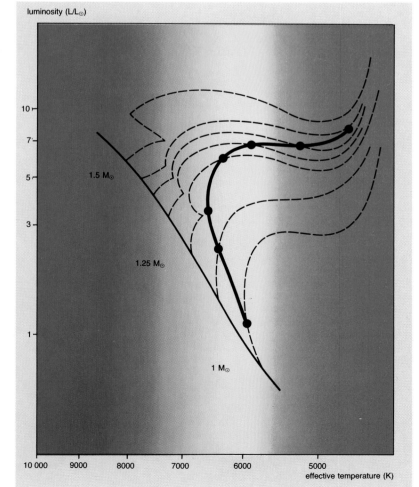

Observational tests of the validity of the models: cluster isochrones (opposite). The stars in a galactic cluster were formed simultaneously, a few million years ago, from a single cloud of interstellar matter. Their chemical compositions are therefore identical, but their masses may be very different. After a given time, their positions in the Hertzsprung–Russell diagram define an isochronal curve, a curve where all the stars are the same age. The most massive stars, which evolve rapidly, have become red giants, while the main sequence is populated by low mass stars. An important observational test of the validity of the models consists of comparing HR diagrams of star clusters with a sequence of theoretical isochrones. These are obtained from the evolutionary tracks of stars of various masses (represented by the dotted lines in the diagram) by joining points corresponding to a given age. The full curve, for example, is the isochrone for four billion years. The agreement between a particular isochrone and the HR diagram of a cluster gives the age of the cluster, and also confirms the validity of the models of internal structure used.

mass ($M_\odot$)	luminosity ($L_\odot$)	effective temperature (K)	central density	central temperature (million K)	radius ($R_\odot$)	lifetime on the main sequence (in millions of years)
0.8	0.25	4940	84	11.4	0.68	20×10^3
1.0	0.77	5730	90	13.5	0.88	10×10^3
1.5	5.2	7290	87	18.5	1.43	1.8×10^3
2.0	16.9	9250	67	20.9	1.60	800
5.0	515	17 020	20	26.8	2.61	78
9.0	3900	23 900	10	30.5	3.64	24
15.0	19 230	31 100	6.3	34.3	4.77	11
30.0	120 000	40 200	3.3	37.3	7.13	5.9

Remark: model at the start of the main sequence phase for a chemical composition of H: 70%; He: 27%; and 3% heavy elements, typical of the young population of our Galaxy (Population I). (After A. Maeder, *Histoire de l'Univers*, Hachette)

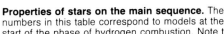

Properties of stars on the main sequence. The numbers in this table correspond to models at the start of the phase of hydrogen combustion. Note the values of central density: they may appear surprising at first sight, the least massive stars being the most dense. At the centre of the Sun the density is one hundred times that of water, but for a 10 solar mass star it is hardly ten times that of water.

via convection. Convection is a means of energy transport with which we are all familiar: in a pan where water is heated, the heat is transferred from the bottom to the surface by the movement of the liquid, which moves from bottom to top.

It is essentially by radiation – radiative transport – that energy is transported towards the surface of a star. Conduction can always be neglected in the case of normal stars. It plays a role only when the gas is degenerate. Convection is a very efficient means of transport but it only occurs in regions where energy transport by radiation is difficult. In convective zones the convection ensures homogeneity. Radiative transfer can be described as a succession of interactions between photons and matter. These interactions modify the motion of the photon which follows a random walk that could be compared to that taken by a drunken man. Between two collisions with matter the photon travels no more than one centimetre. The re-emitted photon could have the same energy as the initial photon, or could well have a different energy. Thus, a photon takes approximately ten million years to move from the centre to the surface. This long period shows us that any change in the energy produced by the thermonuclear reactions at the centre of the star will only be observed millions of years later. The radiation emitted by a star, therefore, does not give any direct information about what is happening in the central reactor. Only neutrinos, which interact very weakly with matter and which, therefore, leave the star as soon as they are created, can convey fundamental information on the thermonuclear reactions taking place. There are no other direct observational tests of the state of the central regions of stars.

Determination of the structure of a star

Knowing the structure of a star, it is possible to determine the temperature, energy flux, pressure and density at all points interior to the star. These quantities are related mathematically by the equations which describe the following: the equilibrium conditions of a gas sphere with equality between pressure and gravitational forces, the production of energy by thermonuclear reactions, the means of energy transport from the centre to the surface, the conservation of mass taking into account possible mass loss by a stellar wind, and the state of the gas.

These equations make use of numerous disciplines within physics from dynamics to atomic physics.

They describe the structure of a star at a given instant. If the mass of gas and the chemical composition are known, they have a unique solution. This basic result, established in 1926, shows that the entire history of a star depends ultimately on only one principal parameter: its initial mass.

Stellar models, evolutionary sequences

The chemical composition of a star is modified in proportion to the rate at which thermonuclear reactions take place in the core of the star, and changes in the internal structure are therefore

expected in the course of time. These changes are described by a series of mathematical models. For each of these models the effective temperature and luminosity of the model star are calculated, and the theoretical point thus obtained may be plotted on the Hertzsprung–Russell diagram. The set of all the points traces out an evolutionary track in the diagram for a star of a given initial mass. This evolutionary track only makes sense if one can deduce the effective temperature and

luminosity of an observed star starting from values attributed by hypothesis to the initial mass and chemical composition.

If there is disagreement between the calculation and the observation, the initial parameters will be modified until the two are in agreement. Although in principle the construction of evolutionary sequences is simple in its methodology, its realisation encounters a number of obstacles. Of these we shall mention only one: how can we take

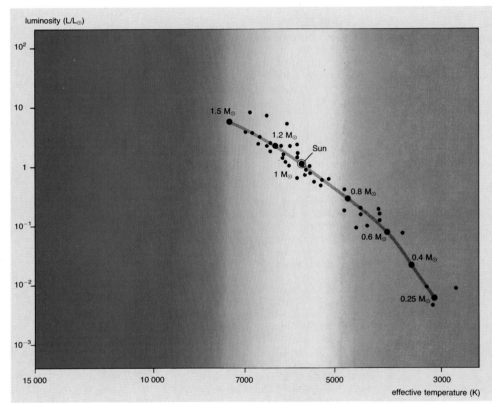

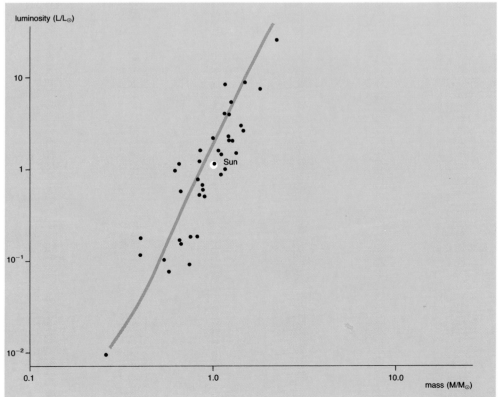

Theoretical and observed main sequences on the Hertzsprung–Russell diagram. The models of stellar structure allow the tracing of a theoretical main sequence in the HR diagram: that is, the sequence of models which are in the phase of hydrogen combustion in the core. This sequence has a certain width since, modifications to the structure in the course of time give rise to a slow variation in luminosity and effective temperature.

The points represent observations of visual binary stars which are on the main sequence and for which values of the effective temperature and luminosity have been determined as accurately as possible. The solid line represents the mean position of the theoretical main sequence, attached to which are the masses of the corresponding model stars. The comparison is, as one can see, satisfactory: the models reproduce the observations well on the whole. However this conclusion must be tempered with the fact that there are no very luminous (i.e. massive) stars on this graph. For these objects the agreement is not so good because mass loss by stellar winds, which plays an important role, has not yet been taken into account correctly in the models. The chemical composition adopted is that of Population I stars.

Theoretical and observed mass–luminosity relation. The mass and luminosity in solar units appear on the abscissa and ordinate respectively.

The observational sequence was provided by observations of visual binary stars belonging to the main sequence, for which masses and luminosities have been determined.

The theoretical relationship is shown by the solid line: it represents a mean value, since the luminosity of a star varies slowly during its time on the main sequence. The assumed chemical composition corresponds to that of Population I.

account of the mass loss that affects some stars in a model of stellar structure? The correct estimate of the change in mass is very important: for example a star of 8 solar masses could, depending on how much mass it loses, end its life as a white dwarf or as a supernova. We must recall also that in the case of close binary systems the stellar wind has a very important role to play in the history of the system. This domain of study has only recently been tackled following observational evidence of the frequency of this phenomenon.

Choice of fundamental parameters: chemical composition and mass

The initial parameters of the models, mass and chemical composition, are relatively easy to ascertain from the observations. Nevertheless the observations only give us information on the chemical composition of the external regions of the star and in no case do we have direct means of testing the chemical modification of the central regions.

It is believed that no convective motion can occur in a star as a whole, ruling out the transport of matter from the centre to the surface of the star. Therefore the chemical composition of the surface cannot be changed in the course of the evolution of the star. Nevertheless, there are some exceptions to this simple scheme. Under certain conditions the envelope, rich in hydrogen since it has not been involved in thermonuclear reactions, can be totally dissipated into the interstellar medium by a very strong stellar wind, leaving the core of the star on view with its modified chemical composition. This is the case for Wolf–Rayet type stars, objects rich in helium because they have been practically reduced to their cores where the hydrogen has undergone fusion.

The chemical composition of stellar atmospheres in the Galaxy shows a certain uniformity, with some variations in the content of hydrogen, helium and heavy elements depending on which population of the Galaxy the star belongs to. It is the same in neighbouring galaxies. In contrast, the determination of the range of initial masses still poses problems. The upper and lower limits of fragments of clouds of interstellar matter that can give birth to stars are not well known. It is estimated that at least 0.04 solar mass is necessary for the central regions to attain temperatures of the order of a few million kelvin, required for the ignition of helium fusion reactions.

On the other hand, the higher the mass of the interstellar gas cloud from which a star forms, the higher the temperature will be as a result of the gravitational contraction of the cloud, because the temperature rise depends on the potential energy of the cloud. Now, at high temperatures, the photons, which possess a high energy, exert a considerable radiation pressure on the outermost layers, preventing them from accreting onto the already formed protostar, whose envelope ionises and is then ejected. The star surrounds itself with an H II region. Thus, from a fragment of 150 solar masses, a star of only 35 solar masses will be formed, the rest being dissipated again into the interstellar medium. The discovery of the very great mass loss to which these objects are subject leads one to think that it will compensate for an instability that was originally expected: the physical parameters defining the stability of models of internal structure with regard to vibrations imply a stellar mass limit of 60 solar masses. Masses larger than the theoretical limit have been measured in observations of binary stars. Taking the uncertainties in the determinations into account, we find that the most massive stars are between 60 and 140 solar masses. Stars whose initial mass is between 150 and 200 solar masses can now be envisaged.

The life of a star

We now describe the subtle mechanisms that govern the structure of a star. In the course of time the star will adjust itself so that the equilibrium between gravitation and the pressure forces due to the thermal agitation of the gas and due to the photons is constantly maintained.

The opacity of the stellar medium plays a fundamental role in the transport of radiant energy: if the medium is transparent, the energy escapes rapidly; if it is opaque, the energy takes a long time to leave. In some way, then, the opacity of the medium controls the flow of energy and thus the luminosity of the star.

A question then arises: since the flow of energy of a star is not governed directly by the rate of nuclear reactions, but also by the opacity of the medium, what is the mechanism that prevents the system from getting carried away? Let us suppose, for example, that the nuclear reactions produce more energy than can be radiated away by the star: there will then be a disequilibrium between pressure and gravitation, and an expansion of the star will result. This in turn will reduce the opacity of the outer regions of the star, enabling the photons to leave more easily. This expansion also affects the central regions, where the temperature drops, consequently reducing the rate of nuclear reactions and therefore finally stabilising the system.

Suppose on the other hand that the nuclear reactions produce too little energy: the star will then begin to contract, the temperature of the gas will rise, the energy flow from thermonuclear reactions will increase, and the equilibrium will be re-established.

This mechanism is of fundamental importance because it will regulate the various stages of the life of a star by adjusting the rate of nuclear reactions through expansion or contraction.

The life of a star, besides its birth and death, consists of two main phases: the period on the main sequence, and the red giant phase. The processes of formation and the subsequent evolution of stars are not yet well understood in detail; in contrast, the main sequence phase is understood rather well.

The graphs of the theoretical evolutionary tracks in the Hertzsprung–Russell diagram enable us to understand the observed distribution of stars in the diagram. The mass is the fundamental parameter that determines the destiny of a star and the regions of the Hertzsprung–Russell diagram it occupies during its life, that is, during the period when it derives its energy from nuclear fusion.

The stars on the main sequence

Nuclear reactions burning hydrogen start at the end of the phase of contraction. When all of the star's energy is supplied by nuclear reactions, the star appears on the main sequence. A star remains on the main sequence during the whole period of combustion of hydrogen in the core.

The lifetime of a star on the main sequence depends on the initial parameters of mass and chemical composition. As the chemical composi-

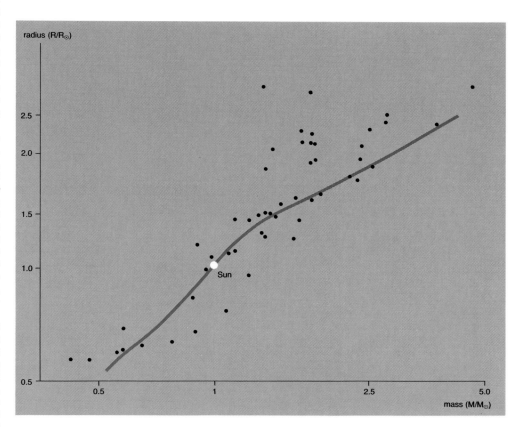

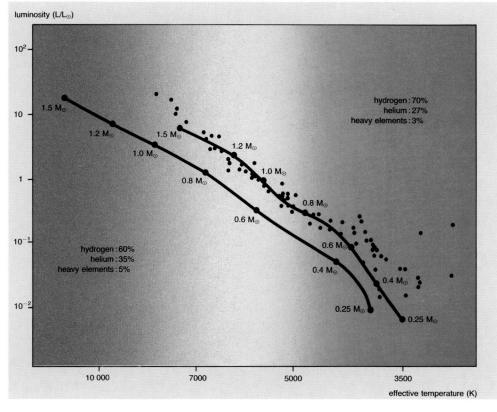

Theoretical and observed mass–radius relation. The mass and radius are plotted in solar units as abscissa and ordinate respectively. Masses and radii of stars on the main sequence are determined by analyses of visual binary stars. The theoretical relation is shown by a continuous line; it represents a mean value since the radius varies slightly during the time the star is on the main sequence. The chemical composition adopted corresponds to Population I.

Sometimes relatively large deviations from the theoretical line are found on plotting observations; in fact, determinations of radius are relatively inaccurate, particularly in the case of close binaries. Moreover, the variations in radius that occur for stars during the main sequence phase also tend to increase the dispersion of the points in the diagram.

Influence of the chemical composition on the position of the main sequence in the Hertzsprung–Russell diagram. In this diagram are plotted observations of stars belonging to the Hyades Cluster. The theoretical main sequence which best fits the observations was obtained with the following initial chemical composition: hydrogen 70 per cent, helium 27 per cent and heavy elements (elements other than hydrogen and helium) 3 per cent. These values correspond to those measured by spectroscopy in the atmospheres of the stars of the cluster. A different theoretical main sequence, calculated with another chemical composition – 60 per cent hydrogen, 35 per cent helium and 5 per cent heavy elements – is incompatible with the observations.

The masses of stars on the two main sequences are indicated at a few particular points.

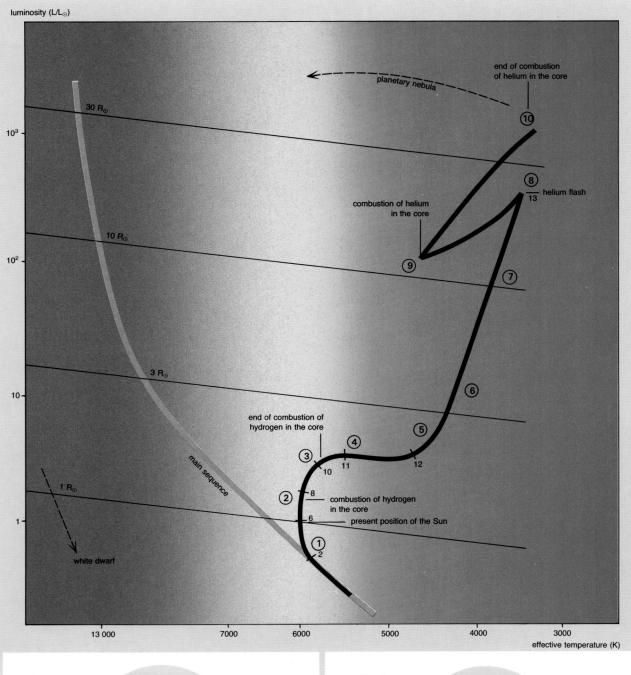

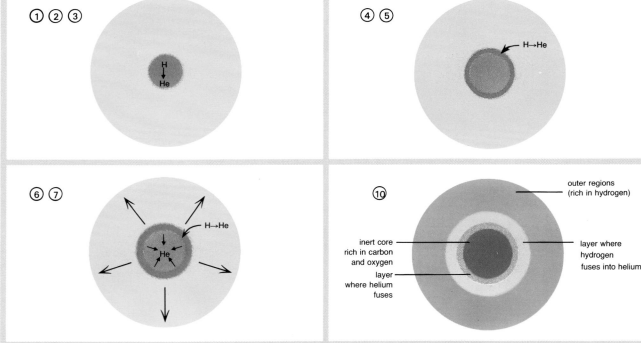

Evolution of a star of 1 solar mass in the Hertzsprung–Russell diagram. The top diagram represents the track of a star of 1 solar mass in the HR diagram. The abscissa is its effective temperature in kelvins and the ordinate its luminosity expressed in solar luminosities. The sloping lines are the isoradius lines whose intersection with the evolutionary track gives the radius of the star at the given point. The numbers in small type, placed along the curve, indicate the age of the star in billions of years, that is, the time elapsed since the ignition of nuclear reactions at the centre of the star. Below the HR diagram are very schematic illustrations of the internal structure at various stages of evolution. (The scales are not constant.)

1. The protostellar phase ends when the nuclear reactions that fuse hydrogen begin at the centre of the star. They continue for about ten billion years. This long period corresponds to the sojourn of a star on the main sequence, the greater part of its lifetime.
2–3. The core, which is gradually enriched in helium, begins to contract. Its temperature increases and the flow of energy becomes more rapid, the rate of nuclear reactions being very sensitive to the temperature. The luminosity of the star increases and the internal structure of the star adjusts itself to maintain its equilibrium and its

energy transport: the radius increases slowly.
4–5. Hydrogen is exhausted in the central regions. The contraction becomes more marked and causes an increase in the temperature sufficient to ignite hydrogen in a thin layer around the core.
6–7. The contraction of the core of the star continues, during which the envelope expands considerably. The radius reaches a value fifty times that on the main sequence. The luminosity increases and the effective temperature decreases; the star becomes a red giant. All of its energy comes from the fusion of hydrogen around the core.
8. When the central temperature passes 10^8 kelvins, the reactions fusing helium start violently: this is the 'helium flash'.
9. Helium combustion stabilises in the core, which expands while the envelope contracts. The star leaves the domain of the red giants.
10. The central regions are now mainly composed of carbon and oxygen, products of the fusion of helium. The output of energy diminishes and a new period of contraction begins, followed by an expansion of the inner layers. The star becomes a red giant again. Helium and hydrogen then burn in a double layer around the core. The subsequent evolution is very rapid. The envelope is ejected during the planetary nebula phase. The star ends its life as a white dwarf.

tion does not vary very much, it does not play an important role. This is not so for the mass. The more massive the star, the more the gravitational contraction during the formation of the star raises the temperature of the gas. Now the rate at which nuclear energy is produced is a function of the temperature of the central regions, implying that the more massive stars have a faster rate of hydrogen combustion than the less massive stars. They are, therefore, also the most luminous and are located high on the main sequence. Their energy is essentially produced by the CNO chain; the proton–proton chain is practically negligible, in contrast to the least massive stars where it predominates. (These reaction chains are discussed on page 280).

Stars spend most of their lives on the main sequence, whatever their mass. This is the reason why there are a large number of stars on the main sequence in the observed Hertzsprung–Russell diagram for most samples apart from the oldest clusters.

What tests do we have of the validity of our models? We know that there is no possibility of observing the internal structure of a star directly. The only possible tests, therefore, aim at comparing the observed properties of stars on the main sequence with the same properties deduced from models. These comparisons test values of the effective temperature, the mass, the radius and the luminosity. The results are quite satisfactory and therefore enable us to interpret the Hertzsprung–Russell diagram in another way apart from a simple 'zoological' description, that is, in terms of stellar structure and evolution.

As we have just seen, the models of internal structure only enable us to describe the bulk properties of stars: mass, radius and luminosity. A more exact description, giving account simultaneously, for example, of the mass loss by stellar wind, the rotation of the star on its axis and the existence of magnetic fields, is extremely difficult. Actually, each of these parameters is treated more or less independently.

Giant stars

A star leaves the main sequence at the end of the combustion of the hydrogen in the central regions. A period of gravitational contraction begins, hydrogen continuing to burn in a layer around the core. The contraction supplies extra energy to the star, increasing its luminosity. The radiation pressure then increases, pushing the envelope outwards. The radius increases so as to reach fifty times its initial value. The surface of a star is thus so large at that time that, despite the great luminosity, the energy radiated per unit surface area, and therefore the effective temperature, decrease. The emitted light is redder than during the main sequence phase: the star has become a red giant.

During this phase, the properties of the core are very different from those of the envelope in the case of a low mass star. The electron gas is degenerate and so conducts heat very well: the temperature becomes uniform in the central regions. When it reaches 100 million kelvin, the density being in the neighbourhood of 10 kilograms per cubic centimetre, fusion reactions between nuclei of helium begin, producing nuclei of carbon and of oxygen. This new period in the life of a star begins rather violently. The degenerate gas is in fact very insensitive to variations in the temperature, which leads to a poor control over nuclear reactions, which escalate: this is the 'helium flash', which affects the core of the star, without being transmitted to the exterior by a sudden change of luminosity. It is of short duration, scarcely a few hundred years.

A new equilibrium state is then established, with a shell where hydrogen burns by the CNO reactions surrounding the core where helium is being fused. The luminosity of the star decreases, and it leaves the red giant branch region. A complicated series of changes to the structure then occurs to adapt the energy lost in the form of radiation to the output of nuclear reactions. There is first an expansion of the core together with a contraction of the regions surrounding it. The reverse mechanism then develops. These oscillations cause variations in the observed luminosity, then are dampened out by viscous forces which dissipate their energy. The star, variable for a while, becomes stable again.

Furthermore, helium, whose combustion con-

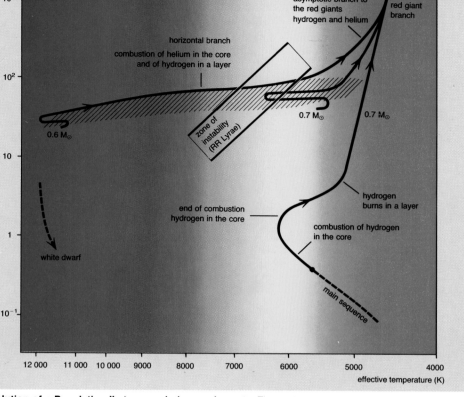

luminosity (L/L)

Evolution of a Population II star poor in heavy elements. The evolution in the HR diagram (above) of a low mass star poor in heavy elements has some peculiar aspects. A star of 0.7 $M_\odot$ containing 0.1 per cent heavy elements remains on the main sequence for sixteen billion years. Its radius and luminosity are then 0.7 $R_\odot$ and 0.7 $L_\odot$ respectively. It reaches the red giant region in three billion years. Until then, few things distinguish it from a Population I star of the same mass (i.e. containing 2 to 3 per cent of heavy elements). However, after the helium flash, it evolves differently by moving towards the left of the diagram in a region called the horizontal branch (a star of lower mass such as 0.6 $M_\odot$ for example, will move further). This phase corresponds to the combustion of helium in the core and lasts about 10 per cent of the time spent on the main sequence. Instabilities may then develop in the envelope: the object becomes a variable RR Lyrae star. Variable stars occupy a well-defined region of instability in the HR diagram, and the star may cross it several times. At the end of the period of combustion of helium in the central regions, a period of helium and hydrogen burning begins in a layer surrounding the core. The point representing the star in the diagram moves asymptotically along the red giant branch. After having (probably) passed through the planetary nebula stage, the star, whose core is rich in carbon, and degenerate, evolves towards the final white dwarf state.

Theoretical evolutionary track of a star of 5 solar masses. The tracks to and from the red giant region are very strongly dependent on the rate of mass loss in the model. The isoradius lines are given for 10 and 100 solar radii. The various phases located by the figures on the diagram are briefly described, with their duration, below.
1–2. Evolution on main sequence until core combustion hydrogen ends (6.5×10^7 years).
2–3. Phase of contraction of the core (2.2×10^6 years).
3–4. Beginning of combustion hydrogen in a layer around the core (1.4×10^5 years).
4–5. Combustion of hydrogen in a thick layer (1.2×10^6 years).
5–6. Contraction of the core and expansion of convective envelope (8×10^5 years).
6–7. First 'red giant' phase (5×10^5 years); point 7 corresponds to start of helium fusion.
7–8. Combustion of helium in the core (10^7 years).
8–9. The structure of the envelope changes (10^6 years).
9–10. Active combustion of helium in the core (10^7 years).
10–11. The outer regions become unstable: this phase corresponds to variables of the Cepheid type: the star is then a blue giant (10^6 years).
11–12. End of the combustion of helium in the core, then very rich in carbon and in oxygen, followed by a new phase of contraction of the core.
12–13. Combustion of helium in a thick layer.
13–14. Contraction of the core and then helium combustion in a thick layer: the star again becomes a red giant and probably a variable of the Mira type. The envelope where helium is burning then becomes thinner. How the evolution continues depends on the mass loss that the star undergoes towards the end of its life.

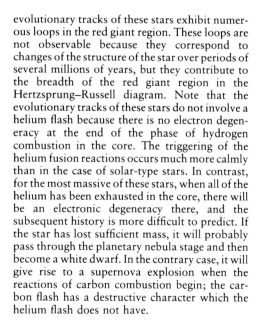

luminosity (L/L$_\odot$)

---

tinues, is gradually exhausted in the central regions. The pressure then decreases until it cannot prevent gravitational contraction. This new phase of evolution resembles that which the star underwent at the point of leaving the main sequence. The temperature increases sufficiently because of this contraction to start nuclear reactions in the layer of helium surrounding the core of carbon and oxygen. This helium layer is itself surrounded by a layer of hydrogen in combustion. The radius and luminosity of the star increase again and the star once more becomes a red giant.

While the star is in the red giant region, a process of mass loss develops, at a rate which varies between 10^{-9} and 10^{-5} solar mass per year. The causes of this ejection of matter into the interstellar medium are not yet well understood. It is generally thought that the radiation pressure becomes sufficiently great to oppose gravity in the outer regions of the star; according to another hypothesis, the ejection is the result of turbulent motions.

When energy is no longer extracted by the star from nuclear reactions in the core, but from reactions taking place in the surrounding layers, the star again enters a phase of instability. Long-period pulsations of some hundreds of days occur, accompanied by changes in luminosity. The star becomes a variable of the Mira type. If these variations are amplified, they can lead to the ejection of an envelope, which is rich in hydrogen, at a velocity of some tens of kilometres per second. This matter begins to seep gradually into the interstellar medium; the process may even be repetitive. The ejected mass is large and may reach 10 to 30 per cent of the initial mass of the star. The observation of planetary nebulae corresponds to this stage. The remaining object

will evolve into a white dwarf. A new chapter opens in the life of the star.

The case of other stars

A low mass star ($M < 0.5$ solar mass) will, in the course of its life, only know the phase of combustion of hydrogen, its mass being insufficient for the subsequent gravitational collapse to ignite the reactions involving helium fusion. It stays in the neighbourhood of the main sequence without ever becoming a giant and ends its life as a white dwarf when its hydrogen is exhausted in the central regions. The lifetimes of these low mass stars are several billion years. They are not very luminous, but they are numerous and constitute, for example, the main sequence of the Hertzsprung–Russell diagram of globular clusters. If the star is more massive than 0.5 solar mass but less than 2.25 solar masses, reactions of helium fusion occur after the fusion of hydrogen. As the core of the star is then partially degenerate, the start of the helium burning reactions is very violent. This is the helium flash. The star remains in the red giant region until the end of its combustion of helium in the core and then becomes a white dwarf, after passing through the planetary nebula phase. The evolutionary track of these stars in the Hertzsprung–Russell diagram is comparable to that of the Sun.

Stars whose mass exceeds 2.25 solar masses but is less than 4 (or even 6) solar masses will also attain the red giant stage. From this stage, any new contraction of the core will only serve to increase the degeneracy without increasing the temperature sufficiently to start new nuclear reactions following the fusion of helium. The

evolutionary tracks of these stars exhibit numerous loops in the red giant region. These loops are not observable because they correspond to changes of the structure of the star over periods of several millions of years, but they contribute to the breadth of the red giant region in the Hertzsprung–Russell diagram. Note that the evolutionary tracks of these stars do not involve a helium flash because there is no electron degeneracy at the end of the phase of hydrogen combustion in the core. The triggering of the helium fusion reactions occurs much more calmly than in the case of solar-type stars. In contrast, for the most massive of these stars, when all of the helium has been exhausted in the core, there will be an electronic degeneracy there, and the subsequent history is more difficult to predict. If the star has lost sufficient mass, it will probably pass through the planetary nebula stage and then become a white dwarf. In the contrary case, it will give rise to a supernova explosion when the reactions of carbon combustion begin; the carbon flash has a destructive character which the helium flash does not have.

The supergiant stars

The most massive stars of all ($M > 8$ solar masses) are those populating the top of the main sequence. These stars are very prodigious in their consumption of nuclear fuel. In a few million years, they can pass through all the stages of nuclear fusion.

Note that the luminosity of a massive star varies little in the course of its evolution. The creation of neutrinos becomes important when the temperature of the core rises, and the

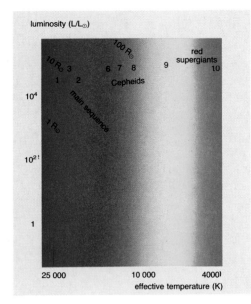

phase	duration (years)	main sequence
1-2	10^7	combustion of hydrogen in the core: the core is convective;
2-3	2×10^5	contraction of the core, com-
3-6	7×10^4	bustion of hydrogen in a layer, causing an expansion of the envelope; start of helium combustion in the core;
6-7	7×10^5	expansion of outer regions follows;
7-9	8×10^5	instabilities of the Cepheid type develop;
9-10	3×10^4	the inert core, rich in carbon, nitrogen and oxygen, is sur- rounded by a first layer where helium burns, and by a second layer where hydrogen burns; the star is a supergiant.

Theoretical evolution of a star of 15 solar masses. This diagram represents the theoretical evolutionary track of a star of 15 $M_\odot$ in the HR diagram. The isoradius lines are drawn for the values 1, 10 and 100 $R_\odot$. Variables of the Cepheid type and supergiants are also indicated. The evolutionary tracks of these stars are much more simple than those of less massive stars. At the end of their evolution, the reactions of carbon fusion can lead to a supernova explosion.

proportion of energy radiated in the form of photons is reduced. On the other hand, the radius varies greatly until it reaches a thousand times that of the Sun, which places these objects in the supergiant region. In these stars, the carbon fusion reactions, after those of hydrogen and of helium, start calmly in a non-degenerate core.

Increasingly heavy elements – oxygen, silicon and magnesium – are then formed. If the star is sufficiently massive, the synthesis continues as far as iron, which is the final product of fusion. At this stage of evolution, the star has a structure like an onion, the iron core being separated from the outer mantle of hydrogen and helium by succes- sive layers of carbon, oxygen, silicon and so on, products of intermediate combustion.

The final stages of the life of a massive star will be dramatic and spectacular. The nuclear fusion reactions are accompanied by a very strong emission of neutrinos, which carry their energy away very quickly. A sudden contraction of the core results, followed by a rise in temperature up to five billion kelvin. In these extreme conditions reactions of photodisintegration of iron can occur, which consume energy in absorbing photons. To satisfy this demand, the core

becomes unstable and then collapses. The matter of the star is then subjected to great changes: it is enriched in neutrons and becomes more and more degenerate while the nuclear reactions become explosive in some outer layers. The energy transport occurs catastrophically. The star becomes a supernova.

The horizontal branch

The Hertzsprung–Russell diagrams of globular clusters are quite different from those of open clusters. As well as the main sequence and the red giant branch, there is a third region called the horizontal branch, which is occupied by stars less massive than the Sun, but of different chemical composition: these stars have a very low content of heavy elements, i.e. elements other than hydrogen and helium. They evolve differently from the Sun. The horizontal branch corresponds to the phase of helium fusion in the core, which happens after the star has been in the red giant phase during the helium flash.

The analysis of the position and the extent of

the horizontal branch of the Hertzsprung– Russell diagram of globular clusters enables us to deduce the initial helium content of the stars in the cluster. This analysis is of fundamental importance in following the evolution of the chemical composition of our Galaxy.

Stellar evolution and stability

The structure of a star undergoes considerable changes in the course of its evolution. It passes through stages, among others, where the state of ionisation of hydrogen and of helium in the envelope can permit the development of instabili- ties giving rise to pulsations, which are man- ifested as variations in luminosity. The compari- son between the properties of real variable stars, for example RR Lyrae stars and Cepheids, and the results of theoretical stellar models is very informative. Cepheids are stars of intermediate mass, from 3 to 16 solar masses. When the instabilities develop, these stars are in the phase of combustion of helium in the core. In contrast, variables of the δ Scuti type have masses greater

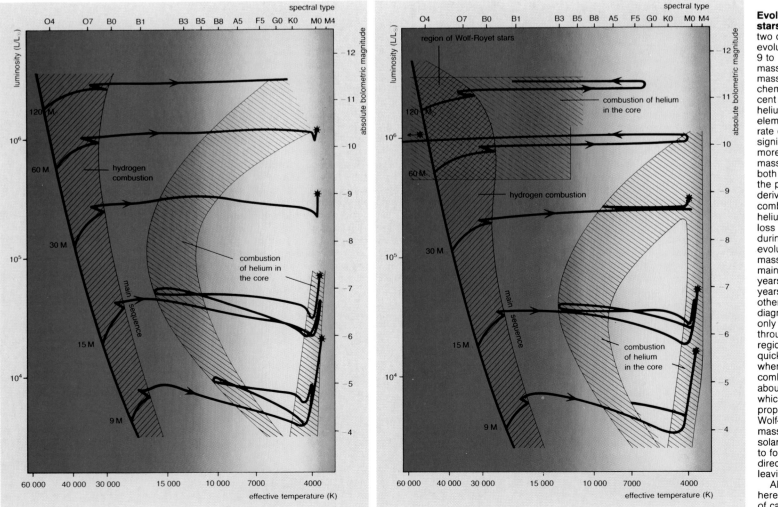

Evolution of very massive stars with mass loss. These two diagrams represent the evolutionary tracks of stars from 9 to 120 solar masses, with mass loss (right), and without mass loss (left), for an initial chemical composition of 70 per cent hydrogen, 27 per cent helium and 3 per cent heavy elements. The mass loss, at the rate chosen here, only has a significant effect if the stars are more massive than 15 solar masses. The hatched areas in both diagrams correspond to the phases where the star derives its energy from the combustion of hydrogen or helium in the core. The mass loss changes relatively little during the various stages of evolution. A star of 60 solar masses thus remains on the main sequence for 4.2 million years instead of 3.8 million years without mass loss. On the other hand, its track in the HR diagram is profoundly altered. It only makes a short passage through the red supergiant region and then evolves very quickly towards the left. This is when the phase of helium combustion takes place, lasting about 200 000 years, during which time the star has properties similar to those of Wolf–Rayet stars. The most massive stars, from 150 to 200 solar masses, that it is possible to form in this context evolve directly towards the left on leaving the main sequence.

All of the tracks represented here end with the combustion of carbon in the core.

than 1.3 solar masses and are still in the hydrogen burning phase. Variables of the β Cephei type are massive stars where the core is beginning its contraction after the hydrogen in the core has been exhausted. The RR Lyrae stars are not very massive, 0.5 to 0.8 solar mass, and undergo their phase of instability during the combustion of helium.

Studies of Cepheids and RR Lyrae stars have given rise to most important theoretical developments. The evolutionary tracks of the relatively massive stars to which the Cepheids correspond show numerous loops in the red giant region during the helium burning phase. This stage is very short and does not represent more than 10 or 20 per cent of the lifetime of the star, 80 per cent being spent on the main sequence. During this helium burning phase, modifications to the structure take place very rapidly and can give rise to instabilities. However, numerous questions remain to be answered, such as the effect of the chemical composition on the models of internal structure and on the development of pulsations; this point is important for the comparison of the Cepheids in the Magellanic Clouds with the Cepheids in our Galaxy. RR Lyrae stars are poor in heavy elements (i.e. elements other than hydrogen and helium). Their instabilities develop during the horizontal branch phase. The analysis in terms of stellar evolution of the instabilities that tend to occur in cool stars of the Mira type is

not as advanced as for the Cepheids or even for the RR Lyrae stars. The main difficulty in explaining the pulsations and the mass ejection is in building convection into the models.

Mass loss and evolution. The very massive stars

The existence of very massive stars of more than 100 solar masses is now acknowledged but it was only in 1981 that the properties of these objects began to be known. This was linked to the discovery, from 1975 onwards, of the importance of the phenomenon of mass loss, thanks to ultraviolet observations obtained by telescopes mounted on satellites. The extent of the mass loss is such that it profoundly alters stellar evolution. Thus, a star of 30 solar masses loses about 10^{-6} solar mass per year, which implies that at the end of the period of combustion of hydrogen its mass is reduced by 24 solar masses.

Before the importance of mass loss was realised, the classical idea of the evolution of massive stars stopped at 60 solar masses. Beyond this limiting value the star would become unstable with respect to any oscillations. But these models did not correctly represent the upper part of the Hertzsprung–Russell diagram. In particular, they did not explain the observation of stars some millions of times more

luminous than the Sun and belonging to the main sequence.

The theoretical diagrams of stellar evolution with mass loss agree well with the observations, the mass loss opposing in some way the development of instabilities.

The evolutionary tracks are totally altered compared with a static situation: for stars of more than 50 solar masses, they no longer pass into the supergiant region. If these ideas are correct, the Wolf–Rayet stars will then be formed from massive stars that have lost enormous quantities of matter, which would lay bare the central regions where the nuclear reactions have profoundly altered the chemical composition. This scenario is not, however, the only possible explanation of the existence of the Wolf–Rayet stars: a mechanism of exchange of matter in a close double star could also produce this type of object.

The hypothesis of mass loss enables the observational facts to be explained in the simplest and most straightforward way. The value of the parameter generally adopted is around 2×10^{-5} solar mass per year. This result is still very much debated, but the tangible fact of mass loss remains just as the profound alterations which it entails in the life of a star.

Michèle GERBALDI

NGC 6164–NGC 6165. This nebulosity, in appearance similar to a planetary nebula, is in fact a very different object. The central star HD 148937 is of spectral type O7 and has a high rate of mass loss: 8×10^{-6} solar masses per year. It belongs to a triple system and is the brightest member. The mass loss probably occurs sporadically and has been going on for a few tens of thousands of years. This mass loss has formed the two nebulosities around the star. It is because two nebulosities can be distinguished that the ensemble has received two catalogue numbers. It is unusual for hot stars to lose sufficient mass to form an observable nebulosity. It is for this reason that HD 148937 is usually thought to belong to the group of Wolf–Rayet stars. (D. F. Malin, Anglo-Australian Telescope Board, © 1981)

Planetary nebulae

Planetary nebulae are gaseous envelopes, more or less circular in shape, surrounding certain hot stars. They were given this name by William Herschel during the eighteenth century because of the resemblance of their disk-like appearance to that of planets. Radiation from the central star excites the gas of the envelope, causing fluorescence as in bright nebulae. The atoms of the envelope absorb the ultraviolet radiation from the star and re-emit it in radio, infrared and visible wavelengths.

The effective temperature of the central star is high (hence the ultraviolet flux). It is usually of the order of 30 000 K and can sometimes reach 100 000 K. On the other hand, these stars are not very luminous (some of them emit only slightly more energy than the Sun). This implies that they must have small radii, the majority looking very much like white dwarfs. Planetary nebulae are therefore always associated with evolved Population II stars or old Population I stars.

The mass of the nebula is between 0.10 and 0.20 solar mass. The gas is very tenuous – 10^{-20} grams per cubic centimetre. Dust is also present. The envelope expands with a velocity of a few tens of kilometres per second. The true size of a planetary nebula is difficult to determine. Although it is relatively easy to measure its angular diameter, its distance must first be known to obtain its absolute dimensions. The central star is in fact peculiar, its properties not matching any normal stars found on the Hertzsprung–Russell diagram. Hence, its distance cannot readily be estimated. The distance of a planetary nebula can be obtained, however, if it is a member of a cluster. For the few cases known one finds a radius between 0.5 and 1 light year. The gas of a planetary nebula is not continuously re-supplied by the star, hence its expansion leads to a complete dissipation into the interstellar medium in about 100 000 years. Nearly a thousand planetary nebulae are known.

The planetary nebula phenomenon must, however, be extremely common. It is, after all, just a normal phase in the life of low mass stars (1 to 5 solar masses), taking place between the red giant phase (where the star would be a Mira variable) and the final phase, the white dwarf. The transition from red giant to planetary nebula would last from 1000 to 10 000 years, and likewise for the next transition from planetary to white dwarf.

A more massive star can also become a white dwarf if during its life it loses enough mass to end up with less than 1.4 solar masses. During the red giant phase stellar winds are an efficient mechanism to lose mass. Thus, it is possible that massive stars could reach the white dwarf state. Once the outer layers are lost the inner zones of the star, which are rich in helium, carbon and nitrogen, become visible to the observer. This is the reason why the central star of a planetary nebula shows a chemical enrichment.

The mechanisms responsible for the expulsion of the envelope are still not well known today. Some similarities have been noted between the spectra of the central stars of planetary nebulae and those of old novae where nebulosities in expansion have also been observed. DQ Her 1934 is one example. It is improbable, however, that all novae become planetary nebulae.

In another case, in 1964, a red star showed a 5 magnitude increase in brightness. Since that time, emission lines indicating the presence of a shell in expansion (with a velocity of 34 kilometres per second) have been seen in its spectrum. This star has a strong stellar wind ejecting gas.

These isolated examples are certainly not representative of all possible mechanisms producing planetary nebulae. In particular, if they are formed from red giants, the stellar wind emitted by these stars should form a detectable halo. Astronomers are attempting to detect such haloes. It is quite possible that it is not necessary to invoke a sudden outburst of matter to explain a planetary nebula. The slow, continuous mass loss of a red giant during its advanced evolutionary stages could do the job.

Michèle GERBALDI

The planetary nebula NGC 6302. Although no central star has been found, this object is classified as a planetary nebula because of its characteristic emission.

Its shape is radically different from that of the Ring Nebula. In fact, only 10 per cent of planetaries are in the shape of a ring; 70 per cent are elongated like this one with two well-distinguished areas, suggesting that matter was ejected in two opposite directions. The velocity of expansion of the gas is high, 400 kilometres per second, indicating that a much more violent event took place than in most planetary nebulae. Furthermore, a spectroscopic analysis near the centre revealed the presence of highly ionised iron atoms (six times ionised). This implies that the central star has a flux much more energetic than is usually found. This object is certainly not a classical planetary nebula. (D. F. Malin, Anglo-Australian Telescope Board, © 1977)

The Helix planetary nebula (NGC 7293). This photograph was obtained with the 150 inch (3.9 metre) Anglo-Australian Telescope at Siding Spring (Australia). The apparent diameter of this object is about 30 arc minutes, roughly the same as the angular diameter of the Moon, and corresponding to a linear extent of 4 light years. The central star is a white dwarf. Various fluorescence processes taking place among the atoms are responsible for its colours in the visible. The blue-green of the inner part is due to oxygen and nitrogen, whereas the overall pink colour is due to the cascade re-emission of hydrogen atoms.

The helical shape of the nebula is really apparent only in certain photographs. The general aspect of the nebulosity, where inner zones seem to point towards the central star, which shows clearly that the gas was expelled from this star.

The helical shape could have other explanations, however, such as rotation of the star, a binary star ejecting matter along its orbital motion, etc. The mechanism is still unknown. (D. F. Malin, Anglo-Australian Telescope Board, © 1979)

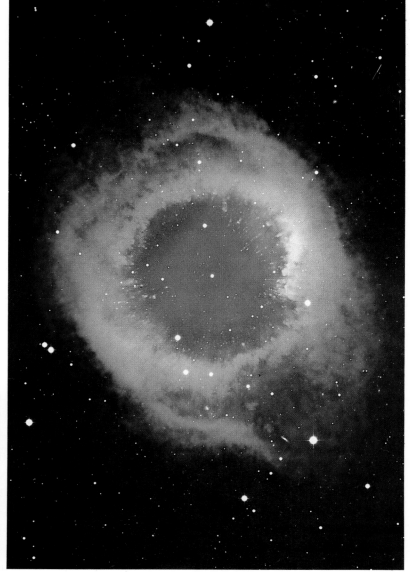

The planetary nebula NGC 7354. This photograph shows a radio picture of the planetary nebula NGC 7354. The intensity of the radio flux of this planetary nebula has been measured at a wavelength of 6 centimetres using a radio interferometer consisting of six of the antennae of the Very Large Array in New Mexico. The angular resolution achieved with these techniques was 1.6 by 2.2 arc seconds.

The radio intensities measured at different points in the nebula were converted into a video signal in order to produce an image which can be seen on a television screen. The colours are arbitrary, corresponding to different intensity levels of the emitted radiation. Strongly emitting regions thus appear in red, while weak ones are blue. This 6-centimetre flux is emitted by ions in the nebula when free electrons are recaptured. These same regions also emit visible light and, as a result, the visible and radio images of the nebula must coincide. This is usually what happens, except when a certain amount of dust is present in the nebula. When this occurs the visible emission is partly absorbed within the nebula. Then only the radio image yields the true size of the nebula because the radio waves are not affected by the dust.

The image in the box in the upper left corner has nothing to do with the nebula; it is only used as a reference to determine the instrumental resolution at the time of observation. (NRAO)

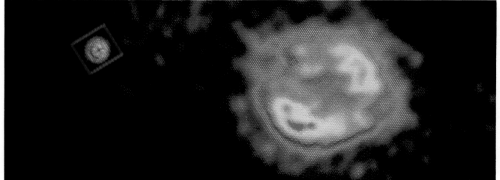

The Ring Nebula in Lyra. Not all planetary nebulae have such regular forms. A few are, for example, hour-glass shaped. No adequate theory exists that can explain all shapes of planetary nebulae. It is possible to understand the shapes of ring-shaped nebulae, however, as shells of expanding gas. The expansion velocity of the envelope of the Ring Nebula in Lyra is estimated to be 19 kilometres per second.

This particular nebula, situated at some 700 parsecs, has a diameter of about 0.5 light years. If its expansion velocity has always been constant, then the nebula would have required 5500 years to reach its present size. The effective temperature of the central star is 70 000 K. Its flux is very rich in ultraviolet radiation which is able to ionise the hydrogen and part of the helium of the nebula. The free electrons thus produced recombine with the ionised atoms, principally with the protons because they are the most numerous. These re-formed hydrogen atoms are in an excited state and the electrons tend to cascade down through the energy levels towards the fundamental state. At each transition from one energy level to another, a photon is emitted. Transitions between highly excited levels produce radio waves, those between the lowest levels visible and ultraviolet light. The shell surrounding the star emits radiation by converting ultraviolet photons into visible light. This light is mostly concentrated into the Balmer series, particularly the Hα line, which gives the nebula its characteristic pinkish colour.

Thus, the image of the planetary nebula we see does not necessarily correspond to all the matter around the central star, but only to the envelope of ionised hydrogen.

The radiation emitted by these nebulae also contains the 658.4 nanometre forbidden line of ionised nitrogen and the doubly ionised oxygen lines at 500.7 and 495.9 nanometres. These last two lines are green.

The forbidden lines are formed in the following way. We have seen that after the electrons are stripped from the atoms by ultraviolet radiation, they are then recaptured by the ions. The free electrons can collide with ions such as those of oxygen, nitrogen or sulphur. These collisions populate excited levels just above the fundamental level. The atoms thus excited will not lose their energy by further collisions because the density of the nebula is very low. The lifetime of these energy levels is very long. The atoms, therefore, stay in their excited states for a long time before they spontaneously return to their fundamental level. These levels are called metastable. It is only after several minutes (instead of 10^{-8} seconds for normal excited levels) that the ion de-excites itself by emitting a photon. Such transitions give rise to spectral lines called forbidden lines because they cannot be detected in the laboratory. This is because vacuums produced on Earth have a much higher density than those found in the interstellar medium, so a collision always takes place before the atom has the time to de-excite itself naturally. This collision changes the energy level of the atom and thus the transition from the previous level is never observed.

The nebula is shown here in four spectral bands corresponding to: blue (upper left), green (upper right), yellow (lower left) and red (lower right). A different region of the nebula appears on each picture.

Near the central star, where the ultraviolet flux is stronger, atoms of oxygen and nitrogen are easily excited, giving a predominently blue-green colour. Further out, when the number of available ultraviolet photons has decreased considerably, only hydrogen is susceptible to being ionised, thus leading to the observed red edge.

The gas of the nebula does not stop there, but extends much farther away. It is neutral, however, and cannot be observed directly in the visible part of the spectrum. The nebula revealed by observations in the visible represents only the gas excited by the radiation from the star. Observations in the radio part of the spectrum enable astronomers to detect the extent of the neutral gas. (Hale Observatories)

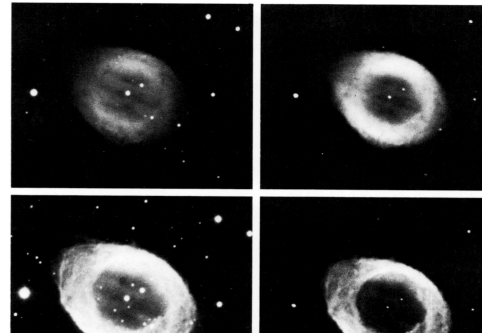

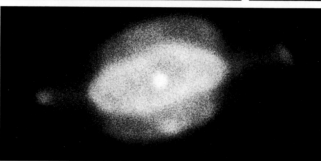

The Saturn planetary nebula (NGC 7009). The curious shape of this nebula (left) is responsible for the nickname Saturn. It looks like a blurred image of Saturn and its rings. The different zones of this nebula are probably layers ejected at different times by the central star. (Kitt Peak National Observatory)

The Dumbbell planetary nebula. Not all the planetary nebulae have a symmetric appearance like the Ring Nebula. This planetary nebula is located in the constellation of Vulpecula at 220 parsecs; its diameter is approximately 0.3 parsec. An analysis of its radiation shows that its structure is similar to that of the Ring Nebula and that it too is excited by the ultraviolet flux of a very hot, central star. (Hale Observatories)

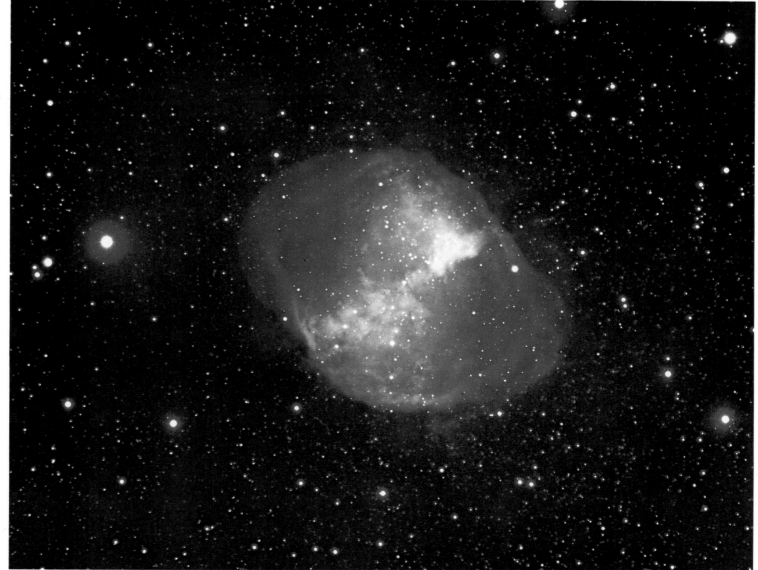

Variable stars

The detection of variable stars

A star is classified as a variable if time variations of its magnitude have been observed. This definition covers all sorts of stars, even though their variability can have quite different origins. These variations can take place not only in visible light but also in the radio or X-ray parts of the spectrum. The detection of variable stars is strongly dependent on the type of instrument and the methods used. If, for example, when two photographs of a star are compared, the exposure times are longer than the time-scale of its variation, the star will appear constant in brightness. Pulsars are a perfect example. They are objects showing periodic brightness variations in the range of a few hundredths of a second to a few seconds. Normally a photographic detector cannot detect such fast fluctuations. It was in fact a radio receiver sensitive to rapid variations that enabled astronomers to discover pulsars. At the other extreme, if the time-scale of the variation is very long (some years) it will be very difficult to reveal these variations, especially since all stars are not observed systematically all of the time.

The principle behind the detection of variable stars is independent of their type: magnitude estimates of a star, obtained on different dates, are compared to magnitudes of stars believed to be constant. The measures of magnitudes are obtained using techniques of photographic photometry or photoelectric photometry. Stars showing slight magnitude variations are difficult to detect. Sometimes motions of the stellar atmosphere take place during the brightness variations, however, and they could be revealed by the Doppler shift of the spectral lines.

If the brightness variation is accompanied by changes in the physical properties of the outer layers, these changes could be analysed by taking spectra simultaneously with the photometric observations. Thus, certain stars believed to be constant could in fact be low-amplitude variables with a very long period.

Classes of variable stars

Variable stars are classified according to the type of their variability. Among the different classes, one comes across stars with periodic light variation of constant amplitude. Other variables show irregular variations and for some others only one change of magnitude was observed – a very important change – so extreme that the structure of the star was drastically changed.

Variable stars can be divided into two groups: stars for which the variations are due to a geometrical effect and the intrinsic variables.

Stars for which the luminosity variation is due to a purely geometrical occultation are members of eclipsing binary systems. In a close binary system transfers of matter between the two stars could also lead to variations of the apparent magnitude of the system.

Intrinsic variables can have quite different origins. Let us suppose that the surface temperature of a star is not uniform. If the star rotates, it will successively present regions of different brightness to the observer, producing variations in the apparent magnitude of the star.

Pulsars and stars with strong magnetic fields are examples of this type of variability. The latter class includes stars of spectral types B and A which are either still on the main sequence or close to it. The structure of their atmospheres is inhomogeneous. Probably because of the surface magnetic field and diffusion processes, some elements accumulate in large areas of the extreme outer atmosphere. These areas have a temperature and chemical composition different from the rest of the outer atmosphere. When these areas are facing the observer, not only a variation of magnitude takes place, but also a change in the appearance of the spectrum can be seen. Stars of this type are called peculiar stars. Their variation is always periodic since it is associated with their rotation. The periods range from half a day to a few hundred days. The variability can be observed from the ultraviolet to the infrared, an interval corresponding to the spectral range where the stellar emission is important. In the visible part of the spectrum the amplitude of the light variation is of the order of a few tenths of a magnitude, but in the ultraviolet it can reach one magnitude. It is, however, out of phase with the visible variation. When the periods are long or the amplitudes too small, the variability cannot normally be detected. These stars can then only be distinguished by their peculiar spectra.

In all other cases, the variability is due to changes of the stellar structure that take place during the course of time.

Three processes can lead to a modification of the structure: instabilities can grow in the stellar atmosphere; matter can transfer between members of a close binary system; or, finally, an instability can occur in the core of the star, thereby modifying all of its internal structure.

Variable stars have been grouped into classes usually designated by the name of the first variable of its kind discovered or the most thoroughly studied example of its kind. These classes are far from homogeneous and all the causes of variability are not yet completely understood.

Supernovae

These objects show a sudden, fantastic increase in brightness, often more than 18 magnitudes, which corresponds to an increase in luminosity of several million times the Sun's luminosity. These are the supernovae.

After these fireworks, the structure of the star is completely modified. An implosion in the central regions of the star, where the thermonuclear fusion reactions are taking place, is responsible for the brightness variation.

Cataclysmic variables

This class includes stars whose brightness increases, in a few hours or a few days, by a factor reaching sometimes a few hundred thousand times their original brightness, a magnitude difference of 7 to 16 magnitudes. After the outburst the brightness decreases more or less regularly during the following few months, until the star returns to its original luminosity. This large brightness variation is accompanied by spectral changes. Certain stars show recurrent variations with ill-defined periodicities that can be as high as several tens of years.

These abrupt variations of brightness can almost certainly be explained by the binary nature of these stars. Nearly all of them have been found to be members of close binary systems.

Belonging to this class of variables are novae and other stars that have light curves very similar to novae. They are of the U Geminorum and Z Camelopardalis types, often called dwarf novae. The R Coronae Borealis stars are almost novae in reverse; they suddenly decrease in brightness by

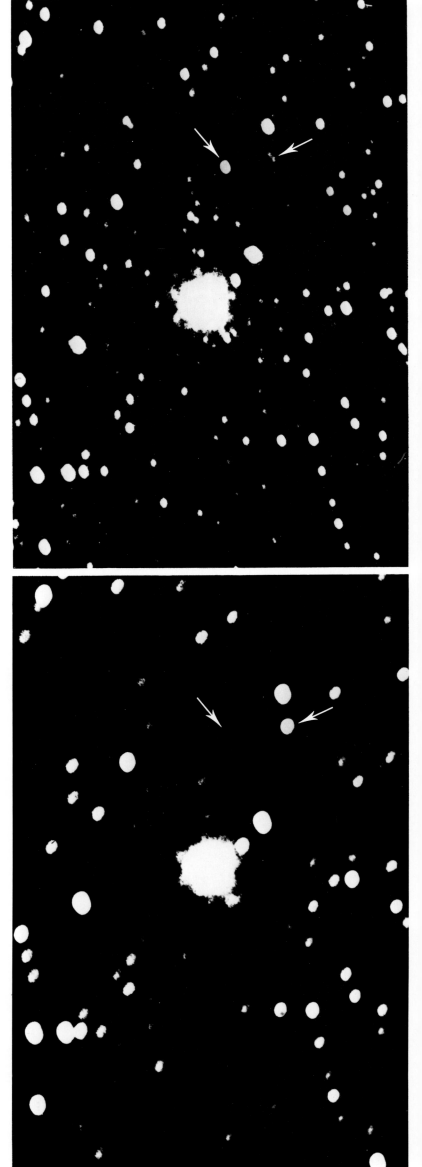

R and S Scorpii. These two stars show quite remarkable light variations. In the top picture R Scorpii is at maximum brightness while S Scorpii is almost invisible. In the bottom picture R Scorpii has disappeared while S Scorpii is bright. These variations in brightness are characteristic of old, red galactic disk stars of spectral class M. The amplitude of the variations is of the order of 4 magnitudes. Their periods are, respectively, 222 days and 177 days. (Yerkes Observatory)

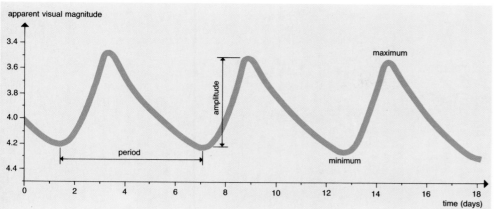

The light curve of a periodic variable (opposite left). The graphical representation of the variation of the apparent magnitude of a star as a function of time is called a light curve. This curve is established from a series of observations distributed (as far as possible) regularly in time.

The 'maximum' is a point on the light curve where the star reaches its maximum brightness. Its magnitude has at that point its minimum numerical value. The 'minimum' is a point on the light curve where the star reaches its faintest luminosity. Here its magnitude has a maximum numerical value. The period is defined as the time interval between two successive maxima or two successive minima. The amplitude is the difference between the magnitudes at minimum and at maximum.

Light curves. Four types of variables, Cepheids, W Virginis, RR Lyrae and δ Scuti stars, show common observational characteristics.

Light curves (below) for a typical star of each group are presented. Because they pulsate, not only the luminosity of these stars varies, but also their radius and their effective temperature. The simultaneous study of these three parameters has led to a better understanding of the nature of these objects. The periodic displacement of the spectral lines, explained by the Doppler effect, shows that the radius of these stars varies with time. The amplitude of the radial velocity variation is of the order of a few tens of kilometres per second. This comes from a radius variation of only a few per cent. The variation in the effective temperature produces changes in the appearance of the spectrum: the spectral type varies periodically.

up to 10 magnitudes, then slowly return to normal. They are not in fact cataclysmic variables; the sharp fading is caused by the formation of solid carbon in the cooler parts of the atmosphere. The star is thus surrounded by soot, which is later blown away by radiation from the star.

Eruptive variables with nebulosity

This class includes variable stars showing unforeseen and abrupt increases in brightness. The amplitude of the luminosity increase is about 3 magnitudes. Sometimes pseudo-periods of the order of a few days are observed, but in general their variability is very irregular.

Most of the stars in this class are young: 'pre-main-sequence stars', which have not yet reached the position of the main sequence in the Hertzsprung–Russell diagram. They are always associated with nebulosity. The main types are: RW Aurigae, T Tauri, UV Ceti and flare stars. Flare stars are found in young open clusters. Their brightness may flare up several times during any 24-hour period. We also find in this class stars called Herbig–Haro objects, which are stars in the process of being formed.

Periodic pulsating variable stars

In this case we are dealing with stars showing periodic variations of their luminosity. These variations are caused by instabilities in the internal structure of the star. A star is a gravitationally stable sphere, but it can oscillate, like a spring, about a mean position. Pulsations are due to modifications of the structure of the outer layers of the star, which are alternately compressed and expanded. The mechanisms responsible for the instability are related to the ionisation of helium in the stellar envelope. Variable stars of this class are located in a well-defined zone on the Hertzsprung–Russell diagram, called the instability strip. During the course of its evolution every star probably passes through this strip. After the disappearance of the causes of instability the star soon returns to its normal, stable state.

Most periodic pulsating stars belong to one of four categories having similar observational characteristics: Cepheids, W Virginis, RR Lyrae and δ Scuti stars. Not only their magnitude varies, but also their radius (because they pulsate) and their effective temperature. The study of these three parameters has enabled astronomers to understand more fully the nature of these variables. Radial variations are detected from the periodic displacement of spectral lines due to the Doppler effect. The amplitude of the observed radial velocity variation is of the order of a few tens of kilometres per second, which corresponds to a change of radius of a few per cent.

The variation of the effective temperature produces changes in the appearance of the spectrum, that is, the spectral type of the star varies periodically.

We shall discuss in some detail the properties of the two most important classes of pulsating variables: the Cepheids and the RR Lyrae stars.

Cepheids: the period–luminosity relation.

The name Cepheid comes from δ Cephei, the first known star of this class, discovered in 1784 by the English astronomer John Goodricke.

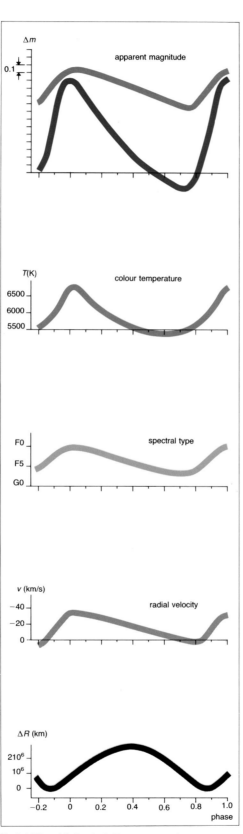

Variability of δ Cephei. The period of δ Cephei is 5.366 days. This diagram displays not only the magnitude variation (Δ*m*) of the star at two wavelengths, 422 and 1030 nanometres, but also the colour temperature variation, the spectral type variation, the radial velocity curve and, finally, the radius variation (Δ*R*). The amplitude of the colour temperature variation is 1500 K while that of the radius is approximately 10 per cent, the mean radius being 15 solar radii. The luminosity variations of Cepheids are due to variations of their effective temperature and not to variations of their radius. Note that the radius variation and the temperature variation are out of phase.

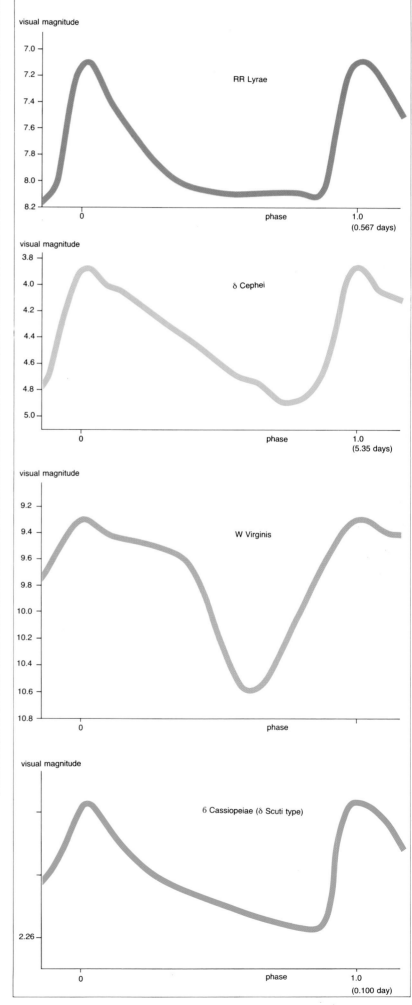

Cepheids, although not very common among variable stars, play a very important role in the study of the structure of the Galaxy. About 700 Cepheids are known in the Galaxy. The Pole Star is a Cepheid whose apparent magnitude varies between 2.5 and 2.6 in about four days. The great importance of these variables is attributed to the fact that there is a linear relation between the period of their luminosity variation and their mean luminosity. This relation was discovered by the American astronomer Henrietta Leavitt in 1912 while observing a few hundred Cepheids in the Magellanic Clouds. The Magellanic Clouds are nearby galaxies, but in 1912 this fact had not been recognised. The period–luminosity relation is an important tool for the determination of distances. Cepheids, being very luminous stars, are still observable at very large distances.

The use of Cepheids as distance indicators in our Galaxy is difficult because most of them are situated in the galactic plane, where they are obscured by the interstellar matter. They have, however, been observed in thirty other galaxies, and it is in the determination of the extragalactic distance scale that the period–luminosity relation plays a major role. Miss Leavitt's observations in the Magellanic Clouds revealed the existence of a correlation between the periods of Cepheids and their apparent visual magnitudes. But in 1912 the distance of the Magellanic Cloud was unknown; one had to consider that all the Cepheids of the Clouds were at the same distance, thus the difference between their apparent magnitude and their absolute magnitude was taken to be a constant. The calibration of the period–luminosity relation is one of the most interesting chapters in the history of twentieth-century astronomy, and one cannot consider the story finished. The calibration in terms of absolute magnitude requires, firstly, the direct determination of the distances of a few Cepheids. This is a difficult project because none of them are sufficiently near the Sun to use the method of trigonometric parallaxes. The problem can be solved by obtaining observations of Cepheids in a few open clusters with known distances. The luminosity of Cepheids ranges from 300 to 26 000 times the luminosity of the Sun, and their radii range from 14 to 200 solar radii.

W Virginis stars

These stars, members of Population II, are found in globular clusters, in the galactic halo and near the nuclear bulge of the Galaxy. They have variations similar to Cepheids and we think that an analogous mechanism is responsible for their pulsation. But W Virginis stars belong to a stellar population quite different from Cepheids. They were formed at least ten billion years ago, at the very beginning of the history of the Galaxy.

W Virginis stars also have a period–luminosity relation, but it is different from that of Cepheids.

RR Lyrae

RR Lyrae stars are always observed among Population II stars. They are then found near the galactic centre, in the galactic halo and in globular clusters. There is no period–luminosity relation for RR Lyrae stars. This is because they constitute a much more homogeneous class than the Cepheid and W Virginis variables. Since they all have nearly the same mean absolute magnitude and are found nearly everywhere in the Galaxy, RR Lyrae stars are used to study the structure of the Milky Way.

More than 4500 RR Lyrae stars have now been detected. The observations of RR Lyrae stars not in globular clusters has been used to determine the extent of the galactic halo.

The radii of these stars are of the order of 8.3 solar radii. They occupy a very well-defined zone in the Hertzsprung–Russell diagram, especially in the Hertzsprung–Russell diagrams of globular clusters, where they are located in an area relatively poor in stars (implying a very brief evolutionary phase) at the level of the horizontal branch. The observation of RR Lyrae stars in globular clusters, in particular their

number, gives clues to the evolutionary stage of the cluster.

Other types of pulsating variables

We will not discuss in detail the properties of the other pulsating variables, the δ Scuti and the AI Velorum stars. The β Canis Majoris stars, sometimes called β Cephei stars, occupy an unusual place among pulsating variables. Their position on the Hertzsprung–Russell diagram is quite different from the other pulsating variables. They are slightly evolved early-type stars located close to the main sequence.

The mechanism put forth to explain the pulsation of these stars, which are not on the instability strip, cannot be explained here. The interpretation of the observations is still quite controversial and we do not know yet the origin of the pulsations of these stars. In contrast Cepheid variables, β Canis Majoris variables reach their maximum brightness when their radius is a minimum.

Variable giant stars

To the right of the instability strip in the Hertzsprung–Russell diagram, one finds several types of variable stars, which can be divided into two main classes.

In the Hertzsprung–Russell diagram the RV Tauri and the yellow semi-regular variables are located between the instability strip and the region occupied by the red variables. They are of spectral types G and K. Their very irregular light curves show fairly unpredictable amplitudes of up to 3 magnitudes. Their periods, also variable, are usually between thirty and fifty days. This class includes variables with a wide range of properties from the point of view of their spectroscopic variations as well as their photometric variations. The class is very different from that of the rather homogeneous Cepheids. Population I and Population II stars are found in this class, and it is probable that a pulsation phenomenon is responsible for their variations.

The red variable class includes giants and supergiants of spectral types M, R, N and S, independently of their light-curve properties: regular, semi-regular or irregular.

The exact classification in one or the other subclass is often difficult. The spectroscopic and photometric variations here are also due to a pulsation of the outer layers of the star. Stars with the largest amplitude belong to a class called the Mira variables, the prototype being the star Mira (o Ceti) in the constellation Cetus (the Whale).

Mira variables have periods ranging from thirty days to more than one thousand days, their amplitudes being from one magnitude up to 10 or 11 magnitudes in the visible. Their light curves are not very regular, that is, they do not repeat exactly from cycle to cycle.

Mira varies in eleven months from a magnitude of 2.5 (once it reached a magnitude of 1.2) to a magnitude of 10. During that time its spectral type changes from M6 to M9. It is one of the very few stars whose diameter has been measured by interferometry. For its estimated distance of 70 parsecs, it has a mean radius equivalent to 420 solar radii. During one cycle this radius varies by about 20 per cent.

Stars of this subclass can be found among Population I as well as Population II stars. Their spectra are rich, in that molecular absorption bands as well as emission lines are observed. A few spectra show lines of short-lived isotopes of technetium. This implies that these isotopes must have been formed in the atmosphere of the star, and this poses interesting questions concerning the structure of these unusual variables.

It is believed that their variation in brightness is caused by pulsational instabilities of the same types as those encountered in Cepheids. But here molecular hydrogen plays the role of helium. These stars have very extended atmospheres with important convective motions and mass loss. For this reason they are sometimes considered progenitors of planetary nebulae. Mira variables, constitute a useful tool for investigations of galactic structure.

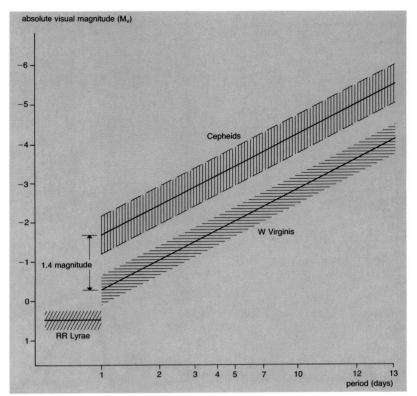

The period–luminosity relation. This empirical relation, discovered from observations, is now interpreted in terms of stages of evolution of stars of different masses and of different chemical composition. It is a very powerful tool for estimating distances. From the known period, the relation allows us to determine the absolute visual magnitude of the star. One must of course know if the star is a Cepheid, a RR Lyrae or a W Virginis variable. The absolute visual magnitude can then be compared to the apparent visual magnitude to obtain the distance to the variable.

Supergiant stars

It is generally acknowledged that all of the brightest stars (i.e. the ones whose luminosity is at least ten thousand the solar luminosity) are pulsating variables. They are located in the upper part of the Hertzsprung–Russell diagram, and they show irregular variations of their brightness, of their spectrum and of their radial velocity. This group of supergiants includes all sorts of stars, such as Cepheids and Mira variables; which have already been discussed, and Wolf–Rayet stars (which are discussed later in this Atlas – page 276) and many more. The amplitude of their light curve is only a few tenths of a magnitude. The range of their periods is very large, from a few hours to dozens of days. The origin of the instability of these objects raises many unanswerable questions. For example, what is the relation between the mechanisms giving rise to the pulsation and the strong rate of mass loss by stellar wind? The current interpretation of the oscillations suggests the presence of non-radial pulsation in the outer layers of the star.

The study of variable stars is one of the keys to understanding the structure of stars. The analysis of pulsations gives us much more information

The number of observed variables. The number of known variables is always changing because of new discoveries. The numbers presented here are taken from the *General Catalogue of Variable Stars* and its supplements published up to 1976. These numbers do not represent the true numbers of variables in the Galaxy because their detection is strongly influenced by interstellar absorption.

variable stars	number observed
Cepheids	670
W Virginis	110
RR Lyrae	5920
δ Scuti – AI Velorum	160
RV Tauri	110
Mira	5200
red variables	5370
β Canis Majoris	50
cataclysmic variables (novae)	380
eruptive variables with nebulosity	1980

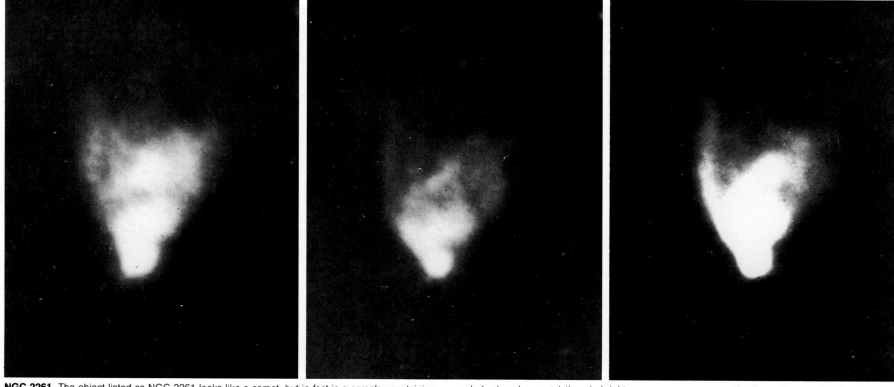

NGC 2261. The object listed as NGC 2261 looks like a comet, but in fact is a complex containing one star and a cloud of gas and dust. The star R Monocerotis located at the tip of the nebula is a T Tauri star, a young variable star. The cloud from which the star was born still surrounds it and now reflects its radiation. The peculiar shape of this nebula most probably comes from irregularities in the density of the interstellar medium in its vicinity.

R Monocerotis, like most T Tauri stars, shows a characteristic radiation, stellar wind and appreciable spectral variations. The amplitude of its irregular variations reaches 4 magnitudes. The nebula also shows variations in brightness, as we can see on the three photographs (Jordan, 1908; Curtis, 1913; Hubble, 1916) but they are not in phase with the variations of the star. This implies that the radiation from the nebula is not just due to reflected light and much more complicated interactions must be taking place. Recent radio observations have revealed the presence of a large molecular cloud associated with this object. This cloud could have an influence on the shape of the nebula. (Yerkes Observatory)

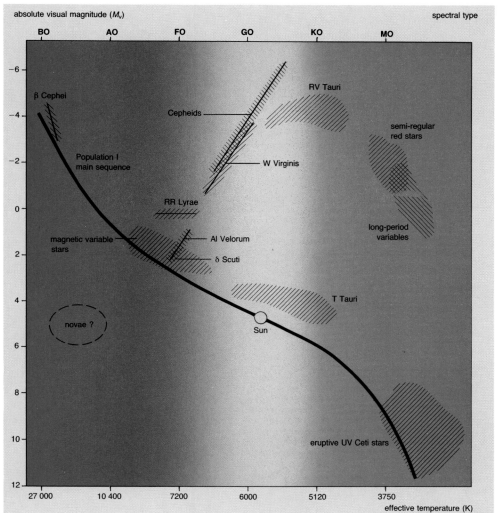

Main properties of pulsating variables. The Hertzsprung–Russell diagram on the left shows the Population I theoretical main sequence, where the Sun is located, and the approximate positions of various classes of variable stars. Several areas of the diagram are occupied by variables. This shows that different kinds of stars can be variable.

The table below summarises the principal properties of each class.

about the inner parts of a star than all of the observations of a static star.

It is important to note that the types of variations we observe probably correspond to a small sample of all the possible instabilities. The best proof of this is our own Sun, which simultaneously exhibits different types of variations. It is, however, often difficult to apply to other stars the fine analysis that can be done on the Sun.

Michèle GERBALDI

class	period (days)	amplitude of the light curve (in magnitude)	spectral type	absolute visual magnitude	population	$M/M_\odot$
δ Scuti	from 0.02 to 0.2	from 0.01 to 0.3	from A2 to F2	from +2 to +3	I	2 } similar
Al Velorum	from 0.02 to 0.2	from 0.03 to 0.8	from A2 to F2 (?)		I	0.5 } properties
W Virginis	from 2 to 45	from 0 to 5	from F2 to G6	from 0 to −3	II	
Cepheids	from 1 to 50	from 0.1 to 2	from F6 to K2	from −6 to −0.5	I	from 3.7 to 14
RR Lyrae	from 0.1 to 1	from 0.3 to 2	from A2 to F2	from 0.0 to +1.0	II	1
β Canis Majoris	from 0.1 to 0.3	from 0.02 to 0.25	from B0 to B3	from −5 to −3	I	
RV Tauri	from 20 to 150	from 3 to 4	from F5 to K5	−3	II	
red variables	from 30 to 1000	from 0.5 to 4.5	M, R, N, S	from −2 to +1	II and I	

White dwarfs

Until the beginning of the twentieth century, physicists had not considered the possibility of physical states in which matter would be much denser than the densities observed in the Solar System of a few grams per cubic centimetre. During the 1920s the progress of quantum mechanics led to a better understanding of matter; atoms were found to be composed of electrons bound to nuclei by electrostatic forces and continuously in motion, producing pressure which prevents further contraction of matter. More precisely, the exclusion principle led to the existence of elementary cells, which can contain no more than two particles. When all the cells are occupied by electrons, matter is said to be degenerate and its density may reach one tonne per cubic centimetre. Degenerate matter is much more packed than ordinary matter for which most of the elementary cells are empty.

The self-gravitational forces of a massive celestial body can compress its matter to a state of electronic degeneracy. This is what has happened in stars called white dwarfs. The interior of a white dwarf is 'cold' (even though the temperature may actually reach one million degrees) in the sense that to keep the star in equilibrium the self-gravitational forces are compensated not by thermal motions as is found in main sequence stars, but by the pressure exerted by degenerate electrons. The interior of a white dwarf is, then, not in a gaseous state. The interior of the star is like a gigantic crystal slowly cooling off.

Subrahmanyan Chandrasekhar, who developed the theory of white dwarfs, predicted that their masses cannot be greater than 1.4 solar masses. Above this value electrons would have to have velocities close to the velocity of light and would be unable to exert a pressure high enough to balance the gravitational forces.

White dwarfs were observed as early as 1910 although their origin was not understood. Considerable information can be gained from the position of stars on a diagram linking their colour (or their surface temperature) and their apparent magnitude. With the discovery of 40 Eridani B astronomers realised that at least one star had a high surface temperature (17 000 K, whereas the surface of the sun is at 6000 K) and an abnormally low intrinsic luminosity. This could be explained only if the radius of 40 Eridani B were very small, comparable to the radius of the Earth. The existence of such objects was rapidly confirmed – two other examples were discovered in 1917. The two newly discovered white dwarfs were Sirius B (the companion of Sirius A, the brightest star in the sky) and van Maanen's star. This latter object has a visual magnitude of 12.4, a mass of 0.68 solar mass and a radius 78 times smaller than that of the Sun. Its average density is, therefore, 300 kilograms per cubic centimetre. The list of known white dwarfs has been increasing ever since, and today a few hundred have been positively identified. Several thousand more are serious candidates.

Even though the central temperature of a white dwarf is less than one million degrees (compared to ten million degrees for the Sun), its atmosphere is, as a rule, hotter than that of an ordinary star. The spectrum of a white dwarf differs from the spectrum of main sequence stars, in that the spectrum of a white dwarf shows very broad spectral lines. Several effects may be responsible for this broadening: the atmospheric pressure, a magnetic field (the magnetic field of a white dwarf may reach one billion times the field of the Sun), or rapid rotation (the rotational

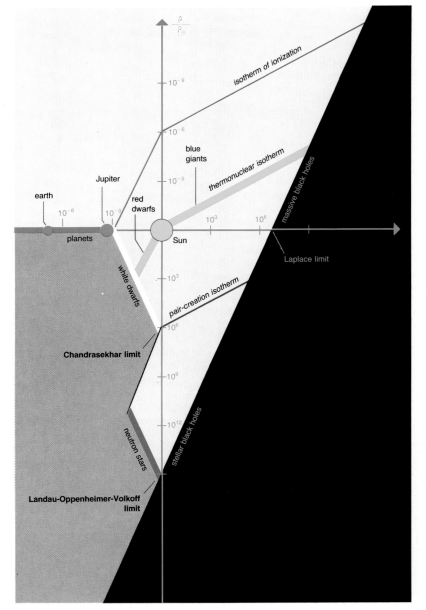

Sirius and its dwarf companion. Sirius is, to the naked eye, the brightest star in the sky. Situated in the constellation Canis Major, it is known as the Dog Star. It is not visible from the northern hemisphere throughout the year. In Ancient Egypt the first appearance of Sirius served to forecast the flood of the Nile, the summer solstice and the warm days ahead. This is the origin of the expression 'dog days'.

Situated at 2.6 parsecs from the Sun, Sirius is a blue star of spectral type A0, twice as massive and twice as big as the Sun. From a study of the perturbations of its proper motion, Friedrich Bessel deduced in 1834 that Sirius was a member of a binary system. Its companion, Sirius B, was, however, not seen until 1862, by Alvan Graham Clark. Sirius B is ten thousand times fainter than Sirius A, and, although having a mass comparable to the mass of the Sun, its diameter is barely five times the Earth's diameter. Sirius B is indeed a white dwarf. Sirius B is the small spot to the right of Sirius A on the picture below. The image of Sirius A is considerably enlarged by the diffusion of light in the photographic emulsion. (Lick Observatory)

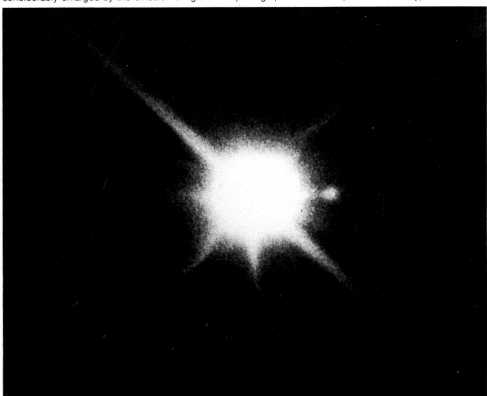

The mass–density diagram of celestial bodies. A celestial object is normally in a state of equilibrium because the forces trying to expand it are balanced by the forces compressing it. Expansive forces can be due to the exclusion principle from quantum mechanics (which keeps the electron or neutron density below a certain critical value) or due to pressure forces when the central temperature of the body is high (for example, the radiation pressure from nuclear reactions). Compressive forces are due in some cases to the electrostatic attraction between electrons and protons, which make up atoms and molecules, or, more generally, due to gravitational forces, which tend to compress objects. Each type of celestial body can be characterised by a relation between the mass and the average density, depending on the forces playing a role in its structure. The mass–density diagram is thus divided into zones, which are defined by the positions of different types of objects. The mass, and average densities are expressed in Solar units.

In the cream area the objects are hot, in the sense that thermal expansion forces dominate. Inside this area the coloured lines delimit three important ranges of temperature. The ionisation isotherm corresponds to a temperature of 10^5 K; it characterises protostars. The thermonuclear isotherm corresponds to 10^7 K, a temperature at which the nuclear fusion of hydrogen into helium starts. This is the source of energy in stellar cores. The pair-creation isotherm corresponds to 10^9 K, above this temperature instabilities become predominant (notably the electron–positron pair creation). Most stars (including the Sun) are located on the thermonuclear isotherm. They are in the main sequence phase. For cold objects the main expansion forces are not thermal in nature but quantum mechanical. The grey area of the diagram is a forbidden area, corresponding to a violation of the exclusion principle. Cold objects are thus located at the border of the cream and grey areas. Below 10^{-3} solar masses (about the mass of Jupiter) the main force of compression is due to electrostatic attraction. Planets are in a state of equilibrium where the density is independent of the mass and is of the same order of magnitude as the density of ordinary matter (1 g/cm³). Above 10^{-3} solar masses, gravitation becomes the main force and gives rise to states much denser than those encountered in the Solar System. White dwarfs, for example, have densities reaching one tonne per cubic centimetre. In white dwarfs the pressure resisting the gravitational contraction is due to electrons which have a peculiar energy distribution imposed by the exclusion principle. This is the electron degeneracy phenomenon. White dwarfs cannot have a mass much larger than 1.4 solar masses, because above this value electrons become relativistic, and their pressure becomes much less, so much so that above the Chandrasekhar limit there is no possible equilibrium between gravitation and degenerate electrons. Another important kind of compact star is the neutron star. In this class, gravity is so strong that it pushes electrons inside atomic nuclei, neutralising protons to produce neutrons. The states of matter that are obtained in this process are even denser, reaching densities comparable to the density of the atomic nucleus: 10^{15} g/cm³. Like electrons, neutrons obey an exclusion principle which allows them to resist the gravitational contraction. As with white dwarfs, however, there is an upper mass limit for neutron stars: the Landau–Oppenheimer–Volkoff limit, corresponding to about 3 solar masses. Above this value degenerate neutrons begin to turn relativistic.

The black area of the diagram is also forbidden, because here gravitation dominates all the possible expansion forces, thermal or quantum forces. No equilibrium is possible in this zone. The border line is occupied by black holes. This line cuts the mass axis at the Laplace limit, which corresponds to black holes with properties predicted by Laplace in 1796: 10^7 solar masses and the same density as the Sun. It cuts the density axis at 10^{15} g/cm³ corresponding to the upper limit for the stability of neutron stars.

During its evolution, a star moves down towards the left of the diagram, towards higher densities and lower masses. A low mass star such as the Sun will, at the end of its main sequence phase (when its nuclear fuel will be exhausted), become a white dwarf. A more massive star leaves the thermal isotherm to enter the pair-creation isotherm. The instabilities trigger its explosion: it becomes a supernova. During this explosion the stellar core collapses. If the core mass is smaller than the Landau–Oppenheimer–Volkoff limit, then the quantum pressure of the degenerate neutrons will stop the gravitational collapse: the core will then become a neutron star. If the mass of the core is larger, nothing can stop the effect of gravitation – the final state will then be a black hole.

period of a white dwarf could be ten seconds while the Sun has a period close to one month). Like main sequence stars, which have stellar spectra classified by the letters B, A, F and G, according to their surface temperature, white dwarf spectra are classified into DB, DA, DF and DG (D stands for dwarf), corresponding to temperatures ranging from 100 000 to 4000 K. Cooler white dwarfs exist, but are not observable. They are the black dwarfs. There is also a spectral type called DC, which represents white dwarfs with a continuum without spectral lines. Such spectra are not found among main sequence stars.

The fundamental technique used for the detection of white dwarfs is based on the analysis of their spectra. Because these stars are of low intrinsic luminosity, only those that are relatively close to the Solar System can be observed. Such nearby stars have a high proper motion. Once a sample of high proper motion stars is selected, a study of their spectra yields their surface temperature and allows us to place them on the colour–magnitude diagram. This can establish without ambiguity if they are in the white dwarf region.

White dwarfs are numerous – it is estimated that they amount to 10 per cent of the stars in our Galaxy – because they represent the final evolutionary phase of low mass stars. When a main sequence star starts to exhaust its nuclear fuel, its core contracts under the effect of gravitation. This contraction is generally stopped by the pressure of degenerate electrons. The star then begins a long cooling process, leading it from the phase of hot white dwarf (surface temperature of 10 000 K) to the phase of a cold, unobservable, black dwarf. Indeed, the Sun, which is now in the middle of its main sequence phase, will, in about five billion years time, reach its old age in the form of a white dwarf. A low mass star on its way to becoming a white dwarf can pass through the intermediate stage of being a planetary nebula, one of the most spectacular celestial events. A planetary nebula is a star 'at the end of its career', very hot and very bright. It expels its atmosphere continuously, thus forming a gaseous nebula in the form of a ring around it. At the end of 30 000 years the star has cooled and shrunken to reach planetary size. Planetary nebulae account for approximately 20 per cent of all white dwarfs.

Jean-Pierre LUMINET

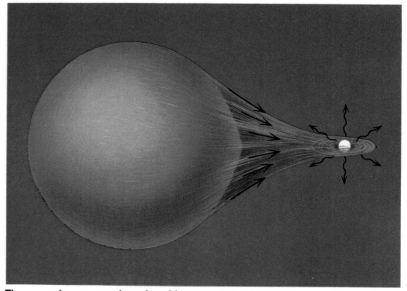

The nova phenomenon in a close binary system. Among the known white dwarfs some belong to binary systems, one component of which is a normal star. When the two components are very near each other (such systems are called close binaries) mass transfer can take place. Shown here is a normal star which has evolved to become a red giant. The atmosphere of the normal star, made up mainly of hydrogen, is sucked up by the strong gravity of the white dwarf. Hydrogen falls on the hot surface of the white dwarf. As the hydrogen accumulates, the pressure and the temperature increase and soon this gas is so hot that nuclear reactions involving hydrogen start just below the white dwarf surface. Suddenly its luminosity increases and a nova is born. Dwarf novae, which include U Geminorum are close relatives of ordinary novae, but their increase in brightness is much smaller; it is believed that the white dwarf captures the gas from the companion in a more irregular fashion, thus giving rise to sporadic increases of brightness.

Nova Aquilae 1918. The most spectacular nova observed during the last hundred years was Nova Aquilae 1918 (a nova is designated by the name of the constellation within which it is seen, followed by the year of its appearance.) On the top photograph, taken before its explosion, the future nova appears as a faint star of visual magnitude 10.6. The lower photograph shows Nova Aquilae 1918 near its climax: its apparent magnitude was then −1.1 (its absolute magnitude was then −8.4). It was for a few days the brightest star in the sky after Sirius.

The brightness of Nova Aquilae 1918 increased by nearly 12 magnitudes – corresponding to a factor of 60 000 – but within a few days it declined rapidly while showing periodic fluctuations. Six months later its apparent visual magnitude was 6; today it is 10.9. (Yerkes Observatory)

A precursor of a white dwarf: the Ring Nebula in Lyra. The Ring Nebula M57 (NGC 6720) located in the constellation Lyra, is one of the most beautiful objects to observe in the sky with a small telescope. It was discovered in 1779 by Antoine Darguier, who, in describing it, compared it to a planet. It was probably for this reason that William Herschel named this class of objects 'planetary nebulae'. It is somewhat unfortunate, because they are actually stars which have reached the end of their main sequence evolution and are shedding their atmospheres. The surface of the blue central star is very hot. Its spectrum tells us that the star is on the verge of starting the slow cooling process leading to a white dwarf.

The gaseous ring surrounding the star is slightly elliptical, indicating that we are not looking at it 'face on', but at a small angle. Moreover, the ring expands at a speed of 19 kilometres per second, and since its present radius is 20 000 astronomical units, one can deduce that the expansion of the nebula started about 6000 years ago.

Note the colour differences across the ring, from blue-green in the inner parts to red on the outer edge. (© Association of Universities for Research in Astronomy, Inc., Kitt Peak National Observatory)

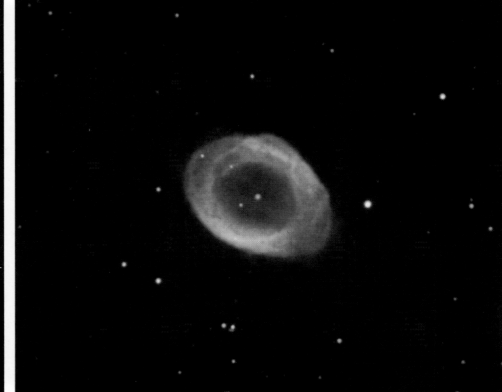

Novae

Novae are stars that show a sudden large increase in brightness, their name coming from the Latin for new star 'nova stella'. Afterwards their luminosity decreases slowly. As in the case of supernovae, novae are eruptive or explosive stars. Novae are, however, easily distinguished from supernovae: the amount of energy they emit is much less. The most powerful ones, like Nova Cygni 1975, shine one thousand times less brightly than supernovae. In contrast to the supernova class, we group under novae quite different types of stars. We classify as novae stars like Nova Cygni, which gain more than ten magnitudes, as well as dwarf novae (also called U Geminorum or Z Camelopardalis stars, after the first stars of this type identified), which increase their brightness only a few magnitudes (two or three). Dwarf novae have recurrent explosions every few months, showing that the event does not drastically affect the star as in the case of a supernova. Novae like Nova Cygni have exploded only once in known history, but are thought to recur about every ten thousand to one hundred thousand years. Recurrent novae, on the other hand, which are somewhat less energetic and intermediate between the two extremes described above, experience explosions every ten to one hundred years. We must note that a few dozen novae are observed per year per galaxy, while a supernova occurs only once every thirty to fifty years in a galaxy. Even if novae are much more numerous than supernovae the total amount of energy liberated by novae is much less; a supernova explosion produces, on average, one million times more energy than a nova explosion.

In the case of a nova explosion only the outer layers of the star seem to be affected. The amount of matter ejected is of the order of 10^{-4} solar mass, with velocities between a few hundred and a few thousand kilometres per second.

Even though novae include quite different objects, they all have one important common property: they are all members of close binary systems. This fact must be responsible for the mechanism leading to the explosion.

The binary system most likely to give birth to a nova consists of a white dwarf and a cool star. When the outer layers of the cool star pass a certain radius, called the Roche limit, they are attracted towards the white dwarf. This phenomenon leads to the formation of an accretion disk around the white dwarf. This disk can be observed in the ultraviolet and X-ray regions of the spectrum because of its high temperature. For dwarf novae, such as the U Geminorum and Z Camelopardalis, the eruption is due to the formation of hot regions on the accretion disk. The transfer of matter from one star to the other does not take place in a steady fashion! The matter being transferred is heated up because it reaches supersonic velocities. Observations of the chemical composition of nova ejecta show that they have a high abundance of elements such as carbon, nitrogen and oxygen (the overabundance relative to the Sun is, depending on the individual case, of the order of 10 to 50). In the case of 'ordinary' novae, the explosion is attributed to nuclear reactions taking place on the surface of the white dwarf buried under the matter from its companion.

The explosion mechanism can in general terms be described in the following way: matter from the cool star, made up mostly of hydrogen and helium, falls on the outer layers of the white dwarf. When sufficient matter has accumulated on the surface the base of the layer becomes compressed and heats up. This triggers reactions of the carbon-nitrogen-oxygen cycle between itself and the outer layers of the white dwarf, which are already rich in these elements. Theoretical models that attempt to predict the effects of these reactions show that the temperature of the outer zones can surpass 100 million degrees. At such temperatures explosive nuclear reactions involving hydrogen, helium, carbon, nitrogen and oxygen are triggered, liberating in a short time – of the order of one hour – sufficient energy to produce an explosion of the outer layers. This leads to the ejection of a significant amount of matter into space.

These reactions also modify the isotopic ratios of carbon, nitrogen and oxygen. The theory predicts that the ratios between the abundances of carbon 13 and carbon 12 and between the abundances of nitrogen 15 and nitrogen 14 increase. Observations appear to confirm this prediction. Other nuclear reactions could also take place during the nova explosion: beryllium 7, which decays into lithium 7, could be synthesised by a reaction of helium 3 with helium 4. Novae do not play as important a role as supernovae in the nucleosynthesis of the matter from which we originate. They could, however, be responsible for the origin of lithium 7 and nitrogen 15 isotopes, whose formation leads to their explosion.

Jean AUDOUZE

Nova Cygni 1975. During the night of 29 August 1975, a new bright star appeared in the constellation of Cygnus (its discovery was announced by a Japanese amateur astronomer). It reached its maximum brightness on 31 August, its apparent magnitude then being 1.8 (top photograph). At this time it was one million times more luminous than the Sun.

Prediscovery photographs do not show the object, which must have been 20th magnitude or fainter; its brightness must, therefore, have increased by a factor of at least twenty million. This puts Nova Cygni 1975 among the most spectacular novae ever observed. Three months after maximum, the nova had dimmed to 11th magnitude (lower photograph). The light curve of Nova Cygni 1975 is shown below. (Lick Observatory)

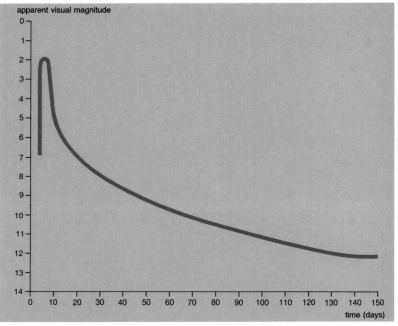

Light curve of the dwarf nova SS Cygni in 1974 (below). This star is a typical dwarf nova: every fifty days or so its brightness increases by four magnitudes (from 12 to 8). This corresponds to its original luminosity multiplied by a factor of forty at maximum light. In the case of ordinary novae the luminosity increases much more (the factor sometimes reaching one million) but the time interval between two successive eruptions could reach a century or more. Dwarf novae liberate relatively little energy because of the small amount of matter accreted. Ordinary novae, having a much longer interval between explosions, can accumulate a much larger amount of external matter.

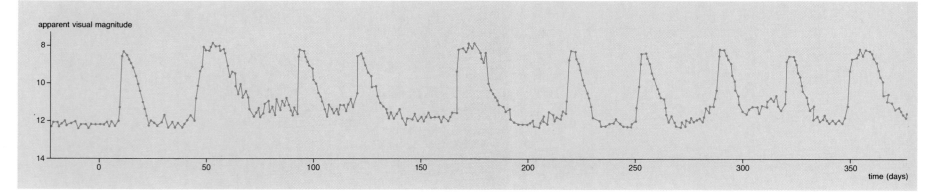

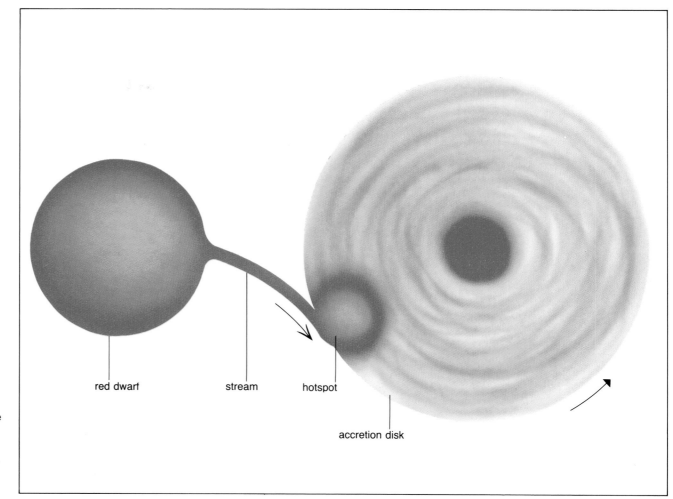

An accretion disk surrounding a nova. The energy liberated by a nova originates from the material drawn away from its companion. This material never falls directly onto the nova; it first forms an 'accretion disk'. A disk is formed because of the combined effects of the gravitational forces produced by the star, the orbital motion of the system and the rotation of the two stars. The velocities of attraction of the material are very high (they are supersonic). The accretion disk is therefore heated to temperatures greater than 10 000 K. The accretion disk emits in the ultraviolet and (sometimes) in X-rays. The place where the stream of transferred material strikes the edge of the disk is sometimes visible as a luminous hot spot. Dwarf novae draw their energy from the radiation flux of the accretion disk.

red dwarf   stream   hotspot

accretion disk

Nebula in expansion around the nova GK Persei 1901. This nebula was ejected by a fairly powerful nova. It is remarkable for its size (it is the largest of all the nebulae observed around novae), for its temperature of the order of 30 000 K and for its great inhomogeneity. These last two properties are explained by the interaction between the nebula in expansion and the interstellar gas. The nebula is rich in nitrogen, which must have been produced by the nuclear reactions responsible for the explosion of GK Persei. (Hale Observatories)

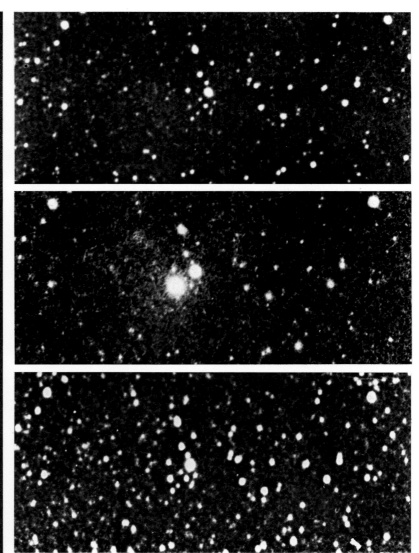

Three photographs of Nova Lacertae 1911 showing its evolution. These three pictures were taken on 7 August 1907, on 31 December 1910, and on 29 September 1911, respectively, by the astronomer Barnard. On the first photograph, taken before the explosion, the star is of magnitude 13. On the 1910 photograph, following the explosion, its magnitude is 7, and, finally, in 1911, the magnitude is 11. It is from old photographs of this type that we can evaluate the mean number of nova explosions taking place per year per galaxy. This number is of the order of twenty to forty. (Yerkes Observatory)

283

Supernovae

A supernova is a phenomenal stellar explosion involving most of the material from a star. It marks the end of the evolution of certain stars. Such events are thought to take place in our Galaxy about once in thirty years. Most supernovae in our Galaxy are undetected because interstellar dust makes them nearly invisible.

Chinese chronicles report the observation of a supernovae in July 1054 in the constellation Taurus. The next two recorded supernovae were reported by Johannes Kepler in 1572 and Tycho Brahe in 1604. Since then the few hundred seen have all been in nearby galaxies. Supernovae surveys are being carried out at several observatories equipped with wide-field Schmidt telescopes. At the time of the explosion the star – which is often previously undetected – increases in brightness by at least 15 magnitudes.

The evolution of a star depends on its mass and its chemical composition. There are two types; Type I supernovae result from the explosion of old stars of relatively low mass that are poor in heavy elements (Population II stars): Type II are the explosions of massive young stars, rich in heavy elements (Population I stars). At maximum light Type I supernovae are nearly three times more luminous than Type II. The luminosity decreases by 3 or 4 magnitud during the first few days after it has reached its maximum. The luminosity then decreases more or less exponentially for several months.

The energy released in a few moments by the explosion is of the order of 10^{44} joules. This fantastic energy corresponds to the total radiated by the Sun in 9 billon years (the Sun is about 4.5 billion years old). Matter ejected by the explosion, is hydrogen poor in the case of Type I supernovae and hydrogen rich for Type II. Depending on the type, they eject between 1 and 10 solar masses of gas. This can correspond to the total mass of the pre-supernova, nothing being left after the explosion. We know, however, since the discovery of pulsars (rapidly rotating neutron stars) in 1968, that after the explosion an extremely dense object can be left. This object, which is the core of the star, is made up exclusively of neutrons almost in contact with each other. These neutron stars can be considered to be like giant atomic nuclei because their proper-

ties depend directly on the interaction of nuclear forces, assisted by gravitational forces.

The search for mechanisms leading to a supernova explosion is very active today. In most theories the energy is primarily of nuclear origin. The theories must explain how a dense core – the pulsar – is formed, and how the nuclear energy is transformed into kinetic energy to eject the outer layers of the star into space.

The mechanism proposed by Fred Hoyle and William A. Fowler thirty years ago to explain the source of energy is still attractive today. It is the photodisintegration of iron, the end of the chain of nuclear reactions during the lifetime of stars of roughly 10 solar masses. The pre-supernova star has a shell structure like an onion. From the surface layer of hydrogen, as we descend we find layers made up of elements having increasing atomic masses. These shells are the products of the different nucleosynthesis phases during the life of the star. The reactions giving birth to the heavier elements and ordered according to increasing temperatures. These temperature increases took place in alternation with successive gravitational contractions. The centre of the star is then composed of a mixture of iron and of nuclei having atomic masses between 50 and 60. Those elements have the highest binding nuclear energy (approximately 8.7 megaelectron volts (MeV) per nucleon). When the central temperature reaches five billion degrees kelvin, matter and radiation are in equilibrium. Gamma-ray photons (γ) have sufficient energy to disintegrate nuclei and produce neutrons (n), in reactions such as:

$$\gamma + {}^{56}\text{Ni} \rightarrow 14\,{}^{4}\text{He}$$
$$\gamma + {}^{54}\text{Fe} \rightarrow 13\,{}^{4}\text{He} + 2n$$
$$\gamma + {}^{56}\text{Fe} \rightarrow 13\,{}^{4}\text{He} + 4n, \text{ etc.}$$

Each of these photodisintegrations robs the gas of some 100 MeV of energy. These extremely endothermic processes break the thermal and hydrostatic equilibrium in the stellar centre and the star collapses. The gravitational energy liberated increases the temperature even more, until the alpha particles are also photodisintegrated. At the moment all of the products of the nucleosynthesis process have been anni-

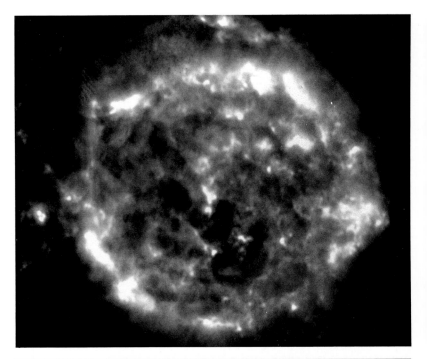

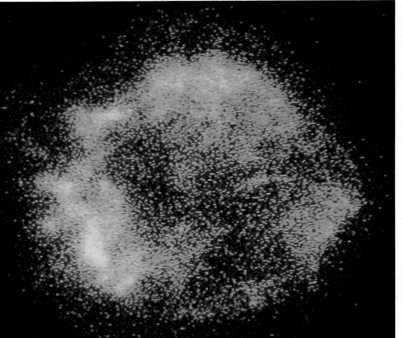

Cassiopeia A. The supernova remnant Cassiopeia A (Cas A; 3C 461) is very young. The explosion, which went unnoticed at the time, took place 300 years ago. This object emits gamma rays, X-rays, radio waves and visible light.

Cassiopeia A is, after the Sun, the strongest radio source in the sky. The top figure shows a map of its radio emission at a frequency of 5 gigahertz (GHz), produced by the very hot gas in the nebula. One can see that the strongest emission comes from the periphery of the remnant in the region of interaction between the expanding shell and the interstellar matter at rest. The radial velocity of the visible filaments in the outer parts is close to 9000 kilometres per second. Cassiopeia A, unlike the Vela and Crab remnants, does not contain a pulsar. An X-ray image (below) obtained with the HEAO-2 satellite does not in fact show any central condensation. (Mullard Radio Astronomy Observatory and Harvard Smithsonian Center for Astrophysics)

The Veil Nebula in Cygnus (NGC 6992). The Veil Nebula represents an advanced stage of the evolution of a supernova remnant which is now dissolving into the interstellar medium. The age of this remnant is very uncertain, but is probably more than 30 000 years. The velocity of expansion is now only 120 kilometres per second. The dense shell formed by the expansion of the hot gas bubble in the interstellar space is now cooling. This shell contains a large amount of dust and can bulldoze its way through space. The gas of this supernova remnant of Type II is very poor in hydrogen, but oxygen and neon, which were produced by nucleosynthesis in the pre-supernova star, are abundant. Eventually the dust shell will break up into small cold clouds leaving behind a bubble of thin gas which will have cooled to 500 000 K. (Hale Observatories)

The Crab Nebula (M1, NGC 1952). The supernova that produced this nebula was most probably of Type I. This remnant is also a strong radio source known as Taurus A. A pulsar in rapid rotation (its period is only 33 milliseconds) is associated with this nebula. The structure of the remnant is dominated by the nebula, whose whitish emission is due to synchrotron radiation. This type of radiation is produced by electrons spiralling along the magnetic field lines originating from the surface of the pulsar. The orange filaments, whose mass is close to 1 solar mass, have a velocity of expansion of 1500 kilometres per second. They are heated and ionised by radiation from the rest of the nebula, which is actually of much lower mass. These filaments are the remains of the explosion of the envelope. (D. F. Malin, Anglo-Australian Telescope Board)

A possible structure for a twenty solar mass supernova. The photodisintegration of iron in the centre of a massive star can lead to the explosion of its outer envelope and the implosion of its core to produce a neutron star or a pulsar. At that moment the star would have a stratified structure. The chemical composition of each successive layer is the result of nucleosynthesis taking place at temperatures (T) and densities (ϱ) increasing as we go to deeper and deeper layers. Each layer is annotated with the chemical symbols representing the elements produced in the layer. Following the supernova explosion, these elements enrich the interstellar medium where new generations of stars will eventually be born.

hilated, the gas now being made up of free neutrons, protons and electrons. But the electrons, being fermions, cannot be compressed strongly without experiencing a large increase in their kinetic energy. The energy of the electrons very rapidly becomes greater than the energy required to transform a proton into a neutron. At that point the electrons are absorbed by the protons. Deprived of one of its components, which was contributing a significant fraction of its pressure, the stellar core collapses at an accelerated rate. The collapse finally stops when the repulsive nuclear force between neutrons comes into action, when the central density is comparable to nuclear densities. The distances between neutrons is then of the order of one fermi (10^{-13} centimetres). The star is now a neutron star. From the onset of the collapse, only a few minutes are required to reach this state. When the core collapse starts, the outer layers of the star, where some nuclear reactions are taking place, fall downwards. The gases become suddenly compressed and overheated. Reaction rates increase tremendously, leading to instabilities and, ultimately, to the explosion of these outer layers.

Other supernova models involve the explosive burning of carbon in stars between 4 and 8 solar masses. In these models no pulsar is formed by the core before the explosion of the envelope. The energy necessary to blow off the envelope can be extracted from the pulsar rotational energy. The transfer of energy can be accomplished by the strong magnetic fields on its surface.

During the neutronisation of the core, some neutrons can be rapidly captured by the nuclei in the envelope. This mechanism is responsible for the production of elements of very large atomic masses, such as uranium.

Neutrinos are released during neutronisation in enormous numbers (approximately 19^{58} of them) in the high temperatures in the supernova (10^{10} K). The first neutrinos from outside the Solar System were detected in a pulse lasting a few seconds on 23 February 1978 when a supernova occurred in the Large Megellanic Cloud, 170 000 light years from Earth. The supernova, SN 1978A, was discovered in a routine photograph by an astronomer at Las Campanas Observatory in Chile. SN 1978A reached magnitude 2.8. It is the brightest supernova for 383 years and the only one whose progenitor star has been identified: a massive star known as Sanduleak-69 202.

The flux of neutrinos from the collapsing core of Sk-69 202 implies that 3×10^{46} joules was carried off by the neutrinos. This is equivalent to the annihilation of one-tenth of the mass of the Sun. For a second or so the luminosity of SN 1978A in neutrinos equalled the luminosity of the Universe.

Amongst the heavy elements produced in the explosion was 0.07 solar masses of ^{56}N. This unstable isotope decays to ^{56}Co to ^{56}Fe, with a mean life of 114 days, producing gamma-radiation, Gamma-rays in the spectral line at 847 KeV from the decay of cobalt in SN 1978A were detected by the Solar Maximum Mission, modulated on and off as the Earth interrupted the direct line of sight to the Large Magellanic Cloud. At the end of 1987 the light from SN 1978A declined exponentially with a time scale of 114 days, powered by the radioactivity. These two observations were direct confirmation that supernovae create the chemical elements.

Supernovae explosions scatter a large quantity of nucleosynthesis products. These have a strong influence on the chemical evolution of galaxies. Long after the explosion, a supernova still reveals itself by its effects on the interstellar medium. Young supernova remnants, such as the 300-year-old Cassiopeia A, appear as vast bubble emitting radiation throughout the spectrum, from radio waves to gamma-rays. The envelope expands at a speed close to 10 000 kilometres per second. As it does so, it pushes the interstellar gas and slows down. Within the remnant, the still-hot gas has a temperature near one million kelvin. After a few hundred thousand years the shell cools suddenly and the remnant dissolves into the surrounding medium. The Veil Nebula in Cygnus is a supernova remnant in this stage.

Supernovae are one of the most important contributors to galactic matter. They not only transmit to the interstellar medium kinetic and thermal energy but, most importantly, they also enrich it with heavy elements from nucleosynthesis. Since nowadays it is generally believed that supernovae trigger formation of new stars, interest in them has been enhanced for astronomers studying stellar evolution or the interstellar medium. The violent shock waves that they liberate could be the catalysts responsible for the formation of stars on a galactic scale.

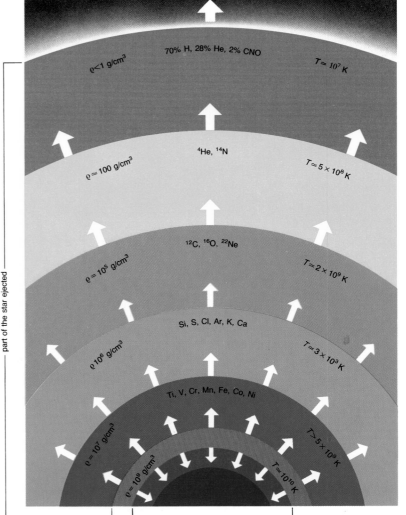

70% H, 28% He, 2% CNO — $\varrho < 1$ g/cm³ — $T \simeq 10^7$ K

^{4}He, ^{14}N — $\varrho \simeq 100$ g/cm³ — $T \simeq 5 \times 10^8$ K

^{12}C, ^{16}O, ^{22}Ne — $\varrho \simeq 10^5$ g/cm³ — $T \simeq 2 \times 10^9$ K

Si, S, Cl, Ar, K, Ca — $\varrho \simeq 10^6$ g/cm³ — $T \simeq 3 \times 10^9$ K

Ti, V, Cr, Mn, Fe, Co, Ni — $\varrho \simeq 10^7$ g/cm³ — $T \simeq 5 \times 10^9$ K

$\varrho \simeq 10^9$ g/cm³ — $T \simeq 10^{10}$ K

part of the star ejected

remaining core: the neutron star

Supernova 1987A. These two photographs show the same star field in the Large Magellanic Cloud, near the Tarantula Nebula (30 Doradus), before and after the explosion. The star Sanduleak-69 202 is marked by an arrow. (D. Malin, Anglo-Australian Telescope Board)

The supernova in the Large Magellanic Cloud: SN 1987A

We had to wait for the night of 23–24 February 1987 to observe a nearby supernova from the beginning. This was quite visible to the naked eye in the Southern hemisphere. This supernova, called 1987 – the first of that year – exploded in the Large Magellanic Cloud, a satellite galaxy of our own, at a distance of 170 000 light-years (about 50 kiloparsecs), in a direction free of interstellar dust. The first picture, of course fortuitous, of the supernova was taken on February 23.443 UT (i.e. at 10 hours 38 minutes UT on 23 February), where it appeared at magnitude 6.5 at the edge of the star formation region 30 Doradus, on a photograph taken by Robert M. McNaught. But Ian Shelton was the first to see it, on a photograph taken on February 24.23 UT from the Las Campanas Observatory in Chile. At almost the same moment it was seen with the naked eye by Oscar Duhalde, working in the same observatory. A little later, at February 23.37 UT the New Zealand amateur astronomer Albert Jones reported a similar observation, mentioning that he had

not seen the supernova on the previous evening, at least down to magnitude 8 at February 23.39 UT. The first photons from the supernova thus arrived at Earth, after a journey of 170 000 years, between 1987 February 23.39 and February 23.443 1987. But the precise instant of the explosion could be determined, in a quite unexpected way, because of a major first in astronomy, namely the detection of neutrinos produced at the centre of the star, at the instant the explosion started, even before the supernova became luminous. The first neutrinos were simultaneously detected at February 23.316 UT by the underground detectors Kamiokande II in Japan and IBM in the U.S. From the day after the announcement of the discovery, the IUE (International Ultraviolet Explorer) satellite was repointed to get the best ultraviolet spectra of the supernova, showing evidence for a rapidly-expanding envelope at a temperature of 14 000 kelvin at the beginning of the observation, followed by a rapid decline of the ultraviolet emission over about a week. Over this time, optical spectra acquired from the ground, notably at the Inter-American Observatory at Cerro Tololo in Chile, and in South Africa, revealed the presence of hydrogen and helium, being expelled at

30 000 kilometres per second: this gave the proof that this was a Type II supernova, whose complex explosion mechanism normally produces a neutron star or black hole.

This region of the Large Magellanic Cloud has been studied for a long time. It is a region of active star formation. Thus it was possible, after some initial hesitation, to identify which star had exploded. The precise site of the explosion, of right ascension 5 h 35 min 49.992 s and declination −69° 17′ 50.08″, coincides with the position of the star Sanduleak-69 202. This was a blue supergiant, of spectral type B3I, which had not shown any signs of particular activity for at least a century. As the distance to the Large Magellanic Cloud is fairly well known at about 50 kiloparsecs, the physical properties can be determined quite precisely from the photometric data. Sanduleak-69 202 was a star of about 20 solar masses, 100 000 times more luminous than the Sun. Its diameter, deduced from its surface temperature, of order 16 000 kelvins, and the luminosity, was about 43 solar radii. The fact that that it was a blue rather than red supergiant surprised most astrophysicists, as supernovae of this type usually result from the explosion of red supergiants, whose envelopes are much more extended and dilute. This paradox was resolved by carefully considering the evolution of massive stars whose chemical composition is intially poor in heavy elements. This was probably the case with Sanduleak-69 202, as the chemical composition of the matter of the Large Magellanic Cloud is three times poorer in heavy elements than the Sun. This has two effects: there is a significant decrease in the energy production rate in the hydrogen-burning phase, which in this type of star is catalysed by carbon, nitrogen, and oxygen (the same energy loss requires higher density and temperature); also a low heavy element content reduces the opacity of the gas inside the star. These two effects, added to the possible action of a stellar wind, which could have removed part of the star's mass during its evolution, make it understandable that the star that exploded was blue, although it would have been a red supergiant about 40 000 years ago.

The event of 23 February 1987 also led to the first detections of neutrinos produced neither on the Earth nor in the Sun. Three detectors, Kamiokande II in Japan, IBM in the US, and Baksan in the Soviet Union, detected the arrival of neutrinos produced at the moment of core collapse of the supernova, 170 000 years earlier. Unfortunately the simultaneous arrival cannot be confirmed to an accuracy of better than a minute, because of the lack of synchronisation of the clocks in the three laboratories. A fourth detector, sheltered from cosmic rays in a tunnel under Mont Blanc, detected a signal almost 5 hours earlier that seems to be unconnected with the supernova. These detectors were not at all designed to detect supernovae, but to set a lower limit to the lifetime of the proton. The Kamiokande II and IBM underground detectors are 1000 and 600 metres below ground respectively, and are reservoirs of several tonnes of pure water in which one measures the recoil electrons or positrons produced by the scattering of a neutrino on an electron or the absorption of an antineutrino by a proton. The Kamiokande detector registered twelve neutrinos (particles or antiparticles) in a space of 12.439 seconds, and IBM, with a higher detection threshold, registered eight in 5.58 seconds. The Baksan detector registered six events, with an uncertainty of nearly a minute in the arrival time.

Analysis of these events gives an estimate of around 4.2 megaelectron volts (48 billion degrees!) for the temperature of the matter producing the neutrinos. The decrease of their energy as a function of arrival time is compatible with a model in which the neutron star cools in a characteristic time of 4.5 seconds. This is a remarkable confirmation of the models of Type II supernovae. However, we should bear in mind the risks of drawing conclusions from such a small statistical sample, and wait for the plentiful detection of neutrinos from a nearer galactic supernova before accepting or rejecting the uncertain parts of the theory.

The evolution of the light from the supernova, or more precisely its debris, has provided much information about the progress of the explosion. The decline of the luminosity has been followed and analysed from the first ultraviolet and optical observations. As the ejected material is

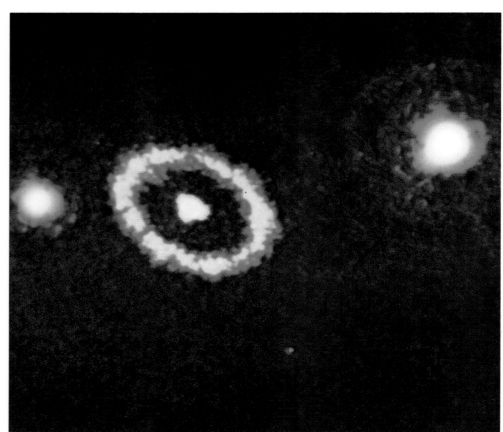

The ring of SN 1987A. A rapidly-expanding luminous ring (velocity of order 10 kilometres per second) was detected around the supernova.

This image was obtained on 23 August 1990 by the Faint Object Camera of the Hubble Space Telescope, at a wavelength of 500.7 nanometres, corresponding to the emission of doubly-ionised oxygen. The central object is the supernova. The plane of the ring is inclined at about 43 degrees to the sky, explaining its elliptical shape; its angular size is 0.88 arcseconds, or about 0.2 parsecs (6200 billion kilometres) for a distance to the Large Magellanic Cloud of 50 kiloparsecs.

The ring is too far from the supernova to consist of material ejected from it. It existed before the explosion and is intepreted as a nebula formed several tens of thousands of years ago by Sanduleak-69 202, as it passed from a red to a blue supergiant. The nebula resulted from the interaction of the rapid stellar wind of the blue supergiant with the slow stellar wind of the earlier red giant phase. The gas in the ring was subsequently heated and ionised by the intense radiation from the supernova. (NASA)

rapidly expanding, it cools very quickly. Gradually the maximum of the emission moves from the ultraviolet to the visible, reaching the infrared a year later. The differences between SN 1987A and other Type II supernovae, especially the high expansion velocity of its hydrogen envelope and the very rapid decline of its ultraviolet emission, can be understood from the fact that the star that exploded was a blue supergiant, which is much more compact than a red supergiant. Because of this it was able to retain its internal energy for longer, to the detriment of its initial luminosity, but to the benefit of its kinetic energy of expansion.

The maximum ultraviolet luminosity, corresponding to the arrival of the shock wave at the surface of the star, could not be observed, only its very rapid decrease, by a factor 1000 in three days. Only twenty days after the first observation by the IUE satellite – made after the shock wave had emerged from the star's surface at about 40 000 kilometres per second – the photospheric temperature decreased from 14 000 to 5500 kelvins. After about four weeks the luminosity of the supernova was dominated by the energy released in the radioactive decay of nickel 56, with a lifetime of six days, and cobalt 56, with a lifetime of 77 days. The energy released by the decay of nickel did not escape the envelope, which was still too dense, but was basically converted into kinetic energy. As the expansion proceeded, the energy liberated from radioactivity as gamma-ray photons was gradually able to escape more and more rapidly, after multiple scattering. This effect resulted in a bump in the supernova light curve about 100 days after the explosion. After this the exponential decay of the supernova light closely followed the radioactive decay of cobalt 56. Interpretation of the light curve shows that 0.07 solar masses of nickel was ejected in the explosion. Comparing this result with calculations of the heavy element abundances produced in the core before and during the explosion, we can conclude that the collapse of the core of Sanduleak-69 202 produced a neutron star of 1.45 solar masses.

A hundred and twenty days after the explosion, the two gamma-ray lines at 847 and 1238 kiloelectron volts from the decay of cobalt 56 became directly observable by the Ginga and SMM (Solar Maximum Mission) satellites and several balloon-borne instruments. This observation, in itself comforting, was astonishing in its speed. Similarly, the Ginga satellite and various instruments on the orbiting space station Mir, detected the first high-energy X-ray emission from the supernova remnant at the end of summer 1987, about twice as quickly as expected. All these observations show that the nucleosynthesis products of the supernova, normally shrouded in the deep layers, had in fact mixed with the outer envelope. This interpretation, appealing to hydrodynamical instabilities of the shock wave propagation through the envelope, has been corroborated by the observation in the infrared of ionised nickel, moving at velocities of more than 3500 kilometres per second, much higher than if it had remained confined to the deep layers of the star.

The exponential decay of the luminosity, dominated by the infrared, seems to have slowed since day 1000 after the explosion. Various interpretations of this phenomenon have been proposed, using the energy input of a central object, the neutron star which has not yet been observed, or of minor radioactive species, such as ^{44}Ti, with no definite conclusion as yet.

The fact that the explosion of SN 1987A was accompanied by the formation of a neutron star of mass 1.4 solar masses remains probable, but it is unlikely that we will observe this in the near future. To achieve this requires the star to become a pulsar. This depends strongly on the density of surrounding matter. For a periodic signal to reach the Earth, it must be neither absorbed nor scattered too much. Besides, not all neutron stars necessarily become pulsars, as this requires amongst other things a strong magnetic field and rapid rotation.

Jean-Pierre CHIÈZE

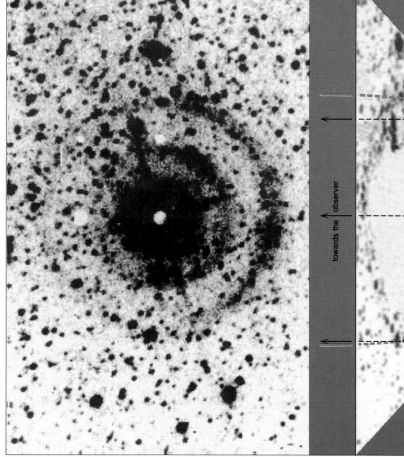
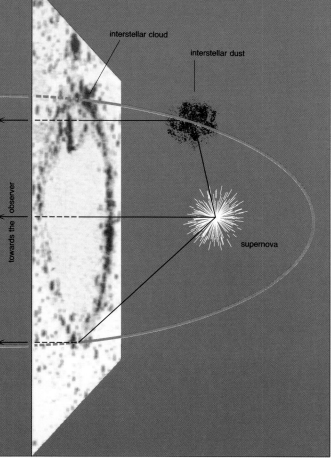

Light echos. SN 1987A lit up the interstellar medium. Two near-circular 'echos', centred on the supernova (overexposed) are visible on the negative obtained on 13 February 1988, one year after the explosion; the angular radii of these echos were then of order 30 and 50 arcseconds.

As shown in the diagram, these 'phantom' nebulae result from the reflection of light emitted at the moment of the explosion by dust grains in the surrounding interstellar medium. At a given epoch after the arrival of the first photons, which travel in straight lines, an observer on Earth sees the photons which have travelled the same distance ct after reflecting from the interstellar matter, where c is the speed of light and t is the time since the explosion. These photons thus appear to come from an ellipsoid of revolution whose foci are the supernova and the Earth. (ESO)

Light curve of SN 1987A. The figure on the left shows the variation of the supernova luminosity over the 180 days following the explosion (which was not observed), i.e. up to 20 August 1987. This is an enlargement of the right-hand figure, which goes up to day 1444, or mid February 1990 ($L_\odot$ is the solar luminosity).

The evolution of the total (or bolometric) luminosity of the supernova gives vital clues about the nature and abundances of the radioactive nuclei whose fission supplies the supernova remnant with energy. The exponential decline of the luminosity reflects the decay of these elements. Until day 120, most of the energy was supplied by the decay of nickel 56 into cobalt 56, with a total mass of 0.07 solar masses. Later, upto about day 500, the luminosity was supplied by the decay of cobalt 56 into iron 56. At present, the supernova luminosity is supplied by the decay of about 0.01 solar masses of the isotope cobalt 57. The presence of a pulsar in the centre of the remnant would be revealed by the stabilisation of the light curve. (after N.B. Stuntzeff et al., 1992)

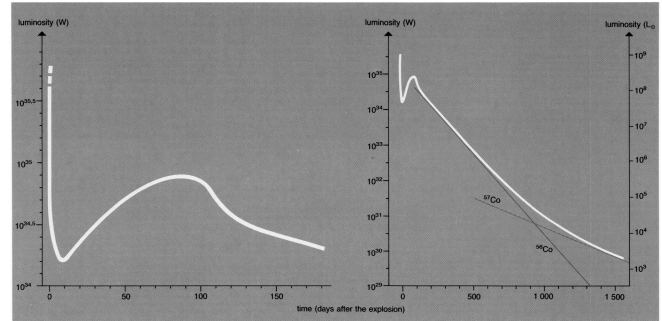

Neutron stars and pulsars

When a massive star has exhausted its thermonuclear fuel, it generally produces a supernova explosion, during which the core of the star collapses onto itself under the force of its own gravity. If the contracting mass is less than 1.4 solar masses, the collapse is halted by the pressure of degenerate electrons and the final state is a white dwarf. But what is the fate of a more massive star undergoing gravitational collapse?

As early as 1934, theorists used quantum mechanics to predict the existence of neutron stars: when gravity becomes too strong for a white dwarf to resist its pull, the electrons are 'pushed' into the interior of the atomic nuclei, converting protons to neutrons. But, just like electrons, neutrons obey an exclusion principle, according to which each neutron must occupy a single elementary cell. When all these cells are full, the neutrons are completely degenerate, and they exert a pressure capable of stopping gravitational collapse.

A neutron star is in many respects an extreme version of a white dwarf: for about the same mass (roughly that of the Sun), a neutron star has a much smaller radius – only 15 kilometres – and a fantastic density – a billion tonnes per cubic centimetre. The temperature of a neutron star is about ten million degrees but, because of its very small size, such an object is usually impossible to detect optically. The mass of a neutron star cannot exceed about three solar masses: above this value, gravity wins against the pressure of degenerate neutrons and the only possible final state is a black hole.

Two important properties of these ultradense stars are their rapid rotation and their strong magnetic fields. We know that all stars rotate, most of them slowly. When a star collapses, its rotation rate increases rapidly (for the same reason that a skater can turn more quickly by bringing her arms in towards her body); a neutron star can thus rotate several times a second. Equally, all stars possess a weak magnetic field, similar to the Earth's. As the star collapses, the strength of the field increases in proportion because it is concentrated over a smaller and smaller surface. Neutron stars have magnetic fields of a million million gauss, that is, a million million times that of the Earth. It is these two properties that allow the detection of neutron stars, in the form of pulsars.

The first pulsar, CP 1919 (this means Cambridge pulsar, at right ascension 19 hours and 19 minutes) was discovered in 1968 by radio astronomers at the Mullard Radio Astronomy Observatory of the University of Cambridge, England. It appears as an object emitting radio pulses of variable intensity but spaced at very regular time intervals: the period, incredibly precise, is 1.337 301 13 seconds. The explanation of this phenomenon soon followed: a pulsar is probably a neutron star whose magnetic field lines accelerate electrons along the magnetic axis, causing the emission of a beam of radio waves rotating with the star and producing a pulse when the beam intercepts the line of sight of an observer. Since 1968, numerous other pulsars have been discovered, and some of them have been found to emit not only in the radio, but also at higher frequencies, up to X-rays and gamma-rays. All frequencies are modulated in the same way by the rotation of the star. More than 300 pulsars are known, mostly situated in the galactic plane within a few kiloparsecs of the Sun. Obviously the most promising places to look for pulsars are supernova remnants. The famous Crab Nebula, the remnant of a pulsar recorded on 4 July 1054, does indeed contain the Crab Pulsar PSR 0532 (PSR denotes pulsar), which, because of its recent formation, is one of the most rapidly rotating known; it turns thirty-three times a second. It is easy to predict that the rotation rate of a pulsar will decrease slowly in the course of time according to the rate at which it dissipates energy. Young pulsars therefore rotate more rapidly than old ones. The measured periods lie between 1.56 milliseconds and 4.3 seconds. However, some ultrarapid pulsars with periods of milliseconds do not seem particularly young, and theoreticians think that they may have been spun up by transfer of mass from a companion star. Pulsar spindown rate lie between 10^{-18} and 10^{-12} seconds per second. These are extremely small values, but they are quite measurable over a period of years. When the rotation rate becomes too small, the pulsar mechanism fails: the average lifetime of a pulsar is less than a few million years.

Another effect contributes to the modification of the rotation rate of a pulsar, but in a more abrupt way. These are 'glitches', which decrease the period of a pulsar by one part in a million in the space of a few days, after which the pulsar settles down into a new stable state. Glitches may be interpreted as 'starquakes', due to certain instabilities in the crust or the core of a neutron star, which cause sharp changes of angular velocity. Glitches are very useful for studying the internal structure of neutron stars, but only a few pulsars, including those in the Crab and Vela, show them.

Although detection of pulsars in supernova remnants has turned out to be rare and difficult, a more widespread phenomenon exists which

Pulsar mechanisms. This compound diagram shows both the mechanism for X-ray pulsars and the mechanism for radio pulsars. In radio pulsars there is a strongly magnetised, rapidly rotating neutron star, with the synchrotron mechanism producing radio emission from regions less than 1000 kilometres from the surface. Note that the magnetic and rotation axes are not in general aligned. In X-ray pulsars, the emission comes directly from the polar caps of the neutron star. In both radio and X-ray pulsars, the rotation of the star produces the observed periodic modulation by a 'lighthouse' effect; a pulse is seen every time the beam intercepts the line of sight of the observer.

1 hot spots (magnetic poles at 10^7 K)
2 flux of particles
3 radio emission
4 magnetic force lines
5 accretion column

The structure of a neutron star. Although it is impossible to reproduce in the laboratory the physical conditions that occur in the interior of a neutron star, and although the properties of ultradense matter are still not well known, theorists predict that the interior of a neutron star (of typical radius 16 kilometres) is divided into five distinct zones:
– because of the extremely strong gravity, the *photosphere* is only about 10 centimetres thick. The properties of matter here are very strongly affected by the temperature (10^7 K) and the magnetic field;
– the *external crust*, about 1 kilometre thick, is a crystalline solid of nuclei immersed in a sea of degenerate relativistic electrons;
– in the *internal crust*, which is about 4 kilometres thick, neutrons are able to exist outside the atomic nuclei, so that matter is in a crystalline form immersed in a sea of electrons and neutrons. The density varies from 4×10^{11} to 2×10^{14} grams per cubic centimetre;
– the *neutron fluid* region, about 10 kilometres thick, constitutes the most important part of the star. Under the effect of gigantic pressures, the internal crust dissolves into a liquid composed mostly of neutrons, though with some protons and electrons too. This liquid is probably a superfluid, which means that its viscosity is close to zero. (Superfluidity is a strange phenomenon, but it has been observed, principally in helium at low temperatures);
– the *solid core*, about 1 kilometre in radius, is much more hypothetical because, at densities of 5×10^{14} grams per cubic centimetre, which occur here, practically nothing is known of the possible states of matter. Nonetheless, there have been speculations on the nature of the elementary particles here: the core may be composed of crystalline neutrons, of a pion condensate, of quark matter, or of hadronic soup.

We stress that this illustration which shows radio and X-ray emission simultaneously, conflates two phenomena which do not occur at the same time in the same objects; in fact, radio pulsars are not usually X-ray pulsars or vice versa.

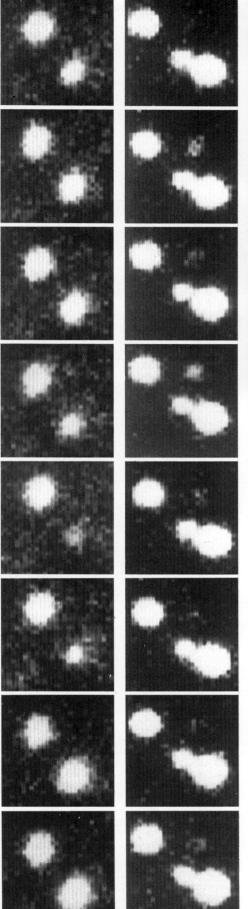

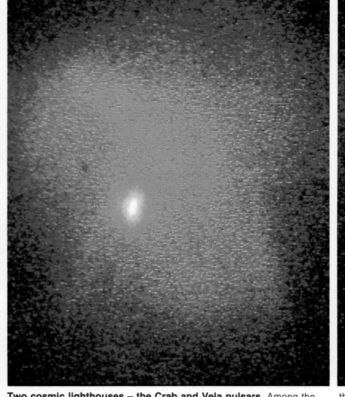

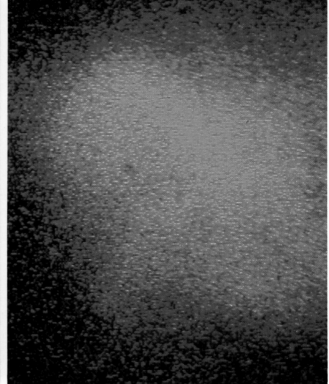

Two cosmic lighthouses – the Crab and Vela pulsars. Among the several hundred known pulsars, all characterised by periodic pulses of radio emission, only two have been detected optically: these are the Crab and Vela pulsars, each associated with supernova remnants. Both these pulsars are particularly young: the Crab supernova is known to have occurred in 1054, and it has been estimated that the Vela supernova happened about 11 800 years ago. Because of their youth, these two pulsars rotate very rapidly – the Crab Pulsar 30 times a second and the Vela Pulsar 11 times a second – and each rotation is accompanied by a pulse of electromagnetic radiation. Optical pulses from the Crab, synchronised with the radio pulses, were discovered in 1969. The Vela Pulsar was detected optically only in 1977, even

though it was discovered in radio emission in 1968.

On the left, the two vertical series of photographs show eight snapshots of the two pulsars, which can be distinguished from the field stars by their variable brightness. Sometimes the Vela Pulsar (right column) disappears altogether: although the interval between pulses is fixed, the intensity of each burst is very variable. (Anglo-Australian Telescope Board, © 1977)

The two pictures above show X-ray images of the Crab Nebula. Superposed on the diffuse X-ray emission of the nebula (in blue) is the much more intense periodic emission of the pulsar itself (the bright spot). (Centre for Astrophysics, Cambridge, Mass.)

The binary pulsar PSR 1913+16 (below right). In 1974 an unusual object, the binary pulsar PSR 1913+16, was discovered. (The numbers indicate the celestial coordinates of the object: right ascension 19 h 13 m, declination +16 degrees.) It is in the constellation of Aquila, 4600 parsecs from the Earth. The binary pulsar consists of a 1.4 solar mass neutron star, pulsing 16.94 times a second, and a 'hidden' companion of the same mass which is probably also a neutron star. These two compact stars rotate about each other in a very tight orbit (a few million kilometres in radius) in 7 hours 45 minutes.

The binary pulsar is a perfect test of the general theory of relativity, which says that an accelerating mass radiates energy in the form of gravitational waves. Dissipation of energy in this way causes the orbit to shrink, and therefore causes a slow diminution, over a period of time, of the orbital period of the binary pulsar. The predictions of

Einstein's theory are in very good agreement with observation: the orbital period of PSR 1913+16 is decreasing by 76 milliseconds per year.

Two other binary pulsars are also now known: PSR 0820+02, with an orbital period of 1710 days, and PSR 0655+64, with a period of 24 hours 41 minutes. Their separations are too large to give rise to observable relativistic effects, but it is possible to measure the masses of the neutron stars in them.

In the photograph, obtained by J. A. Tyson with a video camera mounted on the 4-metre telescope at Kitt Peak, the position of PSR 1913+16 is shown by a cross. It coincides with an optically visible star, probably an old helium star which has ejected its outer layers. It has been suggested by some researchers that this could be the companion of the pulsar. (Bell Laboratories)

Electromagnetic pulses from the Crab Pulsar (below). A pulsar emits electromagnetic radiation not only in the radio region (which allows its detection) but also in higher energy parts of the spectrum, at optical, X-ray and even gamma-ray frequencies. The curves show that the detailed intensity structure of the pulses differs noticeably according to the wavelength at which they are observed. There is also a gradual shift in the arrival time of pulses at different wavelengths. This latter observation leads to an independent estimate of the distance of the pulsar. In contrast, the period of the pulsar, that is, the interval between successive pulses, is exactly the same in all wavelength regions because it is simply the rotation period of the star itself.

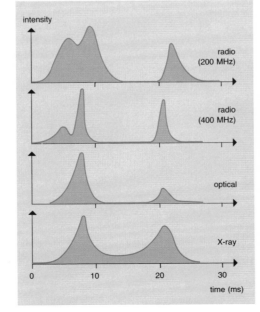

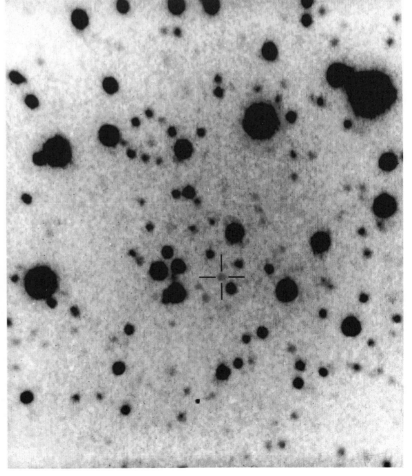

allows the discovery of numerous pulsars: this is the compact X-ray source. In 1971 with the launch of the astronomical satellite Uhuru certain galactic sources were discovered to emit a very strong periodic flux of X-rays. The source Centaurus X-3, for example, has an intrinsic X-ray luminosity ten thousand times the total luminosity of the Sun. Its flux varies periodically every 4.8 seconds and it is eclipsed every 2.087 days, demonstrating that the X-ray source is in orbital motion about a more massive object. The X-ray source is part of a binary system in which a neutron star attracts the stellar wind of a giant star and converts the gravitational energy of the gas into X-rays. Several systems similar to Centaurus X-3 are known and these have allowed, in particular, the measurement of the masses of neutron stars and reasonable confirmation of various theoretical predictions.

Jean-Pierre LUMINET

X-ray and gamma-ray bursters

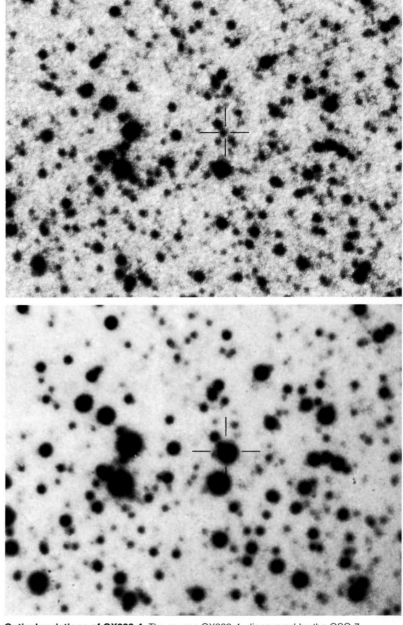

High-energy astrophysics is arguably the branch of astronomy that has most advanced our understanding of the Universe over the past twenty-five years: it deals with the very short wavelength part of the electromagnetic spectrum. It is usual to distinguish X-ray astronomy, concerned with wavelengths shorter than 0.5 nanometre, or equivalently energies greater than 1 kiloelectron volt (keV), from gamma-ray astronomy, which deals with wavelengths shorter than 0.001 nanometer, and therefore energies greater than 500 keV, the annihilation energy of an electron–positron pair. A 500-keV gamma-ray photon has a million times more energy than a photon of visible light.

The development of detection techniques for X-ray and gamma-ray photons took place after the Second World War, when rockets became available and satellites were launched. X-ray and gamma-ray photons interact very strongly with the upper atmosphere and do not reach the ground.

The first satellite dedicated to X-ray astronomy was Uhuru. It was launched from Kenya on 12 December 1970, the anniversary of the country's independence (Uhuru means 'liberty' in Swahili). It was already known that some stars such as the Sun emit X-rays. A notable discovery, in 1962, was Scorpio X-1, the most brilliant X-ray source in the sky. By revealing a large number of powerful new X-ray sources, Uhuru radically transformed our conception of the Universe: it showed us events much more violent than had previously been imagined. Since Uhuru, other X-ray satellites have been launched, among them the well-known HEAO-2, rechristened Einstein to commemorate the 1979 centenary of the great physicist's birth. Most of the permanent

galactic X-ray sources discovered have been identified with systems containing compact stars – white dwarfs, neutron stars or black holes. The high energies involved in such sources require high surface temperatures (10 million K) and gravitational fields sufficiently strong to convert the gravitational energy of matter efficiently into X or gamma-radiation. A number of pulsars, including those in the Crab and in Vela, emit X-rays and gamma-rays. The largest class of bright X-ray sources is that of the X-ray binaries (or X-ray pulsars), in which a compact star pulls onto itself the gas of its companion, which is usually an ordinary star. (This process is called mass transfer.) The luminosities of X-ray sources are mostly between 10^{28} and 10^{31} watts, which is 100 to 100 000 times the total luminosity of the Sun. Such high emission rates require a mass transfer of between 10^{-8} and 10^{-11} solar masses per year, values which are quite compatible with theoretical models.

There are two categories of binary X-ray sources. Massive binaries consist of a very hot and very luminous giant star, of 10 to 20 solar masses, and a less massive compact companion. The atmosphere of the giant star is so extended that some of it falls directly onto the compact star, which is travelling through it. In low mass binaries, the components have comparable masses, so that the capture of gas by the compact star from the normal star is probably accomplished by the intermediate means of an accretion disk, a ring of hot, luminous gas.

One of the most astonishing X-ray sources is undoubtedly AO 538–66. It is a pulsar of period 0.069 seconds, situated in the Large Magellanic Cloud at a distance of 55 000 parsecs. It is optically identified as a binary system containing

Optical variations of GX339-4. The source GX339-4, discovered by the OSO 7 satellite, shows variations of X-ray flux on timescales of several hundred days, as well as smaller amplitude fluctuations over milliseconds. In this it is very similar to the binary X-ray source Cygnus X-1, whose compact component may be a black hole, which alone allows such short-term fluctuations.

The visible component of the binary is identified as a faint star with a magnitude which usually varies between 16 and 18. On 6 March 1981, GX339-4 suddenly became invisible, in the optical as well as X-rays. Taking account of instrumental limitations, this means that its magnitude became larger than 21, corresponding to a brightness decrease by a factor of 100. On 24 March, GX339-4 became visible again and reached a maximum brightness of magnitude 15.4. We may have seen directly the beginning of mass transfer from the visible star on to its compact companion and the rebuilding of an accretion disc. The upper photograph shows the object (marked by a cross) at minimum. The lower photograph shows the object on 1 June 1989, near its maximum. North is up, East to the left. The faint star situated immediately to the north-east of GX339-4 has magnitude 19.5; the brightest star to the South has magnitude 15.9 comparable with that of GX339-4 at its maximum. (H.E. Schuster and G. Pizarro, ESO)

photons/0.8314 s

3 March 1976

32 s

hours (Universal Time)

photons/0.8314 s

4 March 1976

hours (Universal Time)

Flux variations of the rapid X-ray burster MXB 1730–335. The X-ray burster MXB 1730–335, situated in a globular cluster, is called a rapid burster because the average interval between outbursts is only a few tens of seconds, as against a few hours or days for ordinary bursters. It is interesting to note that the strength of an outburst is greater, the longer the quiet period which precedes it. This suggests a process of accumulation then detonation, of material on the surface of a compact star. (Taken from the Encyclopédie Bordas, *La Galaxie, l'univers extragalactique*, p. 263)

a normal star of 12 solar masses and a compact companion, with an orbital period of 16.6 days. This is therefore a massive binary, but it poses a problem for theorists because it emits more X-rays than the entire Milky Way! At first sight, only a black hole seems capable of liberating enough energy continuously, but a neutron star is needed to explain the pulsations.

A universal property of X-ray sources is variability of their flux, on time-scales of a few milliseconds to a few years. Some of these variations are periodic, and can be attributed to the orbital motion of two companions, to the intrinsic rotation of a compact star, or to the precession of an accretion disk. Other X-ray sources, dubbed X-ray bursters, show instead violent and random changes. Some of these are permanent X-ray sources, but others are transitory in the sense that they are only visible in X-rays during an outburst. Outbursts are sudden increases in flux, which may last several days; they may recur but are not periodic. The new astronomy of X-ray and gamma-ray bursters is currently developing and it will raise many fascinating problems in the future. Some tens of X-ray bursters are now known, most of them situated in the plane of the Galaxy, but some are found in globular clusters (dense associations of stars in the halo of the Galaxy). In the latter group, the source MXB 1730–335 has exceptionally rapid bursts, with an average interval of no more than about 10 seconds. Three X-ray binaries are particularly noteworthy: the massive binaries Cygnus X-1 and LMC X-3 (the latter situated in the Large Magellanic Cloud, one of two satellite galaxies to our own) and the dwarf binary A0620-00, as it seems likely that their X-ray emission comes from an accretion disk around a black hole rather than a neutron star.

Theoretical models of X-ray bursters show considerable similarities to models of ordinary novae. The latter are explained by the progressive accumulation of hydrogen onto the surface of a compact star (in this case a white dwarf), followed by a sudden thermonuclear explosion when the density and temperature of the surface layers have been sufficiently increased. In X-ray bursters, the accumulation of gas is onto the surface of a neutron star and the material that undergoes thermonuclear fusion is helium or perhaps even heavier elements; this leads to a greater liberation of energy than in novae.

In contrast, gamma-ray bursters seem to be independent of X-ray bursters: to date, no association has been observed. Gamma-ray bursters were discovered at the end of the 1960s by the American Vela military satellites, whose mission was the detection of experimental thermonuclear explosions by the Soviet Union. Since then, about a hundred gamma-ray events have been detected. Except in three cases the bursts do not clearly repeat (one of these 'repeaters' has produced one hundred and eleven bursts in six years). The duration of outbursts varies from a few milliseconds to a few tens of seconds, and generally there is no periodicity. They are observed only up to a few megaelectron volts, that is in low energy gamma-rays. These energies correspond to temperatures of a few billion degrees, which suggests a new explanation, namely intermittent heating of the surface of a neutron star by the infall of material. The major problem in gamma-ray astronomy is the lack of angular resolution of the detectors used, which are not directional. It is therefore very difficult to localise the sources and to identify them with objects emitting in longer wavelength domains. It is, however, possible to estimate their position by combining observations from a number of satellites (at least three). This allows the definition of 'error boxes', parts of the sky a few arc minutes across, which contain the source responsible for the outburst.

In contrast, unlike the X-ray bursts, their distribution in the sky is uniform, and does not show any particular concentration towards the centre or the plane of the Galaxy. There are therefore two hypotheses: if the gamma-ray bursts are actually situated in the galactic disk, their apparent distribution does not reflect their real distribution (which would be flattened) because of the low sensitivity of the detectors, which only see the closest (within 500 parsecs). If, on the other hand, the bursts are more distant, situated in the halo of the Galaxy or external galaxies, their distribution is really uniform. Theoreticians can only speculate around these two hypotheses. Under the Galactic hypothesis the energies involved in a gamma-ray burst suggest the effects of the intense gravitational field of a neutron star. The absence of optical counterparts shows that these neutron stars must be either isolated or members of binary systems with faint dwarf companions. Three models can explain the physical origin of the bursts themselves. One invokes a high-speed collision between a neutron star (or a white dwarf) and a comet or asteroid. Another appeals to the gradual spindown of the pulsar through

the emission of X-rays: below a certain threshold, the neutron star has to readjust its structure in a kind of 'starquake', releasing enough energy to explain the gamma-ray bursts. Another model, adapted from that for X-ray bursts, invokes a thermonuclear explosion of helium at the suface of a neutron star, triggered by the progressive accumulation of matter either from the interstellar medium or from a companion; the accretion rate would be lower and the magnetic field stronger than in X-ray bursters. This model, constructed by a team from the Paris Observatory, is currently the most complete. If the bursts turn out to be extragalactic, the energy liberated would be as large as in supernovae; a daring idea invokes a gravitational lens effect, which would amplify and multiply the individual events associated with neutron stars (coalescence of binaries, etc). It may, or course, be that none of these explanations is correct: gamma-ray bursts remain one of the great enigmas of contemporary astronomy.

Jean-Pierre LUMINET

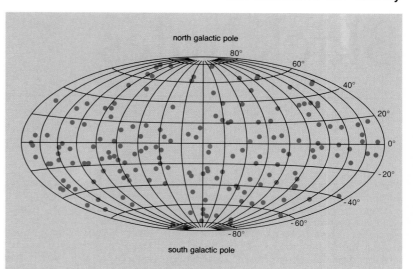

Distribution of gamma-ray bursts on the celestial sphere. The distribution of gamma-ray bursts in galactic coordinates shows no concentration towards the galactic plane, unlike X-ray bursts. The three repeating gamma-ray bursters are shown in red. (from K. Hurley, 1990)

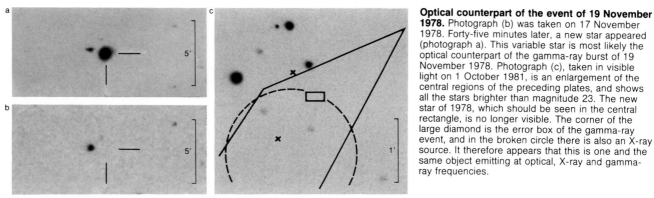

Optical counterpart of the event of 19 November 1978. Photograph (b) was taken on 17 November 1978. Forty-five minutes later, a new star appeared (photograph a). This variable star is most likely the optical counterpart of the gamma-ray burst of 19 November 1978. Photograph (c), taken in visible light on 1 October 1981, is an enlargement of the central regions of the preceding plates, and shows all the stars brighter than magnitude 23. The new star of 1978, which should be seen in the central rectangle, is no longer visible. The corner of the large diamond is the error box of the gamma-ray event, and in the broken circle there is also an X-ray source. It therefore appears that this is one and the same object emitting at optical, X-ray and gamma-ray frequencies.

Error box of the event of 5 March 1979. The error box of the gamma-ray burst of 5 March 1979, determined by simultaneous observations from nine satellites, is superposed on a digitised image of the supernova remnant N49 in the Large Magellanic Cloud. (Original plate from the ESO Quick Blue Sky Survey; digitised by K. Hurley, S. Motch, S. A. Ilovaisky and C. Chevalier.) The very short duration of the pulse requires the emitting object to be smaller than 75 kilometres in size. The burst was followed by 3 minutes of less intense emission,

which showed pulses with a period of 8 seconds; this suggests emission from a hot spot on the surface of a neutron star. If the burster really is in N49, and therefore at a very large distance, its intrinsic luminosity must be a million times greater than that of a gamma-ray burst in our own Galaxy: this raises a number of theoretical problems. Instead of invoking a thermonuclear flash on the surface of a neutron star, a sudden change in the interior might lead to a brief and very intense combustion of the surface layers.

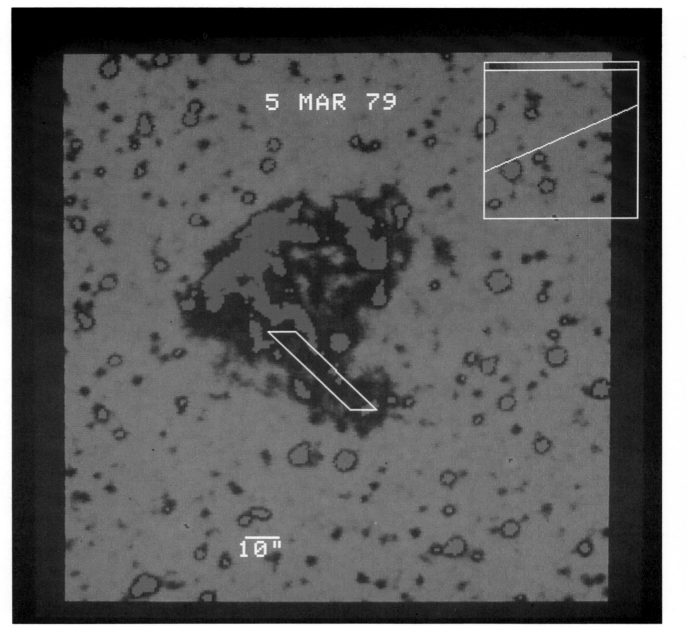

Black holes

When all the thermonuclear fuel such as hydrogen and helium is exhausted in the core of stars, a gravitational collapse of the core results. Stellar evolution ends in the formation of extremely condensed compact objects. White dwarfs and neutron stars belong to this peculiar variety of stars, but their mass cannot exceed 3 solar masses ($M_\odot$). For higher masses, the gravitational compression is no longer compensated by the forces of repulsion of the electrons or of the degenerate neutrons, but continues to crush matter upon itself indefinitely: black holes form.

Since the publication of the general theory of relativity by Einstein in 1915, it has been known that gravitation deforms space. This deformation may be visualised as a gravitational well dug by the bodies in the very structure of space–time. The more massive and dense a body is, the deeper its gravitational well. The final stage of gravitational collapse, the black hole, is characterised by a well so deep that nothing can escape from it, neither particle nor light ray; moreover, all matter falling into the well of a black hole must disappear from the observable Universe.

The physical properties of black holes are so spectacular that for a long time such properties detracted from the credibility of the theory. The theory predicts in particular the possible existence of black holes of all sizes and masses: mini-black holes would have the mass of a mountain concentrated within the volume of an elementary particle; a black hole of one centimetre radius would be as massive as the Earth; stellar black holes would have masses comparable to those of stars within a radius of only a few kilometres; finally, giant black holes would possess a mass equivalent to several hundreds of millions of stars within a radius comparable to that of the Solar System.

Though such concentrations of matter have not yet been directly observed, there is nevertheless strong evidence for the existence of black holes. Note that black holes were envisaged at the end of the eighteenth century by the Englishman

John Mitchell and by Laplace. However, astronomers only really began to be interested in black holes in the 1960s, when phenomena were discovered that are the most energetic ever observed; on the scale of stars, X-ray binaries, and on the extragalactic scale, the radio sources associated with active galactic nuclei and quasars. X-ray binaries are double star systems in which one very compact component, which is optically invisible, emits a considerable flux of X-rays. The detection of these binaries, made possible thanks to satellites like Uhuru and HEAO-1 and -2, is one of the major discoveries of recent years. On a much larger scale, the superactive galaxies such as Seyferts, quasars and BL Lac objects emit far greater amounts of energy than a normal galaxy, at all wavelengths from the radio region through to X-rays and gamma-rays. The important point is that all of these violent phenomena appear to be associated with the presence of very compact massive bodies: neutron stars or stellar black holes in the case of X-ray binaries, supermassive stars or giant black holes for the active galactic nuclei.

It is easy to understand how a black hole can trigger very energetic phenomena in its vicinity. Any object which falls to the surface of the Earth in free fall gives off heat; if the same object were to fall onto the surface of a white dwarf or neutron star where the gravitational field is much stronger, far more energy would be liberated, as light or even as X-rays. Finally, if the object falls into a black hole, the gravitational field is so great that a significant fraction of all of the energy that the object is capable of yielding is liberated: this energy is the energy of the rest mass, given by Einstein's famous equation: $E = mc^2$. A black hole is thus the most efficient device for converting the mass–energy of a body into electromagnetic radiation. In the case of X-ray binary sources, the compact star 'swallows' the atmosphere of its companion and the falling gas produces the observed X-rays. However, the compact star could be either a neutron star or a black hole; the

The formation of a black hole. As a star collapses, its surface gravity becomes greater and greater. The light emitted by the star is consequently more and more curved by gravity. The diagram shows a classical representation of this on the left, with successive states of the star from bottom to top. A space–time diagram is shown on the right (note the time axis, with time increasing upwards). There is a critical moment, t_H, in the course of the collapse when the star forms a black hole on collapsing within its Schwarzschild radius. (This critical radius is equal to 3 kilometres for a star of 1 solar mass.) From that moment, the light rays are bent so much that they cannot leave the star, and a surface known as an event horizon is formed. This surface divides space–time up into two regions: an outer region which includes observers like ourselves, and a region inside the black hole, which becomes totally inaccessible to observation since nothing can escape from it. Once the star has crossed the horizon, it theoretically continues to collapse upon itself to form a singularity where density, pressure and temperature become infinite.

The black hole maelström. The drawing represents the gravitational well of a rotating black hole. The well may be likened to a whirlpool in which the possible ways in which a vessel can navigate are represented by circles of navigation. In the upper part of the well (light grey) the vessel can move in any direction, that is, it can move away from, or towards, or remain at a fixed distance from, the centre of the whirlpool. Inwards of the static limit (dark grey region), whatever the speed of the vessel, it can no longer remain at a fixed distance from the centre of the whirlpool. Between the static limit and the horizon, the vessel can still escape from the central well, but it must move into the interior at an angle determined by the tangents to the circles of navigation. This angle decreases and is more inclined towards the interior as the vessel penetrates more deeply; the vessel can, however, still climb out. (Enough momentum to escape is gained by moving inwards.) Beyond the horizon, the vessel no longer has any chance of escaping even if it moved at the speed of light. It is inexorably attracted to the bottom. This very striking analogy shows that space–time is not frozen around a rotating black hole as in the classical Newtonian concept, but is instead carried along by the rotation much as water is carried along in a whirlpool.

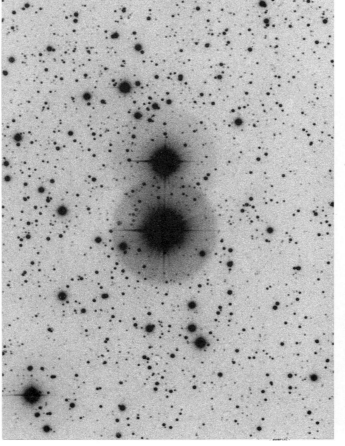

The HDE 226868–Cygnus X–1 system. Cygnus X–1 is one of the brightest X-ray sources discovered in our Galaxy by the Uhuru satellite. The source is associated with an optically visible star numbered HDE 226868, which is a blue giant of 20 solar masses. This star has an invisible companion of 10 solar masses. The X-rays come from 'breathing' of the very extended atmosphere of HDE 226868 by its compact companion. One of the most interesting properties of Cygnus X–1 is the extremely rapid variability of its X-ray luminosity over periods shorter than a thousandth of a second. This shows that the diameter of the region emitting X-rays is less than 300 kilometres, reinforcing the hypothesis of a black hole. Cygnus X–1 is at present the best observational evidence for the existence of stellar black holes. (J. Kristian, Hale observatories)

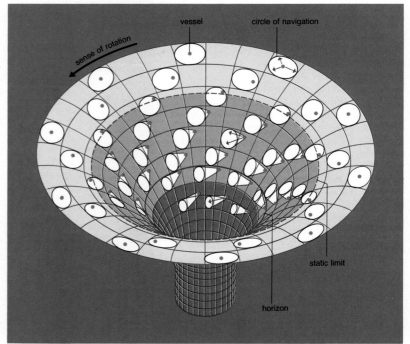

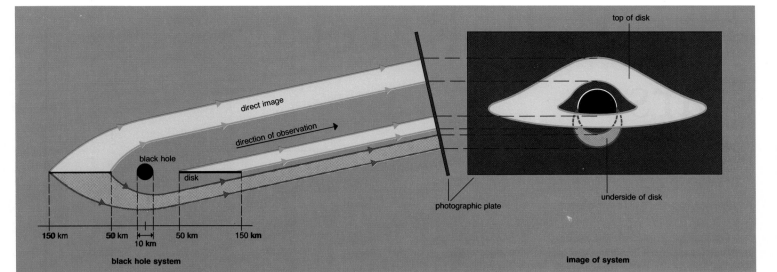

top of disk

direct image

direction of observation

black hole

disk

underside of disk

photographic plate

150 km   50 km   10 km   50 km   150 km

black hole system

image of system

The curvature of space–time in the neighbourhood of a black hole. The drawing above illustrates how the gravitational field of a black hole deforms the structure of space–time, and shows the deflections in the light paths that result. The black hole has a radius of 10 kilometres and is surrounded by a bright disk in the equatorial plane, whose inner and outer edges are 50 and 150 kilometres respectively from the centre of the black hole. The observer is some distance away, in a direction inclined by 10 degrees to the plane of the disk. The light emitted by the disk is received by a photographic plate. The image which results is very different from that which would have resulted if a similar 'ordinary' celestial body had been photographed, such as the ringed Saturn. The curvature of space–time gives rise to two images: one direct image formed by light emitted by the top side of the disk (this is obviously distorted in its upper part) and an indirect image which comes from the underside of the disk! The curvature of space–time is so great close to the black hole that it produces an infinite number of images because the light has to travel an arbitrary number of times around the hole before escaping from the gravitational fields to the photographic plate. However, images of order higher than two are in practice unobservable because they are literally stuck to the edge of the black central disk which represents the black hole itself (and which is larger than the true section of the hole).

A black hole and its image. This simulation of the photographic appearance of a black hole surrounded by a disk of luminous gas was made by J.-P. Luminet. The paths of the light rays were calculated by computer. As in the top figure on this page, the system is viewed from some distance away, from a direction inclined by 10 degrees to the plane of the disk. The simulated image is realistic in the sense that it takes account of the physical properties of the gaseous disk. There are a number of astrophysical situations in which a black hole would be surrounded by a disk of hot gas emitting radiation. Note that only the top half of the indirect image (that emitted by the underside of the disk) is visible, as the bright circle around the black hole is hidden by the image (the gas is assumed to be opaque). The asymmetry of the image is caused by two effects: firstly, the regions of the disk nearest to the hole are the hottest and therefore the brightest (this effect is intrinsic to the hole); secondly, the frequency of the light is altered both by the gravitational field (Einstein effect) and by the rotation of the disk around the hole (Doppler effect). The disk is rotating such that the left-hand part is approaching the observer and the right-hand part receding; this explains why the brightness of the image falls off very rapidly towards the right.

best way to settle the question is to measure the mass of the object, to see whether it exceeds 3 solar masses, the limiting mass of a neutron star; if so, it is probably a black hole. There are currently three serious candidates: Cygnus X-1 and A0620-00 in our galaxy, and LMC X-3 in the Large Magellanic Cloud. We can estimate that our Galaxy contains about a million black holes left over at the end of the evolution of massive stars.

However, the most striking applications of black holes to astrophysics concern the active nuclei of galaxies. The basic idea is the same as before: that energy is produced by heated gas during its free fall into the black hole, but the scale is very much greater. Here the hole has the mass of a billion stars; a typical quasar would have to absorb an amount of gas equivalent to several stars each year. These enormous quantities of gas would be derived from the total rupture of stars, caused either by collisions at high speed or by tidal effects in the gravitational field of the giant black hole.

Another means of proving the existence of black holes in the more or less distant future is the detection of gravity waves (advanced technology would be needed). Such waves are to gravitation what light waves are to electromagnetism. The collapse of a star or the fall of a massive body onto a black hole would give rise to the emission of gravitational waves which could be detected on Earth with sufficiently sensitive antennae. There is no doubt that this technique will be one of the major tools of astronomy in the twenty-first century.

Jean-Pierre LUMINET

Giant black holes in the centres of galaxies?

Astrophysicists think that supermassive black holes of a million to a billion solar masses may form in the nuclei of galaxies by coalescence and accretion. Present direct observational proofs are still insufficient, as although most galactic nuclei do indeed show great concentrations of bright matter, the resolving power of present instruments does not allow us to see regions as small as those occupied by a black hole.

The colour photograph shows the Seyfert galaxy NGC 1566. Its extremely bright nucleus is very similar to those of quasars, although it is less energetic. As its luminosity varies in less than a month, it must be very compact. Moreover, spectra show that hot gas is moving at abnormally high speeds, suggesting that the gas may be in orbit around a black hole of a hundred million solar masses. (D. Malin, Anglo-Australian Telescope Board)

The four black and white photographs show the nucleus and the central regions of the Andromeda galaxy M31: with a very short exposure time (30 s) only the nucleus appears (1); as the exposure time is lengthened (2: 1 min 30 s; 3: 4 min 30 s) the nucleus begins to be masked by the bulge; with an exposure time of 20 min only the bulge appears (4). M31 is a large spiral galaxy whose nucleus is not particularly active; measurements of stellar velocities near its centre nevertheless reveal the presence of a large dark mass of several tens of millions of solar masses, perhaps a black hole which is being 'starved'. (Observatoire de Haute-Provence du CNRS)

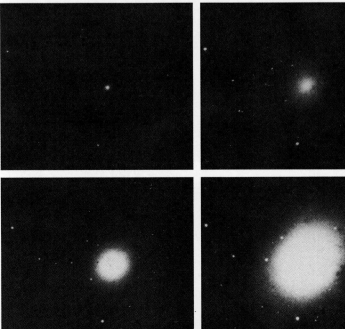

The Wolf–Rayet stars

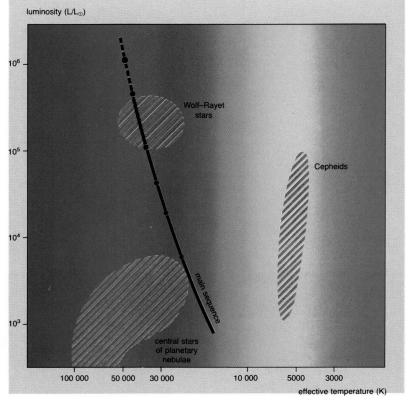

The properties of this group of stars were described for the first time in 1867 by Charles J. Wolf and Georges A. Rayet. The brightest star of this class is the second magnitude star γ^2 Velorum. Approximately 164 Wolf–Rayet stars are known in the Galaxy. They are very luminous stars of spectral type O or B, with a mean absolute visual magnitude of -5. They are thus hot stars, with effective temperatures between 30 000 and 50 000 K, although the precise values are difficult to determine. Their spectra are very peculiar. Along with absorption lines corresponding to spectral type O or B, many emission lines belonging mostly to helium, nitrogen, oxygen, silicon and carbon are seen. These emission lines show that a stellar wind is ejecting matter into space at velocities between 1000 and 3000 kilometres per second, and indicate that an envelope of ejected matter exists around the star. The rate of mass loss by the stellar wind is significant, of the order of 10^{-4} solar masses per year.

Wolf–Rayet stars also show erratic variations in brightness. They often belong to close binary systems, the companion being a main sequence O or B star, less luminous than the Wolf–Rayet.

The atmospheres of Wolf–Rayet stars are hydrogen poor, but rich in carbon and nitrogen. Helium is the most abundant element, but its abundance is difficult to determine accurately.

One sometimes refers to these stars as 'helium stars'. Their masses are estimated to be approximately 10 solar masses or less. Their radii are several times larger than that of the Sun.

The most plausible hypothesis concerning their origin suggests that Wolf–Rayet stars are massive stars that have lost their atmospheres during the course of their evolution. The lack of an atmosphere allows us to observe the central regions where thermonuclear reactions have strongly modified the chemical composition. The hydrogen has been transformed into helium, carbon and nitrogen. Wolf–Rayet stars would then be in the helium-burning phase with a central temperature of the order of one hundred million kelvin.

Wolf–Rayet stars would then simply represent a phase in the evolution of massive stars. This phase should last a few hundred thousand years. Wolf–Rayet stars are young Population I objects, short lived because massive stars evolve so rapidly. Stars with masses in the range of 60 to 100 solar masses have very strong stellar winds. Their mass loss during their hydrogen burning phase could be high enough to allow us to see their helium-rich nucleus. Alternatively, the outer layers may have been lost to a close binary companion.

Michèle GERBALDI

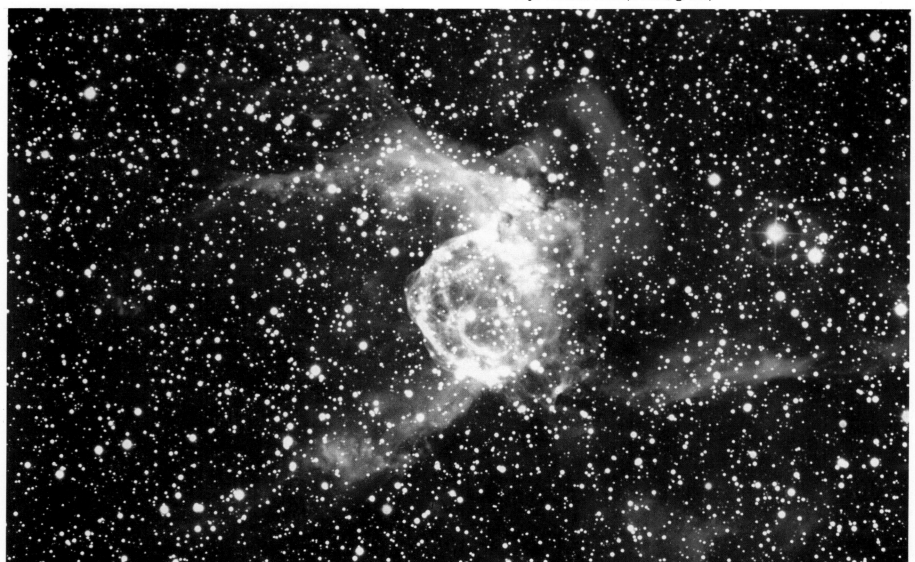

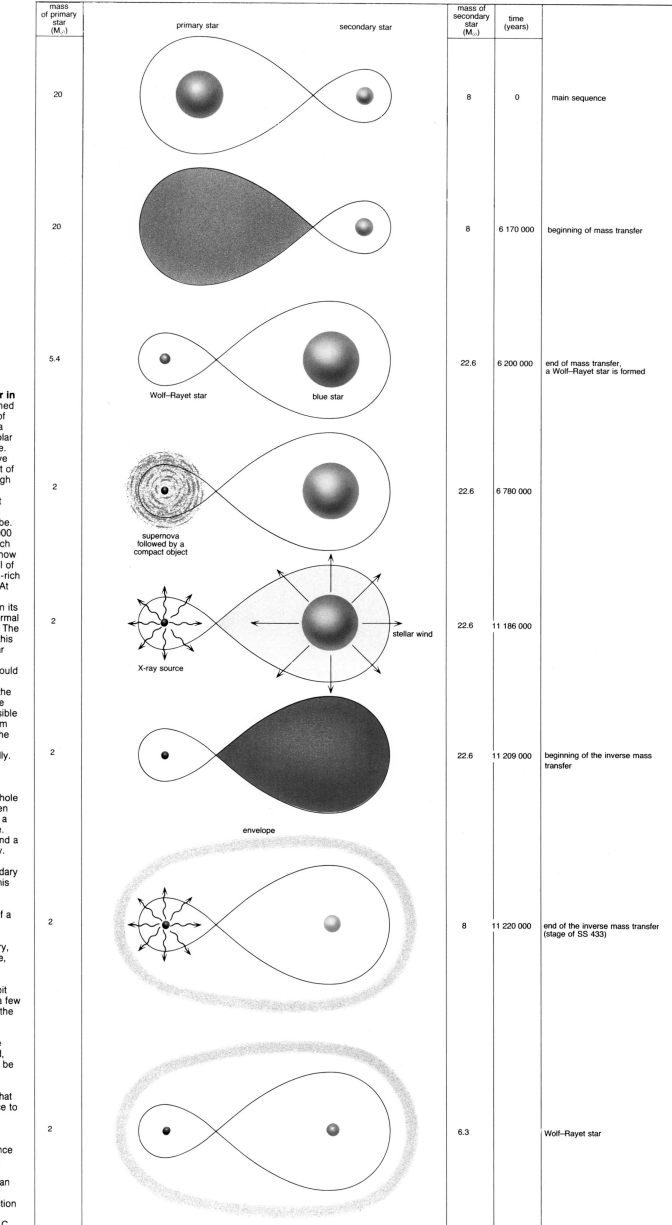

mass of primary star ($M_\odot$)			mass of secondary star ($M_\odot$)	time (years)	
	primary star	secondary star			
20			8	0	main sequence
20			8	6 170 000	beginning of mass transfer
5.4	Wolf–Rayet star	blue star	22.6	6 200 000	end of mass transfer, a Wolf–Rayet star is formed
2	supernova followed by a compact object		22.6	6 780 000	
2	X-ray source	stellar wind	22.6	11 186 000	
2			22.6	11 209 000	beginning of the inverse mass transfer
2	envelope		8	11 220 000	end of the inverse mass transfer (stage of SS 433)
2			6.3		Wolf–Rayet star

Formation and evolution of a Wolf–Rayet star in a binary system. A Wolf–Rayet star can be formed from a massive double star. The orbital periods of these systems are a few days. The evolution of a system made up of a pair of stars of 20 and 8 solar masses, respectively, is represented in the figure. The hydrogen-burning phase of the more massive star gives rise to an appreciable stellar wind. Part of this matter is passed to the secondary star through the inner Lagrangian point. At the end of the hydrogen-burning phase, six million years after it began, the outer layers of the massive star have expanded and are more than filling the Roche lobe. A transfer of mass takes place during some 20 000 years, the mass of the secondary growing to reach more than 22.6 solar masses. The primary star, now with a mass of only 5.4 solar masses, has lost all of its hydrogen-rich envelope and shows its helium-rich core. This helium star is also a Wolf–Rayet star. At this stage the radius of the primary is only a few solar radii, and the helium-burning phase starts in its core. This star is much more luminous than a normal star of identical mass with a hydrogen envelope. The stellar wind continues to be present throughout this phase. The secondary is now a massive blue star burning hydrogen and the orbital period of the system has increased. This Wolf–Rayet phase would last about 580 000 years, then the system will modify itself profoundly. Further evolution turns the primary into a supernova. The collapse of its core will form a neutron star or a black hole. It is possible that at the time of the explosion the binary system would be disrupted, the secondary star leaving the system with a high velocity; in any case the supernova explosion modifies the orbits drastically.

The companion continues its evolution in its hydrogen-burning phase, producing increasingly stronger stellar winds. This matter, strongly accelerated near the neutron star – or the black hole – gives rise to X-ray emission. The system is then called an X-ray binary. The companion becomes a red giant which fills the volume of its Roche lobe. Matter then starts to flow in the other direction and a certain amount of mass goes back to the primary. The opacity around the primary increases, preventing the observation of X-rays. The secondary star also becomes a helium star. At the end of this second mass transfer X-ray emission is again observable. The strange object SS 433 could possibly correspond to this evolutionary phase of a massive binary system. When the second mass transfer is completed the system appears to the observer as a single Wolf–Rayet star. The primary, which is now a compact object, is not observable, since it has received very little mass. The characteristics of this binary system are quite different from the first Wolf–Rayet stage. The orbit has changed, the orbital period now being only a few hours long. A large fraction of the mass has left the system, and the binary is now surrounded by a common gaseous envelope.

Later, another X-ray emission phase may take place: the secondary becoming a supernova and, finally, a compact object. The system could then be destroyed, the two neutron stars ceasing to be bound to each other. Another outcome is also plausible: the helium star and the neutron star, that is, the secondary and the primary, could coalesce to form a single neutron star.

This scenario is a simplified model of the real evolution of a massive binary system. But it can explain the major observational facts: the existence of Wolf–Rayet stars and of binary X-ray sources. Other observations, however, show that real systems are much more complicated than one can imagine.

From an observational point of view the detection of the binary nature of Wolf–Rayet stars is very difficult because of their complex spectra. (After C. de Loore, 1981)

Stellar nucleosynthesis

Nuclear processes

Supernova explosion in the galaxy NGC 4725. In the lower photograph, taken on 2 January 1941, the supernova can be seen, while in the upper photograph, taken on 10 May 1940, it is invisible. This sort of event happens about once every thirty to fifty years in each galaxy. It is the way a high mass star (more than 5 solar masses) finishes its evolution. The importance of this event, as far as nucleosynthesis is concerned, is enormous: most heavy nuclei (all those with masses higher than that of sulphur and lower than that of iron) are synthesised during these explosions, which contribute to the enrichment of the surrounding interstellar medium with the products of stellar nucleosynthesis. (Hale Observatories)

Nucleosynthesis is the transformation of chemical elements into others by means of nuclear reactions. We distinguish nucleosynthetic processes according to their nature and where they occur, and we can classify them into two main categories: those which transform relatively light elements into heavier elements, and those which result in the reverse transformation. Reactions in the first category are of two types: firstly, reactions between charged nuclei and secondly, neutron absorption by charged nuclei. Two sorts of nuclear process partially or completely destroy heavy charged nuclei: these are photodisintegration reactions and spallation ('pulling-out') reactions. Reactions between charged nuclei are also called thermonuclear fusion reactions. Because the nuclei that are interacting all have a positive electric charge, these reactions can occur only in very hot media (the temperature must be greater than several million degrees in the case of the fusion of hydrogen into helium), so that the thermal energy of the nuclei can overcome their electrostatic repulsion. The temperature at which a reaction cycle can occur depends strongly on the nature of the reacting nuclei: the greater their atomic number the greater the electrical repulsion between them and, consequently, the higher the temperature needed. The probability of a thermonuclear reaction, moreover, increases greatly with temperature. The second characteristic of these reactions is that they are exothermic (i.e. they liberate energy) for the elements lighter than iron in atomic weight. The reason is simple: the total mass of the individual nuclei that are susceptible to fusion is slightly more than the mass of the nucleus produced by the reaction. For example, in the fusion of four nuclei of hydrogen into one of helium, the mass loss is 0.7 per cent. The well-known relation $E = mc^2$, established by Albert Einstein, relates the mass lost to the energy released, which is transformed into radiation: 1 gram of hydrogen would change into 0.993 grams of helium and liberate 630 billion joules. Thermonuclear fusion reactions are therefore responsible for the transformation of light elements into heavier elements, but they are also energy sources. They play a fundamental role in the evolution of stars.

Fusion reactions can liberate energy only when they involve elements lighter than iron, which is the most stable element. However, heavy nuclei can be created from lighter nuclei where there is a source of neutrons, particles whose mass is almost equal to that of a proton, and which are electrically neutral. Because they are neutral, the probability of a reaction between a neutron and a nucleus is considerably greater than that of thermonuclear fusion. Neutron–nucleus reactions could, in principle occur at temperatures as low as we like; but only in principle, because neutrons have a limited lifetime – of the order of ten minutes – so neutron absorption reactions can occur only in regions where these particles are formed, that is, where thermonuclear reactions liberate a sufficient neutron flux. Neutron absorption reactions are responsible for the synthesis of elements heavier than iron.

Photodisintegration reactions are the reverse of thermonuclear fusion reactions: in a fusion reaction, light nuclei are turned into heavier nuclei with the emission of photons; in a photodisintegration reaction, photons are absorbed by heavy nuclei and split them into lighter nuclei. While fusion reactions liberate energy, these reactions absorb it. They need temperatures of three or four billion degrees for the photon fluxes to be intense enough. They are proportional to the fourth power of the temperature (if the temperature is multiplied by two, the flux is multiplied by sixteen). As an example, at a temperature of more than three billion degrees, silicon 28 can be split into a nucleus of magnesium 24 and a nucleus of helium 4.

Spallation reactions are induced by very light particles – hydrogen nuclei (protons) or helium 4 nuclei (alpha particles) – moving extremely rapidly and hitting heavier target nuclei (or the reverse, a fast heavy nucleus meeting a slow nucleus of hydrogen or helium). The collision between the very fast particle and the heavy nucleus results in some of the nucleons (protons or neutrons) gaining enough energy to leave the heavy nucleus. These endothermic reactions (i.e. reactions which absorb energy) allow us to explain the formation of lithium (atomic number 3), whose abundance is difficult to account for

type of reaction	reaction sites	reacting chemical species	species produced
thermonuclear fusion	central regions of stars (calm or exploding), the early Universe	from hydrogen up to iron	from helium up to iron
neutron absorption	central regions of stars and exploding stars	iron peak and also heavier nuclei	nuclei heavier than iron
photodisintegration	exploding stars	elements between sulphur and iron	some stable nuclei between neon and iron, and helium
spallation	interstellar medium bombarded by cosmic rays, surfaces of active stars	carbon, nitrogen, oxygen and eventually heavier nuclei	lithium, beryllium and boron

Summary of the principal nucleosynthetic processes

Schematic representation of the four nucleosynthetic processes involved in the formation of the chemical elements. The upper left-hand diagram (a) represents thermonuclear fusion between two charged nuclei. This fusion leads to the formation of a heavier and more highly charged nucleus. It needs a high temperature for the two incident nuclei to overcome their repulsion, but the fusion generally liberates energy which is radiated by the star.

The lower left-hand diagram (b) represents the process of neutron absorption (responsible for the nucleosynthesis of elements heavier than iron). This exothermic process (i.e. one which liberates energy) can occur at any temperature because neutrons are not affected by electrical repulsion during their absorption. In fact, the probability of this reaction is highest when the temperature is very low.

The upper right-hand diagram (c) represents the photodisintegration process. It is an endothermic reaction (it absorbs energy) and is the reverse of the fusion reaction. Many photons are present when the gas is very hot, their number being proportional to the fourth power of the temperature according to the Planck law. A nucleus may split into smaller fragments under the bombardment of these photons.

The lower right-hand diagram (d) represents other endothermic reactions, the spallation reactions. A very high energy particle strikes a target nucleus and rips out nuclei of smaller masses: lithium, beryllium and boron are formed by this process from carbon, nitrogen and oxygen.

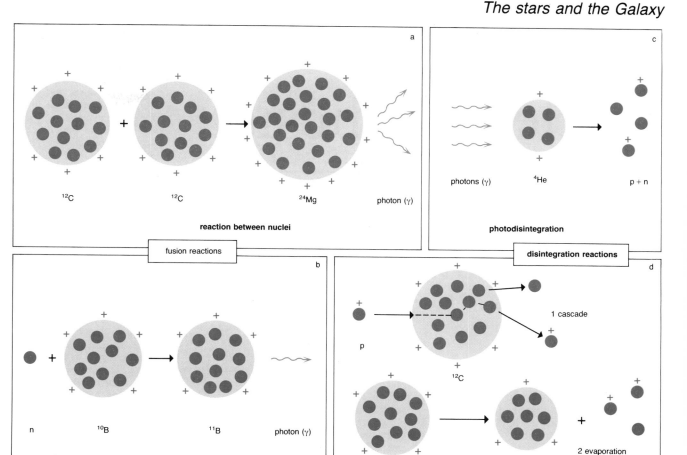

reaction between nuclei

fusion reactions

photodisintegration

disintegration reactions

absorption of a neutron by a nucleus

spallation

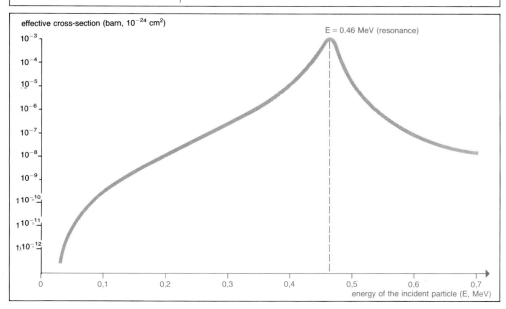

otherwise, from carbon (atomic number 6). They happen mostly during the bombardment of the interstellar medium by cosmic rays, but they can, to a lesser extent, also take place at the surface of certain stars which can create a flux of fast particles.

To sum up, thermonuclear fusion reactions occur in the central regions of stars, which are heated by the energy thus liberated and radiate the energy they produce. They can happen at any time when material that can undergo these reactions is subjected to very high temperatures; this was the situation during the very first stages in the evolution of the Universe, according to the 'Big Bang' theory. Neutron absorption reactions occur in the central regions of stars, where there are reactions that produce enough neutrons. Photodisintegration reactions are important in the last stages of the evolution of massive stars and the onset of supernova explosions.

Jean AUDOUZE

A typical thermonuclear fusion reaction: the reaction between a proton and a carbon 12 nucleus. This figure presents three diagrams giving data concerning the fusion reaction proton + carbon 12 → nitrogen 13 + photons.

The upper diagram represents this reaction between two charged nuclei, the proton and the carbon 12 nucleus, which exert an electrical repulsion on each other. This reaction can occur only at temperatures of the order of ten million degrees at which this repulsion is effectively overcome. It gives rise to a nitrogen 13 nucleus and to energy liberated in the form of photons.

In the middle diagram, we have a schematic diagram of the potential energy between the incident proton and the carbon 12 nucleus. At distances greater than the radius of the carbon nucleus (equal to a few fermis, or a few 10^{-13} cm) the potential is repulsive and decreases as the reciprocal of the distance between the two particles, that is, it becomes higher and higher as the incident particle approaches its target. Inside the nucleus (i.e. in the central few fermis), the potential becomes strongly attractive because of the effect of the short-range nuclear forces which replace the long-range electrical forces). Despite the repulsive potential, the incident particle has a low, but non-zero, probability of penetrating into the nucleus.

At certain energies, the nitrogen 13 nucleus has excited states somewhat like those of atomic electrons. When the energy of the incoming particle corresponds to one of these energies, the probability of a reaction is a maximum. We say then that there is a resonance.

The lower diagram shows the probability of a reaction as a function of energy. This probability at first grows exponentially and very strongly because of the electrical repulsion which resists the reaction between the two nuclei if their relative velocity is too low. Notice a saturation of this probability when the energy is greater than the repulsive potential and a maximum corresponding to a resonance (a coincidence with the energy level of an excited state) of the incident particle with an excited level of the nucleus of nitrogen 13.

Stellar nucleosynthesis

Nucleosynthesis and stellar evolution

The various nuclear processes that can occur govern the evolution of a star and allow us to explain the composition of the matter present in the universe. In fact, a star has only two sources of energy: gravitational energy, liberated when the star contracts (i.e. when it changes its size) and nuclear energy, liberated mostly by fusion reactions.

When a star evolves rapidly, it takes its energy mostly from gravitational contraction; it is inactive from the point of view of nucleosynthesis. On the other hand, the energy liberated when nucleosynthetic processes occur in the central regions stabilises the star since it does not need to contract, or to change its appearance, in order to radiate.

The principal reaction cycles linked to the evolution of stars are the fusion of hydrogen into helium, the fusion of helium into carbon, the fusion of carbon into neon and magnesium and of oxygen into silicon and sulphur, and last, the slow absorption of neutrons, called the s-process. During the explosive phases which characterise the end of massive stars or which set off nova explosions, nuclear fusion reactions, photodisintegration reactions and rapid neutron capture can occur during the passage of the explosion's shock-wave.

The fusion of hydrogen into helium takes place in the central regions of main sequence stars like the Sun and in the intermediate zones of more evolved stars. Two basic reaction chains can take place, the proton–proton (PP) chain, and the carbon–nitrogen–oxygen (CNO) cycle. (See table at top left of opposite page.) The proton–proton chain is divided into three branches, called the PP I, PP II and PP III chains. In the PP I chain, the combustion of hydrogen into helium takes place directly. Helium itself acts as the catalyst in the PP II and PP III chains, and nuclei of carbon, nitrogen and oxygen act as catalysts in the CNO cycle, which was discovered in 1939 by Hans Bethe and Viktor von Weizsäcker. For a star like the Sun, 70 per cent of the energy liberated comes from the PP I cycle, 29 per cent from the PP II cycle, 0.1 per cent from the PP III cycle and less than 1 per cent from the CNO cycle.

The relatively large difference between the mass of the helium nucleus and the mass of four nuclei of hydrogen, as well as the extreme slowness of the fusion of two protons in the $H + H \rightarrow D$ reaction explain why the luminosity of a star like the Sun remains almost constant for a time as long as ten billion years; it is also the reason why so many stars are on the main sequence. In catalytic cycles like the CNO cycle, the elements corresponding to the slowest reaction, like nitrogen 14, have their abundance increased at the expense of the other elements in the cycle.

Fusion of helium into carbon occurs at temperatures of the order of a hundred million degrees and at densities of about 10 000 grams per cubic centimetre, much greater than in the case of the fusion of hydrogen into helium. This cycle occurs in the central regions of red giant stars; it stabilises the stars for periods ten to a hundred times shorter than the preceding cycle for stars on the main sequence, which explains the difference between the number of main sequence stars and the number of red giants. We must note that this cycle does not explain the formation of elements between helium and carbon.

Fusion of carbon and oxygen must occur in the centres of supergiant stars, the last stages of the evolution of stars more massive than the Sun. Stars can be stabilised for relatively short times by these cycles. There are three reasons: firstly, the mass difference between the 'parent' and 'daughter' nuclei is small: the fusion of a given mass of carbon or oxygen liberates at most a hundredth of the amount of energy liberated by the fusion of the same mass of hydrogen; secondly, the quantity of fuel (carbon and/or oxygen) is much less than the amount of hydrogen at the beginning; and lastly, at the temperatures (equal to or greater than a billion degrees) necessary to start these reactions, energy is also liberated in the form of neutrinos, particles of zero or very small mass, which (as opposed to photons) react only very weakly with matter.

Carbon can be transformed into neon or magnesium, and oxygen into silicon or sulphur. Increasingly heavy elements are created in the more massive stars which can provide the necessary central pressures and temperatures.

Explosive phases, like those at the onset of supernovae, can set off the mechanisms responsible for the formation of iron and other elements close to iron in atomic number. Because of the very large electric charges of the nuclei involved, we must call on neutron absorption mechanisms to explain the synthesis of elements heavier than iron. For heavy nuclei, the number of stable isotopes of a given element is generally larger than in the case of light elements. (As an example, xenon has nine stable isotopes while another inert gas, neon, which is lighter, has only three.) We can distinguish three types of isotopes of heavy elements. The first includes those that have a relatively very low number of neutrons and are formed, for example, from other heavy nuclei by proton absorption reactions (these elements are for this reason called 'p'). The second class is formed by slow neutron absorption (the s-process), and the third is those formed by rapid neutron absorption (the r-process). During the red giant stage, a partial mixing of the com-

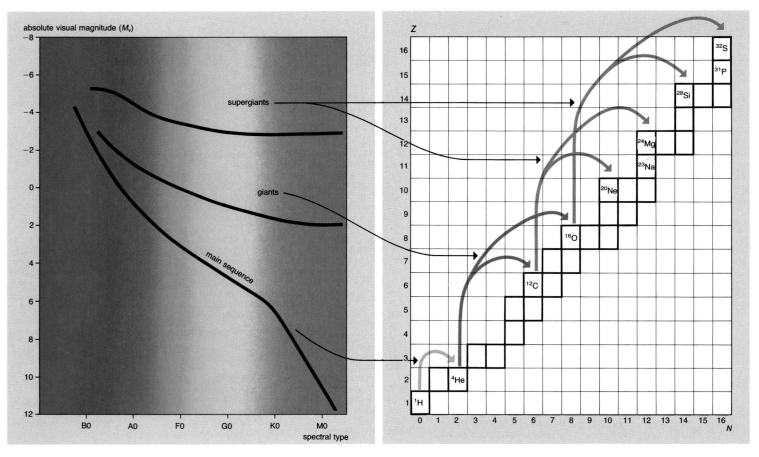

Juxtaposition of a Hertzsprung–Russell (HR) diagram with a Mendeleyev-type (N, Z) diagram. Points representing stable atomic nuclei are plotted as a function of their numbers of neutrons N and protons Z. (The latter is known as the atomic number.) The point representing a star on the HR diagram moves relatively rapidly when gravitational contraction takes place in the star, but nucleosynthesis does not occur. On the other hand, when nucleosynthetic processes are taking place in its central regions, the position of a star in the HR diagram is fixed while its nuclear composition is changing. We can regard the main sequence as associated with the transformation of hydrogen into helium, the red giant stage with the transformation of helium into carbon and oxygen, and some supergiants with the fusion of carbon into neon, magnesium and sodium, and the fusion of oxygen into silicon, phosphorus and sulphur, etc.

PP I cycle	PP II cycle	PP III cycle	CNO cycle
$H + H \rightarrow D + e^+ + \nu$ $D + H \rightarrow {}^3H + \gamma$ ${}^3He + {}^3He \rightarrow {}^4He + 2\,H + \gamma$	${}^3He + {}^4He \rightarrow {}^7Be + \gamma$ ${}^7Be + e^- \rightarrow {}^7Li + \nu$ ${}^7Li + H \rightarrow {}^8Be \rightarrow 2\,{}^4He$	${}^7Be + H \rightarrow {}^8B + \gamma$ ${}^8B \rightarrow {}^8Be + \nu \rightarrow 2\,{}^4He$	${}^{12}C + H \rightarrow {}^{13}N + \gamma$ ${}^{13}N \rightarrow {}^{13}C + e^+ + \nu$ ${}^{13}C + H \rightarrow {}^{14}N + \gamma$ ${}^{14}N + H \rightarrow {}^{15}O + \gamma$ ${}^{15}O \rightarrow {}^{15}N + e^+ + \nu$ ${}^{15}N + H \rightarrow {}^{12}C + {}^4He$

Main sequence: the mechanisms of hydrogen combustion.

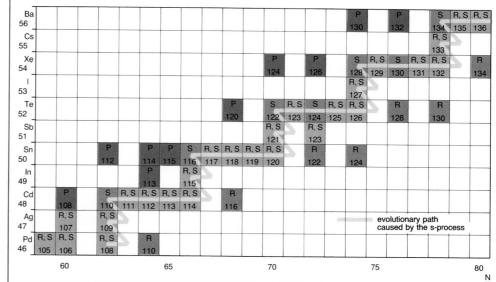

Nucleosynthesis of the heavy elements. This diagram (above) represents a fragment of a nuclear diagram in which the number of neutrons N is plotted as the abscissa with the number of protons Z as the ordinate. Isobars (nuclei with the same atomic mass) are found on the same diagonal from upper left to lower right. We can distinguish three families of stable nuclei: the nuclei marked P (also called p-elements) whose abundance is lower than that of the nuclei of the two other families, and which are relatively rich in protons. These nuclei cannot be formed by neutron absorption; they are formed by proton absorption or neutron emission by partial photodisintegration affecting nuclei of the other two families.

The other two families marked S and R correspond respectively to the nuclei formed by slow absorption and rapid absorption of neutrons. The r-nuclei formed by rapid neutron absorption include nuclei that are relatively rich in neutrons. The s-process of slow neutron absorption follows the path of the continuous line through the middle of the diagram: when neutron absorption leads to the formation of an unstable nucleus, this has enough time to undergo a β-decay by transforming one of its neutrons into a proton, before capturing another neutron; it then goes to the corresponding isobar with an atomic number one higher.

Rapid neutron absorption allows the synthesis of nuclei that are very neutron-rich. When a family of isobars includes a pure r-nucleus (which cannot be synthesised by the s-process) the stable isobar on the s-process path is necessarily a pure s-nucleus because the r-nucleus will have stopped the r-process at its own position. This is not true for all nuclei on the s-process path, most of which can be synthesised by either process. In the Sun, the contributions of the two processes are almost the same.

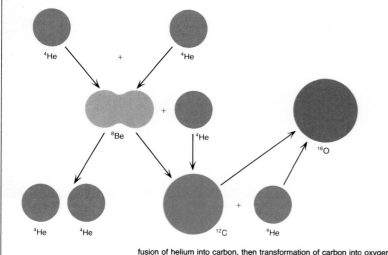

fusion of helium into carbon, then transformation of carbon into oxygen

Combustion of helium into carbon. This process, which occurs at temperatures of the order of one hundred million kelvin in the centres of red giant stars, takes place in two stages in the following way: two helium nuclei undergo an equilibrium reaction, the resultant nucleus being beryllium 8. As this nucleus has an extremely short lifetime (of the order of 10^{-16} seconds), only a very small quantity of beryllium 8 is formed by this equilibrium reaction, which is the first stage of the process. The second stage consists of the fusion of these few beryllium 8 nuclei with helium nuclei to form carbon 12. This second stage is helped by the fact that the fusion reaction between beryllium 8 and helium 4 has a high probability (owing to the presence of an excited level of carbon 12 at an appropriate energy). Some carbon 12 nuclei can then fuse with helium 4 to form oxygen 16. This three-body (or four-body) mechanism was devised by Edwin Salpeter and Fred Hoyle in 1955–56. Fred Hoyle predicted the existence of the excited level of carbon 12 before his prediction could be tested by experiment.

Sketch showing the results of the fusion of carbon and oxygen. The fusion of two carbon nuclei occurs in the normal evolution of massive stars at temperatures of the order of six hundred million kelvin and during supernova explosions at temperatures of the order of two billion kelvin. The non-explosive process forms ${}^{24}Mg$ (very rarely), ${}^{23}Na + H$ or ${}^{20}Ne + {}^4He$. The process also allows the formation of ${}^{25}Mg$ and ${}^{26}Mg$. The fusion of oxygen occurs either at a billion kelvin (non-explosive case) or three billion kelvin (explosive case). The non-explosive case forms ${}^{32}S$ (very rarely), ${}^{31}P + H$ or ${}^{28}Si + {}^4He$. The process allows the synthesis of elements up to the iron peak.

fusion cycle	temperature (10^6 K)	density (g/cm³)	principal elements produced	phase of stellar evolution	mass of the stars undergoing the cycle (in solar masses)
hydrogen	1–20	1–100	${}^4He$ (${}^{14}N$)	main sequence	>0.1
helium	100	10^5	${}^{12}C$, ${}^{16}O$, ${}^{18}O$, ${}^{20}Ne$	red giant	>0.4
carbon	500–800	10^5–10^6	${}^{20}Ne$, ${}^{23}Na$, ${}^{24}Mg$	supergiant	>0.7
oxygen	1000	10^5–10^6	${}^{28}Si$, ${}^{31}P$, ${}^{32}S$	supergiant	>1.0

Thermonuclear fusion cycles occurring during the evolution of stars.

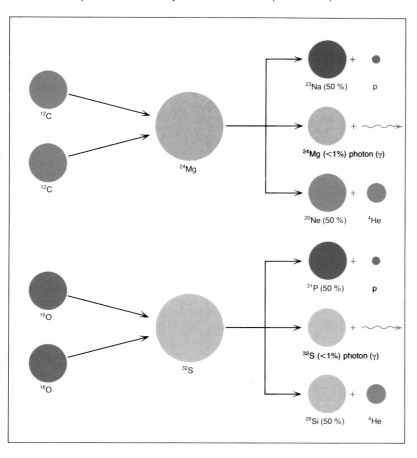

bustion zones of hydrogen and helium leads to reactions (such as ${}^{13}C + \alpha \rightarrow {}^{16}O + n$ or ${}^{22}Ne + \alpha \rightarrow {}^{25}Mg + n$) which provide the necessary neutrons for the process of slow absorption.

If the formation of the light elements lithium, beryllium and boron is explained by bombardment of the interstellar medium by cosmic rays, the nucleosynthesis that takes place in the calm phases of stellar evolution appears not to be able to give a quantitative explanation of the formation of elements between sulphur and iron, or of the heavy elements whose nuclei are rich in neutrons. Furthermore, certain isotopes of relatively light elements, like nitrogen 15 and the heavy isotopes of magnesium, are not formed during the preceding phases in the proportions observed today. We now believe that all of these elements are formed during explosive stellar phases (especially supernovae), either by fusion reactions taking place at temperatures much higher than those reached in calm phases, or by

photodisintegration reactions or by rapid neutron captures.

Photodisintegration reactions first occur at temperatures of three to four billion degrees, starting with silicon; they allow the conversion of silicon into heavier elements by quasi-equilibrium reactions (a succession of photodisintegrations and fusions of the fragments). At temperatures of four to five billion degrees, they affect elements close to iron in mass; this explains the higher abundance of this group compared with the surrounding elements, the stability of the nucleus here being at its greatest.

Since many heavy isotopes cannot be synthesised by slow neutron capture (this is in any case true for elements heavier than lead), we have an indirect proof that rapid neutron capture processes must occur during explosive stellar phases, where very large neutron fluxes can be liberated.

Jean AUDOUZE

Binary stars

Two stars can appear to be very close together in the sky, either because they both happen to be seen in almost the same direction while they are actually a considerable distance apart, or because they really are close together in space. Double stars of the first type are called optical binaries. There are in fact very few examples. Most stars that look double are indeed physically bound: they are true binary stars, or double stars. Each star moves in an ellipse around the other, as the planets do around the Sun. The movements of stars forming an optical binary are obviously unrelated. When we observe two stars close to each other, the characteristics of their motion allow us to decide whether they are an optical binary or a true double star.

There also exist multiple systems, that have at least three stars sufficiently close to each other that their mutual gravitational attraction dominates other gravitational forces. Such systems have a tendency to split into pairs of stars, with the other components perturbing the orbital motion of each pair.

A star also feels gravitational interactions from all the stars in the Galaxy and it has orbital motion around the centre of the Galaxy itself. It is only when two (or more) stars are very close to each other that their mutual gravitational interaction is preponderant over the other interactions. We can then study the relative movements of the components of the system in isolation.

The discovery of binary stars

The term 'double star' was, it seems, used for the first time by Claudius Ptolemy about the two stars v_1 and v_2 Sagittarii, whose angular separation is 14 arc minutes, but which are in fact independent. The first observation of binary stars with a telescope appears to have been made in around 1643 by the Italian, Giambattista Riccioli, who discovered that Mizar (ζ Ursae Majoris) has two components, (ζ_1 and ζ_2). In 1761, Johann Heinrich Lambert thought that such pairs could not all be purely fortuitous: it was already impossible to explain statistically the number of these pairs as chance optical doubles. The idea of real stellar companions was envisaged in 1767 by John Mitchell and in 1779 by Christian Mayer, but at that time no observation would allow the verification of this hypothesis. It was William Herschel who, in 1803, made the first observations that showed the apparent motion of one star of a pair with respect to the other. He concluded that this movement resulted from the mutual interaction of the two stars. However, it was not until 1827 that Félix Savary showed that the relative orbit of the binary star system ζ Ursae Majoris was an ellipse completed in sixty years. This result was an important milestone in the evolution of scientific ideas: for the first time, it had been verified that the terrestrial laws of physics apply irrefutably to the Universe of stars. Until then, the universality of those laws was a matter of speculation, although from a purely philosophical point of view it was generally accepted.

Systematic research into binary stars developed very rapidly, benefiting from the improvements in instruments. Contrary to the position in numerous other branches of astrophysics, observations made during the last century are neither useless nor obsolete: indeed, numerous visual binary stars have not yet completed an orbital revolution since their discovery (i.e. they have not been observed again in their original relative position).

The binary star phenomenon is very common; it has been estimated that more than half of the stars in our Galaxy belong to binary or multiple systems, which poses the problem of their origin.

Parameters measured from observation of binary stars

Analysis of the shapes of the light curves of eclipsing binary stars allows us to find out certain physical parameters of the atmospheres of the stars of the system. When there is a total eclipse, we can determine the ratio of the effective temperatures of the two components; it is also possible, in the case of certain pairs, to deduce the diameters of the stars. This determination requires the knowledge not only of the light curve, but also of the spectra of the two components, which allows us to establish the variation of their radial velocities; the diameters can then be determined as absolute values. It has been possible to calculate diameters in this way for only a small number of stars, but such measurements are important because, along with interferometry, they provide a way to measure directly the diameter of stars.

Certain systems allow a more detailed analysis of the atmosphere of one of the components. For example, the star ζ Aurigae is a system made up of a giant star with a radius 245 times that of the Sun and of spectral type K, and a main sequence star of spectral type B. We observe only the radiation from the B-type star, which is the more luminous and which is periodically eclipsed by the giant star. This giant star has a very extended,

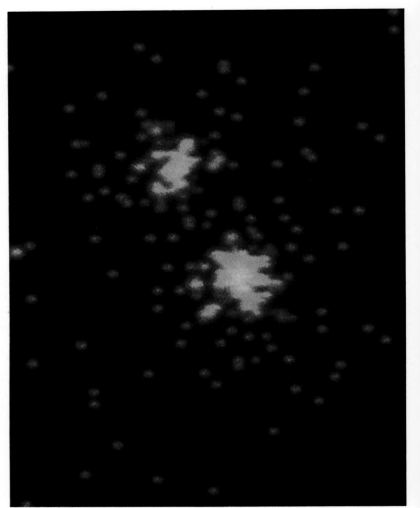

The binary star Alpha Centauri. Alpha Centauri is a triple system, of which we see here the two principal components, A and B, which are dwarfs (luminosity class V), of spectral types G2 and K1, respectively. This X-ray image was made by the Einstein satellite. The third star, Alpha Centauri C or Proxima Centauri, is outside the field, but has an X-ray luminosity comparable with those of the two other stars. These three stars are 1.3 parsecs from the Sun and are our closest neighbours. The apparent separation between the A and B components is 30 arc seconds, corresponding to a real separation of about 40 astronomical units; the system completes a revolution every eighty years. (Einstein, HEAO-2, Harvard-Smithsonian Center for Astrophysics)

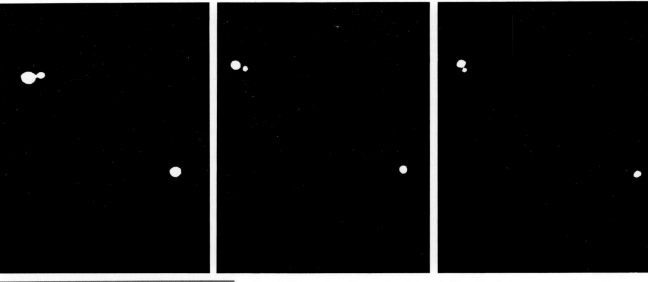

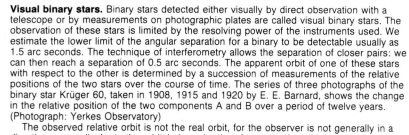

Visual binary stars. Binary stars detected either visually by direct observation with a telescope or by measurements on photographic plates are called visual binary stars. The observation of these stars is limited by the resolving power of the instruments used. We estimate the lower limit of the angular separation for a binary to be detectable usually as 1.5 arc seconds. The technique of interferometry allows the separation of closer pairs: we can then reach a separation of 0.5 arc seconds. The apparent orbit of one of these stars with respect to the other is determined by a succession of measurements of the relative positions of the two stars over the course of time. The series of three photographs of the binary star Krüger 60, taken in 1908, 1915 and 1920 by E. E. Barnard, shows the change in the relative position of the two components A and B over a period of twelve years. (Photograph: Yerkes Observatory)

The observed relative orbit is not the real orbit, for the observer is not generally in a direction perpendicular to the orbital plane; he therefore sees the orbit in projection onto a plane perpendicular to the line of sight. Geometrical considerations allow us to reconstruct the real relative orbit. The diagram shows the apparent orbit of the companion of the star ξ Bootis. (After Hartmann, *Astronomy*, 'The Cosmic Journey', fig. 18–6, p. 334)

The absolute orbit (i.e. the orbit of each component about the centre of mass) can be determined only if we know from elsewhere the position of each component with respect to an 'absolute' reference system defined by distant stars in the Galaxy. The characteristic separation of such pairs varies from 1.2 to several hundred astronomical units. The periods of such systems are always very long: from 1.7 years to more than a century. Furthermore, the time since discovery is not enough to let us determine the periods of several centuries associated with very large orbits. The catalogue published in 1963 by Lick Observatory included 64 246 visual binary star systems, but the orbits of only 700 systems have been completely determined: in general, we have only very fragmentary observations, which are not sufficient to calculate the geometrical parameters of the orbit, but only to show that the stars are physically associated.

Some pairs of binary stars cannot be discovered by their closeness in a telescope, because their angular separation is too large, given their closeness to us and the size of their orbits. Such systems are detected only because the two stars have the same proper motion.

diffuse atmosphere forming its outermost regions. Because this atmosphere is partially transparent, when the eclipse occurs, the B star disappears a little at a time behind the K star. Spectroscopic analysis of the B star seen through the atmosphere of the K star allows a detailed analysis of the various layers of the atmosphere of the latter star. We know of four other systems similar to this one. Furthermore, detailed analysis of the light curves of certain binary stars allows us to find the speed at which the eclipsed star rotates on its axis.

The study of binary stars is fundamental: it is only by observing them that we can find out the mass of stars by direct estimation without relying on any theory *a priori*, about their structure.

It is from these measurements that the mass–luminosity relation was established; this relation plays a fundamental role in the testing of theoretical models describing the internal structure of a star at the beginning of its life.

The masses of the components cannot be determined for all systems. Observations allow us to find the shape of the orbit and the period of revolution; the law of gravitation gives a relation between these observed parameters and the masses of the interacting stars, from which it is possible to find the mass of each star in the system in certain cases.

The classes of binary stars

Binary stars may be detected by various means of observation; their classification is, therefore, based on the technique by which they were discovered. We can distinguish three classes of binary stars: visual binary stars, astrometric binary stars and spectroscopic binary stars.

If the distance of a visual binary star is known, the relative orbit of one of the stars with respect to the other and the orbital period distance can be used to calculate the sum of the masses of the two components. If the motion of each of these components can be determined with respect to the centre of gravity of the system by other means, we can then calculate the mass of each component.

In this way, the masses of Sirius and its companion were calculated: they are respectively 2.27 and 0.98 solar masses. The very different apparent fluxes – the ratio is about 40 000 – of the two stars correspond to a great difference in their luminosities, since they are at the same distance from us. However, the two components of the system have similar colours which allows us to estimate their effective temperature; from the luminosity of Sirius, which is a main sequence star, we have deduced a value for the radius of each star, taking account of the relation linking the luminosity with the radius and the effective temperature. The radius of Sirius A is estimated as 2.3 solar radii and that of Sirius B as a hundredth of a solar radius. Thus, the mean density of Sirius B is a million times higher than that of the Sun; it is a white dwarf. The reasoning applied here can be reversed. If we estimate the masses of the stars from their spectroscopic characteristics by using the relation linking the sum of the masses in a system, its parallax and the orbital period and the semi-major axis of the relative orbit we can obtain a value for the parallax – called the dynamic parallax – from which the distance to the system can be estimated.

Astrometric observations offer the possibility of discovering stars of low mass – less than 0.06 solar masses – causing only minor irregularities (or perturbations) in the proper motion of a more massive star. These low mass stars are unobservable directly because their luminosity is too low. Astrometric measurements also may lead to the discovery of planetary systems perturbing the proper motion of the central star.

Observations of the perturbations in the motion of a star with an invisible companion are not sufficient for a direct determination of the masses. However, these observations can lead to an estimation of the mass of each star in the system, on condition that the parallax is known and the sum of the masses of the components is assumed. The masses thus calculated are only

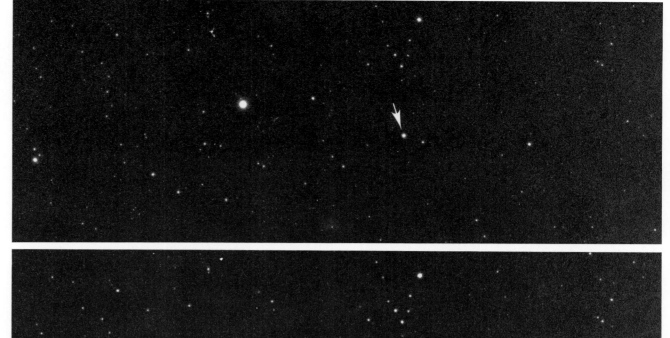

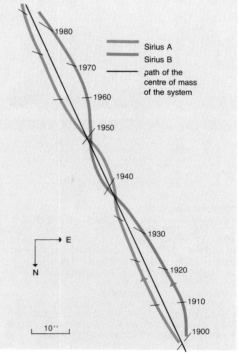

Astrometric binary stars. Repeated measurements of the positions of certain stars have permitted the discovery of irregularities in their proper motions indicating the presence of an invisible companion, either so faint or so close that its image cannot be separated from that of the principal star. Thus in 1844 F. W. Bessell discovered that Sirius shows a non-linear proper motion, whose oscillations resemble those of a visual binary star: he deduced that the proper motion of Sirius is affected by the gravitational interaction with a second star whose luminosity is low. Sirius B, has a magnitude of 8.7 while the magnitude of Sirius A is −1.4. The diagram on the right shows the paths of Sirius A and Sirius B, plotted with respect to an absolute reference system.

F. W. Bessel made a similar observation of the star Procyon, whose companion was detected in 1896. The companion stars are, however, not always observable for some are beyond the limiting magnitudes of even the best-performing instruments. Nowadays we know more than twenty astrometric binary systems. This method is particularly good for the detection of long-period systems (tens of years).

Irregularities are sometimes detected in the orbital motions of a binary star, indicating the presence of an invisible third component. It is then a triple system. The star 61 Cygni is an example.

The best known observed astrometric system is probably Barnard's star, which is the third closest star system to us. The star of spectral type M5 V has a visual magnitude of 9.5; the period of the system – 11.5 years – implies that the companion has a mass similar to the mass of Jupiter. Elaborate calculations suggest that Barnard's star actually has two companions, the second having an orbital period of 26 years and a mass of 1.15 times that of Jupiter. The distance of these objects from the central star would be several astronomical units; these results lead us to think that the perturbations of Barnard's star are due to a planetary system.

In the context of star formation, the importance of such a result is considerable and research into astrometric binary systems among the closest stars to the Sun aims at the discovery of low-mass objects (of the order of 0.01 solar mass), which might be planets. The two photographs above show the movement of Barnard's star in ten years. (Photograph: Lick Observatory)

plausible guesses, but the choice is very limited given the constraints deduced from observations of the system, and notably from the fact that the invisible companion is not very luminous.

Spectroscopic binaries appear as single stars, even in large telescopes, but periodic shifts in the positions of the spectrum lines, or doubling of the lines, reveals the presence of two stars. Observations of a single-lined spectroscopic binary star allow us only to find the sum of the masses of the components, weighted by the inclination of the plane of the orbit to the line of sight. In the case of a double-lined spectroscopic binary, analysis of the two light curves allows us to calculate the mass of each component, but still weighted by the inclination of the orbit. For a sample of many pairs, we can obtain a statistical value of the masses by taking an average value for the inclination. However, if the pair is also a visual

binary star or an eclipsing binary star, the inclination of the orbit can be determined and we then obtain a real value for the mass of each of the components.

The smallest mass values thus measured are those of the visual binary system Ross 614, whose components have masses of 0.114 and 0.062 solar mass. The most massive star detected in this way is part of the system V729 Cygni; it is an eclipsing contact binary and also a double-lined spectroscopic binary, whose principal component has a mass of 58.7 ± 9 solar masses. Even if this mass has been determined correctly from a formal point of view, there are numerous problems in the analysis of the light curve and the spectra leading to the uncertainty. The second component has a mass of (13.7 ± 6.3) solar masses.

continued on p. 305

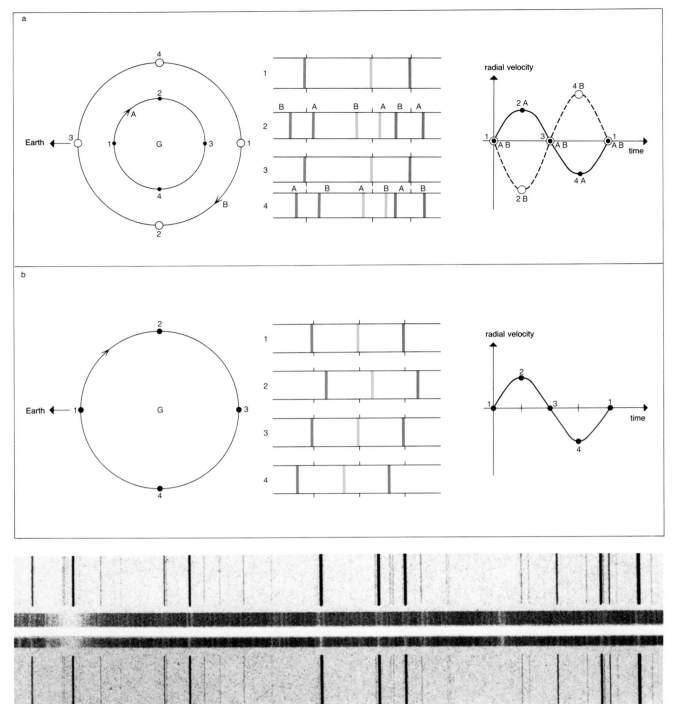

Spectroscopic binary stars. A new way of detecting binary stars was discovered in 1889 by Antonia Maury, who noticed that the spectrum of the star β Aurigae sometimes showed a doubling of the lines. This is interpreted as the spectra of two stars on the plate at once. The positions of the lines vary on the wavelength scale in the course of time because of the orbital motion of each component. The variation in radial velocity is periodic, the period being that of the orbital motion. Such stars are called spectroscopic binary stars.

Diagram (a) shows how the periodic motion of the stars can produce a change in the radial velocity, supposing that the orbits are circular about the centre of gravity (G) of the system and that the Earth is in the orbital plane.

Since the stars are very close together, we will observe them both simultaneously and their two spectra appear on the same photographic plate. When the stars are in positions 1 or 3, the radial velocities due to the orbital motion are zero, and the lines of the two stars are merged. During their movement from position 1 to position 2, star A goes away from the Earth, while star B approaches the Earth. There is therefore a shift of the lines of each star due to the Doppler effect, towards the red for star A and towards the blue for star B. During the movement from position 3 to position 4, the reverse phenomenon occurs. The respective lines of the two stars are not shifted by the same amount; star B must move faster in its orbit than star A to have the same orbital period, since it is further from the centre of gravity. The result of this periodic motion of the stars in their orbits is a sinusoidal variation of their respective radial velocities, represented in the diagram on the right.

We say that such stars are double-lined spectroscopic binaries. In some cases, we observe only one spectrum on the photographic plate, because the second star in the system is too faint to show up. We say it is a single-lined spectroscopic binary star. Figure (b) shows this situation. The theories are the same as in the preceding case. In the general case, where the relative orbits are not circular but elliptical, the radial light curves are still periodic but not sinusoidal.

There are a very great number of spectroscopic binary stars. We estimate that on average one star in every three of four is a spectroscopic binary star. The observed periods vary from several hours up to several years. From the study of the radial velocity curves, it is possible to calculate the parameters defining the elliptical orbit of one star relative to the other, but not to find out the orientation of the orbit in space. However, stars can show periodic variations in their radial velocities without being spectroscopic binaries. In fact, if we calculate the characteristics of their orbits, we find that the radius would be smaller than a plausible radius for a star! These variations are attributed to movements of the whole of the atmosphere of the star: these are variable stars, Cepheids for example. A spectroscopic binary star can be discovered only if it is a close system, where the orbital velocity of the components is high: then the variations in the radial velocity are large enough to be measured. The efficiency of this method is a function of the accuracy of measurement of the Doppler effect. Visual binary systems and astrometric binary systems generally have radial velocity variations too small to be detected, because the orbital motion of stars is much slower when the orbit is larger.

The spectra shown above are those of the spectroscopic binary star HD 6980, consisting of two stars of spectral type G0, similar to the Sun, whose orbital period is 26 days. The radial velocities oscillate between −48 and +48 kilometres per second. The upper spectrum corresponds to the maximum separation of the lines, which are merged in the lower spectrum. The spectra of HD 6980 have on either side the spectrum of an iron arc, which is used for reference. (Photographs, observatoire de Haute Provence du CNRS)

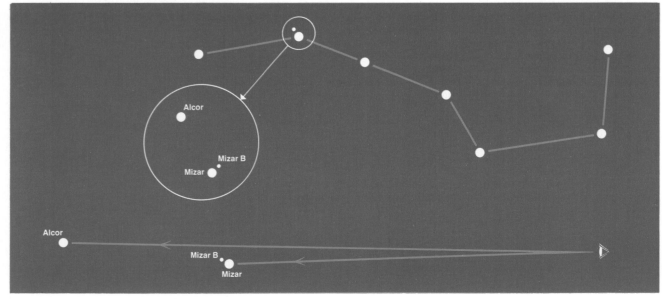

Alcor and Mizar. Separated by an angular distance of 12 arc minutes, Alcor and Mizar (ζ Ursae Majoris) form an optical pair, although their proper motions show that they are linked by gravitational interactions. Mizar is a visual binary system, easily observable in a small telescope, and has a separation of 14.4 arc seconds. Each component is also a spectroscopic binary. It is even possible that Mizar B is a triple system.

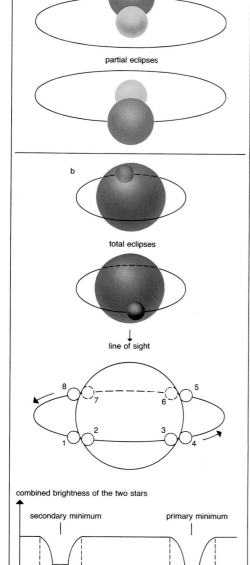

partial eclipses

total eclipses

line of sight

combined brightness of the two stars

secondary minimum   primary minimum

t_1 t_2 t_3 t_4 t_5 t_6 t_7 t_8 time

light curve

Eclipsing binary stars (left). When the orbit of a binary star is seen edge-on or at a very small angle, one of the stars comes periodically into the line of sight between the observer and the other star. Eclipses – or more precisely occultations – occur. The occultation will be partial (a) or total (b) according to the relative diameters of the stars and the inclination of the plane of the orbit to the line of sight.

These binary stars are detected by the periodic variation of the apparent magnitude of the system. They are called eclipsing binaries or photometric binary stars.

The first star of this type to be observed was Algol (β Persei). Its variations in brightness were observed in 1670 by Geminiano Montanari. It seems that a mention of the variability of Algol has even been found in the ancient Chinese annals. It has often been thought that the variability of Algol had been observed by the Arabs because of the meaning of the word Algol: 'the demon's head'. It is more probable that the name represents only a translation of 'the Medusa's head', a name given by the Greeks to the constellation in which Algol lies. The first systematic observations of Algol were made in 1783 by John Goodricke, who observed the variations in its brightness and deduced a period of 2d 20h 49m 3s with an uncertainty of 15 seconds. John Goodricke proposed two models to interpret this variation: one supposed the existence of giant spots on the surface of the star; the other model suggested the existence of eclipses due to a giant planet in orbit around Algol.

We now know more than 4000 eclipsing binary systems. The periods of these systems are generally short: from a few hours up to about ten days. The shortest period yet measured is that of WZ Sagittae: 1 hour 22 minutes, and the longest is that of ε Aurigae: 9883 days, or about 27 years. Measurements of the variations in brightness of the system generally are made by photoelectric photometry. The diagram representing the variations in magnitude as a function of time is called the light curve. Study of the shape of this curve lets us reconstruct the parameters of the orbit.

A schematic light curve is shown here. In reality, the light curve is frequently distorted as a result of interactions between the close components: for example, reflection effects occurring between the components, variation in brightness across the disk of one or the other star, or the deformation of the stars by tidal forces.

Lagrangian points and equipotential surfaces in a binary star system. The two stars (S_1 and S_2) in a binary system revolve around their common centre of gravity. This gives rise to a centrifugal force. At certain points, this centrifugal force is exactly balanced by the combined gravitational attractions of the two stars. An object placed at one of these five equilibrium points, called Lagrangian points, would not move in a frame of reference that rotates with the binary system.

The diagram shows these 'L' points. Only the L_4 and L_5 points correspond to positions of stable equilibrium: if we slightly displace the object from the position L_4 or L_5, the resultant force will push it back towards the equilibrium position. The other three points, L_1, L_2 and L_3, correspond to unstable equilibrium positions: if an object is slightly displaced from one of those points, it will move away from that point.

The diagram also shows a section through the equipotential surfaces of the system.

Near either of the stars, these equipotentials are spherical, because the field of the central star is preponderant. As one goes further away, the effect of the second star becomes larger and larger and the equipotentials are less and less spherical. There is a limiting surface, which defines a volume around each star, called the Roche lobe.

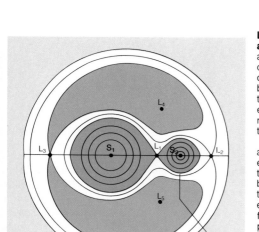

L_4

L_3 S_1 L_1 S_2 L_2

L_5

Roche lobe

Shapes of the stars in a binary system (right-hand diagrams). If the volume of each star is less than that of its Roche lobe, we say that the binary system is detached. Each star has a more or less spherical shape, since this shape is defined by an equipotential surface.

During its evolution, each star expands. If during this expansion, the volume of the star exceeds that of its Roche lobe, there will be mass transfer from the larger star to the smaller star through the inner Lagrangian point, L_1. The system is then semi-detached. The matter might not land directly on the companion, but might form a spiral around the star before falling on its atmosphere. Such matter generally forms an accretion disk around the star.

If both stars expand so that they both fill their Roche lobes, we say it is a contact system; if this expansion continues, the matter of the two stars occupies a larger and larger volume and is mixed, forming an envelope around the two stellar nuclei where thermonuclear reactions continue to occur. It is even possible for the system to lose matter through the L_2 or L_3 points if these are reached. When mass transfer is possible between the two components of a binary system, we say we have a close binary system.

The Algol system has passed through the first stages in the evolution of a close binary system. Initially the system was composed of detached stars with masses of 3 and 1.5 solar masses ($M_\odot$). The more massive star rapidly burnt its hydrogen and while combustion was still taking place in its centre, the star filled its Roche lobe by undergoing a slow expansion. Mass was then transferred via the inner Lagrange point and the roles of the two stars in terms of mass were exchanged. The initially more massive star became a red star of 0.8 solar mass and its companion a blue star of 3.7 solar masses. The system is now semi-detached, with a very small flow of mass between the two components.

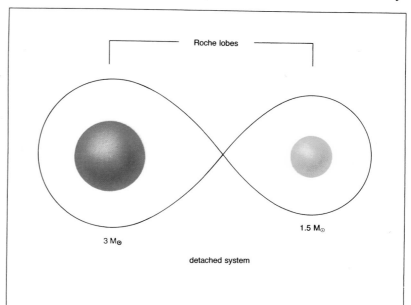

Roche lobes

$3\ M_\odot$ $1.5\ M_\odot$

detached system

formation of the Algol system

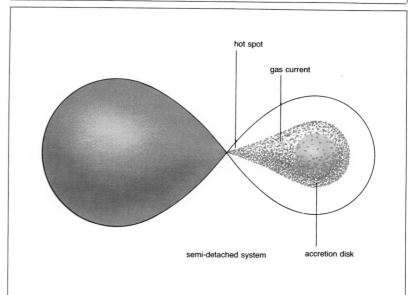

hot spot

gas current

semi-detached system   accretion disk

start of mass transfer

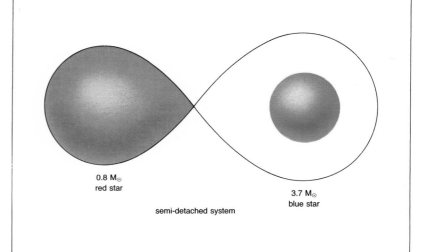

$0.8\ M_\odot$ red star $3.7\ M_\odot$ blue star

semi-detached system

current situation of the Algol system

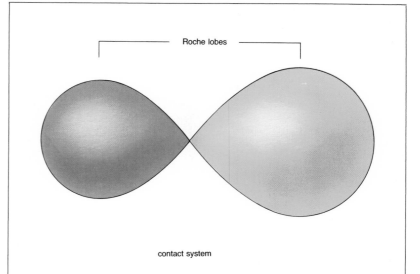

Roche lobes

contact system

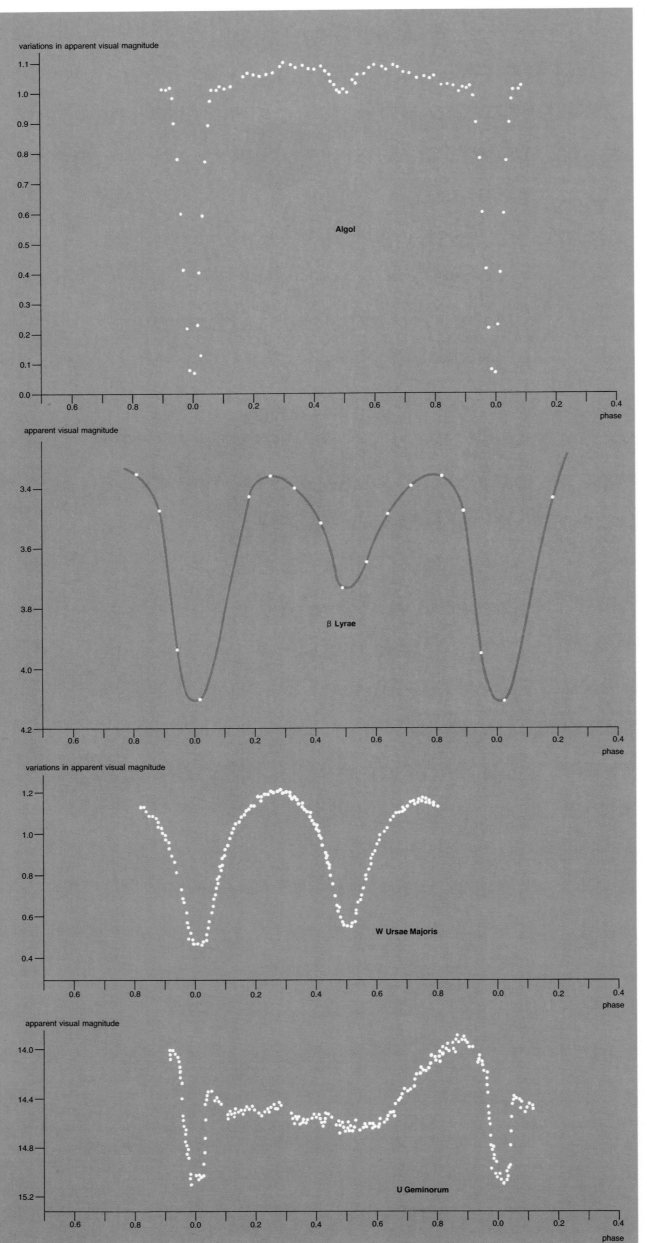

Light curves of close eclipsing binary systems.
Close binary systems are frequently also eclipsing
systems. The light curve of such a system is
strongly affected by effects of the non-spherical
shapes of the stars, and sometimes by the reflection
of the radiation of each component by the other.

In the case of Algol (β Persei), the more massive
star of the system is also the brighter. As it is well
inside its Roche lobe, its surface is very little
deformed and the light curve of such a system is not
changed by the fact that it is a semi-detached
system. In the case of β Lyrae, the brighter star fills
its Roche lobe. The light curve is strongly deformed.
Moreover, in this semi-detached system, a very
strong current of matter exists between the two
components around the interior Lagrange point. The
period is 12.9 days.

The system W Ursae Majoris is probably two stars
in contact which, following mass transfer, have a
common envelope. The light curve shows two
minima of the same depth and between these
minima the light curve is very deformed by the
effects of a strong interaction.

U Geminorum is another eclipsing system
composed of a luminous star that fills its Roche lobe
and a white dwarf, around which there is a large
accretion disk. Strong hydrodynamic phenomena in
this disk make it luminous. Where the gas current
coming from the inner Lagrange point hits the disk
of matter near the white dwarf, a large amount of
energy is liberated in shock waves by the
transformation of the kinetic energy of the gas
current into radiation. This region – the hot spot –
can be more luminous than the stars in the system.

The diversity between these light curves of close
binary systems accounts for the difficulty in
interpreting them. Each system is almost a special
case.

Peculiar stars in binary systems

Metallic-line stars are characterised by a strong magnetic field, which shows up in the splitting of the spectral lines. The especially strong lines of certain metals like manganese or europium are probably the result of a diffusion process in the atmosphere causing some ions to rise to the surface and others to fall. Some of these stars are in binary systems.

Symbiotic stars are objects whose spectra show absorption lines characteristic of a cool star together with emission lines of strongly excited atoms. This fact suggests that they are double-lined spectroscopic binary stars of which one component is a giant star of type K or M and the other is of unknown type but very hot and with tenuous gas surrounding it. These two objects form a close binary.

Lastly, a very high proportion of novae are known to belong to binary stars. Probably all are binaries in fact.

Shapes of the components of a binary system

In the eighteenth century, the French mathematician Joseph Lagrange studied the behaviour of a test particle of negligible mass subjected to the gravitational attraction of two bodies. He showed the existence of a critical surface, made of two lobes, one around each star, called the Roche lobes (shown in the bottom right diagram on page 285). Particles that cross that surface no longer 'belong' to the star within the lobe they have just left, but are subject to the gravitational attraction of both stars, and may eventually leave the binary system. The size and shape of the critical surface is totally determined by the masses and separation of the stars.

Lagrange also demonstrated the existence of a point located between the two stars at which the two lobes join, called the inner Lagrangian point. Particles may transfer from one star to the other by passing through this point. Thus if one or other of the stars should expand to fill its Roche lobe in the course of its evolution, mass transfer will take place.

Binary stars are classified into three groups: detached, semi-detached, or contact systems, according to how many of the stars (none, one, or two) have expanded to fill their Roche lobes.

It may seem that models of the internal structure of a star, which assume that it is spherical, may be inadequate for close binary stars, where the shape of the star can be changed and look pear-shaped. In fact, models for the internal structure describe the configuration of the central regions of the star, where the density of the matter is high enough for it not to be affected by an external gravitational field. The mutual gravitational attraction of the stars affects only the shape of their outermost regions, which we can observe. To which of the three categories a binary belongs depends on the evolution of each of its components.

Evolution of binary stars

Our present knowledge about stellar formation makes it very improbable that binary systems result from the chance meeting of two single stars. The two stars were formed at the same time, but they evolve at different rates because their masses are different. It is the more massive star that is likely to fill its Roche lobe first, thus forming a semi-detached system. However, not all binary systems necessarily pass through this phase: if the two components are initially very far apart, the sizes of the Roche lobes are so large that the stars may never fill them during their expansion. The system will always be detached. The evolution of each component will not be affected by the binary character of the system and will be that of single stars. There is, however, an exception. If one of the components suffers a large mass loss, the mass can be captured by the second star through the inner Lagrangian point. If the binary system is sufficiently close, of such a kind that the Roche lobe can be filled by one of the components during its expansion, there will then be a mass transfer onto its companion. The

way the mass transfer happens, the quantity of mass transferred and the duration of the phenomenon depend essentially on the initial mass of the star and the evolutionary phase in which it is at the time of the transfer. In a close binary system, the change to the masses of the individual components during the evolution implies a profound transformation of their evolutionary scheme compared with isolated stars, but observation of binary systems gives an extra test of the theory of stellar evolution even though it is complicated by this gravitational interaction.

Observational tests of the evolution of close binary stars

Various observations of a binary system can allow us to estimate the region of the Hertzsprung–Russell diagram where the components belong. We have, therefore, some knowledge of the state of the evolution of each star and of its principal properties. We can thus calculate the critical surface for the system and find out whether the system is detached or not.

Sirius is an example of a detached binary star; the more massive component, Sirius A, is still on the main sequence while Sirius B is in a very advanced stage of evolution since it is a white dwarf. Such an observation, contrary to the predictions of the theory of stellar evolution, can be explained thus: Sirius A, initially the less massive star, has gained mass at the expense of Sirius B, initially the more massive star. The fact that Sirius B is a white dwarf necessarily implies that after the mass transfer the remaining mass was no more than 1.4 solar masses. Mass transfer can occur several times during the life of the pair, and not necessarily in the same direction, as the masses change.

Algol (β Persei) is another paradox: it is a triple system, where the eclipsing system, with a period of 2.87 days, is itself in orbital motion with a period of 1.86 years around a third component, much further away. In the eclipsing system, Algol A is apparently a blue main sequence star of type B8, while its companion, Algol B, is a red star, (of type G or K) which has filled its Roche lobe. This system is semi-detached. The star Algol B, a

subgiant, which is the most evolved in the system, has a mass slightly less than that of Algol A. This fact, apparently contradicting stellar evolution theories, results from a large mass transfer from the B component to the A component. This mass transfer is still happening, but very slowly: gaseous streams have been detected around Algol A. Radio emission with erratic variation was detected in 1971 and interpreted as being produced by the gaseous stream from the B component to the A component. Small changes in the orbital period are also attributed to this mass transfer.

The interpretation of light curves of eclipsing stars is very complex for semi-detached systems or contact systems, since the stars are not spherical, and the non-uniform luminosity of their surfaces introduces distortions into the light curve. The curve is also modified by the effect of the gas stream, the gas stream being capable of reflecting the light of the nearer component and thus producing a radiation whose lines are superposed on those of the spectrum of the star. The confusion between these various elements can be almost total and great precautions must be taken in the interpretation of the observations.

Calculations of the evolution of close binary systems show that the final result of the evolution is determined by the mass of the initially more massive star. If this is less than 15 solar masses, the primary star finishes its life as a white dwarf. If the mass is greater than this limit, the evolution is very different. An example of such evolution is given by Wolf–Rayet stars. Binary systems made of massive stars may result in exotic objects: neutron stars or black holes which can be sources of very energetic radiation and of X-rays in particular. The X-rays are generated during the accretion of matter onto a compact object. The binary system is then called an X-ray binary.

This domain of astrophysics has grown considerably over the last twenty years through the development of observations in new spectral ranges, in particular in X-rays, which has deepened our knowledge of the last stages of stellar evolution. Observation of binary stars gives the only possibility of the direct detection of a black hole.

Michèle GERBALDI

star too faint for us to observe its flux

orbit

flux

time

Flux from a non-spherical ellipsoidal variable star. When a star loses its spherical shape because of the gravitational effect in a close binary system, and becomes elongated, the intensity of the radiation at the surface of the star is no longer constant. The most elongated parts of the star are the least luminous. Such a star, because of its orbital motion, shows different parts of its surface to the observer; as these have different luminous intensities and present different light-producing areas as seen from Earth, the observer sees the flux from the star vary, independently of any eclipses that may occur. The flux varies with the same period as the orbital period. Such a system is called an ellipsoidal variable.

Young star clusters

Most of the stars in the Galaxy are now distributed more or less at random in space. The shapes of constellations as seen from the Earth are simply an effect of perspective. There are also, however, obvious groups of stars with a common origin, one of the best known being the Pleiades. These clusters contain a few hundred young stars and are easily distinguishable from globular clusters, which are compact groups of hundreds of thousands of old stars. For this reason the young star clusters are called open clusters.

Studies of the proper motions of open cluster stars show that all of the stars move together. This proves that the stars form a physical group. Because of the effect of perspective, the proper motion trajectories seem to be converging towards a point on the celestial sphere. The location of this point along with the radial velocities of the cluster stars (obtained from the Doppler shift of their spectral lines) permits the determination of the distance of a few open clusters. The most accurate distance determined by this method is that for the Hyades, a loose cluster 43 parsecs from the Sun forming part of the constellation Taurus. Distances of clusters are usually established by comparing the magnitudes and colours of their stars to those of the Hyades stars. This method has a certain weakness, in that it assumes that the properties of the Hyades stars are 'normal' and truly representative of most galactic stars.

Open clusters are concentrated in the disk of the Galaxy where interstellar matter and young Population I stars are found. For this reason open clusters are sometimes called galactic clusters. Over one thousand are known within 3 kiloparsecs of the Sun. At greater distances it becomes increasingly difficult to distinguish clusters against the densely populated background of the Milky Way. One of the closest of the open clusters is the Ursa Major Cluster, which includes several stars of this constellation. Very few open clusters are found away from the galactic plane, and this characteristic is actually used to define the position of the plane of the Galaxy. It is found that the ratio between the thickness and the diameter of the volume occupied by open clusters around the Sun is less than one thousandth.

Other loose groups of stars with common space motions are also found in the Galaxy. These groups, containing O or B stars, are called OB associations. They are nearly always found near complexes of interstellar gas and dust. One well-known association is in the constellation of Orion.

Clusters and associations have a great importance in astronomy because they provide the opportunity to observe groups of stars having the same age and the same origin. It is then reasonable to assume that the chemical composition of the gas cloud from which they originate

The double cluster h and χ Persei. h and χ Persei are twin open clusters visible to the naked eye in the constellation Perseus. They are situated at 2 kiloparsecs from the Sun at the end of a galactic spiral arm and at a high galactic latitude. The lack of nebulosity around them indicates that this region of the Galaxy contains very little gas. These two clusters are located at 15 parsecs from each other. They were probably born out of the same gas cloud. Their ages, determined from the position of the main sequence turn-off in the Hertzsprung–Russell diagram, are, respectively, ten and fifteen million years. Their total mass is of the order of 7000 solar masses. The HR diagram of these clusters is more spread out than that for the Pleiades. (Yerkes Observatory)

The open cluster NGC 3293. The open cluster NGC 3293 is approximately five million years old. It is situated near the Carina Nebula. A wisp of interstellar dust extending from the centre of the cluster may be associated with it. The youngest stars are found on one edge of the cluster while the oldest ones are on the other side, suggesting a progressive wave of star formation. (D. F. Malin, Anglo-Australian Telescope Board, © 1977)

A comparison between the Hertzsprung–Russell diagrams of open and globular clusters. In general the HR diagram of a collection of stars populates three regions: the main sequence, the giant branch and the horizontal branch. Blue giants, if they exist, are located at the upper end of the main sequence (see figure below).

Diagrams of open clusters are composed of a main sequence extending, for the youngest clusters, to the blue giants, which are the most massive stars. They also contain a sprinkling of red giants. The older an open cluster is, the more numerous are its red giants. Sometimes the least massive stars of a very young cluster have not yet reached the zero age main sequence. This sequence is reached by a star once all of its radiation comes from thermonuclear hydrogen burning. Prior to the main sequence stage, a part of the stellar radiation comes from the stellar gravitational contraction. The age of clusters can be determined from the position of the main sequence turn-off (towards the red giant branch). The HR diagrams of globular clusters contain relatively many more red giants which spill towards the horizontal branch. The upper part of the main sequence is missing, since these stars have now become red giants.

The star cluster NGC 2264. This cluster (above) is one of the youngest known open clusters. It is associated with the Cone Nebula, which is an H II region in a giant molecular cloud. This cluster contains a large number of T Tauri stars and pre-main-sequence variables. T Tauri stars are stars in the process of being formed. The least massive stars are still contracting and have not reached the main sequence of the Hertzsprung–Russell diagram where eventually they will burn hydrogen in their core. This cluster is approximately two million years old. (D. F. Malin, Anglo-Australian Telescope Board, © 1981)

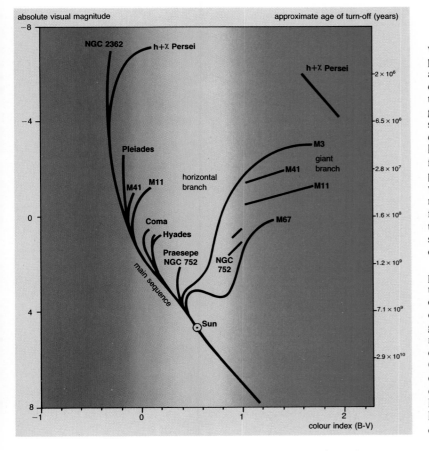

was also homogeneous. Two of the three independent parameters determining the evolutionary state of stars – age, original chemical composition and mass – are thus fixed. Observations allow us to see the effects of evolution in a group of stars of different masses but having the same composition and age. The problem of cluster evolution is actually very complex. It is known, for example, that certain stars lose a fraction of their mass through a stellar wind process. This introduces another parameter which must be taken into account. This phenomenon may explain why white dwarfs are found in a few young open clusters even though the normal evolution of a star would require several thousand million years to produce a white dwarf.

Two independent arguments permit us to pin-point the age of open clusters and associations. First of all, one must remember that stars in open clusters are, contrary to the case of globular cluster stars, only weakly linked to each other by gravitational forces. Because they have small motions relative to each other, eventually they can all escape the gravitational pull of the cluster. Computations show that the mean escape rate is one star per hundred thousands years. From these dynamical arguments one must conclude that the lifetime of a cluster should not exceed one hundred million years, the time required to disperse all the cluster stars into the galactic disk.

Methods based on the theory of stellar evolution lead to similar conclusions regarding ages of open clusters. In fact, they give even more accurate estimates. It was while studying cluster stars that Ejnar Hertzsprung in the Netherlands and Henry Norris Russell in the United States discovered simultaneously that the colour (temperature) and the magnitude (absolute luminosity) were related. The diagram that bears their names synthesises some of the fundamental observations we can collect on stars in what is, today, still one of the most useful tools of astronomy and astrophysics. In this diagram, points for stars in an open cluster are distributed along a narrow diagonal band, the main sequence, which curves upwards at the left towards the region of the red giant. The position of the turn-off from the main sequence is used to estimate the age of clusters.

Associations and open clusters can be considered as genuine nurseries of young stars. For this reason they are important systems for the study of the formation and evolution of stars. The masses of clusters are close to the maximum mass that an interstellar cloud can have if it is not to become gravitationally unstable and begin to contract. Nonetheless, even today, the successive stages leading from this instability to the formation of a star cluster are still not well understood.

Jean-Pierre CHIÈZE

Globular clusters

Globular clusters get their name because they look like giant, spherical swarms of stars. M13 (NGC 6205), a globular cluster in Hercules can be seen with the naked eye, but a large telescope is needed to resolve it into stars.

A typical globular cluster contains between one hundred thousand and one million stars. The stars stay together in space because of their mutual gravitational attraction. The spherical shape of these clusters shows the isotropic properties of these attractive forces. The number of stars per unit volume increases from the outer parts towards the centre of the cluster. Near the centre the stellar density may reach a few thousand stars per cubic parsec, a density a thousand times higher than the stellar density in the solar neighbourhood. A few globular clusters have a slightly ellipsoidal shape, which can be attributed to a slow rotation of the cluster. Omega Centauri (NGC 5139) is one of the more flattened clusters, its minor axis being roughly 80 per cent of its major axis. Diameters of clusters range from 7 to 120 parsecs. The diameter of Omega Centauri is 20 parsecs.

Globular clusters are found around galaxies of all types and sizes. The clusters associated with our Galaxy are spherically distributed about its centre, which is situated in the direction of Sagittarius at some 10 kiloparsecs from the Sun. Our clusters are enclosed in a sphere approximately 60 kiloparsecs in diameter; however, many more clusters are found near the centre of the Galaxy than in its outer parts. About two hundred globular clusters are known. Each cluster travels in an elongated orbit with a period of revolution of the order of two hundred million years. They cross the plane of the Galaxy twice during one of their revolutions. Such crossings produce gravitational tidal effects on the cluster, which loses to the Galaxy its outer stars and any interstellar gas and dust that it may contain. The orbital velocity of a cluster depends on the total mass of the parent galaxy. A cluster situated at a distance R from the centre of its galaxy has a velocity proportional to $(M/R)^{1/2}$, where M is the total mass of the galaxy within a sphere of radius R. From such measures we can get an idea of the contribution of the stars of low luminosity (which are not directly observable) to the total mass of the galaxy. The study of the dynamics of globular clusters is one of the means at our disposal to 'weigh' a galaxy.

Globular clusters are among the most distant observable galactic objects from the Sun. It is not possible to determine their distance directly using, for example, the trigonometric parallax technique used for stars closer than 30 parsecs from the Sun. We must call upon what is called the 'standard candle' technique. We know of several classes of objects in our Galaxy with certain properties – their absolute luminosity or their size – which are sufficiently constant to serve as standards. If we can distinguish and recognise these objects in far away systems, then the comparison of their apparent brightness or their apparent size with those of standard objects of known distance allows us to determine their distance. For globular clusters this comparison is done using a type of variable star called RR Lyrae. These are red giants of Population II found in quite large numbers – a few hundred – in most globular clusters. These variables are easily identifiable from the shape of their light curve. They are very useful because they all have an absolute magnitude near $M = +0.5$ or -0.2. Two factors contribute to decrease their apparent brightness: their distance from us, and the amount of interstellar matter along our line of sight to the variables. The observed apparent magnitude of these variables must be corrected for the effect of interstellar extinction before their distances are determined.

The first distances of globular clusters were obtained by Shapley following the discovery of the period–luminosity relation in 1908 by Henrietta Leavitt. There is a linear relation between the logarithm of the period and the luminosity of the sample of Cepheid variables in the Large Magellanic Clouds. Harlow Shapley assumed that the few Cepheids observed in a few globular clusters shared these same properties. He thus discovered the spherical distribution of the clusters about the galactic centre, and proposed a diameter of 80 kiloparsecs. By the same token he placed the Sun more than 10 kiloparsecs from the galactic centre. The quantitative results of Shapley had to be revised twice: the first time when astronomers realised the importance of the interstellar extinction, and a second time, in the 1940s, when Walter Baade discovered that there are two populations of stars in the Galaxy. Population II stars are old with a surface chemical composition characterised by a low abundance of heavy elements (carbon, nitrogen, oxygen, iron, etc.). Population I stars, on the other hand, are young and are composed of material that has been enriched by nucleosynthetic processes in the stellar interiors of several previous stellar generations. Contrary to what Shapley had assumed, the absolute luminosity of Cepheids in globular clusters is four times less than Population I Cepheids of the same period.

The globular clusters of the Galaxy are very old, being more or less contemporary with its formation. For this reason they are of great

The globular cluster M13. This cluster is located in the constellation Hercules, appearing to the naked eye as a fuzzy little patch of light. M13 is 9000 parsecs from the centre of the Galaxy. It contains several hundred thousand stars which are attracted to each other by gravitational forces stronger than the force exerted by the Galaxy on the cluster as a whole. The stellar density increases strongly towards the centre of the cluster. Stars are so numerous and densely packed that the centre cannot be resolved, that is, stars cannot be seen individually. The brightest stars of a globular cluster are red giants ten billion years old. These stars contribute most of the visible light emitted by the cluster. (US Naval Observatory)

Globular clusters in the Large Magellanic Cloud. The globular clusters of our own Galaxy are all older than ten billion years. Seventeen globular clusters have also been identified in and around the Large Magellanic Cloud, a small irregular galaxy that is a satellite of our own Galaxy. One of these clusters can be seen on the left of the picture. Some of them are no older than ten million years, an age comparable to the age of open clusters. This indicates that stellar formation in the Magellanic Clouds is slower and more regular than in our Galaxy. This difference shows how important globular clusters can be for our understanding of the formation and evolution of galaxies. (Royal Observatory, Edinburgh, © 1981)

interest. Their age is determined by comparing their Hertzsprung–Russell diagrams with computations based on the theories of stellar evolution. Theoretical evolution of stars of different masses but with the same chemical composition as the cluster stars are done to match the Hertzsprung–Russell diagram of the cluster. Taking into account all of the uncertainties, the ages determined are between ten and fifteen billion years, to a precision of the order of 20 per cent. The globular clusters do not all have the same abundance of heavy elements. They are, however, all poorer in heavy elements than the oldest open clusters. Because globular clusters were formed first in the Galaxy, they have benefited very little, or not at all, from the enrichment of interstellar matter by stellar nucleosynthesis and supernovae explosions. About two-thirds of the globular clusters have stars which have one hundred times less heavy elements than the Sun. The stars in the other third have fifty times more heavy elements than those in the first group. This type of observation yields important information on the first stages of the chemical evolution of the Galaxy.

Observations of globular clusters in other galaxies are also of great interest. Globular clusters around the Andromeda Galaxy, M31, are richer in heavy elements than the clusters of our Galaxy. Their space distribution is less concentrated, a fact that may suggest that their early evolution was slower and more irregular. The Large Magellanic Cloud, on the other hand, shows globular clusters of all ages, the youngest being barely ten million years old. The rate of star formation has been slow in this galaxy.

It is generally believed that globular clusters in our Galaxy were formed following the frag-

mentation of the protogalactic cloud. The most distant clusters, being in low density regions, were probably formed over a longer time-scale than the ones close to the galactic centre. The high gas density near the centre could have accelerated their formation. Because clusters evolve very slowly, we can nowadays simulate with computers the successive stages of their life. For a long time stars in a cluster move in fairly regular orbits which criss-cross each other. But when two stars pass close to each other, their mutual gravitational forces perturb their orbits. Gradually, some stars leave the cluster while others fall towards the cluster centre. When the stellar density near the centre reaches several thousand stars per cubic parsec, stellar collisions lead to the formation of close binary systems and possibly to the formation of a black hole. Computations and certain observational data seem to indicate that eventually the outer envelope of the cluster expands and disappears into space. The observation of X-ray emission associated with many globular clusters also suggests such a scenario.

Historically, globular clusters have played a leading role in astronomy. The determination of their galactic distribution at the beginning of the twentieth century led to a more accurate picture of the structure of the Galaxy. Globular clusters were thus used as a first step in the exploration of extragalactic space. The study of globular cluster stars has been, without a doubt, instrumental in the development of the theory of stellar evolution. Globular clusters, because of their ages, give us an insight into the origin and the evolution of galaxies.

Jean-Pierre CHIÈZE

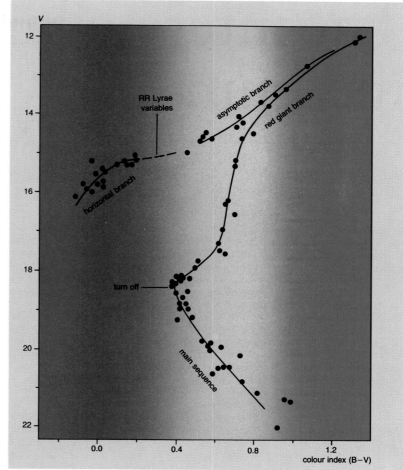

The age of globular clusters. About eighty stars belonging to M92 have been used to plot its Hertzsprung–Russell diagram. Each star is represented by a dot. Its coordinates are: the colour index, $B-V$, which is a measure of the surface temperature of the star, and its apparent visual magnitude, V. Because the cluster dimensions are so small compared to its distance from us, differences in apparent magnitude correspond to differences in absolute magnitude. The effect of interstellar extinction must also be taken into account. Stars populate three areas of the diagram: the main sequence, the giant branch and the horizontal branch. The main sequence stops at a $B-V$ of about 0.4. The position of this 'turn-off' depends on the age of the cluster. Massive stars, which are brighter and hotter than less massive ones, are the first to leave the main sequence. As time passes the main sequence thus loses stars, starting from the upper left part of the diagram. Stars that have left the main sequence have terminated their hydrogen core burning phase and are now red giants. Stellar evolution computations give the mass and the age of these stars as a function of their chemical composition. In fact, only about ten clusters have fairly well-determined ages.

Globular clusters in M87. The M87 galaxy is located at the heart of the Virgo Cluster. It is estimated that this giant elliptical galaxy is surrounded by about 15 000 globular clusters. Several dozen can be seen in this photograph, appearing as small fuzzy points surrounding the overexposed core of M87. It is informative to compare the way in which the clusters are distributed with the distribution of the stars in the parent galaxy. In the case of M87, the clusters extend much further out than the stars. The cluster distribution is also less concentrated towards the centre. The stars of the body of the galaxy were, in all probability, formed much later than those in the clusters, when the diameter of the protogalactic gas was smaller. (Strom and Strom, Kitt Peak National Observatory)

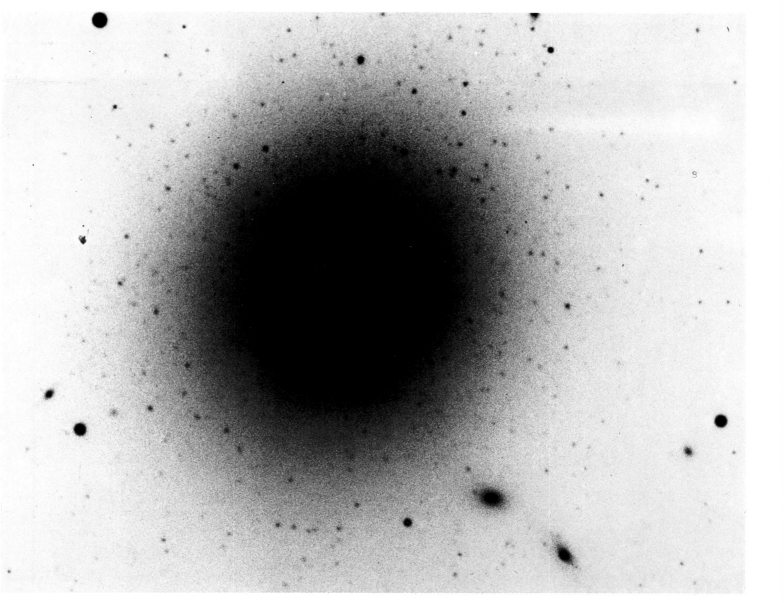

The interstellar medium

In a galaxy like our own, stars have an average separation of 2 to 3 parsecs. The space between the stars is occupied by diffuse matter whose total mass, in our Galaxy, represents about a tenth of that of the stars. The interstellar medium and stars are not two distinct entities coexisting without interaction; on the contrary, the interstellar medium is where stars are born and evolve.

The interstellar medium varies from place to place and is observed in many forms. Invisible to the eye, dark clouds show up when they are seen against the bright Milky Way. There are enormous masses of gas and dust obscuring the dense star clouds in these regions of the sky.

The gaseous nebulae, which have been known for a long time, are another manifestation of the presence of interstellar gas. In these, there are generally hot stars which excite fluorescence in the gas (H II region) or, more simply, illuminate the dust associated with them (reflection nebula).

Observations show that interstellar gas is, for the most part, condensed in the form of interstellar clouds, which have a fairly turbulent appearance. Interstellar matter can also be detected in a more indirect way by the presence of absorption lines in the spectra of hot stars, principally due to sodium and potassium atoms. This sort of observation reveals that interstellar gas is moving in a disordered way: clouds move among themselves with an average speed of some 10 kilometres per second. This constant agitation leads to numerous collisions as a result of which clouds can combine or fragment.

One of the most abundant sources of information on the interstellar medium continues to be observation of the radio waves emitted by the gas. Hydrogen emits at a wavelength of 21.11 centimetres when it is at the low temperatures and densities occurring in space. Hydrogen is by far the most abundant element in interstellar space, representing more than 90 out of every 100 atoms found there. Observation at 21.11 centimetres wavelength allows one to map the regions of the Galaxy rich in hydrogen. By this means it was discovered that hydrogen does not uniformly cover the galactic disk, but rather is concentrated in the central region and in the four spiral arms. A few general facts emerge from these observations. Neutral hydrogen – designated by the symbol H I – is distributed in numerous clouds whose masses range from 0.1 to about 1000 solar masses. Their density is very low, around fifty particles per cubic centimetre. The temperature in the interior of these clouds is very low. The values deduced from the radio power of the emitting gas are generally around 80 K (about −200 °C).

However, there is a population of more massive, denser and even colder clouds, probably produced by the coalescence of a large number of smaller clouds. These are molecular clouds, or molecular complexes. It is principally observations at millimetre wavelengths that have revealed their existence and allowed study of them on a large scale. They are vast aggregations of interstellar gas containing molecules in large quantities. To date, more than sixty types of molecule and radical have been detected, the most abundant being, of course, the hydrogen molecule H_2. The masses of some complexes exceed 500 000 solar masses and their densities, in the central regions, may be more than 10 000 particles per cubic centimetre; the temperatures observed in them can be lower than 10 K. Their

Barnard's Loop. This photograph, taken with a filter that transmits only light in the Hα line of hydrogen, shows a region of the sky around the constellation of Orion. The belt and the quadrilateral formed by Betelgeuse, Bellatrix, Rigel and Saiph can be recognised. The Orion Nebula M42 (NGC 1976) is easily visible in the lower third of the photograph. Barnard's loop, which surrounds it, is seen as a sphere of ionised hydrogen about 80 parsecs in radius. This interstellar bubble is probably the result of the action of radiation pressure from young stars (including the Trapezium group, which ionises the M42 nebula) in its interior; this pressure pushes out the ambient interstellar material. There is as much interstellar dust to the west as to the east of the loop, but there is less hydrogen and the Hα emission is therefore weaker. The age of Barnard's loop is estimated to be three million years, an age comparable to that of the oldest stars in Orion. This represents a large-scale interaction between stars and the interstellar medium. (Photograph by Syuzo Isobe, Tokyo Astronomical Observatory)

The Orion Nebula (M42, 43 NGC 1976, 1982). The first mention of this object as a nebula was made by Nicolas Claude Fabri of Peiresc in 1611; Christiaan Huygens gave a description of it in his *Systema Saturnium*, published in 1659. Situated 450 parsecs from the Sun, the Orion Nebula is today without doubt the best studied H II region. The red colour is due to emission in the Hα line of hydrogen. The centre of the nebula is occupied by the Trapezium, a group of O and B stars. The ultraviolet radiation emitted by these stars ionises the nebula. Just below the Trapezium is a dark region – called the Gulf – due to neutral hydrogen. Observation of emission due to carbon monoxide (CO) and formaldehyde (HCHO) has revealed the presence, just behind the nebula, of a huge molecular cloud. This cloud shows two emission maxima associated with two strong infrared sources, one of which, called the Becklin–Neugebauer object, is actually a star in the process of formation; evidence for this lies in the presence of two masers, due to OH and H_2O. These regions, the densest parts of the molecular cloud, are contracting gravitationally. It is estimated that the nebula visible at present will become dispersed in about ten thousand years, by which time it will be replaced by a new region of hydrogen ionised by the stars currently forming. (D. F. Malin, Anglo-Australian Telescope Board, © 1981)

NGC 6559 and IC 1274–5. The section of sky shown to the left is near the Lagoon (M8) and Trifid (M20) nebulae, in the constellation of Sagittarius. This region, about 1.4 kiloparsecs from the Sun in the direction of the galactic centre, is occupied by a huge complex of interstellar matter, composed mostly of hydrogen (ionised, atomic and molecular) and dust. Ionised hydrogen is revealed here by the red–orange luminosity characteristic of hydrogen Hα emission. In front of this nebula a dark tongue stands out in relief; this is due to a high density of dust, which prevents the transmission of light. These dust grains are about a micrometre in size. The diffuse, bluish regions are reflection nebulae, veils of dust close to bright stars reflecting a part of their light. Their colour comes about because blue light is reflected more efficiently than red by small particles. This very complex nebula is surrounded by a halo of young stars which were born a few million years ago in the heart of this complex of interstellar matter. (Royal Observatory Edinburgh, 1979)

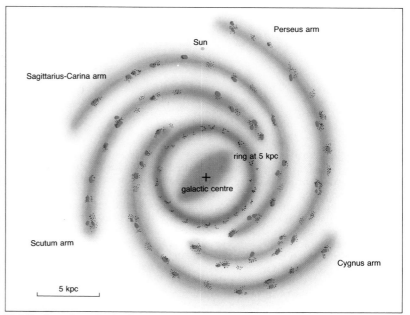

Distribution of gas in the Galaxy. Neutral hydrogen, observed by radio waves at a wavelength of 21 centimetres, is aggregated mostly along the four large spiral arms (in blue). The galactic centre, which contains numerous expanding regions of ionised hydrogen and giant molecular complexes, is surrounded by a huge ring, about 5 kiloparsecs in radius, where a great quantity of atomic and molecular hydrogen is concentrated.

The spiral arms are emphasised by the presence of emission nebulae, or H II regions (in red), ionised by massive O-type stars. Massive molecular clouds (in black), the scenes of active star formation, are also associated with the arms.

Viewed from the side, this structure has the appearance of a very thin disk, about 300 parsecs thick and with a radius of 30 kiloparsecs.

Structure of the interstellar medium. On intermediate scales, over distances of the order of kiloparsecs, the structure of the interstellar medium is probably dominated by the coalescence of supernova remnants of various ages. The medium that results is very hot and very dilute. The temperature (T) is about 500 000 K and the density (n) about 0.003 particles per cubic centimetre. These particles result chiefly from the evaporation of cold clouds of neutral hydrogen – of temperatures about 50 K – within the supernova remnants, which envelop and distort them. The supernova remnants are surrounded by a thick shell which, after roughly five hundred million years, cools rapidly and probably condenses into a series of small dense clouds. This mechanism guarantees the permanence of the cold component of the interstellar medium. On a small scale, the structure of the interstellar medium is dominated by the interactions of dense clouds among themselves and with the supernova remnants. According to the conditions under which these clouds collide, they can fragment into several pieces or, alternatively, coalesce. Small clouds are usually absorbed by large ones; clouds of similar sizes suffering oblique collisions usually fragment.

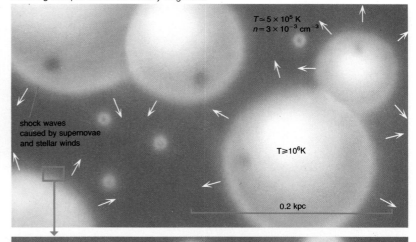

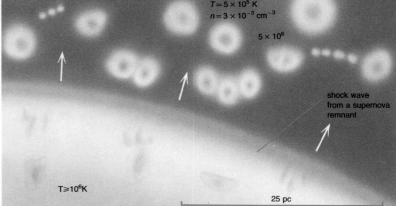

---

ages and lifetimes are not known. Their high masses suggest, however, that they will contract under the effect of their own gravity. Infrared observations in particular have shown that it is precisely these regions where star formation is occurring today. It is estimated that half the mass of the interstellar medium is condensed in the form of molecular clouds.

There are also smaller molecular formations, called Bok globules. Their sizes are generally smaller than a parsec, and their masses are around 200 solar masses. It is known that stars are currently forming in some of these.

Even together, molecular clouds and neutral hydrogen clouds far from fill the total interstellar volume. Quite recent observations in the ultra-violet and X-rays (emission from five, six or seven times ionised oxygen) have 'filled the gap' by providing evidence for the existence of gas which is very hot, with a temperature of the order of 500 000 to one million kelvins, but very dilute, so that a cubic metre of this medium only contains about 3000 particles. In all likelihood, this gas is made up of the residue of supernova explosions which, over all the Galaxy, happen on average once every thirty years. The free electrons in this turbulent interstellar plasma and in the gaseous nebulae cause scintillation of radio sources, in a way analogous to the twinkling of stars due to atmospheric turbulence.

Overall, the interstellar medium contains a magnetic field with an average intensity of a few microgauss, revealed among other things by synchrotron radiation observed in many regions, especially at high galactic latitudes. This synchrotron radiation is produced by the interaction of relativistic electrons, which are travelling at speeds close to the speed of light, with the interstellar magnetic field.

But electrons represent only about one in a hundred of the high-energy particles that streak through the interstellar medium and constitute the galactic cosmic radiation. This is composed of atomic nuclei and elementary particles whose kinetic energy can reach 10^{21} electron volts. The density of cosmic radiation is negligible compared to that of the interstellar gas, which, averaged throughout the Galaxy, reaches 0.3 particles per cubic centimetre. But its importance stems from the kinetic energy it transports, which is comparable to that of the interstellar gas and due to the presence of magnetic fields. Its origin and acceleration mechanisms are still rather uncertain, but there is agreement that supernovae play a dominant role.

In the interstellar medium, material ready to form new generations of stars assembles. Massive stars contribute to it new enriched material which they have synthesised in their interiors during their relatively fleeting existence. Among other things, stellar winds, very strong at the beginning of a star's life, and supernova explosions make a very important contribution of mechanical and thermal energy. The interstellar medium may be regarded as the scene of the physical and chemical life of an evolving galaxy. The gas content of a galaxy varies according to its type. Generally speaking, spiral disk galaxies contain the most gas, and star formation is active in them. On the other hand, elliptical galaxies of the same age contain much less gas and show fewer signs of active star formation. Between these two, irregular galaxies are well provided with gas, but gas which is generally poor in the heavy elements found in our Galaxy; the stars that make up these galaxies form only slowly.

Jean-Pierre CHIÈZE

The interstellar medium

Gaseous nebulae

Nebulae are the most obvious sign of the presence of matter in interstellar space. Several hundred can be counted in our own Galaxy, as well as in several neighbouring galaxies. In our system, they are distributed in the neighbourhood of the galactic disk, where most of the interstellar matter is congregated. Nebulae demonstrate in various ways the action of stars, at different stages of their evolution, on the surrounding diffuse matter. Generally speaking, these effects are greatest at the beginning and at the end of a star's life. Four types of nebulosity are distinguished: regions of ionised hydrogen, often called H II regions; reflection nebulae; planetary nebulae; and supernova remnants.

H II regions are produced by young type O stars, which are massive and very hot and emit very intense ultraviolet radiation. O-type stars, seen most often in groups, have not had time to distance themselves from the interior of the cloud in which they were formed. The intense ultraviolet flux heats and ionises a bubble of gas which is raised to a temperature of 10 000 K.

Buried in the heart of a cloud, this compact H II region is initially hidden from view; it can usually be detected only by the radio or infrared emission from hot dust. Later on, the bubble, whose pressure is higher than that of the medium around it and which is ionising an increasing quantity of gas, becomes optically visible in its entirety. Emission from an H II region is dominated by the recombination lines of hydrogen and helium, and by 'forbidden' lines from singly or doubly ionized oxygen. Recombination radiation consists of photons emitted after the capture of a free electron by an ion; this ion thereby regains the electron which was knocked off by a photon

of ultraviolet radiation from the ionising star. 'Forbidden' lines derive their name from the fact that they come from electromagnetic transitions between atomic levels which are ordinarily, under the pressures found in the laboratory, depopulated by multiple collisions. Only the very low densities in H II regions, of the order of 1000 ions per cubic centimetre, allow these very fragile levels to survive. The continuum radiation, for its part, is dominated by intense infrared emission from the dust, and peaks at a wavelength of about 100 micrometres. At longer wavelengths, radio emission is produced by the slowing down of charged particles within the plasma (this is called free–free emission).

Reflection nebulae have a quite different origin. They are regions rich in interstellar dust, which diffuses the light from neighbouring bright stars, usually of type A, but the gas associated with the solid particles is too cold to become ionised. Reflection nebulae are characterised by their bluish colour. This is due to the fact that the dust grains, about a tenth of a micrometre in size, have a greater reflecting power for blue than for red light, which is therefore more easily transmitted. This is the same reason that cigarette smoke appears blue when illuminated indirectly, but redder when seen in light passing through it. The light reflected by reflection nebulae is weakly polarised. The interstellar magnetic field tends to orient the particles, which are made of elongated silicate or graphite grains encased in a shell of ice. The alignment of the grains is sometimes shown by the 'combed' appearance of certain reflection nebulae.

Planetary nebulae derive their name from the disk-like shape they usually show, which, to

The Trifid Nebula (NGC 6514). This nebula centre is at a distance of about 1.4 parsecs. The red emission is caused by hydrogen ionised by a small group of young stars in the interior of the gaseous mass. At the heart of this expanding H II region a network of bands of interstellar dust is developing, as well as globules of cold gas still not ionised by the stellar radiation. Progressively swallowed and compressed by the H II region, these globules can condense and give birth to new stars. The irregularity of the boundary of the H II region indicates inhomogeneity in the ambient interstellar medium. Indentations correspond to denser regions, temporarily skirted by the hot gas of the nebula. The blue region to the north is a reflection nebula surrounding the star HD 164514. (D. F. Malin, Anglo-Australian Telescope Board, © 1977)

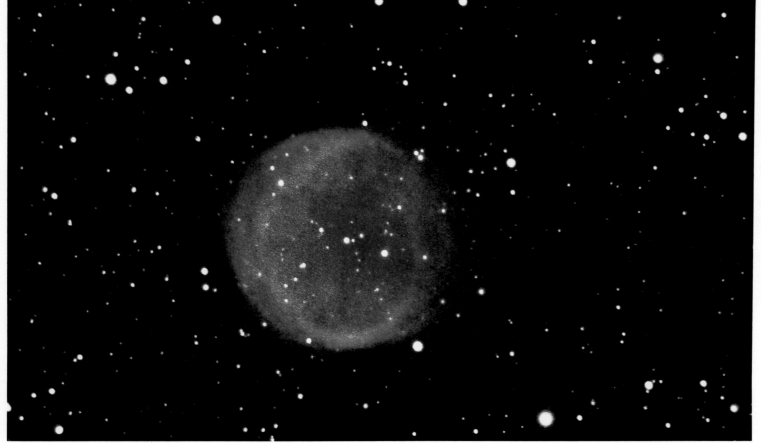

The Aquila Planetary Nebula. This nebula is the result of the ejection of the envelope of a red giant entering an unstable stage of its evolution. The surface temperature of the stellar remnant, visible at the centre of the nebula, exceeds 50 000 K, and its radiation ionises the gas from the ejected layers, which shines by fluorescence; at the same time, the pressure of the wind from the star maintains an expansion speed of about 20 kilometres per second. The expansion of the nebula gradually slows as it becomes more distant from the source of ionising radiation. The gas ejected by the star merges bit by bit with the ambient interstellar medium. It is estimated that planetary nebulae contribute about 15 per cent of the total mass ejected by a group of stars. They therefore play an important role in the redistribution in interstellar space of the chemical elements synthesised in stars. (Hale Observatories)

The Vela supernova remnant.
The supernova explosion that gave birth to this nebula probably dates back more than 10 000 years. In optical emission (from the Hα recombination line of ionised hydrogen), the visible part is at the edge of the shell separating the ambient medium from the gas in the interior of the remnant, which has been heated by the shock wave from the explosion. Overall, the remnant can be detected by its emission in radio waves and X-rays. The centre of the nebula is occupied by a pulsar whose rotation period is about 90 milliseconds. This neutron star is the extremely dense relic of the implosion of the core of the supernova. This and the supernova remnant in the Crab Nebula, are the only two to show high-energy gamma-ray emission. They are also the only ones with which a central pulsar has definitely been associated. (Royal Observatory, Edinburgh, © 1979)

certain early observers, suggested similarity with planets in the Solar System. They are in fact, like H II regions, clouds of gas ionised by a very hot star which is always seen in the centre of the nebula. This time, the luminous gas is the result of the ejection of the surface layers of the ionising central star, which is close to the end of its evolution and whose stripped core is evolving into a white dwarf. Planetary nebulae are not distributed in the Galaxy in the same way as H II regions, which are known to follow the spiral arms. They are scattered throughout the galactic disk, with a marked concentration towards the galactic centre. They appear to follow the distribution of old stars. Planetary nebulae are one of the last stages of evolution of intermediate mass giant stars.

A supernova remnant is the result of the violent explosion of a star. The term remnant is somewhat misleading, because the emitting region contains a mass of gas much greater than the mass of the star which, by exploding, gave birth to the remnant. In fact, the enormous quantity of energy liberated by the supernova causes the formation of a shock wave which propagates at high speed through the ambient interstellar medium, sweeping up an increasing quantity of gas. This, raised to a temperature greater than a million kelvins in a young remnant, emits radiation throughout most of the electromagnetic spectrum, from radio waves to gamma-rays. As they age, supernova remnants cool and finally merge with the ambient medium. Some old remnants, of great size, appear as filamentary shells. Optical emission from filaments accompanies the rapid cooling of that part of the remnant where the density of gas, near the shock, is highest.

Jean-Pierre CHIÈZE

Reflection nebulae around the Pleiades Cluster. The brightest stars of the Pleiades Cluster (right) are each surrounded by blue veils, which are reflection nebulae. The cold gas that bathes these stars is mixed with a large quantity of dust, which appears to consist of elongated grains about 10 micrometres in size. It is the illumination of these grains by the light of the star that gives the characteristic blue colour to this type of nebula. The 'combed' appearance of the nebula around, in particular, the star Merope (below) comes about because the oblong dust grains are aligned by the interstellar magnetic field. (Hansen Planetarium and Kitt Peak National Observatory)

The interstellar medium

Molecular clouds

Molecules were first observed in the interstellar medium in 1937 by the spectroscopist F. Adams, who detected, at wavelengths near 400 nanometres, absorption lines from the radicals CN, CH^+ and CH in the spectra of several bright stars. The lines are very narrow, supporting the idea that the absorbing radicals are at very low temperatures and not a part of the envelope of the observed stars.

The detection of molecules is now largely done using radio astronomical techniques. In 1963, Weinreb and Barrett discovered the hydroxyl radical (OH) in the direction of the galactic centre. The lines of the radical, which are very intense and very narrow, are produced by the stimulated maser emission mechanism. Molecules of SiO and H_2O also emit maser radiation. Maser sources are generally of very small size, less than 3×10^{-4} parsec; they are associated with stars in the process of formation, and are a powerful means for detecting such stars.

Research into interstellar molecules intensified after 1968. Centimetre and millimetre wave observations have brought in a haul of more than fifty interstellar molecules. The most complex organic molecules so far detected in space (including alcohols, acids, aldehydes, amides and amines) are, however, much less complicated than those involved in living material. Despite this, their discovery in a medium previously judged very unfavourable to their synthesis came as a surprise.

Molecules have been detected in abundance in huge interstellar clouds. The giant molecular cloud 'Sagittarius B2' has for a long time been special territory, with a stream of discoveries of new molecules coming from it; most of the known molecules were discovered there. It is estimated that at present half the mass of interstellar gas in the Galaxy is in molecular form. The most abundant of these molecules is of course the hydrogen molecule, because this element is the most widespread in space. Unfortunately, this molecule has no intense line in visible, infrared or radio emission, and is not easily detectable. The asymmetric molecule CO, carbon monoxide, is the one which most easily allows detection of molecular gas, although it is more than 100 000 times less abundant than the H_2 molecule. The strong emission in CO at low temperatures is due to electronic transitions between energy levels caused by the rotation of the molecule about its own axis.

Molecular gas is observed in three broad types of interstellar cloud: dark clouds, molecular clouds and giant molecular complexes.

All these clouds contain a large quantity of interstellar dust. Their central temperature is very low, between 5 and 10 K, and their density, much greater than that of neutral hydrogen clouds, is about 5000 particles per cubic centimetre or more. They are often surrounded by a halo of atomic hydrogen. Dust grains act as a screen against ultraviolet radiation, present throughout interstellar space, which would otherwise destroy the fragile molecules. These same grains catalyse the formation of molecular hydrogen from hydrogen atoms that stick to their surface.

Molecular clouds are characterised by very strong emission from the CO molecule; theoretical arguments have shown that its presence is linked with that of molecular hydrogen in an abundance ratio for CO/H_2 of 8×10^{-6}. The physical conditions in them resemble those in dark clouds. They are distinguished from the latter mostly because their discovery is due to their association with regions of ionised hydrogen and with reflection nebulae. Molecular clouds are special places for the formation of stars. Groups of young stars, of O and B type, are

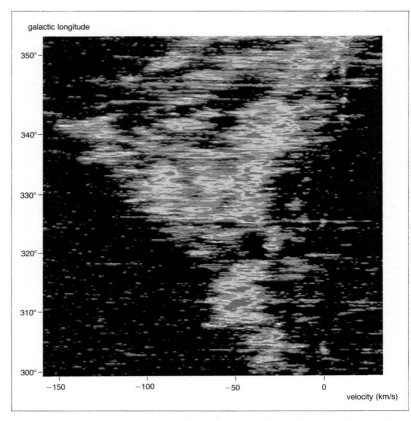

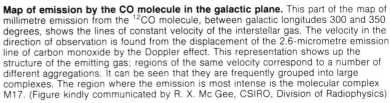

Map of emission by the CO molecule in the galactic plane. This part of the map of millimetre emission from the ^{12}CO molecule, between galactic longitudes 300 and 350 degrees, shows the lines of constant velocity of the interstellar gas. The velocity in the direction of observation is found from the displacement of the 2.6-micrometre emission line of carbon monoxide by the Doppler effect. This representation shows up the structure of the emitting gas; regions of the same velocity correspond to a number of different aggregations. It can be seen that they are frequently grouped into large complexes. The region where the emission is most intense is the molecular complex M17. (Figure kindly communicated by R. X. Mc Gee, CSIRO, Division of Radiophysics)

The dark cloud Rho Ophiuchi. Rho Ophiuchi is a dark cloud of typical size, about 230 parsecs from the Sun. It covers an area of sky 300 times bigger than the Full Moon, in parts of the constellations of Ophichus and Scorpius. This cloud contains a great deal of irregularly distributed dust, which screens the visible light of the stars. Observations at infrared and radio wavelengths nevertheless allow the core of the cloud to be probed. In this way, evidence has been found for the existence of more than sixty young stars, buried in the cloud, as well as six compact H II regions. It is estimated that about 10 per cent of the mass of the cloud has already condensed to form new stars. The densest regions of this cloud contain more than 10 000 molecules of hydrogen per cubic centimetre. The temperature in them is about 10 K. (Royal Observatory, Edinburgh, © 1979)

Interstellar molecules, radicals, and ions detected by the 1990s. The most abundant molecule in space is hydrogen. Others are present only in trace quantities, in molecular clouds and complexes. Some of them represent no more than a millionth of the mass of the clouds. But unfortunately the H_2 molecule cannot be detected directly because its lines in the ultraviolet are strongly absorbed by gas and dust. It is therefore other molecules, most prominently carbon monoxide ^{12}CO or its isotopic analogue ^{13}CO, that allow the detection of molecular gas in space. Some long rectilinear carbon chain molecules and some other unstable molecules, cannot be synthesised in the laboratory, where the conditions prevailing in molecular clouds cannot be simulated. Their identification rests on theoretical calculations of molecular structure. The molecules are classified as a function of the number of atoms in them.
The asterisk shows that the species has only been detected in circumstellar envelopes; the question mark indicates an uncertain identification.

2	3	4	5	6	7	8	9	10	11	13
H_2	H_2O	H_2CO	HCOOH	* C_2H_4	CH_3C_2H	CH_3OCHO	$(CH_3)_2O$	$(CH_3)_2CO$?	HC_9N	$HC_{11}N$
HD	HCO	NH_3	HC_3N	C_6H	CH_3CHO	CH_3C_3N	CH_3CH_2OH	CH_3C_5N ?		
CH	HCO^+	HNCO	CH_2N_2	HCH_2OH	HC_5N		CH_3CH_2CN			
CH^+	CCH	H_2CS	NH_2CN	NH_2CHO	CH_3NH_2		HC_7N			
CN	HCN	H_3O^+ ?	H_2CCO	CH_3CN	CH_2CHCN		CH_3C_4H			
CO	HNC	C_3H	CH_4	CH_3NC	C_6H					
CP	N_2H^+	C_3O	C_4H	CH_3SH						
CS	H_2S	C_3S	CH_2NH	H_2C_4						
OH	OCS	C_3N	* SiH_4	HCC_2HO						
NH	SO_2	* C_2H_2	C_3H_2							
SO	HNO	HCCN	CH_2CN							
NS	HCS^+	$HCNH^+$	C_4Si							
PN	H_2D^+?	$HOCO^+$	H_2C_3							
SiC	SiH_2 ?	HNCS	* C_5							
SiN	* SiC_2									
SiO	C_2S									
SIS	C_2O									
* C_2	* C_3									
CO^+?										
NO										
HCl ?										
* NaCl										
* AlCl										
* KCl										
*AlF ?										

observed in them, as well as strong infrared sources in the interior of the clouds; these are actually stars in the process of formation, with ages of less than a thousand years.

The greater part of molecular gas is assembled in huge molecular complexes, which, with masses of about 100 000 solar masses are the most massive objects in the Galaxy. They are often elongated in shape, and their largest dimension can be as much as 100 parsecs. The total mass of all these giant molecular complexes is about five billion solar masses. They have recently been mapped on a large scale with the help of a small radio-telescope of 1.2 metres diameter, whose large field of view allows detection of CO emission over a large part of the sky. Molecular complexes are most common in the nucleus of the Galaxy, and in a large ring extending from 4 to 8 kiloparsecs, halfway between the Sun and the galactic centre. High-resolution observation of these complexes reveals a very inhomogeneous structure. Neighbouring regions move with very different velocities. Their relative motion, of the order of a few kilometres per second, is supersonic. The various parts of the complex are often themselves composed of an assembly of smaller, but denser, elements.

The formation of giant molecular complexes is still the subject of speculation. The density waves that run through the disk of the Galaxy may cause the condensation and accumulation of numerous small interstellar clouds, whose coalescence leads to the formation of the large complexes. Whatever their origin, the formation of stars within them can only lead to their destruction, and there must be a mechanism that permits their continual renewal.

Many of the questions concerning molecular clouds are still unanswered. It is clear that they play an important role in our Galaxy. As well as being sites for the formation of new generations of stars, their high mass is a factor which must be taken into account in order to understand the structure of the Galaxy itself.

Jean-Pierre CHIÈZE

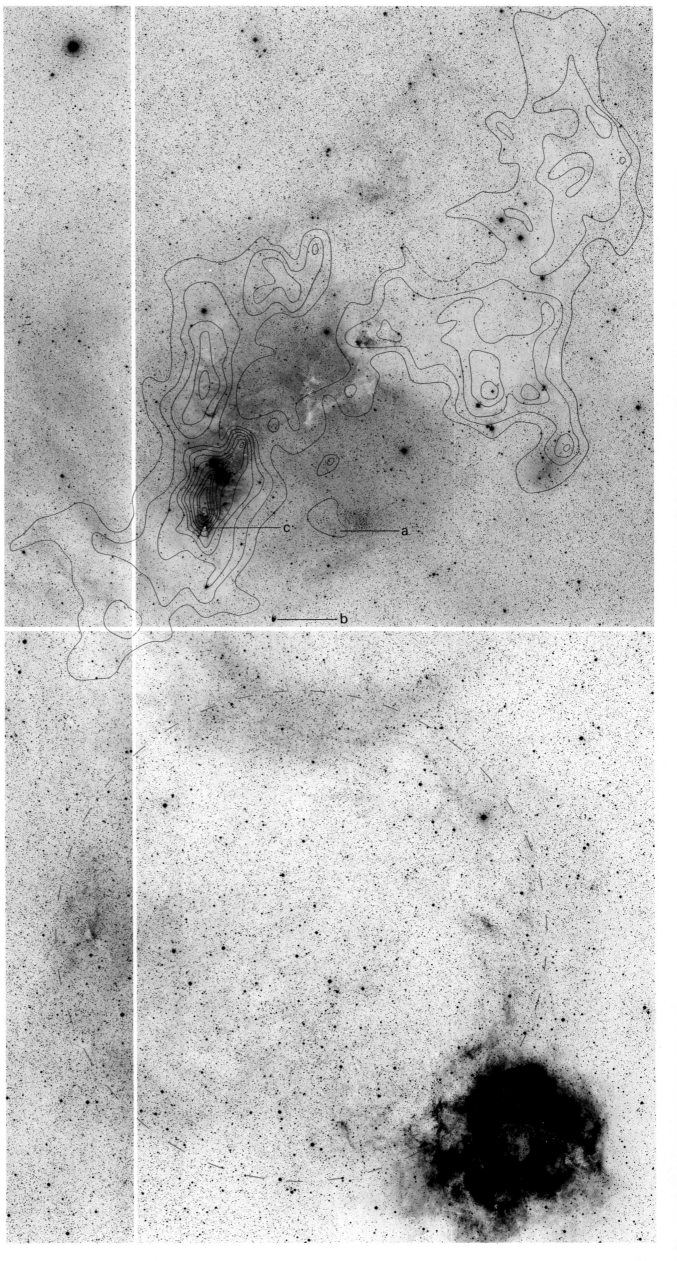

The molecular complex in Monoceros. The molecular complex in Monoceros is a little to the north of the Rosette Nebula. The contours show the places of equal emission intensity in carbon monoxide ^{12}CO. The cloud more to the south is centred on the Cone Nebula (c), itself associated with a cluster of very young stars (NGC 2264) containing many T Tauri variables. The small molecular cloud (a) is associated with the old galactic cluster Trumpler 5. To the south, a part of a supernova remnant (broken line) can be seen. This region of the sky contains several reflection nebulae. To the north of the reflection nebula NGC 2261, the variable Hubble Nebula (b), a Herbig–Haro object is found.

A molecular cloud comprises many condensations, sometimes moving with very different velocities; in this case they are effectively independent. The bridge linking the two clouds is itself composed of a chain of small clouds. The region where CO emission is most intense contains both an infrared source and an H_2O maser. The way in which dust contained in these clouds obscures the sky is very apparent in this montage of negative plates. (POSS)

The interstellar medium

Interstellar masers

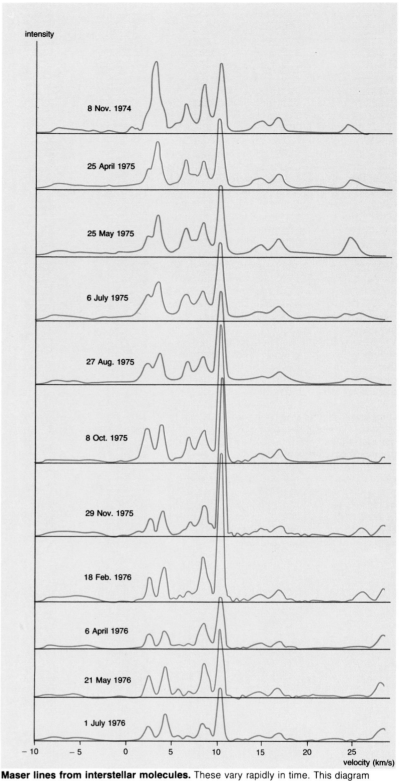

One of the great surprises in astronomy during the 1960s was the discovery of radio emission from a number of previously unsuspected molecules. Even more surprising was the fact that some of these lines were extremely intense and, what is more, variable over a period of time. It was soon understood that this was not normal emission from interstellar molecules, but rather anomalous emission from natural masers. The word maser is an acronym from microwave amplification by stimulated emission of radiation. The possibility of this phenomenon was more or less foreseen by Einstein at the beginning of the century, but masers (and their better-known optical wavelength equivalents, lasers) were not made to work in the laboratory until the 1950s.

The internal energies of atoms and molecules can take only certain specific and well-determined values; we therefore speak of energy levels. Emission of radiation by an atom or molecule comes about when it descends from a higher to a lower energy level, and this radiation, according to the quantum theory, has a wavelength such that each photon emitted has an energy equal to the energy difference between the two levels concerned. Conversely, an atom or molecule in a low energy level, under the influence of radiation at an appropriate wavelength, can be raised to a higher level by absorbing a photon. In addition, Einstein predicted a process called stimulated emission, the exact inverse of such absorption: an atom or molecule in the higher energy level, under the influence of a photon of the appropriate wavelength, can fall to the lower level emitting of course another photon which adds to the first, which stimulated the emission. If this phenomenon dominates, and particularly if stimulated emission is more important than absorption, (i.e. the medium considered has more atoms or molecules in the higher energy level than in the lower), then the medium will act as an amplifier and will emit a very intense spectral line. This condition is difficult to arrange between the very different energy levels corresponding to emission or absorption of optical photons. Consequently the manufacture of lasers is not easy, or at least was not initially. On the other hand, this arrangement is much easier to produce at the close energy levels corresponding to emission or absorption in radio frequencies. Masers exist not only in the laboratory but also in nature. There are many methods of 'pumping' a maser, that is, of raising a majority of its atoms or molecules to the higher energy level: for example, collisions between atoms or molecules, and near or far infrared radiation. Astrophysicists are therefore embarrassed by choice when trying to decide which mechanism pumps a given maser. For some masers, the particular mechanism is still not known.

Several interstellar molecules are capable of

Maser lines from interstellar molecules. These vary rapidly in time. This diagram shows the spectrum of the water vapour (H_2O) masers in the molecular cloud in Orion, obtained at different times separated by a few months. Variations are evident: some components appear or disappear rapidly, others are more stable. The variations result from minor changes in the pumping conditions in the masers. (After Little, White and Riley)

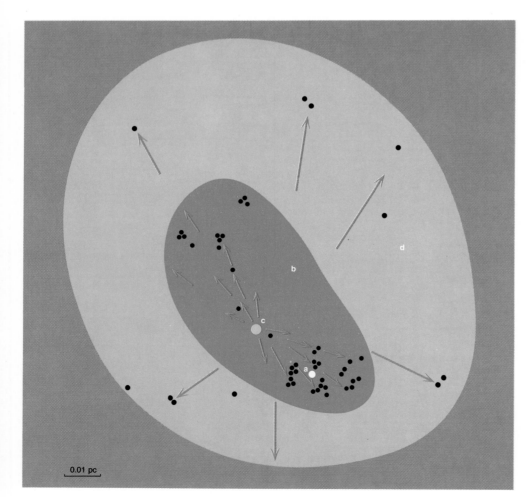

The location of masers. The location of a number of water vapour (H_2O) masers in the molecular cloud in Orion, as observed in 1979, is illustrated in this diagram. Each point represents a different maser source, and the small circle (a) represents a maser which was, for a few months in 1979, the brightest in the sky. Repeated observations at intervals of a few years show that the masers move rapidly. Those in the central region (b) are moving away, at a speed of about 18 kilometres per second, from the presently forming or newly formed central star (c), and appear to have been ejected in two opposite directions. The masers in the exterior region (d) are moving away even more rapidly. (After Genzel, Reid, Moran and Downes)

The Omega Nebula. A gaseous nebula is associated with a molecular cloud which is the site of formation of massive stars; it also contains a number of maser sources. The cloud is observable only in infrared and radio waves. (Kitt Peak National Observatory, Cerro Tololo Inter-American Observatory)

forming natural masers: these are water, H_2O, which emits very intensely at 1.35 centimetres wavelength; hydroxyl, OH (18 centimetres); silicon monoxide, SiO; methyl alcohol, CH_3OH; formaldehyde, HCHO; and many more. Some emissions, especially that of water, are strongly variable in time, with different spectral components appearing and disappearing over periods of a few months.

Very high resolution observations made by intercontinental radio interferometry show, among other things, that each component of the spectrum comes from a different region of the source: these regions, relatively small on the scale of the interstellar medium (their sizes are less than that of the Solar System), approach or recede from the observer at various velocities, producing a shift in the emitted line to wavelengths smaller or greater than that of the molecule at rest. Furthermore, movement of these regions over a period of time is often observed.

Interstellar masers are generally found in the dense parts of molecular clouds, and the emitting regions are always optically invisible because of interstellar absorption; they are almost always associated with star formation. Because the maser sources are small, dense and often hot, it has been suggested that they are protostars, that is, fragments of molecular clouds which are in the process of collapsing onto themselves to form stars. In fact, this is not so, and masers are simply parcels of interstellar matter where conditions are right for pumping; these could be created by infrared radiation, compression or violent motions, and these same conditions appear to be linked with star formation. The motions of maser sources may also indicate the strong winds which accompany the first stages of stellar evolution. For example, many H_2O maser sources in the molecular cloud in Orion are aligned with respect to a point-like object which appears to be a

forming or newly formed massive star, pushing out the nearby material as a strong wind. Recent observations show that such ejection, produced preferentially in two opposite directions, is a quite general phenomenon associated not just with star formation, but also with the nuclei of galaxies, quasars, etc.

James LEQUEUX

Maser emission spectra. The molecular cloud associated with the Orion Nebula contains several molecules showing maser emission. Here the spectra of radio masers in four molecules are shown. The name of the molecule, the frequency of the observed radiation and the date of the observation are shown for each spectrum. The ordinate is intensity (in arbitrary units) and the abscissae show, not the frequency, but the velocity of approach (positive) or recession (negative) of each component of the spectrum, deduced from the frequency according to the Doppler law. Each peak corresponds to a separate emission source and the sources of the different spectra are in general not the same. (After J. M. Moran)

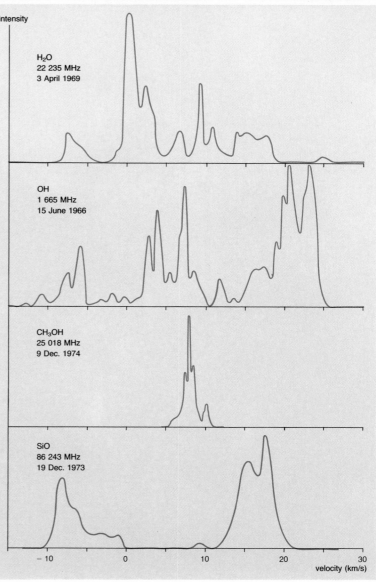

The interstellar medium

Interstellar dust

The interstellar medium contains mostly gas, but there are fine dust particles as well. Although these particles are very small and represent no more than 1 or 2 per cent of the mass of the interstellar medium, they play an extremely important role in astronomy. Indeed, they affect the light from stars in many ways, themselves radiating as much energy as the stars (though in the far infrared, not in visible light), and are also the sites of formation of the most abundant interstellar molecule in the Universe: the hydrogen molecule.

Interstellar dust obscures the light of background stars. This absorption phenomenon, known as interstellar extinction, is partly due to the true absorption of light by dust particles, which are thereby heated, and also to the scattering of light in directions different from the direction of its arrival. This last process can be explained by physical optics: it is due to the diffraction of light. It causes, like absorption, a diminution of the light we receive from a star, because part of it is diverted to different directions and does not reach us. In favourable circumstances, when a star is found just beside a dust cloud, the diffused light illuminates part of the cloud. This is a reflection nebula (although it would be better called a scattering nebula), previously mentioned in connection with gaseous nebulae. In general, starlight attenuated by interstellar dust becomes linearly, and to some extent circularly, polarised; this

The spiral galaxy NGC 891. This photograph of NGC 891, a spiral galaxy like our own seen almost side-on, shows evidence for the existence of a band of dust, which absorbs the light of some of the stars. The dust is in fact mixed with interstellar gas in a very flat disk, flatter than the stellar disk, and the distribution of the gas and dust is fairly chaotic. (D. F. Malin, Anglo-Australian Telescope Board)

occurs because the particles are often aligned by the magnetic field which pervades interstellar space.

Interstellar extinction is a very irksome phenomenon in astronomy. Indeed, dust in the Milky Way prevents us from seeing distant stars near its plane of symmetry. Moreover, the Milky Way appears to us to be divided by a dark band of dust, something which is also seen in external galaxies. Interstellar extinction depends strongly on wavelength: weak in the near infrared and essentially absent in the far infrared and radio. Extinction is noticeable in visible light and important in the ultraviolet. Therefore our Galaxy is much more transparent in infrared than in visible light, and we can 'see', for example, the centre of the Galaxy at wavelengths of about 2 micrometres and longer.

Since starlight is diminished more in the violet than in the blue, and more in the blue than in the green, the colours of stars change through interstellar obscuration and becomes 'reddened'. Measurement of this change of colour, known as interstellar reddening, allows us to estimate the quantity of dust between us and the observed star. This quantity is generally proportional to the amount of interstellar gas measured by other means, implying that gas and dust are well mixed in the interstellar medium.

The energy of the light absorbed by interstellar dust heats it. Clearly, dust cannot be heated indefinitely without emission of heat by radiation in the far infrared. It therefore attains an equilibrium temperature of about 30 K in the Galaxy; however, there are large variations from one location to another. Observations in the far infrared, which have in the past few years become possible with aircraft, balloons and satellites, show, by measurement of thermal emission, that regions where dust is most abundant coincide with dark clouds or molecular clouds. One surprising result of these observations is that our Galaxy radiates half its energy in the infrared and half in visible light, implying that half of all the light from stars is absorbed by interstellar dust. Optical and infrared observations make it

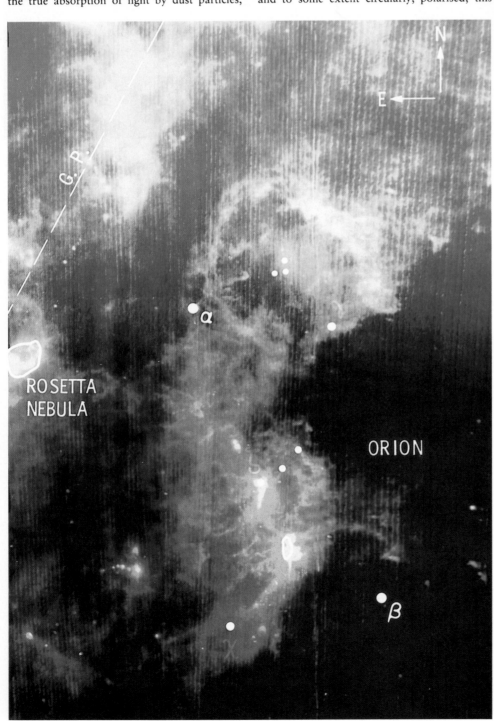

Infrared image of the Orion region. This false-colour image was produced from data obtained by the Infrared Astronomical Satellite (IRAS) and shows a very different view to that obtained by optical telescopes. It actually shows the dust content and not the nebulae or stars. The intensity of infrared radiation is represented by colours: red indicates strong 12 micrometre wavelength radiation; green indicates strong 50 micrometre radiation, and blue shows strong 12 micrometre radiation. Well-known regions of star formation are apparent such as the Orion molecular cloud (the large feature dominating the lower half of the picture), and the Orion and Rosette Nebulae (both marked). These features contain a very great deal of interstellar dust. The Milky Way crosses the upper left hand corner; the galactic plane and the principal stars of Orion are α (Betelgeuse), β (Rigel), γ and δ. The Horsehead nebula is the white spot to the South of the star ζ of Orion's belt. (NASA)

The Horsehead Nebula. This well-known photograph illustrates the effect of interstellar dust on light. The famous Horsehead is a dense dark cloud, where dust is very abundant, which completely absorbs the light from the regions behind it. It stands out from the fairly uniform background, which is a very diffuse gaseous emission nebula. The Horsehead is in fact only a protuberance of an immense dark cloud covering all the lower part of the photograph. Many fewer stars are visible in this region than in the upper part; only the few stars in front of the cloud can be seen, those behind being rendered completely invisible by the extinction. The bright bluish-white nebula to the lower left of the Horsehead is a reflection nebula, produced by the illumination of the near part of the cloud by a star, which is drowned out by the nebula in this overexposed photograph. Finally, the large region at the lower left is a gaseous nebula crossed by lanes of absorbing dust. The circle and cross are artefacts coming from an extremely bright star. (Royal Observatory Edinburgh, 1979)

possible to estimate the sizes and chemical compositions of dust grains, by comparing the observations with the results from theoretical models. Dust grains are very small, with sizes ranging from a few tenths to a few thousandths of a micrometre. They contain various silicates (of aluminium, iron, magnesium, etc.) and carbon (probably in the form of graphite), and also, frozen onto silicate nuclei, molecules of water, ammonia, methane, etc., characteristic traces of which have been detected in the infrared spectrum of the dust. Observations of the chemical composition of interstellar gas, made by the Copernicus satellite, have shown that elements like silicon, aluminium and iron – which appear to partly constitute dust grains – are effectively underabundant in the gas compared to values observed, for example, in the Sun. However, there are regions of hot, dilute interstellar medium where these elements are much less deficient: dust grains are partially destroyed by various processes, such as the passage of a shock wave, thus restoring to the gas part of their material.

Where and how is dust formed? Most of it probably forms in expanding envelopes around cold stars, where initially gaseous elements condense into solid grains. It is believed that small grains can subsequently grow, principally within molecular clouds, by capturing atoms and molecules. Some molecules 'freeze' onto the surface, others form there from atoms and later evaporate: this is the way molecular hydrogen, the principal constituent of dark clouds, is synthesised. Finally, the grains may be abraded or destroyed by their interaction with the surrounding matter. The history of a grain of interstellar dust is probably very complicated.

James LEQUEUX

The infrared emission of galactic interstellar dust. This infrared whole-sky image was obtained by the IRAS satellite. The Milky Way is the bright horizontal band: the direction of the galactic centre is the same as the centre of the image. The blue colour corresponds to intense radiation of about 12 micrometre wavelength, the green to intense radiation of about 60 micrometres, the red to intense radiation of about 100 micrometres. The hottest material appears blue or white, the coolest red. We can distinguish dust emission, which is mixed with interstellar gas in the Milky Way. The diffuse 'S' which crosses the image comes from the weak emission from dust in the plane of the Solar System (the origin of the zodiacal light). The astronomical objects which are also visible on this image are star formation regions in the constellation of Ophiucus (just above the galactic centre) and Orion (the two bright spots below the Milky Way, on the extreme right). The Large Magellanic Cloud is the relatively isolated bright spot, below the Milky Way, to the right of the centre. (NASA)

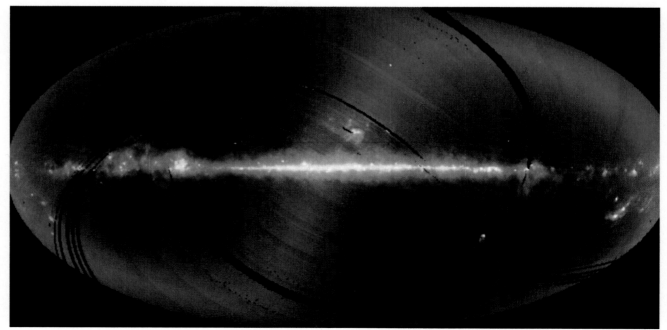

The interstellar medium

The nearby interstellar medium

The interstellar medium, in its diversity and complexity, is strikingly chaotic. The physical conditions within it are extremely varied, and its structures dissolve and re-form continually under the action of supernovae explosions, strong winds blowing from young and massive stars, and ultraviolet radiation from the very youngest stars.

The nearby interstellar medium is no exception. Observations (made in directions well away from the galactic plane) of neutral gas, using the 21-centimetre line of hydrogen, and of ionised gas, using the Hα line, provide evidence for numerous arches or loops and other complex structures indicating a state of great agitation. These observations deal with the medium up to a few hundred parsecs from us.

Paradoxically, the medium very close to us – a few tens of parsecs from the Sun – is difficult to study, particularly because it appears to be very dilute. Only in the past few years have we begun to understand it, especially with the help of observations from space.

The Solar System is not directly immersed in the interstellar medium; the Sun constantly throws off material, as the solar wind, which pushes the interstellar medium well beyond the orbit of Pluto, the most distant planet. However, neutral atoms of hydrogen and helium can penetrate the solar wind, which is completely ionised, without much hindrance. These atoms have been detected by their emission in the far ultraviolet; in addition, their velocities and the direction they come from have been estimated, and even the temperature of the gas they make up has been measured. Their density is about 0.1 atom per cubic centimetre, and their temperature is nearly 10 000 K. They arrive at the Earth at the considerable speed of 25 kilometres per second. Their direction of motion is different from that towards which the Solar System, in comparison with nearby stars, is moving (at 20 kilometres per second). Therefore, the interstellar medium is not at rest, but is in rapid motion with respect to the Sun and also nearby stars. There is nothing surprising in the state of agitation of this medium. In the optical and ultraviolet spectra of nearby stars, very narrow absorption lines from intervening interstellar ions or atoms have been observed. By studying these lines the total number of atoms (or ions) between us and the nearby star can be estimated, and by dividing by the known distance to the star the average density of the medium can be estimated. The result is quite surprising: observations of stars a few parsecs from the Sun imply an average density of interstellar gas of the order of 0.1 atom per cubic centimetre, in agreement with the value obtained from study of the atoms which penetrate the solar wind. Beyond a few parsecs, there is practically nothing, out to considerable distances (about 50 parsecs). We are therefore immersed in a mostly neutral interstellar cloud, which is hot and very dilute, and which is itself immersed in a more dilute medium (less than 0.01 atom per cubic centimetre). Other indications lead us to think that this latter medium is extremely hot, with a temperature of the order of a million degrees. It is therefore identified with the very hot and very dilute component of the interstellar medium which effectively fills the entire Galaxy.

The 'small' cloud which we are in is not symmetric; we appear to be situated close to the edge. Its densest parts, visible in the 21-centimetre line of hydrogen, are found in the direction of the constellations of Scorpio and Ophiuchus. It is possible that the hot 'bubble' was created by the several hot stars found in that direction, and that the cloud was made of shocked gas pushed by the stars, some of which may have recently exploded as supernovae.

James LEQUEUX

The distribution of ionised interstellar hydrogen in the sky (below) was found by combining several photographs taken in the hydrogen Hα line by a small telescope with a large field of view (60 degrees), using a filter which effectively eliminated starlight. This is a negative image. The ionised hydrogen is relatively concentrated along the galactic plane, but, because of interstellar extinction, what is seen is less than 3000 parsecs away.

The direction of the galactic centre (not visible) is to the left. Several rather bright gaseous nebulae can be seen, as well as filaments and loops, as in the distribution of neutral hydrogen. The most notable structures are in the constellations of Vela, Puppis and Carina (towards the middle, a little to the left), and in Orion and Eridanus (towards the middle, to the right and below). These nearby regions are strongly perturbed by supernova explosions and stellar winds from several massive stars. (J.-P. Sivan, Laboratoire d'astronomie spatiale, Marseille)

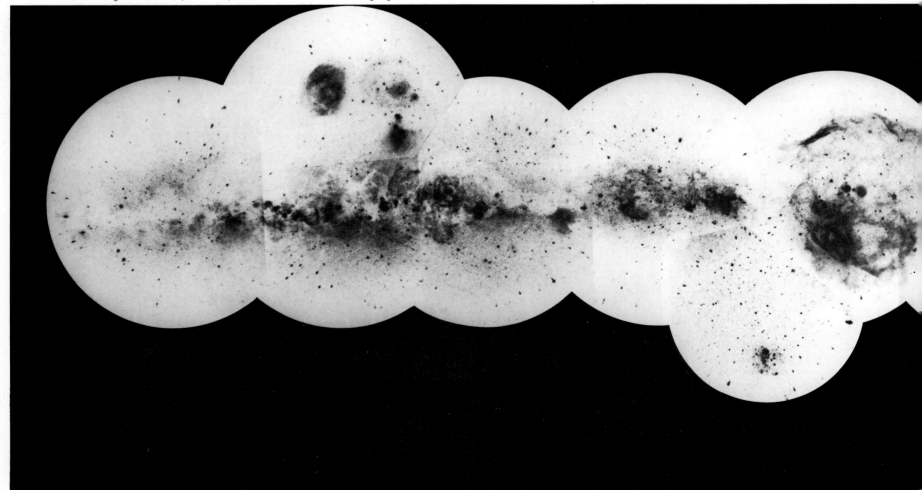

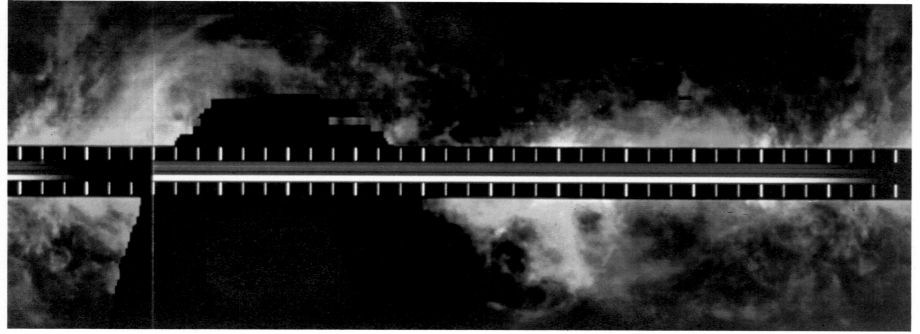

The distribution of neutral interstellar hydrogen in the sky has been obtained from 21-centimetre observations made with the 26-metre radio telescope at the Hat Creek Observatory of the University of California. The galactic plane is in the centre, and the observations do not include either the regions near the plane or those near the pole. The direction of the galactic centre (in the constellation of Sagittarius) is at the centre of the image. The brightness indicates the quantity of gas, and the colour shows velocity of recession (red) or approach (blue). The chaotic appearance of the gas distribution, shown by filaments and loops, etc., can be seen. All the interstellar matter visible here is relatively close, less than a few hundred parsecs from us. (C. Heiles, University of California at Berkeley, and E. B. Jenkins, University of Princeton, Science Photo Library/Cosmos)

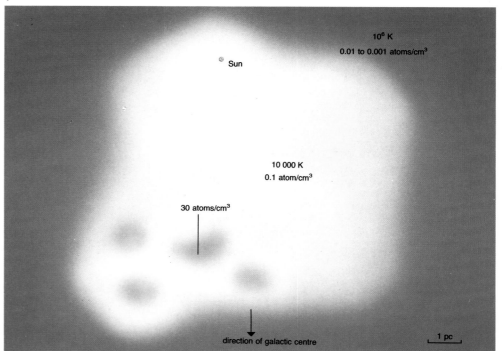

The structure of the very nearby interstellar medium (right) is revealed by observations of interstellar absorption lines in the spectra of neighbouring stars. The Sun (not to scale) is on the edge of a hot, dilute cloud, 5 or 10 parsecs across, in which it moves with a velocity of about 25 kilometres per second. This cloud contains several denser regions, indicated schematically. The whole cloud is immersed in a very hot, very dilute, medium.

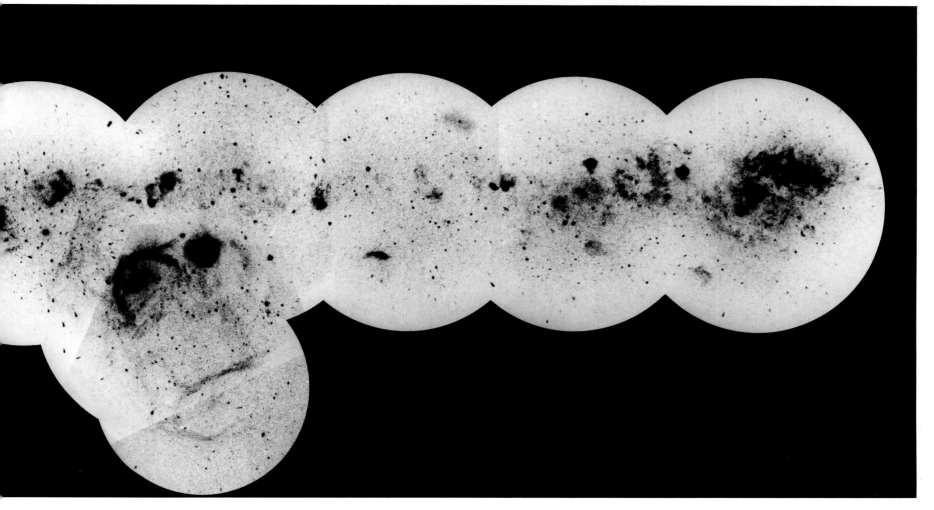

Galactic cosmic rays

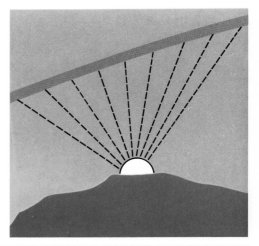

The first experiment in space took place on 7 August 1912; on that day, the Austrian physicist Victor Hess ascended in a balloon with two electroscopes, and noticed that at an altitude of about 5000 metres the electroscopes discharged more rapidly than on the ground, providing evidence for the presence of an ionising agent whose influence increased with altitude. Hess deduced that radiation of extraterrestrial origin was impinging on the atmosphere. It was not until the 1930s that the invention of the Geiger–Muller counter made it possible to be sure that this radiation consisted of charged particles, not photons. Since 1948 it has become known that this radiation consists mostly of nuclei of very high speed, and therefore of very high energy. Cosmic rays also contain fast electrons, but in very small numbers (about one thousandth of the number of nuclei). The energies of these particles range from millions of electronvolts up to 10^{21} electronvolts (the electronvolt is the kinetic energy acquired by an electron whose potential changes by one volt; a particle of 10^{21} electronvolts has enough energy to raise this Atlas by more than ten metres). There are many more particles of relatively low energy (from a million to a billion electronvolts) than of very high energy. Low energy cosmic rays, from a few million to a few hundred million electronvolts, have a very intense flux at energies of some ten million electronvolts; this flux then rises towards a maximum at an energy of about a hundred

million electronvolts. This low-energy part depends very strongly on the level of solar activity, which varies in cycles of about eleven years. When solar activity is at a maximum, a large number of particles of very low energy is ejected as the solar wind. The interaction between this solar wind and cosmic radiation from outside the Solar System strongly reduces the flux of cosmic rays with energies in the range ten to a hundred million electronvolts. This diminution is called the solar modulation. For energies greater than a billion electronvolts, solar modulation has practically no effect. The intensity of cosmic rays decreases sharply with energy, because the production mechanisms tend to accelerate a large number of particles to small and intermediate energies and only a very small number to high energies.

Cosmic rays comprise mostly very energetic atomic nuclei. When these fast particles dive into the atmosphere, they create 'showers' of secondary particles – muons and pions, for example. The relative nuclear composition of cosmic rays is in many ways similar to that of ordinary matter: hydrogen and helium predominate, and iron occupies a relatively important position. But there are some notable differences which help in understanding the origin of cosmic rays. The nuclei of the cosmic rays travel through the interstellar medium, influenced by magnetic fields which continually deflect their trajectories: the resultant distribution of their

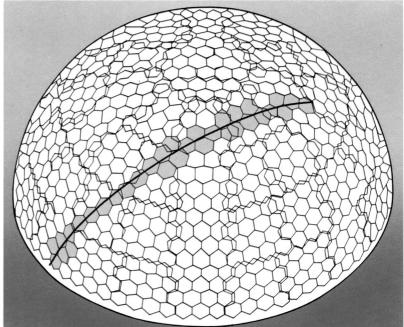

The 'Fly's Eye' Cosmic Ray Observatory. This observatory was planned and constructed to detect cosmic radiation of very high energy (greater than 10^{17} electronvolts) by means of the fluorescent radiation which it induces in the atmosphere. the observatory is sited on a piece of desert-like military land in the state of Utah in the USA, a long way from towns and their atmospheric pollution; it is an assembly of sixty-seven independent mirrors, each 1.60 metres in diameter, arranged on a spherical dome (illustrated above). This resembles a gigantic insect's eye, whence the name. This resemblance is not accidental; the observatory uses the optical properties invented by nature in this kind of arrangement. In the focal planes of each of the sixty-seven mirrors, the Utah University scientists responsible for the operation have installed twelve electronic detectors sensitive to fluorescent traces in the sky. This battery of about a thousand more or less independent detectors allows a precise determination of the trajectories of secondary particles created in the atmosphere by a very-high-energy cosmic-ray particle, and therefore also allows evaluation of the physical parameters of the incident particle. (Illustration kindly communicated by G. L. Cassiday, University of Utah)

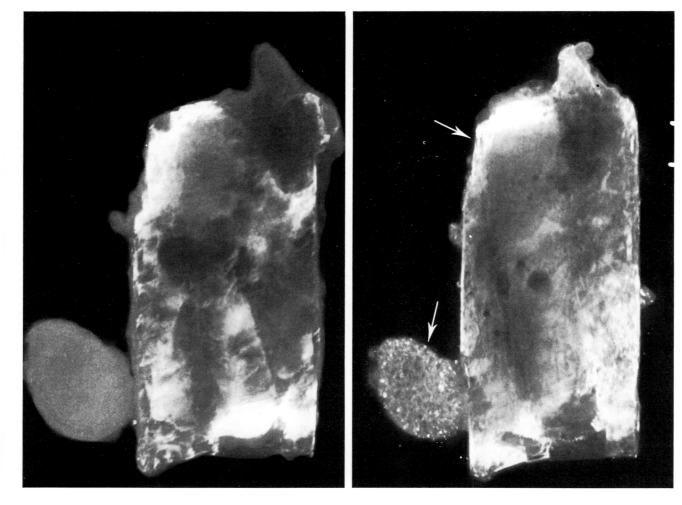

Electron microscope photograph of a meteoritic grain. This photograph shows a silicate grain, a few micrometres in size (1 micrometre equals 3 centimetres on the photograph), extracted from the Orgeuil meteorite, which fell in 1864 near a place of that name in the department of Tarn-et-Garonne in France. This meteorite has a high carbonate content, and is therefore one of the carbonaceous chondrites, the class of meteorites by means of which the chemical composition of the material comprising the Solar System has been determined. This photograph was taken with an electron microscope by the team of M. Maurette, working for CNRS at Toulouse in the Rene Bernas Laboratory, Orsay; it demonstrates the enormous resolving power of this instrument. The plate on the left shows the grain before heating; in the plate on the right, the grain has been raised to a temperature of about 1000 °C. In the latter, a disordered crystallisation can be seen in the places marked by arrows. These places have been subject to bombardment by cosmic rays, favouring the formation of this 'crystallite'. Analysis of grains like these provides evidence of the effects of the flux of cosmic radiation, and in some cases allows one to get an idea of the intensity of the flux.

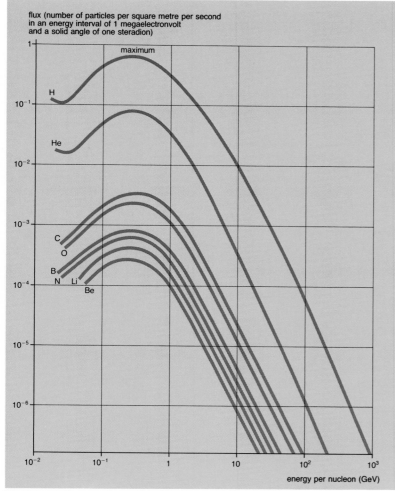

The form of the cosmic-ray flux. Three regions can be distinguished in this representation of the flux of cosmic rays as a function of energy. At energies less than about 10^{-1} gigaelectronvolts (GeV) the flux of cosmic rays decreases strongly with energy. These cosmic rays come from the Sun. There are more particles in this energy range when the Sun is active, that is, when there is a greater number of solar flares, which happens on an eleven year cycle. For energies from 10^{-1} GeV up to a few GeV (1 GeV = 10^9 eV), there is an increase in flux, reaching a maximum at about 1 GeV. These particles are of galactic origin (they come from all the exploding objects and their remnants scattered throughout the galactic disk). In this energy range, cosmic-ray particles are strongly affected by solar modulation: the less energy they have, the more they are prevented from penetrating into the region occupied by the Solar System, hence the maximum at 1 GeV. The maximum flux varies strongly with solar activity. Finally, for energies greater than 10 GeV, solar modulation no longer affects the cosmic radiation. The flux decreases as the energy of the particles to the power 2.6 because it is easier to accelerate particles to low energy than to give them very high energy.

directions of arrival is isotropic; that is to say there is no privileged direction. Unlike rays of light, it is therefore impossible to determine the place of origin of cosmic rays directly.

There are four big differences between the composition of cosmic rays and that of ordinary matter. Firstly, cosmic rays have a very high proportion of heavy elements (carbon, nitrogen, oxygen) compared to the amount of hydrogen and helium. Secondly, elements such as lithium, beryllium and boron, almost absent in ordinary matter, have abundances comparable to those of carbon, nitrogen and oxygen; the ratio of the abundances of these two families is 0.2, but it is 10^{-6} in ordinary matter. Thirdly, nuclei of odd-numbered atomic mass, between silicon and iron in the periodic table of the elements, are more numerous in cosmic rays. Finally, it has been noted that in the case of neon, the ratio between the isotopes neon 22 and neon 20 is four times greater than in the atmosphere.

These facts have been used to help elucidate the mystery of the origin of cosmic rays. Several hypotheses have been advanced to explain the flux of energetic particles. Although it involves only a small number of nuclei compared to the total in galactic matter, it plays a very important role in the overall energy balance of the Galaxy. The energy density carried by cosmic rays is comparable to that of starlight and also to that contained in the 2.7 K cosmological radiation left over from the Big Bang.

Cosmic rays may be of galactic or extragalactic origin. It is generally thought that most cosmic rays do come from objects or regions within our own Galaxy. In fact, recent determinations of the distribution of high energy gamma-rays, made for example by the Cos-B mission, indicate that the sources of cosmic rays must be confined to the galactic disk. There are three types of possible sources: exploding objects, like novae and supernovae; ordinary stars, like the Sun; and interstellar clouds, which may be the scene of particle acceleration. Comparison between abundances in cosmic rays and in

ordinary matter allows the contribution of these different sources to be specified. A very interesting correlation has recently been established between element abundances in cosmic rays at their source and the first ionisation potential of the elements: the most abundant elements in cosmic rays are those most easily ionised. This indicates that ordinary matter, strongly accelerated, could be the origin of part of the cosmic radiation. The overabundance of isotopes like neon 22, synthesised in the exterior zones of novae and supernovae, and the fact that refractory elements are abundant in cosmic rays, are evidence that some cosmic rays must have been formed in exploding stars.

Between their acceleration in exploding sources and their arrival in the Solar System, these particles interact with interstellar gas. The interactions are of two types, nuclear and braking. In the case of nuclear interactions, the energy of the particles is such that they can smash up nuclei of the interstellar medium, and create new elements. Braking reactions are due to interactions between cosmic rays and interstellar electrons. Nuclear reactions due to cosmic rays are the origin of large quantities of elements like lithium, beryllium and boron, which cannot be synthesised in stars. The theory of such production and comparison with observations, particularly the proportion of beryllium 10 whose lifetime is about a million years, shows that the above chemical elements are indeed very likely produced by these interactions in interstellar gas. Furthermore the braking interaction between cosmic rays and interstellar electrons readily accounts for the heating of the interstellar medium.

Despite its low flux, galactic cosmic rays play a very important role in the physics of the Galaxy. It transports large quantities of energy, determines the thermal balance of the galactic disk and is the source of a few chemical elements, such as lithium, beryllium and boron.

Jean AUDOUZE

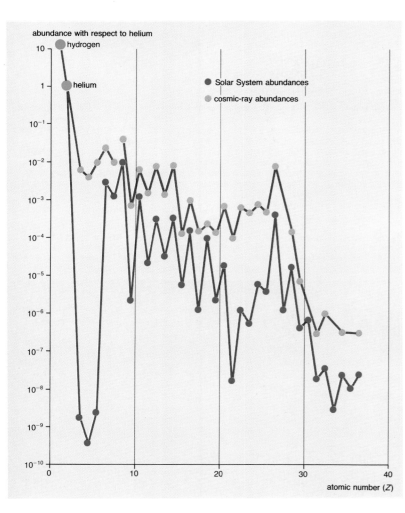

Composition of galactic cosmic rays. The composition of cosmic rays compare with that of 'ordinary' matter (Solar System, stars near the Sun, interstellar medium) in the following ways:
– the lightest and most abundant elements like hydrogen and helium are always the ones found most frequently in cosmic rays. However, their proportion relative to heavier elements (carbon, oxygen and iron, for example) is at least ten times smaller than that observed in ordinary matter;
– elements such as lithium, beryllium and boron, and those with atomic numbers between 21 and 25 (between scandium and manganese), which are very rare in ordinary matter, have a much higher abundance in cosmic rays.

Detailed study of the composition, and comparison with that of the Solar System, has provided the groundwork for hypotheses of the origin of those cosmic rays whose trajectories are 'entangled' with the interstellar magnetic field. It is thought that they are accelerated in the immediate vicinity of supernova remnants or the most massive stars, like Wolf–Rayet stars: elements like hydrogen, helium, carbon, oxygen and iron come from here, and their relative abundances are affected by the acceleration process (leading to the low relative content of hydrogen and helium), which occurs before the formation of grains of interstellar dust. The elements lithium, beryllium and boron, as well as those in the region between scandium and manganese, are formed by nuclear reactions of the cosmic-ray particles with themselves and with interstellar atoms during their 'voyage' through the interstellar medium, characteristics of which can be estimated from the relative composition of these elements.

Our Galaxy

It has been known since the beginning of the seventeenth century that the magnificent tapestry of the Milky Way visible on a clear moonless night is constituted of a very large number of stars. It is now recognised that the Milky Way is what we see of our Galaxy, an assembly of about two hundred billion stars of which the majority are similar to our Sun, which is itself an ordinary member of the Galaxy.

Our Galaxy is a spiral galaxy (see p. 340); it has a morphology wholly similar to the majority of spiral galaxies. Its overall shape is something like that of two fried eggs placed back to back. Thus the shape is that of a disk with a bulge at the centre. The disk is 30 kiloparsecs across, and very thin in relation to its diameter, the thickness of the outer regions being 300 parsecs. The bulge has a diameter of 6 kiloparsecs and a thickness of 1 kiloparsec. Surrounding the bulge is an approximately spherical volume thinly populated by stars; this is called the halo.

Each of the three components of our Galaxy – disk, bulge and halo – has a different density of interstellar gas and a different population of stars. The bulge has a much higher stellar content than the disk with respect to the interstellar medium; the ratio of the mass of matter contained in the interstellar medium to the mass of matter contained in stars is of the order of 0.1 in the disk and less than 0.01 in the bulge. The density of matter in the halo is much less than that of the bulge or that of the disk by a factor of more than ten thousand.

The three types of stellar population are as follows. Firstly, the halo stars which are concentrated in compact, spherical clusters known as globular clusters. Halo stars are very old; they have large velocities out of the plane of the Galaxy, and were formed about 15 billion years ago and are relatively poor in elements heavier than hydrogen and helium. The spherical distribution of the globular clusters can be used to locate the centre of our Galaxy relative to the Sun (see p. 310).

Secondly, the disk stars, known as Population I, are essentially confined to the disk of the Galaxy, having relatively low velocities relative to the galactic plane. They form associations and clusters (known as galactic clusters) and have ages ranging from a few million years up to the age of the oldest stars in the halo. The youngest disk stars are thus very young compared with the 4.5 billion year age of the Solar System, for example. Galactic clusters are looser than globular clusters. In the outer regions of the disk are found relatively more high mass, and therefore young, stars than in the inner regions of the disk. This gives the disk a predominantly blue colour.

Finally, the bulge stars, known as Population II are, in the majority, of low mass and have ages comparable to those of the stars in the halo. It is these stars which give the bulge its orange–yellow colour.

The interstellar medium, which constitutes of the order of 5 to 10 per cent of the mass of the Galaxy, has a very inhomogeneous structure; we distinguish a very dense, cold component (some tens of thousands of particles per cubic centimetre) at a temperature of 100 K, consisting of interstellar clouds, from a second component, a dilute intercloud medium in which the clouds bathe. The interstellar medium is principally constituted of hydrogen (see p. 300). It also

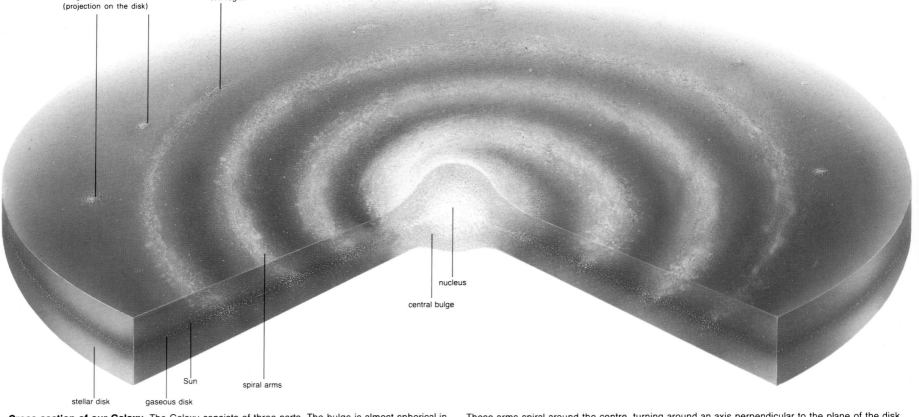

Cross-section of our Galaxy. The Galaxy consists of three parts. The bulge is almost spherical in form and has a radius of about 1 kiloparsec. The densest and most central part of the bulge is called the nucleus. The nucleus has a relatively low content of interstellar gas, less than 10 per cent of the total mass. Nonetheless the molecular clouds Sagittarius A and B are located there; their study gives information on the chemical composition of the nucleus. As in the nucleus of other spiral galaxies, a very small, dense object has been detected in the nucleus; some astrophysicists think that it may be a black hole. The second part of the Galaxy is the disk. Flatter than the bulge, and more extended, it has a radius of about 15 kiloparsecs and a thickness of a few hundred parsecs. The Sun is located about one-third of the way in from the edge of the disk. The structure of the disk is not uniform; from a distance one would see that the density of stars and of ionised hydrogen (seen in H II regions) is rather higher in four structures known as spiral arms.

These arms spiral around the centre, turning around an axis perpendicular to the plane of the disk. The various components of the Galaxy do not, however, rotate with the same angular velocity; the Galaxy does not behave like a solid body. From the centre outwards, the angular velocity increases out to a radius of about 8 kiloparsecs, and then decreases. The Galaxy is thus continuously deformed. While the Sun takes 200 million years to complete a circuit around the Galaxy, a star located 5 kiloparsecs from the centre takes less than half the time. The space between the spiral arms is far from empty; the density of gas and of stars is only a factor of 2 or 3 lower than in the spiral arms. The third component of the Galaxy is the halo, which occupies a spherical volume of radius about 15 kiloparsecs, containing globular clusters and, in general, the oldest stars. The halo contains very little matter compared with the disk and the bulge.

Picture of our Galaxy viewed from outside. This diagram comes from a computer simulation made by J. N. Bahcall. It shows how our Galaxy would look to an exterior observer. At the centre of the galactic disk is the bulge. The 200 million stars in the Galaxy are not uniformly distributed in the disk; the stellar density increases towards the centre and towards the central plane of the disk.

The horizontal band is due to absorption of starlight by interstellar gas and the dust which it contains. Unfortunately the ratio of the mass of dust to the mass of gas is as high as 10 per cent, so the dust absorbs starlight extremely effectively. The distribution of gas is in fact much flatter than that of the stars. (J. N. Bahcall, Institute for Advanced Study, Princeton)

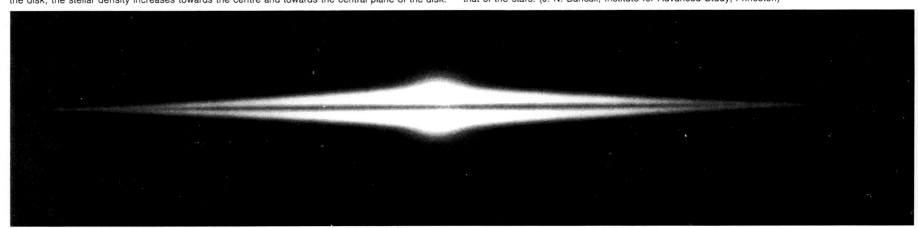

Global view of the night sky. This illustration represents the emission in the visible of the whole of the celestial sphere. This composite plate has been made in galactic coordinates. Our Galaxy is the long whitish trail on the equator of this sphere. It is clearly visible with its stars, dark dust clouds and bright regions of ionised hydrogen. We can also see that the halo of the Galaxy (that is the region outside the equatorial plane) is far poorer in stars than the disk and the bulge. The strong luminosity in the galactic centre, compared to the external regions is also evident. We see that the galactic disk is not completely planar, but that it is slightly 'out of true' due to the internal motion in the Galaxy. The two bright spots in the lower right of the plate (in the southern sky) are the two satellites of our Galaxy, the Magellanic Clouds. (Lund Observatory, Sweden)

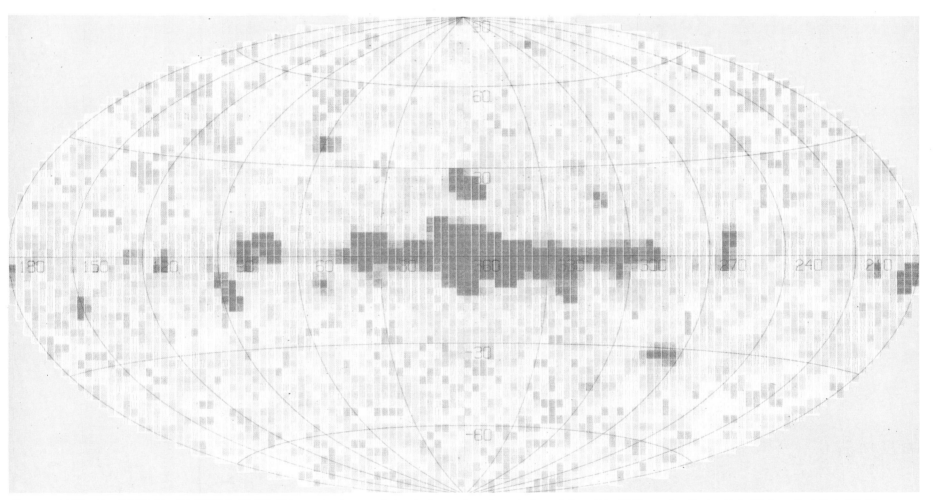

X-ray map of the sky obtained by the HEAO-1 mission. This map has been made in galactic coordinates and shows that the Galaxy is a powerful X-ray emitter: this is because our Galaxy contains supernova remnants like the Crab, which is the dark concentration furthest to the right (next to the number 210). The envelope of gas ejected by a supernova explosion is very hot and therefore emits X-rays.

Also present are X-ray binaries like Cygnus X1 and X3 (near the number 90), Sco X1 (the spot above the galactic centre), and Her X1 (the spot on the longitude 60 above the latitude 30). These binary sources emit X-rays by the exchange of matter between the two stars forming the binary system which causes the liberation of a large quantity of energy. The Large Magellanic Cloud (the spot at lower right) is also a powerful emitter, as are certain globular clusters such as the Virgo Cluster in Coma which is situated at the top of the map. (CFA)

Map of the high energy gamma-ray emission obtained with the satellite Cos B. The different colours represent the different intensities of gamma-ray emission. We see that radiation due to the interaction of very high energy cosmic rays with the interstellar medium has a maximum of intensity in the central regions and in the Galactic disk. The points of intensity correspond to supernova remnants like Vela and the Crab Nebula, which supports the idea that Supernovae are significant sources of cosmic radiation. (ESA photograph).

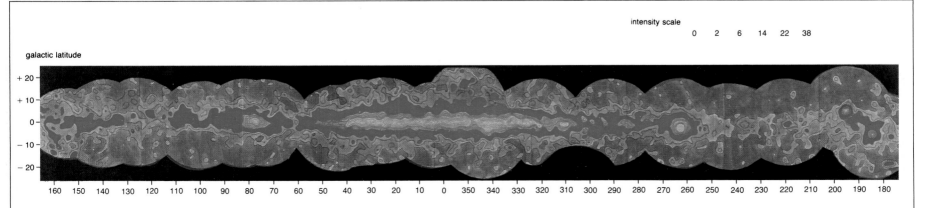

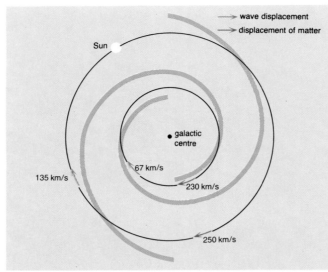

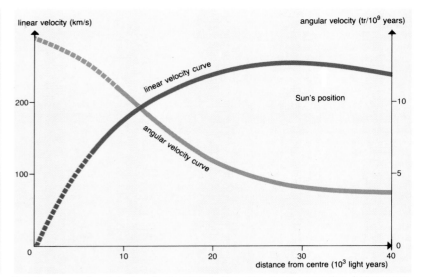

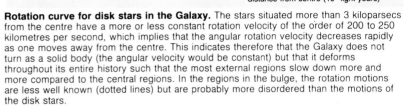

Diagram of the Lin waves capable of explaining the formation and propagation of spiral arms (see p. 340). The two circles correspond respectively to the orbit of the Sun and to that of a star situated at equal distances from the Sun and from the Galactic centre. Unlike the matter, which has a variable angular velocity, the spiral wave moves with a constant angular velocity, which corresponds to linear velocities of 67 kilometres per second on the small circle and 135 kilometres per second on the large circle. In the central regions, the spiral wave therefore moves with a much lower velocity than the matter: this larger velocity difference leads to a gas compression as the strongest wave passes. This gas compression favours star formation whose radiation then contributes to the ionisation of hydrogen. This is the origin of the H II regions (ionised hydrogen) found in the interstellar medium in the arms.

Rotation curve for disk stars in the Galaxy. The stars situated more than 3 kiloparsecs from the centre have a more or less constant rotation velocity of the order of 200 to 250 kilometres per second, which implies that the angular rotation velocity decreases rapidly as one moves away from the centre. This indicates therefore that the Galaxy does not turn as a solid body (the angular velocity would be constant) but that it deforms throughout its entire history such that the most external regions slow down more and more compared to the central regions. In the regions in the bulge, the rotation motions are less well known (dotted lines) but are probably more disordered than the motions of the disk stars.

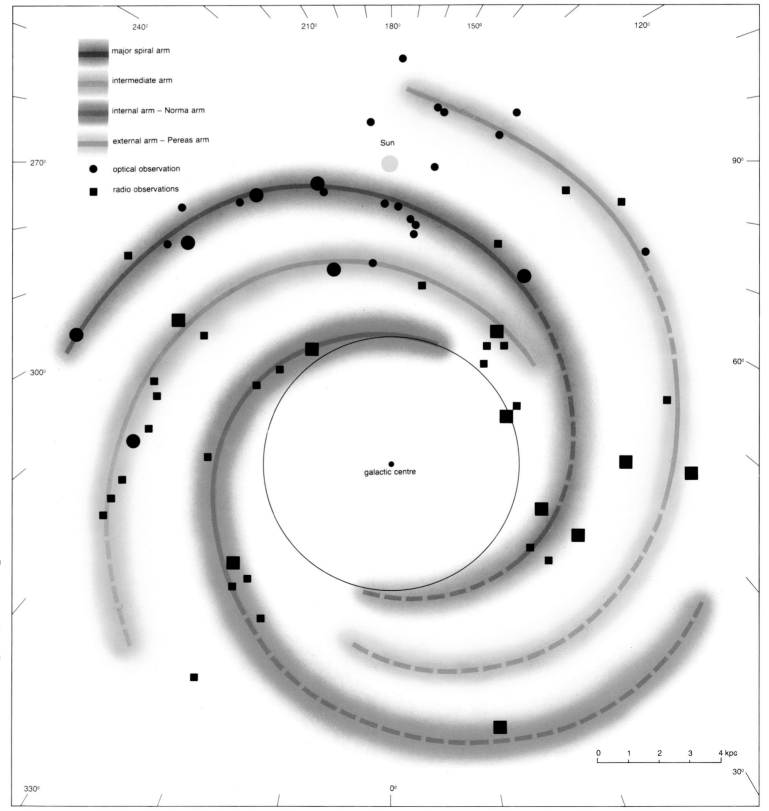

The spiral structure of our Galaxy. This structure is determined from the distribution of regions of ionized hydrogen (H II regions). Circles indicate optical observations. Squares indicate radio observations of recombination lines of hydrogen. The size of the circle or square is proportional to the excitation parameter representative of the size of the H II region. This map enables us to display the four spiral arms in our Galaxy. The dotted parts of the arms are regions which are poorly defined. We see on this map that the Sun lies between the major spiral arm and the Perseus arm. This map is at present the best representation of the spiral structure of our Galaxy. (Map produced by Y. M. and Y. P. Georgelin, Observatory of Marseille)

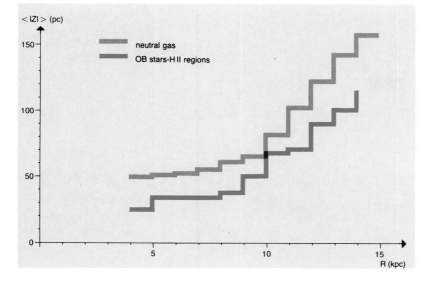

Thickness of neutral gas and of hot stars (and of ionised gas). In this figure is shown the average thickness of the disk of neutral gas (observed through the 21-centimetre line of atomic hydrogen) for galactocentric radii (distance from the galactic centre) from 5 to 15 kiloparsecs. The average thickness of the disk of ionised hydrogen is also shown. This is linked to the distribution of O and B stars because these stars are sufficiently hot to ionise hydrogen. In contrast to the general distribution of stars, which has a tendency to be less thick as one goes from the central regions to the outer regions, the distribution of neutral and of ionised gas (and thus of high mass stars) tends to be thicker on going from the centre to the outer regions. The thickness of the neutral gas goes from 50 to 150 parsecs while that of the ionised gas goes from 30 to 100 parsecs. This result is generally interpreted by saying that the gravitational potential of the outer, less massive regions of the disk is less than that of the more massive inner regions. The gas therefore has a tendency to disperse towards the outer regions; its density there is thus higher, and it is therefore easier for high mass stars to form and ionise a greater thickness of gas. This is very obvious on photographs of spiral galaxies, which are blue in the outer regions and yellow–red in the inner regions.

contains numerous chemical species, including molecules as complex as HC_9N, and an appreciable quantity of tiny solid particles (of size less than a micrometre), known as interstellar dust. The study of the distribution and structure of the interstellar gas, discussed elsewhere in this Atlas, is one of the most active branches of modern-day astrophysics.

Within the disk, there are spiral arms which are concentrations of stars and interstellar gas which spiral out from the ends of the bulge. These arms are characterised by the presence of ionised hydrogen and by the formation of young stars.

The ensemble of the Galaxy – including the spiral arms and the Sun – rotates around the galactic centre, but the Galaxy does not rotate like a solid body: only the innermost central regions of the bulge within 1 kiloparsec from the centre behave in this way, that is, they have the same angular velocity. The rest of the Galaxy appears to have a 'fluid' rotation: the Sun revolves around the centre with a period of about 200 million years, while stars close to the centre take much less time to complete their circuit.

To complete this brief description of the morphology of our Galaxy, it is necessary to point out that the disk is not perfectly flat, but has a certain warp at its extremities. This warp is probably due to tidal effects through gravitational interaction with the nearest galaxies, the Magellanic Clouds and small spheroidal galaxies.

There is a relationship between the relative velocities of the stars situated outside a given radial distance from the centre of the Galaxy and the mass contained within that radius, or again between the latter and the velocity of rotation of the matter as a whole outside the considered region. Contrary to what one would expect, no reduction in rotation velocity is observed at the visible edge of the disk. This observation indicates that the disk extends further out in the form of invisible matter, which increases the ratio of mass to luminosity for the Galaxy. The mass–luminosity ratio, expressed in solar units (about 5 in the disk, and about 10 in the bulge) will overall be of the order of 10 if the invisible matter in the halo is taken into account.

The chemical composition of the Galaxy is also inhomogeneous. The halo is very poor in elements heavier than hydrogen and helium relative to the rest of the Galaxy. On the other hand, the abundance of certain elements such as oxygen and nitrogen increases significantly as one goes from the outer regions of the disk towards the central regions.

As far as the bulge and the halo components are concerned, it is impossible to distinguish our Galaxy from an elliptical galaxy, except that the mass to luminosity ratio is higher for ellipticals. The characteristic spiral structure is very difficult to explain, the spiral arms representing a concentration of stars and interstellar gas 2 to 3 times greater than between the arms. In the later section on spiral galaxies we describe the mechanisms at present invoked to explain the spiral structure: they consist of the differential rotation of a self-maintaining perturbation, spiral in form, or of the contagious effect of supernova explosions upon succeeding generations of stars.

Jean AUDOUZE

Regions of the Milky Way towards the constellations the Southern Cross and Carina. This region of the sky, which the galactic equator crosses from upper right to lower left, shows very dense concentrations of stars as well as numerous ionised hydrogen regions (red). Large quantities of interstellar dust are present, and this partially obscures some of the region by blotting out the light from more distant stars. This effect is most marked in the area of the Coalsack, the dark region at lower left, near the Southern Cross. (D. F. Malin, Anglo-Australian Telescope Board)

The centre of our Galaxy

The centre of the Galaxy in the infrared. To see the central nucleus requires observations in spectral regions such as the infrared, which are unaffected by interstellar absorption. This false colour image is a result of far-infrared observations by the IRAS satellite. It shows the central regions of the Galaxy, within 10 degrees of the centre. Unobscured by interstellar absorption, these regions appear as a bright band along the Galactic plane, with the nucleus near the centre of the picture. The diffuse emission outside the plane arises from local processes: the bright point sources along the plane are giant interstellar clouds of gas and dust heated by hot stars. The intense infrared luminosity of the Galactic centre implies a huge concentration of gas and massive stars. (NASA)

At the beginning of the twentieth century the American astronomer Harlow Shapley determined the position of the dynamical centre of our Galaxy by observing the distribution of globular clusters. He placed it at a distance R_0 about 10 kiloparsecs from the Sun, in the direction of Sagitarius. Today the precise position of the dynamical centre is also established from the distribution of variable stars of RR Lyrae type. The International Astronomical Union currently recognises $R_0 = 8.5$ kiloparsecs, but the most recent work is converging to a value of R_0 of order 8 kiloparsecs. We shall describe here the *nucleus*, which extends 300 parsecs from the Galaxy's dynamical centre, and constitutes the interior of the *bulge*, the vast central thickening of the Galactic disc. The bulge has a very old stellar population, but the nucleus is a turbulent region containing young stars and nonthermal energy sources associated with unusual objects. Another key region of the Galaxy, at similar distances from the centre and the Sun, will also be briefly described.

The recent discovery of activity in the centres of some galaxies, such as Seyfert galaxies, has renewed interest in studying the centre of our own Galaxy. However, telescopes operating in the visible are unable to observe the nucleus because of interstellar dust, microscopic grains found throughout the interstellar medium. Over the distance to the Galactic centre, dust causes about 27 magnitudes of extinction in the visible. Interstellar absorption has much less effect on parts of the electromagnetic spectrum such as radio, infrared, X-rays, and gamma rays. Recent progress in observing in these spectral regions has gradually revealed some astonishing features of the Galactic centre: as it loses some of its mystery the Galactic centre attracts growing interest from astrophysicists.

The total mass in the central nucleus can be estimated from the virial theorem, which gives a simple relation between the potential energy of a system and the kinetic energy of its constituents.

The resulting estimate for the nuclear mass is of order 10^{10} solar masses. The mass of interstellar gas, consisting mainly of hydrogen, is harder to evaluate. Atomic hydrogen, HI, is well observed by radio astronomers because of its 21 cm emission, but is present in only small quantities in the nucleus. Molecular hydrogen, H_2, the main component of the nuclear gas, cannot be directly detected, and its mass is estimated by means of the carbon monoxide molecule CO, which is easy to detect in this region because of a 2.4 millimetre emission line. Assuming that the ratio between CO and H_2 abundances is the same in the nucleus as in the solar neighbourhood, one finds a mass of about 10^8 solar masses. Despite the low accuracy of this estimate, it seems clear that the total mass is much smaller in the central nucleus (about 1 per cent) than in the solar neighbourhood (almost 10 per cent). Most of the matter in the nucleus is thus in the form of stars, which are in general impossible to observe directly because of interstellar absorption. One can, however, pick out some stars in the Galac-

The field of the nucleus in gamma rays. Gamma rays are easily able to penetrate the curtain of interstellar dust. This false colour image results from low-energy gamma-ray observations (40 to 110 kiloelectron volts) made in 1990 by the French telescope SIGMA on the Soviet GRANAT satellite. A grid of Galactic coordinates is superimposed on the gamma-ray image. The dynamical centre has coordinates $l = 0$, $b = 0$ in this grid. The radiation intensity is given by the gradation of colours from blue to white. No gamma-ray emission is detected from the exact direction of the dynamical centre. The gamma-ray source within 1 degree of the centre is certainly situated within the nucleus. Its apparent size actually results from instrumental effects in the telescope rather than representing its real size. The source probably implies the presence of a collapsed star, the endpoint of the evolution of a massive star. This extremely variable gamma-ray source sporadically emits positrons which diffuse out of the source and annihilate in a region with the characteristics of a dense molecular cloud. (CNES/CEA/CESR)

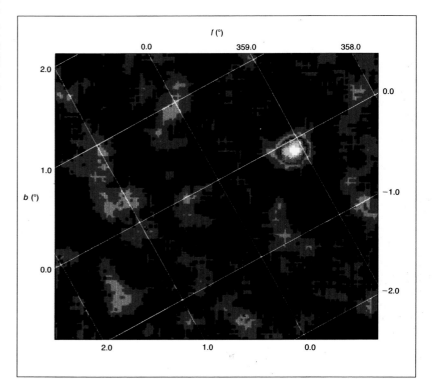

The Scorpio, Ophiuchus and Sagittarius regions of the Milky Way. This photograph was taken in visible light by a wide-field camera at the European Southern Observatory (La Silla, Chile), and covers a field of about 20 degrees × 15 degrees in the Galactic centre. The Galactic nucleus is about halfway between the left-hand edge and the middle of the field, along the centre line. It is hidden by immense absorbing clouds of interstellar gas and dust which appear as dark bands covering field stars. (C. Madsen, ESO)

tic centre, particularly in the infrared. Thus in the near infrared (from 2 to 5 micrometres) one can detect the emission of cool but luminous stars of similar spectral type to the Sun, but near the end of their evolution. One can also detect O and B stars, the youngest, most massive, and hottest. On the one hand they heat the interstellar dust, which then characteristically emits in the far infrared (from 30 to 100 micrometres); their ultraviolet radiation also ionises the interstellar gas where it is less dense than the centres of the densest molecular clouds. The thermal radio emission of these so-called HII regions around hot stars reveals the presence of O- and B-type stars.

The evolution of massive stars usually ends in a supernova, which can be also seen in the radio with a very characteristic spectral shape. A supernova exposion is caused by the collapse of the core of the star to form a compact object, either a neutron star or black hole. The potential wells of these stars may, under certain conditions give rise to the formation of an accretion disc; these become very hot through viscous dissipation and may produce intense X-ray and gamma ray emission. Supernovae and novae, and more generally all the sites of high energy phenomena such as the stellar winds caused by certain massive objects, may inject radioactive products into the interstellar medium. The disintegration of these products produces gamma-ray emission lines which are characteristic of newly-synthesised elements.

The study of all these tracers contributes to an understanding of the shapes of the central nucleus and the physical processes at work there. The most obvious feature is that the stellar density increases very rapidly near the Galactic centre, reaching values a million times larger than in the solar neighbourhood. The interstellar medium of the nucleus is condensed in large molecular cloud complexes. Within the interstellar clouds of the nucleus the isotopic ratios of the most abundant chemical elements such as hydrogen, carbon, nitrogen, and oxygen differ from the ratios measured in the solar neighbourhood. The main differences are a marked underabundance of deuterium with respect to hydrogen and nitrogen 15 with respect to nitrogen 14, as well as overabundances of oxygen 17 and 18 with respect to oxygen 16 and of carbon 13 with respect to carbon 12. These differences are explained by a different chemical evolution because of more rapid transformation of interstellar gas into stars in the nucleus than in the rest of the Galactic disc.

A more advanced chemical evolution in the central nucleus may lead to a lower formation rate for massive stars there than in other active regions of the Galactic disc. This may explain the observed deficit of massive stars in the nucleus. We should, however, note that the deficit might also be explained by postulating a burst of star formation in the nucleus several million years ago. The most massive stars would by now have disappeared because of stellar evolution.

Several gamma-ray sources have been discovered in the Galactic centre. These show the presence of collapsed stars resulting from the evolution of massive stars. One of these sources sporadically emits positrons which annihilate in a medium with the characteristics of a dense molecular cloud after diffusing at least 0.3 parsecs from the source. Observations of the central regions of the Galaxy show the presence of gamma-ray emission at 1809 kiloelectoron volts, the signature of aluminium 26. This radioactive isotope has a lifetime of about a million years and decays into magnesium 26 with the emission of a positron and a gamma-ray photon of energy exactly 1809 kiloelectron volts. This suggests the existence of active nucleosythesis sites in the nucleus.

The most remarkable part of the central region of the Galaxy is inside the nucleus, within a few parsecs of the dynamical centre. This is a rapidly rotating ring of 2 parsecs radius, consisting of unusually hot and dense neutral gas. It surrounds a mysterious radio source, Sagittarius A* (Sgr A*), with a size less than 20 astronomical units, situated precisely at the dynamical centre of the Galaxy. Inside the ring are gas clouds apparently falling into Sgr A*. The virial theorem suggests that there are more than a million solar masses within a parsec of Sgr A*. These observations suggested that Sgr A* might be a supermassive black hole similar to those

invoked to explain observations of active galactic nuclei. However, we would expect such a black hole to reveal itself clearly in the infrared and possibly also in gamma rays, while observations have not confirmed this.

The region at equal distances from the Sun and the Galactic centre, known as the *5 kiloparsec ring*, has very interesting properties which resemble those of the Galactic centre in some respects. Some French astrophysicists have shown that it emits a significant flux in the far infrared, which is a signature of massive stars. We can compare this with the discovery by Japanese astronomers of a near-infrared excess, where low mass stars are prominent. The *5 kiloparsec ring* appears to have relatively more high than low-mass stars. One can explain this in terms of a recent episode of star formation, implying that the formation process is discontinuous. Alternatively it may be that the stars in this region do not have the same mass distribution as in the solar neighbourhood. We see again that understanding the Galaxy requires observations outside the visible region.

Jacques PAUL

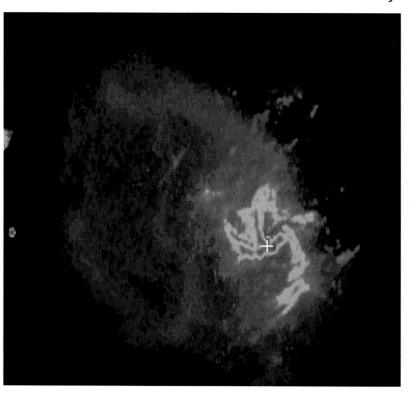

A few parsecs from the dynamical centre. Centimetre radio waves also penetrate the interstellar absorption. This false colour image was produced from 6 centimetre radio observations made at the VLA. The angular resolution is of order 2 arc seconds. A spiral structure is visible within 2 parsecs of the dynamical centre of the Galaxy, with maximum dimension 3 parsecs. The spiral arms may be ionised gas clouds falling in towards the compact radio source Sgr A* (marked by the white cross) which is less than 20 astronomical units in diameter, and lies at the dynamical centre of the Galaxy. These observations suggest that Sgr A* might be a supermassive black hole similar to those invoked to explain observations of active galactic nuclei. (Observers: W.M. Goss, R.D. Ekers, J.H. van Gorkom, U.J. Schwarz; NRAO/IAU)

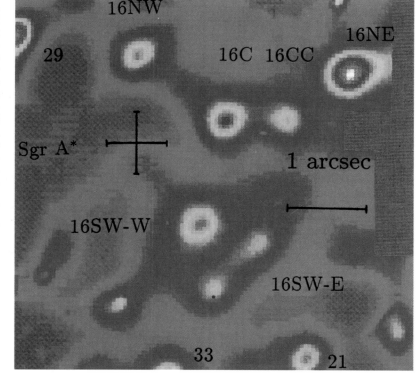

Within a parsec of the dynamical centre of the Galaxy. The latest electronic cameras give infrared images of unprecedented sharpness. This false colour picture of the Galactic centre was obtained by the SHARP system mounted on the European Southern Observatory's 3.5 metre New Technology Telescope. Operating in the K band (2.2 micrometres), this device can reach a resolution of 0.25 arc seconds. In a 6.4 arc second square field (0.25 parsecs at the Galactic centre) one can distinguish about fifteen infrared sources, which are cool luminous stars. The radio source at the dynamical centre of the Galaxy (Sgr A*) (marked by a black cross) appears to have only a weak infrared luminosity, too low to support the supermassive black hole hypothesis. (ESO)

Variation with galactocentric distance of various constituents of the interstellar medium. These diagrams show the dependence of tracers of various constituents with the Galactic disc on R, the distance from the Galactic centre. Atomic hydrogen is only a small fraction of the interstellar gas, whose main constituent is molecular hydrogen (H_2), which cannot be directly detected. By making hypotheses about the CO/H_2 ratio, the 2.4 millimetres emission of the CO molecule allows (a) an estimate of the H_2 distribution, and thus of the whole interstellar gas. The high-energy gamma-ray emission (b) results from the interaction of cosmic rays with the interstellar gas, and thus naturally follows its distribution. We note also a strong similarity with the tracers of young massive stars such as the far-infrared emission (c) coming from interstellar dust heated by such stars, the distribution of the hydrogen recombination line H 166α (d) linked to emission from HII regions around the hottest stars, the distribution of these HII regions themselves (e), and finally those of supernova remnants (f). The similarity between these distributions, all characterised by maximum near $R = 5$ kiloparsecs, is related to the spiral structure of the Galaxy: the spiral arms are the place where the gas from which stars are formed is found. The most massive and thus the hottest stay in the spiral arms because of their rapid evolution, near the gas, which they ionise, and dust, which they heat. For the same reasons one finds supernova remnants in the spiral arms, revealing the final phases of evolution of the most massive stars.

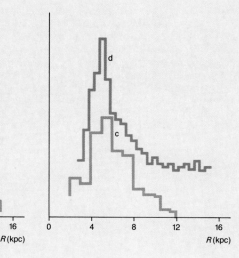

The neighbourhood of the Solar System

The neighbourhood of the Solar System reminds us more of the Great Plains of the Middle West of the USA than a city. If we were to travel at the speed of light, 300 000 kilometres per second, it would take just under six hours to leave the confines of the Solar System. We would then have to wait for four years and four months before arriving at the nearest star, Proxima Centauri. At the risk of disappointing the UFO fanatics, it must be said that we are not unduly troubled by visitors from our neighbourhood! Indeed, it could almost be said that space in the neighbourhood of the Solar System is empty. In fact, interstellar space is filled with atomic gas, molecules and dust grains. They absorb starlight, and so stars appear fainter to us than they would if interstellar space were empty. This absorption has two major consequences. On the one hand, estimates of distance based on luminosity criteria are rather uncertain because the amount of absorption on the line of sight to a star is always badly known. On the other hand, beyond a certain distance, it is impossible to study stars whose intrinsic luminosity is very low. In practice, apart from certain 'holes' in the absorption, stars like the Sun cannot be seen at distances beyond a kiloparsec.

At first sight, the view that we can have of our Galaxy appears to be rather limited; around the Sun, 1 kiloparsec represents a zone with dimensions only one tenth of those of the entire Galaxy. It is possible, however, to observe stars in the neighbourhood of our Sun which come from other regions of the Galaxy because of their having more or less elongated orbits. This has been shown by determining velocities of stars.

When an object moves away from us, its spectrum is shifted towards longer wavelengths by an amount that is proportional to its radial velocity, that is, the component of its velocity along the line of sight. This is the Doppler effect, which, being independent of the velocity perpendicular to the line of sight, allows us to measure only the velocity with which an object is approaching us or receding from us, and not the velocity across our line of sight. In order to measure this other (orthogonal) component, we measure the proper motions of the stars on two photographs of the sky taken, say, ten years apart; the respective positions of the stars will have changed. Measuring these shifts allows us to calculate the velocity. Even for the nearest stars, these shifts are very small; they are measured in thousandths of a second of arc per year. The determinations of the proper motions of the stars are only possible, in fact, for stars located at distances less than a few hundred parsecs. In combining the radial velocities given by the measurement of the Doppler shift and the tangential velocity obtained by measuring the proper motion, we can determine completely the true motion in space of the nearby stars.

The stars in the solar neighbourhood form two large kinematic families. The first, called Population I, has approximately circular orbits around the galactic centre. All of its stars revolve in the same plane, thus forming a disk whose thickness is a hundred parsecs; the components of the velocities of the stars perpendicular to the plane of the disk are relatively low, from 5 to 10 kilometres per second. The stars of Population II in contrast have non-circular orbits and occupy a much greater volume than the disk, known as the halo. Their velocity components perpendicular to the plane of the disk are similarly rather higher, in the region of 50 kilometres per second. This

kinematic structure allows us to determine the distribution of the mass in our Galaxy in a manner which is analogous to the determination of the mass of the Sun from the rotational velocities of the planets. In fact, the radial distribution of mass in our Galaxy has been studied overall from the rotation curve of the gas. The study of stellar kinematics has enabled us to determine the distribution of mass perpendicular to the plane of the disk. It has thus been possible to calculate the mass per unit surface area which must exist in the neighbourhood of the Sun to account for the vertical movements of the stars. Knowing the number of stars in the neighbourhood of the Sun and their luminosity, we can calculate the mass present in the form of stars. This 'luminous' mass is two to three times less than the 'dynamical' mass. There must therefore be another component, massive and of very low luminosity, called the missing mass, which accounts for the difference between the luminous mass and the dynamical mass: the nature of this missing mass has been the subject of numerous discussions. The most probable hypothesis, however, seems to be that there is a population of stars of very low mass (and therefore not very luminous) belonging to the halo; such stars might be black holes or burned out dwarf stars.

Photometric measurements or spectroscopic measurements enable us to determine the chemical abundances of stars in the solar neighbourhood. We term the abundance of heavier elements in stars (such as oxygen, carbon and iron) by the overall term metallicity. Determinations of metallicity have shown that there is a very good correlation between the dynamical and chemical properties of stars. For example, the smaller the component of stellar velocity perpendicular to the disk, the lower on average is the metallicity. The stars in the halo have metallicities ranging from 0.0001 to 0.5 in units of solar metallicity, and stars in the disk have metallicities from 0.1 to 1.5. The halo stars are clearly more deficient in heavy elements than stars in the disk.

The correlations existing between the dynamical and chemical properties enable us to retrace the history of our Galaxy. As we have already said, the distances between the stars are very great, and the interstellar medium is very tenuous. Stars therefore move in their orbits without being deflected by collisions with other stars or being slowed by the interstellar medium. Therefore the mass structure deduced from present-day movements of the stars is virtually identical to that existing when the Galaxy was formed. The existence of old stars, poor in metals, distributed in a roughly spheroidal extended halo, is a reminder of the form of the spheroidal protogalaxy, which must have been a gaseous nebula very poor in heavy elements, composed mainly of hydrogen and helium. The nebula collapsed gravitationally to form stars, which are the present halo stars. Subsequently, the gas remaining after this first phase formed a new, very flattened subsystem in the form of a disk. New stars were formed in this disk, and they continue to form. From the dynamical and chemical properties of the stars in the solar neighbourhood we can estimate the characteristic time-scale for the gravitational collapse of the protogalaxy and the appearance of the disk to be some years. The disk itself appears to evolve with a time-scale of over 10^9 years.

Laurent VIGROUX

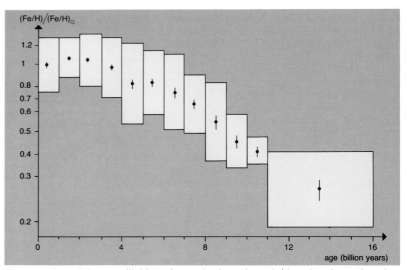

The evolution of the metallicities of stars in the solar neighbourhood as a function of their age. The metallicity is plotted in this diagram as ordinate and normalised to the solar metallicity. The size of the rectangle corresponding to each age represents the dispersion in metallicity corresponding to stars of the same age. The diagram, combined with the distribution of the number of stars as a function of their age, enables us to understand the chemical evolution in the neighbourhood of the Solar System. In the simplest model that could be envisaged, the solar neighbourhood initially consists of gas of zero metallicity. Then stars form and enrich the interstellar medium with heavy elements that they have synthesised. This model does not reproduce the observed properties correctly. Firstly, it predicts too many metal-poor stars. Secondly, it predicts a continuous increase in the metallicity whereas we observe a limit in the metallicity for the youngest stars. These difficulties may be resolved by a model that also attempts to reproduce the dynamical evolution of the stars. The halo was the first part of the Galaxy to be formed. The formation of halo stars enriched the interstellar medium and therefore enriched the gas from which the disk was formed. Even the oldest stars formed in the disk from this already enriched gas will therefore have an enhanced metallicity. If we now suppose that gas still remains in the halo, this gas will continue to fall onto the disk. As this gas has a low metallicity (lower than for the halo stars) it will have the effect of diluting the metals formed in the stars of the disk. This produces a limit to the enrichment of the stars in the disk, just as is observed. (After B. A. Twarog, 1980)

The nebulae NGC 6589 and 6590 (right). Large clouds of gas and dust obscure our view of the Galaxy, thus limiting our field of investigation. This is particularly evident on this photograph, which shows what we see of our Galaxy in the direction of the galactic centre. The distribution of stars is very inhomogeneous. Some regions have a high density of stars; others have virtually none. The latter regions are not actually without stars; light from the stars is absorbed by dust and gas on the line of sight between them and us. The absorption is particularly great in the direction of the galactic centre, which is therefore completely hidden from us. The presence of dust is also shown by the two large blue nebulosities which are due to the diffusion of the light emitted by two very bright stars. The large reddish nebula is a complex of dust and of ionised gas. The red colour results from the emission of the $H\alpha$ line of hydrogen. (D. F. Malin, Anglo-Australian Telescope Board, © 1981)

The nebula in Carina (left). This large gaseous nebula covers a region in our Galaxy which is particularly active in star formation. There are numerous hot young stars inside the nebula whose ultraviolet emission ionises the gas surrounding them, giving rise to an H II region. Although the red colour of this H II region is due to emission of the H_α line of hydrogen, this is not the only emission line of the nebula. Besides other lines of ionised hydrogen, there are lines from various ions present in the nebula, among the most abundant being He, O, N, Ne and S. Measuring the intensities of these lines enables us to determine the abundances of the various elements. H II regions are associated with the young population of our Galaxy, and therefore provide a means of measuring the chemical abundances in this population. The results are in good agreement with those obtained for stars. It is much easier to observe distant H II regions than distant stars. Spectroscopic studies of H II regions have enabled us to confirm over the entire galactic disk the variations of abundance with the distance to the centre of the Galaxy which have been observed in stars of the solar neighbourhood. The nearer one gets to the centre, the higher the abundances of the heavy elements like oxygen and nitrogen. This may be explained by a higher rate of star formation towards the centre, which leads to larger amounts of heavy elements in the interstellar medium. (D. F. Malin, Anglo-Australian Telescope Board, © 1977)

Gould's Belt. This diagram shows the positions, in galactic coordinates, of the brightest young stars in the solar neighbourhood. They are not distributed along the galactic plane, which appears here as the horizontal line at $b = 0°$. They seem, on the contrary, to be distributed within a plane inclined at approximately 20 degrees, represented here by the curved line. They are distributed in a circle, with the Sun at the centre. This circle is called Gould's Belt. A kinematic analysis of the stars shows that the majority belong to a group of stars which is expanding. This expansion also shows up in the motion of neutral gas in the vicinity of Gould's Belt. From the size of the group and the velocity of the expansion, the age of the group is 65 million years. The origin of this structure is undoubtedly a perturbation of the motion of the stars caused by the passage of a density wave corresponding to a spiral arm.

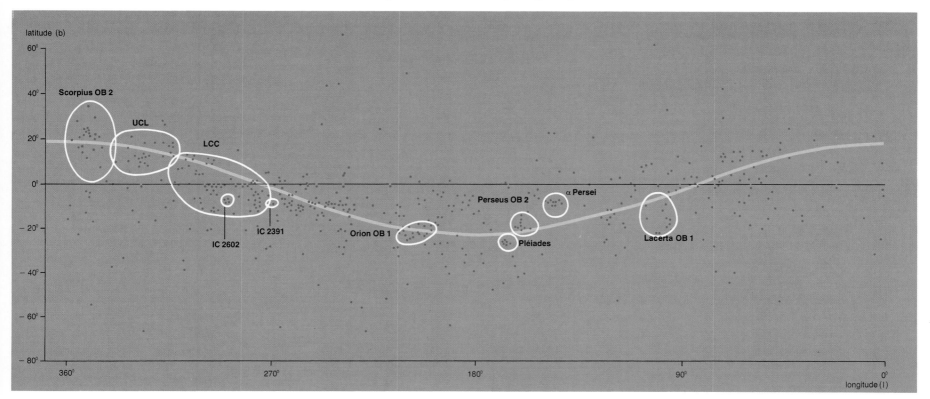

The extragalactic domain

The extragalactic domain

The most fascinating objects for present-day astronomers are the galaxies. Astronomers at last have instruments at their disposal which enable galaxies to be observed in detail. Schematically, galaxies are a bit like bricks which will permit the construction of the Universe in its entirety. At present, more than a billion galaxies are observable; there are perhaps many more, but their detection requires more powerful means of observation such as the Space Telescope. Galaxies are not distributed randomly but form clusters. The enormous distances of galaxies, which generally present great difficulties in their observation, in fact makes it possible to study them in different stages of their evolution. They are receding from one another with velocities that increase proportionally to their distance and which one can measure by the redshift of their emitted lines (Doppler effect). The galaxies having the highest redshifts are thus the most distant from us and therefore those visible at a younger stage of evolution (corresponding to the time necessary for the light to reach us). The majority of astrophysicists today think that quasars (or quasi-stellar objects), at redshifts on average higher than ordinary galaxies, constitute a primitive stage of evolution of galaxies.

A galaxy is an ensemble of stars of various masses, properties and ages, immersed in a more or less dense interstellar medium (in the case of elliptical galaxies, this interstellar medium is practically non-existent). A galaxy can evolve in at least three different ways. The first of these is dynamical evolution. Certain galaxies like our

own have a flattened form, others a more spherical form; these differences in morphology imply differences in the way in which the galaxies were formed and in the way that their various components move and interact. Then from the chemical point of view the composition of a galaxy changes in the course of time. Finally, the galaxy may undergo evolution of its luminosity and of its colour. A galaxy is above all an ensemble of stars which are born, evolve and die (and therefore change luminosity and colour), and which enrich the interstellar medium in chemical elements which they synthesise.

Following the work and observations of astronomers such as Edwin Hubble, we distinguish two main morphological types; elliptical galaxies which are relatively massive, not very luminous, whose colour is rather red and the interstellar gas content low; and spiral galaxies, of which our Galaxy is a prototype, which are themselves subdivided into galaxies with and without bars. Spiral galaxies are formed of three or four components: the central region or bulge which resembles a small elliptical galaxy; the disk which is a more extended and flattened component (as its name indicates), and within which there is a structure of spiral arms. These spiral arms, where stars and interstellar gas are more concentrated, run from the ends of the bulge for the non-barred galaxies and from the ends of the bar for the barred galaxies. In the latter case the bar is a flattened structure which is intermediate in structure between the bulge and the disk. The bulge, the disk (and the bar) are surrounded by a

very tenuous and very extended spherical region within which one finds the globular clusters (clusters of spherical form generally consisting of very old stars). The spiral galaxies are generally less massive and more luminous than the elliptical galaxies. They have an interstellar gas content of the order of 1 to 10 per cent of the total density of matter in the disk. The outer parts of the disk are rather blue, whereas the bulge is more of a yellow–red colour like the elliptical galaxies.

Finally, 10 to 15 per cent of galaxies do not have any particular form, and are termed irregular: these are the least massive galaxies. They are relatively more luminous, and have a higher content of interstellar gas than spiral galaxies and have a dominant blue colour. Hubble gave a classification of these morphological types, just as Linnaeus and Buffon did for vegetable and animal species.

This section of the Atlas closes with an examination of the properties of quasars and of radio sources (the most active galaxies) and with a brief description of the properties of clusters of galaxies. Besides being rather spectacular (we hope the reader will appreciate the beauty of certain pictures), the domain of galaxies is certainly one of those which progresses most at present.

Jean AUDOUZE

On the preceding double page:
The galaxy NGC 253. This galaxy of spiral type, is about ten million light years distant in a group of galaxies in the constellation of Sculptor. NGC 253 appears to be elongated in the photograph because it is viewed edge-on. NGC 253 contains a large quantity of interstellar material. There are two spiral arms running from both sides of the centre; their blue colour comes from the presence of young, massive stars. The nucleus of the galaxy, in contrast, is yellow in colour because of the existence of old, not very massive stars. This nucleus emits the majority of its energy in the infrared wavelength domain. (D. F. Malin, Anglo-Australian Telescope Board, © 1980)

The spiral galaxy NGC 7331 in the constellation of Pegasus. This picture of the spiral galaxy NGC 7331 in the constellation of Pegasus was obtained with the 200-inch telescope at Mount Palomar. We note that the central region or bulge is made of a large number of stars radiating in the yellow (therefore low mass stars predominate). In the outer regions of the arms of this spiral galaxy, we note that the dominant colour is blue, which is an indication of the presence of stars of greater mass (and therefore younger). The density of interstellar gas there is also much greater than in the central regions. On the other hand, the content of heavy elements (carbon, nitrogen, oxygen, iron, etc.) there is less. (Hale Observatories)

The elliptical galaxy NGC 4564. This galaxy was observed with the UK Schmidt telescope, put into service by the Royal Observatory, Edinburgh. This galaxy is completely typical of elliptical galaxies, with its marked ellipticity: there is a great concentration of stars emitting mainly in the yellow and red. (Low mass stars predominate.) Besides its particular morphology, due to the fact that star formation is certainly very rapid during the primordial phase of this galaxy, the ratio of mass to luminosity of this type of galaxy is very high. The most massive galaxies known are elliptical galaxies. We also recall that there is practically no gas in elliptical galaxies. (Royal Observatory, Edinburgh, 1980)

Barred spiral galaxy in the constellation of Fornax. This barred spiral galaxy in Fornax was observed with the 200-inch telescope at Mount Palomar. Three components are visible, the bulge, then a bar which is only distinguished from the bulge by its morphology (and not by its luminosity or colour), and finally a disk, with spiral arms running from the ends of the bar. In the case of this particularly spectacular galaxy, the arms extend a very long way. There are more spiral galaxies in the southern hemisphere than in the northern hemisphere. (Hale Observatories)

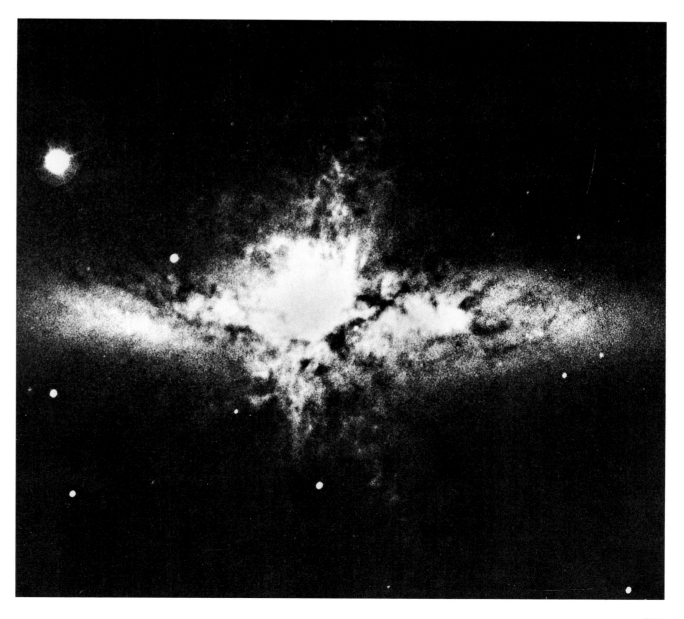

The irregular galaxy M82. This irregular galaxy lies in the constellation of the Great Bear. This picture was taken with the 200-inch telescope at Mount Palomar in light corresponding to the red Hα line of hydrogen. It shows in particular the filaments of hydrogen which extend in all directions from the centre of this galaxy for distances of 3000 parsecs. This galaxy is therefore an active object as is shown by radio astronomical observations: the gas which must have been ejected from the galaxy is detected out to very great distances from the galactic centre. M82 is classified as irregular because it has no particular shape. About 10 to 15 per cent of observed galaxies are in this category. Irregular galaxies are characterised by a very great luminosity relative to their mass, by a predominantly blue colour and by a high density of interstellar gas. The content of heavy elements is lower than in galaxies like our own. (Hale Observatories)

The dynamical evolution of galaxies

The matter in the Universe is not distributed uniformly. It appears to be distributed according to a hierarchical structure. The smallest systems are stars, which have typical masses of about 1 solar mass, then star clusters of 10^6 solar masses, galaxies of 10^{11} or 10^{12} solar masses, clusters of galaxies of 10^{14} or 10^{15} solar masses, and finally the superclusters, more massive again. These systems of matter exist because they are stable. In general, they are subjected to several opposing forces: their self-gravitation which tends to produce a collapse towards the centre; a source of internal pressure which opposes this collapse, and which, if it is great enough, will cause the system to break up; lastly, a force of gravitational

attraction due to the surrounding matter which can tear matter away from the system. Stability results if the self-gravitation is greater than or equal to the disruptive forces of internal pressure and external attraction. In order to explain the formation and evolution of galaxies we need to understand how perturbations that were sufficiently large could have formed in the Universe, that is, regions sufficiently dense that they would, under the influence of their own gravity, evolve in a way independent of the surroundings. Several theories have been proposed to answer this question. Without going too much into detail, these theories predict the formation of perturbations having a certain characteristic mass. In

these hypotheses, the condensations formed have masses between 10^6 solar masses (star clusters) and 10^{12} solar masses (galaxies). The stars and the clusters of galaxies, according to these theories, will be structures formed later. Two possibilities of galaxy evolution can be envisaged depending on whether the typical system formed initially has a mass equal to or less than the mass of present-day galaxies.

If the system has the mass typical of a galaxy, one can, in the first approximation, study its evolution as if it were isolated from its surroundings, and the problem thus reduces to that of understanding a self-gravitating system of gas and stars.

A cluster of galaxies, A 2151, in the constellation of Hercules. In general galaxies are not completely isolated. One finds them preferentially in groups which are very varied in size. The group to which our Galaxy belongs, called the Local Group, contains only three large galaxies, but the majority of galaxies are grouped in vast clusters – like that shown in this photograph – which can contain several thousand galaxies. The evolution of galaxies seems to be connected in some way with their environment. For example, elliptical galaxies are mostly found in large clusters of galaxies. In the same way, the proportion of lenticular galaxies in relation to spiral galaxies varies with the local density of galaxies. (Lenticular galaxies, like spirals, have a bulge and a disk but do not have a spiral structure.) The number of lenticulars increases with the total number of galaxies. At present, it is difficult to say if the influence of the environment shows up in the beginning, in creating initial conditions which favour the appearance of definite galaxy types, or if this influence is felt throughout the evolution of galaxies. By analogy with biological evolution, one could say that the roles played by inheritance and by the environment in the evolution of galaxies are still badly understood. (Hale Observatories)

The dynamical evolution of the gas in a galaxy is very different from that of the stars. The gas in a galaxy is a dissipative system – it can lose its energy. It is constituted of atoms and molecules which are in thermal agitation. It is often gathered into large clouds which have random internal motions analogous to thermal agitation. This internal agitation opposes gravitation and prevents or slows down the collapse of the system. But the collisions between the particles or the clouds cause a dissipation by radiation of the energy associated with their motion. The gas therefore continually loses energy, and is consequently less and less able to resist gravitation. A rapid collapse of the system ensues and the formation of a central condensation. In contrast, the dimensions of the stars are very small in comparison with the distances separating them, and collisions within a stellar system are therefore extremely rare. There is no energy lost by collision and the energy associated with the random motions of the stars will be capable of opposing gravitation more effectively than in a gas. If one looks for an analogy with classical thermodynamics, the gas behaves like an isothermal system and the stars like an adiabatic system.

A stellar system does not remain in equilibrium indefinitely. Even if collisions in the strict sense do not occur, in dense systems a star may pass sufficiently near to another star for their mutual attraction to be much greater than that exerted by the rest of the galaxy. The orbits of the two stars will be modified after their encounter. If such encounters are sufficiently frequent, the motions of the stars can be wholly redistributed. The characteristic time of this phenomenon is called the relaxation time. It is obviously shorter the more frequent the encounters are and the denser the system. The relaxation time is of the order of 10^8 years for a globular cluster, 10^{14} years for a galaxy and 10^{11} years for a cluster of galaxies. As the age of these systems is about 10^{10} years, this relaxation is effective only in clusters of stars. For a system of the size of a galaxy, there is another phenomenon called violent relaxation. The interaction occurs between a star and the mean gravitational potential defined by the other stars as a group. This theory, established by the Cambridge astronomer D. Lynden-Bell shows that a stellar system of the size of a galaxy evolves very rapidly in 10^8 to 10^9 years towards a quasi-stationary configuration under the action of this violent relaxation. Two-body relaxation then transforms this state of quasi-equilibrium into an equilibrium state on a very much longer time-scale of the order of 10^{14} years.

In a galaxy, the interstellar gas and the stars do not form independent systems because the gas is transformed progressively into stars. The dynamical evolution of gas and stars are different, the rate at which gas is transformed into stars will influence the evolution of the galaxy. If the time characteristically taken for star formation is very short compared to the time of dynamical collapse, the galaxy will evolve as a purely stellar system. On the other hand, if this time is very long compared to the dynamical time-scale, the galaxy will evolve as a purely gaseous system. Numerical models have shown that these two hypotheses lead to structures which are very different from the structures actually observed for galaxies. In contrast, a model in which the time of star formation and the dynamical evolution time are of the same order of magnitude produces a final configuration whose density profile corresponds overall to the observed profiles of elliptical galaxies. According to these models, these characteristic time-scales are of the order of 10^8 years.

The appearance of disks, observed in spiral and lenticular galaxies, is connected with another phenomenon: rotation. In a rotating system, the centrifugal force which acts upon the components of the system tends to spread them in the plane perpendicular to the axis of rotation. If the components are stars, the dissipation of energy is negligible and there is no real disk formation. In fact, the stars are distributed in ellipsoids of revolution which are more or less elongated according to their velocities of rotation. It has long been believed that this process is the cause of elliptical galaxies. However, in the 1970s it was observed that the velocities of rotation of these galaxies are too low to explain their flattening. In contrast, in the case of a gaseous component which can lose its energy, a flattening of the system occurs following the formation of a thin disk such as those which are observed.

The following scheme can therefore be imagined. Initially, galaxies are self-gravitating masses of gas, the initial mass being that of the final galaxy. This mass of gas collapses under its own gravitation, while forming stars with a time-scale of some 10^8 years, comparable to the time-scale of the collapse. If star formation takes place over a time long enough to transform all of the gas into stars, the final result is an elliptical galaxy. On the other hand, if for one reason or another star formation is interrupted before all of the gas is transformed into stars, the gas will form a disk under the action of the centrifugal force. There will thus be a system with a central spheroidal part, and a thin disk; that is, a spiral or lenticular galaxy.

Other theories, however, assume that the initial systems have masses much less than galaxies, which are then formed by the coalescence of several systems. If, for example, the mean mass of these objects is 10^8 solar masses, it will be necessary to combine a few thousand of them to form a galaxy. This coalescence of small subsystems is not impossible. Even now about 10 per cent of galaxies are interacting galaxies. These interactions in general lead to deformations in the two galaxies, and not to coalescence. Coalescence is only possible in the case of a very close interaction, when the distance between the two galaxies becomes comparable to their radii. Then there will be mixing and the formation of a new system. This mixing is very effective for the gaseous component, but it occurs just as much in stellar systems. With the present dimensions of galaxies and the distances which separate them, the frequency of collisions is, however, rather low. In contrast, at the origin of the Universe, the distances separating the galaxies or the systems then existing were much smaller. Collisions were therefore much more frequent. Numerical simulations have shown that, in the course of the evolution of the Universe, material systems might undergo between a hundred and a thousand collisions. This process therefore seems to be capable of producing objects at least resembling galaxies, and, even if it is not the only factor, it must play an important role because we observe a dependence of the properties of galaxies as a function of their environment. Depending on whether galaxies occur in a more or less dense galactic medium, galaxies have different characteristics. Their evolution is affected by collisions or sufficiently close encounters to modify their morphology permanently. In the present state of our knowledge, we cannot say that there is a definitive theory of the formation and evolution of galaxies. The various processes described here lead to systems analogous to galaxies, but their relative importance still remains to be clarified.

Laurent VIGROUX

The galaxy NGC 6166. This galaxy is a classic example of a particular type of galaxy called cD which occur at the centres of some dense clusters of galaxies. The colours of this image correspond to different levels of surface brightness. These galaxies are characterised by a very great luminosity – five to ten times that of an elliptical galaxy – and also by a very great size. They are formed in all probability by a progressive capture of cluster galaxies by the most massive galaxy of the cluster. When a cluster galaxy passes near to this supergiant galaxy, the gravitational attraction to which it is subjected can be sufficiently great for it to be swallowed by the supergiant galaxy. This cannibalism would explain the presence of three nuclei at the centre of the galaxy NGC 6166. We would see the digestion of two small galaxies by a larger one. In contrast to normal galaxies, the formation of these cannibal galaxies is only understood in terms of the environment. They only form in clusters that are sufficiently dense for collisions and encounters between galaxies to be numerous. (L. Vigroux)

The interacting galaxies Arp 242. Encounters between galaxies result in the appearance of strange structures; without leading to a coalescence of galaxies like that observed in cD galaxies, these two galaxies are an example of this. Gravitational perturbations of the tidal type tear away the extreme outer stars, which are the least bound, from each galaxy and create peculiar forms; bridges of matter, jets, etc. Numerical simulations on a computer of an encounter between two stellar systems enable us to reproduce the bizarre appearance of these galaxies. (L. Vigroux)

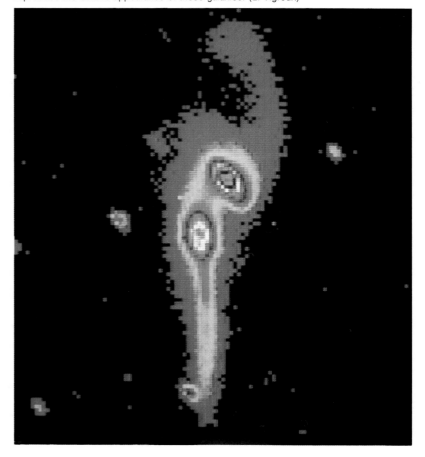

The chemical evolution of galaxies

The youngest stars in our Galaxy, which are localised in its disk, have a proportionately greater content of elements heavier than helium (carbon, nitrogen, etc.) than the oldest stars. This heavy element content is called, slightly inappropriately, the metallicity, and it increases in the course of time in a galaxy such as our own. This is a general property of all galaxies, whether they evolve very rapidly like ellipticals, or rather rapidly like our own, or more slowly like the irregular galaxies (e.g. the *Magellanic Clouds*) or the 'lazy' galaxies, for example blue compact galaxies.

A galaxy transforms the matter of which it was originally formed into stars in the course of its history. These stars evolve more quickly or less quickly, according to their mass: the most massive evolve very rapidly compared with low mass stars because the luminosity increases as the fourth power of the mass, whereas the quantity of nuclear fuel available is only proportional to the mass. A star of 1 solar mass shines for ten billion years and a star of 10 solar masses for only ten million years. At the end of their evolution, the low mass stars (less than 3 or 4 solar masses) lose their outer envelope rather calmly by passing through the planetary nebula stage, whereas more massive stars explode as supernovae. These two processes contribute to returning matter more or less enriched in heavy elements to the interstellar medium. The density and the chemical composition of the interstellar gas from which the different generations of stars are born are thus modified by a succession of mechanisms: the formation of stars, the ejection of stellar material and the explosion of stars.

High mass stars evolve very rapidly, forming heavy elements like carbon, oxygen and iron, for example, and by exploding return the greater part of their matter to the interstellar medium. Less massive stars form much less heavy elements and only return slowly a small fraction of their mass. The relative proportion of high mass and low mass stars is therefore one of the most important factors governing the chemical evolution of a galaxy as well as the rate at which the interstellar material is transformed into stars. Thus in order to study the chemical evolution of galaxies, it is necessary to know (or suppose) the rate of transformation of the interstellar gas into stars, the distribution of stars with respect to their mass at their formation, the nucleosynthetic capacity of every star (i.e. the fraction of its mass which is transformed into a given chemical element) and the fraction of its mass which never returns to the interstellar medium, (i.e. which is left as a white dwarf or neutron star or a black hole). One then compares the results of the analysis to the abundances of chemical elements observed in regions of different ages. The whole of astrophysics is therefore involved in this comparison: observations, principally spectroscopic, enable the abundances of the chemical elements to be found; the construction of models of evolution involves the understanding of the processes of formation, knowledge of nucleosynthesis, and of the way in which stars form, evolve, and end their evolution.

The observational facts to be explained

A model of chemical evolution must simultaneously explain the relative density of interstellar gas and of stars, the relative proportion of stars having a mean abundance in given heavy elements, and the evolution of the abundances of the various chemical elements according to the ages of the objects considered. To give some examples, the Solar System has an age of 4.6 billion years, the halo stars (in particular the stars in globular clusters) have an age of the order of 14 billion years, the stars of the disk have an age between a few thousand years and 10 billion years, and the interstellar medium is presumed to have an age close to zero in that its composition reflects the *present-day* abundances of chemical elements in the particular region being considered.

Models of evolution may be constructed for each galaxy. Obviously most analyses to date have been concerned with our Galaxy because more data have been gathered for it. The abundances of the chemical elements are known in great detail in our Galaxy, not simply in the Solar System through solar spectroscopy and detailed analyses of meteorites, but also in the interstellar medium surrounding our Solar System (thanks to ultraviolet spectroscopy), at the surface of the nearby stars, and in the centre of the Galaxy for numerous elements. The most important observational facts are as follows:
— The relative density of interstellar gas with respect to the total density is 5 to 10 per cent in the solar neighbourhood and less than 1 per cent in the central regions.
— There was a very rapid and large increase of heavy elements during the billion years between the formation of globular clusters and the formation of the oldest disk stars. Stars in globular clusters have abundances of heavy elements which are one-thousandth or one-hundredth of the corresponding abundances of stars in the disk. Note that no stars totally lacking in elements heavier than helium have yet been observed.
— Concerning disk stars, there is a much more moderate relative enrichment (of the order of a factor of three or four which would have been produced during the ten billion years of the existence of the galactic disk).
— When the relative numbers of stars having a given metallicity are compared, few stars are found to have a lower metallicity than the Sun. Other observations are also very important:
— Certain chemical elements, like nitrogen or elements heavier than iron, are even more deficient in halo stars than ordinary heavy elements such as carbon, oxygen and iron.
— The outer regions of the disk are richer in heavy elements than the central regions by factors which can be as much as ten.
— Analyses of interstellar molecules enable the isotopic ratios of certain elements such as carbon, nitrogen, oxygen and sulphur (for example) to be compared. There are very appreciable differences in the ratios found for the nearby interstellar medium, in the interstellar medium in the central regions and in the Solar System.
— Analyses of cosmic rays also give the relative abundances of the elements.

There is much less data for other galaxies. Nonetheless, certain facts readily inspire models. For example, the neighbouring spiral galaxies exhibit the same gradients of metallicity between the outer regions and the inner regions.

The irregular galaxies, like the Magellanic

Diagram of the various processes affecting the evolution of a galaxy. After the contraction and formation of a galaxy which must take place about a few million years after the birth of the Universe, the evolution of the galaxy is affected by the formation, the evolution and the 'death' of stars which leaves a remnant (white dwarf or neutron star) and which partly contaminate the interstellar medium just as some stars accelerate the cosmic rays which bombard them. The physical state of the interstellar gas itself affects the processes of star formation. The various processes are compared with observables such as the luminosity curve of main sequence stars which is linked to the rate of formation of the stars. Stellar evolution may be represented on a Hertzsprung–Russell diagram and affects the colour of the galactic region being considered, etc. The diagram is therefore a collection of the characteristics of formation, dynamical evolution, composition, luminosity and the colours of galaxies.

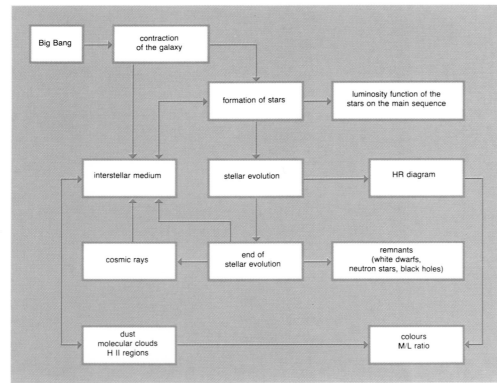

Clouds or the blue compact galaxies, which are relatively rich in gas, also have an appreciably lower metallicity than the mean metallicity of our Galaxy.

These few examples show how we must engage all the observational data of galactic and extragalactic astrophysics in order to understand the evolution of matter in the course of galactic history.

Models of galactic evolution

How galaxies evolve is both complex and not well understood; we can choose only very simple hypotheses as models of evolution.

The present hypotheses are:
– We suppose that the galactic regions which we are studying began at first by being entirely in the form of gas, completely lacking in chemical elements other than those formed by the primordial nucleosynthesis which occurred during the original explosion.
– The rate of star formation is taken to be proportional to the density of interstellar gas.
– The problem is simplified by assuming that the lifetimes of stars contributing to nucleosynthesis are negligible compared with the time-scale of galactic evolution.

Other parameters are also needed, such as the nucleosynthetic capacity of a star and the distribution of the stars as a function of their mass. Note that there are more low mass stars – which evolve slowly and which have a low nucleosynthetic capacity – than high mass stars which have a large capacity and which evolve very quickly. On the other hand, we distinguish the chemical elements which can be synthesised directly from hydrogen and helium – which are called primary elements – from secondary elements which can only be formed from the primary elements. For example, the secondary elements are nitrogen (at least in part) and the elements heavier than iron. The primary elements include carbon, oxygen, and iron.

With these assumptions, models of chemical evolution may be constructed. We do not describe these models in detail here. They account for the increase in metallicity with time,

The Andromeda Galaxy and NGC 2903. A comparison is made here between the central regions of the Andromeda Galaxy (opposite, also known as M31 and NGC 224) observed with the 4-metre telescope at Kitt Peak National Observatory, Arizona, and the spiral galaxy (below) in the constellation of Leo (observed with the 0.9 metre telescope of the McDonald Observatory in Texas). There is a difference between the central regions of these two galaxies, which are made up mainly of low mass stars giving a yellow colour to these regions. The central region of the Andromeda Galaxy is more prominent than that of the galaxy NGC 2903. The arms of the latter show a dominant blue colour due to the presence of a greater proportion of high mass stars. The outer regions are generally poorer in heavy elements than the central regions. This is explained by the fact that these outer regions are less advanced in their evolution than the central regions: enough interstellar matter therefore remains at the outer edges of the galaxies to continue an evolution which has practically ceased at the centre. (Top: Kitt Peak National Observatory; bottom: J. D. Wray, McDonald Observatory)

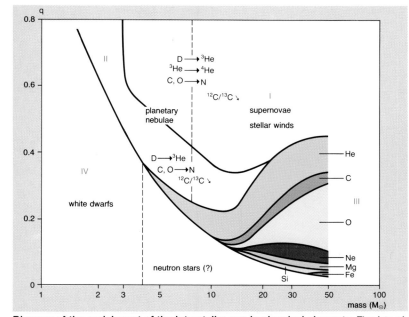

Diagram of the enrichment of the interstellar gas in chemical elements. The fate of the various fractions of the mass of a star at the end of its evolution is represented as a function of the stellar mass expressed in solar masses $M_\odot$. The lower part of the diagram, region IV, represents the matter remaining in the form of stars following the galactic evolution: stars of mass less than 4 $M_\odot$ leave a white dwarf which becomes black, and therefore invisible after a few hundreds of millions of years; stars of higher masses leave a neutron star which is observed as a pulsar for a hundred thousand years. Regions I and II represent the envelopes ejected by planetary nebulae or supernova remnants whose metallicity is not evolved (at least overall) but which have undergone secondary modifications like the transformation of some of their carbon and oxygen to nitrogen, or of deuterium to helium. Region III represents the enrichment itself in helium and in heavy elements. For example, a star of 10 $M_\odot$ leaves 15 per cent of its mass as a neutron star; it ejects 1.5 $M_\odot$ of pure helium and 7 $M_\odot$ as an envelope (70 per cent of its mass) having undergone only secondary nucleosynthetic changes.

Evolution of the abundance of heavy elements in the course of time in our Galaxy (opposite). One notes a sudden increase in the metallicity in the course of the first billion years which must have passed between the formation of the oldest stars which occur in the halo of the Galaxy and the oldest stars in the disk of the galaxy: the metallicity must have increased by a factor of 100 or 1000 during this period. No stars are known today with abundances less than 10^{-4} times that of the Sun. After the formation of the disk, the increase of metallicity is much less (a factor of four between the oldest disk stars and the youngest disk stars). Models of galactic evolution suggest a constant accretion of matter exterior to the disk and the existence of very massive stars at the beginning of the evolution which would enrich the gas very early on.

the gradients in abundance between the outer and inner regions of our Galaxy and the nearby spiral galaxies, the difference in metallicity between the irregular galaxies and the spiral galaxies for example, and lastly the difference in the evolution of the primary elements and secondary elements.

However, models of evolution based upon such simple hypotheses do not adequately explain the fact that metallicity reached saturation a billion years after the beginning of the evolution of our Galaxy. Indeed, a large proportion of stars having a lower metallicity than the Sun is predicted. Astrophysicists are not short of ideas to resolve these difficulties: one theory is that the galactic disk continually accretes matter from outside the Galaxy which has not been enriched by stellar nucleosynthesis – this hypothesis explains the saturation in metallicity well. Also, a burst of massive star formation might occur at the beginning of the evolution of galaxies. This would explain the low but non-zero metallicity of the stars in the halo.

Jean AUDOUZE

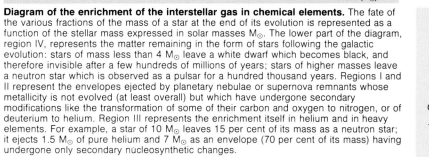

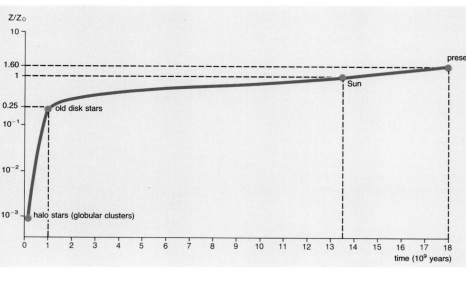

The evolution of the colours of galaxies

A galaxy contains on average a few hundred billion stars. However, most galaxies are too far away for us to be able to resolve the individual stars. Even in the nearest galaxies to us, the Magellanic Clouds, we can only study about 2 per cent of the stars, those which are more massive than 5 solar masses. In other galaxies, even in the best of cases, we can only observe a few very bright stars or small clusters containing more than a dozen stars. Because of the degradation of optical images by the Earth's atmosphere, it is in fact impossible to observe galaxies from the ground with an angular resolution of better than 1 arc second. For a galaxy in the Virgo Cluster, which is the nearest cluster of galaxies to us, this corresponds to a scale length of 60 parsecs. Thus the smallest zones observable will be at least 60 parsecs in diameter, and will contain between a few thousand and a few million stars according to whether the zone is at the centre or at the edge of a galaxy. The light received on Earth results from the addition of the emission of these millions of stars. This is called the integrated emission; the information which we have on the stellar populations in the galaxies is statistical in nature, because it is derived from the study of integrated emissions, and not based on the study of individual stars as in the star clusters of our Galaxy.

There is a fundamental difference between the emission of just one star and the integrated emission of a group of stars. The emission from one star is in the first approximation the same as that from a black body – there is a precise relationship between the emission and the temperature. The hotter a star, the bluer it is and the more luminous it is. No relationship of this kind exists for the integrated emission of many stars. The colours of a galaxy do not correspond to a temperature, but instead depend on the relative proportions of the various types of stars. In a blue galaxy, the light comes essentially from hot stars, whereas in a red galaxy the light from cold stars dominates the emission. The colours of galaxies thus give us information about their stellar content.

In a star cluster in our own Galaxy, the stars are arranged in well-defined sequences in a Hertzsprung–Russell diagram where the magnitude is plotted as a function of the colour. It is technically impossible to establish the Hertzsprung–Russell diagram of a galaxy by studying the individual stars. Instead, we attempt to reconstruct the proportions of stars in the galaxy from the characteristics of the integrated emission, in particular the colours. The principle is simple. Consider a population of stars. At the time of their formation, these stars all lie in the main sequence of the Hertzsprung–Russell diagram. When they age, they leave the main sequence to become red giants. This evolution, which is predicted by models of stellar evolution, occurs more rapidly for more massive stars. Knowing the distribution of stars in this diagram at any moment, we can calculate the integrated luminosity and integrated colours for a number of stars by summing the emissions of all of the stars. We can thus determine the luminosity of a population as a function of its age. To describe a galaxy in terms of a model, it is necessary to introduce populations of various ages. The principle involved is the same but we now add newly-formed stars to the model whose evolution is being followed. Two fundamental parameters govern this type of model: the rate of star formation and the initial mass function, which is the distribution in mass of stars at the time of their formation. In a galaxy where the initial rate of star formation was very high, the majority of the stars today are old, and the light is now

dominated by red giants. In a galaxy where the rate of star formation has remained more or less constant in the course of time, the proportion of young stars is much higher, and there is a large number of very hot, luminous stars which lie on the top part of the main sequence, making the galaxy very much bluer. Variations in the mass function give rise to similar effects, but they also affect the total luminosity; the more the mass function favours massive stars, the more luminous the galaxy. These models have enabled us to derive one fundamental result: the observed colours of galaxies can be explained by all galaxies having the same age and differing only in the characteristic time of star formation.

The use of wide passband filters, 1000 Å in width, such as the U passband (ultraviolet), B passband (blue), V passband (visible), R (red), and J, H, K in the infrared, enables us to measure the approximate flux emitted by a galaxy as a function of wavelength. We can thus calculate the ratios of the intensities measured with the various filters and define colour indices: $U - V$ corresponds to the ratio of the energies passed by the V passband and the U passband. In the colour–colour diagrams, such as $U - B$ plotted against $B - V$, galaxies of various morphological types are

arranged in a regular sequence. The elliptical galaxies are reddest and the spiral and irregular galaxies the bluest. These sequences are explained theoretically if the formation of stars is a continuous process. The ensemble of colours that is observed for a galaxy can only be explained if it contains stars of very varied ages. In particular, all galaxies appear to contain old stars some ten billion years old which are analogous to the Population II stars of our Galaxy. The differences in colour between the various galaxy types can be understood in terms of a greater or lesser rate of star formation. The characteristic time-scale in elliptical galaxies is 100 million years. It is a billion years for the normal spirals and five billion years for the barred spirals and the irregular galaxies.

The comparison of four pictures of galaxies (right-hand page) is very illustrative of the variations of stellar populations that exist within galaxies: there is a continuous transition from elliptical galaxies which are uniformly populated with old stars, to irregular galaxies whose luminosity is dominated by young stars.

Laurent VIGROUX

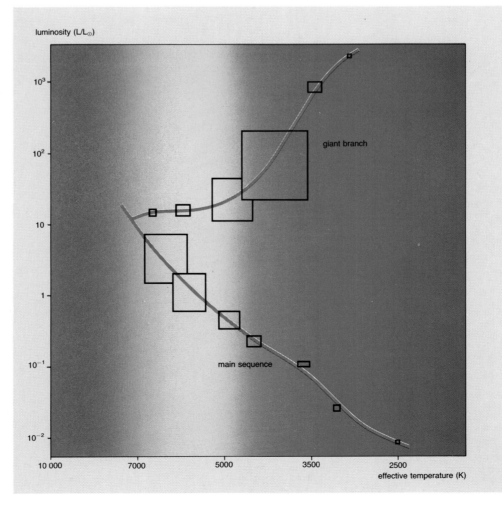

Representation of the stellar population of the elliptical galaxy M32. The positions of the squares in this diagram which plots luminosity as a function of temperature (Hertzsprung–Russell diagram) represent the positions of stars of the galaxy M32. The surface area of the squares corresponds to the relative proportions of these stars at the integrated visual luminosity of the galaxy. This synthetic population has been obtained by a method of population synthesis. One first measures the spectrum, that is, the luminous flux as a function of the wavelength, of the M32 galaxy and of a certain number of stars representative of the stellar population of our Galaxy. A stellar population is then defined as an ensemble of these stars, with a definite proportion of each type. The synthetic spectrum of this population will be the sum of the spectra of each stellar type according to their relative proportions. Classical mathematical algorithms enable the population to be found whose synthetic spectrum most resembles the observed spectrum of the galaxy. It is found that, for M32, most of the luiminosity is contributed by stars of the red giant type, G8–K5, and by stars at the top of the main sequence (F6–G2). From theories of stellar evolution, the age of the system can be determined from the position where the giant branch rejoins the main sequence. Star formation in M32 has thus been estimated to have halted eight billion years ago.

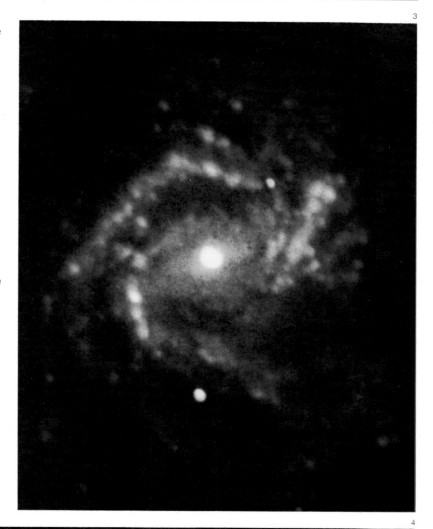

Note on the photographs on this page. The sensation of colours in the human eye comes from the different sensitivities of certain components of the eye to three fundamental colours, red, green and blue. All processes of colour photography and of colour television use this property. The first colour photographs were made by Maxwell in 1862. With the help of a photographer named Sutton, he took photographs of coloured ribbons with filters that were blue, green, yellow and red. The simultaneous projection of these images through the appropriate filters upon a white screen reconstituted the image in colour. This experiment showed that the notion of colour was due to the superposition of three monochromatic images. In the majority of commercial films, three emulsions are superposed: one, sensitive to the blue, gives a yellow image; the second, sensitive to the green, gives a magenta image; and the last, sensitive to the red, gives a cyan image. The colours of the images are the complementary colours of blue, green and red. The superposition of those three images, seen in transmission, or by reflection, reconstitutes the visual appearance of the colours. This procedure can be generalised to give the appearance of colours by the superposition of three ordinary monochromatic images. The four photographs use this process. The colours are due to the projection, on a colour photographic paper, through the appropriate filters, of three negatives made using the U, B and V filters, classical in photometry. The use of colour gives a direct representation of the values of the photometric indices. The correspondence between the colours of these photographs and the stellar population arises through the colours of the stars. One can thus construct a table showing the relationship between star type and colour: a B star appears violet, an A star appears blue–green, and a G star yellow.

1. The elliptical galaxy NGC 4486 (M87). The colours of this image are very uniform. The red colour shows that the light comes essentially from stars of types K to M, that is, cold stars analogous to the red giants in the old population of our Galaxy. Here the population seems to be very uniformly distributed, and, except near the centre, there is not a great variation of colour over the major part of the galaxy. Furthermore, the reddening at the edges is only due to a difference in sensitivity in the UBV photograph used for the picture; this is not a real effect. The small bright white 'tongue' which leaves the centre of the galaxy is not due to the emission of stars but to synchrotron radiation from a jet of relativistic electrons. (J. D. Wray, McDonald Observatory)

2. The Sb-type spiral galaxy NGC 4258. In contrast to M87, two components are easily distinguished. The central part, yellow–white in colour, the bulge, has a population dominated by stars of types G to K. The spiral arms, blue in colour, contain young stars of types O to B. This picture clearly shows the changes in localisation of star formation in spiral galaxies. At the start of its evolution, stars are mainly formed in the central regions. The present colour of the bulge shows that the majority of the stars found there must have been formed twelve billion years ago; then stars were formed in the disk, and the youngest stars are now located in the spiral arms. One may note the difference between the two arms. The top arm is very thin and seems to contain only a very young population. The bottom arm, on the other hand, is more complex. It is much wider and exhibits a transverse variation in colour. The right edge of the arm is marked by a very blue zone containing the youngest stars. Behind the edge, there is a white region dominated by stars of types A to F. The colours of the inner edges of the spiral arms show that star formation must have ceased there a hundred million years ago. This age structure of the stars across a spiral arm can be well understood in a context of the theory of density waves, according to which the formation of stars is due to a shock wave caused by the compression of interstellar gas by the passage of a density wave. The rotation of this density wave gives rise to the basic structure of the spiral arm. One of the edges, where the compression of the interstellar gas occurs, contains all of the youngest stars, and, just behind this, are the stars which had been formed earlier. (J. D. Wray, McDonald Observatory)

3. The SBc-type barred spiral galaxy NGC 4303.
Two components are clearly visible in this picture. The bulge and the central bar are dominated by old stars and the spiral arms by young stars. The relative proportion of young stars is much greater than in NGC 4258. Dust lanes appearing as dark bands run from the nucleus and rejoin the end of the bar and the regions of star formation. This is very common in barred spiral galaxies. These dust lanes embody the density wave (zones where the density of the interstellar medium is high and which rotate around the nucleus) associated with the presence of the bar. In the central regions of this barred galaxy, there is no connection between the density wave and the formation of stars, because the wave rotates more quickly than the interstellar gas. It does not give rise to a shock wave, and there therefore is no star formation. In contrast, beyond the bar, the gas revolves more rapidly than the density wave and a shock wave is produced. The end of the bar corresponds to the point where the gas and the density wave revolve at the same speed. This point is called the point of co-rotation. (J. D. Wray, McDonald Observatory)

4. The Sm-type irregular galaxy NGC 4631. In this galaxy, the old population and the bulge have almost disappeared. Only a close examination will reveal a yellower component at the centre of the galaxy. The bulge is definitely eccentrically placed in relation to the disk. (J. D. Wray, McDonald Observatory)

The Magellanic Clouds

The southern celestial sphere is much more spectacular than the northern hemisphere; the Milky Way is very marked in the south, and it is flanked by two bright, diffuse clouds which are visible to the naked eye. These objects, the Magellanic Clouds, are our nearest neighbours. They are described in the mythologies of the Australian Aborigines, of the Bantu of Southern Africa, and of the islanders of the South Pacific.

The Large Cloud exhibits a very wide luminous bar within a tenuous, scattered disk of clusters of isolated stars that do not appear to be distributed with a well-defined structure such as a spiral. This kind of structure – a bar in a chaotic disk – recurs in numerous galaxies, that by analogy are called Magellanic Irregulars. Certain star clusters in the Large Cloud are extremely prominent. The most remarkable is the 30 Doradus complex, also called the Tarantula Nebula, or, less poetically, NGC 2070. This nebula is the active nucleus of the large Cloud; it has a large number of young stars, in particular Wolf–Rayet stars drowning in a mixture of ionised gas (H II regions) and molecular clouds.

The Large Cloud revolves more slowly than a spiral galaxy: its maximum speed is 70 kilometres per second, whereas values of 200 to 300 kilometres per second are found for spirals. Curiously, the centre of rotation does not coincide with the bar but is a little to the north;

this characteristic occurs in numerous galaxies of this type. This is only possible if the region of the bar – the most luminous part of the Cloud – represents only a small fraction of the Cloud's total mass. From the rotation curve, the total mass comes out at ten billion solar masses, about one twentieth of the mass of our own Galaxy. About 10 per cent of the mass of the Large Cloud is in gaseous form, whereas this proportion is only 5 per cent in the disk of our Galaxy.

The structure of the Small Cloud is more complex than that of the Large Cloud. Although it too has a bar, there is also an enormous outgrowth called the Wing. Measurements of radial velocities of stars in the Small Cloud have shown that the Cloud is made of several parts moving at different speeds. These velocity components are superposed along the line of sight from the Small Cloud; with respect to us, the structure is distributed in depth and it is therefore difficult to determine it accurately. The mass of the Small Cloud is of the order of two billion solar masses – less than that of the Large Cloud – and it is even richer in gas, which comprises about 20 per cent of its mass.

Measuring the distances of the Magellanic Clouds was one of the key events in extragalactic astronomy. In studying Cepheid variables, Henrietta Leavitt of Harvard College demonstrated in 1912 the existence of a relationship between

the period and the mean luminosity of these stars. So period could be used to deduce mean luminosity for this class of star. The subsequent discovery of Cepheid variables in nearby galaxies – the Andromeda Galaxy M31 (NGC 224), the Galaxy in Triangulum, M33 (NGC 598), and NGC 6822 – made it possible to establish the distances of these objects and to demonstrate their extragalactic nature. The exact determination of the period–luminosity relation, as with other distance indicators, has fascinated numerous astronomers. The distance of the Large Cloud is now estimated to be 50 kiloparsecs, and that of the Small Cloud to be 65 kiloparsecs; the uncertainty is 10 per cent. These distances are not large in relation to the size of our Galaxy, whose radius is 15 kiloparsecs. If we imagine the Galaxy to occupy the area of Greater London or New York the Magellanic Clouds would be 100 kilometres from the city and the Andromeda Galaxy, our next nearest neighbour after the Clouds, would be located 1000 kilometres away. This nearness has consequences: the gravitational interaction between our Galaxy and the Clouds shows up as a bridge of gas which is detectable by radio telescopes at a wavelength of 21 centimetres. Between the Magellanic Clouds and the Galaxy there are a series of neutral hydrogen clouds whose total mass is several tens of millions of solar masses. Tidal forces must

Text continued on page 345

The Clouds and their environment. The Clouds take their name from their being mentioned in the accounts of Magellan's voyage around the world (specifically, in the chronicle of Antonio Pigafetta). This wide-field photograph shows the Large Magellanic Cloud at the top, about 25 degrees away from the Small Magellanic Cloud (below). The bright object with a very regular shape in the lower right-hand corner of the picture is a star of our own Galaxy, Achernar (α Eridani), whose image is greatly enlarged by scattering of its light in the photographic emulsion. The Magellanic Clouds, of diffuse shape, are small irregular galaxies very close to our Galaxy, of which they are in fact satellites. Situated at roughly the same distance from us, their visual appearance enables us to account for their respective sizes. Bright spots can be seen in the Large Cloud (in particular the Tarantula Nebula, located just below the bar) which are active regions of star formation. The Small Magellanic Cloud looks like a pear; it too has a bar, and is flanked by a large outgrowth, the Wing. Immediately under the Small Cloud one can see the globular cluster 47 Tucanae (NGC 104) which belongs to our own Galaxy. (Photograph Yerkes Observatory)

The Large Magellanic Cloud.
Apart from the Milky Way, the Large Magellanic Cloud is the most extended object visible in the sky. It extends over more than 50 square degrees. For comparison, the Andromeda Galaxy, M31, occupies an area six times less. The Large Magellanic Cloud is the nearest galaxy to us; its apparently large size is due to its nearness alone because it is intrinsically a small galaxy. Its diameter of 5 kiloparsecs is one sixth of that of our Galaxy, and its mass of 10^{10} solar masses is one twentieth. These values are much less than those found for typical spiral galaxies. This photograph shows the real colours of the Cloud and, although not entirely covering it, enables the most important features to be picked out. The oldest stars, with colours ranging from green to yellow, are distributed all over the Cloud. Many are concentrated in a bar which at first sight appears to be the dominant element of this galaxy. The youngest stars, blue in colour, are distributed here and there, or form small isolated clusters. Reddish filamentary regions are often associated with them. These are regions of ionised gas whose colour derives from the emission of the red Hα line of ionised hydrogen. The largest of these regions, the Tarantula Nebula, is located above the left end of the bar. (D. F. Malin, Anglo-Australian Telescope Board)

The Small Magellanic Cloud.
The Small Magellanic Cloud is a galaxy of the same type as the Large Cloud, located a little further away. The real structure of the Small Cloud remains unknown. The origin of the large structure far down below to the left of the bar, called the Wing, is not well understood. In all probability, the Small Cloud has been deformed by the gravitational interaction with the Large Cloud and with our Galaxy. By their nearness, the two Magellanic Clouds offer numerous possibilities of study, concerning not only the evolution of galaxies but also the evolution of stars. In our Galaxy, the distances of stars are always poorly known. The dimensions of the Clouds being small compared to their distance, all of the stars in one or other of the Clouds may be considered to lie at the same distance from us. The scale of apparent luminosities of the stars of the Clouds therefore correspond in the first approximation to their intrinsic luminosities. Hence a correlation has been shown to exist in the Large Cloud between the mass loss of the massive stars and their luminosities. The correlation is less obvious for the stars in our own Galaxy, simply because the estimation of their intrinsic luminosity was plagued with numerous errors. The comparison of the rate of mass loss of galactic stars with that for the stars of the Clouds has also shown that there is a dependence of this mass loss upon the abundance of heavy elements. In the Clouds, which are underabundant in comparison with our Galaxy, the rate of mass loss is much less. These examples suffice to show the fascination of the Magellanic Clouds for astronomers. (D. F. Malin, Anglo-Australian Telescope Board)

343

The Tarantula Nebula. Also called 30 Doradus, this is the brightest H II region in the Large Magellanic Cloud. It is an enormous complex of ionised gas and young stars. Its diameter is about 250 parsecs for a mass of stars of about 5 million solar masses. This region is particularly active, one third of the young stars in the Cloud are here and it contains most of the stars of the Wolf–Rayet type. Although some people have suggested that it is the centre of the Cloud, 30 Doradus does not occupy a privileged situation and this shows clearly that star formation is not linked to large-scale dynamical phenomena as is the case with spiral galaxies, where there is a relation between density waves, the formation of stars, and spiral arms. In the Clouds, in contrast, the formation of stars is a local phenomenon and seems rather to be due to random phenomena such as collisions between interstellar clouds.

A very bright nucleus 25 parsecs in diameter lies at the centre of 30 Doradus. This nucleus is connected with the presence of a supermassive object called the mystery object R136. This very peculiar object itself consists of three components. The bright nucleus is too small to be resolved from the ground; its size is less than 0.2 parsec. Its precise nature is extremely difficult to discover. If it is a single star, its mass must be several thousand solar masses. Recall that the largest stars attain only 100 solar masses. If it is a cluster of normal stars, it must consist of several tens of stars and thus be very unstable dynamically.

Numerous studies are being undertaken to discover the nature of R136 which is without any doubt a smaller version of the exotic objects which occur in the nuclei of active galaxies. Seyfert galaxies and quasars. (Royal Observatory, Edinburgh, © 1981)

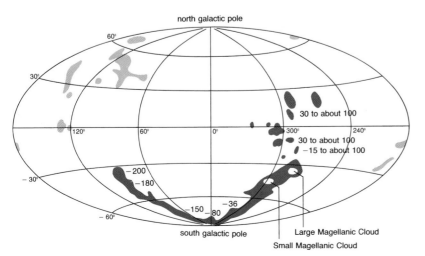

The Magellanic Stream. Systematic investigations undertaken by radio astronomers have revealed the existence of gas clouds falling onto our Galaxy, thanks to their emission of the 21-centimetre line of hydrogen. Their spatial distribution is shown schematically in this map. The largest gaseous complex leaves the Magellanic Clouds and heads towards our Galaxy. The heliocentric velocities of the various parts of this complex are given on the map. The two Magellanic Clouds are connected by a bridge of gas. It would be more precise to say that the Clouds are drowned in an immense gas cloud. This cloud and its continuation towards our Galaxy form what is termed the Magellanic Stream, the origin of which is certainly connected with the gravitational interaction between the Magellanic Clouds and our Galaxy. Numerical simulations have indeed shown that such bridges of gas form between interacting galaxies. However, constructing a detailed model that would reproduce the Magellanic Stream exactly is extremely difficult. The measurement of velocities through the Doppler effect only gives the radial velocity. In other words, we can know the velocity with which the Magellanic Clouds are moving towards us, but not the velocities with which the Clouds are revolving around us. Now the formation of this type of bridge is very sensitive to this velocity of rotation. One must therefore take account of this in numerical models. The study of the kinematics of the system is extremely interesting because it enables a measurement of the total mass of the Galaxy to be made. From our position in the Galaxy, we only know the mass lying within the radius defined by the Sun and the galactic centre. In order to find out the total mass of the Galaxy, one must go further still, that is, as far as the Magellanic Clouds.

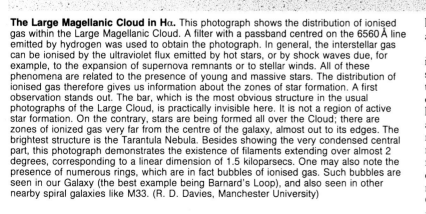

The Large Magellanic Cloud in Hα. This photograph shows the distribution of ionised gas within the Large Magellanic Cloud. A filter with a passband centred on the 6560 Å line emitted by hydrogen was used to obtain the photograph. In general, the interstellar gas can be ionised by the ultraviolet flux emitted by hot stars, or by shock waves due, for example, to the expansion of supernova remnants or to stellar winds. All of these phenomena are related to the presence of young and massive stars. The distribution of ionised gas therefore gives us information about the zones of star formation. A first observation stands out. The bar, which is the most obvious structure in the usual photographs of the Large Cloud, is practically invisible here. It is not a region of active star formation. On the contrary, stars are being formed all over the Cloud; there are zones of ionized gas very far from the centre of the galaxy, almost out to its edges. The brightest structure is the Tarantula Nebula. Besides showing the very condensed central part, this photograph demonstrates the existence of filaments extending over almost 2 degrees, corresponding to a linear dimension of 1.5 kiloparsecs. One may also note the presence of numerous rings, which are in fact bubbles of ionised gas. Such bubbles are seen in our Galaxy (the best example being Barnard's Loop), and also seen in other nearby spiral galaxies like M33. (R. D. Davies, Manchester University)

have pulled this gas from the Clouds and attracted it towards our Galaxy.

The more a galaxy has evolved, the richer it is in heavy elements (such as carbon, oxygen and sulphur) and the poorer it is in gas. We have seen that the Magellanic Clouds are richer in gas than our Galaxy; the corollary is that it is poorer in heavy elements. For example, the oxygen abundances in the Small and Large Clouds are respectively one-third and one-sixth of the corresponding abundances in the Sun. This results from the fact that star formation and the enrichment of interstellar hydrogen take place more slowly in the Magellanic Clouds than in our Galaxy. The precise reasons for this difference are not yet known. The total mass of the Galaxy certainly plays a part since, the more massive a

galaxy, the richer it is in heavy elements, and the more quickly the mechanism works which enriches the galaxy in heavy elements. However, other factors must be involved, and, in particular, the processes of star formation: in our Galaxy, star formation is in the main localised in the spiral arms, whereas in the Clouds star formation seems to occur here and there, in sporadic bursts in zones some hundreds of parsecs in size. Even more curious is the existence in the Clouds of young globular clusters associated with Population I: in our Galaxy, there are only old globular clusters, connected with Population II which precede the formation of the disk.

Laurent VIGROUX

The nearby galaxies

Our Galaxy is not isolated in space. It has satellite galaxies, of which the nearest and the most remarkable are the Magellanic Clouds. In fact, it belongs to a group of about twenty galaxies called the Local Group. The dimensions of this group are of the order of 1 megaparsec. The Local Group forms an entity in the sense that the next nearest galaxies are two or three megaparsecs further away. Three giant galaxies dominate the Local Group: in order of decreasing mass, they are the Andromeda Galaxy (M31), our Galaxy and the galaxy in Triangulum (M33). All three galaxies are spiral. Our Galaxy and the Andromeda Galaxy have further satellite galax-ies. Apart from the Magellanic Clouds, these satellites are all small galaxies called dwarf spheroidals. M31 is also surrounded by a few dwarf spheroidals, but its four principal satellite galaxies are the small elliptical galaxies NGC 221 (M32), NGC 205, NGC 185 and NGC 147. The galaxy in Triangulum does not appear to have any satellites. In addition to these three systems, the Local Group consists of some Magellanic irregular galaxies, NGC 6822 and IC 1613, and ellipticals Maffei I and II.

Laurent VIGROUX

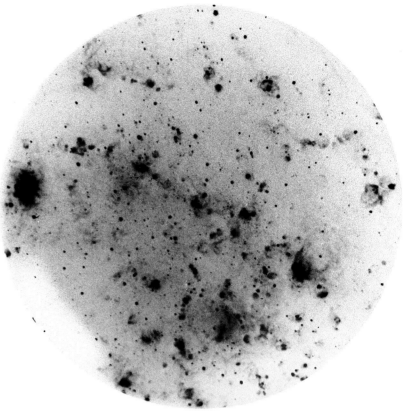

Central region of M33. This picture of the central part of M33 and of its northern arm was obtained using a filter with a passband centred on the Hα line of hydrogen. Regions of ionised gas which emit this line are consequently visible in the photograph. The structure of this hot interstellar gas is very complicated. Apart from the points which are the brightest stars of M33, one sees huge irregular spots. These are the H II regions. In between them the picture reveals the existence of long filaments. Their number increases as one approaches the spiral arms. The origin of these arms is without doubt the agitation of the interstellar medium. The gas would be heated by the passage of a shock wave due to the expansion of a supernova remnant. Some filaments are particularly remarkable for their circular appearance. The rings are in fact hollow spherical bubbles which we see in projection as rings. The sizes of the bubbles vary from 150 to 300 parsecs. They are known also in the Large Magellanic Cloud, and in our Galaxy, where the most celebrated is Barnard's Loop in the constellation of Orion. The origin of these bubbles is still the object of discussion; they might have been formed by the interaction between the particularly intense stellar wind from massive stars and the interstellar medium. (J. P. Sivan, LAS Marseille)

The Galaxy in Triangulum (M33) is a spiral galaxy of the type Sc. It has a mass of 10^{10} solar masses and a diameter of 7 kiloparsecs. This colour photograph enables one to distinguish the various components of this galaxy. The old population, yellow in colour, is distributed uniformly, but the relative proportion of old stars in relation to the young population, blue in colour, is very small. (In the Andromeda Galaxy, the old population is proportionately much greater, and the whole galaxy is more yellow.) At the centre, there is a very bright nucleus which is red in colour owing to Hα emission from hydrogen. This nucleus is only 5 kiloparsecs in diameter, which is very small. From there start spiral arms full of blue stars and H II regions. One of them, located at the end of an arm is particularly bright. Because of its size and the conditions existing there, it closely resembles the Tarantula Nebula in the Large Magellanic Cloud.

The kinematic structure of M33 is complicated by a north–south asymmetry. The southern part of the disk turns at 15 to 20 kilometres per second more slowly than the northern part. This asymmetry is in all probability not due to an excess of mass of the northern part because no asymmetry has been noted in the photometric profile nor in the distribution of neutral hydrogen., It could be caused by the bulge if the latter was slightly displaced with respect to the centre of the disk. Such displacements are often noted in barred spirals but could not be directly verified for the bulges. The asymmetry of M33 also appears in the H II regions. Their excitation is stronger in the northern spiral arms than in the southern arms. This could result from two things: a difference of chemical abundances between the two parts of the galaxy or a difference in the effective temperatures of the stars which ionise the gas in the H II regions. (D. F. Malin, Anglo-Australian Telescope Board)

A dwarf spheroidal galaxy and a globular cluster, NGC 6273. This diffuse cluster of stars (opposite) is a dwarf galaxy, a satellite of our own Galaxy, which lies in the constellation of Sculptor. Our Galaxy has six dwarf satellites of this type, which are the smallest galaxies known. Draco and Ursa Minor, for example, only contain a hundred thousand stars, less than a typical globular cluster. The largest dwarf galaxy contains only twenty million stars. On the other hand, these galaxies are very dispersed and diffuse. The Sculptor galaxy extends over almost 1 kiloparsec; a globular cluster which contains the same number of stars only occupies a diameter of about 230 parsecs, fifty times less. The number of stars per unit volume is about ten thousand times higher in a globular cluster than in a dwarf spheroidal. Below right is shown a globular cluster. A comparison between the two figures is striking; these photographs are not at the same linear scale, with the scale of the globular cluster being a hundred times greater than that of the galaxy. Despite this, the density of stars is much higher in the cluster. The existence of these galaxies is in all probability linked to the formation of our own. Curiously enough, these objects, with the Magellanic Clouds, the Magellanic Stream, and the furthest globular clusters, seem to define a very rough plane. The plane is undoubtedly a remainder of the form of the protogalaxy, something like the way the plane of the ecliptic defined by the planets of the Solar System stems from the disk of the protosolar nebula. The ages of these galaxies, equal to that of the oldest populations of our Galaxy, reinforce this hypothesis. They are contemporaries of the oldest globular clusters, and have comparable chemical compositions: they are underabundant in heavy elements. For example, the metallicity of the dwarf galaxy in Draco is ten times less than that of the Sun. Whereas the chemical composition of the globular clusters is very homogeneous, there is a rather large dispersion in the metallicities of the dwarf galaxies. The metallicities of stars in the same dwarf galaxy vary by a factor of two or three in the smallest dwarf galaxies like Draco, and by a factor of twenty or thirty in the largest, Fornax. This dispersion shows that, in contrast to globular clusters, these galaxies have undergone chemical evolution, which seems to be explainable by a mechanism known as hot wind. When a supernova explodes, it communicates a part of its kinetic energy to the interstellar medium and heats it up. Now if the temperature of a gas increases, it is the velocities of the particles composing it which increase. If the heat given off by the supernova is sufficiently great, it may happen that the velocities of the particles of the gas exceed the escape velocity of the galaxy. In the same way that a projectile having an initial velocity which is greater than the terrestrial escape velocity escapes from the Earth, the particles of this extremely hot gas leave the galaxy. If the mass of the galaxy is small, then the escape velocity is low, one needs only a few supernovae to attain the temperature necessary for the gas to escape; a hot wind (10^6 K) thus blows from the galaxy. When all of the gas has disappeared, there is no more to form new stars, and therefore no more enrichment in heavy elements. Chemical evolution stops and the maximum metallicity of the galaxy will be that which it had when the hot wind began to blow. (Royal Observatory, Edinburgh, © 1979; D. F. Malin, Anglo-Australian Telescope Board, © 1980)

NGC 6822, an irregular galaxy of the Local Group (above). Its structure is very similar to that of the Large Magellanic Cloud. Its mass is lower, 4×10^8 solar masses as opposed to 10^{10} solar masses, as is its size. The stellar populations of the two galaxies resemble each other quite remarkably. NGC 6822 has no HII regions, regions where gas is ionised by the ultraviolet emission of hot stars, mainly O-type and Wolf–Rayet stars. There are two possible explanations:

O-type and Wolf–Rayet stars have very short lifetimes, at most about 10 million years. If there was no star formation in NGC 6822 for a time longer than the life times of these stars, they would all have disappeared, along with the HII regions they ionise. An explanation like this is possible, but our knowledge of the evolution of this type of galaxy suggests that star formation takes place fairly continuously.

The other explanation consists of saying that if there are no O stars this is because none form. The initial mass function (IMF) defines the mass distribution of stars born at a given epoch. Studies have shown that locally stars forming from the same cloud tend to have the same mass. One cloud makes O stars, another B stars, another low-mass stars. . . . In general in galaxies all of this is averaged, and there is no proof of the variation of the IMF from one galaxy to another. The absence of O stars in NGC 6822 may be an example of an IMF which differs from that in our Galaxy or the Large Magellanic Cloud (D.F. Malin, Anglo-Australian Telescope Board, © 1982).

The Andromeda Galaxy

The Andromeda Galaxy (M31) is the most massive of the galaxies in the Local Group: its mass is 3×10^{11} solar masses which is twice that of our Galaxy, and its diameter is 50 kiloparsecs. We see this galaxy practically edge-on – the angle between its plane and the line of sight is only 13 degrees – and, due to this, its spiral structure is difficult to study. Nevertheless, its other characteristics, particularly the size of the bulge compared with the disk, the fraction of mass in the form of gas and its colour associate M31 with Sb-type galaxies; the ultraviolet and radio continuum images confirm the presence of tightly wound arms. The disk contains many H II regions whose chemical composition is revealed by spectroscopic analysis. The abundance of elements in M31 is similar to that in our own Galaxy which suggests comparable evolution. These abundances are not the same across M31: the central regions are about three times richer in heavy elements than the outer regions. This gradient is correlated with other characteristics, in particular the variation of the fraction of mass in gaseous form, which increases from the centre outwards. This correlation indicates that the efficiency of star formation increases as the centre is approached; the gas has been consumed and the process of nucleosynthesis in the centres of stars has increased the abundance of heavy elements. Beyond this simple qualitative interpretation, numerous models have been proposed, but it is still very difficult to account for the existence of gradients and of other radial properties of M31 – or of our Galaxy – if the only explanation is assumed to be an increase in the rate of star formation towards the centre. The origin of the abundance gradients must in part be related to the gas dynamics.

The nucleus of the Andromeda Galaxy is very small; its diameter is of the order of 8 kiloparsecs. Optically, its surface brightness is clearly greater than that of the rest of the galaxy, and it is also a powerful source of radiation at infrared and radio wavelengths. Its kinematic structure is very complex; the movements are non-circular and a dynamical analysis shows that the constituents have mass to light ratios higher than normal stars in the galaxy.

The Andromeda galaxy M31 is interacting with a small elliptical galaxy, M32 (NGC 221) which, in the same way as the companion of the galaxy, M51, may be able to exert a force on the spiral structure of M31. However, the angle of inclination hinders verification of this hypothesis.

The nearness of M32 makes detailed study of its stellar population possible. It has thus been determined that the average metallicity of M32 is that of the Sun and is therefore relatively lower than normal elliptical galaxies. Also, the age of these stars is somewhat less. Models of the synthesis of stellar populations, suggest that the formation of stars in M32 must have ceased eight billion years ago after a duration of several billion years. This result was completely unexpected; it had been thought until then that, in elliptical galaxies, star formation occurred very rapidly, in only a few hundred million years. The stars in M32 should then have been found to have ages in the region of fifteen billion years. Studies of other elliptical galaxies will have to be made to find out if M32 has a pathological behaviour, due, for example, to its interaction with M31, or on the other hand, a behaviour totally typical of elliptical galaxies.

Laurent VIGROUX

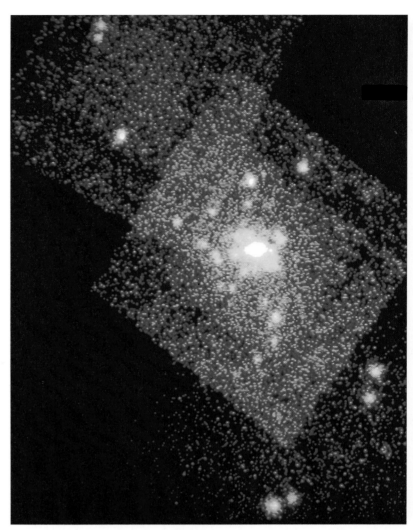

X-ray photograph of the Andromeda Galaxy. This picture was obtained by combining three images taken with the X-ray telescope on board the Einstein satellite, launched in 1978. The energy of incident photons is about 1 kiloelectronvolt. Besides the very bright nucleus, there are several bright spots, which are either globular clusters or stars. Some globular clusters in our own Galaxy were already known to emit X-rays, which were attributed to giant binary stars in the centres of the clusters. Such stars exchange matter during their evolution; this matter, in the form of hot, dense gas emits X-rays. X-ray emission from globular clusters in galaxies other than our own had never been detected before the Einstein satellite was launched (Center for Astrophysics, Cambridge, Mass.)

The Andromeda Galaxy, M31. This spiral galaxy, located 670 kiloparsecs away in the constellation of Andromeda, is the largest of the nearby galaxies forming the Local Group. Its apparent dimensions are very large: 80 by 250 arc minutes. The nucleus of M31 can just be seen with the naked eye, and binoculars are needed to examine it. The exact structure of M31 cannot be determined because of the inclination of the disk, whose size is characteristic of Sb galaxies. The disk contains large numbers of H II regions which are seen as red spots; two thousand have been catalogued. The luminous blue points in the disk are small star clusters known as OB associations; each association is about 100 parsecs across and consists of about a hundred O and B stars. Supernova remnants are also observed. Relative to the mass of M31, the numbers of H II regions, OB associations, and supernova remnants, each associated with recent star formation, shows that the rate of star formation is slightly greater in M31 than in our Galaxy. All around M31, a myriad of small yellow spots can be seen. These are globular clusters analogous to those which populate the halo of our Galaxy. The most distant ones are found at a distance of 30 kiloparsecs from the centre. The halo is thus more extended than the disk containing the young population, which has a radius of only 25 kiloparsecs. The disk also contains large dust zones and dark spots, and therefore contains all the constituents of spiral galaxies.

Two other galaxies are visible on this photograph. The small elliptical galaxy (M32/NGC 221) located in the immediate vicinity of the principal galaxy is interacting strongly with M31. The outer stars in M32 have been torn away by the gravitational attraction of M31. The photometric profile of M32 appears to be truncated in comparison with the profile of a normal elliptic galaxy. For this reason, M32 is often catalogued as a compact galaxy. The other galaxy, NGC 205, is also a satellite elliptical galaxy of M31. The latter has two other satellite elliptical galaxies, NGC 147 and NGC 185, which are also an interacting pair. Around M31 three spheroidal dwarf galaxies of the same type that surround our Galaxy have been found. (Hale Observatories)

Comparison between the optical and ultraviolet images of M31. These two images are at the same scale to facilitate the comparison. The top photograph, taken in the visible region of the spectrum, clearly shows the central bulge, the outer spiral arms and the two satellite galaxies M32 and NGC 205. The ultraviolet photograph (below) was taken at a wavelength of about 200 nanometres using a small telescope carried by a stratospheric balloon. Such observations cannot be made from the ground since the Earth's atmosphere absorbs ultraviolet radiation. All that remains of the bulge in this ultraviolet image is a small circular spot at the nucleus which is probably due to hot, evolved stars, similar to those found in the horizontal branch of the HR diagram for globular clusters. The spiral arms appear very clearly; their ultraviolet emission is due to the presence of very hot stars and H II regions. This image is very representative of the form of the spiral arms in M31. (J. M. Deharveng, LAS, Marseille)

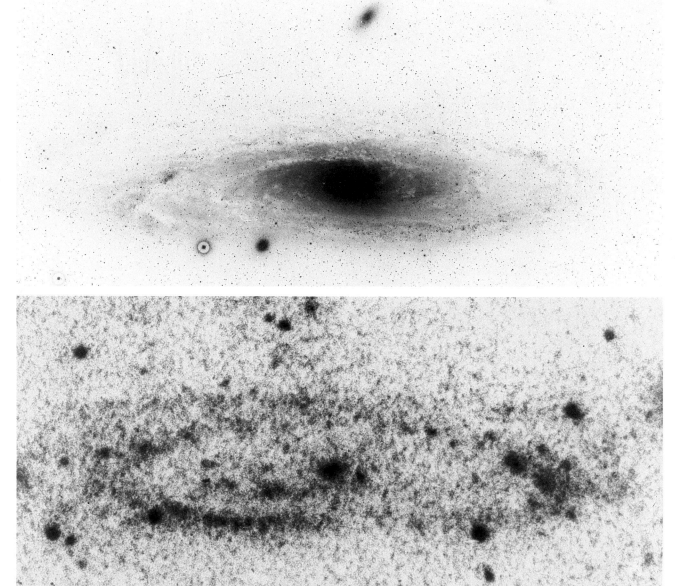

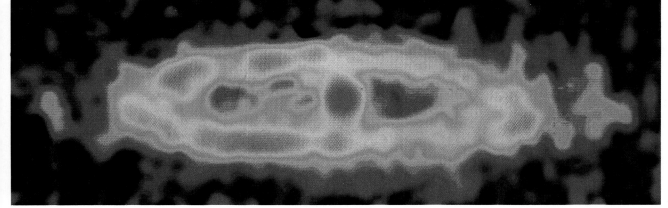

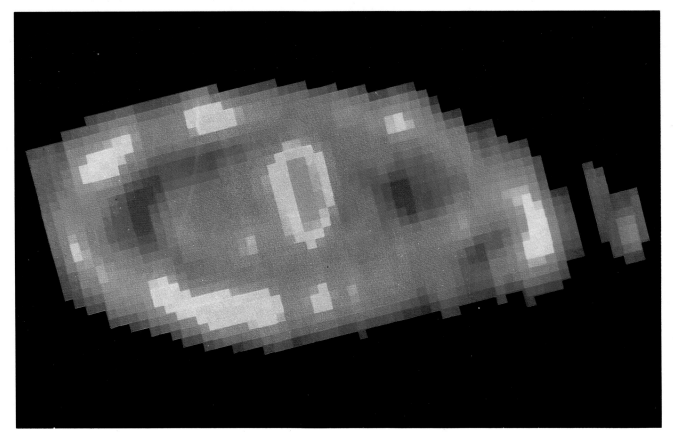

Picture of the Andromeda Galaxy at radio wavelengths. Obtained with a radio telescope, this image (left) shows the appearance of M31 at a wavelength of 11 centimetres. The different colours correspond to different intensities. There is a strong resemblance to the ultraviolet image with the nucleus and the zones of hot ionized gas called H II regions again being visible. At this wavelength, the emission from M31 mainly comes from thermal electrons in the hot, ionised gas. These electrons are in rapid motion. Each time one collides with an ion or another electron, they undergo a deviation in trajectory. This change of direction is accompanied by the emission of a photon. The radiation produced in this way is mainly observable at radio wavelengths. (R. Beck, E. Berkhuigsen, R. Wielebinsky, Max-Planck-Institut für Radioastronomie)

Picture of the Andromeda Galaxy in the infrared. This image was recently processed by computer from observations made by the Infrared Astronomical Satellite (IRAS). The different colours correspond to different intensities from blue (faint) through green, yellow, orange and red (bright). Brighter regions are populated by numerous massive young stars; thus regions in red, orange and yellow are regions of star formation. The observations by IRAS are the first made of the Andromeda Galaxy at infrared wavelengths. (NASA)

The classification of galaxies

Throughout the twentieth century, astronomers have sought to define the fundamental parameters which categorise galaxies and have endeavoured to formulate a method of classification. The one most widely accepted today derives from a morphological classification propounded by Edwin Hubble in 1925. This distinguishes four major classes of galaxies. Elliptical galaxies are characterised by a round or elliptical shape, uniform appearance and a luminosity which is very regularly distributed. On the other hand, spiral galaxies display two components, a central part – the bulge, at first sight similar to elliptical galaxies – and a disk in the plane of which a spiral structure can be observed. Spiral galaxies are themselves divided into two groups, ordinary spirals and barred spirals, depending on whether or not they have a central bar. Lenticular galaxies (or S0) also comprise a bulge and a disk, but the latter does not display a spiral structure. Finally, as their name indicates, irregular galaxies do not have a well-defined structure; no bulge is discernible and they have a chaotic appearance.

Each of these classes is divided into groups. The ellipticals are classified from 0 to 7 according to their ellipticity (E0 for the rounder ones, E7 for the more elliptical ones), the spirals range from Sa to Sd, as well as a type Sm according to the diminution in the size of the bulge in relation to the disk and the opening of the spiral arms. The same applies to the barred spirals, SB.

This classification covers physical differences other than straightforward morphology. Firstly, elliptical galaxies are redder than spiral galaxies, this is interpreted as a difference in their stellar constituents. In elliptical galaxies, the emission of light is dominated by that from cold stars, of the red giant type; the stellar population of elliptical galaxies seems analogous to the old population of our Galaxy. In contrast, there is still a certain amount of activity in spiral galaxies; one ascertains the presence of a young stellar population which appears as blue light. S0 galaxies have a stellar population similar to that of elliptical galaxies. While elliptical and lenticular galaxies have, roughly speaking, the same stellar popula-

Galaxy NGC 4486. This elliptical galaxy appears practically circular. Only its central parts are visible in this short exposure photograph. The true extent of the galaxy is in fact great enough to hide the two small galaxies in the top left hand corner. The distribution of light is very regular in galaxies of this type, which seem to have only one component. The contours of constant surface brightness are called isophotes. The innermost isophotes are almost circular; the outer isophotes corresponding to lower surface brightnesses are more elliptical. Far from the centre, the surface brightness is inversely proportional to the distance from the centre in the first approximation. The small bright points dotted around NGC 4486 are globular clusters; these are particularly numerous in elliptical galaxies. (Kitt Peak National Observatory)

Galaxy NGC 4374. At first sight, this galaxy, although more elongated, appears to be of the same type as the elliptical galaxy in the figure above. However, this is an S0 or lenticular galaxy. While the central part of this galaxy has isophotes which follow a law analogous to that for elliptical galaxies, the outer parts do not follow the same law: the surface brightness decreases exponentially with the radius. These galaxies have two components: a central spherical bulge and a disk. Despite the different morphologies, the stellar populations are, roughly speaking, identical in elliptical and lenticular galaxies. (Kitt Peak National Observatory)

Galaxy NGC 5866. This galaxy is seen edge-on. Thus one sees very clearly the elliptically shaped central bulge and the disk which extends from one side to the other. This structure is characteristic of S0 and spiral galaxies. The disk appears as a dark band due to the presence of dust. Although, due to its inclination, we are unable to see the spiral structure in the plane of the disk, the size of the bulge and the presence of dust suggest that this spiral galaxy is type Sa. (Hale Observatories)

tion, whatever their type, the relative proportion of the young population increases in spiral galaxies of the types Sa to Sm; it is even greater for irregular spirals. The same findings apply to the gaseous content, deduced from radio observations made at the wavelength of the neutral hydrogen transition at 21 centimetres. Whatever their type, elliptical galaxies contain practically no gas at all. On the other hand, spirals have a very large gaseous content; neutral gas represents several per cent of their mass and the fraction of the mass which is in gaseous form increases as one goes from type Sa to type Sm; this fraction ranges from 10 to 20 per cent for irregular spirals. S0 galaxies are intermediate between elliptical and spiral galaxies. Two thirds of them contain no gas, the rest have as much as spirals. Thus the morphological differences between galaxies correspond well with their physical differences. Nevertheless, this classification does not enable one to account for all the phenomena. The dispersion of the physical

Two examples of Sb galaxies: NGC 4736 (left) and NGC 4622 (right). These galaxies are characterised by a bulge which is large in relation to the disk, and by thin, tightly wound arms. Galaxy NGC 4736 is fairly close in morphology to type Sa galaxies. The bulge occupies almost the whole of the galaxy and the arms are very underdeveloped. On the other hand, in NGC 4622 the arms are spread out; they wind round almost twice – this is exceptional. One should also notice their sharpness. (Kitt Peak National Observatory)

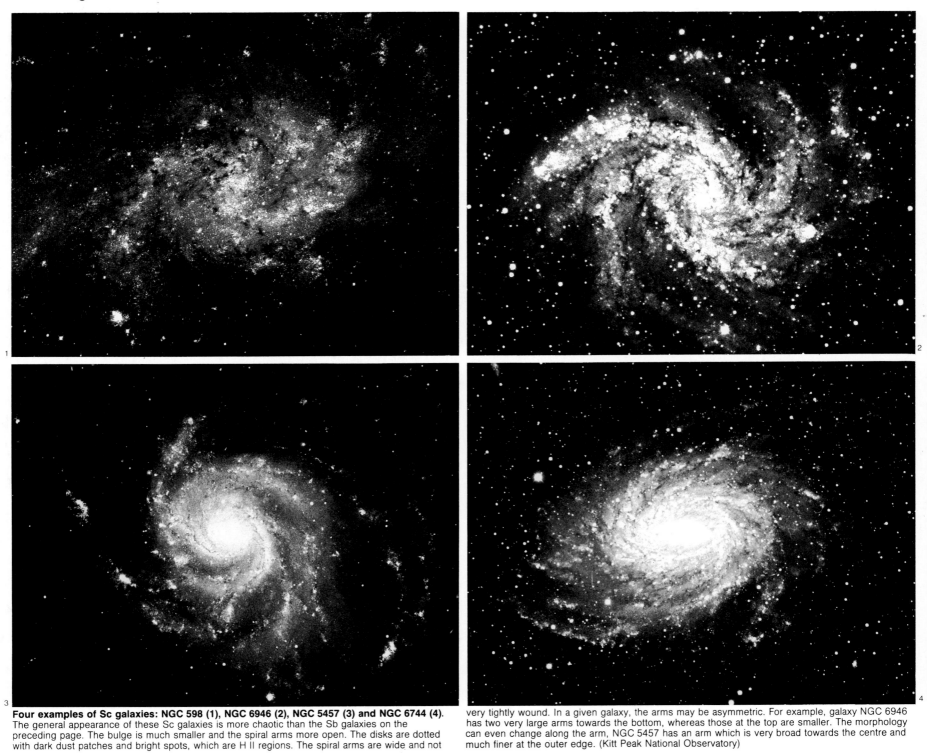

Four examples of Sc galaxies: NGC 598 (1), NGC 6946 (2), NGC 5457 (3) and NGC 6744 (4). The general appearance of these Sc galaxies is more chaotic than the Sb galaxies on the preceding page. The bulge is much smaller and the spiral arms more open. The disks are dotted with dark dust patches and bright spots, which are H II regions. The spiral arms are wide and not very tightly wound. In a given galaxy, the arms may be asymmetric. For example, galaxy NGC 6946 has two very large arms towards the bottom, whereas those at the top are smaller. The morphology can even change along the arm, NGC 5457 has an arm which is very broad towards the centre and much finer at the outer edge. (Kitt Peak National Observatory)

Galaxy NGC 4631. Like NGC 5866 (top of preceding page) this galaxy is seen edge-on, but its appearance is quite different. The bulge has practically disappeared and there is only a very flattened disk. The latter is very contorted and filled in with dust. This galaxy is probably type Sc or even the later Sd or Sm type. In this case, it approximates to an irregular galaxy. (Kitt Peak National Observatory)

properties of galaxies of the same type is very large. Consider for example, in the case of spiral galaxies, the ratio of the mass of hydrogen to the blue luminosity; this ratio is on average 0.3 for Sa spirals and 0.9 for Sd spirals, but we know of Sd spirals which are as gas deficient as Sa spirals. Likewise, the colour dispersion of elliptical galaxies is in general greater than the measurement errors; these dispersions are therefore real.

The preceding paragraphs show that the criteria used to define the types of galaxies are incomplete and that one or more other fundamental parameters must exist which would enable us better to define the galaxies. Finally, it should be mentioned that, even if one does not include active galaxies and interacting galaxies, some galaxies remain unamenable to attempts at classification; it is not unusual to find, according to the catalogues, the same galaxy classified under quite different types.

Initially, this classification was presented as an evolutionary sequence from elliptical galaxies to spirals, from which came the names 'early type'

for ellipticals and 'late type' for spirals, still in current usage in English. Later on, other theories attempted to show that spirals evolved from ellipticals. It now seems to be well established that these ideas were quite wrong. The sequence of galaxies is not one of temporal evolution. On the contrary, the majority of galaxies, if not all, were formed during the same era. Depending upon the initial conditions, they became either elliptical or spiral. Recent studies have shown the importance of star formation in this evolution; a galaxy in which the stars formed quickly, relative to the dynamical evolution time became an elliptical; a galaxy in which the stars formed more slowly became a spiral. At the moment we are still speculating. No theory is yet able to explain the sequence of galaxies in detail. One thing is certain, the differentiation between the different types occurred very early in the history of the universe, at the time of galaxy formation.

Laurent VIGROUX

Barred spiral galaxy NGC 1530 (opposite). The bar of this galaxy is well developed and two thin, very symmetrical arms run from its ends. The ring seen in NGC 3313 (see below) does not appear in NGC 1530. All galaxies with such pronounced bars have very thin symmetrical arms. (Kitt Peak National Observatory)

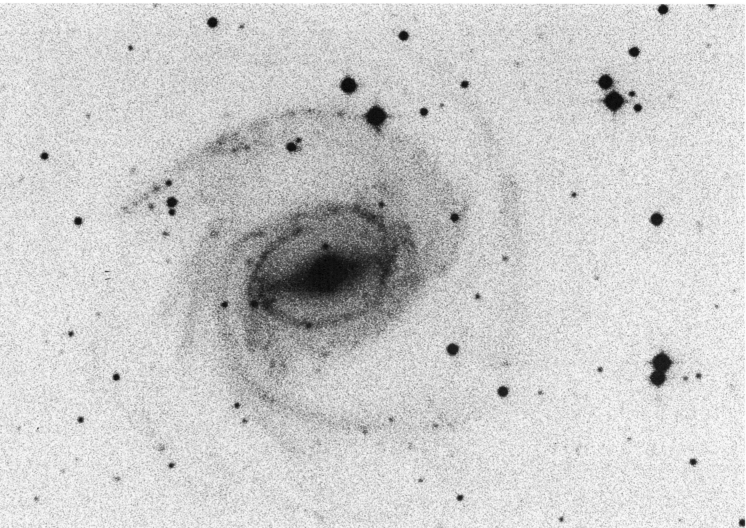

Galaxy NGC 3313. Although this galaxy possesses thin, tightly wound arms like Sb galaxies, its bulge is very elongated, being shaped more like a bar. It is encircled by a very thin ring. The structure of this bulge clearly resembles that of NGC 6744 (facing page). These two galaxies belong to the subgroup of ringed barred spirals RSB; RSBb for NGC 3313 and RSBc for NGC 6744. Their arms are much more symmetrical than the arms of ordinary spiral galaxies, Sb or Sc. Barred spirals are in general much more symmetric than normal spirals. (Royal Observatory, Edinburgh)

Elliptical galaxies

As paradoxical as it may appear, the first question that one could ask about elliptical galaxies is 'what is their real shape?' When we look at the sky, we only see an image in two dimensions, without relief or depth. The absence of relief is particularly troublesome when we observe elliptical galaxies, because their appearance is so regular that it is difficult to know their real shape. Look, for example, at a rugby ball: face-on, it looks circular, but from all other angles it looks like an elongated ellipse.

The shapes of elliptical galaxies in photographs have all shapes from a circle to a slightly elongated ellipse. Since, a priori, it is impossible to know the exact role of projection effects in this apparent distribution of shapes, one may arrive at the following reasoning. Supposing that all elliptical galaxies have the same shape, then, since they are randomly oriented with respect to ourselves, we can calculate the proportion of galaxies which should appear to us as a given shape. Whatever the shape considered, it is impossible to establish the proportion of different types of elliptical galaxies that one observes; as regards the unique shape hypothesis, the Universe contains too many galaxies which appear circular. Elliptical galaxies, therefore, do not all have the same shape. The differences between circular and elliptical galaxies are not solely due to the effect of projection.

It is easiest to imagine that they have the shape of ellipsoids of revolution, that is to say they resemble a ball which would be something between a football and a rugby ball. This idea prevailed for a number of years and has only been disproven very recently. The first doubt came from a study of the kinematics and dynamics of these galaxies. An elliptical galaxy is merely an enormous cluster of billions of stars which remain grouped together simply because of the mutual gravitational attraction. All these stars are moving and they describe orbits, some more

complicated than others, around the centre of the galaxy. When a galaxy rotates the stars tend to move away from the axis of rotation due to the resulting centrifugal force, which manifests itself as a flattening of the galaxy. For each speed of rotation there is an equilibrium configuration such that the galaxy is a flattened ellipsoid with the small axis corresponding to the axis of rotation. This was thought to be the reason for the shape of elliptical galaxies.

The rotation speeds of galaxies can be determined by spectroscopic studies. These galaxies are so far away that we are unable to see the individual stars: the smallest zone that it is possible to study comprises a minimum of several million stars. All these stars are found in different orbits and thus have different speeds. If one takes a spectrum of such a zone, one in fact obtains a superposition of each individual stars' spectrum. The spectra of stars are characterised by a certain number of lines. Because of the Doppler effect, the lines from different stars will no longer be coincident, which produces, in the integrated spectrum of all the stars, a broadening of lines compared to those of individual stellar spectra.

Measurement of the width of the lines thus provides a determination of the dispersion of the velocities of the stars. If the entire galaxy is moving in bulk, for example by rotation, the shift of the spectral lines of all the stars would, in this case, be the same, and would be the same for all stars in the integrated spectrum. Thanks to spectroscopy, we are thus able to measure the velocity dispersion of stars in the galaxies, and the speed of rotation of the galaxies themselves.

These measurements are nevertheless extremely difficult and could not be carried out until the appearance, at the beginning of the 1970s, of new very sensitive detectors. To general surprise, the first sufficiently precise measurements of the speed of rotation and velocity dispersion showed that these galaxies do not turn fast enough for

Galaxy NGC 4472. This elliptical galaxy, about 15 megaparsecs distant, is the largest galaxy in the Virgo cluster. With a diameter of 50 kiloparsecs, and a mass of 10^{12} solar masses, it is twice the size of our Galaxy and ten times more massive. This photograph shows how light is distributed in these elliptical galaxies. The surface brightness decreases from the centre towards the edge. Several more or less empirical laws have been proposed to reproduce the luminosity profiles. A good approximation, far from the centre and the edge, says that the logarithm of the surface brightness decreases as the radius to the power of 1/4. This law, known as de Vaucouleurs' law, applies to nearly all the known elliptical galaxies. Two regions pose problems: the centre and the edge. When one observes a galaxy with a ground-based telescope the light emitted must pass through the Earth's atmosphere before reaching the instrument. This degrades the image which becomes a blur if the object is a point source. Thus it is difficult to know when we observe the centre of a galaxy whether the observed luminosity profile is due to the galaxy itself or whether it is due to atmospheric distortion. The 1986 launch by NASA of the Space Telescope will solve this problem.

Even on a dark night, the sky is a source of light. In particular, it scatters the light emitted by the Sun and the Moon. This emission of the night sky is measured with a certain amount of uncertainty due to the fluctuations of the sky emission. It is clear that one would not be able to detect an object which is fainter than this fluctuation. This is particularly troublesome with regard to elliptical galaxies. We manage to measure them down to a surface brightness which is only 1/100 of the sky's brightness; but beyond that, they disappear. Thus, the measured diameters of elliptical galaxies depend upon the actual observing conditions and so their real size is not known with any certainty. (Kitt Peak National Observatory)

Galaxy NGC 5128. This particular elliptical galaxy in the constellation of Centaurus is identified with a strong radio source. As the source is the strongest in the constellation, it is known as Centaurus A. The centre of the galaxy is occupied by a disk of dust, gas and young stars, and is an active region of star formation. The radio emission is probably linked with the presence of the disk, as all elliptical galaxies with a disk of dust are also radio sources. In general the axis of the radio source is perpendicular to the plane of the disk. NGC 5128 is the most extreme example of this type of galaxy. (D. F. Malin, Anglo-Australian Telescope Board)

their flattening to be accounted for by rotation, least of all if they had the shape of ellipsoids of revolution.

Confirmation that elliptical galaxies had complex shapes was provided by photometric studies. Analysis of surface brightness at each point of the galaxy allows one to trace the lines of equal brightness, or isophotes, which are in general ellipses. But, for certain galaxies, one observes a progressive twisting of these isophotes in proportion to the distance from the centre. Such a twisting is clearly impossible to explain if the galaxy has the shape of an ellipsoid of revolution. The question regarding the shape of an elliptical galaxy is thus neither paradoxical nor stupid: elliptical galaxies do not have a simple shape, they are not ellipsoids of revolution, but rather ellipsoids with three unequal axes.

Most elliptical galaxies do not show signs of much activity. Their stellar populations are dominated by old, cold stars, of the red giant type; contrary to what one observes in our Galaxy, these stars seem to be rich in heavy elements. The absence of young stars and the high metallicity of old stars shows that the process of star formation, chemical enrichment and dynamic evolution toward an equilibrium configuration occurred much more rapidly than in spiral galaxies. Models enable us to estimate the characteristic time of this evolution to be several hundred million years, which is very short compared to the age of the Universe.

Elliptical galaxies were probably formed at the beginning of the history of the Universe, since when they have aged without undergoing any major changes. However, some of them still show, or are just starting to show again, signs of activity. During the 1970s progress made in radio astronomy enabled the presence of neutral hydrogen to be detected in some elliptical galaxies. Others contain a large quantity of dust which collects together to form a disk; very often, these elliptical galaxies with dust are powerful radio sources. Meanwhile the origins of these phenomena remain much debated.

Elliptical galaxies form a very homogeneous family which may possibly be described, to a first approximation, by a single parameter, absolute luminosity. Very strong correlations exist between the luminosity and the other characteristic parameters of these galaxies particularly the metallicity and the velocity dispersion, which, as we have seen, is an indication of the mass of the galaxy. The correlation between luminosity and velocity dispersion shows that all elliptical galaxies have largely the same mass to light ratio, and this indicates that they are made up of the same type of stars.

This uniformity is altogether remarkable. We know of elliptical galaxies with masses as low as a hundred million solar masses ($10^8 \, M_\odot$) and others which are a hundred thousand times more massive ($10^{13} \, M_\odot$). On the other hand, the metallicity is an increasing function of the luminosity (or of the mass). The process of enrichment of these galaxies depends on their mass. Several explanations of this have been proposed, but more detailed models and more precise observations are still necessary before we shall have an exact idea of the evolution of these galaxies.

Laurent VIGROUX

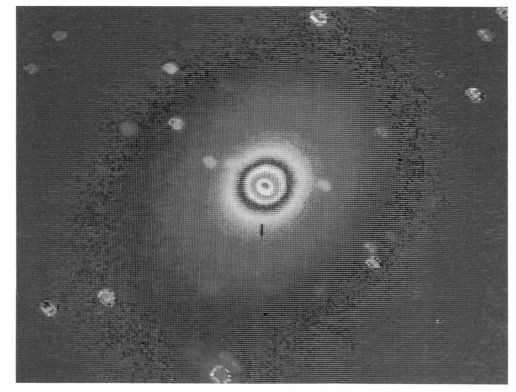

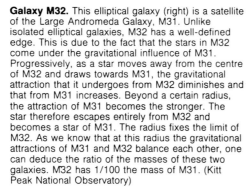

Galaxy NGC 4636. This large galaxy (left) is a member of the Virgo Cluster. The false colours in this picture correspond to different levels of surface brightness. The least luminous regions are blue, and the most luminous are red. As in a normal elliptical galaxy, the lines of equal brightness (isophotes) represented here by lines of the same colour, are ellipses, which rotate going from the centre to the edge of the galaxy. The major axis of the central brightest isophotes corresponds to the minor axis of the outer fainter isophotes. (L. Vigroux)

Galaxy M32. This elliptical galaxy (right) is a satellite of the Large Andromeda Galaxy, M31. Unlike isolated elliptical galaxies, M32 has a well-defined edge. This is due to the fact that the stars in M32 come under the gravitational influence of M31. Progressively, as a star moves away from the centre of M32 and draws towards M31, the gravitational attraction that it undergoes from M32 diminishes and that from M31 increases. Beyond a certain radius, the attraction of M31 becomes the stronger. The star therefore escapes entirely from M32 and becomes a star of M31. The radius fixes the limit of M32. As we know that at this radius the gravitational attractions of M31 and M32 balance each other, one can deduce the ratio of the masses of these two galaxies. M32 has 1/100 the mass of M31. (Kitt Peak National Observatory)

Galaxy NGC 1316. This large elliptical galaxy (below) is found at the centre of the Fornax Cluster of galaxies. It is also a very intense radio source. It is surrounded by a large, very diffuse, shell. The origin and nature of this shell are still a matter of controversy. It might comprise either stars which are escaping from the galaxy, or a gas ionised by relativistic electrons which are being emitted by the galaxy. In the first hypothesis, this type of shell should often be observed in elliptical galaxies. In the second, however, they would only exist around galaxies which are radio sources (Royal Observatory, Edinburgh)

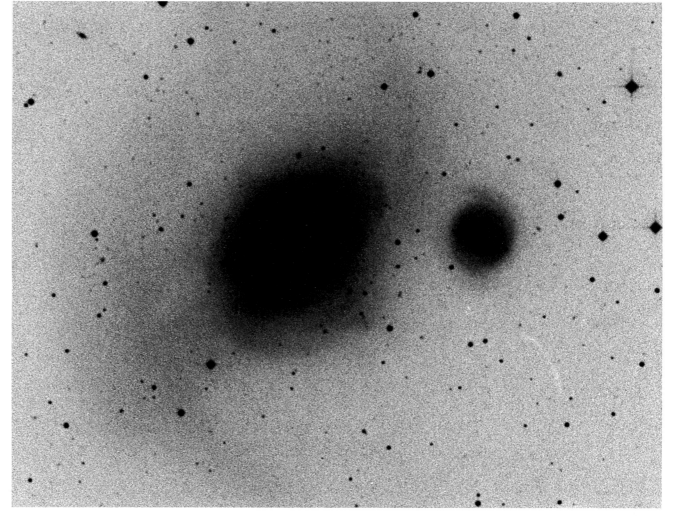

Lenticular galaxies

Lenticular galaxies, often called S0, are those which pose the most problems for astronomers. Their properties are intermediate between those of elliptical and those of spiral galaxies. It now seems to be accepted that they form a separate class and that they are not spiral galaxies changing into elliptical galaxies, or vice versa. However, the evolutionary process which gives birth to lenticular galaxies is still the subject of heated debate and the most contradictory hypotheses are advanced.

Like spirals, lenticulars have a bulge and a disk. The size of the bulge in relation to the disk is greater than in the case of spirals: a spiral disk is roughly ten times bigger than its bulge, whereas the sizes of lenticular bulges and disks are of the same order of magnitude. An S0 disk does not possess any obvious structure, although some do have large bands of dust particles, and one can, in certain cases and with a little imagination, reconstruct irregular faint spiral arms there. The gaseous content of lenticulars is very varied, the richest in gas contain as much as spirals, whereas others, like ellipticals, contain no gas at all; about a third of S0s contain gas. S0 disks, like those of spirals, rotate differentially; the patterns of rotation in the two types are identical, and the masses of lenticular galaxies deduced from their rotation curves are similar to those for spirals.

Morphologically, lenticular galaxies thus resemble spiral galaxies in which the spiral arms have been erased. However, their stellar content brings them closer to elliptical galaxies. Photometric measurements of a large number of S0 galaxies showed that their colours were entirely analogous with those of elliptical galaxies. Light from these galaxies is therefore emitted by the same type of star, old and cold. No trace at all of any recent star formation is observed; they do not possess any H II regions, the regions which, in lenticular galaxies, surround only young, hot stars. The colours, according to numerical models of their evolution, show that star formation stopped at least five billion years ago. This is strikingly different from spiral galaxies where one still sees stars forming.

Why has star formation stopped in S0 galaxies? The answer is probably to be found in the environment of the galaxies. In clusters of galaxies, one finds all types (ellipticals, spirals, lenticulars, etc.) but their relative properties appear to obey certain laws. In particular one can show that the proportion of lenticulars compared to spirals is a function of the total density of galaxies: the higher the local density of galaxies in a cluster, the greater the number of lenticulars found there in relation to the number of spirals. It seems therefore that if a galaxy forms in a dense environment, it is more likely to become an S0 galaxy; if it forms in a more isolated region, it more often becomes a spiral. Several hypotheses have been advanced to explain this. We will only put one forward here, the one that explains the formation of lenticulars by the loss of their interstellar gas. X-ray observations have shown that clusters of galaxies are filled by a very hot intergalactic gas (its temperature is several million degrees) having a low density (a few hundred particles per cubic metre). The galaxies are moving quickly in this gas, which exerts a dynamical force on them, analogous to air resistance.

Although the intergalactic gas has a low density, the movement of galaxies is sufficiently rapid – a few thousand kilometres per second – for the dynamical pressure to be very strong. This pressure is in fact exerted on the interstellar gas in the galaxy; in certain cases, it is sufficiently strong to push the interstellar gas outside the galaxy. Deprived of the gas, the galaxy is no longer able to form stars and it slowly burns out; its spiral arms, which are due to young bright stars, then disappear. One is left with a galaxy with a disk but without spiral structure, that is to say a lenticular galaxy.

However, this explanation is somewhat simplistic. Analysis of the colours shows that this transformation must have taken place more than five billion years ago, and we do not really know what spiral galaxies looked like at that time. Whatever they may have been like, from the dynamic pressure or other effects, like for example tidal effects between galaxies, the environment certainly plays a role in the formation of lenticular galaxies.

Laurent VIGROUX

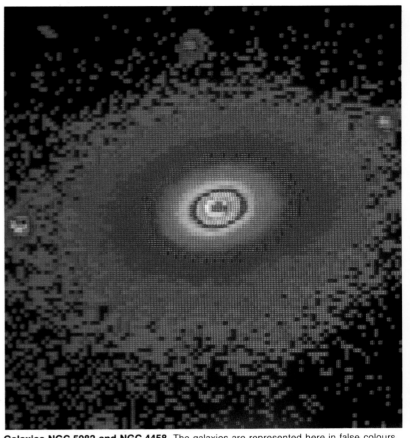

Galaxies NGC 5982 and NGC 4458. The galaxies are represented here in false colours. The lowest luminosities are in blue and the brightest in red. The galaxy NGC 5982 (left) is a rather elongated elliptical galaxy, of type E5. The Galaxy NGC 4458 (right) is a lenticular galaxy. A comparison of the two images allow one to understand the difference between these two types of galaxy. In the former, the lines of equal brilliance, or isophotes (here, zones of the same colour), are quite regular ellipses. A detailed study shows that the

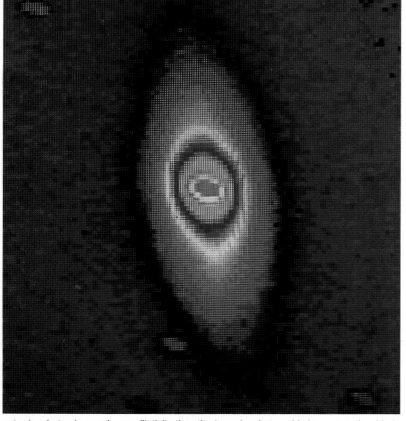

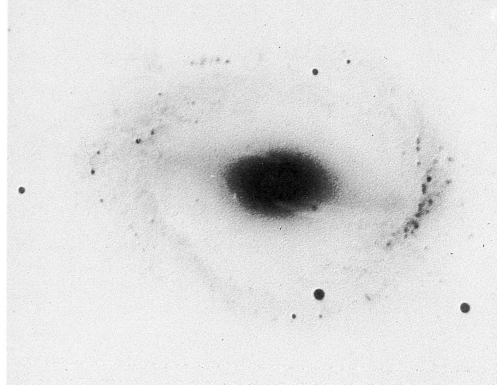

outer isophotes have a lower ellipticity than the inner isophotes; this is more noticeable in the picture of the lenticular galaxy, where the central isophotes are very different from the outer isophotes. The light at the centre is distributed in an almost circular manner, while the isophotes at the edges are extremely elongated ellipses. This variation shows the existence of two components in the lenticular galaxy: a central spheroidal bulge surrounded by a more flattened disk. (L. Vigroux)

Galaxy NGC 1512. This galaxy in the southern hemisphere is a prototype of the class of galaxies called Theta galaxies because of their resemblance to the Greek letter of that name. Note that this is a negative photograph. Theta galaxies are intermediate between barred spirals and lenticular galaxies. The central bar of this galaxy appears to have a double structure: an elliptical bulge, and a disk which is thinner and extended. As accurate photometric studies of this galaxy have not yet been carried out, it is difficult to determine the exact nature of these two components. The bulge contains absorbing bands of dust particles which appear as white bands, and therefore possesses a fairly young stellar population, linked with the presence of the dust.

The bar ends in two coiled spiral arms which close up and give the illusion of forming a ring. Whereas the bar has a morphology similar to lenticular galaxies, the arms, containing H II regions, young stars and bands of dust, resemble those of spiral galaxies.

Theta galaxies have still not been studied sufficiently to give us an accurate idea of their evolution. (ESO)

Galaxy NGC 5102. This southern hemisphere galaxy (left) is typical of lenticular or S0 galaxies. It is difficult to estimate the distance of these objects as they do not contain any standard size objects like the H II regions in spiral galaxies. Nonetheless, the distance of NGC 5102, is estimated as 4 megaparsecs because it belongs to a small group of galaxies of which the most massive is a spiral galaxy, NGC 5128, whose distance is known. This picture clearly shows the existence of two components; the bulge and the disk. The bulge, the brightest part at the centre, resembles an elliptical galaxy. Its extent is 2.4 kiloparsecs (kpc) in its largest dimension and 1.1 kpc in the other. The disk itself is much more elongated with a diameter of about 10 kpc.

This galaxy contains a lot of gas for a lenticular galaxy. The mass of neutral hydrogen has been estimated at 3×10^8 solar masses. Relative to the luminosity of the galaxy, NGC 5102 contains almost as much gas as a normal spiral galaxy. Detailed studies have revealed the existence of bands of dust particles and the presence of some clusters of young stars in the disk.

NGC 5102 thus greatly resembles a spiral galaxy in which the present rate of star formation is very low and where the spiral structure of the disk has disappeared. (D. F. Malin, Anglo-Australian Telescope Board, © 1977)

The galaxy NGC 4594. This galaxy is known as the Sombrero because of its shape. This makes it difficult to assign it a type. At first sight it might appear to belong among the spiral or lenticular galaxies. We can make out a spheroidal bulge containing evolved red stars (the centre is overexposed and appears white) and a thin disk whose blue colour is caused by the very hot yellow stars of which it is composed. The disk is crossed by a dark band revealing the presence of dust. We see this galaxy almost sideways on, and the inclination of the disk is too low for us to see any spiral structure. The ratio between the radii of the bulge and the disk (the latter about 12 kpc) is around 0.3, which would fit a spiral or lenticular structure equally well. This galaxy contains a large amount of gas, but again, some lenticulars such as NGC 5102 have similar amounts. In fact the only discriminating characteristic comes from the presence of H II regions in the disk (they appear as small reddish spots). These H II regions are characteristic of spiral galaxies and are not found in lenticulars. The Sombrero is probably a spiral galaxy, and the ratio of bulge to disk size suggests that it is type Sa.

This example illustrates the difficulties often encountered in classifying a galaxy. There is no obvious difference between a lenticular and a spiral galaxy seen from the side. Conversely, very flattened elliptical galaxies are often classified as lenticular simply because they have an elongated shape. In such cases only very precise photometric studies can decide whether or not a disk is present, and thus whether the galaxy is lenticular or elliptical. (S. Lautsen, ESO)

Galaxy NGC 3718. This very bizarre galaxy illustrates another real difficulty in the classification of galaxies. Some galaxies are classified as S0 simply because it is not known how else to classify them. NGC 3718 is an example. From its appearance, it cannot be a spiral or an elliptical; there is a dominant spheroidal component, barred across its centre by a disk of dust, which is extended by two embryonic arms on each side of the spheroid. The arms are twisted in relation to the plane of the dust disk. Such a galaxy is usually classified as an S0. In catalogues, as a precaution, this galaxy is noted as peculiar.

This inclusion of galaxies which cannot be classified otherwise means that S0 galaxies form a particularly heterogeneous ensemble. The origin of lenticular galaxies will perhaps be easier to understand when a thorough re-classification has been made which keeps only galaxies which really have a lenticular type structure. Using such a homogeneous sample, it will be easier to define the true characteristics of luminosity, colour, mass, etc., of this type of galaxy. (Kitt Peak National Observatory)

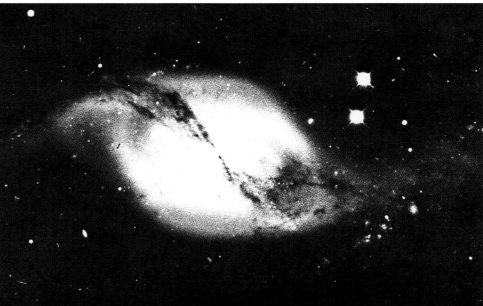

Spiral galaxies

To the amateur astronomer, spiral galaxies are among the most spectacular objects in the sky. The fascination that they exert over the observer, whether professional or amateur, derives from the magnificent spiral arms unfolding with remarkable symmetry. However, the spiral arms represent only a very small part, at the most a few per cent, of the mass of the galaxy. These arms are only visible because they are the active regions of the galaxies. It is here that the majority of stars form, and also where we find the brightest stars.

A spiral galaxy comprises two systems. One almost spherical part – the bulge – occupies the central area; the distance to which it extends depends on the type of galaxy. This bulge has certain analogies with elliptical galaxies; it has the same photometric profile and it seems to be made up of the same type of old stars. The other part is a disk, which contains in particular the spiral arms, and whose luminosity decreases exponentially from the centre towards the edges. This disk is younger than the bulge; it still contains a significant quantity of interstellar gas, which represents a few per cent of its mass, and much activity prevails there: from the standpoint of star formation and chemical enrichment, its evolution is far from finished.

Detailed studies have shown that the formation of these two components must have taken place in two stages. Initially the bulge formed rapidly in a process analogous to that which gave birth to elliptical galaxies. On the other hand, star formation developed more slowly than in elliptical galaxies. Once the bulge was formed, a large quantity of gas remained; this fell into a disk where new stars then formed and where the spiral arms appeared. This picture enables us to explain the general morphological properties of spiral galaxies; moreover, it explains the differences found between the abundances of elements in the bulge and in the disk. Being formed first, the bulge is less rich in heavy elements than the disk, which formed from the already-enriched matter in the bulge. However, this hypothesis does not provide an answer to the question: how are spiral arms formed?

The preliminary to this question is: what are the spiral arms made of? The simplest idea is to say that the stars are arranged in a fixed pattern in these arms. This hypothesis is in fact incompatible with the mode of rotation of disks of spiral galaxies. Like the disk of our Galaxy, these are propelled by differential rotation: that is to say, the angular speed of rotation of a star about the centre of the galaxy varies with the distance of the star from the centre; the further away it is, the lower the angular speed of rotation. If we suppose that all of the stars were initially aligned, the differential rotation would cause the stars near to the centre to turn faster than those further out. This would then produce a winding up effect leading to the formation of a spiral. We can estimate how many turns we would expect for windings due to differential rotation by dividing the age of the disk, approximately ten billion years, by the rotation period of a galaxy, about one hundred million years. Thus windings due to differential rotation ought to consist of something like a hundred turns. However, the spiral arms we observe generally have only one turn, or at most, two. Their existence cannot, therefore, be explained by differential rotation and, conversely, any initial structure in the distribution of stars will be destroyed by differential rotation. This analysis leads to the conclusion that spiral arms are a structural pattern through which stars pass.

One of the first theories proposed to explain spiral arms was that of density waves. A second theory rests on the mechanism of star formation. When stars form, the most massive among them eventually explode: this is the phenomenon of a supernova. When this happens, shock waves propagate through the neighbouring interstellar medium, which can cause gravitational instabili-ties in the interstellar gas and give birth to new stars; this new generation of stars, situated at a certain distance from the first can also form supernovae and thus, step by step, by a sort of contagion, the process of star formation prop-agates, not in a straight line but, due to the differential rotation, along a spiral. Computer simulations based on this second theory repro-duce quite accurately the appearance of spiral arms in galaxies. The presence of young stars in the arms is also explained by this process. The simulations predict relatively broad spiral arms, as observed in most spiral galaxies of the late types, Sc and Sd. In contrast, other galaxies, and in particular barred spirals, have much thinner arms which are best explained by the density wave theory. Both mechanisms certainly co-exist in galaxies, their relative importance explaining properties of the spiral arms such as thickness or thinness, regularity or segmentation.

The density wave theory is one of the big successes of recent astronomy. The idea of applying this theory, already utilised in other areas (hydrodynamics for example), to spiral galaxies, came to the astronomers C. C. Lin and F. Shu in the 1960s. In order to understand what a density wave is, consider the circulation of cars on a two-lane highway. The cars are driving along at about the same speed, say 110 km per hour. A truck driver travelling at 60 km per hour will cause a bottleneck behind him since he forces the faster cars to drive in the single fast lane. The congested traffic travels along with the speed of the lorry, but it is not a solid structure in the sense that one never finds the same cars there. A car can only pass in the bottleneck. At night, an observer in an aeroplane would see a diffuse illumination from the car headlights along the highway. The bottleneck would appear as a more luminous zone because of the higher density of cars there. This bright spot travels with a velocity of 60 km per hour, the speed of the truck, and not at the speed of the cars far from the bottleneck. The congestion is a stable perturbation in the stream of cars which exists as long as the truck is on the highway. According to Lin and Shu, this is analogous to what we observe in spiral galaxies. The role of the truck is played by a perturbation in the gravitational field and that of the cars by the rotating stars. When they cross the perturba-tion – the wave – their motion is slowed down and there is a local increase in density and the appearance of a bright spot. The spiral shape of the wave can explain the spiral arms in galaxies.

This wave also perturbates the motion of gas. According to calculations, this is much more affected than the stars; the contrast between the gas density in and between the arms may be very important. From maps made from radiation emitted by neutral hydrogen in the radio wave-band at a wavelength of 21 centimetres one can obtain information about the distribution of gas in galaxies. This contrast can be strong enough for a shock wave to be produced on the edge of the arm. This shock wave is characterised by a rapid pressure increase and a fall in the velocity. The local increase in density leads to the formation of molecular clouds. These same clouds will give birth to stars by gravitational collapse. The localisation of young O and B stars and regions of ionised gas with which they are associated will therefore be due, according to this theory, to the compression of interstellar gas at the time of its passing through the density wave. We have yet to understand the origin and the evolution of the density wave itself.

Historically, the stability of these waves has been the main theoretical problem. As with all waves, density waves decay with time: if we do not maintain the oscillations of a spring, at the end of a certain time it stops vibrating. In the same way, calculations have shown that without an energy source, density waves must decay in about one billion years. Since the average age of

Text continued on page 361

NGC 1365. This magnificent barred spiral is a member of the Fornax cluster. It is classified SBb. The bar is very well defined although huge dust bands give it a slightly irregular appearance. The two spiral arms running from the ends of the bar are particularly thin and well defined. They contain a large number of regions of ionised gas (the spots). The contrast between the arms and the disk is greater than for normal spirals. (D. F. Malin, Anglo-Australian Telescope Board)

NGC 4565. This galaxy, located in the Virgo cluster, is a good example of a spiral galaxy viewed edge-on. The two components of the galaxy are easily distinguished, the disk, and the spherical bulge at the centre of the galaxy. Although the spiral structure cannot be seen because of the inclination of the disk, it is obvious that the light is less uniformly distributed in the disk than in the bulge. Scattered throughout the disk are dark spots indicating the existence of dust, and bright points which are H II regions. The distribution of light within the bulge resembles that found in elliptical galaxies, with the surface brightness following the law of de Vaucouleurs. In the disk, the surface brightness decreases exponentially from the centre of the galaxy towards the edge. The disk ends abruptly; in contrast to elliptical galaxies, for which it is impossible to define an edge, the edges of the disk in spiral galaxies are easily seen. Perpendicular to the plane of the disk, things are less clear. Recent studies have shown that, even very far out from the centre of the galaxy and well outside the bulge, there are two stellar components. (KPNO)

NGC 2997. The colours in this photograph (above) are the real colours of the galaxy. A picture like this brings out the various components. The centre is yellow. The light there is dominated by Population II stars, i.e. old main sequence stars and red giants. This population is completely analogous to the old population in our Galaxy. The yellow central bulge is relatively small compared to the disk and the spiral arms, both of which start very near the centre of this galaxy. In the bulge are two darker bands due to absorption by dust in the arms. Further out the arms become bluer and are dotted with red spots. The blue colour is caused by the presence of many very hot young stars. The red spots are H II regions, regions of gas ionised by the ultraviolet radiation from hot stars, where the emission is dominated by the Hα radiation from hydrogen which is in the red. Most of the young stars are in the spiral arms, which are regions of enhanced star formation. Note that H II regions are often found on one side of an arm. This is particularly noticeable for the arm near the top of the photograph. The H II regions found there are where the youngest stars are. There seems, therefore, to be an age segregation in the direction perpendicular to the arm. In other words, stars only form on one side of the arm. This observation is naturally explained by the density wave theory. Further out, the situation becomes more chaotic. H II regions seem to be distributed more or less randomly; the arms are not so well delineated and have a more violet colour. Star formation is far less localised than in the central regions, and seems to be due mostly to the contagious type of mechanism. In the same galaxy, the mechanisms of star formation can vary from the centre to the edge. (D. F. Malin, Anglo-Australian Telescope Board)

M83 is a large spiral galaxy in the southern hemisphere. It belongs to the Centaurus group and its distance is 3.7 megaparsecs; it is one of the closest galaxies to us. The arms are tightly wound and rather thick. There is a small bar in the centre. This galaxy represents a category intermediate between normal spirals like NGC 2997 and barred spirals. The colours in this photograph are the real colours of the galaxy. It is on the whole more red than NGC 2997. This is explained by a population which is, on average, older. (D. F. Malin, Anglo-Australian Telescope Board)

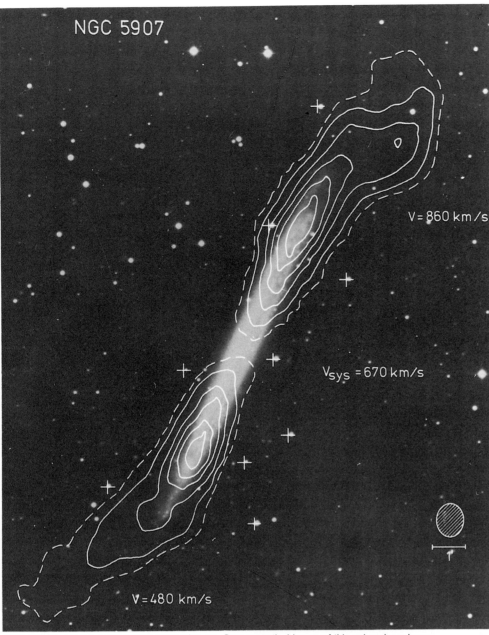

NGC 5907

V = 860 km/s

V$_{sys}$ = 670 km/s

V = 480 km/s

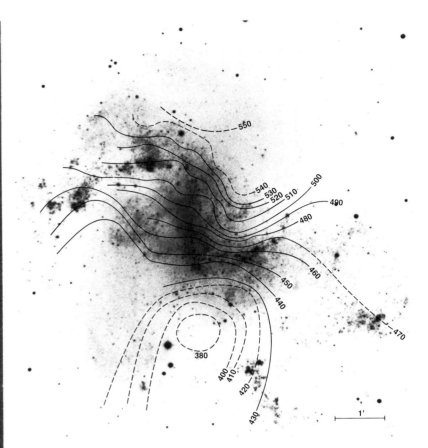

NGC 1313. This galaxy is a barred spiral of the type SBd. Its arms are clearly less pronounced than for SBb spirals. Superimposed on the illustration of this galaxy are lines of equal velocity. These lines have been obtained by measurements of the Doppler shift of the Hα line of ionised hydrogen using the technique of Fabry-Perot interferometry. This method allows us to study the velocities of regions of ionised gas, H II regions. Each constant-velocity line is labelled with its speed in kilometres per second. The velocity of the galactic centre is 465 kilometres per second. The velocity differences between the north and south are due to rotation. Contrary to what one might naïvely assume, the centre of rotation does not coincide with the nucleus which is situated in the centre of the bar, but is located at its southern extremity. The bar rotates around this point, and turns as a solid body as if it were a rigid structure pivoting about an axis. This mode of rotation is totally different from that which we observe in the disks of normal spiral galaxies, which exhibit differential rotation. A large number of bars possess a similar solid body rotation, but this is not the general rule. In other cases, there are gas motions along the bar directed towards the ends. Generally, the velocity field in barred galaxies is more complex than is found in non-barred galaxies. Especially if it is large, a rotating bar causes perturbations of the gravitational field which become apparent as non-circular motions. Numerical simulations have demonstrated that the presence of a bar should increase the velocities perpendicular to the plane of the disk. This leads to a thickening of the disk in barred galaxies relative to non-barred galaxies. (M. Marcelin, Observatoire de Marseille)

NGC 5907 is a large spiral galaxy seen edge-on. Onto an optical image of this galaxy have been superimposed contours of equal intensity obtained from measurements of 21-centimetre emission from neutral hydrogen. These contours are therefore contours of equal gaseous surface density in this galaxy. While the gaseous and luminous (i.e. stellar) disks obviously coincide, the gaseous disk is much more extended. Beyond the stellar disk, it deforms rapidly and symmetrically with respect to the plane of the disk. Similar distortions in the optical disk are observed in some galaxies. Several explanations have been proposed to explain these deformations. One is that there is an interaction with an intergalactic gas having a high enough pressure to distort the external regions of the galaxy where the density is very low. Another explanation may be a tidal interaction with a small satellite galaxy.

Also marked on this photograph are the apparent velocities of the two extremities and the centre of this galaxy. These velocities are interpreted as being the combination of two movements. The whole galaxy is receding from us with a velocity of 670 kilometres per second while simultaneously rotating about its centre. This rotation manifests itself as the difference between the velocities at the two extremities. The galaxy is turning about an axis perpendicular to the plane of the disk. Due to the rotation, the base of the galaxy seems to approach us and the top recedes. The speed of rotation at the extremities is therefore simply the difference between the velocity at the extremities and that at the centre, being 190 kilometres per second. This speed corresponds to one rotation every 300 million years. (R. Sancisi, Westerbork)

Surface density and velocity of the hydrogen gas in M81. The surface density and the velocity of the gas are represented simultaneously in this map of the spiral galaxy M81, of type Sab. The measurements were made with the Dutch Westerbork radio telescope. At any point in the map, the intensity corresponds to the density of the gas and the colour to the velocity, from red through yellow to blue.

The surface density of gas is deduced from measurements of the intensity of 21-centimetre neutral hydrogen emission. The spiral structure is clearly visible. The gas is concentrated in the spiral arms. The shape of the arms is very similar for the neutral gas, young stars and the regions of ionised gas. Perhaps most unexpected is the existence of a hole in the hydrogen at the centre. The gas seems to be concentrated in a ring around the centre, and the surface density of gas diminishes towards the centre, contrary to what we see in optical emission. This hole results from two effects: firstly, a higher rate of star formation which tends to consume the gas in this region, and secondly, for dynamic reasons connected with the viscosity of the gas, there is a tendency for the gas to migrate towards the outer regions and not to remain in the centre. Holes like these seem to exist in the majority of spiral galaxies.

The distribution of velocities in this galaxy is derived by measuring the Doppler shift of the 21-centimetre line of neutral hydrogen i.e. the shift in wavelength of the line owing to the fact that the hydrogen is moving away from us. Blue corresponds to a velocity of 240 kilometres per second, and red to 160 kilometres per second.

These measurements, and other similar measurements of other spiral galaxies, show that, roughly speaking, beyond a certain distance, the linear rotation velocity is constant in the disk. Dynamical models show that, to obtain such a flat rotation curve, the mass of the galaxy must be distributed in the disk with constant surface density. This result, definitively established towards the end of the 1970s, is very surprising. As we have already stated, the surface brightness of a disk decreases exponentially with the distance from the centre. If the mass remains constant, this means that the mass to light ratio of stars varies with radius. This variation is too great to be explicable if the stars in the disk are normal stars analogous to those we find in the disk of our Galaxy. The mass surplus is probably to be found as very low luminosity stars. At present, we believe that these stars are the remainder of first generations of stars formed in the galaxies and distributed in a massive, very low luminosity halo surrounding it. This halo would in fact contain the major fraction of the galactic mass. Numerous studies currently underway may confirm or otherwise this hypothesis. This problem, known as the missing mass problem, remains one of the major problems in the evolution of galaxies. (Westerbork)

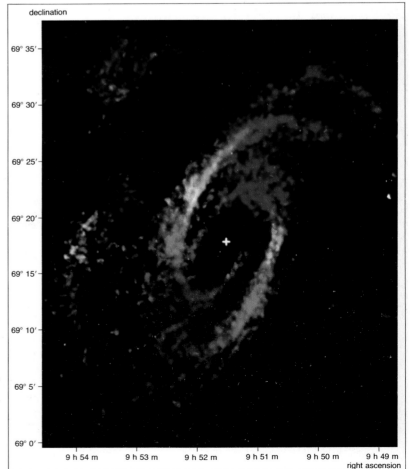

The Whirlpool Galaxy. These interacting galaxies lie in the constellation of Canes Venatici. The larger galaxy, NGC 5194, is known as the Whirlpool Galaxy. It has a magnificent coiled spiral structure, with one of its arms extending out to the galaxy NGC 5195. The latter is also a spiral galaxy, but its arms are clearly less developed; they are partially masked by large filaments which are due to young stars and ionised gas. Note the large dust lane occupying the interior of the arm linking the two galaxies. It is particularly noticeable on NGC 5195 where it carves a dark patch between the arm and the nucleus. The spiral structure of the larger galaxy is caused by its interaction with the smaller. A numerical simulation where two uniform disks of different sizes collide can produce a configuration analogous to that observed in this system. (J. D. Wray, McDonald Observatory)

spiral galaxies is about ten billion years, there must be an 'engine' capable of perpetuating these density waves. The spiral galaxies with the most pronounced arms are in general either interacting galaxies like M51, or barred galaxies like NGC 1365. This observational connection has been confirmed by numerical simulations which have shown that the interaction between two galaxies and the presence of a bar produces strong density waves. But this does not explain why non-barred spirals exist, nor the origin of the bars themselves. Here again, the answer seems to have been supplied by numerical simulations. According to these calculations, a thin disk of stars rotating differentially is not a stable structure. When, in addition to their speed of rotation, the stars have a small component of random velocity, as is the case for stars in the disk of our Galaxy, these random velocities increase with time and de-stabilise the disk. The principal mode of instability leads to the formation of a density wave having the shape of a bar, and not a spiral shape. The disk, assumed to be uniform to start with, begins by acquiring a spiral structure due partly to the differential rotation. This spiral structure only lasts several galactic revolutions. It then disappears and gives way to a bar which, at its maximum intensity, obtained after about ten rotations, can reach 40 per cent of the total mass of the disk. Subsequently, the bar partially disperses and becomes stabilised at a level representing 25 per cent of the mass. The speed at which this sequence progresses depends strongly on the mass of the spherical component, bulge or halo that added to the disk re-creates the true conditions in spiral galaxies. A massive halo acts as a stabiliser. It lowers the progressive increase in random velocities. The more massive the halo, the more stable the disk and the more slowly the sequence of events just described will take place. The time that we have indicated corresponds to a halo having the same mass as the disk. If the halo mass is ten times larger than the disk mass, the evolution will occur three to four times slower. The existence of massive haloes therefore explains the persistence of the spiral structure in non-barred galaxies. According to these ideas, the transition from non-barred to barred spiral galaxies should therefore be due to a variation of the ratio of the mass of the halo to that of the disk. Galaxies with very massive haloes would evolve slowly and would still be in the intermediate phase of a normal spiral. Galaxies having a low mass halo would have developed a large enough bar to generate a wave of spiral density affecting especially the gaseous content of these galaxies. These would become barred spiral galaxies.

Massive haloes allow us to reproduce the morphology of spiral galaxies and to explain their rotation curves. But, at present, they remain hypothetical. They have escaped all attempts to detect them so far. If they exist, they must be made up of very low luminosity stars.

Laurent VIGROUX

NGC 1300 is an SBb galaxy. Its arms are more coiled than those of NGC 1365 on page 340, and the bar is less obscured by dust. However, we see two dark lanes, one from each side of the nucleus, extending into the arms. These dust lanes embody zones where gas is compressed by shock waves, due to the gas passing through density waves. The shock waves start on the edges of the nucleus and continue into the arms. The dust lanes and zones of gas compression lie on the inside edges of the arms; this is particularly visible for the left arm. A naïve explanation of this galaxy is that its structure results from its clockwise rotation, which would leave behind trails which have the appearance of these spiral arms. Stirring a cup of coffee slowly easily gives an analogous structure. However, if this were the true explanation, the compression zones would be found on the outsides of the arms and not on the inside as observed. In fact, contrary to intuition, there is a rotation of the gas in an anticlockwise sense with respect to the density waves, which form a shock wave on the inside edge. (Hale Observatories)

Irregular galaxies

Some galaxies do not have a well-defined structure. Amongst these two groups have been particularly well studied: Magellanic irregular galaxies and blue compact galaxies.

When an astronomer speaks of irregular galaxies, he thinks in fact of the Magellanic type of irregular galaxy, the adjective referring to the Magellanic Clouds, particularly the Large Cloud, typical of these galaxies. Like spirals, these objects have a bulge, although much smaller, and a disk which evidently possesses no spiral structure. In fact, one often clearly sees a bar in the disk, similar to the one observed in the Large Cloud. These galaxies generally have rather low masses – between one and ten billion solar masses, compared with one hundred billion solar masses for a typical spiral galaxy – and they are relatively rich in gas, which accounts for 10 to 20 per cent of their mass. Magellanic irregular galaxies therefore resemble spiral galaxies, but are smaller and less evolved; for that reason they are classed next to spirals. There is however a continuity between spiral galaxies and irregular galaxies.

In the 1950s the Swiss–American astronomer Fritz Zwicky discovered another class of irregular galaxies, blue compact galaxies, which resemble isolated H II regions: if the Tarantula Nebula, the large complex of H II regions and young stars in the Large Magellanic Cloud, was separated from the Large Cloud, it would be a blue compact galaxy. These galaxies generally have much smaller dimensions than irregular galaxies, and their masses are of the order of several hundred million solar masses. They are particularly active regions of star formation; this activity is evident from the large quantity of ionised gas that they contain. We can calculate the number of ultra-violet photons necessary to ionise the gas and from this derive the mass of young stars, as the strength of ionising ultraviolet flux from very hot stars is known. It is found that the total mass of young stars needed for the ionisation is approximately equal to the mass of the galaxy. The rate of star formation is therefore very high. One may draw a first conclusion: blue compact galaxies have not always had this high rate of star formation, since they do not contain enough matter to form stars at this rate for more than a few hundred million years. Two hypotheses may be considered: these galaxies are either young galaxies in the process of forming their stars, or galaxies of normal age in which the star formation occurred in successive bursts. In the first case, they would be the only examples known of young galaxies; in the second case, we would be witnessing a burst of star formation. Numerous studies are underway to investigate these ideas. Whatever the case, young or not, blue compact galaxies are, like the irregular Magellanic galaxies, not highly evolved. As regards chemical composition, these two types of galaxy strongly resemble each other: their heavy element abundance is low compared to that in our Galaxy; they contain for example between three and twenty times less oxygen. We can appreciate their relevance to understanding the chemical evolution of galaxies. When we study our Galaxy, we observe, in some sense, a finished product and it is therefore difficult to reconstruct its evolution. In irregular galaxies and in blue compact galaxies, however, we see intermediate stages, easier to understand. Also, these galaxies contain numerous H II regions of ionised gas whose abundances we can determine relatively easily by measuring the intensity of their emission lines. Much work has been undertaken to study the abundances of the elements in these galaxies and to relate these to other more notable characteristics such as mass, luminosity, gaseous content, and among the most engaging results that have already been obtained, is the knowledge of the rate of enrichment in heavy elements of the interstellar matter by stars. But, without doubt, the most interesting result has been a determination of the primordial helium abundance. We believe that helium,

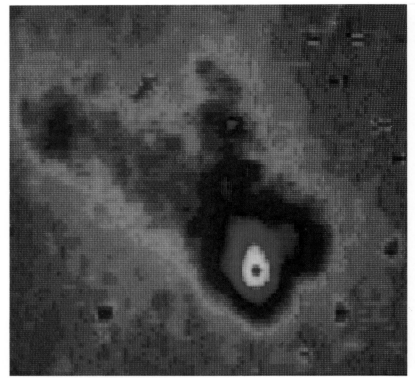

The blue compact galaxy II ZW 40. This galaxy is about 7 megaparsecs distant. It is very small. Its largest dimension is about 1 kiloparsec, i.e. it is thirty times smaller than our Galaxy. This illustration represents, in false colours, the distribution of surface brightness. The outer zones in dark green are the least luminous, and the brightest central part is represented in yellow and blue. This galaxy appears to be particularly contorted. The roughly circular nucleus is surrounded by two large extensions. Each extension itself is subdivided into a number of condensations. Dynamic analyses have shown that each extension has a mass of about 4×10^7 solar masses ($M_\odot$) while the mass of the nucleus is only 2×10^7 $M_\odot$. In spite of this, the nucleus is approximately forty times brighter than the two extensions. The mass to light ratios are thus very different in the extensions and the nucleus. This is due to distinct stellar populations. The nucleus is the centre of an enormous amount of stellar activity. The great majority of the stars here are young and very luminous. In contrast, there is little star formation in the extensions and the stars there are less luminous.

The gas in the nucleus is ionised by the ultraviolet flux from its stars. This nucleus therefore appears like a large H II region of about 500 parsecs in size. Spectroscopic analysis of this gas has shown that the abundance of heavy elements, such as nitrogen or oxygen, is about ten times lower than in our Galaxy.

The mass of the gas contained in this galaxy is estimated at 10^8 $M_\odot$. The mass of young stars formed in the nucleus, 2×10^7 $M_\odot$, therefore represents one fifth of the gas available for forming stars. The oldest stars are about ten million years old. If star formation continued in this galaxy at its present rate, all the remaining gas would be used up in fifty million years. If we recall that the age of these galaxies is normally fifteen to twenty billion years, we see that the rate of star formation in II ZW 40 is too high to be anything other than a transitory period in the history of this galaxy. It is not known whether this high level of activity is happening for the first time, or if it is a periodic phenomenon which has already taken place several times. (L. Vigroux)

The galaxy IC 5152. This object, situated in the southern constellation of Indus, shows the continuity betweeen the irregular Magellanic galaxies and the spirals. Some observers place IC 5152 in the first of these classes (1m); but this galaxy seems to have a small bulge, and is sometimes classified among type Sd spirals.

IC 5152 is close enough for its brightest stars to be resolved (but note that the very luminous blue star at the upper right of the photograph belongs to our Galaxy: this is HD 209142, with apparent magnitude 8). The distance to IC 5152 can be estimated by two methods: first by assuming that its brightest stars have the same intrinsic luminosity as their counterparts in the Milky Way (measurement of their apparent luminosity then gives their distances); second by studying its Cepheid variables (these stars are valuable distance indicators). IC 5152 lies between 1.5 and 4.5 Mpc away, i.e. a little outside the Local Group. (D.F. Malin, Anglo-Australian Telescope Board)

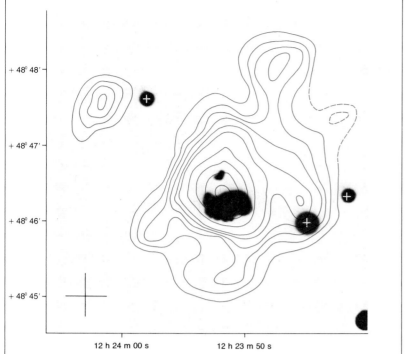

The blue compact galaxy I ZW 36. This galaxy is completely analogous to II ZW 40. It has an irregular shape and possesses a small growth towards the north. The four circular spots are stars in our Galaxy and have nothing to do with I ZW 36. A map of the distribution of neutral hydrogen deduced from radio observations of the 21-centimetre line of hydrogen has been superposed on the optical image. The radio extension is not aligned with the optical extension. The isolated cloud situated north-east of the galaxy at top left does not have an optical counterpart. Apart from this cloud, the distribution of gas is fairly inhomogeneous, with several peripheral condensations. The central condensation does not coincide exactly with the active nucleus of the galaxy.

Studies of the dynamics of the gas seem to show that the mass of the galaxy is much larger than would be deduced from its optical appearance alone. Its 'visible' mass such as gas accounts for only about 10 per cent of its total mass. Several hypotheses have been put forward on the nature of this hidden mass such as interstellar molecules, dust or old stars. So far no definitive solution to this problem has been found. (F. Viallefond, T. X. Thuan)

The irregular Magellanic galaxy NGC 55. This galaxy belongs to the Sculptor group, the closest group of galaxies to the Local Group. Its structure is similar to that of the Large Magellanic Cloud, but it is seen from the side, while we see the Large Magellanic Cloud almost face-on. NGC 55 has a bright reddish region (this colour is mainly produced by cool evolved stars) which is a kind of flattened bulge or smeared-out bar, and a diffuse bluish disk (this colour is caused by very hot young stars and gaseous nebulae). Note that, unlike in spiral galaxies, the bulge is not central, but at one end of the disk. A model of this galaxy shows that its centre of mass does not coincide with the centre of the bulge. This should be compared with the bar of the Large Magellanic Cloud which is displaced from the rotation axis, which here does pass through the centre of mass. The disk has a complex structure. We observe large bright spots, which are H II regions, where the gas is ionised by the ultraviolet flux of hot stars, and also large dark regions. These are interstellar dust clouds, which absorb starlight. The distance to NGC 55 is of the order of 3 megaparsecs. (R.M. West, ESO)

unlike heavier elements, was produced principally at the time of the primordial explosion (Big Bang). During galactic evolution, stars enrich galaxies in heavy elements but also in helium. The current helium abundance is the result of these two phenomena. The study of irregular galaxies has shown that there is a relationship in these galaxies between the abundance of heavy elements and the helium abundance (the richer a galaxy is in oxygen, for example, the richer it is in helium). This relationship throws light on the galactic chemical evolution. If we extrapolate this relationship down to zero oxygen abundance, in other words to the conditions which prevailed before galactic enrichment, just after the Big Bang, we obtain an estimate of the amount of primordial helium. This shows that primordial helium accounted for 22 per cent of the matter formed immediately after the Big Bang, the remainder being principally hydrogen. It is curious to find that the study of small, rather ugly galaxies provides us, by indirect means, with one of the strongest constraints for cosmological models.

Laurent VIGROUX

The galaxy M82. This galaxy is the prototype of another class of irregular galaxies which are often called class II irregulars. They are characterised by a highly chaotic structure and above all by bright filaments which seem to be ejected from the galaxy.

The dynamics of these galaxies are very complicated and do not bear any resemblance to those of a rotating galaxy. In certain cases, such as M82, a radio source is associated with the nucleus of the galaxy; for a long time it was thought that these were exploding galaxies. In the last few years, detailed studies of M82 have shown, more prosaically, that its structure is the result of a very strong interaction with the neighbouring spiral galaxy M81. M82 is not exploding, it is only being manhandled during an encounter with another galaxy. (Hale Observatories)

The discovery of radio galaxies

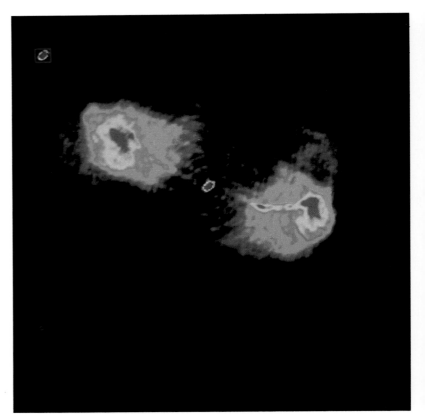

Observations of electromagnetic waves corresponding to the radio waveband are possible from the ground for wavelengths between 30 metres and 1 millimetre, that is for frequencies between 10^7 hertz (10 MHz) and 3×10^{11} hertz (300 GHz). The first observations were made in 1930 by Karl Jansky who detected emission from the centre of our Galaxy and the diffuse emission associated with the Milky Way. During the 1940s other observations were made by Grote Reber, but radio astronomy did not really get off the ground until the late 1940s, when physicists were using radar equipment invented during World War II. In 1949, the Australians John Bolton, Gordon Stanley and O. B. Slee identified two intense radio sources in the southern sky, Centaurus A and Virgo A with the nearby galaxies NGC 5128 and NGC 4486 (M87), respectively. Then, in 1954, Walter Baade and Rudolf Minkowski showed that the radio source Cygnus A is associated with a distant galaxy. These identifications brought to light the extragalactic character of certain radio sources and allowed the discovery of a new class of objects, radio galaxies. In the 1960s progress in radio observations led to the discovery of quasars.

During the 1950s and 1960s research into extragalactic radio sources was hindered by two difficulties. First, the energy received by a radio telescope when observing a radio source is extremely weak; second, the position of a radio source must be known with a precision of a few arc seconds so that its identification with an optical object is possible and its distance may then be estimated. These difficulties were overcome by the construction of radio telescopes with large surfaces, then interferometers, and finally with the introduction of more and more sensitive detectors. For example, the energy that would be received by the 100-metre diameter radio telescope of Effelsburg, near Bonn (FRG), on observing an intense radio source of flux density 1000 janskys (Jy) for a year in a 50 MHz frequency band is only 10^{-3} calories. This amount of energy would only raise the temperature of 1 gram of water by 0.001 °C. Moreover, one should note that a 1000 jansky radio source is an intense radio source and that most radio sources studied have a flux of less than 1 jansky ($10^{-26}\text{Wm}^{-2}\,\text{Hz}^{-1}$).

Since the angular sizes of radio sources are generally less than 1 arc minute, their structures can only be studied with radio telescopes having a resolution better than or equal to 10 arc seconds. The resolution improves as the wavelength of the observation decreases and the diameter of the telescope increases, such that one would need an instrument having a diameter of 15 kilometres to obtain a resolution of 10 arc seconds at a wavelength of 1 metre. Resolutions of less than an arc second have been obtained by interferometers and by making observations at shorter and shorter wavelengths. Only obstacles of a technical nature – it is difficult to construct antennae with surfaces manufactured to a precision better than one-tenth of a wavelength, and to construct very sensitive receivers – have hindered the development of radio astronomy at shorter wavelengths, and in particular millimetre radio astronomy. The largest interferometers in operation today – the VLA (Very Large Array) in the USA and the MERLIN (Multi Element Radio Linked Interferometer Network) in Great Britain – use arrays of antennae for which the distances between the farthest antennae are respectively of the order of 30 and 130 kilometres. The development of intercontinental interferometry enables us to observe sources with arrays of detectors for which the longest

Radio map of 3C 388. The radio galaxy 3C 388 was observed at a wavelength of 6 centimetres (4885 MHz) with the VLA in the United States by J. Burns and W. Christiansen. The resolution of the instrument is one arc second; the beam of the radio telescope is represented at the top left of the illustration. The image presented here was constructed using a computer, the different colours corresponding to different intensities of radio emission. The radio source 3C 388 is a double source consisting of two extended radio lobes. There is also an unresolved radio component coincident with the nucleus. The volume of the ejected radio lobes is much greater than the volume of the optical galaxy. Each lobe contains a zone of intense radio emission and a jet directed towards the nucleus of the galaxy. The right hand jet is much more obvious. These jets correspond to jets of plasma originating from the galactic nucleus, which keep these zones of intense radio emission and the extended lobes supplied with a magnetic field and ultrarelativistic electrons. (NRAO)

Map of the sky at 408 MHz. This map (below) is a representation, in galactic coordinates, of the sky as observed at 408 MHz (a wavelength of 73 centimetres). The centre of our Galaxy occupies the centre of the map and the plane of the Galaxy, which in the sky corresponds to the Milky Way, is the horizontal axis of symmetry in the figure. The intensity of radio emission increases from blue through red. This map shows the intense and diffuse emissions of the galactic centre and galactic plane which were discovered by Karl Jansky and Grote Reber. The radio emission perpendicular to the galactic plane and directed towards the north galactic pole is called the north galactic spur; it is associated with a supernova remnant located in the neighbourhood of the Solar System. A few radio sources outside the galactic plane have been marked: (1) Centaurus A – associated with the galaxy NGC 5128; (2) Virgo A – associated with the galaxy M87 and the Virgo cluster of galaxies; (3) the quasar 3C 273; (4) and (5) the Large and Small Magellanic Clouds, which are satellite galaxies of our own Galaxy. (kindly communicated by R. Wielebinski/Max-Planck-Institut für Radioastronomie)

north galactic pole

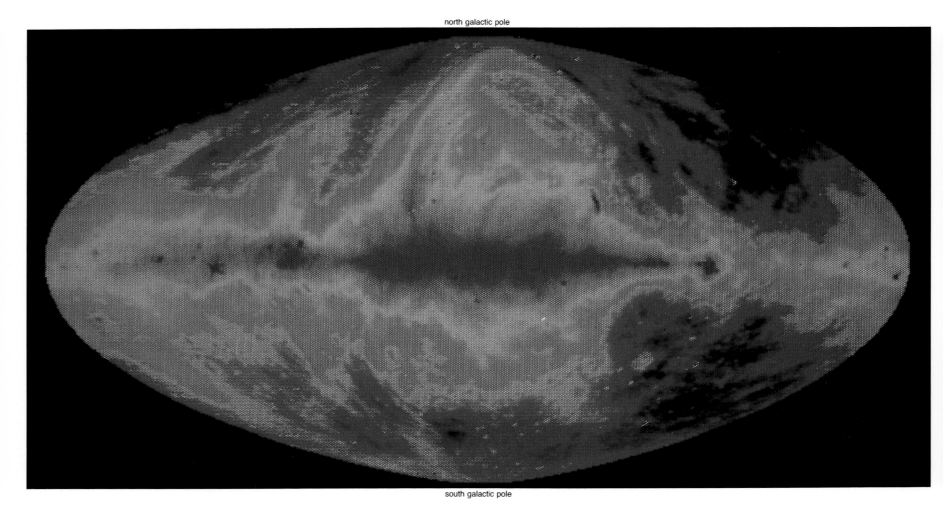

south galactic pole

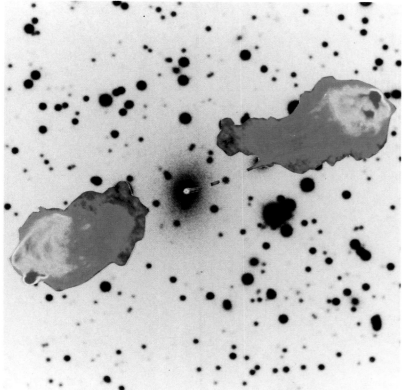

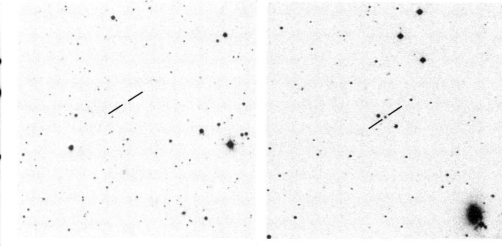

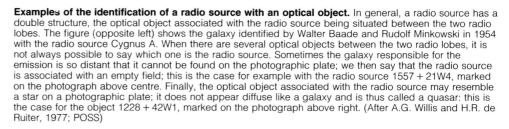

Examples of the identification of a radio source with an optical object. In general, a radio source has a double structure, the optical object associated with the radio source being situated between the two radio lobes. The figure (opposite left) shows the galaxy identified by Walter Baade and Rudolf Minkowski in 1954 with the radio source Cygnus A. When there are several optical objects between the two radio lobes, it is not always possible to say which one is the radio source. Sometimes the galaxy responsible for the emission is so distant that it cannot be found on the photographic plate; we then say that the radio source is associated with an empty field; this is the case for example with the radio source 1557 + 21W4, marked on the photograph above centre. Finally, the optical object associated with the radio source may resemble a star on a photographic plate; it does not appear diffuse like a galaxy and is thus called a quasar: this is the case for the object 1228 + 42W1, marked on the photograph above right. (After A.G. Willis and H.R. de Ruiter, 1977; POSS)

baseline may be as much as 9000 kilometres. Nevertheless, in the 1960s, before large interferometers had been developed, we were able to determine precisely the position of a radio source by observing its occultation by the Moon; the variation of the radio flux as a function of time enabled the profile of the radio source in the direction of the occultation to be obtained with a resolution of the order of an arc second, which is independent of the wavelength of the observation. In 1963, this method provided a determination of the position of the radio source 3C 273 with a sufficiently accurate precision to identify it with a stellar-like object which was the first quasar ever detected.

The discovery of the sky's diffuse radio emission and of radio galaxies posed the problem of the origins of this radiation. In the radio waveband, astronomers can observe localised radiation at certain wavelengths – this being either line radiation or continuous radiation. In radio astronomy different types of line radiation can be distinguished: lines from transitions between hyperfine structure levels (such as the line emitted at 21 centimetres by the hydrogen atom), transition lines between the rotational levels in radicals or in molecules (certain lines have enabled the discovery of several tens of radicals and organic molecules) and, finally, the

recombination lines of ions (characteristic of hydrogen, helium, carbon atoms, etc.) present in ionised regions or H II regions.

Continuous radiation also has various origins. One can discriminate the 2.7 K thermal background radiation of the sky – which is the fossil radiation characteristic of the beginning of the explosion of the Universe (Big Bang); the 'bremsstrahlung', or radiation from the free electrons being decelerated in the field of atomic nucleii (this radiation is observed in H II regions); and finally synchrotron radiation, which is radiation from decelerating ultrarelativistic electrons spiralling in a magnetic field. The study of the polarisation and the distribution, as a function of frequency, of the received radiation enables one to identify these various types of continuous radiation. The theorists Vitaly Ginzburg and Josef Shklovsky showed in 1950 that the radiation from radio galaxies must be synchrotron radiation. This theory has been confirmed by the observation of the polarisation of radiation from radio galaxies. It has been established that the radio flux observed, $S(\nu)$, follows a power law in frequency ν, that is of the form $S(\nu) = S_0\nu^{-\alpha}$, where α is called the spectral index, usually with a value from 0 to 2. One can show that the observed flux $S(\nu)$ is produced by a distribution of ultrarelativistic electrons of the form: $N(\varepsilon) = $

$K\varepsilon^{-\gamma}$, where ε is the electron energy and $N(\varepsilon)$ the number of electrons with that energy, with $\gamma = 2\alpha + 1$. This result demonstrates that radio emission from radio galaxies is from a plasma containing very energetic particles (ultrarelativistic electrons) and magnetic fields. The existence of such plasmas shows that radio galaxies are the sites of extremely violent phenomena; the aim of studies of radio galaxies is to understand the origin of these phenomena and their evolution with time.

One of the major problems is therefore the determination of the physical characteristics of the plasma responsible for the radio emission. In particular it is important to know the structure of the magnetic field in the interior of the plasma, as well as its intensity. The plasma is confined by the pressure of the intergalactic medium, so that radio observations enable us to probe this medium. A question of great interest at the moment is the origin of the plasma jets in the radio galaxy nucleus: it appears these are accelerated by a spinning, magnetised black hole with a mass of around 10^8 solar masses.

Jacques ROLAND and Florence DURRET

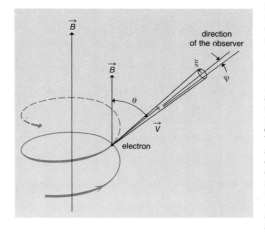

Synchrotron radiation. Synchrotron radiation is radiation from the braking of an electron travelling close to the speed of light in a magnetic field. Such electrons are said to be ultrarelativistic; their kinetic energy, defined by $E_k = mc^2\left(\dfrac{1}{\sqrt{1-v^2/c^2}} - 1\right)$ is large, compared to their rest energy, given by $E_r = mc^2$ (m is the mass of an electron, v its velocity and c the velocity of light). The motion of an electron in a magnetic field $\vec{B}$ is helical; the synchrotron radiation emitted by the moving electron is concentrated in a cone of very small half angle ξ. The result is that this radiation can only be seen by an observer for whom the angle ψ between the direction of motion of the electron and the direction of the observer is smaller than ξ. In general, synchrotron radiation from an ensemble of electrons is only emitted uniformly in all directions if this ensemble has a distribution of velocities such that all angles θ are possible between the magnetic field $\vec{B}$ and the velocity vector of the electron $\vec{V}$. (After Moffet, 1975)

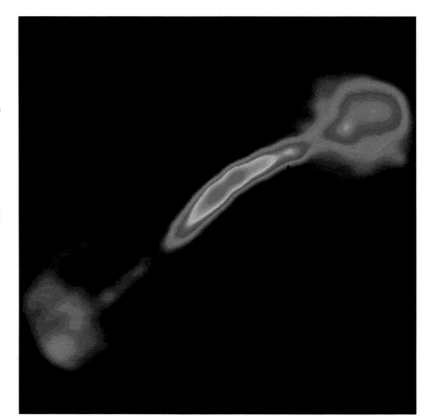

Radio map of the Source 1333–33 (right). This source is associated with the eleventh magnitude galaxy IC 4296. It is one of the closest known radio galaxies at a distance of 36 megaparsecs. The total extent of the source is 350 kiloparsecs. Given its small distance, a relatively large region of the sky 40 arc minutes square had to be observed to make this complete map. The VLA was used in its C configuration by N. Killen and R. Ekers at a wavelength of 21 centimetres. The resolution is 60 arc seconds. The nucleus of the galaxy, located in the centre of the map, has ejected two very bright radio jets which feed the extended components. This radio map is very different from that of 3C 388 (opposite page). In particular, note the structure of the extended radio lobes, the shape of the radio jets and the compact component associated with the galaxy. These structural differences indicate differences in formation of these radio galaxies, i.e. differences in the activity of the nuclei and differences in the properties of the plasmas ejected from them. (NRAO)

Radio galaxies

After the identification of the first radio sources (Cygnus A, Centaurus A and Fornax A) with galaxies, the study of the structure of these objects has shown that the radio emission originates in two extended regions, the lobes, situated on opposite sides of the galaxy. This double structure makes the peculiar character of radio galaxies apparent; these are galaxies capable of ejecting large amounts of plasma comprising magnetic fields and relativistic particles responsible for the radio emission. The central galaxies are therefore the sites of extremely violent events. Recent observations have shown that between the two radio lobes is a compact radio component which coincides with the nucleus of the galaxy. The fundamental peculiarity of radio galaxies is thus the existence, at the centre of the galaxy, of an active nucleus responsible for ejecting the two extended radio components. The study of extended radio sources therefore consists of determining their physical characteristics (such as volume of the radio lobe, average magnetic field, and energy of relativistic particles) and then understanding their evolution in time and their formation from the active nucleus. At present, most of the physical parameters necessary to understand them are only known to an order of magnitude, which explains the absence of a satisfactory theory enabling them to be studied in detail.

The principal characteristic of extended radio sources is the large volume occupied by the plasma ejected from the galaxy. The radio lobes of extended radio sources are the largest objects known in the universe. In the extreme case of 3C 236, the most extended radio source known, the volume of the radio lobes is 10^{67} cubic metres, which is more than six thousand times the volume of the galaxy itself. Whereas 3C 236 extends to 5 megaparsecs, the average extent of extended radio sources is several hundred kiloparsecs. In comparison the average radius of a galaxy is of the order of 20 kiloparsecs. By convention, radio sources having linear dimensions greater than or equal to 0.1 kiloparsec are called extended radio sources. The average spectral index of extended radio sources is about 0.75 and one of their characteristics is that their radio flux is constant in time. The energy radiated in the radio region (between 10 MHz and 100 GHz) by the large majority of extended radio sources is between 10^{35} and 10^{36} watts; it can reach 10^{39} watts in extreme cases. (By way of comparison, the energy radiated in the optical region by a typical galaxy is 10^{37} watts.) Nevertheless, if the energy radiated in the radio region is not exceptional compared to the energy radiated by the nuclei of galaxies in the infrared or in X-rays, the continuous energy in the form of magnetic field and relativistic and non-relativistic particles is gigantic; it is on average 10^{53} joules. (This energy is equivalent to that which would be obtained if one million solar masses were converted into energy with 100 per cent efficiency.) The mass contained in the radio lobes may reach 100 million solar masses. Note that, in the case of the most powerful radio sources, these values are multiplied by a factor of ten. It is estimated that the average velocity of separation of the radio lobes from the nucleus is less than or equal to 30 000 kilometres per second; taking into account the dimensions of extended radio sources, it is concluded that their age is between 1 and 100 million years. The galaxies responsible for the formation of extended radio sources are the sites of the most violent events known in the Universe. The nucleus of a galaxy producing 10^{53} joules in ten million years has to liberate in this period between 10^{38} and 10^{39} watts, that is ten times more than the energy emitted in the optical

region by the whole galaxy. It is interesting to note that the objects associated with extended radio sources are elliptical galaxies or quasars.

The study of the polarisation of radio radiation at various wavelengths enables one to determine the direction of the magnetic fields in the lobes, as well as the density of the thermal component of the plasma. It is found that, on a large scale, the magnetic field is not disordered inside the lobe, but is oriented along the axis joining the two extended components. Radio observations alone do not enable us to obtain the average magnetic field inside the lobes, but, with certain assumptions, one can estimate its order of magnitude: in the most extended sources (like 3C 236) the magnetic field is of the order of 10^{-6} gauss, whereas inside the zones of intense emission (hot spots) of sources similar to 3C 452, it may reach 5×10^{-4} gauss. The lifetime of ultrarelativistic electrons radiating at 1000 MHz is ten million years in a magnetic field of 10^{-6} gauss whereas it would only be ten thousand years in a magnetic field of 5×10^{-4} gauss. For the different types of extended radio sources, an ultrarelativistic electron moving at the speed of light does not have time to cross a radio lobe. Consequently, inside radio lobes, a re-acceleration of ultrarelativistic electrons is necessary to explain the age (1 to 100 million years) of extended radio sources.

A model of extended radio sources must be able to account for the general properties of these sources. Advances towards a comprehensive model of extended sources are being made by two complementary strategies, which provide a good illustration of how physics research is done. The first involves making observations – and interpreting them – of radio sources and their associated galaxies in several wavebands: for example, optical and ultraviolet observations can constrain nuclear activity and the rate at which fuel falls into the central black hole; X-ray observations (such as from the EINSTEIN satellite) provide information on the bremsstrahlung from the surrounding medium and hence its density; and of course there are the radio observations themselves. In the second approach, theoretical physicists make computer simulations of the dynamics of the radio-emitting plasma, computing the flow patterns produced when a jet hits intergalactic medium and predicting the distribution of radio synchrotron emission. Many puzzles remain, but the combination of these two strategies has produced an enormous improvement in recent years in our understanding of extended radio sources.

Jacques ROLAND and Florence DURRET

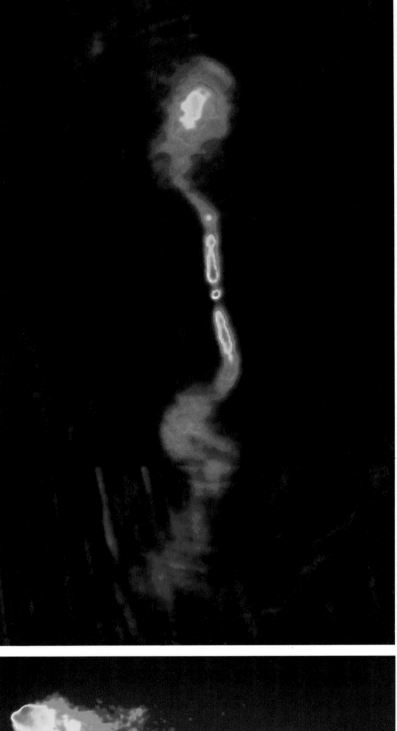

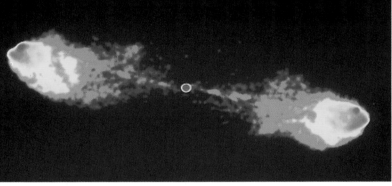

Two types of double extended radio sources. Radio sources have a characteristic double structure, the only exceptions being trailing radio galaxies, which are observed in clusters of galaxies. Two extreme types of extended radio sources are illustrated here by 3C 449 (top) and 3C 452 (bottom). The radio map of 3C 449 was made at 1.465 GHz using the VLA (after R. A. Perley, A. G. Willis and J. S. Scott 1979) and that of 3C 452 at 8 GHz (A. R. S. Black 1993) also using the VLA.

There is an unresolved compact radio source, which coincides with the nucleus of the optical galaxy in both cases. The two types of radio sources differ primarily in their global properties. Sources like 3C 452 are the most powerful, their luminosities at a wavelength of 21 centimetres being of the order of 10^{26} watts per hertz whereas the luminosities of radio sources like 3C 449 are only 10^{24} watts per hertz. The volumes of the two types of sources are also very different. Sources like 3C 449 are the most extended radio sources known whereas sources similar to 3C 452 have volumes more than 100 times smaller. The spectral indices of sources of the 3C 452 type are about 0.9 whereas those for sources like 3C 449 are about 0.6. Great differences in small-scale structure in the two types of source are revealed by high resolution mapping. Sources like 3C 452 are characterised by two zones of intense emission at the extremities of the extended lobes. The average magnetic field in these zones is about 5×10^{-4} gauss (G) and the spectral index is smaller than that for the surrounding extended radio emission. In contrast, radio sources like 3C 449 are characterised by an absence of such zones in the extended components and also by the existence of two brilliant radio jets directed from the nucleus towards the extended lobes. The magnetic field is less than or equal to 10^{-5} G and the spectral index increases along the jet from the nucleus to the lobes. This change in spectral index indicates the ageing of the ultrarelativistic electrons, responsible for the synchrotron emission. In sources like 3C 452, the nucleus ejects matter in the form of a jet containing no relativistic particles and a magnetic field less than 10^{-6} G. This jet is extremely thin and its speed is close to that of light. This interaction of the jet with the intergalactic medium which causes the formation of regions of intense emission, with the appearance of strong turbulence, amplifies the magnetic field to 5×10^{-4} G by the dynamo effect and the formation of a relativistic component in the plasma. After crossing these zones, the diffuse plasma gives birth to the extended component. In comparison, in the case of sources like 3C 449, the nucleus ejects matter in the form of a wide jet containing relativistic particles and magnetic fields of the order of 10^{-5} G. This jet feeds the extended components whose spectral index is larger than that in the jet. (NRAO)

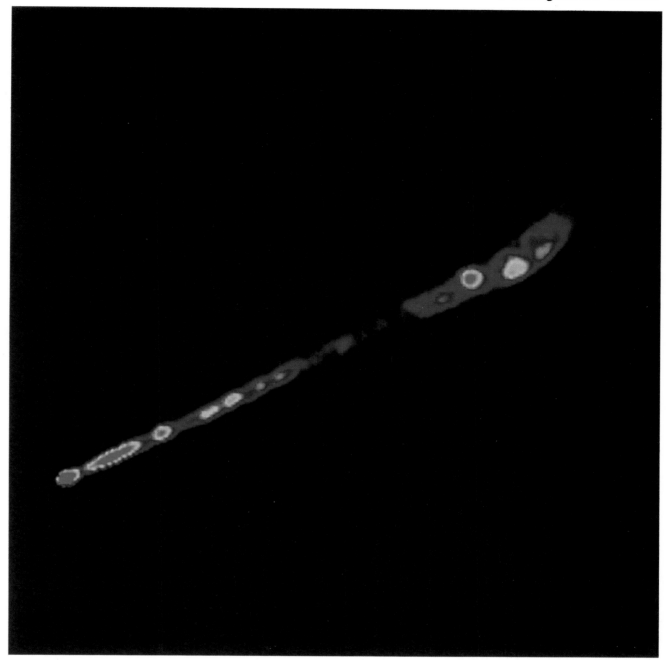

Two examples of the transfer of energy from the nucleus to the extended components. For the extended radio components resulting from activity in the nucleus of the galaxy, one of the problems to resolve is that of the transfer of energy from the nucleus out to the extended components. Radio maps made at high resolution and high sensitivity have indicated the existence of jets of plasma which enable energy to be re-supplied to the extended components. The alignment of the observed jets with the extended components shows that the direction of ejection from the nucleus can remain constant for periods between 1 and 100 million years. In certain cases, for example NGC 6251 (opposite, observed at 20 centimetres wavelength by A. G. Willis, R. A. Perley and A. H. Bridle using the VLA) we see one single jet leaving the nucleus (bottom left), which shows that the ejection can be sporadic while its direction remains constant. In the case of the radio source 4C 26.03 (below, observed at 20 centimetres wavelength by E. B. Fomalont, C. Lari, P. Parma, R. Fanti and R. Ekers using the VLA) the direction of ejection is not constant but the two jets have undergone a precession around the nucleus; the phenomenon, clearly visible on this enlargement of the central part of the radio source, explains the S shape seen in certain extended radio sources. The continuous or almost continuous ejection from the nucleus enables us to estimate the energy necessary, in the course of time, to make the extended radio components (NRAO)

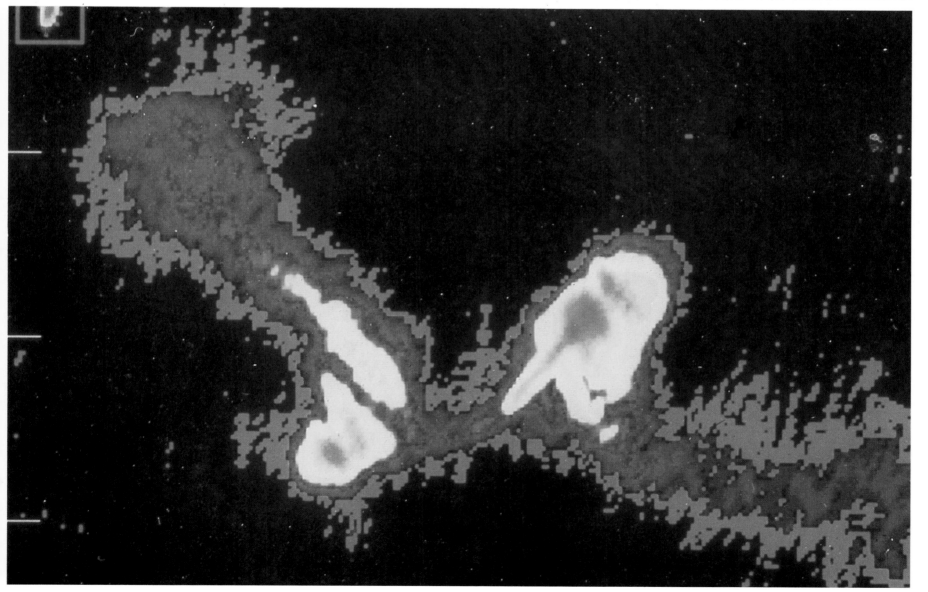

Compact radio sources and jets

Observations of extended radio sources have indicated the existence of unresolved radio components – of angular dimensions less than an arc second – which coincide with the nucleus of the galaxy. The study of their structure has shown that these radio components, in fact, often have a diameter less than 0.01 arc second; for this reason they have been given the name compact radio sources.

These radio sources are primarily characterised by a cut-off (sharp decrease of intensity) in their radio spectrum due to the synchrotron self-absorption of long wavelength radiation; this happens at frequencies around 100 MHz. Another characteristic of these radio sources is the variability of their radio flux, which is observed on time-scales of the order of weeks to a year. Their radio spectra are often flat: the spectral index is about zero compared to about 0.75 for extended radio sources. If we assume that the emission from compact radio sources is isotropic, the energy radiated in the radio region is of the order of 10^{34} to 10^{35} watts.

Even if compact radio sources are characteristic of the nuclei of galaxies associated with extended radio sources, we also see them in the active nuclei of galaxies which are not associated with extended radio sources, and also in the closest 'normal' galaxies. In the latter, the energy radiated by the compact radio sources is only 10^{26} watts. The characteristic angular dimension of compact radio sources is 0.001 arcsecond, which corresponds to a dimension of 0.01 parsec for the closest galaxies possessing a weakly active nucleus (for the closest galaxy with an active nucleus, NGC 1275, this dimension corresponds to 0.4 parsec) and to several tens of parsecs for the most distant compact radio sources which constitute quasars.

Compact sources afford a particular opportunity to help us understand the dynamics of jets, because very long baseline interferometry often reveals components of their jets moving away from the nucleus at a speed apparently faster than light. This superluminal motion is only apparent, and arises because the jet is in fact pointing almost directly at us (see p 369), but it proves *directly* that, at least near the nucleus of a radio source, jets do travel at relativistic speeds.

Some compact sources have X-ray and optical properties just like the nuclei of large, double radio sources, and we infer that these compact sources will grow into large doubles. Others have a range of exotic properties and understanding them is an exciting challenge.

Jacques ROLAND and Florence DURRET

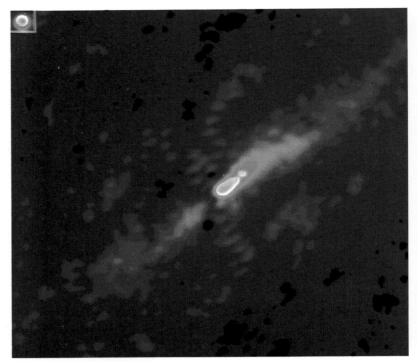

Observation of the jet associated with the galaxy NGC 315. The radio galaxy associated with the galaxy NGC 315 is a very extended double radio source. The cartography of the central regions of the radio source shows that it comprises a unique jet and that it possesses an asymmetric structure around the nucleus. The map presented here was obtained by E. B. Fomalont, A. H. Bridle, A. G. Willis and R. G. Strom using the VLA then under construction. Only the first eight antennae installed were used and the resolution obtained at a wavelength of 21 centimetres was 11 arc seconds. The radio galaxy NGC 315 is one of the best studied at various resolutions. On the following page is a study of its extended and its compact structures. (NRAO)

Variations in the flux of the compact radio source 3C 454.3. Studies of the time variation of the flux from compact radio sources are very important as they show that these sources are the sites of very violent events characterised by time-scales of the order of between a month and a year. Given that these variations are linked to the nucleus of the galaxy responsible for the formation of the extended components, their observation enables us to obtain information about the activity in the nucleus, and on the mode of ejection of the plasmas responsible for the extended radio sources. Variations are observable for wavelengths between 1 millimetre and 1 metre, but in certain exceptional cases, we have been able to observe correlated flux variations in the radio, infrared and optical regions. In the case of the quasar 3C 454.3, variation in the observed flux between 1966 and 1969 was detected firstly at short wavelengths and then at longer wavelengths with a smaller and smaller amplitude. This type of variation is characteristic of the adiabatic expansion of an initially opaque synchrotron source, which becomes transparent at longer and longer wavelengths. When the source expands, the synchrotron re-absorption becomes less and less efficient at large wavelengths and the source can radiate at longer and longer wavelengths. On the other hand, when a cloud of plasma expands, the average magnetic field diminishes and the observed radio flux decreases. One can also observe another type of variation characteristic of radio sources of the BL Lac type; these variations have an amplitude which is independent of the wavelength and appears simultaneously at various wavelengths. This type of variation is that of a transparent radio source whose increase in flux is due to an abrupt increase in the number of ultrarelativistic electrons responsible for the synchrotron radiation; the drop in flux from the source is therefore due to the expansion of the source and to the loss of energy of the electrons. (After K. I. Kellermann and I. I. K. Pauliny-Toth, 1981)

Spectra of compact radio sources. Compact radio sources are usually variable and their spectra have to be determined by simultaneously measuring their flux at different frequencies using several radio telescopes. The spectra of 2134 + 00, 3C 345, 0007 + 10 and 3C 84 are presented here (after K. I. Kellermann and I. I. K. Pauliny-Toth, Compact radio sources, *Annual Review of Astronomy and Astrophysics*, vol. 19, p. 373).

Compact radio sources have a characteristic cut-off at low frequencies (clearly visible in the case of 2134 + 00) which is due to the re-absorption of the synchrotron radiation by the radio source itself. For a compact source, if the density of the synchrotron radiation becomes sufficient, the electrons can re-absorb it, and this re-absorption becomes more efficient as the radio frequency goes down; it becomes negligible for frequencies greater than the frequency corresponding to a maximum in the observed spectrum. The spectrum of the radio source 3C 84 is a power law at low frequencies (less than 1 GHz). Such a power law spectrum is characteristic of extended radio sources. At higher frequencies, the spectrum is similar to that of 2134 + 00. Thus simply knowing the spectrum of 3C 84 shows that the radio source has an extended and a compact component.

The spectrum of 3C 345 is similar to that of 3C 84, but it stays flat at higher frequencies. This flat spectrum is characteristic of the presence of several compact components; as the frequencies corresponding to the maximum in the spectrum of each component are different, the superposition of the spectra of these different components produces a flat spectrum.

The components whose maximum flux is observed between 10 and 100 GHz are the most compact components. Their dimensions are less than 0.1 parsec. This is the case for the radio source 0007 + 10 where the maximum in the spectrum occurs at 100 GHz and for which the part of the spectrum observed corresponds to the synchrotron re-absorption.

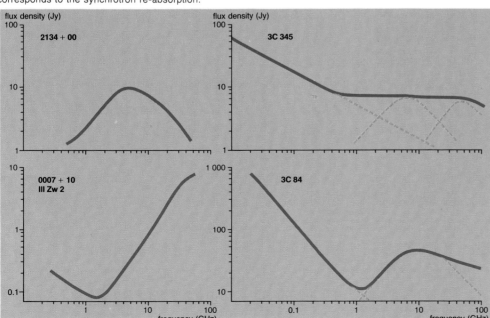

Structure of the radio galaxy NGC 315. The study of the structure of compact radio sources requires high resolution observations. To obtain a resolution of 10^{-3} arc seconds observing at a wavelength of 6 centimetres, it is necessary to use an interferometer with a separation between the most distant antennae of about 10 000 kilometres. Radio telescopes situated on different continents such as America and Europe are therefore used. This observational technique, called intercontinental interferometry, poses delicate problems and the first reliable maps were only obtained in the 1980s. These maps are limited to a dynamic range of the order of ten (the dynamic range is the ratio of the strongest intensities observed in the map to the weakest). Compact sources may have a complex structure, consisting of several components whose flux vary with time. The mapping of such sources with a dynamic range limited to ten does not enable us to study unambiguously the variation of their structure with time. In fact, a dynamic range of 100 would be necessary for this type of study. With the various interferometers (Westerbork, Cambridge, VLA, MERLIN) capable of mapping radio sources with resolutions which reach 0.1 arc second and the intercontinental interferometry allowing us to observe compact radio sources with resolutions of the order of 0.001 arc second, it is possible to study continuously how a galactic nucleus ejects plasma on scales of 1 parsec up to extended components which measure 1 megaparsec. In general, compact components are double and appear as an unresolved nucleus and a jet aligned with the direction of the extended components. Some have a much more complex structure. A composite radio map of the radio galaxy NGC 315 is shown on the right and can be compared with the 21-centimetre radio map on the facing page. (After K. I. Kellermann and I. I. K. Pauliny-Toth, 1981).

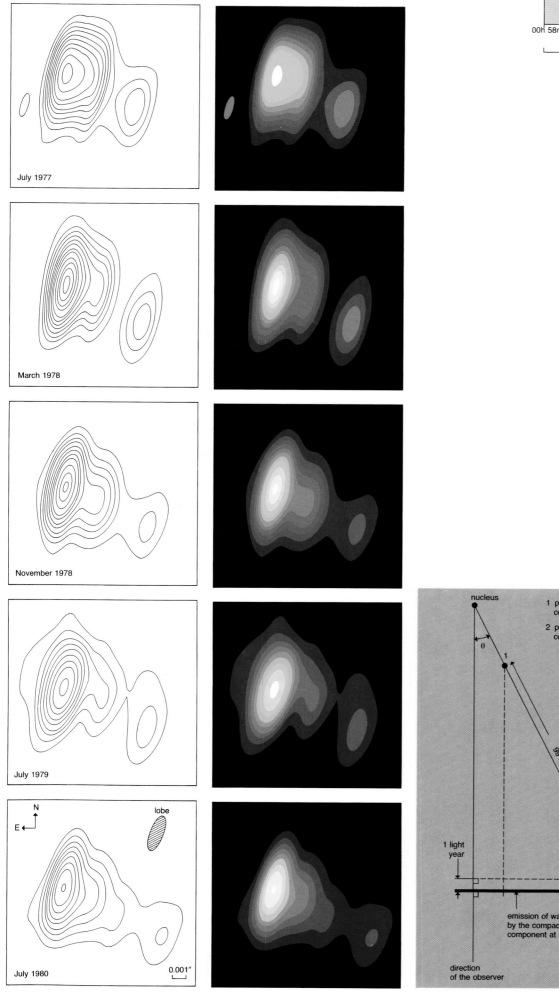

July 1977

March 1978

November 1978

July 1979

July 1980

N
E

lobe

0.001"

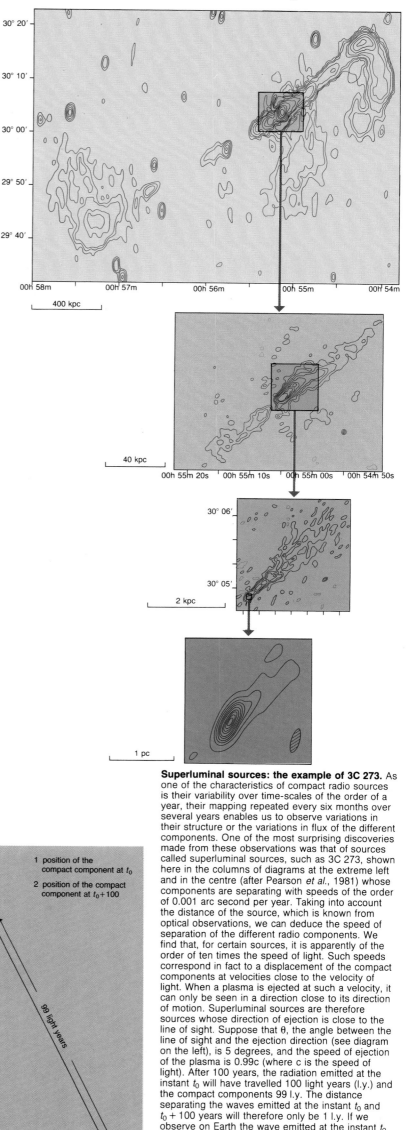

1 position of the compact component at t_0

2 position of the compact component at $t_0 + 100$

nucleus

θ

1

99 light years

1 light year

emission of waves by the compact component at t_0

direction of the observer

ejection direction

2

Superluminal sources: the example of 3C 273. As one of the characteristics of compact radio sources is their variability over time-scales of the order of a year, their mapping repeated every six months over several years enables us to observe variations in their structure or the variations in flux of the different components. One of the most surprising discoveries made from these observations was that of sources called superluminal sources, such as 3C 273, shown here in the columns of diagrams at the extreme left and in the centre (after Pearson *et al.*, 1981) whose components are separating with speeds of the order of 0.001 arc second per year. Taking into account the distance of the source, which is known from optical observations, we can deduce the speed of separation of the different radio components. We find that, for certain sources, it is apparently of the order of ten times the speed of light. Such speeds correspond in fact to a displacement of the compact components at velocities close to the velocity of light. When a plasma is ejected at such a velocity, it can only be seen in a direction close to its direction of motion. Superluminal sources are therefore sources whose direction of ejection is close to the line of sight. Suppose that θ, the angle between the line of sight and the ejection direction (see diagram on the left), is 5 degrees, and the speed of ejection of the plasma is 0.99c (where c is the speed of light). After 100 years, the radiation emitted at the instant t_0 will have travelled 100 light years (l.y.) and the compact components 99 l.y. The distance separating the waves emitted at the instant t_0 and $t_0 + 100$ years will therefore only be 1 l.y. If we observe on Earth the wave emitted at the instant t_0 in 1980 for example, the wave emitted at the instant $t_0 + 100$ years will be observed in 1981. The separation of the two components on the celestial sphere observed in 1980 is $d = D \sin \theta = 9$ l.y. with $D = 99$ l.y. The speed of separation (V) observed on the celestial sphere therefore equals 9c. Knowing the observed speed of separation and the angular distance of the components, one can deduce the year of the start of the separation. We find that the year deduced corresponds to a period of great flux variation in the compact radio source.

Active galactic nuclei

In radio astronomy, intercontinental interferometry has shown that the compact radio sources found at the centres of radio galaxies have angular diameters of about 0.001 arc second. The primary source giving rise to the energy released by radio galaxies or active galaxies is called the active nucleus. We now define this more precisely.

In a normal galaxy the density of stars increases towards the centre. We thus define the centre of the galaxy as a nucleus of stellar origin (i.e. constituted of stars) which does not have the characteristics of an active nucleus. We can define an active galaxy morphologically as an object which looks like a star and which is not resolved (i.e. having an angular diameter of less than one arc second) on short-exposure photographs. The luminosity of an active nucleus is generally much less than that of the galaxy but for very active nuclei the nuclear luminosity may exceed that of the rest of the galaxy. Spectroscopically, the nucleus of a normal galaxy has a spectrum

characterised by the presence of absorption lines; for spiral galaxies, there may also be narrow emission lines. In contrast, an active nucleus has a spectrum with a continuum originating not from stars but from synchrotron or inverse Compton emission, and, except for BL Lac objects, strong emission lines, often with wide lines of the Balmer series. Finally, the nucleus of an active galaxy is optically variable. To summarise, the nucleus of an active galaxy is of stellar appearance on short-exposure photographs, being unresolved, is optically variable, and its spectrum contains a continuum of non-thermal origin, often with strong, wide emission lines. There is probably some weak activity in the nuclei of normal galaxies, but the galaxies are too far away for it to be seen. Our definition of an active galaxy does not preclude the existence of weakly active nuclei at the centre of normal galaxies. The concept of an active galaxy can be morphologically defined: a galaxy is said to be an N galaxy if it has an unresolved nucleus with a luminosity compara-

ble to that of the galaxy. Note that this definition says nothing about the particular character of the nuclear spectrum.

In 1943, well before the discovery of radio galaxies and quasars, Carl Seyfert noted that certain galaxies exhibited a very bright nucleus and had spectra that differed from those of other galaxies. It was only twenty years afterwards, with the development of new optical detectors, that the study of Seyfert galaxies developed. A Seyfert galaxy is a galaxy with a bright nucleus, unresolved, with an optical spectrum containing emission lines, generally wide in the case of the Balmer series and narrow in the case of the

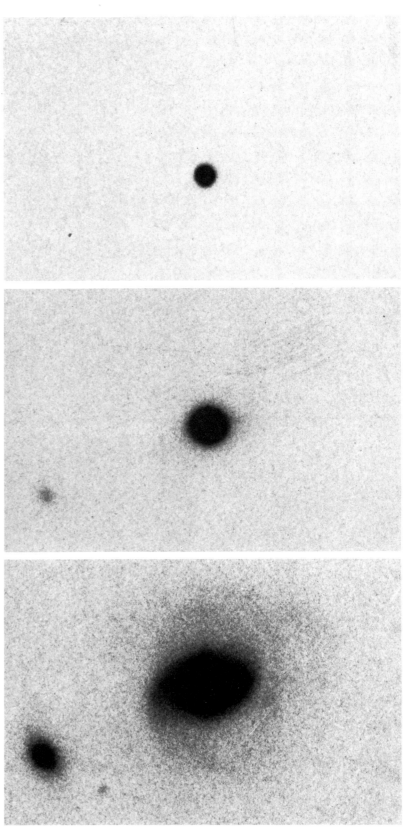

An example of a galaxy with an active nucleus. The photograph below shows the galaxy NGC 1275, taken with the 4-metre telescope at Kitt Peak National Observatory, USA. NGC 1275 is the brightest member of the Perseus cluster of galaxies, numbered A 426 in the Abell catalogue of clusters of galaxies. The optical extent of the galaxy is 80 arc seconds. Optical and spectroscopic studies have shown the galaxy to be very peculiar. It appears that the nucleus is of the BL Lac type, and that the filaments visible around the nucleus are due either to matter being ejected from parts of the nucleus, as in the case of Centaurus A, or to the accretion of cold intergalactic gas. NGC 1275 is one of the nearest known galaxies with an active nucleus and Very Long Baseline Interferometry (VLBI) has been used to map the central radio components with a resolution of better than one light year. The extent of the various compact radio components forming the nucleus is 1/80 000 of the size of the galaxy, showing that the active galactic nucleus occupies a volume many times smaller than the galaxy itself. In the case of NGC 1275, the volume occupied by the nucleus is over 10^{14} times smaller than that of the galaxy. Finally, we note that the luminosity of a very active nucleus may be comparable with the luminosity of the rest of the galaxy, which shows the extreme violence of the phenomena associated with the activity of the nuclei of galaxies. (Kitt Peak National Observatory)

A galaxy with an active nucleus: ESO 113–IG 45. The short-exposure (several minutes) photographs on the right show galactic nuclei. In such photographs, a galactic nucleus resembles an unresolved star (of angular diameter less than 1 arc second): thus it is said to have a stellar appearance. Part of the light emitted by an active nucleus is of synchrotron or inverse Compton origin. The ultraviolet continuum is particularly evident for nearby active galaxies. On long-exposure photographs the stellar envelope is readily visible (this is only the case with nearby galaxies). The luminosity of very active nuclei may be comparable with that of the galaxy itself, as is the case with the galaxy ESO 113–IG 45. During periods of extreme nuclear activity, its luminosity can exceed that of the galaxy, and the stellar envelope surrounding the galaxy can hardly be seen. (After R. M. West, A. C. Danks and G. Alcaino)

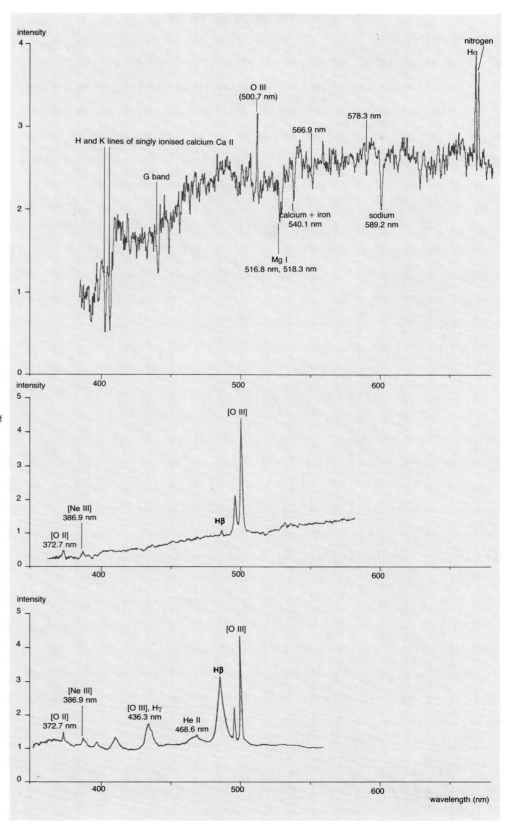

The nature of BL Lac. BL Lac was for a long time thought to be a variable star. Then it was identified as a radio source, also found to be variable. Its optical spectrum showed no lines and so it took a long time to determine its real nature. If BL Lac is observed at minimum luminosity, an elliptical nebulosity can be seen surrounding it. The spectrum of this nebulosity is typical of elliptical galaxies, and BL Lac therefore turns out to be a peculiar type of active galactic nucleus. Until now, every observation of the underlying galaxy associated with a BL Lac object has revealed an elliptical galaxy. Distances of BL Lac objects can only be measured if the underlying galaxy can be observed, which is rarely the case. (After Wlerick *et al.*, 1973)

Examples of spectra: a normal spiral galaxy, a narrow emission line galaxy (Markarian 176) and a Seyfert galaxy (Markarian 509). The spectrum of a normal galaxy (upper right, after Bergeron) is characterised by the existence of absorption lines. Spiral galaxies, being the galaxies richest in gas, have spectra with narrow emission lines derived from the ionisation of interstellar clouds by young stars. In contrast, elliptical galaxies are poor in interstellar gas, and have only stellar absorption lines in their spectra. There are two types of lines, permitted lines and forbidden lines. The nomenclature is as follows: the line [O III] 5007, for example, is the line of doubly ionised (III) oxygen (O) at a wavelength of 500.7 nanometres. The square brackets simply indicate that the line is a forbidden line. In the spectra of normal galaxies, the principal absorption lines are as follows: the H and K lines of singly ionised calcium Ca II, of wavelengths 393.37 nm and 396.85 nm; the G band; the atomic magnesium doublet Mg I; a line due to a blend of calcium and iron; two unidentified lines at 540.1 nm and 578.3 nm; sodium at 589.2 nm, and a few lines of iron. Most of the lines are of stellar origin but some are due to a mixture of gas and stars (H and K lines of calcium, D lines of sodium).

The spectrum of the spiral galaxy presented here also has emission lines due to ionised clouds in the galaxy; these lines are the line [O III] 5007, the Hα line, and a nitrogen line. We note that all of these lines are observed at wavelengths (λ) greater than the wavelengths (λ0) of the same lines observed on the Earth. The galaxy spectrum is said to be displaced towards the red and the quantity $z = (\lambda - \lambda_0)/\lambda_0$ is called the redshift of the galaxy. The redshift of galaxies is a corollary from the law of the expansion of the universe (Hubble's law) which enables us to calculate the distance to a galaxy from the relation $D = cz/H_0$, where c is the velocity of light (300 000 kilometres per second), H_0 the Hubble constant (50 kilometres per second per megaparsec is an acceptable value) and D the distance, expressed in megaparsecs. This law is valid for galaxies of redshift less than 0.2.

Certain galaxies, such as Markarian 176 (centre) have characteristic spectra with numerous emission lines (compare the Hβ line with that of the galaxy Markarian 509, below). These emission lines are due to a sudden increase in the rate of star formation in the galaxy. The newly formed stars ionise the interstellar medium and give rise to emission lines. These galaxies do not have very active nuclei comparable to Seyfert nuclei, but are merely in a peculiar evolutionary phase, characterised by rapid star formation for a short time. The spectrum of Markarian 509 is typical of Seyfert galaxies. It can be seen that the Balmer lines Hβ, Hγ are very wide while the forbidden lines such as [O III] are narrow and of comparable width to the Hα line of the normal spiral galaxy spectrum.

Notes on line formation. In general, an atom, an ion, or a molecule can radiate energy if it passes from an excited state to a less excited state. An atom may be in an excited state after a collision or after absorption of a photon. The lifetime of a given state is defined as the average time the atom is in this excited state. Typical lifetimes for atoms are less than 10^{-6} seconds. Certain states can have lifetimes exceeding 1 second; they are called metastable. Transitions from an excited state to a less excited state may be of various types, depending on the states of the atom and of the emitted photon. Two types of transitions may be distinguished: the most frequent, called dipole transitions, and the rarest, called quadrupole transitions. When an atom can radiate both types of transitions, it will emit practically all of the photons corresponding to the dipole transition, and the line corresponding to the quadrupole transition will be almost unobservable. In contrast, if the atom is in a metastable excited state, emission of a dipole photon is impossible and, if the atom is not perturbed for a time exceeding the lifetime of the metastable state, it will emit a quadrupole photon. The lines corresponding to these quadrupole transitions are the forbidden lines; those corresponding to the dipole transitions are the permitted lines. The interstellar medium particularly favours the emission of quadrupole photons. Certain conditions must be fulfilled for this: atoms must be excited to a metastable state by absorbing radiation from a star situated in a cloud; the medium must be sufficiently rarefied for the time between atomic collisions to exceed the lifetime of the metastable state, and the density of radiation in the cloud must be low. These conditions are generally fulfilled in the interstellar medium and the forbidden lines are observable, corresponding to quadrupole transitions, and can be stronger than the permitted lines.

forbidden lines. Note that galaxies do exist with both permitted lines and narrow forbidden lines, but containing a much less active nucleus. These latter galaxies are distinguished from Seyfert galaxies by the fact that their nuclei are resolved and that their spectra do not contain the non-thermal continuum characteristic of active nuclei. Reflecting the degree of nuclear activity, there is a continuum of galaxies ranging from galaxies with narrow emission lines to Seyfert galaxies. Seyferts are not strong radio galaxies (not extended double sources), and they do not radiate a major part of their energy at X-ray frequencies, but they are on the other hand strong infrared sources. An important part of their infrared radiation is of synchrotron origin, proving the presence of an active nucleus, but the rest of their infrared emission is due to radiation from dust surrounding the nucleus.

Although Seyfert galaxies are by definition N galaxies, the converse is not true! N galaxies do exist for which the spectrum of the nucleus does not show strong emission lines. These N galaxies, with spectra that do not show strong emission lines, are called BL Lac objects. They are galactic nuclei with extremely variable flux, and strong, rapidly varying polarisation. The typical time-scales of the observed variations in flux and in polarisation are of the order of a week to a day or even shorter. It is difficult to study these objects because the lack of emission lines in their spectra makes it harder to determine their distance and hence their physical characteristics. However, stellar lines are observed in many cases and this gives information on the distance to these objects. In other cases, when absorption lines are observed in their spectra, these may arise from a galaxy on the line of sight, and a minimum distance can be deduced for the object. The lack of emission lines in the spectra of BL Lac objects is not clearly understood; it may be due to a cut-off in the continuum ionising radiation, or to a lack of the gas producing emission lines, or the emission lines may be so broad that they cannot be distinguished from the continuum.

Jacques ROLAND and Florence DURRET

The discovery of quasars

During the 1950s, the efforts of radio astronomers were directed towards mapping the radio sky in detail, and then towards identifying the optical objects corresponding to the radio sources. The first radio surveys of the sky, made at Cambridge in England, resulted in various catalogues, of which the most important was the Third Cambridge Catalogue (abbreviated: 3C). The survey for this catalogue covered all parts of the sky north of declination $-5°$, and contains all the sources (about 400) for which the radio flux at 178 megahertz is greater than 9 janskys (Jy). Two radio sources of the Third Cambridge Catalogue, 3C 48 and 3C 273, were to be the genesis of the discovery of quasars. Although the first radio galaxies were identified at the beginning of the 1950s, it was more than ten years before radio positions were obtained that were sufficiently accurate to identify new objects. In 1961, the radio astronomer Thomas Matthews, having obtained a precise position for the source 3C 48, realised that the object associated with it seemed to be a star. The optical spectrum was peculiar; it contained numerous emission lines, but none of their wavelengths corresponded with those of known lines. The radio astronomers believed they had found a peculiar radio star, and there the matter rested for the next two years.

Then, in 1963, on observing the occultation of the radio source 3C 273 by the Moon, Hazard, Mackey and Shimmins obtained the position of the source to a precision of one arc second. Using the 100-inch telescope at Mount Palomar, Maarten Schmidt identified this source with an object of star-like appearance, similar to that identified with 3C 48. He obtained a spectrum of 3C 273 which also showed emission lines, but which were recognisable as known lines. The lines were those of atomic hydrogen, shifted towards the red. The observed wavelength (λ) was related to the wavelength (λ_0) obtained in the laboratory by the equation $\lambda = \lambda_0 (1 + z)$, where z is the redshift. In the case of the radio source 3C 273, the redshift was 0.158. Greenstein and Matthews re-examined the spectrum of 3C 48 obtained previously, and realised that the emission lines of 3C 48 were displaced towards the red by a factor of 0.367. These two newly discovered radio sources identified with objects of stellar appearance, were called quasars: this name is a contraction of the expression 'quasi-stellar radio sources'. These redshifts implied that both objects were located far beyond our local group of galaxies.

The discovery of the quasars revolutionised our knowledge of the Universe. The majority of the galaxies that can be seen with the Schmidt telescope at Mount Palomar, which are brighter than magnitude 19 or 20, exhibit spectra with redshifts of under 0.2. Almost all of the galaxies observable with present-day telescopes are located within a sphere of radius (D) of 1200 megaparsecs. The two newly discovered quasars, although much brighter than the most distant galaxies, turned out to be extremely distant objects. Indeed, they were the most distant objects known at that time. Quasars were then considered, a priori, to be the best objects for probing the Universe beyond the most distant galaxies. But, if the quasars are as distant as their redshifts would appear to indicate, then their bright apparent magnitude imply that they are the most luminous objects in the Universe. Their luminosity, being 100 times that of the brightest galaxies known, therefore poses two problems: first, what is the nature of such luminous objects; second, what is their source of energy? Twenty-five years after their discovery, we know the answer to the first question, and we hope to be able to answer the second.

Radio studies of quasars show that many have the extended double radio structure characteristic of radio galaxies. Other quasars consist of a single compact source. The quasars with extended radio structures often also have a compact radio component coinciding with the quasar itself. Radio variability is shown by this component, in the same way as the quasar varies optically. The time-scales of the variability for the optical and radio components are of the order of one year. Quasars have optical angular sizes less than one arc second. In fact, their angular sizes are of the order of 0.001 arc second, as is shown by intercontinental interferometric observations. If the optical variability of the quasar is due to changes in the quasar itself, considered as a single object, then the dimensions of the component which is varying are of the order of the characteristic time-scale of the variability expressed in light years. We therefore deduce that the quasars must be objects with dimensions of the order of a parsec or less, which shows that their volume is 10^{12} times smaller than that of a normal galaxy. Although they were discovered by radio astronomy, optical research has shown that only 10 per cent of the quasars are in fact radio emitters. It is difficult to know whether this figure corresponds to a difference in activity between the radio quasars and the radio-quiet quasars, or if it corresponds to anisotropy of radio emission in quasars: it may be that there are relativistic radio jets which are not observable unless the jets are more or less pointed at the observer. The optical spectra of quasars are similar to those of Seyfert galaxies. Both emit a large part of their energy in the infrared, but quasars are much more intense at X-ray and gamma wavelengths than Seyfert galaxies.

The radio morphology of quasars, their spectroscopic properties, their compact structure of the order of 0.001 arc second, and finally their variability, show the quasars to be yet another form of the active galactic nucleus phenomenon already known in radio galaxies, N galaxies, Seyfert galaxies and BL Lac objects. The only difference between the quasars and nuclei of nearby galaxies is the far greater luminosity of the former, although the most powerful radio galaxies, such as 3C 295, have luminosities comparable with some quasars. It therefore seems that the quasars are the very extreme case of the active galaxy nuclei. Many kinds of observations can confirm this hypothesis. In particular, for the nearest quasars, the galaxy underlying the quasar may be studied. This type of observation is especially difficult because the luminosity of the quasars is much greater than that of a typical galaxy; the light coming from the stellar envelope of the galaxy is completely swamped by that coming from the quasar. Nonetheless, twenty years after the discovery of quasars, it was shown that the nebulosity around the nearest quasar 3C 273 has a size, profile and luminosity comparable with those of giant elliptical galaxies. In the case of the quasar 3C 48, it has been shown that the nebulosity around the quasar partially consists of stars. It was known for some years that it was partly composed of ionised gas; the stellar component has been shown to exist by the presence of absorption lines typical of galaxies. The redshift of the stellar envelope is the same as that of the quasar, which shows that the latter is an active nucleus of a galaxy. At least 200 quasars have now been shown to be embedded in galaxies. To sum up, even if a model of the quasar remains to be constructed, we can nevertheless be certain of the nature of quasars, which are galactic nuclei responsible for the most violent events known at present in the Universe.

Jacques ROLAND and Florence DURRET

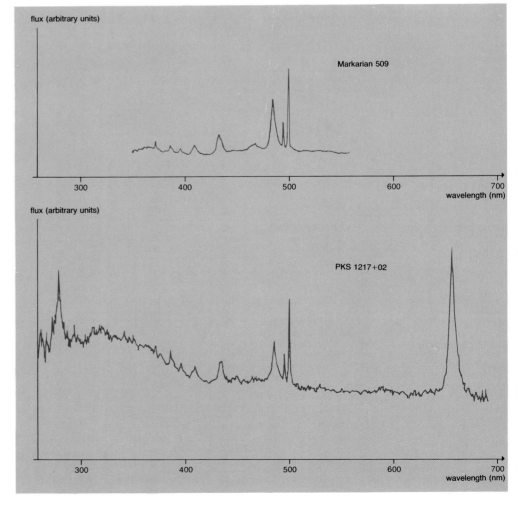

Spectra of the quasar PKS 1217 + 02 and of the Seyfert galaxy Markarian 509. The quasar PKS 1217 + 02 is one of the nearest known quasars. Its redshift is therefore comparable with those of distant galaxies (of redshift 0.2). We remark that the spectra of PKS 1217 + 02 and Markarian 509 are similar in all respects, noting, in the case of the quasar, the intensity of the blue continuum. From the point of view of spectroscopy, quasars cannot be distinguished from Seyfert galaxies, and so they are considered to be the extreme case of active nuclei. In fact, the similarity of the spectra implies that the same phenomena occur in quasars and in the nuclei of Seyfert galaxies, and therefore shows the analogous nature of these two types of object. (After P. Véron)

	Seyfert galaxies (galaxies for which the Balmer lines are broad)	BL Lac objects (galaxies practically devoid of observable emission lines)
N galaxies (nucleus less luminous than the galaxy)	NGC 4151	BL Lac
quasars (nucleus more luminous than the galaxy)	3C 273	OJ 287

The various types of galaxies with extremely active nuclei.

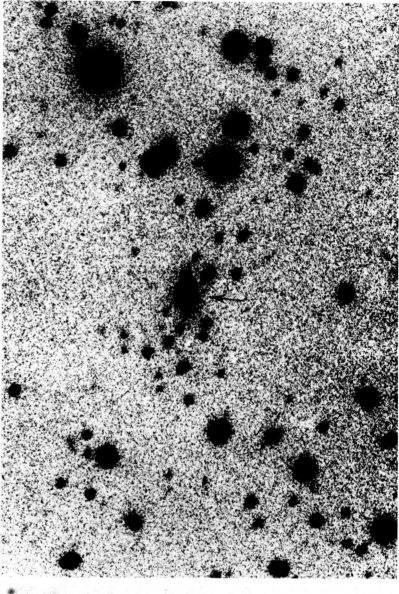

Optical investigation of a quasar using an objective prism. After the discovery of the first two quasars, 3C 48 and 3C 273, astronomers noticed that they were much bluer than normal galaxies. This is due to continuum synchrotron radiation, easily observable in the blue. Because it was difficult to obtain accurate radio positions at that time in order to identify new quasars, a systematic investigation of blue stars on photographs taken with the Schmidt telescope at Mount Palomar was begun, in the hope of discovering new quasars optically. Unfortunately, a large proportion of the objects turned out to be stars, making it impossible to construct a homogeneous sample of quasars.

Nowadays the optical investigations of quasars involve an objective prism, used to obtain the spectra of all of the objects in the field in a single exposure. The Schmidt telescope can be used in this manner to explore a zone 6 degrees square, but the magnitude limit is not very deep. In order to reach faint quasars a 4-metre telescope is used instead of a Schmidt. The field is then much smaller, and only a small part of the sky has been examined in this way, but the number of quasars found now exceeds 1500. As can be seen from the two photographs, quasars cannot be distinguished, a priori, from stars on photographs of the sky, whereas, on objective prism plates, the strong emission lines show up the quasars. Spectra of stars in the field are uniform, unlike the quasar spectrum shown, which has a redshift of 2.06, and is made conspicuous by the presence of the Lyman-alpha line of laboratory wavelength 121.6 nm, observed here shifted to 372.1 nm. (After P. S. Osmer)

The nature of quasars. The photograph shows a nearby quasar, 3C 206, located in a cluster of galaxies. The fact that the redshifts of the galaxies in the cluster and the emission spectrum of the quasar are identical, demonstrates that the redshift of quasar emission lines is indeed indicative of the distance of the quasar. (Wyckoff, ESO)

In the case of the quasar 3C 48, it was shown recently that the surrounding nebulosity partially consists of stars, demonstrating the nature of quasars as active galactic nuclei. Even if satisfactory models of quasars have not yet been constructed, their nature is no longer in doubt. (Hale Observatories)

Quasars, cosmology, and gravitational lenses

There is one very peculiar thing about astronomical observations: when we observe a distant galaxy at, say, 1000 megaparsecs, we observe it as it was approximately three billion years ago. It is therefore obvious that studying very distant galaxies gives us knowledge of their evolution in the course of time. As the great majority of the observable galaxies are located within a sphere of radius of 1200 megaparsecs, it is impossible (with present techniques) to investigate the evolution of galaxies over times longer than four billion years. Now it is generally believed that the galaxies were formed when the Universe was very young and that their formation has long since ceased. With the age of the Universe being at most twenty billion years, the epoch of galaxy formation and of rapid evolution of galaxies is not accessible by observing galaxies four billion light years away. On the other hand, quasars are much further away than galaxies, the most distant quasars being of the order of fifteen billion light years away. (Note that it is impossible to give an exact age for the quasars because it depends on which model of the Universe is adopted, and present-day astronomical observations do not allow us to determine the appropriate model). The study of the quasars is obviously of interest because we can investigate the evolution in time of the activity in the nuclei of galaxies. Very few quasars are observed with redshifts over 3.5. Why this should be so is not clear at present; it may be that the first very active nuclei only appeared after five billion years. Given the prodigious output from the nuclei of active galaxies, their active phase probably does not last for more than 100 million years. We can ask whether this activity is recurrent. There is no observational evidence of

such a possibility. On the contrary, the numbers of very luminous nuclei have considerably decreased with the passage of time. It is estimated that the density of active nuclei at the present day is a thousand times less than it was ten billion years ago. This spectacular reduction in the activity of the nuclei is not yet understood. The study of the structure of the Universe, and the determination of its characteristics (whether it is closed or open) require observations of high redshift objects (redshifts much greater than 0.2). After their discovery, the quasars were considered to be excellent probes for determining the characteristics of the Universe. However, those expectations were not fulfilled, because quasars are not good 'standard candles' – there is no reason why their luminosities should remain constant over periods as long as ten billion years. It is very difficult to see how we can find objects showing little evolution with time which would enable us to probe the Universe over times of ten billion years. On the other hand, studies of the optical spectra of quasars have revealed the existence of a very large number of absorption line systems, permitting analysis both of the distribution of matter on the line of sight, and of its evolution in time. In fact, a galaxy with a redshift of 2 cannot be observed directly, but its presence on the line of sight to a quasar can be indicated through the heavy element absorption lines observed in a quasar spectrum. This has already allowed us to demonstrate the existence of diffuse haloes surrounding galaxies as well as certain properties of the gas distribution for times between ten and fifteen billion years ago.

Since quasars are extremely distant and compact objects, besides opening up the greater part

of the observable Universe, they can be used to define useful systems of reference for astronomical problems and to verify the general theory of relativity. Tests of the theory of relativity are important because several other theories of gravitation have been formulated since general relativity which predict different results at higher precision.

One experiment carried out involving quasars is the measurement of the gravitational bending by the Sun of electromagnetic waves emitted by quasars. A ray of light passing near a massive body is deflected. The confirmation of this deflection of light by the Sun in 1917 was one of the first crucial tests of general relativity. During a total solar eclipse, the stars in the neighbourhood of the Sun were photographed, and their positions measured. Then the measured positions were compared with the known positions; it was found that the stars in the neighbourhood of the solar disk were displaced by just over an arc second. In 1924 the experiment resulted in the angle of deflection being measured with a precision of only 25 per cent.

It would be possible to repeat the experiment with quasars of angular size up to 0.001 arc second for which the absolute position is known to 0.001 arc second. The angle of deflection could be measured to a precision of 0.05 per cent using the intercontinental network of very long baseline interferometers, which will allow the various theories of general relativity to be tested.

The second verification of general relativity must remain qualitative at present in view of the difficulties in the relevant calculations. It consists of obtaining two images of a single quasar in the sky, arising from a gravitational lens effect

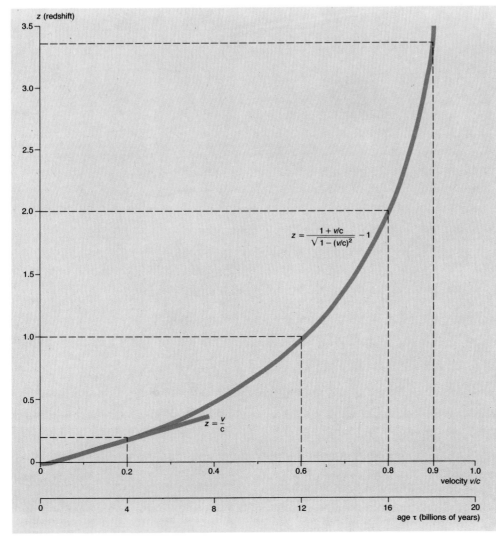

The redshifts of quasars and their epoch of formation. The great majority of the observable galaxies have redshifts of 0.2 or less and are thus located within a sphere of radius of 1200 megaparsecs (assuming $H_0 = 50$ kilometres per second per megaparsec). Galaxies 1200 megaparsecs away are seen as they were four billion years ago. The much more luminous quasars can be seen at much greater redshifts, although very few are known with redshifts over 3.5. The velocity of recession v of a given object is related to its redshift z by the formula $z = [(1 + (v/c)] \ [(1 - (v/c)^2]^{-1/2} - 1$ which is well approximated by $z = v/c$ for redshifts up to 0.2. Thus quasars of redshift 3.5 have velocities of recession 0.9c or 270 000 kilometres per second. Although the age of objects of low redshift (under 0.2) can be derived straightforwardly, for objects of higher redshift the age depends very much on which model of the Universe, open or closed, is adopted. Nonetheless, an upper limit τ for the age of an object in years can always be calculated: $\tau < 10^{12}/H_0 \ (v/c)$. For example, a quasar of redshift 3.5 is less than eighteen billion years old. The age of the Universe itself is less than twenty billion years (obtained by substituting $v = c$). It thus follows that at least three-quarters of the time the Universe has been in existence can be probed with quasars. Why there are so few with redshifts over 3.5 is not known; the simplest explanation is that nuclei of galaxies as active as the quasars had not been formed in the first billion years of the Universe.

caused by a galaxy located on the line of sight between the quasar and the observer.

The other fundamental application of quasars is the possibility of defining a system of reference on the celestial sphere. This system would be made up of about 200 quasars with absolute positions known to around 0.001 arc seconds. Being very distant, quasars do not show any apparent motion on the celestial sphere. They are also very compact, and can be regarded as points: they are therefore ideal objects to define a system of reference which could be the basis of studies of the rotation of the Earth to a greater precision than can be obtained at present. In particular, it would be possible to observe the variation in the rotation of the Earth due to atmospheric motions or movements inside the Earth and to measure the deformations of the crust as well as continental drift. Finally, we note that observations of the apparent motion of the nucleus of our own Galaxy would demonstrate the motion of the Sun around the galactic centre.

Jacques ROLAND and Florence DURRET

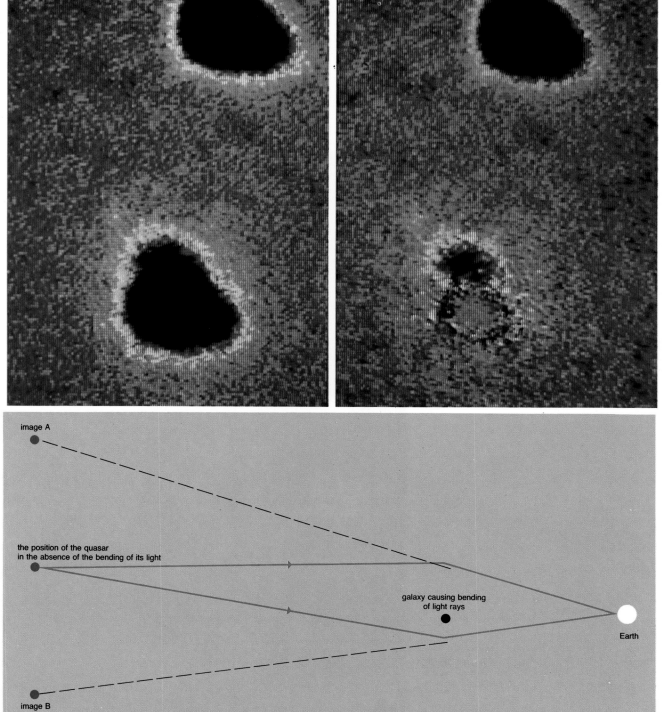

An example of a gravitational lens: the quasar 0957 + 561 A, B, (opposite right). The photographic sky survey made with the Schmidt telescope at Mount Palomar shows that the image of the quasar 0957 + 561 is double. The spectra of the two barely resolved images are in fact so similar that the possibility is excluded of a galaxy nucleus with two active and distinct components (as happens sometimes in supergiant galaxies of the cD type). The only possible interpretation is that the first gravitational lens has been found. Rays of light from the quasar are deflected when passing near a galaxy on the line of sight to the quasar, forming two images of the original quasar (shown here in false colour). When a computer is used to analyse the images by making the intensities of the northern and southern images equal, and then subtracting the northern image from the southern, the image of a distant galaxy is seen 1 arc second to the north of the southern image. This is the galaxy which has given rise to the two quasar images, themselves six arc seconds apart. (Photograph kindly communicated by Alan N. Stockton, Institute for Astronomy and Planetary Geosciences Data Processing Facility, University of Hawaii)

Absorption spectrum of high-redshift quasar (below). The optical spectra of quasars may show a large number of absorption lines in addition to the emission lines. This is particularly the case for the most distant quasars at redshifts above 2. There are three types of absorption line systems in quasars. Firstly, there are broad lines with redshifts very close to the emission-line redshift, which originate from absorbing clouds surrounding the quasar. These clouds may fall towards the quasar or move away from it. For clouds falling onto the galactic nucleus, the absorption line has a redshift greater than that of the nucleus; for ejected clouds the reverse is the case. Thus the redshift measured depends not only on the velocity of recession of the quasar due to the expansion of the Universe, but also on the motion of the clouds around the quasar.

The second kind of absorption line system consists of narrow lines whose redshift is generally very much less than the emission-line redshift of the quasar. Such lines are due to absorbing matter located somewhere on the line of sight between the quasar and the observer, such as a diffuse galactic halo. Although most such galaxies are too distant to be observed directly, this effect

has been clearly shown to occur in the case of the apparent association between 3C 232 and the galaxy NGC 3067 which is close enough to us to be photographed (photograph below right, from the Palomar Observatory Sky Survey). Although one object is much further away than the other, the quasar (indicated by the lines) and the galaxy (centre) are very close on the celestial sphere. The quasar spectrum contains absorption lines which have the same redshift as the absorption lines of the galaxy and which are due to absorption of the light from the quasar by the (invisible) galactic halo lying on the line of sight.

The third kind of absorption lines are seen, for example, in the spectrum of the quasar PKS 2126–158 (below left). These make up what is called a 'Lyman-alpha forest' and consist of dozens of narrow absorption lines attributed to the Lyman-alpha line of atomic hydrogen. There is much evidence to show that they are due to multiple clouds of intergalactic hydrogen lying on the line of sight, which are more than ten billion years old. (After P. Young *et al.*, 1979)

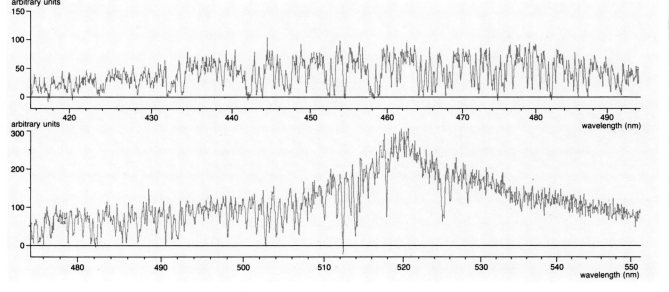

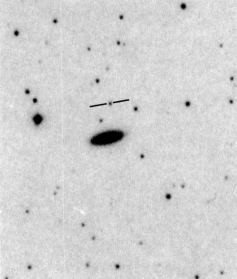

Quasars and the large scale structure of the Universe

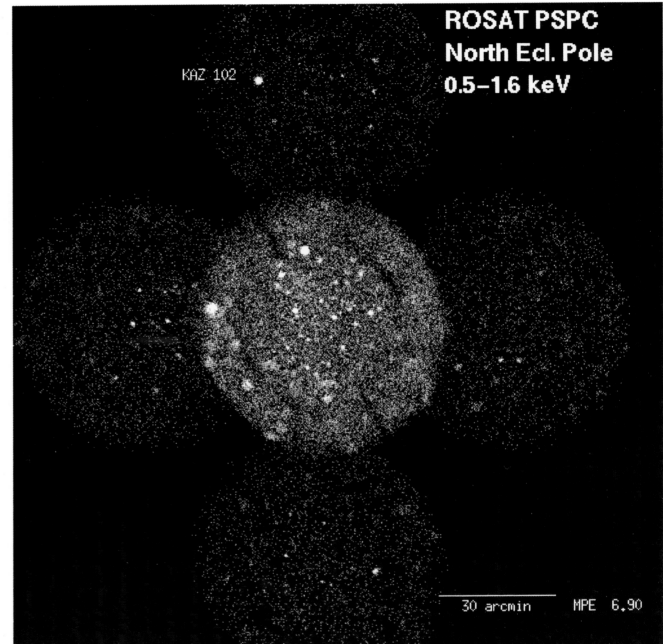

ROSAT PSPC
North Ecl. Pole
0.5–1.6 keV

KAZ 102

30 arcmin   MPE 6.90

Several decades after the identification in 1963 of the first quasar, 3C 273, searches for these still mysterious objects continue. Numerous past and current sky surveys have been performed with the aim of finding as many as possible, determining their distances and studying their spatial distribution in the Universe as well as their evolution in time.

Despite these efforts the number of known quasars – several thousand – is much smaller than the potentially observable number, which is several million if we assume that every quasar brighter than 22nd magnitude is detectable. However, there is no need to catalogue all observable quasars if we can establish complete samples of objects representative of the whole population. Quasar searches are slow and complicated, and it has been possible to explore only tiny parts of the sky systematically. The main difficulty is to pick out the quasars from all the other astronomical objects: on a photographic plate or a digital image from a CCD detector a quasar is indistinguishable from a star (remember that quasar is the contraction of *quasi-stellar object*). Methods of finding quasars are thus based on certain properties such as the presence of intense emission lines, excess radiation in the ultraviolet and X-ray regions, intensity variations or radio emission. This provides a list of 'quasar candidates' from quite rapid imaging observations of the sky. Confirming that a given object is a quasar requires a detailed spectrum and is much more costly in telescope time. The use of objective prisms allows one to acquire images and spectral information simultaneously: we get a low-resolution spectrum for all objects in the field, and quasars reveal themselves by the presence of intense emission lines which can be recognised by a computer on digital images. In this way about ten very distant quasars have been found with redshifts greater than 4. Further, fibre optic systems now allow one to obtain detailed spectra of several objects in a field simultaneously and make possible statistical investigation of faint and distant quasars. The population of unidentified or undetectable quasars is revealed by its collective contribution to various radiation backgrounds, particularly in the X-ray region. The results of the ROSAT X-ray satellite support the idea that much of the diffuse X-ray background may come from quasars.

The search for high-redshift quasars involves the study of the edges and past of the Universe. The now-classic interpretation is that the quasar redshift $z = \Delta\lambda/\lambda_0$ is cosmological, and related to the distance d between the quasar and Earth by the formula $z = dH_0/c$ (to first order), where H_0 is the Hubble constant characterising the expansion of the Universe and c the speed of light. Observing a quasar at high redshift means seeing it as it was when the light reaching us was emitted, i.e. several billion years ago. From a representative sample of quasars at various redshifts we can hope to develop a typical evolutionary sequence for quasars and answer several questions. When did the quasar phenomenon

The X-ray background. This composite image of the sky in X-rays (with energies between 0.5 and 1.6 kiloelectron volts) in the direction of the North ecliptic pole was obtained by the PSPC (Position-Sensitive Proportional Counter) instrument on the ROSAT satellite, with exposures of 49 000 seconds for the central field and 5000 seconds for the adjacent fields; the angular resolution is of order 25.

The point X-ray sources visible in the central field account for about 30 per cent of the total detected flux. The optical identification of these sources is not yet complete, but similar studies for completely analysed fields suggest that the majority of them (about 80 per cent) are quasars and other active galactic nuclei. The remainder of the total flux (about 70 per cent) comes from

unresolved emission.

This image also allows the detection of a fairly close cluster of galaxies, revealed by weak extended emission whose surface brightness is less than 50 per cent of the background (a little above and to the left of the centre of the central field).

Deeper observations in a direction where X-ray absorption by the interstellar medium of our Galaxy is weaker have shown that about 45 per cent of the diffuse background can be resolved into discrete sources. The source Kaz 102 is a previously-known quasar belonging to the Kazanian catalogue. (Max-Planck-Institut für Physik und Astrophysik)

appear in the history of the Universe? What is a quasar's lifetime? Can a quasar be reactivated by interaction with galaxies? Are there different types of quasars (e.g. radio-loud quasars, about 10 per cent of known quasars, and radio-quiet quasars) or do the observed differences result from different evolutionary stages or differing orientations of the quasar rotation axis to the line of sight? A long-debated question is whether the quasar number decreases with redshift. We now know that quasars exist out to redshifts greater than 4, since about ten have been identified. Present data suggest that only the density of faint quasars decreases above redshift 3.5. The identification of quasars out to redshifts of order 5 shows that these objects were already formed

only several billion years after the Big Bang, implying a significant constraint on models of the evolution of the Universe.

The role of quasars as markers of the Universe connects the study of their distribution to a crucial problem of modern cosmology: how to reconcile the near-perfect global isotropy of the early Universe deduced from observations of the 2.7 K background radiation (particularly by the COBE satellite) with the undeniable presence of matter inhomogeneity at various scales (galaxies or quasars, clusters and superclusters of galaxies, voids and filaments). The fluctuations giving rise to these structures should have left their traces on the 2.7 K background radiation, but have not been detected. The existence of inhomo-

376

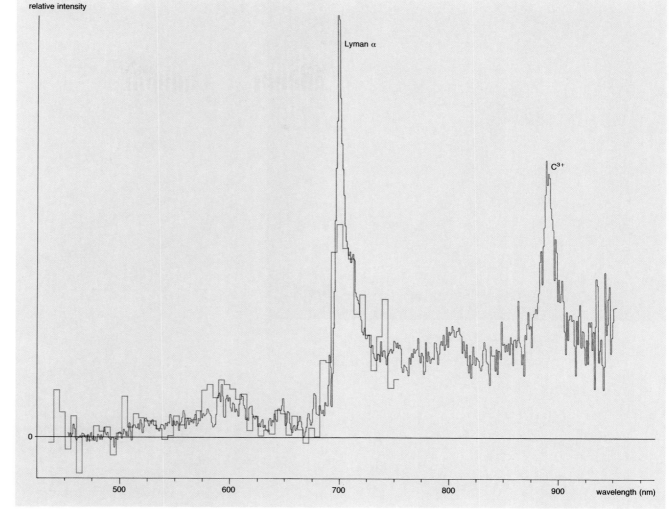

relative intensity

Lyman α

C³⁺

Spectra of a distant quasar, PC 1158+4635. In blue, a low resolution (13 nanometres) spectrum of this object obtained in 50 seconds during a sky survey showed the presence of a very strong and asymmetric emission line as well as an abrupt drop in emission to the blue side of the line (towards shorter wavelengths). This strongly-redshifted Lyman α line of hydrogen (it is observed at 700 nanometres while the laboratory wavelength is 121.6 nanometres), shows that the object is a very good quasar candidate.

This was confirmed by detailed spectroscopic study: shown in red, a spectrum with high signal to noise, obtained with about 3600 seconds integration time (resolution 2.5 nanometres) showing several spectral lines, in particular the emission line of the carbon ion C^{3+}. These allow a precise determination of the quasar redshift: at 4.733 this was the largest redshift known at the beginning of 1991 (in April of that year a quasar with even higher redshift – 4.897 – was discovered; this is PC 1247+3406).

The lower intensity observed at wavelengths less than 700 nanometres is caused by a large number of narrow Lyman α lines (the Lyman α forest); a spectral resolution of 0.1 nanometres would be required to pick them out individually. (after P. Schneider, M. Schmidt and J.E. Gunn, 'PC 1158+4635: An optically selected quasar with a redshift of 4.73', in *The Astronomical Journal*, vol XCVIII, no 6, pp 1951–8, 1989)

Distribution of quasars over the sky. This negative, obtained in the visible, shows the distribution of quasars (shown by circles) over the sky in a field of 1.3°×1.6° around the nearby galaxies NGC 470, NGC 474 and NGC 520. The *apparent* alignments in projection on the plane of the sky might suggest a physical connection between these galaxies and the quasars, despite the high redshifts of the latter (indicated beside each one). However, statistical studies of a larger number of objects argue that the alignments are accidental. At present only the cosmological interpretation of the redshifts gives a coherent framework for describing and understanding the distant Universe. (after H. Arp and O. Duhalde, 'Quasars near NGC 520', in *Publications of the Astronomical Society of the Pacific*, vol XCVII, no 593, pp 1149–57, 1985)

geneities in the observed galaxy or quasar distributions imposes tight constraints on allowable cosmological models. Thus looking for irregularities in the quasar distribution, particularly at scales greater than a few hundred megaparsecs, is fundamental in view of the fact that only quasars are detectable in the remote Universe. Several significant clumps have been detected at scales from 50 to 200 megaparsecs (with $H_0 = 100$ km s^{-1} Mpc^{-1}) and for redshifts from 0.4 to 1.3. We should also mention highly speculative searches for 'exotic' structures in the quasar distribution (double fields, etc) which might signal the presence of a cosmic string or unorthodox structure (multiply-connected, foamlike...) in the Universe.

Quasars are also interesting in that they reveal the presence of matter along the line of sight. They act as probes of the intergalactic space between themselves and the observer. Several effects result from intervening objects. The first is gravitational: if a galaxy (or cluster of galaxies) of large enough mass is aligned with the quasar, the emitted light is deflected before reaching us, which can lead to distortion or amplification of the quasar image. We sometimes even observe several images: this is a gravitational lens. Another effect is produced not by the entire galaxy but by those of its stars which pass very close to the quasar line of sight: the mass of a star is clearly not enough to distort the images formed by the galaxy appreciably, but it is able to produce an intensity variation lasting several months as it passes across the quasar. Detection of this effect requires accurate photometry of the images over several months and is therefore difficult. It is nevertheless extremely important, as the speed of the variation depends among other things on the small-scale structure of the quasar: passing in front of it, the star acts as a magnifying glass and allows

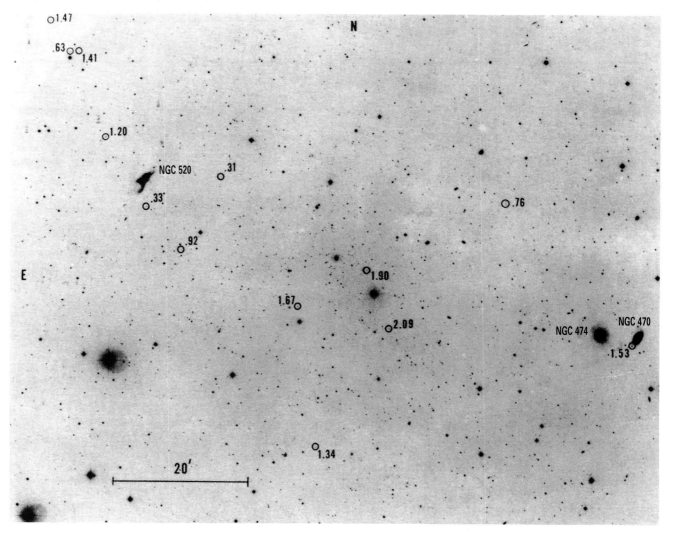

N

○1.47

.63 ○○ 1.41

○1.20

NGC 520   ○.31

.33 ○

○.76

.92 ○

○1.90

1.67 ○

○2.09

NGC 474   NGC 470

1.53 ○

E

○1.34

20'

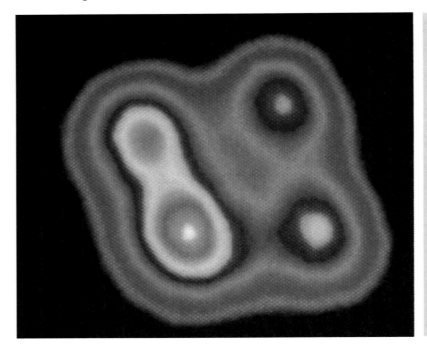

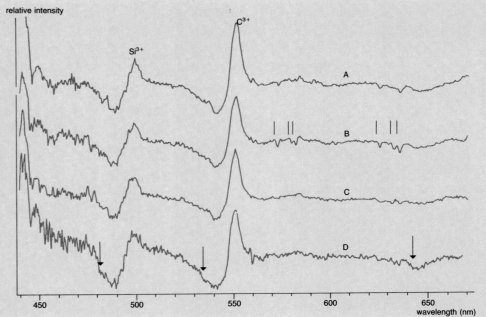

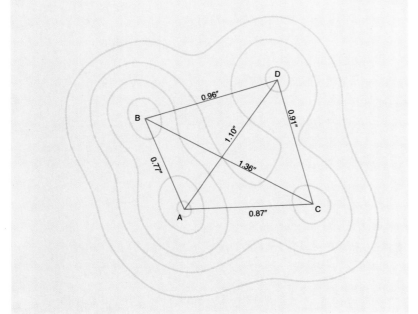

A remarkable gravitational mirage: the 'four leaf clover' (H 1413+117). The false-colour CCD image was obtained at the CFH 3.6 metre telescope through a red filter. It shows the four components A, B, C and D of the quasar H 1413+117, separated by about 1 arc second on the sky (the figure shows the exact angular distances).

The spectra of these components were taken with the two-dimensional spectrograph SILFID (French acronym for *direct fibre imaging integrated linearising spectrograph*). This uses a system of optical fibres to allow simultaneous observation of the four images. Their spectra are very similar, showing that these are multiple images of a single source produced by gravitational deflection. The emission lines at 500 and 550 nanometres arise from ions of silicon Si^{3+} and carbon C^{3+} (laboratory transitions at 140 and 155 nanometres, implying an emission redshift of 2.55).

H 1413+117 is a broad absorption line quasar, as shown by the characteristic shape of the spectrum to the left of the two visible emission lines. The absorption lines (marked by vertical lines) reveal a significant difference between the spectra: they are much more noticeable in component B. They belong to two systems whose redshifts are 1.4382 and 1.6603. It is very likely that at least one of them is caused by the same galaxy which acts as the lens. The slightly different paths through the halo of the intervening galaxy taken by the light of images A, B, C and D explain the differing intensities of the observed narrow absorption lines. One can also see subsidiary features (marked by arrows) in the broad absorption line profiles of component D; these features suggest that the source has intrinsic temporal variations (appearing in A, B, C, and D at different times because of the unequal light paths), or that an individual star (as well as its host galaxy) is involved in the light deflection. (after M.-C. Angonin, M. Remy, J. Surdej and C. Vanderriest, 'First spectroscopic evidence of microlensing on a BAL quasar? The case of H 1413+117', in *Astronomy and Astrophysics* vol CCXXXIII, no 1, pp 5–8, 1990)

us to measure the size of the quasar emission region.

A second signature of the intergalactic medium involves the presence of gas along the line of sight, whereas the preceding phenomenon depended only on the mass of the intervening matter. This second signature was discovered immediately after the identification of the first quasars: as well as the broad emission lines which characterise them, quasar spectra often show narrow absorption lines. One can identify the atoms or ions involved (carbon, oxygen, or magnesium for example) and thus the transitions between energy levels, and deduce the redshift of the absorbing gas. Most of the lines are in the ultraviolet, which is much richer in transitions than the visible region: they are redshifted into the visible region and are thus observable from the ground. We should stress that only a small amount of gas (much less than that encountered in crossing the disk of a galaxy) is needed to produce these lines. Detecting these absorption-line systems is therefore a very sensitive means of probing intergalactic space.

We can distinguish three categories of absorption lines. Some have a similar redshift to the quasar and result from matter ejected from it. This is not surprising, as the jets of radio sources (some of which are associated with quasars) clearly show that ejection can occur. In fact radio-loud quasars show an excess of these absorption line systems. We also encounter very broad lines in this group: the ejection velocity

deduced from the emission and absorption redshifts may reach 60 000 kilometres per second; unlike the narrow lines, these systems are only observed towards radio-quiet quasars.

Narrow lines of elements other than hydrogen detected at redshifts much lower than that of the quasar can be explained as resulting from absorption in galactic halos. For low-redshift systems (around 0.5) it has been possible to identify directly the galaxy whose halo produces the absorption lines. The halo sizes can be estimated as about 40 kiloparsecs. These are therefore much more extended than galactic disks (about 10 kiloparsecs) and seem to be present around most galaxies. They may result from supernova explosions in the disk, the matter ejected at high velocity carrying interstellar gas out with it to large distances. Gas from satellite dwarf galaxies might also be torn out by tidal forces and progressively dispersed. However they result, the gaseous halos revealed by quasar absorption lines are ideal objects for cosmology. Unlike the galaxies they surround they can be observed out to redshifts as high as those of distant quasars ($z=4.897$ in 1992). The parameters characterising them (size, ionisation degree, element abundances) probably vary with redshift (i.e. with cosmic time). If we can decipher these variations we may be able to discover the conditions holding in the distant Universe (ambient radiation, gas density). Moreover, the study of these line systems may allow us to discover if the observed galaxy distribution in the nearby Universe (walls, fila-

ments and voids) is also present at redshifts of order 2–3. In some rare cases the line of sight passes so close to the centre of an intervening galaxy that it crosses its disk. We can then expect to find gas with properties similar to the well-studied properties of interstellar cloud in our Galaxy. Atomic hydrogen has been detected towards several quasars through the 21 centimetre radio absorption line, and in one case molecular hydrogen has been detected.

Finally, studies of quasars with quite good spectral resolution (several tens of nanometres) reveal a third category of lines, and thus a new class of objects, the Lyman α clouds. Their name comes from the fact that their only signature is a Lyman α absorption line; elements such as carbon, oxygen and silicon, visible in the earlier cases, are too underabundant to be detected here. The exact nature of these objects remains enigmatic: we do not know whether they are dwarf galaxies, forming galaxies, or primordial clouds. Nevertheless several features are clear: the number of lines is very large (hence the name Lyman α forest for the spectral region where they appear), and is about thirty times greater than the number of systems resulting from galactic halos at redshift 2. Moreover, this number increases rapidly with redshift. We can also see that the spatial distribution of Lyman α clouds is much more uniform than that of galaxies in the nearby Universe (there are, in particular, no clear voids). We would need detailed study of low-redshift objects of this type in order to understand these surprising properties.

If we assume that the Lyman α clouds represent the discrete component of the intergalactic medium, we can ask if there is a diffuse component distributed throughout space. Analyses of quasar spectra give a partial answer to this question by showing that the density of atomic hydrogen must be extremely low. Consequently any diffuse intergalactic medium must be very strongly ionised. Further, the absence of distortion from the theoretical spectrum of a black body in the 2.7 K background shows that the density of any such medium also must be very low.

Quasars thus give us precious information about matter along their lines of sight. But there is a reverse side to the coin: the physical effects revealing the presence of this material may affect our determination of the properties of quasars and their distribution in the Universe. Gravitational lensing makes some quasars brighter for example, leading to overestimates of their luminosities, and perhaps making them detectable at all. On the other hand, if the gas producing the absorption lines is associated with absorbing and scattering dust, like the interstellar medium in our Galaxy, a quasar situated behind a galaxy will appear fainter than in reality. These two effects can profoundly influence our determination of the quasar luminosity and distribution functions as they affect near and distant objects differently; only the latter have a non-negligible probability of having a galaxy on their line of sight. The quasar population must be studied in close comparison with other objects which can affect the light they emit.

Patrick BOISSÉ and Hélène SOL

A quasar with low-redshift absorption lines: PKS 1127-14. The spectrum of this object clearly shows an emission line of the carbon ion C^{3+} (at 155 nanometres in the laboratory, implying an emission redshift of 1.18) as well as three absorption lines of the magnesium ion Mg^+ (a doublet at 279.6 and 280.3 nanometres) and magnesium itself (at 285.2 nanometres), characterised by an absorption redshift of 0.3130.

The image of the field around the quasar is shown in false colours giving the intensity received through a red filter; the pixel size is 0.67 arc seconds by 0.67 arc seconds and the breadth of the image is 34 arc seconds. The image shows two non-stellar objects (2 and 3) near the quasar; the nearer is at 9.6 arc seconds and its spectrum is that of a galaxy with a redshift of 0.313, identical to the absorption redshift. This galaxy thus has a halo large enough to cover the quasar image: its radius is at least 30 kiloparsecs. (data obtained at the 3.6 metre ESO telescope; after J. Bergeron and P. Boissé, 'A sample of galaxies giving rise to Mg II quasar absorption systems', in *Astronomy and Astrophysics* vol. CCXLIII, p. 344 1991)

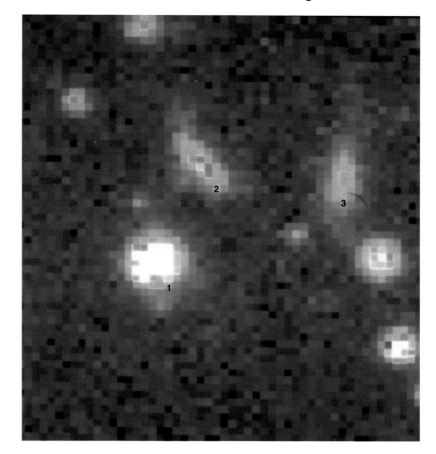

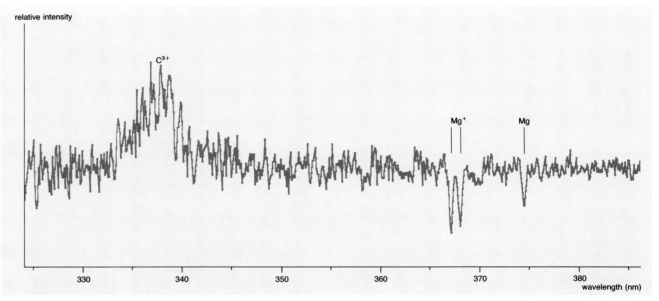

Intergalactic hydrogen clouds and the Lyman α forest. This shows a portion of the blue spectrum of the quasar 2206 – 199 N (emission redshift 2.55) at high resolution (0.04 nanometres). The very strong absorption at 374 nanometres is a Lyman α line (121.6 nanometres in the laboratory) with absorption redshift 2.07625 (the dashed curve is a theoretical fit to the line profile). Its noticeable broadening does not result from the usual turbulence of the absorbing gas but from the very large number of hydrogen atoms along the line of sight, which produce a high opacity. Although this amount of gas is completely compatible with the presence of a normal galactic disk, the abundances of elements such as carbon, oxygen and silicon are extremely low (less than 1 per cent of solar abundances). We are thus probably seeing a recently formed galaxy which has undergone little chemical evolution. Most of the other narrow lines are also Lyman α without any associated lines of heavy elements: they constitute the Lyman α forest (after M. Rauch, R.F. Carswell, J.G. Robertson, P.A. Shaver and J.K. Webb, 'The heavy element abundances in the z = 2.076 absorption system towards the QSO 2206 – 199 N', in *Monthly Notices of the Royal Astronomical Society*, vol. CCXLII, no 4, pp 698–703, 1990)

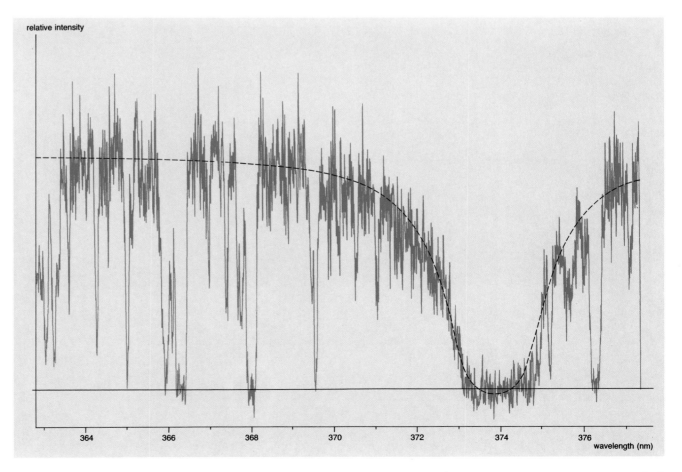

Energy and black holes in galactic nuclei

Although observations have made it possible to show that quasars are extremely active galactic nuclei and to discover the general properties of galactic nuclei, as yet, no wholly satisfactory model of active galactic nuclei exists which will take account of those general properties. Nonetheless, it is now possible to distinguish the general constraints which must be satisfied, in a model of an active galactic nucleus. For the model to be acceptable, it must not only be coherent, and explain the observed properties, but it must also predict new properties which can be experimentally verified. The main features of such a model must be mapped out using the accumulation of observational data before the model can be elaborated.

The first characteristic of active galactic nuclei is the colossal energy which they liberate. Very active nuclei give out 10^{53} to 10^{54} joules in 100 million years. To give some idea of how great these values are, the Sun, whose luminosity is 4×10^{26} watts, will give out 10^{44} joules if it shines for ten billion years, corresponding to the conversion of 10^{27} kilograms (from the equation $E = mc^2$). The mass of the Sun is 2×10^{30} kilograms, so it will be seen that the Sun annihilates one two-thousandth of its mass in ten billion years. A normal galaxy, therefore, consisting of 100 billion stars like the Sun, consumes only 0.005 solar mass per year to maintain its luminosity. An active galactic nucleus, which liberates 10^{54} joules, consumes 5×10^6 solar masses in only 100 million years, or 0.05 solar mass per year. We thus conclude that the process of conversion of mass into energy in an active galactic nucleus must be more efficient than nuclear reactions. Astrophysicists have therefore tended to think that the ultimate energy source of active galactic nuclei must be gravitational energy because it is the only form of energy which can be converted with an efficiency higher than that of nuclear reactions. Nonetheless, even though various processes are known which convert gravitational energy into radiant energy,

exactly what goes on in active galactic nuclei is not clear. The energy liberated by an active nucleus is extraordinarily great; in the case of radio galaxies it is possible to estimate how much mass is ejected by the nucleus to form the extended radio lobes. Although it is difficult to make an estimate from radio observations, values between 10^9 and 10^{11} solar masses have been found for various radio galaxies. The mass of a normal galaxy is 10^{11} solar masses; masses of as much as 10^{13} solar masses are found for supergiant galaxies of the cD type. The mass ejected by an active nucleus may thus be 1 per cent or more of the galactic mass.

The colossal energies liberated by an active galactic nucleus, together with the enormous masses which may be ejected, show active nuclei to be extraordinary objects within galaxies. Yet two other parameters of active nuclei, their volume and mass, make them entirely unique. Defining and determining the radius of an active nucleus is not easy, but a nucleus defined as containing the primary energy source is smaller than the most compact sources observed in radio astronomy. The nearby active nuclei have sizes less than 1 light-year. For models of nuclei involving a supermassive black hole, this size corresponds to that of the accretion disk around the black hole. For galaxies with only weakly active nuclei (like our Galaxy) the size of the nucleus is less than a light-month.

Given the order of magnitude of the size of the nucleus, it remains to determine the other fundamental parameter, the mass. It is difficult to estimate the mass of active nuclei by optical means; direct observation of the region of the neighbourhood of the nucleus will only be possible with resolutions of 0.001 arc second or less. As the limit of resolution for Earth-bound telescopes is one arc second, the regions at present observed in fact correspond to volumes rather greater than those containing an actual supermassive black hole and its accretion disk. Nonetheless, optical observations show that for

nearby active galaxies a volume not more than 100 parsecs in radius contains a mass of 10^8 to 10^9 solar masses. There are several methods of determining the mass of a nucleus, based on the effects of the mass on its environment. The rotation curve of the velocities of stars around the nucleus can be studied – it can be shown that the form of the curve requires the presence of about 10^9 solar masses in a sphere of radius of 100 parsecs (about 300 light-years). Studies of profiles of broad emission lines also define the mass of the central object. Unfortunately, results obtained by the latter method may be interpreted in many totally different ways, such as if the profiles were due to gas being ejected from the nucleus. It may be that optical observations with the Space Telescope will show that a mass of 10^8 to 10^9 solar masses is contained in a sphere of radius less than or equal to 1 light-year. Taken overall, the properties of active nuclei – colossal masses within small volumes – tend to favour a model involving a supermassive black hole of 10^9 solar masses.

The most active galactic nuclei are not observed in the neighbourhood of our own Galaxy but at distances of about ten billion light years. This shows that the numbers of active nuclei have decreased in the course of time. The lifetime of an active nucleus is about 100 million years and it is remarkable that the part of the Universe containing the most active nuclei is about ten billion light years away. Whether every galaxy has at some time had a nucleus in an active phase is not known, but the majority of galaxies, such as our own, have very weakly active nuclei.

Active galactic nuclei are, with their parameters such as volume and mass, very peculiar objects. Galactic nuclei are the seat of exceptionally violent phenomena over very short times compared to the age of a galaxy, and their activity in the majority of cases probably has no influence on the rest of the galaxy.

Jacques ROLAND and Florence DURRET

Accretion disk around a black hole and formation of jets from the nucleus (below). (a) The gravitational field of a black hole sucks in the gas in its vicinity. This gas is flattened into a thin disk by the combined effect of gravity and the centrifugal force resulting from the rotation of the gas around the hole (region I). The gas is warmed by its own viscosity and accelerated by the field towards the black hole. When the temperature rises, the disk becomes thicker, and a bulge is formed (region II). Close to the black hole, the temperature reaches many millions of degrees, and the flow is turbulent (region III). About 80 per cent of the energy radiated by the gas before it falls into the hole is in the form of X-rays. Region I is 20 times larger than region II, which is 100 times larger than region III.
(b) The gas surrounding the black hole and the accretion disk contains magnetic fields (B). The magnetic field of the disk creates an electric field (E) perpendicular to the disk. Electrons and protons of the gas surrounding the black hole are accelerated in two directions perpendicular to the disk, giving rise to two jets of relativistic electrons responsible for the synchrotron emission. (After B. Carter and J.-P. Luminet [left] and Lovelace [right])

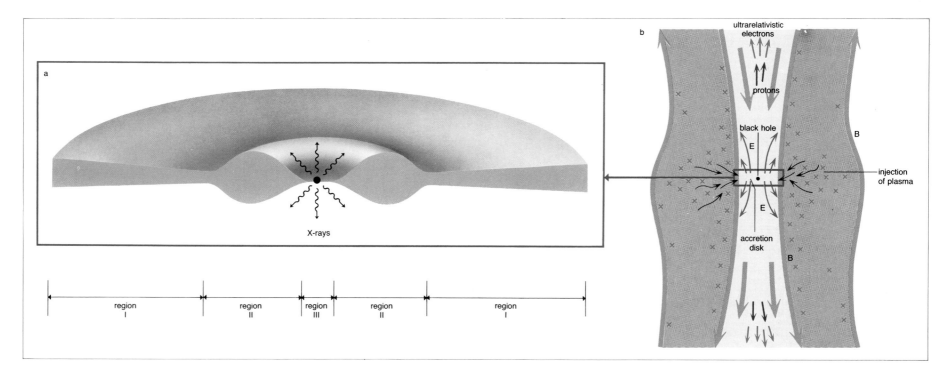

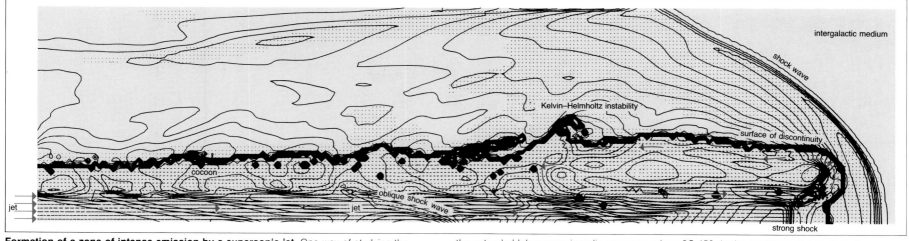

Formation of a zone of intense emission by a supersonic jet. One way of studying the formation of the extended components of radio sources is to make numerical simulations of hydrodynamic systems. In the model of Norman and his colleagues, for example, the nucleus ejects a supersonic jet which behaves hydrodynamically; the effects of the magnetic fields and relativistic particles are assumed to be negligible, with the predominant influence being exerted by the thermal component of the ejected plasma. A strong shock forms at the tip of the jet. The backflow is subsonic and turbulent. Electrons from the flow cross and re-cross the strong shock and become ultrarelativistic, giving rise to intense synchrotron emission from the zone. Such zones are seen in sources such as 3C 452 and Cygnus A. In the turbulent subsonic backflow just behind the shock the magnetic field of the jet (10^{-6} gauss) is amplified by a dynamo effect up to 5×10^{-4} gauss. Then the plasma returns towards the galaxy, giving rise to a cocoon which corresponds to the extended lobes seen in radio sources such as 3C 452. In the most powerful sources, the velocity of the jet is twenty per cent of the speed of light, i.e. 60 000 kilometres per second. The jet cannot consist of electrons and positrons (as is supposed by some authors) because it is the kinetic energy of the protons which, just behind the shock, amplifies the magnetic field up to 5×10^{-4} gauss. A model such as the above can be used to explain radio sources as powerful as Cygnus A if the jet injects about 2 solar masses per year, which represents a total mass loss of 10^8 solar masses for a radio source of this type. In the model of Norman and his colleagues it is the presence of the very thin intergalactic medium (with 10^{-6} electrons per cubic centimetre) which gives rise to the frontal shock. The presence of the cocoon increases the stability of the jet through Kelvin–Helmholtz instabilities but the overall state of the jet is not clear at present. (After Norman *et al.*, 1982)

Jet associated with the galaxy M87 (Virgo A). The radio source Virgo A was identified as early as 1949 by Bolton, Stanley and Slee with the galaxy M87, also known as NGC 4486. The galaxy is the brightest in the Virgo Cluster. The bright jet was discovered when a short exposure was made in order to study the optical nucleus of the galaxy. It was much later that the jet was found to emit at radio wavelengths; its optical and radio emission are both of synchrotron origin. The jet, which is bright compared with those of Cygnus A, cannot be described by a model of the Norman type; the effects of the magnetic field and of the relativistic particles are not negligible within the jet. Other radio galaxies are known to have jets like that of M87. (J.-L. Nieto, Canada-France-Hawaii)

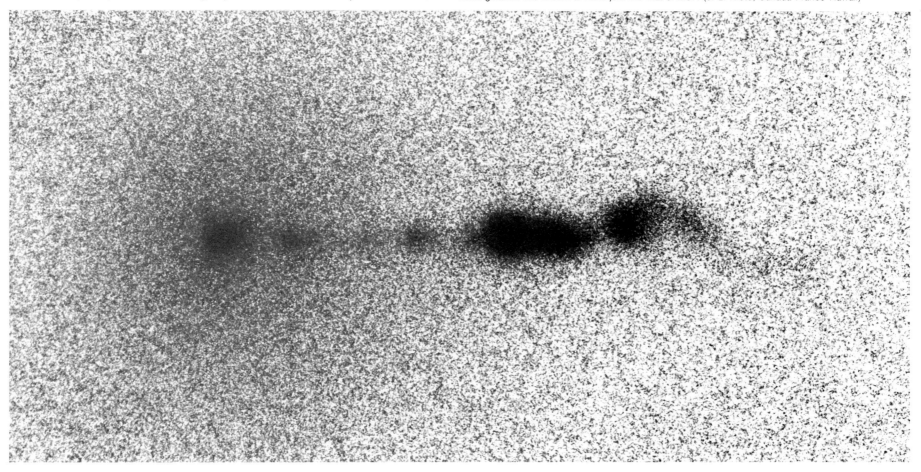

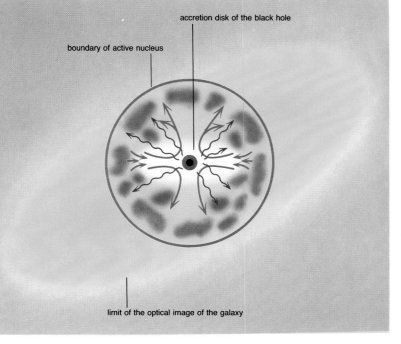

Example of a model of a galactic nucleus (left). At the beginning of the 1950s, radio galaxies were believed to result from collisions between galaxies, gas in the one galaxy colliding with gas in the other, dissipating the kinetic energy of the galaxies by forming plasma, giving rise to radio emission. After the discovery of quasars, many models of active nuclei were proposed; it was already known that the activity of Seyfert galaxies and radio galaxies resulted from activity in the galactic nucleus. In the first type of model, supernova outbursts were involved. As the centre of a galaxy contains a dense concentration of stars, the supernova rate there is higher than anywhere else in the galaxy; also, collisions between stars favour the formation of supernovae. However, such models were not able to account for the general properties of active nuclei – in particular why matter should be ejected in preferred directions – and were therefore abandoned. The next type of model involved a supermassive object of 10^6 to 10^8 solar masses at the centre of the galaxy, the gravitational collapse of this 'spinar' having been halted by an intense magnetic field. The rotation of the spinar would then accelerate the particles in its neighbourhood in such a way as to form the ultrarelativistic components responsible for the radio emission. Such models resemble the current models of pulsars in supernova remnants, apart from the great difference in mass. As the evolution of spinars results in the long term in gravitational collapse to form black holes, the models for active galactic nuclei that are now most favoured involve a supermassive black hole of mass 10^6 to 10^8 solar masses at the centre of the galaxy. The edge of the hole (region III in the diagram on page 358) is a source of X-rays and of intense ultraviolet radiation (violet colour) which ionises the gas clouds around the hole. The broad emission lines arise from dense clouds (green colour) which lie within a spherical volume of a radius equal to that of the accretion disk. The great width of the lines arises from the motions of the clouds, which, besides orbiting around the hole, are subject to radiation pressure from the central source. Clouds which are not very dense (red–brown colour) give rise to narrow lines. Note that the diagram is not drawn to a constant scale; the radius of the zone where X-rays and ultraviolet radiation originate is 0.001 parsec; the radius of the zone containing the clouds originating the broad lines is between 0.01 and 0.1 parsec; the zone of narrow line formation extends to 1 parsec; and finally, the radius of the galaxy is 10 kiloparsecs. (After Suzy Collin)

Interactions between galaxies

All populations have their oddities – and the world of galaxies is no exception. Galaxies can be classified morphologically into four well-defined types: ellipticals, lenticulars, spirals and irregulars. However, one to two per cent of observed galaxies are abnormal in the sense that it is impossible to include them in this classification: they are termed peculiar. These peculiar galaxies are generally found in small groups of galaxies, and often in pairs which immediately suggests that the reason for their deformity is gravitational interaction at short range or even collisions between galaxies.

Galaxies are more or less scattered throughout the Universe. Studies of the nearby bright galaxies have shown that the average distance between two typical galaxies, each 50 kiloparsecs in diameter, with a mass of 10^9 solar masses and a velocity of recession of a few hundred kilometres per second, is of the order of 3000 kiloparsecs. Consequently, the probability of galaxies meeting one another seems wholly negligible: only one collision in ten thousand billion years would occur, or one collision in a time 100 times the age of the universe and thus 100 times the age of the galaxies themselves. In contrast, galaxies in a cluster are much closer together; they have individual velocities exceeding 1000 kilometres per second, and their average distance apart is less than 500 kiloparsecs. It thus works out that there is a good chance that every cluster galaxy will have had at least one short-range gravitational interaction with another galaxy since the formation of galaxies fifteen billion years ago.

This interaction usually shows up through the existence of filaments, bridges of matter, deformed galactic disks, and other exotic structures. However, such structures are of very low luminosity; it was only in the 1950s that the characteristic symptoms of interacting galaxies appeared frequently on photographic plates. In 1959, a first atlas of 356 specimens was published by B. A. Vorontsov-Velyaminov. The superb *Atlas of Peculiar Galaxies* (1966) of Halton C. Arp also shows the great diversity of structures associated with these objects.

The gravitational interaction between neighbouring galaxies produces tidal forces. A body subjected to such forces by a nearby body tends to align itself towards the other body and be compressed perpendicular to that direction; this is why the oceans of the Earth are lifted up both where the pull of the Moon is the strongest, but also at the diametrically opposite point. Tidal forces acting on a rigid body such as a star or planet generally produce only small deformations. On the other hand, galaxies are made up of gas and stars and being feebly bound may be considerably deformed by tidal forces arising from a nearby galaxy.

Our Galaxy is itself subject to gravitational tides caused by its two satellite galaxies, the Large and Small Magellanic Clouds: a long trail of gas, the Magellanic Stream, circulates from the

The Seyfert Sextet VV 115 (left). This curious ensemble is numbered 115 in the Vorontsov–Velyaminov catalogue of peculiar galaxies. In actual fact, the sextet consists of only five galaxies and a large cloud of gas (top left). This cloud was ejected by the principal galaxy in the group, NGC 6027, which is a tidally distorted spiral galaxy. There was a controversy at one time as to whether all of the five galaxies in the group were physically associated. Four of the galaxies are definitely interacting gravitationally – the spiral galaxy NGC 6027, and the three lenticular galaxies which form a vertical chain on its right. This is verified by the velocities of recession measured for the galaxies; according to the theory of the cosmological expansion, measuring the velocity of recession of a galaxy (itself derived from the redshift of its spectrum) gives the distance of the galaxy directly. All four galaxies have a velocity of recession of 4480 kilometres per second, corresponding to a distance of 81 megaparsecs. The fifth member of the group (the large blob below NGC 6027) is a spiral galaxy whose velocity of recession of 19 930 kilometres per second puts it five times further away. It therefore happens to be in the field of the other four galaxies by chance, and they alone are interacting gravitationally. (Arp, Hale Observatories)

Clouds to our Galaxy. Another effect of the proximity of the Clouds is the warping of the plane of the Galaxy, discovered in 1957. The Sun is known to occupy an off-centre position in the very plane of the Galaxy; the galactic disk is turned up on one side of the position of the Sun and down on the other and if it were possible to observe the Galaxy edge-on, its disk would probably have the appearance of a much flattened S in the neighbourhood of the Sun. It is thought that this distortion was triggered off by the Magellanic Clouds recently passing very close to the disk (the Clouds orbit around the Milky Way Galaxy in about 500 million years). Moreover, the distortion is not rigid; it oscillates slowly around an equilibrium position over a period calculated to be at least one billion years. The disk of M31, the Andromeda Galaxy, is just

Text continued on page 386

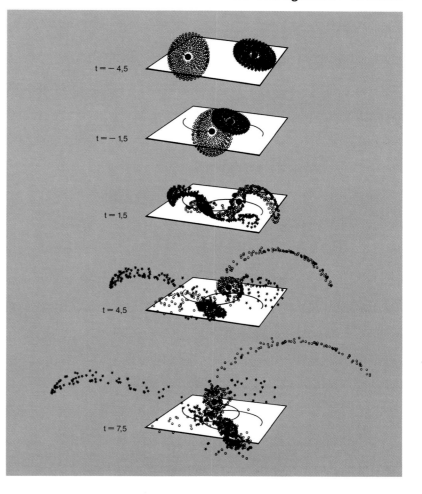

The pair of galaxies NGC 4038 and NGC 4039: the antennae. The photograph (below) of this close pair of interacting galaxies shows the remarkable distortion which has developed as a result of the tidal effects between the two bodies. Even though the distance between the centres of the galaxies is less than 20 000 parsecs, the two ends of the 'antenna' are separated by more than 150 000 parsecs, five times the diameter of the Milky Way Galaxy. In the central regions of each galaxy is a complex mixture of dust, ionised gas and young stars. At the tip of the longest 'antenna' is a region of numerous recently formed stars. What has happened is that a shock wave resulting from the encounter of the two galaxies has considerably compressed the interstellar gas, which has given rise to a new generation of stars.

Whereas a physicist who wants to study collisions of elementary particles can set up the appropriate experiments and vary the initial conditions in order to derive the general laws, the astrophysicist cannot possibly do this sort of thing with galaxies. Instead, he has recourse to what is termed a simulation in which an elaborate computer program recreates the physical phenomena artificially. Until the 1960s, the hypothesis that tidal effects would explain the deformations in galaxies was extremely controversial; it was then that the Toomre brothers, Alar and Juri, constructed a very elaborate program which simulated a close approach between two identical spiral galaxies. The result of the simulations (shown on the right) was a definite confirmation of the theory of gravitational tides.

The Toomre model is necessarily simplified because it is out of the question to store the positions and velocities of the tens of billions of stars in real galaxies. The model assumes the mass of each galaxy to be concentrated at the centre, around which gravitates a disk of 350 points for which the differential rotation simulates that found in spiral galaxies. The disks are initially angled at 60 degrees with respect to one another. The galaxies then approach each other in an elliptical orbit in accordance with the laws of celestial mechanics.

The evolution is obviously speeded up in the figure; the unit of time is 100 million years. The instant zero (not shown) which is the moment of closest approach of the two nuclei, and of greatest mutual influence, is between frames 2 and 3. The particles subsequently leave the disks and give rise to filaments, which develop in opposite directions and are bent by their differential rotations. After 750 million years, the system attains a configuration very similar to that now observed for the NGC 4038/4039 pair. (D. F. Malin, Anglo-Australian Telescope Board)

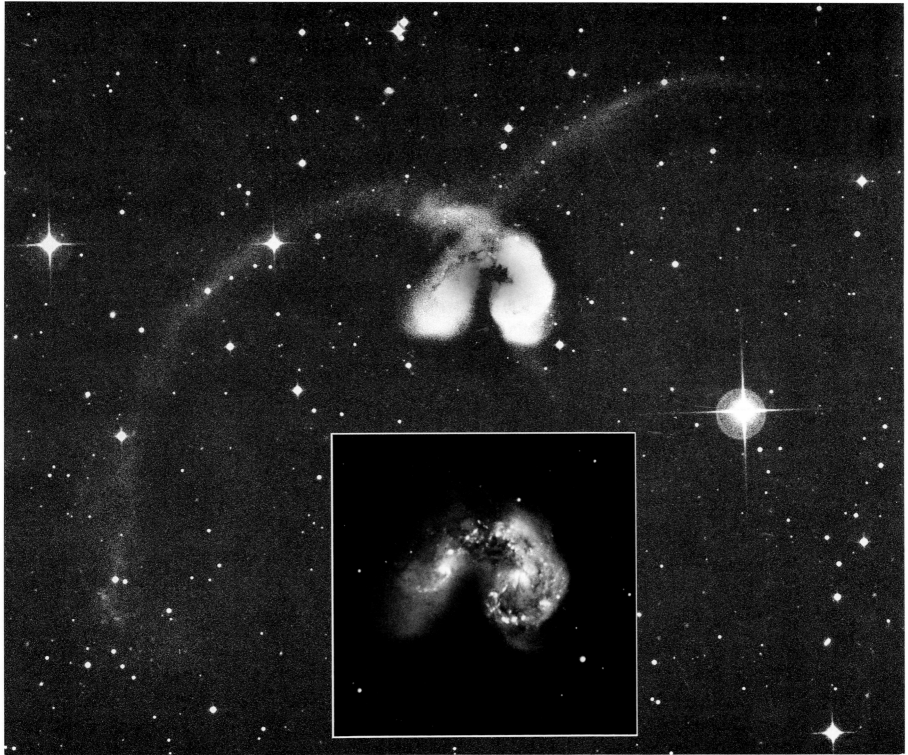

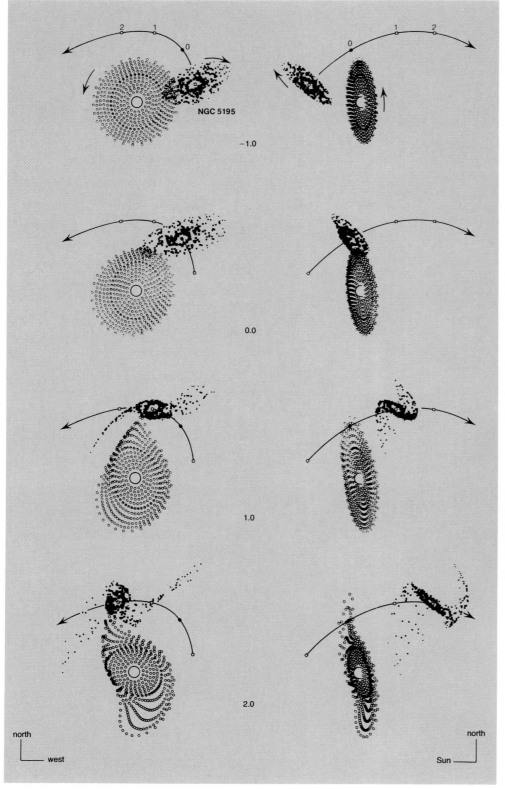

The Whirlpool Galaxy in Canes Venatici, M51. The spiral structure of this very celebrated object was observed by Lord Rosse in 1845. The structure is in fact the result of a gravitational iteraction between the galaxies NGC 5194 and NGC 5195. NGC 5194 is a large spiral galaxy of 10^{11} solar masses. The nucleus contains very hot gas, dust and many young, hot stars. The gas is in rapid motion and gives off radio emission. The rather blue companion galaxy NGC 5195 is probably much younger than NGC 5194, and three times less massive. Of irregular shape, it appears to be evolving towards a barred spiral structure, the bar being induced by the gravitational interaction. NGC 5195 revolves slowly around the large spiral galaxy in several hundreds of millions of years. (Photograph, above, Kitt Peak National Observatory)

Alar and Juri Toomre have carried out computer simulations (see left) showing the evolution of the pair of interacting galaxies. Two sequences are shown, each of four frames, showing how the galaxies would look from two different directions. The left-hand sequence shows the view from the Earth. The trajectory of NGC 5195 is shown; the time-scale given is in units of 100 million years. Note the similarity of the simulation and the reality. Such simulations are particularly interesting as they enable us to go back in time.

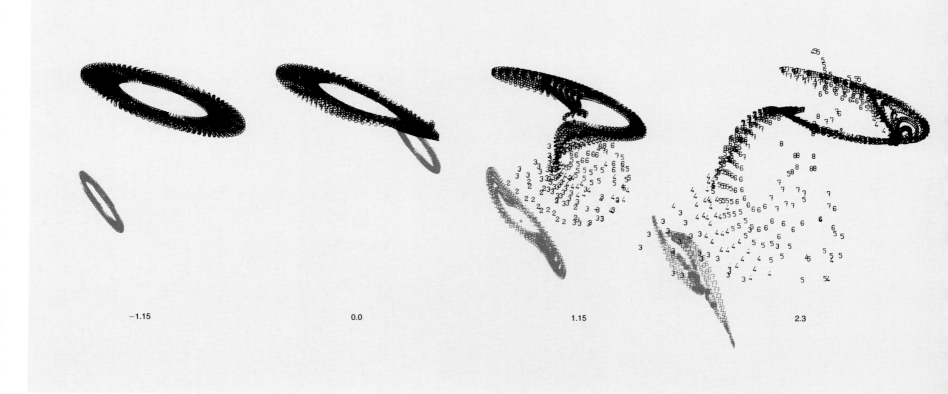

The galaxy pair NGC 5426 and NGC 5427. This is a remarkable example of interacting spiral galaxies. The distortion of their disks, caused by the mutual effects of gravitational tides, is not static, but dynamic. The disks probably oscillate around a mean position with a period of the order of 100 million years, just as the disk of our own Galaxy slowly oscillates under the influence of the Magellanic Clouds. (ESO)

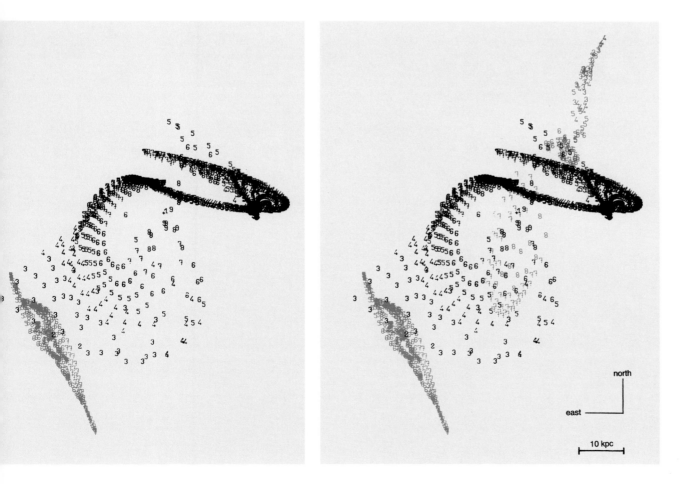

The NGC 4631/4656 galaxy pair. A numerical model of these interacting galaxies was constructed by Françoise Combes. Six phases in their evolution are shown in the figure, with 100 million years between each phase. Initially, the two galaxies are normal spiral galaxies with different masses, NGC 4631 being the more massive of the two. The close approach of their disks ends in the creation of a bridge of matter connecting the two galaxies as well as a tail escaping from NGC 4631 towards the north. Other extensions are also present which cannot be explained by a model of just two interacting galaxies. There is in fact a third galaxy, NGC 4627, not very far away which is probably gravitationally bound to the NGC 4631/4656 pair. Too small to perturb its two companions, it will have lost all of its gas through the interaction and given rise to new filaments. (F. Combes)

as warped; in this case, however, the satellite galaxies, which are very distant, do not appear to be responsible. More generally, nearly 80 per cent of spiral galaxies are observed to have a deformed outer disk, without these galaxies necessarily having very close companions. They will have suffered a gravitational tide over a billion years ago, triggering a slow undulation of their disks.

All tides cause kinetic energy to be lost from the body subject to the tides. In this way, the tides of the Earth–Moon system lead to a slow decrease in the distance of the Moon. The same kind of thing happens when galaxies interact: the creation of filaments, the bridges of matter and other distortions necessitate a capture of energy at the expense of the kinetic energy of the interacting galaxies; this is called dynamical friction. Two galaxies orbiting one another will lose their orbital energy bit by bit through dynamical friction, and will approach each other and end up merging after billions of years.

During interactions between galaxies, the interstellar gas is strongly compressed in certain regions. The compression of the gas in interstellar clouds leads to the formation of stars: strongly compressed clouds succumb to their own self-gravitation to form protostars. These are very luminous, and contract slowly as their energy is dissipated in the form of radiation; their central temperature increases until it reaches the temperature of ten million degrees, which is the threshold at which thermonuclear reactions are triggered, stabilising the stars. Thus we can predict that interacting galaxies must possess active regions of accelerated star formation, or else have a younger population of stars overall than isolated galaxies of the same type. That is what is actually observed; the spheroidal and elliptical galaxies in a cluster are, for example, extremely luminous because they are populated with numerous young stars.

The probability of two galaxies colliding head-on is less than the probability of a gravitational interaction, which will produce tidal effects. However, a number of ring galaxies are known, including the celebrated Cartwheel, which exhibits an amazing structure: the rim, the spokes and the hub are present! Such ring galaxies are the result of head-on collisions; in fact, the nuclei of galaxies, which are populated with stars and which are regions of high density, can easily pass through the more diffuse and gaseous disks, leaving behind an annular wave analogous to the wave produced when a pebble is thrown into water.

In rich clusters of galaxies dynamical friction plays a major role – it appears to be responsible for the 'sweeping up' of gas: the interstellar gas in spiral galaxies may be wrung out during gravitational interactions with other galaxies in the cluster. Dynamical friction will feed the intergalactic medium, which will tend to be gravitationally confined to the interior of the cluster; it is known that clusters of galaxies possess a very hot intracluster gas, detectable by its X-ray radiation. On the other hand, in the course of the gas being swept up, the spirals will be reduced to spheroidal galaxies, lenticulars and ellipticals, which are very poor in gas. The 'regular' clusters, that is, those which have evolved and have members that have undergone numerous interactions, show a strong predominance of spheroidals and ellipticals,. especially in the central regions, where the frequency of encounters is very high. Such clusters also have one or several supergiant 'cannibal' galaxies which might well result from the merging of many normal galaxies. One suspects more and more that elliptic = spiral + spiral is the basic equation for galactic alchemy. Under this hypothesis, galaxy collisions produce not only 'monsters', but create the morphology of a large fraction of galaxies.

Jean-Pierre LUMINET

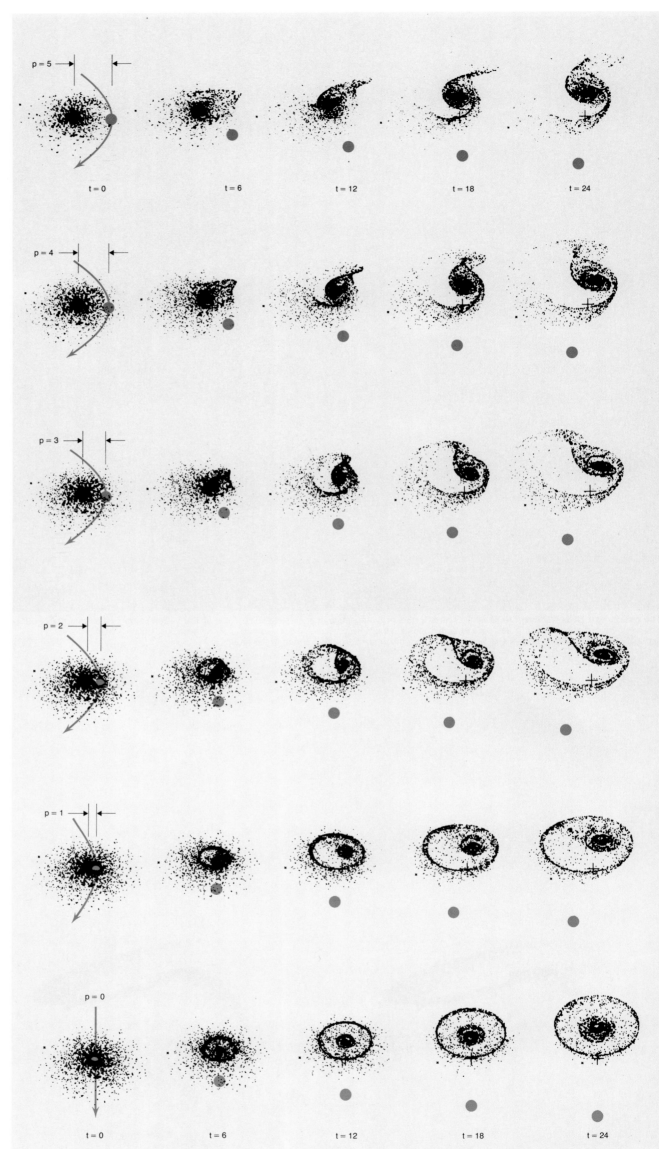

Simulations of direct collisions between a large and a small galaxy. This series of computer simulations, made by Alar Toomre in 1978, shows how the initial conditions affect the evolution of the system and hence its final configuration. The large galaxy is represented by a disk of 2000 particles; a companion galaxy of half its mass (represented by a filled circle) follows a parabolic orbit passing more or less close to the centre of the large galaxy and perpendicular to its disk. The evolution of the system is then determined by the sole datum of the minimum distance of approach of the two systems, or 'impact parameter'. This impact parameter (*p*) is relatively large in the top sequence (five times a certain unit of length) in which the companion galaxy brushes the edge of the disk. The impact parameter is reduced from sequence to sequence until, in the bottom sequence, its value is zero, corresponding to a collision which is exactly head-on. The unit of time (*t*) increasing from left to right in a given sequence) is 100 million years. The structures obtained after 2.4 billion years range from barred spirals for large impact parameters to ring galaxies for head-on collisions. Collisions with low impact parameters are obviously less likely to occur than collisions of high impact parameters; consequently, ring galaxies are rare. (Document kindly communicated by A. Toomre)

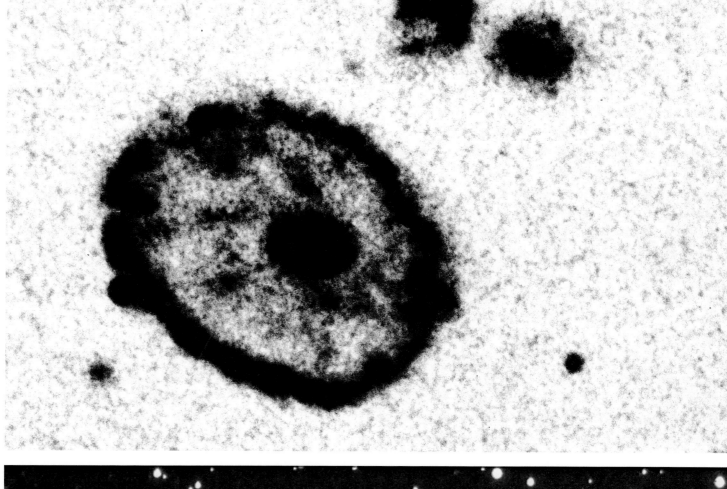

The Cartwheel Galaxy, A 0035. This very appropriately named ring galaxy is 200 megaparsecs distant. The rim, which is continuing to expand, has a diameter equal to that of our own Milky Way Galaxy. In the interior of the Wheel, the hub and spokes consist mainly of old, red stars. As is suggested by the simulations on the left-hand page, this galaxy was originally a large spiral galaxy through which a smaller galaxy passed about 300 million years ago. A circular shock wave gave rise to the gaseous, expanding rim; the initial compression of the gas has produced numerous massive bright young stars. The supernova rate observed there is 100 times the mean rate for normal galaxies. The little galaxy intruder has continued its trajectory and is now 100 kiloparsecs from the hub, on the axis of the Wheel. In the photograph, the intruder is the galaxy without spiral arms. (Royal Observatory, Edinburgh)

The galaxy pair IG 29 and 30. This surprising pair of interacting galaxies is a perfect example of a head-on collision. A ninety-minute exposure with the 1-metre ESO Schmidt telescope reveals a luminous bridge connecting the two galaxies and a diffuse ring around the larger galaxy. Whether this ring is made of gas or stars is unknown. (ESO)

Groups of galaxies

Galaxies are not uniformly distributed in the sky but appear to be in groups. Galaxies which lie in the same direction may of course be unconnected with one another, just as projection effects for the stars in our Galaxy give rise to the constellations, which are not genuine groups. A more accurate idea of the positions of galaxies can be gained by measuring their distances. If many galaxies lie in the same direction of the sky and are found to be at the same distance from us, they are indeed located in the same limited region of space. Analysis of the distribution of the nearby galaxies reveals that certain groups are real; galaxies do have a tendency to conglomerate in more or less rich 'clumps'.

The nearest example is the Magellanic Clouds, two irregular, diffuse galaxies in the southern hemisphere which are small satellites of our Milky Way Galaxy. The Clouds can be seen with the naked eye. The gravitational interaction between the Milky Way Galaxy and its two satellites gives rise in particular to the Magellanic Stream, which is a stream of neutral hydrogen pulled out from the Magellanic Clouds by the tidal forces exerted by our own Galaxy, forming a bridge of matter linking the Galaxy to its satellites.

Giant galaxies are often accompanied by a whole procession of smaller satellites gravitating around them. Thus, the Andromeda galaxy M31 (NGC 224), a beautiful spiral about 700 kiloparsecs from us, and which can be regarded as the twin sister of the Milky Way in view of its proximity and similar size and structure, has a number of satellites such as M32 (NGC 221), NGC 147 and NGC 205, small elliptical galaxies gravitating about M31 in about 500 million years. M31 itself is a beautiful spiral galaxy about 700 kiloparsecs from us.

The mass of a giant galaxy can be calculated by applying the laws of celestial mechanics to the motions of satellite galaxies around a giant galaxy. The values derived are comparable to those obtained from rotation curves (the method is based on studying the speed of galactic rotation of the stars as a function of their distance from the centre). The Andromeda Galaxy, for example, is estimated to have a mass of around 300 billion solar masses, about one and a half times that of our own Galaxy. Another beautiful spiral galaxy, the Triangulum Nebula, is located not far from M31 and, even though it is not a satellite of the galaxy, it is a close companion. There are no surprises about these small groups with a few galaxies around a more sizeable galaxy – such groups are easily understood in terms of the gravitational force exerted by the dominant galaxy on its near neighbours. Less predictable is the distribution of galaxies on a slightly larger scale within a spherical volume of radius 15 megaparsecs centred on the Milky Way Galaxy. This volume is modest compared to that of the observable Universe. Of course, a great quantity of galaxies of all types and sizes can be found there, ranging from pygmy galaxies resembling the globular star clusters, to giant galaxies like the Andromeda Galaxy. But the remarkable fact is that thirty or so galaxies can be seen out to 1.3 megaparsecs, and several thousand between 2.4 and 15 megaparsecs, but *none* between 1.3 and 2.4 megaparsecs. The thirty nearby galaxies in fact form a group isolated from other galaxies, called the Local Group.

On taking an inventory of the Group, we find the following: two giant spiral galaxies, which are our own Galaxy and the Andromeda Galaxy (M31, otherwise known as NGC 224); two average spirals, the Triangulum Nebula (M33/NGC 598) and the Large Magellanic Cloud; an elliptical galaxy with a nucleus (M32/NGC 221); half-a-dozen small irregular galaxies; a dozen dwarf elliptical galaxies, and several very feeble objects resembling isolated globular clusters. Additional objects are added from time to time as astronomical techniques improve. However, the list is rather incomplete; dwarf galaxies are almost impossible to detect at distances greater than that of the Andromeda Galaxy. It is estimated, using the observed density in the immediate neighbourhood of our Galaxy to make the extrapolation, that the Local Group

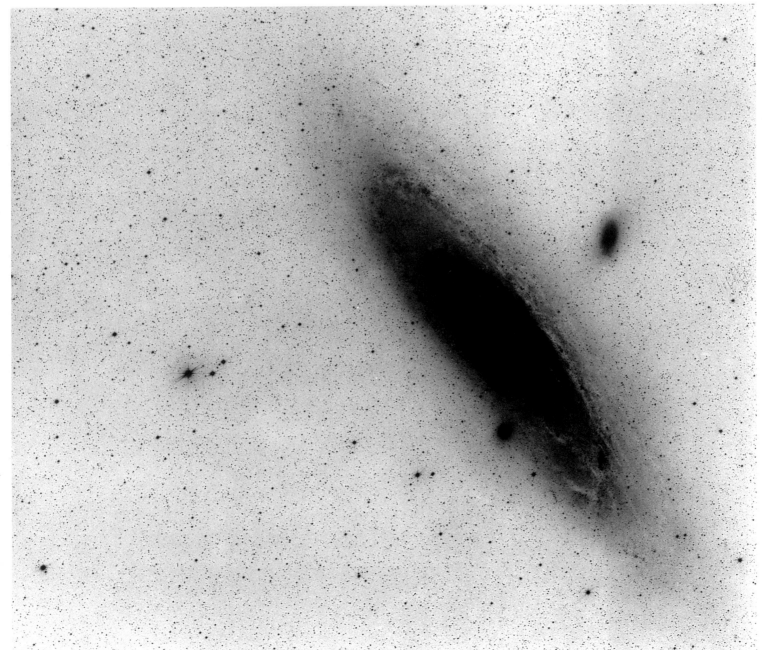

The Andromeda Galaxy and two of its satellites. The Great Galaxy in Andromeda, M31, is visible to the naked eye in the northern hemisphere. It extends over 3 degrees in the sky, or six times the angle subtended by the moon. Historically, it was the first object whose extragalactic nature was recognised, paving the way for all modern astronomy and cosmology. The Andromeda Galaxy is a spiral galaxy containing 400 billion stars. It is the dominant member of the Local Group, and has a veritable court of smaller satellites gravitating around it, ranging from globular clusters to dwarf elliptical galaxies.

Two of its close satellites can be seen in the photograph, the small galaxy NGC 205 very close to the disk, and an elliptical galaxy, M32, a little further away, which is resolved into stars, of which the brightest are red giants. (POSS)

Fornax I. This group of galaxies lies in the constellation of the Furnace in the southern hemisphere at a distance of 16 megaparsecs. The group is centred on the large elliptical galaxy NGC 1399; a close inspection of the photograph reveals that NGC 1399 is surrounded by a swarm of tiny images of stellar appearance. These are thought to be globular clusters, each containing many tens of thousands of stars, just like the globular clusters in the halo of our own Galaxy.

Another remarkable galaxy in the photograph, the large galaxy NGC 1365 at lower left, is a perfect prototype of the barred spiral; it proves, moreover, to be a galaxy with a very active nucleus. It is not certain whether this galaxy belongs to the Fornax I group. The other members of the group are NGC 3245, NGC 3254, NGC 3274 and NGC 3277. (D. F. Malin, Anglo-Australian Telescope Board)

A group of galaxies in Leo. This small group of several galaxies is 16.5 megaparsecs away and belongs to the Local Supercluster. Four galaxies are visible in the photograph. The dominant galaxy is the beautiful spiral galaxy NGC 3190. The two other spirals NGC 3185 and NGC 3187 are in fact barred spirals. The elliptical galaxy at the upper left is NGC 3193. (Hale Observatories)

The extragalactic domain

Cartography of the Local Group. The Local Group consists of twenty or so listed galaxies (see table page 370) together with many tens of 'pygmy' galaxies and isolated globular clusters within a spherical volume of 1 megaparsec radius. Only the most important objects are shown in the figure; the position of each galaxy is given in a system of coordinates termed 'supergalactic', defined by the supergalactic plane passing through the Sun, the centre of the Milky Way, and the centre of the Virgo cluster of galaxies. The Virgo Cluster is in fact the very heart of a vast system of several tens of groups of galaxies, called the Local Supercluster, and it is convenient to give the positions of nearby galaxies in supergalactic coordinates. It just so happens that the direction of the north galactic pole is sufficiently close to the direction of the Virgo Cluster from the plane of the Milky Way Galaxy to be perpendicular to the supergalactic plane, making it possible to define all of the coordinate axes without ambiguity. On the three-dimensional 'map' we immediately notice the groups of galaxies around the two dominant members, which are our own Galaxy and the Andromeda Galaxy, M31. (The centre of mass of the Local Group is near to the mid-point of these two galaxies and thus does not coincide with the origin of supergalactic coordinates.) Note also the chain of galaxies from Wolf–Lundmark through the M31 subgroup to IC10.

Intergalactic distances are quite small relative to the dimensions of galaxies, in contrast to the situation with stars. For example, the distance from the Milky Way to the nearest large galaxy is only 20 galactic diameters, whereas the distance from the Sun to the nearest star, Alpha Centauri, is thirty million times the diameter of those stars. To illustrate the extraordinary way galaxies populate cosmic space, we can imagine our Galaxy to be reduced to the dimensions of Paris. The stars will then be tiny spheres one hundredth of a micrometer in diameter. The Andromeda Galaxy will be found 200 kilometres away, and the Local Group will be entirely contained within a spherical volume 300 kilometers in radius. Within a radius of 100 000 kilometres there will be many hundreds of millions of galaxies.

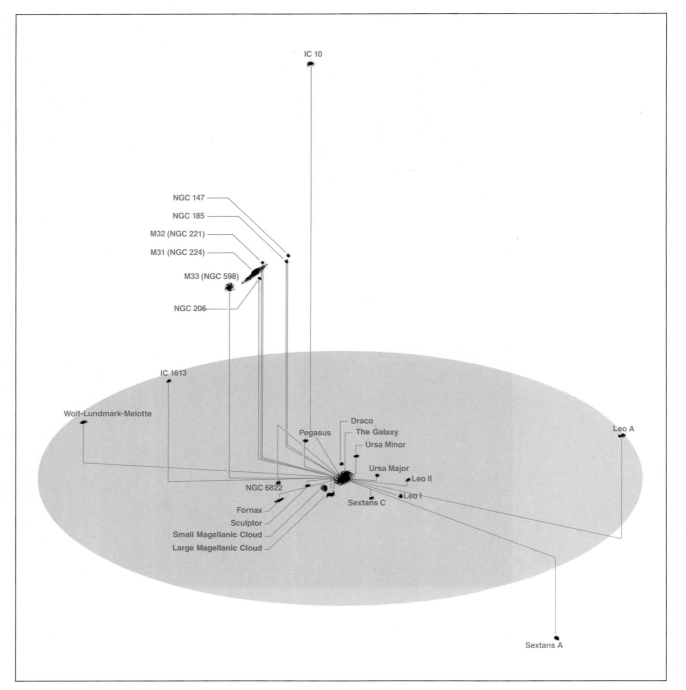

could well contain 100 of these pygmy galaxies. (A galaxy is defined as a dwarf galaxy if it is less luminous than absolute magnitude −16. For comparison, the Andromeda Galaxy is a hundred times more luminous at its absolute magnitude of −21.)

The Local Group does not have a central condensation, but instead has two subgroups centred around the two dominant galaxies, which are our own Galaxy and M31. However, there is a narrow zone of sky corresponding to the plane of our Galaxy, in which it is extremely difficult to make extragalactic observations because of the absorption of radiation by accumulated gas and dust. This again prevents an exhaustive list being drawn up for the Local Group.

In the case of the Local Group, the important question then is whether the association is merely fortuitous or whether it is a stable group of objects 'cemented' together by the self-gravitation of the ensemble. No definite answer can be given, but the fact that there are a large number of similar groups like the Local Group beyond 2.4 megaparsecs suggests that such groups are in fact relatively isolated structures bound by gravity. The astronomer Gérard de Vaucouleurs has listed more than fifty groups of galaxies within a radius of 16 megaparsecs, each containing several tens of members. It is very much a matter of there being a new kind of object which in some sense plays the role of basic 'building blocks' in the medium- and large-scale structure of the Universe. A group is defined as an association of galaxies, of which a dozen are more luminous than absolute magnitude −16 within a volume of about a cubic megaparsec. This is equivalent to a numerical density of galaxies ten times the mean

galaxies	galaxy type	diameter (kpc)	mass ($M_\odot$)	distance (kpc)	radial velocity (km/s)
M31	Sb	50	3×10^{11}	670	−275
NGC 221	E0	1	3×10^{9}	660	−210
NGC 205	E5	2	10^{10}	640	−240
NGC 185	E5	1	10^{9}	660	−300
NGC 147	E5	1	10^{9}	660	−250
Our Galaxy	Sc	30	1.5×10^{11}	—	—
Large Magellanic Cloud	Irr	7	10^{10}	50	+270
Small Magellanic Cloud	Irr	3	2×10^{9}	65	+168
Sculptor	dwarf spheroidal	1	3×10^{6}	85	
Fornax	dwarf spheroidal	2	2×10^{7}	170	+ 40
Leo I	dwarf spheroidal	0.7	3×10^{6}	230	
Leo II	dwarf spheroidal	0.7	10^{6}	230	
Draco	dwarf spheroidal	1	10^{5}	67	
Ursa Minor	dwarf spheroidal	—	10^{5}	67	
M33	Sc	8	10^{10}	730	−190
IC 1613	Irr	1	3×10^{8}	740	−240
NGC 6822	Irr	2	4×10^{8}	470	− 40
Ursa Major	dwarf spheroidal			120	
Sextans A	Irr			1000	
Sextans C	dwarf spheroidal			140	
Leo A	Irr			1100	
IC 10	spiral			1260	
WLM	Irr			870	

Some galaxies of the Local Group.

The galaxies of the M83 group. Gérard de Vaucouleurs has identified a group of half a dozen galaxies around the spiral galaxy M83, including in particular the spirals NGC 4945 and NGC 5068, the lenticular galaxies NGC 5128 and NGC 5102 (see page 338) and the irregular galaxy NGC 5253. The group lies in the southern hemisphere, at a distance of 4 megaparsecs. It is a sparse group, 2 megaparsecs in diameter, and it may well not be gravitationally bound.

M83 (NGC 5236) is a beautiful spiral galaxy seen face-on (right), and in luminosity and size is very much like our own Galaxy. There is a marked difference between the rather old stars of the nucleus and the young stars in the spiral arms. Many of these young stars, called Population I, are massive and evolve rapidly; as a result the supernova rate is the highest known for any galaxy, with a supernova every ten or fifteen years.

NGC 5128 (below) is crossed by a wide band of absorption due to obscuring dust. The galaxy is a radio source, known as Centaurus A. The radio structure is complex. Two radio components are inside the optical galaxy, while two further components extend beyond it by over a megaparsec or so. (ESO)

density outside groups. The nearest group of galaxies to the Local Group is the Sculptor Group 2.4 megaparsecs away, consisting of half a dozen bright spirals more or less distributed in a circle of diameter 1 megaparsec. The Ursa Major–Camelopardalis Group, at a distance of 3 megaparsecs, is more spread out. Its members lie within a spherical volume of diameter 2.1 megaparsecs, and are concentrated around two dominant galaxies, like the Local Group. The group includes the Maffei galaxies, discovered in 1968, which were thought at the beginning of the 1970s to belong to the Local Group in the neighbourhood of IC 10, whereas they are in fact located in the plane of our Galaxy, making it difficult to observe them and to measure their distance.

Bearing all this in mind, we can ask whether all galaxies in the Universe are members of groups or of even larger structures such as the rich clusters of galaxies. A partial answer to this question is provided by studies of nearby galaxies; within a radius of 20 megaparsecs, it seems that 10 to 20 per cent of galaxies do not belong to a group. Such galaxies are called field galaxies.

It is gravitation which is responsible for galaxies being in groups. Studies of the dynamics of a group based on measuring the velocities and masses of its members enable us to deepen our understanding of such systems. The Local Group is well known and the most studied. It is clear that its dwarf elliptical galaxies play an entirely negligible role in the overall motions of the Group as their masses are very small. The dynamics of the Local Group are in fact dominated by the Andromeda Galaxy (M31), our Galaxy and the Triangulum Nebula, M33. The centre of mass of the Local Group is located quite close to the mid-point of our Galaxy and M31. Our Galaxy is moving away from this centre of mass at 170 kilometres per second. The radial velocities of M31 and M33 relative to our Galaxy can also be measured (i.e. the components of the velocities in the line of sight) giving their intrinsic radial velocities relative to the centre of mass of the Local Group. If the masses and velocities of galaxies in a group are known, their kinetic energies can be calculated, and a total mass for

The group of galaxies around M81. This group lies in the constellation of the Great Bear. At a distance of 2.5 megaparsecs, it is one of the nearest groups to the Local Group, and it is a member of the Local Supercluster. It contains spiral and irregular galaxies. At the centre of the photograph is Bode's Nebula, a large spiral galaxy. The size of its bulge indicates that it is in an advanced state of evolution.

To its right, the galaxy M82 appears to be an irregular galaxy but it is probably an edge-on spiral deformed by gravitational interaction with M81. The group almost certainly extends into the constellation of the Giraffe in an obscured region near the galactic plane: it bears the name Ursa Major–Camelopardalis, and includes, notably, the two Maffei galaxies. (USIS)

Stephan's Quintet. This spectacular group comprises the galaxies NGC 7317 (furthest to the right), NGC 7318 a and b (strongly interacting), NGC 7319 to the top left, and the elliptical galaxy NGC 7320 to the south. This group illustrates well how misleading projection effects can be. When the velocities of recession of the galaxies are measured (and hence their distances can be deduced from Hubble's Law of the expansion of the universe) it is found that NGC 7317 to NGC 7319 have the same velocity of recession of 6000 kilometres per second, while the velocity of recession of NGC 7320 is only 800 kilometres per second. NGC 7320 is thus by far the nearest galaxy and is not related to the four interacting galaxies.

At upper left, part of a sixth galaxy can be seen; its velocity of recession is the same as that of the NGC 7317–9 quartet, and on long exposures there seems to be a bridge of matter connecting the galaxy to the quartet. There is indeed a Stephan Quintet, although it is not quite the one that we thought! (H. Arp, Kitt Peak National Observatory)

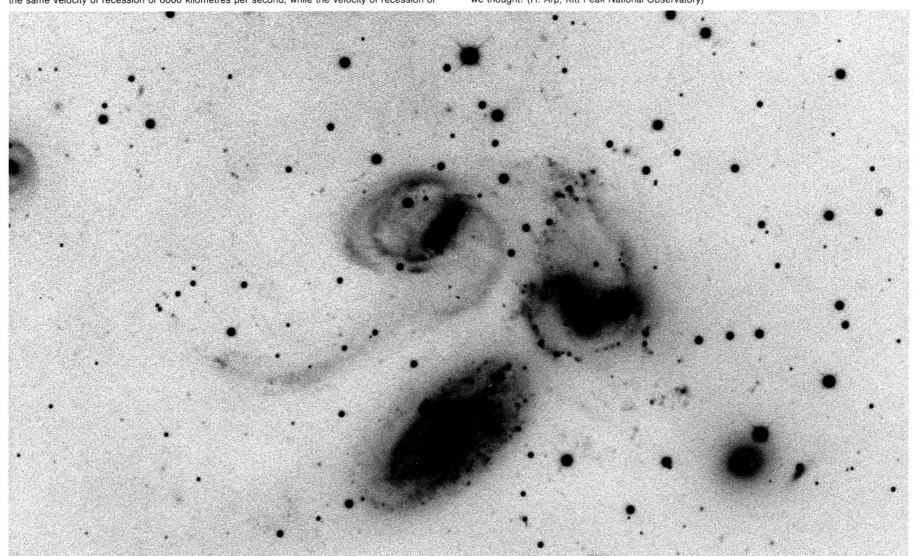

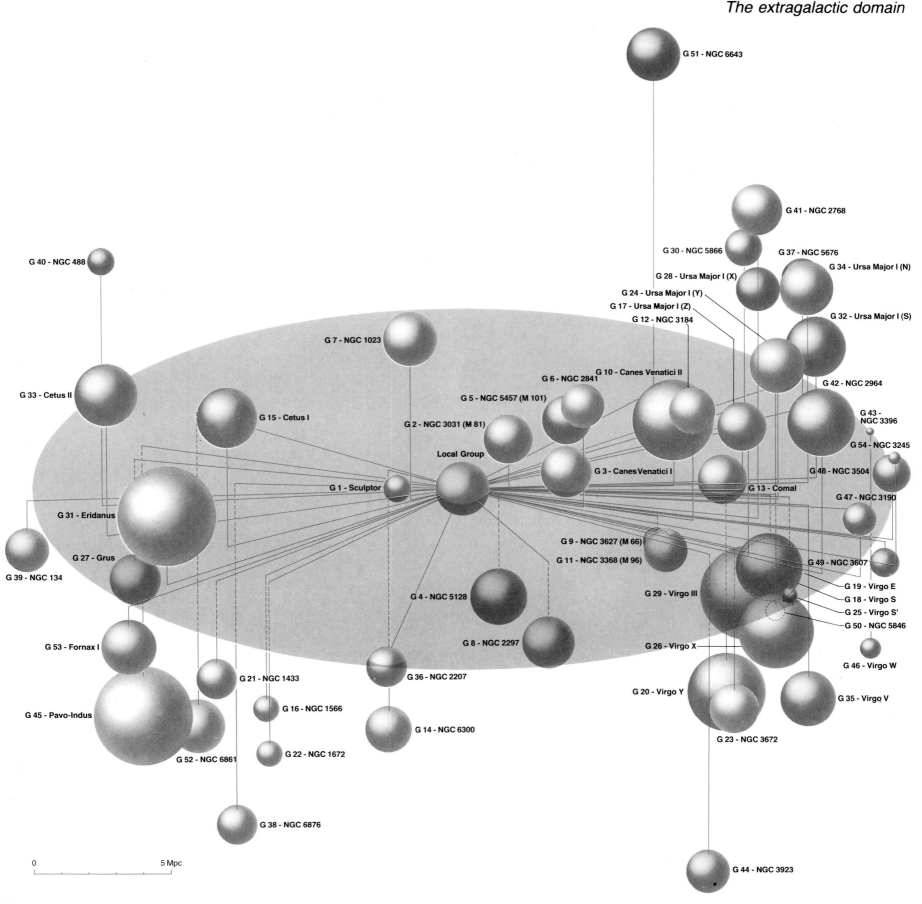

The spatial distribution of the nearby groups of galaxies. Within a radius of 16 megaparsecs of our Galaxy, Gérard de Vaucouleurs has catalogued over fifty groups of galaxies analogous to our own Local Group. These groups are here illustrated schematically by spheres with sizes that are proportional to their actual dimensions. The supergalactic coordinate system is used in the figure (already defined in the legend to the map of the Local Group on page 368). Each group is labelled with a group number and with the name of the constellation in which it lies (e.g. Cetus,

Pavo-Indus) or the (Messier or NGC) catalogue number of the dominant galaxy in the group. The Local Group is at the very centre of the drawing. Note the greater number of groups in the right half; these groups are in fact members of the Local Supercluster, centred on the Virgo Group; the Local Group lies on the periphery. The lighter coloured spheres are those located above the plane. The two spheres shown with red dotted lines are presented thus because they are masked by their neighbours.

the group may be obtained. The mass estimated in this way is called the dynamical mass. We then compare this with the luminous mass derived by adding up the masses of each of the members of the group (of which only a fraction is observable, but the most luminous galaxies are the most massive).

For the Local Group the luminous mass is about 650 billion solar masses, of which 70 per cent is concentrated in M31 and in our own

Galaxy. The dynamical mass, however, is four times as much. Similar discrepancies are found with other groups – the dynamical mass can even be ten times the luminous mass. We are therefore led to postulate the existence of non-luminous mass, called the missing mass. The intergalactic medium can account for some of the difference.

Using rotation measures of individual galaxies, many astronomers believe that these could be much more extended, and thus much more

massive, than currently thought: they might possess superhalos consisting of very faint low mass stars, so numerous that they contribute at least as much mass as the luminous part of the galaxy, or more massive but extinct stars.

The missing mass problem recurs for much greater structure such as clusters and superclusters of galaxies; it is one of the basic theoretical questions of extragalactic astronomy.

Jean-Pierre LUMINET

Clusters and superclusters of galaxies

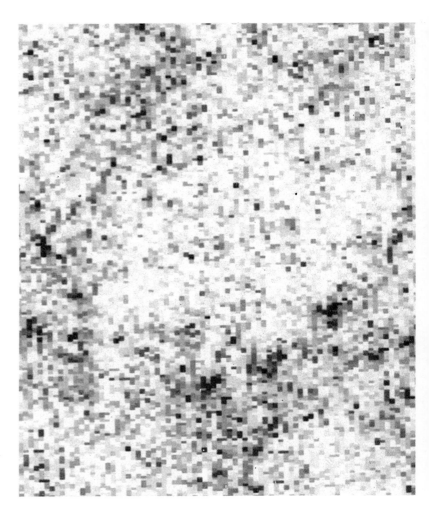

Well before galaxies were recognised as independent stellar systems analogous to our Milky Way, astronomers had noticed their tendency to cluster in certain regions of the sky rather than being distributed randomly. This property already appears clearly on a sky atlas giving the positions of the thousands of galaxies among the 13 226 objects catalogued by the Dane Johan Ludvik Emil Dreyer in the *New General Catalogue of Nebulae and Clusters of Stars* (1888) and its two supplements, the *Index Catalogue . . .* (1895) and the *Second Index Catalogue . . .* (1908).

It appears that there is a basic tendency in nature to make the objects of a given class form the elements of a new class of higher order:

elementary particles assemble to form atoms, which assemble in molecules, which in turn form nebulae and stars (and possibly planets), which are themselves parts of much vaster galaxies, which in turn form clusters. The extremes of the hierarchy are at the frontiers of human knowledge: elementary particles are made from even smaller units, quarks; while clusters of galaxies form parts of gigantic structures called superclusters.

Beginning in 1933, Harlow Shapley published a catalogue of twenty-five clusters and put forward the idea that these associations of galaxies are not fortuitous but result from evolutionary processes. The same year, Fritz Zwicky studied a particularly rich cluster of galaxies in

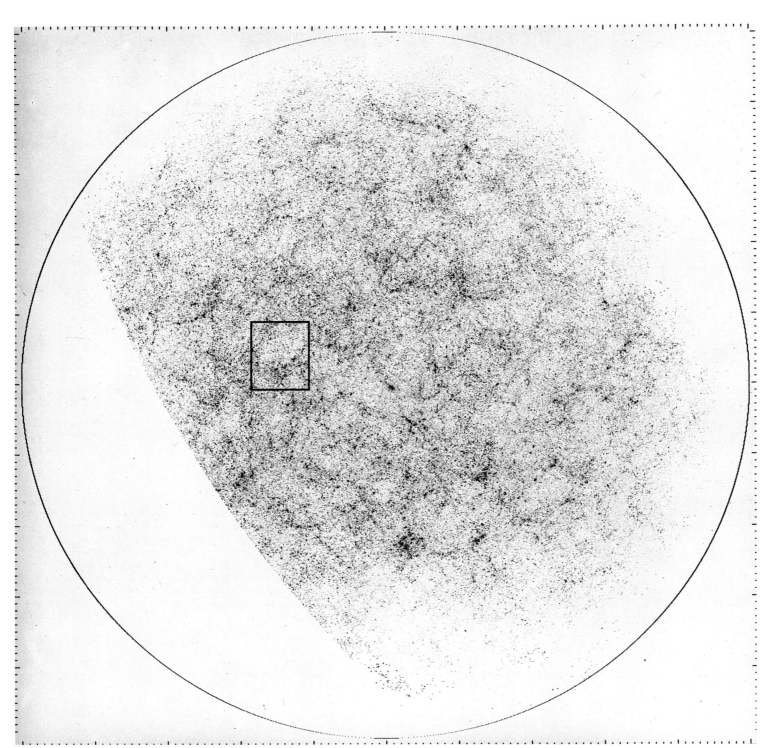

The cosmic tapestry This map of the sky was created by astronomers from Princeton University using the catalogue of galaxies assembled by D.C. Shane and C.A. Wirtanen at Lick Observatory, California. It shows the two-dimensional distribution of more than a million galaxies brighter than apparent magnitude 19. These galaxies are at an average distance of 430 megaparsecs. The sky is cut into tiny squares with side 10 minutes whose greyness is proportional to the number of galaxies to be found there, from zero (lightest squares) to ten galaxies (darkest squares). Each square contains on average one galaxy but 1600 of them (out of about a million) contain at least ten. The tendency to cluster is clear, while long filaments suggest that the clusters themselves form part of very extended superclusters.

The North Galactic Pole is at the centre of the map; we can see a region very rich in galaxies there, which is the Coma cluster. The map is bounded by the Galactic equator, where the dust and stars of the Milky Way prevent any counting of galaxies; the empty region at lower left is the part of the sky which is too far south to be visible from Lick Observatory. An enlargement of a small region of the 'tapestry' (above) shows the meticulous work required to create this map. (M.Seldner, B.Siebers, E.J.Groth, and P.J.E.Peebles)

The Virgo cluster. Extending over about one hundred square degrees of sky near the North Galactic Pole, the Virgo cluster is the closest rich cluster: it probably contains 10 to 20 thousand galaxies (15 of these galaxies appear in the famous Messier catalogue, containing one 109 objects). The cluster diameter is about 2.7 megaparsecs. The average velocity of its galaxies, caused by the cosmological expansion, is about 1140 kilometres per second, indicating an average distance of 15 megaparsecs (for a Hubble constant of 75 kilometres per second per megaparsec). The Virgo cluster is a typical irregular cluster, without a central condensation or marked segregation between the various types of galaxy. Elliptical galaxies form about 30 per cent of its population. The Virgo cluster is in the centre of a vast system, the local Supercluster, which includes our own Local Group of galaxies.

X-ray emission from the Virgo cluster. Clusters of galaxies contain an intracluster gas of very high temperature (from 10 to a 100 million degrees) radiating X-rays. This image from the Einstein satellite shows the X-ray emission of the intracluster gas around M87 (the diameter of the central region, shown in yellow, is about 10 arc minutes; for comparison the diameter of M87 in the visible is about 2 arc minutes). This giant elliptical galaxy itself emits X-rays from its active nucleus, and is the hot spot at the centre of the image. (Harvard-Smithsonian Center for Astrophysics)

The centre of the Virgo cluster. The two largest galaxies, M84 (right) and M86 (left) are surrounded by a retine of spiral and elliptical galaxies. On the far left a spiral, NGC 4438, is being deformed by a tidal interaction with its companion elliptical galaxy, NGC 4435. (David F. Malin, ROE and AATB).

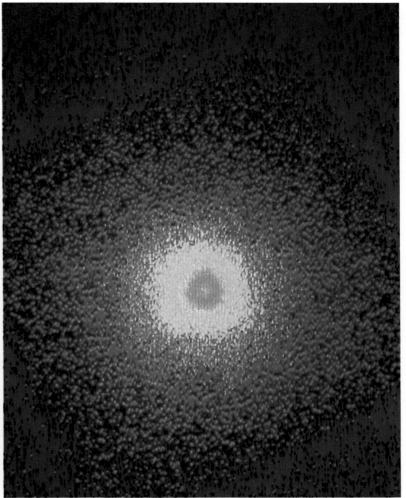

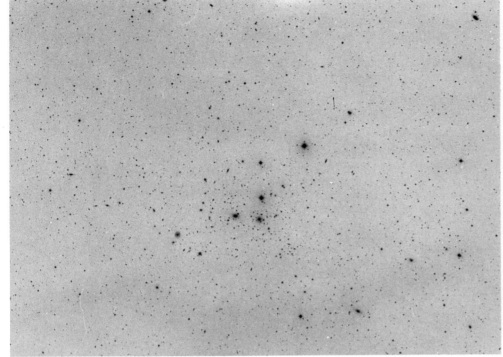

The Coma cluster (Abell 1656). Situated (hence the name) in the constellation of Coma Berenices, only 2 degrees from the North Galactic Pole, the Coma cluster thus occupies an ideal astronomical position, as at this latitude there is no obscuration from the dust and clouds of the Milky Way. The Coma cluster extends over several square degrees of sky and contains more than a thousand bright galaxies. It is one of the nearest rich clusters. Its average distance is measured by the velocities of its galaxies, themselves deduced from the redshift of their spectra resulting from the expansion of the Universe. Several hundred redshifts have been measured; they dive an average recession velocity of 6900 kilometres per second, putting the centre of the Coma cluster at 90 megaparsecs (for a value of the Hubble constant H_0 of 75 kilometres per second per megaparsec).

Coma is a typical regular cluster, i.e. it is spherically symmetric and has a strong central concentration, resembling the distribution of stars in a globular cluster or elliptical galaxy, on a much larger scale. It is thus an evolved cluster, in the sense that the galaxies of its core have undergone so many gravitational interactions that they have shared out their kinetic energies, and are distributed in a 'relaxed' state.

The total diameter of the cluster is estimated at 2.7 megaparsecs, while the core is five times smaller. The most massive galaxies have had time to reach the centre of mass of the cluster and are thus almost all in the core; this is populated mainly by elliptical and spheroidal galaxies, whose average mutual distance is three times smaller than the distance between our Milky Way and the Andromeda galaxy, giving an idea of the richness of the cluster. The spiral galaxies are in the periphery.

It is possible to detect a slow rotation of the cluster, which is itself only a small part of a vast supercluster containing another rich cluster, Abell 1367. At right, the photograph shows the centre of the Coma cluster. The colours are reconstructed from several black and white pictures taken in the red, green and blue regions of the spectrum at the Schmidt telescope of Mount Palomar by Laird A. Thompson. One can count about 300 elliptical and spheroidal galaxies. The brightest object is a blue star in our own Galaxy. The others are galaxies of the cluster, each containing several tens of billions of stars. The two most important are NGC 4889, a supergiant elliptical (left), and NGC 4874 (right), a spheroidal whose gas has been swept out by close interactions with other galaxies of the cluster. (left, POSS; right L.A. Thompson and Kitt Peak national Observatory)

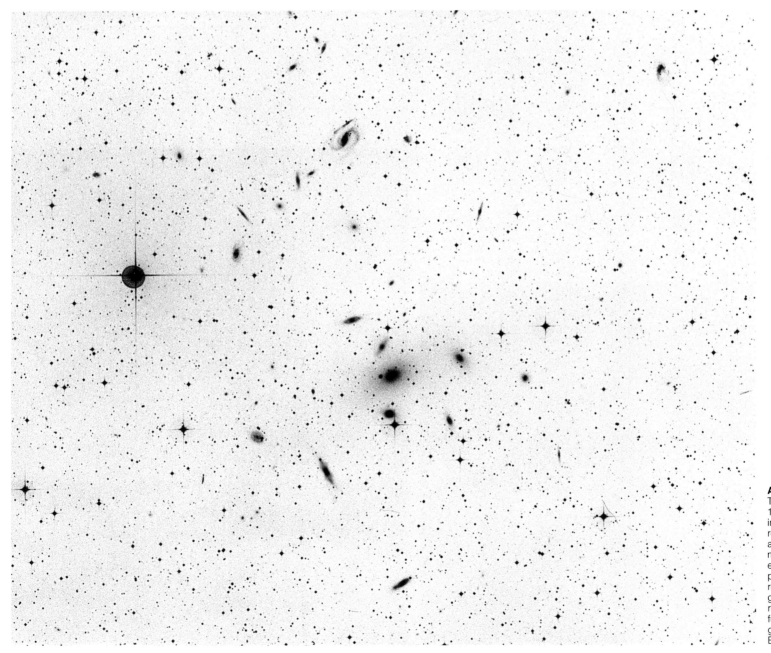

A cluster of galaxies in Pavo. This cluster is at a distance of 100 megaparsecs. We immediately recognise it as a regular cluster, centred around a giant elliptical galaxy. It has a mixed population of spirals and ellipticals, but spirals predominate in the outer regions. The ellipticals and giant spheroidals of the central regions may result from the fusion of several smaller galaxies. (Royal Observatory Edinburgh)

the constellation of Coma Berenices and raised the problem of the missing mass.

Even the definition of a cluster of galaxies is still debated. Until the beginning of the 1950s, astronomers believed that most galaxies were isolated (they were called 'field galaxies') and that only a minority belonged to groups of varying richness. With the development of observing techniques and methods of statistical analysis, it became possible to show that while a few galaxies appear really isolated, the majority are members of double, triple, or multiple systems; the convention is now reversed, and has gone as far as discussing clusters with a single member! Without falling into this excess of zeal, we are forced to recognise that it is not simple to decide if a set of galaxies in a given volume is a cluster or not. A mathematical tool borrowed from statistics, the correlation function, is very useful for clarifying the problem. The correlation function is an objective measure of any tendency to cluster that the galaxies may have, and starts from the simple idea that if the galaxies belong to a cluster the distances between them are generally smaller than if they are uniformly distributed. Calculating correlation functions is a long, arduous, but fruitful task. One first has to 'unpack' a catalogue of galaxies giving their apparent positions in the sky; the catalogue of Donald C. Shane and Carl A. Wirtanen gives, for example, all the galaxies brighter than apparent magnitude 18.7 observable from Lick Observatory in the USA, and represents about a million galaxies (and about 12 years work!). One then cuts the celestial sphere into small squares of side one degree and counts the number of catalogued galaxies in each square and compares it with the average. Analysis of the Shane–Wirtanen catalogue shows the presence of a uniform background of about fifty galaxies per square degree on which are superimposed a large number of clusters containing more than one hundred galaxies per square degree. It shows in particular that one has a better chance of finding a galaxy in the sky by looking near an already known galaxy, and an even better chance by looking near a pair of galaxies, and so on. The tendency to cluster appears to decrease strongly over distance scales greater than 20 megaparsecs, which corresponds to the observed size of the most extended clusters.

The correlation functions, however, do not say at what level of organisation of galaxies we can speak of clusters. The choice is, in fact, arbitrary. We know, for example, that our Galaxy is a member, along with the Andromeda galaxy and about thirty smaller nearby galaxies, of a relatively isolated system known as the Local Group. Tens of similar groups have been observed, but these clusters of galaxies are far from constituting the summit of hierarchical structure in the Universe: there are thousands of much richer groups each containing several hundred to several tens of thousands of members in a volume comparable to that of the Local Group (i.e. within a radius of several megaparsecs). It is these very rich and dense groups that we shall call clusters of galaxies here; the number density of galaxies at the centre of rich clusters can be a thousand to a million times the mean number density of galaxies in the Universe. The two nearest clusters to the Milky Way are those of Virgo, at 20 megaparsecs in the direction of the North Galactic Pole, and Coma, at 100 megaparsecs in the constellation of Coma Berenices. To find the distance of a cluster of galaxies we have to measure the mean distance of several of its members using the Hubble law. At such distances the expansion of the Universe shows itself as a recession of all galaxies, which causes a shift of their spectra to lower frequencies; measuring this *redshift* gives the recession velocity of the galaxies of the cluster, which, in turn, gives their distance by dividing by the Hubble constant. The mean distance of the cluster can thus be found. The most distant clusters observed at present have recession velocities of 100 000 kilometres per second.

The expansion velocity of the Universe gives the galaxies a purely radial motion; on this are superimposed the proper motions of the galaxies in the cluster's gravitational field. Analysis of this velocity dispersion gives valuable information about the internal dynamics of clusters. It is generally thought that clusters are stable, gravitationally bound systems, but the question is still debated. What appears clear is that their central cores are gravitationally bound, as otherwise the

The Perseus cluster (Abell 425). This highly irregular cluster is only 17 degrees above the Galactic plane; because of this, many stars of the Milky Way appear in front of its galaxies.

The distance of the Perseus cluster, deduced from measurements of redshifts of twenty-four of its galaxies (average recession velocity 5490 kilometres per second), is about 72 megaparsecs (for $H_0 = 75$ kilometres per second per megaparsec). This, with Virgo and Coma, is one of the nearest rich clusters. The velocity dispersion of the galaxies is very high, of order 1400 kilometres per second.

The core of the cluster is dominated by a supergiant elliptical galaxy, NGC 1275 (the brightest galaxy, at the extreme left of the photograph), which is a galaxy with an active nucleus, called a Seyfert galaxy, associated with a powerful radio source, Perseus A.

The cluster contains other radio sources, in particular NGC 1265 and IC 310, which have 'head–tail' structures characteristic of radio galaxies moving at high speed within a rich intracluster gas. The Perseus cluster is extremely rich in very hot gas, making it one of the brightest extragalactic X-ray sources, with the emission concentrated around NGC 1275, also catalogued as Perseus X-1. (Kitt Peak National Observatory)

Absolute motion of the Local Group and the Great Attractor. By measuring the anisotropy of the cosmological background in two opposing directions in the sky (dipole effect) astronomers on the Earth can determine the motion of the Local Group with respect to this distant absolute rest frame. But they have first to take account of relative motions on a small scale: the Earth gravitates around the Sun at 30 kilometres per second: the Solar System moves around the centre of our Galaxy at 230 kilometres per second, and the Galaxy itself moves at 40 kilometres per second towards the centre of mass of the Local Group, in the direction of its large neighbour, the Andromeda galaxy.

Correcting for these motions internal to the Local Group, there remains a velocity of 600 kilometres per second in the opposite direction to the Galactic rotation, perpendicular to the line joining the Solar System to the Galactic centre, and about 27 degrees above the Galactic plane. The gravitational attraction of the Virgo supercluster can explain only one of these velocity components (green vector). The other velocity component (blue vector), which is larger, is directed towards the Hydra–Centaurus supercluster, itself under the influence of a yet stronger and more distant overdensity. Called the Great Attractor and hypothetically situated at twice the distance of the Hydra–Centaurus supercluster, this overdensity, formed by an enormous concentration of galaxies, is supposed to be responsible for the general drift of the Local Group, the Virgo supercluster, and the Hydra–Centaurus supercluster with respect to the absolute reference frame constituted by the cosmic background radiation. (from an illustration by Hank Iken.)

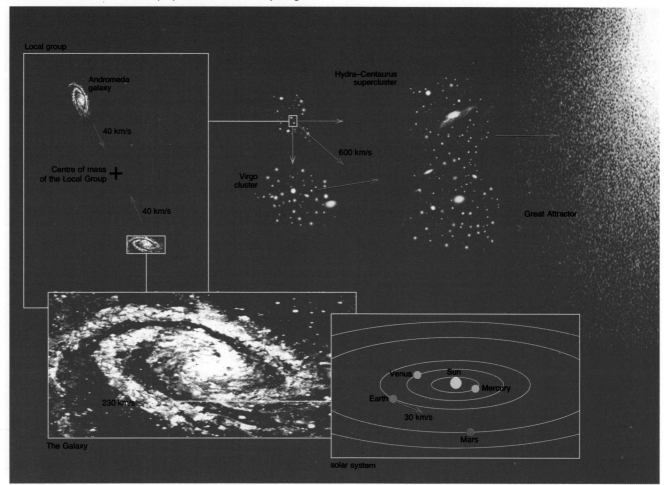

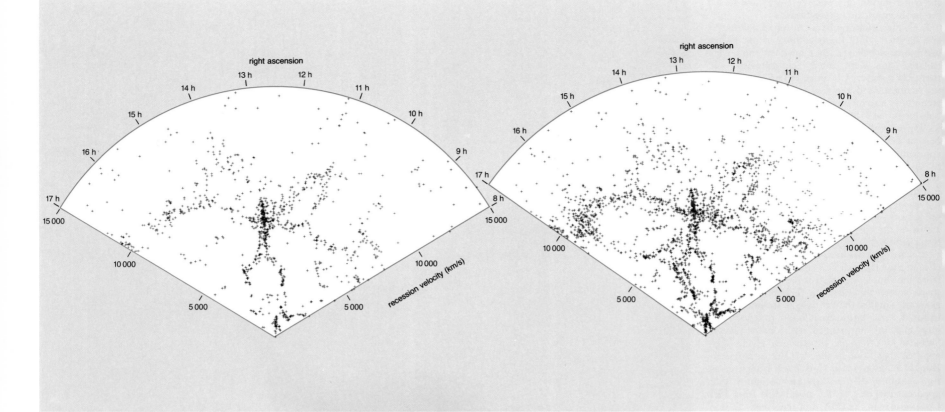

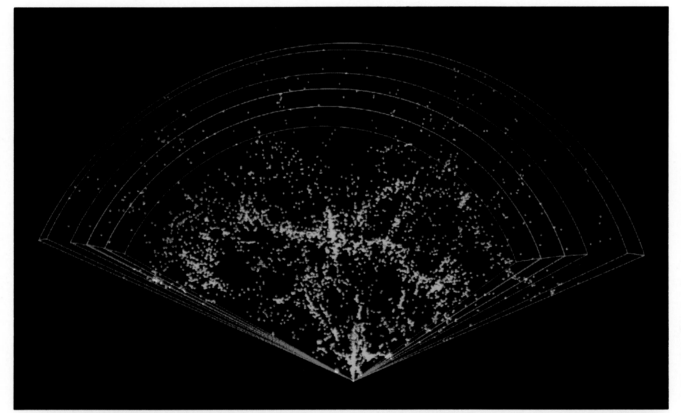

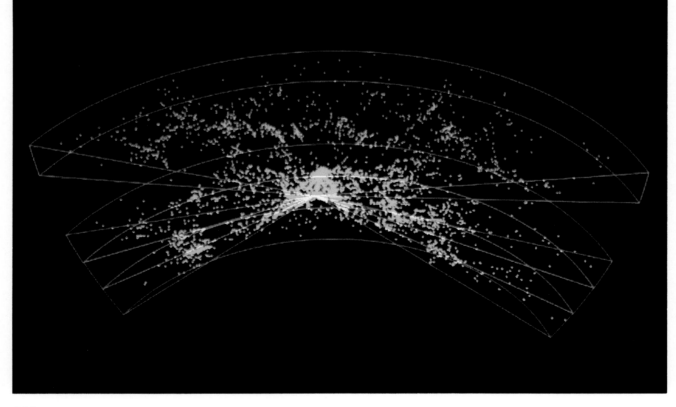

The large-scale Universe: filaments and voids.
The distribution of matter in structures larger than clusters of galaxies is a lively area of research. Observations accumulate rapidly because of new detectors and progress in data handling. Analyses of the angular distribution of galaxies on photographic plates are now complemented by the three-dimensional information supplied by redshift measurements (the redshift is related to the distance by Hubble's law). In the 1950s it took several hours to acquire a spectrum. Today this takes only a few minutes and deep studies of the Universe can be carried out systematically. At the beginning of the 1990s, for example, we have measured more than 30 000 galaxy redshifts, representing about one hundred-thousandth of the volume of the visible Universe. New maps, though fragmentary, have provided big surprises. Galaxies seem to be distributed on surfaces like flakes around quasi-spherical voids, forming a foam-like structure. A team of astronomers from the Harvard-Smithsonian Center for Astrophysics (Cambridge, Massachusetts) has mapped a vast set of nearby galaxies (brighter than magnitude 15.5) in this way, in slices of sky between 8h and 17h right ascension and 8.5 degrees to 44.5 degrees declination (each slice has a width of 6 degrees in declination), up to a recession velocity of 15 000 kilometres per second (a depth of 200 megaparsecs for $H_0 = 75$ kilometres per second per megaparsec). The Sun is at the tip of each slice.

The illustrations above show as examples, left, the slice between 26.5 degrees and 32.5 degrees declination (containing 1074 galaxies), right, the superposition of three slices between 26.5 degrees and 44.5 degrees declination (containing 2500 galaxies).

Bubbles and filaments appear distinctly. The voids have diameters of about 50 megaparsecs. A strange structure with the shape of a human silhouette seen from the front with outspread arms and legs, is visible. The 'torso' of this silhouette (along the central line of sight, near 13 h right ascension) corresponds to the Coma cluster. The most striking structure (the 'arms' of the silhouette) is called the Great Wall. Situated at a distance of about 100 megaparsecs. It is about 60 megaparsecs by 170 megaparsecs in area, with a 'thickness' of only 5 megaparsecs (hence the name). Its mass is ten times higher than that of the Local Supercluster. The illustrations opposite show four superimposed slices seen from the Solar System. The separate slice (orange dots) lies between declinations +8.5 degrees and +14.5 degrees, while the three other slices (yellow dots) are between +26.5 degrees and 44.5 degrees.

Surveys performed over other slices of the sky also reveal the existence of large voids and filaments or walls. They place tight observational constraints on models of the formation and evolution of galaxies. (M.J.Geller and J.P. Huchra, Harvard-Smithsonian Center for Astrophysics)

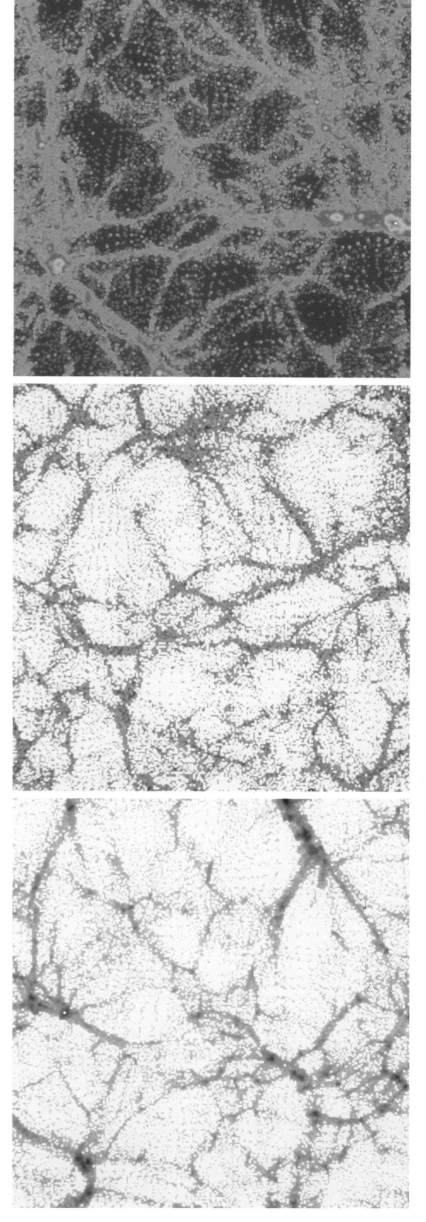

The Universe simulated by computer. To try to understand the formation processes of galaxies, clusters, and even larger structures such as 'filaments', 'voids' and 'bubbles', astrophysicists perform numerical computer simulations of the evolution of the matter density in an expanding universe over tens of billions of years, starting from a given initial distribution. Here the initial conditions are close to those of the so-called cold dark matter model, in which most of the matter in the Universe is assumed nonluminous. Gravity alone is responsible for the clumping of matter and its organisation into characteristic structures. The result reproduces fairly accurately the structure currently observed, from galaxies to superclusters of galaxies.

These three images show the matter density in three spatial slices each of side 70 megaparsecs and thickness 4 megaparsecs, projected on three perpendicular planes. The (arbitrary) colours have been chosen so as to bring out various aspects of the distribution. The upper image emphasises the densest regions: filaments, bubble boundaries. The middle image emphasises instead the low-density regions, and clearly shows the large 'voids' which run through the luminous matter, actually containing dark matter; the lower image is a compromise. (J.-M. Alimi, Observatoire de Meudon)

cores would 'evaporate' in a time of the order of that for a galaxy to cross the core; this is easily estimated by dividing the core diameter by the average dispersion velocity. This time is generally only about 100 million years, much shorter than the age of the Universe, which is more than 10 billion years; thus the cores of clusters must be bound. The dynamics of the peripheral galaxies (constituting the halo) is by contrast much less well understood; the outer layers of a rich cluster could collapse gravitationally. If the cluster cores are stable there must be equipartition of energy among the core galaxies; this means that because of the many gravitational interactions between the galaxies the kinetic energy is equally distributed among the galaxies. The time required to achieve this – called the relaxation time – is actually comparable with the cluster age for the cores of rich clusters, but much larger for the peripheral galaxies. Because of this equipartition of energy, the more massive galaxies tend to move more slowly and gradually to spiral towards the centre of gravity of the cluster (the kinetic energy of a cluster is proportional to its mass and the square of its velocity). Consequently an 'evolved' cluster will have had time to develop a high concentration of massive bright galaxies, while an unevolved cluster will not have a dominant condensation. These two types of configuration and the various intermediate stages are actually observed in clusters of galaxies.

There are several catalogues of clusters of galaxies. Notably, George Abell catalogued 2712 rich clusters from the sky maps produced by the Schmidt telescope of the Mount Palomar Observatory down to photographic magnitude 21. There are two large morphological classes: regular and irregular clusters; these are analogous to the very concentrated globular star clusters and the much sparser and more dispersed open (or galactic) star clusters.

Regular clusters are spherically symmetric and have a strong central condensation, the centre itself frequently being marked by a supergiant elliptical cD galaxy. These are evolved clusters with a very deep central potential well; the dispersion velocities of the galaxies are therefore large, of the order of 1000 kilometres per second. Also, all the bright galaxies are elliptical (or S0), and so have a depleted gas content. This may have been swept out of the galaxies in gravitational interactions between near neighbours. The diameters of regular clusters vary between 1 and 10 megaparsecs, and their masses are of order 10^{15} solar masses, equivalent to 10 000 giant galaxies like the Milky Way. The nearby Coma cluster is the prototype regular cluster.

Irregular clusters have neither symmetry nor any unique concentration in the distribution of their galaxies; they are little evolved dynamically. All types of galaxies are found in them. The velocity dispersions are lower here, and the diameters vary from 1 to 10 megaparsecs with the masses running from 10^{12} to 10^{14} solar masses. The Virgo cluster is a typical irregular, the closest rich cluster to the Local Group. The latter is irregular itself, as are all poor clusters of galaxies.

Rich clusters of 10 megaparsecs diameter are not the largest structures in the Universe. In fact the distribution of rich clusters consists of a roughly uniform background of clusters on which are superimposed clusters of clusters, called superclusters. Our Local Group and several tens of groups of nearby galaxies are, for example, part of a much larger system, about 30 megaparsec in diameter, centred on the Virgo cluster: this is the Local Supercluster. However, the latter is relatively poor as it has only one rich cluster at its centre, plus several poor clusters. Many other superclusters are known, such as Hydra–Centaurus, Perseus–Pisces, or the 'Great Wall', each one containing on average about ten rich clusters. The largest size of these structures is of the order of 100 megaparsecs, but unlike clusters they are usually filamentary in shape.

Recent discoveries, both observational and theoretical (using numerical simulations) about the organisation of superclusters into lines, filaments, flakes or walls, as well as the existence of vast regions practically devoid of luminous galaxies, lead to the idea that the Universe is composed of 'bubbles' of about 100 megaparsecs in diameter. Inside a 'bubble', space is almost empty; galaxies and clusters are found at the boundaries, while particular concentrations of superclusters are found along lines of intersection between boundaries.

The theoretical and cosmological impact of such a discovery is immense; structures as vast as superclusters and voids reflect the original distribution of matter, unlike structures at smaller scales, which have been perturbed by mixing and by earlier interactions between galaxies which have removed all trace of the initial conditions. These structures can therefore tell us about the processes which form galaxies, which are still not understood.

These very partial views of the large scale structure of the Universe demand deeper observations. Through new ultrasensitive photographic plates we can now reach apparent magnitude 26, allowing the cataloguing of more than 100 million galaxies in the sky. But only the systematic measurement of their distance (using their redshifts) will give their true distribution in space rather than their projected distribution. There will soon be enough data to tell us if we really live in a 'cellular' universe.

It is thought that the hierarchy of structures in the Universe stops at the superclusters: it is observed that on scales larger than 1000 megaparsecs (i.e. at distances of more than three billion light years), variations in the number density of galaxies become smaller than half of their average density. Consequently the Universe is homogeneous and isotropic on the very large scale, as is spectacularly shown by the perfectly uniform distribution of the cosmic background radiation at 2.7 K. This marks the true beginning of the realm of relativistic cosmology

Jean-Pierre LUMINET

The intergalactic medium and dark matter

In models of the long-term evolution of the Universe based on general relativity the mean density plays a central role: if it is less than a critical value which today is about 10^{-29} grams per cubic centimetre, the Universe will continue to expand indefinitely. If the mean density is greater than the critical value the expansion will stop and give way to contraction. It is therefore natural to try to catalogue every possible form of matter and estimate its contribution to the mean density of the Universe.

The most obvious contribution is that of the luminous matter, i.e. that of the stars and galaxies; its density is fifty times less than the critical value. However, this visible mass cannot account for the strength of the gravitational field at the edges of galaxies, as shown by measurements of the rotation velocity of stars and hydrogen clouds in spiral galaxies, and the detection of vast amounts of plasma around elliptical galaxies. It is generally thought that galaxies are much more extended and massive than indicated by their luminosities.

The luminous mass is also too small to explain the dynamics of clusters of galaxies. If we measure the velocity dispersion of the galaxies in a cluster we can work out its mass; this *dynamical* mass is always greater than the luminous mass.

Dark matter exists on all scales. The discrepancy between what is directly observed as luminous matter and what is inferred from dynamics varies from a factor of two in the neighbourhood of the Solar System to a factor of about five in galaxies and thirty or more in clusters of galaxies. This is called the *missing mass* problem.

It is not surprising that non-luminous matter exists: just as the interior of each galaxy has its content of gas and dust, the space between the galaxies is not completely empty. More surprising is the fact that the dark matter appears to dominate luminous matter quite strongly. Besides the dynamical measurements, a cosmological proof of the importance of the dark matter is provided by observations of the abundances of light elements, deuterium, helium, and lithium, made in the first seconds of the Universe. These abundances depend on the density of baryons (protons and neutrons) present in the Universe at that epoch, and their accurate determination allows us to calculate by extrapolation the present density of the Universe. The value found in this way corresponds perfectly to what is deduced from the dynamics of luminous matter, and is five times less than the critical value. An intergalactic medium therefore clearly exists, and while not deciding between open and closed models of the Universe, remains a subject of immense interest because of the influence it may have on the evolution of galaxies.

Another way of detecting dark matter appeared in the 1980s after the discovery of the first gravitational lens. When a very massive body, a galaxy or cluster of galaxies lies between the Earth and a very distant object – a galaxy or quasar – its gravitational field amplifies and focusses the light from the distant object. This effect is predicted by general relativity, and is the same as that by which the Sun bends the light rays from a distant star. This was first observed by Arthur Stanley Eddington in 1919 from the island of Principe, during a solar eclipse. In 1931 Fritz Zwicky realised that if the deflector is not a star but an entire galaxy, the same source could appear in several different positions. We had to wait until 1979 to observe finally these cosmic optical illusions: the double quasar Q 0957+561 A, B, in the constellation of Ursa Major, made of two objects separated by about 6 seconds of arc, whose extremely similar spectra showed that they were two images of the same quasar caused by the gravitational action of a deflecting galaxy. The lens effect was confirmed a little later by the discovery of the lensing galaxy, then by radio observations.

Such alignments are not as rare as one might think: we already know about twenty gravitational lenses. These objects are very interesting for cosmologists, as the detailed analysis of the light rays reveals the mass distribution of the lens. Comparing the visible size with its real size deduced from the light deflection one can calculate the proportion of dark matter it contains. Further, the lensed light is delayed, partly because of the increased path length and the variation of proper time of the photon as it passes around the deflector. Measuring the difference between the optical paths gives us the distance of the lensed objects, and thus the age of the Universe and its rate of expansion. The time delay is a measurable quantity for a quasar, as its intrinsic flux can vary on a timescale less than the delay. For Q 0957+561, constant monitoring carried out over eight years from 1980 to 1987 by a team of researchers from the Observatoire de Meudon showed that there was a delay of 415 days.

In 1987 a spectacular discovery made by another French team increased the importance of gravitational lenses: giant structures in circular arcs were detected in several distant clusters of galaxies. The effect was predicted by Einstein in 1936 for the case of perfect alignment between the source and the deflector: the arc is the image of a distant galaxy formed by the presence of potential wells of the cluster along the line of sight. We know about ten arcs. Analysis of the images allows determination of the amount of dark matter in the deflecting galaxies. The resulting values are in perfect agreement with the dynamical masses, confirming the presence of dark matter of about five to ten times the mass of the luminous matter.

What is this dark matter? In clusters the existence of an intra-cluster gas has been well established since 1972, with the discovery of the first 'head–tail' radio source. This is a radio galaxy moving in the intra-cluster and leaving a trail behind it which resembles the trails left by comets in the hot plasma of the solar wind. A second major discovery in 1977 was the observation of lines of strongly ionised iron in certain clusters, such as those in Virgo, Coma and Perseus, showing that the temperature of the intra-cluster gas had to be of the order of ten million degrees. Further, the heavy-element abundances of the intra-cluster gas show that it is not primordial but has undergone stellar nucleosynthesis: put another way, the intra-cluster gas has passed at least once through the galaxies of the cluster to undergo an entire cycle of 'astration' leading to the production of heavy elements.

Gas at ten million degrees must emit X-rays. The Uhuru, Einstein and ROSAT satellites confirmed that clusters are strong extended X-ray sources, with a great variety of sizes and luminosities. Several tens of clusters have been detected in X-rays; their luminosities range from 10^{36} to 10^{38} watts, representing surface brightnesses between five and five hundred times the sky background. There is a correlation between the X-ray emission of a cluster and its richness, which specifies the degree of central condensation.

In some clusters, such as the Perseus cluster, much of the X-ray emission comes from a region around an active galaxy which is itself an X-ray

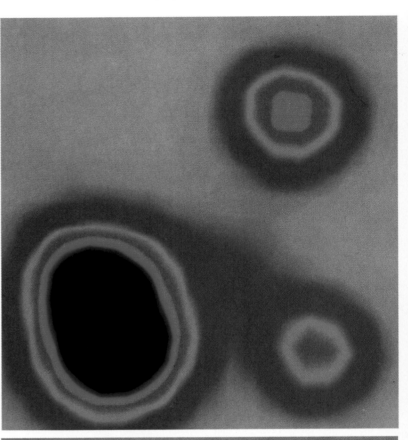

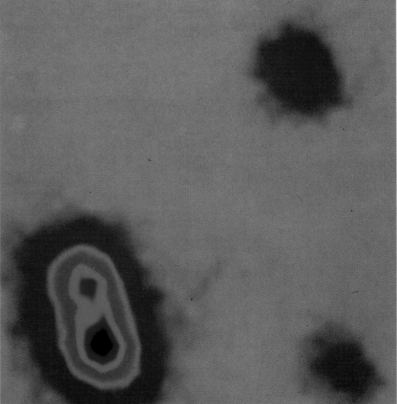

The triple quasar PG 115+080. In May 1980, one year after the discovery of the first gravitational lens Q 0957+561, a second lens was found, also by accident. On a false colour CCD image with angular resolution of 0.9″ (top) PG 115+080 looks like a triple quasar whose very close components (separations of order 2″) have identical spectra and redshifts ($z=1.723$).

The most plausible gravitational lens model is based on the hypothesis that the lensing galaxy is either an edge-on spiral or a very flattened spiral. This model requires the existence of five images: the brightest, slightly elongated, is actually a very close double image, as clearly shown in the lower illustration, whose resolution is an exceptional 0.3: the fifth component must be very faint and has not been detected, like the lensing galaxy.

Unlike what is found for the double quasar Q 0956+561, significant time variations of the 'three' components occur practically in phase. The quasar–lens–observer configuration is such that there is partial cancellation between geometrical delays and those caused by the potential well created by the lensing galaxy. (G. Lelièvre, J.-L. Niéto, J. Arnaud, CFH)

emitter; however, the sum of the X-ray luminosities of the individual galaxies is only a fraction of the total X-ray luminosity of the cluster, showing that the hot gas is diffuse and not localised. The radiation is *bremsstrahlung* (braking radiation), i.e. the emission of X-ray photons by the electrons of a hot plasma deflected by the electric fields of the ions.

An interesting consequence of the existence of hot gas in clusters is the Sunyaev–Zeldovich effect: this is the distortion of the spectrum of the cosmological radiation at 2.7 K when this is observed through a cluster: the cosmological photons gain energy by interacting with the very fast electons of the intra-cluster gas, causing a slight deviation of the temperature of the cosmological radiation of about a ten-thousandth of a degree.

The most recent studies of the intra-cluster gas lead to mass estimates of the same order as the luminous mass; it therefore cannot account for the whole of the dark matter. This may be in the form of invisible stars. One possibility is stars at the end of their evolution which no longer radiate: white dwarfs, neutron stars, or

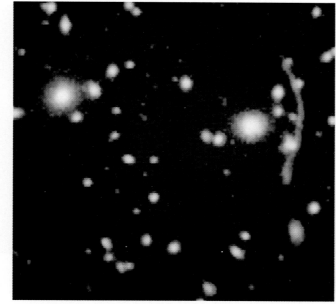

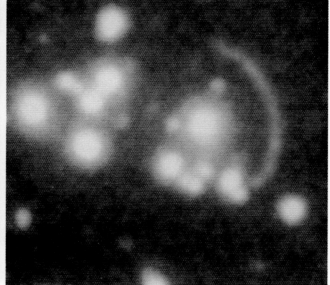

Einstein arcs. The image on the left was obtained in September 1985 by Bernard Fort and Yannick Mellier of the Laboratoire d'Astrophysique de Toulouse, using the CFH telescope. It shows the central region of the galaxy cluster Abell 370; a large blue arc extends over more than 30″. In April 1987 a team including among others Bernard Fort, Yannick Mellier and Geneviève Soucail gave the first irrefutable proof of the gravitational origin of the arc, by means of spectroscopic studies carried out at the large European Southern Observatory telescope at La Silla, Chile: while the redshift of the cluster is 0.37, that of the arc is 0.72. The interpretation is clear: the arc is the image of a galaxy situated behind the cluster, at twice the distance, of the order of 7 billion light years; this image is deformed by the gravitational lens effect of the total mass, visible and invisible, of the cluster.

The Einstein arc of the galaxy cluster Cl 2244-02 (right) shows a near-circular curve of about 110° (its apparent size is 15″). It consists of two gravitational images of a background source which is probably a spiral galaxy seen face on whose redshift is 2.24 (the cluster is much nearer, at z=0.329). It is not uniformly bright but shows six brightness peaks. This is caused by the association of two images of the galaxy: in each of them, the bulge and spiral arms produce three brightness peaks. This gravitational lens gives an estimate of the total mass of the cluster Cl 2244-02 as two hundred billion solar masses, much higher than its luminous mass. (left, B. Fort, CFH; right, F. Hammer, CFH)

The Einstein cross. This magnificent gravitational lens is a quasar, Q 2237+0305, about 8 billion light years away, observed through a galaxy five times closer. The lens effect caused by the nucleus of this galaxy on the light from the quasar produces four components separated by 1–2″.

These false colour images were obtained at the CFH telescope using the pupil segmentation technique and a photon-counting camera: the upper one shows the whole galaxy; the lower one, whose resolution is 0.3″ shows more closely the four components of the image as well as the galactic nucleus, visible at the centre of the cross. (G. Lelièvre, J.-L. Niéto, J. Arnaud, CFH)

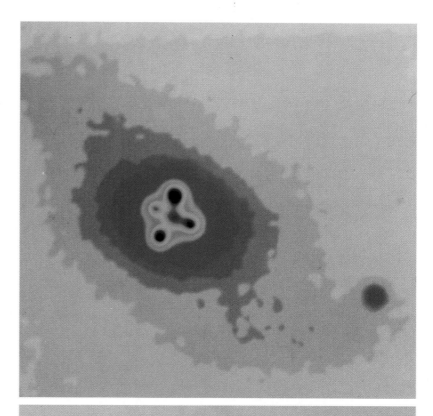

Head–tail radio galaxies. A radio galaxy has an active nucleus which continually ejects plasma at high velocity: this plasma is projected very far outside the optical galaxy and forms extended lobes emitting radio waves through the synchrotron effect. Rich clusters of galaxies often contain particular radio galaxies called head–tail, which show the existence of intergalactic intra-cluster gas and the influence this can have on the galaxies immersed in it. When a radio galaxy moves inside a rich cluster its velocity in the cluster's gravitational field can reach 1000 kilometres per second; the radio lobes are swept back because of the intra-cluster gas pressure (at rest in this gravitational field), forming a sort of tail which can be single or double. This radio image (at 6 centimetre wavelength) shows the radio galaxy NGC 1265, moving in the gas of the Perseus cluster at a velocity of 2000 kilometres per second. (NRAO, AUI, Observers C.P. O'Dea, F.N. Owen)

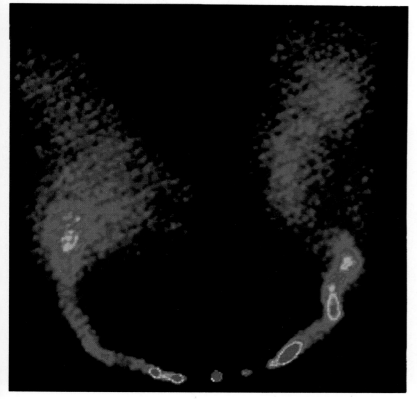

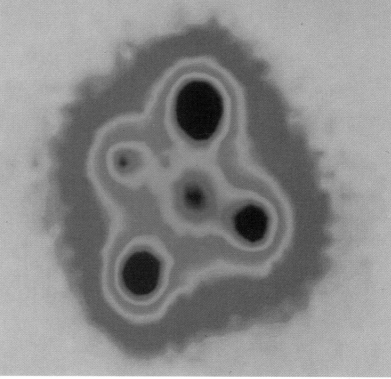

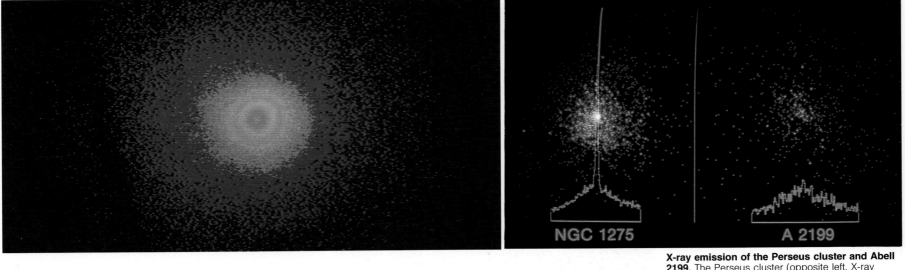

X-ray emission of the Perseus cluster and Abell 2199. The Perseus cluster (opposite left, X-ray observation by the Einstein satellite) has an X-ray luminosity of 4×10^{37} watts (one hundred billion times the Sun's luminosity) of which one quarter comes from the active galaxy NGC 1275 (centre) and the rest from the hot intra-cluster gas. Emission lines of iron XXIV (twenty-three times ionised) have been discovered there, indicating a gas temperature of around 100 million degrees. Perseus is a good example of a regular cluster with a strong central condensation, in contrast with irregular clusters like Abell 2199 which do not have such condensations. The images at right allow a comparison of the two clusters through their X-ray luminosity distributions. While the Perseus profile shows a very sharp central peak (corresponding to the emission of NGC 1275), that of Abell 2199 has none. (Harvard–Smithsonian Center for Astrophysics)

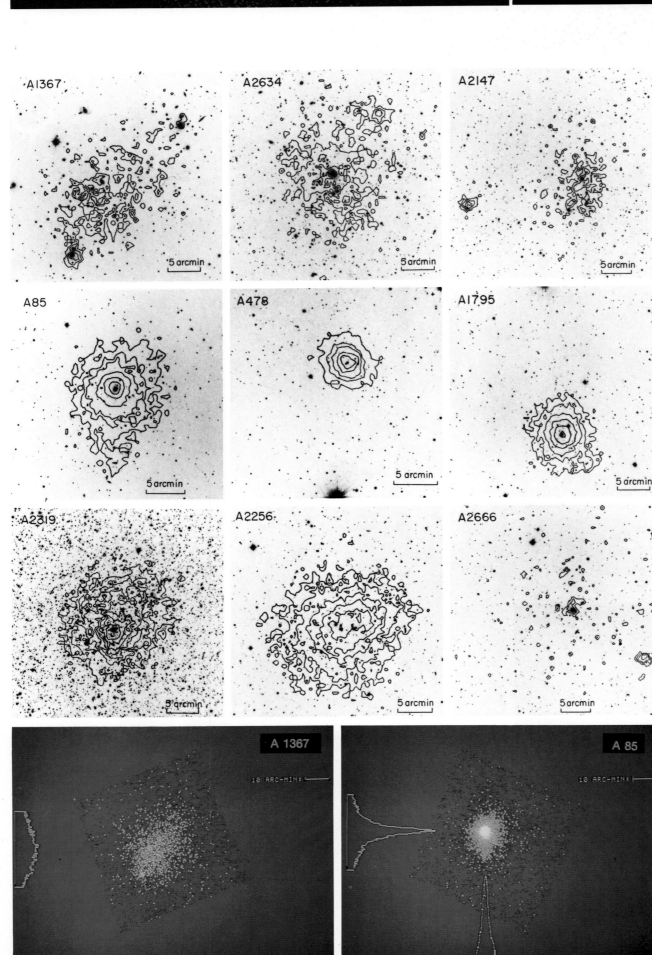

X-ray structure of clusters of galaxies. The pictures opposite are a superposition of the optical image and the X-ray emission contours for nine rich clusters in the Abell catalogue. The IPC (Imaging Proportional Counter) of the Einstein satellite observed the X-ray isophotes of each cluster and allowed comparison with the optical images demonstrating correlations between the distribution of the galaxies (radiating in the optical) and the gas (radiating in X-rays). The X-ray structure of rich clusters is grouped in three categories related to the stage of dynamical evolution of the cluster. The clusters A 1367, A 2634 and A 2147 have X-ray emission in large dispersed clumps; these are 'young' clusters, little evolved in the sense that the galaxies have not yet formed a clear central condensation and their velocity dispersions are small. These properties are seen even more clearly on the X-ray colour image of A 1367 on which is marked the distribution of X-ray luminosity (opposite left). These clusters still contain many spiral galaxies rich in gas, which have not been swept clean by close interactions with other galaxies, while the intra-cluster gas is relatively underabundant and only a weak X-ray emitter. The clusters A 85, A 478 and A 1795 show, by contrast, an X-ray structure continuously distributed around a very strong central peak. The luminosity profile of A 85 (colour image opposite right) shows this very clearly. These are evolved clusters which have developed a deep potential well at their centres, often around a giant cD galaxy (central and Dominant). These clusters have few spirals, as the gas has already been swept out; the velocity dispersion, gas temperature and X-ray flux are high.

The clusters A 2319, A 2256 and A 2666 represent an evolutionary stage intermediate between little-evolved and highly-evolved clusters; the X-ray flux distribution is fairly continuous but the central peak is much less pronounced. (monochrome images: Smithsonian Institution Astrophysical Observatory, after C. Jones *et al.* 1979; colour images, Harvard–Smithsonian Center for Astrophysics)

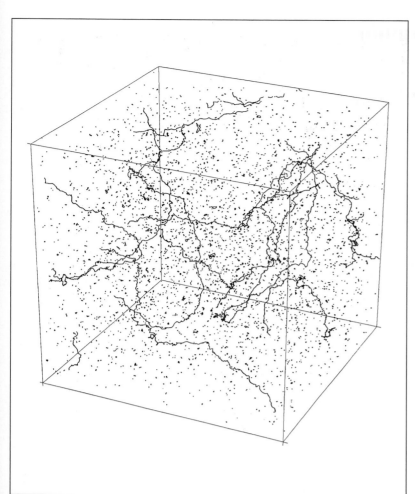

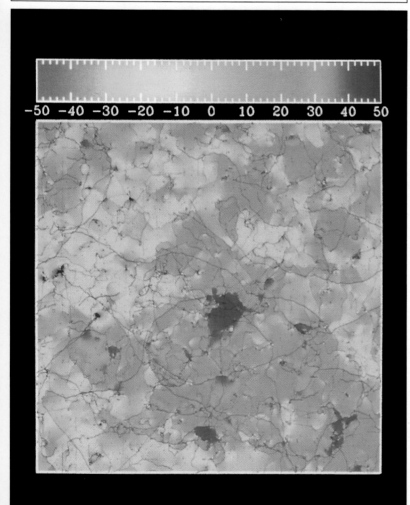

-50 -40 -30 -20 -10 0 10 20 30 40 50

Cosmic strings. A candidate for the dark matter is not a particle but a structure called a cosmic string. Similar to crystalline defects forming, for example, between regions of a pond which froze at different times, cosmic strings would be large topological defects caused by symmetry-breaking in the cooling of the early Universe. They would have the form of very thin tubes with a very high line density of mass-energy (1 million billion tons per centimetre); they could therefore be detected by gravitational lens effects, or by the gravitational waves they would emit as they decay in the course of their evolution. However, their total energy density would not be enough to close the Universe.

The upper figure shows a computer simulation of the evolution of a network of cosmic strings. As the strings spread out in the Universe and cross each other they form closed loops several hundred light years in diameter, large enough to cause accretion of ambient matter and trigger the formation of clusters and galaxies.

The lower figure shows the results of a calculation of the fluctuations of the cosmic background radiation at 2.7 K which would be caused by the evolution of a network of cosmic strings. The size of this map is 7°×7°; the upper scale gives the values of the temperature fluctuations ($\Delta T/T_{average}$). The motion of the strings across the sky should produce a characteristic pattern of anisotropies in the cosmological background radiation through the gravitational lens effect. But these anisotropies are not observed, imposing constraints on the theories in which galaxy formation is caused by cosmic strings. (upper figure, D. Bennett and F. Bouchet; lower figure, F. Bouchet, D. Bennett and A. Stebbins)

black holes. It is, however, very implausible that such stars could have formed in large numbers in the halos of galaxies, even more so outside these galaxies. Symbolic of invisible matter, black holes play a different role: besides those which are the residue of stellar evolution – it seems that we have detected several among the X-ray sources in our Galaxy – one suspects that much more massive black holes, from a million to a billion solar masses exist – in the centres of active galaxies and quasars. An even more interesting idea is that black holes of intermediate mass down to a billion tons could also exist in abundance and evade detection (those of lower mass would have 'evaporated' through quantum-mechanical effects). These so-called primordial black holes would have been produced by local condensations of the very early Universe. But their formation mechanism remains very speculative, and their possible contribution to the dark matter remains completely unknown. The most likely class of invisible stars are those of low mass. Small stars ($0.1M_\odot$) invisible in the optical are rather unlikely candidates as galactic halos do not show any excess brightness at infrared wavelengths, where these objects emit. By contrast, smaller objects, brown dwarfs or 'jupiters', intermediate between planets and stars, might populate galactic halos. Detecting them would be impossible without their gravitational lensing effects. In this case, the 'microlens' would be a brown dwarf situated in our Galaxy or a nearby galaxy, with the distant source being a quasar. The lensing would produce a luminosity variation and a distortion of the quasar's spectrum (in fact, an entire class of galaxies with active nuclei which vary extremely rapidly has been interpreted as resulting from microlensing).

Besides offering an explanation of the discrepancy between the luminous and dynamical masses, dark matter has a second facet which is even more intriguing. Theoretical arguments suggest that the mass density of the Universe might be *exactly* equal to the critical value. In particular, the theory of inflation, deduced from elementary particle physics applied to the very early Universe and developed to explain a number of things not accounted for in the standard Big Bang model, predicts that the Universe, after a phase of exponential expansion, is flat (and thus, through general relativity, has the critical density). Now calculations of Big Bang nucleosynthesis forbid the baryonic mass density (including small stars, black holes and 'jupiters') to exceed one fifth of the critical value. The flatness of the Universe imposed by inflation therefore demands study of non-baryonic dark matter.

Another theoretical constraint is set by the formation of galaxies and the development of large-scale structure in the Universe. We know so little about this question that we still have to decide whether the stars or the large galaxy clusters that formed first. The galaxies and clusters must have formed from small density fluctuations in the early Universe. If we assume that all the matter is baryonic, the natural growth of inhomogeneities under gravity requires that the density contrast should have been of the order of one thousandth at the era of decoupling between radiation and baryonic matter (a million years after the Big Bang) in order to produce the structure we see today. But searches for anisotropy in the cosmological 2.7 K radiation, emitted exactly at this epoch, which must therefore bear the imprint of these matter inhomogeneities, show that on angular scales corresponding to large galaxy clusters (a few minutes of arc) there are no fluctuations of more than one hundred thousandth in density contrast. There is therefore a contradiction, unless we invoke dark matter consisting of weakly interacting massive particles (WIMPs), already condensed into fluctuations at an epoch before baryon–radiation decoupling, and weakly coupled to radiation so as not to leave an imprint on the 2.7 K radiation. After recombination baryonic matter will fall into the potential wells of the dark matter to form the visible parts of the galaxies, immersed in the extended halos of dark matter.

Neutrinos of non-zero rest mass are a possible source of 'hot' dark matter, i.e. with speed so large that it cannot condense into fluctuations except of very large scale. The advantage of neutrinos over other WIMPs is that we are certain that they exist and that they are as numerous as the photons of the cosmic back-

ground radiation. They would close the Universe if their real mass was between a ten thousandth and a hundred thousandth of that of the electron. They present a difficulty, however; they cannot explain the small fluctuations, or account for the distribution of galaxies and individual clusters. The key point is, of course, to know whether neutrinos really have a rest mass. Laboratory experiments, performed particularly in the Soviet Union, are very controversial, but they have stimulated intense theoretical activity directed at finding even more elusive particles.

A plethora of other WIMPs predicted by various unified interaction theories but not observed in the laboratory (perhaps because their masses are too high to be created by current particle accelerators) provide potential candidates for 'cold' dark matter, i.e. of low enough velocity to form small-scale fluctuations very early. For example, axions are light particles produced in the quark–hadron transition which preceded primordial nucleosynthesis, and accumulations of them could 'seed' galaxy formation. Supersymmetry theories supply known particles (photon, graviton etc) with massive partners (photino, gravitino etc). Calculations show that, for example, the Universe would contain enough photinos of mass between one and fifty times that of the proton to make it closed. In general, cold dark matter accounts well for the distribution of matter on all scales: WIMPs would be distributed much more uniformly than baryonic matter, in particular populating the 'voids' in the distribution of galaxies.

The practically inexhaustible richness of elementary particle theories at very high energies provides many candidates for the dark matter. Cosmology has therefore today reached a watershed, becoming an experimental science through elementary particle physics. The way in which Nature arranges itself to hide a large fraction of its matter fascinates a growing number of physicists. The search for cool stars by their gravitational lens effect and for weakly interacting particles using ionisation detectors or bolometers may elicit the ultimate Copernican revolution, of knowing that the matter we are made of is marginal to the Universe.

Jean-Pierre LUMINET

The scientific perspective

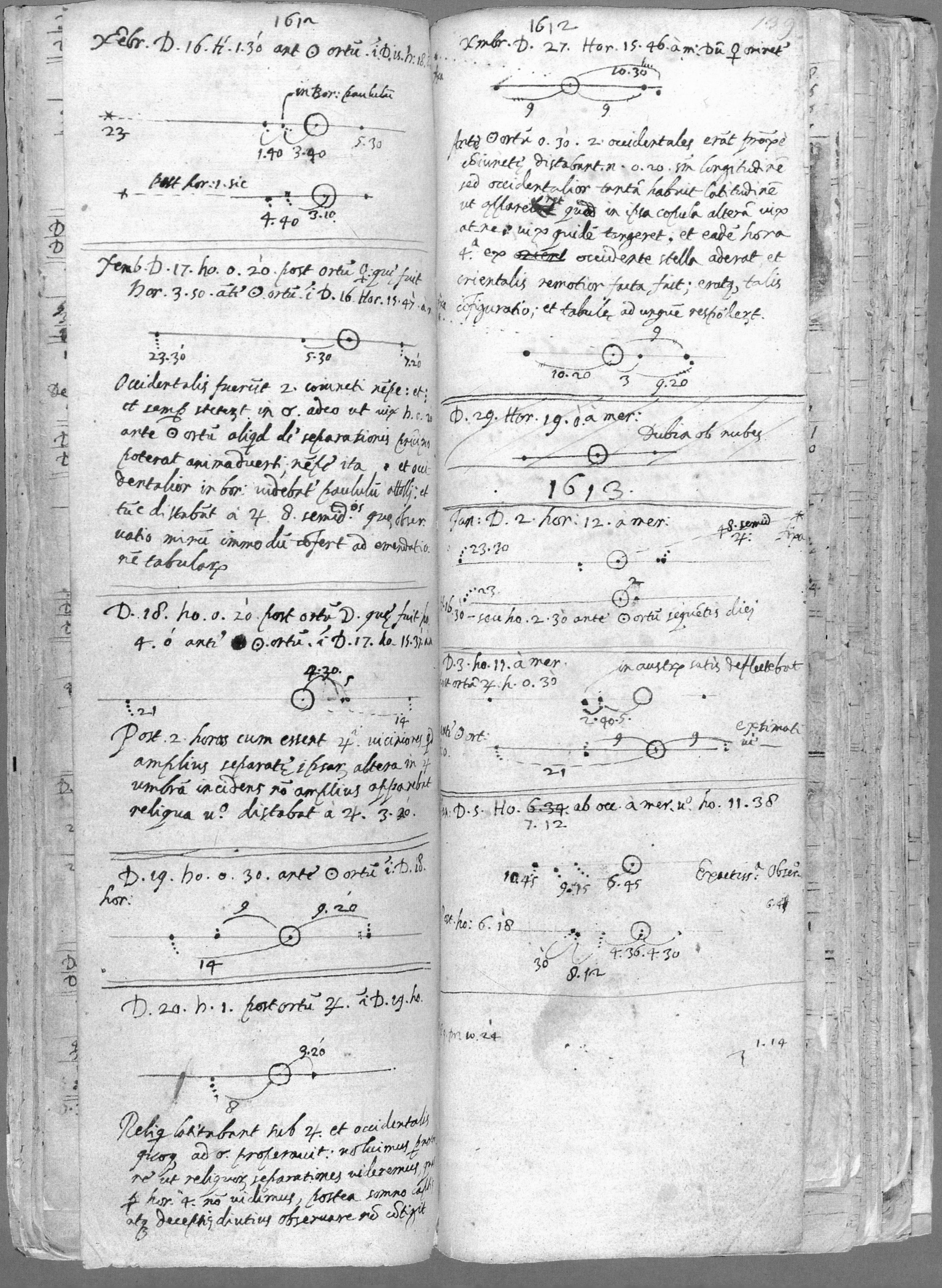

Cosmology

Cosmology is the domain of astronomy – and of physics – which attempts to describe the properties and evolution of the Universe in its entirety. As one of the fundamental questions of this discipline is that of the origin of the Universe, an important part of this section will be devoted to explaining the model of sudden creation, better known as the Big Bang theory, whose principal founders were Georges Lemaître and George Gamow. As the name indicates, the theory attributes an explosive origin to the whole Universe; this simple model explains the properties of the Universe observed today, and because of this is generally favoured by specialists. However, caution is needed in this matter as we only have a very fragmentary ensemble of observational facts at our disposal, of which the majority have only been known since a very recent date. It is therefore unfortunately impossible to predict whether the hypotheses at present favoured in cosmology will remain valid in the long term. It is necessary at this point to emphasise the fact that the situation of cosmology is unique and complex. It is unique since its ultimate goal is to include all the other sciences, which are only particular applications of cosmology. Note that our own existence is in itself a cosmological fact which must be integrated into any theory. Cosmology is complex because the majority of observations which have a cosmological significance are difficult: they are concerned with the furthest objects and therefore the faintest. Because the velocity of light is finite, the astronomer, in observing the further objects, in fact has the possibility of observing a larger volume of the Universe and, above all, to go back in time in studying numerous objects – galaxies and stars – in very different phases of their evolution. Although the task of cosmology is difficult, it is not hopeless as use can be made as it were of observations of the past.

The principles of cosmology

In order to simplify their work, astronomers have established a number of propositions, which they in the main agree to adopt and which, by a sort of misuse of language, they call principles of cosmology. Five basic principles may be enumerated: the principle of uniformity, the cosmological principle, the anthropic principle, the principle of equivalence and Mach's principle.

The principle of uniformity states that the laws of physics which govern the behaviour of objects in our immediate environment, and which can be verified in the laboratory, apply to the whole of the observable Universe.

The cosmological principle states that the Universe is the same everywhere except for local irregularities; it therefore assumes that the Universe is homogeneous and isotropic, which considerably simplifies the models used to describe it.

Cosmology must explain our own existence; certain cosmologists attach little importance to this principle, called the anthropic principle; others consider it to be fundamental.

The principle of equivalence states that all bodies – whether constituted of matter or of antimatter – have a positive or zero mass which explains the ensemble of their dynamic properties: in particular, this principle includes the equivalence of gravitational mass and inertial mass.

From Mach's principle, the inertia of every body is determined by the distribution of matter in the Universe; as a corollary, this distribution of masses determines the geometry of the Universe. Mach's principle is therefore one of the principles on which the

general theory of relativity is founded; this theory gives matter the property of acting on all kinds of motion, including the motion of light.

These rather simple principles therefore guide the construction of cosmological models which are moreover supported by observations.

Cosmology and observations

The only approach which would enable cosmology to progress consists of establishing simple physico-mathematical models to represent the geometry and evolution of the Universe, and confronting predictions made from the models with observations. There is obviously an infinite number of possible models for the Universe, even if one keeps strictly within the context of traditional physics, that is, general relativity. Therefore one only keeps the simplest models, that is, those specified by the minimum number of parameters, which is a way of applying a principle of simplicity. None could claim that the Universe is so constructed that it could really be described by one of the simplest possible models, but it would be satisfying for a first step to succeed in constructing a simple model which agreed with the ensemble of observations and whose free parameters, reduced in number, would be determined by those observations. The simplest models at our disposal are the zero pressure models of Friedmann, based on general relativity, on the principles mentioned earlier, and on the fact that the pressure seems to be negligible in the actual Universe. In these models, the geometry and the evolution of the Universe on the very large scale are wholly determined by the values at a given time of just two parameters: these parameters are themselves defined in terms of a characteristic dimension of the Universe R. This dimension could, for example, be the distance between two distant galaxies. The velocity with which this distance is expanding at a given time is then the derivative of R with respect to time, that is $\dot{R} = dR/dt$, and the acceleration is the second derivative, that is $\ddot{R} = d^2R/dt^2$. The first parameter expresses the velocity with which the Universe is expanding, and is called the Hubble constant, defined by $H = \dot{R}/R$. The second parameter is the deceleration parameter $q = -\ddot{R}/RH^2$, which measures how rapidly the expansion is slowing down. The observational determination of the values of these parameters at the present epoch would suffice to completely specify the model and thus the past and future of the Universe it is supposed to describe. This is the goal of observational cosmology. The values of the Hubble constant and of the deceleration parameter at the present epoch are denoted by H_0 and q_0 respectively. Sometimes a third parameter called the cosmological constant Λ which is invariable with time is also added; it suffices here to say that this parameter was introduced for purely mathematical reasons, and that its physical significance is unclear. However, it is difficult to construct models without it, and, though there have been in the recent past several attempts to confront models with a non-zero cosmological constant with observations which seem to be difficult to explain otherwise, the tendency at present is not to bring in this third parameter.

We must now review the main observational facts which enable us, at least in principle, to set limits to the parameters of the models of the Universe, if not to determine their values. With the exception of the first – the Olbers–Chéseaux paradox – all of these observations are barely half a century old.

The Olbers–Chéseaux paradox

To say that the night sky is dark appears to be stating the obvious. Nevertheless, this simple observational fact has profound cosmological implications. These implications, undoubtedly suspected for a very long time, were first stated explicitly by the Swiss Jean Philippe Loys de Chéseaux and then by the German Heinrich Olbers at the end of the eighteenth century. Suppose that the Universe is uniform, static and infinite, and that its geometry is Euclidean; every line of sight would then end on a star and the sky would thus have a surface brightness equal to the mean surface brightness of the stars: this is clearly not the case. We can attempt to elucidate this paradox by postulating the presence of absorbing matter which limits how far we can see. However, this matter would in time itself be heated up; when it reached equilibrium, it too would have the same surface brightness as the stars. Clearly at least one of our original assumptions must be incorrect. The paradox is in fact resolved by the finite size of the Universe and the fact that it is continually

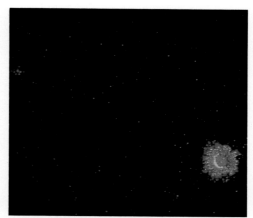

The quasar 3C 273 observed in X-rays. A careful study of observations made in the X-ray region by the satellite HEAO–2 has revealed that over a hundred quasars are very powerful X-ray emitters and contribute to the diffuse X-ray radiation observed in all directions in the sky. This X-ray picture shows 3C 273 at lower right, and another source at top left which is another quasar of a much higher redshift. Its distance is 4 megaparsecs in contrast to 0.9 megaparsecs for 3C 273. Quasars, which have the point-like appearance of a star, are very important in cosmology; radio-emitting quasars have the same radio properties as radio galaxies. Quasars are galaxies in a primitive phase of their evolution, which will therefore have been very energetic objects when young. The extremely large distances of quasars enable us to measure the expansion and therefore the age of the Universe as about 15 billion years. (Centre for Astrophysics)

The elliptical galaxy M87. This elliptical galaxy, which is a member of the Virgo Cluster, has been identified with a very powerful radio source called Virgo A. This source emits a very great amount of energy at radio wavelengths of the order of 10^{53} joules. Radio galaxies sometimes show some peculiarities at visible wavelengths. In this photograph can be seen a jet of gas emitted by the centre of the galaxy which derives from the very intense activity in the galactic centre. Indirect evidence for the presence of a black hole at the centre is provided by the high velocity of the surrounding gas. The intense radio emission would be produced by matter being heated as it fell into the black hole. This mechanism would also give rise to the very great quantities of energy emitted in the radio domain by quasars. (Kitt Peak National Observatory)

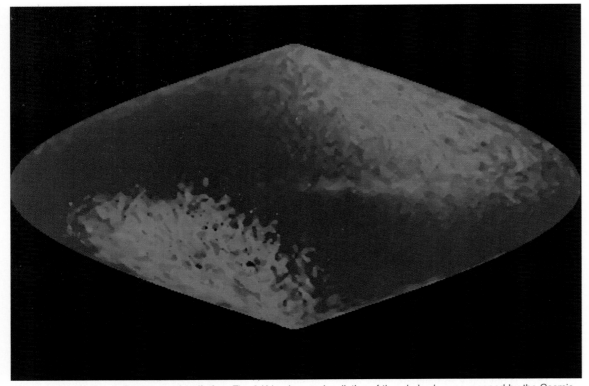

The cosmic microwave background radiation. The 3 K background radiation of the whole sky was mapped by the Cosmic Background Explorer (COBE). The temperature is that of a perfect black body at 2.730 K. Small fluctuations in the temperature, of just a few millionths of a degree, are thought to be due to the formation of structure in the universe. (NASA-GSFC).

only small uncertainties compared with the uncertainties in the determination of galaxy distances.

The determination of distances, which is very difficult, is made in stages. Firstly, it is necessary to calibrate the relationship between period and luminosity for variable stars (RR Lyrae stars and Cepheids) using stars in our own Galaxy whose distances are known. This calibration already introduces an uncertainty of about 20 per cent. Secondly, the same types of variable stars are identified in other galaxies; obtaining the luminosities of these stars using the period–luminosity relation enables the distance of the stars and thus the galaxies to be calculated. This method can be used out to distances of about 5 megaparsecs. However, it introduces additional random observational errors, besides a possible systematic error arising from the fact that the variable stars involved do not necessarily have the same properties in all galaxies because of differences in chemical composition. Thirdly, in the galaxies whose distances have thus been determined, it is necessary to identify other more luminous objects which can be used as secondary distance criteria in order to go out to greater distances. One can for example study the luminosity of the brightest stars or of globular clusters, or study the diameter of the brightest H II regions, to establish a relationship between the morphology and size of the galaxies themselves. These criteria may be used with galaxies out to distances of 20 megaparsecs. Finally, using the thus enlarged sample of galaxies, one can attempt to identify tertiary distance criteria based on global properties such as morphology, diameter, luminosity and radio properties. The use of these criteria enables us to go out to 100 megaparsecs; at present the determination of distances further out is impossible.

This step-by-step method of determining distances is precarious because the random errors and the systematic errors are cumulative. Also, the universality of the distance criteria is an unverifiable assumption. One should not therefore be surprised that the most recently determined values of H_0 are still not better known than the range 50 to 100 kilometres per second per megaparsec. One looks for different methods of measuring H_0. We shall give a typical example of a method later.

expanding (because of this expansion, the light from stars further and further out is shifted more and more to the red until it no longer contributes to the sky brightness). The paradox is therefore no longer a difficulty for modern cosmology. Note that the upper limits of intensity of the integrated visible light of the furthest galaxies (called the sky background) derived from present-day observations are still too high to derive interesting constraints on models of the Universe. The same is true in most wavelength domains.

The expansion of the Universe and the Hubble constant

It was only at the beginning of the 1920s that the American Edwin Hubble, who had the new 100-inch telescope at Mount Wilson at his disposal, definitively established that the Andromeda Nebula, M31, was located beyond our Galaxy, and defined extragalactic distance criteria which enabled him to estimate the distances of increasingly distant galaxies. In 1916 and later, Vesto M. Slipher and Francis Pease succeeded in obtaining spectra for numerous galaxies and measured the often large shift in wavelength of spectral lines present in the spectra with respect to the same lines observed in laboratory spectra (if λ_0 is the laboratory wavelength, and $\Delta\lambda$ the shift in wavelength, the spectral shift is $\Delta\lambda/\lambda_0$). In 1919, Harlow Shapley noticed that the great majority of the shifts were towards the red. He attributed these shifts to the Doppler effect and deduced that most of the galaxies were receding from us. However, it was only in 1929 that Hubble made the fundamental discovery that the spectral shift was proportional to the distance of the galaxy, which implied that the Universe was expanding if the effect was indeed a Doppler effect. This discovery is now known as Hubble's Law. The velocity of recession of a galaxy is $v = cz$ if z is small, c being the velocity of light. (The exact formula is $1 + z = \sqrt{(c + v)/(c - v)}$; as v approaches c the spectral shift tends to infinity.) The present Hubble constant H_0, at the time t_0 is defined by $H_0 = \dot{R}(t_0)/R(t_0) = v/d = cz/d$, d being the distance of the galaxy. In his first investigation, Hubble had estimated H_0 to be 530 kilometres per second per megaparsec; his distances were in fact underestimated, and present-day estimates for H_0 are five to ten times smaller.

Hubble's Law and the determination of H_0 pose three problems. First of all, is the spectral shift a cosmological effect arising from the expansion of the Universe or is it due to some other cause? This was immediately an extremely controversial question, and the debate raged again with greater intensity than before with the discovery of quasars (the first in 1963) some of which had redshifts as large as 3.5. However, a convincing proof to the contrary is lacking and it is very difficult to find a plausible physical mechanism to explain the spectral shifts. One is forced to accept the reality of the expansion of the Universe.

Next, how far out is Hubble's Law verified? Distances can only be measured out to about 100 megaparsecs; on this scale Hubble's Law is well verified, but there are smaller regions where there are deviations from the average undoubtedly owing to local motions of matter.

Finally, how accurately can H_0 be measured? The determination of redshifts and local deviations from Hubble's Law result in

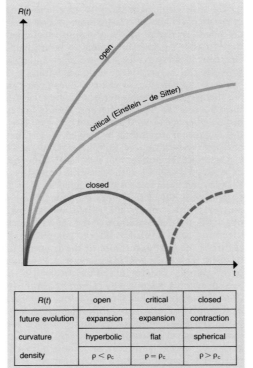

$R(t)$	open	critical	closed
future evolution	expansion	expansion	contraction
curvature	hyperbolic	flat	spherical
density	$\rho < \rho_c$	$\rho = \rho_c$	$\rho > \rho_c$

Evolution of the geometry of space in a Universe undergoing a primordial explosion. This evolution depends on the value of the density of matter in the Universe; if the density is high, the matter exerts a self-gravitational force on the Universe which is sufficient for it to recontract after its initial expansion. The Universe is said in this case to be closed, that is, it may undergo a large number of expansions and contractions and therefore pass through many critical phases similar to the Big Bang. The Universe adopts this behaviour if its present density is greater than a critical value of the order of 10^{-29} grams per cubic centimetre. If the density is equal to or less than this critical value, the self-gratification is not great enough to prevent the universe continuing to expand. If the present density is equal to the critical density, the model is known as an Einstein–de Sitter model. The Universe is said to be open in the last two cases.

Geometric cosmological tests

Once the expansion of the Universe is accepted, and H_0 determined as accurately as possible, one can go some steps further. In the Friedmann models of zero pressure and zero cosmological constant, the deceleration parameter q_0, which is always positive or zero, is related to the geometry of the Universe and to its local density ϱ_0 in a very simple way. The relation between q_0 and ϱ_0 is $\varrho_0 = 2q_0\varrho_{\text{crit}}$, where ϱ_{crit} is what is known as the critical density. This critical density is equal to $3H_0^2/8\pi G$ where G is the gravitational constant. The value of ϱ_{crit} is of the order of 10^{-29} grams per cubic centimetre. Putting aside the stationary Universe model of non-zero pressure, there are three kinds of model:
– if q_0 is greater than 0.5, the geometry of the Universe is elliptic in the Riemann sense (it is impossible to explain this fully here but the sum of the angles of a triangle would exceed the Euclidean value of 180 degrees). The local density ϱ_0 exceeds the critical density ϱ_{crit}, and the expansion of the Universe will come to a halt after a certain time and be followed by a phase of contraction. The Universe is said to be closed. The present age of the Universe is less than $2/3H_0$.
– if q_0 is equal to 0.5, the geometry of the Universe is Euclidean, and $\varrho_0 = \varrho_{\text{crit}}$; the expansion slows down continually but never halts. The present age is equal to $2/3H_0$;
– if q_0 is positive but less than 0.5 the geometry of the Universe is hyperbolic and ϱ_0 is less than ϱ_{crit}. (The sum of the angles of a triangle would be less than the Euclidean value of 180 degrees.) The expansion slows without halting. The age of the Universe is between $1/H_0$ and $2/3H_0$. The Universe is said to be open.

Two different approaches to determining q_0, and thus the appropriate model, are possible. In the first approach, effects of the geometry of the Universe on the observed properties of the furthest objects are studied. Such tests are known as geometric cosmological tests. In the second, physical quantities such as the local density ϱ_0 or the age of the Universe are measured in order to find q_0. These are physical cosmological tests.

Counts of galaxies and of extragalactic radio sources

In a Euclidean Universe uniformly populated with identical objects it can be shown that the number N of objects emitting an apparent flux (light, radio, etc.) greater than a given flux S is proportional to $S^{-1.5}$. The regions of the Universe nearest to us out to a few hundred megaparsecs approximate a Euclidean Universe. At large' distances, the number of sources with flux greater than S, $N(>S)$, increases less quickly with increasing flux S; how quickly depends on the geometry and therefore upon q_0. In principle, then, q_0 can be determined by counting sources. The first attempt to do this was made by Hubble, who counted galaxies in a particular region as a function of their apparent magnitude m up to magnitude 19. (The relation between m and S is $m = -2.5 \log S + \text{constant}$.) Hubble showed that, up to the limit of his counts, that $N(>S)$ was indeed proportional to $S^{-1.5}$. This is not surprising as his galaxies are in fact relatively nearby objects; his result does no more than support the hypothesis that the Universe is homogeneous. After the advent of radio astronomy, the same test was used on extragalactic radio sources by radio astronomers at Cambridge under the direction of Martin Ryle. The surprising result was that the number of objects with flux greater than S varies as $S^{-1.8}$. Later counts have confirmed this, at least at relatively low frequency. No model of the Universe predicts such behaviour, which can only be explained by evolutionary effects, radio sources at large distances being more luminous and more numerous than they are today. We recall that the more distant the objects which we are studying, the less time has elapsed since the beginning of the Universe at the moment when they emitted the radiation that we observe today.

Studies since then have shown that the radio sources are galaxies and quasars, and that quasars show the greatest effects of evolution with cosmic time; the numbers of quasars at redshifts of 2.5 are thousands of times greater than at the present day. This is not surprising: being very powerful radio emitters, quasars can be seen at greater distances than galaxies and evolutionary effects are best seen for quasars. Evolution is also present in the optical for quasars and even for galaxies; such effects have recently been detected in much deeper galaxy counts than those made by Hubble. These evolutionary effects are of fundamental importance in cosmology in that they show that the Universe is not immutable, definitively ruling out the model of a stationary Universe with identical properties at all points and all times which was introduced in the 1950s. The expansion of the Universe in this model is compensated for by the continuous creation of matter at every point in the Universe. Thus the local density remains constant. The pressure in this stationary Universe is negative; the physical interpretation of this is unclear. Besides ruling out this model, the evolutionary effects all but destroy any hope of using galaxy counts and radio source counts to determine q_0.

The magnitude–redshift relation

The magnitude–redshift relation, applied to the brightest galaxies in clusters, whose luminosities are supposed identical, has been used to establish the Hubble Law at quite large distances. At even larger distances – several hundreds of megaparsecs – one can expect that the geometry of the Universe, by affecting the propagation of light, will modify this relationship in a manner which will depend on q_0. One can therefore envisage using the magnitude–redshift relation of the very distant galaxies to determine q_0. Various attempts have been made to do this. Unfortunately, as we have seen, galaxies, and their luminosities, evolve with cosmic time. One obviously hope to calculate the magnitude of this effect from our theoretical notions of the evolution of galaxies and hence correct the observations, but these ideas are still too uncertain for a reliable result to be derived.

The angular diameter–redshift relation

The angular diameter–redshift relation is a geometric test similar to the preceding one: the geometry of the Universe affects the relation between linear dimensions, angular diameters, distances and redshift for the distant stars in, an important manner which would make the test very sensitive if it could be used. Unfortunately, attempts to use the apparent dimensions of galaxies and of clusters of galaxies are inconclusive, and tests involving radio sources have only succeeded in demonstrating the existence of evolutionary effects, the linear dimensions of the farthest objects being smaller than the nearest objects.

The physical cosmological tests

The use of geometric cosmological tests, as we have seen, has up to now only given ambiguous results because of the effects of evolution with cosmic time which have been shown to exist. They will therefore only permit q_0 to be determined if one succeeds in constructing a definite theory of this evolution, which has not yet been done. Therefore physical tests have gained or regained much interest for researchers.

The age of the oldest objects in the Universe

The age of the oldest objects in the Universe is obviously smaller than the age of the Universe; the upper limit to the age of the Universe thus obtained is interesting. As we saw above, the age of the Universe is less than the inverse of the Hubble constant, $1/H_0$ (from 10 to 20 billion years for values of H_0 between 50 and 100 kilometres per second per megaparsec), and is smaller the larger the value of q_0. The oldest objects in our Galaxy are the globular clusters. Recent determinations of their age, unfortunately still uncertain, are between 14 and 18 billion years. Although these values are approximate, they do at least rule out values of H_0 greater than 100 kilometres per second per megaparsec. These approximate values compare well with ages of 15 to 20 billion years already determined from the analysis of lunar and terrestrial meteoritic crystals containing uranium 238, thorium 232 and rhenium 187. These techniques, known as cosmochro-

A galaxy at the limit of the Universe. This image, obtained by the 10-metre Keck Telescope, shows in the centre the galaxy 4C 41.17, which is one of the most distant galaxies known. The redshift is 3.8, which corresponds to a distance of about 12 billion light years. The image, shown here in false colour, was obtained in the infrared at a wavelength of 2200 nm. (W.M. Keck Observatory/Keith Matthews and James Graham).

Hubble diagram for clusters of galaxies. Each point in this diagram represents a cluster of galaxies. The relative velocity of each cluster is plotted as a function of its apparent magnitude corrected for interstellar absorption; the apparent magnitude is thus an indicator of distance. The second derivative (acceleration or deceleration) of the expansion of the universe is not necessarily zero. This plot, which includes not only the nearest clusters of galaxies but also some distant clusters, enables the deceleration parameter of the universe to be determined. A Universe for which the deceleration parameter is low, between 0 and 1/2, can expand continually and is open. For a deceleration parameter greater than 1/2, the Universe is closed, that is, it may pass through a succession of expansions and contractions. The diagram shows the theoretical curves for four values of q_0, including two extreme models – the model of continuous creation known as the steady state, and also a very high density model with $q_0 = 5$. Even if the greatest weight is given to the points representing the furthest clusters of galaxies, the data do not enable us to make a choice between values for q_0 of 0 and 1, but merely exclude the extreme cases. (After a diagram by A. Sandage)

nology, enable us to estimate the time since the formation of these elements within stars. Because our Galaxy has had to form more stars at the very beginning of its evolution than later and because the beginning of its history must have taken place a few million (at most a billion) years after the birth of the Universe, one finds again, by this technique, that the ages determined by nucleosynthesis are completely comparable with those deduced from the expansion of the Universe or the study of stars in globular clusters.

The density of the local Universe

The value of q_0 may be obtained by determining the density of the local Universe. In principle, it is simple; it suffices to locate and count the galaxies within a given volume, to evaluate their masses, and to divide the total mass of the galaxies by the volume. The first attempts to estimate the density gave values of around 10^{-31} grams per cubic centimetre, considerably less than the critical density (10^{-29} grams per cubic centimetre). However, various recent observations have led to a significant increase in the estimated value of the local density. Firstly, it has been realised, from dynamical considerations, that elliptical and even spiral galaxies possess a more or less spherical halo, invisible optically but containing much matter; although the nature of this halo is not yet known, its existence could only be doubted with difficulty. Furthermore, similar dynamical considerations, applied this time to clusters of galaxies, suggest that the mass of the clusters is greater than the combined masses of the member galaxies. At least part of the mass responsible has been detected as a very hot gas emitting in the X-ray region in some large clusters; the mass of this gas is of the order of the mass of the galaxies themselves. The combination of these two factors could well lead to the true density of the Universe being a factor of ten higher than that originally estimated, the new value thus being about 10^{-30} grams per cubic centimetre. This value still implies an open Universe (q_0 less than 0.5) but even then other matter may not have been taken into consideration. For example, if neutrinos have a small non-zero mass, as appears to be suggested by certain still much debated laboratory experiments, they could contribute considerably (by some unknown amount) to the mass of the Universe; values of q_0 equal to or less than 0.5 might not be excluded.

The electromagnetic background radiation of the Universe

The discovery of the universal electromagnetic background radiation in 1965 by Arno Penzias and Robert Wilson was as important for cosmology as the discovery of the expansion of the Universe. This radiation was immediately identified as the relic of the radiation which filled the Universe in its first phase, predicted by George Gamow in 1949. The radiation was in fact produced during the phase of the recombination of hydrogen at an epoch when the temperature was of the order of 4000 K; the spectrum

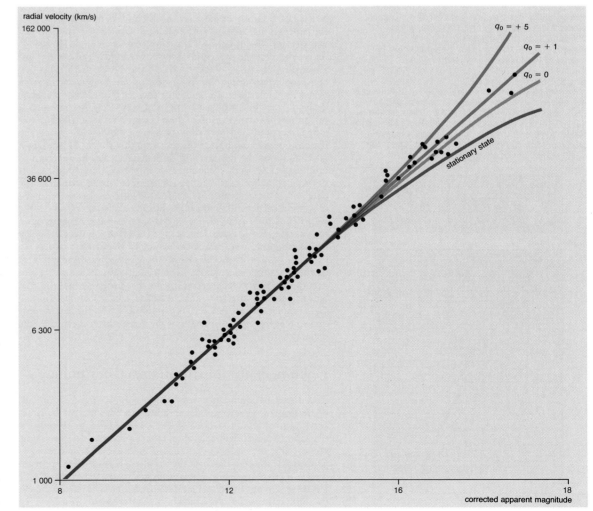

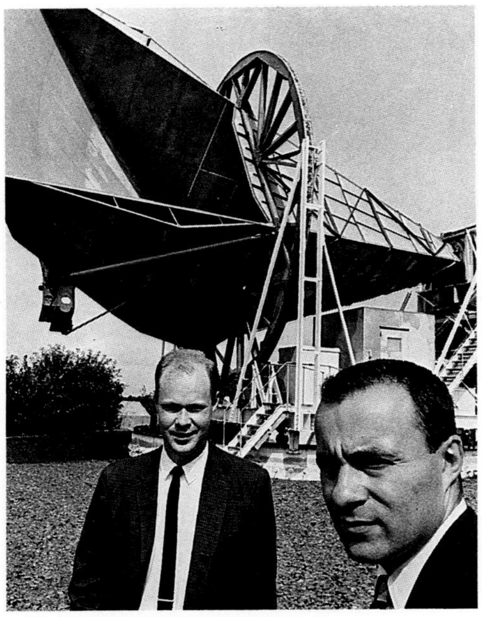

The discovery of the fossil radiation at 3 K. The large antenna of the Holmdel station of the Bell Research Laboratories of the Bell Telephone Company in New Jersey enabled Arno A. Penzias (right) and Robert W. Wilson to discover the 3 K fossil radiation in 1965; they understood the importance of their discovery thanks to conversations with astrophysicists of the University of Princeton who had predicted that there would be significant radiation at 7 centimetres if the Universe began with an explosion. This discovery struck a fatal blow to the theory of continuous creation. (Bell Laboratories)

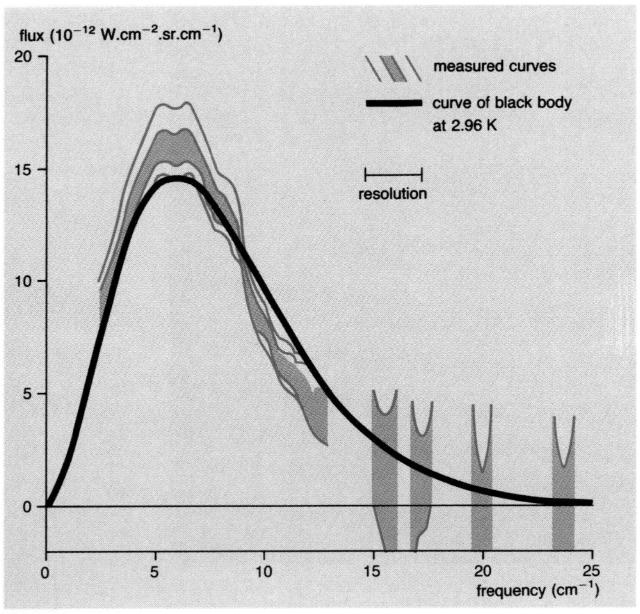

Distribution of the flux of the cosmological radiation as a function of the frequency (upper right). The flux precisely follows the Planck distribution for a black body at a temperature of 2.96 K. Measurements in the infrared were made using photometers cooled by liquid helium on board a balloon. The distribution of flux (which is isotropic) is due to the cooling through expansion of a hotter fossil radiation, then at some tens of thousands of kelvin, which was emitted at the end of the radiative era when primordial hydrogen ceased to be ionised.

Evolution of the temperature and density as a function of time for a Big Bang model of the Universe involving a sudden and explosive creation. Four eras of extremely different duration are distinguished in the evolution of the Universe. The hadronic era ceases when the age of the Universe is of the order of 10^{-4} second. During this era, the physical laws of the Universe are determined by nucleons (protons and neutrons) and their constituents (quarks, etc.). The leptonic era lasts about 10 seconds, during which electrons and positrons determine the equilibrium between the protons and neutrons. Neutrons cannot disintegrate during this era. The radiation era is dominated by the energy of the photons and lasts about a million years. At the start of the radiation era, when neutrons begin to disintegrate, primordial nucleosynthesis begins. This era ends when the energy of the photons is no longer enough to ionise the hydrogen. At this point, the Universe is no longer opaque and the stellar era, in which we are now, begins. The cosmological radiation is emitted at the moment when the radiation era ends and the stellar era begins. The stellar era is characterised by the hierarchical organisation of matter into clusters of galaxies, stars and so on. The photonic radiation produced during the radiation era is no longer the dominant source of energy.

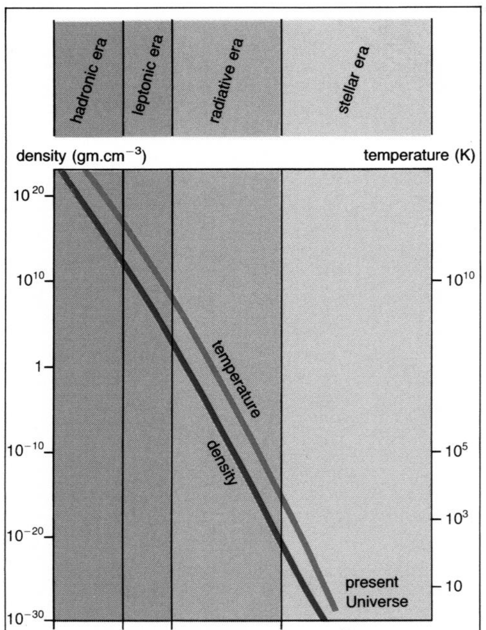

of this radiation was thus that of a black body, and the expansion of the Universe has not altered its properties, the characteristic temperature of the radiation being inversely proportional to the spectral redshift z. Compared to the initial 4000 K, the present temperature of 2.7 K suggests that this radiation was produced at the epoch corresponding to a redshift of the order of 1500. The existence of this radiation is important for two reasons. Firstly, it definitively rules out the model of a stationary Universe, already weakened by counts of radio sources. The universal radiation could not in fact be produced in such a model. Despite this, certain workers did try to show that the radiation could result from the superposition of contributions from extragalactic radio sources of unknown nature, but the idea did not stand up to a critical examination, nor to the numerous observations which the model engendered. Secondly, knowing the value of the radiation temperature enables us to derive a relationship between the temperature and density of the primordial Universe; combined with other observations this relationship can be exploited by cosmology. The most recent observations, in particular those made from aeroplanes or by balloon, have confirmed the black-body character of the emission and shown a high degree of isotropy. This uniformity in all directions and the absence of structure at various angular scales is wholly remarkable; those characteristics also set very strong constraints upon the relative size of fluctuations of density and temperature in the primordial Universe. Some cosmologists think that fluctuations are necessary for the formation of galaxies and clusters of galaxies. However, there are a few minor deviations from these characteristics. The best established is a slight anisotropy, with the radiation being slightly more intense in a particular direction in space than in the opposite direction. This anisotropy is simply interpreted as arising from motion of the Earth in the direction where the radiation is most intense. After correcting for the various known movements of the Earth with respect to the centre of the Local Group of galaxies to which we belong, a velocity of the order of 600 kilometres per second with respect to the mean velocity of the objects in the Universe is implied. It seems, moreover, that the spectrum of the cosmological radiation is not precisely that of a black body. If the latter distortions are confirmed, they imply that the radiation has undergone interaction with something, perhaps dust formed from heavy elements derived from the initial generations of stars. Another interesting phenomenon involving interaction between radiation and matter is an effect predicted by the Soviets Sunyaev and Zeldovich; when the cosmological radiation passes through a cluster of galaxies it is scattered by the hot gas within the cluster and its spectrum is modified. The slight reduction in intensity thus produced at centimetre wavelengths appears to have been convincingly detected. This reduction depends upon the temperature of the gas, upon its density, and on the Hubble constant H_0. Now the X-ray emission from the same intracluster gas also depends upon the gas temperature and density, and on another power of H_0; by combining the two observations it is therefore possible in principle to measure H_0 directly without requiring measurements of the distances of galaxies. The present observations are not accurate enough to be interesting, but it is probable that the goal will be reached in a few years. Another method of determining H_0, completely independent of the above, is by studying double quasars which are two images of the same object created by the deflection of light rays by an intervening galaxy.

Helium, deuterium, lithium and cosmology

The observed abundances of helium (atomic mass 3 and 4), deuterium (atomic mass 2) and lithium (atomic mass 7) set constraints on what kind of model is chosen for the primordial explosion; the present-day abundances predicted by the model chosen for the subsequent evolution must also match. The processes leading to the formation of these lightest elements must have taken place only a few minutes after the birth of the Universe. Deuterium is in fact destroyed, not synthesised, by the thermonuclear reactions which occur in the centres of stars. Even if the bombardment of the interstellar medium by cosmic rays is able to produce the quantities observed of the elements lithium 6, beryllium 9 and boron 11, it can synthesise at the most only a few per cent of the quantity of deuterium observed. Helium 4, and in a lesser amount, helium 3, is produced by the thermonuclear fusion of hydrogen which constitutes the main source of energy for the ensemble of stars in a galaxy. A comparison between the luminosity of a galaxy integrated over its lifetime and the present content of helium 4 shows that at the most a few per cent of the helium observed is of stellar origin. This observation is corroborated by almost direct observations of the primordial abundance of helium in certain 'lazy' galaxies which are just beginning to undergo star formation: nucleosynthesis is little active and the amount of heavy elements is very low while the helium content is large. Finally, although lithium 7 can be produced during the evolution of galaxies in small quantities (less than 10 per cent) by galactic cosmic rays or in certain red giants, novae and supernovae, the element could have been formed completely naturally after the first minutes of the existence of the Universe.

Deuterium is observed in the Solar System and in the interstellar medium. Unfortunately, determining its abundance is difficult. For example, deuterium occurs as heavy water in meteorites and terrestrial rocks; the relative proportion of heavy water to ordinary water depends on the temperature at which the rock or meteorite solidified. In the interstellar medium, use must be made of ultraviolet astronomy; certain physical processes such as the pressure exerted by light rays upon deuterium atoms can affect the results. Despite these difficulties, it is found that the ratio of deuterium to primordial hydrogen is $(2 \pm 1) \times 10^{-5}$.

Helium 3 can only be observed in the solar wind or when certain meteorite crystals rich in occluded gas are heated. Radio-astronomical observations of the interstellar medium merely give an upper limit of the ratio of helium 3 to primordial hydrogen, which, coincidentally is again $(2 \pm 1) \times 10^{-5}$.

The most recent observations of 'lazy' galaxies enable a primordial abundance of helium 4 of 0.08 ± 0.01 to be deduced, which does not appear to depend on the relative heavy element content observed in those galaxies deficient in heavy elements.

The primordial content in lithium 7 is at present the subject of a very interesting debate based on spectroscopic observations made in 1981 and 1982 by French astronomers of old stars in the halo of our Galaxy. These observations appear to indicate a primordial ratio of lithium 7 to hydrogen of 10^{-10}. However, this value is ten times less than that deduced by observing young stars. Because the observations of the very old halo stars involve only a very small number of stars, and because these stars might have destroyed part of their lithium 7, the primordial abundance of lithium 7 remains an open question. Happily, all this does not much affect the nature of the model of the Universe responsible for these observed relative contents of heavy elements.

The theory of sudden creation

In the context of the theory of sudden creation, usually known as the Big Bang, it is assumed that the Universe was extremely hot and dense when it was born, as corroborated by all the observations already mentioned, at an epoch about 15 to 20 billion years ago. In every model of the Big Bang the temperature was greater than 10^{13} K during the first instants of the Universe, since when the Universe has continued to expand and cool.

The history of the Universe falls classically into four eras in this theory. The first lasts about 10^{-4} second, and corresponds to when the temperature was greater than 10^{12} K and the density greater than 10^{14} grams per cubic centimetre. This first era is called the hadronic era because the elementary particles governed by the nuclear interaction, that is, the hadrons (pi-mesons, protons, neutrons and heavier elementary particles) determined the physics of the Universe. In the following it will be necessary to use concepts of elementary particle physics to describe the physical processes which take place in this era. The reader can, however, skip this description without great loss.

In the light of recent theories of elementary particle physics which attempt to unify the nuclear interaction (also called the strong interaction, governing the physics of hadrons), the weak interaction (which governs the physics of leptons – electrons and neutrinos), and the electromagnetic interaction (which governs the physics of radiation), this hadronic era is itself subdivided into three phases. The first, called the Planck time, lasts 10^{-43} second; during this phase, the physics of the Universe is indeterminate by the simple application of the Heisenberg uncertainty principle. The temperature of the particles during this phase is of the order of 10^{32} K. The first of the phases which has a physical meaning is a phase during which the three interactions (nuclear, weak and electromagnetic) cannot be distinguished; this phase ends when the temperature of the Universe is of the order of 10^{27} K and its age of the order of 10^{-30} second. At the end of this period, it is possible to distinguish the nuclear force from the other two which remain mixed up together. Also at the end of this phase, elementary particles appear, such as quarks, which will be the basis for the formation of hadrons, and also all the other particles which will be necessary to bring about the physical equilibrium of the Universe (like gluons – which play the same role in quark physics as pi-mesons in nuclear physics or as photons in electromagnetic emission – and Higgs' particles, to mention two examples).

The transition between this phase and the next, corresponding to the appearance of quarks, is where one generally places the baryosynthesis arising from the creation of the elements constituting the nucleons (protons and neutrons). This is accompanied by a kind of loss of symmetry in the Universe: the new theory of the unification of the fundamental interactions in fact implies that the total electric charge of the Universe cannot be conserved, and that, in consequence, a particle which was previously thought to be external, the proton, will have a finite lifetime. The theory predicts a value of 10^{30} years to within a factor of ten. As would be imagined, numerous teams of researchers have attempted to measure this lifetime. To give some idea, this lifetime would imply that one of the 2×10^{27} protons of which your body is formed would disintegrate during your lifetime. The second prediction of this theory of unification is that there exists a large excess of matter over antimatter. At the beginning of the 1970s, a group of particle physicists led by Roland Omnes, who were interested in cosmology, had constructed a symmetry model in which the Universe consisted of equal quantities of matter and antimatter. This theory, attractive because it adopted an apparently very simple and therefore reasonable hypothesis, met with two fatal flaws. The first followed from the size of the globules of matter or of pure antimatter which could persist because of the repulsive pressure exerted by photons arising from the matter–antimatter annihilation: these globules would have to be much smaller in size than galaxies and they would not be able to persist over lifetimes comparable to the age of the Universe. The second pitfall arose from the fact that, as we have seen previously, the ratio of the number of nucleons (protons or neutrons) to the number of photons constituting the Universe is of the order of 10^{-10}, certainly greater than 10^{-11}. The theories of the unification of fundamental interactions lead very naturally to such values, while the models of a symmetric Universe lead to values at least 100 to 1000 times smaller (and therefore incompatible with observations) because of the annihilation which will have had to take place during the primordial phases of the Universe. To sum up these considerations, models of a symmetric Universe in which matter and antimatter occur in equal quantities are not compatible with observation, while unification theories which predict a high finite lifetime for the proton are supported by the observed ratio between the density of matter and that of the residual radiation.

After the appearance of quarks, the electromagnetic interaction and the weak interaction separate at an epoch where the age of the Universe is of the order of 10^{-6} to 10^{-8} second and the temperature between 10^{13} and 10^{17} K.

The second of the four eras describing the history of the Universe is called the leptonic era: it ends when the Universe is about 10 seconds old and has a mean temperature of the order of 10^{10} K and a density of the order of 10^4 grams per cubic centimetre. During this epoch, the physics of the Universe was governed by the electrons and positrons which were in equilibrium with the radiation. At the end of this era is located the primordial nucleosynthesis, which begins when the lepton–radiation equilibrium ceases and because of that, the neutrons which were stable particles until then are no longer in equilibrium with the protons and become subject to beta disintegration (neutron → proton + electron + antineutrino) without the reverse reaction taking place.

The third era is therefore dominated by radiation; it is called the radiative era. Its duration is of the order of a million years, and ends when the energy contained in radiation becomes less than that contained in matter, that is, at an epoch when the temperature is of the order of 10 000 K and the density of the order of 10^{-21} grams per cubic centimetre. At the end of this era hydrogen ceases to be ionised: the protons and the electrons combine to form hydrogen atoms. Then the present era begins,

called the stellar era. This era, now about 15 billion years old, is characterised today by a radiation temperature of 2.7 K and a density of the order of 10^{-31} to 10^{-30} grams per cubic centimetre. As the gas constituting the Universe was completely ionised during the radiative era, the stars that we see today could not then have existed and it is therefore impossible to go back beyond that epoch by observing more and more distant objects at larger and larger redshifts. The transition between the radiative era and the stellar era therefore constitutes an insuperable observational barrier which can only be overcome by our knowledge of nuclear physics (for the primordial nucleosynthesis at the beginning of the radiative era) and our knowledge of particle physics (for all the events occurring during the hadronic and leptonic phases).

The period during which the galaxies and clusters of galaxies were isolated and formed was in all probability at the beginning of the stellar era, when the Universe was immersed in radiation of temperature between 100 and 10 000 K. The formation of galaxies is attributed to the birth at this epoch of fluctuations which are of two kinds. The first, called adiabatic, would have masses up to 10^{15} solar masses, corresponding to the masses of clusters of galaxies. These fluctuations would themselves be fragmented by other smaller fluctuations, giving birth to individual galaxies. These fluctuations, called isothermal, would give birth to globules with masses of the order of 10^6 solar masses, similar to masses of large globular clusters; galaxy formation would take place through coalescence of such globules, and then the formation of clusters of galaxies. The very recent discovery of large regions of the sky in which the most powerful telescopes are unable to detect any objects constitutes an argument in favour of the birth of large adiabatic fluctuations. Since such fluctuations would affect the isotropy and homogeneity of the universal background radiation, the physical parameters of the fluctuations must of course be consistent with the observed small distortions of the background radiation, which are limited to one ten-thousandth of the temperature of the radiation.

We have given a rough description of the various events and phases marking the birth of the Universe; we will now show how the processes of primordial nucleosynthesis enable us to add important constraints on models of the Universe.

The primordial nucleosynthesis and its cosmological consequences

The preceding considerations have already set a number of conditions which must be fulfilled by an expanding Universe: we know that there must have been a particularly hot primordial phase with temperatures greater than 10^{12} K, and that the Universe must be asymmetric, that is, with the quantity of matter greatly in excess of the quantity of antimatter of which only traces can exist. Models of the Big Bang which are even more precise can be constructed; they are called standard models because they straightforwardly explain the formation of the elements deuterium, helium 3 and 4, and lithium 7 in the porportions now observed. These models assume that the Universe is homogeneous and isotropic (i.e. that it obeys the cosmological principle) and that its expansion is well described by the general theory of relativity (which has received support from a large number of recent observations). The models also assume that the density of leptons (electrons and neutrinos) is strictly less than that of photons, which works out at less than 400 leptons per cubic centimetre in the entire Universe.

In these simplified models, it is easy to evaluate the velocity of expansion of the Universe: this velocity is inversely proportional to the square root of the density of the Universe and is the same in every direction.

Nucleosynthesis begins as soon as neutrons are no longer in

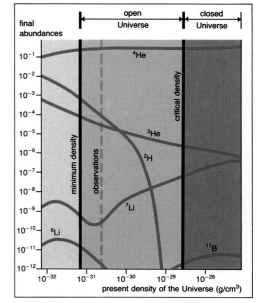

The final abundances of the light elements as a function of the present density of the Universe. Because the expansion of the Universe took place in a quasi-adiabatic fashion, there is a relation between this present density and the density of matter at the moment of nucleosynthesis, that is, about 3 minutes after the primordial explosion: the higher the present density, the higher the matter density at that moment. As can be seen from the figure, the abundances of deuterium and helium 3 decrease as the present density of the Universe increases, while helium 4 and lithium 7 behave in the opposite fashion: when the density increases, the rate of reactions involving charged particles increase, explaining the partial destruction of deuterium and of helium 3 in favour of the formation of heavier nuclei. These models of primordial nucleosynthesis reproduce the observed abundances well for a value of the present density of the Universe of the order of 2 to 3 times 10^{-31} grams per cubic centimetre, that is, a value almost a hundred times less than the critical density. This is one of the best arguments for an open Universe which expands continuously.

Cycles of nuclear reactions at the origin of the primordial nucleosynthesis. The primordial nucleosynthesis took place about 3 minutes after the original explosion at the moment when the neutrons, previously in equilibrium with the protons, began to be subject to beta disintegration transforming them into protons with the emission of electrons and antineutrinos. The first reaction of neutron capture by the proton is very rapid since neutrons are neutral in charge and do not repel the charged protons. The formation of deuterium at this epoch is therefore extremely easy, in contrast to what happens in the interior of stars. The nuclei thus formed can themselves interact with a proton or a neutron to give a nucleus of helium 3 or of tritium, or two deuterium nuclei may interact to give a nucleus of tritium and a proton or a nucleus of helium 3 and a neutron. A deuterium nucleus may collide with a tritium nucleus to give a nucleus of helium 4 and a neutron, and so on. Lithium 7 is formed, for example, by the interaction of a nucleus of helium 3 and a nucleus of helium 4, to give a nucleus of beryllium 7, which, on capturing an electron, gives lithium 7. Other reactions involving these few nuclei lead to the formation of deuterium, helium 3 and 4, and lithium 7, to the practical exclusion of all other nuclei (tritium and beryllium 7 end by giving birth to helium 3 and lithium 7).

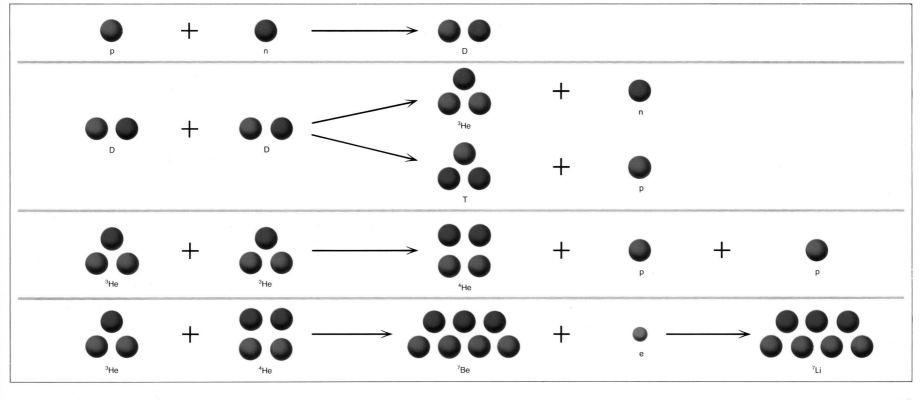

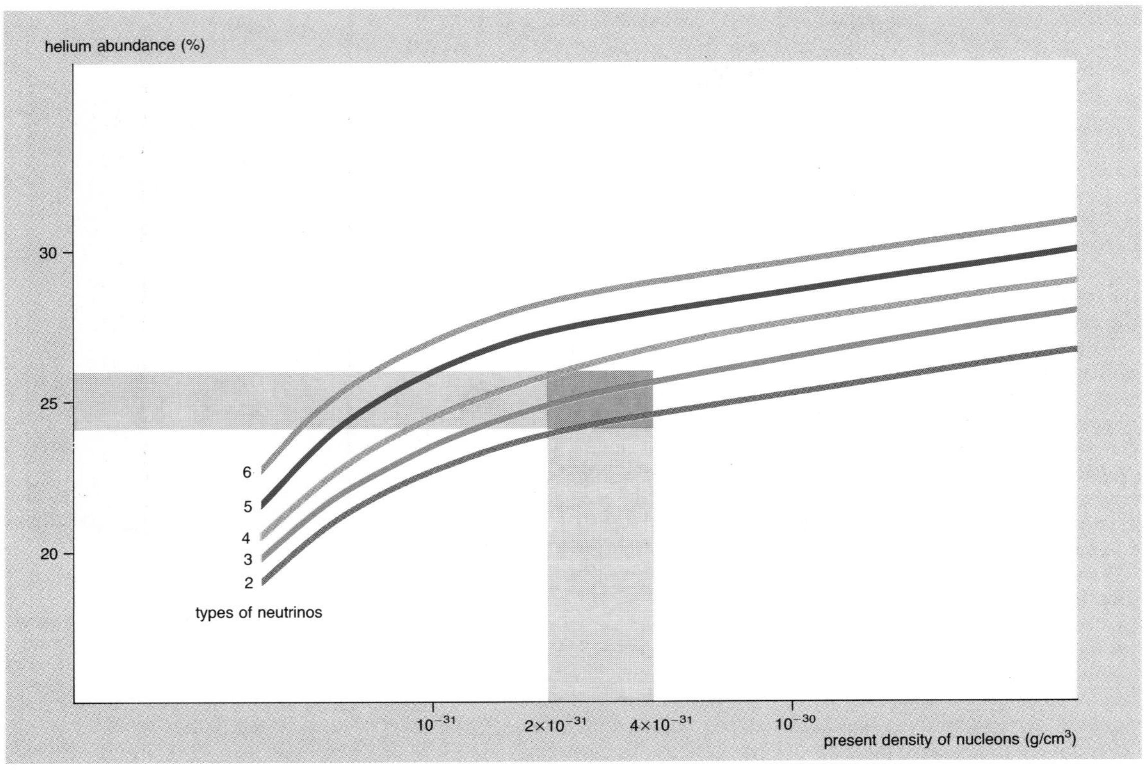

helium abundance (%)
30
25
20
6
5
4
3
2
types of neutrinos
10^{-31}
2×10^{-31}
4×10^{-31}
10^{-30}
present density of nucleons (g/cm³)

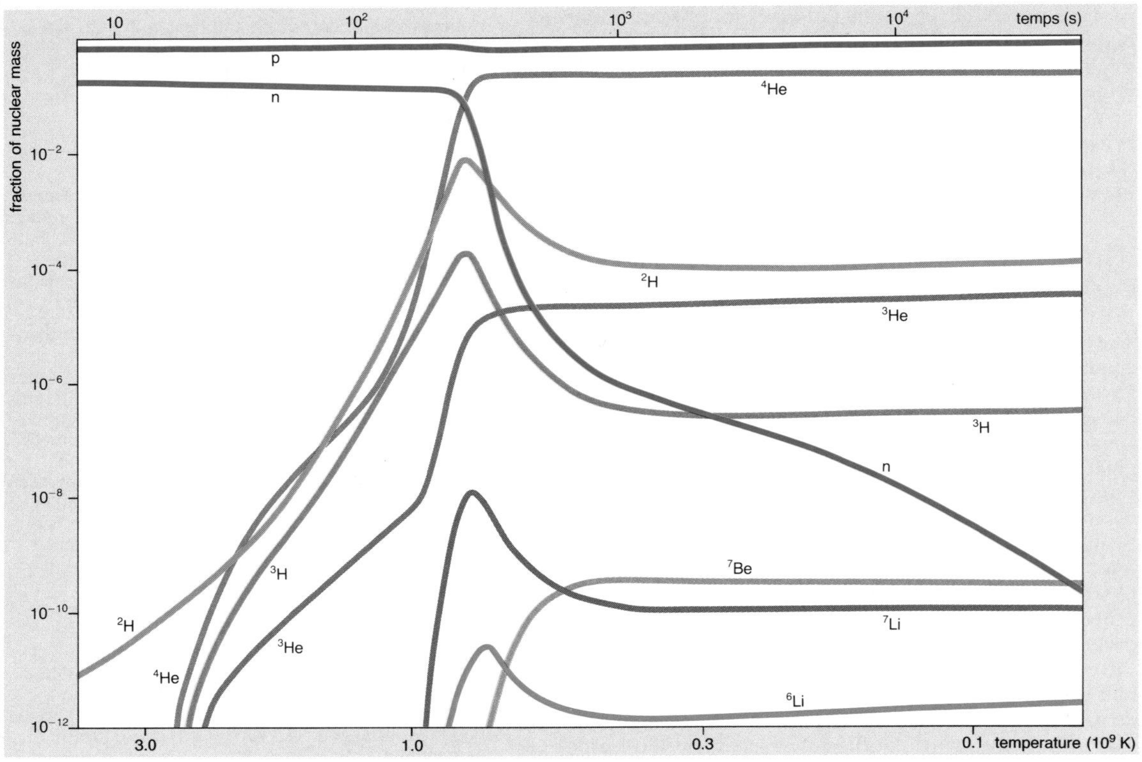

10
10^2
10^3
10^4
temps (s)
fraction of nuclear mass
10^{-2}
10^{-4}
10^{-6}
10^{-8}
10^{-10}
10^{-12}
p
n
⁴He
²H
³He
³H
n
⁷Be
⁷Li
⁶Li
²H
⁴He
³H
³He
3.0
1.0
0.3
0.1 temperature (10^9 K)

The number of families of neutrinos that can exist, deduced from the abundance of helium 4. The top figure (above) represents the helium abundance as a function of the present density of the Universe; the abundance is calculated for various numbers of neutrino families, from 2 to 6. As the number of neutrino families increases, the total density (including the rest mass and the energy present in the Universe) increases because the energy transported by every family of neutrinos is less. This energy is, however, comparable in its order of magnitude to the energy carried by photons. When this density increases, the velocity of expansion of the Universe also increases, and this advances the time when the neutrons cease to be in equilibrium with the protons, that is, the ratio of neutrons to protons is increased. As the resulting abundance of helium is a function of this ratio, it can be understood why the helium abundance is proportional to the number of families of neutrinos. From a present density of the Universe between 2 and 4 times 10^{-31} grams per cubic centimetre, one notes that the primordial abundance of helium (determined from the less evolved galaxies) sets an upper limit of three to the number of neutrino families. As three different types of neutrino are already known (muonic, electronic and tauic) one must not in principle discover any more neutrino families. (From *La Recherche*, no. 137, October 1982)

Evolution of the abundance (in mass) of various chemical species. In the lower diagram is shown the order in which elements were synthesised during the first few minutes of the Universe, when the temperature was between 100 million and 3 billion K. Primordial nucleosynthesis truly began when the neutrons began to disintegrate. Significant amounts of deuterium (^{2}H), then helium 3 (^{3}He) and tritium (^{3}H), followed by helium 4 (^{4}He), lithium 7 (^{7}Li) and beryllium 7 (^{7}Li) are synthesised when the temperature is of the order of 500 to 800 million K.

equilibrium with protons and begin to disintegrate, that is, when the temperature is about 10^9 K. Neutrons and protons then fuse to give birth to deuterium, which in its turn can capture a proton to produce a nucleus of helium 3. The fusion of two helium 3 nuclei gives a nucleus of helium 4; a nucleus of helium 3 fused to a nucleus of helium 4 gives rise to beryllium 7 which, by capturing an electron, leads to lithium 7. These processes of nucleosynthesis last a few minutes, and the present density of the Universe is very dependent on the final result. Thus knowledge of this present density enables us to estimate the density of the Universe at the moment of the primordial nucleosynthesis: if this density was relatively low, the reactions which follow the absorption of neutrons and which lead to the destruction of deuterium and (in lesser measure) of helium 3 are much less efficient since their rate is lower than in the case of high density. One notes that the primordial nucleosynthesis of deuterium and of helium 3 and lithium 7 set an upper limit to the present density of the Universe. In the standard model of the Big Bang, with the observed abundances of these elements, this upper limit is of the order of 5×10^{-31} grams per cubic centimetre, namely about ten times more than the density deduced from the distribution of visible matter in the Universe. However, this density is about five times less than that derived from the distribution of the *total* mass of the Universe (deduced from the dynamics of clusters of galaxies, for example). It is about forty times less than the critical density (about 2×10^{-29} grams per cubic centimetre) which corresponds to the frontier between models in which the Universe continually expands (called open models of low density) and models in which the Universe may undergo a succession of phases of expansion and contraction (closed models). The primordial nucleosynthesis of the light elements deuterium, helium 3 and lithium 7 therefore has two important consequences: firstly, it implies that the Universe is in continual expansion; secondly, in order to reconcile the low density of ordinary nuclear matter with the higher densities necessary to explain the dynamics of large structures of matter such as the clusters of galaxies, it leads to the idea that the difference between these high 'dynamic' densities and the low 'nuclear' densities is made up by neutrinos, which would exist in numbers comparable with the numbers of photons and would have a very small but non-zero mass.

The nucleosynthesis of helium does not supply any significant constraint on the present density of the Universe. On the other hand, it does depend very strongly on the rate of expansion of the Universe, that is, the velocity at which it expands. In fact, the helium 4 content is greater for a higher ratio of neutron density to proton density at the moment the neutrons cease to be in equilibrium with the protons and begin to be subject to beta disintegration. As for this relative content of neutrons with respect to protons, the higher the temperature of the Universe when the neutrons begin to disintegrate, the higher the ratio of neutrons to protons. If this proton–neutron decoupling took place in conditions of high temperature the Universe was in rapid expansion.

The velocity of expansion of the Universe is obviously very dependent on the laws describing the gravitational interaction between its different components; in the case of the standard model, the relevant theory is the theory of general relativity, which predicts that the rate of expansion of the Universe is inversely proportional to the square root of the total density of matter. When the nuclear density (more precisely, the density of matter in the form of nucleons) is fixed, various parameters can influence this rate of expansion. Firstly there is the degree of isotropy of the Universe; if the matter in the Universe is more concentrated in certain directions than in others (contrary to what is assumed in the standard Big Bang model), the velocity of expansion is greater in those directions, and could therefore enhance the formation of helium 4. Secondly, the number of leptons, and therefore of neutrinos relative to the number of photons could affect the velocity of expansion. The number of leptons is an important parameter which directly connects the world of cosmology with elementary particle physics; it merits some discussion. Classically, there are only two families of leptons, the family of mu-mesons and the family of electrons and positrons. Each of these two families has a distinct family of neutrinos associated with it, namely the electronic neutrinos and muonic neutrinos, respectively. Classically, during the second era in the history of the Universe – the leptonic era – one would expect the energy in the Universe to be distributed more or less equally between the photons and the two families of neutrinos. In fact, photons and neutrinos do not obey the same laws of statistical physics because one can in principle create an infinite number of photons while the number of neutrinos (like that of the nucleons or the electrons) is limited by the Pauli exclusion principle. This leads to the result that the photons carry almost twice as much energy as each of the families of neutrinos. If other families of leptons were to be discovered, this would imply the existence of further families of neutrinos, which would result in a higher total density of the Universe and therefore would have contributed to accelerating the expansion. The particle physicists indeed very recently experimentally discovered a new family of leptons called τ leptons.

Relationships therefore exist between the abundance of primordial helium, the relative content of photons and of nuclear matter, and properties of particle physics such as the lifetime of the neutron in relation to its beta disintegration, or the number of different families of neutrinos. The most recent studies appear to show that a primordial abundance of helium 4 of 0.24 ± 0.01 derived from observations of 'lazy' galaxies implies that the maximum number of different families of neutrinos (and therefore of families of leptons) is three. These studies therefore predict in principle that no more families of leptons can be discovered; or the classical models of the Big Bang will be in serious difficulty. Hence despite the small number of observable parameters involved in cosmology, the standard Big Bang models are capable of being confirmed or ruled out by developments in particle physics, which gives the models, besides their simplicity and usefulness in making predictions, as great a scientific value as the majority of physical theories.

Primordial nucleosynthesis is therefore of great interest, not only as the mechanism for explaining the presence of the lightest elements in a simple and straightforward way, but also it reveals some of the most important aspects of the history of the Universe and also of particle physics: the presence of deuterium, helium 3 and lithium 7 sets an upper limit to the density of nuclear matter, which implies that the Universe must expand continually and must be rather old, between 15 and 20 billion years old. The rather high primordial abundance of helium 4 is directly linked to the expansion of the Universe and in principle sets a limit to the number of lepton types that can exist in nature.

The models of the Big Bang are therefore at present able to explain the expansion of the Universe and thus its large-scale motion, the origin of the 2.7 K radiation, the presence of deuterium, helium and at least part of the lithium in the observable matter, and the excess of matter over antimatter. The models imply a coherent age for the Universe whatever method of estimating it is used (evolution of stars in globular clusters, the determination of H_0, long-lived radioactive nuclei) and predict that the number of lepton families is limited to those already known; they are in accord with the recent theories which attempt to unify the fundamental interactions of physics. An important question must now be considered; this is whether neutrinos have a finite mass.

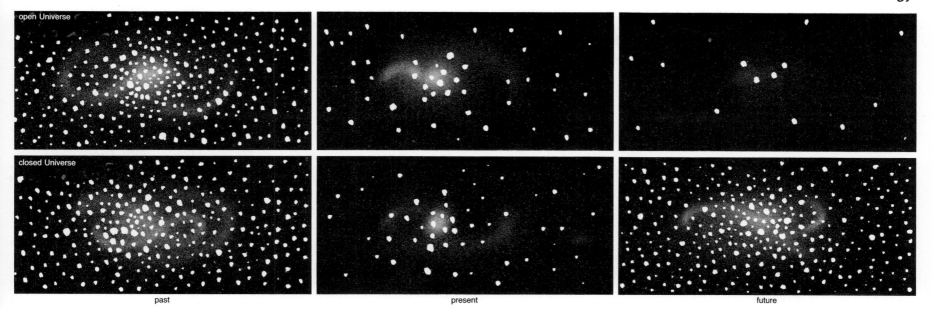

open Universe

closed Universe

past   present   future

Do neutrinos have a mass?

The existence of neutrinos was proposed for the first time by Paul Dirac on a theoretical basis to explain the conservation of the angular momentum, or spin, of the ensemble of particles produced by the beta disintegration of the neutron. These neutrinos, whose existence was subsequently confirmed experimentally by numerous physicists including the American Frederick Reines, were supposed to travel at the velocity of light and therefore to have zero rest mass (like photons).

For some years now, physicists have been less sure that neutrinos have zero rest mass. Numerous cosmological and astrophysical problems would in fact have simple and elegant solutions if neutrinos were to have a very low mass equivalent to an energy of about 30 electron volts (eV) – 17 000 times less than the mass of the electron, i.e. 5×10^{-32} gram.

The first such problem has already been mentioned. It concerns the difference between the mass density deduced from the nucleosynthesis of the light elements and the much larger mass density deduced from the dynamics of clusters of galaxies of between 20 and 50 per cent of the critical density. The Universe would then remain in a state of continuous expansion. If the density of neutrinos (comparable to that of photons because of the equi-partition of energy between the various particles) is substantially higher than that of the ordinary nuclear matter, with a neutrino mass of (at least) 4 eV the mass of neutrinic matter will exceed that of ordinary matter. A mass of 30 eV for the neutrino corresponds to the density deduced from the dynamics of the clusters. A mass of 100 eV would give the Universe a density greater than the critical density, and it would then contract again. Masses of neutrinos greater than 150 eV are not compatible with the lower limit of the age of the Universe, which is of the order of 12 billion years. If this hypothesis of low mass neutrinos, which is at present neither confirmed nor ruled out by experiment, is correct, the dynamics of the Universe is dominated by neutrinos and not by the matter of which we are formed. This would be a great blow to our deeply rooted feeling that man is (or should be) at the centre of everything.

The existence of massive neutrinos would lead to other interesting consequences. For example, because the photon flux has a temperature of 2.7 K, the neutrino flux would have a temperature of 2 K: the mean kinetic energy of neutrinos is only 1/100 000 of their mass (in the case where they have a mass of 30 eV) that is, that their velocity is low (three-thousandths of the speed of light) which confines them within clusters of galaxies or even to the interior of galaxies or within their haloes. This implies that the distribution of matter in the Universe remains discontinuous even if low mass neutrinos dominate the dynamics.

However, if the mass of the neutrino is zero, the low nuclear density and thus low total matter density renders the formation of protogalactic perturbations extremely difficult, and they would therefore appear relatively late, hardly compatible with the very small distortions observed for the 2.7 K background radiation. The presence of neutrinos of low mass facilitates the formation of these perturbations in making them compatible with the small distortions in this radiation. Finally, in another domain of astrophysics, the Sun does not emit the flux of neutrinos theoretically predicted; among the various possible explanations, the existence of massive neutrinos capable of disintegrating into particles of smaller mass would contribute to an explanation of the relatively small observed flux of solar neutrinos, and assist in resolving a particularly complex problem.

As yet, no experiment has provided a sufficiently precise determination of the mass of neutrinos. Perhaps astronomers themselves will contribute part of the answer to the problem by discovering in the ultraviolet certain characteristic emissions resulting from the transformation of neutrinos into lighter neutrinos and into radiation (photons).

An open Universe and a closed Universe. This diagram shows schematically an open Universe, which expands without ceasing, and a closed Universe in which the present expansion will be followed by a contraction. An open Universe corresponds to relatively low present densities of the Universe; it is favoured by the results of the primordial nucleosynthesis. However, if neutrinos have a mass of a few tens of electron volts, the possibility cannot be totally excluded that the Universe has a much higher density. The Universe would undergo a new phase in which it would contract if the mass contained in neutrinos dominates that contained in ordinary matter. (From *La Recherche*, no. 136, September 1982)

Velocity of recession of galaxies as a function of their distance. The velocity of recession of galaxies is in the first approximation proportional to their distance. The slope of the line is between 50 and 100 kilometres per second per megaparsec. The simplest interpretation of this diagram, constructed by Edwin P. Hubble, is that the Universe exploded around 15 billion years ago. The scatter of the points representing the furthest galaxies in this diagram is a result of the difficulty of accurately measuring the distances of the furthest galaxies. The Hubble Law appears to be well established today, confirming the theory of the Big Bang.

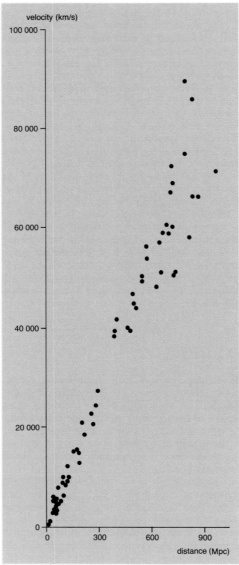

Some rival theories to the Big Bang

Before concluding by speculating on the future of the Big Bang theory, it is convenient to recall briefly alternative theories which are at present much less favoured by astrophysicists. The models of continuous creation and of a symmetric Universe with equal quantities of matter and of antimatter have already been mentioned. Many other models have been proposed, among them a model involving spectral redshifts which are not cosmological in origin, and Dirac's model involving variable gravity.

The model of spectral redshifts which are not cosmological in origin assumes that the redshifts of quasars and of galaxies do not indicate their velocities of recession relative to us, but are derived from the interaction of the photons with matter, with the interaction varying from one source to another. The model is static and no origin is necessary. This theory implies that the laws of physics are not the same everywhere and that there are interactions between photons and matter which have not yet been established either experimentally or theoretically.

The variable gravity model proposed by Dirac and his successors is inspired by the following interesting fact. The ratio between the electric and gravitational forces between a proton and a neutron is of the order of 10^{40}, as is the ratio of the hypothetical radius of the Universe to the classical radius of the electron. (This hypothetical radius of the Universe is calculated by dividing the velocity of light by the age of the Universe, and is therefore proportional to c/H_0.) Dirac therefore suggested that the force of gravity must be decreasing with time since the Hubble constant has itself increased during the evolution of the Universe. This idea is consistent with observations except at one point; another theory of gravity due to Carl Brans and Robert Dicke would replace the theory of general relativity, and the most recent observational tests in fact favour Einstein's theory rather than the variable theory of gravity.

Given the various observational cosmological tests (radio source counts, distributions of magnitudes and of redshifts, dynamics of clusters of galaxies, etc.), the Hubble Law, the discovery of the 2.7 K radiation, and the existence of light elements, the theory of a singular origin or Big Bang appears to be well established today. The theory has numerous interesting implications in the field of particle physics: predominance of matter over antimatter, plausibility of theories unifying the fundamental interactions of physics and therefore a finite lifetime for the proton (estimated at 10^{30} years), maximum number of lepton families, finite mass of neutrinos (estimated to be a few electron volts). The model predicts that the Universe must be continually expanding and that nuclear matter must turn into radiation in about 10^{30} years. However, it is clear that cosmology will again in the future undergo a revolution as important as the establishment of the Big Bang model by Gamow and Lemaître for two reasons. Firstly, progress will continue to be made in particle physics; secondly, the advent of the ensemble of modern observational techniques will enable fainter and fainter objects to be observed which are more and more distant and closer to the singular origin.

Jean AUDOUZE and James LEQUEUX

The extraterrestrial life debate

The idea of extraterrestrial life is very ancient. However, it was almost always opposed by religious faiths and was refuted, especially by the Christian Church, for a long time. In its intellectual and ideological omnipotence, the Church could not accept the existence of any place other than the territories known to Christianity and the abstract vision of some place beyond the grave. It took the discovery of new continents by Christopher Columbus and Ferdinand Magellan to show that inhabited regions were not so restricted as some people wished to believe – and would have others believe. Nevertheless, only the extent of the Earth was in question. More than anyone else it was Copernicus who, through his profound conceptual revolution that revealed the Earth as similar to other celestial bodies, finally persuaded many philosophers of the possible existence of life elsewhere in the Universe.

Towards the end of the nineteenth century, Camille Flammarion made no secret of his conviction of the plurality of worlds and he used his observations to try to characterise the populations that he thought inhabited the different planets of the Solar System. He explained it thus in an article published in 1880 in *Astronomie populaire* where he wrote, not without a certain amount of caution, 'All we can believe is that since a relationship exists between mental faculties and physical conditions, the harsher conditions are on a planet the poorer must sensibility be there, so that clearly the inhabitants of Mercury and Venus may well be less intellectual than we are. On the other hand intelligent life matures with time and since Mars was formed before the Earth and cooled more quickly it must be more advanced in every way. There can be no doubt that life there was at its zenith when we were still in our infancy!'

There is no question that such faith in the natural and universal emergence of life was influenced by the dogma of spontaneous generation, which Louis Pasteur was opposing so vehemently at that time. As a result of numerous observations showing that life was not so widespread as had been believed, at the beginning of the twentieth century scientists began to consider the problem in terms of two questions that were much less ambitious but better considered than hitherto. No longer was it a question of whether life exists elsewhere but one of how life came to appear on Earth and following from that – where else could it develop?

For the past fifty years a new discipline, exobiology, the study of extraterrestrial life, has been endeavouring to answer these two questions. These efforts have involved the cooperation of researchers in many very different fields: astrophysicists, physicists, chemists, biologists and even sociologists. The activities and the basic theories of this new science are the subject of this chapter.

The history of matter

The currently accepted laws of physics enable us to unveil the history of the matter of the universe. This matter has been subjected to a continuous recycling and two processes are clearly evident, one of high and one of low energy.

The high-energy process. Matter was partly formed in the few minutes that followed the enormous initial explosion known as the Big Bang. However, only the light elements such as hydrogen, deuterium and helium were formed in this way. Much later, when matter was rearranged in galaxies, by gravitational forces, great clouds formed deep inside the galaxies and sometimes collapsed inwards on themselves. In this way the first stars were born but these earliest stars probably had no planetary systems because heavy elements did not yet exist. However, thermonuclear reactions in the centres of these stars were soon creating heavier elements, especially carbon, nitrogen and oxygen. When a massive star has exhausted its nuclear fuel it explodes, launching its material into the interstellar medium where the heavy elements formed in the star amalgamate with the gas clouds. When a cloud, thus seeded with heavy elements, collapses to form new stars, the new stars will probably have planetary systems. A cycle of star creation proceeds in this way inside the galaxy; eventually the interstellar gas is progressively locked up in the inert residue of stellar systems; planets, red and white dwarfs, neutron stars and black holes. Throughout this cycle of production of heavy elements, considerable amounts of energy are involved and, in general, matter lacks the time to fashion itself into structures more complicated than atoms.

The low-energy process. By contrast, events are different in two parts of the process just described: in the 'residue', notably planets, and in the interstellar medium where more than fifty kinds of organic molecule have already been discovered. In these sites the energy conditions and the elapsed time appear to have favoured different evolution of the matter. This process is still very poorly understood in almost every case, except one, the evolution of the Earth itself.

Since its formation, about 4.5 billion years ago, the Earth must have lost its original atmosphere; it was probably swept away by the violent solar wind coming from the newly-formed Sun. Later, gases escaped from the Earth's crust as it became heated by the numerous meteorite impacts that frequently occurred in the Solar System at that epoch. These gases formed a new atmosphere, made up chiefly of hydrogen, nitrogen, methane, ammonia, water vapour and carbon dioxide. The temperature was low enough to allow the water vapour to condense and form the oceans. Since the atmosphere contained neither oxygen (whose appearance followed that of life; it was photosynthesis in plant life that manufactured all the oxygen that we breathe) nor ozone, ultraviolet radiation from the Sun could freely penetrate the atmosphere, right down to the surface of the Earth. This energy flux, combined with that obtained from other sources, such as storms and volcanoes, led to the formation of more complex organic molecules that were dissolved in the relatively fresh-water oceans of that epoch. About 1930, John Haldane in England and Alexander Oparin in the USSR suggested that life was born in this 'warm dilute soup', but our knowledge of this phase is somewhat hazy because no trace of it survives.

Fossils dating from about a billion years after the formation of the Earth, have been discovered in very ancient sedimentary rocks. They are microscopic unicellular organisms, similar to bacteria. However, a further two billion years elapsed before the first multicellular organisms that we know of developed. This time lapse suggests that it is easier to proceed from the inert to unicellular life than from unicellular to multicellular life.

The theory of evolution that was put forward by Charles Darwin is based on the reorganisation of matter among living beings. The process, which only involves small amounts of energy, could be interrupted at any time if the conditions that foster its progress become unfavourable, so evolution on Earth may well have been arrested on many occasions. Indeed, it is known that very small variations in solar radiation can cause important climatic changes; that the Solar System (including the Earth) has passed through upwards of a hundred dense interstellar clouds in its motion around the Galaxy; that supernovae may have occurred sufficiently close to the Earth to smother it with high-energy radiation; that approximately once in every hundred million years – that is more than forty times in the Earth's lifetime – a very large meteorite or comet strikes the Earth, an event that can affect for months, or even years, the equilibrium of the whole planet, by raising the temperature by some tens of degrees, by changing the albedo and by poisoning the waters and so on. Nevertheless we continue to exist.

Is our survival due to extraordinary good luck or is life itself an extremely stable system which, once established, is capable of adapting to new and sometimes very difficult conditions? This query provokes a more fundamental question: What is life?

What is life?

In attempting to answer this question the only available data that we can use to assist us are derived from a single source – life on Earth. It is, therefore, worth discussing again how life first appeared before going on to attempt to draft, in very general terms, a definition of life in the Universe.

Returning to the first billion years of the Earth's existence to study the processes that then operated, we can identify five important phases.

The first of these was rendered intelligible by a laboratory experiment carried out in 1952 by Stanley Miller and Harold Urey. They showed that organic molecules can be synthesised when an atmosphere without oxygen but containing hydrogen, nitrogen, methane, ammonia, water vapour and carbon dioxide is subjected to ultraviolet radiation and electrical discharges. It was subsequently demonstrated that such synthesis is equally

The oldest fossils (unicellular organisms), found in sedimentary rocks. (David I. Groves, University of Western Australia)

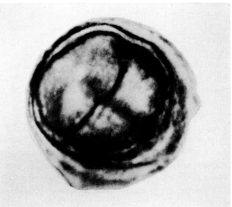

Precambrian multicellular organisms, 850 million years old. (J. W. Schopf, University of California, Los Angeles)

possible through the agency of heat or of particle bombardment or even of resonant vibrations following a shock. The first experiment proved remarkable in that it resulted in the synthesis of some of the amino acids found in living matter; these are the acids of which the proteins in living organisms are composed. By varying the composition of the original mixture – always kept free of oxygen – and by changing the experimental conditions, it was possible to show that, under conditions that must actually have existed at some epoch in the history of the Earth, the twenty known biological amino acids can be synthesised.

With regard to the second phase, many investigators, and especially F. Raulin and G. Toupance in France, have succeeded in showing that only a few atmospheric precursors such as formaldehyde or hydrogen cyanide acid are formed in the gas itself and that it is only after dissolution in water that the more complex molecules known as monomers are formed (amino acids, heterocyclic bases, sugars).

The third stage leads to the synthesis of polymers of biological interest from these monomers. Here also the possibility of such synthesis has been confirmed by experiment.

The fourth stage is an organic phase, separate from the aqueous one. Vladimir Fok and Alexander Oparin have demonstrated that micro-drops, enclosed within a membrane-like envelope, are capable of catalysing elementary reactions such as those that govern the exchanges of cells with the exterior medium.

Finally, with the increasing complexity of the medium, organic structures begin to appear; these have a new and very important property, that of auto-reproduction. In fact they are primitive living species. This final stage still remains totally incomprehensible, which explains why it has become the focus of intensive laboratory research.

Although schematic, this review of the processes leading up to the appearance of life on Earth enables us to have an inkling of a solution to the problem of going from the inert to life and, at the same time, to have a better understanding of the remarkable uniformity, on the microscopic scale, of all living matter on Earth. All organisms are formed from the same 'bricks' – twenty amino acids and five nucleotides – that are 'cemented' together by only six different chemical liaisons.

In all life there is a continuous utilisation of material – the consumption of food – from which energy is obtained, through metabolic reactions, for the maintenance of the life of the organism. Therefore the difference between the inert and the living may be defined in terms of the following two characteristics:

– in the short term, the survival of a living creature depends on a system of auto-regulation that enables it to maintain constancy in its interior in spite of variations in the constraints imposed on it by the exterior medium.

– in the long term the survival of life depends on its capability of reproducing itself; continuity in the life–death cycle is maintained by DNA, which transmits to the next generation information on its structure. Furthermore, this substance is capable of alteration in the course of transmission so that it has the ability to adapt to constraints or to evolution in the exterior medium.

With these ideas of what life is we are prompted to ask how can we discover evidence of extraterrestrial life?

Direct research on extraterrestrial life

This research commenced with investigations of meteorites, extraterrestrial objects that abound on the Earth. The first results were very encouraging; as early as 1961 they furnished evidence of the presence of unicellular organisms in the interior of some meteorites. In fact, as was later to be proved, it was terrestrial contamination of the samples that generated these organisms and the realisation of this rendered all such investigations suspect. However, more recent analyses and particularly those of the Allende meteorite, have revealed the presence of numerous amino acids that are not found in living organisms on Earth. Moreover, some of these amino acids have an optical asymmetry opposite to that of their biological counterparts. It seems that the existence of complex organic compounds of extraterrestrial origin is now well established.

The advent of the space age enabled direct research to be undertaken into life elsewhere in the Solar System. Since 1969 the exploration of the Moon has confirmed a number of previously held beliefs: there is no atmosphere there, the 'night' temperature is much lower than that of the 'day' and the bombardment by the solar wind gives conditions unsuitable for the emergence of life. No trace of organic compounds was detected in the lunar samples.

These results justified the formulation of hypotheses for other places in the Solar System. It is very probable that life has not appeared on Mercury since conditions there are similar to those on the Moon. Venus is no longer a propitious site either; there is no water there now and the soil temperature is about 500 °C – conditions that chemical and energy considerations indicate are very unfavourable for the appearance of life. However, we should be cautious in dismissing the whole of the planet because there are protected regions where temperatures are lower and conditions could be conducive to the emergence of life.

Beyond the orbit of the Earth, temperature conditions become more favourable for the appearance of life. From this point of view Mars is particularly well situated; its soil temperature falls

No trace of any kind of life was found on our satellite, the Moon, but thanks to its enterprise and resourcefulness, intelligent life arrived there. (NASA)

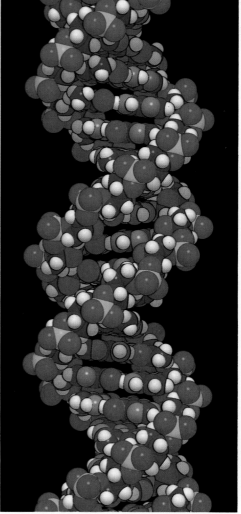

Synthetic image of the double helix of the DNA molecule. (Nelson L. Max, Computer Graphics Group, Lawrence Livermore National Laboratory)

The original experiment by Stanley Miller and Harold Urey (left) and the more recent experiment devised by Didier Mourey of the University of Paris, Val-de-Marne (right).

Micro-drops with envelope and rich in polymers, first studied by Alexander Oparin. (Sidney W. Fox, University of Miami)

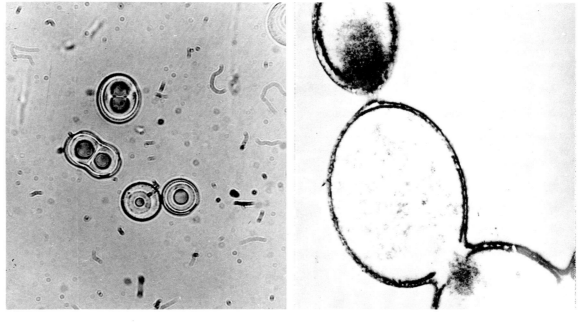

What form of life?

In discussing research on extraterrestrial life, the kind of life that we normally have in mind is similar to that found on Earth, which is based on chemical reactions with the carbon atom. As a working theory this is quite logical because, on the one hand, the only form of life that we know of derives from carbon chemistry and, on the other hand, carbon is the only element from which very complex chemistry can be developed. This fact is confirmed by observations of the interstellar medium. Out there, in very cold and almost empty space, where conditions are so different from those on Earth, the only complex molecules that have been detected are carbon compounds.

Nevertheless, some scientists have freely speculated on other possible forms of life. Thus, a form of life based on silicon chemistry has been suggested because this element is chemically similar to carbon. However, silicon is much less abundant than carbon and its capacity for chemical combination is much smaller (laboratory attempts to develop an 'organic chemistry' based on silicon failed) so that any form of life founded on this element seems highly improbable.

Another approach is to suppose that life elsewhere, even though founded on carbon, could assume forms very different to terrestrial life. And so, aware that this chemistry is prevalent in the interstellar medium, the astrophysicist Sir Fred Hoyle conceived a very original form of life in his science fiction novel *The Black Cloud* in which he depicts life out there as having evolved to take the form of an organised and thinking cloud that is nourished by light from the stars.

Other scientists, forsaking carbon chemistry and abandoning conventional notions about living matter, have suggested even more astonishing life forms. Jean Schneider of Meudon Observatory has propounded a crystalline form of life based on the complex structure that dislocation networks can adopt in the interior of crystals. Such networks can be stable, very elaborate and capable of producing other networks (by a process analogous to the synthesis of proteins) inside the same crystal. This form of life is nourished by mechanical energy – such as that furnished by shock vibrations. The objection to this idea is that the capacity for reproduction in such networks appears very limited. But if such a life form exists, it could develop inside solid bodies subject to mechanical stresses – as in the interior of planets, in white dwarf stars or in neutron stars.

Freeman Dyson and Frank Drake have proposed an even more bizarre form of life, based on the nuclear force that bonds protons to neutrons inside atomic nuclei; it is not related to the electromagnetic forces involved in chemical reactions. This kind of life could emerge on the surface of a neutron star where the nuclear force is capable of fashioning stable structures constructed from a number of nucleons, ranging from 1 (the nucleus of the hydrogen atom) to 238 (comprising the 92 protons and 146 neutrons in the nucleus of uranium 238). This force might also build much larger structures. (The largest nucleus formed in the laboratory contains 265 nucleons but its lifespan is less than 10^{-9} second – i.e. a nanosecond!) But whether or not a thing is unstable depends on one's point of view. In fact, bearing in mind their size and their agitation speeds on the surface of a neutron star, these nuclei are displaced by amounts comparable with their size in about 10^{-21} second. A man moving on the surface of the Earth covers a distance equivalent to his size in about one second. Consequently that which to us appears unstable might seem extremely stable to a 'nuclear being' (scale: 10^{-21} second). Imagine, in particular, a nucleus formed by nuclear forces and containing tens of thousands of nucleons, with a lifetime of the order 10^{-15} second. In terms of the number of interactions it experiences on the surface of a neutron star, this nucleus will last a long time and suffer millions of collisions before disappearing. It might be supposed then that, amid this wealth of interaction and through some kind of selection mechanism, life could appear in the form of very complex and heavy nuclei, capable of reproduction and of interacting methodically with their environment. If such life ever appeared its evolution should be proportionally more rapid than ours because its time-scale is so much shorter than ours (10^{-21} compared with 1 second). Since about a billion years were required for the generation of life on Earth, it should take only about 1/30 second for the transformation of inert matter into 'nuclear' life on the surface of a neutron star. Members of such ephemeral civilisations should measure about 10^{-11} centimetres and live approximately 10^{-15} second.

Communication with these civilisations appears difficult, not only because they would probably use gamma photons (characteristics of nuclear reactions) rather than radio photons but more especially because of the incredible brevity of their messages. In fact on our time-scale a civilisation like theirs should only last for a nanosecond.

within a comparatively clement temperature range (0 to $-100\,°C$) and it has a tenuous atmosphere composed of carbon dioxide and a little nitrogen. Moreover, it has been known since 1971, thanks to the Mariner 9 probe which orbited the planet, that Mars was once a site of great physical activity; there is evidence of enormous volcanoes, of extraordinary tectonic activity and of the remains of vast river beds and their tributaries. It really seems that water once flowed freely on Mars. It was partly in the hope of discovering how life could have commenced elsewhere, in conditions different to ours, and perhaps with the aim of finding out how matter is transformed from the inert to the living state, that the two Viking missions to Mars were organised by NASA.

The Viking 1 probe landed on 20 July 1976 at Chryse Planitia and Viking 2 at Utopia Planitia on 3 September 1976. They carried several instruments capable of detecting even a germ of life but the cameras only revealed a desert-type landscape utterly devoid of any sign of life. Using an arm that could be manipulated, samples of the soil were scooped into three separate analysing devices: the first observed gaseous exchanges that could possibly be produced by micro-organisms, the second investigated the soil's capacity for assimilating nourishment and the third sought to discover photosynthetic properties in it. The results established beyond doubt that the soil is very active but the observed activity is chemical rather than biological. It should also be stressed that the chemical analysis of the samples failed to discover any organic compounds, not even very simple ones. It may be concluded from this that the hypothesis of life on Mars has been seriously discredited.

Further out in the Solar System the asteroids are not, a priori, auspicious sites for the generation of life because they are so much like the Moon. It is to be noted, however, that some of the meteorites collected on Earth, in a number of which amino acids have been discovered, may well have come from the asteroid belt. Only exploration will settle this question decisively.

Going out still further in the Solar System we encounter the giant planets, Jupiter and Saturn. For a long time it has been known that their atmospheres, which are essentially composed of hydrogen and helium, also contain significant amounts of methane and ammonia. Also, it has been demonstrated that organic compounds are formed easily in such atmospheres – indeed the rich colours observed there are probably due to these compounds. Although Jupiter and Saturn are far from the Sun, there are in the atmospheres of these gaseous planets zones of favourable temperature where life may possibly have evolved. That is why Carl Sagan is prepared to imagine life forms coasting in their atmospheres, rather like jellyfish in our oceans.

The richness of the world of the giant planets and of their satellites was disclosed by the Voyager 1 and 2 missions. Some of these satellites are solid rock, others are covered by ice, yet on Io there are active volcanoes. Such variety can only be propitious for the appearance of life; it may be compared with an immense laboratory in which an experiment is carried out many times,

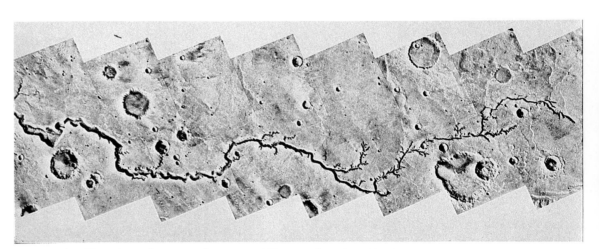

A river bed on Mars shows that water, the source of life, once flowed on this planet. (NASA)

Hollows scooped out in the Martian soil during the Viking experiments to discover evidence of life. (NASA/NSSDC)

always with slightly different initial conditions. The atmosphere of Titan, the largest of Saturn's satellites, is very similar to the primitive terrestrial atmosphere; it is composed of hydrogen, nitrogen and ammonia and the atmospheric pressure is about that on Earth. Moreover, some organic compounds – among them hydrogen cyanide – have been detected there and some observers believe that the reddish hue of this satellite is due to polymers of such atmospheric precursors. Yet, however extraordinary Titan may appear, it should be remembered that it is much further from the Sun than the Earth is and that its surface temperature is about −180 °C, so that water cannot exist in the liquid state there. However, Tobias Owen describes it as a place where there are probably oceans of liquid methane and where showers of this liquid fall to form streams on the surface. This bizarre world therefore seems a promising place for the emergence of some form of life. It should in any case be able to reveal the first stages in the transition from inert to living matter.

The greater their distance from the Sun, the colder objects are, but this does not in itself prevent the emergence of a complex chemistry. Indeed it has been suggested that in comets, in favourable conditions, the most complex organic chemistry may develop and the first 'bricks' of life may well have been deposited by comets in favourable sites, such as, for example, the Earth. This brings to mind the theory of panspermatism, a popular nineteenth century explanation of the appearance of life. This theory states that life, formed elsewhere, propagates itself from one planetary system to another through the agency of micro-organic spores. Although such a concept only shirks the very problem of the origin of life it still remains perfectly plausible: it is not possible that, in spite of all the precautions taken, the probes sent into the Solar System may already have deposited on the Moon, Mars or Venus some germs of an alien life? Moreover, the use of micro-organisms for the transformation of the atmospheres of Mars and Venus to render them habitable by man has already been proposed. The hypothesis of organised panspermatism has even been advanced by Francis Crick and Leslie Orgel who suggest that the Earth, like other sterile planets, could have been seeded by intelligent beings from another star system. Such an idea takes us outside the relatively narrow confines of the Solar System and leads on to a much more general investigation of the appearance and evolution of life.

Indirect research on extraterrestrial life

Where should we search for life? What kind of life are we looking for? As we have seen, the most favourable places for its appearance are on planets and in the interstellar medium. In the interstellar medium all the basic elements of carbon chemistry are found together: hydrogen (and helium), carbon itself, nitrogen and oxygen. There is no longer any doubt that this carbon chemistry develops in the interstellar medium which, with its few hundred atoms per cubic centimetre and a temperature of −250 °C, is very different from planetary environments. More than fifty compounds have recently been detected in the interstellar medium, one of which, hydrogen cyanide, is an atmospheric precursor whose importance has already been emphasised. These compounds are found deep within relatively dense clouds that are full of interstellar dust and are liable to collapse inwards under gravity, thus forming new stars and planets. Some researchers believe that this process does not necessarily result in the destruction of the organic compounds and that it is in the interstellar medium that the true origin of life is to be found.

However, just considering the only form of life that we know of in the Solar System, let us try to estimate the number of other sites propitious for the appearance of life in other stars of our Galaxy. We can at once eliminate some stars since it is necessary for planetary systems to have been formed about stars with sufficiently long lifetimes for the development and evolution of life; these are the cooler stars, which take more than a billion years to consume their fuel – essentially they are stars of types F, G and K (the Sun is a type G star). If all double and multiple star systems are rejected as well, because their radiation is not stable enough, there still remain about twenty billion stars in our Galaxy which are stable for long enough for life to emerge.

How many of these stars have a planetary system? For this we need to observe planetary systems around stars other than the Sun. There are eight stars where it is thought that there are dark companions, detectable by the gravitational perturbation of the central star. In particular, Barnard's star possesses companions of masses 1.1 and 0.8 times the mass of Jupiter. This suggests that the formation of planetary systems is relatively frequent.

The observation of disks around very young stars of T Tauri type over the last few years seems to support this idea: when an interstellar cloud collapses to form a star, it must form a disk in which dust grains quickly concentrate (as in the rings of Saturn), then adhere and gradually form bodies of a few kilometres in size, and in certain cases planets. A spectacular example of such a protoplanetary disk has recently been discovered around the star β Pictoris by the infrared satellite IRAS, and then detected from the ground by Bradford A. Smith and Richard J. Terrile. Seen from the side, the disk shows that dust gravitates out to more than 1000 AU from the central star, showing the huge size of the system, twenty times more extended than our Solar System. This is a very young system, but we do not yet know its exact evolutionary state: in particular we do not know if planets have already formed there. However, the presence of such a disk

so close to the Solar System confirms again the idea that stars must fairly generally be surrounded by a planetary system. We can estimate that about 20 billion stars in our Galaxy each have a planetary system, and it is reasonable to suppose that at least one planet in each of these systems is at the right distance from the star – i.e. in the ecosphere of the star – so as to be neither too hot nor too cold (like the Earth in the Solar System).

Suppose that in our Galaxy there are twenty billion planets on which life may have appeared, what means have we of detecting them? As has just been explained, it is almost impossible to observe these planets directly. Besides, when we reflect on the difficulty of detecting life in the other planets of our own Solar System, without the aid of space probes, we realise how hopeless the task is.

Still, if over the past fifty years, a radio astronomer in some other planetary system had observed the Solar System at television wavelengths he would have found the Earth more radiant than the Sun and from that he would have had no difficulty in deducing that the Earth is inhabited – and by intelligent life. That is why, if life has managed to evolve elsewhere and to develop an advanced technology, we can envisage the possibility not just of detecting it but of communicating with it.

The evolution of life towards civilisation

By using our knowledge of cosmic evolution and by invoking the law of averages – we are on an average or *ordinary* planet in orbit about an *ordinary* star that is situated in an *ordinary* part of an *ordinary* galaxy and consequently the appearance and evolution of life on Earth are examples of 'ordinary' events – we can attempt to calculate *N*, the number of civilisations contemporaneously present in our Galaxy. To achieve this, a famous formula – once described as the contraction of a vast amount of ignorance into a tiny space – was proposed by Frank Drake:

$$N = N_* f_p n_e f_v f_i f_c T$$

where N_* is the number of stars formed annually in the Galaxy that are stable for long enough for the appearance of life; f_p is the fraction of these stars with a planetary system; n_e is the number of planets of each system that are at an opportune distance from the central star (within its ecosphere); f_v is the fraction of such planets on which life has appeared; f_i is the fraction of these different kinds of life that evolve to intelligence and civilisation; f_c is the

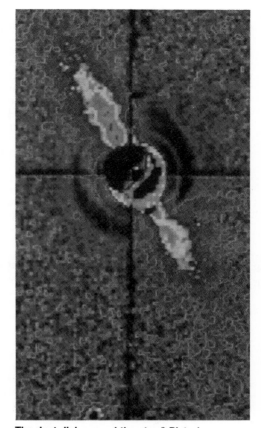

The dust disk around the star β Pictoris.
Bradford A. Smith and Richard J. Terrile used a stellar coronagraph to show the existence of an immense dust disk of more than 1000 AU radius gravitating around the central star (here hidden behind an occulting mask). The system is seen from the side and might represent a planetary system in formation. Recent spectroscopic observations by Roger Ferlet, Lew Hobbs, Anne-Marie Lagrange and Alfred Vidal-Madjar seem to indicate that bodies of a few kilometres diameter may be present in this disk, and that if they collide, some of them may fall onto the central star. (B.A. Smith and R.J.Terrile)

Titan, a satellite of Saturn. One of the places in the Solar System where life is perhaps being born. (NASA)

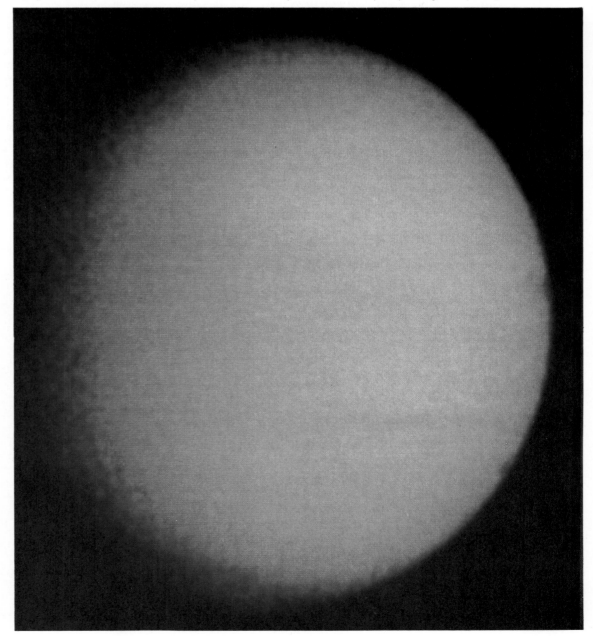

fraction of such civilisations endeavouring to communicate with other civilisations; and finally T is the length of time such endeavours are continued.

The following numerical values are possible: $N_* = 20$; $f_p = 0.5$; $n_e = 1$; $f_v = 0.2$; $f_i = 1$; $f_c = 0.5$. According to these figures twenty stars similar to the Sun are born in the Galaxy every year and ten of these have planetary systems, all with one well-situated planet; on one in five of such conveniently sited planets life appears, and always evolves to develop intelligence and civilisation: half such civilisations try to communicate with others.

With these assumptions, N is equal to T, that is, the number of civilisations in the Galaxy trying to communicate simultaneously is comparable with their average life time, expressed in years. The most optimistic estimates, a billion civilisations, or the most pessimistic, just one – our own, are therefore admissible.

In fact, it is very difficult to estimate N without theorising on the evolution of a technologically advanced civilisation. Many scientists, like Josef Shklovsky, Carl Sagan, Sir Fred Hoyle, Freeman Dyson and Nikolai Kardashev, have envisaged various possible cases. Associating evolution with the consumption of energy they classify civilisations in three main categories. Type I civilisations operate on a planetary scale; they are capable of changing the environment but they are satisfied with the resources of the planet itself; type II civilisations require all the energy emitted by the central star (and Dyson imagines that to contain this they construct a sphere around the star); type III civilisations depend on groups of stars – they operate on a galactic scale. Civilisations of types II and III are obviously too different from our own for communication with them. Nevertheless, direct evidence of them could be obtained, on a galactic scale for type III and, for type II by the detection of a stellar type source re-emitting residual thermal energy in the infrared.

The lifespan of these civilisations is widely debated. There is a view that it may be very brief (if civilisation automatically destroys itself once it has discovered nuclear energy) or, on the other hand it may be very long, practically unlimited, if some kind of artificial life – robots for example – succeeds the organic life.

To these views Michael Hart contributed a weighty argument that enlivened the debate. He showed that it should not be difficult for an advanced civilisation to achieve interstellar travel with speeds approaching one tenth that of light. Thus the propagation front of such a civilisation should spread through the galaxy at nearly the same speed and consequently the whole galaxy could be colonised in less than a million years – an extraordinarily short period in the cosmic time-scale. If just one civilisation had attained this level it should have already manifested its presence; but we know this has not happened and we may therefore conclude that such civilisations do not exist and consequently, that we are alone in the Galaxy. Michael Papagiannis has tried to counter this argument by asserting that such a civilisation is actually present but concealed – hiding in the asteroid belt, for example. Such considerations have led Pierre Connes to develop for civilisations the equivalent of Olbers's paradox: why is the manifestation of intelligence not evident?

The debate continues and recently William Newman and Carl Sagan resumed an argument of Sebastian von Hoerner which claims that even a civilisation capable of propagating with the speed of light but with a population growth like ours – 2 per cent per year – could not overcome the problems of demographic explosion. Such a rate should therefore be rapidly reduced by any civilisation mindful of its survival, even, and especially if, it is travelling in interstellar space. Newman and Sagan show that with zero growth rate the speed of propagation of a civilisation could be relatively slow and that consequently a number of civilisations could coexist in the Galaxy without contacting one another.

Another theory, that of the 'galactic zoo' formulated by J. A. Ball seeks to explain why contact has never been realised by suggesting that one civilisation would dominate the Galaxy and prevent communication between the planetary 'cages'.

We should be aware that all this is highly speculative and that all these ideas can only be compared with a single observation – that of our own civilisation. From this stems the need for further observations and especially for direct communication with other civilisations.

Communication

If other technologically advanced civilisations exist how can we contact them?

Although possible, direct communication seems very difficult. However, we should bear in mind that the vast interstellar spaces are not necessarily an insurmountable obstacle; new generations can be born in spacecraft, methods of suspension of life (hibernation, for example) will probably be discovered and much faster craft will be constructed. And, as Paul Langevin predicted, a traveller will age much more slowly as his speed approaches that of light. As an example of this Carl Sagan has calculated that a traveller on board a vessel capable of an acceleration of 1 g (the same as that experienced on the surface of the Earth) could reach M31, the Andromeda Galaxy, in twenty-five years of his own time frame. Needless to say there could be no return journey although, strictly speaking, a return to Earth would be possible after a few million years, when the traveller would be completely forgotten.

Once again we are in the realm of speculation and it would be dangerous to associate this with the question of UFOs (unidentified flying objects), however tempted one might be to do so.

The question of UFOs is a delicate one because the problem has never been precisely defined and, more often than not, the available information has been provided by untrained observers. Because of that, some scientists refuse to discuss the subject on the grounds that it is irrational, but one may ask whether it is not possible that a blunt refusal to examine a question that deeply interests so many people is itself irrational.

It is a very complex subject. Several official investigations conducted both in the USA and in the USSR reported that, while they were satisfied that UFOs do not exist, there were incidents they could not explain. Similarly, those scientists who have seriously studied the records admit that there are aspects of the UFO phenomenon that they do not understand. As a result, in several countries there are associations examining this problem.

If we accept that we have never been visited by extraterrestrial life, all that remains is the possibility of communication by means of signals that are propagated with the speed of light – and this, as Giuseppe Cocconi and Philip Morrison demonstrated in 1959, seems a simple proposition. But here also there is a dilemma: in which direction should the signal be beamed and what kind of signal should be used? Effectively we are searching for a needle in a haystack without knowing exactly what a needle is. To give but one illustration of our predicament: if a thousand civilisations were at this moment communicating in our Galaxy, finding out the direction in which to point our aerial to hear one of them would be like trying to find a particular grain in a cubic metre of sand.

But research is not deterred by adversity so Giuseppe Cocconi and Philip Morrison were soon advocating the use of the highly transparent 'radio window' for the transmissions, and particularly the 21-centimetre wavelength because this corresponds to the emission frequency of hydrogen, the most abundant element in the Universe. If we can expect other civilisations to be motivated by such logical and thrifty reasoning then we should commence listening at once, without further theorising.

The first attempt, Project Ozma (named after the King of Oz) was directed by Frank Drake and William Waltman, in 1960, when they listened, at this wavelength, to two nearby stars τ Ceti and ε Eridani. The result was not a success but the experiment stimulated fresh thinking on the problem and contributed to the development of research strategy.

Later, some new objects that behaved peculiarly were discovered; these were, in 1961, quasars, which could have been associated with a type III civilisation, and in 1967, pulsars, which emit a signal so regularly that for a while astrophysicists thought they were listening to 'Little Green Men' (which is why, for a short time, pulsars were humorously designated LGM).

In 1971, a gigantic listening project was devised, under the direction of Bernard Oliver and John Billingham. Entitled Project Cyclops, it envisaged the deployment of more than a thousand orientatable dish aerials of a hundred metres diameter; an array capable of radio communication with a civilisation anywhere in the Galaxy. But the project never materialised.

For their part the Soviet Union initiated Project CETI (Communication with Extra-Terrestrial Intelligence), in accordance with the decisions of the Soviet–American Conference organised by Sagan in 1973. Up to now the results of this ambitious programme have been negative; they only serve to underline how difficult the undertaking is. Thus while 'listening' to 600 nearby stars on 21-centimetre wavelength Shklovsky was unable to pick up a single signal emanating from another civilisation. However, it should be pointed out that he only

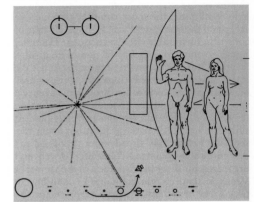

Messages sent in space probes, in Voyager 1 and 2 (above) **and in Pioneer 10 and 11** (below). In the Voyagers the message is in a modern medium, a video disc, but it contains basically the same information as the more 'intelligible' diagrams on plaques in the Pioneers. In the latter the 21-centimetre line of neutral hydrogen is used as a Rosetta Stone. The 21 centimetre line provides a standard of both length and time and is illustrated (top left) by a hydrogen atom that is shown rotating. It is possible to check the standard of length since the probe itself is drawn to this scale on the plaque. The spider-like mark and the lines radiating from it give the directions of the principal pulsars as seen from the Earth, and knowing the scale of time, their periods also.

Any civilisation with the ability to find the probes in interstellar space would know the pulsars well and would certainly have included them in detailed charts of the Galaxy. This intelligence would then realise that there is only one point in space and time within our Galaxy that corresponds to the spider. Having thus located the Sun, members of this civilisation would recognise the planets that surround it and find the one from which the probe was sent (the third, as indicated at the bottom of the diagram). What they would probably find most difficult to comprehend are the drawings of a man and a woman – especially if they themselves resemble the spider in the centre! (NASA)

The mythological papyrus of Nespakashouty. The god Geb, representing the Earth, reclining under the celestial vault formed by the figure of Nut. (Musées nationaux, Paris)

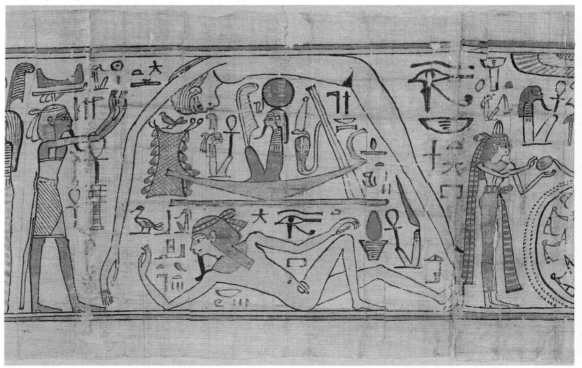

'listened to each star for a few minutes and the probability that a civilisation should be transmitting in our direction in so short a time is obviously very small. The difficulty in all these investigations is that one must find not only the right frequency on which to communicate, but also guess the right direction in the sky. Recently Jean-Louis Basdevant and Alfred Vidal-Madjar suggested that civilisations probably show themselves in two stages. First, they show where they are and demonstrate in a simple way the radio frequency on which communication can be established, and used in the second stage. Thus 'they' signal their presence in the ultraviolet, at the wavelength of the hydrogen Lyman α line. The interstellar medium is so opaque at this wavelength that the sky is uniformly dark and without stars. However, optical solitons can propagate without attentuation in this medium. The first contact would thus be made by means of emission of solitons at the wavelength of Lyman α, which is very costly in energy. For the observer these would appear as 'abnormal' bright points in the sky. But we would still have to communicate by radio!

If we are to establish contact one day, we too must transmit since we cannot communicate while we remain silent. So, in 1974, a message was sent by the Arecibo radio telescope in the direction of the cluster of stars in M13. This was the message that Drake had composed and shown to his colleagues several years earlier to see if they could decipher it: the supposed interstellar message was interpreted by many radio astronomers. The great adventure of interstellar communication had effectively commenced.

There is of course another means of communication – the bottle in the sea, or more precisely, in space. Prompted by Sagan, our civilisation has already released four of these, two in Pioneer 10 and 11 and two in the Voyager 1 and 2 spacecraft: these four messengers will leave the Solar System and go on through space, covering a distance equivalent to our separation from the nearest star in 80 000 years. Information on our civilisation is conveyed in engraved plaques or in video disks.

A science has just been born – exobiology. Its object is the study of life, in all its forms, in the Universe. In the laboratory an attack is being made on the exciting problem of how life emerged from inert matter. In our Solar System it appears that one site, Titan, should be of value in solving some of the unknowns.

The study of biological aspects of life is well underway. Life exists – of that there can be no doubt. It evolves, but we do not know exactly how and the existence of other civilisations seems plausible if not demonstrable. It has been said that 'the absence of evidence is not evidence of absence'. Still, it is worth repeating that the study of these civilisations is very speculative and our enthusiastic quest for communication may be nothing more than a desire for eternity. Indeed what could be more alike, spiritually, than the engraved plaque in Pioneer 10 and the messages addressed to the afterlife that Ancient Egyptians placed inside their tombs? If the question of eternity were to be resolved, civilisation might no longer seek to communicate.

All that and many other things are possible. But one thing is certain – we experience a profound desire to communicate with the unknown. But supposing effective contact were established, what effects would it have? Exchanging information with beings that evolved independently of ourselves could hardly be anything other than a considerable bonus and should in any case result in radical changes in our evolution and in our way of thinking.

For the moment, faced with this hidden ferment, one can only echo the famous words of Enrico Fermi: 'But where are they?'

Alfred VIDAL-MADJAR

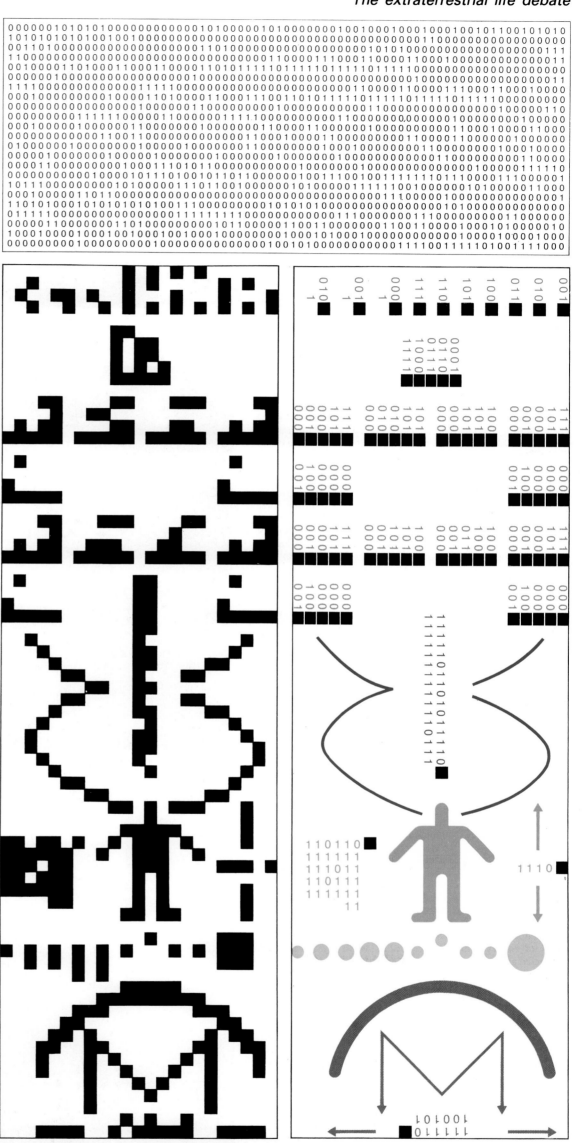

Message transmitted by the Arecibo radio telescope (above). This coded message in a form similar to Morse, but with the digits 0 and 1 instead of dots and dashes (see top). The message may be translated in the form of an image (lower left) where the white squares represent the zeros and the black squares the ones. The information is contained in 1679 characters; this figure is the product of two prime numbers (73 × 23) and there is only one way of composing an image from the message thus coded, in 73 rows and 23 columns. The picture on the right is a more intelligible translation of the message: the Arecibo radio telescope is recognisable at the bottom (at least to those who know it!); above it are, first, the Sun and planets, with the Earth offset, next, a human form, then, the double helix of the DNA molecule and finally, chemical formulae for the compounds of this molecule. (After Carl Sagan and Frank Drake, 1975)

The globular cluster M13 (left). The message transmitted from Arecibo in 1974 will reach M13 around the year 30 000. But will we ever receive a reply? (Hale Observatories)

Astronomical observations

Observation of the sky began in prehistory. The splendours of the starry sky and of the Sun and Moon have always stimulated dreams and poetry, bringing to life an aesthetic appreciation that has driven us to look further and to discover the marvels of the Universe. Until the beginning of the twentieth century information reached us only by means of visible light. Nevertheless, light carries information of different kinds. For example, it is characterised by both its intensity and its colour. Having established methods for charting the positions of the stars in the sky, it became possible to analyse the light and derive the information concealed in it. This objective started the trend towards bigger and better telescopes, which is still in progress today.

Radiation other than visible light also reaches us from the sky. This discovery has led to the development of new instruments and techniques. Meteorites and cosmic rays bring us direct material evidence of what there is in the neighbourhood of the Earth. Radio waves can tell us about the most distant parts of the Universe.

By means of satellites and space telescopes it has become possible to gather information much more widely; probes have been sent to the boundaries of the Solar System and observatories have been placed in orbit around the Earth to study the parts of the electromagnetic spectrum that are absorbed by the Earth's atmosphere: the infra-red, ultraviolet, X-rays and gamma rays, etc.

This account is concerned mainly with the history and recent development of astronomical instrumentation.

The quest for light

Only a tiny part of the electromagnetic spectrum can pass through the Earth's atmosphere and be easily detected; that is visible light. It is in this part of the spectrum that the Sun emits most of its energy and, because we are on the Earth, the human eye has evolved to respond best precisely in this region.

The human eye is a remarkable instrument for collecting photons, the 'particles' of light; when they pass through the lens in the eye, an image is formed on the retina, which detects the light; the information is then transmitted to the brain, where it can be analysed. The human eye is among the most sensitive detectors of light that exist. Each of the cells of the retina can 'see' a few photons arriving in an interval of the order of one-tenth of a second; being very adaptable, the eye can also observe beams of light a billion times brighter than the faintest signal it can detect! This is an enormous dynamic range, and it can also distinguish colours.

The angular resolution of the human eye, that is the angle between two objects which it can distinguish, is of the order of one minute of arc, which corresponds to an object an inch across at a distance of about 300 feet (or a millimetre at about 3.5 metres). This resolution is the result of the size of the cells of the retina, which is of the order of a few micrometres and corresponds to the diffraction limit defined by the aperture of the pupil of the eye. Interference occurs when light waves are diffracted as they pass through any aperture. This effect blurs the intrinsic quality of the image which could be produced by an optical instrument if diffraction did not occur, thus limiting its resolving power. In the case of the eye, the blur is about one minute of arc across, which corresponds to a diameter of a few micrometres on the retina. The larger the aperture an instrument (such as a telescope) has, however, the smaller the blurring effect. The brain, by processing all the picture elements that constitute the image on the retina (about six million), gives us the impression of a perfect and continuous image. The brain corrects the inherent faults of the optical system (we have a blind spot on the retina, which is difficult to recognise) and gives the impression of movement as well as stereoscopic vision. The eye and the brain working together are capable of remarkable feats; for example, in order to maintain good contrast in an image the eye operates in rapid jerks and slow drifts, which allow the brain to maintain an image of optimum quality, even at low light intensity. This method has been used in instruments, particularly in infrared astronomy, to detect weak signals against a background of intense noise.

At the same time, however good the eye may seem, it is limited in several respects: first in sensitivity and resolution because of the size and the efficiency of the detection system formed by the cells of the retina, second in the colours it can

The temple of Cape Sounion. The gift of vision enables our minds to take in the whole Universe. (P. Boulat, Cosmos)

The observatory at Stonehenge. Begun more than 4000 years ago, Stonehenge is probably one of the oldest astronomical observatories. The precise alignments of some megaliths with the directions of the rising and setting Sun at the solstices, and of the Moon when at its extreme declinations, have provoked even the astrophysicist Fred Hoyle to suggest that Stonehenge must be the oldest computer on Earth. This enormous computer was used to predict eclipses. (A. Hasarth, Explorer)

The interferometer near Grasse, France. This interferometer consists of two telescopes, each with an aperture of 1.5 metres, placed inside spherical mountings; the spheres can be pointed in any direction. (A. Labeyrie)

distinguish and, finally, by the exposure time – about one-tenth of a second. Moreover, it does not retain a precise record of what has been observed and the processing by the brain can lead to errors; it can create optical illusions or, more serious, it can even create observations of things that do not exist, except in the imagination of the observer: a celebrated example is that of the 'canals' of Mars. The aim of the quest for photons is to overcome all these limitations.

The earliest instruments merely marked the positions of the Sun and stars. The oldest of these is probably the deep well dug near Syene, in Upper Egypt, in which one can see the Sun's light reflected by the water at midday on the summer solstice. This well allowed Eratosthenes of Cyrene, with the help of a traveller who walked to Alexandria, to show that the Earth was not flat, and even to measure its circumference reasonably accurately. Sundials track the changing position of the Sun, whose apparent movement allows time to be measured.

Light-gathering power and resolution remained limited to those of the eye until the true quest for photons was eventually started by the invention of the telescope. Galileo, hearing of the invention, was the first to apply it to astronomy. The principle of his first telescope was simple; an objective lens collected the light falling on an area larger than that of the eye, and formed, in its focal plane, an image of distant objects. An eyepiece allowed the observer to magnify this image. Thus, the resolution was improved and fainter objects could be observed.

Throughout the evolution of astronomical instrumentation, improvements have been made progressively to overcome the imperfections of earlier instruments. Defects become a serious problem when the aperture of a simple refracting telescope is made large, particularly because of chromatic aberration. In passing through the lens, photons of different colours are deviated unequally, creating coloured fringes round the image of an object. This problem was overcome in the invention of reflecting telescopes, which use mirrors as objectives. The principal types of optical system were devised by Newton and Cassegrain.

The sizes of conventional reflecting telescopes have increased considerably: the diameter of the well-known Palomar telescope, commissioned in 1948, is 5.08 metres. The maximum reached 6 metres in 1978 with the construction of the telescope at Zelenchukskaya, USSR but the sizes of telescopes have not increased as much as one might have expected in the last seventy years. Very large instruments are the exception rather than the rule. There are many reasons for this. Above all, it is difficult to grind and polish very large surfaces to the highest optical quality. Large telescopes must also be positioned and oriented with precision. Such telescopes are not limited by diffraction but by the Earth's atmosphere, the disturbances of which always blur the images so that these large telescopes have a best resolution which is only that of a telescope of about 10 centimetres aperture (about 1 second of arc, or the angle covered by a penny at a distance of 4 kilometres). Because the sky is not totally dark, even at night (it scatters light and emits certain colours like a faint, permanent aurora) it is futile simply to increase the sizes of the mirrors in order to see fainter objects because, then, the contrast between the objects and the sky background is no better; even in good conditions for observation, the atmospheric emission within a patch the size of the seeing blur is equivalent to the light received from a star of about the 20th magnitude. It will be difficult to observe from the ground any objects fainter than the 25th magnitude without trying to compensate for the effects of atmospheric turbulence. If one succeeds in reducing the image blur, fainter stars will be observable. Some progress has already been made in this direction.

A modern telescope should, thus, certainly be large but, above all, it should be very well situated. The Canada–France–Hawaii telescope (CFHT), for example, although slightly smaller than the Anglo-Australian Telescope (3.9 metres) or the 4-metre telescopes at Kitt Peak and Cerro Tololo, with its situation at 4200 metres on Mauna Kea, is one of the best modern instruments for ground-based astronomy.

The tendency towards large telescopes has, however, revealed some weaknesses. In particular, as one increases the size of telescopes, the field of view is reduced. Bernard Schmidt invented a type of telescope with a large aperture and a large field of view. A telescope of this type of 48 inches (1.22 metres) aperture, installed at Mount Palomar, was used to make the well-known Palomar Observatory Sky Survey, on which objects down to the 21st magnitude are recorded. However, this survey does not cover the whole sky because, as it is situated in the northern hemisphere, the telescope cannot observe the most southerly regions. In fact, the geographical distribution of large observatories is concentrated in the northern hemisphere. This situation began to change at the start of the 1980s.

Progress has also been made in improving detectors. In fact, the human eye, even when aided by the largest telescope, cannot see very faint objects. The invention of the photographic plate, although much less sensitive than the eye (about 100 photons are needed to make any impression on a plate) led to significant progress in astronomy. Photography allowed very long exposure times, up to a whole night in length, which compensated for poor sensitivity. Furthermore, it has a great potential for storing information but, at the same time, certain features of the photographic plate remain as limitations: it lacks sensitivity; it can be saturated by a bright light; there is a threshold below which it cannot reach, whatever the exposure time may be.

It is clear that much can be gained, especially if detection of individual photons is possible. Modern detectors aim to do this in various ways. Andre Lallemand at Meudon Observatory pioneered the construction of a camera in which a quarter of the incident photons liberated electrons which were accelerated and refocussed by an electron-optical system onto a photographic plate capable of detecting each fast-moving electron. Although it reaches within a factor of about four of the ultimate limit – the noise-free detection of every photon – this camera has been little used because of the difficulty of operating it. It requires an ultra high vacuum and a cryogenic system.

Other electronographic cameras, in which the light-sensitive photocathode was protected, by a thin sheet of mica, from damaging gases liberated at the photographic plate; were developed by McGee at Imperial College, London, and later by McMullan at the Royal Greenwich Observatory.

Meanwhile, other improvements made detectors easier to handle and allowed the photographic plate to be discarded. Image intensifiers giving up to 10^{10} photons output for a single input photon were produced; coupling these intensifiers to TV cameras allowed Boksenberg, at University College, London, to construct the IPCS (Imaging Photon Counting System) in which the theoretical limit would be reached if every incident photon generated a burst of photons at the output; in fact only about 20 per cent of them do so. For some time now the type of TV camera that uses a scanning electron beam has, little by little, been replaced by small electronic detectors called Charge Coupled Devices, (CCDs) which allow a much better performance (they respond to about 80 per cent of the incident photons at some wavelengths and their information storage capacity is larger) and more convenient usage. The performance of the CCD is so good that small observatories equipped with telescopes of 1 metre aperture are now capable of work the same as the Palomar 200-inch (5-metre) telescope could do in 1948. If we take account of the improvement in detectors and as the norm the capability of the Palomar telescope at the time of its commissioning, the largest existing telescopes are each equivalent to one of more than 20 metres aperture in those days.

The ultimate limit in detection, the observation of a single photon, has been achieved. Is the quest therefore finished? The answer is no, because battles against other problems have been carried on in parallel, in particular that of improving spectroscopy.

Spectroscopy allows us to 'test' the photons for their colour, that is to say, their wavelength or their energy. The radiation characteristic of atoms or molecules that are well known in the laboratory can be detected in the light from astronomical objects and so it is possible to deduce the composition of the stars. Furthermore, the shifts in the wavelengths of features in a spectrum from the wavelengths observed in the laboratory allow us to measure the velocity, relative to ourselves, of the object emitting; this phenomenon is called the Doppler effect. In this way Edwin Hubble discovered the expansion of the Universe. He found that the most distant objects showed a greater displacement of the spectrum towards the red than nearer objects.

At first prisms were used to disperse light into its component colours. Objective prisms placed in front of telescope objectives are still used to examine the light of all the objects in the field of view of the instrument simultaneously. Today, spectroscopy uses very varied methods to separate photons of different wavelenths. The majority of these methods depend on the wave nature of light and cause constructive interference for a chosen wavelength, the other wavelengths being eliminated by destructive interference. The system most frequently used nowadays is the grating, an optical surface on which up to a few thousand straight parallel grooves are rules in each millimetres. This surface reflects light of a given wavelength in a particular direction. It is then sufficient to place a detector at the right position to measure, in the spectrum of the source being studied, the intensity of light received of the wavelength considered. Such instruments allow radial velocities to be measured with an accuracy better than 1 kilometre per second. This may seem to be very inaccurate in relation to velocities experienced in everyday life but very often quite accurate enough in the case of the observation of astronomical objects. In fact, the velocities of recession of the galaxies are expressed in thousands of kilometres per second and movements within our own Galaxy in hundreds of kilometres per second; even the movements in our own Solar System amount to tens of kilometres per second (the Earth's velocity in its orbit around the Sun is about 30 kilometres per second).

It therefore appears that all the limits have been reached: it is difficult to build larger telescope mirrors; it is impossible to detect anything fainter than a single photon. However, a further gain can be obtained in a new way, by optical interferometry. By this technique, the resolving power of a telescope can be increased from about one arc second to a few thousandths of an arc second! With this resolving power, stellar diameters can be measured directly. We know that diffraction prevents a telescope of 10-centimetre aperture from resolving objects separated by less than an arc second. A telescope of 1-metre aperture can, theoretically, separate objects 0.1 arc second apart, and a 10-metre telescope should separate objects 0.01 arc second apart, and so on. However, the information is blurred by the continuous movement of the atmosphere. In fact, as Antoine Labeyrie has shown, the information is not completely lost; if a 'snapshot' is made of the star image one can see, instead of one large patch of light, a multitude of little ones, called speckles, showing the

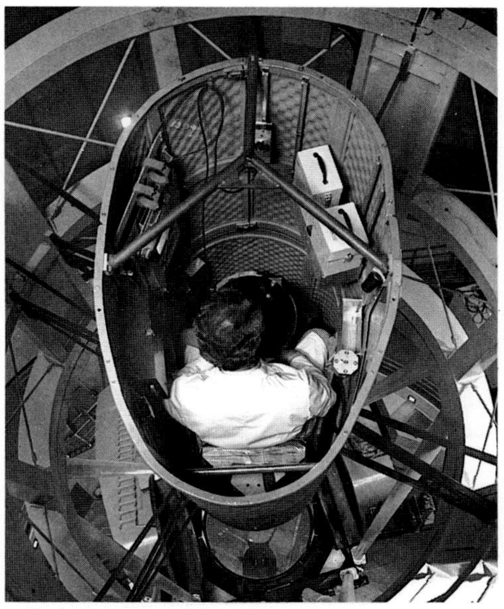

The observer's cage at Lick Observatory. The observer must sometimes ride in a small cage at the focus of the primary mirror in order to carry out observations. An observer may sometimes have to spend whole nights in this minute capsule (made as small as possible to avoid obscuring too great a part of the incoming starlight). (Lick Observatory)

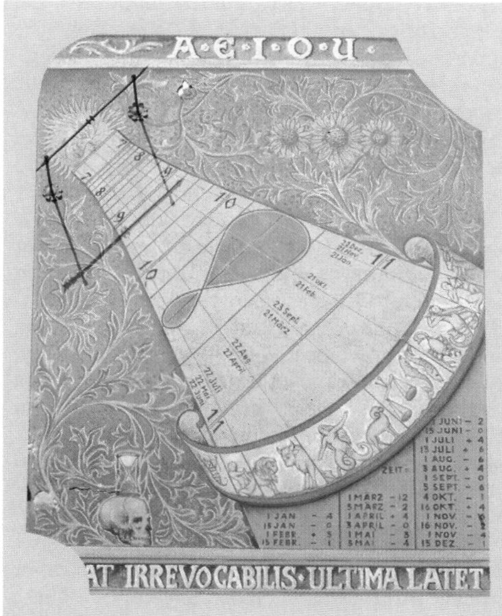

Sundial at Merano, Italy. One of the most aesthetically pleasing astronomical instruments. Here the study of the sky is reunited with art and poetry. (Fiori, Explorer)

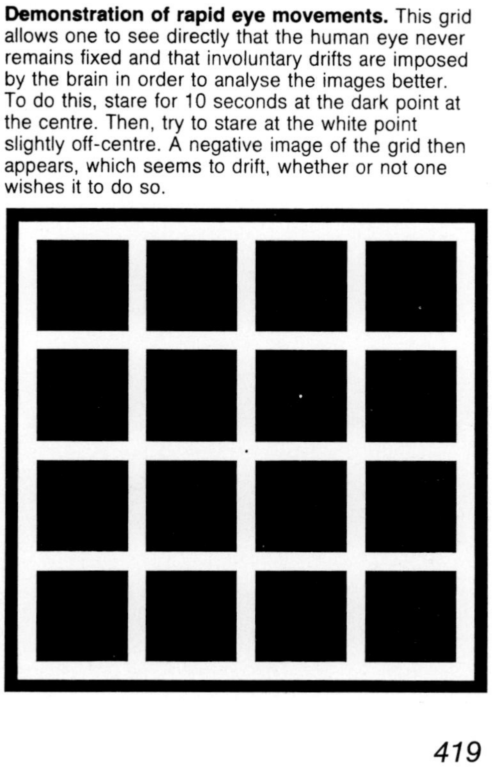

Demonstration of rapid eye movements. This grid allows one to see directly that the human eye never remains fixed and that involuntary drifts are imposed by the brain in order to analyse the images better. To do this, stare for 10 seconds at the dark point at the centre. Then, try to stare at the white point slightly off-centre. A negative image of the grid then appears, which seems to drift, whether or not one wishes it to do so.

The Canary Islands Observatories of the IAC

The Canary Islands were discovered astronomically in 1856 by the Astronomer Royal for Scotland, Charles Piazzi Smyth, who site-tested on peaks on Tenerife, particularly Guajara (2717 metres). Tenerife was the base for several astronomical expeditions during the next hundred years, including the solar eclipse expeditions of 1959 and Spain set up a permanent observatory in 1970. As a result, growing ever more aware of the excellent conditions for astronomical observation to be found on the heights of the islands of Tenerife and La Palma, Spain decided 'to make the observatories of the Instituto de Astrofísica de Canarias (IAC) available to the international scientific community' and the International Agreement on Cooperation in Astrophysics was signed on 26 May 1979. In this way Spain has internationalized the observatories of the IAC, 'affording the Signatory Bodies an effective voice in the decision making', through an International Scientific Committee (CCI).

Since then development has been so rapid that the Teide Observatory (OT), on Tenerife, and the Roque de los Muchachos Observatory (ORM), on La Palma, now take their place amongst the best in the world and without doubt combine as Europe's natural Northern Hemisphere Observatory, representing as of 1993 scientific interests of Belgium, Denmark, Eire, Finland, France, Germany, Italy, Norway, Spain, Sweden, The Netherlands and The United Kingdom. They were formally inaugurated in 1985 by six European Heads of State and the entire Spanish Royal Family. The international astronomical community representatives were headed by five Nobel Laureates.

Apart from the telescopes mentioned elsewhere in this section, there are several other optical, infrared and solar telescopes. These include, at the Roque de los Muchachos Observatory, the 2.6 metre Nordic Optical Telescope (Denmark, Finland, Norway and Sweden) which is an altazimuth telescope with Cassegrain focus, equipped with active optics, and used mainly for optical and infrared observations of celestial objects and the Carlsberg Automatic Meridian Circle. The Instituto de

Astrofísica de Canarias operates at the Teide Observatory the 1.5 metre infrared Carlos Sanchez Telescope, which is used for general infrared observations and mapping of the galactic centre, and a new 80 centimetre general purpose optical telescope.

For both observatories Spain provides the infrastructure, including roads, power, communications, etc., and receives 20 per cent of the observing time on each instrument. A further 5 per cent of the observing time is available for collaborative projects among the countries which have signed the international agreements. The remaining time is divided between the countries which participate in the operation of each telescope.

Cosmic Microwave Background Experiment. The Nuffield Radio Astronomy Laboratory of Manchester University has installed three radio telescopes working at 10, 15 and 33 gigahertz which measure the anisotropy of the microwave background radiation. An important requirement for the understanding of galaxy formation, and in fact any structure in the present observable universe, is a knowledge of the initial distribution of matter in the very early universe. The only way to measure this is through the temperature variations produced by the seed density perturbations present when the hot dense universe changed to a cooler transparent one. In this way these perturbations in temperature were frozen into this opaque wall with a temperature of about 3000 kelvins at a red-shift of about 1000. The resulting distribution in structure covers a large range of angular scales with an amplitude of typically 30 microkelvins within the now apparent temperature of 3 kelvins of the microwave background radiation. The reason that measurements of anisotropy are important is that the amplitude of these fluctuations is dependent on the density and the material content of the universe, therefore theories of galaxy formation can be tested by the requirement that they produce present structure from the observed anisotropy.

Solar telescopes at the Teide Observatory. The 45 centimetre Gregory Coude Telescope – left – (Germany) is a solar telescope with a focal length of 28 metres and is equipped with a spectrograph. It is used for studies of fine-scale solar surface features with high spectral resolution and high precision polarimetry. The 40 centimetre Vacuum Newton Telescope – centre – (Spain) provides images of the whole solar disk and is often used in combination with the Gregory Coude and Vacuum Tower Telescopes. The 60 centimetre Vacuum Tower Telescope – right – (Germany) is the largest solar telescope in the Canary Islands (1993). It has a focal length of 46 metres and a high resolution Echelle spectrograph and is mainly used for studies of fine-scale solar surface features with very high spectral resolution.

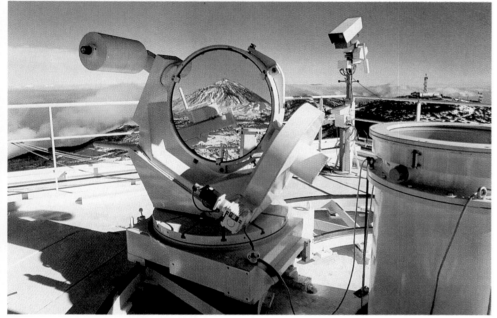

Ceolostat of the Vacuum Tower Telescope.

The factors which have led to the installation of so many optical telescopes at ORM represent a compromise between communication, infrastructure and technical base (all of which imply proximity to a sufficiently large population) and height, seeing, dark sky and absence of pollution (all of which imply distance from people). In fact the geographical location of La Palma makes it competitive with the best sites at the same altitude in the world. The weather of the western Canaries is oceanic, dominated by the cold Canary Current, flowing southwards in the continuation of the clockwise circulation of the Gulf Stream in the North Atlantic Ocean. As the weather systems drift from the west the cold current produces a contraction of the convection at the ocean surface, and the inversion layer below which the cloud is contained lies at approximately 1500 metres. Thus, both the Roque de los Muchachos and the Teide Observatories (both at 2400 metres) protrude into clear air. In order to avoid possible deterioration of these privileged conditions, the Spanish Parliament passed a Bill on 31 October 1988, the 'Law to Protect the Astronomical Quality of the Observatories of the Instituto de Astrofísica de Canarias', generally known as the 'Law of the Sky', which is the first national law in the world of its kind. The decrees which implement it follow the recommendations of the International Astronomical Union and will guarantee that the excellent quality of these observatories remains unspoilt.

Future

France is building a solar telescope (THEMIS) at the Teide Observatory. It will be a Ritchey–Chretien vacuum tube telescope with altazimuth mounting and a primary mirror 90 centimetres in diameter, its main instruments will be spectrographs. First light is expected in 1995. Italy is building its National Telescope 'Galileo' at the

Roque de los Muchachos Observatory. It will be a Ritchey–Chretien telescope with altazimuth mounting, a primary mirror of 3.5 metres in diameter and active optics. Its main instruments will be imaging cameras and spectrographs for the visible and the infrared. First light is expected in 1995. The BET (Big European Telescope) is at present being negotiated between Spain and

Solar Seismology. A number of spectrophotometers for resonant scattering have been installed at the Solar Laboratory and in other installations of the Teide Observatory. Helioseismology is at present one of the few techniques for the study of the Sun's interior. Different types of wave, resonant modes of oscillation, are generated in the interior of the Sun and these produce small movements in the solar atmosphere which can only be detected from the Earth with very precise instrumentation.

other interested European countries. It will be an optical–infrared telescope with a primary mirror of between 8 and 10 metres to be sited at the Roque de los Muchachos Observatory. The LEST (Large Earth-based Solar Telescope) is at an advanced stage of design and is only awaiting the successful outcome of funding negotiations between the participating countries so that construction can be started at the Roque de los Muchachos Observatory. It will have a 2.5 metre primary mirror, making it by far the largest and most important solar telescope in the world. Its design specifications include live optics components, with a wavefront sensor, a fast correlation tracker and imaging Stokes polarimeters.

Franciso SANCHEZ

High Energy Gamma-Ray Array. This rapidly expanding installation has a unique combination of scintillation detectors, air cerenkov counters, cerenkov telescopes, muon detectors etc. The photograph shows one of the cerenkov telescopes. Its main activity is the search for extended air showers originating from Ultra High Energy (UHE) cosmic rays ($E > 10^{13}$ electron volts) in order to understand the origin of cosmic rays of the highest energy.

The Teide Observatory (Tenerife)

diameter	name	organization	date
25 cm	Razdow Heliograph (solar)	IAC (Sp.)	1969
40 cm	Vakuum Newton (solar)	KIS (DFG, G), IAC	1972
45 cm	Gregory Coudé (solar)	USG (DFG, G)	1986
50 cm	MONS (optical)	MU (B)	1972
60 cm	Vacuum Tower (solar)	KIS (DFG, G)	1989
75 cm	Auto SN Patrol (optical)	LBL (USA), IAC	1994
80 cm	IAC-80 (optical)	IAC	1993
90 cm	THEMIS (solar)	CNRS (F)	1995
1.55 m	Carlos Sánchez (infrared)	IAC	1972
Other installations			
Solar Laboratory (Pyramid)	Mark-I spectrophotometer	BU (UK)	1976
	Mark-II spectrophotometer	BU (UK)	1989
	Paul-II spectrophotometer	BU (UK)	1989
	Space-III spectrophotometer	GOLF consortium	1989
	IRIS-T spectrophotometer	NU (F)	1989
	LOI	ESA, IAC	1990
	TON	NTHU (ROC)	1993
	NIP	NSO (USA), IAC	1986
	GONG Fourrier Tacometer	NSO (USA), IAC	1994
Radio Telescopes	10 GHZ (microwaves)	UM (UK), IAC	1984
	15 GHZ (microwaves)	UM (UK), IAC	1989
	33 GHZ (microwaves)	UM (UK), IAC	1991

The Roque de los Muchachos Observatory (La Palma)

diameter	name	organization	date
18 cm	CAMC	CUO (D), RGO (SERC, UK), RIOA (Sp.)	1984
50 cm	Solar Tower	ROYAC, Stockholm Observatory (S),	1985
60 cm	60-cm (optical)	KVA (S)	1982
1.00 m	JACOBUS Kapteyn (optical)	RGO, NWO, DIAS (I)	1984
2.54 m	Isaac Newton (optical)	RGO, NWO	1984
2.56 m	NOT (optical)	NOTSA	1990
3.5 m	GALILEO (optical)	CRA (It.)	1995
4.20 m	William Herschel (optical)	RGO, NWO (NL)	1987
Other installations Gamma-ray detectors	HEGRA	MPIfPA, Münich (G), Univ. of Kiel (G), Univ. Complutense de Madrid (Sp.)	1988

Date: Year operation commenced.

Abbreviations:
B – Belgium;
BU – Birmingham University;
CAMC – Carlsberg Automatic Meridian Circle;
CNRS — Centre National de Récherche Scientifique;
CRA – Consiglio per le Ricerche Astronomiche;
CUO – Copenhagen University Observatory, Denmark;
D – Denmark;
DFG – Deutsche Forschungsgemeinschaft;
DIAS – Dublin Institute for Advanced Studies, Ireland;
ESA – European Space Agency;
F – France;
G – Germany;
GOLF – Global Oscillations at Low Frequencies;
GONG – Global Oscillations Network Group;
HEGRA – High Energy Gamma-Ray Array;
I – Ireland;
It. – Italy;
IAC – Instituto de Astrofísica de Canarias;
ING – Isaac Newton Group of Telescopes;
KIS – Kiepenheuer Institut für Sonnenphysik;
KVA – Kungliga Vetenskapsakademien;
LBL – Lawrence Berkeley Laboratory;
LOI – Luminosity Oscillations Imager;
MPIfPA – Max-Planck-Institut für Physik und Astrophysik;
MU – Mons University;
NIP – Normal Incidence Pyrheliometer;
NL – The Netherlands;
NOT – Nordic Optical Telescope;
NOTSA – Nordic Optical Telescope Scientific Association, for Research Councils of Denmark, Finland, Norway, Sweden;
NTHU – National Tsing Hua University;
NU – Nice Université;
NWO – Nederlands Wetenschappelijk Onderzoek;
RGO – Royal Greenwich Observatory of the SERC, UK;
RIOA – Real Instituto y Observatorio de la Armada;
ROC – Republic of China;
ROYAC – Royal Swedish Academy of Sciences;
SERC – Science and Engineering Research Council;
S – Sweden;
Sp. – Spain;
TON – Taiwan Oscillations Network Project;
UK – United Kingdom;
UM – University of Manchester;
USA – United States of America;
USG – Universitäts-Sternwarte Göttingen;

inhomogeneous structure of the air above the telescope. If the objects observed could theoretically be resolved by the telescope, it would seem that the structure of these patches is not quite random. If, for example, a double star is examined, a pattern of doubled speckles will be seen, revealing the presence of two objects and their separation. The distribution of brightness across the surface of Betelgeuse, a red supergiant star which should theoretically be resolved by the telescope employed, has been successfully reconstructed in this way. Likewise, the CFHT has been used to measure the diameters of Pluto and its satellite Charon and their separation; this result cannot be obtained by any telescope used in the classical manner.

We are always limited by the aperture of even the largest telescopes. In fact, the resolving power of a large telescope is equivalent to that of two small telescopes separated by the diameter of the large one. With a pair of telescopes, instead of obtaining speckles, interference fringes are observed directly but these disappear if the object has a diameter greater than the resolution of the large telescope. This effect can be seen directly; it is only necessary to make two small holes about 0.5 millimetres apart in a piece of aluminium foil, using a needle, and put this about 15 metres from an electric light bulb. When you look at the bulb through the two holes you will see a blurred patch consisting of bright and dark fringes; if you move towards the bulb the fringes will eventually disappear. The explanation is as follows: you have effectively a small interferometer with a diameter of 0.5 millimetres, whose resolving power is a few minutes of arc. When the bulb is 15 metres or more from the interferometer its apparent diameter is smaller than that resolving power, and all of the light from the bulb that passes through the holes contributes to the observed interference fringes. When the bulb is closer, it is resolved by the system, that is to say, for example, that the left-hand part of the bulb gives one system of fringes and the right-hand half another system, but displaced just so that the bright fringes of the first system fall on the dark fringes of the second. The systems of fringes then disappear, and the interferometer can now resolve the object.

H. Fizeau had proposed this method as early as 1868, and in 1890 A. A. Michelson used it to measure the apparent diameters of the four Galilean satellites of Jupiter. Later, A. A. Michelson and F. G. Pease built an interferometer with movable mirrors which allowed them to obtain the effect of a variable separation. While observing Betelgeuse they saw the fringes disappear when the separation of the mirrors was increased to 3 metres, which showed that they had then completely resolved the disk of the star, which is 0.05 arc seconds in diameter.

A. Labeyrie built an interferometer, near Grasse, in France, that is based on the same principle but is formed of two small movable telescopes, whose maximum separation is 67 metres. This system therefore has the same resolving power as a telescope of 67 metres aperture, that is to say a few thousandths of a second of arc, which corresponds to an object a few metres across at the distance of the Moon! The apparent diameters of a number of stars have been measured with this system. An interferometer incorporating larger telescopes is under construction and it will allow fainter objects to be measured.

Ground-based astronomy has thus made considerable progress during the twentieth century. Numerous improvements have been made, particularly in the development of photon detectors.

However, whatever effort we make, it will always be impossible to retrieve information which is not carried by visible photons. For this reason astronomy has evolved towards collecting other data, which is how radioastronomy was born.

The search for other open windows on the Universe: radio astronomy

Radio astronomy owes its origin to pure chance. K. G. Jansky, an engineer at Bell Telephone Laboratories, had been sent in 1931 to a telecommunications station in New Jersey to solve a problem of radio interference. He then discovered that some of the disturbance came from the direction of the Milky Way, our Galaxy. Radio astronomy was born, but Jansky did not pursue his research in this new field. It was only after 1945 that radio astronomy truly sprang into life. In fact, the development of radar has led to considerable progress in electronics and the detection of weak radio signals. The quest for photons in the radio region of the spectrum then began in the same way as in optical astronomy.

The principles are the same: as large a surface as possible should be used to collect the photons and concentrate them on a detector system. In radio astronomy, the equivalents of the mirrors of optical astronomy are large, smooth metal surfaces or metallic mesh, which reflect the radio waves. Photographic plates or other detectors of optical astronomy are replaced by antennae that transform the radio waves into electrical signals.

In radio astronomy the difficulties arise from the weakness of the energy collected: the total energy collected by all the radio telescopes in fifty years is less than is needed to turn a single page of this Atlas!

The buildings of large collectors began very quickly: as early as 1957 the first of the giants, the well-known radio telescope at Jodrell Bank in England, the brain-child of Bernard Lovell, was completed. Its reflecting surface is 76 metres (250 feet) in diameter. Also at this time, the first artificial satellites were launched, and this radio telescope was found to be the only tracking station capable of detecting the faint signals sent by the space probes. The first photograph of the back of the Moon was published by the UK and not by the USSR who had sent the probe!

The early construction of Jodrell Bank was a master-stroke, since even today there is only one larger, fully-steerable, radio telescope, that at Effelsburg in the Federal Republic of Germany, 100 metres in diameter, which was brought into use in 1971.

Some radio telescopes are not steerable in all directions; they use the rotation of the Earth to sweep the sky. Thus, by slightly changinging their orientation each day, a radio map of the sky can be created line by line, in the same way as television pictures. The largest radio telescope in the world is of this type, a gigantic bowl of wire netting 305 metres in diameter but near Arecibo in Puerto Rico. Its focus, where the detectors are placed, is 130 metres above the reflecting surface.

Other giant radio telescopes have been built with several reflecting surfaces, the movable surface being flat and therefore easier to adjust. In this way the radio telescope at Nançay in France allows observations of the sky with a very large collecting area.

But, as in optical astronomy, the resolution of a radio telescope is limited by the size of its collector. Because radio waves have a much longer wavelength than visible light, the resolution that can be obtained is considerably worse – typically tens of seconds of arc – and very large collecting areas are needed to achieve this. It is increasingly difficult to build larger and larger instruments, but resorting to interferometry allows the attainment of a resolving power equivalent to that of a large telescope with a diameter equal to the distance separating two smaller radio telescopes.

The principle is exactly the same as that developed in optical astronomy, except that it was generally used much earlier in radio astronomy. One of the principal difficulties in interferometry is that the two telescopes need to be stable to about one wavelength. In optics, the precision needed is of the order of one thousandth of one millimetre, while for working in the radio region it is from several centimetres to several metres. This is why it was much easier to create radio interferometers than optical ones. One of the first instruments was conceived by Martin Ryle at Cambridge, UK, using two fixed antennae 800 metres apart, and a third antenna that could be moved along rails 800 metres long. The greatest separation of the antennae is thus 1.6 kilometres. The signals received by the different antennae are fed into a computer, which is able to reconstruct the shape of the radio source being observed with the resolution of a radio telescope 1600 metres in diameter.

This method was rapidly developed by connecting antennae further and further apart. In 1972, eight new antennae, distributed along a baseline 5 kilometres long, were brought into use beside the first interferometer at Cambridge. In the Netherlands, an interferometer consisting of fourteen antennae on a baseline of 3 kilometres has been installed at Westerbork. But the largest interferometer is the Very Large Array (VLA) near Socorro in New Mexico; twenty-seven movable dishes provide the resolving power that could be obtained with a telescope 27 kilometres in diameter!

It is possible to connect widely separated radio telescopes to create a very long baseline, simply by recording, simultaneously with the astronomical radio signals, the 'pips' of a very precise clock. Later the signals can be mixed and the interferance created as if the measurements had been made at the same observatory.

Radioastronomers at Jodrell Bank have interconnected six dishes, obtaining the resolution of an instrument 137 kilometres in diameter. For the present, only the size of the Earth limits this method. Antennae on different continents are connected in the Very Long Baseline Interferometry system (VLBI). The resolution that would be obtained by a radio telescope the size of the Earth (a few ten-thousandths of a second of arc) is thus obtained. This would allow us to see an object of a few tens of centimetres in size on the Moon!

As in the optical, radio spectroscopy has also been developed, thanks to frequency-selecting filters, which allows us to receive chosen wavelengths. The familiar radiation of atomic hydrogen at 21 centimetres wavelength can be observed with spectral resolution equivalent to a Doppler shift due to a velocity of a few kilometres per second. Numerous other line emissions have been observed, especially towards short wavelengths between a few centimetres and a few millimetres. At this boundary between the radio region and the infrared region, an entirely new technique is being developed, aimed particularly at the observation of complex molecules, of which more than 50 different species have already been detected in the interstellar medium.

In this region of the spectrum there are observational difficulties associated with the shorter wavelengths being used. To observe at 1 millimetre wavelength, the surface of the reflector must be accurate to better than one-tenth of one millimetre. Nevertheless, such radio telescopes have been built, with diameters between 5 and 45 metres. The Ratan-600, built in the Caucasian mountains, near the site of the optical telescope at Zelenchukskaya, is in the form of a gigantic ring 600 metres across.

Also in this region of the spectrum, interferometry seems very promising. A project of the Institute of Millimetre Radioastronomy, established by France and West Germany, involves the construction of a large millimetre-wave antenna, 30 metres in diameter, in Spain, in the Sierra Nevada and an interferometer consisting of three movable dishes each 15 metres in diameter on a base line 400 metres in length.

Lastly, another type of data has been gathered in a much more

The galaxy M81 in the visible and the ultraviolet. These two false colour pictures were taken (top) in the red by a ground-based 0.9 metre telescope, and (bottom) in the ultraviolet (160–220 nanometre) by the UIT (Ultraviolet Imaging Telescope) of the Astro-1 mission, which took place from 2–11 December 1990 on the space shuttle *Columbia*.

The latter image is much richer in information than any picture of M81 from a ground-based observatory: it shows groups of very young hot stars in the spiral arms of the galaxy, where new stars are forming at a high rate (Kitt Peak National Observatory and NASA).

Speckles in the image of a double star. A speckle image of a star observed at the focus of the Mount Palomar 5-metre telescope. The turbulence of the Earth's atmosphere causes interference to occur between different parts of the wavefront, and the effect is either constructive or destructive at different places in the focal plane. Thus, in the absence of the Earth's atmosphere, a perfect telescope would concentrate the star light into a single speckle. By detailed analysis of the small patches it is possible to recover the real resolving power of the telescope. (A. Labeyrie, CERGA)

Radio telescope at Nançay (France). A view of the stationary spherical dish, which focusses the radio waves from the sky, which have been reflected to it by a movable flat mirror.
The spherical dish is covered by a fine wire mesh which allows it to reflect radio waves just as a normal mirror reflects visible light. (De Sazo, Rapho and H. Veiller, Explorer)

The radio telescope at Arecibo. The largest radio telescope (and/or radar) in the world is in Puerto Rico. It is stationary and has a diameter of 305 metres. Scanning of the sky is carried out by the rotation of the Earth and displacements of the detectors suspended by cables high above the ground. This observatory has been used to transmit a powerful radio signal to any possible civilizations which might live in the globular cluster M13. (Arecibo Observatory; kindly provided by F. D. Drake)

The Very Large Array (VLA) near Sorocco in New Mexico is the largest radio interferometer. Its resolving power is equivalent to that of a radio telescope more than 20 kilometres in diameter, with its twenty-seven dishes each 25 metres in diameter. (G. Heisler, Liaison Gamma)

systematic manner in the radio region than in the visible region: the polarisation of radiation. The emission mechanisms of radio radiation are often much more coherent than typical sources of visible light and, as a result, many radio sources emit polarised radiation, in which the directions of vibration of the radio waves are not random. One of the causes of this polarisation of radio waves is the presence of magnetic fields. It is possible, by means of this kind of observation, to detect magnetic fields in the Universe from 10^8 down to 10^{-9} tesla.

In 40 years, radio astronomy has not only caught up with its elder cousin, optical astronomy, but has even overtaken it in some fields. It allows us to observe some of the most distant and the largest sources (radio sources with an extent of some millions of parsecs) as well as the smallest (neutron stars a few tens of kilometres in diameter). But the advent of radioastronomy has done much more; it has shown us how a new sphere of observations can be rich in discoveries, and in discoveries as fundamental as quasars, pulsars, and the comological black-body background radiation. Because of the nature of the data collected, computers have to be used to process the results but we can then see pictured what our eyes can never see directly.

The foundations of 'invisible' astronomy have been laid by radio astronomy. Astronomers and astrophysicists were ready to start the search for information in yet more regions of the spectrum, a hope that was realised, when instruments were able to operate above the Earth's atmosphere in the 1960s.

The quest for information wherever it may be found

The greater part of the information that comes to us from the stars is in the form of photons, of which the majority are absorbed by the Earth's atmosphere and never reach the ground. The arrival of the space age has progressively opened windows in the electromagnetic spectrum: the first observations in the X-ray region were made in 1963, in the infrared in 1965, in the ultraviolet from 1968, gamma radiation from 1972, the sub-millimetre region since 1973 and, finally, the extreme ultraviolet region from 1975. Now, all the windows are more or less open and the gaps that appear to exist between these regions arise because different techniques must be used in changing from one region to another.

However, in all the regions, it is always photons that are observed, and the same fundamental process must be undergone. It is necessary to try to collect as many photons as possible and then to focus them so as to detect them as efficiently as possible. The ultimate aim is to select the photons more and more precisely and so to read more accurately the positional and spectral information they carry.

Each region of the spectrum presents its own difficulties so the methods in some are more advanced technically than in others.

In the infrared and sub-millimetre regions the difficulties arise because everything on the Earth emits these wavelengths, simply because their temperature is of the order of 300 degrees above absolute zero. In particular, both the Earth's atmosphere and the mirrors of the telescopes emit radiation and the detectors are sensitive to all the radiation coming from the materials surrounding them. Because of this, the detectors are enclosed in chambers kept at a temperature of a few kelvin, which emit very little themselves and protect the detectors from radiation from their surroundings. Nevertheless, even these precautions are insufficient since a detector views space through the Earth's atmosphere and by means of reflections from telescope mirrors: it receives all their unwanted radiation directly. The ideal solution is to place such an observatory in space and to cool the mirrors to a few kelvin. The Infrared Astronomy Satellite (IRAS), which was operating for most of 1983, soon made a number of unexpected discoveries, including some comets! But while astronomers were waiting for this satellite, they found cunning ways to detect a faint source against a very bright background (as if you wished to observe the stars in full daylight). To do this, one mirror is vibrated rapidly so as to observe alternately the sky in the direction of the source and another nearby area of the sky. These oscillations, which resemble the involuntary movements of the human eye, make it possible to detect a faint source against a bright background. indeed, with a detector tuned to the frequency with which the mirror is oscillated, it is possible to measure just the modulated part of the signal to recover the required data; this technique is called synchronous demodulation.

A large infrared telescope, 0.91 metres in diameter (36 inches), was built by NASA to use this method of observation and has been used since 1975 mounted on a C-141 aircraft. When flying at an altitude of 12 kilometres it is above more than 99 per cent of the atmospheric water vapour, which is the principal cause of absorption at these wavelengths. This observatory is known as the Kuiper Airborne Observatory (KAO) after the planetary astronomer Gerard P. Kuiper. Many observations that are impossible from ground level can be made with it. An important discovery, the rings around Uranus, was shared by the KAO because of its ability to observe anywhere on Earth. An occultation of a star by the planet Uranus was observable mainly over the oceans, and at very few ground-based observatories; the dark rings around the planet were detected by their effect on the starlight.

Generally speaking, the principal observations made in the

The VLA

The VLA – the *Very Large Array* – is a radio-telescope which can map the sky with a resolution comparable with that obtainable by a large optical telescope.

The resolving power of a telescope is proportional to its aperture and inversely proportional to the wavelength of the observed radiation. As radio waves have wavelengths much longer than optical light, gigantic telescopes would be needed to obtain resolving powers comparable to those of optical telescopes. Fortunately it is unnecessary that the apertures of the telescope should be in the ratio of the wavelengths. Radio interferometers, made up of several linked antennae, are built precisely to observe radio sources at high resolution. Although first developed in the optical, it is in radioastronomy that interferometry has made its most important contributions. It is easy to transport radio signals by means of a waveguide, without losing their information content; these signals can then be easily handled.

The VLA is an interferometer employing the technique called aperture synthesis, which uses the Earth's rotation; its resolving power is equivalent to that of an antenna – clearly impossible to build – of 27 kilometres diameter. Its best angular resolution is of the order of a few tenths of an arc second. The observable wavelengths run from centimetres to decimetres.

The first studies for the VLA began in 1962–3. An interferometer with two antennae was built at Green Bank, in West Virginia, with the aim of testing various instruments and technologies. In 1967, the VLA project was submitted to the National Science Foundation, and in 1972 the US Congress authorised contruction of the instrument. The first antenna became operational in 1975 and the last one in 1979, although observations had been carried out since 1977. The instrument was completed in 1980 and officially inaugurated in 10 October 1980, thus becoming the third station of the National Radio Astronomy Observatory, a research organisation run by a private consortium of universities – Associated Universities Inc. – under contract to the National Science Foundation.

About 100 people are responsible for running and maintaining this scientific complex. Proper use of it requires close collaboration between observers and technicians. The astronomer prepares an observing programme, which he discusses with the operator; the instructions are then programmed into a computer, which then executes the programme automatically. The data acquisition system has two main parts. The first works in real time, controlling the the antennae and the data collection for the whole of an observation (1 min to 3 days). When the observation is finished, the data are sent to a second system, which analyses them and prepares a high-quality radio map.

Its flexibility and power make the VLA unique in resolution and sensitivity. Because of this it is vital to many areas of astrophysical research: planetology, stellar evolution, interstellar gas clouds, physics of radio sources, structure of the Universe. ... It has allowed us to discover the chemical composition of the giant planets and to examine solar flares in minute detail. The VLA is a vital instrument for the study of protostars and the remnants of novae and supernovae. The research carried out on active galactic nuclei and quasars is of great importance for understanding the physical processes by which these objects liberate vast quantities of energy, and for the possible detection of giant black holes.

Elisabeth VANGIONI-FLAM

The architecture of the VLA. The VLA has seventy-two observing stations distributed over three arms in the form of a near-perfect three-pointed star. The diagram at upper right shows the directions of these three arms with respect to geographical north, and the angles between them. The east and west arms are 21 kilometres long (the distance between the geometric centre and station number 72); the northern arm is shorter (18.9 kilometres): making it 21 kilometres long would have required construction work, and the terrrain is very difficult to the north of the site.

There are nine antennae on each arm. Twenty-four stations able to take these antennae are distributed along each arm (the antennae are moved to these stations using a complex vehicle which runs on rails). The distribution of the stations has been calculated so as to give the best quality observations.

The system can adopt four standard configurations. In configuration A the nine antennae are distributed over 21 kilometres; one may thus observe bright radio sources at high resolution, but with low throughput. Configuration D, the most compact, is confined within a circle of 600 metres radius; this allows observation of extended sources at low resolution but high sensitivity. Configurations B and C are intermediate (radii 6.4 and 1.95 kilometres respectively). It requires between one-and-a-half and five days to move from one configuration to another by moving the antennae. These movements can be used to give observations in non-standard configurations. The VLA repeats a given configuration every fifteen months or so. The observing stations are coded as follows: one letter denotes the arm (N, E or W), and is followed by a number equal to the product $\beta \times n$, where β is 8, 4, 2 and 1 for configurations A, B, C, and D, and n is the number of the antenna position at the station. Certain stations can be used in several configurations. For example station W16 can receive antenna 8 in configuration C (2×8), or antenna 4 in configuration B (4×4), or antenna 2 in configuration A (8×2).

The data collected by the antennae are channelled to the control centre by underground waveguides consisting of steel tubes covered on the inside by a mesh of copper wires. No amplifier is required between the antennae and the control centre. The general configuration of the instrument is not frozen; it is continually improved in response to developments in receiver technology and data handling, as well as the needs of the scientific community.

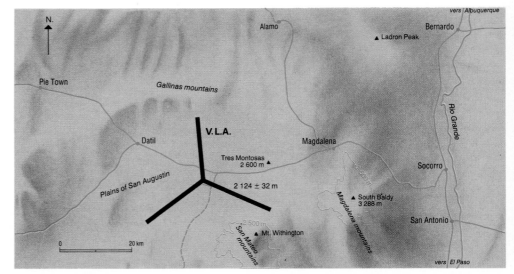

The site of the VLA. The construction of the VLA was preceded by painstaking studies, as the site had to satisfy a number of conditions. A huge, flat, isolated desert site was required to reduce radio interference. A relatively high altitude and a dry climate were required to reduce the effects of the atmosphere. Finally the latitude had to be low enough for the instrument to be able to observe as large a part of the celestial sphere as possible.

Eighty sites were originally considered: a minute examination of their characteristics reduced the number to three, which were then studied in detail. The VLA was finally built in New Mexico, 80 kilometres west of Socorro, in the San Augustin plains. These plains were the floor of an ancient lake which dried up several thousand years ago. The altitude of the centre of the VLA is 2124 metres, and the whole instrument lies between 2092 and 2156 metres altitude (2124 ± 32 metres). The mountains bordering the valley form an effective screen against radio interference. The climate is semi-desert: the annual rainfall is about 270 millimetres and the extreme temperatures are −23 and +35 °C.

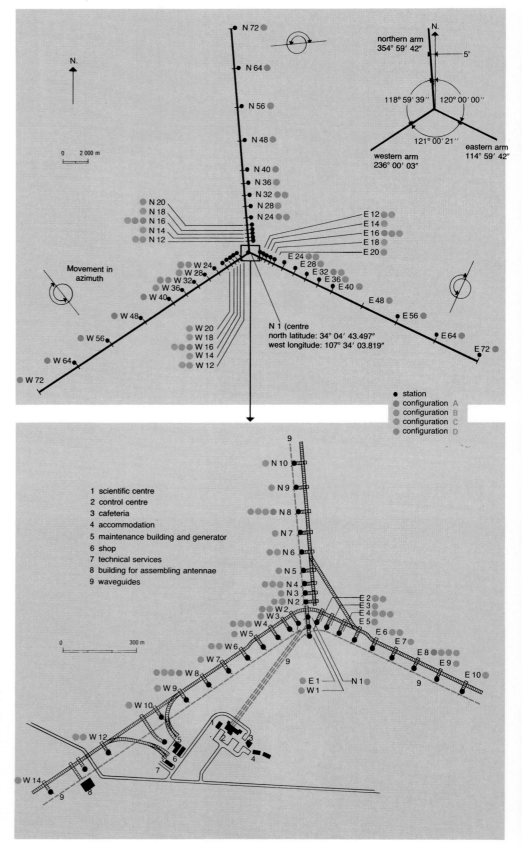

Aerial view of the VLA. The VLA is seen here in configuration C; the west arm stretches out to the horizon. The antennae were assembled in the large yellow building, which is now used for maintainance. Twenty-seven antennae at most are necessary for an observation, but twenty-eight were built, as each one is overhauled every three years while the VLA remains in continuous operation.

The railway lines carrying the antenna transport vehicle are visible. This machine weighs 73 tonnes and can carry 300 tonnes. A complex system of hydraulic jacks allows it both to pivot so as to reach the observing stations on sections of rail perpendicular to the main one, and to lift and place the antennae. (NRAO)

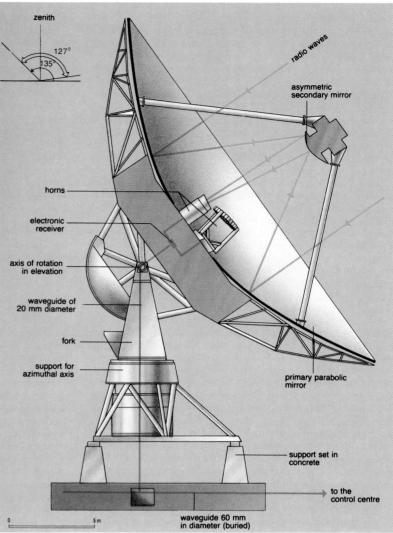

Structure of an antenna (figure opposite and photograph below). Each antenna, with a total weigth of 193 tonnes, has a horizontal mount allowing an azimuthal rotation of ±270 degrees; the elevation varies between 8 and 135 degrees. The antennae are of Cassegrain type. The aluminium primary mirror has a diameter of 25 metres – corresponding to a geometric aperture of 491 square metres – and has focal length 9 metres. Its structure is extremely rigid: the displacement between the actual and theoretical surface is of order 0.5 millimetres. The secondary mirror focusses the radio waves onto the detecting horns at the centre of the primary mirror. Changing the observing frequency is done by rotating the asymmetric secondary mirror about an axis aligned with that of the primary mirror; the focus moves along a circle of radius 97 centimetres along which various receivers are placed. Below the primary mirror a chamber maintained at constant temperature (±1 °C) contains a receiver which amplifies the radio waves by a factor of a million. The waves are then funnelled along waveguides to the principal waveguide. Another chamber, at the base of the antenna, contains the equipment required for pointing the antenna at the radio source, whose direction in the sky varies throughout the observation. (NRAO)

Data handling. The data collected during an observation are analysed by a computer which produces a radio map of the source.

This map can be in one of three forms. Using a printer or VDU it can be produced as a grey-scale map (centre) of contours of equal intensity (isophotes) (256 shades of grey are available), or mapped in a range of 64 false colours (below).

The observer can change the brightness or colours so as to bring out details of the map. The colour map is that of the radio galaxy 2354 + 471, situated in a poor cluster of galaxies. The red corresponds to maximum intensity, dark blue to minimum intensity. The various parts of the active galaxy are visible, including two jets indicating the presence of high-energy particles ejected from the nucleus. These jets feed two diffuse lobes.

The contrast between the narrowness of the jets and the wide lobes can be understood as the passage from a dense medium (interstellar gas) to a diffuse one (intergalactic gas). This observation was performed at a wavelength of 20 centimetres (1460 gigahertz) by J.O. Burns and S.A. Gregory, who used the VLA in a nonstandard configuration. The field is 3 minutes square and the angular resolution is around 2 arcseconds (NRAO).

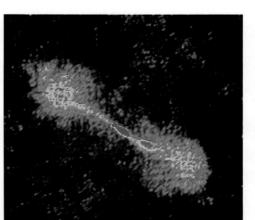

wavelength in cm	angular resolution ("): extent of observable radio source (")			
	configuration			
	A	B	C	D
400	25:720	75:2400	250:7800	800:18,000
90	6:170	17:540	56:1800	200:4200
20	1.4:38	3.9:120	12.5:420	44:900
6	0.4:10	1.2:36	3.9:120	14:300
3.6	0.24:7	0.7:20	2.3:60	8.4:180
2	0.14:4	0.4:12	1.2:40	3.9:90
1.3	08:2	111	0.9:25	2.8:60

The VLA's performance. Observations are made in four frequency domains; these contain atomic or molecular emission lines of particular importance for the study of the Universe, as for example the neutral hydrogen line at 1420 gigahertz (21.1 centimetres) or the water vapour line at 22.235 gigahertz (1.35 centimetres). The two main ways of using the VLA are observations in a given frequency band and spectroscopy; one can also study the polarisation of the incident radiation. It is possible to change from one band to another in a few seconds, or to observe in two bands simultaneously.

The angular resolution and throughput of observable radio sources depend on the frequency range and the chosen configuration. The angular resolution in the east–west direction is quasi-independent of the declination of the radio source. By contrast the resolution is strongly dependent on the declination in the north–south direction. For a declination of 30° south the resolution in the north–south direction is three times worse than in the east–west direction; for a declination of 40° south – the limit for the VLA – it is four times worse.

infrared region are: studies of the thermal emission from cool objects, studies of regions that are made opaque at shorter wavelengths by the presence of dust, such as the sites of star formation, analysis of the chemical composition of cold and dense clouds, research into complex organic molecules, and analysis of the 3-kelvin black-body radiation remaining from the early Universe.

Turning to shorter wavelengths, the ultraviolet has also become accessible through the use of satellites. It could be said that the optics and the detectors for this region are quasi-classical, the principal problem being to place the equipment in orbit and to make it work reliably there. As the wavelengths become shorter, the quality of the mirrors must be even better than for visible light, but they can be made comparatively easily up to the extreme ultraviolet region, at about 100 nanometres. Beyond this threshold, towards even shorter wavelengths, serious difficulties arise because all materials are opaque and mirrors reflect very inefficiently. This arises simply from the nature of matter, which strongly absorbs radiation of these wavelengths. It is also for this reason that the ultraviolet is especially interesting because almost all atoms or ions can be observed through their characteristic lines in this region of the spectrum. For example, atomic hydrogen can be detected at these wavelengths with a sensitivity about a million times as great as that needed to detect the 21-centimetre radiation in the radio region.

A number of early experiments were launched on sounding rockets. Subsequently two astronomical satellites have made their mark on observations in this spectral region. These are the satellite Copernicus, a telescope 80 centimetres in diameter that worked for more than eight years in orbit, and the International Ultraviolet Explorer (IUE) which is still working. As ultraviolet radiation is strongly absorbed by matter it is particularly suitable for studying the less dense regions of the universe, the intergalactic medium, stellar winds and the outer layers of stars, the upper atmospheres of planets and the interplanetary medium. In all of these regions, ultraviolet observations can be used to measure the atomic and ionic composition of the medium, and its temperature and physical state in the range of temperature from a few tens of degrees to about 10^6 kelvin.

At even shorter wavelengths is the X-ray region, which provides data on extremely hot media and places where very violent processes are taking place. The objects observed are quasars, pulsars, nuclei of galaxies, the intergalactic medium in clusters of galaxies and the remnants of supernovae. Many experiments had been flown, but the High Energy Astrophysical Observatory (HEAO-2, renamed Einstein) was the first true X-ray observatory. The technical difficulties stem mainly from the fact that all classical optical concepts are useless, X-rays being capable of penetrating all matter. Mirrors can only be used at grazing incidence, which results in the need for enormous instruments that collect very little radiation. For the same reason, spectroscopy is very difficult. X-ray photons can undergo diffraction by crystals but the resolution is relatively poor.

Gamma-rays are even more energetic than X-rays but it is impossible to deflect them in any type of optical system whatever. Photons are collected one by one and their direction of arrival is estimated from the direction of the electron–positron pairs that they create in the detector. This is the technique that has been used in the satellite Cos-B which, little by little, established a gamma-ray map of the sky. These photons are so scarce that it is often necessary to spend some months on one source before being able to make a relatively detailed map of it. Since April 1991 the Americans have been receiving data from the Compton Gamma Ray Observatory.

Gamma-ray astronomy is concerned with the most violent events in the Universe and it allows us to investigate quasars, pulsars and sources of X-rays up to the highest energy levels. Strangely, these observations also allow us to investigate the interstellar medium because, in collisions between high energy cosmic rays and atoms, gamma-ray photons are often emitted. Thus, the map of the sky in gamma-rays made by Cos-B shows the disk of our Galaxy very clearly through the presence of the diffuse interstellar medium.

By observing in space, we can also detect high energy cosmic-ray particles, which are stopped by the Earth's atmosphere. They have been studied with equipment quite unlike any optical instrumentation. These instruments can collect the cosmic

rays and sort them according to their mass, charge and energy. The results obtained show that some cosmic rays come from the Sun and others from the disk of the Galaxy. In the latter case it is impossible to determine the source of the particles but, by analysing their composition, it is possible to formulate hypotheses concerning the places of their formation.

Within the Solar System, it has proved possible to gather data on the spot using interplanetary probes. Development of these probes have produced a new type of astronomical instrumentation.

Even if the techniques of observation evolve, many of the basic methods remain unchanged. One of the important processes is that of image formation. Thus, very often there are small telescopes on these probes to form an image of the planet to be surveyed. One of the pioneers in this respect was the instrument that sent back the first picture of the far side of the Moon. A photograph was taken, developed on board, then analysed point by point. This data was transmitted to the ground rather in the way used sometimes by journalists to send a picture by telegraph but, very soon, this system was replaced by television cameras. Several planets have been observed from close quarters, and the Moon and Mars from their surfaces. Pictures of our own planet from space allow us to understand more clearly its structure and evolution.

A large number of devices have been developed to make more accurate observations from space probes. It would take too long to describe all of these so we will try only to distinguish the principal types of instruments devised for observations of the Solar System.

From probes designed to fly by, or go into orbit around, a planet, one can observe the exterior surface and examine the very low density media that may extend very far into space, such as belts of charged particles. It is possible to measure the magnetic field, or electric field (by measuring the voltage between two spheres), determine what neutral or charged particles are present (using mass spectrometers), detect the impact of micrometeorites (by studying the puncturing of gas-filled cells), map the elevation or topography of the surface (using radar and probe the composition of the surface (by the observation of induced emission of X-rays and gamma-rays); very tenuous atmospheres can be detected (by observation of ultraviolet emission) and the thermal balance and composition of atmospheres studied (by ultraviolet and infrared observations). The satellite itself can become an analysing instrument. Changes in its orbit reveal details in the gravitational field (for example, this is how the lunar 'mascons', regions where the lunar material appears to be denser, were discovered); braking of the satellite reveals the existence of an atmosphere, and studying this can lead to knowledge of the density and temperature of the upper atmosphere (on the Earth, models describing the upper atmosphere have been based for a long time on this type of data). It is interesting to note that the braking can be measured so precisely by accelerometers, in which the movements of a ball in equilibrium in the interior of the satellite are measured in relation to the satellite itself, that it is now possible to detect the acceleration of the satellite caused by the pressure of light from the Sun! Finally, the radio transmissions of a satellite can become a measuring instrument, especially when the satellite passes behind the atmosphere of a planet. By analysing the disturbances of the radio signals it is possible to determine the variation of the temperature of the atmosphere with altitude.

A great variety of instruments has been flown around or past planets; others have been landed on their surfaces. On planetary surfaces, some of the instruments used in orbit, such as cameras and magnetometers, may also be used but instruments, to measure the wind, the pressure and temperature of the atmosphere and vibrations of the surface are employed as well. Some equipment is relatively simple (for example scoops, drills and hammers); a charged sheet of aluminium foil was used to collect particles of the solar wind reaching the surface of the Moon. An example of more sophisticated apparatus is the gas chromatograph (a long, porous, very fine tube, through which a mixture of gas is passed), which can detect, in a mixture of gases, a component which is only one billionth (1 part in 10^9) of the principal gas. However, some would argue that human astronauts are more useful and versatile than instruments alone, with their unique ability to note, to sort and to select. The activities of

The Einstein Satellite. The Einstein Observatory is a satellite that carries an X-ray telescope. It weighs 3175 kilograms and is 6.7 metres long and 2.4 metres in diameter. This instrument has made it possible to investigate the hottest and most active regions and bodies in the Universe. (CFA)

One of the mirrors of the X-ray telescope in the Einstein Satellite. This 'cylinder' receives the X-rays at grazing incidence and focusses them on detectors placed on its axis at a distance of several metres. The successful manufacture of mirrors of this type is a considerable technological achievement. (CFA)

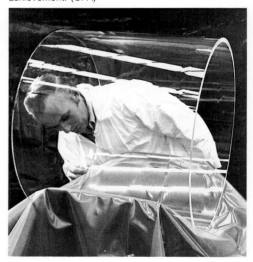

Aircraft flown by NASA and named the Kuiper Airborne Observatory (KAO). This aircraft carries an infrared observatory. When it is flying at 12 500 metres (41 000 feet) the aircraft is above 99 per cent of the atmospheric water vapour (the chief cause of infrared absorption). The telescope (aperture 0.91 metre) is pointed through the opening in the top of the fuselage. (NASA, Sygma)

Cos-B and astronomy

Gamma-ray astronomy is the best means of studying sites of large energy transfer associated with violent interactions between particles, with nuclear processes and extreme electromagnetic phenomena. These violent episodes are usually linked to the late stages of stellar evolution: supernovae, pulsars, neutron stars in close binary systems, and black holes.

The Cos-B satellite was a detector for radiation between 30 megaelectron-volts and 10 gigaelectron-volts. It was operational between 1975 and 1982. This mission, the first of its type in astronomy, produced a sky map unexpected in its details and its implications for astrophysics.

In 1969 several proposed scientific projects were examined. Two of them, Cos-A and Cos-B, were concerned with the radiation of the Universe. Cos-A was designed to observe X-rays and gamma rays, and Cos-B only gamma rays. After various consultations with the scientific community Cos-A was eliminated, and in May 1969, five European institutions and countries (European Space Agency, France, Germany, The Netherlands and Italy) decided to build it together.

The satellite was launched in 9 August 1975 from California. At that time a mission of two years was expected; it in fact lasted until the middle of 1982.

A gamma-ray detector must achieve two aims: first, to detect a photon and deduce its energy and direction, and second, to reject every incident charged particle. The second task is extremely difficult as cosmic rays, made up of protons, nuclei and electrons, are 10 thousand times more abundant than the desired photons. The telescope on board Cos-B is a good example of this type of detector. The development and calibration of the instrument were performed using large particle accelerators.

The mission programme mainly involved a survey of the Milky Way; a gamma-ray map of the Galactic plane was indeed obtained. A third of the observation time was used for observations at high Galactic latitude, particularly in regions populated by objects such as pulsars, quasars, and Seyfert galaxies.

The full data set confirms that on a large scale, most of gamma radiation comes from within our Galaxy. It is possible that this radiation results from the interaction of high-energy cosmic rays with interstellar matter.

On smaller scales, a detailed analysis of the data revealed the existence of point sources emitting gamma rays, which are associated either with interstellar clouds or pulsars (such as those of the Crab and Vela) along with certain extragalactic sources (such as the quasar 3C 273).

Gamma rays are the most energetic part of the electromagnetic spectrum. The data set from Cos-B has provided important information on the physical processes producing violent events in the Universe.

The success of Cos-B only emphasised the need for more observation in this domain. The European telescope SIGMA (Système d'Imagerie Gamma par Masque Aléatoire), launched in 1990 on board the Soviet GRANAT satellite, took over from Cos-B. It operates in the energy domain between 20 kiloelectron-volts and 2 megaelectron-volts. The use of a randomly coded mask gives it an angular resolution of 3 arc minutes. This mask projects a certain shadow pattern as a function of the position of the source in the sky; this image is analysed by 61 photomultipliers and then deconvolved mathematically. The field of view is 4 degrees by 4 degrees. Compared with Cos-B, the energy resolution is improved, making the detector a true imaging spectrometer. Its mission is to observed localised sources rather than diffuse gamma-ray emission. The gamma camera results from a collaboration of the Comissariat à l'Energie Atomique (Saclay) and the Centre

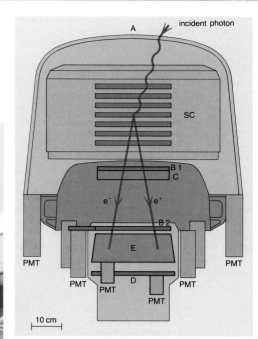

Cross-section of the Cos-B telescope. The main component of the gamma-ray detector is the spark chamber (SC). It contains gas (neon with a small percentage of ethane at a pressure of 2000 hectopascals) and 17 thin plates. The scintillator A is a sort of protector. When charged particles (cosmic rays) strike the counter, a system operates to prevent the spark chamber from working. An incident gamma photon penetrates the spark chamber and interacts with plates to convert itself into an electron–positron pair. The charged particles continue their tracks, ionising the gas, towards the detectors B1, C, and B2. Almost instantaneously the spark chamber is activated. A high voltage is applied between the plates, and sparks appear along the tracks ionised by the charged particles. These sparks reveal the trajectories of the charged particles, allowing one to deduce that of the incident photon. Finally a measurement of the energy of the electron, which represents almost all the energy of the incident photon, is performed by a calorimeter. This is composed of detectors. The first (E) is able to absorb totally electrons of 300 megaelectron-volts. Below this energy, electrons are detected by D. The various photomultipliers (PMT) associated with the detectors transform the photoelectrons into an electrical signal. The whole instrument gives an angular resolution of 2 degrees by 5 degrees.

Cos-B. This photograph shows the cylindrical shape of the satellite, 1.4 metres in diameter and 1.13 metres in height. The observing apparatus is in the centre of the cylinder. The electronics is mounted in the lower part so as not to reduce the field of view. The total weight is 278 kilograms, of which 119 kilograms is actually part of the experiment. A system of wires (in the base of the satellite) is used to point the satellite in the desired direction. The satellite is stabilised by rotation about its axis.

The orbit was chosen to allow long uninterrupted observations. The sky was observed for a month, at fixed orientation. The parts of the orbits corresponding to crossing the Van Allen belts reduced the observation time to 30 hours (for an orbital periof of 37 hours). If they had been in operation, the detectors could have been damaged by the powerful radiation emitted in this region. The data obtained by the satellite (about 25 hours per orbit) were collected by ground stations in Belgium and Alaska, then recorded on magnetic tape. These were then processed by the participants in the Cos-B project, the various scientific results being published jointly (ESA).

d'Etude Spatiale des Rayonnements (Toulouse). Its main objectives are the observation of Galactic sources, with priority for those in the Galactic centre, and, above all, extragalactic sources (nearby galaxies, active galactic nuclei, quasars), which are active in the spectral range covered by SIGMA. A map of the Galactic centre has been made. A single gamma-ray source dominates: at 45 arc minutes from the centre itself, this strange object intrigues more than theoreticians.

Cos-B and SIGMA have inaugurated the era of European gamma-ray astronomy. NASA has launched a large space observatory, the Gamma Ray Observatory (GRO), renamed Compton, placed in orbit by the space shuttle *Atlantis* on 7 April 1991, which has four large instruments studying a very large range of gamma-ray frequencies. The sensitivity of the spark chamber is ten times higher than that of Cos-B. Several research institutes which worked on Cos-B are participating in the American experiments.

Elisabeth VANGIONI-FLAM

Observation of the Galactic plane in gamma rays. This map shows on the one hand the region of sky observed by Cos-B, and on the other hand the twenty-five point sources with energies above 100 megaelectron-volts it detected. The map is given in galactic coordinates (l = longitude, b = latitude). Note that the observations were principally made along the Galactic disc (latitudes between +20 degrees and −20 degrees). The concentration of sources along the disc is noteworthy, and very significant (detecting objects at high latitude is easier because the background radiation is lower). The distribution shows that most gamma-ray sources are within our Galaxy, and the distribution in longitude suggests a concentration towards the central regions. Because of the mediocre angular resolution of the instrument it is very difficult to identify them securely with known objects, as several candidates appear possible. The SIGMA satellite was designed to get round this shortcoming. Two Galactic objects giving an unmistakable signature are the Vela (1) and Crab (2) pulsars, while another source has been associated with a nearby cloud complex known as Rho Ophiuchi (4). Finally, an extragalactic objects was identified in the quasar 3C 273 (3). This is the most distant source currently known (about 2.5 billion light years). The colour of the sources is related to the intensity of the gamma-ray emission. The Vela pulsar is the strongest source observed at this wavelength.

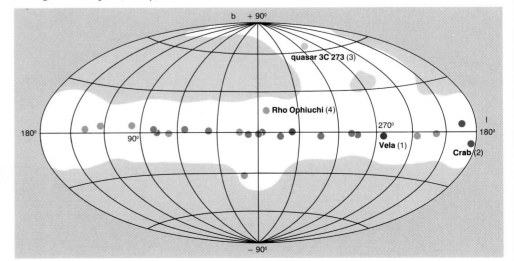

astronauts certainly results in an enormous variety of data being returned from the Moon.

Automatic instruments and astronauts both have their advantages and disadvantages. Trends in instrumentation are towards more sophisticated automatic and independent systems. Developing successful instruments is fraught with many difficulties. Not only are they necessarily complex but they must work for a long time (flight durations of several years have been achieved), sometimes in extreme conditions (the Venera probes, after surviving the interplanetary vacuum, had to work at 500 °C and under a pressure of 90 atmospheres. Though difficult these objectives have been achieved. It is predicted that the Viking spacecraft, which landed on Mars, will continue to send us pictures until 1994 (which will be 20 years of operation) and the Pioneer and Voyager probes will send us information from the boundaries of the Solar System probably for more than 10 years after their launch. Space exploration allows us to retrieve unique data, (impossible to obtain from the Earth) which may completely transform our understanding of the astronomical object studied: the close-up view of Saturn's rings is a striking example.

Less classical instrumentation

There is one kind of astronomical data that falls from the sky – in the form of meteorites. At the start of the last century their origin was unknown and the idea that they could come from the sky seemed strange and improbable. Nevertheless, these stones bring us data about the formation of the Solar System. To recover the information contained in meteorites a sophisticated laboratory capable of analysing their chemical composition is required. Some of the material could have had an interstellar origin; meteorites may be an important source of information about a medium that will certainly remain inaccessible directly for a very long time. However, the origin of these stones is not known for certain. As a result of future missions to comets or asteroids their origin may be clearly identified.

Cosmic rays (high energy atomic particles) present the same problem since it is not easy to discover their source. When they arrive in the upper atmosphere, they collide with the atoms or molecules and generate a shower of secondary particles which, after many collisions, finally reach the ground. They were discovered in 1912 by Victor Franz Hess during a balloon flight, but it has been necessary to wait for the use of large charged particle accelerators to be able to analyse them correctly. However, these machines can only generate collisions with energies up to 10^{15} electronvolts while the energy of some cosmic rays can reach 10^{20} electronvolts, which corresponds to a particle possessing as much energy as a well-served tennis ball! Such particles are extremely rare, though and the whole of the energy that they bring to the Earth is no more than the whole of the energy that reaches us as star light. Thus, for example, a particle with energy greater than 10^{15} electronvolts is received only once per hour on a surface of 100 square metres. Pierre Auger and his collaborators succeeded in detecting such cosmic rays by means of Geiger counters only, by using as the detector the atmosphere itself, which produces a multitude of secondary particles, giving a signal from the primary cosmic ray that is very well spread out and easy to detect.

Showers are now observed when they reach the ground by means of arrays of detectors distributed over several square kilometres. The base of a shower is generally spread over a circle with a radius of a kilometre. The largest of these arrays of

detectors are situated in New Mexico, Australia, and the USSR.

The detectors themselves have evolved substantially; currently they consist of large tanks filled with very pure water. The particles emit a measurable amount of light as they pass through the water faster than the local velocity of light. The number of showers and the energy of the primary particles can be determined. One more item of information may possibly be recovered for very high energy particles (10^{20} electronvolts): that is the direction of arrival, which, as in the case of photons, indicates the position of the source in the sky. The interval between the times of arrival of a shower at different parts of the array of detectors is used to derive the direction of the primary cosmic ray. The results have shown that the most energetic cosmic rays do not come from within our own Galaxy, though their source is still unknown. A new and more powerful detection system is in the course of construction in the USA, which will use the principle of measuring the light emitted by the shower in its passage through the atmosphere. This system will be ten times as efficient as the older arrays. The detector acts rather like the eye of an insect (it has been called the 'fly's eye'): a large number of mirrors with very sensitive photomultipliers monitor all directions of the sky. When a shower occurs, the flash of light is recorded by a succession of these detectors, which makes it possible to determine the position of the shower. The direction of arrival can be found from the intervals between the arrival times of the flash on the different detectors.

In this instrument, the telescope objective is replaced by the atmosphere, which collects the information, while a 'fly's eye' acts as detector.

There is yet another range of instruments that have had such an impact on our understanding of the universe that it would be a grave omission to ignore them. Computers are indisputably important astronomical instruments. Without them, a number of discoveries would never have been possible. Although John Couch Adams and Urbain Jean Joseph Le Verrier independently predicted the position of Neptune without computers (but with some good luck, for at least one of them made some numerical errors), this achievement was probably at the limit of what could be done by human calculation. The discovery of Uranus' rings was made because a computer was used to predict the date and place where an occultation of a star by Uranus would be visible.

Computers are, of course, essential for space missions; a computer guides the rocket, corrects the orbit of the probe, organises a rendezvous in space and controls the state of all the instruments to ensure the complete success of a mission; without computers on board, interplanetary probes would not be able to land because instructions sent from the Earth would take tens of minutes, if not hours, to reach the probe and an hour in the course of a landing is certainly too much. The computer's capacity for work is utilised on Earth, too. The quantity of data collected by all the observatories is such that in future it will be impossible to analyse the data without selecting and processing them with the help of large computers. Radio astronomy with an interferometer is an example where the computer plays an indispensible role. It would be impossible to make high resolution radio maps of the sky without it. Gamma-ray astronomy also offers a good example: only one event in a hundred is significant and should be recorded; who would ever do the unrewarding work of sorting if it were not the computer?

Now used in all parts of astronomy, the computer not only carries out tasks such as pointing the telescope, it also helps in the understanding of the phenomena that are observed. Modern theoretical models in astrophysics could not have been developed without computers All the models, from the best known, such as the Big Bang or that of stellar evolution, to the least well-known, such as those that describe physico-chemical equilibria in

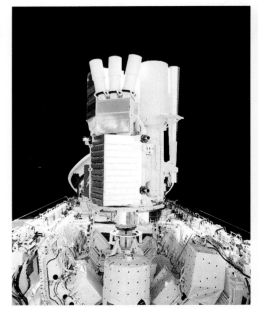

The Astro-1 payload on the space shuttle *Columbia*. The Astro-1 mission took place from 2–11 December 1990 and acquired 820 ultraviolet and X-ray images of 66 different objects; it is impossible to observe at these wavelengths from the ground.
Astro-1 had four instruments: the Hopkins Ultraviolet Telescope, the Wisconsin Ultraviolet Photopolarimetry Experiment and the Broad Band X-ray Telescope. (NASA)

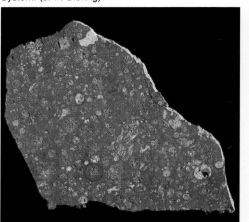

A piece of the Allende meteorite which fell on 8 February 1969. This is a carbonaceous chondrite and it contains a great wealth of information on its history. This stone is one piece of the jig-saw puzzle that, one day, will reveal the origin of the Solar System. (J.-P. Bibring)

The 'fly's eye' cosmic-ray observatory built at Little Granite Mountain in Utah, USA. It consists of a large number of mirrors, each 1.5 metres in diameter. Mounted at the bottom of a metal drum, each mirror examines a part of the sky. When a cosmic ray releases a shower of particles and light near the observatory, the light is detected and the direction of arrival and the energy of the cosmic ray are reconstructed. This observatory will study high energy cosmic rays whose origins are still unknown. (Photograph kindly provided by G. L. Cassiday, University of Utah)

Arguably the most sophisticated astronomical instrument to be sent into space: Man. On the Moon, he carried out essential tasks such as searching for and selecting specimens of as great a variety as possible, on or under rocks, etc. (NASA)

Cerro Tololo Observatory in Chile. Optical telescopes can sometimes be used, like radio telescopes, in an active mode (equivalent to radar). A very powerful laser beam can be transmitted and then received back at the observatory after bouncing off reflectors placed on the Moon. These observations have enabled us to measure continental drift as well as to test the theory of general relativity. (D. Kirkland, Contact, Cosmos)

planetary atmospheres, or on the surfaces of grains of interstellar dust, represent hours and hours of calculations on the most powerful computers. One of the greatest uses is made in the preparation of dynamical meteorological forecasts.

If all the experimental equipment used to measure physical quantities essential to astrophysical models were mentioned here, the list would be very long. Thus, for example, particle accelerators give the products of collisions between atoms at high energy, which ultimately will permit a better understanding of the evolution of stars. In laboratories, experiments are appearing which, by simulation, provide very valuable information on the phenomena we would like to understanding in the astrophysical context. Hot or cold vacuum chambers, high pressure vessels, plasma retorts, have all been built with this aim. In some experiments, the conditions for the appearance of life have been simulated in glass vessels in which electric discharges occur and the contents are illuminated with ultraviolet light and bombarded by particles. All of these experiments, sometimes very specialised, in some sense form part of the whole of astrophysical instrumentation, since they undeniably contribute to the progress of that science.

Nevertheless, if physics brings much to astrophysics, the inverse is also true. An experiment that is impossible in the laboratory sometimes exists ready set up in an astrophysical situation where the state and evolution of matter under the most extreme conditions of density, temperature and pressure can be observed. This close interaction between physics and astrophysics finds its culmination in the theory of relativity. As one of the conceptual foundations of modern astrophysics the theory of relativity was too important to remain untested, so we include here a description of the equipment that has been put into operation to verify this theory, and so may be said to contribute significantly to modern astronomical instrumentation.

One of the predictions of the theory of general relativity is the slowing of clocks in a gravitational field, also manifested in the gravitational red-shift. In 1965, an experiment mounted between the top and bottom of a tower 25 metres high showed a small shift of the frequency of gamma-ray photons emitted in the disintegration of a radioactive isotope of iron, Fe^{57}. The experiment confirmed the theoretical prediction to almost 1 part in 100. In 1977 Carroll O. Alley took some atomic clocks in an aircraft up to an altitude of 10 kilometres. The retardation of clocks that remained on the ground, and were thus in a stronger gravitational field than those on the plane, again confirmed the theory to about 1 part in 100. In 1977 Vessot and Levine carried out the most convincing experiment. They flew an atomic clock in a rocket to an altitude of more than 10 000 kilometres and were able to confirm the theory with an accuracy better than 1 part in 10 000.

Another prediction of the theory of general relativity is the deviation of light in a gravitational field. Observations of the displacement of the apparent positions of stars during total eclipses of the Sun were carried out between 1919 and 1976. Unfortunately, the displacement cannot be measured very accurately but the observations confirm the theory to within 10 per cent. This test has become much more accurate thanks to the method of radio interferometry, the VLBI method giving a much more accurate position of the source observed. After the data had been recorded on nearly 10 000 kilometres of magnetic tape, the analysis of the results by a computer required more than five months to confirm the prediction to nearly 1 part in 100.

A third prediction of the theory is that the time taken by a light or radio signal to travel between two points should increase if a massive body comes close to the path taken by that signal. Two planets situated on opposite sides of the Sun can be used in this experiment. The first measurements were made by radar echoes, but soon the tracking of probes throughout the Solar System was made more acccurate. The most spectacular result was obtained in 1976 with the Viking probes landed on Mars. The time for transmission and return of a radio signal could be measured so accurately that it was possible to determine the distance between a point on the Earth and a point of Mars with an uncertainty of only about 2 metres! The result obtained, however, could only confirm the prediction of the theory within 2 parts in 1000 because of disturbances to the propagation time of the signal caused by the interplanetary plasma. The result is nonetheless remarkable.

Another prediction concerns the precession of the periastron of one star orbiting in the gravitational field of a second star. In 1859, Leverrier noted a significant difference between the observed and predicted positions of Mercury. He determined the displacement of the perihelion of Mercury as 36 arc seconds per century. The value recently determined from two hundred years of observations is 43 arc seconds per century, with an uncertainty of about 0.4 arc seconds. In 10 years, radar measurements have increased their accuracy to 0.2 arc seconds, but there is a problem in the interpretation of this displacement in terms of relativity. It could be explained equally well by the fact that the Sun is not exactly spherical, especially if the interior is rotating faster than the solar surface. The recently-discovered binary pulsar, could in principle provide a test of this effect. The two pulsars are so close to each other that advance of the periastron is 4.2 degrees per year, or 35 000 times as fast as in the case of Mercury. However, too little is known of this system for it to be used to test the theory of general relativity rigorously; on the contrary one can understand the system better by using the theory.

Finally, another feature of the theory of general relativity is that the constant of gravitation should indeed be constant and not change in the course of time as was suggested by P. A. M. Dirac in 1938 in his own theory of gravitation. This change, if it exists, could be observed from its effect on the distances of the planets. By using radar echoes on the planets, it can be shown that the constant of gravitation does not vary by more than 1 part in 10^{10} per year.

Using reflectors left on the Moon by different Apollo craft, it has been possible to send a laser beam from an observatory on Earth and obtain a reflection; although the intensity of the reflection is no more than a few photons, this technical feat has been successful. The distance from Earth to Moon has been measured in this way to within a few centimetres. This is accurate enough to say that the constant of gravitation does not vary by more than a few parts in 10^{11} per year. This result cannot be improved significantly because the loss of matter from the Sun through the solar wind would give rise to a change in the Earth–Moon distance of about that order of magnitude.

These measurements on the moon are so convincing that they allow us to reject the Brans–Dicke theory of gravitation, which gives predictions differing by more than a metre from those obtained by observation.

Thus, in a general way, all the observations confirm Einstein's theory of general relativity but do not always cause us to reject other theories, simply because these, having more free parameters, can always be adjusted to fit the observations. The simplest theory that remains in agreement with all the observations is that of general relativity.

Observation of the invisible

We shall see how it is possible to 'observe' regions or events that appear to be totally beyond any direct observation.

The first of these inaccessible regions is the past. Is an event that which has long since taken place definitely lost to observation? Not always. In particularly, there are traces on the Earth of its own past. Some records of the recent past can obviously be recovered from written history, but it is the much more distant past that concerns us here. Let us take some examples: the change in the length of the day can be found imprinted in some fossilised shells, likewise, the variations of the solar cycle produce, through the collisions of solar wind particles with atoms in the upper atmosphere, more or less of the carbon isotope carbon-14 in the Earth's atmosphere. This carbon is absorbed by trees and enters the trunk and branches, so it is possible to follow changes in the solar cycle through the years as far back as some tens of thousands of years. Thus it has been found that the Sun has experienced periods of inactivity which appear to coincide with very distinct cold periods on the Earth, the latest of which corresponded with the reign of the French King Louis XIV, the Sun King!

More surprising yet are the traces of very rare elements found in layers of sedimentary rocks on the scale of the whole planet. Iridium is found concentrated in a layer scarcely a centimetre thick in the middle of sedimentary rocks hundreds of metres thick. This could be explained by the fall of a meteorite or a comet about 70 million years ago, which would have released a catastrophe on a global scale: giant waves reaching 8 kilometres in height may have ravaged the continents, destroying almost all the larger living species. This provides a plausible explanation of the end of the dinosaurs.

Other difficult regions to observe are the interiors of celestial bodies. Yet, even there, some methods exist. In the case of planets, seismology, either passive, or active (for example, recording the effects of the impact of a Lunar Module when it was deliberately crashed on the Moon allows us to probe their interiors. To be able to observe its interior directly it would be necessary to break up a planet. This may, in fact, have already occurred. Some asteroids may be fragments of a larger object, thus revealing its interior to full view. To study them, though, it would be necessary to go to the asteroid belt.

It is possible to probe the interior of a star by observing its natural vibrations; this has already been done in the case of the Sun, using instruments able to record movements of its surface with a velocity of a few millimetres per second! The results appear to show that the interior is much as predicted.

There is still another way to examine the interiors of stars. Neutrinos are subatomic particles that interact so rarely with matter that they can pass straight through a star. In the case of the Sun, theory predicts a certain rate of production of neutrinos in the course of the thermonuclear reactions taking place in its centre. These neutrinos should, therefore, be detectable on Earth, though this is not done easily, as the great majority of them pass through the Earth without difficulty. R. Davis Jr has, nevertheless, measured the flux of neutrinos with an instrument made of an enormous tank filled with 400 000 litres of tetrachloroethylene, which contains many chlorine atoms. This tank is installed at the bottom of a disused gold mine; only neutrinos are able to reach it. Once a day, on average, a chlorine atom is changed to argon by a collision with a solar neutrino. After several months, the atoms of argon thus produced in the tank can be flushed out and their number estimated from the rate at which they decay radioactively; the flux of neutrinos through the tank can then be estimated. A serious problem is apparent in the case of the Sun, the observed flux being about three times smaller than that which has been predicted.

Neutrino astronomy is only in its infancy, and it will be

necessary to wait for larger neutrino observatories before probing the interiors of other celestial bodies where powerful thermonuclear reactions are in progress. The problem of determining the direction of a neutrino remains; R. Davis was obliged to assume that all of those he detected had come from the Sun.

However, some regions of the universe are so dense that they remain opaque even to neutrinos! Is there any way to investigate these regions? There is in fact one way, using gravitational waves.

The theory of general relativity predicts the emission of gravitational waves when coherent movements of large masses occur near the speed of light. Thus, for example, the nucleus of a supernova during its explosion, or matter close to a black hole could be sources of gravitational waves. To detect these waves, at least two particles are required, in different places. The particles are not displaced in exactly the same way by the passage of a gravitational wave; their separation changes. It is therefore sufficient to measure their separation to detect the passage of a gravitational wave. This is, however, not an easy task for the expected effects are incredibly small. Let us take as an example the source that is supposed to be the most powerful emitter of gravitational waves: a collision between two giant black holes occurring in a distant galaxy or quasar. This source would produce about 10 joules per square metre at the Earth. An aluminium block weighing a tonne would be set into vibration by such a wave, but the movements of its ends would be less than the diameter of an atomic nucleus!

Incredible as it seems, Joseph Weber of the University of Maryland has been developing a gravitational wave detector since 1960. This consists of a bar of aluminium weighing 1400 kilograms, 1.53 metres in length and 0.66 metres in diameter. The difficulty is to isolate the bar from all possible sources of noise. This has been done, however. Weber had then to face another obstacle inherent in the system: vibrations produced by the thermal agitation of the molecules of the bar itself! He overcame this difficulty by making two identical bars and mounting them several hundred kilometres apart, one near Washington, the other near Chicago. Then, examining the displacements of the bars, he looked for coincidences. In fact, only gravitational waves passing through the Earth at the speed of light would seem to be able to cause a simultaneous signal in both bars. Since 1967, coincident events have been recorded. It was then that a long period of difficulty began for Weber; theoreticians considered that his detector was not sensitive enough to record waves from foreseeable cosmic events; at the same time other experimenters threw themselves into this type of observation, but only the Munich–Frascati group was able to record such events. Since then, the situation has not changed. However, by refining his measurements even more, Weber again found, in 1974, some very clear coincidences between his two detectors. One possible explanation is a low-frequency electromagnetic disturbance, which could certainly propagate over long distances and could cause a response in both highly sensitive detection systems, but it is only a hypothesis.

Weber has, moreover, discovered anisotropy linked to sidereal time by analysing about 150 recorded events. The Earth turns once relative to the stars in a sidereal day (23 h 56 m 4 s solar time), and no terrestrial activity can explain the observed correlation. On the other hand, a distant source outside the Solar System, would account for it. Furthermore, Weber has found that the increase and decrease of events occurs twice in each sidereal day. This is exactly what one would expect: the bar is most sensitive when it is perpendicular to the source, which occurs twice in each sidereal day as it is carried round by the Earth's rotation.

Impressive as this result is, it still has not been confirmed by other experimenters, who are using more sensitive antennae, made of very pure materials and cooled to near absolute zero temperature in order to minimise the background noise. Although Weber's results are not confirmed, a number of other researchers are making a start in the detection of gravitational waves.

The future

Before concluding, we will take a brief look at the future of astronomical instrumentation.

For ground-based optical and infrared astronomy, two main courses of action seem obvious: the building of very large light collectors, and of long base-line interferometers. But why, one may ask, continue to observe from the ground, since the atmosphere puts a serious constraint on resolution and, furthermore, emits light? The reasons are financial. The cost of a ground-based telescope with an aperture of 15 metres would be only one-tenth of the cost of the Space Telescope which, it is hoped, will be launched in 1986, and would collect forty times as much light. The need for light is so great, and the financial arguments so important that ground-based observations are the most realistic short-term solution. Already, a telescope has been built in Arizona that may become a prototype for giants of the future, called the Multiple Mirror Telescope (MMT). It is difficult and certainly very expensive to build ever larger telescopes. It is easier to join several together. The MMT, with its six mirrors each 1.8 metres in diameter, has demonstrated that the method works. The principal difficulty is that of controlling the positions of the mirrors with respect to each other throughout

The Gamma Ray Observatory. On 5 April 1991 one of the heaviest scientific satellites ever launched (17.5 tons) was placed in orbit by the shuttle *Atlantis*. This space observatory, renamed Compton, is devoted entirely to astronomy. It contains two imaging telescopes operating at energies between 1 and 30 megaelectron-volts. (NASA)

an exposure, in spite of all the vibrations, atmospheric turbulence and thermal distortions. This feat has been achieved, originally using laser beams, but these are also affected by atmospheric turbulence and later it was found that if the mirrors were aligned by observing a comparatively bright star near the faint object, the mirrors would stay in alignment for 15 to 30 minutes while the fainter object was observed. A multiple mirror telescope with an aperture equivalent to 15 metres, is under consideration in the USA. With such a giant, the spatial resolution and the sensitivity will be sufficient to probe the Universe five times as far out as it is known today and, probably, to detect planetary systems around stars near our Solar System.

In radioastronomy, the problem is the same as that in optical astronomy. The only way to collect more energy is to increase the number of antennae. The Very Large Array (VLA) in New Mexico has twenty-seven mobile antennae each 25 metres in diameter, together equivalent to a radio telescope over 20 kilometres in diameter. Only arrays designed to communicate with extraterrestrial civilisations will be larger in size. In radio interferometry, the limits imposed by the size of the Earth have already been reached by the VLBI system. Only radio interferometry carried out with one radio telescope on Earth and another in orbit in the Solar System will bring any marked increase in resolution. Apart from the problem of placing a radio telescope in an orbit that takes it far from the Earth, such a project is feasible with modern technology since we know already how to measure the distance between two objects in the Solar System with an error of only about a metre, the accuracy that would be needed for interferometry. The improvement in resolution with such a system could be of the order of a million, which would allow us to see radio signals of the strength emitted by our civilization through the whole Galaxy.

Even larger and heavier instruments may be placed in orbit, either with the Ariane rocket or the Space Shuttle. The Salyut flights suggest that habitable space stations may soon be in orbit. That would make it possible to assemble even larger instruments. All the non-optical telescopes launched into space until the present time have suffered, above all, from the problem of inadequate size. There are projects to launch large telescopes for all regions of the electromagnetic spectrum, from the infrared to gamma rays, including X-rays and ultraviolet. Some instruments tackle the problem of improving performance in ways other than making the telescope as large as possible.

As an example we mention Sigma, the gamma-ray observatory of the Centre National d'Etudes Spatiales, launched in 1989, whose primary aim was to improve the angular resolution; this had never been better than a degree in this region. No focussing system can be used, so the plan is to place a coded grid far in front of a detector. A point source could immediately be recognised because it would reproduce the pattern of the grid on the detector. A resolution of a minute of arc could be reached, which would allow the identification of many of these sources.

Another example is the European Space Agency's astrometric satellite Hipparcos, also launched in 1989, will enable us to measure the distance of a hundred thousand stars, whereas at present the distances of only a few hundred of them are known accurately. This instrument will use a double mirror, which will simultaneously reflect onto a grid the images of two widely separated parts of the sky. Since the satellite will rotate at a known speed, by measuring the time interval between the passages of two stars in front of the grid, the angle between the two objects can be measured with great accuracy. Some objects are very distant, so it will be possible to determine the parallax of

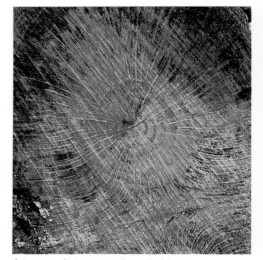

A tree trunk can sometimes hold an astronomical record of the past. Solar activity causes variations in the percentage of carbon-14 in the atmosphere. Measurement of the percentage of carbon-14 a tree's annual growth rings reveals past variations in solar activity. (R. Tixador, Top)

Supernova 1987A: when a star itself becomes an astrophysical instrument. This explosion of a star of 20 solar masses revealed its core collapse in a burst of neutrinos detected at the Earth. The star also behaves as a giant cosmic lighthouse: the light flash it produced, which lasted about a year, illuminated interstellar dust grains which in turn reflected this light. Two 'echos' are visible on this negative image (taken on 13 February 1988, a year after the explosion, observed on 13 February 1988). They are almost circular, centred on the supernova remnant. As they are only a mirage, it is not at all paradoxical that they are seen to move at twenty-five × the speed of light. (ESO)

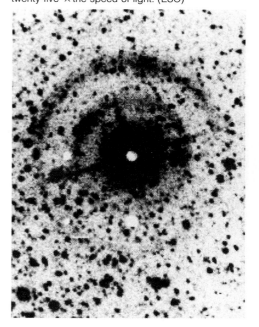

others that are nearer by comparison with these. The results may change our perception of the distance scale of the Universe itself, since at present it is only based on indirect measurement.

So as to harmonise its long-term projects with those of the other large space agencies, the European Space Agency (ESA) has defined a long-term programme called Horizon 2000. This programme contains large projects called 'cornerstones', designed to give Europe the possibility of breakthroughs in certain important areas of astrophysics. Thus, as the first cornerstone, Soho–Cluster, which is already decided upon, will involve a solar satellite Soho (Solar and Heliospheric Observatory), placed at the Lagrange point (i.e. in permanent equilibrium between the Sun and the Earth, at a million kilometres from the latter), one of whose main aims will be to probe the Sun's interior by seismology. The other component, Cluster, will be formed from several satellites in orbits in the Earth's magnetosphere, arranged so that they can observe phenomena three-dimensionally. Later cornerstones will be an X-ray satellite, also decided upon, another in infrared astronomy, and a cometary sample mission. Among these cornerstones there will be other more modest missions; after the remarkable success of the infrared survey satellite IRAS, which provided a new view of the sky, one of these will be an infrared satellite for pointed observations, ISO, involving a telescope and instruments cooled to very low temperature.

There are also more long-term projects for orbital platforms and to build very large telescopes with collecting areas reaching 100 metres in diameter, and for interferometer systems up to 30 metres. Some plans for interferometers propose flying several separate telescopes separated by even larger distances. The solar radiation pressure could be used to control the distance between the telescopes.

There are proposals to detect gravitational waves in space by a principle very different from that used by Weber. The masses, whose separations are measured accurately, are formed of mirrors mounted at the ends of long arms, their separations being controlled by an optical interferometer system using the two mirrors. The effects being sought are so small that they are in danger of being disturbed by the noise introduced by the photons when they are reflected from the mirrors.

As far as Solar System astronomy is concerned, there are several projects to visit various objects. For example, some probes have been planned to meet Halley's comet in 1986. There are also projects for flybys of some of the small bodies in the belt of asteroids, for balloons floating in the atmosphere for Venus, for a probe to fall into the atmosphere of Jupiter, for vehicles on Mars or even balloons rolling on the surface of Mars, solar observatories especially to study oscillations of the Sun and, in the more distant future, a module to descend into the atmosphere of Titan, a satellite of Saturn that could contain a primitive form of life. Nearer home, there are proposals with the aim of measuring our own environment ever more accurately. In particular, there are proposals for swarms of satellites to study the interaction of the solar wind with the magnetosphere, for lasers in orbit to cause fluorescence in the atmosphere in order to understand its structure more thoroughly, for large radar satellites to monitor the positions of icebergs, for ever more refined meteorological satellites. This list is obviously not complete but it shows that scientists are not running short of ideas.

Finally we return to the Earth, to conclude this survey of the future with a project for a neutrino observatory which is remarkable both for its originality and for the impact it could have on our understanding of the Universe. Neutrinos are produced in collisions between highly energetic protons and other matter. The sites of these collisions are sometimes observable but are more often hidden from us by large quantities of matter, and might remain unsuspected for a long time. Let us take, as an example, the strange object SS433, recently discovered, which emits a thousand times as much energy as the brightest star in our Galaxy. Its discovery was delayed because it is hidden by the matter which, by falling onto it, releases this energy. To see directly what is happening in this object, one is obliged to turn to neutrinos. These interact very weakly with matter (more than 10^{14} of them pass through our bodies every second) so vast quantities of matter are required to detect any of them. In the Dumand project, planned by a consortium of American physicists, it is proposed to use as the detector a cube of sea water with sides 1 kilometre in length! The passage of a neutrino that interacted with the matter in this cube would be detected either by a flash of light produced by the shower of particles that it would produce, or by a sound wave, if it had enough energy. The proposal consists of installing photon detectors and microphones every 50 metres along vertical cables spaced at 50 metres from each other, all at more than 5 kilometres' depth in the ocean, which is deeper than the showers of particles from cosmic rays can reach. The whole volume of the 1-kilometre cube will thus be filled with detectors able to record the passage of a neutrino that has suffered an interaction and to determine its direction of arrival to about half a degree. The importance of this form of telescope is that it would be able to operate continuously and would be sensitive to neutrinos coming from any direction, even from below, i.e. those that have passed through the whole Earth.

This project is not a Utopian dream. Some tests have already been carried out in the sea and the site of the observatory has been chosen, on the sea bed near the Hawaiian islands. Its cost is equivalent to that of a modern large telescope, so it is a real possibility.

One of the first of a new generation of large ground-based telescopes: the Multiple Mirror Telescope (MMT). Installed on Mount Hopkins in Arizona, it brings the light collected by six mirrors, each 1.8 metres in diameter, to a common focus. (Smithsonian Observatory)

Astronomical instrumentation, with an extraordinary variety of forms, allows us to extend our sensory abilities. However, let us not forget that, in the last resort, it is with our own senses that we receive information, and that it is our brains that process it, with all their analytical power but with weaknesses as well. The desire for aesthetic and intellectual satisfaction drives us to seek ever further discoveries.

Martin Harwit, Professor of Astronomy at Cornell University, has tried to analyse, in a statistical manner, the progression of discoveries in astronomy. With this aim, he has made a list of forty-three great discoveries in date order. Straight away the spectacular increase in the number of discoveries made in the 1960s can be seen. Other conclusions can also be drawn from these analyses.

First, the most important discoveries flow from a distinct technical advance. This was the case with Galileo's telescope and also the orbiting observatories.

Second, when a new technique has been developed, important discoveries follow rapidly (usually in less than 5 years). Galileo built his telescope only 18 months after the first known description of a telescope. As a corollary, a new instrument rapidly exhausts its possibilities for dramatic discoveries, becoming an instrument for deep study and thorough investigation.

Third, new cosmic phenomena are often discovered by physicists or engineers originally trained in disciplines other than astronomy. This is especially true for radio astronomy and X-ray and gamma-ray astronomy.

Many discoveries have been made with equipment originally developed for military use. Radar opened the door to radio astronomy. Infrared detection had initially been developed by the military and the gamma-ray bursts were discovered by the military Vela satellites.

Instruments that have made discoveries have often been built by teams of observers who have then had exclusive use of them. This was the case for the first infrared, X-ray and gamma-ray observations, as well as the discovery of the 3-degree black-body radiation by A. A. Penzias and Robert W. Wilson, or that of pulsars by Anthony Hewish and Jocelyn Bell.

New discoveries are often made by chance, thanks to a mixture of good luck and the wish to understand a surprising result; this applies to the discovery of radio emission from the Galaxy by Karl Jansky in 1931 when he was investigating the noise on a radio telephone circuit, and to Victor Hess who flew in a balloon to study atmospheric electricity and discovered cosmic rays in 1912.

The Ariane rocket. The European rocket, Ariane, is capable either of putting an observatory in orbit, or of launching an interplanetary probe. (CNES)

The Hubble Space Telescope

The result of an international collaboration between NASA and ESA, the Hubble Space Telescope was placed in orbit on 25 April 1990. From the beginning of the project, it had been decided that shuttles would visit it regularly to maintain it, with astronauts replacing defective parts in addition to their other jobs. These plans assumed special importance because of the problems that appeared after launch.

The advent of the Space Telescope had been eagerly awaited as a space observatory is free of the limitations that the atmosphere imposes on ground-based instruments (absorption of radiation, degradation of images by turbulence . . .). The Hubble Space Telescope therefore carried many of the hopes of the scientific community. This can be judged by the expected performance: an angular resolution of 0.1 arc seconds (making it a quite exceptional instrument for the epoch when it was planned, at the beginning of the 1970s: since then, however, ground-based telescopes have made such progress that they rival this), a pointing accuracy of 0.007 arc seconds, and a limiting magnitude of 29. Armed with this advanced technology astronomers hoped to be able to study objects fifty times fainter than those observable from the ground.

Three priority subjects had been selected: the determination of the distances of galaxies and measurement of the Hubble constant, the study of quasar absorption lines, and the detection of faint objects in particular regions of the sky. Through these observations it might be possible to choose between various cosmological models.

Quasars and active galaxies, studied in great precision, would certainly yield some of their secrets. Observation of the most distant parts of the Universe promised to show galaxies at the beginning of formation. Besides these things, it was hoped to discover other planetary systems and understand stellar evolution better. In short, every area of astronomy would benefit from this instrument.

One can imagine the consternation of astronomers when they realised that the telescope suffered from spherical aberration – traceable to defective checking during the polishing of the primary mirror – and could not complete all of its mission. Mechanical problems also appeared: the solar panels began to vibrate because of temperature changes during passages into the Earth's shadow, causing incorrect pointing of the telscope; two of the six attitude control gyroscopes failed, and a third works only intermittently.

The essential advantage of the Hubble Space Telescope over ground-based telescopes should have been its resolving power; this was considerably altered, and half of the scientific programme had to be postponed. The instrument most affected is the Wide Field Camera (from 80 to 90 per cent of the observations were postponed). The Faint Object Camera lost about 50 per cent of its programme. The high resolution spectrograph and the photometer were less affected. Only the ultraviolet observations remained relatively unaffected.

Fortunately the mission of the Hubble Space

Placement in orbit. The space shuttle *Discovery* was launched from Cape Canaveral on 24 April 1990. The following day the Hubble Space Telescope was placed in a circular orbit at 610 kilometres altitude. It is seen here slowly leaving the cargo bay of the shuttle (whose disconnected remote arm is visible). The protective shutter is closed; the solar panels and the high-gain antenna are deployed.

Communication between the ground and the telescope uses the TDRS system of geostationary satellites. The command and control centre of the satellite is at Goddard Space Flight Center, Greenbelt, about 40 kilometres from Baltimore, where the Space Telescope Science Institute develops the scientific programme and collects the scientific data (NASA).

Telescope could last at least fifteen years, which should allow the initial objectives to be achieved. Many studies have been carried out so that the instrument can recover at least in part its expected performance. NASA plans to install corrective optics in 1994.

However, these ups and downs should not obscure the fact that the Hubble Space Telesope has already fulfilled part of its mission: the data collected already show that significant scientific results can be obtained despite the spherical aberration, thanks partly to computer techniques for correcting images.

Elisabeth VANGIONI-FLAM

Architecture and instruments of the Hubble Space Telescope. With a mass of 11.6 tonnes, a length of 13.16 metres and a maximum diameter of 4.26 metres, the space observatory has three parts: the Support System Module (SSM), the Optical Telescope Assembly (OTA) and the scientific instruments.

The SSM is the structure containing the two other assemblies. Its main functions are mechanical protection, the electrical power supply through the solar panels, the storage and transmission of data, thermal regulation and the attitude and pointing control systems.

The OTA (shown in section at right) includes the optical telescope itself, its structure and the startrackers of the pointing system. The deflectors, principal and central, eliminate stray light (solar, terrestrial, lunar). The two mirrors are the main elements of the optical telescope. The 2.4 metre primary mirror is 30 centimetres thick and weighs 740 kilograms. It is made of special glass (92.5 per cent silica SiO_2, 7.5 per cent titanium dioxide TiO_2) with a very low expansion coefficient; the polishing of this, which lasted about two years, was the origin of the spherical aberration. The secondary mirror has a 30 centimetre diameter. The telescope is of Cassgrain type. Arrows show the paths followed by the light rays. These fall on the primary mirror, which sends the beam towards the secondary mirror. This mirror reflects the light, which is brought to

a focus behind the primary mirror (1.5 metres from its front surface) through a 60 centimetre diameter hole in its centre. The five scientific instruments (two cameras, two spectrographs and a photometer) then convert the images into useful data. The two cameras differ in their fields, spatial resolutions and wavelength ranges. The Wide Field Camera covers the blue, ultraviolet, and particularly red and infrared regions of the spectrum. This instrument – which is situated in a radial module – should have been used mainly for cosmological tests. The Faint Object Camera, built by the European Space Agency, has a very small field, but was designed to obtain the largest possible spatial resolution. An image of the observed source is obtained by summing the exposures made over several orbits. The two spectrographs have a complementary role: they cover a spectral domain impossible to study with a single instrument. The high resolution spectrograph analyses the visible and ultraviolet parts of the spectrum, and studies the composition of the interstellar medium as well as the abundance of the chemical elements at the surface of stars. The faint object spectrograph determines the composition, charactersitics and dynamics of distant sources. Finally, the photometer is for following rapid flux changes in compact objects (its time resolution is 10 microseconds).

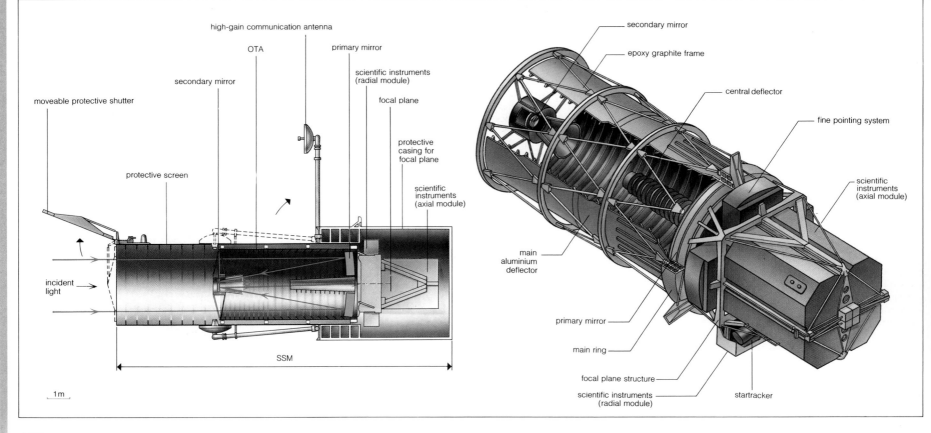

Considering these different points, the importance of instrumentation in astronomical discovery is clear. However, can one attempt to predict possible future discoveries that new instruments may bring? To do this, Martin Harwit studied the improvements in resolution over the entire electromagnetic spectrum, reached in 1939, 1959 and 1979, and showed how, each time there was an improvement, the possibility of a discovery appeared. Thus, for example, in the radio region, three discoveries are noted, each of which required a particular resolution that only one type of instrument could give. These were the discovery of the isotropic black-body radiation, which required an instrument with a large field; the discovery of quasars, needing a resolution of 1 minute of arc to identify the sources, and finally that of the superluminal velocities of some sources, which required a resolution of a milli-arc second. In each case, when the instrumentation allowed it, the discovery was made.

Harwit has also shown that, in the sciences in general, progress is made through conflicts between theory and observation. This is rarely the case is astronomy. In general, theory does not anticipate discovery and is only a poor guide. On the contrary, when a new phenomenon has been discovered, theoreticians produce several different explanations, though not more than one of these could be correct.

Theory is the tool that allows better analysis or interpretation of the observations, and forms an integral part of our 'instrumentation'. Sometimes, if no theory is able to explain the accumulated data, a search for other observations may allow us to understand the phenomenon better, and thus to fit a model to it. If there are several models, they can help us to choose a new observation by means of which we could select the best of the models.

On the other hand, there are also theoretical discoveries that have resulted from the use of a model pushed to the limit of its logic. In this way, black holes, neutron stars, and the presence of degenerate matter at the centre of stars, were discovered. But, strangely, the theoretical and observational discoveries were quite independent. The first white dwarf star was observed twenty years before a correct theoretical explanation of it was given. On the other hand, neutron stars were predicted forty years before the discovery of pulsars. Quasars are not fully understood more than twenty years after their discovery, whereas forty years after the first theoretical studies of black holes, there is not yet one indisputable observational candidate. But this is not a recent phenomenon. Copernicus predicted that the movement of the Earth should produce an effect of parallax, which was not observed until three hundred years later.

One can thus conclude that astronomy is a rich area for discoveries, and that these are made mainly by means of new instrumentation, which explains the extraordinary leap forward that has taken place during the last twenty years.

One might gain the impression that, in some areas, an intrinsic limit of detection has been reached, though progress has by no means come to a halt. In others, a first detection has not even yet been successful. The Universe may be sending us information in a form still unsuspected, for example, perhaps through the intermediary of other intelligent beings. So keep looking.

Alfred VIDAL-MADJAR

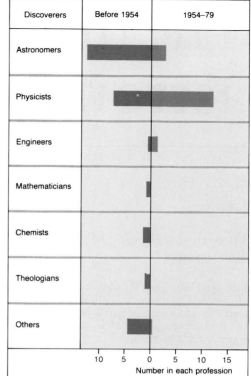

Discoverers	Before 1954	1954–79
Astronomers		
Physicists		
Engineers		
Mathematicians		
Chemists		
Theologians		
Others		

Number in each profession

Professions of the discoverers of forty-three phenomena (see the list in the table at the foot of the page). It is interesting to note that, after 1954, discoveries have not usually been made by professional astronomers, but rather by specialists who have developed a new technique. Even before 1954, more than half of the astronomical discoveries have been made by non-astronomers. (After M. Harwit, 'Physicists and Astronomy. Will you join the dance?' in *Physics Today*, vol. 34, p. 172, 1981)

Dates of the discovery of forty-three cosmic phenomena. (after M. Harwit, 1981)

Date	Discovery	Date	Discovery	Date	Discovery
Antiquity	Stars	1910	Giants/main sequence stars	1963	Quasars
Antiquity	Planets	1912	Cosmic rays	1965	Microwave background
Antiquity	Novae	1912	Pulsating variables	1965	Masers
1577	Comets	1915	White dwarfs	1965	Infrared stars
1610	Moons	1917	Galaxies	1966	X-ray galaxies
1655	Rings	1929	Cosmic expansion	1968	Pulsars
1754	Galactic clusters	1930	Interstellar dust	1968	Gamma-ray background
1785	Clusters of galaxies	1934	Novae/supernovae	1970	Infrared galaxies
1798	Interplanetary matter	1939	Galaxies with/without gas	1971	Superluminal sources
1801	Asteroids	1942	Supernova remnants	1973	Gamma-ray bursts
1803	Multiple stars	1946	Radio galaxies	1974	Unidentified radio sources
1861	Variable stars with nebulosity	1947	Magnetic variables		
1864	Planetary nebulae	1949	Flare stars		
1864	Globular clusters	1957	Interstellar magnetic fields		
1865	Ionised gas clouds	1962	X-ray stars		
1903	Cold interstellar gas	1962	X-ray background		

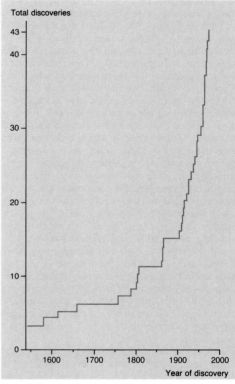

Cumulative graph of forty-three major astronomical discoveries (see table below). The increase in the last few years is spectacular and is mainly due to technological progress. When a technical innovation appears, a discovery often follows. (After M. Harwit, 1981)

The history of astronomy

Sumerian Zodiac. Representation of the map of the world. (British Museum, London)

From the awakening of thought to the Renaissance

Humanity has existed for over a million years. Yet for thousands of centuries, an almost total silence reigns. Very little trace of early man's industrial activity is left – chipped stones, fragments of tools, scrapers, axes and harpoons. Of his mental activity, no material evidence remains.

Suddenly, in the last part of the paleolithic age about 20 000 years ago, evidence of human thought appears: tombs, engravings and sculptures abound. One recognises constellations of stars on the first gravestones. Astronomy and arithmetic were certainly the oldest of the sciences. The attention of mankind must have been drawn to the great astronomical phenomena very early on. And firstly to the most obvious phenomenon of all, the alternation of days and nights.

Well before the invention of writing, mankind was acquainted with the phases of the Moon, the diurnal movement of the stars bringing the constellations back each night in an unchangeable pattern, and the periodic return of the seasons. The first notions of cosmography arose to meet the needs of people who were navigators, migrators and farmers.

From prehistory, the study of the sky must have given rise to two trains of thought: the realisation of the existence of immutable natural laws and the temptation to locate all-powerful supernatural beings in the sky. These ideas must surely have given rise to the first myths and stellar cults which have an important place in the various primitive religions. As far back as one goes in the history of man, one encounters cosmogonies. Mankind has wondered about his origins in all ages, in every latitude and at every level of civilisation. The formation of the Earth and of the heavens, the appearance of life, and the emergence of man are the essential questions in our search for our origins. The stakes, so to speak, are not only scientific; driven by far more than simple curiosity, our ancestors tried to formulate answers before even being able to express the questions correctly. The links between certain celestial and terrestrial phenomena were noted very early on, in particular between the cycle of the seasons and the Sun's passage through the Zodiac. This was an introduction to the principle of causality, the irrational application of which led to astrology, which is now considered a pseudo-science.

Besides the great periodic phenomena, there are the exceptional events such as eclipses and comets, which were interpreted as the signs of (more often than not) malevolent intentions on the part of the gods. In the same way, the irregular behaviour of the planets was found to be intriguing. Thus a mixture of astronomy, astrology and religion developed, the first often in the service of the other two.

Before the Greek era, in the great empires of Mesopotamia, Egypt, and of course China, mythical cosmologies saw the light of day, together with the first division of the sky into constellations and the first measurements of time and of the directions of the stars. Astronomy had little in the way of instruments for its first faltering beginnings, just the human eye. In order to measure the time, astronomers used the gnomon, a simple rod stuck in the ground, and the polos, a hollow half-sphere on which one observes the shadow of the point of a style fixed at the centre, as well as the hour-glass and the clepsydra (water clock). Among the instruments for taking aim at the stars a simple straight rod, the alidade, was used. Then came the compass, the double-jointed alidade, and other devices for measuring angular separations on the sky. Lastly came the great quadrants, on which the shadow of a rod indicated the noon altitude of the Sun, and the armillary spheres, jointed concentric circles capable of turning with respect to each other and therefore able to simulate the celestial motions. And, born of the desire to know the time at night; came the planispheric astrolabe. This instrument, invented by the Greeks, passed on by the Arabs to the West, served to determine the hour of day or night throughout the Middle Ages. Only with the invention of the pendulum clock did the astrolabe disappear; its pedagogical and calculatory functions were equally important.

With these rudimentary instruments, that astronomers only abandoned at the end of the seventeenth century, naked-eye observations of the sky gave birth to the first calendars, sky maps and ephemerides. However, just as with the other natural sciences, the rise of Greek culture marked the first great note in the history of astronomy. It all started at the beginning of the sixth century BC with the creation of the Ionian school by Thales. Science and religion separated, and the supernatural became remote from the explanation of phenomena. Thales, his pupils, and his successors constructed a theory of the Universe which was atomic and kinetic. Being materialists, they arrived at a concept of the world where the gods played a secondary role.

Just as the Ionian school flourished at Miletus and Samos, in southern Italy the School of Pythagoras was born. Founded on a mathematical concept of the world, it attributed mystic geometrical properties to numbers and figures. Its dogmatism resulted in

Map of the heavens. Board I of the work of Peter Apian (1495–1552) also called Petrus Apianus, *Astronomicum Caesareum*, printed in Ingolstadt in 1540. (Photograph J. Counil, observatoire de Paris)

a dualist and religious concept of the Universe.

These two very different schools of thought gave birth to the great astronomical discoveries of Antiquity: the Earth recognised as an isolated heavenly body in space; its sphericity, without a top or bottom; the fall of massive bodies towards the centre of the Earth; the first measurement of the dimensions of the Earth, Moon and Sun, and of the Earth–Moon distance; the first attempt to measure the distance of the Sun, albeit fruitless; the theories of motion of the Moon and of the Sun; the theory of eclipses; and the ingenious geometric combinations able to account for the apparent wanderings of the planets on the celestial sphere.

However, Antiquity also bequeathed a series of three postulates to astronomers which blocked the developmnent of celestial science. First of all, there was the geocentric concept which held that the Earth, utterly immobile, occupied the centre of the Universe. The second postulate was that the Universe was divided into two worlds: the cosmos – a world of purity where nothing could change, world of the ether and of circular motion – and the sublunar world – a world of impurity and of change, world of the Earth and of the four elements, and of rectilinear motion, up and down. The final postulate was that uniform circular motion or a combination of such motions were the only possible motions for the heavenly bodies.

These cosmological postulates, closely linked to the physics of Aristotle, reigned almost without challenge for twenty centuries. The exception was the system of Aristarchos (290 BC). This placed the Sun at the centre of the Universe and attributed a double motion to the Earth, a rotation upon its axis and a revolution around the Sun. The views of Aristarchos, the first and last heliocentric astronomer of Antiquity and the only true precursor of Copernicus, were quickly swept away by the flood of geocentrists who had both immediate understandability and common sense on their side. Ptolemy could only triumph. His *Grand Mathematical Syntax* (AD 140) which was passed on by the Arabs to the West under the name of the *Almagest*, was both the crowning achievement and the sum of ancient astronomy. It presents an overall concept of the Universe which is harmonious and geometric, and also a complete treatise of practical astronomy, accompanied by the indispensable concepts of geometry and a basic form of early trigonometry, which one needs to be acquainted with to understand the work. It depends totally on the physics of Aristotle which preceded it by six centuries, except for the equant – a veritable fudge! At the centre, of course, the Earth is immobile. The Moon revolves around it first, then Mercury, Venus, the Sun, Mars, Jupiter and Saturn. The stars are fixed to the eighth sphere; they are thus all at the same distance from the Earth.

Ptolemy constructed ingenious combinations of uniform circular motion in order to explain the apparently extremely complex motions of the planets. From an Aristotelian point of view, two of these combinations, the epicycle system and the eccentric system, can be considered orthodox; the third, the equant point system, in contrast could not be tolerated! There one was no longer speaking of uniform circular motion.

Nonetheless, Ptolemaic astronomy functioned for fourteen centuries without fundamental upheaval, even if the Arab astronomers who passed it on to the Christian West improved upon it at some points, such as the motion of the Moon. Not until the Renaissance did a handful of people guide astronomy onto a new path, and they did so even before the progress of observations forced a shattering revision.

'De Revolutionibus'

The publication in 1543 of the treatise *De revolutionibus orbium caelestium* by Nicholas Copernicus was an event of paramount importance. It was the end of the geocentric era. Copernicus fixed the Sun at the centre of his system. The Earth was given the rank of a planet, rotating on its axis once every twenty-four hours, and performing a revolution around the Sun in one year. Of the three postulates which blocked the development of astronomy, the geocentric one was explicitly denied. However, although the dichotomy of the Universe was also implicitly denied in consequence (what would remain of the purity of the sublunar world with the Earth wandering around there?) the principle of circular motion was not rejected, but reinforced. In fact, Copernicus replaced the equant circles, a clever device of Ptolemy, by the postulate of uniform circular motion.

In appearance, therefore, there was little change. In the improbable combination of circles which constituted the machinery of the Universe, two parts exchanged their places, the Earth and the Sun. But this exchange had at least two immediate consequences. On the one hand, the motion of the Earth around the Sun opened up a new strategic field for astronomy. Using this new possibility, Kepler discovered that the planets move in ellipses with the Sun at one of the foci; the principle of uniform circular motion thus joined the geocentric concept in being rejected. On the other hand, Copernican cosmology was fundamentally incompatible with Aristotelian physics, even without Copernicus proposing a new physics. This gap created a new, dynamic situation: all physicists who wanted to adopt the cosmology of Copernicus had to reject the physics of Aristotle and attempt to establish a new physics. Galileo, the founder of modern physics, understood this straight away. It was he who fashioned a revolutionary tool from the work of Copernicus.

The Copernican system is simpler than Ptolemy's. However, it

View of the site of Carnac in Brittany.
(Photograph E. V. Leroy, Pix)

Tablet representing an astrological calendar. Low-Mesopotamia, third to first centuries BC. The tablet is divided into twelve parts corresponding to the twelve signs of the Zodiac. (Musée de Louvre; picture from the Musées nationaux, Paris)

Solar sundial of the Forbidden City of Peking.
(Photograph G. Gester, Rapho)

Frontispiece of the 'Almagestum novum' published in 1651 by Riccioli. Uranus weighs the systems of the world. Ptolemy and his geocentric world system, in the lower part of the frontispiece, are not in contention. It is the system of Tycho Brahe, much favoured by Riccioli, in which the Sun revolves around the Earth, which gets the verdict over the system of Copernicus. (J. Counil)

is still necessary to define what sort of simplicity we are talking about. Although it can be argued that Copernican astronomy uses as many circles in total as the astronomy of Ptolemy and is just as complicated in that respect, that is not the problem. The problem is cosmological. Even in the overall simplified presentation of Ptolemy's system we cannot do without the first epicycle. The first rather large epicycle replaces the circuit of the Earth around the Sun; if this is forgotten, we cannot understand the behaviour of the planets on the background of the heavens. In the Copernican system, the first epicycle is small; it explains the deviations between the mean motion and the true motion (in reality between uniform circular motion and non-uniform elliptical motion). This epicycle can be forgotten for a short time without major consequences. Another simplicity of the Copernican system is that although the inferior and superior planets have different movements, they have an identical cosmological status. In the Ptolemaic system, it is necessary to give them a different cosmological status, and in so doing to exchange the roles of the epicycles and the deferents. Lastly, and above all, in the Copernican cosmos there is a simple connection between the order of the planets and the periods of their revolution: the further a planet is from the Sun the more slowly it circles the Universe. Without such a simple connection, Kepler would never have looked for, nor found, his third law of planetary motion. None of this could have happened in Ptolemaic cosmology.

For all that, the game between Ptolemy and Copernicus was not yet over. Even Tycho Brahe, the first astronomer to perfect the instruments of Antiquity and to improve the precision of observations by a factor of ten, was not a Copernican. He opted for an intermediate system, the most practical, kinematically speaking. The Earth was again considered immobile at the centre of the world, which was very practical for describing the motions of the other bodies of the system: the Sun revolved around the Earth but Mercury and Venus revolved around the Sun, the orbits of Mars, Jupiter, and Saturn enclosing them all, which explained the different motions of the inferior and superior planets. This clever system was not the true system of the world; Brahe's merits were different. He bequeathed excellent planetary observations, notably of the course of Mars, which brought the young Kepler to Prague to see Brahe, just before his death.

Until the beginning of the seventeenth century, the Copernican ideas were cautiously propagated across Europe particularly by the Lutheran scholar Rheticus, the sole disciple of Copernicus during his life. Contrary to a commonly-received idea, the Catholic Church did not stir at first. The first condemnations came from the Protestant world. Even before *De Revolutionibus* was published, Luther, informed by public rumour, violently and strongly condemned the fool who set the Earth in motion. Even the gentle Melanchthon considered that the attempt to move the Earth and stop the Sun was absurd and that the propagation of such ideas should not be tolerated by a wise government. Then there was indifference, especially since *De Revolutionibus* was prefaced by an anonymous 'Introduction' which many scholars believed to be by Copernicus, and which was in fact by the theologian Andreas Osiander, who had supervised the editing of the work. This clever 'Introduction' presented the Copernican system as a mathematical theory among so many others, permitting mathematicians to use the Copernican data without accepting Copernican cosmology.

It was only later, in Italy, that history took a more serious turn. It came out that *De Revolutionibus* addressed itself not to those drawing up astronomical tables but to physicists such as Galileo and to philosophers like Giordano Bruno.

Bruno, going further than Copernicus, stated that, if the Earth is a planet like the others, the division of the Universe into cosmos and sublunar world makes no sense. He proclaimed the unity of the heavens and the Earth, the identical nature of the Sun and the stars, the infinity of the Universe and the plurality of worlds.

At the end of 1609 and at the beginning of 1610, Galileo aimed a telescope towards the sky. A whole new world was revealed to his gaze. The Milky Way, was resolved into millions of stars; the rings of Saturn, greatly distorted by his telescope, and the surface of Jupiter appeared; the measurement of shadows on the Moon enabled him to estimate the height of the lunar mountains; the phases of Venus became evident; and, above all, the discovery of the satellites of Jupiter demonstrated that there was not, as affirmed by the physics of Aristotle, a unique centre of motion in the Universe. Whether one's preferred theory was heliocentric or geocentric, there were at least two centres of revolution: the Earth, around which revolves the moon, and Jupiter, around which revolves four satellites! From simple kinematic hypothesis, the system of Copernicus became a physical reality.

The reaction of the Catholic Church was as brutal as ineffective. The advocacy of Bruno, burnt alive in Rome in 1600 for theological heresies, brought the theses of Copernicus into disrepute. They were solemnly condemned in 1616, more than 70 years after their publication. Galileo was tried and abjurated in 1633. But the century did not end before Newton established the identical nature of the forces of attraction which govern the movements of the planets, and gravity which governs the fall of terrestrial bodies. The distinction between the sublunar world and the rest of the Universe collapsed for good. A unique law was given for the whole Universe.

But let us return to the young Kepler. When he arrived in Prague on 4 February 1600, he had already published his *Mysterium Cosmographicum*. He showed himself to be a staunch Copernican, displaying an incredible energy in explaining the structure of the world by successive nestings of regular convex polyhedra. But the astronomical contribution therein was small, apart from the fact that it signalled an inconsistency in the Copernican system. He affirmed that the orbital planes of the planets, close without being merged, pass through the Sun, and that their inclinations to the ecliptic remain constant, while Copernicus had them passing through the centre of the Earth!

In 1601, making the most of his opportunities, Kepler got hold of the observational journals of Tycho Brahe and in particular those dedicated to the orbit of Mars. On getting the post of Imperial Mathematician at Tycho's observatory, thus benefiting from job security, Kepler attacked the delicate problem of the orbit of Mars. He expressed the positions of the planet in relation to the centre of the Sun, and then perceived that it is impossible to explain the motion by a combination of uniform circular motions. So he returned to the equant, rejected by Copernicus. He indeed found a circle and an equant point satisfactory for four judiciously chosen positions of Mars. But the other points did not fit the orbit thus determined. The equant, like the epicycles and the eccentrics, could not explain the observations of Tycho Brahe. The discrepancies were of the order of ten arc minutes. Kepler's confidence in the qualities of Tycho Brahe as an observer was such that he bowed to this and reconsidered the whole problem, going back to a careful study of the Earth's orbit. By an unbelievable chance, because he made two errors in this calculation which cancelled each other out, he discovered his first law: the radius vector joining the planet to the Sun sweeps out equal areas in equal times. Then he discovered that the orbit of Mars is oval in form. Kepler's attempts to construct orbits were numerous, but he hounded the problem and found a path long known to geometers, but which no one had dared to suggest for the planets: the ellipse. This is Kepler's second law: the planets move in ellipses of which the Sun occupies one focus.

Calculators of the motions of the Galilean satellites

In 1609, the first two laws of planetary motion appeared in *Astronomia nova*. At this time, Galileo observed the moons of Jupiter, and was soon imitated by Kepler himself who gave them the name of satellites. However, Kepler's first two laws, even though important, did not satisfy the curiosity of their author. Kepler's aim, was to find another law: he persisted with Tycho Brahe's notebooks in order to find a mathematical expression for the connection shown to exist by Copernicus between the distances of the planets from the Sun and their orbital periods. He put the same relentlessness into this as into his researches into the structure of the world. In 1618, after nine years of work upon the data of Tycho Brahe, Kepler stated in his *Harmonice mundi*, his third law: the squares of the periods of revolution of the planets are proportional to the cubes of their mean distances from the Sun.

Copernicus was already largely superseded; the axiom of uniform circular motion had been abolished the laws of planetary motion opened the way to a much deeper understanding than the invention of a hierarchy of circles designed to account for the apparent motions of the planets. Copernicus became even more outmoded; in Italy, Galileo delivered another mortal blow to Aristotelian physics. His experiments on falling bodies, experiments rendered in mathematical language, not only led to the law of inertia (an essential part of the understanding of stellar dynamics) but laid the foundations of modern physics.

In fact, between 1600 and 1610, Galileo established that the trajectories of projectiles whose initial velocities are horizontal are parabolae. It was Torricelli who generalised this result for any initial velocity. These results gave rise to the idea of a

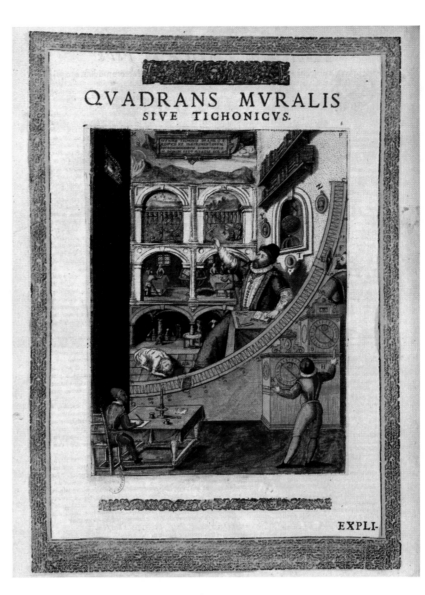

The great wall quadrant of Uraniborg Observatory, used by Tycho Brahe and his assistants. The person on the right-hand edge of the picture (representing Tycho Brahe) is aiming at the Moon, which is at upper left. In the foreground, another person is noting the time, while a third (on the left) transcribes the observation into a logbook. The large figure at centre right depicts a lifesize painting of Tycho Brahe which was made on the actual wall of the quadrant. The illustration is taken from *Tychonis Brahe astronomiae instauratae mechanice* (work preserved at the Bibliothèque nationale, picture Bibl. nat. Paris)

The world system of Kepler. These engravings, taken from the *Mysterium cosmographicum* by Johannes Kepler (1596), show the instruments of Tycho Brahe and, at lower left, the spheres corresponding to the orbits of the six planets known at that time; the spheres are inscribed within the five Platonic solids, the cube, tetrahedron, dodecahedron, icosahedron and octahedron. (Photograph J. Counil, observatoire de Paris)

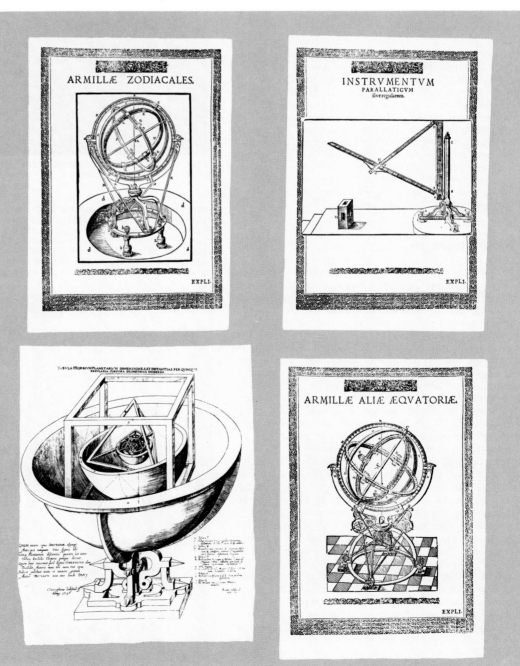

conservation of the horizontal component of velocity for massive bodies, leading to the concept of perpetual motion, without which something would be needed to push the stars continually. Towards 1630, Descartes showed these discoveries to be general to matter and postulated an axiom that the natural motion is in a straight line, and not circular. Simultaneously and independently celestial mechanics and terrestrial mechanics progressed in giant strides. Kepler and Galileo analysed and noted facts without understanding the causes and without unifying these two new fields of knowledge, but the ground was prepared for the genius of Newton.

The birth of modern astronomy

It is told that Newton meditated in a garden in Lincolnshire one evening in 1665. He had fled there from Cambridge because the London plague had closed the University. There was a full Moon. A slight sound was heard: an apple had just fallen to the ground. Newton asked himself the question: why does the Moon not fall to the Earth like this apple? A paradoxical answer came to him, which could only have come to a genius: at every instant, the Moon falls towards the Earth! The force which attracts the apple to the Earth and the force which holds the Moon to the Earth are the same.

The story may be legendary but it has a ring of truth. It is above all exemplary, not because it teaches that it suffices to dream in a garden to make in a few seconds one of the greatest discoveries of all time, but because it illustrates the fact that some fundamental ideas in the sciences are born from elementary questionings about everyday events. It was thus the same with special relativity, descended from the commonplace question that Einstein had the guile to ask, and which he had the genius to answer: what are two *simultaneous* events?

The intuition that the force which makes apples fall and that the gravity which governs the motions of the stars are one and the same force is one thing; to verify it by calculation and putting this intuition into mathematical form is another. The history of the development of the thoughts of Newton is not known for certain. Documents on this precise matter are rare. It has been claimed that the value of the radius of the Earth that was accepted in 1665 which was the value involved in the estimation of the fall of the Moon, was too inaccurate to verify the calculation, and that Newton would therefore only have undertaken his calculations again when in possession of a new measurement of an arc of meridian made by the Abbé Picard. However, that does not explain why Newton only took up the subject again in 1682 – Picard's measurement was completed in 1670 – and the improvement in accuracy of 15 per cent does not appear to be decisive in the circumstances. Newton had other far more formidable obstacles on his route. He had to establish that the attraction diminished as the square of the distance. But above all gravity results from the attraction upon a body, such as the moon or an apple, of *all* points of the Earth. It was not self-evident that gravity behaved as if all of the mass of the Earth was concentrated at its centre. Without this being demonstrated, the uncertainty in the distance of the apple from the Earth was of much greater importance than that connected with estimating the length of an arc of meridian.

Newton had to create an adequate mathematical tool in order to resolve the problem – showing that simply having a reverie under an apple tree is not enough – this tool was the calculus of fluxions, forerunner of our differential calculus. Newton's correspondence between 1670 and 1680 shows, by the way, that hesitations and false starts were numerous. In 1679, for example, Hooke proposed to study the effect of the rotation of the Earth upon falling bodies. Newton gave the wrong solution: for him, the resulting trajectory had to be a spiral. Hooke himself gave the correct solution: the trajectory is an ellipse. Newton recognised his error in a letter to Halley and it is probable that Hooke's problem boosted Newton's work on gravity. From 1683 to 1685, Newton wrote the great work which truly inaugurates modern astronomy and which appeared in 1687, thanks to Halley's financial assistance: the *Philosophiae naturalis principia mathematica*.

Newton founded mechanics upon three elementary principles
– all bodies not subject to an external force remain at rest or in uniform motion at constant velocity in a straight line;
– the change of motion is proportional to the motive force impressed; also, the acceleration is in the direction of the force;
– to every action there is an equal and opposite reaction.

In making the grand synthesis of the works of Kepler on the motions of the planets, the works of Galileo on falling bodies, and those of Huygens on the centrifugal force, and relying on the previous principles, Newton stated the universal law of gravitation: any two bodies attract one another with a force proportional to their masses and inversely proportional to the square of the distance between their centres of gravity. The last bolt which locked the door to progress in the development of astronomy was removed! The Universe does not consist of two parts; the same law governs the fall of terrestrial bodies (including the trajectory of cannon balls) and the celestial motions. For the first time, astronomy had at its disposal a differential law which could be used to deduce from the state of a system at a given instant its state at the instant immediately following. Before Newton, the science of celestial motions only had at its disposal numerical tables relying on arbitrary and artificial geometrical models, and on empirical and integral laws such as those of Kepler, only describing the overall motions.

The other works which made Newton famous are concerned with the nature of light. On 6 February, 1672, Newton wrote to Oldenburg: 'To perform my late promise to you, I shall without further ceremony acquaint you, that in the beginning of the Year 1666 (at which time I applyed my self to the grinding of Optick glasses of other figures than *Spherical*,) I procured me a Triangular glass-Prisme, to try therewith the celebrated *Phaenomena* of *Colours*. And in order thereto having darkened my chamber, and made a small hole in my window-shuts, to let in a convenient quantity of the Suns light, I placed my Prisme at its entrance, that it might be thereby refracted to the opposite wall. It was at first a very pleasing divertisement, to view the vivid and intense colours produced thereby'. This first moment of aesthetic emotion having passed, Newton undertook a series of experiments on this decomposition of white light into light of various colours. His experiments were the basis of spectroscopy with which the physical study of the stars truly began. Thanks to it, astronomers know the chemical composition of the stars, their temperatures and the phenomena that occur in their midst. Thus Newton was the pioneer of two great routes by which astronomy now began to progress: celestial mechanics and stellar physics.

Three significant dates had occurred within one hundred and fifty years: 1543, publication of *De Revolutionibus*; 1609, first observations of the sky and publication of the *Astronomia nova*; 1687, publication of the *Principia*. The year 1667 was also important. It was in that year that Auzout and Picard invented the filar micrometer and equipped the eye-pieces of telescopes with this new improvement. The telescope, which previously merely increased the resolution light-gathering power of the eye, now began to introduce precise angular measurements to astronomy. For example, the positions of the satellites of Jupiter in relation to their mother planet, previously estimated approximately, were from then on determined to a few arc seconds or so.

A few years later, the application of the pendulum to clocks brought precision to the measurement of time, and astronomy, the descriptive science, became a science of precision. The first of its areas to benefit was stellar astronomy. From the eighteenth century onwards, the important stages in this quest for precision in astronomical measurements were the catalogues of Flamsteed drawn up at Greenwich for the northern sky, and those of the Abbé Lacaille at the Cape of Good Hope for the southern sky, then those of Bradley, Piazzi and Bessel. Then came the compilation of the great catalogues of Argelander – the *Bonner Durchmusterung* – and of Gould – the *Uranometria argentina*. Both are fundamental in contemporary astronomy.

The eighteenth century also saw the triumph of celestial mechanics. It was such a triumph that the discipline became the model for all the exact sciences and it seemed for a while that all the domains of physics were destined to be reduced to mechanics. Initially, however, the greatest resistance to the theories of Newton was in France under the pressure of numerous powerful Cartesians. Yet, paradoxically, it was the French astronomers who took up the challenge again. With the Swiss Euler, Clairaut, d'Alembert and Mapertuis were the engineers of this extraordinary development which culminated in the works of Lagrange and Laplace, reaching its zenith in the computation of the position of Neptune by Le Verrier. (The position was also independently calculated by the English mathematician John Couch Adams.)

Lagrange, born in Turin but French by adoption, produced considerable mathematical works and new results in celestial mechanics. He studied the motions of the satellites of Jupiter and the secular and periodic variations of planets undergoing perturbations. It was to resolve these problems that he introduced the method of the variation of arbitrary constants. His great work remains *Méchanique analytique*. As for Laplace, his work was immense and famous. Born in 1749, he showed, in 1773, the invariability of the major axes of the planets and established, in 1784, the stability of the Solar System. After many ups and downs in his work, his cosmogonic hypothesis on the formation of the Solar System (*Exposition du système du monde*, 1796) was the order of the day. His *Traité de méchanique céleste* (1798–1825) and his *Théorie analytique des probabilitiés* (1812) made his name immortal.

While the French continued the work of Newton at the highest level, the English did not remain idle. Two fundamental discoveries are credited to Bradley. In 1727, intending to demonstrate the existence of the annual parallax of the fixed stars, he discovered aberration (annual displacement of the stars where the velocity of the Earth in its orbit and the velocity of light are combined). Bradley thus confirmed both the finite velocity of light, discovered by 1676 by Roemer, and the Copernican system. At last, two centuries after Copernicus a physical phenomenon was observed which confirmed the revolution of the Earth around the Sun. Then in 1748, Bradley described nutation: the axis of the Earth oscillates under the gravitational influence of the moon with a period of 18 years. It was Newton's laws of gravitation which were this time confirmed by observation.

In less than two centuries, astronomy had totally changed its character. Before the seventeenth century, to practise astronomy, one took a cosmological model, generally that of Ptolemy, slightly amended, or exceptionally replaced (as in the case of Copernicus). Then, equipped with the eye and three or four rudimentary instruments, one observed the principal heavenly bodies from time to time and thus corrected the predictions of astronomical tables. The stars were considered as points of light with no apparent importance except that of defining positions at

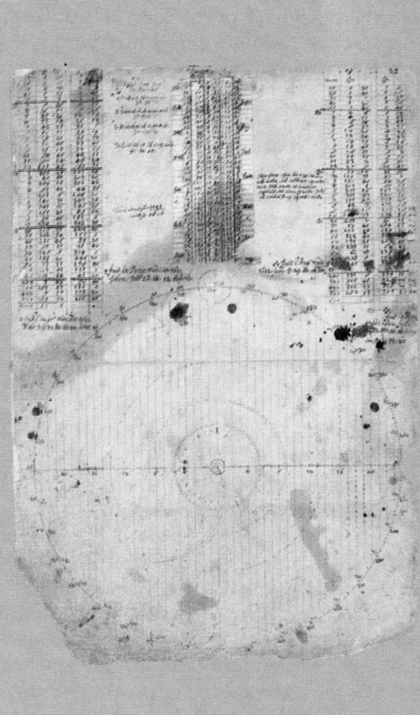

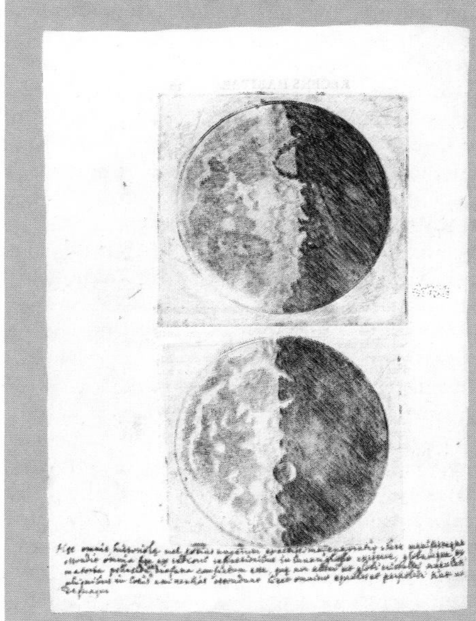

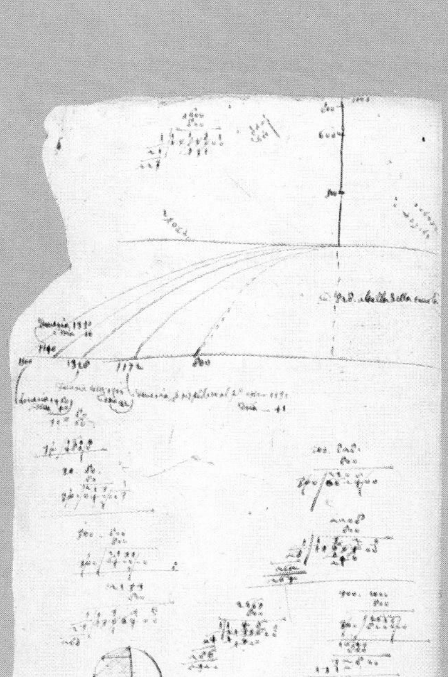

Pages of the journals of Galileo. From top to bottom: calculations of the positions of the Galilean satellites; drawings of the moon; discovery of parabolic motion. (Biblioteca nazionale centrale di Firenze, pictures G. Sansoni, UFPI)

known instants. The tasks and techniques were therefore simple, as was the mathematical tool: elementary trigonometry. There was not much physics.

From the beginning of the seventeenth century, the invention of the telescope revolutionised astronomy. In a few nights of observation with the telescope, Galileo contributed more new knowledge than all of his predecessors put together. A trend was begun which has continued ever since, in which the development of techniques was inextricably linked with theoretical knowledge. The construction of more and more powerful telescopes enabled astronomers to probe further and further into the Universe. The name of William Herschel shone brightly in this domain. Hitherto associated with mathematics alone, from this time on astronomy was linked with all the sciences of nature.

Jean-Pierre VERDET

The development of traditional astronomy

The Industrial Revolution which began in England at the end of the eighteenth century, spread to Europe and to the United States during the nineteenth century. Manufacturers and entrepreneurs arose, taking control over industry and commerce and imposing a new social and economic order. The mathematical sciences quickly became a principal basis of technical development. The new bourgeosie assisted in the development of these sciences, convinced that technical progress and the study of nature were necessarily good and beneficial for their own advancement. The development of astronomy shows that practical utility is not the only motive for interest in science: after mathematics, astronomy was the most highly regarded domain of research because it was considered to be the most appropriate to deepen knowledge of the mechanisms governing nature. Numerous observatories were founded in the most advanced countries, notably within the universities (Cambridge University Observatory was founded in 1823) and even with private funds, like Harvard Observatory in the United States, founded in 1844 by citizens of Boston.

As is well known, traditional astronomy benefited from the considerable progress made in the construction of machines and notably in precision engineering. Until the beginning of the nineteenth century, the best instrument constructors were found in England. Then it was the turn of Germany. Foremost among the new successes was the meridian circle, which enables two coordinates of a star to be measured simultaneously. Bessel (1784–1846), at Konigsburg, in Germany, was the pioneer of this type of instrument which enabled the precision of observations to be considerably improved. Fraunhofer (1787–1826) perfected the construction of optical components which, together with excellent mechanisms, constituted instruments of outstanding quality. Fraunhofer and others, like Ramsden in England, improved or invented auxiliary measuring instruments. The design of astronomical clocks also made great progress. Gauss, at Göttingen in 1804, devised the method of least squares which enabled an objective analysis of observations to be made for the first time (before this, the astronomer was satisfied with more or

less arbitrarily selecting the observations which appeared to him to be the best. With all these improvements, astronomers not only improved the classical measurements (for example, those of the distances within the Solar System) but also tackled new problems. Bessel in Germany, Wilhelm Struve at Dorpat (today called Tallin), then Otto Struve at Pulkovo in Russia succeeded in measuring for the first time in 1840 the distances of certain stars thanks to their apparent displacement in the course of the year with respect to more distant stars.

It was also in the nineteenth century that systematic astronomical observations went beyond the work of isolated and individual astronomers to become part of large international enterprises. For example, astronomers in Germany organised the first of these large programmes in 1871, carried out by sixteen observatories in various latitudes, each charged with observing all of the brightest stars occurring in a given declination zone in the sky. The triumphant completion of the resultant catalogue, the *Bonner Durchmusterung*, which contains more than 100 000 positions of stars, took many years. This was not the only such work: for example, an astronomical conference convened at Paris in 1889 took a decision to construct a photographic atlas of the sky, using a rather large number of identical telescopes made by the brothers Prosper and Paul Henry. An enormous amount of work was involved in the preparation of this photographic map of the sky called The Carte du Ciel, and it took much longer than expected. These projects were the forerunners of the massive international collaborations of today.

The zenith of celestial mechanics

The term 'celestial mechanics' was coined by Laplace, who made important improvements in the methods of calculating the movements of celestial bodies. Further improvements continued throughout the nineteenth century. As a union of the twin sciences, mathematics and astronomy, celestial mechanics was considered in some sense to be the supreme science, the triumph of the human intellect. Improved by Gauss, Cauchy, Bessel and Jacobi, to mention only the most famous mathematical astronomers of the first half of the nineteenth century, celestial mechanics reached its zenith as Le Verrier (1811–77), studying the mysterious perturbations in the motions of the planet Uranus, upon which Laplace had worked in vain, attributed them to a new planet, the position of which he predicted more or less correctly. J. C. Adams carried out an analogous work at Cambridge in England, but it was from the information of Le Verrier that Neptune was found at Berlin Observatory by Galle on 23 September 1846.

This discovery was considered by the general public to be almost miraculous: that was in fact the case, because the problem had several possible solutions for the orbit of the perturbing planet. Le Verrier did not choose the correct one and it was a happy coincidence that Neptune was found near to the predicted position. This story was published at the time by two Americans, Peirce and Walker, whose behaviour, although perfectly conforming to scientific ethics, was considered to be an insult to Le Verrier. Le Verrier continued to improve the predictions of the motions of the planets and had remarkable success; however, he conducted himself like a veritable dictator at the Paris Observatory, scorning all that was not celestial mechanics, and prevented very promising developments in other areas of research.

The establishment of a theory of motion for the moon was an even more difficult undertaking than that of the motions of the planets. However, it was carried out very well in the middle of the nineteenth century by Hansen at Gotha and independently by Delaunay at Paris at the price of twenty years of tedious and laborious efforts. Nevertheless, certain aspects of the motion of the Moon resisted all rational explanation, just like certain features of the motion of Mercury. That may seem unimportant to us, but it greatly irritated astronomers at that time, and also the navigators who used the motion of the moon to determine their longitude at sea. These anomalies were not to be explained until about 1920 in a totally unexpected way: those in the motion of the Moon were in fact due to irregularities in the rotation of the Earth (Chandler had discovered in 1885 that the rotation axis of the Earth was not fixed), and those in the motion of Mercury were due to an effect explained by Einstein in his famous theory of general relativity. Thus the science of celestial mechanics gave rise to two very important discoveries: one discovery on the practical level, because it demonstrated the existence of irregularities in what was believed to be the best possible clock, the rotation of the Earth, and the other on the theoretical level.

The beginnings of astrophysics

Until the beginning of the nineteenth century, astronomy was devoted almost entirely to the description of the motions of celestial bodies and to the causes of these motions. For the rest astronomy contented itself with simple descriptions. The idea was firmly fixed that it would be impossible to know the chemical composition of the stars and their physical conditions since one could not go and look. Some astronomers resigned themselves to this. Others used their imagination, occasionally backed up by fantasies of observation (like those of the very famous canals of Mars supposed to have been constructed by living beings). However, some physicists began to think that it might not be impossible to gain objective knowledge of the nature of stars.

The telescope of Lord Rosse with which he discovered the spiral structure of certain nebulae. (Yerkes Observatory)

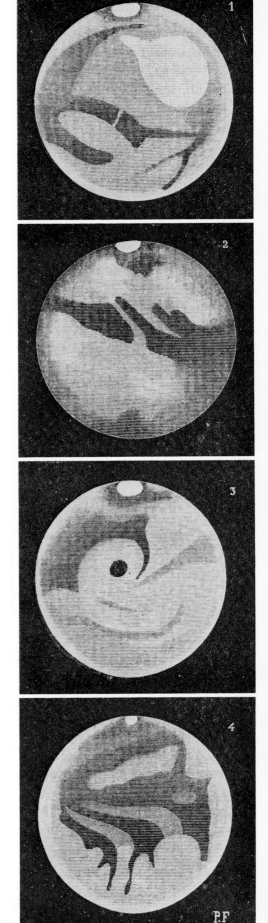

'Telescopic views of the planet Mars in 1877.'
Drawings taken from the book *Popular Astronomy* by Camille Flammarion.

Several physicists, such as Ångström, Foucault and Stokes, realised in the 1850s that a pair of lines observed by Fraunhofer in the spectrum of the Sun – one of the first applications of the spectrograph – coincided in wavelength with an identical pair of lines observed in the laboratory in the spectrum of sodium. The small step to the realisation that the Sun contains sodium was taken immediately. But it was Kirchhoff who laid the foundations of the chemical analysis of bodies through spectroscopy: when he measured the wavelengths of several thousand lines in the solar spectrum, he recognised that a large number of them were due to elements such as hydrogen, iron, sodium, magnesium and calcium. From 1887, Rowland, an American who had considerably improved the spectrograph, observed both in the Sun and in the laboratory a much larger number of lines. On measuring their wavelengths precisely, he recognised not less than thirty-six terrestrial elements in the Sun. This was in 1896; by 1928, he had increased the number to fifty-one. Towards the middle of the nineteenth century, interest was growing in the structures of the Sun, and in particular in its outer atmosphere, the chromosphere and the corona, visible during total eclipses. In 1868, the first spectra of the chromosphere were obtained during an eclipse, by Janssen. Besides the classical lines of hydrogen, the spectra contained an intense line which did not correspond to anything observed on Earth. This was attributed to an unknown element called helium. Helium was subsequently discovered on Earth, but it is interesting to see that astronomy thus made its first substantial contribution to physics. The links between physics (and chemistry) and astronomy became closer and closer: one could truly begin to speak of astrophysics.

In 1870, the Englishman, Young, discovered new lines, this time in the corona, which were attributed to a further new element, coronium. However, laboratory research was in vain this time, and it was only in 1941 that Edlén, a Swede, was able to explain how the lines originated; they came from iron, calcium, nickel and other elements that were very strongly ionised in conditions practically impossible to obtain in the laboratory.

The second half of the century was also marked by a large number of studies, both morphological and descriptive, of the surface and atmosphere of the Sun. The astrophysical laboratory founded in 1876 by Janssen at Meudon was notably dedicated to these studies. It was, however, necessary to wait until laws governing radiation had been stated by Kirchhoff, Boltzmann and Wien before a reasonable idea of the temperature of the solar surface, about 6000 K, could be obtained in 1893. This was before the physical law governing radiation was finally established by Max Planck in 1906. A further interesting point in the history of the sciences concerns evolution of ideas for the origin of energy emitted by the Sun. After Robert Mayer had stated the principle of the conservation of energy in 1842, he proposed the first 'modern' explanation: the solar energy was derived from a supply of kinetic energy from meteorites continually falling upon the Sun's surface. However, this idea met insurmountable difficulties and was soon abandoned. The explanation put forward by Helmholtz was much neater. He noticed that, if the Sun contracts, its shrinking releases gravitational energy which is converted into thermal energy, the Sun's present gravitational energy enabling it to radiate at its present rate for twenty-five million years. At that time no one had any idea of time-scales in the Universe and this value appeared to be totally acceptable until well into the twentieth century. Although Helmholtz's idea is not at present relevant to the Sun, it does in fact apply to certain phases of stellar evolution. It anticipated the appearance in scientific thought of the idea of evolution which was resoundingly stated by Charles Darwin in 1859 in the *Origin of Species*. Like living species, the Sun was no longer considered to be unchanging in its properties. Cosmogony, that is the study of the formation and evolution of the heavenly bodies and particularly of the Solar System made a triumphal entry onto the astronomical scene at the same time.

In 1871, the American Lane advanced a theory of the formation of the Sun which anticipated modern ideas, and also foresaw its death as a small cold body after the exhaustion of its gravitational energy. Charles Darwin's son George, made important contributions to evolutionary cosmology by showing that tidal forces exerted on the planets by their satellites and reciprocally (for example by the Moon on the Earth) are able to slow down their rotation in the long run.

While this research into cosmology was developing it is remarkable that no work was done on the problem which interests us most of all: the origin of life. No doubt this was a consequence of the materialism which was deeply engrained in nineteenth-century philosophy. In contrast, much was said about the plurality of inhabited worlds, an idea clearly rather antagonistic to religious beliefs at the time, and adopted enthusiastically by people like Camille Flammarion, whose philosophy was so characteristic of the late nineteenth century. In the early twentieth century people began to doubt the dogma of the plurality of worlds. Certain astronomers, particularly Eddington and Jeans, declared themselves persuaded that our planetary system and life on Earth have a good chance of being unique in the Galaxy and perhaps in the Universe. This debate continues today.

The applications of spectroscopy to astronomy were of course not limited to the Sun. From 1823, Fraunhofer described in detail the spectra of some bright stars, but his untimely death prevented him from continuing with this work, which was only much later taken up again, after Kirchhoff had demonstrated the interest in observing astronomical spectra and comparing them with spectra obtained in the laboratory. William Huggins, a wealthy Englishman, who had constructed a private observatory, and Father Secchi in the Vatican were the main pioneers of astronomical spectroscopy. Huggins recognised in 1863 that similar elements are present upon Earth, in the Sun and in numerous bright stars. From 1863 to 1868, Secchi laid the foundations of the classification of stars by means of the features of their spectra. Huggins did not content himself with studying the spectra of stars with the eye, but succeeded in taking photographs of spectra, at almost the same time as Draper, an American. In 1864, Huggins also recognised emission lines in the spectra of certain nebulae similar to those produced in a rarefied gas subjected to an electric discharge. These were gaseous nebulae. However, other nebulae, notably the Great Nebula in Andromeda, did not show emission lines, and it was correctly surmised that these nebulae were star clusters which were too distant for the individual stars to be resolved. Huggins also studied comets, in which he discovered and identified molecules. Above all, he was the first to determine the velocity of recession of a star from the displacement of its lines due to the Doppler effect.

Following these discoveries, the compilation of large star catalogues commenced, containing star positions, apparent magnitude (measured with rudimentary photometers), spectral type (classified according to an improved version of Secchi's system, still in use), radial velocity, and so on. There was also much interest in the many varieties of double stars and peculiar stars. Not much was yet understood about these objects (they are much discussed in nineteenth-century astronomical literature).

There were only rudimentary ideas about stellar evolution: it was thought that either stars were continuously cooling, or that they were being heated up and later cooling down during their evolution. These incorrect ideas paralysed progress for some time.

To study fainter and fainter stars and nebulae, more and more powerful instruments were constructed. This quest for big telescopes, begun by Herschel towards the end of the eighteenth century, resulted in huge refracting telescopes about 1 metre in diameter, which after their heyday in about 1900 were hardly used any more except for observations of binary stars. The use of reflecting telescopes, invented by Newton, increased in importance. By 1845, William Parsons (Lord Rosse) had constructed a telescope six feet in diameter with which he discovered spiral structure in several nebulae. The quality of this outstanding instrument was considerably improved by the French physicist and astronomer Foucault, who constructed the first glass mirrors. This paved the way for the great telescopes which were to open the era of contemporary astronomy in the twentieth century.

Despite the enormous effort of nineteenth-century astronomers, in constructing instruments and making observations and astronomical calculations, knowledge of the Universe was still sketchy at the dawn of the twentieth century. Although it was known that the Sun and stars contain the same chemical elements as the Earth, the prevalent ideas about their atmospheres, internal constitutions, and evolution were, to say the least, highly speculative. In planetary astronomy, fantasy supplied what was lacking in knowledge. Lastly, there was no idea of the size of the Milky Way, and the existence of other galaxies besides our own was not known. The concept of an infinite Universe, although commonly accepted, remained solely theoretical; and these concepts remained very anthropocentric. In short, the nascent astrophysics was still in the same state as the old astronomy before Kepler and Newton: an abundance of data and of recently accumulated observational facts had not been integrated into sound theories. This was not the fault of astronomy, but of physics: for example, physics had not yet produced a theory of radiation enabling the spectra of stars to be interpreted correctly. However, great hopes were raised as links were forged between astronomy and modern physics: the history of astronomy in the twentieth century is a progression parallel to that of physics. Advances in astronomy have depended very greatly on technological progress, enabling large telescopes to be built, and the advancement of radio astronomy and of space astronomy. The progress of physics and of technology has been so rapid that we need not be surprised that astronomy has known a veritable revolution in eighty years, leading to a complete change of our concept of the Universe. Nor is there any reason to think that this

The Milky Way galactic system of William Herschel, 1784. (Yerkes Observatory)

A large refractor at Meudon Observatory. (Picture G. Servajean, Meudon Observatory)

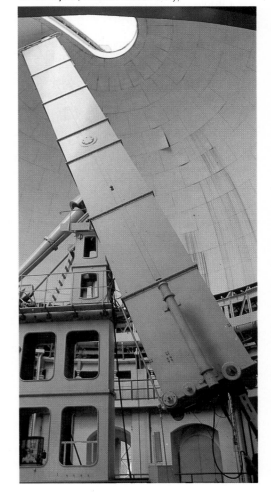

Karl Jansky in front of the antennae which enabled him to discover radio waves from space. (Bell Laboratories)

revolution is over: astronomy shows no sign of stagnation or even breathlessness.

It is difficult to write an objective history of contemporary astronomy. Nonetheless, with a few examples, we can illustrate the evolution of thought and of methods of work in astronomy, and the mutual influences of astronomy, of physics and of technology. We can distinguish three phases. From the beginning of the century up to about 1950, astronomy was still essentially optical astronomy, even though large instruments enabled spectacular discoveries to be made, and theory progressed considerably, laying the foundations used today. From 1950 to 1970, the development of radio astronomy – a new observational technique born outside traditional astronomy – led to spectacular discoveries of several unexpected classes of objects. After 1970, the astronomical scene was further enriched with observations made from space, and techniques of observation and calculation were revolutionised by the computer.

1900–1950. The birth of contemporary astronomy

We will content ourselves here with a few general insights into the relationships of astronomy to atomic and nuclear physics, and an examination of how, thanks to instrumental development, we have begun to form some idea of the true scale of distances within the Universe.

At the beginning of this century, the spectrum of the Sun, of stars and of gaseous nebulae were well known. What was lacking was a theory of such spectra. Empirical rules were discovered describing the relationships between wavelengths of lines of a given element in some simple cases. Balmer's name is particularly remembered in this connection, but Huggins also contributed. In 1913, Bohr constructed a model of the atom which gave the key enabling the coded message in the spectra to be deciphered: this model included Planck's quantum theory, and enabled the interpretation of numerous spectral lines to be reduced to the knowledge of a few atomic energy levels. From then on, progress in analysing astronomical spectra was closely linked to progress in theoretical and laboratory spectroscopy, and even contributed to them. Among the more noteworthy discoveries, we must mention the identification of lines observed in various astronomical objects which do not correspond to any lines detected in the laboratory: Bowen in the United States and Edlén in Sweden were, respectively, the authors of the first identifications of mysterious lines in gaseous nebulae (1927) and in the solar corona (1941). The lines in the solar corona were used to study qualitatively its properties and evolution (notably by Bernard Lyot in France) well before they were identified. It was on these occasions that it was realised how physical conditions in the Universe can differ from those realisable in terrestrial laboratories, and how the study of the stars is consequently useful for studying physics in general. In the United States, England and Germany the theory of the production and transport of radiation was developed. This resulted at least in a partial understanding of the physics of the Sun and the stars. Work on the astronomical aspects of the problem was undertaken by Karl Schwartschild (1905–10), Milne, and above all Eddington (1923). A fundamental by-product of this research was the possibility of determining the qualitative abundances of the elements in the stars from the intensities of their spectral lines. It was only then that astronomers realised that hydrogen is by far the principal constituent of the Universe, followed by helium and the other elements.

The relationship between astronomy and nuclear physics in the first half of the twentieth century was also close. The thermonuclear nature of the Sun's energy is now so well known

that we can scarcely imagine that this idea is relatively recent and that the idea was not at all obvious. We saw previously that, following Helmholtz, it was believed at the beginning of this century that the Sun derives its energy from gravitational contraction. However, the relatively short life that the Sun would have had in this hypothesis was gradually revealed to be incompatible with the very much longer time-scales revealed by the progress of geology. Another hypothesis had to be found. Radioactivity was first considered, then the discovery of the possibility of nuclear transmutation by Rutherford in 1919 revealed more promising sources of energy. In 1929, Eddington proposed, among other possibilities, the conversion of hydrogen to helium: the problem was to know how and where this transformation occurred. This required the collaboration of astronomers – to study the internal structure of stars and there estimate pressure and temperature – and nuclear physicists. Eddington worked out the essential astronomy but it was only in 1938 that C. F. von Weizsäcker and Hans Bethe were able to establish the detailed outlines of the nuclear reactions by which hydrogen is transformed into helium. The details of stellar evolution were still very far from being understood, in particular, why stars are distributed as they are in the surface temperature–luminosity diagram, constructed between 1905 and 1915 by E. Hertzsprung and H. N. Russell. However, it seems clear that astronomy has lent impetus to the development of nuclear research: why should we not one day manage to control it, as the Sun and stars do naturally?

The last important area of astronomical research during the first half of the twentieth century concerned the establishment of a scale of distance for the universe. At the end of the nineteenth century, only the distances of the nearest stars had been measured, by the geometrical method of parallaxes. The Dutchman Kapteyn thought of a statistical method which could go further, and which was the first to give, in 1908, a reasonable idea of the true dimensions of our Galaxy. Just after this, the Henrietta Leavitt showed that certain classes of variable stars have a period of variation which depends on their intrinsic luminosity. In comparing this intrinsic luminosity with the flux received on Earth, one can therefore obtain their distance if the period–luminosity relation has been calibrated by observations of stars of the same type whose distance is known in some other way. After much trial and error, H. Shapley, of Harvard College Observatory applied this method to our Galaxy and showed in 1916–17 that the Sun is not located at the centre. This was the start of a brilliant series of works on galactic structure, dominated by the Swede Lindblad and Jan Oort; for their part, the Americans Trumpler and Adams laid the foundations of our knowledge of interstellar matter. However, the most spectacular discoveries of the whole half-century were definitely the discovery of the extragalactic nature of the numerous nebulae (the first was the Nebula in Andromeda) which we now call galaxies, and the discovery of the expansion of the Universe. These two discoveries were made with the new giant 2.5 metre diameter telescope at Mount Wilson in California, and were due to Edwin Hubble in the period 1924 to 1929. The discoveries were based on an application of the distance criterion of Miss Leavitt to the individual stars in the nearest of these galaxies, on the establishment of new distance criteria and, lastly, on spectroscopic measurements, difficult at that time, of the velocities of recession of the galaxies. The impressive dimensions and time-scales of the observable Universe seemed extraordinary. Paralleling these observational results, theoretical cosmology developed rapidly upon the foundation of the general theory of relativity formulated by A. Einstein in 1916, and soon confirmed by observational results; the whole harmonious ensemble is without doubt one of the greatest scientific achievements of the period 1900–1950.

The first solar tower in the world. This tower, dating from 1925, is situated at Mount Wilson in the United States. Its height is 45 metres and its objective is 30 centimetres in diameter. (Picture Serge Koutchmy, Institut d'astrophysique de Paris CNRS)

1950–1970. Radio astronomy and the new astronomical objects

The period 1940–45 was one of stagnation for astronomy. However, the technical developments realised in pursuit of military goals, particularly in the domain of radar and electronics, had repercussions in astronomy. That was how radio astronomy developed. Of course Karl Jansky had discovered radio emission from the Milky Way in 1931, but this discovery was practically ignored by optical astronomers, who did not know what to make of it. After the war, radio and radar engineers began to construct increasingly sensitive radio antennae and receivers for radio observations of the sky. Radio astronomy was thus born outside traditional optical astronomy, within physics laboratories. A series of striking discoveries soon showed the importance of this new means of making observations and attracted the attention of astronomers. From the start of the 1950s, some groups of radio astronomers joined observatories and regular collaborations were established. Nevertheless, consequences of the initial separation still remain.

The first series of discoveries was to do with the Solar System. It was quickly perceived that solar activity is much more spectacular at radio wavelengths than at optical wavelengths, and that the radio emission comes, at least at metre and decametre wavelengths, from the upper regions of the solar corona, which it is very difficult to observe by any other means. A thorough revision of our knowledge of the outer layers of the Sun resulted. Similarly, radio observations of the planets led to unexpected discoveries such as the high temperature at the surface of Venus and the powerful sporadic radio emissions from Jupiter, recognised much later to come from a magnetosphere somewhat akin to that of the Earth. These observations of the planets were more or less the prerogative of the United States. Whereas Europe and Australia were the principal countries interested in solar radio astronomy.

Two schools of radio astronomy in Britain made crucial contributions to the development of radio telescopes. At Jodrell Bank, near Manchester, a huge single-dish radio telescope was constructed under the inspired leadership of Sir Bernard Lovell. Completed in 1955, the 76-metre dish was for many years the world's largest fully-steerable radio telescope. It was the fore-runner of 90-metre telescopes later constructed in the USA and West Germany. At Cambridge the technique of aperture synthesis was invented. This enabled small radio telescopes to be linked in pairs. By using the Earth's natural rotation, and observing the same object for twelve hours, the Cambridge astronomers could synthesise the angular resolution of large dishes. Synthetic apertures several kilometres across were used to construct a whole series of catalogues of radio sources in the sky. This same technique is employed in the Very Large Array (VLA), in New Mexico, which is currently the world's best interferometer telescope. By linking radio dishes on separate continents, a technique known as Very Long Baseline Interferometry (VLBI) it is possible to resolve to one-thousandth of a second of arc, something even the Space Telescope cannot achieve in the visible region. Sir Martin Ryle was awarded the Nobel Prize in Physics for this amazing technique; he was the first astronomer to get the award.

In 1951, totally new possibilities were opened up for galactic astronomy by the almost simultaneous discovery, in the United States, The Netherlands and Australia, of the 21-centimetre line of hydrogen. Radio waves, unlike light, are not absorbed by interstellar dust: the whole of the Galaxy suddenly became accessible to observations, resulting in the first evidence for the spiral structure of our Galaxy. However it was only towards the end of the 1960s that the theory of density waves, developed by Lin and Shu in the United States, succeeded in giving a satisfactory explanation of the origin of the spiral structure. Another advantage of making observations using the 21-centimetre line is that, with hydrogen being the main constituent of interstellar gas, the intensity of the line gives a measurement of the quantity of gas present in the Galaxy and in those more distant galaxies where it can be observed. These measurements are, however, incomplete because the interstellar medium not only contains atoms, but also molecules. Little was known about these molecules before 1963 when, through the influence of the American physicist Townes, several were discovered through their radio emission. Not only was the existence demonstrated of clouds of dust and gas scarcely suspected beforehand, constituted entirely of molecules, and which are the site of star formation, but some molecules were found to give natural maser emission, of interest to the physicists as well as the astronomers.

Classical astronomy has not gained a great deal from radio astronomy, because the stars are weak radio sources. It is in the theoretical domain that the period 1950–1970 was particularly fruitful. Thanks to the progress of nuclear physics, and also, in the last phase, the use of the first computers which enabled us to bring under control the complex problems posed by the internal structure of stars, it has been possible for the first time to understand the main features of their evolution and show how they synthesise the heavy elements. It is a vast field, where many astronomers have become famous. The chance discovery of pulsars in 1967 by Hewish and Bell at Cambridge in England demonstrated the existence of objects which theorists had predicted from before the war: neutron stars.

The 1950s saw a new expansion of extragalactic astronomy and of cosmology, certainly stimulated by the 200-inch telescope at Mount Palomar, but also connected with three completely unexpected discoveries in radio astronomy: the discovery of intense radio emission from some galaxies (1954), then from quasars (1963), and lastly (perhaps the most important of all) the discovery of radiation from the primaeval Universe at millimetre wavelengths by the Americans Penzias and Wilson in 1967 which earned them the Nobel Prize. (Other Nobel prizes have been awarded for astronomy were to A. Hewish (pulsars) and recently to S. Chandrasekhar and W. Fowler for their work on stellar structure and nuclear physics.) Quasars, the most luminous objects in the Universe, enable us to probe it very deeply. There were great hopes of using this to discover its structure and its geometry. Unfortunately, however, quasars evolve with time, making their use very tricky. The interpretation of the radiation from the primaeval Universe is easier and has confirmed, in a manner difficult to refute, the old idea of the 'primaeval atom' of the Belgian Abbé Lemaître, which dates from about 1930.

We cannot leave the period 1950–1970 without saying a few words about space astronomy which has known a difficult birth in those years: much effort, but also much failure and few results. Nevertheless, space dominates the following period in the same way that radio astronomy dominated the 1950s and the 1960s.

After 1970: space and computers: astronomy tomorrow

It is very difficult to write the history of contemporary astronomy especially because it is as rich as that of astronomy itself, a science whose vitality never ceases to surprise us. We will therefore find in these lines only a short survey of trends in astronomy today.

The most striking trend of all is certainly the prodigious development of space astronomy. It has enabled the Solar System to be explored directly (the spectacular and unexpected results are atill far from being understood), and also the whole electromagnetic spectrum to be observed. Very sophisticated satellites have observed or are currently observing in gamma rays, X-rays, the ultraviolet and the infrared. The classical subject of astrometry can now be extended by observations from space by the European satellite Hipparcos, which can improve the accuracy of stellar position and distance measurements by several orders of magnitude. Space technology is now fully mature: setbacks are comparatively rare, as the hardware has become remarkably reliable. Astronomers today can make observations from space observatories (by sending commands via telecommunications) just as they use a telescope or radio telescope on the ground. The space community, initially isolated like that of radio astronomers, is being integrated with classical astronomers.

Data processing has invaded astronomy like everything else. Computers open up huge possibilities of calculations and the modelling of natural phenomena (this is their most well-known aspect) but they can also pilot very complex instruments from the ground into space, and finally permit an infinitely more precise and advanced analysis of observational data than was possible before. From this results not only an improvement in performance, in flexibility and in facility of use of traditional instruments, but also the possibility of constructing extremely complex instruments, unimaginable without data processing, such as large radio interferometers. Finally, the astronomer can now rapidly use observational techniques with which he is not familiar without special training thanks to the standardisation of the modes of control of instruments. Although this has the disadvantage that the astronomer tends to lose direct contact with the instrument and even with the sky – as far as the public at large are concerned, what does have the romance of traditional astronomy is the possibility that a given astronomer can use several complementary techniques to study the object which he wants to understand.

The contemporary period is characterised finally, by a tendency (not at all new) towards building very large instruments. It is curious that this tendency has been most spectacular in the new domains of observation, particularly in radio astronomy. However optical astronomy is also in a state of flux. The construction of giant telescopes about 10 metres in diameter is envisaged, or even a network of large interconnected telescopes. Obviously, such projects are only rarely possible for a single country and usually require international cooperation. Astronomers have, very happily, practice in such collaborations and often get the projects completed despite economic and political difficulties. Yet is there a place for small, clever, projects on the scale of a nation, or an observatory, or an individual? The answer is yes. It is certain that there will always be a need for small, well-conceived, specialised instruments besides the monsters, just as space exploration of the planets with its immense means will not render obsolete well-conceived observations of the planets from the ground. Getting the balance right, however, is not easy, and this will be one of the main problems of forthcoming decades – giving way neither to the tendency to build gigantic instruments nor to that of dissipation of effort.

James LEQUEUX

200-inch (5-metre) telescope at Mount Palomar in the United States (Hale Observatories)

Observation satellite in the ultraviolet domain (IUE [International Ultraviolet Explorer], NASA)

SKY MAP
See caption page 444

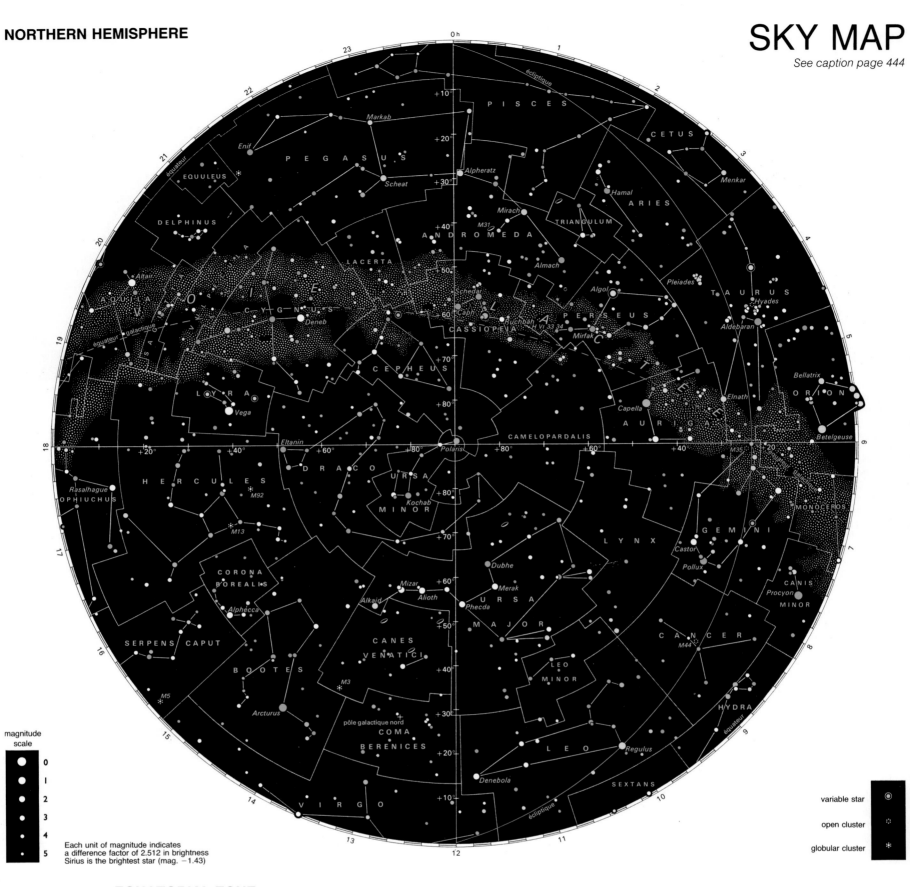

PISCES

CETUS

PEGASUS

Markab

Enif

EQUULEUS

DELPHINUS

Scheat

Alpheratz

ARIES

Hamal

Menkar

équateur

+10°

+20°

+30°

+40°

ANDROMEDA

M31

Mirach

TRIANGULUM

Almach

Algol

Pleiades

TAURUS

Hyades

Aldebaran

écliptique

LACERTA

+50°

Scheda

CASSIOPEIA

NGC VI 33 34

Mirfak

PERSEUS

Elnath

Capella

ORION

Bellatrix

V O I E

CYGNUS

Deneb

Ruchbah

+60°

L A C T É E

AURIGA

Betelgeuse

Altair

AQUILA

équateur galactique

+70°

CEPHEUS

M35

MONOCEROS

LYRA

Vega

CAMELOPARDALIS

+80°

+80°

+60°

+40°

+20°

Eltanin

Polaris

+80°

Rasalhague

OPHIUCHUS

HERCULES

M92

DRACO

URSA

Kochab

MINOR

GEMINI

Castor

Pollux

CANIS

Procyon

MINOR

M13

+70°

LYNX

Dubhe

Merak

Mizar

Alkaid

Alioth

URSA

CORONA

BOREALIS

Alphecca

+60°

Phecda

MAJOR

CANCER

M44

SERPENS CAPUT

CANES

VENATICI

+50°

+40°

LEO

MINOR

HYDRA

BOOTES

M3

équateur

M5

Arcturus

+30°

pôle galactique nord

COMA

BERENICES

+20°

LEO

Regulus

+10°

Denebola

SEXTANS

VIRGO

écliptique

magnitude scale

0
1
2
3
4
5

variable star

open cluster

globular cluster

Each unit of magnitude indicates
a difference factor of 2.512 in brightness
Sirius is the brightest star (mag. −1.43)

EQUATORIAL ZONE

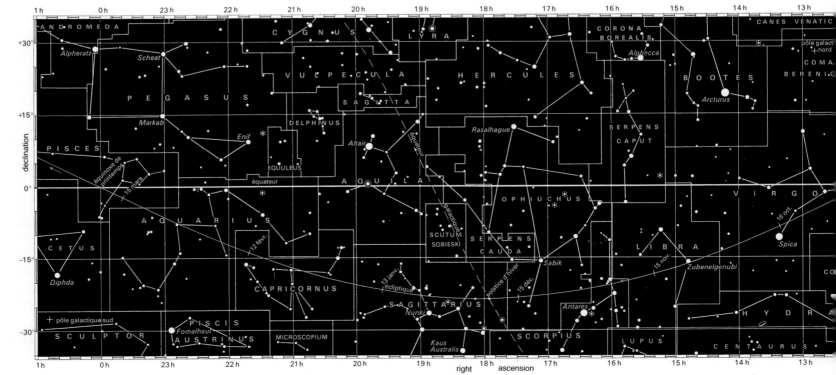

ANDROMEDA

CYGNUS

LYRA

CANES VENATIC.

CORONA
BOREALIS

pôle galact
nord

Alpheratz

Scheat

+30°

VULPECULA

HERCULES

Alphecca

COMA
BERENIC.

PEGASUS

SAGITTA

BOOTES

Markab

DELPHINUS

Arcturus

+15°

Enif

EQUULEUS

Altair

Rasalhague

SERPENS
CAPUT

PISCES

équateur

équateur

AQUILA

0°

VIRGO

15 mars

AQUARIUS

OPHIUCHUS

CETUS

SCUTUM
SOBIESKI

SERPENS
CAUDA

LIBRA

−15°

12 févr.

Sabik

Zubenelgenubi

15 nov.

Diphda

CAPRICORNUS

13 janv.

écliptique

Antares

16 oct.

Spica

solstice d'hiver

15 déc.

HYDRA

pôle galactique sud

SCULPTOR

PISCIS

Fomalhaut

AUSTRINUS

MICROSCOPIUM

SAGITTARIUS

Nunki

SCORPIUS

LUPUS

CENTAURUS

−30°

Kaus
Australis

declination

right ascension

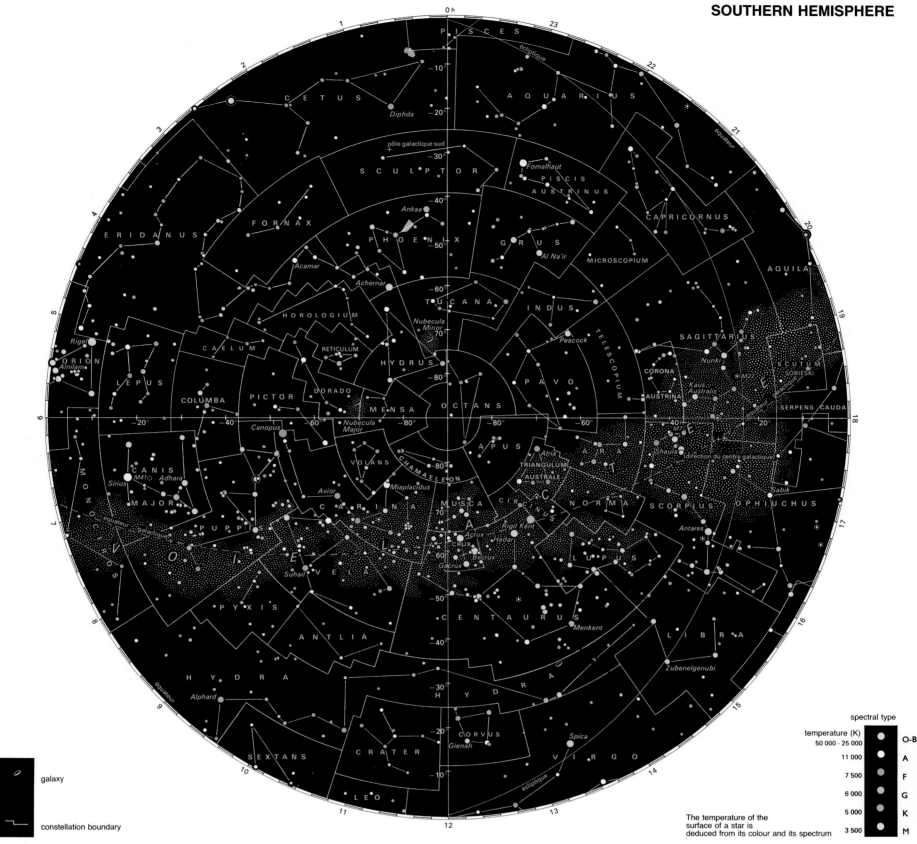

SOUTHERN HEMISPHERE

galaxy

constellation boundary

spectral type

temperature (K)

50 000 - 25 000	O-B
11 000	A
7 500	F
6 000	G
5 000	K
3 500	M

The temperature of the
surface of a star is
deduced from its colour and its spectrum

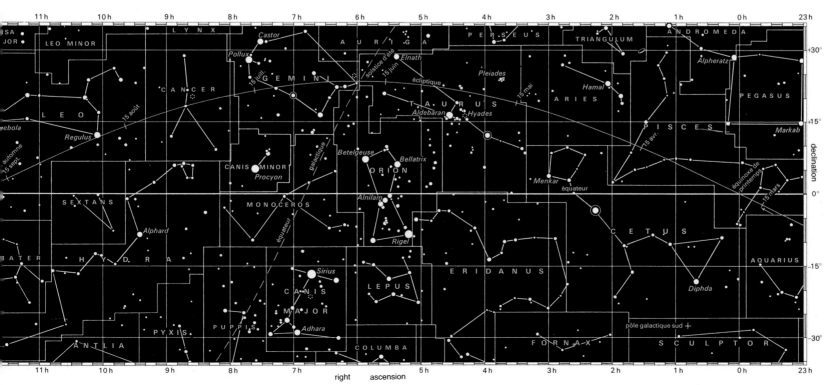

right   ascension

Map of the heavens. The ancients believed that the stars were fixed to a sphere circling the Earth. This concept of a celestial sphere is used in drawing up maps of the heavens today. A system of coordinates very similar to latitude and longitude on terrestrial maps is used to define the positions of stars on this sphere. Remember that latitude measures how far a point is north or south of the Earth's equator, while longitude is defined as the angle between the great circle passing through the point and the two poles; and the great circle passing through Greenwich and the poles. The latter great circle is known as the Prime Meridian.

On the celestial sphere, the north and south poles are defined by the Earth's axis of rotation, and the celestial equator is the projection of the Earth's equator upon the sky. The celestial equivalents to latitude and longitude are declination and right ascension. Declination is the angle between the star and the equatorial plane; as with latitude, declination runs from $-90°$ to $+90°$. Right ascension is measured in hours, from 0 to 24 hours, and is the angle between the star and the position of the Sun at the vernal equinox (on about 21 March) when day and hight are equal. This position of the Sun is known as the First Point of Aries; the celestial equator and the ecliptic (the Sun's path) intersect there, and define the origin point for right ascension.

Our map of the heavens is presented in three parts, the northern hemisphere (left), the southern hemisphere (right), and the part of the sky near to the celestial equator (bottom). The stars are marked on this map by circles whose radii are proportional to their brightness, and their colour varies with the temperature of the star's surface.

The apparent path of the Sun through the sky is called the ecliptic; the Sun makes a complete circuit each year. Besides the vernal equinox, the autumnal equinox and summer and winter solstices are also marked.

The galactic equator follows the bright band of stars known as the Milky Way, which consists of stars of our own Galaxy. The most densely populated region of the Milky Way is in the southern hemisphere, in Sagittarius, and is in the direction of the centre of our Galaxy.

The heavens are divided up into arbitrary regions within which the brightest stars (having the smallest magnitudes) form patterns called constellations. The constellations in the northern sky are of great antiquity; the twelve constellations through which the Sun passes are known as the Zodiac.

One of the stars in the Little Bear (Ursa Minor) lies within a degree of the north celestial pole. The star is for this reason known as the Pole Star, North Star, or Polaris. The South celestial Pole lies in the constellation of the Octant (Octans).

In the southern constellations Doradus and Toucan there are two masses of stars visible to the naked eye, Nubecula Major and Nubecula Minor. They are two irregular galaxies, satellites of our Galaxy, known as the Large and the Small Magellanic Clouds. The map shows some galaxies like M31 in Andromeda, which are visible to a good eye or with binoculars. Also shown are globular clusters like M13 in the constellation of Hercules, and certain galactic clusters such as M35 in Gemini. In Taurus for example, the Crab Nebula is the remnant of the best known supernova.

This map of the sky can therefore be used to pinpoint the brightest stars.

Anyone with a small telescope can use it to try to find particularly interesting objects like nebulae or globular clusters.

Bibliography

GENERAL REFERENCE

Stuart Ross Taylor, *Solar System Evolution - A New Perspective*, Cambridge Univ. Press, 1993

Erika Böhm-Vitense, *Introduction to Stellar Astrophysics*, Cambridge Univ. Press, 1992

Jean Audouze, Guy Israel and Jean-Claude Falque, *The Cambridge Atlas of Astronomy*, Cambridge Univ. Press, 1994

William H. Donahue, *Johannes Kepler, New Astronomy*, Cambridge Univ. Press, 1992

Nigel Henbest, Heather Couper and M. Marten, *The Guide to the Galaxy*, Cambridge Univ. Press, 1994

David F. Malin, *A View of the Universe*, Cambridge Univ. Press, 1993

William Liller and Ben Mayer, *The Cambridge Astronomy Guide*, Cambridge Univ. Press, 1992

Stephen P. Maran, *The Astronomy and Astrophysics Encyclopaedia*, Cambridge Univ. Press, 1992

Michael Rycroft, *The Cambridge Encyclopaedia of Space*, Cambridge Univ. Press, 1990

O. Abell, D. Morrison and S. C. Wolfe, *The Realm of the Universe*, Saunders, 5th ed., 1992

THE SUN

Peter R. Wilson, *Solar and Stellar Activity Cycles*, Cambridge Univ. Press, 1994

Manfred Schüssler and Wolfgang Schmidt, *Solar Magnetic Fields*, Cambridge Univ. Press, 1994

Judit Pap, *The Sun as a Variable Star*, Cambridge Univ. Press, 1994

John K. Hargreaves, *The Solar-Terrestrial Environment*, Cambridge Univ. Press, 1992

J. H. K. Phillips, *Guide to the Sun*, Cambridge Univ. Press, 1992

P. V. Foukal, *Solar Astrophysics*, John Wiley and Sons, 1990

A. N. Cox, W. C. Livingston and M. S. Matthews eds, *Solar Interior and Atmosphere*, Univ. Arizona Press, 1991

K. Haubauer, *Exploring the Sun, Solar Science Since Galileo*, New Series in NASA History, The Johns Hopkins Univ. Press, 1991

THE SOLAR SYSTEM

Patrick Moore, *New Guide to the Planets*, Sidgwick and Jackson, 1993

N. Henbest, *The Planets: Portraits of New Worlds*, Penguin Books, 1994

N. Henbest and H. Couper, *The Planets*, Pan Books, 1985

G. J. Consolmagno and M. W. Schaffer, *Worlds Apart: A Textbook in Planetary Sciences*, Prentice Hall US, 1993

T. Padmanabhan, *Structure Formation in the Universe*, Cambridge Univ. Press, 1993

A. S. Eddington, *The Internal Constitution of the Stars*, Cambridge Univ. Press, 1988

J. K. Beatty, B .O'Leary and A. Chaikin (eds.) *The New Solar System*, Cambridge Univ. Press-Sky Publishing Corporation, 3rd ed, 1990

W. K. Hamblin and E. H. Christiansen, *Exploring the Planets*, Macmillan, 1990

K. R. Lang and C. A. Whitney, *Wanderers in Space*, Cambridge Univ. Press, 1991

Mercury

R. G. Strom, *Mercury: The Elusive Planet*, Cambridge Univ. Press, 1987

F. Villas, C. R. Chapman and M. S. Matthews (eds.), *Mercury*, Univ. Arizona Press, 1988

Venus

G. Hunt and P. Moore, *The Planet Venus*, Faber and Faber, 1982

E. Burgess, *An Errant Twin*, Colombia Univ. Press, 1985

P. Cattermole, *Venus: The New Geology*, UCL Press, 1994

'Magellan at Venus', *Journal of Geophysical Research*, vol. XCVII, E8, p. 13063, 1992

The Earth

Paul Hodge, *Meteorite Craters and Impact Structures of the Earth*, Cambridge Univ. Press, 1994

G. C. Brown, C. J. Hawkesworth and R. L. C. Wilson (eds.), *Understanding the Earth*, Cambridge Univ. Press, 1985

P. Cattermole and P. Moore, *The Story of the Earth*, Cambridge Univ. Press, 1985

J. K. Hargreaves, *The Solar-Terrestrial Environment*, Cambridge Univ. Press, 1992

The Moon

P. D. Spudis, *The Geology of Multi-Ring Impact Basins – The Moon and other Planets*, Cambridge University Press, 1993

P. H. Cadogan, *The Moon: Our Sister Planet*, Cambridge Univ. Press, 1981

D. Baker, *The Moon: Physics, Geology and Evolution*, Ellis Horwood, 1993

I. Ridley, *Understanding the Moon*, Chapman and Hall, 1994

Mars

J. Blunck, *Mars and its Satellites*, Exposition Press Inc., 1982

M. H. Carr, *The Surface of Mars*, Yale Univ. Press, 1981

H. Kieffer *et al.* (eds.), *Mars*, Univ. Arizona Press, 1992

'Mars Observer', *Journal of Geophysical Research*, vol. XCVII, E5 p. 7663, 1992

J. Oberg, *Mission to Mars* , Stackpole Books, 1982

Viking Orbiter Views of Mars, NASA SP-441, NASA, 1982

The Asteroids

J. Erickson, *Target Earth!: Asteroid Collisions Past and Future*, Tab Books, 1991

R. P. Binzel, T .Gehrels and M. S. Matthews (eds.), *Asteroids II*, Univ. Arizona Press, 1989

C. R. Chapman and D. Morrison, *Cosmic Catastrophes*, Plenum Press, 1989

J. C. Cunningham, *Introduction to Asteroids*, Willmann-Bell, 1988

Jupiter

John H. Rogers, *The Giant Planet Jupiter*, Cambridge Univ. Press, 1995

E. Burgess, *By Jupiter, Odysseys to a Giant*, Colombia Univ. Press, 1982

D. Morrison (ed.), *Satellites of Jupiter*, Univ. Arizona Press, 1982

D. Morrison and J. Samz, *Voyage to Jupiter*, NASA SP-439, NASA, 1980

'Ulysses at Jupiter', *Science*, vol. CCLVII, p. 1503, 1992

C. M. Yeates *et al.*, *Galileo: Exploration of Jupiter's System*, NASA SP-479, NASA, 1985

Saturn

T. Gehrels and M. S. Matthews (eds.), *Saturn*, Univ. Arizona Press, 1984

D. Morrison, *Voyages to Saturn*, NASA SP-451, NASA, 1982

A. F. O'D. Aleksander, *The Planet Saturn : A History of Observation, Theory and Discovery*, Dover, 1980.

'Pioneer Saturn', *Science*, vol. CCVII, p. 400, 1980

'Voyager 1', *Science*, vol. CCXII, p. 159, 1981

'Voyager 1', *Science*, vol. CCXCII, p. 675, 1981

'Voyager 2', *Science*, vol. CCXV, p. 499, 1982

Uranus

Garry Hunt and Patrick Moore, *Atlas of Uranus*, Cambridge Univ. Press, 1989

E. D. Miner, *Uranus: Planets, Rings and Satellites*, Ellis Horwood, 1990

E. Burgess, *Uranus and Neptune: The Distant Giants*, Colombia Univ. Press, 1988

J. T. Bergstrahl, E. D. Miner and M. S. Matthews (eds.), *Uranus*, Univ. of Arizona Press, 1991

'Voyager 2', *Science*, vol. CCXXXIII, p. 39, 1986

'The Voyager 2 Encounter with Uranus', *Journal of Geophysical Research*, vol. XCII, p. 14873, 1987

Neptune

Garry E. Hunt and Patrick Moore, *The Atlas of Neptune*, Cambridge Univ. Press, 1994

E. Burgess, *Far Encounter: The Neptune System*, Colombia Univ. Press, 1992

P. Moore, *The Planet Neptune*, Ellis Horwood, 1988

M. Grosser, *The Discovery of Neptune*, Dover, 1979

'The Neptune System: a Post-Voyager Perspective', *Geophysical Research Letters*, vol. XVII, pp. 1643–1776, 1990

'Triton', *Science*, vol. CCL, pp. 410–443, 1990

'The Voyager 2 Encounter with the Neptunian System', *Science*, vol. CCXLVI, pp. 1417–1501, 1989

Pluto

R. P. Binzel, 'Pluto', *Scientific American*, vol. CCLXII, no. 6, p. 50, 1990

A. White, *The Planet Pluto*, Pergamon Press, 1980

Comets

Stephen J. Edberg and David H. Levy, *Observing Comets, Asteroids, Meteors, and the Zodiacal Light*, Cambridge Univ. Press, 1994

M. E. Bailey, S. V. M. Clube and W. M. Napier, *The Origin of Comets*, Pergamon Press, 1990

J. C. Brandt and R. D. Chapman, *Introduction to Comets*, Cambridge Univ. Press, 1981

J. C. Brandt and R. D. Chapman, *The New Comet Book*, W. H. Freeman, 1991

'Encounters with Comet Halley. The first results', *Nature*, vol. CCCXXI, p. 259, 1986

M. Grewing, F. Praderie and R. Reinhard (eds.), *The Exploration of Halley's Comet*, Springer-Verlag, 1988

B. G. Marsden, *Catalogue of Cometary Orbits*, Smithsonian Astrophysical Observatory, 6th ed., 1989

R. L. Newburn, M. Neugebauer and J. Rahe (eds.), *Comets in the Post-Halley Era*, Kluwer, Dordrecht, 1990

F. L. Whipple, *The Mystery of Comets*, Smithsonian Institutional Press, 1985

L. L. Wilkening (ed.), Comets, Univ. Arizona Press, 1982

THE STARS AND THE GALAXY

The Hertzsprung–Russell diagram

Erika Böhm-Vitense, *Introduction to Stellar Astrophysics*, Cambridge Univ. Press, 1991

Systems of Reference and Determining Distances

L. V. Morrison and G. F. Gilmore, *Galactic and Solar System Optical Astrometry*, Cambridge Univ. Press, 1994

J. Franco, L. Aguilar, S. Lizano and E. Daltabuit, *Numerical Simulations in Astrophysics*, Cambridge Univ. Press, 1994

S. Van Den Bergh and C. J. Pritchet (eds.), *The Extragalactic Distance Scale*, Kluwer, 1988

A. Van Helden, *Measuring the Universe*, Univ. Chicago Press, 1986

The Evolution of the Stars

Lawrence Aller, *Atoms, Stars and Nebulae*, Cambridge Univ. Press, 1991

J. Franco, F. Ferrini and G. Tenorio-Tagle, *Star Formation, Galaxies and the Interstellar Medium*, Cambridge Univ. Press, 1993

N. Prantzos, E. Vangioni-Flam and M. Cassé, *Origin and Evolution of the Elements*, Cambridge Univ. Press, 1993

Roger J. Tayler, *The Stars: Their Structure and Evolution*, Cambridge Univ. Press, 1994

R. Weinberger and A. Acker, *Planetary Nebulae*, Kluwer, 1993

D. C. Black and M. S. Matthews (eds.), *Protostars and Planets II*, Univ. Arizona Press, 1985

A. Maeder and A. Renzini (eds.), *Observational Tests of the Stellar Evolutionary Theory*, Reidel, 1984

T. X. Thuan, T. Montmerie and J. Tran Thanh Van (eds.), *Starbursts and Galaxy Evolution*, Edition Frontières, 1987

Novae and Supernovae

Richard McCray, *Supernovae and Supernova Remnants*, Cambridge Univ. Press, 1994

A. Jevicki and C. Tan, *Particle Strings and Supernovae*, World Scientific Publishing, 1989

P. Murdin, *Supernovae*, Cambridge Univ. Press, 1985

R. A. Chevalier, 'Supernova 1987A at five years of age', *Nature*, vol. CCCLV, p. 691, 1992

M. Livio and G. Shaviv (eds.), *Cataclysmic Variables and Ralted Objects*, Reidel, 1983

A. G. Petschek (eds.), *Supernovae*, Springer-Verlag, 1990

V. Trimble, 'Supernovae', *Review of Modern Physics* part 1, vol. LIV, no. 4, 1982; part #2, vol. LV, no. 2, 1983

S. E. Woosley and T. A. Weaver, 'The Great Supernova of 1987', *Scientific American*, vol. CCLXI, no. 2, p. 32, 1989

Neutron Stars and Pulsars

Vassily Beskin, Alexander Gurevich and Yakov Istomin, *Physics of the Pulsar Magnetosphere*, Cambridge Univ. Press, 1993

K. A. Van Riper, R. J. Epstein and C. Ho, *Isolated Pulsars*, Cambridge Univ. Press, 1993

James M. Nemec and Jaymie M. Matthews, *New Perspectives on Stellar Pulsation and Pulsating Variable Stars*, Cambridge Univ. Press, 1993

J. M. Irvine, *Neutron Stars*, Oxford Univ. Press, 1978

V. M. Lipunov, *Astrophysics of Neutron Stars*, Springer-Verlag, 1992

A. G. Lyne and F. Graham-Smith, *Pulsar Astronomy*, Cambridge Univ. Press, 1990

S. L. Shapiro and S. A. Teukolsky, *Black Holes, White Dwarfs and Neutron Stars*, Wiley, 1983

Black Holes

Igor Novikov, *Black Holes and the Universe*, Cambridge Univ. Press, 1990
Kip S. Thorne, Richard H. Price and Douglas A. MacDonald, *Black Holes the Membrane Paradigm*, Yale Univ. Press, 1986
Stephen W. Hawking, *Hawking on the Big Bang and Black Holes*, World Scientific Publishing, 1992
J. A. H. Flutterman, F. A. Handler and R. A. Matzner, *Scattering from Black Holes*, Cambridge Univ. Press, 1988

Wolf-Rayet Stars

C. W. De Loore and A. J. Willis (eds.), *Wolfe Rayet Stars: Observations, Physics, Evolution*, Reidel, 1982
M. C. Lortet and A. J. Willis (eds.), *Wolfe Rayet Stars: Progenitors of Supernovae?*, Observatoire de Paris, 1983
K. Van Der Hucht and B. Hidayat, *Wolfe-Rayet Stars and Interrelations with other Massive Stars in Galaxies*, Kluwer, 1991

Stellar Nucleosynthesis

N. Prantzos, E. Vangioni-Flam and M. Cassé, *Origin and Evolution of the Elements*, Cambridge Univ. Press, 1993
R. E. S. Clegg, W. P. S. Meikle and I. R. Stevens, *Circumstellar Media in Late Stages of Stellar Evolution*, Cambridge Univ. Press, 1994
Juhan Frank, Andrew R. King and Derek J. Raine, *Accretion Power in Astrophysics*, Cambridge Univ. Press, 1992
J. Audouze and S. Vauclair, *An Introduction to Nuclear Astrophysics: The Formation and Evolution of Matter in the Universe*, Reidel, 1980

Binary Stars

Antoine Duquennoy and Michel Mayor, *Binaries as Tracers of Stellar Formation*, Cambridge Univ. Press, 1992
Walter H. G. Lewin, Jan Van Paradijs and Edward P. J. Van Den Heuvel, *X-ray Binaries*, Cambridge Univ. Press, 1995
J. E. Pringle and R. A. Wade, *Interacting Binary Stars*, Cambridge Univ. Press, 1985

Cosmic Rays

R. B. Partridge, *3K: The Cosmic Microwave Background Radiation*, Cambridge Univ. Press, 1995
V. S. Berezinski, S. V. Bulanov, V. A. Dogiel and V. L. Ginzburg, *Astrophysics of Cosmic Rays*, North-Holland, 1990
M. S. Longair, *High Energy Astrophysics*, Cambridge Univ. Press, 1994
P. Sokolsky, *Introduction to Ultrahigh Energy Cosmic Ray Physics*, Addison-Wesley, 1989

THE EXTRAGALACTIC DOMAIN

The Galaxies

Roger J. Taylor, *Galaxies: Structure and Evolution*, Cambridge Univ. Press, 1993
Jay M. Pasachoff, Hyron Spinrad, Patrick Osmer and Edward S. Cheng, *The Farthest Things in the Universe*, Cambridge Univ. Press, 1994
A. C. Fabien, *Clusters and Superclusters of Galaxies*, 1992
W. Oegerle, M. Fitchett and L. Danly, *Clusters of Galaxies*, Cambridge Univ. Press, 1990
B. J. T. Jones and J. E. Jones, *The Origin and Evolution of Galaxies*, Reidel, 1982
J. L. Sersic, *Extragalactic Astronomy*, Reidel, 1982
H. Verenberg, *Atlas of Deep-Sky Splendours*, Cambridge Univ. Press–Sky Publishing Corporation, 4th ed., 1983

Radio Sources – Quasars

R. J. Davis and R. S. Booth, *Sub-arcsecond Radio Astronomy*, Cambridge Univ. Press, 1993
K. Kellermann and I. I. K. Pauliny-Toth, 'Compact radio Sources', *Annual Review of Astrononomy and Astrophysics* vol. XIX, p. 373, 1981
J. S. Miller (ed.), *Astrophysics of Active Galaxies and Quasi-Stellar Objects*, Oxford Univ.Press, 1985
H. R. Miller and P. J. Whita (eds.), *Variability of Active Galactic Nuclei*, Cambridge Univ. Press, 1991
J. Roland, H. Sol and G. Pelletier (eds.), *Extragalactic Radio Sources. From Beams to Jets*, Cambridge Univ. Press, 1992
E. Valtaoja and M. Valtonen (eds.), *Variability of Quasars*, Cambridge Univ. Press, 1991
R. J. Weyman, R. F. Caswell and M. G. Smith, 'Absorption lines in the spectra of quasi stellar objects', in *Annual Review of Astronomy and Astrophysics*, vol.XIX, p. 41, 1981

COSMOLOGY

G. G. Fazio and R. Silberberg, *Currents in Astrophysics and Cosmology*, Cambridge Univ. Press, 1993
Jayant V. Narlikar, *Introduction to Cosmology*, Cambridge Univ. Press, 1993
Mark Srednicki, *Particle Physics and Cosmology*, North-Holland, 1990
John McLeish, *Cosmology: Meanings of the Universe*, Bloomsbury, 1993
S. Weinberg, *The First Three Minutes*, Basic Books, 1993
Matts Roos, *Introduction to Cosmology*, John Wiley and Sons, 1994
G. O. Abell and G. Chincarini (eds.), *Early Evolution of the Universe and Its Present Structure*, Reidel, 1983

R. Harrison, *Cosmology*, Cambridge Univ. Press, 1982
J. N. Islam, *The Ultimate Fate of the Universe*, Cambridge Univ. Press, 1981
D. W. Sciama, *Modern Cosmology and the Dark Matter Problem*, Cambridge Univ. Press, 1993
J. Silk, *The Big Bang: The Creation and Evolution of the Universe*, W. H. Freeman and Co., 1988

LIFE IN THE UNIVERSE

Ben Zuckerman and Michael H. Hart, *Extraterrestrials, Where are They?*, Cambridge Univ. Press, 1995
Joseph A. Angelo, *The Extraterrestrial Encyclopaedia: Our Search for Life in Outer Space*, Facts on File Inc., 1985
D. Goldsmith and T. Owen, *The Search for Life in the Universe*, Addison-Wesley, 1992
A. Hansson, *Mars and the Development of Life*, Ellis Horwood, 1991
J. Billingham (ed.), *Life in the Universe*, MIT Press, 1981
F. Dyson, *Origins of Life*, Cambridge Univ. Press, 1986
M. D. Papagiannis (ed.), *The Search for Extraterrestrial Life: Recent Developments*, Reidel, 1985

THE HISTORY OF ASTRONOMY

Alexander S. Sharov and Igor D. Novikov, *Edwin Hubble, the Discoverer of the Big Bang Universe*, Cambridge Univ. Press, 1993
M. Hoskin (eds.), *The General History of Astronomy, Vol.IV, Astrophysics and Twentieth-Century Astronomy to 1950*, Part A, O. Gingerich (ed.), Cambridge Univ. Press, 1984
O. Pedersen, *Early Physics and Astronomy*, Cambridge Univ. Press, 1993

OBSERVATIONAL ASTRONOMY

Wil Tirion, *Cambridge Star Atlas 2000*, Cambridge Univ. Press, 1991
Robert C. Smith, *Observational Astrophysics*, Cambridge Univ. Press, 1995
Jack Newton and Philip Teece, *The Guide to Amateur Astronomy*, Cambridge Univ. Press, 1994
Patrick Martinez, *The Observer's Guide to Astronomy*, Cambridge Univ. Press, 1994
William Liller, *The Cambridge Guide to Astronomical Discovery*, Cambridge Univ. Press, 1992
David Levy, *The Sky*, Cambridge Univ. Press, 1993
Alan Hirshfeld, Roger W. Sinnott and François Ochsenbein, *Sky Catalogue 2000.0*, Cambridge Univ. Press, 1992
C. Sutton, *Spaceship Neutrino*, Cambridge Univ. Press, 1992

About the Authors

Monique Arduini-Malinovski. Director of astronomy and solar physics programmes in the Scientific Programmes Division of the Centre National d'Etudes Spatiales.

Solar and space physics; analysis of X-ray spectroscopy of the solar corona; atomic physics related to solar physics.

Jean Audouze. Directeur de Recherche in the CNRS, former Director of the Institut d'Astrophysique, Paris, lecturer at the Ecole Polytechnique.

Nucleosynthesis and galactic evolution: primordial and explosive nucleosynthesis; evolution of galaxies; origins of cosmic rays and the solar system.

Jean-Loup Bertaux. Directeur de Recherche in the CNRS, Service D'Aeronomie du CNRS (Verrièes-le-Buisson).

Specialist in the physics and composition of the interplanetary medium, comets, and planetary atmospheres; principal investigator and coinvestigator of many space missions (Vega, Venera, Voyager, Mars–94/96 . . .).

Jean-Pierre Bibring. Professor at Université de Paris–XI (Orsay), Labaratoire René-Bernas (Orsay).

Study of irradiation of small lunar and meteoritic grains (implantation effect of solar-system silicates); collection and study of interplanetary grains in the upper atmosphere.

André Boischot. Astronome Titulaire at the Observatoire de Paris-Meudon.

Solar physics; radioastronomical study of planetary magnetospheres and the origin of their electromagnetic emission; director of large network of decametric antennae of the radioastronomical observatory at Nancay; coinvestigator of the Voyager mission.

Patrick Boissé. Deputy Director of the Department of Physics at the Ecole Normale Superieure.

Absorption-line systems in quasars; study of the structure of interstellar clouds.

André Brahic. Professor at Université de Paris-VII, Astrophysicist at the Observatoire de Paris-Meudon.

Dynamics of stars and the Solar System; formation of the rings of the giant planets; coinvestigator of the Voyager mission.

Michel Cassé. Astrophysicist at the Commissariat à l'Energie Atomique. Conseiller Scientific at the Grande Bibliothèque de France.

Stellar evolution and nucleosynthesis; cosmic rays; gamma rays.

Anny Cazenave. Ingénieur at the Centre Nationale d'Etudes Spatiales, director of the research group for spacy geodesy (CNES, Toulouse).

Specialist in dynamical problems of Solar System formation in general, and the Earth in particular.

Jean-Pierre Chièze. Ingénieur at the Commissariat á l'Energie Atomique (Military applications, Bruyères-le-Chatel).

Physics of the interstellar medium (interstellar clouds); formation and evolution of galaxies; supernova explosions.

Florence Durret. Agrégée de l'Université, Doctor of Astrophysics, Assistant at the Observatoire de Paris-Meudon, Researcher at the Institut d'Astrophysique (CNRS, Paris).

Active galatic nuclei, Seyfert galaxies, giant HII regions: optical, ultraviolet and X-ray observations.

Daniel Gautier. Directeur de Recherche in the CNRS, director of the Infrared Astronomy Laboratory at the Observatoire de Paris-Meudon.

Specialist in the physics and composition of the atmospheres of giant planets; coinvestigator of the Voyager and Cassini missions.

Michèle Gerbaldi. Maître de Conferences at the Université de Paris-XI (Orsay), Researcher at the Institut d'Astrophysique (CNRS, Paris).

Specialist in stellar atmospheres; spectrophotometric study of the atmospheres of hot stars with peculiar chemical composition, founder member of the association Comité de Liaison Enseignants et Astronomes.

Guy Israël. Directeur de Recherche in the CNRS, Service d'Aeronomie of the CNR (Verriéres-le-Buisson).

Specialist in the physics of planetary atmospheres; principal investigator for the Vega mission to Venus and the Cassini/Huygens mission.

Yves Langevin. Chargé de Recherche in the CNRS, Laboratoire René-Bernas (Orsay).

Planetologist; specialist in the irradiation of the lunar surface; physics of the regolith.

James Lequeux. Astronome de Premiére Classe, Observatoire de Paris-Meudon.

Galactic and extragalactic astrophysics; physics of the interstellar medium; chemical evolution of galaxies; research into the stellar populations of the Magellanic Clouds and unevolved galaxies (irregular and blue compact galaxies).

Jean-Pierre Luminet. Chargé de Recherche in the CNRS, Astrophysicist at the Observatoire de Paris-Meudon.

Application of General Relativity to astrophysics; giant black holes in galactic nuclei, primordial structure and cosmological models.

Philippe Masson. Professeur at the Université de Paris-XI, Laboratoire de Géologie Dynamique Interne (Orsay).

Geologist and planetologist; associated with the NASA planetary geology programme; specialist in comparative planetology and the study of large-scale tectonic structure.

Pierre Mein. Astronome at the Observatoire de Paris-Meudon.

Specialist in solar physics (detailed studies of the Sun's atmosphere).

Jacques Paul. Ingénieur at the Commissariat á l'Energie Atomique (Direction de Sciences de la Matière, Service d'Astrophysique, Saclay).

Gamma ray astronomy; coleader of the SIGMA (imaging gamma-ray satellite) project.

Jacques Roland. Chargé de Recherche in the CNRS, Institut d'Astrophysique, Paris.

Quasars and active galaxies; optical and radio observations; determination of inter-galactic magnetic fields.

Hélène Sol. Chargée de Recherche in the CNRS, Observatoire de Paris-Meudon.

Active galactic nuclei and quasars; extragalactic jets; extragalactic magnetic fields.

Pierre Thomas. Professeur at the Ecole Normale Supérieure de Lyon.

Tectonics of the telluric planets and satellites; geological influence of large impact basins; fracture phenomena of the Earth's crust and the connection with vulcanism.

Catherine Turon. Astronome at the Observatoire de Paris-Meudon.

Leader of the Inca consortium, charged with producing the imput catalogue for the Hipparcos astrometric satellite; space astrometry.

Elisabeth Vangioni-Flam. Ingénieur de Recherche in the CNRS, Institute d'Astrophysique, Paris.

Explosive nucleosynthesis; evolution of cosmic rays; primordial nucleosynthesis; chemical evolution of galaxies; origin of the light elements.

Jean-Pierre Verdet. Astronome at the Observatoire de Paris-Meudon.

Historian of astronomy, mainly up to the eighteenth century.

Jean-Claude Vial. Directeur de Recherche in the CNRS, Laboratoire de Physique Stellaire et Planetaire du CNRS (Verriéres-le-Buisson).

Specialist in solar physics from space (chromosphere and transition zone).

Alfred Vidal-Majdar. Directeur de Recherche in the CNRS (Institut d'Astrophysique, Paris), Maître de Conferences at the Ecole Polytechnique.

Abundances of the interstellar medium with cosmological applications; study of the hot phases of the interstellar medium; ultraviolet space physics.

Laurent Vigroux. Ingénieur in the Commissariat á l'Energie Atomique (Institut de Recherche Fondamentale, Service d'Astrophysique, Saclay).

Observation and evolution of galaxies; design, building and use of a charge-coupled device camera for extragalatic astronomy; nucleosynthesis.

Acknowledgements. Many researchers and organisations have allowed us to use photographs and other material: it is impossible to cite all of them, but we should particularly like to thank Raymond M. Batson, of the US Geological Survey, who supplied almost all of the planetary maps; James I. Vette, Director of the World Data Center A for Rockets and Satellites, who supplied the images from the National Space Science Data Center (referred to as NASA/NSSDC); the Jet Propulsion Laboratory (images referred to as JPL/NASA); Philippe Masson; S. Ichtiaque Rasool; Margaret B. Weems, of the National Radio Astronomy Observatory, who supplied the VLA images (referred to as NRAO), Kurt Kjär, of the European Southern Observatory (ESO photographs). Photographs referenced as POSS are part of the Palomar Sky Survey, copyright of the National Geographic Society.

Index

This index is selective. It does not duplicate the list of contents, or cover every single item in the book, or every appearance of those entries included. We have tried to provide a guide which will be useful on several levels, and be wide-ranging but still intelligible.

To avoid duplication, a glossary of astronomical terms is combined with the index; glossary entries naturally have no page references.

Page references in Roman type refer to the text: those in italics refer to illustrations and captions.

Aananin Rhea crater *207* (40 N, 330)
Abbe Lunar crater *111* (58 S, 175 E)
Abbe de la Caille catalogue 437
Abbot, Charles (1872–1973) 49
Abell catalogue *399*; **Abell 85** *402*; **Abell 370** *401*; **426** (Perseus Cluster) *397, 400, 401, 402*; **Abell 478** *402*; **Abell 1367** *402*; **Abell 1656** (Coma Berenices) *13, 325, 394, 396, 397, 398, 399*; **Abell 1795** *402*; **Abell 2147** *402*; **Abell 2151** (Hercules Cluster) *336*; **Abell 2199** *402*; **Abell 2256** *402*; **Abell 2319** *402*; **Abell 2634** *402*; **Abell 2666** *402*
Abel Lunar crater *111* (36 S, 85 E)
Abenezra Lunar crater *111* (21 S, 12 E)
Absorption Decrease in the intensity of radiation as a consequence of part of its energy being used to excite or ionise matter (gas, dust) between the emitter and the observer.
Abu Nuwas Mercury crater *75* (17.5 N, 21)
Abul Wafa Lunar crater *111* (2 N, 117 E)
Abulfeda Lunar crater *111* (14 S, 14 E)
Abundances of the chemical elements The abundance of a chemical element is the ratio of the number of atoms (or ions) of this element per unit volume of the object or region considered to the corresponding number of hydrogen atoms. The mass abundance of a given element is the ratio of its mass density to the total mass density.

We can thus define the abundances of all observable chemical elements from hydrogen to uranium in regions and objects as diverse as the Sun and Solar System, stars, the interstellar medium, cosmic rays, and other galaxies.

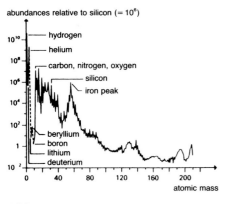

Acamar θ Eridani. Star *433*. See **Eridanus**
Acceleration parameter 406
Accretion. Gravitational capture of matter by an astronomical body. Accretion is the opposite of mass loss. For example, in a close binary system, one of the stars may increase its mass by accreting matter from its companion star. *60, 61, 62, 62, 63, 66, 70, 82, 121, 121, 162, 264, 283.*
Accretion disc *197, 283, 290, 291, 303, 380, 380*
Achelous Ganymede crater *183* (66 N, 4)
Achernar (α Eridani). Star *253, 342, 443*. See **Eridanus**
Achilles Trojan asteroid *162*
Achondrites. Rocky meteorites devoid of chondrites 130, *130, 162*
Acidalia Planitia. Mars *137* (from 14 to 55 N, and 0 to 60).
Acrux (α Crucis). Star *253, 443*. See **Crux**
Active galaxies 370–371, 386, *395, 401, 403*
Adad Ganymede crater *183* (62 N, 352)
Adal Callisto crater *178* (77 N, 79)
Adams, John Couch (1819–92) 218, 438, 440
Adams Lunar crater *111* (32 S, 69 E)
Adams Martian crater *139* (31 N, 227).
Adapa Martian crater *183* (83 N, 22)
Adhara (ε Canis Majoris). Star *443*. See **Canis Major**
Adiabatic A change of a system (change of physical state, temperature, pressure or density variation. . .) is called adiabatic if there is no exchange of energy with the surroundings.
Adjua Rhea crater *207* (48 N, 127)
Adlinda Callisto crater *178* (58 S, 20)
Adonis Asteroid *162*
Adonis Linea Europa *181*
Adrastea Jovian satellite *57, 174, 175, 175*
Adrastus Dione crater *206* (62 S, 35)
Advection of heat 96
Adventure Rupes Mercury *75* (64 S, 63)
Aeneas Trojan asteroid *162*
Aeneas Dione crater *206* (28 N, 47)
Aerosols 68 *68, 78, 80, 81, 96, 210, 211, 211*
Aestum Sinus or **Torrid** Gulf, Moon, *109, 110* (12 N, 9 W)
Africanus Horton Mercury crater *75* (50.5 S, 42)
Agamemnon Trojan asteroid *162*
Agatharchides Lunar crater *110* (20 S, 31 W)
Agenor Linea Europa *181*
Aglaonice Venus *87*
Agrippa Lunar crater *111* (4 N, 11 E)
Ägröi Callisto crater *178* (42 N, 12)
Agunua Rhea crater *207* (70 N, 66)
Ahmad Baba Mercury crater *74, 75* (58.5 N, 127)
Aino Planitia Venus *86* (from 35 S to 70 S, from 55 to 115)

Airy Lunar crater *111* (18 S, 6 E)
Airy Martian crater *138* (0.5 N, 0)
Aitken Lunar crater *111* (17 S, 173 E)
Ajax Trojan asteroid *162*
Akna Montes Venus *86* (68 N, 315)
Akycha Callisto crater *178* (74 N, 325)
Al-Biruni Lunar crater *111* (18 N, 93 E)
Al-Hamadhani Mercury crater *75* (39 N, 89.5)
Al-Jahiz Mercury crater *75* (1.5 N, 22)
Al Naiir (α Gruis). Star. *443*
Al-Qahira Vallis Mars *139* (from 14 to 19 S, from 194 to 200)
Alauda Asteroid *161*
Alba Fossae Mars *137* (from 38 to 49 N, from 109 to 117)
Alba Patera Martian volcano 133, *137* (40 N, 110), 140, *141, 144, 145*
Albategnius Lunar crater *111* (12 S, 4 E)
Albor Tholus Mars *139* (19 N, 210)
Alcor (ζ Ursae Majoris). Star, *302*. See **Ursa Major**
Alcor Tholus Mars *139* (19 N, 210)
Aldebaran (α Tauri). Star *252, 253, 256, 442, 443*
Alden Lunar crater *111* (68 S, 111 E)
Alekhin Lunar crater *111* (68 S, 130 W)
Alembert, Jean le Rond d' (1717–83) 104, 437
Alencar Mercury crater *75* (63.5 S 104)
Alexander Lunar crater *111* (40 N, 14 E)
Alfr Callisto crater *178* (9 S, 222)
Alfraganus Lunar crater *111* (6 S, 19 E)
Algol (β Persei). Star *303, 304, 305, 442, 443*. See **Perseus**
Alhazen Lunar crater *111* (18 N, 70 E)
Ali Callisto crater *178* (57 N, 58)
Aliacensis Lunar crater *111* (31 S, 5 E)
Alinda Asteroid *163*
Alioth (ε Ursae Majoris). Star *422*. See **Ursa Major**
Alkaid (η Ursae Majoris). Star *442*. See **Ursa Major**
Allende Meteorite *61, 129, 131, 413, 428*
Almach (γ Andromedae). Star *442*. See **Andromeda**
Almanon Lunar crater *111* (17 S, 15 E)
Alnilam (ε Orionis). Star *443*. See **Orion**
Alpes (Vallis) Moon *110* (49 N, 2 E)
Alpetragius Lunar crater *110* (16 S, 4 W)
Alpha Regio Venus *86* (28 S, 3)
Alphard (α Hydrae). Star *442, 443*. See **Hydra**
Alpheca Star *442*. See **Corona Borealis**
Alpheratz Star *442, 443*. See **Andromeda**
Alphonsus Lunar crater *110* (13 S, 3 W)
Altai (Rupes) Moon *111* (24 S, 22 E)
Altaïr (α Aquilae). Star *253, 442, 443*. See **Aquila**
Altazimuth Mount 425
Alter Lunar crater *110* (19 N, 108 W)
Altitude Angular distance *h* between the horizontal and the position of the observed star. *h* is measured in degrees (from 0 to +90 to the north

and to −90 to the south). Altitude can be replaced by the complementary angle ζ = (90° −*h*), known as the zenith distance (horizontal coordinates).
Amalthea Jovian satellite *57, 163, 174, 174, 175*
Amantis Asteroid *163*
Amata Dione crater *206* (0, 290)
Amaterasu Patera Io *187* (38 N, 307)
Amazonis Planitia Mars *133, 136* (from 0 to 40 N, and from 140 to 168)
Ameta Rhea crater *207* (58 N, 13)
Amici Lunar crater *110* (10 S, 172 W)
Amirani Io volcano *187* (27 N, 119)
Ammonia 192, 210, 218
Ammura Ganymede crater *183* (36 N, 337)
Amor Asteroids 130, 160, *162, 163*
Amru Al-Qays Mercury crater *75* (13 N, 176)
Amundsen Lunar crater *111* (83 S, 103 W)
Ananke Jovian satellite *175*
Anarr Callisto crater *178* (43 N, 3)
Anaxagoras Lunar crater *110* (76 N, 10 W)
Anaximander Lunar crater *110* (66 N, 48 W)
Anaximenes Lunar crater *110* (75 N, 45 W)
Anchises Trojan asteroid *162*
Anchises Dione crater *206* (36 S, 63)
Andal Mercury crater *75* (47 S, 38.5)
Andel Lunar crater *111* (10 S, 13 E)
Anders Lunar crater *110* (42 S, 144 W)
Anderson Lunar crater *111* (16 N, 171 E), *112*
Andromeda Constellation *442, 443* (from 22h 56m to 2h 36m, and from 21.4° to 52.9°); Alpheratz, star *442*; β **Andromedae** (Mirach), star; γ **Andromedae** (Almach), star *442*
Andromeda (Galaxy) M31, NGC 224, *293, 309, 339, 342, 343, 346, 346, 348–9, 355, 386, 388, 388, 390, 390 393, 397, 406, 416, 439, 440*
Andromedids Meteor shower 127
Ångström, Anders Jonas (1814–74) 438
Ångström Lunar crater *110* (30 N, 42 W)
Anguis (Mare) Moon *111* (23 N, 69 E)
Aningan Callisto crater *178* (51 N, 11)
Ankaa Star *433*. See **Phoenix**
Annihilation Process in which a particle and its antiparticle (e.g. electron and positron, proton and antiproton etc) interact and entirely transform their mass into radiation.
Anomalous Month *105*
Anorthosite Group of intrusive rocks of granular appearance and composed almost entirely of anorthosic feldspar 121, *121*
Ansgarius Lunar crater *111* (14 S, 82 E)
Anshar Sulcus Ganymede *183* (15 N, 200)
Antares (α Scorpii). Star 18, *252, 253, 256, 443*. See **Scorpius**
Antenna 21, *20, 422–5*

isotopes of rubidium and strontium in meteorites and terrestrial samples. Study of the ratio ^{87}Sr/^{86}Sr compared with that of ^{87}Rb/^{86}Sr in various minerals shows that the points lie on a straight line with slope inversely proportional to the age of the Solar System. The age of cosmic rays, i.e. the time which elapses between production and the moment of detection, is of order several tens of thousands of years (it is the absence of beryllium 10, a radioactive element with a lifetime of 1.6×10^6 years, which disintegrates into boron 10, which gives us a good idea of this age).

The age of the stars of a cluster is determined by what is called Sandage's method. This uses the Hertzsprung–Russell diagram of stars of the cluster to find the relative length of the main sequence. The longer its main sequence, the younger a cluster is (to within a few thousand years). Globular clusters have the shortest main sequences, with ages between 12 and 20 billion years. It is thought that all galaxies must have formed in the first billion years after the beginning of the Universe. The age of the Universe can be determined by three different methods giving almost the same answer: the age of the globular clusters, the age of certain radioactive elements such as rhenium 187, and the age deduced from the inverse of the Hubble constant. These three methods determine the age of the Universe at about 15 billion years. *405, 406–11*

Balmer Lunar crater *111*(20 S, 70 E)
Balmer Radiation 242
Balmer Series 250, 256, 275, 370, 371, *371*
Balzac Mercury crater *75*(11 N, 14 S)
Bamberga Asteroid *161*
Ban Mimas crater *203* (46 N, 150)
Banachiewicz Lunar craters *110* (51 N, 135 E) and *111* (6 N, 80 E)
Barabashov Martian crater *137*(47 N, 69)
Barbier Lunar crater *111*(24 S, 158 E)
Barnard, Edward (1857–1924) 174, 283, 300
Barnard Lunar crater *111*(30 S, 86 E)
Barnard Regio Ganymede *183* (22 N, 10)
Barnard's Belt 310, *310*
Barnard's Star Faint low-mass star, about 2 parsecs from the Solar System in the direction of Ophiuchus. Because of its large proper motion it is thought that this star may be surrounded by a planetary system similar to the Solar System. *253*, 254, 301, 345, 346, 415
Barnard–5 Interstellar cloud *260*
Barocius Lunar crater *111*(45 S, 17 E)
Barringer Lunar crater *110*(29 S, 151 W)
Barrow Lunar crater *110*(73 N, 10 E)
Bartels Lunar crater *110*(24.5 N, 90 W)
Bartók Mercury crater *75*(29 S, 135)
Baryons 400, 403
Basalt Dark rock with four main constituents; labrador, pyroxene, olivine and magnetite; the most widespread volcanic rock. 120, *120*, 121
Bashō Mercury crater *75*(32 S, 170.5)
Bavorr Callisto crater *178*(48 N, 23)
Bayer Lunar crater *111*(51 S, 35 W)
Beaumont Lunar crater *111*(18 S, 29 E)
Becklin–Neugebauer Object Very strong infrared source discovered in the Orion Nebula in 1966 by Eric E. Becklin and Gerry Neugebauer; it may be a protostar undergoing gravitational contraction.
Becquerel Lunar crater *111*(41 N, 129 E)
Becquerel Martian crater *137*(22 N, 8)
Becvar Lunar crater *111*(2 S, 125 E)
Bedivere Mimas crater *203* (10 N, 141)
Beer Martian crater *137*(15 S, 8)
Beethoven Mercury crater *75*(20 S, 124)
Behaim Lunar crater *111*(16 S, 79 E)
Beijerinck Lunar crater *111*(13 S, 152 E)
Beli Callisto crater *178*(61 N, 79)
Belinda Uranus satellite *216*
Bell Lunar crater *110*(22 N, 96 W)
Bell Regio Venus *86* (32 N, 48)
Bellatrix (γ Orionis). Star *310*, *318*, *442*, *443*. See Orion
Bellingshausen Lunar crater *111*(61 S, 164 W)
Bello Mercury crater *75*(18.5 S, 120.5)
Bellot Lunar crater *111*(13 S, 48 E)
Belopolsky Lunar crater *110* (18 S, 128 W)
Belus Linea Europa, *181*
Belyayev Lunar crater *111*(23 N, 143 E)
Bennett Comet, 226, 227, 228
Bergstrand Lunar crater *111*(19 S, 176 E)
Berkner Lunar crater *110*(25 N, 105 W)
Berlage Lunar crater *111*(64 S, 164 W)

Bernini Mercury crater *75*(79.5 S, 136)
Bernoulli Lunar crater *111* (34 N, 60 E)
Berosus Lunar crater *111*(33 N, 70 E)
Berzelius Lunar crater *111*(37 N, 51 E)
Bessel, Friedrich (1784–1846) 234, 438
Bessel Lunar crater *111*(22 S, 18 E)
Bessel catalogue 437
Beta Decay Weak interaction in which a neutron is transformed into a proton and an electron (or a proton into a neutron and a positron). The process involves an uncharged particle of zero (or extremely small) mass, called a neutrino. 409–10
Beta Regio Venus 65, 84, 85, *86* (27 N, 285), 87, 88
Beta Taurids Meteorite shower *127*
Betelgeuse (α Orionis). 18, 246, 252, 253, 256, *256*, 310, 318, 422, *442*, *443*. See Orion
Bethe, Hans Albrecht (born 1906) 267, 298, 440
Bettina Asteroid *161*
Bettinus Lunar crater *111*(63 S, 45 W)
Bhabha Lunar crater *111*(56 S, 165 W)
Bianca Uranus satellite *216*
Bianchini Lunar crater *110*(49 N, 34 W)
Biblis Patera Martian volcano *136* (2 N, 124), *141*
Biela Lunar crater *111*(55 S, 52 E)
Biela Comet *127*
Biermann, Ludwig F. (born 1907) 37, 229
Big Bang Today the generally accepted cosmological theory, according to which the Universe began in a primordial explosion about fifteen billion years ago. 165, 166, *166*, 297, 363, 365, 403, 405, 406, 408–11, 412
Big Bertha Saturn spot 193
Big Joe Martian rock 156
Billy Lunar crater *110*(14 S, 50 W)
Binary Pulsar PSR 1913+16 289
Binary Stars 300–5, *422*
Binary System System of two stars moving around their common centre of mass. Each star follows an elliptical orbit. About two-thirds of the stars in our Galaxy are members of binary or multiple systems. Close binary systems can transfer mass via the process of accretion; **binary stars** *252*, *264*, *276*, *283*, 300, *348*; **astrometric binary stars** 301, *301*; **spectroscopic binary stars** 301, *302*; **visual binary stars** 310
Binary X-Ray Sources 290, 292–3
Binda Meteorite *130*
Biot, Jean-Baptiste (1774–1862) 126
Biot Lunar crater *111*(23 S, 51 E)
Birkeland, Olaf Kristian (1867–1917) 36
Birkeland Lunar crater *111*(30 S, 174 E)
Birkham Rhea crater *207*(70 N, 303)
Birkhoff Lunar crater *110*(59 N, 148 W), 113
Birmingham Lunar crater *110*(64 N, 10 W)
Birt Lunar crater *110*(22 S, 9 W)
Bjerknes Lunar crater *111*(38 S, 113 E)
Bjerknes Martian crater *139*(43 S, 189)
Black Body Ideal body which absorbs and re-emits all the radiation incident upon it.
  A black body is in thermodynamic equilibrium with the radiation within it. The intensity of the radiation emitted as a function of wavelength depends only on the temperature.
Black Dwarf 27, 281, *338*

Black Holes 244, 280, 290, 292–3, 295, 309, 324, 329, 338, 380, *380*, 403, *405*, 433
Blancanus Lunar crater *111*(64 S, 21 W)
Blanchinus Lunar crater *111*(25 S, 3 E)
Blazhko Lunar crater *110*(31 N, 148 W)
Blue Compact Galaxies 362–3
Blue Giants Stars 306, *307*
Bobone Lunar crater *110*(29 N, 131 W)
Boccaccio Mercury crater *75*(80.5 S, 30)
Bochica Patera Io *187* (61 S, 22)
Bode Lunar crater *110*(7 N, 2 W)
Boeddicker Martian crater *139*(15 S, 197)
Boethius Mercury crater *75*(0.5 S, 74)
Boguslawsky Lunar crater *111*(75 S, 45 E)
Bohnenberger Lunar crater *111*(16 S, 40 E)
Bohr, Niels, (1885–1962) 250, 440
Bohr Lunar crater *110*(13 N, 86 W)
Bohr Atom 250
Bok Globules Small (10^3 to 10^4 AU) dark clouds found in dusty gas-poor regions in the Milky Way. They are named after the American astronomer Bart J. Bok, who first studied them. 60, 261, 263, 311
Bolometric Correction Produces the bolometric magnitude when added to the visual magnitude
Boltzmann, Ludwig (1844–1906) 439
Boltzmann Lunar crater *111*(78 S, 99 W)
Boltzmann Constant [symbol: *k*]. Proportionality constant relating the absolute temperature *T* to the average kinetic energy of a particle: $k = 1.38066 \times 10^{-23}$ J K^{-1}
Bolyai Lunar crater *111*(34 S, 125 E)
Bond Martian crater *137*(33 S, 36)
Bond (G.) Lunar crater *111*(32 N, 36 E)
Bond (W.) Lunar crater *110*(64 N, 3 E)
Bonpland Lunar crater *110*(8 S, 17 W)
Boole Lunar crater *110*(64 N, 87 W)
Boötes (The Herdsman) Constellation *442* (from 13h 33m to 15h 47m, and from 7.6° to 55.2°); α Boötis (Arcturus), star, *253*, 256, *442*
Borda Lunar crater *110*(25 S, 47 E)
Borealis Planitia Mercury *75*(70 N, 80), 76
Borman Lunar crater *110*(37 S, 142 W)
Bors Mimas crater *203*(45 N, 163)
Boscovich Lunar crater *111*(10 N, 11 E)
Bose, Satyendranath (1894–1974) 14
Bose Lunar crater *111*(54 S, 170 W)
Bosons Class of elementary particles whose spin is an integer multiple of ℏ (examples: photons, π and K mesons, helium atoms) 14
Boss Lunar crater *110*(46 N, 90 E)
Botticelli Mercury crater *75*(64 N, 110)
Bouguer Lunar crater *110*(52 N, 36 W)
Bouguer Martian crater *138*(19 S, 333)
Boussingault Lunar crater *111*(70 S, 50 E)
Bouvard Lunar crater *110*(36 S, 78 W)
Bowell Comet 232
Bowen, Ira (1898–1973) 440
Boyer, Charles 82
Boyle Lunar crater *111*(54 S, 178 E)
Bradfield Comet 230
Bradley, James (1693–1762) 91, 437
Bradley Catalogue 437
Bragg Lunar crater *110*(42 S, 103 W)
Bragi Callisto crater *178*(77 N, 69)

Brahe, Tycho (1546–1601) 266, 284, 435, 436, *436*
Brahms Mercury crater *75*(58.5 N, 177)
Bramante Mercury crater *75*(46 S, 62)
Brami Callisto crater *178*(26 N, 18)
Bran Callisto crater *178*(25 S, 207)
Brashear Lunar crater *111*(74 S, 172 W)
Brashear Martian crater *137*(54 S, 120)
Brayley Lunar crater *110*(21 N, 37 W)
Bredikhin Lunar crater *110*(17 N, 158 W)
Breislak Lunar crater *111*(48 S, 18 E)
Bremsstrahlung Radiation emitted or absorbed when an electron is slowed or accelerated by the electric field of a nucleus without being captured. This radiation is thus not quantised, and can in principle be emitted or absorbed at any wavelength. 42, 403

BREMSSTRAHLUNG

Brenner Lunar crater *111*(39 S, 39 E)
Brianchon Lunar crater *110*(77 N, 90 W)
Briault Martian crater *138*(10 S, 270)
Bridgman Lunar crater *111*(44 N, 137 E)
Briggs Lunar crater *110*(26 N, 69 W)
Brightness temperature Temperature of a black body which would radiate the same power per unit surface area at the same wavelength as the object considered.
Brisbane Lunar crater *111*(50 S, 65 E)
Brontë Mercury crater 74, *75*(39 N, 126.5)
Brouwer Lunar crater *110*(36 S, 125 E)
Brown Lunar crater *111*(47 S, 16 W)
Brown Dwarf 264, 265, 403
Brownian Motion Random motion of microscopic particles resulting from continuous irregular collisions with other particles of the medium. An example is the motion of smoke particles in air, which can be observed with a microscope.
Brunelleschi Mercury crater *75*(8.5 S, 22.5)
Brunner Lunar crater *111*(10 S, 91 E)
Bruno, Giordano (1548–1600) 435, 436
Buch Lunar crater *111*(39 S, 18 E)
Budh Planitia Mercury *75*(18 N, 148)
Buffon, Georges de (1707–88) 60
Buffon Lunar crater *110*(41 S, 134 W)
Buga Callisto crater *178*(22 N, 326)
Buisson Lunar crater *111*(1 S, 113 E)
Bulagat Rhea crater *207*(35 S, 12)
Bulialdus Lunar crater *110*(21 S, 22 W)
Bumba Rhea crater *207*(70 N, 40)
Bunsen, Robert (1811–99) 246, 419
Bunsen Lunar crater *110*(41 N, 85 W)
Burckhardt Lunar crater *111*(31 N, 57 E)
Bürg Lunar crater *111*(45 N, 28 E)
Buri Callisto crater *178*(43 N, 44)
Burnham Lunar crater *111*(14 S, 7 E)
Burr Callisto crater *178*(40 N, 136)
Bursters (X-Ray) 290–1
Bursters (Gamma-Ray) 290–1, *291*
Burton Martian crater *136*(14 S, 156)

Büsching Lunar crater *111*(38 S, 20 E)

Butlerov Lunar crater *110*(9 N, 110 W)

Buys-Ballot Lunar crater *111*(21 N, 175 E)

Byrd Lunar crater *110*(83 N, 0)

Byrgius Lunar crater *110*(25 S, 65 W)

Byron Lunar crater *111*(8 S, 40 W)

Cabannes Lunar crater *111*(61 S, 171 W)

Cabeus Lunar crater *111*(85 S, 40 W)

Cadmus Linea Europa *181*

Caelum (The Chisel) Constellation *443* (from 4h 18m to 5h 3m, and from −27.1° to −48.8°)

Ceasar (Julius) Lunar crater *111*(9 N, 15 E)

Cajori Lunar crater *111*(48 S, 168 E)

Calippus Lunar crater *111*(39 N, 11 E)

Callicrates Mercury crater *75*(65 S, 32)

Callisto Satellite of Jupiter. *57, 58, 175, 176, 176, 177, 177, 178, 178, 179*

Calorimeter *427*

Caloris Basin on Mercury *73, 74, 74, 76, 77, 118*

Caloris Montes Mercury *75* (from 22 to 40 N, 180)

Caloris Planitia Mercury *75* (30 N, 195)

Camelopardalis (The Giraffe) Constellation *442* (from 3h 11m to 14h 25m, and from 52.8° to 85.1); **Z Camelopardalis**, star *277, 282*

Cameron, Alastair G.W. (born 1927) *265*

Camichel, Henri, (born 1917) *82*

Camille Asteroid *160, 161*

Camões Mercury crater *75*(70.5 S, 70)

Campanus Lunar crater *110*(28 S, 28 W)

Campbell, Bruce T.E. *264, 265*

Campbell Lunar crater *111*(45 N, 152 E)

Campbell Martian crater *139*(54 S, 195)

Cancer (The Crab) Constellation *442* (from 7h 53m to 9h 19m, and from 6.8° to 33.3°)

Candor Chasma Mars *53, 137* (5 S, 75)

Canes Venatici (The Hunting Dogs) Constellation *442* (from 12h 4m to 14h 5m, and from 28° to 52.5°)

Canes Venatici Groups of galaxies *393*

Canis Major (The Greater Dog) Constellation *278, 279, 443* (from 6h 9m to 7h 26m, from −11° to −33.2°): α **Canis Majoris** (Sirius), star *248, 252, 253, 256, 280, 301, 443*; **Sirius A** *280, 305*; **Sirius B**, *256, 280, 305*; ε **Canis Majoris** (Adhara), star *443*

β **Canis Majoris** Group of variable stars *278, 278, 279*

Canis Minor (The Lesser Dog) Constellation *442, 443* (from 7h 4m to 8h 9m, and from −0.1° to 13.2°); α **Canis Minoris** (Procyon), star *253, 256, 256, 442, 443*

Cannizzaro Lunar crater *110*(55 N, 100 W)

Cannon Lunar crater *111*(20 N, 80 E)

Canopus (α Carinae). Star *253, 256 443*. See **Carina**

Cantor Lunar crater *111*(38 N, 118 E)

Canyon Diablo Meteorite *127, 128*

Capella Lunar crater *111*(8 S, 36 E)

Capella (α Aurigae). Star *253, 256, 442*. See **Auriga**

Caph (β Cassiopeiae). Star *442*. See **Cassiopeia**

Capri Chasma Region of Mars *137* (18 S, 50), *142*

Capricorn (The Sea Goat) Constellation *442, 443* (from 20h 4m to 21h 57m, and from −8.7° to −27.8°)

Capuanus Lunar crater *110*(34 S, 26 W)

Carbon 14 *47, 48*

Carbon Monoxide *314, 315, 328, 329*

Carbonaceous Chondrites Meteorites with the highest carbon content. Of all the matter in the Solar System, their composition is thought to be closest to that of the protosolar nebula. They are used in drawing up the table of abundances of chemical elements *128–30, 158, 158, 162*

Cardanus Lunar crater *110*(13 N, 73 W)

Carducci Mercury crater *75*(36 S, 90)

Carina (The Keel) Constellation *320, 443* (from 6h 2m to 11h 18m, and from −50.9° to −75.2°); α **Carinae** (Canopus), star *253, 256, 443*; β **Carinae** (Miaplacidus), star *443*; ε **Carinae** (Avior), star *443*

Carina Galaxy *390*

Carlini Lunar crater *110*(34 N, 24 W)

Carme Satellite of Jupiter *175*

Carnot Lunar crater *110*(52 N, 144 W)

Carpatus Montes Moon *110*(15 N, 24 W), *111*

Carpenter Lunar crater *110*(70 N, 50 W)

Carrington, Richard (1826–75) *38*

Carrington Lunar crater *111*(44 N, 62 E)

Carthage Meteorite *128*

Carthage Linea Dione *206* (12 N, 325)

Cartwheel Galaxy (A 0035) *386, 387*

Carver Lunar crater *111*(43 S, 127 E)

Casatus Lunar crater *111*(73 S, 35 W)

Cassandra Dione crater, *206* (42 S, 242)

Cassegrain (XVII century) *419*

Cassegrain Lunar crater *111*(40 S, 113 E)

Cassegrain Telescope *420, 420, 425, 432*

Cassini, Jean Dominique (1625–1712) *104, 196, 208, 238*

Cassini Lunar crater *111*(40 N, 5 E)

Cassini Martian crater *138*(24 N, 328)

Cassini Division *52, 57, 190, 196, 197, 198, 199, 199*

Cassini's Law *104*

Cassiopeia Constellation *442* (from 22h 56m to 3h 6m, and from 46.4° to 77.5°); Schedar, star *442*; β **Cassiopeiae** (Caph), star *442*; δ **Cassiopeiae** (Ruchbah), star *442*

Cassiopeia A Radiosource *244, 284, 285*

Castalia Asteroid *160, 160*

Castor (α Geminorum) Star *442*. See **Gemini**

Catalán Lunar crater *111*(46 S, 87 W)

Catalogues Galaxies, nebulae, stars, radiosources and other objects have been catalogued for many years. The oldest catalogues are the most famous, such as that of Charles Messier (1730–1817), who listed the brightest objects in the sky (such as the Crab Nebula, and nearby galaxies). Lists of galaxies and stars were also produced by Johan Ludvig Dreyer in the NGC (*New General Catalogue of Nebulae and Clusters*), and by Henry Draper in the HD catalogue. Radiosources (radiogalaxies and quasars) can be found in catalogues compiled by radio observatories such as Cambridge (e.g. 3C stands for the 3rd Cambridge catalogue of

radiosources). There are many other catalogues: for example the Uhuru catalogue (from the Swahili word for freedom) of X-ray sources, the Abell catalogue of clusters of galaxies, the Markarian catalogue of galaxies with active nuclei, etc. *254, 394, 397 339, 437, 438*

3C Catalogue *372*; **3C 48** *372*; *373*; **3C 65** *407*; **3C 84** *368*; **3C 206** *373*; **3C 232** *375*; **3C 236** *366*; **3C 256** *407*; **3C 273** *364, 365, 379, 372, 373, 405 427, 427*; **3C 295** *372*; **3C 338** *365*; **3C 345** *368*; **3C 354.3** *368*; **3C 388** *364*; **3C 449** *366*; **3C 452** *366, 366, 381*; **3C 461** *284*

4C Catalogue; **4C 2603** *367*

Catharina Lunar crater *111*(18 S, 24 E)

Catilus Dione crater, *206* (3 S, 275)

Caucasus Montes Moon, *111* (36 N, 8 E)

Cauchy, Augustin (1789–1857) *438*

Cauchy Lunar crater *111*(10 N, 39 E)

Cavalerius Lunar crater *110*(5 N, 67 W)

Cavendish Lunar crater *110*(25 S, 54 W)

Cayley Formation Moon *113*

CCD (charge coupled device) *12, 419*

Celestial Mechanics *57, 58, 121, 218, 292, 383, 388*

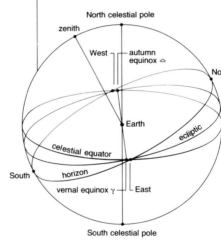

CELESTIAL SPHERE

Celestial Sphere Imaginary sphere of very large radius, on which are projected all celestial objects, with the Earth at the centre. The main reference circles of this sphere are the ecliptic, the celestial equator and the horizon. Any point is specified with respect to the cardinal points and the equinoxes.

Censorinus Lunar crater *111*(0, 32 E)

Centaurus (The Centaur) Constellation *309, 336, 419* (from 11h 3m to 14h 59m, and from −29.9° to −64.5°); α **Centauri** (Rigil Kentaurus), double star *54, 232, 253, 300, 443*; β **Centauri** (Hadar), star *253, 443*; ν **Centauri** (Menkent), star *443*; ω **Centauri**, globular cluster *308*; **Proxima Centauri**, star *18, 253*

Centaurus A Radiosource *354, 364, 364, 366, 370, 390*

Centaurus X-3 X-ray source *289*

Cepheids Very luminous periodically variable stars. There are two types: type I Cepheids (thought to be the youngest) have periods between 5 and 10 days;

type II Cepheids (less luminous) have periods between 10 and 30 days.

Cepheids are particularly useful

distance indicators; there is a relation between their period and their intrinsic luminosity, which through comparison with their apparent luminosity gives an estimate of their distance.

Cepheus Constellation *442* (from 20h 1m to 8h 30m, from 53.1° to 88.5°); α **Cephei**, star *91*; β **Cephei**, star *278*; γ **Cephei**, star *264*, δ **Cephei**, star *198, 201, 277, 278*; η **Cephei**, star *264*

Cepheus Lunar crater *111*(41 N, 46 E)

Ceraski Lunar crater *111*(49 S, 141 E)

Ceraunius Fossae Mars *137* (from 20 to 38 N, from 105 to 112)

Ceraunius Tholus Martian volcano *137*(24 N, 97), *141*

Cerenkov Radiation Radiation produced by the motion of a charged paticle in a medium of refractive index differing from 1, and having an initial velocity greater than that of electromagnetic waves. This type of radiation is used in some cosmic ray detectors.

Ceres Asteroid *160, 161, 162, 163*

Cerulli Martian crater *138*(32 N, 338)

Cervantes Mercury crater *75*(75 S, 122)

CETI (Communication with Extra-Terrestrial Intelligence) *416*

Cetus (The Whale) Constellation *442, 443* (from 23h 55m to 3h 21m, and from −25.2° to 10.2°); α **Ceti** (Menkar), star *442, 443*; β **Ceti** (Diphda), star *443*; o **Ceti**, star *252, 278*; τ **Ceti**, star *253, 416*; **UV Ceti**, star *277*

Cetus Group Group of galaxies *393*

CFHT Canada-France-Hawaii Telescope, *11, 419, 420–1*

Chacornac Lunar crater *111*(30 N, 32 E)

Chaffee Lunar crater *110*(39 S, 155 W)

Challenger Space shuttle *23, 53*

Challis Lunar crater *110*(78 N, 9 E)

Chalybes Regio Io *187* (55 N, 85)

Chamaeleon (The Chamaeleon) Constellation *443* (from 7h 32m to 13h 48m, and from −75.2° to −82.8°)

Chamberlin Lunar crater *111*(59 S, 96 E)

Champollion Lunar crater *111*(37 N, 175 E)

Chandler, Seth (1846–1913) *438*

Chandler Lunar crater *111*(44 N, 171 E)

Chandrasekhar, Subrahmanyan (born 1910) *280*

Chandrasekhar Limit Maximum possible mass for a white dwarf: if the mass of the star exceeds the critical value of 1.44M$_\odot$ the weight of the outer layers becomes too large for the degeneracy pressure of the gas to support; there is thus no stable equilibrium configuration *289*

Chang Heng Lunar crater *111*(18.5 N, 111.5 E)

Chant Lunar crater *110*(41 S, 110 W)

Chao, Meng-Fu Mercury crater *75* (87.5 S, 132)

Chaplygin Lunar crater *111*(6 S, 150 E)

Chapman, Sydney (1888–1970) *36*

Chapman Lunar crater *110*(50 N, 101 W)

Charlier Lunar crater *110*(36 N, 132 W)

Charon Satellite of Pluto *57, 224–5*

Chaucer Lunar crater *110*(3 N, 140 W)

Chauvenet Lunar crater *111*(11 S, 137 E)

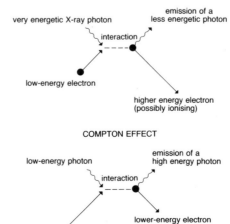

COMPTON EFFECT

INVERSE COMPTON EFFECT

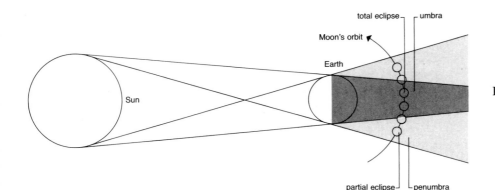

LUNAR ECLIPSE

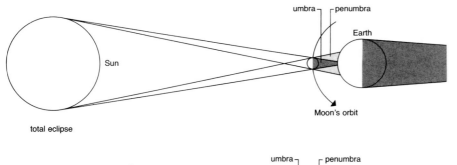

total eclipse

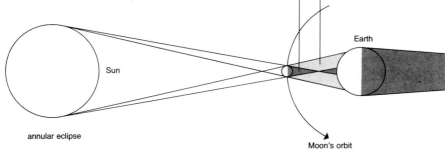

annular eclipse

SOLAR ECLIPSE

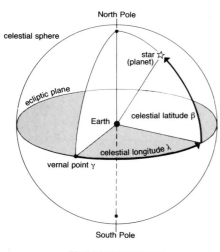

ECLIPTIC COORDINATES

the celestial equator. The celestial longitude λ, which runs from 0° to 360°, and the celestial latitude β, running from −90° to +90°, are the coordinates in this system.

Eddie Martian crater *139*(12 N, 218)

Eddington, Arthur (1882–1944) *266*, 440

Eddington Lunar crater *110*(22 N, 72 W)

Edison Lunar crater *111*(25 N, 99 E)

Edlén, Bengt (born 1906) *439*, 440

Effective temperature of a star Temperature of a black body producing the same luminosity per unit area as the star. *248*, *256–9*

Egdir Callisto crater *178*,(31 N, 35)

Egede Lunar crater *110*(49 N, 11 E)

Ehrlich Lunar crater *110*(41 N, 172 W)

Eichstädt Lunar crater *110*(23 S, 80 W)

Eijkman Lunar crater *111*(62 S, 141 W)

Eimmart Lunar crater *111*(24 N, 65 E)

Einstein, Albert (1879–1955) *14*, *292*, *296*, *316*, *411*, *429*, *438*, *441*

Einstein Artificial satellite. See **HEAO 2**

Einstein Lunar crater *110*(18 N, 86 W)

Einstein arc *401*

Einstein effect *293*, *400*

Einthoven Lunar crater *111*(5 S, 110 E)

Eisila Regio Venus *86* (from 5 N to 20 N, and from 0 to 50)

Eitoku Mercury crater *75*(21.5 S, 157.5)

Ejriksson Martian crater *138*(19 S, 174)

Elaine Mimas crater *203*(40 N, 102)

Élara Jovian satellite *175*

Electromagnetic radiation Emission of photons: the least energetic are those of radio waves, and the most energetic correspond to gamma rays, with visible in between.

radio: radiation of very long wavelengths (longer than 5 x 10^{-4}m)
infrared: wavelengths from 8 x 10^{-7} to 5 x 10^{-4}m;
visible: wavelengths from 4 to 8 x 10^{-7}m, corresponding to the range of sensitivity of the human eye;
ultraviolet: wavelengths from 10^{-8} to 4 x 10^{-7}m;
X-rays: wavelengths from 10^{-11} to 10^{-8}m;
gamma rays: the most energetic radiation, with wavelengths shorter than 10^{-11}m.

Only the radio, a small part of the infrared, and visible radiation penetrate the absorption of the Earth's atmosphere. Other methods, involving the use of rockets, balloons, or satellites, are needed to detect radiation at all other wavelengths.

Electron *14*, *25*, *25*, *42*, *44*, *44*, *285*, *408–10*

Electron Volt Energy unit used in atomic and nuclear physics. It is the energy acquired by an electron passing through a potential drop of 1 volt:
    1 eV = 1.6 x 10^{-19} J.

Elementary particles There are three categories of elementary particles. The photon is the vector particle of the electromagnetic interaction; it has zero rest-mass and by definition moves at the speed of light. Leptons are the particles which interact via the weak nuclear force. These include the muon (heavy electron), the electron and their neutrinos. The hadrons are the particles which interact through the strong nuclear force; these include the pions (π mesons) which are the vector particles of the strong interaction; the kaons (K mesons), the nucleons (proton and neutron) and the hyperons (also called strange particles). All of these particles (particularly the hadrons) are probably formed from more elementary particles called quarks. As these particles govern three of the four fundamental forces their behaviour influences the evolution of the entire Universe. *13*, *14*, *15*, *42*, *44*, *403*, *408–11*

Elger Lunar crater *110*(35 S, 30 W)

Ellerman Lunar crater *110*(26 S, 121 W)

Elliptical Galaxies *311*, *334*, *335*, *336*, *337*, *340*, *340*, *346*, *348*, *350*, *350*, *351*, *353*, *354–5*, *356*, *356*, *357*, *366*, *371*, *382*, *386*, *388*, *389*, *390*, *395*, *396*, *399*, *400* *401*, *407*

Ellison Lunar crater *110*(55 N, 108 W)

Ellyay Rhea crater *207*(76 N, 95)

Elnath (β Tauri). Star *442*, *443*. See **Taurus**

Eltanin (γ Draconis). Star *442*. See **Draco**

Elvey Lunar crater *110*(9 N, 101 W)

Elysium Fossae Mars *139* (from 21 to 30 N, and from 217 to 224)

Elysium Mons Mars *139* (25 N, 213), *140*, *145*

Elysium Planitia Mars *133*, *133*, *139* (from 10 S to 30 N, and from 180 to 260), *141*

Emakong Patera Io *187* (0, 110)

Emden Lunar crater *110*(63 N, 176 W)

Emitted flux Energy flux emitted per unit area of a surface (unit: watts per square metre).

Enceladus Saturnian satellite *57*, *57*, *58*, *190*, *194*, *202*, *202*, *204*

Encke Comet *127*, *226*, *228*, *230*, *233*, *238*, *239*

Encke Lunar crater *110*(5 N, 37 W)

Encke's Division *52*, *197*, *198*

Endeavour Rupes Mercury *75*(38 N, 31)

Endymion Lunar crater *110*(55 N, 55 E)

Energy balance *70*, *81*, *95*, *96*, *97*, *97*

Engelhardt Lunar crater *110*(5 N, 159 W)

Enif (ε Pegasi). Star *442*, *443*. See **Pegasus**

Enlil Ganymede crater *183*(52 N, 301)

Eos Chasma Mars, *137* (from 7 to 17 S, and from 30 to 53)

Eötvös Lunar crater *111*(36 S, 134 E)

Ephemerides List of calculated positions of an astronomical object at various epochs

Epidemiarum (**Palus**) Moon *110*(31 S, 26 W)

Epigenes Lunar crater *110*(73 N, 4 W)

Epimenides Lunar crater *110*(41 S, 30 W)

Epimethea Saturnian satellite *202*, *209*

Equatorial coordinates Spherical coordinates with the celestial equator as the fundamental plane. The reference point is the vernal equinox, which is the intersection of the ecliptic and the celestial equator. This point moves because of the precession of the Earth's axis, so it is necessary to specify the epoch, currently either 1950 or 2000. Right ascension α, varying from 0 to 24 h, and declination δ, varying from 0° to ±90°, are the coordinates of a point in this system. The latter is equivalent to geographic coordinates.

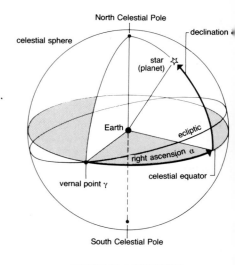

EQUATORIAL COORDINATES

Equatorial Mount *421*

Equiano Mercury crater *75*(39 S, 31)

Equilibrium temperature Theoretical temperature attained by the surface of a body under the effect of the Sun's radiation and the reflection of the surface (Bond albedo) alone.

Equinoxes The two points of the celestial sphere where the celestial equator intersects the ecliptic. The vernal equinox is the point at which the Sun passes from the south to the north. By definition, this point is taken as the reference point for equatorial coordinates (its right ascension and declination are zero).

Equuleus (The Little Horse) Constellation *442* (from 20h 54m to 21h 23m, and from 2.2° to 12.9°)

Eratosthenes Lunar crater *110*(15 N, 11 W)

Eratosthenes of Cyrenia (circa 276–circa 195 BC) *419*

Eridanus (The River) Constellation *443* (from 1h 22m to 5h 9m, and from 0.1° to −58.1°; α Eridani (Achernas) star, *253*, *342*, *443*; **Eridani B**, star *253*, *280*, ε Eridani star *253*, *264*, o² **Eridani**, star *253*; θ **Eridani** (Acamar), star *443*

Eridanus Cluster Group of galaxies *393*

Erlik Callisto crater *178*(66 N, 358)

Eros Asteroid *161*, *162*, *163*

Erosion *87*, *89*, *146*, *147*

Erro Lunar crater *111*(6 N, 98 E)

Escalante Martian crater *138*(0, 245)

Eshmun Ganymede crater *183*(22 S, 187)

Esnault-Pelterie Lunar crater *110*(47 N, 142 W)

ESO catalogue; ESO 113-1G 45 *370*; **ESO 138-1G 29/30** *387*

Espin Lunar crater *111*(28 N, 109 E)

Etana Ganymede crater *183*(78 N, 310)

Euclid Lunar crater *110*(7 S, 29 W)

Euctemon Lunar crater *110*(77 N, 30 E)

force	carrier particle*	range	relative intensity	examples
strong nuclear	gluon, meson π	10^{-13}cm (1 fermi)	~ 10	nuclear fusion or fission
weak nuclear	intermediate boson	$\leq 10^{-16}$cm	$\sim 10^{-13}$	beta decay
electromagnetic	photon	very large as $1/r^2$	1/137	magnetism emission of light
gravitational	graviton (?)	very large as $1/t^2$	$\sim 10^{-38}$	universal atraction motion of the planets*

* particle exchanged in the interaction

FUNDAMENTAL FORCES OF PHYSICS

interaction is mediated by the exchange of photons of all energies, from the least energetic radio photons up to gamma rays.

The two other interactions are short- range, and act only over distances of a few fermi (1 fermi = 10^{-15}m);

– the strong interaction holds the atomic nucleus together against the electrical repulsion of the protons and involves the exchange of mesons between all nucleons (protons and neutrons);

– the weak interaction involves the leptons, such as the electron, positron, muons and neutrinos; beta decay is an example of the weak interaction.

Recent work, recognised in the award of the Nobel Prize to S. Weinberg, A. Salam and S. Glashow, has shown that the weak and electromagnetic forces can be unified, i.e. described as two aspects of the same force. Physicists are currently attempting to construct Grand Unified theories, which would unify the strong, weak and electromagnetic interactions. 14, 15, *15,* 25, 408–9, 411, 414

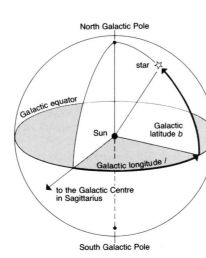

GALACTIC COORDINATES

Helberg Lunar crater *110*(22 N, 102 W)

Helen Planitia Venus *86*(58 S, 260)

Helicon Lunar crater *110*(40 N, 23 W)

Helios A & B Interplanetary probes 23

Helioseismology 24, *26*

Heliosphere. See **Solar Cavity**

Helium Chemical element making up 8–10% of the nuclei in the universe (in the form of ⁴He) and 22–30% of the observed mass. Its two main isotopes, ³He and ⁴He, are mainly formed in primordial nucleosynthesis. 24, *25*, *25*, 27, 29, *29*, 218, *218*, 363, 408–10

Helium Flash 27

Helix Nebula Planetary nebula (NGC 7293) 274

Hell Lunar crater *110*(32 S, 8 W)

Hellas Planitia Mars basin *65*, *133*, *134*, *138* (from 30 to 60 S, and from 272 to 313), 149, 157

Heller Rhea crater *207*(9 N, 310)

Hellespontus Montes Mars *138* (from 35 to 50 S, and from 310 to 319)

Helmholtz, Hermann von (1821–94) *439*, 440

Helmholtz Lunar crater *111*(72 S, 78 E)

Helmholtz Martian crater *137*(46 S, 21)

Henderson Lunar crater *111*(5 S, 152 E)

Hendrix Lunar crater *111*(48 S, 161 W)

Hengo Chasma Venus *86*(8 N, 355)

Heno Patera Io *187*(57 S, 312)

Henry, Paul (1848–1905) 438

Henry (**Paul**) Lunar crater *110*(24 S, 57 W)

Henry, Prosper (1849–1903) 438

Henry (**Prosper**) Lunar crater *110*(24 S, 59 W)

Henry Martian crater *138*(11 N, 336)

Henyey Lunar crater *110*(13 N, 152 W)

Haphaestus Fossae Mars *138* (from 18 to 25 N, and from 233 to 242)

Haphaestus Patera Io *187* (2 N, 290)

Hepti Callisto crater *178*(64 N, 27)

Heraclitus Lunar crater *111*(49 S, 6 E)

Herbig, George H. 264

Herbig – Haro Objects Bright concentrations of gas and dust exhibiting intense emission lines which are thought to contain stars in their first stages of evolution. They are strong infrared sources, and show strong mass loss. *261*, 277

Hercules Constellation *336*, *442* (from 15h 47m to 18h 56m, and from 3.9° to 51.3°)

Hercules Lunar crater *110*(46 N, 39 E)

Hercules Cluster. Abell 2151 Cluster of galaxies *336*

Hercules X-1 X-ray source *325*

Herculina Asteroid *161*

Herigonius Lunar crater *110*(13 S, 34 W)

Hermann Lunar crater *110*(1 S, 57 W)

Hermione Asteroid *161*

Hermite Lunar crater *110*(85 N, 90 W)

Hero Rupes Mercury *75*(57 S, 173)

Herodotus Lunar crater *110*(23 N, 50 W)

Herschel, William (1738–1822) 210, 211, 12, 216, 255, 267, 274, *275*, *281*, 300, 437, 439

Herschel Lunar crater *109*, *110*(6 S, 2 W)

Herschel (**C.**) Lunar crater *110*(34 N, 31 W)

Herschel (**J.**) Lunar crater *110*(62 N, 41 W)

Herschel 36 Star *60*

Herschel Martian crater *139*(14 S, 230)

Herschel Mimas crater *203*(0, 105)

Hertz Lunar crater *111*(14 N, 104 E)

Hertzsprung, Ejnar (1873–1967) 256, 257, 440

Hertzsprung Lunar crater *110*(0, 130 W), *113*

Hertzsprung–Russell Diagram 18, 27, *27*, 242, 252, 256–9, *260*, 261, 267, 268, 269, 270, *270*, 271, 272, *272*, 273, 274, 277, 278, 279, *279*, 294, 298, *298*, 306, *307*, *307*, 309, 338, 340, *340*

Hesiod Mercury crater *75*(58 S, 35.5)

Hesiodus Lunar crater *110*(29 S, 16 W)

Hesperia Planum Mars *138* (from 10 to 35 S, and from 242 to 258)

Hess, Victor (1883–1946) 322, 428, 433

Hess Lunar crater *111*(54 S, 174 E)

Hestia Rupes Venus *86*(5 N, 65)

Hevelius, Johannes (1611–87) 111

Hevelius Formation Moon *108*, *109*, *110*(2 N, 67 W)

Hewish, Anthony (born 1924) 433, 441

Heymans Lunar crater *110*(75 N, 145 W)

Hidalgo Asteroid *54*, *162*

Hilbert Lunar crater *111*(18 S, 108 E)

Hildas Asteroid *162*

Himalia Jovian satellite *175*

Hind Lunar crater *111*(8 S, 7 E)

Hippalus Lunar crater *110*(25 S, 30 W), *114*

Hipparchus (2nd century BC) 91, 246, 254

Hipparchus Lunar crater *111*(6 S, 5 E)

Hipparchus Martian crater *136*(45 S, 151)

Hipparchus Catalogue 246, 254

Hipparcos Artificial satellite 254–5, 441

Hippocrates Lunar crater *110*(71 N, 146 W)

Hirayama Lunar crater *111*(6 S, 93 E)

Hiroshige Mercury crater *75*(13 S, 27)

Hiruko Patera Io *187* (65 S, 331)

Hitomaro Mercury crater *75*(16 S, 16)

Hodr Callisto crater *178*(69 N, 87)

Hoenir Callisto crater *178*(36 S, 261)

Hoffmeister Lunar crater *111*(15 N, 137 E)

Hogg Lunar crater *111*(34 N, 122 E)

Hogni Callisto crater *178*(14 S, 5)

Hohmann Lunar crater *110*(18 S, 94 W)

Holbein Mercury crater *75*(35.5 N, 29)

Holberg Mercury crater *75*(66.5 S, 61)

Holden Lunar crater *111*(19 S, 63 E)

Holden Martian crater *137*(26 S, 34)

Holetschek Lunar crater *111*(28 S, 151 E)

Homer Mercury crater *75*(1 S, 36.5)

Hommel Lunar crater *111*(54 S, 33 E)

Hooke, Robert (1635–1703) *166*, 437

Hooke Lunar crater *111*(41 N, 55 E)

Hooke Martian crater *137*(45 S, 44)

Horace Mercury crater *75*(68.5 S, 52)

Horizontal coordinates These are defined on a sphere with the observer at the centre. The equatorial plane is the observer's horizon. The straight line passing through the observer perpendicular to the horizon plane defines the zenith in the north and the nadir in the south. The position of an object in the sky is then defined by two coordinates:

– the altitude *h* given by the angle between the object and the horizon; *h* is measured in degrees and varies from 0 to 90 degrees at the zenith and 0° to –90° at the nadir; the altitude is often replaced by its complement ζ, called the zenith distance;

– the azimuth is the angle between the vertical passing through the object and a reference point, the north; the azimuth varies from 0° to 360°. This coordinate system is purely local, as the apparent position of an object varies with time, and depends on the place of observation on the Earth.

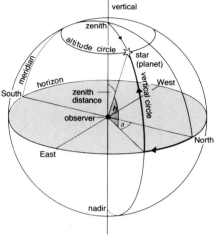

HORIZONTAL COORDINATES

Horologium (The Clock) Constellation *443* (from 2h 12m to 4h 18m, and from –39.8° to +67.2°)

Horrebow Lunar crater *110*(59 N, 41 W)

Horrocks Lunar crater *111*(4 S, 6 E)

Horsehead Nebula *318*, *319*

Horus Patera Io *187*(10 S, 340)

Hour Angle Angle between the meridian at a point and the hour circle passing through the specified object. The angle is expressed in hours, minutes and seconds, hence the name. Hour angle is equivalent to terrestrial longitude.

Houzeau Lunar crater *110*(18 S, 124 W)

Howard–Koomens–Michels Comet 232

Howe Venus crater *87*

Hoyle, Fred (born 1915) 284, 414, 416

Hubble, Edwin (1889–1953) 334, 350, 405, 406, 407, 419, 441

Hubble Lunar crater *111*(22 N, 87 E)

Hubble Variable nebula *315*

Hubble Diagram *407*, *411*

Hubble's Law Empirical law stating that the recession velocity of a galaxy is proportional to its distance: this distance is determined using several indicators such as Cepheids (variable stars whose periods are directly related to their intrinsic luminosities), HII regions, supernova remnants, etc. The further a galaxy is from us, the greater its recession velocity. The constant of proportionality is called the Hubble constant and has a value between 50 and 100 km s⁻¹ Mpc⁻¹. The inverse of the Hubble constant defines an expansion age for the Universe of between 10 and 20 billion years. *371*, *376*, *396*, *397*, *398*, *399*, 405, 406, 407, 408, 411, *411*, 432

Hubble Space Telescope *52*, 265, 432, *432*

Huggins, William (1824–1910) 439, 440

Huggins Lunar crater *110*(41 S, 2 W)

Huggins Martian crater *139*(49 S, 204)

Hugo Mercury crater *75*(39 N, 47.5)

Humason Comet 228

Humboldt Lunar crater *111*(27 S, 81 E)

Humboldt Sea Moon *110*(55 N, 75 E), *113*

Humorum Mare Moon *52*, *110*(23 S, 38 W), *113*, *114*, *118*, 119

Hun Kal Mercury crater *75*(0.2 S, 20)

Huo Hsing Vallis Mars *138* (from 28 to 34 N, and from 292 to 299)

Hussey Martian crater *136*(54 S, 127)

Hutton Lunar crater *111*(37 N, 169 E)

Huygens, Christiaan (1629–95) 136, 196, 197, 210, *310*, 437

Huygens Martian crater *138*(14 S, 304)

Hyades Cluster 269, 306, *442* (4h 20m, + 17°)

Hydra (The Water Snake) Constellation *442*, *443* (from 8h 8m to 14h 58m, and from 6.9° t –35.3°; α **Hydrae** (Alphard) star *442*, *443*

Hydra-Centaurus Supercluster of galaxies *397*, *399*

Hydrogen The lightest and the most abundant chemical element. It constitutes about 90% of existing nuclei in observed matter. Heavier nuclei are synthesised from hydrogen either in primordial nucleosynthesis, or in the interior of stars. Hydrogen has a heavy isotope, deuterium, which is a hundred thousand times less abundant and synthesised primordially. Its abundance is used to determine if the Universe will continue to expand forever or not. 24, *25*, 27, 28, 29, *29*, 71, 82, *211*, 218, 284, 286, 291, 314, 315, 318, 319, 328, 329, *329*, 378–9 408–9

Hydrus (The Lesser Water Snake) Constellation *443* (from 0h 2m to 4h 33m, and from –58.1° to –82.1°)

Hygeia Asteroid *161*, *162*

Hyginus Lunar crater *111*(8 N, 6 E)

Hyginus Rima Moon *111*(8 N, 6 E)

Hypatia Lunar crater *111*(4 S, 23 E)

Hyperion Saturnian satellite *57*, *57*, 58, *202*, 203, 209

Hyperons Elementary particles heavier than nucleons (protons and neutrons), which are unstable in the laboratory but which probably exist in the core of neutron stars

Iapetus Saturnian satellite *57*, 58, *202*, *202*, 203, 208, 216, *216*

Ibn Yunus Lunar crater *111*(14 N, 91 E)

Ibsen Mercury crater *75*(24 S, 36)

Icarus Asteroid *162*

Icarus Lunar crater *110*(6 S, 173 W)

Ictinus Mercury crater *75*(79 S, 165)

Ideler Lunar crater *111*(49 S, 22 E)

Idelson Lunar crater *111*(81 S, 114 E)

Iduna Asteroid *163*

IG 29 & 30 Galaxies *387*

Igaluk Callisto crater *178*(5 N, 315)

Igraine Mimas crater *203*(40 S, 225)

Iha Dione crater *206*(2 N, 242)

Ikeya–Seki Comet 232

IMB (Irvine Michigan Brookhaven) Neutrino detector 286, 287

Imbrium Mare Moon *109*, *109*, *110* (36 N, 16 W) 112, *113*, 114, 116, 117, 118, *118*, 119, *120*, 121, *121*

Imdr Regio Venus *86* (45 S, 212)

Imhotep Mercury crater *75*(17.5 S, 35.5)

Inaehus Tholus Io *187* (29 S, 354)

Inclination Angle between the orbital plane of a planet and the ecliptic or equivalently the angle between the rotation axis of a star and the line of sight. *56*, *57*

Indus (The Indian) Constellation *443* (from 20h 25m to 23h 25m, and from –45.4° to –74.7°); ε **Indi**, star *253*

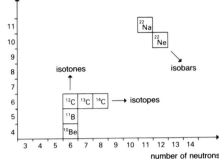

ISOBARS, ISOTONES, ISOTOPES

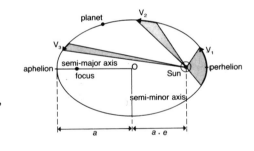

cooler than any radiating object behind it.

Kirkwood Lunar crater *110*(69 N, 157 W)

Kirkwood Gaps *160, 162, 163*

Kishar Ganymede crater *183*(78 N, 330)

Kishar Sulcus Ganymede crater *183*(15 S, 220)

Klaproth Lunar crater *111*(68 S, 22 W)

Kleimenov Lunar crater *110*(33 S, 141 W)

Klein Lunar crater *111*(12 S, 3 E)

Klute Lunar crater *110*(37 N, 142 W)

Knobel Martian crater *139*(6 S, 226)

Kobayashi–Berger–Milon Comet *231*

Koch Lunar crater *111*(43 S, 150 E)

Kochab (β Ursae Minoris). Star *442*. See **Ursa Minor**

Kohlschütter Lunar crater *111*(15 N, 154 E)

Kohoutek Comet *227*

Kolhörster Lunar crater *110*(10 N, 115 W)

Komarov Lunar crater *111*(25 N, 153 E)

Kondratyuk Lunar crater *111*(15 S, 115 W)

König Lunar crater *110*(24 S, 25 W)

Konstantinov Lunar crater *111*(20 N, 159 E)

Kopff Lunar crater *110*(17 S, 90 W)

Korolev Lunar crater *110*(5 S, 157 W), *113*

Kostinsky Lunar crater *111*(14 N, 118 E)

Kovalevskaya Lunar crater *110*(31 N, 129 W)

Kovalsky Lunar crater *111*(22 S, 101 E)

Krafft Lunar crater *110*(17 N, 73 W)

Kramers Lunar crater *110*(53 N, 128 W)

Krasnov Lunar crater *110*(31 S, 89 W)

Krasovsky Lunar crater *110*(4 N, 176 W)

Kreutz Group of comets *232*

Krieger Lunar crater *110*(29 N, 46 W)

Krusenstern Lunar crater *111*(26 S, 6 E)

Krüger 60 Double star *253, 300*

Kuan Han–Ch'ing Mercury crater *75*(29 N, 53)

Kuanja Chasma Venus *86*(13 S, 104)

Kugler Lunar crater *111*(53 S, 104 E)

Kuiper, Gerard Pieter (1905–73) *210, 216, 223*

Kuiper Martian crater *136*(57 S, 157)

Kuiper Mercury crater *72, 75*(11 S, 31.5)

Kuiper Airborne Observatory (KAO) *426, 426*

Kukarkin, Boris (born 1909) *282*

Kulik, Leonid (1883–1942) *233*

Kulik Lunar crater *110*(42 N, 155 W)

Kumpara Rhea crater *207*(10 N, 320)

Kun Lun Chasma Rhea crater *207*(45 N, 290)

Kunowsky Lunar crater *110*(3 N, 32 W)

Kuo Shou Ching Lunar crater *110*(8 N, 134 W)

Kurchatov Lunar crater *111*(39 N, 142 E)

Kurosawa Mercury crater *75*(52 S, 23)

Labeyrie, Antoine (born 1943) *418, 422*

La Caille Lunar crater *111*(24 S, 1 E)

Lacaille Stars *253*

Lacchini Lunar crater *110*(41 N, 107 W)

Lacerta (The Lizard) Constellation *442* (from 21h 55m to 22h 56m, and

from 34.9° to 56.9°); **10 Lacertae**, star *256*

BL Lacertae Galaxy with active nucleus *256, 368, 370, 371, 371, 372*

Lacertids (BL Lac Objects) Galaxies with the following properties:
– rapid intensity variations in the radio, infrared and visible;
– absence of spectral lines in low-dispersion spectrograms;
– significant polarisation.
These objects may be quasars at very low redshift, and emit most of their radiation in the visible. BL Lacertae is the prototype of this class and is the closest quaser-like object to us. *370–1*

La Condamine Lunar crater *110*(53 N, 28 W)

Lacroix Lunar crater *110*(38 S, 59 W)

Lacus Veris Moon *117*

Lada Terra Venus *86*(from 45 S to 65 S, and from 330 to 90)

Lade Lunar crater *111*(1 S, 10 E)

Laertes Tethys crater *205*(50 S, 60)

Lagalla Lunar crater *110*(44 S, 23 W)

Lagoon Nebula M8 (NGC 6523), *60, 311*

Lagrange, Joseph (1736–1813) *104, 203, 305, 437*

Lagrange Lunar crater *110*(32 S, 69 W)

Lagrange Points Points in the orbital plane of two bodies gravitating about their common centre of mass, where a third body of negligible mass could remain in relative equilibrium. *174, 303, 304, 305*

Lakshmi Planum Venus *53, 84, 85, 86*(67 N, 330), *87*

Lalande Lunar crater *110*(4 S, 8 W)

Lalande 21185 Star *253*

Lallemand, André (born 1904) *419*

Lamarck Lunar crater *110*(23 S, 69 W)

Lamb Lunar crater *111*(43 S, 101 E)

Lambert, Johann Heinrich (1728–77) *300*

Lambert Lunar crater *110*(26 N, 21 W)

Lambert Martian crater *138*(20 S, 335)

Lame Lunar crater *111*(15 S, 65 E)

Lamèch Lunar crater *111*(43 N, 13 E)

Lamerok Mimas crater *203*(65 S, 285)

Lamont Lunar crater *111*(5 N, 23 E)

Lampland Lunar crater *111*(31 S, 131 E)

Lampland Martian crater *137*(36 S, 79)

Landau Lunar crater *110*(42 N, 119 W)

Landau–Oppenheimer–Volkoff Limit *280*

Landsat–1 Artificial satellite for studying earth resources *94, 100, 101*

Lane Lunar crater *111*(9 S, 132 E)

Langemak Lunar crater *111*(10 S, 119 E)

Langevin Lunar crater *111*(44 N, 162 E)

Langley Lunar crater *110*(52 N, 87 W)

Langmuir Lunar crater *110*(36 S, 129 W)

Langrenus Lunar crater *106, 107, 111*(9 S, 61 E)

Lansberg Lunar crater *110*(0, 26 W)

La Pérouse Lunar crater *111*(10 S, 78 E) (Perugia)

Laplace, Pierre Simon de (1749–1827) *58, 60, 104, 196, 197, 292, 437, 438, 439*

Laplace Limit *280*

Laplace Relations *58*

Large Magellanic Cloud Galaxy *286, 286, 390, 443* (5h 26m, −69°)

Larissa Neptune satellite *220, 223*

Larissa Chasma Dione *206*(30 N, 75)

Larmor Lunar crater *110*(32 N, 180 W)

Lassell, William (1799–1880) *216, 222*

Lassell Lunar crater *110*(16 S, 8 W)

Lassell Martian crater *137*(21 S, 63)

Latagus Dione crater *206*(15 N, 26)

Latium Chasma Dione *206*(20 N, 65)

Laue Lunar crater *110*(29 N, 96 W)

Launcelot Mimas crater *203*(9 S, 317)

Lauritsen Lunar crater *111*(27 S, 96 E)

Lausus Dione crater *206*(37 N, 27)

Lavinia Planitia Venus *86*(47 S, 350), *87*

Lavoisier Lunar crater *110*(36 N, 80 W)

Leavitt, Henrietta (1868–1921) *278, 308, 342, 440, 441*

Leavitt Lunar crater *110, 111*(46 S, 140 W)

Lebedev Lunar crater *111*(48 S, 108 E)

Lebedinsky Lunar crater *110*(8 N, 165 W)

Leda Jovian satellite *175*

Leda Planitia Venus *86*(45 N, 65)

Lee Lunar crater *110*(31 S, 41 W)

Leeuwenhoek Lunar crater *110*(30 S, 179 W), *113*

Legendre Lunar crater *111*(29 S, 70 E)

Legentil Lunar crater *110*(73 S, 80 W)

Lehmann Lunar crater *110*(40 S, 56 W)

Leibnitz Lunar crater *111*(34 S, 178 E)

Leighton, Robert B. *26*

Lemaître, Georges (1894–1966) *405, 411, 441*

Lemaître Lunar crater *111*(62 S, 150 W)

Le Monnier Lunar crater *111*(26 N, 31 E)

Length of the day *59, 59, 68*

Lenticular Galaxies *336, 337, 350, 350, 351, 356–7, 382*

Lenz Lunar crater *110*(3 N, 102 W)

Leo (The Lion) Constellation *255, 339, 389, 442, 443* (from 9h 18m to 11h 56m, and from −6.4° to 33.3°); α Leonis (Regulus), star *257, 442, 443*, β Leonis (Denebola) star *442, 443*; AD Leonis, star *254*

Leo A Galaxy *390*

Leo I Galaxy *390*

Leo II Galaxy *390*

Leo Minor (The Lesser Lion) Constellation *442, 443* (from 9h 19m to 11h 4m, and from 23.1° to 41.7°)

Leonids Meteorite shower *127*

Leonov Lunar crater *111*(19 N, 148 E)

Leopardi Mercury crater *75*(73 S, 180)

Lepaute Lunar crater *110*(33 S, 34 W)

Lepton Era *408, 409, 410*

Leptons *14, 408–11*

Lepus (The Hare) Constellation *443* (from 4h 54m to 6h 9m, and from −11° to −27.1°)

Lermontov Mercury crater *75*(15.5 N, 48.5)

Lerna Regio Io *187*(65 S, 300)

Letronne Lunar crater *110*(10 S, 43 W)

Leucippus Lunar crater *110*(29 N, 116 W)

Leuschner Lunar crater *110*(1 N, 109 W)

Le Verrier, Urbain Jean Joseph (1811–77) *57 218, 429, 437, 438*

Le Verrier Lunar crater *110*(40 N, 20 W)

Le Verrier Martian crater *138*(38 S, 343)

Levi-Civita Lunar crater *111*(24 S, 143 E)

Lewis Lunar crater *110*(19 S, 114 W)

Lexell Lunar crater *110*(36 S, 4 W)

Ley Lunar crater *111*(45 N, 154 E)

Leza Rhea crater *207*(22 S, 902)

Liang K'ai Mercury crater *75*(39.5 S, 183.5)

Liapunov Lunar crater *111*(27 N, 88 E)

Libra (The Scales) Constellation *243, 442, 443* (from 14h 8m to 15h 59m, and from −0.3° to −29.9°); α Librae (Zubenelgenubi), star *442, 443*

Libration Oscillations of the Moon making 60% of its surface visible even though it always presents the same face to the Earth

Libya Linea Europa *181,*

Licetus Lunar crater *110*(47 S, 6 E)

Li Ch'ing Chao Mercury crater *75*(77 S, 73)

Lichtenberg Lunar crater *110*(32 N, 68 W)

Lick Lunar crater *111*(14 N, 53 E)

Liebig Lunar crater *110*(25 S, 48 W)

Li Fan Martian crater *136*(47 S, 153)

Light Bending *374, 375, 377, 378, 400*

Light Year Unit equal to the distance travelled by light (or any other electro-magnetic radiation) *in vacuo* in one (tropical) year:
1 light year = 9.4605 x 10^15 m = 0.3066 pc = 63,240 AU.

Lilius Lunar crater *111*(55 S, 6 E)

Limb Apparent edge of the disc of an object seen in projection on the sky.

Lin, Chia Chiao (born 1916) *358, 441*

Lindblad, Bertil (1895–1965) *440*

Lindblad Lunar crater *110*(70 N, 97 W)

Lindenau Lunar crater *111*(33 S, 25 E)

Linné Lunar crater *111*(25 N, 12 E)

Li Po Mercury crater *75*(17.5 N, 35)

Lithium *408–10*

Lithosiderites Meteorites made of a mixture of iron and silicates *128*

Lithosphere *65, 76, 90, 91, 93, 102*

Littrow Lunar crater *111*(22 N, 31 E)

Liu Hsin Martian crater *136*(53 S, 172)

LM Lunar module *12, 119*

Lobachevsky Lunar crater *111*(9 N, 112 E)

Local Group *346, 347, 348, 388, 390, 390, 392, 393, 393, 395, 397, 399, 408*

Local Supercluster *390, 393, 395, 399,*

Lockyer Lunar crater *111*(46 S, 37 E)

Lockyer Martian crater *139*(28 N, 199)

Lodurr Callisto crater *178*(52 S, 270)

Lodygin Lunar crater *110*(18 S, 147 W)

Loewy Lunar crater *110*(23 S, 32 W)

Lohrmann Lunar crater *110*(1 S, 67 W)

Lohse Lunar crater *111*(14 S, 60 E)

Lohse Martian crater *137*(43 S, 16)

Loki Io *187*(19 N, 305)

Loki Io Patera *187*(13 N, 310)

Lomonosov Lunar crater *111*(28 N, 98 E)

Longomontanus Lunar crater *111*(50 S, 21 W)

Loni Callisto crater *178*(4 S, 215)

Loreley Asteroid *161*

Lorentz Lunar crater *110*(34 N, 100 W), *113*

Losy Callisto crater *178*(68 N, 329)

Lot Mimas crater *203*(29 S, 226)

Louville Lunar crater *110*(45 N, 45 W)

Love Lunar crater *111*(6 S, 129 E)

Lovelace Lunar crater *110*(82 N, 107 W)

Lovell, Alfred Charles Bernard (born 1913) *422*

Lovell Lunar crater *110*(39 S, 149 W)

Lowa Rhea crater *207*(45 N, 10)

Lowell, Percival (1855–1916) *224*

Lowell Lunar crater *110*(13 S, 103 W)

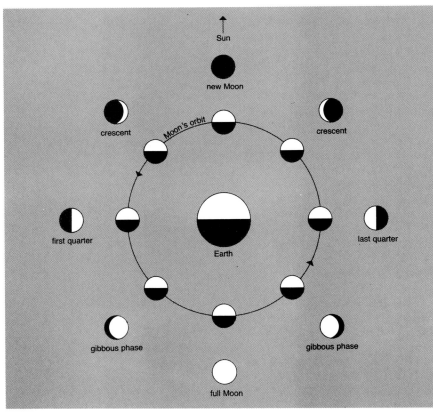

PHASES OF THE MOON

Ophir Chasma Mars *137*(from 3 to 9 S, and from 64 to 75)

Ophiuchus (The Serpent Holder) Constellation *319, 442, 443* (from 15h 58m to 18h 42m, and from 14.3° to −30.1°); α **Ophiuchi** (Rasalhague), star *442, 443*; η **Ophiuchi** (Sabik), star *442, 443*; ϱ **Ophiuchi**, star *245*; ϱ **Ophiuchi**, cloud *260, 314, 427*

Oppenheimer Lunar crater *110*(35 S, 166 W)

Oppolzer Lunar crater *110*(2 S, 1 W)

Opposition Alignment of the Sun, Earth and a superior planet in that order.

Orbital Elements Quantities specifying the position and orbit of an astronomical object about another (such as the Sun). There are six elements:

– *a*, the semi-major axis of the elliptical orbit;

– *e*, the eccentricity of the ellipse; *e* = *c/a*, where *c* is the distance between the centre of the ellipse and one of the foci. If *c* vanishes, so does *e* and the ellipse is a circle. Conversely as *e* increases the ellipse becomes more elongated. The upper limit to *e* is 1;

– *i* is the angle between the orbital plane and the ecliptic (for an object in orbit about the Sun) or the plane of the sky (for an object orbiting about another star);

– Ω is the longitude of the ascending node; it is the angle in the plane of the ecliptic (or the sky) between the direction of the vernal equinox (or the line of sight) and the line of nodes;

– ω is the argument of perihelion (periastron), i.e. the angle in the orbital plane between the line of nodes and the direction of perihelion (periastron); perihelion (periastron) is the point of closest approach to the Sun (star);

– *T* (the epoch) is the last orbital element: it is not geometrical, but specifies the instant of perihelion (periastron) passage (or any other convenient point such as the centre of an eclipse for an object outside the Solar System).

The elements *a* and *e* characterise the trajectory; *i* and Ω specify the position of the orbital plane with respect to the ecliptic (sky) plane, and ω gives the orientation of the ellipse in this plane. *56, 57, 160, 234*

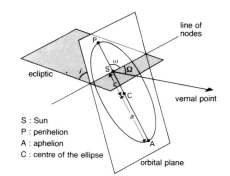

S : Sun
P : perihelion
A : aphelion
C : centre of the ellipse

Orbital Period *57, 58*

Orbital Resonance *56, 57, 58, 58, 224, 224*

Orbits (Planets and natural satellites) *54, 54, 56, 57, 57, 58, 58, 78, 79, 158, 159, 218, 222, 223, 224, 225;* Comets *54, 226, 231, 234, 234;* Asteroids *160, 161, 162*

Orcus Patera Mars *139*(14 N, 181)

Ordinary Chondrites Meteorites *128–30, 129*

Oresme Lunar crater *111*(43 S, 169 E)

Orgeuil Meteorite *129*

Orientale Mare Moon *52, 108, 109, 110*(19 S, 95 W), *112, 113, 117, 120, 121, 121*

Orion (Hunter) Constellation *310, 318, 319, 320, 346, 442, 443* (from 4h 41m to 6h 23m, and from −11° to 23°); α **Orionis** (Betelguese), star *18, 246, 252, 253, 256, 256, 310, 318, 422, 422, 443;* β **Orionis** (Rigel), star *253, 253, 256, 256, 257, 310, 318, 443;* γ **Orionis** (Bellatrix), star *310, 318, 442, 443;* ε **Orionis** (Alnilam), star *443;* χ **Orionis** (Saïph), star *310, 318, 442, 443;* FU **Orionis**, star *264*

Orion Cloud *245, 260, 261, 316, 317*

Orion A. Molecular cloud *262, 318*

Orion Nebula M42 (NGC 1976) *310, 317, 318*

Orionids Meteorite shower *127*

Orlov Lunar crater *110*(26 S, 175 W)

Ormazd Rhea crater *207*(62 N, 55)

Orontius Lunar crater *110*(40 S, 4 W)

Osiris Ganymede crater *182, 183*(39 S, 161)

Oski Callisto crater *178*(56 N, 266)

OSO (Orbital Solar Observatory). Artificial satellites for solar observations; OSO–1 *22*; OSO–7 *23, 290*; OSO–8 *43*

Ossa Chasma Mimas *203*(25 S, 290)

Ostwald Lunar crater *111*(11 N, 122 E)

Ottar Callisto crater *178*(60 N, 100)

Oudemans Martian crater *137*(10 S, 92)

Outer Planets *56*

Outgassing Process in which gas trapped in the interior of a planet escapes and becomes part of its atmosphere. *66, 67, 70, 82*

Ovda Regio Venus *85, 86*(5 S, 92)

Ovid Mercury crater *75*(69.5 S, 23)

Ozma Project *416*

Padua Linea Dione *206*(15 S, 225)

Palatine Chasma Dione *206*(70 S, 330)

Palatine Linea Dione *206*(50 S, 312)

Palisa Lunar crater *110*(9 S, 7 W)

Palitzsch, Johann Georg (1723–88) *234*

Palitzsch Lunar crater *111*(28 S, 65 E)

Pallas Asteroid *160, 161, 162, 163*

Pallas Lunar crater *110*(5 N, 2 W)

Palmieri Lunar crater *110*(29 S, 48 W)

Pan Amalthea *174*

Paneth Lunar crater *110*(63 N, 95 W)

Pan Ku Rhea crater *207*(72 N, 120)

Pangea Chasma Mimas *203*(40 S, 315)

Pannekoek Lunar crater *111*(4 S, 140 E)

Papagiannis, Michael *416*

Papaleksi Lunar crater *111*(10 N, 164 E)

Paracelsus Lunar crater *111*(23 S, 163 E)

Parallax The apparent movement of an object viewed against a distant background as the observer changes position. Also the angle subtended by some conventionally chosen baseline. The parallax gives the distance of the object from the baseline.

Paraskevopoulos Lunar crater *110*(50 N, 150 W)

Parenago, Pavel Petrovitch (1906–60) *282*

Parenago Lunar crater *110*(26 N, 109 W)

Parga Chasma Venus *86*(from 5 to 35 S, and from 220 to 300)

Parker, Eugene Newman (born 1927) *37*

Parkhurst Lunar crater *111*(34 S, 103 E)

Parrot Lunar crater *111*(15 S, 3 E)

Parry Lunar crater *110*(8 S, 16 W)

Parsec (symbol: pc). The parsec (contraction of *parallax second*) is a unit of length equal to the distance of an object whose annual parallax is one second of arc. It is usually used for objects outside the Solar System: 1pc = 3.0857 x 10^{16} m = 3.2616 light years = 206 265 AU.

Parsons, William, Lord Rosse (1800–67) *384, 438, 439*

Parsons Lunar crater *110*(37 N, 171 W)

Pascal Lunar crater *110*(74 N, 70 W)

Paschen Lunar crater *110*(14 S, 141 W)

Pasiphae Jovian satellite *175*

Pasteur Lunar crater *111*(11 S, 105 E)

Pasteur Martian crater *138*(19 N, 335)

Patientia Asteroid *161, 162*

Patroclus Trojan asteroid *162*

Pauli Lunar crater *111*(45 S, 137 E)

Pauli Exclusion Principle In its restricted form: no two electrons of an atom can have the same values of the four quantum numbers. The exclusion principle applies quite generally to fermions but not bosons. *280, 280*

Pavlov Lunar crater *111*(29 S, 142 E)

Pavlova Venus *86*(14 N, 40)

Pavlovka Meteorite *130*

Pavo (The Peacock) Constellation *13, 396, 443* (from 17h 37m to 21h 30m, and from −56.8° to −75°); α **Pavonis** (Peacock), star *443*

Pavo-Indus Group of galaxies *393*

Pavonis Mons Martian volcano *134, 137*(1 N, 113), *141, 142*

Pawsey Lunar crater *111*(45 N, 144 E)

Peacock (α Pavonis). Star *433*. See **Pavo**

Peary Lunar crater *110*(89 N, 50 E)

Pease, Francis Gladhem *422*

Pease Lunar crater *110*(13 N, 106 W)

Pedn Rhea crater *207*(50 N, 340)

Pegasus Constellation *334, 442, 443* (from 21h 6m to 0h 13m, and from 2.2° to 36.3°); α **Pegasi** (Markab), star *442, 443*; β **Pegasi** (Scheat), star *252, 442, 443*; ε **Pegasi** (Enif), star *442, 443*

Pegasus Galaxy *390*

Peirce, Charles Sanders (1839–1914) *438*

Peirce Lunar crater *111*(18 N, 53 E)

Peiresc, Nicolas Claude Fabri de (1580–1637) *310*

Peirescius Lunar crater *111*(46 S, 71 E)

Pekko Callisto crater *178*(17 N, 6)

Pele Io volcano *186, 187*(19 S, 257), *188*

Pelion Chasma Mimas *203*(25 S, 215)

Pellinore Mimas *203*(35 N, 128)

Pelorus Linea Europa *181*

Pentland Lunar crater *111*(64 S, 12 E)

Penzias, Arno (born 1933) *407, 408, 433, 441*

Percivale Mimas crater *203*(1 S, 171)

Perelman Lunar crater *111*(24 S, 106 E)

Perepelkin Lunar crater *111*(10 S, 128 E)

Perepelkin Martian crater *137*(52 N, 65)

Peridie Martian crater *137*(26 N, 276)

Perigee Closest point to the Earth of a satellite orbit. *105*

Perihelion Closest point to the Sun of a planet's orbit. This point evolves in time, a phenomenon called precession. The precession of the perihelion of Mercury is one of the best indications in favour of general relativity. *56, 224, 234, 234*

Perihelion Precession *57, 429*

Period–Luminosity Relation *255, 278, 406*

Permafrost Ground permanently frozen below a shallow depth. On the Earth, the typical depth is of order 50 cm to 1 m. The surface is frozen only at the lowest temperatures.

Perrin, Jean (1870–1942) *266*

Perrine Lunar crater *110*(42 N, 131 W)

Perrine Regio Ganymede *183*(40 N, 30)

Perseids Meteorite shower *127*

Perseus Constellation *442, 443* (from 1h 26m to 4h 46m, and from −30.9° to 58.9°); α **Persei** (Mirfak), star *442, 443*; β **Persei** (Algol), star *303, 304, 305, 442, 443*; h and χ **Persei**, double cluster *306*

Perseus Cluster (Abell 426) *397, 400, 401, 402*

Petavius Lunar crater *111*(25 S, 61 E)

Petermann Lunar crater *110*(74 N, 69 E)

Peters Lunar crater *110*(69 N, 31 E)

Petrarch Mercury crater *75*(30 S, 26.5), *77*

Petrie Lunar crater *110*(46 N, 108 E)

Petropavlovsky Lunar crater *110*(37 N, 115 W)

Petrov Lunar crater *111*(61 N, 88 E)

Pettit Martian crater *136*(12 N, 174)

Pettit Lunar crater *110*(28 S, 86 W)

Petzval Lunar crater *111*(63 S, 113 W)

Phase Angle Angle formed by the Sun, a planet, and the Earth. For an inferior planet, situated between the Earth and the Sun, the maximal phase angle corresponds to quadrature. For a superior planet, outside the Earth's orbit about the Sun, this angle is 180° at inferior conjunction

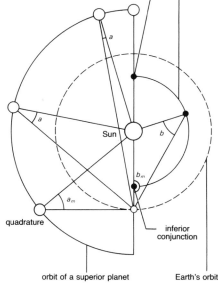

a : phase angle of a superior planet (a_m : maximum value)

b : phase angle of an inferior planet (b_m : maximum value)

PHASE ANGLE

Phecda (γ Ursae Majoris). Star *442*. See **Ursa Major**

Phidias Mercury crater *75*(9 N, 150)

Phillips Lunar crater *111*(26 S, 78 E)

Philolaus Lunar crater *110*(75 N, 33 W)

Philoxenus Mercury crater *75*(8 S, 112)

Philus Sulci Ganymede *183*(37 N, 215)

Phineus Linea Europa *181*

Phlegra Montes Mars *139* (from 30 to 46 N, 195)

Phobos Martian satellite *57*, 136, 158–9, 163, *163*

Phobos Missions 26, 27, 158, *158*

Phocylides Lunar crater *111*(54 S, 58 W)

Phoebe Saturnian satellite *57*, 202, 203, 209

Phoebe Regio Venus 84, *86*(12 S, 280), 87, 88, *89*

Phoenix (The Phoenix) Constellation *443* (from 23h 24m to 2h 24m, and from −39.8° to −58.2°); Ankaa, star *442*

Photochemistry 68, *68*, 80

Photodisintegration Nuclear reactions induced by the intense photon fluxes in the very hot (>10⁹ K) central regions of massive stars. Heavy nuclei (e.g. silicon or iron) are fragmented into lighter nuclei. These reactions absorb energy (endothermic). 284, 285, *285*

Photon Elementary particle of zero rest-mass and charge, the quantum of the electromagnetic field; the photon has spin 1 and is thus a boson. The photon's energy is proportional to the frequency of the wave with which it is associated: $E = h\nu$, with h = Planck's constant: $h = 6.626176 \times 10^{-34}$ J s. *24*, *24*, 25, 28, 285, *287*, 290, 408–11, 419, 422, 423, 426, 427

Photosphere *18*, 19, 28–9, 30, *30*, 34, 38, *45*, 288

Phrygia Sulcus Ganymede *183*(20 N, 5)

Piazzi, Giuseppe (1746–1826) 160

Piazzi Lunar crater *110*(36 S, 68 W)

Piazzi Catalogue 437

Piazzi Smyth Lunar crater *110*(42 N, 3 W)

Picard, Jean (1620–82) 437

Picard Lunar crater *111*(15 N, 55 E)

Piccolomini Lunar crater *111*(30 S, 32 E)

Pickering, William (1858–1938) *224*

Pickering Martian crater *136*(34 S, 133)

Pictet Lunar crater *110*(43 S, 7 W)

Pictor (The Painter's Easel) Constellation *443* (from 4h 32m to 6h 52m, and from −53.1° to −64.1°); β Pictoris 264, 265, *265*, *415*, *415*

Pigalle Mercury crater *75*(37 S, 10.5 W)

Pingré, Alexandre-Gui (1711–96) 232

Pingré Lunar crater *111*(54 S, 78 W)

Pioneer Interplanetary probes 23, *52*, 83, *167*, 172, *173*, 194, *195*, 197, 428; **Pioneer Venus** *54*, 65, 68, 78, 79, 80, *80*, 81, *82*, 83, 84, 85, 87, 88; **Pioneer–10** *52*, 167, 176, 186, 189, *416*, *417*; **Pioneer–11** *52*, 167, 175, 176, 186, 190, *194*, 198, *416*, *417*

Pirquet Lunar crater *111*(20 S, 140 E)

Pisces (The Fishes) Constellation *442*, *443* (from 22h 49m to 2h 4m, and from −6.6° 33.4°)

Piscis Austrinus (The Southern Fish) Constellation *442*, *443* (from 21h 25m to 23h 4m, and from −25.2° to −36.7°); α Piscis Austrini (Fomalhaut), star *442*, *443*

Pitatus Lunar crater *110*(30 S, 14 W)

Pitiscus Lunar crater *111*(51 S, 31 E)

Pizzetti Lunar crater *111*(35 S, 119 E)

Plagioclase Family of feldspars varying from sodium-rich to calcium-rich and including aluminium silicate

Plana Lunar crater *111*(42 N, 28 E)

Planck, Max (1858–1947) 439, 440

Planck Lunar crater *111*(58 S, 138 E), *113*

Planck Distribution *408*,

Planck Rima Moon *111*(65 to 54 S, and 129 to 125E)

Planck's Constant. See **Photon**

Planck's Law. Relation giving the intensity of thermal equilibrium (black body) radiation as a function of the wavelength and temperature of the emitter. The cosmic background radiation follows Planck's law. 248, 251, 296, 297

Planetary Atmospheres 56, 66–7, 68–71, 414, 428

Planetary Nebula 13, 27, 243, *244*, 259, 271, 274–5, 281, *281*, 338

Planetary Tides 54, 57, 58, 59, *59*, 62, 64, 80, 99, 101, *153*, *154*, 155, 158, *188*, 222

Planetesimal or Planetoid Theoretical object growing by accretion of matter from the primitive protoplanetary nebula up to a size between a metre and a kilometre. 60, 61, 63, *63*, 70

Planets 52, 54, *54*, 55, 56, *56*, 57, 62–3, 68, 70, 231, 232, 254, 255, 264–5, 412

Plaskett Lunar crater *110*(82 N, 175 E)

Plasma 42, 44, *45*, 365, 366, *367*, 368, *368*

Plasmapause Region of the terrestrial ionosphere at an altitude of 4 to 7 earth radii where the plasma density decreases very sharply. It marks the transition between a cold dense plasma of terrestrial origin and a hot tenuous plasma arising from both the Earth and the solar wind.

Plate Tectonic Theory describes the evolution of the Earth's rigid crust constituting the lithosphere. This is assumed to be formed of a mosaic of rigid plates of various sizes, which move with respect to each other. 90–3

Plato Lunar crater *110*(51 N, 9 W)

Playfair Lunar crater *111*(23 S, 9 E)

Pleiades (Cluster) *238*, 251, 306, 313, *442*, *443* (3h 44m, 24°)

Plinius Lunar crater *111*(15 N, 24 E)

Plummer Lunar crater *110*(25 S, 155 W)

Plutarch Lunar crater *111*(25 N, 79 E)

Pluto 54, *54*, 56, *56*, 57, 222, 224–5

Plutonic Rock Eruptive rock crystallising at depth within the Earth

Po Chü-I Mercury crater *75*(6.5 S, 165.5)

Pogson, Norman (1809–91) 246

Pogson Lunar crater *111*(42 S, 111 E)

Poincaré, Henri (1854–1912) 196, 197

Poincaré Lunar crater *111*(57 S, 161 E)

Poinsot Lunar crater *110*(79 N, 147 W)

Poisson Lunar crater *111*(30 S, 11 E)

Polar Aurorae 102, *173*

Polaris (α Ursae Minoris). Star *442*. See **Ursa Minor**

Polarisation The direction of the electric field associated with a light wave. Unpolarised light has all directions perpendicular to the wave direction equally represented. In polarised light, the electric field may oscillate in a plane (linearly polarised light) or move around a circle or ellipse (circularly and elliptically polarised light). In astronomy polarised light can be produced by scattering from dust, or directly from synchrotron or cyclotron emission, with varying degrees of polarisation. 423

Pollack, James B. 70, 71

Pollux (β Geminorum). Star *442* See **Gemini**

Polybius Lunar crater *111*(22 S, 26 E)

Polygnotus Mercury crater *75*(0, 68.5)

Polzunov Lunar crater *111*(26 N, 115 E)

Poncelet Lunar crater *110*(76 N, 55 W)

Pons Lunar crater *111*(25 S, 22 E)

Pontanus Lunar crater *111*(28 S, 15 E)

Pontécoulant Lunar crater *111*(59 S, 65 E)

Popov Lunar crater *111*(17 N, 100 E)

Porter Lunar crater *111*(56 S, 10 W)

Porter Martian crater *137*(50 S, 114)

Portia Uranian satellite 216

Posidonius Lunar crater *111*(32 N, 30 E)

Positron 328, 329, *408*, 409, 427

Pourquoi-Pas Rupes Mercury *75*(58 S, 156)

Powers of Ten Astronomy is a domain of both very small and very large numbers: distances in the Solar System are millions of kilometres, and hence in billions of metres (the Earth–Sun distance is around 150,000,000,000 m); the power received from a radio source is of order a billionth of a billionth of a watt (0.000,000,000,000,000,001 W). These numbers are difficult to write accurately and to visualise; manipulating them in this form can easily produce errors. Physicists and astronomers therefore prefer to use the notation in powers of ten, and this is in general used in the present volume. The power n of ten indicates that the figure 1 is followed by n zeros, if n is positive. If n is negative, $-n - 1$ zeros appear after the zero and decimal point before the figure 1 appears. As examples:
$10^3 = 1,000$
$10^6 = 1,000,000$
$10^1 = 10$
$10^{-1} = 0.1$
$10^{-4} = 0.000,1$
$10^{-8} = 0.000,000,01.$

Po Ya Mercury crater *75*(45.5 S, 21)

Poynting Lunar crater *110*(17 N, 133 W)

Poynting–Robertson Effect Loss of angular momentum by orbiting particles because of their absorption and re-emission of solar radiation. 238, *239*

Prager Lunar crater *111*(4 S, 131 E)

Prandtl Lunar crater *111*(60 S, 141 E)

Praxiteles Mercury crater *75*(27 N, 60)

Precession Slow periodic motion of the rotation axis of a body. For example, the Earth rotates around an axis making an angle of 23.5° with respect to the normal to the ecliptic. The Earth's axis describes a cone in space whose base is called the circle of precession. The precession of the equinoxes corresponds to a slight movement to the west (50.26 seconds per year); the period of the equinoctial precession is about 25,800 yr. The tropical year, i.e. the time between two passages of the vernal equinoxes, is 365.242 days. Thus the sidereal year is 20 minutes longer. The precession of the planets corresponds to a slow periodic motion induced by the interactions of the planets with each other. 57, 91.

Precession of Perihelion 57, 429

Prefixes Various names and symbols

have been adopted to denote decimal multiples and sub multiples in SI units.

Priam Trojan asteroid 162

Priestley Martian crater *139*(54 S, 228)

Priestly Lunar crater *111*(57 S, 108 E)

Prinz Lunar crater *110*(26 N, 44 W)

Procellarum Oceanus Moon *52*, 108, 110(10 N, 47 W), *113*, 114, 117, *120*

Proclus Lunar crater *111*(16 N, 47 E)

Proctor, Richard A. (1837–88) 136

Proctor Lunar crater *111*(46 S, 5 W)

Proctor Martian crater *138*(48 S, 330)

Procyon (α Canis Minoris). Star 253, 256, *256*, *442*, *443*. See **Canis Minor**

Prometheus Io volcano 186, *187*(3 S, 153), 188

Prominence Structure of hot hydrogen projecting from the Sun. These vertical tongues can reach heights of 200,000 km. They can be seen on the surface of the Sun during total solar eclipses. 30, *30*, 31, 40, *40*, 41, 44, 46

Proper Motion Apparent angular motion of a star on the celestial sphere, i.e. in a direction perpendicular to the line of sight. 254, *254*

Protagoras Lunar crater *110*(56 N, 7 E)

Proteus Neptunian satellite 223, *223*

Proton Elementary particle with positive charge equal and opposite to that of the electron. Its mass is 1836.152 times that of the electron; it is a fermion (spin 1/2). 14, *24*, 25, *25*, 44, 285, 408–10

Protonilus Mensae Mars *138*(from 38 to 49 N, and from 303 to 325)

Protoplanet 264, *265*

Protosolar Nebula 60–1, 62, 66, *66*, 67, 70, 131, 165, 166, 190, 237

Protostar Gas cloud at the end of its contraction and about to become a star.

Proust Mercury crater *75*(20 N, 47)

Proxima Centauri Star 18, 253. See **Centaurus**

Psyche Asteroid *161*

Ptolemaeus Lunar crater *110*(14 S, 3 W), 111

Ptolemaeus Martian crater *136*(46 S, 158)

Ptolemy, Claudius (c.90–c.168 AD) 300, 435, *435*

Puccini Mercury crater *75*(64.5 S, 46)

Pu Chou Chasma Rhea *207*(35 S, 100)

Puiseux Lunar crater *110*(28 S, 39 W)

Pulsar Extremely dense collapsed star emitting rapid pulses detected in the radio (and sometimes also in the visible and X-rays). The pulse periods vary from a few milliseconds to several seconds: pulsars are definitely established as neutron stars. 276, 284, *285*, *285*, 287, 288–9, 290, *338*, *416*, 426, 427, 429, 433, 441

multiplying factor	prefix	symbol
10¹⁸	exa	E
10¹⁵	peta	P
10¹²	tera	T
10⁹	giga	G
10⁶	mega	M
10³	kilo	k
10²	hecto	h
10	deca	da
10⁻¹	deci	d
10⁻²	centi	c
10⁻³	milli	m
10⁻⁶	micro	μ
10⁻⁹	nano	n
10⁻¹²	pico	p
10⁻¹⁵	femto	f
10⁻¹⁸	atto	a

467

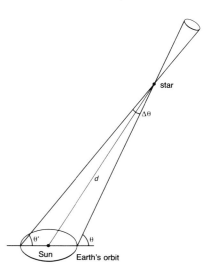

Viking – The Biological programme

The landing of the two Viking spacecraft on Mars allowed the first exobiological experiments. The main aim of these missions was to discover if the Martian surface matter had the properties required to support life. Each station had four instruments which operated as automatic mini-laboratories studying samples of the Martian surface (less than one gram of material for each sample).

The first experiment, designed by NASA engineers under the name of the Pyrolytic Release Experiment (cf fig.1), looked for evidence of photosynthesis by possible bacteria in the sample. A light source with the ultraviolet filtered out was used. If organisms able to perform photosynthesis were present in the sample, it should have been revealed by the absorption of carbon from the small amounts of carbon dioxide and carbon monoxide in the apparatus. After incubation for 120 hours the sample was heated to liberate by pyrolysis the organic compounds possibly synthesised. To identify them, small amounts of the same gases containing the radioactive isotope carbon 14 were added to the Martian samples, thus acting as a tracer.

The second experiment, called the Labelled Release Experiment (cf fig.2) tested the possible metabolic burning of nutrients. This involved checking the ability of any Martian micro-organisms to nourish themselves in this way from a mixture of typical nutrients. The method was to try to detect the carbon dioxide rejected into the atmosphere after oxidation of the organic components of the mixture. As in the preceding experiment, the carbon atoms in the organic molecules were radioactive carbon 14 isotopes. The results followed from analysis of the radioactivity of the gas liberated by the putative micro-organisms after nutrition.

The third experiment, called the Gas Exchange Experiment (cf fig.3), checked for the changes that might occur in a small sample of the Martian atmosphere in the presence of biological activity. Nutrients, in the form of a water solution very rich in organic compounds, were supplied to any bacteria existing in the Martian soil. This was done in two distinct stages. In the first – humidification – phase enough nutrients were placed in a container to make the atmosphere very rich in water vapour from the organic mixture. The object of this phase was to 'wake up' the organisms, which would be revealed by changes to the environment of the sample. In the second – saturated – phase, the

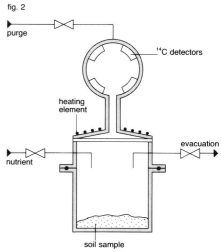

fig. 2

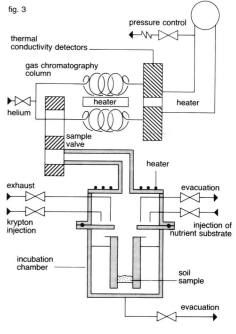

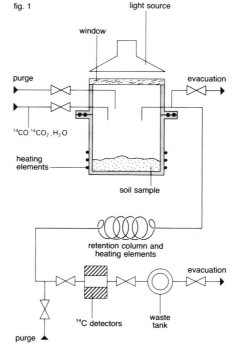

fig. 1

fig. 3

vapour from the mixture was in contact with the sample. The experiment was designed to test the metabolic activity of any Martian bacteria through changes in the oxygen, carbon and nitrogen constituents. For example, if the process was a fermentation, the atmosphere in the container should have become enriched in carbon dioxide. The detector was an extremely sensitive gas-phase chromatograph, used in the laboratory to detect traces of very heavy constituents. The principle of this apparatus is to adsorb the constituents to be measured on to a metallic surface. They remain there for a time determined by their composition. The analysis is made by desorbing the constituents through heating the surface. Because their retentions differ one can separate them by cycles of varying length. The apparatus can detect even small changes in the atmospheric composition above the sample. The experiment has to last several weeks.

The fourth experiment (cf fig.4) was only indirectly biological. The mass spectrometer analysed as many molecules as possible in the Martian atmosphere. When connected to a vessel containing a sample of Martian soil heated to 220° C and 500° C, this apparatus could detect the outgassing of any organic molecules. In front of this measuring instrument was a gas-phase chromatograph, which would only detect puffs of gas of different composition at well-separated intervals. To achieve the precision required to demonstrate the existence of very small concentrations of organic compounds the entry to the apparatus had a system designed to eliminate very abundant constituents such as carbon dioxide and carbon monoxide.

The results of the four experiments did not allow the conclusion that any biochemical activity exists on Mars.

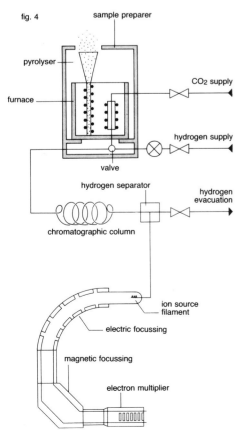

fig. 4

sample preparer

pyrolyser

CO₂ supply

furnace

hydrogen supply

valve

hydrogen separator

hydrogen evacuation

chromatographic column

ion source filament

electric focussing

magnetic focussing

electron multiplier

The open cluster NGC 6520 and the dark cloud Barnard 86. NGC 6520, an open cluster of young stars, is shown in the left-hand part of this photograph superimposed upon the myriads of old stars in our Galaxy. On the right is a dark cloud, Barnard 86; this type of object does not radiate visible light but blocks out the light of the stars located behind the cloud. (Photograph D. F. Malin, Anglo-Australian Telescope Board © 1980)